Discovering Advanced Algebra

An Investigative Approach

SECOND EDITION

Jerald Murdock

Ellen Kamischke

Eric Kamischke

Project Editor
Elizabeth DeCarli

Project Administrator
Tamar Wolins

Consulting Editors
Heather Dever, Joan Lewis, Nathaniel Lewis, Sharon Taylor

Editorial Consultants
David Rasmussen

Editorial Assistants
Christa Edwards, Juliana Tringali

Mathematical Content Reviewers
Larry Copes, Inver Grove Heights, Minnesota
James Wilson, Ph.D., University of Georgia, Athens, Georgia

Multicultural and Equity Reviewer
Swapna Mukhopadhyay, Ph.D., Portland State University, Portland, Oregon

Teacher Reviewers
Rhonda Ball, Troy High School, Troy, Ohio
Joan Conway, Waseca High School, Waseca, Minnesota
Judy Hicks, Ralston Valley High School, Arvada, Colorado
Paul Mlakar, St. Mark's School of Texas, Dallas, Texas
Jennifer North Morris, Santa Rosa High School, Santa Rosa, California
William Putnam, John Marshall High School, Rochester, Minnesota

Scientific Content Reviewer
Art Fortgang, Mills High School, Millbrae, California

Accuracy Checkers
Dudley Brooks, Cavan Fang

Editorial Production Supervisor
Kristin Ferraioli

Production Editor
Holly Rudelitsch

Copyeditor
Margaret Moore

Production Director
Christine Osborne

Production Coordinator
Ann Rothenbuhler

Cover Designers
Jill Kongabel, Jeff Williams

Text Designer
Marilyn Perry

Art Editor
Jason Luz, Maya Mélenchuk

Photo Researcher
Marcy Lunetta

Illustrators
Juan Alvarez, Robert Arnow, Pamela Hobbs, William Pasini, Sue Todd

Technical Art
ICC Macmillan Inc., Matt Perry, Jason Luz

Compositor and Prepress
ICC Macmillan Inc.

Printer
Webcrafters, Inc.

Textbook Product Manager
Tim Pope

Executive Editor
Josephine Noah

Publisher
Steven Rasmussen

This material is based upon work supported by the National Science Foundation under award number MDR9154410. Any opinions, findings, and conclusions or recommendations expressed in this publication are those of the authors and do not necessarily reflect the views of the National Science Foundation.
Key Curriculum Press
1150 65th Street
Emeryville, CA 94608
editorial@keypress.com
www.keypress.com
Printed in the United States of America
10 9 8 7 6 5 4 3 2 14 13 12 11 10 09
ISBN 978-1-55953-984-5

Acknowledgments

Creating a textbook and its supplementary materials is a team effort involving many individuals and groups. We are especially grateful to thousands of *Discovering Advanced Algebra* and *Discovering Algebra* teachers and students, to teachers who participated in the summer institutes and workshops, and to manuscript readers, all of whom provided suggestions, reviewed material, located errors, and most of all, encouraged us to continue with the project.

Our students, their parents, and our administrators at Interlochen Arts Academy have played an important part in the development of this book. Most importantly, we wish to thank Carol Murdock, our parents, and our children for their love, encouragement, and support.

As authors we are grateful to the National Science Foundation for supporting our initial technology-and-writing project that led to the 1998 publication of *Advanced Algebra Through Data Exploration.* This second edition of *Discovering Advanced Algebra* has been developed and shaped by what we learned during the writing and publication of earlier versions of this text and *Discovering Algebra* and by our work with so many students, parents, and teachers who were searching for a more meaningful algebra curriculum.

Over the course of our careers, many individuals and groups have been instrumental in our development as teachers and authors. The Woodrow Wilson National Fellowship Foundation provided the initial impetus for involvement in leading workshops. Publications and conferences produced by the National Council of Teachers of Mathematics and Teachers Teaching with Technology have guided the development of this curriculum. Individuals such as Ron Carlson, Helen Compton, Frank Demana, Arne Engebretsen, Paul Foerster, Christian Hirsch, Glenda Lappan, Richard Odell, Heinz-Otto Peitgen, James Sandefur, James Schultz, Dan Teague, Charles VonderEmbse, Bert Waits, and Mary Jean Winter have inspired us.

The development and production of *Discovering Advanced Algebra* has been a collaborative effort between the authors and the staff at Key Curriculum Press. We truly appreciate the cooperation and valuable contributions offered by the Editorial and Production Departments at Key Curriculum Press. Finally, a special thanks to Key's publisher, Steven Rasmussen, for encouraging and publishing a technology-enhanced *Discovering Mathematics* series that offers groundbreaking content and learning opportunities.

Jerald Murdock
Ellen Kamischke
Eric Kamischke

A Note from the Publisher

For more than 30 years, Key Curriculum Press has developed mathematics materials for students and teachers. The mathematics that we learn and teach in school has changed to reflect advances in technology, new discoveries in mathematics and science, and changing societal needs. However, in spite of the changes in mathematics and mathematics education, we have found one truth that has not changed: Students learn mathematics best when they understand the concepts behind it. *Discovering Advanced Algebra* asks you to discuss mathematics, engage in hands-on activities, analyze and develop mathematical models for real-world situations, and make connections between different mathematical representations.

This diverse approach to teaching provides access for students of all learning styles. Kinesthetic learners make connections between their movement and graphs and equations representing motion. Visual learners appreciate the engaging graphics and art as well as the emphasis on multiple representations of relations and functions. Auditory learners enhance their understanding as they explore and discuss concepts with other students. Whatever your learning style, you'll benefit from deeply exploring mathematical concepts, discussing your ideas with other students, and applying your discoveries to solving real-world problems.

Investigations are at the heart of this book. Through the investigations, you will concretely explore interesting problems and generalize concepts. And if you forget a concept, formula, or procedure, you can always re-create it—because you developed it yourself the first time! You'll find that this approach allows you to form a deep and conceptual understanding of advanced algebra topics. As you progress through the text, you'll see that graphing calculators and other technologies are used to explore patterns and to make, test, and generalize conjectures. The use of technology allows you to explore more interesting problems and to move easily among different representations.

As Glenda Lappan, mathematics professor at Michigan State University and former president of the National Council of Teachers of Mathematics, said about the first edition of this book, "Students coming out of a year with this text . . . will know the mathematics they know in deeper, more flexible ways. They will have developed a set of mathematical habits of mind that will serve them very well as students or users of mathematics. They will emerge with a sense of mathematics as a search for regularity that allows prediction."

If you are a student, we believe that what you learn this year will serve you well in life. If you are a parent, we believe you will enjoy watching your student develop mathematical confidence. And if you are a teacher, we believe *Discovering Advanced Algebra* will greatly enrich your classroom. The professional team at Key Curriculum Press wishes you success and joy in the lifetime of mathematics ahead of you. We look forward to hearing about your experiences.

Steven Rasmussen, Publisher
Key Curriculum Press

Contents

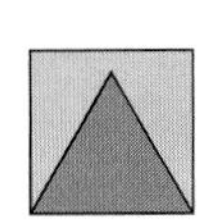
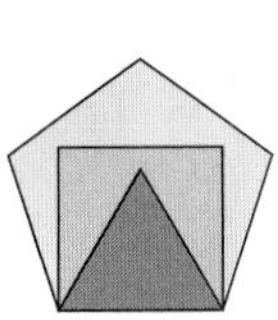
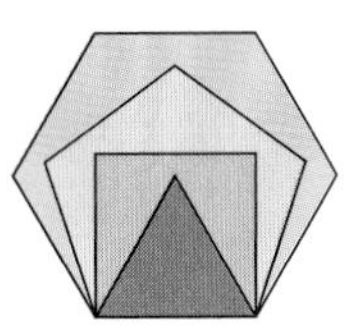

A Note to Students from the Authors

Jerald Murdock

Ellen Kamischke

Eric Kamischke

The goal of this stage of your mathematical journey is to develop advanced algebraic tools and the mathematical power that will help you participate fully in life beyond school. On this journey you will make connections between algebra and the world around you.

Important decision-making situations will confront you in life, and your ability to use mathematics and algebra can help you make informed decisions. You'll need skills that can evolve and be adapted to new situations. You'll need to interpret numerical information and use it as a basis for making decisions. And you'll need to find ways to solve problems that arise in real life, not just in textbooks. Success in algebra is a gateway to many varied career opportunities.

Hopefully, you've already found out that learning algebra is more than memorizing facts, theories, and procedures. With your teacher as a guide, you'll learn algebra by doing mathematics. You'll make sense of important algebraic concepts, learn essential skills, and discover how to use algebra.

During the **Investigations,** you will talk about algebra, share ideas, and learn from and with your fellow group members. You will also work and communicate with your teammates to strengthen your understanding of the mathematical concepts. To enjoy and gain respect in your role as a learner, honor differences among classmates, listen carefully when others are sharing, stay focused during the process, be responsible and respectful, and share your own ideas and suggestions.

The right technology can help you explore new ideas and answer questions that come up along the way. Using a graphing calculator, you will be able to manipulate large amounts of data quickly so that you can see the overall picture. Throughout the text you can refer to **Calculator Notes** for information that will help you use this tool. Technology will play an important role in your life and future career. Learning to use your graphing calculator efficiently and being able to interpret its output will prepare you to use other technologies successfully in situations to come.

The book itself will be a guide, leading you to explore ideas and ponder questions. Read it carefully—with paper, pencil, and calculator close at hand—and take good notes. Concepts and problems you have encountered before can help you solve new problems. Work through the **Examples** and answer the questions that are asked along the way. Some **Exercises** require a great deal of thought. Don't give up. Make a solid attempt at each problem that is assigned. Sometimes you'll make corrections and fill in details later, after you discuss a problem in class. Features called **Project, Improving Your . . . Skills,** and **Take Another Look** will challenge you to extend your learning and to apply it in creative ways.

Just as this book is your guide, your notebook can be a log of your travels through Algebra 2. In it you will record your notes and your work. You may also want to keep a journal of your personal impressions along the way. And just as every trip results in a photo album, you can place some of your especially notable accomplishments in a portfolio that highlights your trip. Collect pieces of work in your portfolio as you go, and refine the contents as you make progress on your journey. Each chapter ends with **Assessing What You've Learned.** This feature suggests ways to review your progress and prepare for what comes next: organizing your notebook, writing in your journal, updating your portfolio, and other ways to reflect on what you have learned.

You should expect hard work, interesting activities, and occasional frustration. Yet, as you gain more algebra skills, you'll overcome obstacles and be rewarded with a deeper understanding of mathematics, an increased confidence in your own problem-solving abilities, and the opportunity to be creative. From time to time, look back to reflect on where you have been. We hope that your journey through *Discovering Advanced Algebra* will be a meaningful and rewarding experience.

And now it is time to begin. You are about to discover some pretty fascinating things.

Jerald Murdock
Ellen Kamischke
Eric Kamischke

CHAPTER

0 Problem Solving with Algebra

The California Academy of Sciences building, designed by Pritzker Prize–winning architect Renzo Piano (b 1937), has a living roof covered with nine species of California native plants. The planning and implementation of the roof required the collaboration of architects, botanists, living roof experts, and contractors. The living roof provides insulation for the building as well as habitat for native birds, butterflies, and insects in Golden Gate Park, San Francisco.

OBJECTIVES

In this chapter you will

- solve problems both on your own and as a member of a group
- use pictures and graphs as problem-solving tools
- learn a four-step process for solving problems with symbolic algebra
- practice strategies for organizing information before you solve a problem

LESSON

0.1 Pictures, Graphs, and Diagrams

A whole essay might be written on the danger of thinking without images.

SAMUEL COLERIDGE

In this textbook there are many problems that ask you to look at situations in new and different ways. This chapter offers some strategies to approach these problems. Although some of the problems in this chapter are fictitious, they give you a chance to practice skills that you will use throughout the book and throughout life.

This first lesson focuses on using a sketch, graph, or diagram to help you find a solution.

EXAMPLE A

Allyndreth needs to mix some lawn fertilizer with 7 liters of water. She has two buckets that hold exactly 3 liters and 8 liters, respectively. Describe or illustrate a procedure that will give exactly 7 liters of water in the 8-liter bucket.

▶ Solution

There is more than one solution to this problem. The picture sequence below shows one solution.

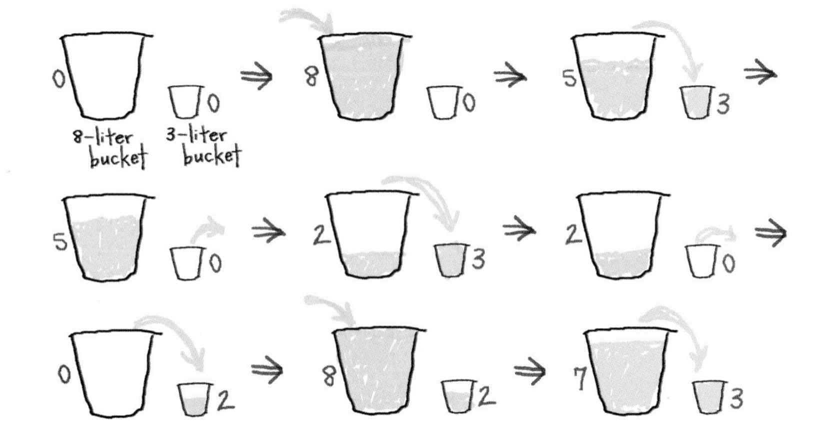

A written description of the solution to Example A might be complex and hard to understand. Yet the pictures help you keep track of the amount of water at each step of the solution. You also see how the water is poured into and out of the buckets. Can you think of a different solution? If so, does your solution take more or fewer steps?

Using pictures is one way to visualize a problem. Another problem-solving strategy is to use objects and act out the problem. For instance, in Example A you could use paper cups to represent the buckets and label each cup with the amount of water at each step of the solution. When you act out a problem, it helps to record positions and quantities on paper as you solve the problem so that you can recall your own steps.

Problem solving often requires a group effort. Different people have different approaches to solving problems, so working in a group gives you the opportunity to hear and see different strategies. Sometimes group members can divide the work based on each person's strengths and expertise, and other times it helps if everyone does the same task and then compares results. Each time you work in a group, decide how to share tasks so that each person has a productive role. You might find that working together, your group can solve challenging problems that would have stumped each of you individually.

The following investigation will give you an opportunity to work in a group and an opportunity to practice some problem-solving strategies.

Investigation
Polar Bear Crossing the Arctic

A polar bear rests by a stack of 3000 pounds of fish he has caught. He plans to travel 1000 miles across the Arctic to bring as many fish as possible to his family. He can pull a sled that holds up to 1000 pounds of fish, but he must eat 1 pound of fish at every mile to keep his energy up.

What is the maximum amount of fish (in pounds) the polar bear can transport across the Arctic? How does he do it? Work as a group and prepare a written or visual solution.

Pictures are useful problem-solving tools in mathematics but are not limited to diagrams like those in Example A. Coordinate graphs are some of the most important problem-solving pictures in mathematics.

EXAMPLE B

A line passes through the point (4, 7) and has slope $\frac{3}{5}$. Find another point on the same line.

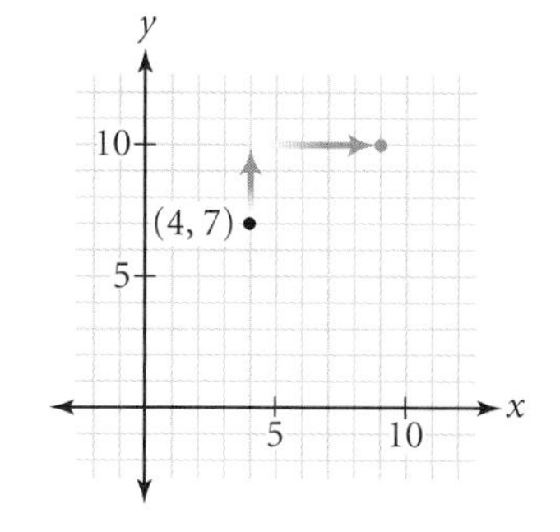

▸ Solution

You *could* use a formula for slope and solve for an unknown point. But a graph may be a simpler way to find a solution.

Plot the point (4, 7). Recall that slope is $\frac{\text{change in } y}{\text{change in } x}$ and move from (4, 7) according to the slope, $\frac{3}{5}$. One possible point, (9, 10), is shown. Can you find another point on the line?

René Descartes

Mathematics CONNECTION

Coordinate graphs are also called Cartesian graphs, named after the French mathematician and philosopher René Descartes (1596–1650). Descartes was not the first to use coordinate graphs, but he was the first to publish his work using two-dimensional graphs with a horizontal axis, a vertical axis, and an origin. Descartes's goal was to apply algebra to geometry, which today is called analytic geometry. Analytic geometry in turn laid the foundations of modern mathematics, including calculus.

Although pictures and diagrams have been the focus of this lesson, problem solving requires that you use a variety of strategies. As you work on the exercises, don't limit yourself. You are always welcome to use any and all of the strategies that you know.

History CONNECTION

George Pólya (1887–1985) was a Hungarian-American mathematician often recognized for his contribution to the study of problem solving. In his 1945 book, *How to Solve It,* he describes four steps used in the process of solving a problem:

- Understand the problem
- Devise a plan
- Carry out the plan
- Look back

Solving a challenging problem often involves moving between these steps. For example, a solution plan that fails may lead to deeper understanding of a problem.

You can learn more about Pólya and his contributions to mathematics and problem solving by using the Internet links at www.keymath.com/DAA .

EXERCISES

Practice Your Skills

1. The pictures below show the first and last steps of solutions to bucket problems similar to Example A. Write a statement for each problem.

a. @

0 0 ⇒ . . . ⇒ 4 0

10-liter bucket 7-liter bucket

b.

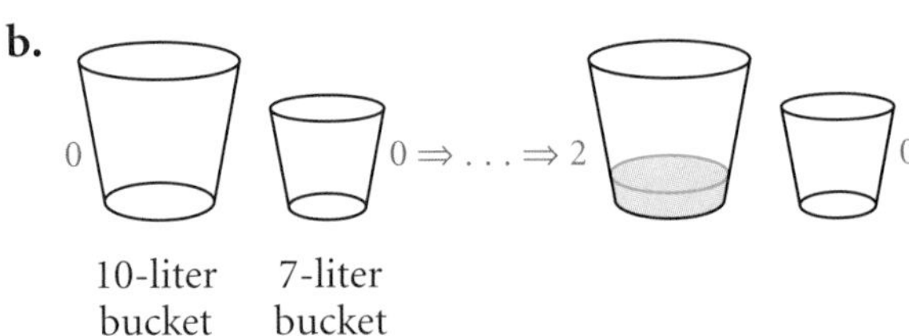

2. Find the slope of each line.

a.

y 5 5 x

b.

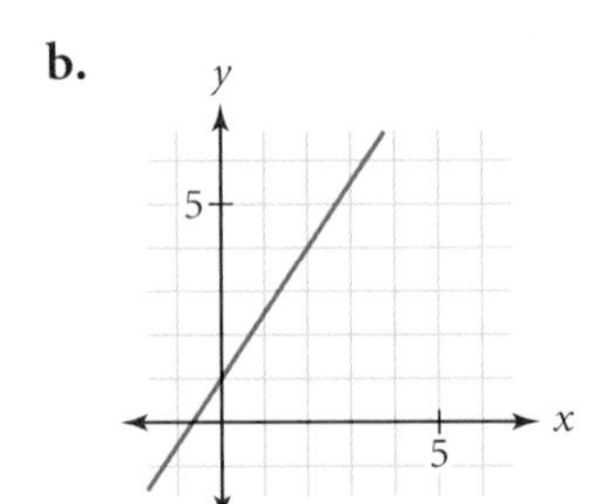

keymath.com/DAA

3. A line passes through the point (4, 7) and has slope $\frac{3}{5}$. Find two more points on the line other than (9, 10), which was found in Example B.

4. Find the slope of the line that passes through each pair of points.

a. (2, 5) and (7, 10) ⓐ **b.** (3, −1) and (8, 7)

c. (−2, 3) and (2, −6) **d.** (3, 3) and (−5, −2)

5. The following problem appears in *Problems for the Quickening of the Mind,* a collection of problems compiled by the Anglo-Saxon scholar Alcuin of York (ca. 735–804 C.E.). Describe a strategy for solving this problem. Do not actually solve the problem. ⓗ

A wolf, a goat, and a cabbage must be moved across a river in a boat holding only one besides the ferryman. How can he carry them across so that the goat shall not eat the cabbage, nor the wolf the goat?

Reason and Apply

6. For each situation, draw and label a diagram. Do not actually solve the problem.

a. A 25 ft ladder leans against the wall of a building with the foot of the ladder 10 ft from the wall. How high does the ladder reach? ⓐ

b. A cylindrical tank that has diameter 60 cm and length 150 cm rests on its side. The fluid in the tank leaks out from a valve on one base that is 20 cm off the ground. When no more fluid leaks out, what is the volume of the remaining fluid?

c. Five sales representatives e-mail each of the others exactly once. How many e-mail messages do they send?

7. Use this graph to estimate these conversions between grams (g) and ounces (oz).

a. 11 oz ≈ _?_ g ⓐ **b.** 350 g ≈ _?_ oz ⓐ

c. 15.5 oz ≈ _?_ g **d.** 180 g ≈ _?_ oz

e. What is the slope of this line? Explain the real-world meaning of the slope.

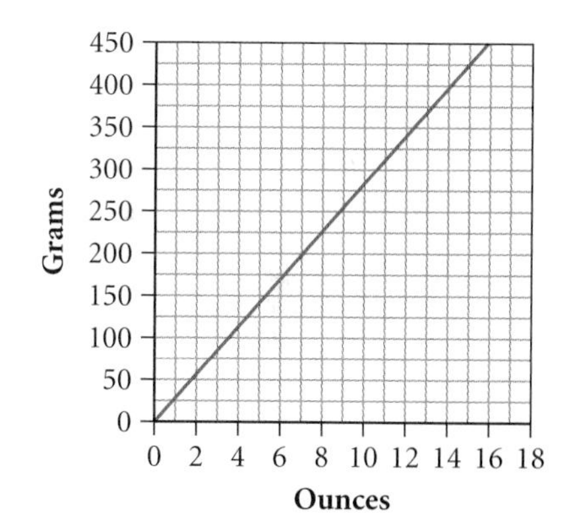

History CONNECTION

The unit of measure today called the pound comes from the Roman unit *libra*, abbreviated "lb," a weight of around 5000 grains (a very small unit of weight). It was subdivided into ounces, or *onzas*, abbreviated "oz." In England the Saxon pound was based on a standard weight of 5400 grains kept in the Tower of London. Now many countries have adopted the Système Internationale (SI), or metric system, as the standard form of measurement.

8. Explain how this graph helps you solve these proportion problems. Then solve each proportion.

a. $\frac{12}{16} = \frac{9}{a}$ ⓐ **b.** $\frac{12}{16} = \frac{b}{10}$

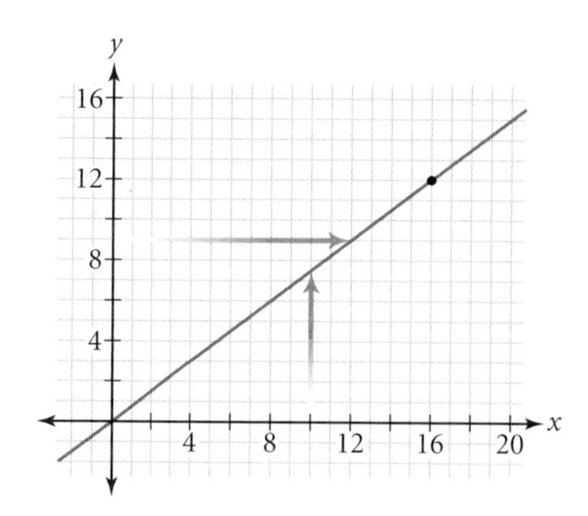

9. APPLICATION Kyle has a summer job cleaning pools. He needs to measure exact amounts of chlorine but has only a 10-liter bucket and a 7-liter bucket. Assuming an unlimited supply of chlorine, describe or illustrate a procedure that will ⓗ

a. Give exactly 4 liters of chlorine in the 10-liter bucket.

b. Give exactly 2 liters of chlorine in the 10-liter bucket.

Pool on a Cloudy Day with Rain (1978) by David Hockney (b 1937)

10. You can use diagrams to represent algebraic expressions. Explain how this rectangle diagram demonstrates that $(x + 2)(x + 3)$ is equivalent to $x^2 + 2x + 3x + 6$.

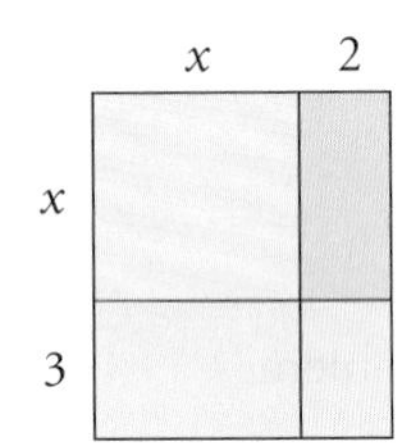

11. Draw a rectangle diagram to represent each product. Use the diagrams to expand each product.

a. $(x + 4)(x + 7)$ ⓐ **b.** $(x + 5)^2$ ⓗ **c.** $(x + 2)(y + 6)$ **d.** $(x + 3)(x - 1)$

12. Solve the problem in Exercise 5. Explain your solution.

Review

13. Translate each verbal statement into a symbolic expression.

a. Three more than a number

b. Four less than twice a number

c. Two-thirds of a number

14. Match these values to the points on the number line.

a. $\frac{3}{8}$ **b.** $\frac{13}{9}$ **c.** $\frac{16}{25}$ **d.** $\frac{5}{14}$

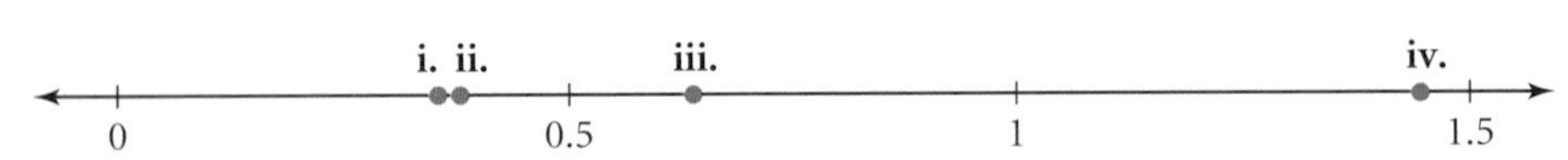

15. Convert these decimal values to fraction form.

a. 0.375 **b.** 1.42 **c.** $0.\overline{2}$ ⓐ **d.** $0.\overline{35}$

16. Match each percent with an equivalent fraction or decimal.

a. 115% **i.** $\frac{3}{20}$

b. 15% **ii.** 0.0015

c. 1.5% **iii.** 1.15

d. 0.15% **iv.** $\frac{3}{200}$

LESSON

0.2

Symbolic Representation

On the second day, the customer service team had planned to double the number of calls answered on the first day, but they exceeded that by three dozen. Seventy-five dozen customer service calls in two days set a new record.

You can translate the paragraph above into an algebraic equation. Although you don't know how many calls were answered each day, an equation will help you figure it out.

For many years, problems like this were solved without writing equations. Then, around the 17th century, the development of symbolic algebra made writing equations and finding solutions much simpler. Verbal statements could then be translated into symbols by representing unknown quantities with letters, called variables, and converting the rest of the sentence into numbers and operations. (You will actually translate this telephone call problem into algebraic notation when you do the exercises.)

History CONNECTION

Muhammad ibn Mūsā al-Khwārizmī (ca. 780–850 C.E.), an Iraqi mathematician, wrote the first algebra treatise. The word *algebra* comes from the treatise's title, *Kitāb al-jabr wa'al-muqābalah,* which translates to *The Science of Completion and Balancing.* Al-Khwārizmī wanted the algebra in this treatise to address real-world problems that affected the everyday lives of the people, such as measuring land and trading goods. You can learn more about al-Khwārizmī and the history of algebra by using the links at www.keymath.com/DAA .

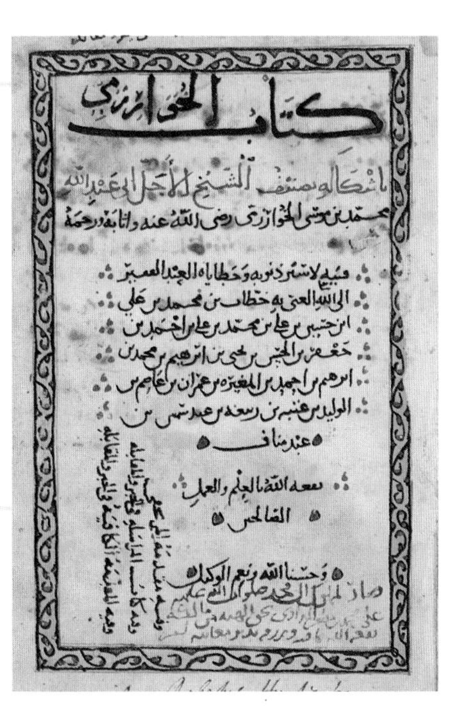

The postage stamp above depicts al-Khwārizmī and was issued in the former Soviet Union. At left is the title page of al-Khwārizmī's treatise on algebra.

Language is designed for describing qualities rather than quantities. Mathematical symbols are better for describing quantities. A very helpful problem-solving technique is to use symbols to represent numbers that quantify data such as time, weight, or distance. These numbers might vary or remain constant. Their values might be given in a problem or unknown.

EXAMPLE A

When Adam and his sister Megan arrive at a party, they see that there is 1 adult chaperone for every 4 kids. Right behind them come 30 more boys, and Megan notices that the ratio is now 2 boys to 1 girl. However, behind the 30 boys come 30 more girls, and Adam notices that there are now 4 girls for every 3 boys. What is the final ratio of adult chaperones to kids?

▶ *Solution*

To represent this problem with symbolic algebra, you first need to determine which quantities are unknown. Assign a variable to each unknown.

Let B represent the original number of boys.

Let G represent the original number of girls.

Let A represent the number of adults.

Next, write equations for the information in the problem.

$$\frac{A}{B+G} = \frac{1}{4}, \text{ or } A = \frac{1}{4}(B+G)$$

When Adam and Megan arrive, there is 1 adult for every 4 kids, so the ratio of adults to kids is 1:4, or $\frac{1}{4}$. The total number of kids is represented by $B + G$. Call this Equation 1.

$$\frac{B+30}{G} = \frac{2}{1}, \text{ or } B + 30 = 2G$$

After 30 more boys come, the ratio of boys to girls is $\frac{2}{1}$. Call this Equation 2.

$$\frac{G+30}{B+30} = \frac{4}{3}, \text{ or } G + 30 = \frac{4}{3}(B+30)$$

After 30 more girls come, the ratio of girls to boys is $\frac{4}{3}$. Call this Equation 3.

You want to know the final ratio of adults to kids: $\frac{A}{B+30+G+30} = \underline{?}$

Now solve the equations to find values for the variables. Equations 2 and 3 involve only variables B and G, so you might eliminate B and solve for G.

$G + 30 = \frac{4}{3}(2G)$	Using Equation 2, substitute $2G$ for $B + 30$ in Equation 3.
$3(G + 30) = 4(2G)$	Multiply both sides by 3.
$3G + 90 = 8G$	Multiply and distribute.
$90 = 5G$	Add $-3G$ to both sides of the equation.
$G = 18$	Divide both sides by 5.
$B + 30 = 2(18)$	Substitute 18 for G in Equation 2.
$B = 36 - 30 = 6$	Subtract 30 from both sides.

Now that you have found values for B and G, you can use Equation 1 to find A.

$$A = \frac{1}{4}(6 + 18) = 6$$

Substitute for B and G in Equation 1.

Finally, interpret the solution.

$$\frac{A}{B+30+G+30} = \frac{6}{6+30+18+30} = \frac{6}{84} = \frac{1}{14}$$

The final ratio of adults to kids is 1:14.

You may have noticed that Example A used a four-step solution process. What are the four steps? Let's apply this process to another problem.

EXAMPLE B

Three friends went to the gym to work out. None of the friends would tell how much he or she could leg-press, but each hinted at their friends' leg-press amounts.

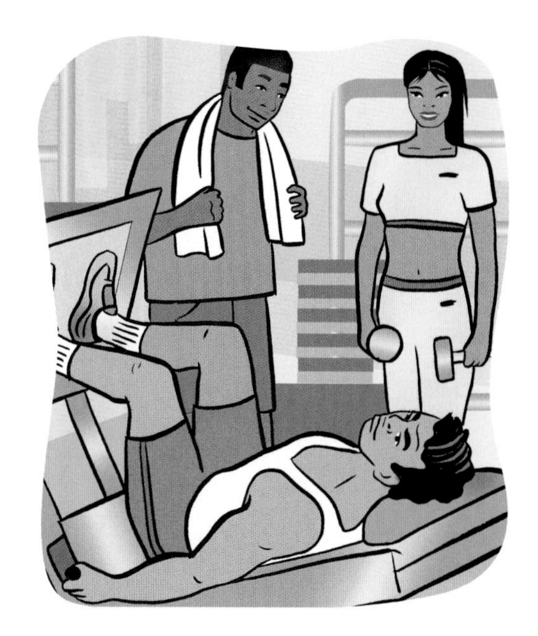

- Chen said that Juanita and Lou averaged 87 pounds.
- Juanita said that Chen leg-pressed 6 pounds more than Lou.
- Lou said that eight times Juanita's amount equals seven times Chen's amount.

Find how much each friend could leg-press.

▶ Solution

First, list the unknown quantities and assign a variable to each.

Let C represent Chen's leg-press weight in pounds.

Let J represent Juanita's leg-press weight in pounds.

Let L represent Lou's leg-press weight in pounds.

Second, write equations from the problem.

$$\begin{cases} \dfrac{J+L}{2} = 87 & \text{Chen's statement translated into an algebraic equation. Call this Equation 1.} \\ C - L = 6 & \text{Juanita's statement as Equation 2.} \\ 8J = 7C & \text{Lou's statement as Equation 3.} \end{cases}$$

Third, solve the equations to find values for the variables.

$J + L = 174$	Multiply both sides of Equation 1 by 2.
$C - L = 6$	Equation 2.
$J + C = 180$	Add the equations.
$7J + 7C = 1260$	Multiply both sides of the previous equation by 7.
$7J + 8J = 1260$	Equation 3 allows you to substitute $8J$ for $7C$.
$15J = 1260$	Add like terms.
$J = 84$	Divide both sides by 15.
$8(84) = 7C$	Substitute 84 for J in Equation 3.
$C = 96$	Solve for C.
$96 - L = 6$	Substitute 96 for C in Equation 2.
$L = 90$	Solve for L.

Last, interpret your solution. Chen leg-presses 96 pounds, Juanita leg-presses 84 pounds, and Lou leg-presses 90 pounds.

The investigation will give you a chance to try the four-step solution process on your own or with a group.

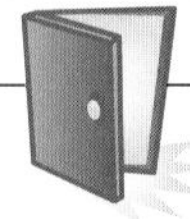

Investigation
Problems, Problems, Problems

Select one of the problems below, and use these four steps to find a solution.

Step 1 List the unknown quantities, and assign a variable to each quantity.

Step 2 Write one or more equations that relate the unknown quantities to conditions of the problem.

Step 3 Solve the equations to find a value for each variable.

Step 4 Interpret your solution according to the context of the problem.

When you finish, write a paragraph answering this question: Which of the four problem-solving steps was hardest for you? Why?

Problem 1

On Monday the manager scheduled two clerks and one supervisor for 8-hour shifts. She was pleased because she met the daily salary budget.

On Tuesday the manager needed two clerks for 8 hours, a third clerk for 4 hours, and one supervisor for 4 hours. She was very pleased to be \$10 under budget.

On Wednesday she needed three clerks for 8 hours and one supervisor for 4 hours. This day she was over budget by \$20.

All of the clerks make the same hourly wage. What is the daily salary budget?

Problem 2

Sandy uses a $\frac{1}{2}$-ton pickup (a truck that can carry 1000 pounds) to transport ingredients for mortar. Sand comes in 50-pound bags and cement comes in 40-pound bags. Mortar is made from five bags of sand for every bag of cement. How many bags of each should Sandy load to make the most mortar possible?

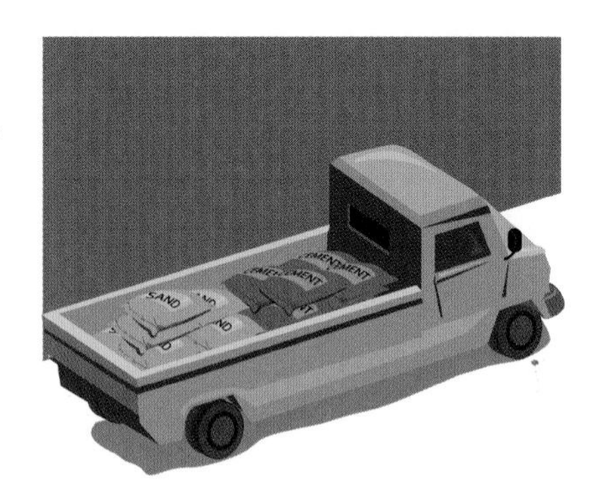

The four problem-solving steps in the investigation help you organize information, work through an algebraic solution, and interpret the final answer. As you do the exercises in this lesson, refer back to these four steps and practice using them.

Exercises

Practice Your Skills

1. Explain what you would do to change the first equation to the second.

a. $a + 12 = 47$
$a = 35$ (@)

b. $5b = 24$
$b = 4.8$

c. $-18 + c = 28$
$c = 46$

d. $\frac{d}{-15} = 4.5$
$d = -67.5$

2. Which equation would help you solve the following problem?

Each member of the student council made three copies of the letter to the principal. Adding these to the 5 original letters, there are now a total of 32 letters. How many members does the student council have?

A. $5 + c = 32$ **B.** $3 + 5c = 32$ **C.** $5 + 3c = 32$

3. Solve each equation.

a. $5 + c = 32$ **b.** $3 + 5c = 32$ (@) **c.** $5 + 3c = 32$

4. Which equation would help you solve this problem? What does x represent?

On the second day, the customer service team had planned to double the number of calls answered the first day, but they exceeded that by three dozen. Seventy-five dozen customer service calls in two days set a new record.

A. $x + 3 = 75$ **B.** $x + 2x + 3 = 75$ **C.** $2x + 3 = 75$

5. Solve each equation.

a. $x + 3 = 75$ **b.** $x + 2x + 3 = 75$ **c.** $2x + 3 = 75$

Reason and Apply

6. Julie's math teacher likes to give her students "mathemagical" number tricks. Try the steps of her favorite number trick below.

Step 1	Write down the first three digits of your seven-digit phone number (don't include the area code).
Step 2	Multiply by 80. (@)
Step 3	Add 1.
Step 4	Multiply by 250.
Step 5	Add the last four digits of your phone number to your results from Step 4.
Step 6	Add the last four digits of your phone number again.
Step 7	Subtract 250.
Step 8	Divide your number by 2.

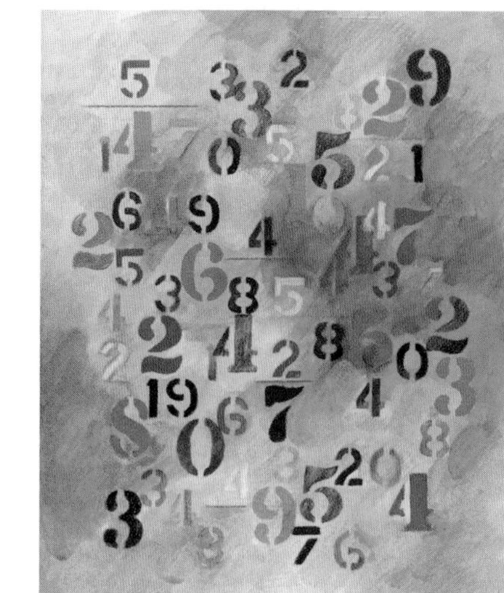

a. How does your answer to Step 8 compare to your phone number?

b. Now you'll represent the number trick to explain why it works. Write an algebraic expression for each step of the process. For each step you should modify your answer from the previous step. Be sure to combine like terms and do the arithmetic for each step. Use x to represent the first three digits of your phone number, and use n to represent the last four digits of your phone number.

c. Will the number trick work for all phone numbers? Use your algebraic expression for Step 8 to explain your answer.

7. Here is a problem and three related equations.

Anita buys 6 large beads and 20 small beads to make a necklace. Ivan buys 4 large and 25 small beads for his necklace. Jill selects 8 large and 16 small beads to make an ankle bracelet. Without tax, Anita pays \$2.70, and Ivan pays \$2.85. How much will Jill pay?

$$\begin{cases} 6L + 20S = 270 & \text{(Equation 1)} \\ 4L + 25S = 285 & \text{(Equation 2)} \\ 8L + 16S = J & \text{(Equation 3)} \end{cases}$$

a. What do the variables L, S, and J represent?

b. What are the units of L, S, and J? ⓗ

c. What does Equation 1 represent?

This necklace was crafted around the 3rd to 2nd millennium B.C.E. in the southwest Asian country of Bactria (now in Afghanistan).

8. Follow these steps to solve Exercise 7.

a. Multiply Equation 1 by -2.

b. Multiply Equation 2 by 3.

c. Add the resulting equations from 8a and b.

d. Solve the equation in 8c for S. Interpret the real-world meaning of this solution.

e. Use the value of S to find the value of L. Interpret the real-world meaning of the value of L.

f. Use the values of S and L to find the value of J. Interpret this solution.

9. The following problem appears in *Liber abaci* (1202), or *Book of Calculations,* by the Italian mathematician Leonardo Fibonacci (ca. 1170–1240).

If A gets 7 denarii from B, then A's sum is fivefold B's. If B gets 5 denarii from A, then B's sum is sevenfold A's. How much has each?

$$\begin{cases} a + 7 = 5(b - 7) & \text{(Equation 1)} \\ b + 5 = 7(a - 5) & \text{(Equation 2)} \end{cases}$$

a. What does a represent?

b. What does b represent?

c. Explain Equation 1 with words.

d. Explain Equation 2 with words.

10. Use Equations 1 and 2 from Exercise 9.

a. Explain how to get

$$b + 5 = 7[(5b - 42) - 5]$$ ⓐ

b. Solve the equation in 10a for b.

c. Use your answer from 10b to find the value of a.

d. Use the context of Exercise 9 to interpret the values of a and b.

Review

11. APPLICATION A 30°-60°-90° triangle and a 45°-45°-90° triangle are two drafting tools used by people in careers such as engineering, architecture, and drafting. The angles in both triangle tools are combined to make a variety of angle measures in hand-drawn technical drawings, such as blueprints. Describe or illustrate a procedure that will give the following angle measures.

a. 15°  **b.** 75° **c.** 105°

12. Rewrite each expression without parentheses.

a. $3(x + 7)$ **b.** $-2(6 - n)$ **c.** $x(4 - x)$

13. Substitute the given value of the variable(s) into each expression and evaluate.

a. $47 + 3x$ when $x = 17$ **b.** $29 - 34n + 14m$ when $n = -1$ and $m = -24$

14. Find the slope of the line that passes through each pair of points.

a. (4, 7) and (8, 7) **b.** (2, 5) and (−6, 3)

15. Rudy is renting a cabin on Lake Tahoe. He wants to make a recipe for hot chocolate that calls for 1 cup of milk. Unfortunately, he has only a 1-gallon container of milk, a 4-cup saucepan, and an empty 12-ounce soda can. Describe or illustrate a process that Rudy can use to measure exactly 1 cup of milk. (There are 8 ounces per cup and 16 cups per gallon.)

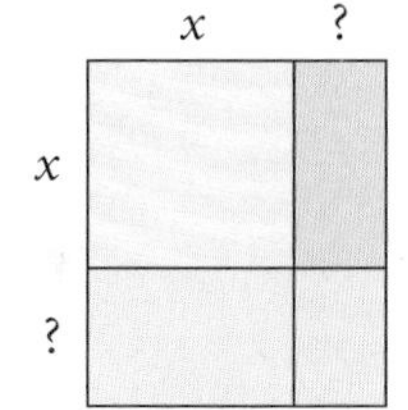

16. Draw a rectangle diagram to represent each product. Use the diagrams to expand each product.

a. $(x + 1)(x + 5)$ **b.** $(x + 3)^2$ **c.** $(x + 3)(x - 3)$

IMPROVING YOUR REASONING SKILLS

Fair Share

Abdul, Billy, and Celia agree to meet and can tomatoes from their neighborhood garden. Abdul picks 50 pounds of tomatoes from his plot of land. Billy picks 30 pounds of tomatoes from his plot. Unfortunately, Celia's plants did not get enough sun, and she cannot pick any tomatoes from her plot.

They spend the day canning, and each has 36 quarts of tomatoes to take home. Wanting to pay Abdul and Billy for the tomatoes they gave to her, Celia finds \$8 in her wallet. How should Celia divide the money between her two friends?

LESSON

0.3

Organizing Information

If one and a half chickens lay one and a half eggs in one and a half days, then how long does it take six monkeys to make nine omelets?

Everybody gets so much information all day long that they lose their common sense.

GERTRUDE STEIN

What sort of problem-solving strategy can you apply to the silly problem above? You could draw a picture or make a diagram. You could assign variables to all sorts of unknown quantities. But do you really have enough information to solve the problem? Sometimes the best strategy is to begin by organizing what you know and what you want to know. With the information organized, you may then find a way to get to the solution.

EXAMPLE A | How many seconds are in a calendar year?

▶ Solution

First, identify what you know and what you need to know.

Know	Need to know
1 year	Number of seconds

It may seem like you don't have enough information, but consider these commonly known facts:

1 year = 365 days (non–leap year)

1 day = 24 hours

1 hour = 60 minutes

1 minute = 60 seconds

This fragment of an ancient Roman calendar shows months, days, and special events. You can learn how to read Roman calendars, or *fasti,* with an Internet link at www.keymath.com/DAA .

You can write each equality as a fraction and multiply the chain of fractions such that the units reduce to leave seconds.

$$1\ \cancel{\text{year}} \cdot \frac{365\ \cancel{\text{days}}}{1\ \cancel{\text{year}}} \cdot \frac{24\ \cancel{\text{hours}}}{1\ \cancel{\text{day}}} \cdot \frac{60\ \cancel{\text{minutes}}}{1\ \cancel{\text{hour}}} \cdot \frac{60\ \text{seconds}}{1\ \cancel{\text{minute}}} = 31{,}536{,}000\ \text{seconds}$$

There are 31,536,000 seconds in a non–leap-year calendar year.

In Example A, you discovered there are 31,536,000 seconds in 1 year. If you want to know how many seconds there are in 5 years, you can just multiply this number by 5. When quantities are related like this, they are said to be in **direct variation** with each other. Another example of direct variation is the relationship between the number of miles you travel and the time you spend traveling. In each case the relationship can be expressed in an equation of the form $y = kx$, where k is called the **constant of variation.** [▶ You can use the **Dynamic Algebra Exploration** at www.keymath.com/DAA to review direct variation.◀]

keymath.com/DAA

EXAMPLE B

To qualify for the Interlochen 470 auto race, each driver must complete two laps of the 5-mile track at an average speed of 100 miles per hour (mi/h). Due to some problems at the start, Naomi averages only 50 mi/h on her first lap. How fast must she go on the second lap to qualify for the race?

▶ Solution

Sort the information into two categories: what you know and what you might need to know. Assign variables to the quantities that you don't know. Use a table to organize the information, and include the units for each piece of information.

Know	Might need to know
Speed for first lap: 50 mi/h	Speed for second lap (in mi/h): s
Average speed for both laps: 100 mi/h	Time for first lap (in h): t_1
Length of each lap (in mi): 5	Time for second lap (in h): t_2

Use the units to help you find connections between the pieces of information. Speed is measured in miles per hour and therefore calculated by dividing distance by time. You might also remember the relationship *distance* = *rate* · *time,* or $d = rt$. Because you know the distance and rate for the first lap, you can solve for the time: $t = \frac{d}{r}$.

$$t_1 = \frac{5 \text{ miles}}{50 \text{ miles per hour}} = \frac{1}{10} \text{ hour, or 6 minutes}$$

It takes one-tenth of an hour, or 6 minutes, to do the first lap. You know the distance and speed for the first and second laps together, so solve for the time for both laps. Then you can subtract to find what you are looking for, the time for the second lap.

$$(t_1 + t_2) = \frac{10 \text{ miles}}{100 \text{ miles per hour}} = \frac{1}{10} \text{ hour, or 6 minutes}$$

Note that the time for both laps is the same as the time for the first lap. This means the time for the second lap, t_2, must be zero. It's not possible for Naomi to complete the second lap in no time, so she cannot qualify for the race. Would the solution be different if the laps were 10 miles long? d miles long?

Some problems overwhelm you with lots of information. Identifying and categorizing what you know is always a good way to start organizing information.

EXAMPLE C

Lab assistant Jerry Anderson has just finished cleaning a messy lab table and is putting the equipment back on the table when he reads a note telling him *not* to disturb the positions of three water samples. Not knowing the correct order of the three samples, he finds these facts in the lab notes.

- The water that is highest in sulfur was on one end.
- The water that is highest in iron is in the Erlenmeyer flask.
- The water taken from the spring is not next to the water in the bottle.
- The water that is highest in calcium is left of the water taken from the lake.
- The water in the Erlenmeyer flask, the water taken from the well, and the water that is highest in sulfur are three distinct samples.
- The water in the round flask is not highest in calcium.

Organize the facts into categories. (This is the first step in actually determining which sample goes where.)

▶ Solution

Information is given about the types of containers, the sources of the water, the elements found in the samples, and the positions of the samples on the table. You can find three options for each category.

Containers: round flask, Erlenmeyer flask, bottle

Sources: spring, lake, well

Elements: sulfur, iron, calcium

Positions: left, center, right

Now that the information is organized and categorized, you need to see where it leads. You will finish this problem in Exercise 9.

Career CONNECTION

Event planners make a career organizing information. To plan a New Year's Eve celebration, for example, an event planner considers variables such as location, decorations, food and beverages, number of people, and staff. For the celebration of the year 2000, event planners around the world worked independently and cooperatively to organize unique events for each city or country and to coordinate recording and televising of the events.

Berlin, Germany (left), and Seattle, Washington (right), celebrate the new millennium at midnight, January 1, 2000.

Use this investigation as an opportunity to practice categorizing and organizing information as you did in Example C.

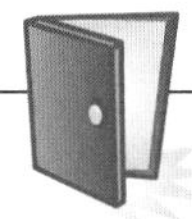

Investigation
Who Owns the Zebra?

There are five houses along one side of Birch Street, each of a different color. The home-owners each drive a different car, and each has a different pet. The owners all read a different newspaper and plant only one thing in their garden.

- The family with the station wagon lives in the red house.
- The owner of the SUV has a dog.
- The family with the van reads the *Gazette.*
- The green house is immediately to the left of the white house.
- The *Chronicle* is delivered to the green house.
- The man who plants zucchini has birds.
- In the yellow house they plant corn.
- In the middle house they read the *Times.*
- The compact car parks at the first house.
- The family that plants eggplant lives in the house next to the house with cats.
- In the house next to the house where they have a horse, they plant corn.
- The woman who plants beets receives the *Daily News.*
- The owner of the sports car plants okra.
- The family with the compact car lives next to the blue house.
- They read the *Bulletin* in the house next to the house where they plant eggplant.

Who owns the zebra?

Organizing the known information and clarifying what you need to find out is a very useful strategy. Whether it involves simply keeping track of units or sorting out masses of information, organizing your data and making a plan are essential to finding a solution efficiently.

Malaysian-American architect and artist Daniel Castor (b 1966) created this pencil drawing of the Amsterdam Stock Exchange. Castor's work—which he calls "jellyfish" drawings—organizes lots of information and several perspectives into one drawing. He describes his art as "[capturing], in two dimensions, the physical power of the spaces yielded by [the] design process, a power that cannot be adequately described by word or photograph."

EXERCISES

Practice Your Skills

1. Use units to help you find the missing information.

   a. How many seconds would it take to travel 15 feet at 3.5 feet per second? ⓐ

   b. How many centimeters are in 25 feet? (There are 2.54 centimeters per inch and 12 inches per foot.)

   c. How many miles could you drive with 15 gallons of gasoline at 32 miles per gallon?

2. Emily and Alejandro are part of a math marathon team on which they take turns solving math problems for 4 hours each day. On Monday Emily worked for 3 hours and Alejandro for 1 hour. Then on Tuesday Emily worked for 2 hours and Alejandro for 2 hours. On Monday they collectively solved 139 problems and on Tuesday they solved 130 problems. Find the average problem-solving rate for Emily and for Alejandro.

   a. Identify the unknown quantities and assign variables. What are the units for each variable? ⓐ

   b. What does the equation $3e + 1a = 139$ represent? ⓗ

   c. Write an equation for Tuesday.

   d. Which of these ordered pairs (e, a) is a solution for the problem?

      i. (34, 37) ii. (37, 28) iii. (30, 35) iv. (27, 23)

   e. Interpret the solution from 2d according to the context of the problem.

3. To qualify for the Interlochen 470 auto race, each driver must complete two laps of the track at an average speed of 100 mi/h. Benjamin averages only 75 mi/h on his first lap. How fast must he go on the second lap to qualify for the race?

4. Use the distributive property to expand and combine like terms when possible.

a. $7.5(a - 3)$

b. $12 + 4.7(b + 6)$

c. $5c - 2(c - 12)$ ⓗ

d. $8.4(35 - d) + 12.6d$

5. Solve each equation.

a. $4.5(a - 7) = 26.1$ ⓐ

b. $9 + 2.7(b + 3) = 20.7$

c. $8c - 2(c - 5) = 70$

d. $8.4(35 - d) + 12.6d = 327.6$

6. Chase rides his bike 5 miles to school in 25 minutes. The relationship between the distance he rides and the time it takes is a direct variation.

a. Write a direct variation equation between d, the distance Chase rides, and the time it takes, t. Use k for the constant of variation.

b. Use the information about Chase's ride to school to find the value of k and its units.

c. If Chase always rides at the same rate, how long will it take him to ride 8 miles?

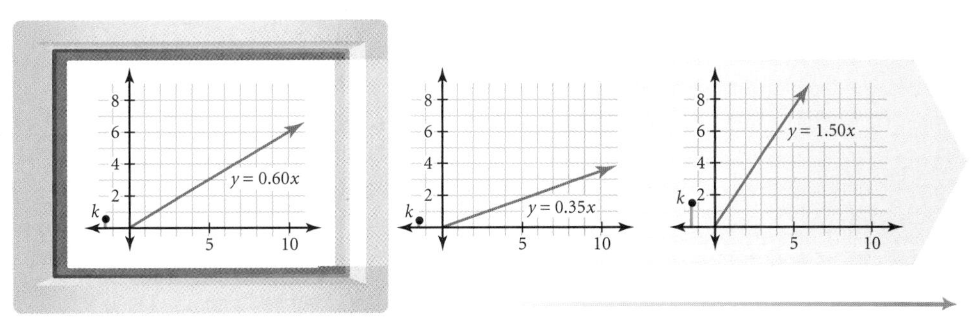

[▶ You can use the **Dynamic Algebra Exploration** at www.keymath.com/DAA to review direct variation and to help you solve this exercise.◀]

keymath.com/DAA

Reason and Apply

7. APPLICATION Alyse earns \$15.40 per hour, and she earns time and a half for working past 8:00 P.M. Last week she worked 35 hours and earned \$600.60. How many hours did she work past 8:00 P.M.? ⓗ

8. The dimensions used to measure length, area, and volume are related by multiplication. Find the information about each rectangular box. Include the units in your solutions.

a. A box has volume 12,960 cm^3, height 18 cm, and length 30 cm. Find the width. ⓐ

b. A box has volume 486 in^3 and height 9 in. Find the area of the base.

c. A box has base area 3.60 m^2 and height 0.40 m. Find the volume.

d. A box has base area 2.40 ft^2 and volume 2.88 ft^3. Find the height.

Nicaraguan artist Federico Nordalm (b 1949) created this painting, titled *Box of Apples.*

9. Lab assistant Jerry Anderson has just finished cleaning a messy lab table and is putting the equipment back on the table when he reads a note telling him *not* to disturb the positions of three water samples. Not knowing the correct order of the three samples, he finds these facts in the lab notes.

- The water that is highest in sulfur was on one end.
- The water that is highest in iron is in the Erlenmeyer flask.
- The water taken from the spring is not next to the water in the bottle.
- The water that is highest in calcium is left of the water taken from the lake.
- The water in the Erlenmeyer flask, the water taken from the well, and the water that is highest in sulfur are three distinct samples.
- The water in the round flask is not highest in calcium.

Determine which water sample goes where. Identify each sample by its container, source, element, and position. ⓗ

10. APPLICATION Paul can paint the area of a 12-by-8 ft wall in 15 min. China can paint the same area in 20 min.

a. In which equation does t represent how long it would take Paul and China to paint the area of a 12-by-8 ft wall together? Explain your choice. ⓗ

i. $\frac{15t}{96} + \frac{20t}{96} = \frac{1}{96}$ **ii.** $(96)15t + (96)20t = 96$ **iii.** $\frac{96t}{15} + \frac{96t}{20} = 96$

b. Solve all three equations in 10a.

c. How long would it take Paul and China to paint the wall together?

11. Kiane has a photograph of her four cats sitting in a row. The cats are different ages, and each cat has its own favorite toy and favorite sleeping spot.

- Rocky and the 10-year-old cat would never sit next to each other for a photo.
- The cat that sleeps in the blue chair and the cat that plays with the rubber mouse are the two oldest cats.
- The cat that plays with the silk rose is the third cat in the photo.
- Sadie and the cat on Sadie's left in the photo don't sleep on the furniture.
- The 8-year-old cat sleeps on the floor.
- The cat that sleeps on the sofa eats the same food as the 13-year-old cat.
- Pascal likes to chase the 5-year-old cat.
- If you add the age of the cat that sleeps in a box and the age of the cat that plays with a stuffed toy, you get the age of Winks.
- The cat that sleeps on the blue chair likes to hide the catnip ball, which belongs to one of the cats sitting next to it in the photo.

Who plays with the catnip-filled ball?

Review

12. Rewrite each expression using the properties of exponents so that the variable appears only once.

a. $(r^5)(r^7)$ @ b. $\frac{2s^6}{6s^2}$ c. $\frac{(t)(t^3)}{t^8}$ d. $3(2u^2)^4$

13. Joel is 16 years old. His cousin Rachel is 12.

a. What is the difference in their ages?

b. What is the ratio of Joel's age to Rachel's age?

c. In eight years, what will be the difference in their ages?

d. In eight years, what will be the ratio of Joel's age to Rachel's age?

14. Draw a rectangle diagram to represent each product. Use the diagrams to expand each product.

a. $(3x + 1)(x + 5)$ b. $(x + 3)^2$ c. $(x + y)(2x + 7)$

15. APPLICATION Iwanda sells African bead necklaces through a consignment shop. At the end of May, the shop paid her \$100 from the sale of 8 necklaces. At the end of June, she was paid \$187.50 from the sale of 15 necklaces. Assume that the consignment shop pays Iwanda the same amount of money for each necklace sold.

a. Write a direct variation equation for the amount Iwanda is paid in terms of the number of necklaces sold.

b. If the shop sells only 6 of her necklaces in July, how much money will Iwanda get?

c. If the materials for each necklace cost \$8, how much profit did Iwanda make each month?

IMPROVING YOUR REASONING SKILLS

Internet Access

A nationwide Internet service provider advertises "1000 hours free for 45 days" for new customers. How many hours per day do you need to be online to use 1000 hours in 45 days? Do you think it is reasonable for the Internet service provider to make this offer?

CHAPTER 0 REVIEW

There is no single way to solve a problem. Different people prefer to use different problem-solving strategies, yet not all strategies can be applied to all problems. In this chapter you practiced only a few specific strategies. You may have also used some strategies that you remember from other courses. The next paragraph gives you a longer list of problem-solving strategies to choose from.

Organize the information that is given, or that you figure out in the course of your work, in a list or table. **Draw a picture,** graph, or diagram, and label it to illustrate information you are given *and* what you are trying to find. There are also special types of diagrams that help you represent algebraic expressions, such as rectangle diagrams. **Make a physical representation** of the problem. That is, act it out or make a model. **Look for a pattern** in numbers or units of measure. Be sure all your measures use the same system of units and that you compare quantities with *the same* unit. **Eliminate some possibilities.** If you know what the answer cannot be, you are partway there. **Solve subproblems** that present themselves as part of the problem context, or **solve a simpler problem** by substituting easier numbers or looking at a special case. Don't forget to **use symbolic representation!** Assign variables to unknown quantities and write expressions for related quantities. Translate verbal statements into equations, and solve the equations. **Work backward** from the solution to the problem. For instance, you can solve equations by undoing operations. **Use guess-and-check,** adjusting each successive guess by the result of your previous guess. Finally, but most importantly, **read the problem!** Be sure you know what is being asked.

EXERCISES

@ Answers are provided for all exercises in this set.

1. You are given a 3-liter bucket, a 5-liter bucket, and an unlimited supply of water. Describe or illustrate a procedure that will give exactly 4 liters of water in the 5-liter bucket.

2. Draw a rectangle diagram to represent each product. Use the diagrams to expand each product.

a. $(x + 3)(x + 4)$ **b.** $(2x)(x + 3)$ **c.** $(x + 6)(x - 2)$ **d.** $(x - 4)(2x - 1)$

3. Use the Pythagorean Theorem to find each missing length.

a.

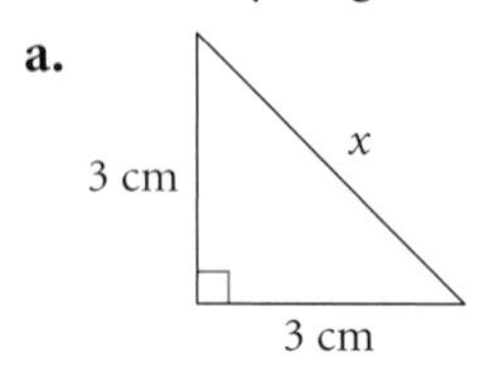

b.

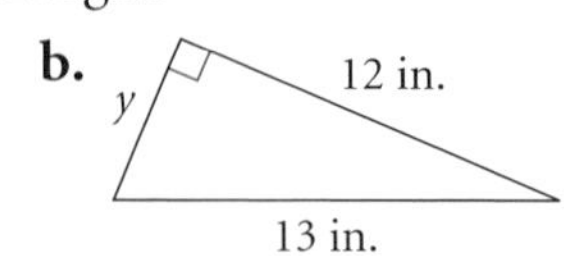

4. This graph shows the relationship between distance driven and gasoline used for two cars going 60 mi/h.

a. How far can Car A drive on 7 gallons (gal) of gasoline?

b. How much gasoline is needed for Car B to drive 342 mi?

c. Which car can drive farther on 8.5 gal of gas? How much farther?

d. The graph shows the direct variation between the distance driven and the gasoline used. What is the constant of variation for each car?

e. What is the slope of each line? Explain the real-world meaning of each slope.

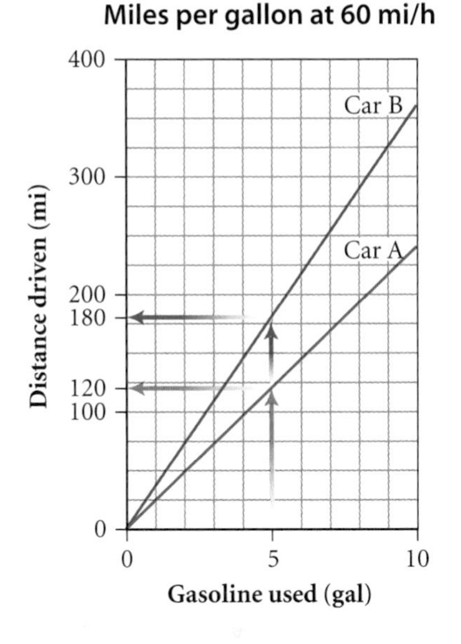

Consumer CONNECTION

The U.S. Environmental Protection Agency (EPA) obtains fuel-economy estimates each year on car manufacturers' new models. In a controlled laboratory setting, professional drivers test the cars on a treadmill-type machine with engines and temperature conditions that simulate highway and city driving. Hybrid cars—a cross between fuel-powered cars and electric cars—are some of the highest-rated vehicles for fuel economy.

This diagram illustrates the components of a parallel hybrid car.

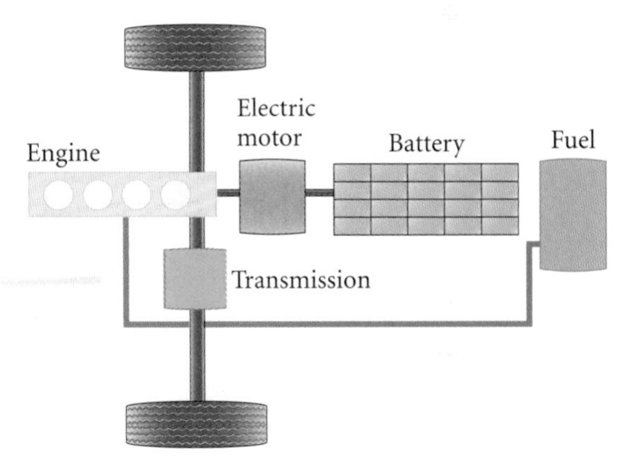

5. Solve each equation. Check each answer by substituting into the original equation.

a. $3(x - 5) + 2 = 26$

b. $3.75 - 1.5(y + 4.5) = 0.75$

6. In 6a and b, translate each verbal statement to a symbolic expression. Combine the expressions into an equation to solve 6c.

a. Six more than twice a number

b. Five times three less than a number

c. Six more than twice a number is five times three less than the number. Find the number.

7. APPLICATION Keisha and her family are moving to a new apartment 12 miles from their old one. You-Do-It Truck Rental rents a small truck for \$19.95 per day plus \$0.35 per mile. Keisha's family hopes they can complete the move with five loads all on the same day. She estimates that she will drive the truck another 10 miles for pickup and return.

a. Write an expression that represents the cost of a one-day rental with any number of miles.

b. How much will Keisha pay if she does complete the move with five loads?

c. How much can Keisha save if she completes the move with four loads?

8. Toby has only a balance scale, a single 40 g mass, and a stack of both white and red blocks. (Assume that all white blocks have the same mass and all red blocks have the same mass.) Toby discovers that four white blocks and one red block balance two white blocks, two red blocks, and the 40 g mass. He also finds that five white blocks and two red blocks balance one white block and five red blocks.

a. List the unknown quantities and assign a variable to each.

b. Translate Toby's discoveries into equations.

c. Solve the equations to find a value for each variable.

d. Interpret the solution according to the context of the problem.

Science CONNECTION

Calculating how much food a human being needs and planning how much food can be carried are critical factors for space travel. Weight and volume of food must be limited, trash must be minimized, and nutritional value and variety must be provided. Current U.S. space shuttle flights can carry 3.8 pounds of food per person per day, providing 3 meals a day for up to 14 days.

Astronaut Linda Godwin handles food supplies aboard the space shuttle *Endeavour* in December 2001.

9. Amy says that it is her birthday and that six times her age five years ago is twice as much as twice her age next year. How old is Amy?

10. Scott has 47 coins totaling $5.02. He notices that the number of pennies is the same as the number of quarters and that the sum of the number of pennies and quarters is one more than the sum of the number of nickels and dimes. How many of each coin does Scott have?

11. The height of a golf ball in flight is given by the equation $h = -16t^2 + 48t$, where h represents the height in feet above the ground and t represents the time in seconds since the ball was hit. Find h and interpret the real-world meaning of the result when

a. $t = 0$ **b.** $t = 2$ **c.** $t = 3$

12. Rewrite each expression using the properties of exponents.

a. $(4x^{-2})(x)$ **b.** $\frac{4x^2}{8x^3}$ **c.** $(x^3)^5$

13. Consider the equation $y = 2^x$.

a. Find y when $x = 0$. **b.** Find y when $x = 3$.

c. Find y when $x = -2$. **d.** Find x when $y = 32$.

14. Use units to help you find the missing information.

a. How many ounces are in 5 gallons? (There are 8 ounces per cup, 4 cups per quart, and 4 quarts per gallon.)

b. How many meters are in 1 mile? (There are 2.54 centimeters per inch, 12 inches per foot, 5280 feet per mile, and 100 centimeters per meter.)

15. Bethany Rogers just started working as a sales assistant. She learns that all three sales representatives have important meetings with clients, but she uncovers only these clues.

- Mr. Bell is a sales representative although he is not meeting with Mr. Green.
- Ms. Hunt is the client who will be meeting in the lunch room.
- Mr. Green is the client with a 9:00 A.M. appointment.
- Mr. Mendoza is the sales representative meeting in the conference room.
- Ms. Phoung is a client, but she will not be in the 3:00 P.M. meeting.
- Mrs. Plum is a sales representative, but not the one meeting at 12:00 noon.
- The client with the 9:00 A.M. appointment is not meeting in the convention hall.

Help Bethany figure out which sales representative is meeting with which client, where, and when.

TAKE ANOTHER LOOK

1. You have seen that the multiplication expression $(x + 2)(x + 3)$ can be represented with a rectangle diagram in which the length and width of the rectangle represent the factors and the area represents the product. Find a way to represent the multiplication of three factors, such as $(x + 2)(x + 3)(x + 4)$. Explain how the geometry of the diagram represents the product.

2. Recall Jerry Anderson's problem with the water samples (Exercise 9 in Lesson 0.3). In the table below, each of the six subsections compares two of the characteristics. For example, the upper-left subsection compares the position and the type of container. Each statement given in the problem translates into yeses (Y) or noes (N) in the cells of the table. When you have two noes in the same row or column of one subsection, the third cell must be a yes. And when you have a yes in any cell of a subsection, the other four cells in the same row and column must be noes. The table will eventually show you how the characteristics match up. Use this table to solve Jerry's problem.

	Erlenmeyer flask	Round flask	Bottle	Calcium	Sulfur	Iron	Lake	Well	Spring
Left									
Center									
Right									
Lake									
Well									
Spring									
Calcium									
Sulfur									
Iron									

3. Find a way to use the distributive property to rewrite $(x + 2)(x + 3)$ without parentheses. Compare and contrast this method to the rectangle diagram method. Look back at Exercise 11 in Lesson 0.1, and explain how you could use the distributive property for each multiplication.

4. Use your graphing calculator or geometry software to explore the slopes of lines. What is the slope of a horizontal line? Of a vertical line? What slopes create a diagonal line at a 45° angle from the x-axis? For which slopes does the line increase from left to right? For which slopes does it decrease? Can you estimate a line's slope simply by looking at a graph? Write a short paper summarizing your findings.

keymath.com/DAA

Assessing What You've Learned

WRITE IN YOUR JOURNAL Recording your thoughts about the mathematics you are learning, including areas of confusion or frustration, helps point out when you should seek assistance from your teacher and what questions you could ask. Keeping a journal is a good way to collect your ideas and questions, and if you write in it regularly, you'll track the progress of your understanding throughout the course. Your journal can also remind you of interesting contributions to make in class or prompt questions to ask during a review period.

Here are some questions you might start writing about.

- How has your idea about what algebra is changed since you finished your first-year course in algebra? Do you have particular expectations about what you will learn in advanced algebra? If so, what are they?
- What are your strengths and weaknesses as a problem-solver? Do you consider yourself well organized? Do you have a systematic approach? Give an everyday example of problem solving that reminds you of work you did in this chapter.

GIVE A PRESENTATION In the working world, most people will need to give a presentation once in a while or contribute ideas and opinions at meetings. Making a presentation to your class gives you practice in planning, conveying your ideas clearly, and adapting to the needs of an audience.

Choose an investigation or a problem from this chapter, and describe the problem-solving strategies you and your group used to solve it. Here are some suggestions to plan your presentation. (Even planning a presentation requires problem solving!)

- Work with a partner or team. Divide tasks equally. Your role should use skills that are well established, as well as stretch your abilities in new areas.
- Discuss the topic thoroughly. Connect the work you did on the problem with the objectives of the chapter.
- Outline your talk, and decide what details to mention for each point and what diagrams, graphs, or pictures would clarify the presentation.
- Speak clearly and loudly. Reveal your interest in the topic you chose by making eye contact with listeners.

CHAPTER

1 Sequences

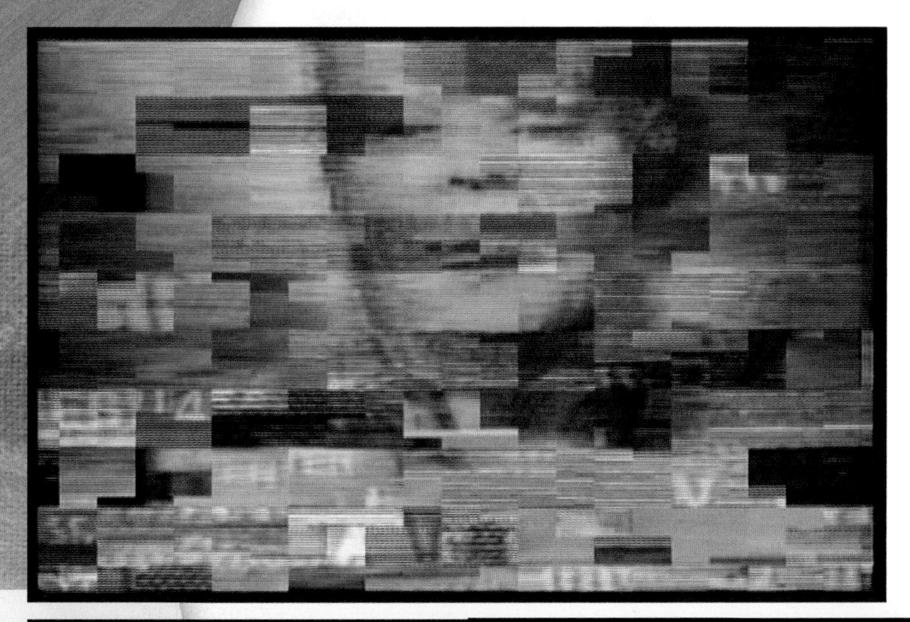

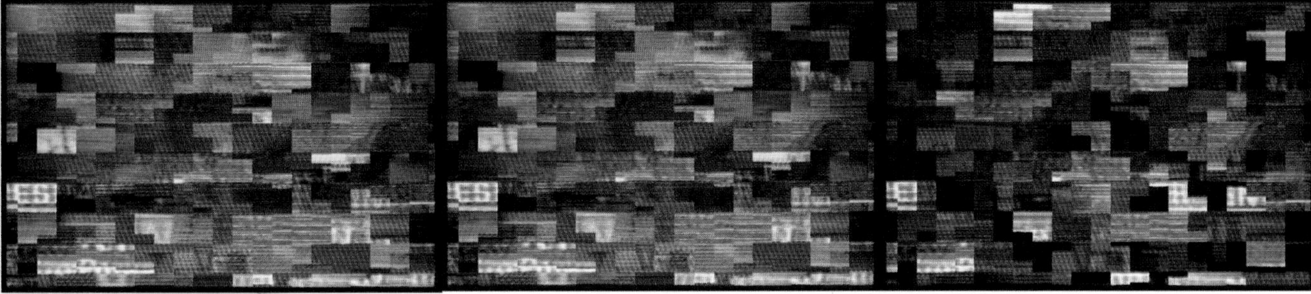

To create the video piece *Residual Light,* experimental video artist Anthony Discenza (American, b 1967) recorded 3 hours of commercial television by filming the TV screen while continuously channel surfing. This 3-hour sample was then compressed in stages by recording and re-recording the material on analog and digital tape while controlling the speed of the playback. Through this recursive process, the original 3 hours was gradually reduced to just 3 minutes. This 3-minute sequence was then slowed back down, resulting in a 20-minute loop.

OBJECTIVES

In this chapter you will

- recognize and visualize mathematical patterns called sequences
- write recursive definitions for sequences
- display sequences with graphs
- use sequences to model growth and decay
- investigate what happens to sequences in the long run

Differences and Ratios

In Chapter 1, you will study number patterns and how quantities like populations and bank balances change over time. In this lesson you will review some concepts and skills that will help prepare you for your work in the rest of the chapter.

Two important words that describe how a quantity has changed are *difference* and *ratio*. You subtract *now* − *previous* to find the difference. You divide $\frac{now}{previous}$ to find the ratio.

EXAMPLE A

Find the difference and ratio for each situation and interpret their meanings.

a. Last year's college tuition was \$20,000. This year it is \$23,000.

b. One hundred fifty acres of land were flooded last week, but now only 120 acres are flooded.

▶ Solution

a. $difference = now - previous$

$$= 23{,}000 - 20{,}000 = 3{,}000$$

$$ratio = \frac{now}{previous}$$

$$= \frac{23{,}000}{20{,}000} = \frac{23}{20} = 1.15$$

The difference tells you that tuition has increased by \$3,000. The ratio is 1.15, or 1 + 0.15, which means that tuition increased 15% from last year to this year.

b. $difference = now - previous$

$$= 120 - 150 = -30$$

$$ratio = \frac{now}{previous}$$

$$= \frac{120}{150} = \frac{4}{5} = 0.8$$

The difference tells you that 30 fewer acres are flooded now. Because the ratio, 0.8, is less than 1, you can think of it as 1 − 0.2. This means that the amount of flooded land decreased by 20% from last week to this week.

EXAMPLE B Find the new quantities described.

a. 80 increased by 12% **b.** 50 decreased by 4%

▶ Solution

a. $\textit{previous} + \textit{change} = 80 + 0.12 \cdot 80$
$= 80(1 + 0.12) = 80(1.12) = 89.6$

80 increased by 12% is 89.6.

b. $\textit{previous} - \textit{change} = 50 - 0.04 \cdot 50$
$= 50(1 - 0.04) = 50(0.96) = 48$

50 decreased by 4% is 48.

EXAMPLE C Sujata earned $40,000 this year. Next year her salary will increase by 3% and she will get an additional $1,000 bonus. What will her new salary be?

▶ Solution

This calculation involves both a percent change and an addition.

$\textit{previous}(1 + \textit{rate}) + \textit{bonus} = 40{,}000(1 + 0.03) + 1{,}000$
$= 40{,}000(1.03) + 1{,}000$
$= 41{,}200 + 1{,}000 = 42{,}200$

Sujata's new salary will be $42,200.

EXERCISES

1. Find the difference and ratio for each (*previous, now*) pair of numbers.

a. 100, 120 @ **b.** 50, 53 **c.** 200, 140 @ **d.** 12, 9

2. Match each description to its formula.

a. 100 increased by 5% @ **i.** 100(1 + 0.5)
b. 100 decreased by 50% **ii.** 100(1 − 0.05)
c. 100 increased by 50% **iii.** 100(1 − 0.5)
d. 100 decreased by 5% **iv.** 100(1 + 0.05)

3. Find each new quantity described.

a. 20 increased by 15% @ **b.** 60 increased by 20% **c.** 300 decreased by 18% @
d. 40 decreased by 30% **e.** 110 increased by 20 **f.** 250 decreased by 40

4. The price of a $30,000 automobile is reduced by 15%, and then the price is cut by $1,200. What is the new price? @

5. Kevin has $400 in his savings account. This amount will increase by 2% by the end of the year because of the interest the account earns. Kevin plans to withdraw $150 on January 1 to purchase a new speaker system for his MP3 player. What will his balance be?

6. Describe a single pattern of change in each list of numbers.

a. 10, 12, 14, 16 **b.** 5, 15, 45, 135

LESSON

1.1

Recursively Defined Sequences

For every pattern that appears, a mathematician feels he ought to know why it appears.

W. W. SAWYER

Look around! You are surrounded by patterns and influenced by how you perceive them. You have learned to recognize visual patterns in floor tiles, window panes, tree leaves, and flower petals. In every discipline, people discover, observe, re-create, explain, generalize, and use patterns. Artists and architects use patterns that are attractive or practical. Scientists and manufacturing engineers follow patterns and predictable processes that ensure quality, accuracy, and uniformity. Mathematicians frequently encounter patterns in numbers and shapes.

The arches in the Santa Maria Novella cathedral in Florence, Italy, show an artistic use of repeated patterns.

Scientists use patterns and repetition to conduct experiments, gather data, and analyze results.

You can discover and explain many mathematical patterns by thinking about recursion. **Recursion** is a process in which each step of a pattern is dependent on the step or steps that come before it. It is often easy to define a pattern recursively, and a recursive definition reveals a lot about the properties of the pattern.

EXAMPLE A

A square table seats 4 people. Two square tables pushed together seat 6 people. Three tables pushed together seat 8 people. How many people can sit at 10 tables pushed together? How many tables are needed to seat 32 people? Write a recursive definition to find the number of people who can sit at any linear arrangement of square tables.

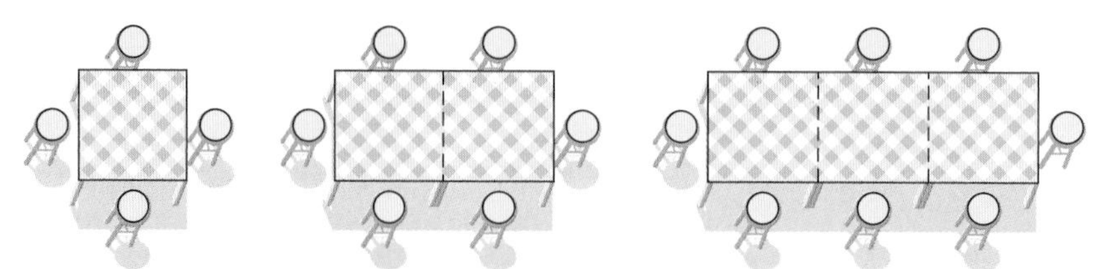

▶ Solution

Sketch the arrangements of four tables and five tables. Notice that when you add another table, you seat two more people than in the previous arrangement.

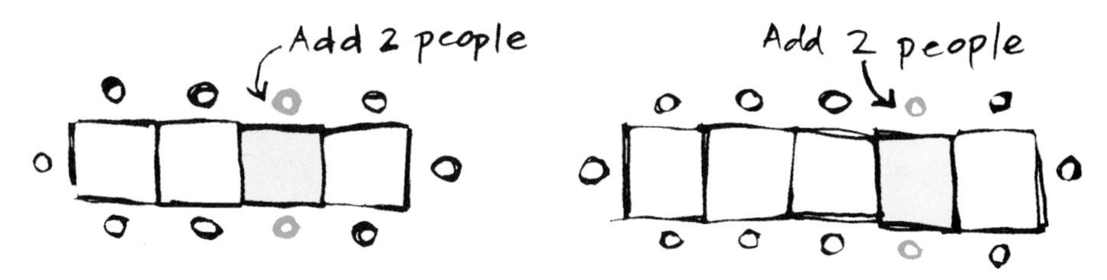

You can put this information into a table, and that reveals a clear pattern. You can continue the pattern to find that 10 tables seat 22 people, and 15 tables are needed for 32 people.

Tables	1	2	3	4	5	6	7	8	9	10	. . .	15
People	4	6	8	10	12	14	16	18	20	22	. . .	32

You can also organize the information like this:

number of people at 1 table = 4

number of people at 2 tables = number of people at 1 table + 2

number of people at 3 tables = number of people at 2 tables + 2

If you assume the same pattern continues, then

number of people at 10 tables = number of people at 9 tables + 2.

In general, the pattern is

number of people at n tables = number of people at $(n - 1)$ tables + 2.

This rule shows how to use recursion to find the number of people at any number of tables. In recursion, you use the previous value in the pattern to find the next value.

A **sequence** is an ordered list of numbers. The table in Example A represents the sequence

4, 6, 8, 10, 12, . . .

Each number in the sequence is called a **term.** The first term, u_1 (pronounced "u sub one"), is 4. The second term, u_2, is 6, and so on.

The nth term, u_n, is called the **general term** of the sequence. A **recursive formula,** the formula that defines a sequence, must specify one (or more) starting terms and a **recursive rule** that defines the nth term in relation to a previous term (or terms).

You generate the sequence 4, 6, 8, 10, 12, . . . with this recursive formula:

$u_1 = 4$

$u_n = u_{n-1} + 2 \quad \text{where } n \geq 2$

Because the starting value is $u_1 = 4$, the recursive rule $u_n = u_{n-1} + 2$ is first used to find u_2. This is clarified by saying that n must be greater than or equal to 2 to use the recursive rule.

This means *the first term is 4* and *each subsequent term is equal to the previous term plus 2.* Notice that each term, u_n, is defined in relation to the previous term, u_{n-1}. For example, the 10th term relies on the 9th term, or $u_{10} = u_9 + 2$.

EXAMPLE B

A concert hall has 59 seats in Row 1, 63 seats in Row 2, 67 seats in Row 3, and so on. The concert hall has 35 rows of seats. Write a recursive formula to find the number of seats in each row. How many seats are in Row 4? Which row has 95 seats?

An opera house in Sumter, South Carolina.

▶ ***Solution***

First, it helps to organize the information in a table.

Row	1	2	3	4	. . .
Seats	59	63	67		. . .

Every recursive formula requires a starting term. Here the starting term is 59, the number of seats in Row 1. That is, $u_1 = 59$.

This sequence also appears to have a common difference between successive terms: 63 is 4 more than 59, and 67 is 4 more than 63. Use this information to write the recursive rule for the nth term, $u_n = u_{n-1} + 4$.

Therefore, this recursive formula generates the sequence representing the number of seats in each row:

Row	1	2	3	4	. . .
Seats	59	63	67		
		+4	+4	+4	

$u_1 = 59$

$u_n = u_{n-1} + 4 \quad \text{where } n \geq 2$

You can use this recursive formula to calculate how many seats are in each row. [▶ See **Calculator Note 1B** to learn how to do recursion on your calculator. ◀]

$u_1 = 59$ — The starting term is 59.

$u_2 = u_1 + 4 = 59 + 4 = 63$ — Substitute 59 for u_1.

$u_3 = u_2 + 4 = 63 + 4 = 67$ — Substitute 63 for u_2.

$u_4 = u_3 + 4 = 67 + 4 = 71$ — Continue using recursion.

$\vdots$

Because $u_4 = 71$, there are 71 seats in Row 4. If you continue the recursion process, you will find that $u_{10} = 95$, or that Row 10 has 95 seats.

You can graph the sequence from Example B by plotting (*row, seats*) or, more generally, (n, u_n).

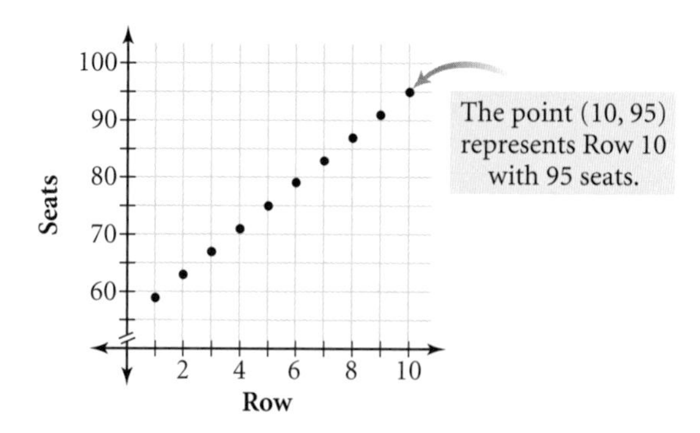

In Examples A and B, a constant is added to each term of the sequence to generate the next term. This type of sequence is called an **arithmetic sequence.**

Arithmetic Sequence

An **arithmetic sequence** is a sequence in which each term is equal to the previous term plus a constant. This constant is called the **common difference.** If d is the common difference, the recursive rule for the sequence has the form

$$u_n = u_{n-1} + d$$

The key to identifying an arithmetic sequence is recognizing the common difference. If you are given a few terms and need to write a recursive formula, first try subtracting consecutive terms. If $u_n - u_{n-1}$ is constant for each pair of terms, then you know your recursive rule must define an arithmetic sequence.

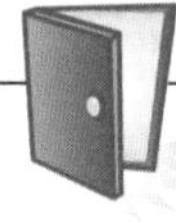

Investigation
Monitoring Inventory

Art Smith has been providing the prints of an engraving to FineArt Gallery. He plans to make just 2000 more prints. FineArt has already received 470 of Art's prints. The Little Print Shoppe also wishes to order prints. Art agrees to supply FineArt with 40 prints each month and Little Print Shoppe with 10 prints each month until he runs out.

Step 1 As a group, model what happens to the number of unmade prints, the number of prints delivered to FineArt, and the number delivered to Little Print Shoppe in a **spreadsheet** like the one below. [▶ See **Calculator Note 1C** for different ways to create this table or spreadsheet on your calculator. ◀]

Month	Unmade prints	FineArt	Little Print Shoppe
1	2000	470	0
2			

Step 2 Use your table from Step 1 to answer these questions:

a. How many months will it be until FineArt has an equal number or a greater number of prints than the number of prints left unmade?

b. How many prints will have been delivered to the Little Print Shoppe when FineArt has received twice the number of prints that remain to be made?

Step 3 Write a short summary of how you modeled the number of prints and how you found the answers to the questions in Step 2. Compare your methods with the methods of other groups.

Career CONNECTION

Economics is the study of how goods and services are produced, distributed, and consumed. Economists in corporations, universities, and government agencies are concerned with the best way to meet human needs with limited resources. Professional economists use mathematics to study and model factors such as supply of resources, manufacturing costs, and selling price.

The sequences in Example A, Example B, and the investigation are arithmetic sequences. Example C introduces a different kind of sequence that is also defined recursively.

EXAMPLE C

The geometric pattern below is created recursively. If you continue the pattern endlessly, you create a **fractal** called the Sierpiński triangle. How many red triangles are there at Stage 20?

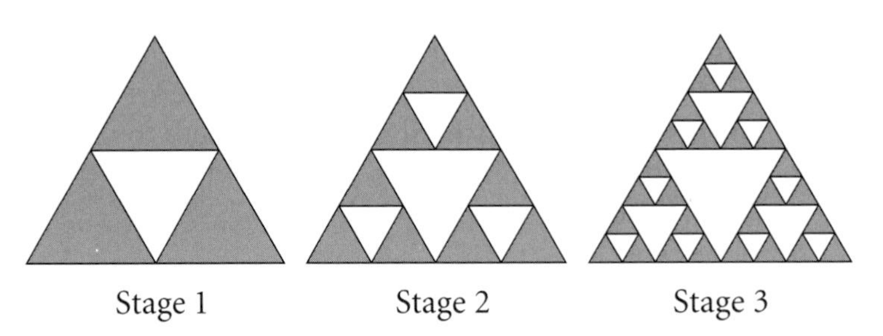

This stamp, part of Poland's 1982 "Mathematicians" series, portrays Waclaw Sierpiński.

Mathematics CONNECTION

The Sierpiński triangle is named after the Polish mathematician Waclaw Sierpiński (1882–1969). He was most interested in number theory, set theory, and topology, three branches of mathematics that study the relations and properties of numbers, sets, and points, respectively. Sierpiński was highly involved in the development of mathematics in Poland between World War I and World War II. He published 724 papers and 50 books in his lifetime. He introduced his famous triangle pattern in a 1915 paper.

▶ Solution

Count the number of red triangles at each stage and write a sequence.

3, 9, 27, . . .

The starting term, 3, represents the number of triangles at Stage 1. You can define the starting term as term one, or u_1. In this case, $u_1 = 3$.

Starting with the second term, each term of the sequence is 3 times the previous term, so that 9 is 3 times 3 and 27 is 3 times 9. Use this information to write the recursive rule and complete your recursive formula.

$u_1 = 3$

$u_n = 3 \cdot u_{n-1}$ where $n \geq 2$

Using the recursive rule 19 times, you find that $u_{20} = 3{,}486{,}784{,}401$. There are over 3 billion triangles at Stage 20!

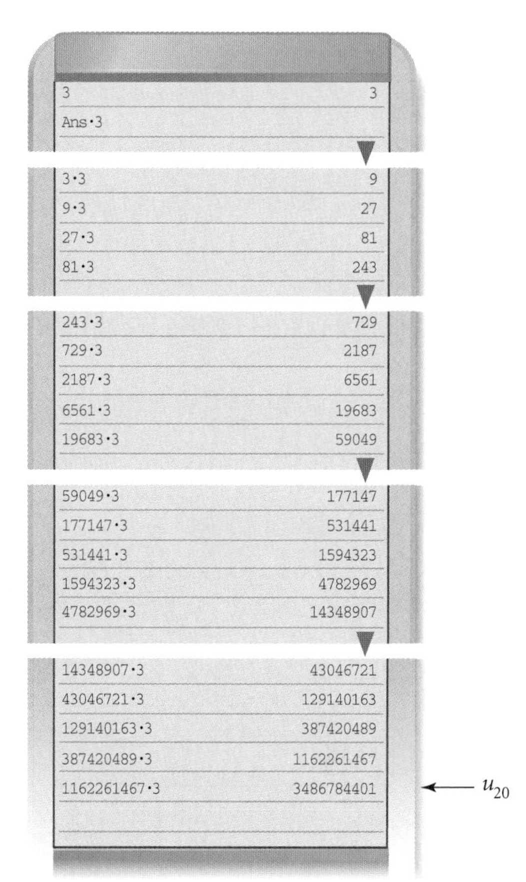

In Example C, each term is multiplied by a constant to generate the next term. This type of sequence is called a **geometric sequence.**

Geometric Sequence

A **geometric sequence** is a sequence in which each term is equal to the previous term multiplied by a constant. This constant is called the **common ratio.** If r is the common ratio, the recursive rule for the sequence has the form

$u_n = r \cdot u_{n-1}$

You identify a geometric sequence by dividing consecutive terms. If $\frac{u_n}{u_{n-1}}$ has the same value for each pair of terms, then you know the sequence is geometric.

Arithmetic and geometric sequences are the most basic sequences because their recursive rules use only one operation: addition in the case of arithmetic sequences, and multiplication in the case of geometric sequences. Recognizing these basic operations will help you easily identify sequences and write recursive formulas.

EXERCISES

Practice Your Skills

1. Match each description of a sequence to its recursive formula.

a. The first term is −18. Keep adding 4.3.

b. Start with 47. Keep subtracting 3.

c. Start with 20. Keep adding 6.

d. The first term is 32. Keep multiplying by 1.5.

i. $u_1 = 20$
$u_n = u_{n-1} + 6 \quad \text{where } n \geq 2$

ii. $u_1 = 47$
$u_n = u_{n-1} - 3 \quad \text{where } n \geq 2$

iii. $u_1 = 32$
$u_n = 1.5 \cdot u_{n-1} \quad \text{where } n \geq 2$

iv. $u_1 = -18$
$u_n = u_{n-1} + 4.3 \quad \text{where } n \geq 2$

2. For each sequence in Exercise 1, write the first 4 terms of the sequence and identify it as arithmetic or geometric. State the common difference or the common ratio for each sequence. ⓐ

3. Write a recursive formula and use it to find the missing table values. ⓐ

n	1	2	3	4	5	. . .	
u_n	40	36.55	33.1	29.65		. . .	12.4

4. Write a recursive formula to generate an arithmetic sequence with a first term 6 and a common difference 3.2. Find the 10th term.

5. Write a recursive formula to generate each sequence. Then find the indicated term.

a. 2, 6, 10, 14, . . . Find the 15th term.

b. 0.4, 0.04, 0.004, 0.0004, . . . Find the 10th term.

c. −2, −8, −14, −20, −26, . . . Find the 30th term.

d. −6.24, −4.03, −1.82, 0.39, . . . Find the 20th term. ⓐ

History CONNECTION

Hungarian mathematician Rózsa Péter (1905–1977) was the first person to propose the study of recursion in its own right. In an interview she described recursion in this way:

> The Latin technical term "recursion" refers to a certain kind of *stepping backwards* in the sequence of natural numbers, which necessarily ends after a finite number of steps. With the use of such recursions the values of even the most complicated functions used in number theory can be calculated in a finite number of steps.

In her book *Recursive Functions in Computer Theory,* Péter describes the important connections between recursion and computer languages.

Rózsa Péter

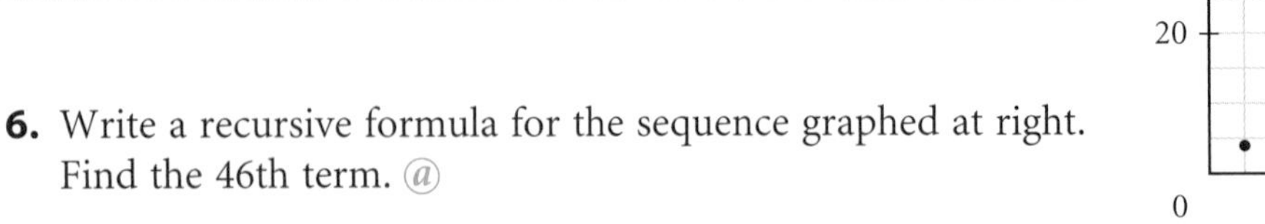

6. Write a recursive formula for the sequence graphed at right. Find the 46th term. ⓐ

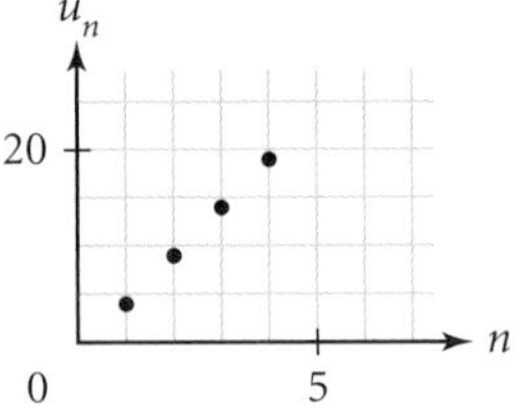

Reason and Apply

7. Write a recursive formula that you can use to find the number of segments, u_n, for Figure n of this geometric pattern. Use your formula to complete the table.

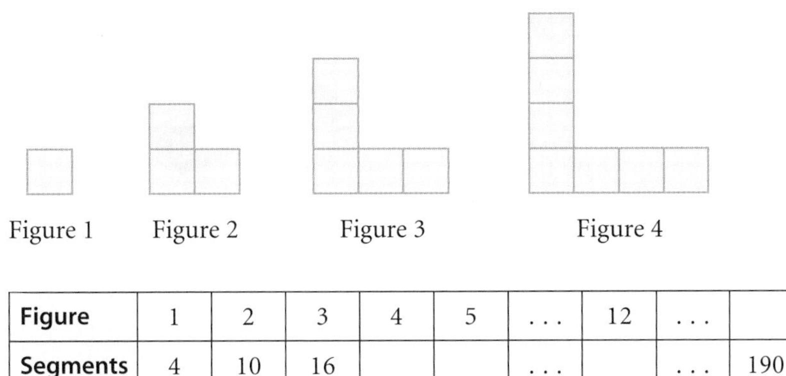

Figure	1	2	3	4	5	. . .	12	. . .	
Segments	4	10	16			. . .		. . .	190

8. A 50-gallon (gal) bathtub contains 20 gal of water and is filling at a rate of 2.4 gal/min. You check the tub every minute on the minute.

a. Suppose that the drain is closed. When will you discover that the water is flowing over the top? @

b. Now suppose that the bathtub contains 20 gal of water and is filling at a rate of 2.4 gal/min, but the drain is open and water drains at a rate of 3.1 gal/min. When will you discover that the tub is empty?

c. Write a recursive formula that you can use to find the water level at any minute due to both the rate of filling and the rate of draining.

9. A car leaves town heading west at 57 km/h.

a. How far will the car travel in 7 h?

b. A second car leaves town 2 h after the first car, but it is traveling at 72 km/h. To the nearest hour, when will the second car pass the first?

10. APPLICATION Inspector 47 at the Zap battery plant keeps a record of which AA batteries she finds defective. Although the battery numbers at right do not make an exact sequence, she thinks they are close to an arithmetic sequence.

a. Write a recursive formula for an arithmetic sequence that estimates which batteries are defective. Explain your reasoning.

b. Predict the numbers of the next five defective batteries.

c. How many batteries in 100,000 will be defective?

11. The week of February 14, the owner of Nickel's Appliances stocks hundreds of red, heart-shaped vacuum cleaners. The next week, he still has hundreds of red, heart-shaped vacuum cleaners. He tells the manager, "Discount the price 25 percent each week until they are gone."

   a. On February 14, the vacuums are priced at \$80. What is the price of a vacuum during the second week? @

   b. What is the price during the fourth week?

   c. When will a vacuum sell for less than \$10?

12. Carl and David's teacher asked them to write the first four terms of a sequence that starts with 100 and has a common difference of −8. Carl says the first four terms are 100, 108, 116, 112. David says the first four terms are 100, 92, 86, 78. Who is correct? Write a clear explanation of how to use the common difference to build the sequence.

13. Taoufik picks up his homework paper from the puddle it fell in. Sadly he reads the first problem and finds that the arithmetic sequence is a blur except for two terms.

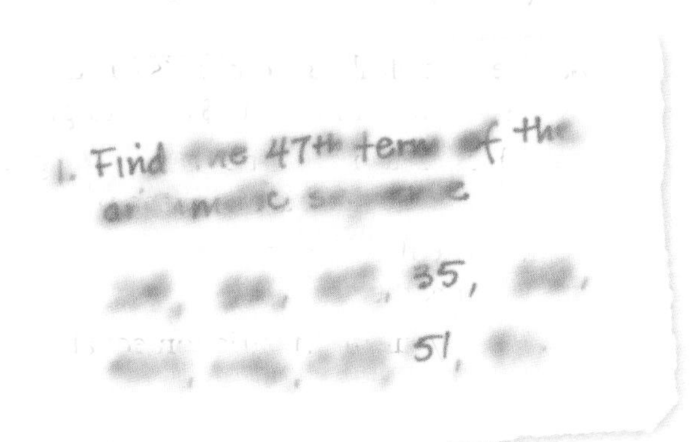

   a. What is the common difference? How did you find it? (h)

   b. What are the missing terms?

   c. What is the answer to Taoufik's homework problem?

Review

14. Ayaunna starts 2.0 m from a motion sensor. She walks away from the sensor at a rate of 1.0 m/s for 3.0 s and then walks toward the sensor at a rate of 0.5 m/s for 4.0 s.

   a. Create a table of values for Ayaunna's distance from the motion sensor at 1-second intervals. @

   b. Sketch a time-distance graph of Ayaunna's walk.

15. Write each question as a proportion and then find the unknown number.

a. 70% of 65 is what number? @

b. 115% of 37 is what number?

c. 110 is what percent of 90?

d. What percent of 18 is 0.5?

16. Find the area of this triangle using two different strategies. Describe your strategies.

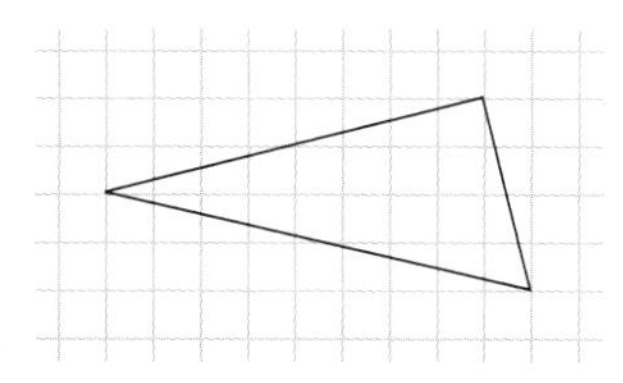

17. APPLICATION Sherez is currently earning $390 per week as a store clerk and part-time manager. She is offered either a 7% increase or an additional $25 per week. Which offer should she accept?

18. The cost of filling your car's gas tank varies directly with the number of gallons of gas you put in. It costs $41.88 to put 12 gallons in the gas tank.

a. How much does it cost to put 1 gallon of gas in the tank? This is the constant of variation for this relationship.

b. Write the equation for the direct variation between c, the cost of filling the tank, and g, the number of gallons put in the tank.

c. Use your direct variation equation to find the cost of filling a tank with 22 gallons of gas.

IMPROVING YOUR REASONING SKILLS

Fibonacci and the Rabbits

Suppose a pair of newborn rabbits, one male and one female, is put in a field. Assume that rabbits are able to mate at the age of one month, so at the end of its second month a female can produce another pair of rabbits. Suppose that each female who is old enough produces one new pair of rabbits (one male, one female) every month and that none of the rabbits die. Write the first few terms of a sequence that shows how many pairs there will be at the end of each month. Then write a recursive formula for the sequence.

This sequence is called the **Fibonacci sequence** after Italian mathematician Leonardo Fibonacci (ca. 1170–1240), who included a similar problem in his book *Liber abaci* (1202). How is the Fibonacci sequence unique compared with the other sequences you have studied?

Two arctic hares blend with the white tundra in Ellesmere Island National Park in northern Canada.

LESSON

1.2 Modeling Growth and Decay

Each sequence you generated in Lesson 1.1 was either an arithmetic sequence with a recursive rule in the form $u_n = u_{n-1} + d$ or a geometric sequence with a recursive rule in the form $u_n = r \cdot u_{n-1}$. You compared consecutive terms to decide whether the sequence had a common difference or a common ratio.

In most cases you have used u_1 as the starting term of each sequence. In some situations, it is more meaningful to treat the starting term as a zero term, or u_0. The zero term represents the starting value before any change occurs. You can decide whether it would be better to begin at u_0 or u_1.

EXAMPLE A

An automobile depreciates, or loses value, as it gets older. Suppose that a particular automobile loses one-fifth of its value each year. Write a recursive formula to find the value of this car when it is 6 years old, if it cost \$23,999 when it was new.

Consumer CONNECTION

The *Kelley Blue Book*, first compiled in 1926 by Les Kelley, annually publishes standard values of every vehicle on the market. Many people who want to know the value of an automobile will ask what its "Blue Book" value is. The *Kelley Blue Book* calculates the value of a car by accounting for its make, model, year, mileage, location, and condition.

▸ *Solution*

Each year, the car will be worth $\frac{4}{5}$ of what it was worth the previous year. Therefore the sequence has a common ratio, which makes it a geometric sequence. It is convenient to start with $u_0 = 23{,}999$ to represent the value of the car when it was new so that u_1 will represent the value after 1 year, and so on. The recursive formula that generates the sequence of annual values is

$u_0 = 23{,}999$ — Starting value.

$u_n = 0.8 \cdot u_{n-1}$ where $n \geq 1$ — $\frac{4}{5}$ is 0.8.

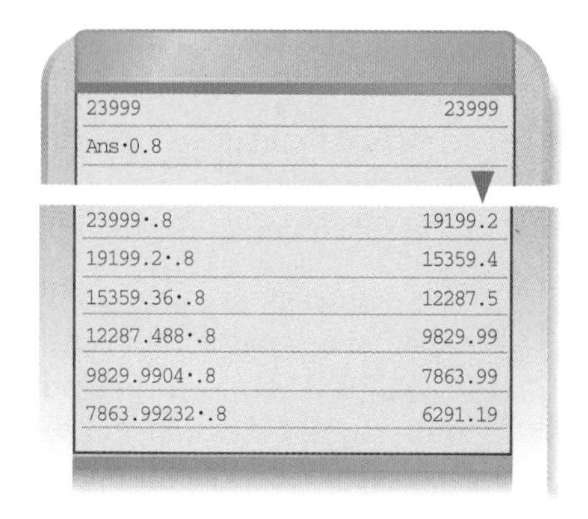

Use this rule to find the 6th term.

After 6 years, the car is worth \$6,291.19.

In situations like the problem in Example A, it's easier to write a recursive formula than an equation using x and y.

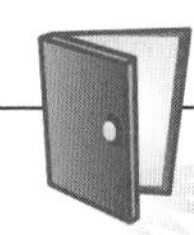

Investigation
Looking for the Rebound

You will need

- a ball
- a motion sensor

When you drop a ball, the rebound height becomes smaller after each bounce. In this investigation you will write a recursive formula for the height of a real ball as it bounces.

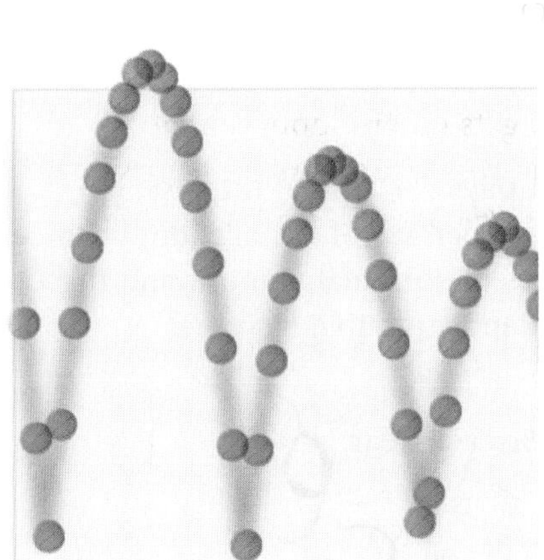

Procedure Note

Collecting Data

1. Hold the motion sensor above the ball.
2. Press the trigger, then release the ball.
3. If the ball drifts, try to follow it and maintain the same height with the motion sensor.
4. If you do not capture at least 6 good consecutive bounces, repeat the procedure.

Step 1 Set up your calculator and motion sensor and follow the Procedure Note to collect bouncing-ball data. [► See **Calculator Note 1D** for calculator instructions on how to gather data. ◄]

Step 2 The data transferred to your calculator are in the form (x, y), where x is the time since you pressed the trigger, and y is the height of the ball. Trace the data graphed by your calculator to find the starting height and the rebound height after each bounce. Record your data in a table.

Step 3 Graph a scatter plot of points in the form (*bounce number, rebound height*). Record the graphing window you use. [► See **Calculator Notes 1E, 1F, 1G,** and **1H** to learn how to enter, plot, trace, and share data. ◄]

Step 4 Compute the rebound ratio for consecutive bounces.

$$\textit{rebound ratio} = \frac{\textit{rebound height}}{\textit{previous rebound height}}$$

Step 5 Decide on a single value that best represents the rebound ratio for your ball. Use this ratio to write a recursive formula that models your sequence of *rebound height* data, and use it to generate the first six terms.

Step 6 Compare your experimental data to the terms generated by your recursive formula. How close are they? Describe some of the factors that might affect this experiment. For example, how might the formula change if you used a different kind of ball?

You may find it easier to think of the common ratio as the whole, 1, plus or minus a percent change. In place of r you can write $(1 + p)$ or $(1 - p)$. The car example involved a 20% (one-fifth) loss, so the common ratio could be written as $(1 - 0.20)$. Your bouncing ball may have had a common ratio of 0.75, which you can write as $(1 - 0.25)$ or a 25% loss per bounce.

In Example A, the value of the car decreased each year. Similarly, the rebound height of the ball decreased with each bounce. These and other decreasing geometric sequences are examples of **decay.** The next example is one of **growth,** or an increasing geometric sequence.

Interest is a charge that you pay to a lender for borrowing money or that a bank pays you for letting it invest the money you keep in your bank account. **Simple interest** is a percentage paid on the **principal,** or initial balance, over a period of time. If you leave the interest in the account, then in the next time period you will receive interest on both the principal and the interest that were in your account. This is called **compound interest** because you are receiving interest on the interest.

EXAMPLE B

Gloria deposits $2,000 into a bank account that pays 7% annual interest compounded annually. This means the bank pays her 7% of her account balance as interest at the end of each year, and she leaves the original amount and the interest in the account. When will the original deposit double in value?

▶ *Solution*

The balance starts at $2,000 and increases by 7% each year.

$u_0 = 2000$

$u_n = u_{n-1} + 0.07 \cdot u_{n-1}$ where $n \geq 1$ — The recursive rule that represents 7% growth.

$u_n = (1 + 0.07)u_{n-1}$ where $n \geq 1$ — Factor.

Use your calculator to compute year-end balances recursively.

Term u_{11} is 4209.70, so the investment balance will more than double in 11 years.

2000	2000
Ans·(1+0.07)	
2000·(1+.07)	2140.
2140.·(1+.07)	2289.8
2289.8·(1+.07)	2450.09
2450.086·(1+.07)	2621.59
2621.59202·(1+.07)	2805.1
2805.1034614·(1+.07)	3001.46
3001.460703698·(1+.07)	3211.56
3211.5629529569·(1+.07)	3436.37
3436.3723596639·(1+.07)	3676.92
3676.9184248404·(1+.07)	3934.3
3934.3027145792·(1+.07)	4209.7

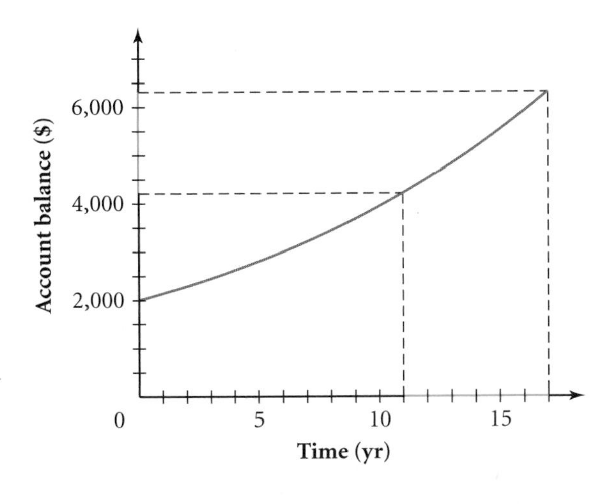

Compound interest has many applications in everyday life. The interest on both savings and loans is almost always compounded, often leading to surprising results. This graph and spreadsheet show the account balance in Example B.

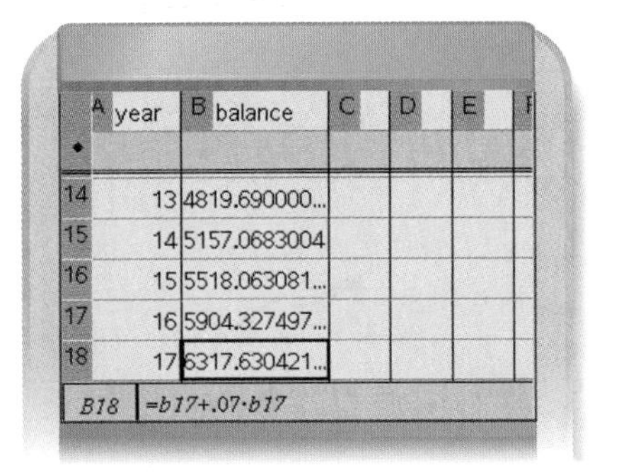

	A year	B balance	C	D	E
14	13	4819.690000...			
15	14	5157.0683004			
16	15	5518.063081...			
17	16	5904.327497...			
18	17	6317.630421...			

B18 =b17+.07·b17

Leaving just $2,000 in the bank at a good interest rate for 11 years can double your money. In another 6 years, the $2,000 will have tripled.

Some banks will compound the interest monthly. You can write the common ratio as $\left(1 + \frac{0.07}{12}\right)$ to represent one-twelfth of the annual interest, compounding monthly. When you do this, n represents months instead of years. How would you change the rule to show that the interest is compounded 52 times per year? What would n represent in this situation?

EXERCISES

Practice Your Skills

1. Find the common ratio for each sequence, and identify the sequence as growth or decay. Give the percent change for each.

a. 100, 150, 225, 337.5, 506.25, . . . @

b. 73.4375, 29.375, 11.75, 4.7, 1.88, . . .

c. 80.00, 82.40, 84.87, 87.42, 90.04, . . .

d. 208.00, 191.36, 176.05, 161.97, . . .

2. Write a recursive formula for each sequence in Exercise 1. Use u_0 for the first term and find u_{10}. @

3. Write each sequence or formula as described.

a. Write the first four terms of the sequence that begins with 2000 and has the common ratio 1.05. @

b. Write the first four terms of the sequence that begins with 5000 and decays 15% with each term. What is the common ratio?

c. Write a recursive formula for the sequence that begins 1250, 1350, 1458, 1574.64, ...

Films quickly display a sequence of photographs, creating an illusion of motion.

4. Match each recursive rule to a graph. Explain your reasoning.

A. $u_0 = 10$
$u_n = (1 - 0.25) \cdot u_{n-1}$ where $n \geq 1$ (@)

B. $u_0 = 10$
$u_n = (1 + 0.25) \cdot u_{n-1}$ where $n \geq 1$

C. $u_0 = 10$
$u_n = 1 \cdot u_{n-1}$ where $n \geq 1$

i.
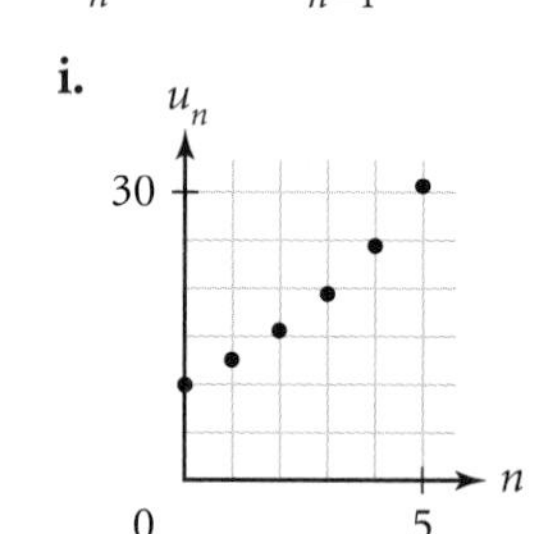

ii.
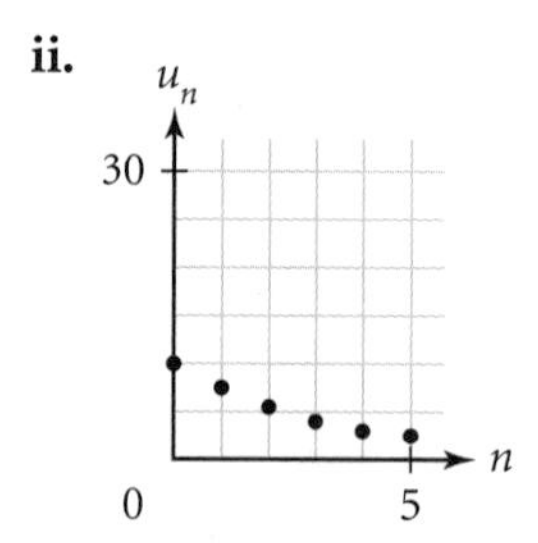

iii.
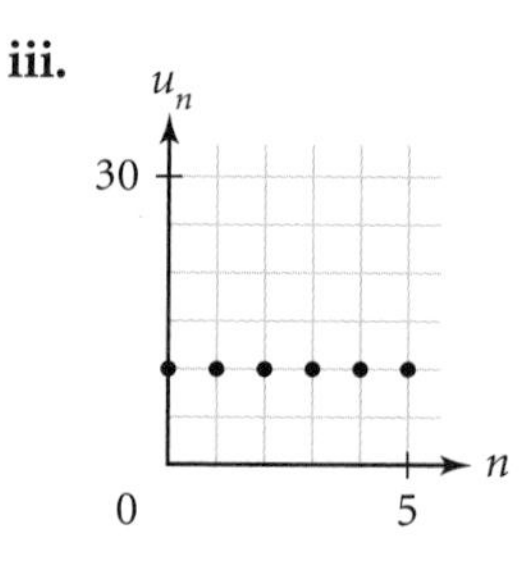

5. Factor these expressions so that the variable appears only once. For example, $x + 0.05x$ factors into $x(1 + 0.05)$.

a. $x + Ax$ (@)
b. $A - 0.18A$ (@)
c. $x + 0.08125x$
d. $2u_{n-1} - 0.85u_{n-1}$

Reason and Apply

6. Suppose the initial height from which a rubber ball drops is 100 in. The rebound heights to the nearest inch are 80, 64, 51, 41, ...

a. What is the rebound ratio for this ball? (h)
b. What is the height of the tenth rebound?
c. After how many bounces will the ball rebound less than 1 in.? Less than 0.1 in.?

7. Suppose the recursive formula $u_0 = 100$ and $u_n = (1 - 0.20)u_{n-1}$ where $n \geq 1$ models a bouncing ball. Give real-world meanings for the numbers 100 and 0.20.

8. Suppose the recursive formula $u_{2008} = 250{,}000$ and $u_n = (1 + 0.025)u_{n-1}$ where $n \geq 2009$ describes an investment made in the year 2008. Give real-world meanings for the numbers 250,000 and 0.025, and find u_{2012}. (@)

9. **APPLICATION** A company with 12 employees is growing at a rate of 20% per year. It will need to hire more employees to keep up with the growth, assuming its business keeps growing at the same rate.

a. How many people should the company plan to hire in each of the next 5 years?
b. How many employees will it have 5 years from now?

10. **APPLICATION** The table below shows investment balances over time.

Elapsed time (yr)	0	1	2	3	. . .
Balance ($)	2,000	2,170	2,354.45	2,554.58	. . .

a. Write a recursive formula that generates the balances in the table. (@)
b. What is the annual interest rate?
c. How many years will it take before the original deposit triples in value?

11. APPLICATION Suppose you deposit $500 into an account that earns 6% annual interest. You don't withdraw or deposit any additional money for 3 years.

a. If the interest is paid once per year, what will the balance be after 3 years? ⓐ

b. If the interest is paid every six months, what will the balance be after 3 years? This is also referred to as 6% compounded semiannually. Divide the annual interest rate by 2 to find the semiannual interest rate.

c. What will the balance be if you receive 6% compounded *quarterly* for 3 years? ⓗ

d. What will the balance be if you receive 6% compounded *monthly* for 3 years?

12. APPLICATION Suppose $500 is deposited into an account that earns 6.5% annual interest and no more deposits or withdrawals are made.

a. If the interest is compounded monthly, what is the monthly rate?

b. What is the balance after 1 month?

c. What is the balance after 1 year?

d. What is the balance after 29 months?

13. Suppose Jill's biological family tree looks like the diagram at right. You can model recursively the number of people in each generation.

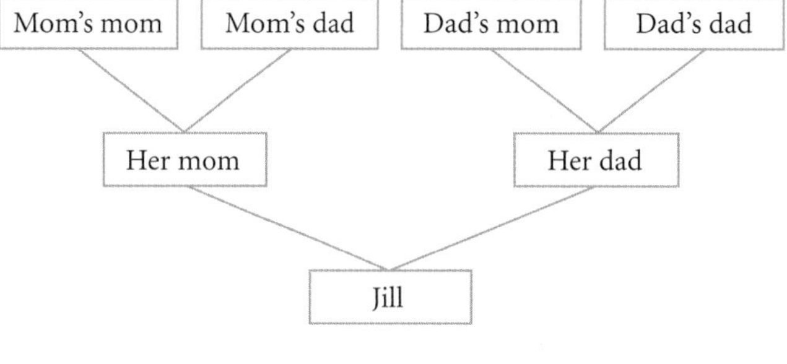

a. Make a table showing the number of Jill's ancestors in each of the past five generations. Use u_0 to represent Jill's generation.

b. Look in your table at the sequence of the number of ancestors. Describe how to find u_n if you know u_{n-1}. Write a recursive formula.

c. Find the number of the term of this sequence that is closest to 1 billion. What is the real-world meaning of this answer?

d. If a new generation is born every 25 years, approximately when did Jill have 1 billion living ancestors in the same generation?

e. Your answer to 13c assumes there are no duplicates, that is, no common ancestors on Jill's mom's and Jill's dad's sides of the family. Look up Earth's population for the year you found in 13d. You will find helpful links at www.keymath.com/DAA. Describe any problems you notice with the assumption of no common ancestors.

Cultural CONNECTION

Family trees are lists of family descendants and are used in the practice of genealogy. People who research genealogy may want to trace their family's medical history or national origin, discover important dates, or simply enjoy it as a hobby. Alex Haley's 1976 genealogical book, *Roots: The Saga of an American Family,* told the powerful history of his family's prolonged slavery and decades of discrimination. The book, along with the 1977 television miniseries, inspired many people to trace their family lineage.

Alex Haley (1921–1992)

14. APPLICATION Carbon dating is used to find the age of ancient remains of once-living things. Carbon-14 is found naturally in all living things, and it decays slowly after death. About 11.45% of it decays in each 1000-year period of time.

Let 100%, or 1, be the beginning amount of carbon-14. At what point will less than 5% remain? Write the recursive formula you used.

At an excavation site in Alberta, Canada, these scientists uncover the remains of an Albertosaurus (a relative of the Tyrannosaurus) about 65–70 million years old.

15. APPLICATION Between 1970 and 2000, the population of Grand Traverse County in Michigan grew from 39,175 to 77,654.

a. Find the percent increase over the 30-year period. ⓐ

b. What do you think the *annual* growth rate was during this period? ⓗ

c. Check your answer to 15b by using a recursive formula. Do you get 77,654 people after 30 years? Explain why your recursive formula may not work.

d. Use guess-and-check to find a growth rate, to the nearest 0.1% (or 0.001), that comes closest to producing the 30-year growth experienced.

e. Use your answer to 15d to estimate the population in 1985. How does your estimate compare with the average of the populations of 1970 and 2000? Explain your results.

16. Taoufik looks at the second problem of his wet homework that fell in a puddle.

a. What is the common ratio? How did you find it?

b. What are the missing terms?

c. What is the answer he needs to find?

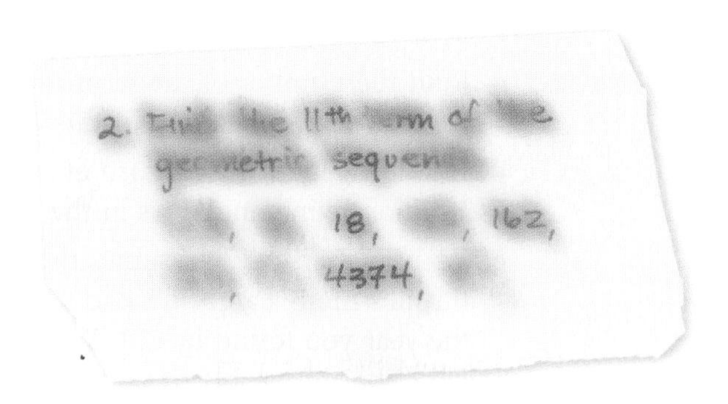

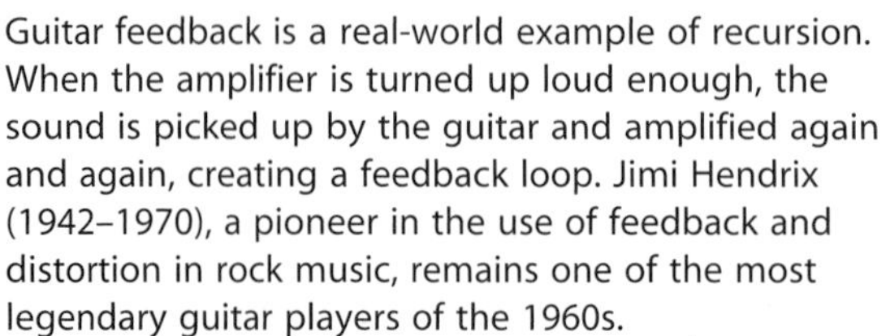

Guitar feedback is a real-world example of recursion. When the amplifier is turned up loud enough, the sound is picked up by the guitar and amplified again and again, creating a feedback loop. Jimi Hendrix (1942–1970), a pioneer in the use of feedback and distortion in rock music, remains one of the most legendary guitar players of the 1960s.

Review

17. The population of the United States grew 13.20% from 1990 to 2000. The population reported in the 2000 census was 281.4 million. What population was reported in 1990? Explain how you found this number.

18. An elevator travels at a nearly constant speed from the ground level to an observation deck at 160 m. This trip takes 40 s. The elevator's trip back down is also at this same constant speed.

a. What is the elevator's speed in meters per second?

b. How long does it take the elevator to reach the restaurants, located 40 m above ground level? @

c. Graph the height of the elevator as it moves from ground level to the observation deck.

d. Graph the height of the elevator as it moves from the restaurant level, at 40 m, to the observation deck.

e. Graph the height of the elevator as it moves from the deck to ground level.

The CN Tower in Toronto is one of Canada's landmark structures and one of the world's tallest buildings. Built in 1976, it has six glass-fronted elevators that allow you to view the landscape as you rise above it at 15 mi/h. At 1136 ft, you can either brace against the wind on the outdoor observation deck or test your nerves by walking across a 256 ft^2 glass floor with a view straight down.

19. Consider the sequence 180, 173, 166, 159, ...

a. Write a recursive formula. Use $u_1 = 180$.

b. What is u_{10}?

c. What is the first term with a negative value?

20. Solve each equation.

a. $-151.7 + 3.5x = 0$ @

b. $0.88x + 599.72 = 0$

c. $18.75x - 16 = 0$

d. $0.5 \cdot 16 + x = 16$

21. For the equation $y = 47 + 8x$, find the value of y when

a. $x = 0$

b. $x = 1$ @

c. $x = 5$

d. $x = -8$

LESSON

1.3 A First Look at Limits

A mathematician is a machine for turning coffee into theorems.

PAUL ERDÖS

Increasing arithmetic and geometric sequences, such as the number of new triangles at each stage in a Sierpiński triangle or the balance of money earning interest in the bank, have terms that get larger and larger. But can a tree continue to grow larger year after year? Can people continue to build taller buildings, run faster, and jump higher, or is there a limit to any of these?

How large can a tree grow? It depends partly on environmental factors such as disease and climate. Trees have mechanisms that slow their growth as they age, similar to human growth. (Unlike humans, however, a tree may not reach its full growth until 100 yr after it starts growing.) This giant sequoia, the General Sherman Tree in Sequoia National Park, California, is considered to be the world's largest tree. The volume of its trunk is over 52,500 ft^3.

The women's world record for the fastest time in the 100 m dash has decreased by about 3 s in 66 yr. Marie Mejzlíková (Czechoslovakia) set the record at 13.6 s in 1922, and Florence Griffith-Joyner (USA), shown at right, set it at 10.49 s in 1988. In the 1998 article "How Good Can We Get?" Jonas Mureika predicts that the ultimate performance for a woman in the 100 m dash will be 10.15 s.

Decreasing sequences may also have a limit. For example, the temperature of a cup of hot cocoa as it cools, taken at one-minute intervals, produces a sequence that approaches the temperature of the room. In the long run, the hot cocoa will be at room temperature.

In the next investigation you will explore what happens to a sequence in the long run.

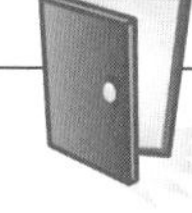

Investigation

Doses of Medicine

You will need

- a bowl
- a supply of water
- a supply of tinted liquid
- measuring cups, graduated in milliliters
- a sink or waste bucket

Our kidneys continuously filter our blood, removing impurities. Doctors take this into account when prescribing the dosage and frequency of medicine.

In this investigation you will simulate what happens in the body when a patient takes medicine. To represent the blood in a patient's body, use a bowl containing a total of 1 liter (L) of liquid. Start with 16 milliliters (mL) of tinted liquid to represent a dose of medicine in the blood, and use clear water for the rest.

Step 1 Suppose a patient's kidneys filter out 25% of this medicine each day. To simulate this, remove $\frac{1}{4}$, or 250 mL, of the mixture from the bowl and replace it with 250 mL of clear water to represent filtered blood. Make a table like the one below, and record the amount of medicine in the blood over several days. Repeat the simulation for each day.

Day	Amount of medicine (mL)
0	16
1	
2	
3	

Step 2 Write a recursive formula that generates the sequence in your table.

Step 3 How many days will pass before there is less than 1 mL of medicine in the blood?

Step 4 Is the medicine ever completely removed from the blood? Why or why not?

Step 5 Sketch a graph and describe what happens in the long run.

A single dose of medicine is often not enough to treat a patient's condition. Doctors prescribe regular doses to produce and maintain a high enough level of medicine in the body. Next you will modify your simulation to look at what happens when a patient takes medicine daily over a period of time.

Step 6 Start over with 1 L of liquid. Again, all of the liquid is clear water, representing the blood, except for 16 mL of tinted liquid to represent the initial dose of medicine. Each day, 250 mL of liquid is removed and replaced with 234 mL of clear water and 16 mL of tinted liquid to represent a new dose of medicine. Complete another table like the one in Step 1, recording the amount of medicine in the blood over several days.

Step 7 Write a recursive formula that generates this sequence.

Step 8 Do the contents of the bowl ever turn into pure medicine? Why or why not?

Step 9 Sketch a graph and explain what happens to the level of medicine in the blood after many days.

The elimination of medicine from the human body is a real-world example of a dynamic, or changing, system. The cleanup of a contaminated lake is another real-world example. Dynamic systems often reach a point of stability in the long run. The quantity associated with that stability, such as the number of milliliters of medicine, is called a **limit.** Mathematically, we say that the sequence of numbers associated with the system approaches that limit. Being able to predict limits is very important for analyzing real-world situations like these. The long-run value helps you estimate limits.

Environmental CONNECTION

The Cuyahoga River in Cleveland, Ohio, caught fire several times during the 1950s and 1960s because its water was so polluted with volatile chemicals. The events inspired several clean water acts in the 1970s and the creation of the federal Environmental Protection Agency. After testing the toxic chemicals present in the water and locating possible sources of contamination, environmental engineers established pollution control levels and set standards for monitoring waste from local industries. Cleanup efforts are ongoing, and the river has been designated an Area of Concern by the Environmental Protection Agency.

This photograph shows a fire on the Cuyahoga River in 1952.

Each of the sequences in the investigation approached different long-run values. The first sequence approached zero. The second sequence was shifted and it approached a nonzero value. A **shifted geometric sequence** includes an added term in the recursive rule. Let's look at another example of a shifted geometric sequence.

EXAMPLE

Antonio and Deanna are working at the community pool for the summer. They need to provide a "shock" treatment of 450 grams (g) of dry chlorine to prevent the growth of algae in the pool. Then they add 45 g of chlorine each day after the initial treatment. Each day, the sun burns off 15% of the chlorine. Find the amount of chlorine after 1 day, 2 days, and 3 days. Create a graph that shows the chlorine level after several days and in the long run.

▶ ***Solution***

The starting value is given as 450. This amount decays by 15% a day, but 45 g is also added each day. The amount remaining after each day is generated by the rule $u_n = (1 - 0.15)u_{n-1} + 45$, or $u_n = 0.85u_{n-1} + 45$. Use this rule to find the chlorine level in the long run.

$u_0 = 450$ The initial shock treatment.

$u_1 = 0.85(450) + 45 = 427.5$ The amount after 1 day.

$u_2 = 0.85(427.5) + 45 \approx 408.4$ The amount after 2 days.

$u_3 = 0.85(408.4) + 45 \approx 392.1$ The amount after 3 days.

To find the long-run value of the amount of chlorine, you can continue evaluating terms until the value stops changing, or see where the graph levels off. From the graph, the long-run value appears to be 300 g of chlorine.

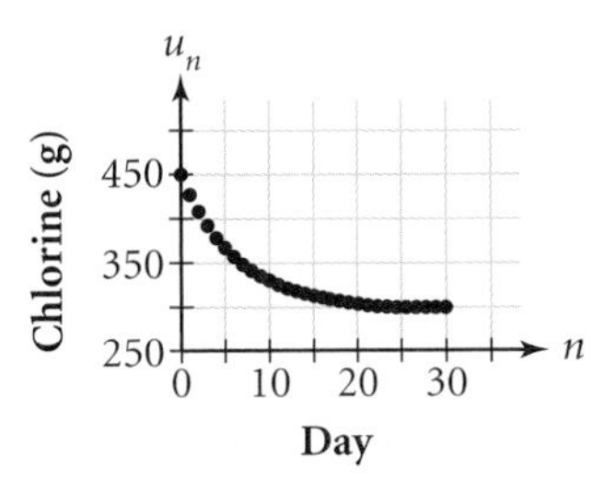

You can also use algebra to find the value of the terms as they level off. The key is to assume that the terms stop changing. Then you can set the value of the next term equal to the value of the previous term and solve the equation.

$u_n = 0.85u_{n-1} + 45$	Recursive rule.
$c = 0.85c + 45$	Assign the same variable to u_n and u_{n-1}.
$0.15c = 45$	Subtract $0.85c$ from both sides.
$c = 300$	Divide both sides by 0.15.

The amount of chlorine will level off at 300 g, which agrees with the long-run value estimated from the graph.

The study of limits is an important part of calculus, the mathematics of change. Understanding limits mathematically will give you a chance to work with other real-world applications in biology, chemistry, physics, and social science.

EXERCISES

Practice Your Skills

1. Find the values of u_1, u_2, and u_3. Identify the type of sequence (arithmetic, geometric, or shifted geometric) and tell whether it is increasing or decreasing.

a. $u_0 = 16$
$u_n = (1 - 0.05)u_{n-1} + 16$ where $n \geq 1$ ⓐ

b. $u_0 = 800$
$u_n = (1 - 0.05)u_{n-1} + 16$ where $n \geq 1$

c. $u_0 = 50$
$u_n = (1 - 0.10)u_{n-1}$ where $n \geq 1$

d. $u_0 = 40$
$u_n = (1 - 0.50)u_{n-1} + 20$ where $n \geq 1$ ⓐ

2. Solve each equation.

a. $a = 210 + 0.75a$ **b.** $b = 0.75b + 300$ ⓐ **c.** $d = 0.75d$ ⓐ **d.** $c = 210 + c$

3. Find the long-run value for each sequence in Exercise 1. ⓗ

4. Write a recursive formula for each sequence.

a. 200.00, 216.00, 233.28, 251.94, . . .

b. 0, 10, 15, 17.5, 18.75, . . . ⓐ

c. The first term is 400. The next term is the previous term increased by 5% and then decreased by 20.

Reason and Apply

5. The Osbornes have a small pool and are doing a chlorine treatment. The recursive formula below gives the pool's daily amount of chlorine in grams.

$u_0 = 300$

$u_n = (1 - 0.15)u_{n-1} + 30 \quad \text{where } n \geq 1$

a. Explain the real-world meanings of the values 300, 0.15, and 30 in this formula. ⓐ

b. Describe what happens to the chlorine level in the long run.

6. APPLICATION On October 2, 2008, Sal invested \$24,000 in a bank account earning 3.4% annually, compounded monthly. On November 1, one month's interest was added to his account. The next day Sal withdrew \$100. He continued to withdraw \$100 on the second day of every month after that. He records his account balance in a table.

a. Find the missing values in the table.

b. Write a recursive formula for the balances on the second day of each month. Start with $a_0 = 24{,}000$. ⓐ

c. What is the value of a_4? What is the meaning of this value? ⓐ

d. What will the balance be on October 2, 2009? On October 2, 2011?

Date	Balance (\$)
Oct 2, 2008	24,000.00
Nov 1, 2008	24,068.00
Nov 2, 2008	23,968.00
Dec 1, 2008	
Dec 2, 2008	23,935.91
Jan 1, 2009	
Jan 2, 2009	

7. APPLICATION Consider the bank account in Exercise 6.

a. What happens to the balance if the same interest and withdrawal patterns continue for a long time? Does the balance ever level off?

b. What monthly withdrawal amount would maintain a constant balance of \$24,000 in the long run?

8. APPLICATION The Forever Green Nursery owns 7000 white pine trees. Each year, the nursery plans to sell 12% of its trees and plant 600 new ones.

a. Find the number of pine trees owned by the nursery after 10 years.

b. Find the number of pine trees owned by the nursery after many years, and explain what is occurring.

c. What equation can you solve to find the number of trees in the long run? ⓐ

d. Try different starting totals in place of the 7000 trees. Describe any changes to the long-run value.

e. In the fifth year, a disease destroys many of the nursery's trees. How does the long-run value change?

9. APPLICATION Jack takes a capsule containing 20 milligrams (mg) of a prescribed allergy medicine early in the morning. By the same time a day later, 25% of the medicine has been eliminated from his body. Jack doesn't take any more medicine, and his body continues to eliminate 25% of the remaining medicine each day. Write a recursive formula for the daily amount of this medicine in Jack's body. When will there be less than 1 mg of the medicine remaining in his body? ⓐ

10. Consider the last part of the Investigation Doses of Medicine. If you double the amount of medicine taken each time from 16 mL to 32 mL, but continue to filter only 250 mL of liquid, will the limit of the concentration be doubled? Explain.

11. Suppose square *ABCD* with side length 8 in. is cut from paper. Another square, *EFGH,* is placed with its corners at the midpoints of *ABCD,* as shown. A third square is placed with its corners at midpoints of *EFGH,* and so on.

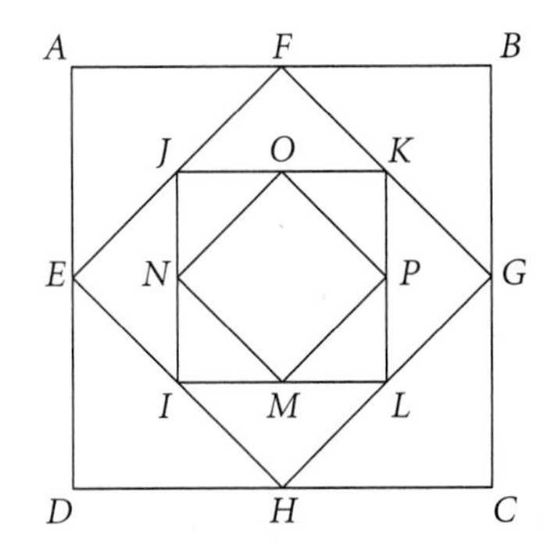

a. What is the perimeter of the ninth square? ⓗ

b. What is the area of the ninth square?

c. What happens to the ratio of perimeter to area as the squares get smaller?

Review

12. Assume two terms of a sequence are $u_3 = 16$ and $u_4 = 128$.

a. Find u_2 and u_5 if the sequence is arithmetic. **b.** Find u_2 and u_5 if the sequence is geometric.

13. A park biologist estimates the moose population in a national park over a four-year period of mild winters. She makes this table.

Year	Estimated number of moose
2004	760
2005	835
2006	920
2007	1010

a. Write a recursive formula that approximately models the growth in the moose population for this four-year period.

b. The winter of 2008 was particularly severe, and the park biologist has predicted a decline of 10% to 15% in the moose herd. What is the range of moose population she predicts for 2008?

14. If a rubber ball rebounds to 97% of its height with each bounce, how many times will it bounce before it rebounds to half its original height? ⓐ

IMPROVING YOUR VISUAL THINKING SKILLS

Think Pink

You have two 1-gallon cans. One contains 1 gallon of white paint, and the other contains 3 quarts (qt) of red paint. (There are 4 quarts per gallon.) You pour 1 qt of white paint into the red paint, mix it, and then pour 1 qt of the mixture back into the can of white paint. What is the red-white content of each can now? If you continually repeat the process, when will the two cans be the same shade of pink?

LESSON

1.4 Graphing Sequences

By looking for numerical patterns, you can write a recursive formula that generates a sequence of numbers quickly and efficiently. You can use graphs to help you identify patterns in a sequence.

Investigation

Match Them Up

Match each table with a recursive formula and a graph that represent the same sequence. Think about similarities and differences between the sequences and how those similarities and differences affect the tables, formulas, and graphs.

1.

n	u_n
0	8
1	4
3	1
6	0.125
9	0.015625

2.

n	u_n
0	0.5
1	1
2	2
3	4
4	8

3.

n	u_n
0	−2
1	1
2	2.5
4	3.625
5	3.8125

4.

n	u_n
0	−2
2	2
5	8
7	12
10	18

5.

n	u_n
0	8
1	6
3	2
5	−2
7	−6

6.

n	u_n
0	−4
1	−4
2	−4
4	−4
8	−4

A. $u_0 = 8$
$u_n = u_{n-1} - 2 \quad \text{where } n \geq 1$

B. $u_0 = 8$
$u_n = 0.5u_{n-1} \quad \text{where } n \geq 1$

C. $u_0 = 0.5$
$u_n = 2u_{n-1} \quad \text{where } n \geq 1$

D. $u_0 = -2$
$u_n = u_{n-1} + 2 \quad \text{where } n \geq 1$

E. $u_0 = -4$
$u_n = u_{n-1} \quad \text{where } n \geq 1$

F. $u_0 = -2$
$u_n = 0.5u_{n-1} + 2 \quad \text{where } n \geq 1$

i.

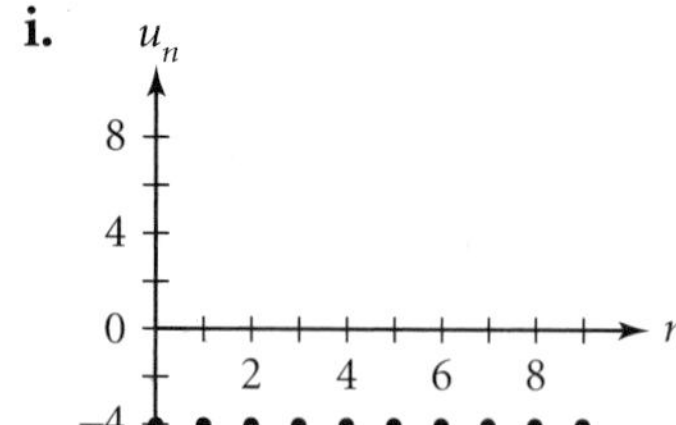

ii.

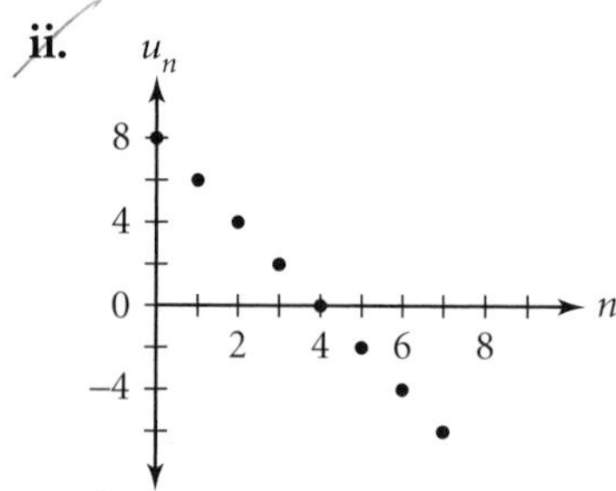

iii.

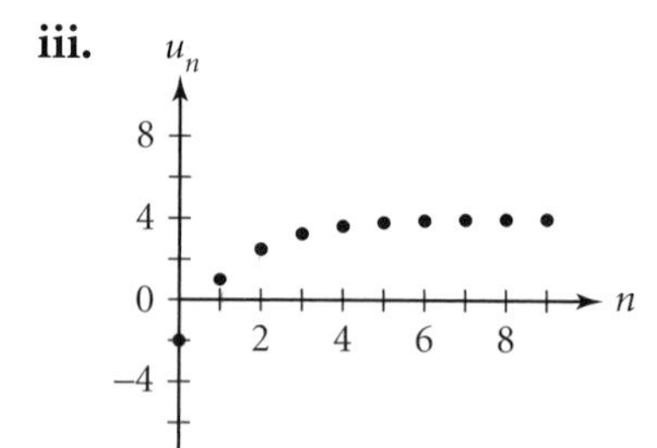

iv.

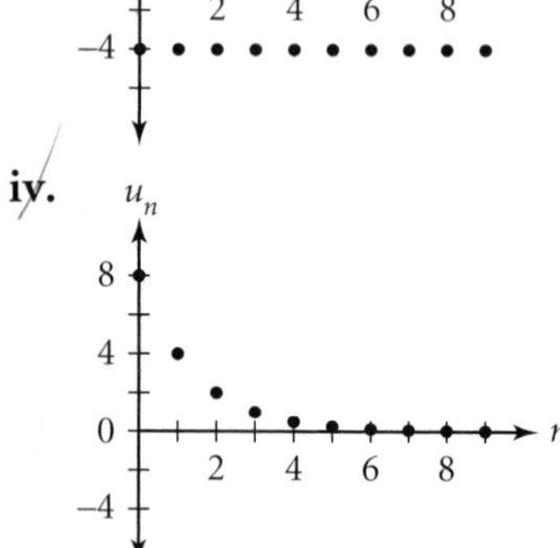

v.

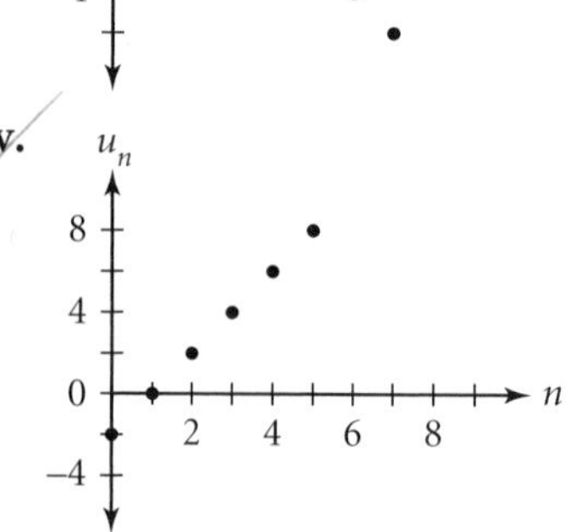

vi.

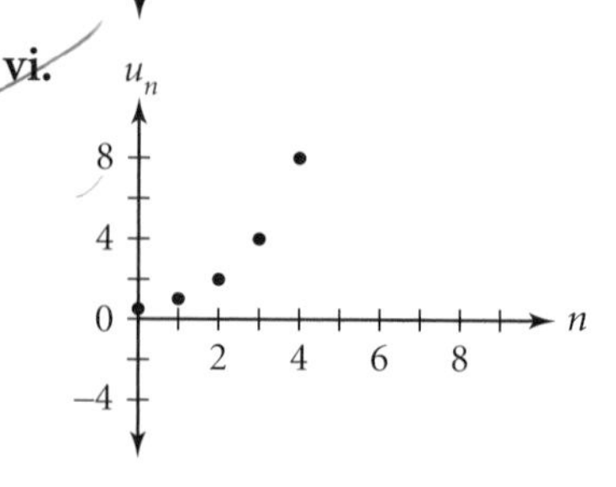

Write a paragraph that summarizes the relationships between different types of sequences, recursive formulas, and graphs. What generalizations can you make? What do you notice about the shapes of the graphs created from arithmetic and geometric sequences?

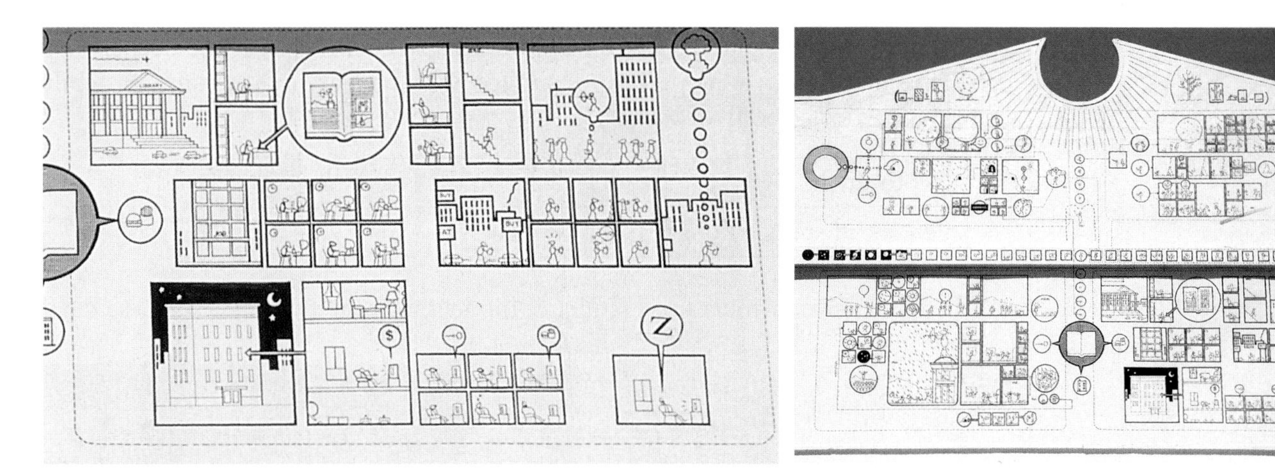

Many cartoons and comic strips show a sequence of events in linear order. The artwork of American artist Chris Ware (b 1967) breaks the convention by showing many sequences intertwined. This complex mural by Ware, at 826 Valencia Street in San Francisco, depicts human development of written and spoken communication.

The general shape of the graph of a sequence's terms gives you an indication of the type of sequence necessary to generate the terms.

The graph at right is a visual representation of the first five terms of the arithmetic sequence generated by the recursive formula

$$u_1 = 1$$
$$u_n = u_{n-1} + 2 \quad \text{where } n \geq 2$$

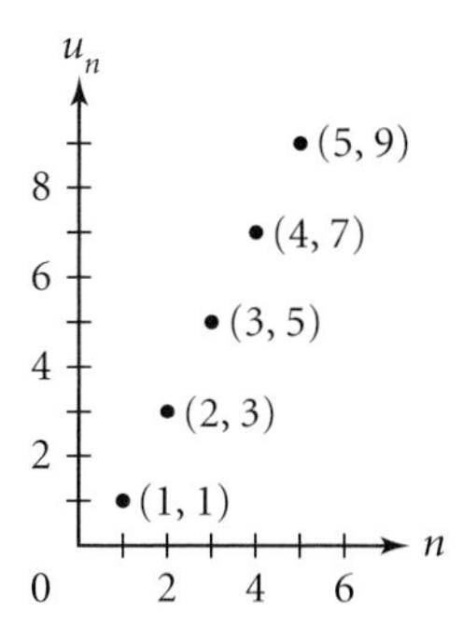

This graph, in particular, appears to be **linear,** that is, the points appear to lie on a line. The common difference, $d = 2$, makes each new point rise 2 units above the previous point.

Graphs of sequences are examples of **discrete graphs,** or graphs made of isolated points. It is incorrect to connect those isolated points with a continuous line or curve because the term number, n, must be a whole number.

The general shape of the graph of a sequence allows you to recognize whether the sequence is arithmetic or geometric. Even if the graph represents data that are not generated by a sequence, you may be able to find a sequence that is a **model,** or a close fit, for the data. The more details you can identify from the graph, the better you will be at fitting a model.

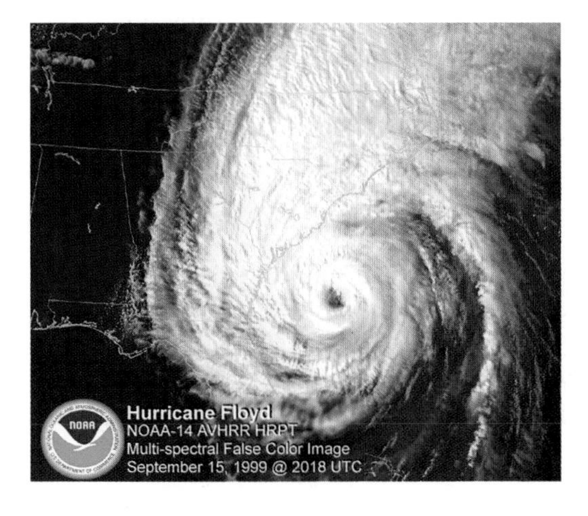

Weather forecasting is one career that relies on mathematical modeling. Forecasters use computers and sophisticated models to monitor changes in the atmosphere. Trends in the data can help predict the trajectory and severity of an impending storm, such as a hurricane.

EXAMPLE

In deep water, divers find that their surroundings become darker the deeper they go. The data here give the percent of surface light intensity that remains at depth n ft in a particular body of water.

Depth (ft)	0	10	20	30	40	50	60	70
Percent of surface light	100	78	60	47	36	28	22	17

Write a recursive formula for a sequence model that approximately fits these data.

Science CONNECTION

Marine life near the ocean's surface relies on organisms that use the sun for photosynthesis. But lifeforms at the bottom of the ocean, where sunlight is virtually absent, feed on waste material or microorganisms that create energy from chemicals released from Earth's crust, a process called chemosynthesis. Squat lobsters and galatheid crabs are among many deep-sea lifeforms that thrive near hydrothermal vents, where Earth's crust releases chemical compounds.

▶ Solution

A graph of the data shows a decreasing, curved pattern. It is not linear, so an arithmetic sequence is not a good model. A geometric sequence with a long-run value of 0 will be a better choice.

The starting value at depth 0 ft is 100% light intensity, so use $u_0 = 100$. The recursive rule should have the form $u_n = (1 - p)u_{n-1}$, but the data are not given for every foot, so you cannot immediately find a common ratio. The ratios between the given values are all approximately 0.77, or $(1 - 0.23)$. Because the light intensity decreases at a rate of 0.23 every 10 feet, it must decrease at a smaller rate every foot. A starting guess of 0.02 gives the model

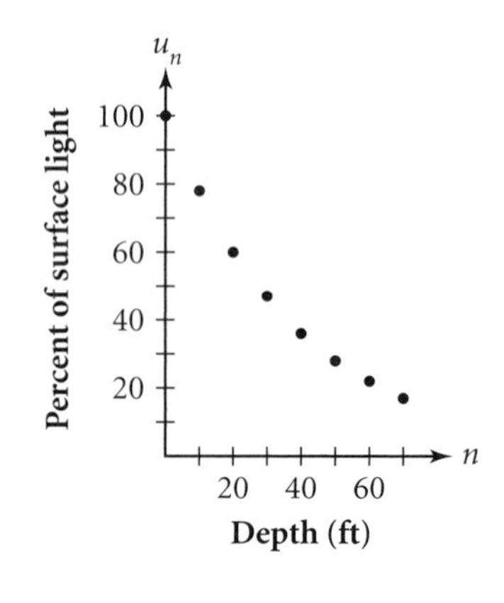

$$u_0 = 100$$
$$u_n = (1 - 0.02)u_{n-1} \quad \text{where } n \geq 1$$

Check this model by graphing the original data and the sequence on your calculator. The graph shows that this model fits only one data point—it does not decay fast enough. [▶ See **Calculator Note 1I** to learn about creating sequences on your calculator and **Calculator Note 1J** to learn about graphing sequences. ◀]

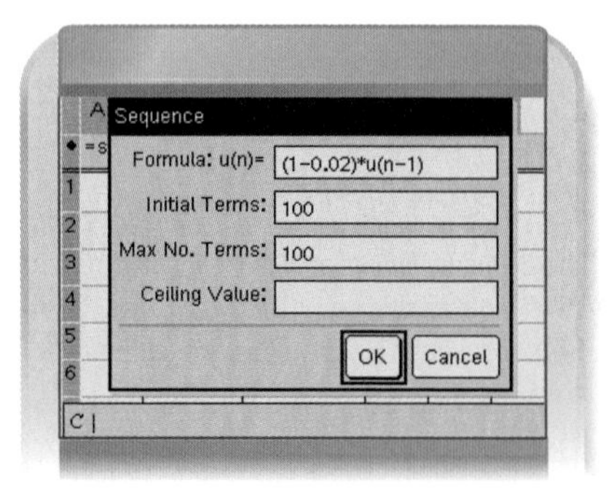

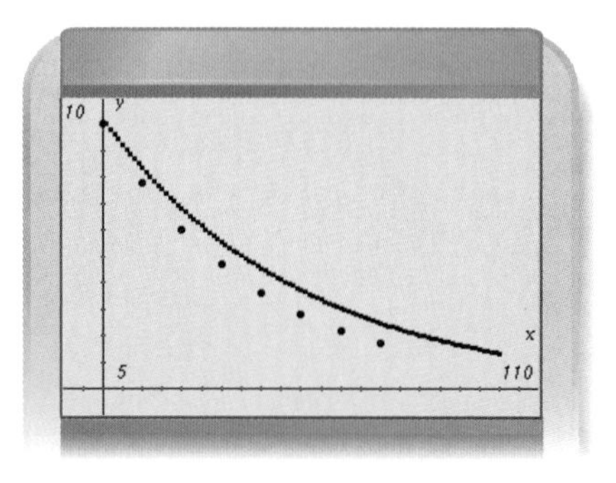

Experiment by increasing the rate of decay. With some trial and error, you can find a model that fits the data better.

$$u_0 = 100$$
$$u_n = (1 - 0.025)u_{n-1} \quad \text{where } n \geq 1$$

Once you have a good sequence model, you can use your calculator to find specific terms. For example, the value of u_{43} means that at depth 43 ft approximately 34% of surface light intensity remains.

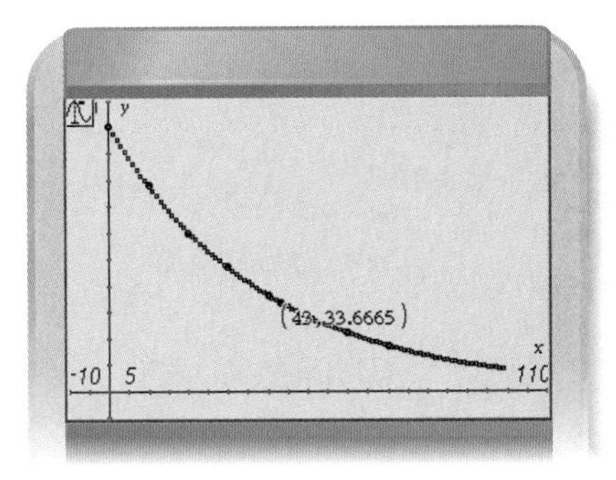

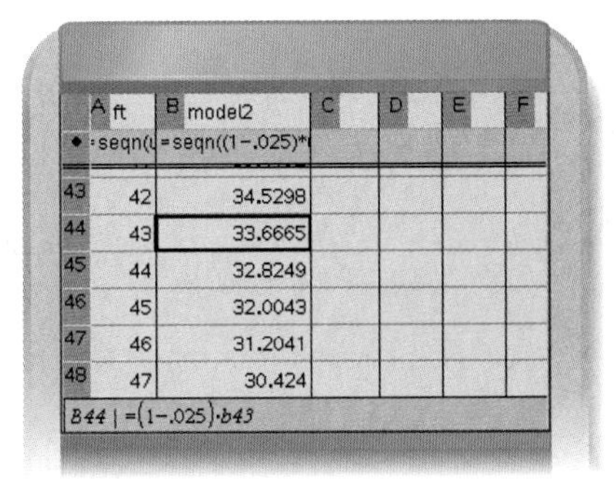

As you see in the calculator screen on the previous page, some calculators use $u(0)$, $u(1)$, $u(2)$, . . . , $u(n-1)$, and $u(n)$ instead of the subscripted notation u_0, u_1, u_2, . . . , u_{n-1}, and u_n. Be aware that $u(5)$ means u_5, not u multiplied by 5. You may also see other variables, such as a_n or v_n, used for recursive formulas. It is important that you are able to make sense of these equivalent mathematical notations and be flexible in reading other people's work.

Being alert also pays off when working with graphs. Graphs help you understand and explain situations, and visualize the mathematics of a situation. When you make a graph or look at a graph, try to find connections between the graph and the mathematics used to create the graph. Consider what variables and units were used on each axis and what the smallest and largest values were for those variables. Sometimes this will be clear and obvious, but sometimes you will need to look at the graph in a new way to see the connections.

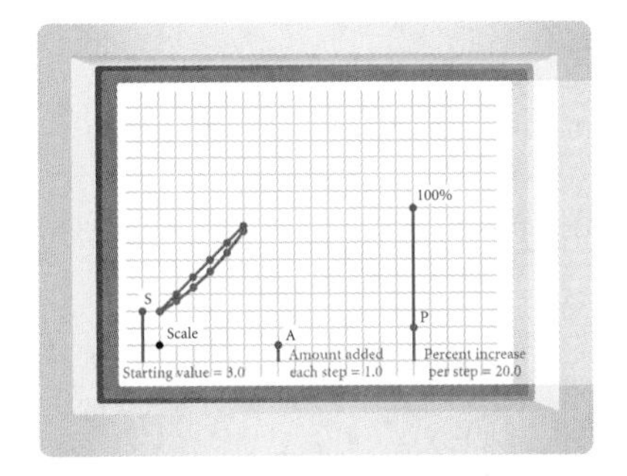

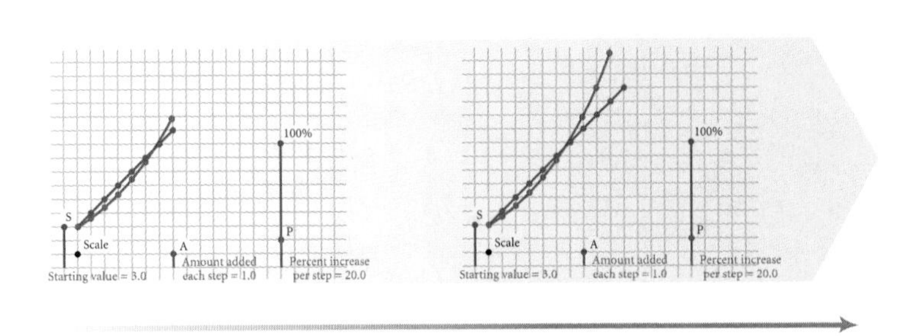

[▶ You can use the **Dynamic Algebra Exploration** Sequence Graphs at www.keymath.com/DAA to explore the graphs of arithmetic and geometric sequences.◀]

keymath.com/DAA

EXERCISES

You will need

A graphing calculator for Exercises **1**, **4**, **6**, **9–13**, and **15**.

Practice Your Skills

1. Suppose you are going to graph the specified terms of these four sequences. For each sequence, what minimum and maximum values of n and u_n would you use on the axes to get a good graph?

a.

n	0	1	2	3	4	5	6	7	8	9
u_n	2.5	4	5.5	7	8.5	10	11.5	13	14.5	16

b. The first 20 terms of the sequence generated by

$u_0 = 400$
$u_n = (1 - 0.18)u_{n-1}$ where $n \geq 1$ @

c. The first 30 terms of the sequence generated by

$u_0 = 25$
$u_n = u_{n-1} - 7$ where $n \geq 1$

d. The first 70 terms of the sequence generated by

$u_0 = 15$
$u_n = (1 + 0.08)u_{n-1}$ where $n \geq 1$ @

2. Match each graph with a formula and identify the sequence as arithmetic or geometric.

A.

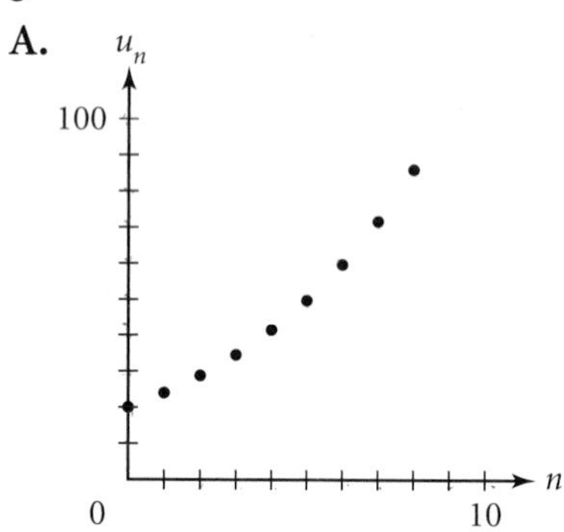

B.

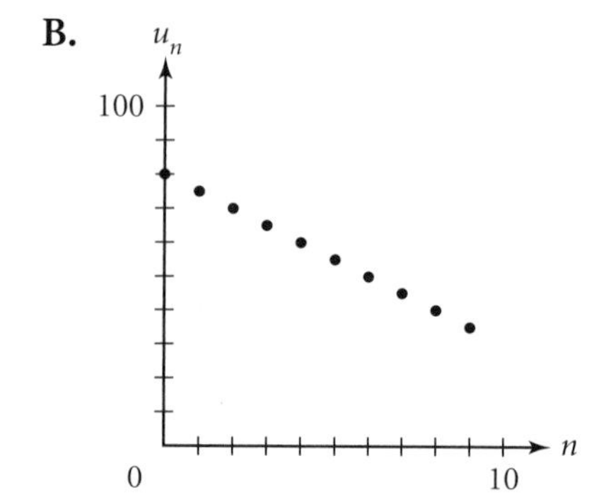

C.

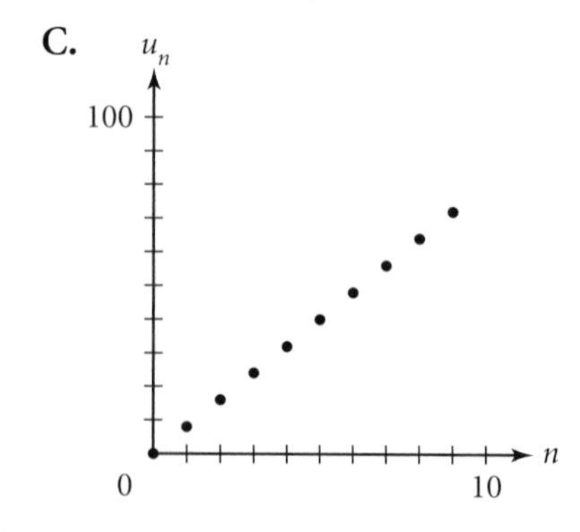

D.

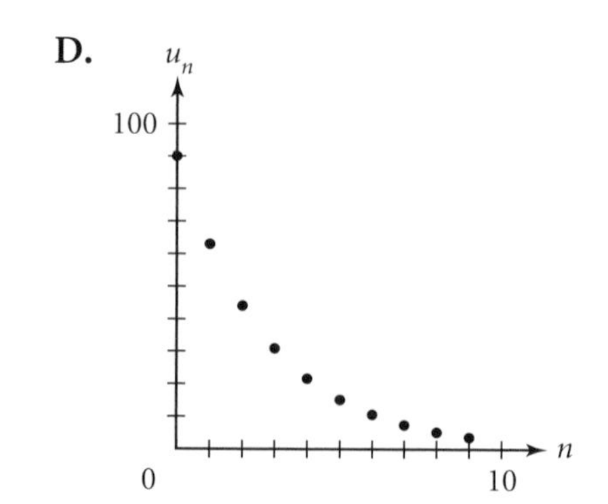

i. $u_0 = 0$
$u_n = u_{n-1} + 8$ where $n \geq 1$

ii. $u_0 = 20$
$u_n = 1.2u_{n-1}$ where $n \geq 1$

iii. $u_0 = 90$
$u_n = 0.7u_{n-1}$ where $n \geq 1$

iv. $u_0 = 80$
$u_n = u_{n-1} - 5$ where $n \geq 1$ @

3. Imagine the graphs of the sequences generated by these recursive formulas. Describe each graph using exactly three of these terms: arithmetic, decreasing, geometric, increasing, linear, nonlinear, shifted geometric.

a. $v_0 = 450$
$v_n = a \cdot v_{n-1}$ where $n \geq 1$ and $0 < a < 1$ @

b. $v_0 = 450$
$v_n = b + v_{n-1}$ where $n \geq 1$ and $b < 0$

c. $v_0 = 450$
$v_n = v_{n-1} \cdot c$ where $n \geq 1$ and $c > 1$

d. $v_0 = 450$
$v_n = v_{n-1} + d$ where $n \geq 1$ and $d > 0$

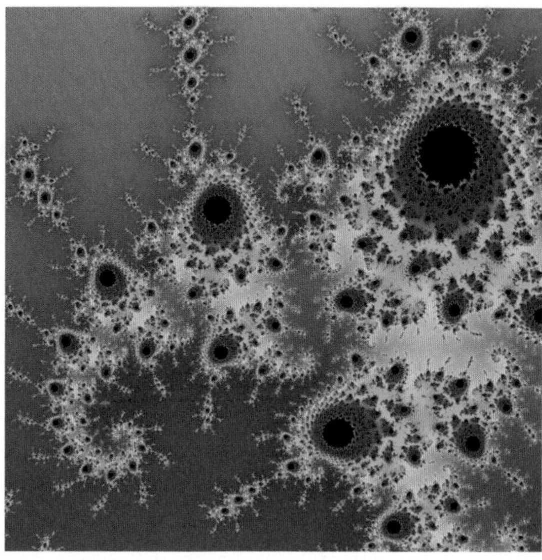
This complex fractal was created by plotting points generated by recursive formulas.

Reason and Apply

4. Consider the recursive rule $u_n = 0.75u_{n-1} + 210$.

a. What is the long-run value of any shifted geometric sequence that is generated by this recursive rule?

b. Sketch the graph of a sequence that is generated by this recursive rule and has a starting value

i. Below the long-run value @
ii. Above the long-run value
iii. At the long-run value

c. Write a short paragraph describing how the long-run value and starting value of each shifted geometric sequence in 4b influence the appearance of the graph.

5. Match each recursive formula with the graph of the same sequence. Give your reason for each choice.

A. $u_0 = 20$
$u_n = u_{n-1} + d$ where $n \geq 1$

B. $u_0 = 20$
$u_n = r \cdot u_{n-1}$ where $n \geq 1$

C. $u_0 = 20$
$u_n = r \cdot u_{n-1} + d$ where $n \geq 1$

i.
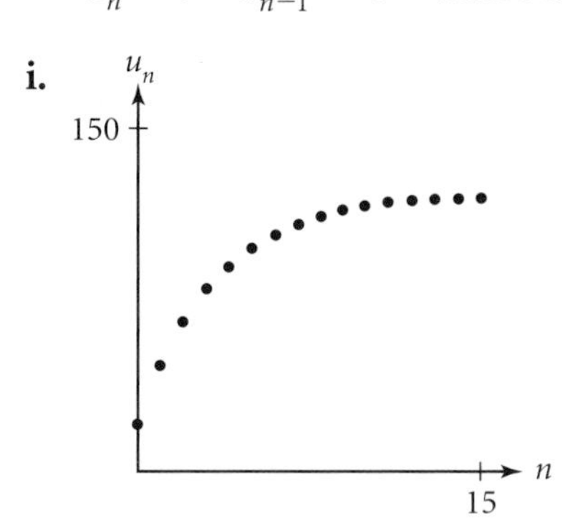

ii.
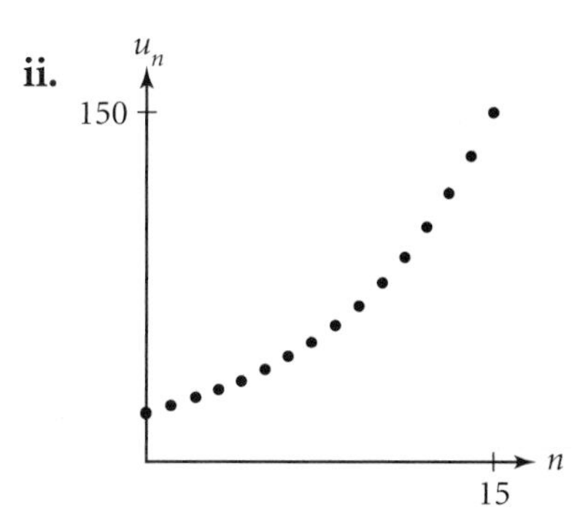

iii.
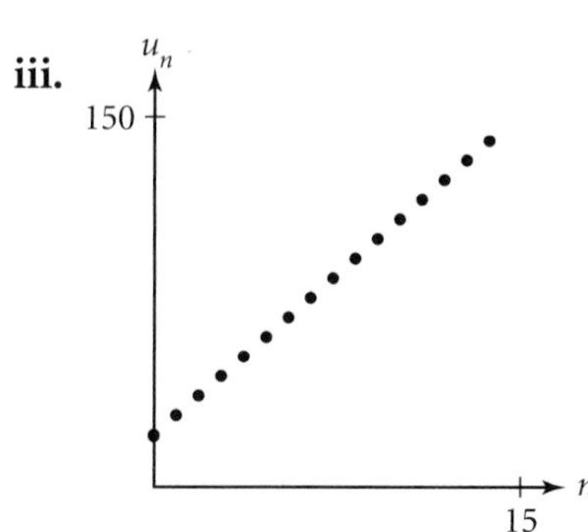

6. Consider the geometric sequence 18, −13.5, 10.125, −7.59375, ... @

   a. Write a recursive formula that generates this sequence.

   b. Sketch a graph of the sequence. Describe how the graph is similar to other graphs that you have seen and also how it is different.

   c. What is the long-run value?

7. Your friend calls on the phone and the conversation goes like this:

   Friend: What does the graph of an arithmetic sequence look like?

   You: Be more specific.

   Friend: Why? Don't they all look the same?

   You: Yes and no.

   Explain for your friend how the graphs of arithmetic sequences are similar and how they vary.

8. Your friend calls back and asks about geometric sequences. Explain how the graphs of geometric sequences are similar and how they vary.

9. APPLICATION The Forever Green Nursery has 7000 white pine trees. Each year, the nursery plans to sell 12% of its trees and plant 600 new ones.

   a. Make a graph that shows the number of trees at the nursery over the next 20 years.

   b. Use the graph to estimate the number of trees in the long run. How does your estimate compare to the long-run value you found in Exercise 8b in Lesson 1.3?

10. APPLICATION The Bayside Community Water District has decided to add fluoride to the drinking water. The ideal concentration of fluoride in drinking water is between 0.7 mg/L and 1.2 mg/L. If the concentration gets higher than 4.00 mg/L, people may suffer health problems. If the concentration is less than 0.7 mg/L, it is too low to promote dental health. Suppose 15% of the fluoride present in the water supply is consumed during a period of one day. Write a sequence and create a graph to analyze each of these scenarios.

A water treatment plant

   a. If the fluoride content begins at 2.00 mg/L and no additional fluoride is added, how long will it be before the concentration is too low to promote dental health? @

   b. If the fluoride content begins at 2.00 mg/L and 0.50 mg/L is added daily, will the concentration increase or decrease? What is the long-run value? Explain your reasoning.

   c. Suppose the fluoride content begins at 2.00 mg/L and 0.10 mg/L is added daily. Describe what happens.

   d. The Water District board members vote that there should be an initial treatment of 2.00 mg/L, but that the long-run fluoride content should be 1.00 mg/L. How much fluoride needs to be added daily for the fluoride content to stabilize at 1.00 mg/L?

11. APPLICATION As the air temperature gets warmer, snowy tree crickets chirp faster. You can actually use a snowy tree cricket's rate of chirping per minute to determine the approximate temperature in degrees Fahrenheit. Use a graph to find a sequence model that approximately fits these data. (h)

Snowy Tree Crickets' Rates of Chirping

Temperature (°F)	50	55	60	65	70	75	80
Rate (chirps/min)	40	60	80	100	120	140	160

Snowy tree crickets are about 0.7 in. long, are pale green, and live in shrubs and bushes. Only male crickets chirp, and they have different chirps for different activities, such as mating and fighting. All species of crickets chirp by rubbing their wings together.

12. When the country of Bhutan was recognized as a monarchy by Britain in 1907, it had a population of 239,000. Modify the population data in the table by subtracting 1907 from the year and 239 from the population. Use a graph to find a sequence that approximately fits the modified data. (h)

Population of Bhutan

Year	Population (thousands)
1960	867
1965	950
1970	1045
1975	1157
1980	1281
1985	1424
1990	1598
1995	1793
2000	2005
2005	2232

(*www.census.gov*)

13. Consider the recursive formula

$$u_0 = 450$$
$$u_n = 0.75u_{n-1} + 210 \quad \text{where } n \geq 1$$

a. Find u_1, u_2, u_3, u_4, and u_5.

b. How can you calculate backward from the value of u_1 to u_0, or 450? In general, what operations can you perform to any term in order to find the value of the previous term?

c. Write a recursive formula that generates the values from 13a in reverse order.

Review

14. Find the value of a that makes each equation true.

a. $47{,}500{,}000 = 4.75 \times 10^a$ **b.** $0.0461 = a \times 10^{-2}$ **c.** $3.48 \times 10^{-1} = a$

15. *Mini-Investigation* For a–c, find the long-run value of the sequence generated by the recursive formula.

a. $u_0 = 50$
$u_n = (1 - 0.30)u_{n-1} + 10 \quad \text{where } n \geq 1$

b. $u_0 = 50$
$u_n = (1 - 0.30)u_{n-1} + 20 \quad \text{where } n \geq 1$ (a)

c. $u_0 = 50$
$u_n = (1 - 0.30)u_{n-1} + 30 \quad \text{where } n \geq 1$

d. Generalize any patterns you notice in your answers to 15a–c. Use your generalizations to find the long-run value of the sequence generated by

$$u_0 = 50$$
$$u_n = (1 - 0.30)u_{n-1} + 70 \quad \text{where } n \geq 1$$

EXPLORATION

Graphs of Sequences

In this chapter you have created sequences of numbers using recursion. In this exploration you will use recursion and **iteration** to create a sketch that allows you to explore the graphs of these sequences. *Iteration* means repeating a process (in this case, arithmetic operations) over and over again.

Activity

Repeat After Me

Step 1 In a new sketch, choose **Graph | Define Coordinate System.** Adjust the scale until the x-axis goes from about -30 to 30 units. To adjust the scale, drag a number on either axis.

Step 2 Use the **Point** tool to construct a point anywhere in the plane, and use the **Text** tool to label it point A. With point A selected, measure its x- and y-coordinates by choosing **Measure | Abscissa (*x*)** and **Measure | Ordinate (*y*).** Choose **Graph | Snap Points** so that point A moves between integer coordinates when you drag it.

Step 3 Without doing any calculations or drawings, imagine adding 1 to both the x- and y-coordinates and plotting the new point. Then imagine adding 1 to both coordinates of the new point to get another new point, and so forth. How would the points be arranged? If you connected them, what shape would they make? You will use Sketchpad to check your predictions.

Step 4 Choose **Measure | Calculate** and click on the measurement of x_A on the screen to enter it into the calculator. Then calculate $x_A + 1$. Follow the same steps to calculate $y_A + 1$. The results will be the coordinates of a new point.

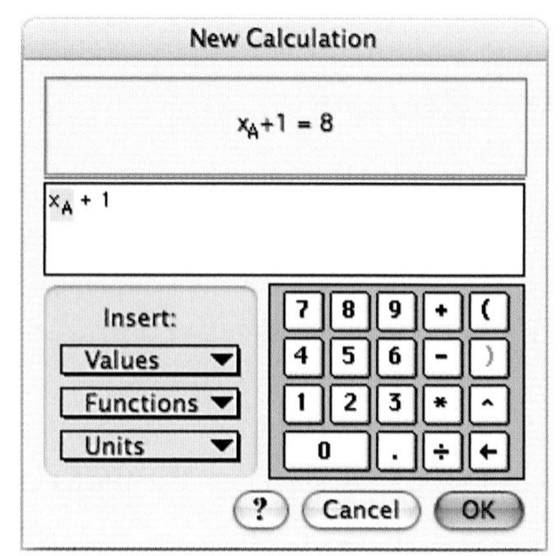

Step 5 Use the **Arrow** tool to select the new x-coordinate $(x_A + 1)$ and the new y-coordinate $(y_A + 1)$ in order. Then choose **Graph | Plot As (*x*, *y*)** to plot the new point. Label the new point A'.

Step 6 Hide the original x_A-coordinate measurement and hide the calculation $x_A + 1$.

Step 7 Now you will use iteration to create more points. Select point A and choose **Transform | Iterate.** You'll get a dialog box that asks you to map the pre-image, point A, to another object. Click point A' in the sketch to indicate that the same operation should next be applied to A'. Click Iterate.

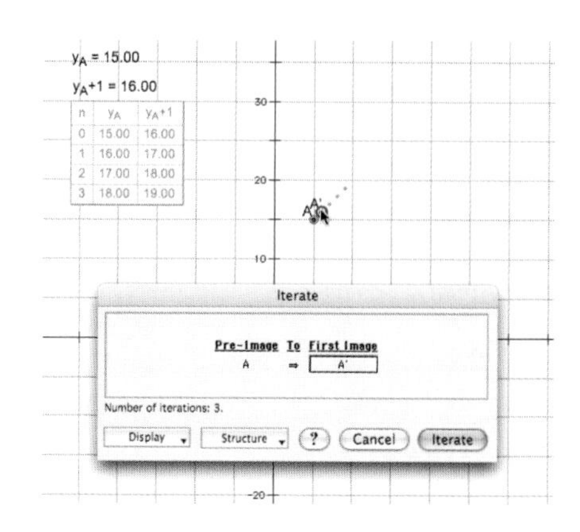

Sketchpad plots several more points by repeating (iterating) the operation that maps point A to point A'. Sketchpad also creates a table showing the calculations for each iterated point.

Step 8 Select one of the iterated points, and press the + key several times. This will increase the number of iterations. Increase it until there are at least ten points. Select point A and choose **Measure | Coordinates.**

Step 9 Drag your initial point A around. Describe the sequence of images generated by this iteration. Is the sequence arithmetic, geometric, or shifted geometric?

Questions

1. Move point A to (0, 2). With $y_0 = 2$, write a recursive formula to generate the y_n-coordinates. *Note:* If you change the scale of the axes or move the axes, you will need to reposition point A.

2. Change the iteration rule $(y_A + 1)$ to $(y_A + 4)$, and then describe how the change affects the table, graph, and sequence formula.

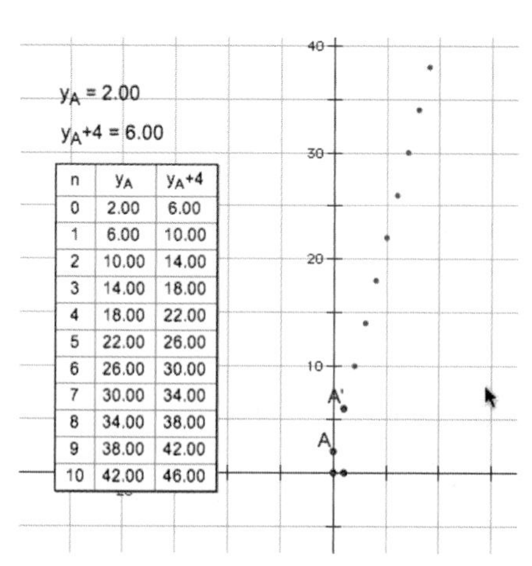

n	y_A	y_A+4
0	2.00	6.00
1	6.00	10.00
2	10.00	14.00
3	14.00	18.00
4	18.00	22.00
5	22.00	26.00
6	26.00	30.00
7	30.00	34.00
8	34.00	38.00
9	38.00	42.00
10	42.00	46.00

3. How would you change your sketch to produce the sequence $u_0 = 1$ and $u_n = 1.5 \cdot u_{n-1}$? What happens to the values in the table? What happens to your graph? What kind of sequence have you created?

4. Using the iteration rule $y_A \cdot 2$, drag point A to the x-axis. Describe the pattern of the iterated points. Drag point A below the x-axis, and then describe the pattern of your sequence.

5. Explore the effect of using different values of B in the iteration rule $y_A \cdot B$. Which values provide geometric sequences with long-range tendency toward the x-axis?

6. Drag point A and describe the pattern of the iterated points generated by the iteration rule $0.5 \cdot y_A + 3$.

LESSON

1.5 Loans and Investments

. . . so little stress is ever laid on the pleasure of becoming an educated person, the enormous interest it adds to life.

EDITH HAMILTON

In life you will face many financial situations, which may include car loans, checking accounts, credit cards, long-term investments, life insurance, retirement accounts, and home mortgages. You will need to make intelligent choices about your money. Fortunately, much of the mathematics is no more complicated than the recursive formula u_0 = principal (starting amount), $u_n = r \cdot u_{n-1} + d$.

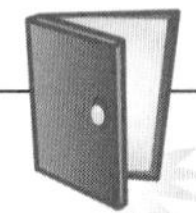

Investigation
Life's Big Expenditures

In this investigation you will use recursion to explore loan balances and payment options. Your calculator will be a helpful tool for trying different sequence models.

Part 1

You plan to borrow \$22,000 from a bank to purchase a new car. You will make a payment every month to the bank to repay the loan, and the loan must be paid off in 5 years (60 months). The bank charges interest at an annual rate of 7.9%, compounded monthly. Part of each monthly payment is applied to the interest, and the remainder reduces the starting balance, or principal.

Step 1 What is the *monthly* interest rate? What is the first month's interest on the \$22,000? If you make a payment of \$300 at the end of the first month, then what is the remaining balance?

Step 2 Record the balances for the first 6 months with monthly payments of \$300. How many months will it take to pay off the loan?

Step 3 Experiment with other values for the monthly payment. What monthly payment allows you to pay off the loan in exactly 60 months?

Step 4 How much do you actually pay for the car using the monthly payment you found in Step 3? (*Hint:* The last payment should be a little less than the other 59 payments.)

Part 2

Use the techniques that you discovered in Part 1 to find the monthly payment for a 30-year home mortgage of \$146,000 with an annual interest rate of 7.25%, compounded monthly. How much do you actually pay for the house?

Consumer CONNECTION

In the third quarter of the year 2007, the National Association of Home Builders reported that the median price for a home in the United States was \$239,000. Mortgage lenders often require monthly payments to be no more than 28% of a family's monthly income. The most affordable place to buy a house was Kokomo, Indiana (median home price of \$97,000), where people with median incomes could afford 90.5% of the homes sold. In contrast, the least affordable houses were in Napa, California, where the median home price was \$585,000, and people earning the median income could afford only 3.3% of the homes sold.

Investments are mathematically similar to loans. With an investment, deposits are added on a regular basis so that your balance increases.

EXAMPLE

Gwen's employer offers an investment plan that invests a portion of each paycheck before taxes are deducted. Gwen gets paid every week. The plan has an annual interest rate of 4.75%, compounded weekly, and she decides to contribute \$10 each week. What will Gwen's balance be after 5 years?

▶ Solution

Gwen's starting balance is \$10. Each week, the previous balance is multiplied by $\left(1 + \frac{0.0475}{52}\right)$ and Gwen adds another \$10. A recursive formula that generates the balance is

$$u_0 = 10$$
$$u_n = \left(1 + \frac{0.0475}{52}\right)u_{n-1} + 10 \quad \text{where } n \geq 1$$

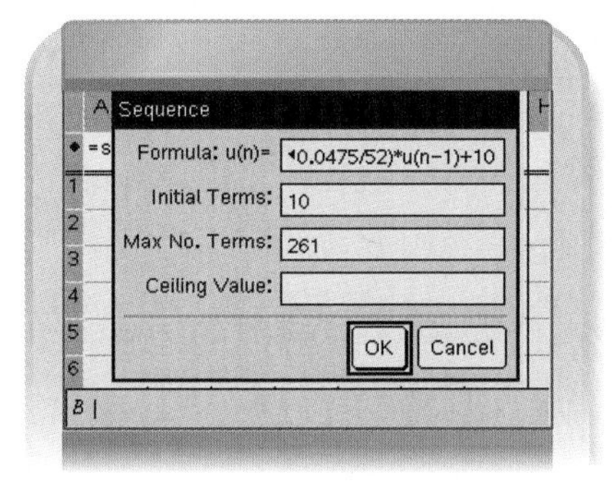

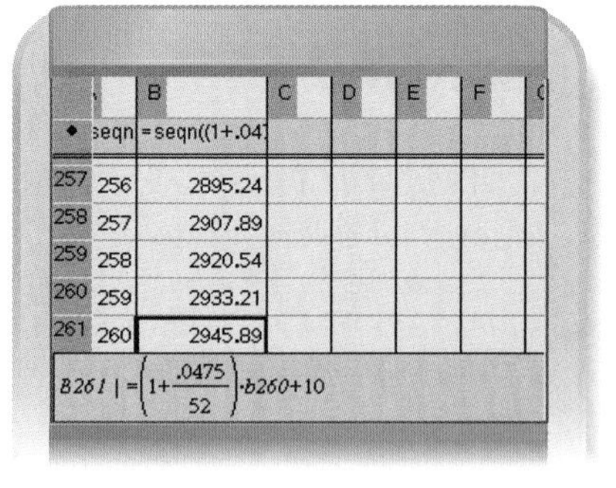

There are 52 weeks in a year and 260 weeks in 5 years. The value of u_{260} shows that the balance after 5 years is \$2,945.89.

As you work on each exercise, look for these important pieces of information: the principal, the deposit or payment amount, the annual interest rate, and the frequency with which interest is compounded. You will be able to solve many financial problems with these values using recursion.

EXERCISES

You will need

A graphing calculator for Exercises **5–12** and **14.**

Practice Your Skills

1. Assume the sequence generated by $u_0 = 450$ and $u_n = (1 + 0.039)u_{n-1} + 50$ where $n \geq 1$ represents a financial situation and n is measured in years. ⓐ

a. Is this a loan or an investment? Explain your reasoning.

b. What is the principal?

c. What is the deposit or payment amount?

d. What is the annual interest rate?

e. What is the frequency with which interest is compounded?

2. Answer the questions in Exercise 1a–e for the sequence generated by $u_0 = 500$ and $u_n = \left(1 + \frac{0.04}{4}\right)u_{n-1} - 25$ where $n \geq 1$. Let n be measured in quarter-years.

3. Find the first month's interest on a \$32,000 loan at an annual interest rate of

a. 4.9% ⓐ **b.** 5.9% ⓐ **c.** 6.9% **d.** 7.9%

4. Write a recursive formula for each financial situation.

a. You borrow \$10,000 at an annual interest rate of 10%, compounded monthly, and each payment is \$300. ⓐ

b. You buy \$7,000 worth of furniture on a credit card with an annual interest rate of 18.75%, compounded monthly. You plan to pay \$250 each month.

c. You invest \$8,000 at 6%, compounded quarterly, and you deposit \$500 every 3 months. (Quarterly means four times per year.)

d. You enroll in an investment plan that deducts \$100 from your monthly paycheck and deposits it into an account with an annual interest rate of 7%, compounded monthly.

Reason and Apply

5. **APPLICATION** Find the balance after 5 years if \$500 is deposited into an account with an annual interest rate of 3.25%, compounded monthly. ⓐ

6. **APPLICATION** Consider a \$1,000 investment at an annual interest rate of 6.5%, compounded quarterly. Find the balance after

a. 10 years **b.** 20 years **c.** 30 years

7. ***Mini-Investigation*** Find the balance of a \$1,000 investment, after 10 years, at an annual interest rate of 6.5% when compounded

a. Annually

b. Monthly

c. Daily (In financial practice, daily means 360 times per year, not 365.)

d. How will the balances compare from investments that have interest rates that are compounded annually, monthly, or daily after the same length of time?

8. APPLICATION Beau and Shaleah each get a $1,000 bonus at work and decide to invest it. Beau puts his money into an account that earns an annual interest rate of 6.5%, compounded yearly. He also decides to deposit $1,200 each year. Shaleah finds an account that earns 6.5%, compounded monthly, and decides to deposit $100 each month. ⓐ

a. Compare the amounts of money that Beau and Shaleah deposit each year. Describe any differences or similarities.

b. Compare the balances of Beau's and Shaleah's accounts over several years. Describe any differences or similarities.

9. APPLICATION Regis deposits $5,000 into an account for his 10-year-old child. The account has an annual interest rate of 8.5%, compounded monthly.

a. What regular monthly deposit amount is needed to make the account worth $1 million by the time the child is 55 years old?

b. Make a graph of the increasing balances.

10. APPLICATION Cici purchased $2,000 worth of merchandise with her credit card this past month. Then she was unexpectedly laid off from her job. She decided to make no more purchases with the card and to make only the minimum payment of $40 each month. Her annual interest rate is 18%, compounded monthly.

a. Find the balance on the credit card over the next 6 months. ⓐ

b. When will Cici pay off the total balance on her credit card?

c. What is the total amount paid for the $2,000 worth of merchandise?

11. APPLICATION Megan Flanigan is a loan officer with L. B. Mortgage Company. She offers a loan of $60,000 to a borrower at 9.6% annual interest, compounded monthly.

a. What should she tell the borrower the monthly payment will be if the loan must be paid off in 25 years? ⓗ

b. Make a graph that shows the unpaid balance over time.

Review

12. The school cafeteria offers a choice of ice cream or frozen yogurt for dessert once a week. During the first week of school, 220 students chose ice cream and only 20 students chose frozen yogurt. During each of the following weeks, 10% of the frozen-yogurt eaters switched to ice cream and 5% of the ice-cream eaters switched to frozen yogurt.

a. Calculate the number of frozen-yogurt eaters who switched to ice cream during the first week.

b. Calculate the number of ice-cream eaters who switched to frozen yogurt during the first week.

c. Assume the cafeteria population remains constant at 240 students. From your answers to 12a and b, find the number of frozen-yogurt eaters after 1 week.

d. How many ice-cream eaters are there after 1 week?

e. How many frozen-yogurt eaters and ice-cream eaters will there be after 3 weeks? After 10 weeks? What will happen in the long run?

Pablo Picasso (1881–1973), *Man with a straw hat and an ice cream cone*, 1938. 61 × 46 cm. Musee Picasso, Paris, France

13. Is this sequence arithmetic, geometric, shifted geometric, or something else? (h)

$$\frac{1}{1}, \frac{1}{2}, \frac{1}{3}, \frac{1}{4}, \ldots$$

14. Consider the geometric sequence generated by

$$a_0 = 4$$
$$a_n = 0.7a_{n-1} \quad \text{where } n \geq 1$$

a. What is the long-run value?

b. What is the long-run value if the common ratio is changed to 1.3?

c. What is the long-run value if the common ratio is changed to 1?

15. Find the value of n in each proportion.

a. $\frac{2.54 \text{ cm}}{1 \text{ in.}} = \frac{n}{12 \text{ in.}}$ **b.** $\frac{1 \text{ km}}{0.625 \text{ mi}} = \frac{n}{200 \text{ mi}}$ **c.** $\frac{1 \text{ yd}}{0.926 \text{ m}} = \frac{140 \text{ yd}}{n}$

Project

THE PYRAMID INVESTMENT PLAN

Have you ever received a chain letter offering you prizes or great riches? The letters ask you to follow the simple instructions and not break the chain. Actually, chain-letter schemes are illegal, even though they continue to be quite common.

Suppose you have just received this letter, along with several quotes from "ordinary people" who have already become millionaires. How many rounds have already taken place? How many more rounds have to take place before *you* become a millionaire?

Your project should include

- A written analysis of this plan.
- Your conclusion about whether or not you will become a millionaire.
- An answer to the question "Why do you think chain-letter schemes and pyramid plans are illegal?"

Join the Pyramid Investment Plan (PIP)!

Become a millionaire! Send only $20 now, and return it with this letter to PIP. PIP will send $5 to the name at the top of the list below. Then the second person will move up to the top of the list, and your name will be added to the bottom of the list. You will receive a new letter and a set of 200 names and addresses. Your name will be in position #6. Make 200 copies of the letter, mail them, and **wait to get rich!**

Each time a recipient of one of your letters joins PIP, your name advances toward the top of the list. When your name reaches the top, each of the thousands of people who receive that letter will be sending money to PIP, and you will receive your share. **A conservative marketing return of 6% projects that you will earn over $1.2 million!**

1. *Chris*
2. *Katie*
3. *Josh*
4. *Kanako*
5. *Dave*
6. *Miranda*

EXPLORATION

Refining the Growth Model

Until now you have assumed that the rate of growth or decay remains constant over time. The terms generated by the recursive formulas you have used to model arithmetic or geometric growth increase infinitely in the long run.

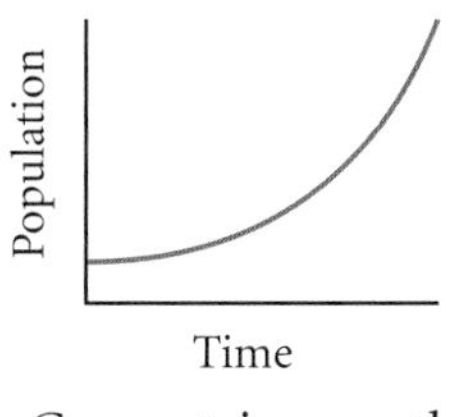

Geometric growth

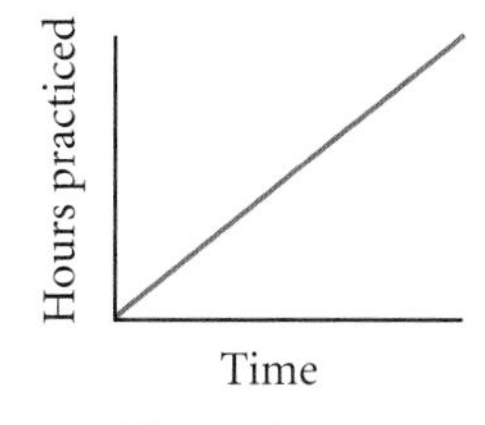

Arithmetic growth

However, environmental factors rarely support unlimited growth. Because of space and resource limitations, or competition among individuals or species, an environment usually supports a population only up to a limiting value.

The graph at right shows a **logistic function** with the population leveling off at the maximum capacity. If a population's growth is modeled by a logistic function, the growth rate isn't a constant value but, rather, changes as the population changes.

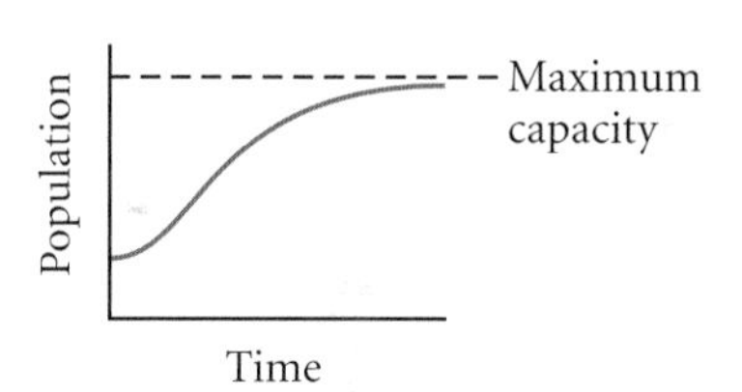

Logistic functions are used to model more than just population. Researchers in many fields apply logistic functions to study things such as the spread of disease or consumer buying patterns. In this exploration you will see an example of logistic growth.

Environmental CONNECTION

Some desert cities in the United States have growing populations and little supply of natural resources. Large populations need, for example, a plentiful water supply for drinking and sewage systems, as well as for luxuries such as watering lawns and filling swimming pools. Population growth reduces groundwater supply, which in turn activates old earthquake faults and cracks in the earth's surface, and damages buildings. Using logistic functions to model population growth can help government agencies monitor natural resources and avoid environmental catastrophies.

The Hoover Dam, located on the border of Nevada and Arizona, helps supply water to desert cities such as Las Vegas.

Activity

Cornering the Market

A company has invented a new gadget, and everyone who hears about it wants one! The company hasn't started advertising the new gadget yet, but the news is spreading fast. Simulate this situation with your class to see what happens when a popular new product enters the market.

Each person in your class will be assigned an I.D. number. At time zero, only one person has bought the gadget, and at the end of every time period, each person who has one tells another person about it, and they go out and buy one (unless they already have one).

Step 1 For about 10 time periods, each person who knows about the gadget generates a random I.D. number and tells that person, who immediately goes out and buys one. [► See **Calculator Note 1L** to learn how to generate random numbers. ◄] Create a table like the one at right, and record the total number of people who have the gadget.

Time period	Number of people who own the gadget
0	1
1	
2	
3	

Step 2 Enter your data into two columns, *time* and *people*. Make a scatter plot of your data. Describe your scatter plot. Explain why the number of people who own the gadget doesn't always double for each time period.

Step 3 Divide each term in *people* by the previous term and enter these ratios into a new column, *ratio*. These ratios show you the rate at which the number of people who own the gadget grows during each time period.

Step 4 In this activity the growth rate depends on the number of people who own the gadget. Turn off the scatter plot from Step 2 and make a new scatter plot of (*people, ratio*). What happens to the growth rate as the number of people who own the gadget increases?

The net growth at each step depends on the previous population size u_{n-1}. So the net growth is a function of the population. This changes the simple growth model of $u_n = u_{n-1}(1 + p)$, in which the rate, p, is a constant, to a growth model in which the growth rate changes depending on the population.

$$u_n = u_{n-1}\left[1 + p \cdot \left(1 - \frac{u_{n-1}}{L}\right)\right]$$

where p is the unrestricted growth rate, L is the limiting capacity or maximum population, and $p \cdot \left(1 - \frac{u_{n-1}}{L}\right)$ is the net growth rate.

Step 5 | What is the maximum population, L, for the gadget-buying scenario? What is u_0? What is the unrestricted growth rate, p? What does $\frac{u_{n-1}}{L}$ represent? Write the recursive formula for this logistic function.

Step 6 | Create a new column, *model,* defined as $1 + p \cdot \left(1 - \frac{people}{L}\right)$, using p and L from Step 5. Plot (*people, model*) and (*people, ratio*).

Step 7 | Turn off the scatter plots from Step 6 and check how well your recursive formula from Step 5 models your original data (*time, people*).

History CONNECTION

At the end of the 18th century, data on population growth were not available, and social scientists generally agreed with English economist Thomas Malthus (1766–1834) that a population always increases exponentially, eventually leading to a catastrophic overpopulation. In the 19th century, Belgian mathematician Pierre François Verhulst (1804–1849) and Belgian social statistician Adolphe Quételet (1796–1874) formulated the net growth rate expression $p \cdot \left(1 - \frac{u_{n-1}}{L}\right)$ for a population model. Quételet believed that limitations on population growth needed to be accounted for in a more systematic manner than Malthus described. Verhulst was able to incorporate the changes in growth rate in a mathematical model.

A "closed" environment creates clear limitations on space and resources and calls for a logistic function model.

EXAMPLE

Suppose the unrestricted growth rate of a deer population on a small island is 12% annually, but the island's maximum capacity is 2000 deer. The current deer population is 300.

a. What net growth rate can you expect for next year?

b. What will the deer population be after 1 year?

c. Graph the deer population over the next 50 years.

▶ Solution

a. There is a maximum capacity, so the population can be modeled with a logistic function. The net growth rate is $p\left(1 - \frac{u_{n-1}}{L}\right)$, where p is the unrestricted growth rate and L is the limiting value, in this case, the island's maximum capacity. Because $p = 12\%$ and $L = 2000$ deer, the net growth rate will be

$$0.12\left(1 - \frac{u_{n-1}}{2000}\right)$$

Using $u_0 = 300$, this gives a growth rate of $0.12\left(1 - \frac{300}{2000}\right) = 0.102$, or 10.2%, for the first year.

b. The recursive rule for this logistic function is $u_n = u_{n-1}\left[1 + 0.12\left(1 - \frac{u_{n-1}}{2000}\right)\right]$

Using $u_0 = 300$,

$$u_1 = 300\left[1 + 0.12\left(1 - \frac{300}{2000}\right)\right]$$
$$= 300(1 + 0.102)$$
$$= 330.6$$

So the deer population after 1 year will be around 331.

c. The graph shows the population as it grows toward the maximum capacity of 2000 deer.

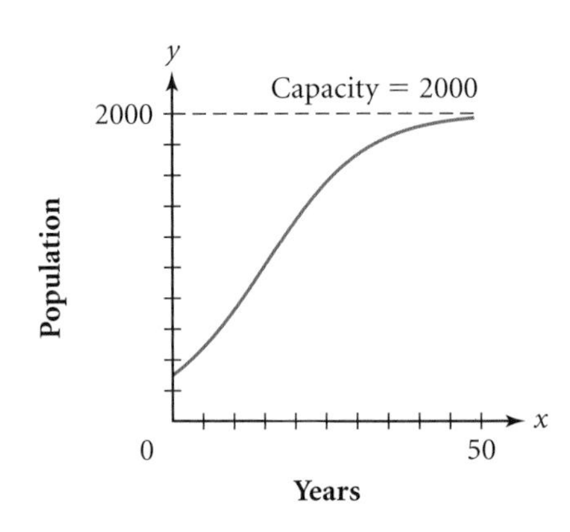

Now try your hand at applying logistic function models with these questions.

Questions

When humans inhabit areas previously dominated by animals, the population growth of the animals may be adversely affected.

1. Bacteria grown in a culture dish are provided with plenty of food but a limited amount of growing space. Eventually the population will become overcrowded, even though there is plenty of food. The bacteria grow at an unrestricted rate of 125% per week initially. The starting population is 50, and the capacity of the dish is 5000. Find the net growth rate and the population at the end of each week for 7 weeks.

2. A large field provides enough food to feed 500 healthy rabbits. When food and space are unlimited, the population growth rate of the rabbits is 20%, or 0.20. Complete each statement using the choices less than 0, greater than 0, 0, or close to 0.20.

a. When the population is less than 500, the net growth rate will be _?_.

b. When the population is more than 500, the net growth rate will be _?_.

c. When the population is very small, the net growth rate will be _?_.

d. When the population is 500, the net growth rate will be _?_.

3. Suppose the recursive rule $d_n = d_{n-1}\left[1 + 0.35\left(1 - \frac{d_{n-1}}{750}\right)\right]$ will give the number of daisies growing in the median strip of a highway each year. Presently there are about 100 daisies. Write a paragraph or two explaining what will happen to the population of daisies. Explain and support your reasoning.

CHAPTER 1 REVIEW

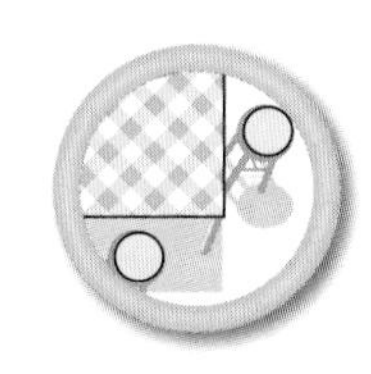

A **sequence** is an ordered list of numbers. In this chapter you used **recursion** to define sequences. A **recursive formula** specifies one or more starting terms and a **recursive rule** that generates the *n*th term by using the previous term or terms. You learned to calculate the terms of a sequence by hand and by using recursion and sequences on your calculator.

There are two special types of sequences—arithmetic and geometric. **Arithmetic sequences** are generated by always adding the same number, called the **common difference,** to get the next term. Your salary for a job on which you are paid by the hour is modeled by an arithmetic sequence. **Geometric sequences** are generated by always multiplying by the same number, called the **common ratio,** to get the next term. The growth of money in a savings account is modeled by a geometric sequence. For some **growth** and **decay** scenarios, it helps to write the common ratio as a percent change, $(1 + p)$ or $(1 - p)$. Some real-world situations, such as levels of medicine in the body, are modeled by **shifted geometric sequences** that use a recursive rule with both multiplication and addition.

Many sequences have a long-run value after many, many terms. Looking at a graph of the sequence may help you see the long-run value. Graphs also help you recognize whether the data are best modeled by an arithmetic or geometric sequence. The graph of an arithmetic sequence is **linear** whereas the graph of a geometric sequence is curved.

EXERCISES

You will need

A graphing calculator for Exercises **2**, **4**, and **7–10**.

@ Answers are provided for all exercises in this set.

1. Consider this sequence:

256, 192, 144, 108, . . .

a. Is this sequence arithmetic or geometric?

b. Write a recursive formula that generates the sequence. Use a_1 for the starting term.

c. What is the 8th term?

d. Which term is the first to have a value less than 20?

e. Find a_{17}.

2. Consider this sequence:

3, 7, 11, 15, . . .

a. Is this sequence arithmetic or geometric?

b. Write a recursive formula that generates the sequence. Use u_1 for the starting term.

c. What is the 128th term?

d. Which term has the value 159?

e. Find u_{20}.

3. List the first five terms of each sequence. For each set of terms, what minimum and maximum values of n and a_n would you use on the axes to make a good graph?

a. $a_1 = -3$
$a_n = a_{n-1} + 1.5$ where $n \geq 2$

b. $a_1 = 2$
$a_n = 3a_{n-1} - 2$ where $n \geq 2$

4. APPLICATION Atmospheric pressure is 14.7 pounds per square inch (lb/in^2) at sea level. An increase in altitude of 1 mi produces a 20% decrease in the atmospheric pressure. Mountain climbers use this relationship to determine whether or not they can safely climb a mountain and to periodically calculate their altitude after they begin climbing.

a. Write a recursive formula that generates a sequence that represents the atmospheric pressure at different altitudes.

b. Sketch a graph that shows the relationship between altitude and atmospheric pressure.

c. What is the atmospheric pressure when the altitude is 7 mi?

d. At what altitude does the atmospheric pressure drop below 1.5 lb/in^2?

These mountaineers climbed to an altitude of 11,522 feet to reach the top of Mt. Clark in Yosemite National Park, California.

Environmental CONNECTION

Humans of any age or physical condition can become ill when they experience extreme changes in atmospheric pressure in a short span of time. Atmospheric pressure changes the amount of oxygen a person is able to inhale which, in turn, causes the buildup of fluid in the lungs or brain. Someone who travels from a low-altitude city like Akron, Ohio, to a high-altitude city like Aspen, Colorado, and immediately ascends a mountain to ski, could get a headache, become nauseated, or even become seriously ill.

5. Match each recursive formula with the graph of the same sequence. Give your reason for each choice.

A. $u_0 = 5$
$u_n = u_{n-1} + 1$ where $n \geq 1$

B. $u_0 = 1$
$u_n = (1 + 0.5)u_{n-1}$ where $n \geq 1$

C. $u_0 = 5$
$u_n = (1 - 0.5)u_{n-1}$ where $n \geq 1$

D. $u_0 = 5$
$u_n = u_{n-1} - 1$ where $n \geq 1$

i.
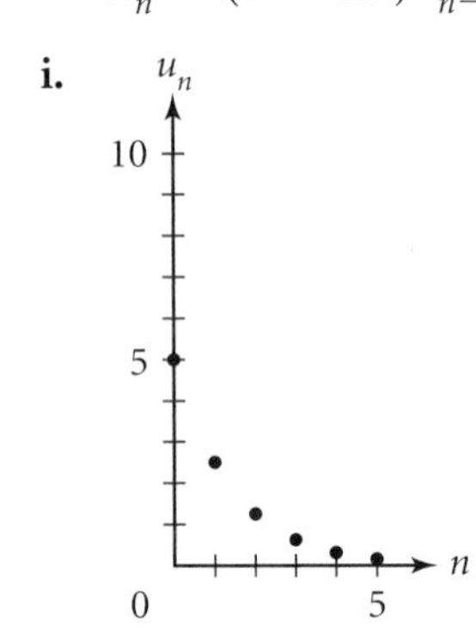

ii.
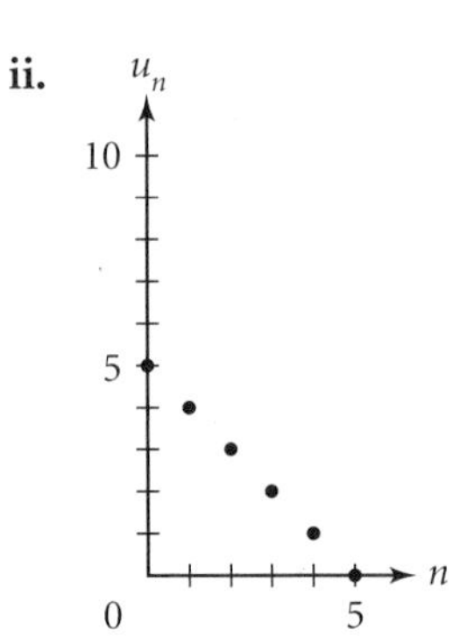

iii.
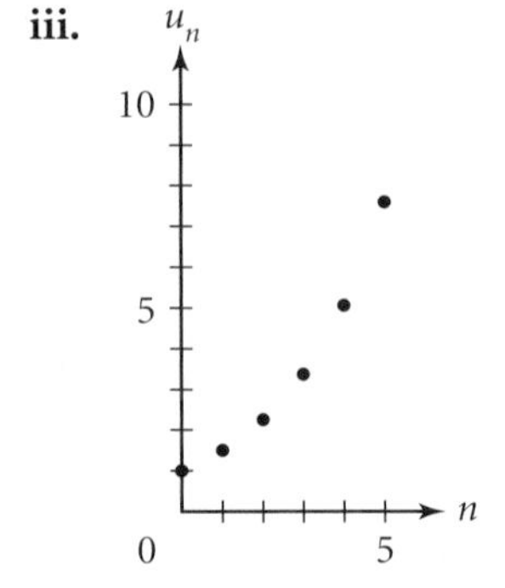

iv.
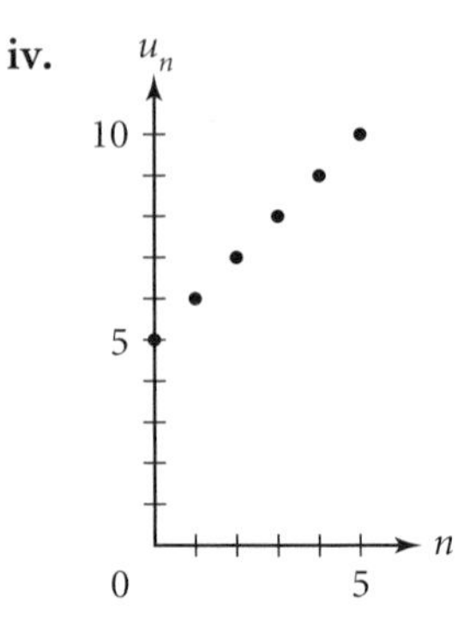

6. A large barrel contains 12.4 gal of oil 18 min after its drain is opened. How many gallons of oil were in the barrel before the drain was opened if it drains at a rate of 4.2 gal/min?

7. APPLICATION The enrollment at a college is currently 5678. From now on, the board of administrators estimates that each year the school will graduate 24% of its students and admit 1250 new students. What will the enrollment be during the sixth year? What will the enrollment be in the long run? Sketch a graph of the enrollment over 15 years.

8. APPLICATION You deposit $500 into a bank account that has an annual interest rate of 5.5%, compounded quarterly.

a. How much money will you have after 5 yr if you never deposit more money?

b. How much money will you have after 5 yr if you deposit an additional $150 every 3 mo after the initial $500?

9. APPLICATION This table gives the consumer price index for college tuition from 1980 to 2005. Use a graph to find a sequence model that approximately fits these data.

U.S. Consumer Price Index for College Tuition

Year	Consumer price index
1980	70.8
1985	119.9
1990	175.0
1995	264.8
2000	331.9
2005	475.1

(U.S. Department of Labor, Bureau of Labor Statistics)

Economics CONNECTION

The consumer price index (CPI) is a measure of the change over time in the prices paid by consumers for goods and services, such as food, clothing, and health care. The Bureau of Labor Statistics obtains price information for 80,000 items in order to adjust the index. The price of specific items in 1982–1984 is assigned an index of 100, and the index for all subsequent years is given in relation to this reference period. For example, an index of 130 means the price of an item increased 30% from the price during 1982–1984. Learn more about the CPI with the links at www.keymath.com/DAA.

10. APPLICATION Oliver wants to buy a cabin and needs to borrow $80,000. What monthly payment is necessary to pay off the mortgage in 30 years if the annual interest rate is 8.9%, compounded monthly?

TAKE ANOTHER LOOK

1. In Lesson 1.1, Example A, you saw an arithmetic sequence that generates from a geometric pattern: the number of seats around a linear arrangement of square tables is 4, 6, 8, ... Notice that the sequence comes from the increase in the perimeter of the arrangement. What sequence comes from the increase in area?

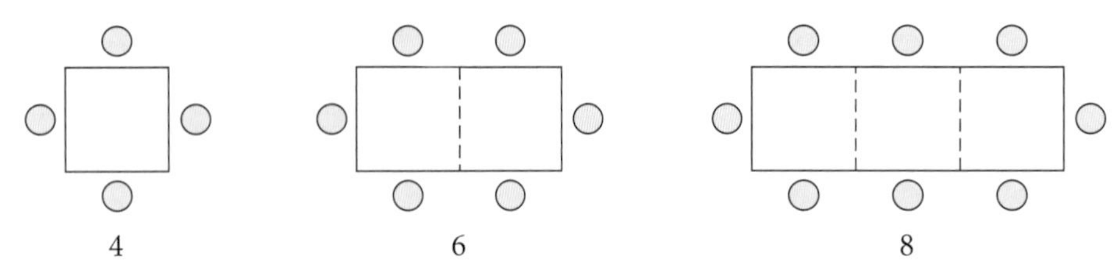

Use these geometric patterns to generate sequences using perimeter, area, height, or other attributes, and give their recursive formulas. Which sequences exhibit arithmetic growth? Geometric growth? Are there sequences you cannot define as arithmetic or geometric?

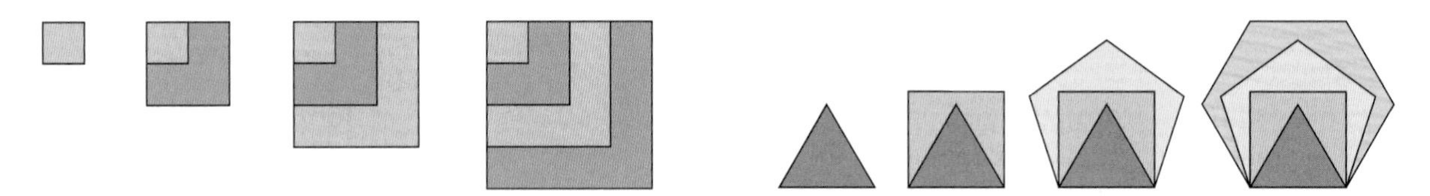

Invent a geometric pattern of your own. Look for and define different sequences.

2. Here's a strategy an algebra student discovered for finding the monthly payment you would need to make to pay off a loan. First, specify the loan amount, the annual interest rate, and the term (length) of the loan. Choose trial monthly payment amounts that form a sequence, say, \$0, \$1, \$2, \$3. Record the final balance remaining at the end of the term for each payment amount. Next, explore the differences in the final balances and find a pattern. Finally, use the pattern to find a monthly payment that results in a zero final balance at the end of the term. Tell how you know that your payment amount is correct. How many trial payment amounts do you need in order to determine the monthly payment? Try this strategy for one of the exercises in Lesson 1.5.

3. Imagine a target for a dart game that consists of a bull's-eye and three additional circles. If it is certain that a dart will land within the target but is otherwise random, what sequence of radii gives a set of probabilities that form an arithmetic sequence? A geometric sequence? Sketch what the targets look like.

4. As you probably realized in Lesson 1.5, your graphing calculator is a useful tool for solving loan and investment problems. With the added help of a special program, your calculator will instantly tell you information such as the monthly payment necessary for a home mortgage. [► See **Calculator Note 1M** to learn how to use your calculator's financial-solver program. ◄] Try using this program to solve some of the exercises in Lesson 1.5 or a problem of your own design.

Assessing What You've Learned

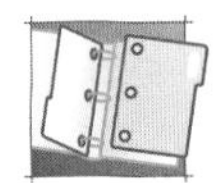

ORGANIZE YOUR NOTEBOOK Wouldn't you like to have a record of what you've learned and just what you are expected to know in this algebra course? Your own notebook can be that record if you enter significant information and examples into it on a regular basis. It is a good place for new vocabulary, definitions and distinctions, and worked-out examples that illustrate mathematics that's new to you. On the other hand, if your notebook is simply a stack of returned homework, undated class notes, and scratch-paper computations and graphs, it is too disorganized to perform that service for you.

Before you get far along in the course, take time to go through your notebook. Here are some suggestions:

- Put papers into chronological order. If your work is undated, use the table of contents in this book to help you reconstruct the sequence in which you produced the items in your notebook. You can number pages to help keep them in order.
- Go through the book pages of Chapter 0 and Chapter 1, and see whether you have a good record of how you spent class time—what you learned from investigations and homework. Fill in notes where you need them while information is fresh in your mind. Circle questions that you need to have cleared up before going on.
- Reflect on the main ideas of this chapter. Organize your notes by type of sequence, and make sure you have examples of recursive formulas that produce different types of graphs. Be sure you can identify the starting value and the common difference or common ratio for each type of sequence, and that you know how to use the information to find a later term in the sequence.

PERFORMANCE ASSESSMENT It's important to know how well you are progressing in this course so that you're encouraged by your gains. Assessing what you've learned also alerts you to get help before you're confused about important ideas that you'll have to build on later in the course.

One way to tell whether or not you understand something is to try to explain it to someone else. Choose one of the modeling problems in Lesson 1.4, such as the population of Bhutan in Exercise 12, and present it orally to a classmate, a relative, or your teacher. Here are some steps to follow:

- Explain the problem context, and give the listener some background on what the problem calls for mathematically.
- Be sure the listener knows what you mean by finding a model. Then describe how you find the model, using the given information, and how you express mathematically what the model is.
- Comment on how well the model fits the original data.
- Show the listener how to use the model. For example, ask a question about the future population of Bhutan, and show the listener how to use the model to answer the question. Show at least one way to check that your result is reasonable. Don't forget to interpret the mathematical result in terms of the problem's real-world context.

CHAPTER

2 Describing Data

Contemporary German artist Andreas Gursky (b 1955) digitally manipulated this photograph so that objects in the background appear as clearly as objects in the foreground. The resulting work of art, *99 Cent* (1999), conveys the vast amount of information that surrounds you in a supermarket. What data about the inventory of this store are helpful to you as a customer? What data are helpful to the management of the store? How would you describe and summarize those data with graphs or numbers?

OBJECTIVES

In this chapter you will

- create, interpret, and compare graphs of data sets
- calculate numerical measures that help you understand and interpret a data set
- make conclusions about a data set and compare it with other data sets based on graphs and numerical values

Measures of Central Tendency and Dot Plots

A sample of the weights of backpacks was collected from 30 students. In describing this sample of numerical data, you might want to give the "typical" value for the set, rather than listing all 30 weights. Values called **measures of central tendency** are used to summarize data into a single value or **statistic.**

The **mean** is the sum of all the data values divided by the number of values.

The **median** is the middle number when the data are arranged in order.

The **mode** is the value that occurs most frequently in the data.

EXAMPLE A

Find the three measures of central tendency for these backpack weights (in pounds):

10, 20, 9, 17, 3, 10, 15, 15, 7, 10, 9, 10, 9, 7, 4, 6, 7, 9, 13, 10, 8, 7, 4, 4, 8, 33, 10, 9, 7, 16

▶ *Solution*

The sum of all 30 values is 306. When you divide by 30, you find 10.2 lb is the mean.

The median is 9 lb, the middle value when the data are arranged in order.

3, 4, 4, 4, 6, 7, 7, 7, 7, 7, 8, 8, 9, 9, 9, 9, 9, 10, 10, 10, 10, 10, 10, 13, 15, 15, 16, 17, 20, 33

$$\frac{9+9}{2} = 9$$

Because there is an even number of values, the median is the mean of the two middle values.

The mode is 10 lb, the weight that occurs most frequently.

You can justify using any of these three statistics as a typical weight. If you wanted to present a statistic that implies backpacks are too heavy, you might use the mean because it is higher than the median or mode due to one very large data value. When a data set has one or more values that are far from the rest, the median often is more representative of the data than the mean.

Health CONNECTION

The weight of a backpack, if carried improperly, can cause physical injuries. The American Chiropractic Association recommends that the items inside a backpack weigh no more than 10% of the person's body weight. To prevent muscle aches, fatigue, and pain in the shoulders, neck, and back, use straps on both shoulders, and adjust the straps so that the backpack falls below the shoulders and rests on the hips.

A **dot plot** is a statistical graph in which each value in the sample is plotted above a number line. The dot plot for backpack data looks like this.

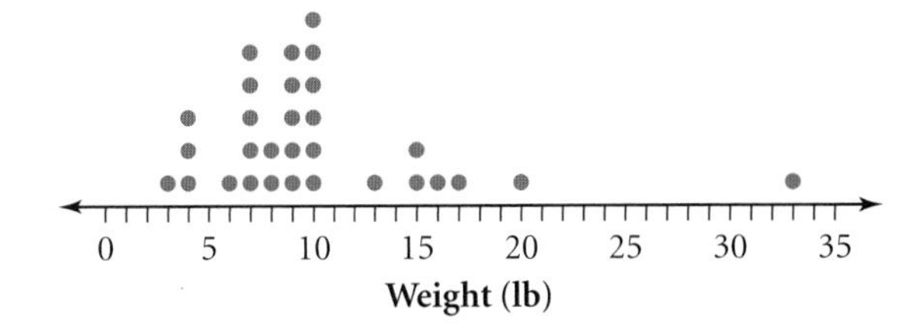

If you have a dot plot of a data set, you can find the median or the mode by counting the dots and noting their position.

EXAMPLE B Use the dot plot to find the three measures of central tendency for this sample.

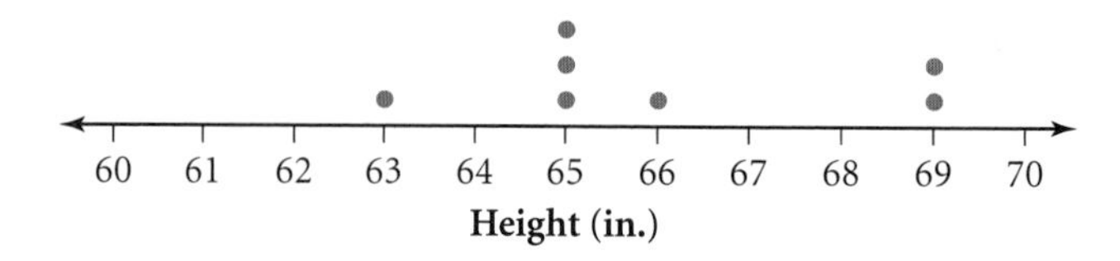

▸ ***Solution*** There are seven data values, so the fourth data value is the median. Counting in four values from either end, the median is 65 in.

The mean is $\frac{63 + 65 + 65 + 65 + 66 + 69 + 69}{7} = \frac{462}{7} = 66$ in.

The mode is 65 in., the height that occurs most frequently.

EXERCISES

1. Find the mean, median, and mode for each data set.

a. Time for pizza delivery (min): {28, 31, 26, 35, 26} ⓐ

b. Yearly rainfall (cm): {11.5, 17.4, 20.3, 18.5, 17.4, 19.0} ⓗ

c. Cost of a small popcorn at movie theaters ($): {2.75, 3.00, 2.50, 3.50, 1.75, 2.00, 2.25, 3.25}

d. Number of pets per household: {3, 2, 1, 0, 3, 4, 1}

2. A data set has a mean of 12 days, the median is 14 days, and there are three values in the data set.

a. What is the sum of all three data values? ⓐ

b. What is the one value you know?

c. Create a data set that has the statistics given. Is there more than one data set that could have these statistics?

3. What are the values of the data set in the dot plot? ⓐ

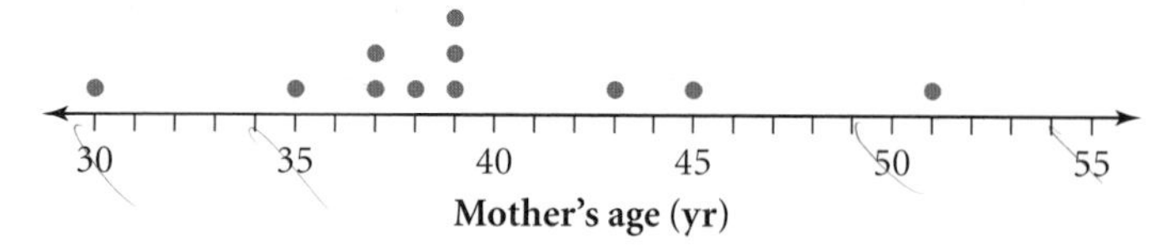

4. Find the mean, median, and mode for the data in Exercise 3.

5. Make a dot plot of these assignment scores for Peter and Amulet, and find the three measures of central tendency for each student's scores.

a. Peter's scores (%): {82, 86, 81, 84, 87, 84, 98}

b. Amulet's scores (%): {80, 86, 74, 84, 85, 81, 84}

c. Usually the mean score is used to summarize. Which student would prefer to use his or her median score? Explain.

d. What other statistics might be useful in describing and comparing the scores? ⓗ

LESSON

2.1

Box Plots

That is what learning is. You suddenly understand something you've understood all your life, but in a new way.

DORIS LESSING

Newspapers, magazines, the evening news, commercials, government bulletins, and sports publications bombard you daily with data and statistics. As an informed citizen, you need to be able to interpret this information in order to make intelligent decisions.

In this chapter you will graph data sets in several different ways. You'll also study some numerical measures that help you better understand what a data set tells you. A good description of a data set includes not only a measure of central tendency, such as the mean, median, or mode, but the spread and distribution of the data as well. This is often done with a set of summary values or a graph.

EXAMPLE A

Owen is a member of the student council and wants to present data about backpack safety to the school board. He collects these data on the weights of backpacks of 30 randomly chosen students. Owen wants to present a graph that shows the distribution and shape of the backpack data. Create a **box plot (box-and-whisker plot)** of the data. [▶ See **Calculator Note 2B** to learn how to create a box plot on your calculator. ◀]

Student	Grade	Weight of backpack (lb)
1	Junior	10
2	Senior	20
3	Junior	9
4	Junior	17
5	Junior	3
6	Junior	10
7	Senior	15
8	Junior	15
9	Senior	7
10	Senior	10
11	Junior	9
12	Senior	10
13	Senior	9
14	Junior	7
15	Senior	4
16	Senior	6
17	Senior	7
18	Senior	9
19	Junior	13
20	Junior	10
21	Senior	8
22	Senior	7
23	Senior	4
24	Senior	4
25	Junior	8
26	Junior	33
27	Senior	10
28	Senior	9
29	Senior	7
30	Junior	16

▸ Solution

A box plot (or box-and-whisker plot) is created from the **five-number summary** of the data.

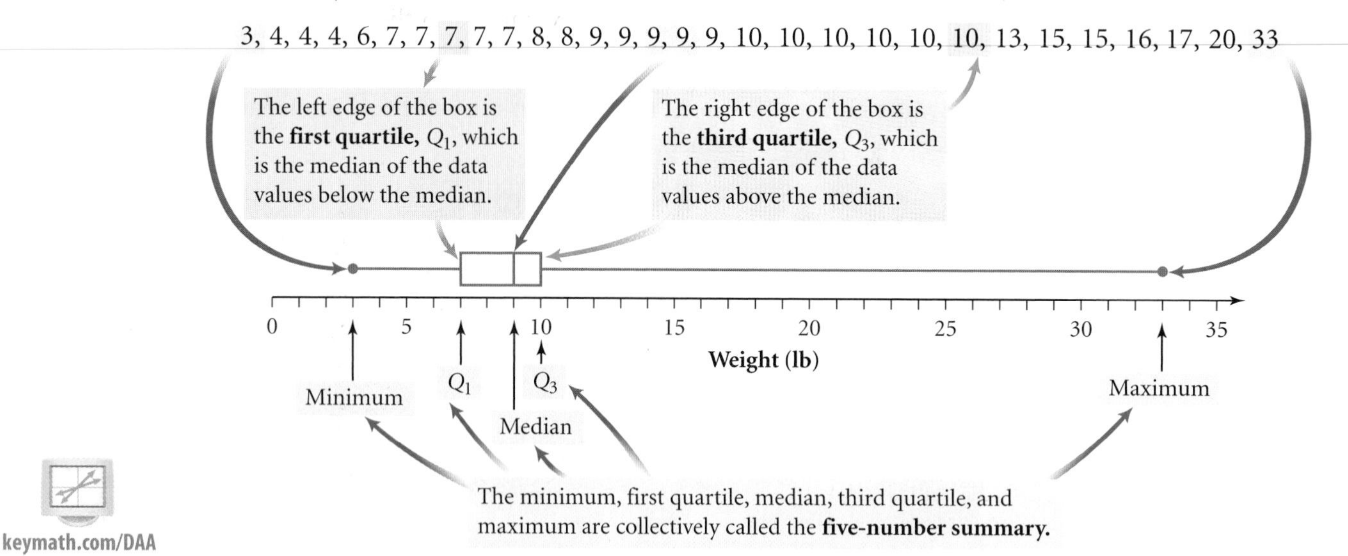

keymath.com/DAA

The lines extending from the "box" are called "whiskers." They identify the minimum and maximum values of the data.

The data set in Example A did not include every student in the school, so it may or may not tell much about all student backpack weights. If Owen took his sample from the first 30 students who arrived to a single class, then the data set might be **biased,** or unfair: It could represent students who hurry to class because their backpacks are too heavy. How might the information be biased if Owen took the sample from the first 30 volunteers?

Assume that Owen's data are from a **simple random sample** of the population. This means that every student is equally likely to be selected. Statistics for a simple random sample are **unbiased estimates** for the population, so the sample is **representative** of the population. This means that you can conclude that results for the sample data, such as a median of 9 lb, apply to all backpacks in the school.

A box plot uses the five-number summary to show how data are spread around the median. Several statistics based on the five-number summary also describe spread. The **range** is the difference between the maximum and the minimum. Another measure of spread is the **interquartile range (*IQR*),** which is defined as the difference between the third quartile (Q_3) and the first quartile (Q_1), or the length of the box in the box plot. The *IQR* is less affected than the range by extreme values in the data. Can you create two data sets with the same range where one has an *IQR* half as big as the other?

You can use a graph of data to look for clusters, gaps, and extreme values in the sample. One backpack in Owen's sample weighed 33 lb, far more than the next largest weight of 20 lb. Would the sample be more representative of the population if that very heavy backpack were omitted?

Extreme values are called **outliers** when there is a gap between them and the rest of the data. A hybrid graph called a **modified box plot** can be used to show these gaps. In a modified box plot, any values that are more than 1.5 times the *IQR* from the ends of the box are plotted as separate points.

EXAMPLE B

Use the backpack data from Example A to answer each question.

a. Find the range and the interquartile range.

b. Create a modified box plot showing the outliers.

▶ *Solution*

Refer to the five-number summary and box plot from the solution of Example A.

a. The range is equal to the maximum minus the minimum: $33 - 3 = 30$ lb. The IQR is $Q_3 - Q_1 = 10 - 7 = 3$ lb.

b. Use the $1.5 \cdot IQR$ rule to draw "fences" that divide the outliers from the rest of the data. The fences are located at $Q_1 - 1.5 \cdot IQR = 7 - 1.5(3) = 2.5$ and at $Q_3 + 1.5(3) = 10 + 1.5(3) = 14.5$. Any value outside the fences, that is, less than 2.5 or greater than 14.5, is plotted as an individual point.

When drawing a modified box plot, draw the whisker from the end of the box to the value that is closest to, but inside, the fence. The minimum value, 3 lb, is inside the fence, so the lower whisker looks the same. Close to the upper fence at 14.5 lb, one backpack weighed 13 lb (<14.5) and the next weighed 15 lb (>14.5). So the upper whisker goes from 10 to 13, and all the values above the fence are marked with individual points.

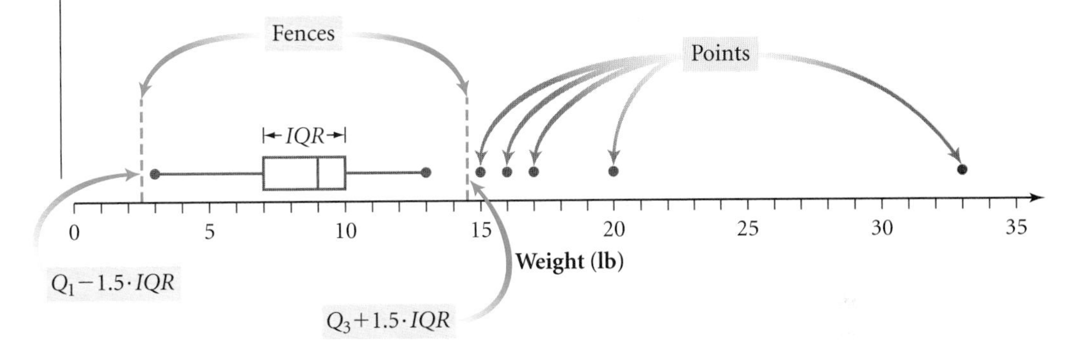

Cultural CONNECTION

During the 15th and 16th centuries, the Inca used *quipus,* a system of knotted cords, to record numerical information, such as population. The number of knots, the number of strings, and variations in color and thickness combined to create a sophisticated method for keeping data and statistics, which scholars have yet to fully decipher.

The illustration at left, from a 17th-century letter by Felipe Guáman Poma de Ayala, shows an Incan treasurer holding a *quipu.* The photo shows an actual Incan *quipu.*

Statisticians often talk about the shape of a data set. **Shape** describes how the data are distributed relative to the center. A **symmetric** data set is balanced, or nearly so, at the center. Note that it does not have to be exactly equal on both sides to be called symmetric. **Skewed** data are spread out more on one side of the center than on the other side. The backpack data provide an example of skewed data. You will learn more about spread in the next lesson. For now, a box plot can be a good indicator of shape because the median is clearly visible as the center.

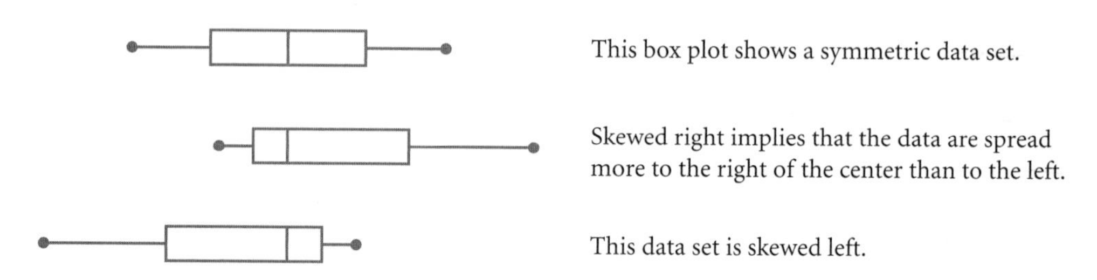

This box plot shows a symmetric data set.

Skewed right implies that the data are spread more to the right of the center than to the left.

This data set is skewed left.

Investigation
Pulse Rates

You will need

- a watch or clock with a second hand

Pulse rate is often used as a measure of whether or not a person is in good physical condition. In this investigation you will practice making box plots, compare box plots, and draw some conclusions about pulse rates.

Step 1 Measure and record your resting pulse for 15 s. Multiply this value by 4 to get the number of beats per minute. Pool data from the entire class.

Step 2 Exercise for 2 min by doing jumping jacks or by running in place. Afterward, measure and record your exercise pulse rate. Pool your data.

Step 3 Order each set of data. Calculate the five-number summaries for your class's resting pulse rates and for your exercise pulse rates.

Step 4 Prepare a box plot of the resting pulse rates and a box plot of the exercise pulse rates. Determine a range suitable for displaying both of these graphs on a single axis.

Step 5 Draw conclusions about pulse rates by comparing these two graphs. Be sure to compare not only centers but also spreads and shapes. Could your conclusion apply to a larger population? Describe the population and explain how your class is representative of that population.

If your sample is representative of a larger population, then the shape and spread of your sample data will be like the shape and spread of the entire population. So in general you can draw conclusions about the population by describing the sample. What factors will influence how confident you are in your conclusions?

EXERCISES

You will need

A graphing calculator for Exercises **15–17** and **20.**

Practice Your Skills

1. Approximate the values of the five-number summary for this box plot. Give the full name for each value. @

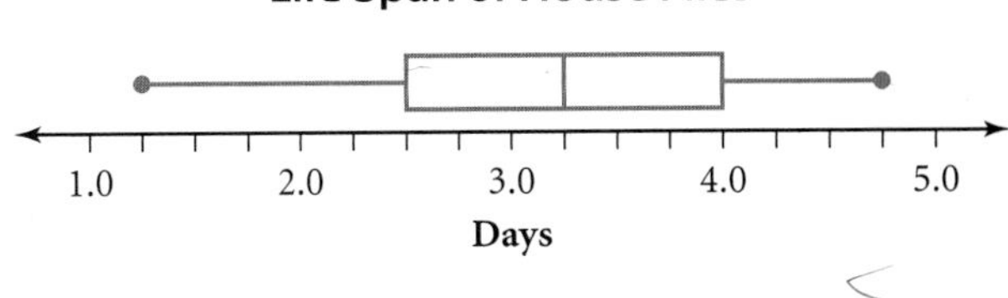

2. Match this data set to one of the four box plots.

Licks to the center of a lollipop:
{470, 510, 547, 558, 561, 574, 593}

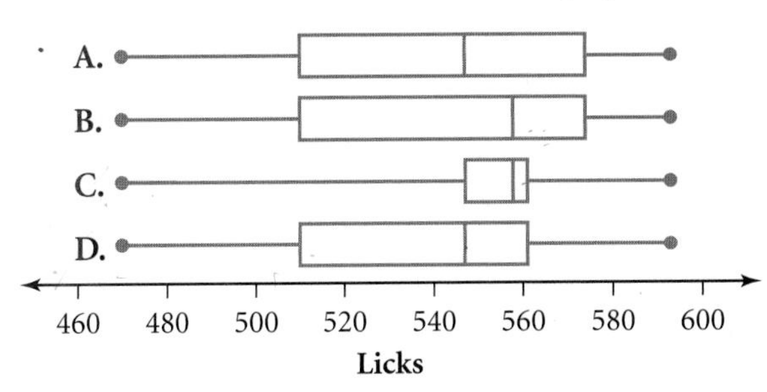

3. Match this box plot to one of the data sets for the number of minutes on the phone spent by 13 customer service representatives in a given hour.

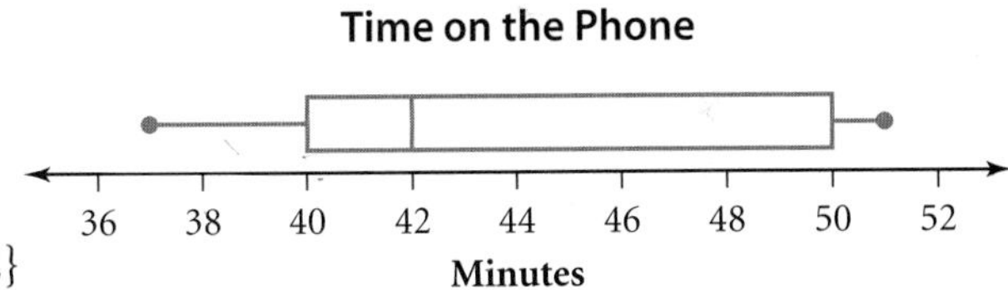

A. {36, 37, 38, 39, 40, 41, 42, 43, 44, 46, 48, 50, 52}

B. {37, 40, 42, 50, 51, 51, 51, 51, 51, 51, 51, 51, 51}

C. {36, 37, 40, 40, 40, 42, 42, 43, 44, 50, 50, 51, 52}

D. {37, 39, 40, 40, 40, 41, 42, 43, 44, 49, 51, 51, 51}

4. Calculate the range and interquartile range for each plot in Exercise 2. @

5. In a classroom experiment, 11 bean plants were grown from seeds. After two weeks, the heights in centimeters of the plants were 9, 10, 10, 13, 13, 14, 15, 16, 17, 19, and 21.

a. Find the five-number summary. @

b. Find the range and interquartile range.

c. What are the units of the range and interquartile range?

Reason and Apply

6. Invent a data set with seven data values such that the mean and the median are both 84, the range is 23, and the interquartile range is 12. (h)

7. Here are the scores on semester assignments for two students.

Connie: {82, 86, 82, 84, 85, 84, 85}
Oscar: {72, 94, 76, 96, 90, 76, 84}

Find the mean and median for each set of scores, and explain why they do not tell the whole story about the differences between Connie's and Oscar's scores.

8. These box plots represent Connie's and Oscar's scores from Exercise 7.

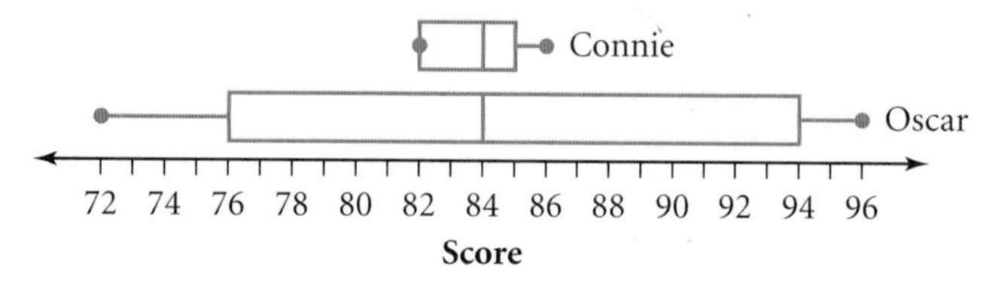

Write a paragraph describing the information pictured in the box plots. Use the box plots to help you draw some statistical conclusions. In your description, include answers to such questions as What does it mean that the second box plot is longer? Where is the left whisker of the top box plot? What does it mean when the median isn't in the middle of the box? What does it mean when the left whisker is longer than the right whisker? What information is conveyed by the range and interquartile range for each student? (h)

9. Homer Mueller has played in the minor leagues for 11 years. His home run totals, in order, for those years are 56, 62, 49, 65, 58, 52, 68, 72, 25, 51, and 64.

a. Construct a box plot showing Homer's data.

b. Give the five-number summary. (a)

c. Find the range and interquartile range.

d. Find the mean.

e. How many home runs would Homer need to hit next season to have a 12-year mean of 60? (h)

Sports CONNECTION

Baseball fans thrive on the "stats" of their favorite players. For example, batting average (AVG or BA) is the number of hits divided by the number of at bats. The slugging average (SA or SLG) is the total bases per at bat, where singles are one base, doubles are two bases, triples are three bases, and homeruns are four bases.

Right fielder Roberto Clemente (1934–1972) played for the Pittsburgh Pirates from 1955 to 1972. He participated in twelve All Star Games, won twelve Gold Glove Awards, and was posthumously elected to the National Baseball Hall of Fame.

10. Invent a data set with seven distinct values and a mean of 12.

11. Invent a data set with seven values and a mode of 70 and a median of 65.

12. Invent a data set with seven values that creates this box plot.

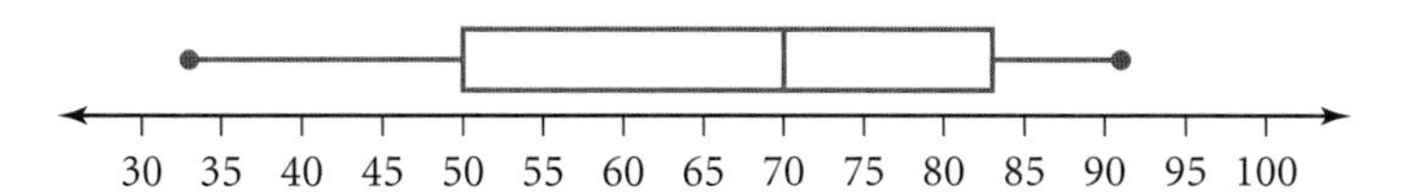

13. In order to monitor weight, a cookie manufacturer samples jumbo chocolate chip cookies as they come off the production line. The weights in grams of 11 cookies are 25, 30, 27, 33, 32, 28, 17, 22, 25, 30, and 28.

a. Find the five-number summary and the interquartile range.

b. Calculate the locations of the outlier fences. ⓐ

c. Create a modified box plot of these data. Describe the shape of the data set.

14. A reporter waited until school was over and then interviewed the first ten students to leave the school grounds about the amount of emphasis the school places on athletics. Would you consider this a biased or an unbiased sample? Explain.

15. APPLICATION Lord Rayleigh was one of the early pioneers in studying the density of nitrogen. (Read the Science Connection below.) The following are data that he collected. Lord Rayleigh's measurements first appeared in *Proceedings of the Royal Society of London* (London, vol. 55, 1894). Each piece of data is the mass in grams of nitrogen filling a certain flask under a specified temperature and pressure.

Mass of nitrogen produced from chemical compounds (g)		
2.30143	2.29890	2.29816
2.30182	2.29869	2.29940
2.29849	2.29889	2.30074
2.30054		

Mass of nitrogen produced from the atmosphere (g)		
2.31017	2.30986	2.31010
2.31001	2.31024	2.31010
2.31028	2.31163	2.30956

a. Calculate the five-number summary for each set of data.

b. On the same axis, create a box plot for each set of data.

c. Describe any similarities and differences in the shapes of the box plots. Do the box plots support Lord Rayleigh's conjecture? ⓗ

d. If Lord Rayleigh performed the experiment exactly the same way each time, why did he get all these different values?

Science CONNECTION

One of the earliest persons to study the density of nitrogen was the English scientist Lord Rayleigh (1842–1919, born John William Strutt). Working with fairly small samples, he noticed that the density of nitrogen produced from chemical compounds was different from the density of nitrogen produced from the atmosphere. On the supposition that the air-derived gas was heavier than the "chemical" nitrogen, he conjectured the existence in the atmosphere of an unknown ingredient. In 1894, Lord Rayleigh isolated the unknown ingredient, the colorless, tasteless, and odorless gas called argon. In 1904, Lord Rayleigh was awarded the Nobel Prize in physics for his discovery. You can learn more about Rayleigh's work by using the links at www.keymath.com/DAA .

Lord Rayleigh

16. Refer to the backpack data listed in Example A. Separate the data by grade level.

a. Compute the mean and median weights for the juniors and for the seniors.

b. Create a modified box plot for each grade level. Put both box plots on the same axis.

c. Based on the information in your box plots, write a brief statement analyzing these two groups. Use the vocabulary developed in this lesson in your statement.

d. Based on your box plots, explain why the means or medians may have been greater in each grade level.

17. Identify the potential outlier(s) in the scores on this 20-question true-false test using the $1.5 \cdot IQR$ rule. Without this rule, would you classify the score of 11 as an outlier? Explain.

16, 18, 15, 16, 11, 15, 19, 16, 18, 14, 6, 18, 15, 17

Review

18. Find the next three terms in each sequence and write a recursive formula.

a. 42, 45, 48, . . .

b. 16, 40, 100, . . .

19. Evaluate each expression. Write your answers both in radical form and in decimal form rounded to one decimal place.

a. $\sqrt{\frac{432}{6}}$ (@)

b. $\sqrt{\frac{782 + 1354}{24}}$

c. $\sqrt{\frac{49 + 121 + 16 + 81 + 100}{4}}$

20. **APPLICATION** Rebecca wants to buy a used drum set that costs \$400. If she buys it now on credit, she will pay an annual interest rate of 15%, compounded monthly, on the unpaid balance. She can also wait until she has saved the money in her bank account, which earns an annual interest rate of 5%, compounded monthly. Either way she can contribute \$40 each month from her weekend job. How long will it take to pay for the drum set if she buys it on credit? How long would she have to save to buy it? Give Rebecca advice on whether to buy it now or later. (h)

21. Solve.

a. $x + 8 = 15$

b. $3x = 15$

c. $3x + 8 = 15$

EXPLORATION

Precision, Accuracy, and Significant Figures

When you make a measurement, how precise are you? How accurate are you? The words "precision" and "accuracy" are often used interchangeably, but they have very different meanings.

Precision refers to the repeatability of a measurement. If you measure something several times and get very similar answers each time, your measurements are precise. This is like throwing darts and having them all land in the same small region, but not necessarily in the center of the target.

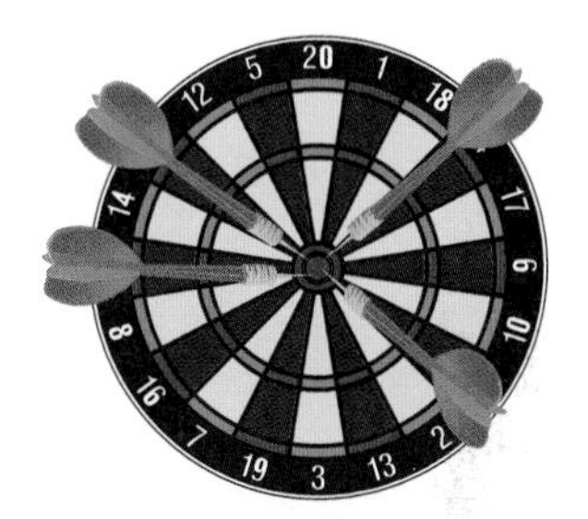

Accuracy refers to how close your measurement is to the actual value. For example, dart throws that all land in the center of the target are accurate.

Your measurements may be very precise but not accurate if your measuring device is not properly calibrated. For example, measuring lengths with a ruler that has been stretched a bit may result in precise measurements, but they won't be accurate.

On the other hand, if all your measurements are accurate (close to the actual value), then they will be precise, because they'll be close to each other.

Activity

Measuring Up

You will need

- a centimeter ruler

Work with a partner or your group to make the following measurements and calculations. Report your results to your teacher or a class recorder.

Step 1 Measure and record the dimensions of this textbook.

Step 2 Your teacher will record all measurements on the board. Compare the length measurements. How close are they? Do the same for the measurements of width and depth.

Step 3 Use your own measurements to calculate the volume of the textbook. Your teacher will record all the volumes on the board. How do they compare?

Step 4 Because all of you measured the same-sized object using the same type of measuring device, it might seem that the measurements should be identical. However, whenever you make a measurement, there is a built-in uncertainty in using the measuring device.

Scientists use **significant figures** (or **significant digits**) to monitor the degree of uncertainty of measurements when they are used in calculations. In general, the last digit of a measurement is uncertain. For example, suppose you use a meterstick, marked in millimeters, to measure the length of an object. If you record a measurement of 24.33 cm, the last digit is not certain. Another person might find the length to be 24.32 cm or 24.34 cm. However, you will agree on the 24.3 because you can read the meterstick accurately to the millimeter, or 0.1 cm. Based on this information, the precision of the measurement is $\pm$ 0.01 cm.

Look at your measurements and those of your classmates. Do they agree for one digit? Two digits? Describe how well the measurements of each dimension (length, width, and depth) agree.

Now you'll investigate the rules for identifying the number of significant figures in a measurement. Study the following chart showing the number of significant figures in various numbers.

Data value	Number of significant figures
125	3
4.7	2
5	1
1.08	3
0.004728	4

Data value	Number of significant figures
20.0	3
0.340	3
500	ambiguous case
5.00×10^2	3
5×10^2	1

Did you discover a pattern?

Nonzero figures always count. The zeros are a bit more complicated. There are three types of zeros: leading, trapped, and trailing.

- *Leading zeros,* such as those in 0.004728, never count.
- *Trapped zeros,* as in 1.08, always count.
- *Trailing zeros,* or those at the end of a number, count only when there is a decimal point. In the numbers 20.0, 300.00, and even 50., the zeros count as significant figures. If there is no decimal point, as in 500, then it isn't possible to tell whether the zeros are significant.

You can use scientific notation to avoid ambiguous cases. The measurement 5.00×10^2 has three significant figures, whereas the measurement 5×10^2 has only one significant figure.

LESSON

2.2

Measures of Spread

Out on the edge you see all kinds of things you can't see from the center.

KURT VONNEGUT

Students Connie and Oscar from Exercises 7 and 8 in Lesson 2.1 had the same mean and median scores, but the ranges and interquartile ranges for their scores were very different. The range and interquartile range both measure the spread of the data.

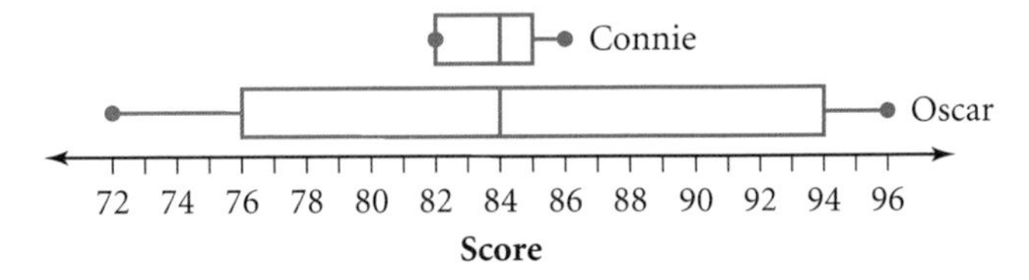

The interquartile range describes the spread of data relative to the median. It can also be useful to look at the spread relative to the mean. One way to measure the spread of data relative to the mean is to calculate the **deviations,** the signed differences between the data values and the mean. Recall the mean score for each student was 84. Here are the deviations for Connie's and Oscar's scores.

Assignment	Connie's score (%)	Deviation	Oscar's score (%)	Deviation
1	82	82 − 84 = **−2**	72	72 − 84 = **−12**
2	86	86 − 84 = **2**	94	94 − 84 = **10**
3	82	82 − 84 = **−2**	76	76 − 84 = **−8**
4	84	84 − 84 = **0**	96	96 − 84 = **12**
5	85	85 − 84 = **1**	90	90 − 84 = **6**
6	84	84 − 84 = **0**	76	76 − 84 = **−8**
7	85	85 − 84 = **1**	84	84 − 84 = **0**

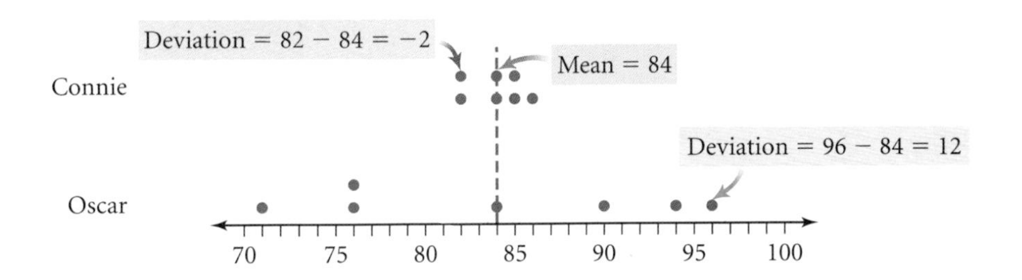

What do the deviations tell you about each student's performance? Certainly Connie's scores are far more predictable than Oscar's scores. Her deviations from the mean are smaller.

In the investigation, you'll perform an experiment and then explore ways to describe the **spread,** or variability, of your results.

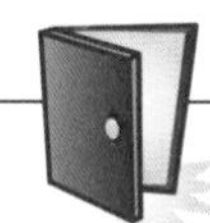

Investigation
A Good Design

You will need

- a measuring tape or metersticks
- paper
- books
- a pad of paper or cardboard
- a rubber band
- a ruler

In a well-designed experiment, you should be able to follow a specific procedure and get very similar results every time you perform the experiment. In this investigation you will attempt to control the setup of an experiment in order to limit the variability of your results.

Select and perform one of these experiments. Make complete and careful notes about the setup of your experiment.

Experiment 1: Rolling Ball

In this experiment you'll roll a ball of paper down a ramp and off the edge of your desk. Build your ramp from books, notebooks, or a pad of paper. Select the height and slope of your ramp and the distance from the edge of your desk, and determine any other factors that might affect your results. Make a ball by crumpling a piece of paper, and roll it down the ramp. Record the horizontal distance to the place where the ball hits the floor. Repeat this procedure with the same ball, the same ramp setup, and the same release another seven or eight times.

Experiment 2: Rubber Band Launch

In this experiment you'll use a ruler to launch a rubber band. Select the height and angle of your launch and the length of your stretch, and determine any other factors that might affect your results. Launch the rubber band into an area clear of obstructions. Record the horizontal distance of the flight. Repeat this procedure as precisely as you can with the same rubber band, the same launch setup, and the same stretch another seven or eight times.

Step 1 Use your data from Experiment 1 or 2. Calculate the mean distance for your trials and then calculate the deviations.

Step 2 In general, how much do your data values differ from the mean? How does the variability in your results relate to how controlled your setup was? Determine a way to calculate a *single* value that tells how accurate your group was at repeating the procedure. Write a formula to calculate your statistic using the deviations.

Step 3 A value known as the **standard deviation** helps measure the spread of data away from the mean. Use your calculator to find this value for your data. [▶ See **Calculator Note 2A** to learn how to calculate the standard deviation using your calculator. ◀]

Step 4 What are the units of the statistic you calculated in Step 2? The units of the standard deviation you found in Step 3 are the same as those of the original measurements. How does your statistic compare to the standard deviation?

Step 5 If you were going to repeat the experiment, how would you change your procedures to minimize the standard deviation?

If you want to measure the spread of data, it is typical to start by finding the mean. Next you find the deviation, or directed distance, from each data value to the mean. When you compare the results from two groups in the previous investigation, you may find that one group's mean is 200 cm and another group's mean is 300 cm. However, if their deviations are similar, then the groups performed the experiment equally well. The deviations let you compare the spread independent of the mean.

Consider Connie's and Oscar's scores and their deviations from the mean score for each student. How can you combine the deviations into a single value that reflects the spread in a data set? Finding the sum is a natural choice. However, if you think of the mean as a balance point in a data set, then the directed distances above and below the mean should cancel out. Hence, the deviation sum for both Connie and Oscar is zero.

Connie's deviation sum: $-2 + 2 + -2 + 0 + 1 + 0 + 1 = 0$

Oscar's deviation sum: $-12 + 10 + -8 + 12 + 6 + -8 + 0 = 0$

In order for the sum of the deviations to be useful, you need to eliminate the effect of the different signs. Squaring each deviation is one way to do this.

Connie			Oscar		
Score	Deviation	$(\text{Deviation})^2$	Score	Deviation	$(\text{Deviation})^2$
82	−2	**4**	72	−12	**144**
86	2	**4**	94	10	**100**
82	−2	**4**	76	−8	**64**
84	0	**0**	96	12	**144**
85	1	**1**	90	6	**36**
84	0	**0**	76	−8	**64**
85	1	**1**	84	0	**0**
$sum = \mathbf{14}$			$sum = \mathbf{552}$		
Variance → $\frac{sum}{6} = \mathbf{2.\overline{3}}$			$\frac{sum}{6} = \mathbf{92}$		
Standard deviation → $\sqrt{\frac{sum}{6}} \approx \mathbf{1.5}$			$\sqrt{\frac{sum}{6}} \approx \mathbf{9.6}$		

Begun in 1992 by Lorraine Serena and Elena Siff, *Women Beyond Borders* is a worldwide art project in which women artists are given identical wooden boxes and asked to transform them. If you consider the original box to be the mean, some of the resulting artworks have large deviations.

When you sum the squares of the deviations, the sum is no longer zero. The sum of the squares of the deviations, divided by one less than the number of values, is called the **variance** of the data. The square root of the variance is called the **standard deviation** of the data. The standard deviation provides one way to judge the "average difference" between data values and the mean. It is a measure of how the data are spread around the *mean.*

From left: **1.** The original pine box; **2.** Darlene Nguyen-Ely, USA, Vietnam, *Journey #17;* **3.** Madoka Hirata, Japan, *The Distance from Time #1;* **4.** Elena Mary Siff, USA, *Narcissism;* **5.** Cirenaica Moreira Diaz, Cuba, *Untitled;* **6.** Alejandra Mastro Sesenna, Guatemala, *Eva's Last Wish;* **7.** Gordana Kaljalovic Odanovic, Yugoslavia, *Model of Intimacy;* **8.** Lilian Nabulime, Kenya, *My Self.* On the web at *www.womenbeyondborders.org.*

In statistics, the mean is often referred to by the symbol $\bar{x}$ (pronounced "x bar"). Another symbol, Σ (capital *sigma*), is used to indicate the sum of the data values. For example, $\sum_{i=1}^{5} x_i$ means $x_1 + x_2 + x_3 + x_4 + x_5$, where x_1, x_2, x_3, x_4, and x_5 are the individual data values. So the mean of n data values is given by

$$\bar{x} = \frac{\sum_{i=1}^{n} x_i}{n}$$

The sum of all data values.

The number of data values.

Sigma notation is useful for defining standard deviation.

Standard Deviation

The **standard deviation,** s, is a measure of the spread of a data set.

$$s = \sqrt{\frac{\sum_{i=1}^{n} (x_i - \bar{x})^2}{n - 1}}$$

where x_i represents the individual data values, n is the number of values, and $\bar{x}$ is the mean. The standard deviation has the same units as the data.

The larger standard deviation for Oscar indicates that his scores generally lie much farther from the mean than do Connie's. A large value for the standard deviation tells you that the data values are not as tightly packed around the mean. As a general rule, a set with more data near the mean will have less spread and a smaller standard deviation.

You may wonder why you divide by $(n - 1)$ when calculating standard deviation. As you know, the sum of the deviations is zero. So, if you know all but one of the deviations, you can calculate the last deviation by making sure the sum will be zero. The last deviation depends on the rest, so the set of deviations contains only $(n - 1)$ independent pieces of data.

In Lesson 2.1, you learned one rule for identifying potential outliers using the five-number summary. Values that are more than $1.5 \cdot IQR$ from either Q_1 or Q_2 are potential outliers. A different rule used by some computer software is to identify any values more than two standard deviations from the mean as potential outliers.

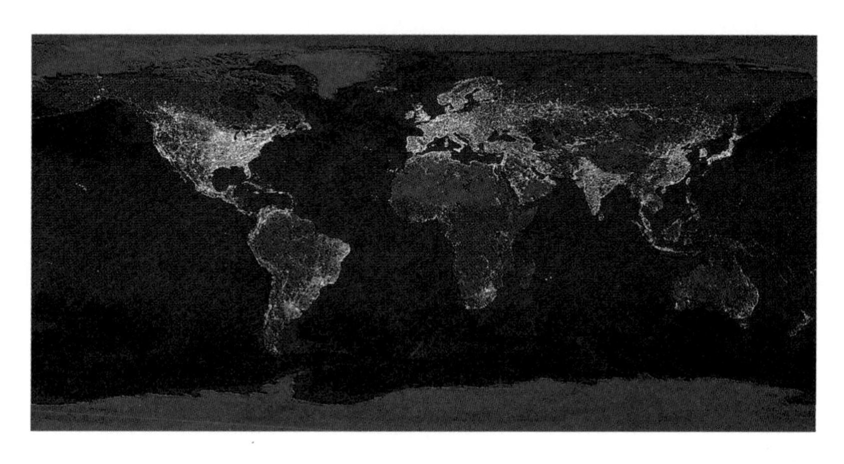

The National Geophysical Data Center created this map of human settlements, based on composite data from many satellites about sources of nighttime light. You could consider isolated points of light to be outliers because they are far removed from dense concentrations of light.

EXAMPLE This table gives the student-to-teacher ratios for public elementary and secondary schools in the United States.

Student-to-Teacher Ratios for Public Elementary and Secondary Schools (2003–2004 School Year)

State	AK	AL	AR	AZ	CA	CO	CT	DE	FL	GA	HI	IA	ID
Ratio	17.2	12.6	14.7	21.3	21.1	16.9	13.6	15.2	17.9	15.7	16.5	13.8	17.9

State	IL	IN	KS	KY	LA	MA	MD	ME	MI	MN	MO	MS	MT
Ratio	16.5	16.9	14.4	16.1	16.6	13.6	15.8	11.5	18.1	16.3	13.9	15.1	14.4

State	NC	ND	NE	NH	NJ	NM	NV	NY	OH	OK	OR	PA	RI
Ratio	15.1	12.7	13.6	13.7	12.7	15.0	19.0	13.3	15.2	16.0	20.6	15.2	13.4

State	SC	SD	TN	TX	UT	VA	VT	WA	WI	WV	WY
Ratio	15.3	13.6	15.7	15.0	22.4	13.2	11.3	19.3	15.1	14.0	13.3

(U.S. Department of Education, National Center for Education Statistics)

a. Calculate the mean and the standard deviation. What do the statistics tell you about the spread of the student-to-teacher ratios?

b. Identify any values that are more than two standard deviations from the mean. What percentage of all values are more than two standard deviations from the mean?

▸ Solution

a. The mean ratio is approximately 15.55 students per teacher. The standard deviation is approximately 2.48 students per teacher. More than half the ratios are within 2.48 students of the mean.

b. The mean minus two standard deviations is $15.55 - 2(2.48) \approx 10.58$. There are no states that average fewer than 10.58 students per teacher. The mean plus two standard deviations is $15.55 + 2(2.48) \approx 20.51$. There are four states that exceed this value: OR (20.6), CA (21.1), AZ (21.3), and UT (22.4). So $\frac{4}{50}$, or 8%, of the states are more than two standard deviations from the mean.

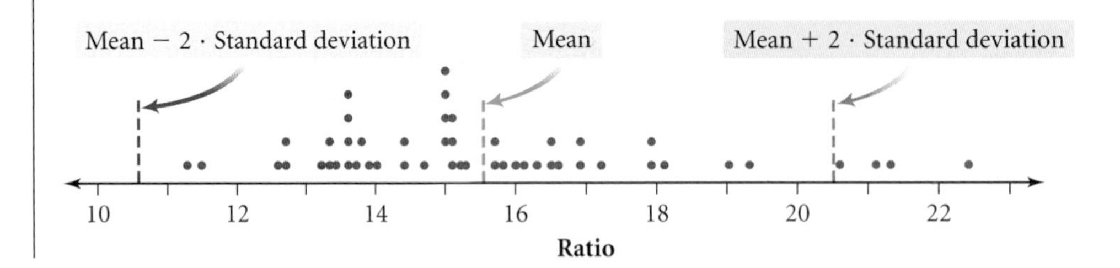

Be careful when you draw conclusions from data. Consider what the data values represent and whether the sample represents a larger population. Would it make sense to look at the student-to-teacher ratio for your school (or any other school) and compare it with these values? Why or why not?

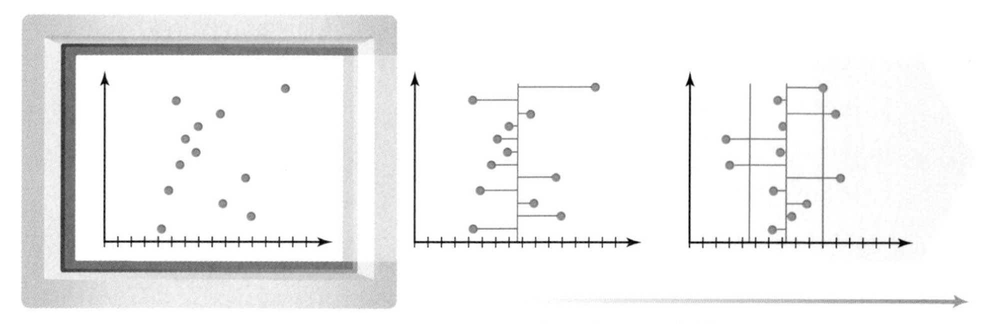

[▶You can use the **Dynamic Algebra Exploration** at www.keymath.com/DAA to explore how changing a data set changes the mean and standard deviation.◀]

As you work the exercises, you may notice that finding a measure of central tendency is usually a necessary step in measuring the spread. In order to calculate the standard deviation, you first need to calculate the mean. The interquartile range relies on the first and third quartiles, which in turn rely on the median.

EXERCISES

You will need

A graphing calculator for Exercises **2, 8, 12, 13, 15,** and **18.**

Practice Your Skills

1. Given the data set {41, 55, 48, 44}: ⓐ

a. Find the mean.

b. Find the deviation from the mean for each value.

c. Find the standard deviation of the data set.

2. The lengths in minutes of nine music CDs are 45, 63, 74, 69, 72, 53, 72, 73, and 50.

a. Find the mean.

b. Find the deviation from the mean for each value.

c. Find the standard deviation of the data set.

d. What are the units of the mean, the deviations from the mean, and the standard deviation?

3. Students collected eight length measurements for a mathematics lab. The mean measurement was 46.3 cm, and the deviations of *seven* individual measurements were 0.8 cm, −0.4 cm, 1.6 cm, 1.1 cm, −1.2 cm, −0.3 cm, and −1.0 cm. ⓗ

a. What were the original eight measurements collected?

b. Find the standard deviation of the original measurements.

c. Which measurements are more than one standard deviation above or below the mean?

4. Invent a data set with six data values such that the mean is 10 and the standard deviation is $\sqrt{\frac{2}{5}} \approx 0.632$. ⓗ

5. While you can't actually find the mean and standard deviation from a box plot, you can match these by estimation and elimination.

a. mean = 28, standard deviation = 6

b. mean = 25, standard deviation = 10

c. mean = 25, standard deviation = 6

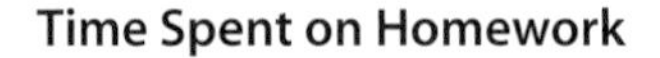

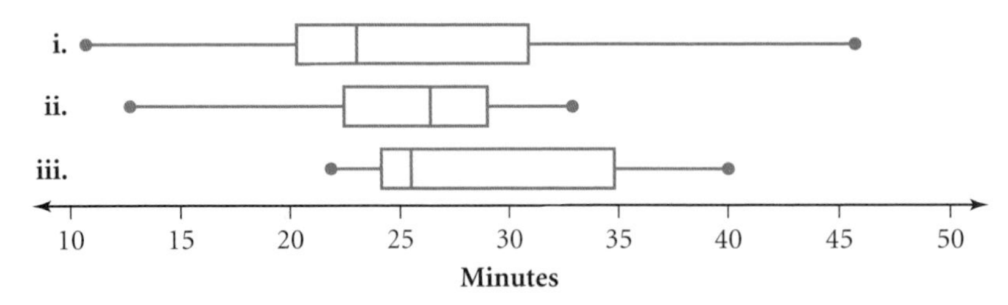

Reason and Apply

6. APPLICATION The mean diameter of a Purdy Goode Compact Disc is 12.0 cm, with a standard deviation of 0.012 cm. No CDs can be shipped that are more than two standard deviations from the mean. How would the company's quality control engineer use those statistics?

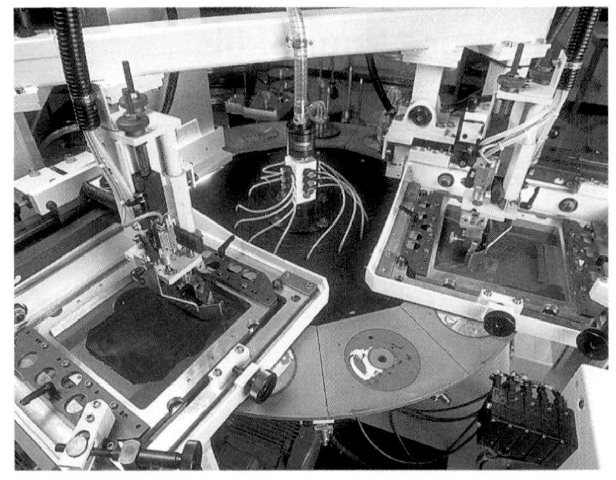

This automated machine paints labels on compact discs.

7. Two groups do the rubber band launch experiment seven times.

Group A distance (cm): {182, 186, 182, 184, 185, 184, 185}

Group B distance (cm): {52, 94, 66, 116, 100, 76, 84}

a. What is the mean for each group? ⓐ

b. What is the standard deviation for each group? ⓐ

c. Based on your answers for 7a and b, compare the results of the two groups.

8. Find the standard deviation and interquartile range of the backpack data from Example A in Lesson 2.1. Which of these two values is larger? Will this value always be larger? Explain your reasoning and find or create another data set that supports your answer.

9. Two data sets have the same range and interquartile range, but the first is symmetric and the second is skewed left.

a. Sketch two box plots that satisfy the conditions for the two sets. ⓐ

b. Would you guess that the standard deviation of the skewed data set is less than, more than, or the same as the first? Explain your reasoning.

c. Invent two data sets of seven values each that satisfy the conditions. ⓗ

d. Find the standard deviation for the two data sets. Do the standard deviations support your answer to 9b?

10. APPLICATION The students in four classes recorded their resting pulse rates in beats per minute. The class means and standard deviations are given at right.

a. Which class has students with pulse rates most alike? How can you tell?

b. Can you tell which class has the students with the fastest pulse rates? Why or why not?

Resting Pulse Rates (beats/min)

Class	Mean	Standard deviation
First period	79.4	3.2
Third period	74.6	5.6
Fifth period	78.2	4.1
Sixth period	80.2	7.6

11. Here are the mean daily temperatures in degrees Fahrenheit for two cities.

a. Find the mean and standard deviation for each city.

b. Draw a box plot for each city. Find each median and interquartile range.

c. Which city has the smaller spread of temperatures? Justify your conclusion.

d. Does the interquartile range or the standard deviation give a better measure of the spread? Justify your conclusion.

Mean Temperatures (°F)

Month	Juneau, Alaska	New York, New York
January	26	32
February	29	34
March	34	41
April	41	50
May	48	60
June	54	69
July	57	75
August	56	74
September	50	67
October	42	57
November	33	47
December	29	37

12. Members of the school mathematics club sold packages of hot chocolate mix to raise funds for their club activities. The numbers of packages sold by individual members are given at right.

a. Find the median and interquartile range for this data set. @

b. Find the mean and standard deviation. @

c. Draw a box plot for this data set. Use the $1.5 \cdot IQR$ definition to name any numbers that are outliers.

d. Remove the outliers from the data set and draw another box plot.

e. With the outliers removed, recalculate the median and interquartile range and the mean and standard deviation.

f. Which is more affected by outliers, the mean or the median? The standard deviation or the interquartile range? Explain why you think this is so.

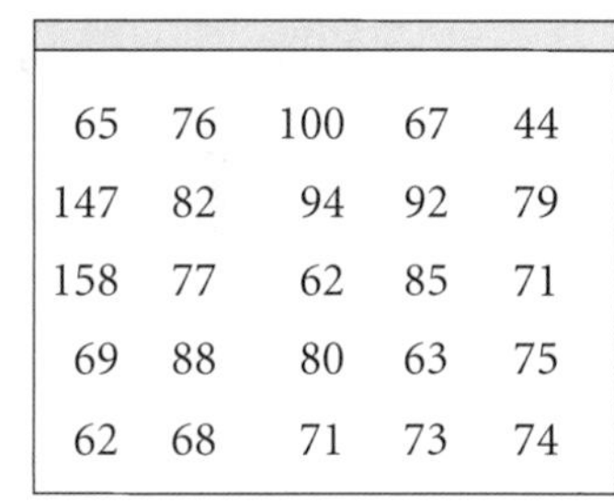

65	76	100	67	44
147	82	94	92	79
158	77	62	85	71
69	88	80	63	75
62	68	71	73	74

13. Your group built a catapult and launched tennis balls 20 times, trying to keep all the conditions the same for each launch. The mean distance traveled was 23.5 m, with a standard deviation of 4.7 m. Would you be surprised if the next launch was 35 m? Why or why not?

14. APPLICATION Matt Decovsky wants to buy a CD/MP3 player for his car at an online auction site. Before bidding, he decides to do some research on the selling price of recently sold players. His search comes up with these 20 prices.

\$54.99	\$89.99	\$120.00	\$89.95
\$129.99	\$282.95	\$50.99	\$116.00
\$72.51	\$62.00	\$42.00	\$96.00
\$158.00	\$83.89	\$157.95	\$194.99
\$106.95	\$149.95	\$175.00	\$289.00

a. Find the mean, median, and mode.

b. Draw a box plot of the data. Describe its shape.

c. Find the interquartile range and determine whether there are any outliers.

d. While looking over the items' descriptions, Matt realizes that the outlier CD/MP3 players contain features that don't interest him. If he removes the outliers, will the mean or the median be less affected? Explain.

e. Sketch a new box plot with the outliers removed. How does this help you support your answer in 14d?

f. If you were Matt, what might you set as a target price in your bidding? Explain your reasoning. ⓗ

15. A sample is biased if it excludes or underrepresents some part of the sampled population. The mean of a biased sample is not representative of the mean of the population. It may be too high or too low.

Suppose you want to describe the height of students at your school, and you use the boys' varsity basketball team as your sample. How might the mean and standard deviation of the sample compare with the mean and standard deviation of the population?

Review

16. Celia lives 2.4 km from school. She misses the school bus and starts walking at 1.3 m/s. She has 20 min before school starts. Write a recursive formula and use it to find out whether or not she gets to school on time.

17. Solve.

a. $\frac{x+5}{4} + 3 = 19$ ⓐ

b. $\frac{3(y-4)+6}{6} - 2 = 7$

18. These data sets give the weights in pounds of the offensive and defensive teams of the 2007 Super Bowl Champion Indianapolis Colts. (*www.superbowl.com*)

Offensive players' weights (lb): {198, 332, 290, 295, 295, 320, 252, 185, 230, 214, 251}

Defensive players' weights (lb): {245, 300, 274, 268, 227, 235, 243, 182, 180, 203, 206}

a. Find the mean and median weights of each team.

b. Prepare a box plot of each data set. Use the box plots to make general observations about the differences between the two teams.

LESSON

2.3

Histograms and Percentile Ranks

You miss 100 percent of the shots you never take.

WAYNE GRETZKY

A box plot gives you an idea of the overall distribution of a data set, but in some cases you might want to see other information and details that a box plot doesn't show. A **histogram** is a graphical representation of a data set, with columns to show how the data are distributed across different intervals of values. Histograms give vivid pictures of distribution features, such as clusters of values, or gaps in data.

The columns of a histogram are called **bins** and should not be confused with the bars of a bar graph. The bars of a bar graph indicate categories—how many data items either have the same value or share a characteristic. The bins of a histogram indicate how many numerical data values fall within a certain interval. You would use a bar graph to show how many people in your class have various eye colors, but a histogram to show how many people's heights fall within various intervals.

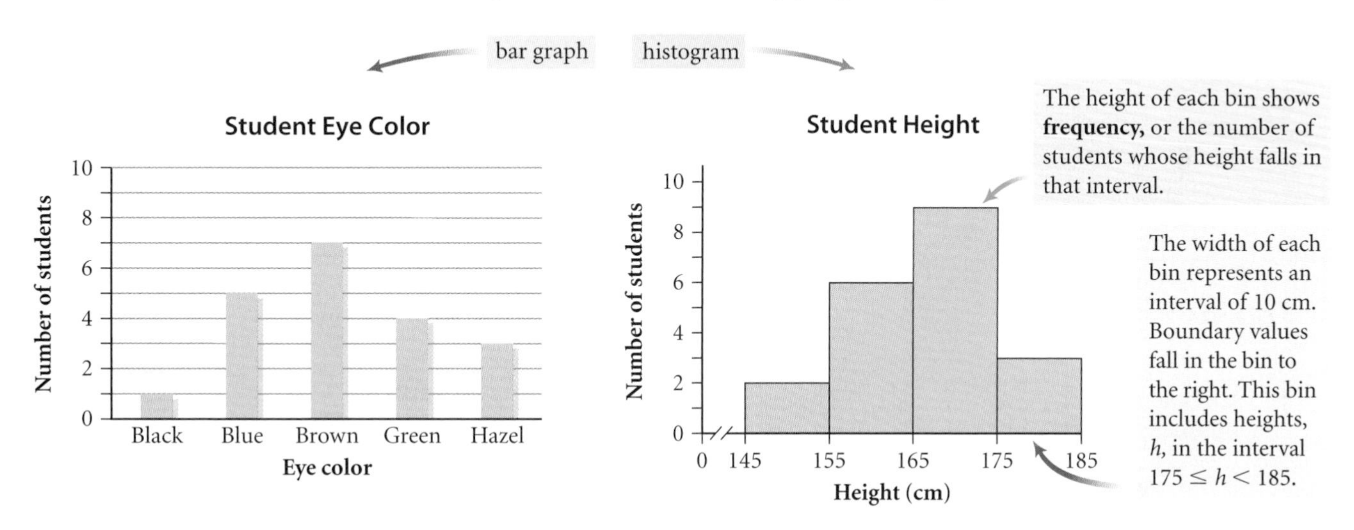

Histograms are a good way to display information from large data sets. Although you can't see individual data values, you can see the shape of the data and how the values are distributed throughout the range. As you will see in Example A, bin width depends on how much detail you want to show, but all the bins should have the same width.

Some stereo equalizers have spectrum displays that resemble histograms. These displays are similar to histograms because they show the output frequencies by intervals, or bands. They are different because the bands may not represent equal intervals.

EXAMPLE A

Shatevia took a random sample of 50 students who own MP3 players at her high school and asked how many songs they have stored. The two graphs were constructed from the data in the table.

Number of Songs Stored in MP3 Player

921	866	943
796	933	900
976	841	901
905	925	863
834	895	1013
987	975	891
933	875	833
898	926	966
885	956	934
935	846	904
975	864	765
863	924	906
806	944	915
864	812	862
927	1004	925
874	974	787
896	794	

Graph A
Number of Songs in MP3 Player

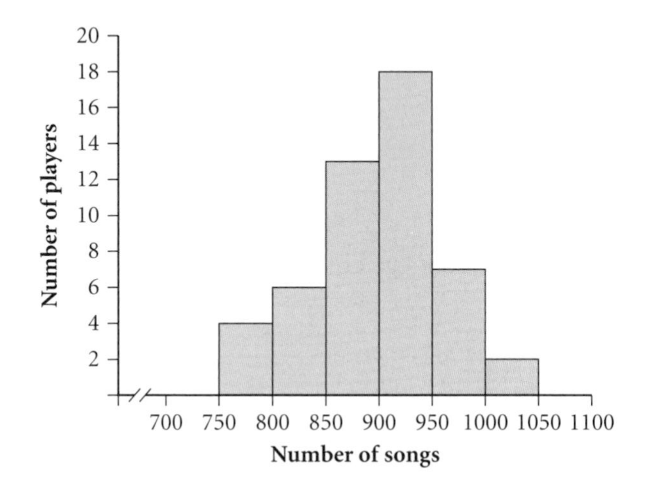

Graph B
Number of Songs in MP3 Player

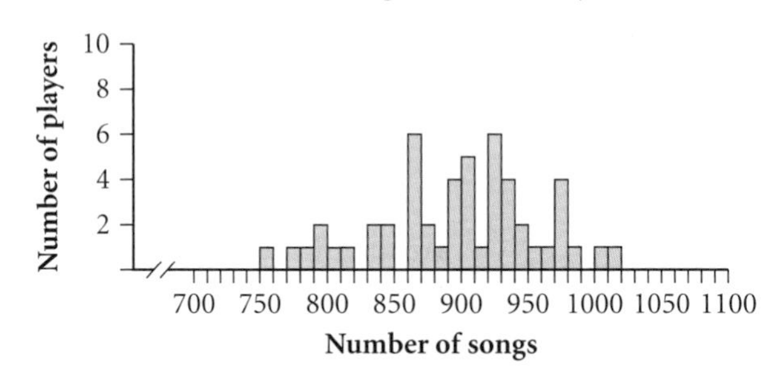

a. What is the range of the data?

b. What is the bin width of each graph?

c. How can you know if the graph accounts for all 50 values?

d. Why are the columns shorter in Graph B?

e. Which graph is better at showing the overall shape of the distribution? What is that shape?

f. Which graph is better at showing the gaps and cluster in the data?

g. What percentage of the players have fewer than 850 songs stored?

▶ Solution

a. The number of songs goes from a low of 765 songs to a high of 1013 songs. The range is 248 songs.

b. The bin width of Graph A is 50 songs, and the bin width of Graph B is 10 songs.

c. The sum of all the bin frequencies is 50 for each of the graphs.

d. The bins in Graph A hold the values of up to five bins from Graph B. With smaller bin widths you will usually have shorter bins.

e. Graph A shows that the distribution is skewed left. This fact is harder to see with all the ups and downs in Graph B.

f. With more bins you can see gaps and clusters in the data. A dot plot is like a histogram with a very small bin width. Graph B is the better graph for seeing gaps and clusters.

g. Add the bin frequencies for the bins below (to the left of) 850 songs. There are 10 data values, so 10 out of 50, or 20% of the sample, had fewer than 850 songs.

The **percentile rank** of a value is the percentage of data values that are below the given value. In the example, 850 songs has a percentile rank of 20 because this value is greater than 20% of the values in the sample.

Suppose a large number of students take a standardized test, such as the SAT. There are so many individual scores that it would be impractical to look at all of the actual numbers. A percentile rank gives a good indication of how one person's score compares to other scores across the country.

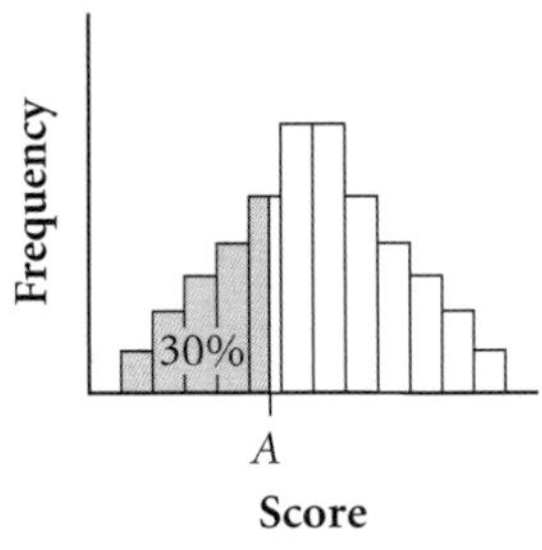

Students with score *A* are at the 30th percentile, because their score is better than the scores of 30% of the tested students.

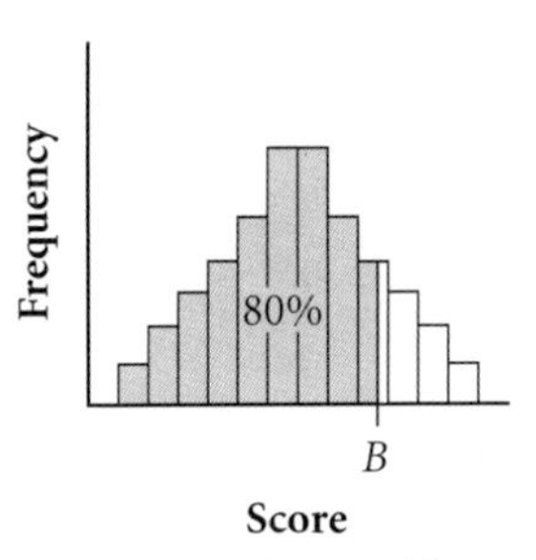

Likewise, students with score *B* are at the 80th percentile, because 80% of the tested students have scores that are lower.

EXAMPLE B

The data used in this histogram have a mean of 34.05 and a standard deviation of 14.68.

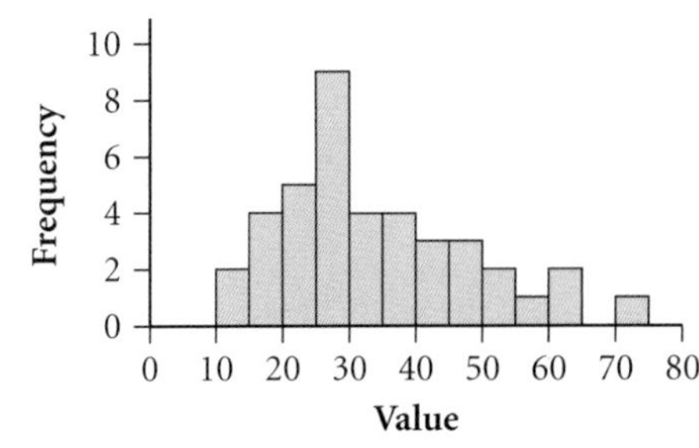

a. Approximate the percentile rank of a value two standard deviations above the mean.

b. Approximately what percentage of the data values are within one standard deviation of the mean?

▶ Solution

Add the bin frequencies to find that there are 40 data values in all.

a. The value of two standard deviations above the mean is $34.05 + 2 \cdot 14.68$, or 63.41. All of the data values in the ten bins up to the value of 60 are less than 63.41. Adding the bin frequencies up to 60 gives 37.

Therefore $\frac{37}{40}$, or approximately 92.5%, of the data lie below 63.41. So 63.41 is approximately the 93rd percentile.

b. One standard deviation above the mean is 48.73, and one standard deviation below the mean is 19.37. This interval includes at least those values in the bins from 20 to 45. So $\frac{25}{40}$, or approximately 62.5%, of the data lie within one standard deviation of the mean.

Combining what you know about measures of central tendency and spread with different displays of data enables you to provide a complete picture of a data set. The following investigation gives you an opportunity to analyze data using all of the statistics and graphs you have learned about.

Investigation
Eating on the Run

Teenagers require anywhere from 1800 to 3200 calories per day, depending on their growth rate and level of activity. The food you consume as part of your diet should include sufficient fiber, moderate levels of carbohydrates and fat, and as little sodium, saturated fat, and cholesterol as possible. The table shows the recommended amounts of carbohydrates and fiber and the maximum amounts of other nutrients in a healthy 2500-calorie diet.

Nutrition Recommendations for a 2500-Calorie Diet

Nutrient	Amount
Total fat	Less than 80 g
Saturated fat	Less than 25 g
Cholesterol	Less than 300 mg
Sodium	Less than 2400 mg
Total carbohydrate	375 g
Dietary fiber	30 g

(U.S. Food and Drug Administration)

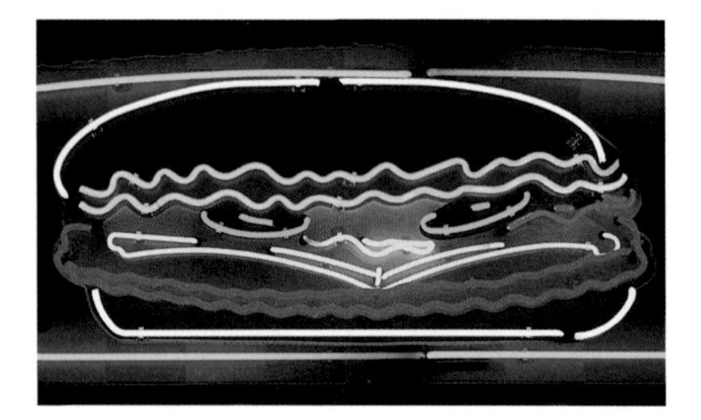

So, how does fast food fit into a healthy diet? Examine the information about the nutritional content of fast-food sandwiches. With your group, study one of the nutritional components (total calories, total fat, saturated fat, cholesterol, sodium, or total carbohydrate). Use box plots, histograms, and the measures of central tendency and spread to compare the amount of that component in the sandwiches. You may want to divide your data so that you can make comparisons between different types of sandwiches or between restaurants. As you do your statistical analysis, discuss how these fast-food items would affect a healthy diet. Prepare a short report or class presentation discussing your conclusions.

[► See **Calculator Note 2C** to learn how to make histograms on your calculator. ◄]

Consumer CONNECTION

Fast food is a popular choice today because it is quick and convenient. Despite being high in fat, calories, sodium, and cholesterol, fast food is not bad, nutritionists say, but should be consumed in moderation with an otherwise healthy diet. Many fast-food restaurants have responded to America's new health consciousness and now offer low-calorie menu items such as salads, lean meats, and chili.

Fast-Food Nutrition Facts

Sandwich	Total calories	Total fat (g)	Saturated fat (g)	Cholesterol (mg)	Sodium (mg)	Carbohydrate (g)
Burger King Double Whopper with Cheese	1060	69	27	185	1540	53
Carl's Jr. Western Bacon Cheeseburger	657	31	12	82	1387	65
Carl's Jr. The Six Dollar Chili Cheese Burger	926	57	27	154	1960	57
Hardee's 1/2-Lb. Grilled Sourdough Thickburger	1100	74	30	155	1430	61
Hardee's 1/2-Lb. Six Dollar Burger	1120	73	30	150	1870	72
Hardee's 1/3-Lb. Cheeseburger	680	39	19	90	1320	51
Jack in the Box Hamburger with Cheese	355	17.5	7	55	770	31
Jack in the Box Sourdough Jack	715	51	18	75	1165	36
McDonald's Quarter Pounder with Cheese	540	29	13	95	1240	39
Sonic No. 1 Sonic Burger	577	36	7	37	753	43
Wendy's Hamburger, Kids' Meal	270	9	3.5	30	610	33
Burger King Chicken Whopper	570	25	4.5	75	1410	48
Carl's Jr. Bacon Swiss Crispy Chicken Sandwich	757	38	11	91	1554	72
Carl's Jr. Carl's Western Crispy Chicken Sandwich	747	28	10	79	1875	92
Carl's Jr. Charbroiled Santa Fe Chicken Sandwich	610	32	7	97	1440	45
Carl's Jr. Southwestern Spicy Chicken Sandwich	624	41	9	63	1642	48
Dairy Queen Grilled Chicken Sandwich	340	16	2.5	55	1000	26
Jack in the Box Chicken Sandwich	390	21	4	35	730	39
Jack in the Box Chicken Sandwich with Cheese	430	24	6	45	880	40
Jack in the Box Sourdough Grilled Chicken Club	505	27	6.5	75	1220	35
McDonald's Crispy Chicken	510	26	4.5	50	1090	47
McDonald's Chicken McGrill	400	16	3	70	1020	37

(*www.foodfacts.info*)

In order to provide a complete statistical analysis of a data set, statisticians often need to use several different measurements and graphs. Throughout this course you will learn about more statistics that help you make accurate predictions and conclusions from data.

EXERCISES

You will need

A graphing calculator for Exercise **7**.

Practice Your Skills

1. The histogram at right shows a set of data of backpack weights.

a. How many values are between 2 kg and 3 kg?

b. How many values are in the data set?

c. Make up a set of data measured to the nearest tenth of a kilogram that creates this histogram. ⓗ

Weight of Students' Backpacks

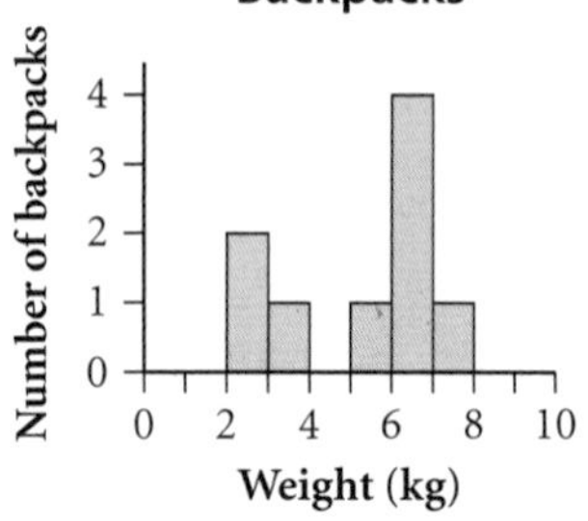

2. Study these four histograms.

i.

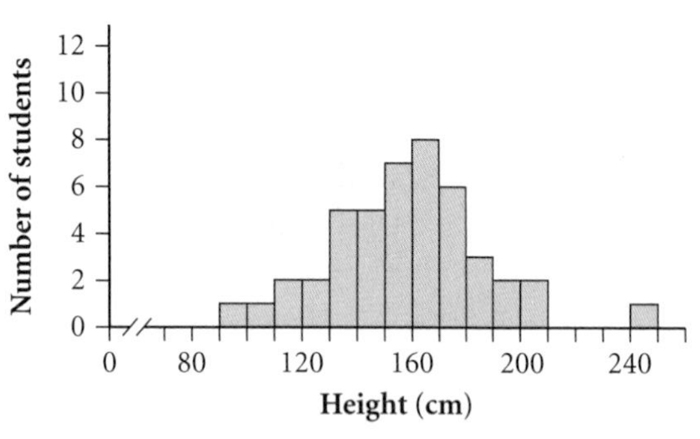

ii.

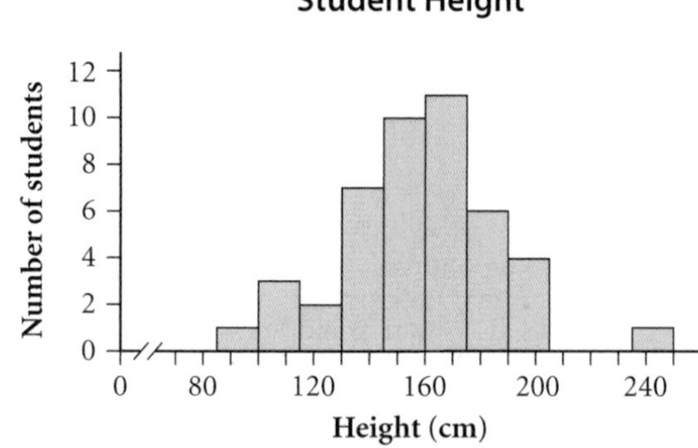

iii.

Student Height

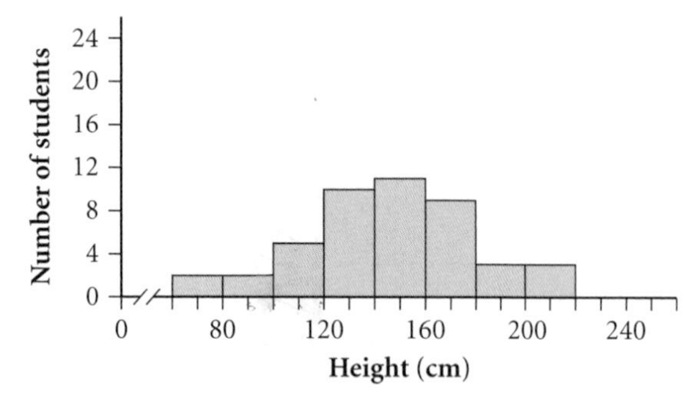

iv.

Student Height

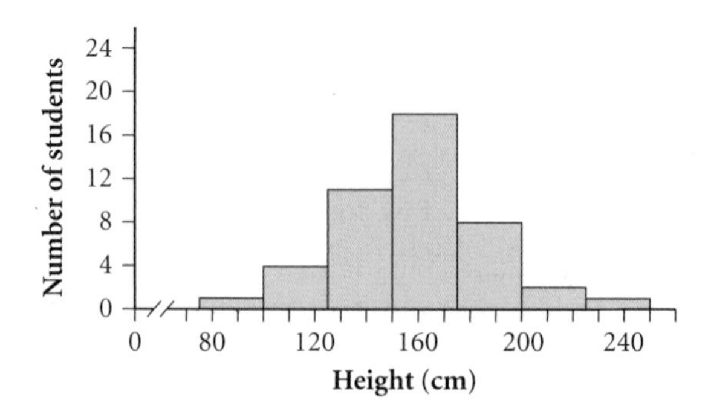

a. What is the bin width of each histogram above?

b. Which histogram could not come from the same data set as the other three? Explain why. ⓗ

3. These data are the head circumferences in centimeters of 20 newborn girls:

{31, 32, 33, 33, 33, 34, 34, 34, 34, 35, 35, 35, 35, 35, 35, 36, 36, 36, 37, 38}

a. How many of the 20 newborn girls have a head circumference less than 34 cm?

b. What is the percentile rank of 34 cm for this sample? (h)

c. What is the percentile rank of 38 cm for this sample?

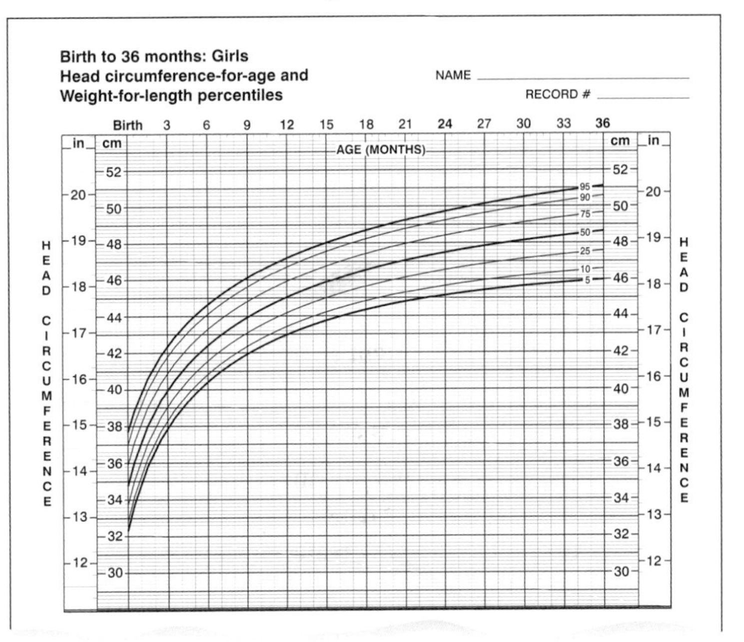

Growth charts, such as the one shown at right, often give a range of data divided into percentiles. This chart shows the 5th, 10th, 25th, 50th, 75th, 90th, and 95th percentiles.

Reason and Apply

4. Carl and Bethany roll a pair of dice 1000 times and keep track of the result of each roll. The frequency of each result is listed below and shown in the histogram.

Result	Frequency
2	26
3	56
4	83
5	110
6	145
7	162
8	149
9	114
10	73
11	61
12	21

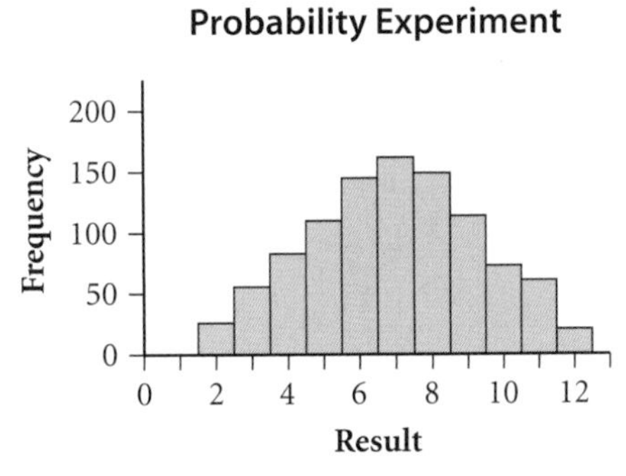

a. Explain why the histogram is mound-shaped.

b. Describe how to find the mean result and the median result for this data set. (h)

5. Rita and Noah survey 95 farmers in their county to see how many acres of sweet corn each farmer has planted. They summarize their results in a histogram.

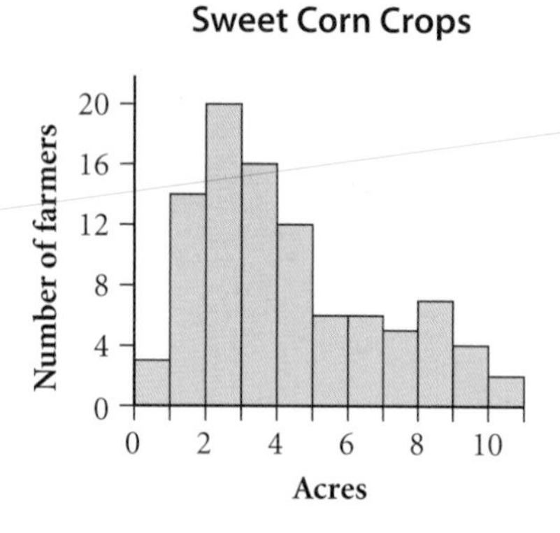

A harvesting vehicle at work in a corn field

a. The distribution is skewed right. Explain what this means in terms of the data set.

b. Sketch a possible box plot of the data. Explain how you estimated the five-number summary.

c. Is it reasonable for Rita and Noah to use their results to draw conclusions about the number of acres of sweet corn in farms across the entire state?

6. Describe a situation and sketch a histogram to reflect each condition named below.

a. mound-shaped and symmetric @

b. skewed left

c. skewed right

d. rectangular

7. Ignacio kept a log of the amount of time he spent doing homework and watching television during 20 school days.

Day	1	2	3	4	5	6	7	8	9	10
Homework (min)	4	10	40	11	55	46	46	23	57	28
Television (min)	78	30	15	72	25	30	90	40	35	56

Day	11	12	13	14	15	16	17	18	19	20
Homework (min)	65	58	52	38	38	39	45	27	41	44
Television (min)	12	5	95	27	38	50	10	42	60	34

a. Draw two box plots, one showing the amount of time spent doing homework, and one showing the amount of time watching television. Which distribution has the greater spread? @

b. Make an educated guess about the shape of a histogram for each set of data. Will either be skewed? Mound-shaped? Check your guess by drawing each histogram.

c. Calculate the median and interquartile range, and the mean and standard deviation, for both homework and television. Which measure of spread best represents the data?

8. At a large university, 1500 students took a final exam in chemistry.

a. Frank learns that his score of 76 (out of 100) places him at the 88th percentile. How many students scored lower than Frank? How many scored higher?

b. Mary scored 82, which placed her at the 95th percentile. Describe how Mary's performance compares with that of others in the class.

c. The highest score on the exam was 91. What percentile rank is associated with this score?

d. Every student who scored above the 90th percentile received an A. How many students earned this grade?

e. Explain the difference between a percent score and a percentile rank. In your opinion, should you be evaluated based on percent or percentile? Why?

Science CONNECTION

How are speed limits determined? Radar checks are performed at selected locations on the roadway to collect data about drivers' speeds under ideal driving conditions. A statistical analysis is then done to determine the 85th percentile speed. Studies suggest that posting limits at the 85th percentile minimizes accidents and traffic jams and that drivers are more likely to comply with the speed limit. You can learn more about how speed limits are determined by using the links at **www.keymath.com/DAA** .

9. APPLICATION Traffic studies have shown that the best speed limit to post on a given road is the 85th percentile speed.

Assume a road engineer collects these data to determine the speed limit on a local street.

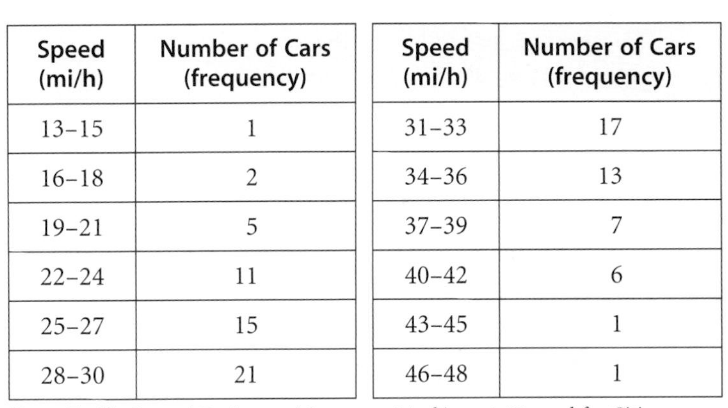

Speed (mi/h)	Number of Cars (frequency)	Speed (mi/h)	Number of Cars (frequency)
13–15	1	31–33	17
16–18	2	34–36	13
19–21	5	37–39	7
22–24	11	40–42	6
25–27	15	43–45	1
28–30	21	46–48	1

(Iowa Traffic Control Devices and Pavement Markings: A Manual for Cities and Counties, Center of Transportation Research and Education, 2001)

a. Draw a histogram for these data. @

b. Find the 85th percentile speed.

c. What speed limit would you recommend based on this traffic study?

d. What other factors should be considered in determining a speed limit?

10. APPLICATION A road engineer studies a rural two-lane highway and presents this histogram to the County Department of Highways.

a. For how many cars was data collected?

b. What is the 85th percentile speed?

c. What speed limit would you recommend for this highway?

Speed Limit Study

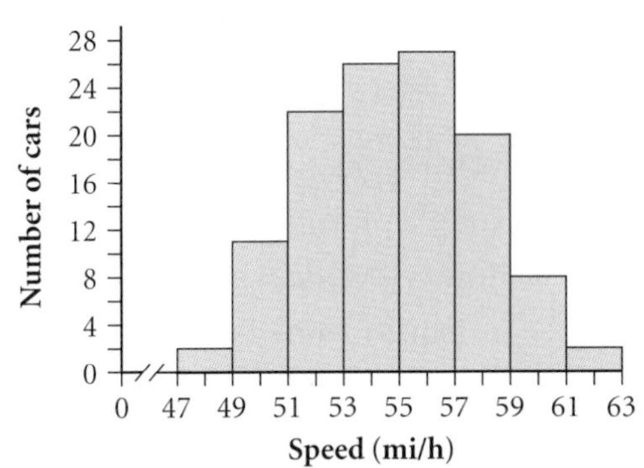

Review

11. Penny calculates that the deviations from the mean for a data set of eight values are 0, −40, −78, −71, 33, 36, 42, and 91.

a. How do you know that at least one of the deviations is incorrect?

b. If it turns out that 33 is the only incorrect deviation, what should the correct deviation be?

c. Use the corrected deviations to find the actual data values, the standard deviation, the median, and the interquartile range if the mean is

i. 747 **ii.** 850

d. Write your observations from 11a–c. (h)

12. At Piccolo Pizza Parlor, a large cheese pizza sells for \$8.99. Each topping costs an additional \$0.50.

a. How much will a four-topping pizza cost? (a)

b. The Piccolo Extra Special has eight toppings and costs \$12.47. How much do you save by ordering this special combination instead of ordering eight toppings of your own choosing?

13. Courtney can run 100 m in 12.3 s. Marissa can run 100 yd in 11.2 s. Who runs faster? (There are 2.54 centimeters per inch, 12 inches per foot, and 3 feet per yard.) (h)

Project

STEM-AND-LEAF PLOTS

John Tukey introduced **stem-and-leaf plots** in 1972 as a form for one-variable data that is appropriately compact and easy to look over.

Something like a sideways histogram, the stem-and-leaf plot is more detailed because individual data values can be found in the graph. Here are Connie's and Oscar's scores from Exercise 7 in Lesson 2.1 displayed in a stem-and-leaf plot.

Connie		Oscar
	7	2
	•	6 6
5 5 4 4 2 2	8	4
6	•	
	9	0 4
	•	6

Key

7 | 2 = 72

How do you think a stem-and-leaf plot works, and why do you need a key? Read about stem-and-leaf plots in a high school or college statistics book and prepare a research project.

Your project should include

- Instructions to make a stem-and-leaf plot from data.
- Sample plots using data from this chapter. Include variations of stem-and-leaf plots.
- How to decide what level of accuracy is needed to communicate the data usefully.
- Insights obtained from your stem-and-leaf plots.

EXPLORATION

Census Microdata

Since the establishment of the United States, the Constitution has required a nationwide population count, or census. The first U.S. census was conducted in 1790, less than a year after George Washington was inaugurated. A full census has been conducted every 10 years since.

The first censuses were primarily concerned with the number of people so that the federal government could make decisions about representation and taxation. Today, however, the U.S. Census Bureau collects a wide variety of data, including age, sex, race, national origin, marital status, and education. You can learn more about the history of the U.S. Census by visiting the Internet links at www.keymath.com/DAA .

A census taker gathers data at a New York City home in 1930.

The most detailed information published by the U.S. Census Bureau is called microdata—data about individuals. These microdata are originally published as an array of numbers, as shown below. You can see that microdata, by themselves, would not be very useful to someone trying to make decisions.

Collecting data is only the first part of the U.S. Census Bureau's job. The Bureau also analyzes these data and publishes reports that help federal, state, and local governments, organizations, businesses, and citizens make decisions. In this exploration you'll use Fathom Dynamic Data™ Software to analyze a set of census microdata. You'll make some conjectures about the data set and use what you've learned about statistics in this chapter to support or refute your conjectures.

```
.........1.........2..
1234567890123456789012
P0089277000001400000130
P0089277011001390000110
P0089277020001104000100
P0089277020001064000100
P0089277020001044000080
P0089277001001190000180
P0089277010001470000180
```

This table shows 1990 U.S. Census microdata about people living around Berkeley, California. Each row represents one individual. Shaded and unshaded rows group individuals that live in the same household. Columns represent specific information about each individual. For example, columns 15–16 indicate age. In the first row the person is 40 years old.

Social Science CONNECTION

In theory, the census counts every person living in the United States. In actuality, despite outreach programs that attempt to count everyone, including people without housing and people who are not able to read or complete the census, the 2000 U.S. Census is estimated to have excluded up to 3.4 million people. Furthermore, the U.S. Census Bureau today uses both a short form and a long form, such that some questions are asked only of a small sample (5% to 20%) of the population, and the results are statistically applied to the whole population. Some statisticians believe that a census founded entirely on random sampling but conducted more thoroughly by tracking down every single person in that sample may be more beneficial in the future.

Activity

Different Ways to Analyze Data

Step 1 Start Fathom. From the File menu choose **Open Sample Document.** Open one of the census data files in the **Social Science** folder. You'll see a box of gold balls, called a collection, that holds data about several individuals, or cases.

Step 2 Click on the collection and then choose **New | Case Table** from the Object menu. You now have a table of your microdata. Scroll through the data. How many people are there? What specific data, or attributes, were collected about each person?

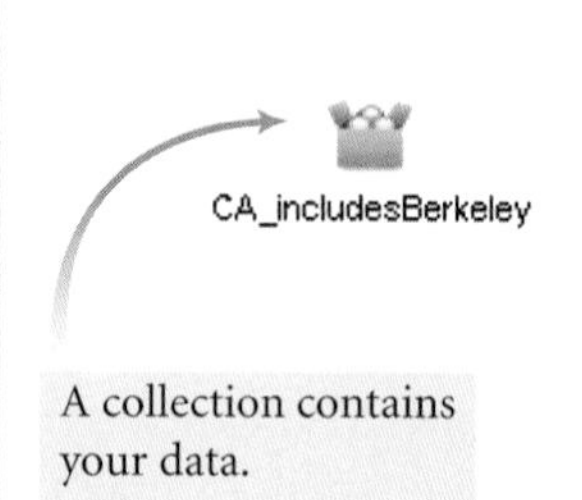

A collection contains your data.

CA_includesBerkeley

	sex	age	race
1	F	38	White
2	M	42	White
3	M	35	Korean
4	F	34	Korean
5	M	3	Korean

A case table lets you see the attributes for each case in the collection.

Step 3 Make a conjecture about the people in your data set. Your conjecture can be about just one attribute, such as "The majority of people have some education beyond high school," or it can be about a combination of attributes, such as "The males are older than the females." Your conjecture should be something that you can test and, therefore, support or refute with statistics.

Step 4 Begin testing your conjecture by calculating summary statistics, such as the mean, median, standard deviation, and interquartile range. Choose **New | Summary Table** from the Object menu, and then drag and drop attributes from your case table. From the Summary menu, choose **Add Basic Statistics** and **Add Five-Number Summary** to see some of the statistics that you have studied in this chapter. Based on these statistics, does your conjecture seem to be true? Why or why not?

CA_includesBerkeley

	sex: F	sex: M	Row Summary
age	267	233	500
	37.127341	34.592275	35.946
	21.6569	19.776365	20.819654
	1.3253808	1.2955927	0.93108324
	0	0	0
	0	0	0
	21	20	21
	33	35	34
	48	47	48
	90	88	90

S1 = count( )
S2 = mean( )
S3 = stdDev( )
S4 = stdError( )
S5 = count(missing( ))
S6 = min( )
S7 = Q1( )
S8 = median( )
S9 = Q3( )
S10 = max( )

This summary table shows statistics for the ages of females and males. For example, the median age of females is 33.

Step 5 Graphs are another way to test your conjecture. Choose **New | Graph** from the Object menu. Drag and drop attributes into your graph and then choose **Box Plot** from the pull-down menu in the corner of the graph window. Do box plots help you support or refute your conjecture?

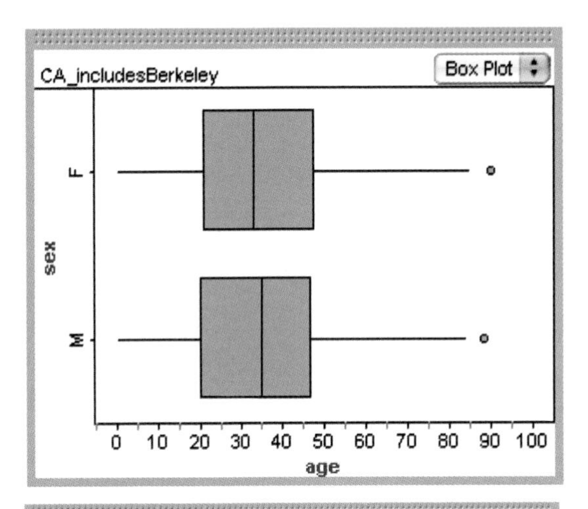

Step 6 Now create histograms to test your conjecture. Either you can follow the process in Step 5 to create a new graph or you can simply use the pull-down menu to change your box plots to histograms. What new information do the histograms give you? Do histograms help support your conjecture? Do the histograms better support your conjecture if you change the bin width?

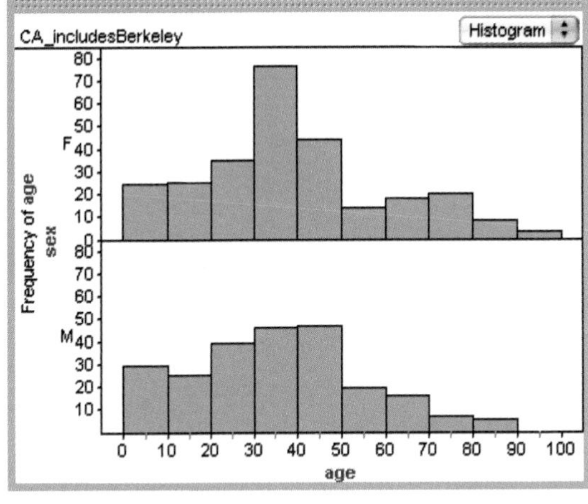

Step 7 Look at all of your analyses, including the summary statistics and graphs. Do you think your conjecture is true? If you aren't sure, you might want to modify your conjecture or think about factors that might also be at work. Write a short paragraph explaining your findings. Think about these questions as you write: What factors might affect your analysis? Can you explain any outliers in your data? Which statistic or graph revealed the most about these data? In what ways might citizens or governments use these data to make informed decisions?

Questions

1. In this exploration, you've seen some ways to determine whether or not a conjecture is true. Is it possible for a conjecture to appear to be true or false, depending on what statistic or graph you select? Make a new conjecture for your microdata and try to find one statistic or graph that supports the conjecture and one that refutes the conjecture.

2. State and federal decision-makers often have to compare data from different regions to make sure they are meeting everyone's needs. Use Fathom to compare microdata for two different geographic regions. Make and test a few conjectures about how the regions compare and contrast. Describe at least one way in which these communities could use your graphs to make decisions.

CHAPTER 2 REVIEW

Sets of data, such as temperatures, the sugar content of breakfast cereals, and the number of hours of television you watch daily, can be analyzed and pictured using tools that you learned about in this chapter. The **mean** and the **median** are **measures of central tendency.** They tell you about a typical value for the data set. But a measure of central tendency alone does not tell the whole story. You also need to look at the **spread** of the data values. The **standard deviation** helps you determine spread about the mean, and the **interquartile range** helps you determine spread about the median. These measures of spread are also frequently used to identify **outliers** in the data set.

To display a data set visually, you can use a box plot or a histogram. A **box plot** shows the median of the data set, the **range** of the entire set, and the interquartile range between the **first quartile** and the **third quartile.** A **histogram** uses **bins** to show how the data are spread throughout the entire range. By changing the width of the bins, you can get a different perspective on the distribution. Neither a box plot nor a histogram shows individual data values, but both help you see whether the data set is **symmetric** or **skewed.**

Percentile ranks are useful to show how one data value compares to the data set as a whole. The percentile rank of one data value tells you the percentage of values that are less than the given data value. Percentile ranks are not the same as percent scores.

By using a combination of **statistics** and graphs, you can better understand the meaning and implications of a data set. A careful analysis of a data set helps you make general conclusions about the past and predictions for the future.

EXERCISES

You will need

A graphing calculator for Exercises **3, 6** and **8.**

@ Answers are provided for all exercises in this set.

1. Which box plot has the greater standard deviation? Explain your reasoning.

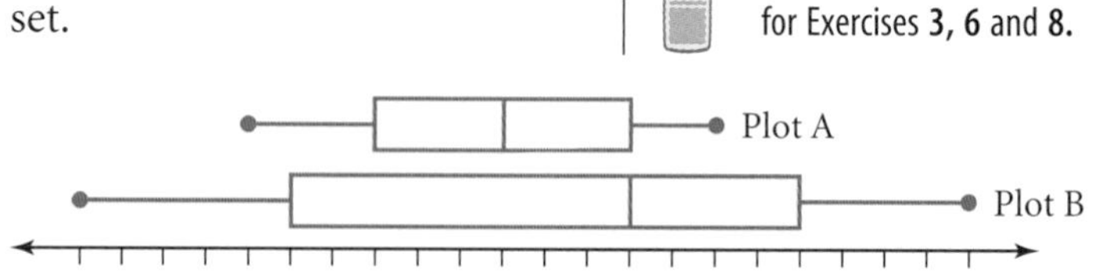

2. Consider these box plots. Group A conducted the rubber band launch experiment 30 times, and Group B conducted the experiment 25 times.

a. How many data values are represented in each whisker of each box plot?

b. Which data set has the greater standard deviation? Explain how you know.

c. Draw two histograms that might represent the information pictured in these two box plots.

Rubber Band Launch

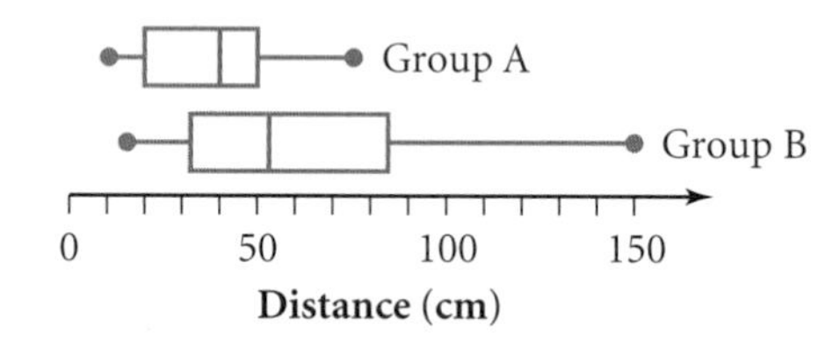

3. The San Antonio Spurs won the 2006 National Basketball Association Championship. This table gives the total points scored by each player during that season.

Total Points Scored by San Antonio Spurs Players (2005–2006 Season)

Player	Points	Player	Points	Player	Points
Brent Barry	635	Francisco Elson	350	Fabricio Oberto	349
Matt Bonner	275	Melvin Ely	89	Tony Parker	1429
Bruce Bowen	510	Michael Finley	740	Beno Udrih	340
Jackie Butler	41	Manu Ginobili	1240	Jacque Vaughn	192
Tim Duncan	1599	Robert Horry	268	James White	50

a. Find the mean, median, and mode for this data set.

b. Find the five-number summary.

c. Draw a modified box plot of the data. Describe the shape of the data set.

d. Calculate the interquartile range.

e. Would you identify any of the point totals as outliers? Explain.

Tim Duncan

4. Invent two data sets, each with seven values, such that Set A has the greater standard deviation and Set B has the greater interquartile range.

5. A random sample of songs was taken from the Hot 100 list in *Billboard Magazine* in 2007. This histogram shows the number of weeks each song had been on the list at the time of the sample.

Use the information in the histogram to summarize the data. Be sure to estimate at least one measure of center and one measure of spread, describe the shape of the data set, and note any other interesting characteristics of the data.

Sample of Songs from Hot 100 List

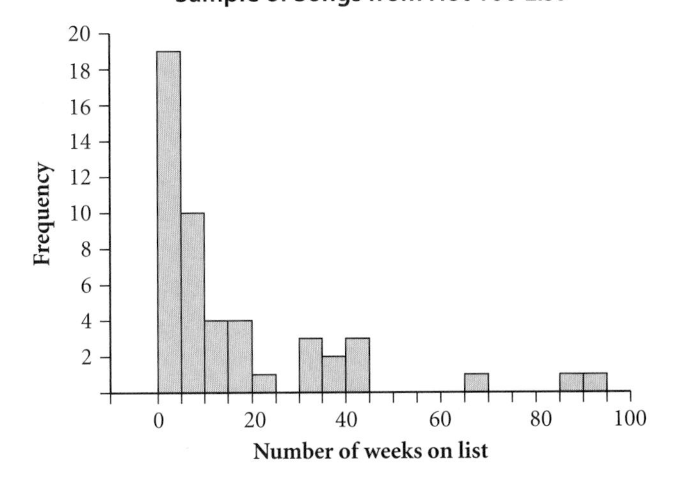

6. Below is a list of Academy Award winners in the Best Actress and Best Actor categories and each person's age when he or she received the award.

Academy Award Winners 1976–2007

Year	Best actress in a leading role, Age	Best actor in a leading role, Age
1976	Faye Dunaway, 36	Peter Finch, 60
1977	Diane Keaton, 32	Richard Dreyfuss, 30
1978	Jane Fonda, 41	Jon Voight, 40
1979	Sally Field, 33	Dustin Hoffman, 42
1980	Sissy Spacek, 31	Robert De Niro, 37
1981	Katharine Hepburn, 74	Henry Fonda, 76
1982	Meryl Streep, 33	Ben Kingsley, 39
1983	Shirley MacLaine, 49	Robert Duvall, 53
1984	Sally Field, 38	F. Murray Abraham, 45
1985	Geraldine Page, 61	William Hurt, 36
1986	Marlee Matlin, 21	Paul Newman, 62
1987	Cher, 41	Michael Douglas, 43
1988	Jodie Foster, 26	Dustin Hoffman, 51
1989	Jessica Tandy, 81	Daniel Day-Lewis, 32
1990	Kathy Bates, 42	Jeremy Irons, 42
1991	Jodie Foster, 29	Anthony Hopkins, 54

(*www.imdb.com*)

Year	Best actress in a leading role, Age	Best actor in a leading role, Age
1992	Emma Thompson, 33	Al Pacino, 52
1993	Holly Hunter, 36	Tom Hanks, 37
1994	Jessica Lange, 45	Tom Hanks, 38
1995	Susan Sarandon, 49	Nicolas Cage, 31
1996	Frances McDormand, 39	Geoffrey Rush, 45
1997	Helen Hunt, 34	Jack Nicholson, 60
1998	Gwyneth Paltrow, 26	Roberto Benigni, 46
1999	Hilary Swank, 25	Kevin Spacey, 40
2000	Julia Roberts, 33	Russell Crowe, 36
2001	Halle Berry, 33	Denzel Washington, 47
2002	Nicole Kidman, 35	Adrien Brody, 29
2003	Charlize Theron, 27	Sean Penn, 42
2004	Hilary Swank, 29	Jamie Foxx, 36
2005	Reese Witherspoon, 28	Philip Seymour Hoffman, 37
2006	Helen Mirren, 60	Forest Whitaker, 44
2007	Marion Cotillard, 32	Daniel Day-Lewis, 50

a. Find the mean age and the median age for the Best Actress winners.

b. Find the mean age and the median age for the Best Actor winners.

c. On the same axis, draw two box plots, one for the age of Best Actress winners and the other for the age of Best Actor winners.

d. Draw two histograms, one for Best Actress and one for Best Actor.

e. Use your graphs from 6c and d to predict which data set has the greater standard deviation. Explain your reasoning. Then calculate the standard deviations to check your prediction.

f. Julia Roberts was 33 years old when she won Best Actress in 2000. What is her percentile rank among all Best Actress winners from 1976 to 2007? Explain what this percentile rank tells you.

At the 74th annual Academy Awards, Halle Berry was the first African-American to win Best Actress. Denzel Washington was the second African-American to win Best Actor.

7. APPLICATION Following the 1998 Academy Awards ceremony, Best Actress nominee Fernanda Montenegro (age 70) said Gwyneth Paltrow (age 26) had won Best Actress for *Shakespeare in Love* because she was younger. This comment caused Pace University student Michael Gilberg and professor Terence Hines to test the theory that younger women and older men are more likely than older women and younger men to receive an Academy Award for Best Actress and Best Actor. Their study was published in the February 2000 issue of *Psychological Reports.*

Assume that you are working with Gilberg and Hines. Use your statistics and graphs from Exercise 6 to confirm or refute the theory that younger women and older men are more likely to win. Prepare a brief report on your conclusions.

8. The 2005 U.S. passenger-car production totals, rounded to the nearest thousand, are shown at right.

a. Make a box plot of these data.

b. Make a histogram using a bin width that provides meaningful information about the data.

c. Suppose a different year has a similar distribution but the total number of cars produced is 400 thousand greater than in 2005. Describe how this could affect the shapes of your box plot and histogram.

d. What is the percentile rank of Acura?

e. What is the percentile rank of Ford?

2005 U.S. Passenger-Car Production

Brand	Number of cars (thousands)
Acura	89
BMW	20
Buick	74
Cadillac	169
Chevrolet	572
Chrysler	97
Dodge	228
Ford	473
Honda	493
Lincoln	66
Mercury	46
Mitsubishi	63
Nissan	383
Pontiac	228
Saturn	110
Subaru	87
Toyota	509

(The World Almanac and Book of Facts 2007)

TAKE ANOTHER LOOK

1. Which measure of central tendency do you think of when someone says "average"? Without clarification, an "average" could be the mean, the median, or the mode. Find newspaper and magazine articles that state an "average," such as "The average American family has 2.58 children." Read the articles closely and analyze any data that are provided. Can you find enough information to tell which measure of central tendency is being used? Do you find that most articles are or are not specific enough with their mathematics? What conclusions can you make?

2. The calculation of mean that you learned in this chapter—the sum of all data values divided by the number of values—is more precisely called the **arithmetic mean.** Other means include the geometric mean, the harmonic mean, the quadratic mean, the trimean, and the midmean. Research one or more of these means, and compare and contrast the calculation to the arithmetic mean.

3. In Lesson 2.3, you learned how to find the median of a data set by looking at a histogram. How would you use the histogram to approximate the mean? The mode?

4. Another measure of spread, the mean deviation, *MD*, uses absolute value to eliminate the effect of the different signs of the individual deviations.

$$MD = \frac{\sum_{i=1}^{n} |x_i - \bar{x}|}{n}$$

Try using mean deviation for some of the exercises in which you calculated standard deviation. How do the values compare? When might standard deviation or mean deviation be the more appropriate measure of spread?

Assessing What You've Learned

BEGIN A PORTFOLIO An artist usually keeps both a notebook and a portfolio. The notebook might contain everything from scratch work to practice sketches to notes about past or future subject matter. The portfolio, in contrast, is reserved for the artist's most significant or best work.

As a student, you probably already keep a notebook that contains everything from your class notes to homework to research for independent projects. You can also start a separate portfolio that collects your most significant work.

Review all the work you've done so far and find your best works of art: the neatest graphs, the most thorough calculations for various statistics, the most complete analysis of a data set, or the most comprehensive project. Add each piece to your portfolio with a paragraph or two that addresses these questions:

- What is the piece an example of?
- Does this piece represent your best work? Why else did you choose it?
- What mathematics did you learn or apply in this piece?
- How would you improve the piece if you redid or revised it?

WRITE TEST ITEMS Writing your own problems is an excellent way to assess and review what you've learned. If you were writing a test for this chapter, what would it include? Start by having a group discussion to identify the key ideas of the chapter. Then divide the lessons among group members, and have each group member write at least one problem for each lesson assigned to him or her. Try to create a mix of problems, from simple one-step exercises that require you to recall facts and formulas, to complex multistep problems that require more thinking. Because you'll be working with data and statistics, you'll need to carefully consider which statistics and graphs are appropriate for the data.

Share your problems with your group members and try out one another's problems. Then discuss the problems in your group:

- Were the problems representative of the content of the chapter?
- Were any problems too hard or too easy?
- Were the statistics appropriate for the data?

CHAPTER

3 Linear Models and Systems

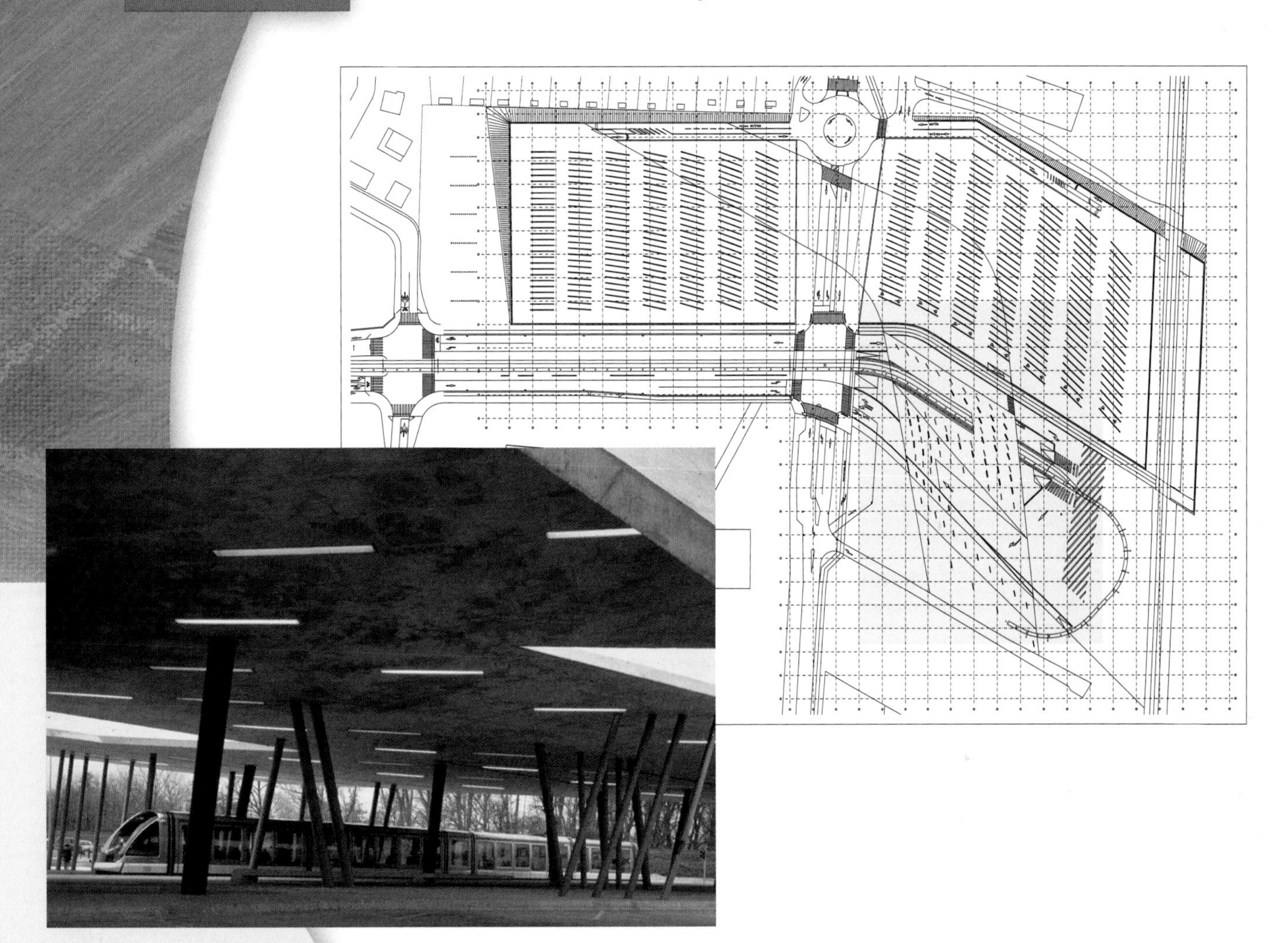

Iraqi-British architect Zaha Hadid (b 1950) designed this commuter train station on the outskirts of Strasbourg, France. She worked with the concept of overlapping fields and lines to represent the interacting patterns of moving cars, trains, bicycles, and pedestrians. The roof of the waiting area in the station is punctured by angled lines that let sunlight shine on the floor, shifting throughout the day. You can see these lines in the photograph, which shows the region of the architectural plan highlighted in blue.

OBJECTIVES

In this chapter you will

- review linear equations in intercept form and point-slope form
- explore connections between arithmetic sequences and linear equations
- find lines of fit for data sets that are approximately linear
- solve systems of linear equations

Linear Relationships

An equation with two variables describes a mathematical relationship between the two variables. In a linear relationship such as $4x + 7y = 42$, a value of one of the variables always corresponds to a single value of the other variable.

EXAMPLE A

Given the relationship $4x + 7y = 42$:

a. Find y when $x = 9$.

b. Find x when $y = 6$.

▶ Solution

a. Substitute 9 for x and then solve for y. Think of undoing each operation that is done to y.

$4(9) + 7y = 42$	Substitute 9 for x in the original equation.
$36 + 7y + -36 = 42 + -36$	Add -36 to both sides to undo adding 36.
$7y = 6$	Add.
$\frac{1}{7} \cdot 7y = \frac{1}{7} \cdot 6$	Multiply both sides by $\frac{1}{7}$ to undo multiplying by 7.
$y = \frac{6}{7}$	

b. Substitute 6 for y and then solve for x.

$4x + 7(6) = 42$	Substitute 6 for y in the original equation.
$4x + 42 + -42 = 42 + -42$	Add -42 to both sides to undo adding 42.
$4x = 0$	Add.
$x = 0$	Multiply both sides by $\frac{1}{4}$ to undo multiplying by 4.

At times it is helpful to rewrite an equation in a different form. For example, to graph the equation $4x + 7y = 42$ on a graphing calculator, you'll need to solve it for y.

EXAMPLE B

Given the relationship $4x + 7y = 42$:

a. Solve for y.

b. Solve for x.

▶ Solution

a. Your steps will be similar to the steps in Example A, part a. However, the final equation will still have a term with x.

$4x + 7y = 42$	Original equation.
$7y = 42 - 4x$	Add $-4x$ to both sides.
$\frac{1}{7} \cdot 7y = \frac{1}{7} \cdot (42 - 4x)$	Multiply both sides by $\frac{1}{7}$.
$y = 6 - \frac{4}{7}x$	Distribute and multiply.

b. The steps will be similar to the steps in Example A, part b.

$4x + 7y = 42$	Original equation.
$4x = 42 - 7y$	Add $-7y$ to both sides.
$\frac{1}{4} \cdot 4x = \frac{1}{4} \cdot (42 - 7y)$	Multiply both sides by $\frac{1}{4}$.
$x = \frac{21}{2} - \frac{7}{4}y$	Distribute and multiply.

Because the instructions asked you to solve for x, but not to leave the equation in a specific form, you might have written the final equation as $x = \frac{1}{4}(42 - 7y)$, or $x = -1.75y + 10.5$, or another equivalent equation. Each form can provide different information about the mathematical relationship between x and y.

EXAMPLE C | Given the relationship $5 - 6(y + 2) = x$, solve for y.

▶ Solution

Remember to think about undoing the order of operations as you solve for y.

$5 - 6(y + 2) = x$	Original equation.
$-6(y + 2) = x - 5$	Add -5 to both sides to undo adding 5.
$y + 2 = -\frac{1}{6}(x - 5)$	Multiply both sides by $-\frac{1}{6}$ to undo multiplying by -6.
$y = -\frac{1}{6}(x - 5) - 2$	Add -2 to both sides to undo adding 2.

Depending on how the equation is going to be used, you might make further changes to the right side of this equation. However, the current form is correct and meaningful. Unless you are instructed to put an equation into a particular form, you can stop after you have isolated the specified variable.

EXERCISES

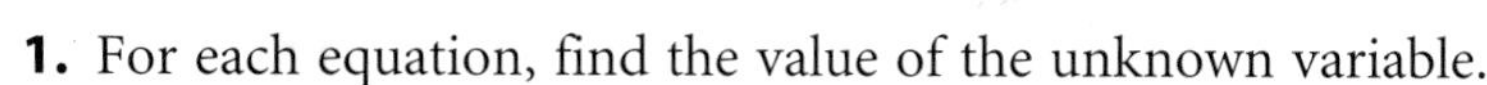

1. For each equation, find the value of the unknown variable.

a. $y = 3x - 11$ when $x = -5$ @

b. $a = 12 - \frac{1}{2}b$ when $b = 30$

c. $4.5s - 3.5t = 19$ when $s = 2.2$ @

d. $\frac{3}{5}x + \frac{7}{3}y = 20$ when $x = 12$

2. Solve for the indicated variable in each equation.

a. $y = 3x - 11$ for x @

b. $\frac{y - 3}{x - 2} = \frac{5}{4}$ for y

c. $4.5s - 3.7t = 18.1$ for s @

d. $a = 7 + 4(b - 3)$ for b

3. Consider the graph of the equation $y = 7 + 2.4(x - 5)$.

a. Find the y-intercept (the y-value when $x = 0$). @

b. Find the x-intercept (the x-value when $y = 0$).

c. The ratio of these intercepts $\left(\frac{y\text{-intercept}}{x\text{-intercept}}\right)$ is similar to what value? Why?

LESSON

3.1 Linear Equations and Arithmetic Sequences

You can solve many rate problems by using recursion.

Matias wants to call his aunt in Chile on her birthday. He learned that placing the call costs \$2.27 and that each minute he talks costs \$1.37. How much would it cost to talk for 30 minutes?

Valparaiso, Chile

You can calculate the cost of Matias's phone call with the recursive formula

$$u_0 = 2.27$$
$$u_n = u_{n-1} + 1.37 \quad \text{where } n \geq 1$$

To find the cost of a 30-minute phone call, calculate the first 30 terms, as shown in the calculator screen.

	A n	B	C	D	E
◆		=seqn(u(n−1)+1.37			
27	26	37.89			
28	27	39.26			
29	28	40.63			
30	29	42.			
31	30	43.37			

B =seqn(u(n−1)+1.37,{2.27},50)

As you learned in algebra, you or Matias can also find the cost of a 30-minute call by using the **linear equation**

$$y = 2.27 + 1.37x$$

where x is the length of the phone call in minutes and y is the cost in dollars. If the phone company always rounds up the length of the call to the nearest whole minute, then the costs become a sequence of discrete points, and you can write the relationship as an explicit formula,

$$u_n = 2.27 + 1.37n$$

where n is the length of the call in whole minutes and u_n is the cost in dollars.

An **explicit formula** gives a direct relationship between two discrete quantities. How does the explicit formula differ from the recursive formula? How would you use each formula to calculate the cost of a 15-minute call or an n-minute call?

In this lesson you will write and use explicit formulas for arithmetic sequences. You will also write linear equations for lines through the discrete points of arithmetic sequences.

EXAMPLE A

Consider the recursively defined arithmetic sequence

$$u_0 = 2$$
$$u_n = u_{n-1} + 6 \quad \text{where } n \geq 1$$

a. Find an explicit formula for the sequence.

b. Use the explicit formula to find u_{22}.

c. Find the value of n so that $u_n = 86$.

▶ Solution

a. Look for a pattern in the sequence.

u_0	2
u_1	$8 = 2 + 6 = 2 + 6 \cdot 1$
u_2	$14 = 2 + 6 + 6 = 2 + 6 \cdot 2$
u_3	$20 = 2 + 6 + 6 + 6 = 2 + 6 \cdot 3$

Notice that the common difference (or rate of change) between the terms is 6. You start with 2 and just keep adding 6. That means each term is equivalent to 2 plus 6 times the term number. In general, when you write the formula for a sequence, you use n to represent the number of the term and u_n to represent the term itself.

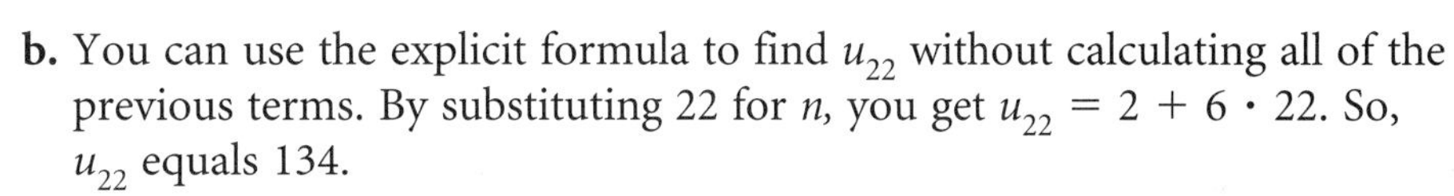

b. You can use the explicit formula to find u_{22} without calculating all of the previous terms. By substituting 22 for n, you get $u_{22} = 2 + 6 \cdot 22$. So, u_{22} equals 134.

c. To find n so that $u_n = 86$, substitute 86 for u_n in the formula $u_n = 2 + 6n$.

$86 = 2 + 6n$	Substitute 86 for u_n.
$14 = n$	Solve for n.

So, 86 is the 14th term.

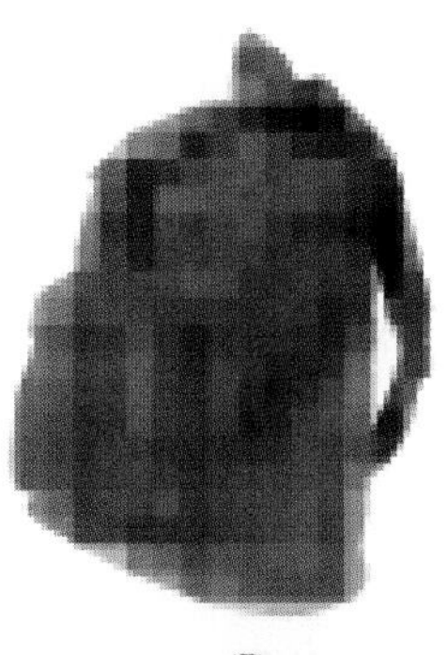

When you download photos from the Internet, sometimes the resolution improves as the percentage of download increases. The relationship between the percentage of download and the resolution could be modeled with a sequence, an explicit formula, or a linear equation.

You graphed sequences of points (n, u_n) in Chapter 1. The term number n is a whole number: 0, 1, 2, 3, ... So, using different values for n will produce a set of discrete points. The points on this graph show the arithmetic sequence from Example A.

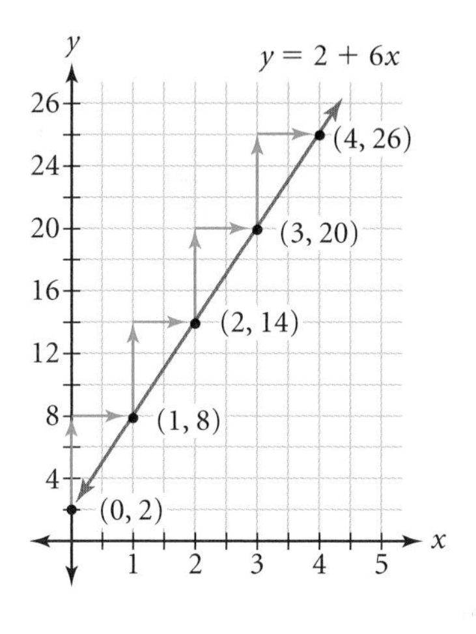

When n increases by 1, u_n increases by 6, the common difference. In terms of x and y, the number 6 is the change in the y-value that corresponds to a unit change (a change of 1) in the x-value. So the points representing the sequence lie on a line with a slope of 6. In general, the common difference, or rate of change, between consecutive terms of an arithmetic sequence is the **slope** of the line through those points.

The pair (0, 2) names the starting value 2, which is the y-intercept. Using the intercept form of a linear equation, you can now write an equation of the line through the points of the sequence as $y = 2 + 6x$, or $y = 6x + 2$.

In this course you will use n when you are writing a formula for a sequence of discrete points. When you write an equation of the line that contains the points of the sequence, you will use x to indicate that all real numbers can be input values.

In the investigation you will focus on the relationship between the formula for an arithmetic sequence and the equation of the line through the points representing the sequence.

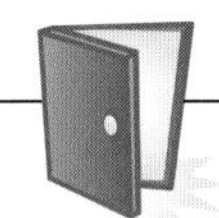

Investigation
Match Point

Below are three recursive formulas, three graphs, and three linear equations.

1. $u_0 = 4$
$u_n = u_{n-1} - 1 \quad \text{where } n \geq 1$

2. $u_0 = 2$
$u_n = u_{n-1} + 5 \quad \text{where } n \geq 1$

3. $u_0 = -4$
$u_n = u_{n-1} + 3 \quad \text{where } n \geq 1$

A.

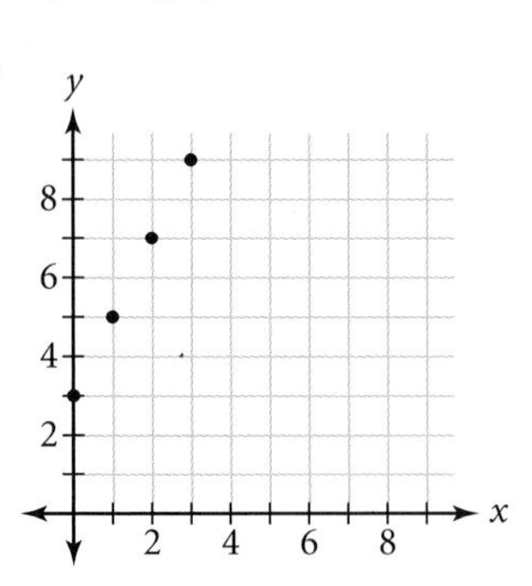

B.

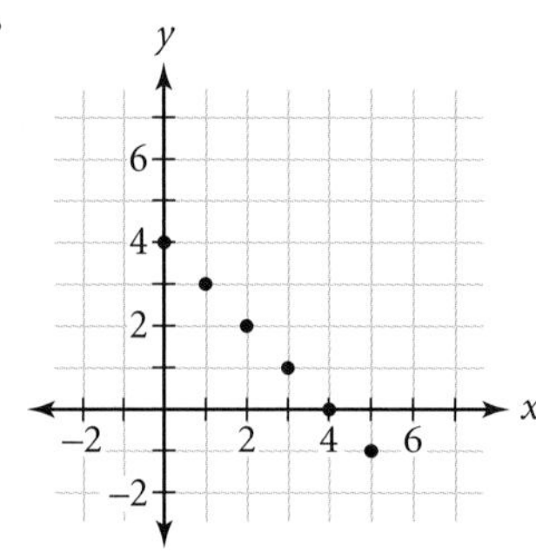

C.

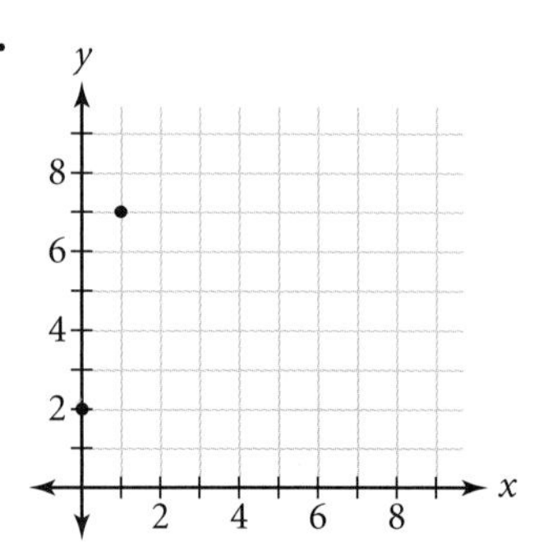

i. $y = -4 + 3x$

ii. $y = 4 + x$

iii. $y = 2 + 5x$

Step 1 Match the recursive formulas, graphs, and linear equations that go together. (Not all of the appropriate matches are listed. If the recursive rule, graph, or equation is missing, you will need to create it.)

Step 2 Write a brief statement relating the starting value and common difference of an arithmetic sequence to the corresponding equation $y = a + bx$.

Step 3 Are points (n, u_n) of an arithmetic sequence always collinear? Write a brief statement supporting your answer.

EXAMPLE B

Retta typically spends \$2 a day on lunch. She notices that she has \$17 left after today's lunch. She thinks of this sequence to model her daily cash balance.

$t_1 = 17$
$t_n = t_{n-1} - 2 \quad \text{where } n > 1$

a. Find the explicit formula that represents her daily cash balance and an equation of the line through the points of this sequence.

b. How useful is this formula for predicting how much money Retta will have each day?

▶ Solution

Use the common difference and starting term to write the explicit formula.

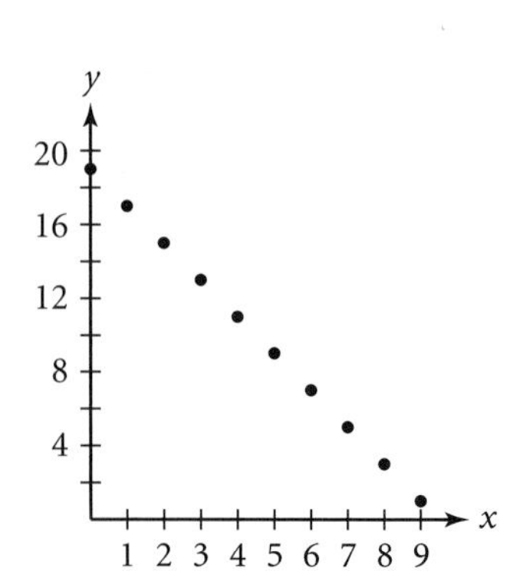

a. Each term is 2 less than the previous term, so the common difference of the arithmetic sequence and the slope of the line are both -2. The term t_1 is 17, so the previous term, t_0, or the y-intercept, is 19.

The explicit formula for the arithmetic sequence is $t_n = 19 - 2 \cdot n$, and the equation of the line containing these points is $y = 19 - 2x$.

b. We don't know whether Retta has any other expenses, when she receives her paycheck or allowance, or whether she buys lunch on the weekend. The formula could be valid for eight more days, until she has \$1 left (on t_9), as long as she gets no more money and spends only \$2 per day.

For both sequences and equations, it is important to consider the conditions for which the relationship is valid. For example, a phone company usually rounds up the length of a phone call to the nearest minute to determine the charges, so the relationship between the length of the call and the cost of the call is valid only for a call length that is a positive integer. Also, the portion of the line left of the y-axis, where x is negative, is part of the mathematical model but has no relevance for the phone call scenario.

EXERCISES

You will need

A graphing calculator for Exercise **15**.

Practice Your Skills

1. Consider the sequence ⓐ

$a_0 = 18$
$a_n = a_{n-1} - 3 \quad \text{where } n \geq 1$

a. Graph the sequence.

b. What is the slope of the line that contains the points? How is that related to the common difference of the sequence?

c. What is the y-intercept of the line that contains these points? How is it related to the sequence?

d. Write an equation for the line that contains these points.

2. Refer to the graph at right.

a. Write a recursive formula for the sequence. What is the common difference? What is the value of u_0?

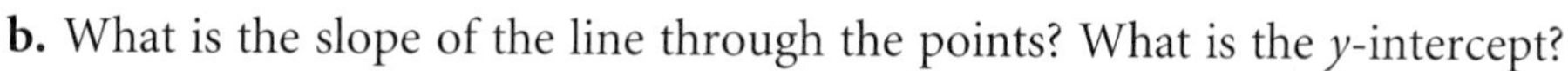

b. What is the slope of the line through the points? What is the y-intercept?

c. Write an equation for the line that contains these points.

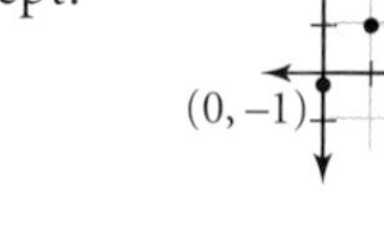

3. Write an equation for the line that passes through the points of an arithmetic sequence with $u_0 = 7$ and a common difference of 3. ⓐ

4. Write a recursive formula for a sequence whose points lie on the line $y = 6 - 0.5x$.

5. Find the slope of each line.

a. $y = 2 + 1.7x$ ⓐ **b.** $y = x + 5$ **c.** $y = 12 - 4.5x$ **d.** $y = 12$ ⓗ

Reason and Apply

6. An arithmetic sequence has a starting term, u_0, of 6.3 and a common difference of 2.5.

a. Write an explicit formula for the sequence.

b. Use the formula to figure out which term is 78.8.

7. Suppose you drive through Macon, Georgia (which is 82 mi from Atlanta), on your way to Savannah, Georgia, at a steady 54 mi/h.

a. What is your distance from Atlanta two hours after you leave Macon? ⓐ

b. Write an equation that represents your distance, y, from Atlanta x hours after leaving Macon.

c. Graph the equation.

d. Does this equation model an arithmetic sequence? Why or why not?

Savannah, Georgia

8. APPLICATION Melissa and Roy both sell cars at the same dealership and have to meet the same profit goal each week. Last week, Roy sold only three cars, and he was below his goal by \$2,050. Melissa sold seven cars, and she beat her goal by \$1,550. Assume that the profit is approximately the same for each car they sell.

a. Use a graph to find a few terms, t_n, of the sequence of sales amounts above or below the goal. ⓗ

b. What is the real-world meaning of the common difference?

c. Write an explicit formula for this sequence of sales values in relation to the goal. Define variables and write a linear equation.

d. What is the real-world meaning of the horizontal and vertical intercepts?

e. What is the profit goal? How many more cars must Roy sell to be within \$500 of his goal?

9. The points on this graph represent the first five terms of an arithmetic sequence. The height of each point is its distance from the x-axis, or the value of the y-coordinate of the point.

a. Find u_0, the y-coordinate of the point preceding those given.

b. How many common differences (d's) do you need to get from the height of $(0, u_0)$ to the height of $(5, u_5)$?

c. How many d's do you need to get from the height of $(0, u_0)$ to the height of $(50, u_{50})$?

d. Explain why you can find the height from the x-axis to $(50, u_{50})$ using the equation $u_{50} = u_0 + 50d$.

e. In general, for an arithmetic sequence, the explicit formula is $u_n = \underline{\ ?\ }$.

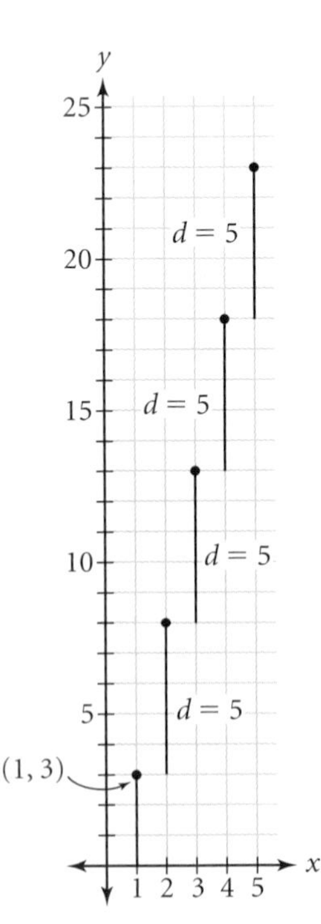

10. A gardener planted a new variety of ornamental grass and kept a record of its height over the first two weeks of growth.

Time (days)	0	3	7	10	14
Height (cm)	4.2	6.3	9.1	11.2	14

a. How much does the grass grow each day?

b. Write an explicit formula that gives the height of the grass after n days.

c. How long will it take for the grass to be 28 cm tall? ⓐ

Heather Ackroyd and Dan Harvey print photographs on grass, such as this one, *Sunbathers 2000.* They position a photographic negative over growing grass, and the chlorophyll reacts to give a temporary image in shades of green and yellow. They use genetically modified grass to make their images last longer.

11. An arithmetic sequence of six numbers begins with 7 and ends with 27. Follow 11a–c to find the four missing terms.

a. Name two points on the graph of this sequence: $(\underline{?}, 7)$ and $(\underline{?}, 27)$. ⓗ

b. Plot the two points you named in 11a and find the slope of the line connecting the points.

c. Use the slope to find the missing terms.

d. Plot all the points and write the equation of the line that contains them. ⓗ

12. APPLICATION If an object is dropped, it will fall a distance of about 16 feet during the first second. In each second that follows, the object falls about 32 feet farther than in the previous second.

a. Write a recursive formula for the distance fallen each second under free fall.

b. Write an explicit formula for the distance fallen each second under free fall.

c. How far will the object fall during the 10th second? ⓐ

d. During which second will the object fall 400 feet?

Science CONNECTION

Leonardo da Vinci (1452–1519) was able to discover the formula for the **velocity** (directional speed) of a freely falling object by looking at a sequence. He let drops of water fall at equally spaced time intervals between two boards covered with blotting paper. When a spring mechanism was disengaged, the boards clapped together. By measuring the distances between successive blots and noting that these distances increased arithmetically, da Vinci discovered the formula $v = gt$, where v is the velocity of the object, t is the time since it was released, and g is a constant that represents any object's downward acceleration due to the force of gravity.

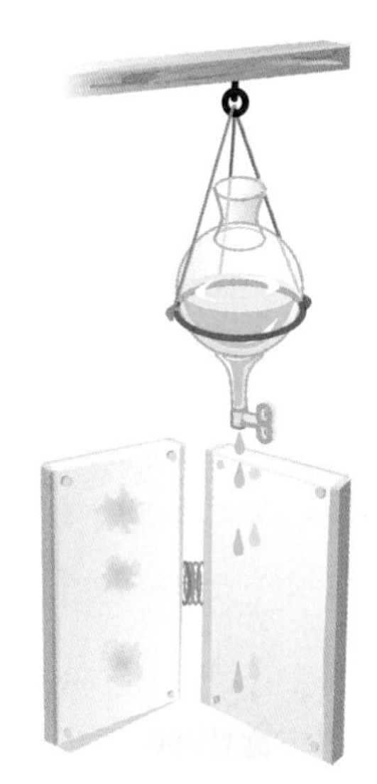

Review

13. **APPLICATION** Suppose a company offers a new employee a starting salary of \$18,150 with annual raises of \$1,000, or a starting salary of \$17,900 with a raise of \$500 every six months. At what point is one choice better than the other? Explain.

14. Suppose that you add 300 mL of water to an evaporating dish at the start of each day, and each day 40% of the water in the dish evaporates.

    a. Write and solve an equation that computes the long-run water level in the dish. @

    b. Will a 1 L dish be large enough in the long run? Explain why or why not.

15. **APPLICATION** Five stores in Tulsa, Oklahoma, sell the same model of a graphing calculator for \$89.95, \$93.49, \$109.39, \$93.49, and \$97.69.

    a. What are the median price, the mean price, and the standard deviation?

    b. If these stores are representative of all stores in the Tulsa area, of what importance is it to a consumer to know the median, mean, and standard deviation? Which is probably more helpful, the median or the mean?

16. Based on this histogram, create a possible box plot of the data. If you assume all of the data values are integers, how could you find the five-number summary?

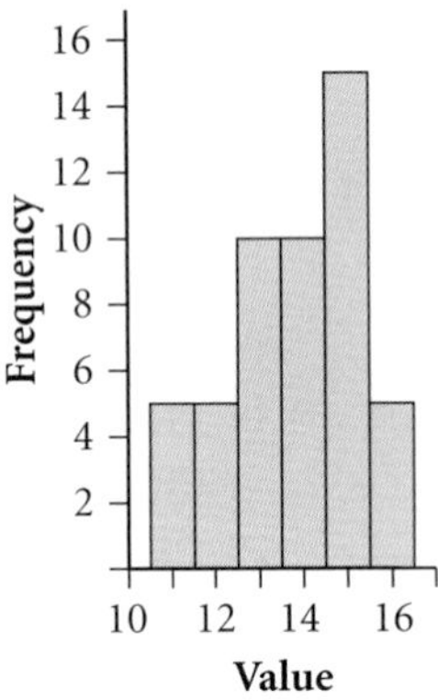

IMPROVING YOUR REASONING SKILLS

Sequential Slopes

Here's a sequence that generates coordinate points. What is the slope between any two points of this sequence?

$$(x_0, y_0) = (0, 0)$$
$$(x_n, y_n) = (x_{n-1} + 2, y_{n-1} + 3) \quad \text{where } n \geq 1$$

Now match each of these recursive rules to the slope between points.

a. $(x_n, y_n) = (x_{n-1} + 2, y_{n-1} + 2)$

b. $(x_n, y_n) = (x_{n-1} + 1, y_{n-1} + 3)$

c. $(x_n, y_n) = (x_{n-1} + 3, y_{n-1} - 4)$

d. $(x_n, y_n) = (x_{n-1} - 2, y_{n-1} + 10)$

e. $(x_n, y_n) = (x_{n-1} + 1, y_{n-1})$

f. $(x_n, y_n) = (x_{n-1} + 9, y_{n-1} + 3)$

A. 0 B. 1 C. 3 D. -5 E. $\frac{1}{3}$ F. $\frac{-4}{3}$

In general, how do these recursive rules determine the slope between points?

LESSON

3.2

Revisiting Slope

There is more to life than increasing its speed.

MOHANDAS GANDHI

Suppose you are taking a long trip in your car. At 5 P.M., you notice that the odometer reads 45,623 miles. At 9 P.M., you notice that it reads 45,831. You find your average speed during that time period by dividing the difference in distance by the difference in time.

$$\text{Average speed} = \frac{45{,}831 \text{ miles} - 45{,}623 \text{ miles}}{9 \text{ hours} - 5 \text{ hours}} = \frac{208 \text{ miles}}{4 \text{ hours}} = 52 \text{ miles per hour}$$

You can also write the rate 52 miles per hour as the ratio $\frac{52 \text{ miles}}{1 \text{ hour}}$. If you graph the information as points of the form (*time, distance*), the slope of the line connecting the two points is $\frac{52}{1}$, which also tells you the average speed.

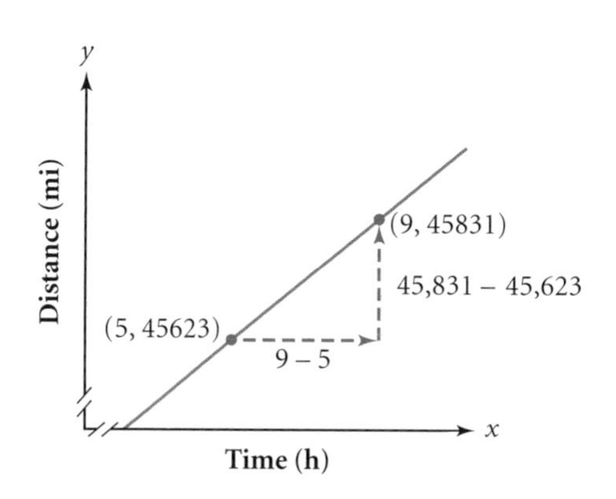

Slope

The formula for the slope between two points, (x_1, y_1) and (x_2, y_2), is

$$slope = \frac{y_2 - y_1}{x_2 - x_1} \quad \text{where } x_1 \neq x_2$$

The slope will be the same for any two points selected on the line. In other words, a line has only one slope. Two points on a line can have the same y-value; in that case, the slope of the line is 0. If they had the same x-value, the denominator would be 0 and the slope would be undefined. So the definition of slope specifies that the points cannot have the same x-value. What kinds of lines have a slope of 0? What kinds of lines have undefined slope?

Slope is another word for the steepness or rate of change of a line. If a linear equation is in **intercept form,** then the slope of the line is the coefficient of x.

Intercept Form of the Equation of a Line

You can write the equation of a line as

$$y = a + bx$$

where a is the y-intercept and b is the slope of the line.

Slope is often represented by the letter m. However, we will use the letter b in linear equations, as in the intercept form $y = a + bx$.

When you are using real-world data, choosing different pairs of points results in choosing lines with slightly different slopes. However, if the data are nearly linear, these slopes should not differ greatly.

Investigation
Balloon Blastoff

You will need

- paper
- tape
- a balloon
- a straw
- string
- a motion sensor

In this investigation you will launch a rocket and use your motion sensor's data to estimate the rocket's speed. Then you will write an equation for the rocket's distance as a function of time. Choose one person to be the monitor and one person to be the launch controller.

Procedure Note

1. Make a rocket of paper and tape. Design your rocket so that it can hold an inflated balloon and be taped to a drinking straw threaded on a string. Color or decorate your rocket if you like.
2. Tape your rocket to the straw on the string.
3. Inflate a balloon but do not tie off the end. The launch controller should insert it into your rocket and hold it closed.
4. Tie the string or hold it taut and horizontal.

Step 1 Hold the sensor behind the rocket. At the same time the monitor starts the sensor, the launch controller should release the balloon. Be sure nobody's hands are between the balloon and the sensor. [▸ See **Calculator Note 3C.** ◂]

Step 2 Retrieve the data from the sensor to each calculator in the group.

Step 3 Graph the data with time as the independent variable, x. What are the domain and range of your data? Explain.

Step 4 Sketch the graph of the data and select four representative points from the rocket data. Mark the points on your sketch and explain why you chose them.

Step 5 Record the coordinates of the four points and use the points in pairs to calculate slopes. This should give six estimates of the slope.

Step 6 Are all six slope estimates that you calculated in Step 5 the same? Why or why not? Find the mean, median, and mode of your slope estimates. With your group, decide which value best represents the slope of your data. Explain why you chose this value.

Step 7 What is the real-world meaning of the slope, and how is this related to the speed of your rocket?

In many cases, when you try to model the steepness and trend of data points, you may have some difficulty deciding which points to use. In general, select two points that are far apart to minimize the error. They do not need to be data points. Disregard data points that you think might represent measurement errors.

When you analyze a relationship between two variables, you must decide which variable you will express in terms of the other. When one variable depends on the other variable, it is called the **dependent variable.** The other variable is called the **independent variable.** Time is usually considered an independent variable.

Next, you need to think about the domain and range. The set of possible values for the independent variable is called the **domain** and the set of values for the dependent variable is called the **range.**

EXAMPLE

Daron's car gets 20 miles per gallon of gasoline. He starts out with a full tank, 16.4 gallons. As Daron drives, he watches the gas gauge to see how much gas he has left.

a. Identify the independent and dependent variables.

b. State a reasonable domain and range for this situation.

c. Write a linear equation in intercept form to model this situation.

d. How much gas will be left in Daron's tank after he drives 175 miles?

e. How far can he travel before he has less than 2 gallons remaining?

▸ *Solution*

The two variables are the distance Daron has driven and the amount of gasoline remaining in his car's tank.

a. The amount of gasoline remaining in the car's tank depends on the number of miles Daron has driven. This means the amount of gasoline is the dependent variable and distance is the independent variable. So use x for the distance (in miles) and y for the amount of gasoline (in gallons).

b. The independent variable is the distance Daron drives. Because his car can go a maximum of 20(16.4), or 328 miles, the domain is $0 \leq x \leq 328$ miles. The dependent variable is the amount of gasoline remaining, so the range is $0 \leq y \leq 16.4$ gallons.

c. Daron starts out with 16.4 gallons. He drives 20 miles per gallon, which means the amount of gasoline decreases $\frac{1}{20}$, or 0.05, gallon per mile. The equation is $y = 16.4 - 0.05x$.

d. You know the x-value is 175 miles. You can substitute 175 for x and solve for y.

$$\begin{aligned} y &= 16.4 - 0.05 \cdot 175 \\ &= 7.65 \end{aligned}$$

He will have 7.65 gallons remaining.

e. You know the y-value is 2.0 gallons. You can substitute 2.0 for y and solve for x.

$$\begin{aligned} 2.0 &= 16.4 - 0.05x \\ -14.4 &= -0.05x \\ 288 &= x \end{aligned}$$

When he has traveled more than 288 miles, he will have less than 2 gallons in his tank.

EXERCISES

You will need

A graphing calculator for Exercises **7**, **10–12**, and **15**.

Practice Your Skills

1. Find the slope of the line containing each pair of points.

a. $(3, -4)$ and $(7, 2)$ ⓐ **b.** $(5, 3)$ and $(2, 5)$ **c.** $(-0.02, 3.2)$ and $(0.08, -2.3)$

2. Find the slope of each line.

a. $y = 3x - 2$ **b.** $y = 4.2 - 2.8x$ **c.** $y = 5(3x - 3) + 2$ ⓐ

d. $y - 2.4x = 5$ **e.** $4.7x + 3.2y = 12.9$ **f.** $\frac{2}{3}y = \frac{2}{3}x + \frac{1}{2}$

3. Solve each equation.

a. Solve $y = 4.7 + 3.2x$ for y if $x = 3$. ⓐ

b. Solve $y = -2.5 + 1.6x$ for x if $y = 8$.

c. Solve $y = a - 0.2x$ for a if $x = 1000$ and $y = -224$.

d. Solve $y = 250 + bx$ for b if $x = 960$ and $y = 10$.

4. Find the equations of both lines in each graph.

a.

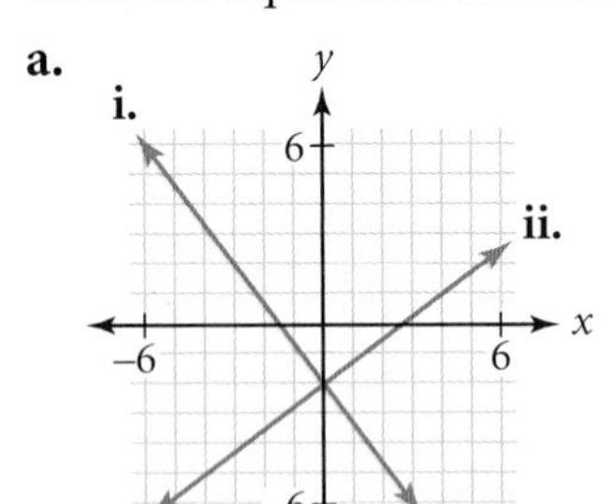

b.

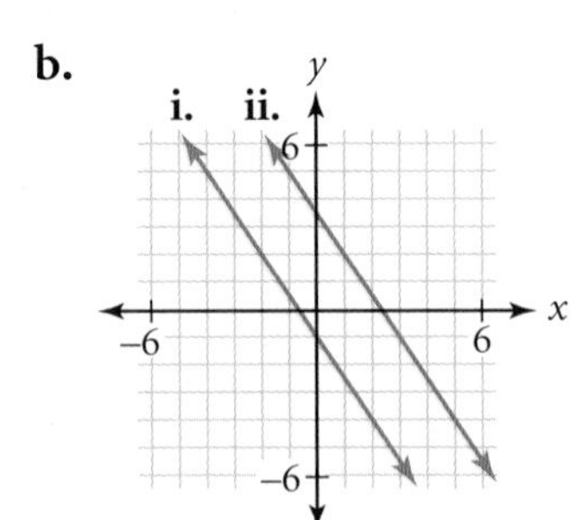

c. What do the equations in 4a have in common? What do you notice about their graphs? ⓐ

d. What do the equations in 4b have in common? What do you notice about their graphs?

5. The equation of line m is $y = -4 + 2.5x$. Line l is parallel to line m. Line n is perpendicular to line m. What are the slopes of lines l and n? ⓗ

Reason and Apply

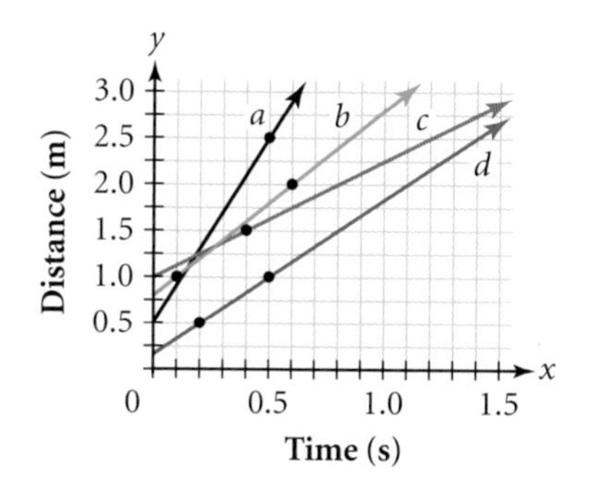

6. Four rocket flights were recorded. Each rocket started at a different distance from the motion sensor and traveled at a different speed. Use the graph to answer the questions.

a. Which rocket is the fastest? Explain how you know.

b. Which rocket is the slowest? Explain how you know.

c. Which rocket is moving at 2 meters per second? Explain how you know.

7. APPLICATION Layton measures the voltage across different numbers of batteries placed end to end. He records his data in a table.

Number of batteries	1	2	3	4	5	6	7	8
Voltage (volts)	1.43	2.94	4.32	5.88	7.39	8.82	10.27	11.70

a. Let x represent the number of batteries, and let y represent the voltage. Find the slope of a line approximating these data. Be sure to include units with your answer. ⓐ

b. Which points did you use and why? What is the real-world meaning of this slope?

c. Is this an example of direct variation? Why or why not?

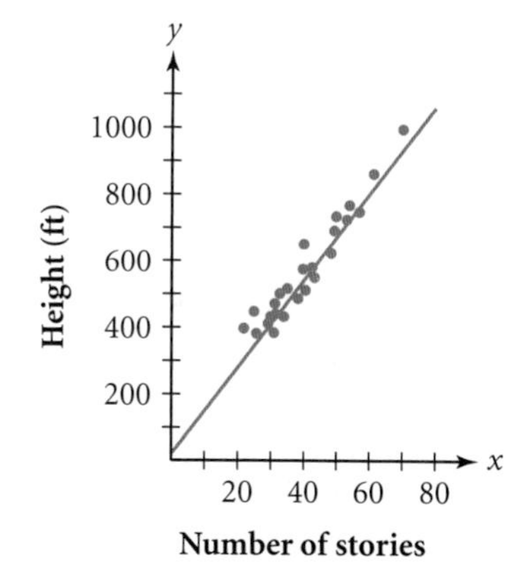

8. This graph shows the relationship between the height of some high-rise buildings in feet and the number of stories in those buildings. A line is drawn to fit the data.

a. Estimate the slope. What is the meaning of the slope? ⓗ

b. Estimate the y-intercept. What is the meaning of the y-intercept?

c. Explain why some of the points lie above the line and some lie below.

d. According to the graph, what are the domain and range of this relationship?

Many earthquake-prone areas in the United States have strict building codes to ensure that high-rise buildings can withstand earthquakes. Some taller buildings have cross-bracing to make them more rigid and lessen the amount of shaking. The extent of damage that an earthquake can cause depends on several factors: strength of the earthquake, type of underlying soil, and building construction. The shaking increases with height, and the period of shaking (in seconds) is approximately equal to 0.1 times the number of stories in the high-rise. For more information on engineering and earthquakes, see the links at www.keymath.com/DAA.

Steel braces will reduce earthquake damage to this concrete building.

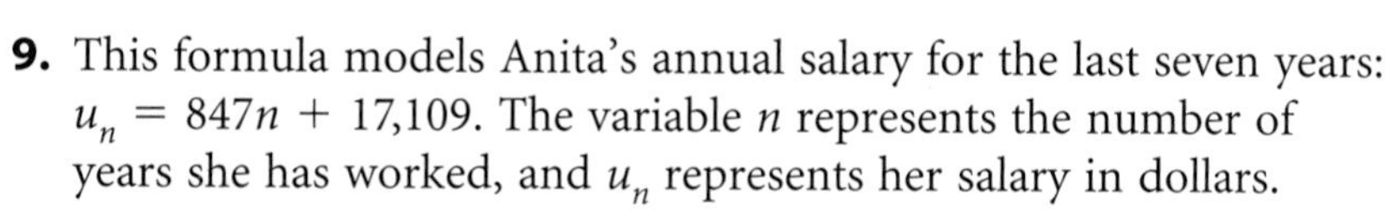

9. This formula models Anita's annual salary for the last seven years: $u_n = 847n + 17{,}109$. The variable n represents the number of years she has worked, and u_n represents her salary in dollars.

a. What did she earn in her fifth year? What did she earn in her first year? (Think carefully about what n represents.)

b. What is the rate of change of her salary?

c. What is the first year Anita's salary will be more than $30,000?

10. APPLICATION This table shows how long it took to lay tile for hallways of different lengths.

Length of hallway (ft)	3.5	9.5	17.5	4.0	12.0	8.0
Time (min)	85	175	295	92	212	153

a. What is the independent variable? Why? Make a graph of the data.

b. Find the slope of the line through these data. What is the real-world meaning of this slope?

c. Which points did you use and why?

d. Find the y-intercept of the line. What is the real-world meaning of this value?

11. APPLICATION The manager of a concert hall keeps data on the total number of tickets sold and total sales income, or revenue, for each event. Tickets are sold at two different prices.

Total tickets	448	601	297	533	523	493	320
Total revenue ($)	3,357.00	4,495.50	2,011.50	3,784.50	3,334.50	3,604.50	2,353.50

a. Find the slope of a line approximating these data. What is the real-world meaning of this slope?

b. Which points did you use and why?

12. APPLICATION How much does air weigh? The following table gives the weight of a cubic foot of dry air at the same pressure at various temperatures in degrees Fahrenheit.

Temp. (°F)	0	12	32	52	82	112	152	192	212
Weight (lb)	0.0864	0.0842	0.0807	0.0776	0.0733	0.0694	0.0646	0.0609	0.0591

a. Make a scatter plot of the data.

b. What is the approximate slope of the line through the data?

c. Describe the real-world meaning of the slope.

Recreation CONNECTION

Hot air rises, cool air sinks. As air is heated, it becomes less dense and lighter than the cooler air surrounding it. This simple law of nature is the principle behind hot-air ballooning. By heating the air in the balloon envelope, maintaining its temperature, or letting it cool, the balloon's pilot is able to climb higher, fly level, or descend. Joseph and Jacques Montgolfier developed the first hot-air balloon in 1783, inspired by the rising of a shirt that was drying above a fire.

Review

13. Rewrite each expression by eliminating parentheses and then combining like terms.

a. $2 + 3(x - 4)$ @ **b.** $(11 - 3x) - 2(4x + 5)$ **c.** $5.1 - 2.7[1 - (2x + 9.7)]$

14. Solve each equation.

a. $12 = 6 + 2(x - 1)$ **b.** $27 = 12 - 2(x + 2)$

15. Charlotte and Emily measured the pulse rates of everyone in their class in beats per minute and collected this set of data.

{62, 68, 68, 70, 74, 66, 82, 74, 76, 72, 70, 68, 80, 60, 84, 72, 66, 78, 70, 68, 66, 82, 76, 66, 66, 80}

a. What is the mean pulse rate for the class?

b. What is the standard deviation? What does this tell you?

16. Each of these two graphs was generated by a recursive formula in the form $u_0 = a$ and $u_n = (1 + r)u_{n-1} + p$ where $n \geq 1$. Describe the parameters a, r, and p that produce each graph. (There are two answers to 16b.)

a.

b.

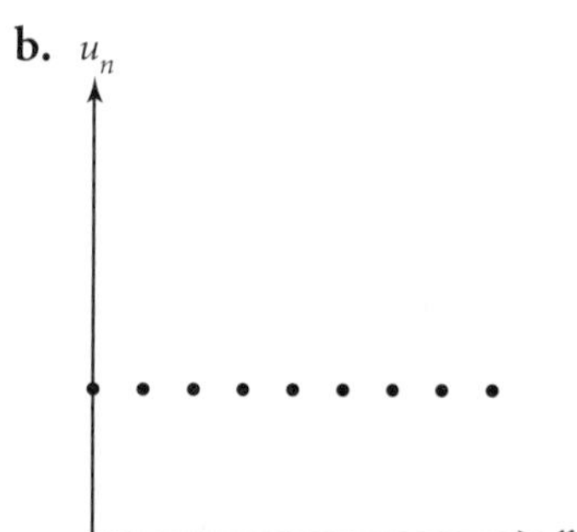

17. Each of these graphs was produced by a linear equation in the form $y = a + bx$. For each graph, tell if a and b are greater than zero, equal to zero, or less than zero.

a.

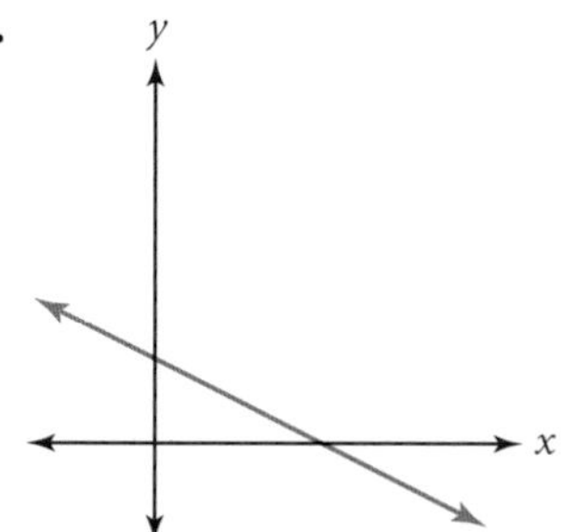

b.

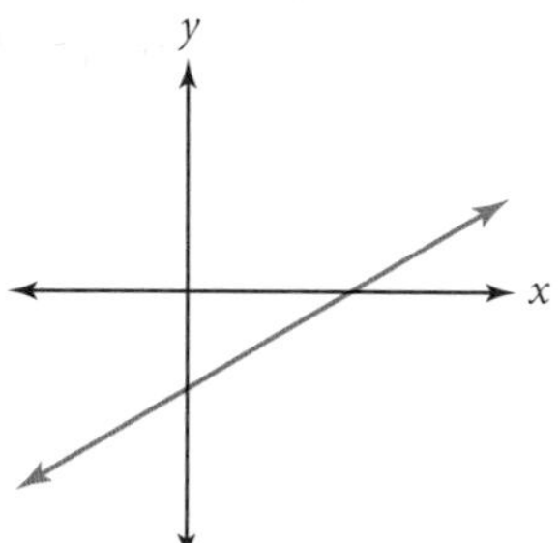

c.

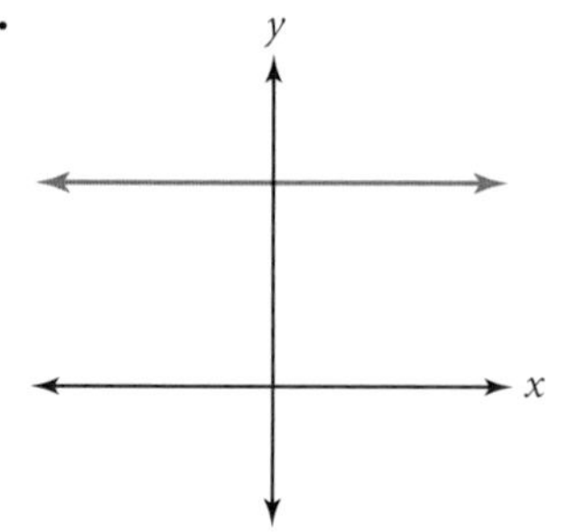

d.

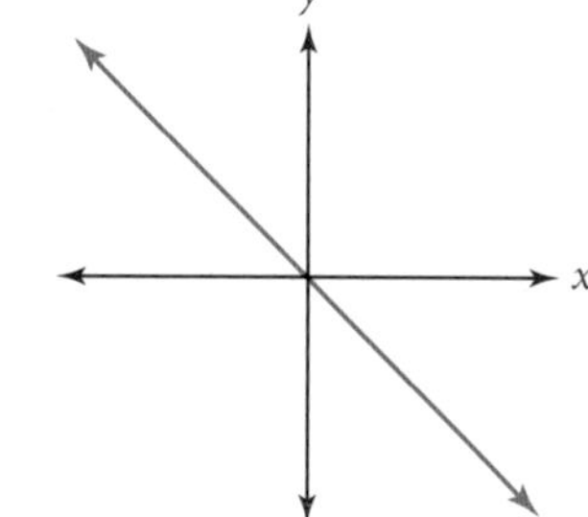

LESSON 3.3

Fitting a Line to Data

Odd how the creative power at once brings the whole universe to order.

VIRGINIA WOOLF

When you graph points from an arithmetic sequence, they lie on a line. When you collect and graph real-life data, the points may appear to have a linear relationship. However, they will rarely lie exactly on a line. They will usually be scattered, and it is up to you to determine a reasonable location for the line that summarizes or gives the trend of the data set. A line that fits the data reasonably well is called a **line of fit.** A line of fit can be used to make predictions about the data, so it is also called a **prediction line.**

There is no single list of rules that will give the best line of fit in every instance, but you can use these guidelines to obtain a reasonably good fit.

Finding a Line of Fit

1. Determine the direction of the points. The longer side of the smallest rectangle that contains most of the points shows the general direction of the line.

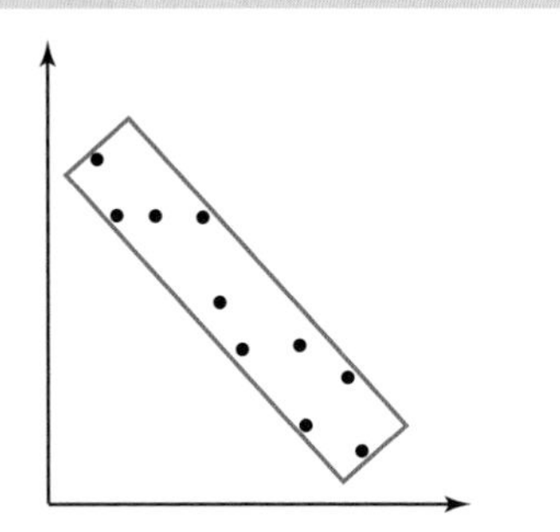

2. The line should divide the points equally. Draw the line so that there are about as many points above the line as below the line. The points above the line should not be concentrated at one end, and neither should the points below the line. The line has nearly the same slope as the longer sides of the rectangle.

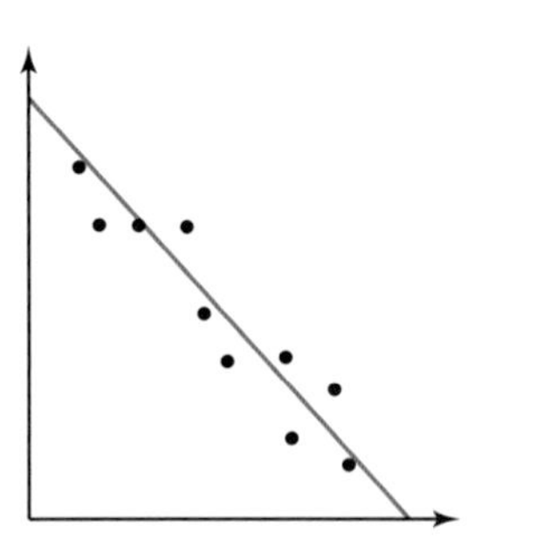

Some computer applications allow you to manually fit a line to data points. You can also graph the data by hand and draw the line of fit.

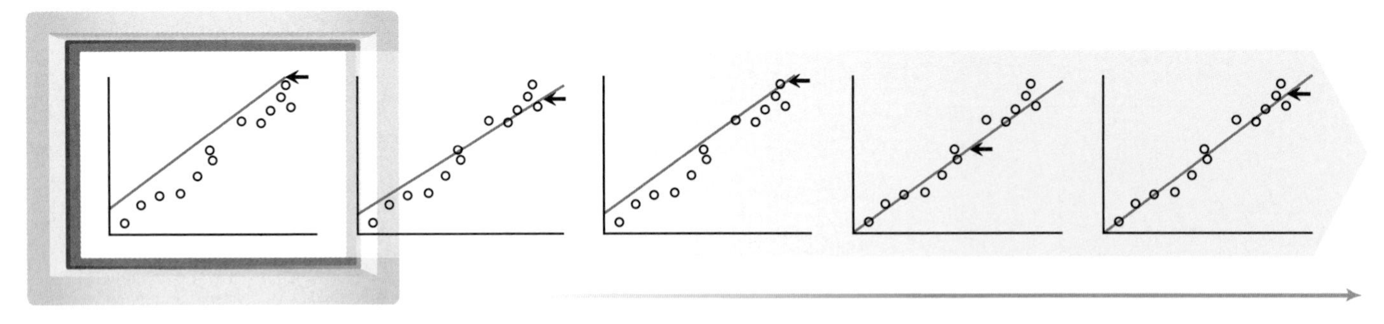

keymath.com/DAA

[► You can explore a movable line of fit using the **Dynamic Algebra Exploration** at www.keymath.com/DAA . ◄]

Once you have drawn a line of fit for your data, you can write an equation that expresses the relationship. To indicate that the line is a prediction line, the variable $\hat{y}$ ("y hat") is used in place of y.

As you learned in algebra, there are several ways to write the equation of a line. You can find the slope and the y-intercept and write the intercept form of the equation. Often, it is easier to choose any two points on the line, use them to calculate the slope, and use the slope and either of the points to write the **point-slope form** of the equation. Either method should give almost exactly the same results. In general, using two points that are farther apart results in a more accurate calculation of the slope than using two points that are closer together.

Point-Slope Form

The formula for the slope is $b = \frac{y_2 - y_1}{x_2 - x_1}$, so you can write the equation of a line with slope b and containing point (x_1, y_1) for any general point (x, y) as $b = \frac{y - y_1}{x - x_1}$ or, equivalently, as

$$y = y_1 + b(x - x_1)$$

This is called the point-slope form for a linear equation.

You can then use this equation to predict points for which data are not available.

EXAMPLE

On a barren lava field on top of the Mauna Loa volcano in Hawaii, scientists have been monitoring the concentration of CO_2 (carbon dioxide) in the atmosphere since 1959. This site is favorable because it is relatively isolated from vegetation and human activities that produce CO_2. The average concentrations for 13 different years, measured in parts per million (ppm), are shown here.

Year	CO_2 (ppm)
1980	338.69
1982	341.13
1984	344.42
1986	347.15
1988	351.48

Year	CO_2 (ppm)
1990	354.21
1992	356.37
1994	358.88
1996	362.64
1998	366.63

Year	CO_2 (ppm)
2000	369.48
2002	373.10
2004	377.38

(Carbon Dioxide Information Analysis Center)

a. Find a line of fit to summarize the data.

b. Predict the concentration of CO_2 in the atmosphere in the year 2050.

Science CONNECTION

With a name meaning "long mountain," Mauna Loa has an area of 2035 mi^2, covers half of the island of Hawaii, and is Earth's largest active volcano. Volcanologists routinely monitor Mauna Loa for signs of eruption and specify hazardous areas of the mountain.

Caldera on Mauna Loa

▸ Solution

Let x represent the year, and let y represent the concentration of CO_2 in parts per million. A graph of the data shows a linear pattern.

a. You can draw a line that seems to fit the trend of the data.

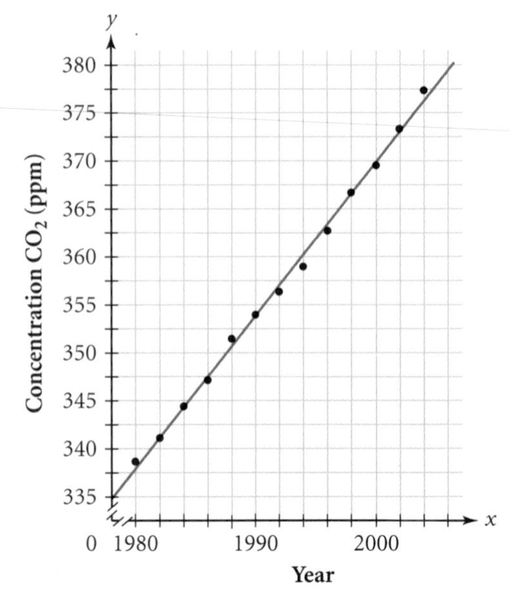

The line shown on the graph is a good fit because many of the points lie on or near the line, the data points above and below the line are roughly equal in number, and they are evenly distributed on both sides along the line. You do not know what the y-intercept should be because you do not know the concentration in the year 0; you have no reason to believe that the line passes through (0, 0).

Next, you need to find the equation of this line of fit. The y-intercept is not easily available, but you can choose two points and use the point-slope form. In this case the line of fit passes through a number of the data points. You might choose the points (1982, 341.1) and (2002, 373.1). They are far enough apart. When you have a data set that is more scattered, you might choose points on the line of fit that are not actually data points.

You can use these points to find the slope.

$$\text{slope} = \frac{373.1 - 341.1}{2002 - 1982} = \frac{32}{20} = 1.6$$

This means the concentration of CO_2 in the atmosphere has been increasing at a rate of about 1.6 ppm each year.

Use either of the two points you chose for the slope calculation, and substitute its coordinates for x_1 and y_1 in the point-slope form of a linear equation.

$$\hat{y} = 373.1 + 1.6(x - 2002)$$ Use point (2002, 373.1), substituting 2002 for x_1 and 373.1 for y_1.

b. You can substitute 2050 for x and solve for $\hat{y}$ to predict the CO_2 concentration in the year 2050. The prediction is 449.9 ppm.

$$\hat{y} = 373.1 + 1.6(2050 - 2002)$$
$$= 449.9$$

Environment CONNECTION

Cars and trucks emit carbon dioxide, methane, and nitrous oxide by burning fossil fuels. These gases trap heat from the sun, producing a "greenhouse effect" and causing global warming. Some scientists warn that if accelerated warming continues, higher sea levels will result. To learn how scientists are studying the greenhouse effect, see the links at www.keymath.com/DAA .

The planet Venus suffers from an extreme greenhouse effect due to constant volcanic activity, which has created a dense atmosphere that is 97% carbon dioxide. As a result, surface temperatures reach 462°C.

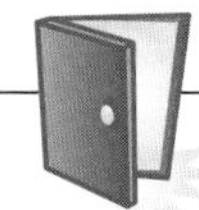

Investigation
The Wave

You will need

- a stopwatch or watch with second hand

Sometimes at sporting events, people in the audience stand up quickly in succession with their arms upraised and then sit down again. The continuous rolling motion that this creates through the crowd is called "the wave." You and your class will investigate how long it takes different-size groups to do the wave.

Step 1 Using different-size groups, determine the time for each group to complete the wave. Collect at least nine pieces of data of the form (*number of people, time*), and record them in a table.

Step 2 Plot the points, and find the equation of a reasonable line of fit. Write a paragraph about your results. Be sure to answer these questions:

- What is the slope of your line, and what is its real-world meaning?
- What are the x- and y-intercepts of your line, and what are their real-world meanings?
- What is a reasonable domain for this equation? Why?

Step 3 Can you use your line of fit to predict how long it would take to complete the wave if everyone at your school participated? Everyone in a large stadium? Explain why or why not.

Finding a value between other values given in a data set is called **interpolation.** Using a model to extend beyond the first or last data points is called **extrapolation.** How would you use the Mauna Loa data in the example to estimate the CO_2 levels in 1960? In 1991?

EXERCISES

You will need

A graphing calculator for Exercises **8–10.**

Practice Your Skills

1. Write an equation in point-slope form for each line shown.

a. ⓐ

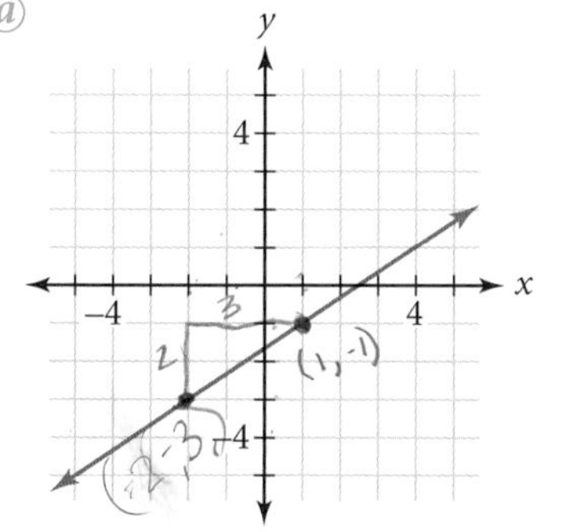

b.

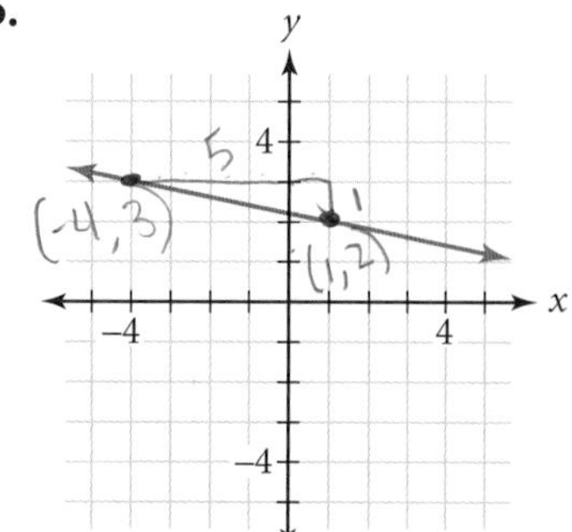

2. Write an equation in point-slope form for each line described.

a. slope $\frac{2}{3}$ passing through $(5, -7)$ ⓐ

b. slope -4 passing through $(1, 6)$

c. parallel to $y = -2 + 3x$ passing through $(-2, 8)$

d. parallel to $y = -4 - \frac{3}{5}(x + 1)$ passing through $(-4, 11)$

3. Solve each equation.

a. Solve $u_n = 23 + 2(n - 7)$ for u_n if $n = 11$. ⓐ

b. Solve $d = -47 - 4(t + 6)$ for t if $d = 95$.

c. Solve $y = 56 - 6(x - 10)$ for x if $y = 107$.

4. Consider the line $y = 5$.

a. Graph this line and identify two points on it.

b. What is the slope of this line?

c. Write the equation of the line that contains the points $(3, -4)$ and $(-2, -4)$.

d. Write three statements about horizontal lines and their equations.

5. Consider the line $x = -3$.

a. Graph it and identify two points on it.

b. What is the slope of this line?

c. Write the equation of the line that contains the points $(3, 5)$ and $(3, 1)$.

d. Write three statements about vertical lines and their equations. ⓗ

Reason and Apply

6. Of the graphs below, choose the *one* with the line that best satisfies the guidelines on page 138. For each of the other graphs, explain which guidelines the line violates.

a.

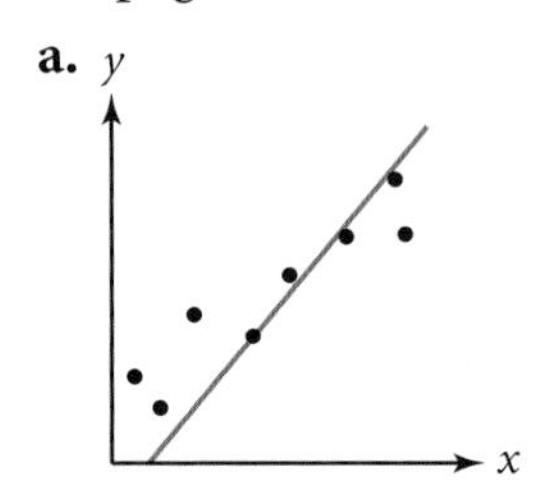

b.

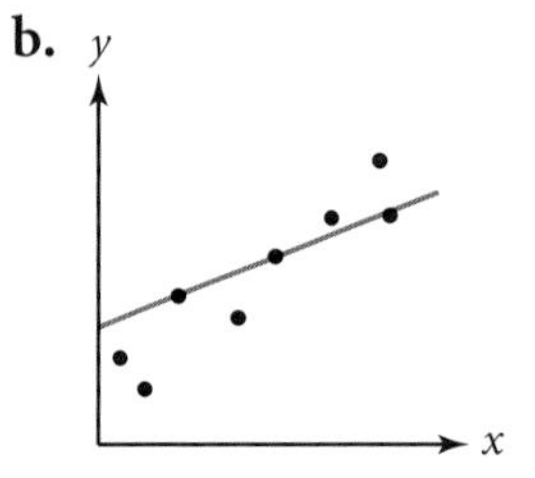

c.

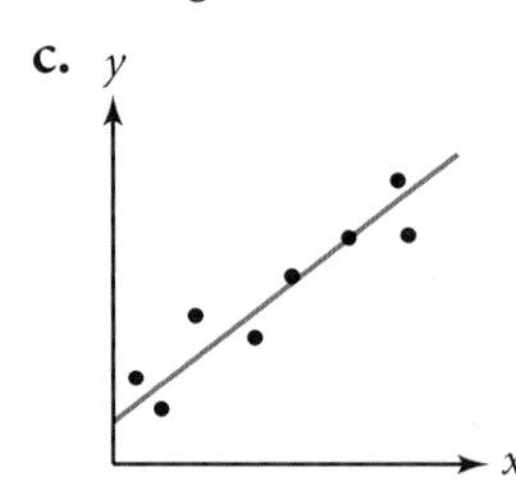

d.

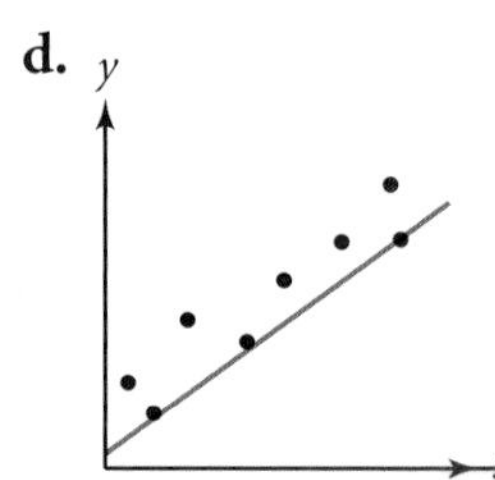

7. For each graph below, lay your ruler along your best estimate of the line of fit. Estimate the y-intercept and the coordinates of one other point on the line. Write an equation in intercept form for the line of fit.

a. ⓐ

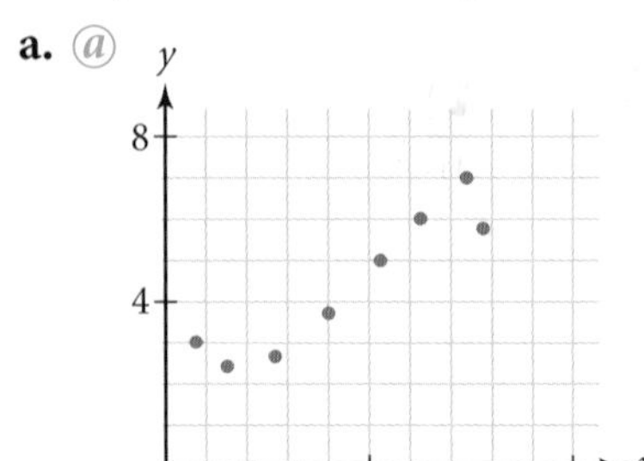

b.

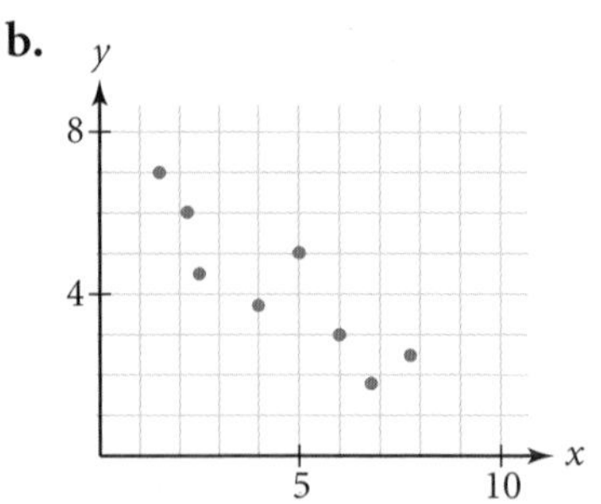

c.

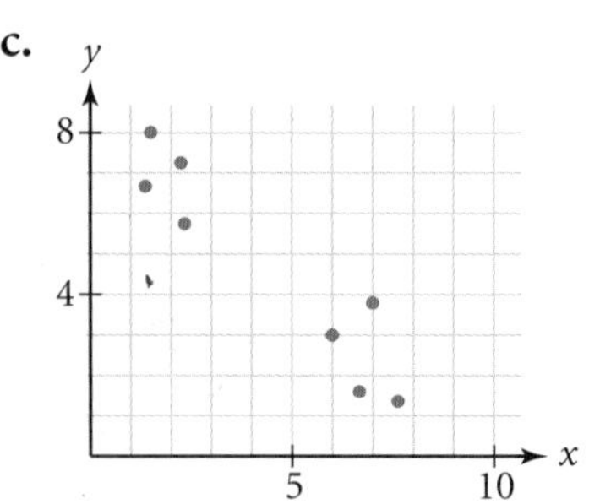

8. APPLICATION A photography studio offers several packages to students posing for yearbook photos. Let x represent the number of pictures, and let y represent the price in dollars.

Number of pictures	30	18	15	11
Price ($)	27.00	20.00	17.00	14.00

a. Plot the data, and find an equation of a line of fit. Explain the real-world meaning of the slope of this line.

b. Find the y-intercept of your line of fit. Explain the real-world meaning of the y-intercept.

c. If the studio offers a 50-print package, what do you think it should charge?

d. How many prints do you think the studio should include in the package for a $9.99 special?

9. Use height as the independent variable and length of forearm as the dependent variable for the data collected from nine students.

a. Name a good graphing window for your scatter plot.

b. Write a linear equation that models the data. @

c. Write a sentence describing the real-world meaning of the slope of your line.

d. Write a sentence describing the real-world meaning of the y-intercept. Explain why this doesn't make sense and how you might correct it.

e. Use your equation to estimate the height of a student with a 50 cm forearm and to estimate the length of a forearm of a student 158 cm tall.

Height (cm)	Forearm (cm)
185.9	48.5
172.0	44.5
155.0	41.0
191.5	50.5
162.0	43.0
164.3	42.5
177.5	47.0
180.0	48.0
179.5	47.5

10. APPLICATION This data set was collected by a college psychology class to determine the effects of sleep deprivation on students' ability to solve problems. Ten participants went 8, 12, 16, 20, or 24 hours without sleep and then completed a set of simple addition problems. The number of addition errors was recorded.

Hours without sleep	8	8	12	12	16	16	20	20	24	24
Number of errors	8	6	6	10	8	14	14	12	16	12

a. Define your variables and create a scatter plot of the data.

b. Write an equation of a line that approximates the data and sketch it on your graph.

c. Based on your model, how many errors would you predict a person to make if she or he hadn't slept in 22 hours?

d. In 10c, did you use interpolation or extrapolation? Explain.

Review

11. The 3rd term of an arithmetic sequence is 54. The 21st term is 81. Find the 35th term. ⓐ

12. Write the first four terms of this sequence and describe its long-run behavior. ⓗ

$$u_1 = 56$$
$$u_n = \frac{u_{n-1}}{2} + 4 \quad \text{where } n \geq 2$$

13. Given the data set {20, 12, 15, 17, 21, 15, 30, 16, 14}:

a. Find the median.

b. Add as few elements as possible to the set in order to make 19.5 the median.

14. You start 8 meters from a marker and walk toward it at the rate of 0.5 m/s.

a. Write a recursive rule that gives your distance from the marker after each second.

b. Write an explicit formula that allows you to find your distance from the marker at any time.

c. Interpret the real-world meaning of a negative value for u_n in 14a or 14b.

Project

TALKIN' TRASH

How much trash do you and your family generate each day? How much trash does that amount to for the entire U.S. population daily? Annually? How much land is needed to dump garbage? Research some data on U.S. population and waste production. Find linear models for your data. Use your equations to predict the population and the amount of waste that you might expect in the years 2020 and 2030. Use your graphs and equations to decide if the amount of waste is increasing because of the increase in population or because of other factors.

Garbage piles up in New York City during a 1982 sanitation workers strike.

Your project should include

- Data and sources.
- Linear models for predicting population and waste, an explanation of how you found them, and real-world meanings for each part of your model.
- Domain and range for each model.
- Population and waste predictions for 2020 and 2030, including amount of waste per person per day.
- A complete analysis of your findings, in paragraph form.

LESSON

3.4

The Median-Median Line

Have you noticed that you and your classmates frequently find different equations to model the same data? Some lines fit data better than others and can be used to make more accurate predictions. In this lesson you will learn a standard method for finding a line of fit that will enable each member of the class to get the same equation for the same set of data.

You can use many different methods to find a statistical line of fit. The **median-median line** is one of the simplest methods. The procedure for finding the median-median line uses three points $(M_1, M_2, \text{ and } M_3)$ to represent the entire data set, and the equation that best fits these three points is taken as the line of fit for the entire set of data.

To find the three points that will represent the entire data set, you first order all the data points by their domain value (the *x*-value) and then divide the data into three equal groups. If the number of points is not divisible by 3, then you split them so that the first and last groups are the same size. For example:

18 data points: split into groups of 6-6-6
19 data points: split into groups of 6-7-6
20 data points: split into groups of 7-6-7

You then order the *y*-values within each of the groups. The representative point for each group has the coordinates of the median *x*-value of that group and the median *y*-value of that group. Because a good line of fit divides the data evenly, the median-median line should pass between M_1, M_2, and M_3, but be closer to M_1 and M_3 because they represent two-thirds of the data. To accomplish this, you can find the *y*-intercept of the line through M_1 and M_3, and the *y*-intercept of the line through M_2 that has the same slope. The mean of the three *y*-intercepts of the lines through M_1, M_2, and M_3 gives you the *y*-intercept of a line that satisfies these requirements.

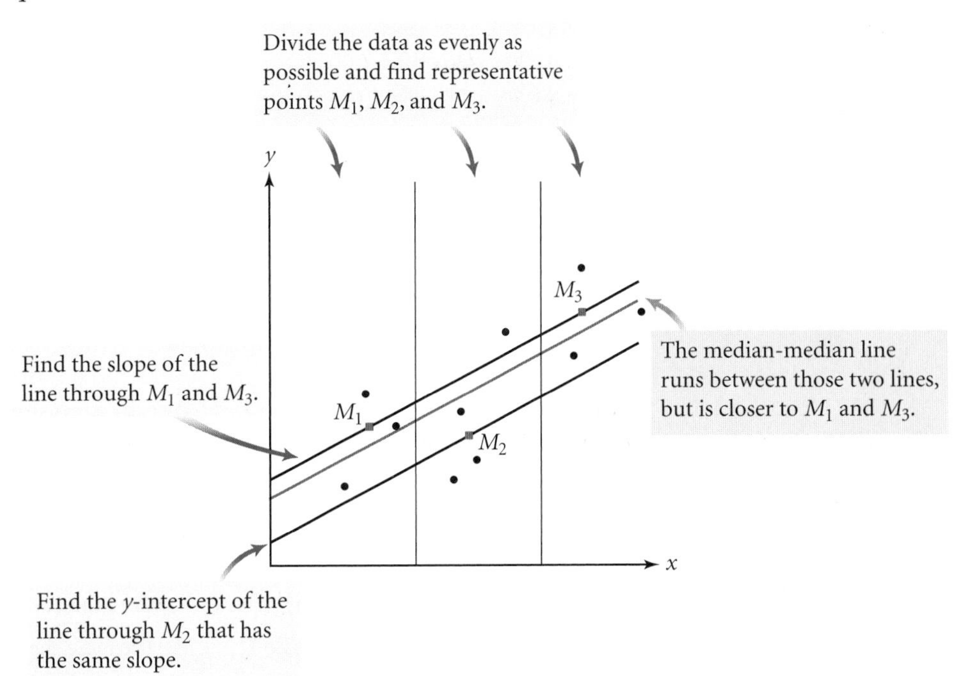

Don't forget to order the y-values in each group when finding the median y-value. Carefully study the following example to see how this is done.

EXAMPLE

Find the median-median line for these data.

x	y	x	y
4.0	23	8.9	50
5.2	28	9.7	47
5.8	29	10.4	52
6.5	35	11.2	60
7.1	35	11.9	58
7.8	40	12.5	62
8.3	42	13.1	64

▸ ***Solution***

First, group the data and find M_1, M_2, and M_3. There are 14 data points, so split them into groups of 5-4-5.

x	y	**(median x, median y)**
4.0 5.2 5.8 6.5 7.1	23 28 29 35 35	$M_1(5.8, 29)$
7.8 8.3 8.9 9.7	40 42 50 47	$M_2(8.6, 44.5)$
10.4 11.2 11.9 12.5 13.1	52 60 58 62 64	$M_3(11.9, 60)$

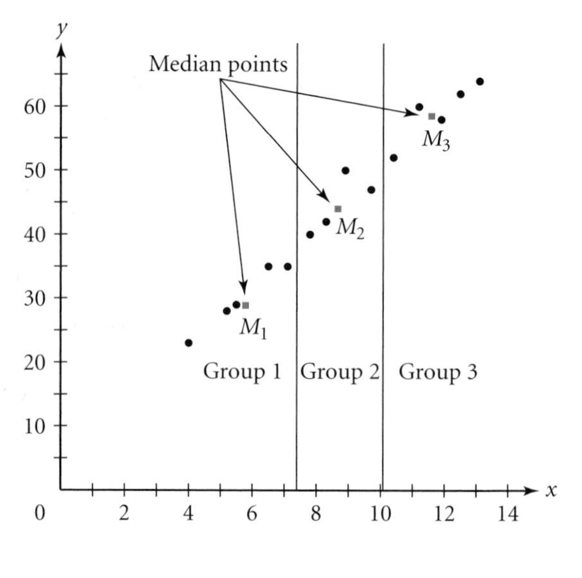

The slope of the median-median line is determined by the slope of the line through M_1 and M_3.

$$\text{slope} = \frac{60 - 29}{11.9 - 5.8} \approx 5.082$$

Next, find an equation for the line containing points $M_1(5.8, 29)$ and $M_3(11.9, 60)$. You have already found that the slope is about 5.082. Using M_1, you can write the point-slope form of the equation. Rewrite the equation in intercept form to find the y-intercept. (The calculations use the unrounded value of the slope.)

$y \approx 29 + 5.082(x - 5.8)$ Write the equation in point-slope form.

$y \approx 29 + 5.082x - 29.4754$ Distribute the 5.082.

$y \approx -0.475 + 5.082x$ Add like terms. This is the intercept form. The y-intercept is about -0.475.

If you use M_3 instead of M_1, you should get the same equation, because your equation is the line through both M_1 and M_3.

Next, find the equation of the line that is parallel to the line through M_1 and M_3 and passes through the middle representative point, $M_2(8.6, 44.5)$.

$y \approx 44.5 + 5.082(x - 8.6)$ Write the equation in point-slope form.

$y \approx 0.795 + 5.082x$ Distribute and add like terms to find the intercept form of the equation.

The median-median line is parallel to both of these lines, so it will also have a slope of 5.082. To find the y-intercept of the median-median line, you find the mean of the y-intercepts of the lines through M_1, M_2, and M_3. Note that the y-intercept of the line through M_1 is the same as the y-intercept of the line through M_3.

$\frac{-0.475 + (-0.475) + 0.795}{3} \approx -0.052$ Find the mean of the y-intercepts.

So, finally, the equation of the median-median line is $\hat{y} = -0.052 + 5.082x$. Note that this line is one-third of the way from the first line to the second line.

Finding a Median-Median Line

1. Order your data by domain value first. Then, divide the data into three sets equal in size. If the number of points does not divide evenly into three groups, make sure that the first and last groups are the same size. Find the median x-value and the median y-value in each group. Call these points M_1, M_2, and M_3.
2. Find the slope of the line through M_1 and M_3. This is the slope of the median-median line.
3. Find the equation of the line through M_1 with the slope you found in Step 2. The equation of the line through M_3 will be the same.
4. Find the equation of the line through M_2 with the slope you found in Step 2.
5. Find the y-intercept of the median-median line by taking the mean of the y-intercepts of the lines through M_1, M_2, and M_3. The y-intercepts of the lines through M_1 and M_3 are the same.
6. Finally, write the equation of the median-median line using the mean y-intercept from Step 5 and the slope from Step 2.

Japanese artist Yoshio Itagaki (b 1967) created *Tourist on the Moon #2* as a commentary on tourists' desire to document their visits to spectacular scenes. He is both amused by and critical of the human appetite for sensation and novelty. This work is a triptych—a piece in three panels.

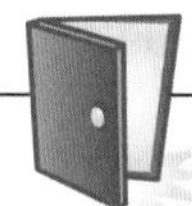

Investigation

Airline Schedules

In this investigation you will use data about airline flights to find a median-median line to model the relationship between the distance of a flight and the flight time. You will use the linear model to make predictions about flight times and distances that aren't in the table.

Destination	Flight time (min)	Distance (mi)
Cincinnati, OH	64	229
Houston, TX	189	1092
Los Angeles, CA	288	1979
Memphis, TN	104	610
Denver, CO	180	1129
Phoenix, AZ	248	1671
Louisville, KY	67	306
San Francisco, CA	303	2079
Omaha, NE	120	658
New Orleans, LA	156	938

Step 1 The flights listed here are morning departures from Detroit, Michigan. Write a complete sentence explaining what the first line of data tells you.

Step 2 Graph the data on graph paper.

Step 3 Show the steps to calculate the median-median line through the data. Write the equation of this line. Use your calculator to check your work. [▶ See **Calculator Note 3D** to learn how to find the median-median line with your calculator.◀]

Step 4 On your graph, mark the three representative points used in the median-median process. Add the line to this graph.

Step 5 Answer these questions about your data and model.

a. Use your median-median line to interpolate two points for which you did not collect data. What is the real-world meaning of each of these points?

b. Which two points differ the most from the value predicted by your equation? Explain why.

c. What is the real-world meaning of your slope?

d. Find the y-intercept of your median-median line. What is its real-world meaning?

e. What are the domain and range for your data? Why?

f. Compare the median-median line method to the method you used in Lesson 3.3 to find the line of fit. What are the advantages and disadvantages of each? In your opinion, which method produces a better line of fit? Why?

Step 6 Summarize what you learned in this investigation and describe any difficulties you had.

There are many ways to find a line of fit for linear data. Estimating the line of fit is adequate in many cases, but different people will estimate different lines of fit and may use their own bias to draw the line higher or lower. A median-median line is a systematic method, accepted by statisticians, that summarizes the overall trend in linear data.

EXERCISES

You will need

A graphing calculator for Exercises **6** and **8–11.**

Practice Your Skills

1. How should you divide the following sets into three groups for the median-median line method?

a. set of 51 elements ⓐ **b.** set of 50 elements **c.** set of 47 elements **d.** set of 38 elements

2. Find an equation in point-slope form of the line passing through

a. (8.1, 15.7) and (17.3, 9.5) ⓐ **b.** (3, 47) and (18, 84)

3. Find an equation in point-slope form of the line parallel to $y = -12.2 + 0.75x$ that passes through the point (14.4, 0.9).

4. Find an equation of the line one-third of the way from $y = -1.8x + 74.1$ to $y = -1.8x + 70.5$. (*Hint:* The first line came from two points in a median-median procedure, and the other line came from the third point.) ⓐ

5. Find the equation of the line one-third of the way from the line $y = 2.8 + 4.7x$ to the point (12.8, 64).

Reason and Apply

6. APPLICATION Follow these steps to find the equation of the median-median line for the data on life expectancy at birth for males in the United States for different years of birth. Let x represent the year of birth, and let y represent the male life expectancy in years.

Year of birth	Male life expectancy (years)
1910	48.4
1920	53.6
1930	58.1
1940	60.8
1945	62.8
1950	65.6
1955	66.2
1960	66.6
1965	66.8
1970	67.1
1975	68.8
1980	70.0
1985	71.2
1990	71.8
1995	72.5
2000	74.1
2003	74.8

(*The World Almanac and Book of Facts 2007*)

a. How many points are there in each of the three groups?

b. What are the three representative points for these data? Graph these points. ⓐ

c. Draw the line through the first and third points. What is the slope of this line? What is the real-world meaning of the slope?

d. Write the equation of the line through the first and third points. Rewrite this equation in intercept form.

e. Draw the line parallel to the line in 6d passing through the second point. What is the equation of this line? Rewrite this equation in intercept form.

f. Find the mean of the y-intercepts and write the equation of the median-median line. Graph this line. Remember to use the intercepts of all three lines.

g. The year 1978 is missing from the table. Using your model, what would you predict the life expectancy at birth to be for males born in 1978?

h. Use your model to predict the life expectancy at birth for males born in 1991.

i. Using this model, when would you predict the life expectancy at birth for males in the United States to exceed 80 years?

7. A set of data has been summarized by the three median-median points (2, 11), (5, 32), and (8, 41).

a. What is the equation of the median-median line? ⓐ

b. The centroid (or balance point) of the triangle formed by the three points can be found by calculating the means of the x- and y-coordinates $(\bar{x}, \bar{y})$. What is the centroid of the triangle formed by these three points?

c. What is the equation of the line, parallel to the median-median line, passing through the centroid?

d. How does this line compare to the median-median line?

e. Use your results to make a conjecture: Write an equation for the median-median line using the slope (b) and the centroid $(\bar{x}, \bar{y})$.

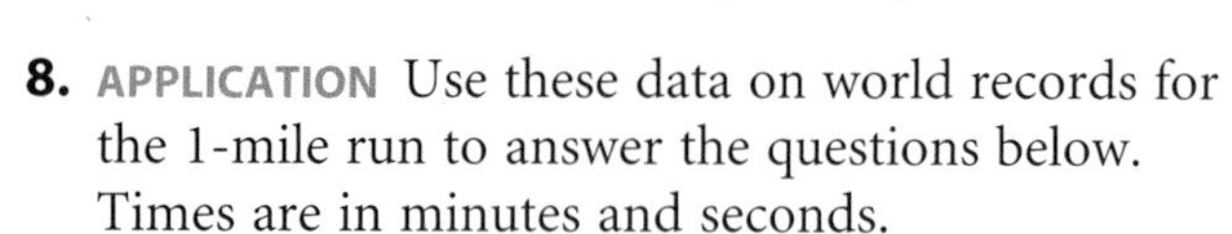

8. APPLICATION Use these data on world records for the 1-mile run to answer the questions below. Times are in minutes and seconds.

World Records for 1-Mile Run

Year	Runner	Time
1937	Sydney Wooderson, U.K.	4:06.4
1942	Gunder Haegg, Sweden	4:06.2
1945	Gunder Haegg, Sweden	4:01.4
1958	Herb Elliott, Australia	3:54.5
1967	Jim Ryun, U.S.	3:51.1
1979	Sebastian Coe, U.K.	3:49.0
1982	Steve Scott, U.S.	3:47.69
1985	Steve Cram, U.K.	3:46.31
1993	Noureddine Morceli, Algeria	3:44.39
1999	Hicham El Guerrouj, Morocco	3:43.13

(*International Association of Athletics Federations*)

a. Let x represent the year, and let y represent the time in seconds. What is the equation of the median-median line?

b. What is the real-world meaning of the slope? ⓗ

c. Use the equation to interpolate and predict what new record might have been set in 1954. How does this compare with Roger Bannister's actual 1954 record of 3:59.4? ⓐ

d. Use the equation to extrapolate and predict what new record might have been set in 1875. How does this compare with Walter Slade's 1875 world record of 4:24.5?

e. Describe some difficulties you might have with the meaning of 1075.7 as a y-intercept.

f. Has a new world record been set since 1999? Find more recent information on this subject and compare it with the predictions of your model.

9. The number of deaths caused by automobile accidents, D, per hundred thousand population in the United States is given for various years, t.

t	1924	1925	1926	1927	1928	1929	1930	1931	1932	1933
D	15.5	17.1	18.0	19.6	20.8	23.3	24.5	25.2	21.9	23.3

(W. A. Wilson and J. I. Tracey, *Analytic Geometry*, 3rd ed., Boston: D. C. Heath, 1949, p. 246.)

a. Make a scatter plot of the data.

b. Would you describe this pattern with a single line? Explain why or why not. ⓐ

c. Create one or more models and give an appropriate domain and range for each model.

d. Would you use this model to predict the number of deaths from auto accidents in 2010? Explain why or why not.

10. Devise a mean-mean line procedure and use it on the data in Exercise 8. What difficulties might arise when using this method? Compare the advantages and disadvantages of this method and the median-median line method.

Review

11. Create a data set of 9 values such that the median is 28, the minimum is 11, and there is no upper whisker on a box plot of the data.

12. What is the equation of the line that passes through a graph of the points of the sequence defined by

$$u_1 = 4$$
$$u_n = u_{n-1} - 3 \quad \text{where } n \geq 2$$

13. The histogram at right shows the results of a statewide math test given to eleventh graders. If Ramon scored 35, what is the range of his percentile ranking? (h)

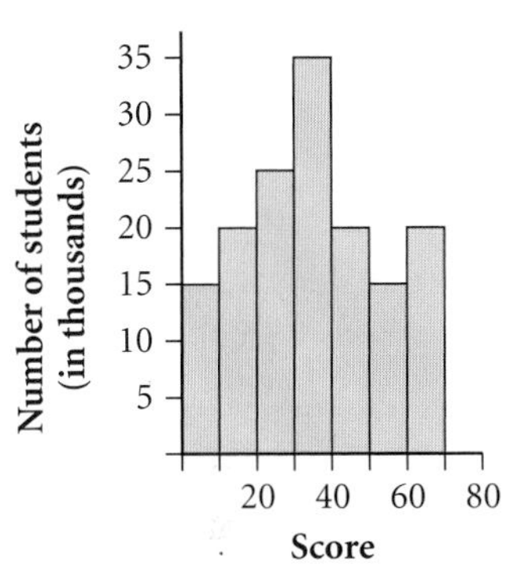

14. Earl's science lab group made six measurements of mass and then summarized the results. Someone threw away the measurements. Help the group reconstruct the measurements from these statistics.

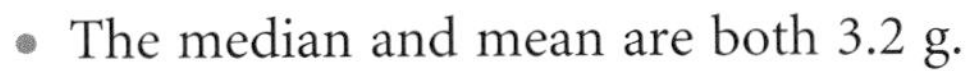

- The median and mean are both 3.2 g.
- The mode is 3.0 g.
- The *IQR* is 0.6 g.
- The largest deviation is −0.9 g.

15. Travis is riding with his parents on Interstate 15 across Utah. He records the digital speedometer reading in mi/h at 4:00 P.M. and every five minutes for the next hour. The mean of the 13 readings is 58.3 mi/h and the standard deviation is 7.4 mi/h.

a. If you assume a symmetric mounded set of data, what are your estimates for the fastest and slowest of his 13 readings?

b. If you assume the data are skewed left, what are your estimates for the fastest and slowest of the readings?

c. Which of your assumptions and estimates seem more likely to you and why?

Project

COUNTING FOREVER

How long would it take you to count to 1 million? Collect data by recording the time it takes to count to different numbers. Predict the time it would take to count to 1 million, 1 billion, 1 trillion, and to the amount of the federal debt.

Your project should include

- ▶ Explanations of the procedure you used to collect these data and how you made your predictions.
- ▶ Your data, graphs, and equations.
- ▶ A summary of your predictions.

In 1992, this sign in New York City showed the increasing federal debt. What is the current debt?

LESSON

3.5 Prediction and Accuracy

The growth of understanding follows an ascending spiral rather than a straight line.

JOANNA FIELD

The median-median line method gives you a process by which anyone will get the same line of fit for a set of data. However, unless your data are perfectly linear, even the median-median line will not be perfect. In some cases you can find a more satisfactory model if you draw a line by hand. Having a line that "looks better" is not a convincing argument that it really is better.

y
Data point
Positive residual
Line of fit
Point predicted by line of fit
Point predicted by line of fit
Negative residual
Data point
x

You need a way to evaluate how accurately your model describes the data. One excellent method is to look at the **residuals,** or the vertical differences between the points in your data set and the points generated by your line of fit.

residual = y-value of data point − y-value of point on line

Similar to the deviation from the mean that you learned about in Chapter 2, a residual is a signed distance. Here, a positive residual indicates that the point is above the line, and a negative residual indicates that the point is below the line.

A line of good fit should have about as many points above it as below. This means that the sum of the residuals should be near zero. In other words, if you connect each point to the line with a vertical segment, the sum of the lengths of the segments above the line should be about equal to the sum of those below the line.

EXAMPLE A

The manager of Big K Pizza must order supplies for the month of November. The numbers of pizzas sold in November during the past four years were 512, 603, 642, and 775, respectively. How many pizzas should she plan for this November?

▸ Solution

Let x represent the past four years, 1 through 4, and let y represent the number of pizzas. Graph the data. The median-median method gives the linear model $\hat{y} \approx 417.3 + 87.7x$.

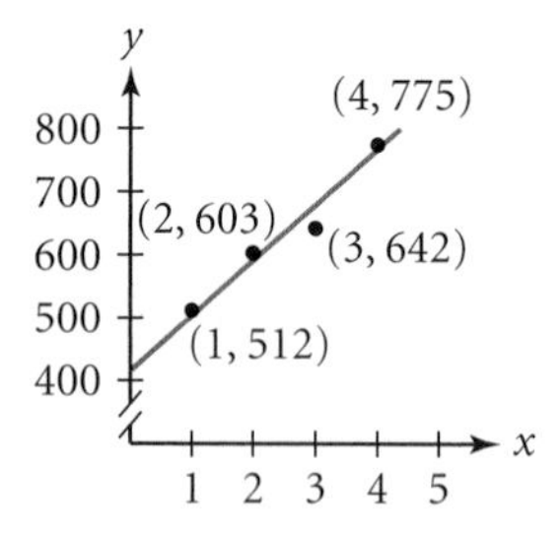

Calculating the residuals is one way to evaluate the accuracy of this model before using it to make a prediction. Evaluate the linear equation when x is 1, 2, 3, and 4. When $x = 1$, for example, $417.3 + 87.7(1) = 505.0$. A table helps organize the information.

Year (x)	1	2	3	4
Number of pizzas (y)	512	603	642	775
y-value from line $\hat{y} \approx 417.3 + 87.7x$	505.0	592.7	680.3	768.0
Residual ($y - \hat{y}$)	7.0	10.3	−38.3	7.0

The sum of the residuals is $7 + 10.3 + (-38.3) + 7.0$, or -14.0 pizzas, which is fairly close to zero in relation to the large number of pizzas purchased. The linear model is therefore a pretty good fit. The sum is negative, which means that all together the points below the line are a little farther away than those above.

If the manager plans for $417.3 + 87.7(5)$, or approximately 856 pizzas, she will be very close to the linear pattern established over the past four years. Because the residuals range from -38.3 to 7, she may want to adjust her prediction higher or lower depending on factors such as whether or not the supplies are likely to spoil or whether or not she can easily order more supplies.

The investigation gives you another opportunity to fit a line to data and to analyze the real-world meaning of the linear model. You'll also calculate residuals and explore their meaning.

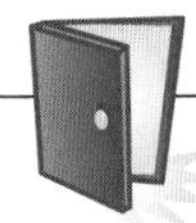

Investigation

Spring Experiment

You will need

- a spring
- a mass holder
- small unit masses
- a support stand
- a ruler

In this investigation you will collect data on how a spring responds to various weights. You will create a model using a median-median line and then you will use residuals to judge the accuracy of any predictions made using your model.

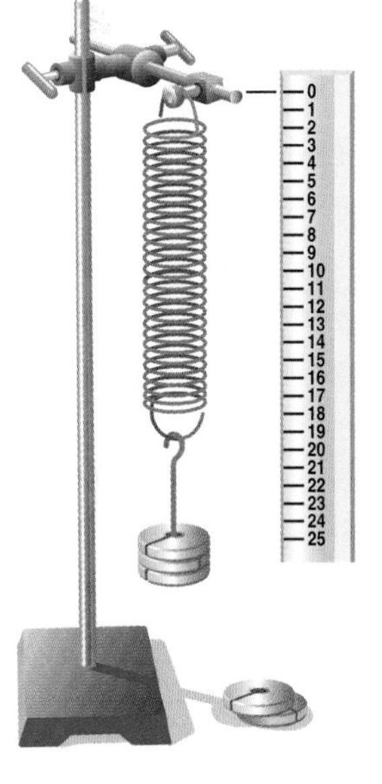

Procedure Note

1. Attach the mass holder to the spring.
2. Hang the spring from a support, and the mass holder from the spring.
3. Measure the length of the spring (in centimeters) from the first coil to the last coil.

Step 1 Place different amounts of mass on the mass holder, recording the corresponding length of the spring each time. Collect about 10 data points of the form (*mass, spring length*).

Step 2 Plot the data and find the median-median line.

Step 3 Give the real-world meanings of the slope and the y-intercept.

Step 4 Use a table to organize your data and calculate the residuals. Then answer these questions.

a. What is the sum of the residuals? Does it appear that your linear model is a good fit for the data? Explain.

b. Find the greatest positive and negative residuals. What could the magnitude of these residuals indicate?

c. If you used your model to predict the length of the spring for a particular weight and then measured the spring for that weight, would you be surprised if the predicted length and the measured length were different? Why or why not?

d. Use your model to predict the length of your spring for a weight 2 units larger than your heaviest weight. Now give an estimate in the form _?_ ± _?_ that you believe would contain the actual measured length.

The sum of the residuals for your line of fit should be close to zero. However, as shown at right, you could find a poorly fitting line and still get zero for the sum of the residuals. Due to the residual sizes, you should realize that this model will not make accurate predictions. A line of good fit should also follow the direction of the points. This means that the individual residuals should be as close to zero as possible.

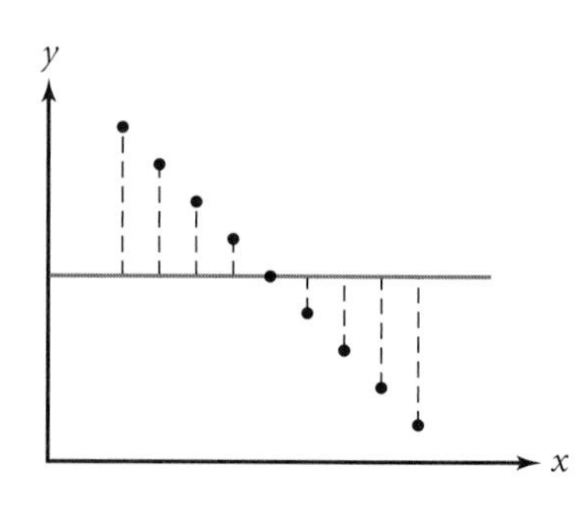

keymath.com/DAA

Because the sum of the residuals does not give a complete picture, you need some other way to judge how accurate predictions from your model will be. One useful measure of accuracy starts by squaring the residuals to make them all positive. For the pizza data and the linear model in Example A, this gives

Year (x)	1	2	3	4
Number of pizzas (y)	512	603	642	775
Residual	7.0	10.3	−38.3	7.0
(Residual)^2	49	106.78	1469.44	49

The sum of the squares is quite large: 1674.22 pizzas^2. Perhaps some kind of average would give a better indication of the error in the predictions from this model. You could divide by 4 since there are four data points. However, this is not necessarily the best thing to do. Consider that it takes a minimum of two points to make any equation of a line. So two of the points can be thought of as defining the line. The other points determine the spread. So you should actually divide by 2 less than the number of data points.

$$\frac{1674.22}{4-2} \approx 837.11 \text{ pizzas}^2$$

The measure of the error should be measured in numbers of pizzas, rather than pizzas^2, so take the square root of this number.

$$\sqrt{837.11} \approx 28.9 \text{ pizzas}$$

This means that generally this line should predict values for the number of pizzas that are within 28.9 pizzas of the actual data. This value is called the **root mean square error.**

Root Mean Square Error

The root mean square error, s, is a measure of the spread of data points from a model.

$$s = \sqrt{\frac{\sum_{i=1}^{n}(y_i - \hat{y}_i)^2}{n - 2}}$$

where y_i represents the y-values of the individual data pairs, $\hat{y}_i$ represents the respective y-values predicted from the model, and n is the number of data pairs.

You should notice that root mean square error is very similar to standard deviation, which you learned about in Chapter 2. Because of their similarities, both are represented by the variable s.

EXAMPLE B

A scientist measures the current in milliamps through a circuit with constant resistance as the voltage in volts is varied. What is the root mean square error for the model $\hat{y} = 0.47x$? What is the real-world meaning of the root mean square error? Predict the current when the voltage is 25.000 volts. How accurate is the prediction?

Voltage (x)	5.000	7.500	10.000	12.500	15.000	17.500	20.000
Current (y)	2.354	3.527	4.698	5.871	7.053	8.225	9.403

▶ Solution

A graph of the data with the linear model shows a good fit. To calculate the root mean square error, first calculate the residual for each data point, take the sum of the squares of the residuals, divide by $n - 2$, and take a square root. [▶ See **Calculator Note 3E** for ways to calculate the root mean square error on your calculator. ◀]

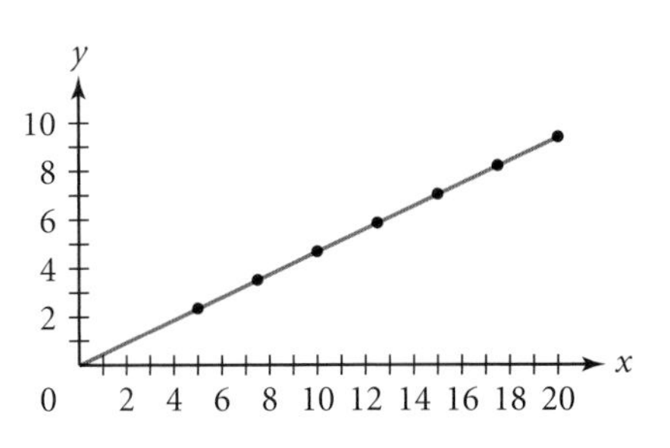

Voltage (x)	5.000	7.500	10.000	12.500	15.000	17.500	20.000
Current (y)	2.354	3.527	4.698	5.871	7.053	8.225	9.403
Prediction from line ($\hat{y}$)	2.350	3.525	4.700	5.875	7.050	8.225	9.400
Residual ($y - \hat{y}$)	0.004	0.002	−0.002	−0.004	0.003	0	0.003
(Residual)^2	1.6×10^{-5}	4.0×10^{-6}	4.0×10^{-6}	1.6×10^{-5}	9×10^{-6}	0	9.0×10^{-6}

$$s = \sqrt{\frac{5.8 \times 10^{-5}}{7-2}} = \sqrt{\frac{5.8 \times 10^{-5}}{5}} = \sqrt{1.16 \times 10^{-5}} \approx 0.0034$$

This means that values predicted by this equation will generally be within 0.0034 milliamp of the actual current.

For a voltage of 25.000 volts, the model gives 0.47(25.000), or 11.750 milliamps, as the current. Considering the root mean square error, the scientist could expect a reading between 11.7466 milliamps and 11.7534 milliamps.

You have seen many ways to find a line of fit for data. Calculating residuals and the root mean square error now gives you tools for evaluating how well your line fits the data.

EXERCISES

You will need

A graphing calculator for Exercises **3, 5, 8, 10,** and **11.**

Practice Your Skills

1. The median-median line for a set of data is $\hat{y} = 2.4x + 3.6$. Find the residual for each of these data points.

a. (2, 8.2) @ **b.** (4, 12.8) **c.** (10, 28.2)

2. The median-median line for a set of data is $\hat{y} = -1.8x + 94$. This table gives the x-value and the residual for each data point. Determine the y-value for each data point. @

x-value	5	8	12	20
Residual	−2.7	3.3	2.1	−1.1

3. Return to Exercise 6 in Lesson 3.4 about life expectancy for males. Use your median-median line equation to answer these questions.

a. Calculate the residuals.

b. Calculate the root mean square error for the median-median line.

c. What is the real-world meaning of the root mean square error?

4. Suppose the residuals for a data set are 0.4, −0.3, 0.2, 0.1, −0.2, −0.3, 0.2, 0.1. What is the root mean square error for this set of residuals? @

Reason and Apply

5. APPLICATION This table gives the mean height in centimeters of boys ages 5 to 13 in the United States.

Age	Height (cm)
5	110
6	116
7	121
8	127
9	132
10	137
11	143
12	150
13	156

(*www.fpnotebook.com*)

a. Define variables, plot the data, and find the median-median line. @

b. Calculate the residuals.

c. What is the root mean square error for the median-median line?

d. What is the real-world meaning of the root mean square error?

e. If you use the median-median line to predict the mean height of boys age 15, what range of heights should be predicted? Are you using interpolation or extrapolation to make your prediction?

6. Consider the residuals from Exercise 5b.

a. Make a box plot of these values.

b. Describe the information about the residuals that is shown in the box plot.

7. With a specific line of fit, the data point (6, 47) has a residual of 2.8. The slope of the line of fit is 2.4. What is the equation of the line of fit? ⓗ

8. The following readings were taken from a display outside the First River Bank. The display alternated between °F and °C. However, there was an error within the system that calculated the temperatures.

°F	18	33	37	25	40	46	43	49	55	60	57
°C	−6	2	3	−3	5	8	7	10	12	15	13

a. Plot the data and the median-median line.

b. Calculate the residuals. You will notice that the residuals are generally negative for the lower temperatures and positive for the higher temperatures. How is this represented on the graph of the data and line?

c. Adjust your equation (by adjusting the slope or y-intercept) to improve this distribution of the residuals.

d. Calculate the root mean square error for the median-median line and for the equation you found in 8c. Compare the root mean square errors for the two lines and explain what this tells you.

e. Use your equation from 8c to predict what temperature will be paired with

i. 85°F ⓐ **ii.** 0°C

9. Alex says, "The formula for the root mean square error is long. Why do you have to square and then take the square root? Isn't that just doing a lot of work for nothing? Can't you just make them all positive, add them up, and divide?" Help Calista show Alex that his method does not give the same value as the root mean square error. Use the residual set −2, 1, −3, 4, −1 in your demonstration. Do you think Alex's method could be used as another measure of accuracy? Explain your reasoning.

10. Leajato experimented by turning the key of a wind-up car different numbers of times and recorded how far it traveled.

Turns	0.5	1	1.5	2	2.5	3	3.5	4
Distance (in.)	33	73	114	152	187	223	256	298

a. Graph the data and find the median-median line.

b. Calculate the root mean square error.

c. Use extrapolation to predict how far the car will go if you turn the key five times. Use the root mean square error to describe the accuracy of your answer.

11. APPLICATION Since 1964, the total number of electors in the electoral college has been 538. In order to declare a winner in a presidential election, a majority, or 270 electoral votes, is needed. The table at right shows the number of electoral votes that the Democratic and Republican parties have received in the presidential elections from 1964 to 2004.

a. Let x represent the electoral votes for the Democratic Party, and let y represent the votes for the Republican Party. Make a scatter plot of the data.

b. Why are the points nearly linear? What are some factors that make these data not perfectly linear?

c. Sketch the line $y = 269$ on the scatter plot. How are all the points above the line related?

d. Find the residuals for the line $y = 269$. What does a negative residual represent?

e. What does a residual value that is close to 0 represent?

Electoral Votes

Year	Democrats	Republicans
1964	486	52
1968	191	301
1972	17	520
1976	297	240
1980	49	489
1984	13	525
1988	111	426
1992	370	168
1996	379	159
2000	266	271
2004	251	286

History CONNECTION

Article II, Section 1, of the U.S. Constitution instituted the electoral college as a means of electing the president. The number of electoral votes allotted to each state corresponds to the number of representatives that each state sends to Congress. The distribution of electoral votes among the states can change every 10 years depending on the results of the U.S. census. The actual process of selecting electors is left for each state to decide.

Election workers in Dade County, Florida, hand-check ballots during the 2000 presidential election.

Review

12. Write an equation in point-slope form for each of these lines.

a. The slope is $\frac{3}{5}$ and the line passes through (2, 4.7). ⓐ

b. The line is a direct variation through (6, 21). ⓗ

c. The line passes through (3, 11) and (−6, −18).

13. Create a set of 7 values with median 47, minimum 28, and interquartile range 12. ⓗ

14. Solve.

a. $3 + 5x = 17 - 2x$ ⓐ **b.** $12 + 3(t - 5) = 6t + 1$

15. David deposits $30 into his bank account at the end of each month. The bank pays 7% annual interest compounded monthly.

a. Write a recursive formula to show David's balance at the end of each month.

b. How much of the balance was deposited and how much interest is earned after

i. 1 year ⓐ **ii.** 10 years **iii.** 25 years **iv.** 50 years

c. What can you conclude about making regular deposits into a bank account that earns compound interest?

Fathom
Dynamic Data™ Software

EXPLORATION

Residual Plots and Least Squares

You have learned a couple of different ways to fit a line to data. You've also used residuals to judge how well a line fits and to give a range for predictions made with that line. In this activity you will learn two graphical methods to judge how well your line fits the data. You will also use these methods to identify outliers.

Activity

A Good Fit?

Step 1 Start Fathom and open the sample document titled **States–CarsNDrivers.ftm.** When the file opens, you will see a collection and a case table. This collection gives you various data about the population, drivers, vehicles, and roadways in each U.S. state and the District of Columbia in the year 1992.

Step 2 Create a new graph. Drag the attribute PopThou (population in thousands) to the x-axis and drag the attribute DriversThou (licensed drivers in thousands) to the y-axis. Your graph will automatically become a scatter plot. Describe the trend in the data and give a possible explanation for the trend.

Step 3 With the graph window selected, choose **Add Movable Line** from the Graph menu. Drag the movable line until it fits the data well. What is the equation of your estimated line of fit? Based on your line, which states are outliers?

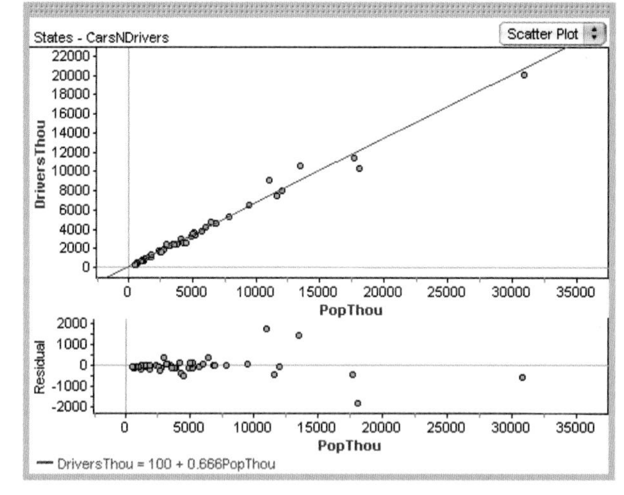

Step 4 Choose **Make Residual Plot** from the Graph menu. What does the residual plot show you? From the residual plot, which states are outliers? Are they the same states you selected in Step 3?

Step 5 Experiment by moving the movable line and observing how the residual plot changes.

Step 6 Return to your scatter plot and residual plot. With the graph window selected, choose **Show Squares** from the Graph menu. Move the movable line and observe how the squares change. Explain how each square is drawn.

Step 7 Move your line back to a position where it is a good fit. (You might want to turn off the squares while you do this.) Notice the size of the squares for the outliers that you identified in Step 4. How do they compare to the other squares?

Step 8 In addition to the movable line, Fathom will graph a median-median line and a least squares line. (You'll learn more about the least squares line in Chapter 11.) Try adding a median-median line and a least squares line to your graph, either alone or with your estimate of a line of fit. How do the equations compare? How do the residual plots compare? How do the squares compare?

Questions

1. How can you identify outliers from the squares? How did you identify outliers from looking at the line of fit? How did you identify outliers from the residual plot? How are the approaches interrelated? Do you find one approach easier than the other?

2. As you change the slope of your line, what happens to the residual plot? As you change the y-intercept of your line, what happens to the residual plot? Explain how you can use the residual plot to adjust the fit of your line.

3. Graphs A, B, and C are residual plots for different lines of fit for the same data set. How would you adjust each line to be a better fit?

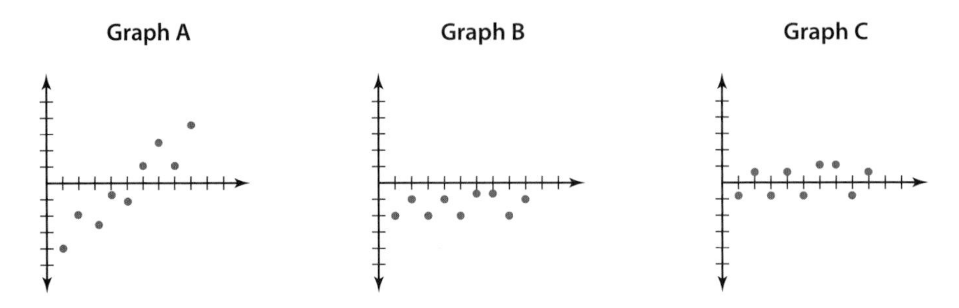

4. Explain how you can use the squares to adjust your movable line to a better fit. Based on your explanation, how do you think the least squares line got its name?

5. When you experimented with all three lines of fit in Step 8, did you get the same equation for all three? Give some reasons why this may or may not have happened.

LESSON

3.6

Linear Systems

The number of tickets sold for a school activity, like a spaghetti dinner, helps determine the financial success of the event. Income from ticket sales can be less than, equal to, or greater than expenses. The break-even value is the intersection of the expense function and the line $y = x$. This equation models all the situations in which the expenses, y, are equal to income, x.

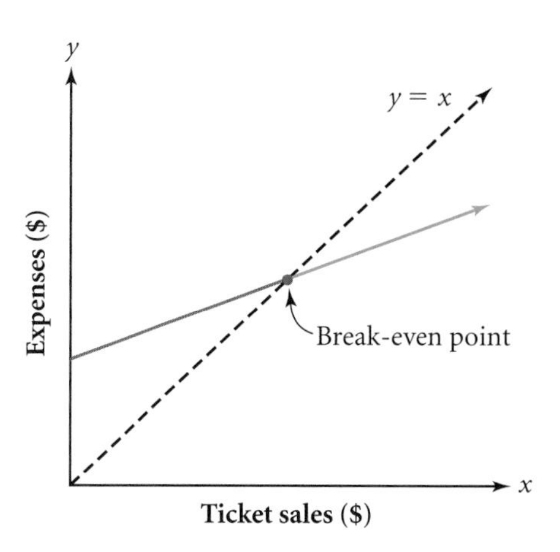

In this lesson you will focus on mathematical situations involving two or more equations or conditions that must be satisfied at the same time. A set of two or more equations that share the same variables and are solved or studied simultaneously is called a **system of equations.**

American painter I. Rice Pereira (1907–1971) explored light and space in her works, characterized by intersecting lines in mazelike patterns. This piece is titled *Green Mass* (1950).

EXAMPLE A

Minh and Daniel are starting a business together, and they need to decide between long-distance phone carriers. One company offers the Phrequent Phoner Plan, which costs 20¢ for the first minute of a phone call and 17¢ for each minute after that. A competing company offers the Small Business Plan, which costs 50¢ for the first minute and 11¢ for each additional minute. Which plan should Minh and Daniel choose?

▶ Solution

The plan they choose depends on the nature of their business and the most likely length and frequency of their phone calls. Because the Phrequent Phoner Plan costs less for the first minute, it is better for very short calls. However, the Small Business Plan probably will be cheaper for longer calls because the cost is less for additional minutes. There is a phone call length for which both plans cost the same. To find it, let x represent the call length in minutes, and let y represent the cost in cents.

Because x represents the length of the entire phone call, $x - 1$ represents the number of minutes after the first minute. Therefore, a cost equation that models the Phrequent Phoner Plan is

$$y = 20 + 17(x - 1)$$

A cost equation for the Small Business Plan is

$$y = 50 + 11(x - 1)$$

A graph of these equations shows the Phrequent Phoner Plan cost is below the Small Business Plan cost until the lines intersect. You may be able to estimate the coordinates of this point from the graph.

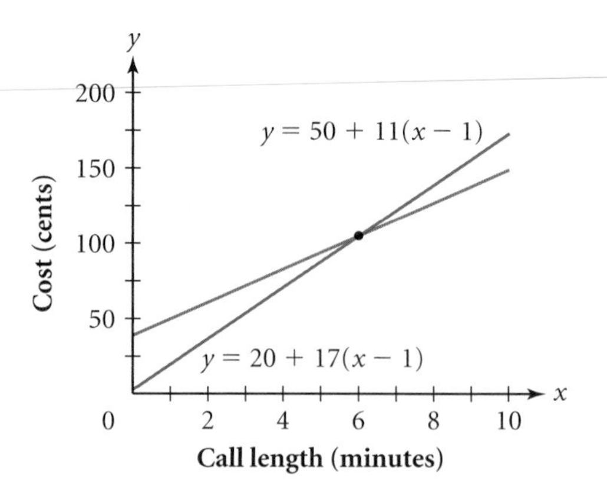

Or you can look in a table to find an answer. The point of intersection at (6, 105) tells you that a six-minute call will cost $1.05 with either plan.

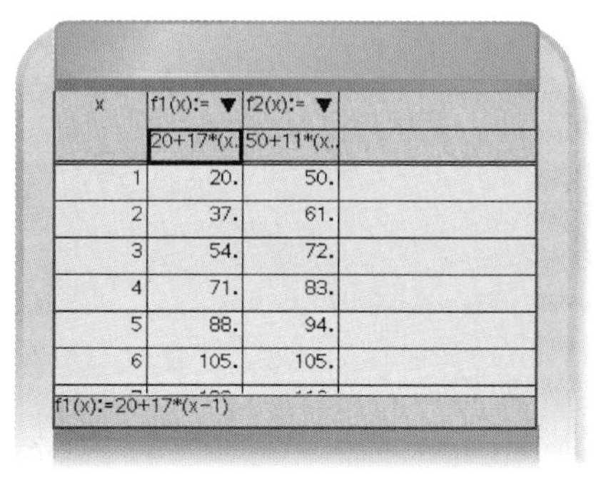

If Minh and Daniel believe their average business call will last less than six minutes, they should choose the Phrequent Phoner Plan. But if they think most of their calls will last more than six minutes, the Small Business Plan is the better option.

You can also use equations to solve for the point of intersection. How would you solve the system of equations symbolically?

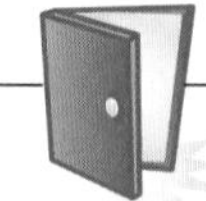

Investigation
Population Trends

The table below gives the populations of San Jose, California, and Detroit, Michigan.

Populations

Year	1950	1960	1970	1980	1990	2000	2005
San Jose	95,280	204,196	459,913	629,400	782,248	894,943	912,232
Detroit	1,849,568	1,670,144	1,514,063	1,203,368	1,027,974	951,270	886,671

(*The World Almanac and Book of Facts 2007*)

Step 1 Estimate the year that the two cities had the same population. What was that population?

Step 2 Show the method you used to make this prediction. Choose a different method to check your answer. Discuss the pros and cons of each method.

In algebra, you studied different methods for finding the exact coordinates of an intersection point by solving systems of equations. One method is illustrated in the next example.

EXAMPLE B

Justine and her little brother Evan are running a race. Because Evan is younger, Justine gives him a 50-foot head start. Evan runs at 12.5 feet per second and Justine runs at 14.3 feet per second. How far will they be from Justine's starting line before Justine passes Evan? What distance should Justine mark for a close race?

▸ Solution

You can compare the time and distance that each person runs. Because the distance, d, depends on the time, t, you can write these equations:

$d = rt$	Distance equals the rate, or speed, times time.
$d = 14.3t$	Justine's distance equation.
$d = 50 + 12.5t$	Evan's distance equation.

Graphing these two equations shows that Justine eventually catches up to Evan and passes him if the race is long enough. At that moment, they are at the same distance from the start, at the same time. You can estimate this point from the graph or scroll down until you find the answer in the table. You can also solve the system of equations.

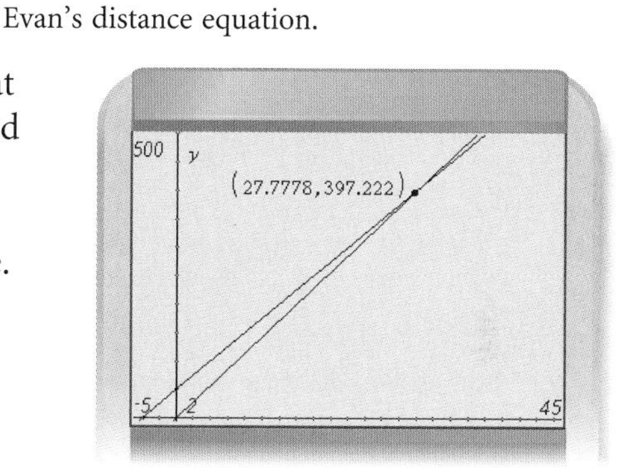

Because you want to find out when Justine's distance and Evan's distance are equal, you can substitute Justine's expression for distance, $14.3t$, for the distance, d, in Evan's equation. Then you'll solve for t, which will give you the time when the distances are equal.

$14.3t = 50 + 12.5t$	Substitute $14.3t$ for d in Evan's distance equation.
$1.8t = 50$	Subtract $12.5t$ from both sides.
$t = \frac{50}{1.8} \approx 27.8$	Divide both sides by 1.8.

So, Justine passes Evan after 27.8 seconds. Now you can substitute this value back into either equation to find their distances from the starting line when Justine passes Evan.

$$d = 14.3t = 14.3 \cdot \frac{50}{1.8} \approx 397.2$$

If Justine marks a 400 ft distance, she will win, but it will be a close race.

The method of solving a system demonstrated in Example B uses one form of **substitution.** In this case you substituted one expression for distance, $14.3t$, for d in the other equation. The resulting equation had only one variable, t. When you have the two equations written in intercept form, substitution is a straightforward method for finding an exact solution. The solution to a system of equations with two variables is a pair of values that satisfies both equations. Sometimes a system will have many solutions or no solution.

EXERCISES

You will need

A graphing calculator for Exercises **1**, **5**, and **8–10**.

Practice Your Skills

1. Use a table to find the point of intersection for each pair of linear equations.

a. $\begin{cases} y = 3x - 17 \\ y = -2x - 8 \end{cases}$ @

b. $\begin{cases} y = 28 - 3(x - 5) \\ y = 6 + 7x \end{cases}$

2. Write a system of equations that has (2, 7.5) as its solution.

3. Write the equation of the line perpendicular to $y = 4 - 2.5x$ and passing through the point (1, 5).

4. Solve each equation.

a. $4 - 2.5(x - 6) = 3 + 7x$ @

b. $11.5 + 4.1t = 6 + 3.2(t - 4)$

5. Use substitution to find the point (x, y) where each pair of lines intersect. Use a graph or table to verify your answer.

a. $\begin{cases} y = -2 + 3(x - 7) \\ y = 10 - 5x \end{cases}$ @

b. $\begin{cases} y = 0.23x + 9 \\ y = 4 - 1.35x \end{cases}$

c. $\begin{cases} y = -1.5x + 7 \\ 2y = -3x + 14 \end{cases}$

Reason and Apply

6. The equations $s_1 = 18 + 0.4m$ and $s_2 = 11.2 + 0.54m$ give the lengths of two different springs in centimeters, s_1 and s_2, as mass amounts in grams, m, are separately added to each.

a. When are the springs the same length?

b. When is one spring at least 10 cm longer than the other?

c. Write a statement comparing the two springs.

7. APPLICATION This graph shows the Kangaroo Company's production costs and revenue for its pogo sticks. Use the graph to estimate the answers to the questions below.

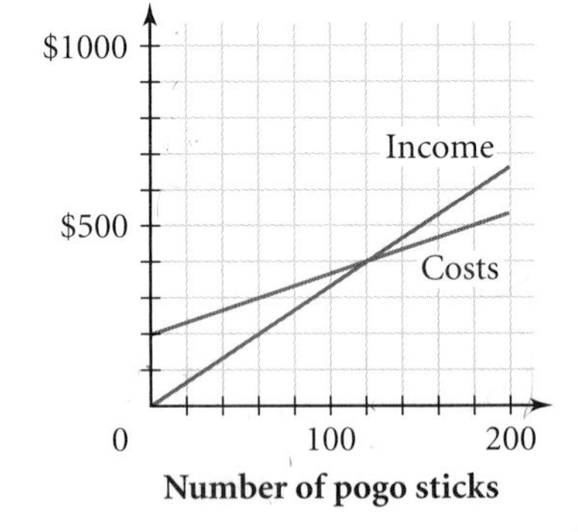

a. If 25 pogo sticks are sold, will the company earn a profit? Describe how you can use the graph to answer this question. @

b. If the company sells 200 pogo sticks, will it earn a profit? If so, approximately how much?

c. How many pogo sticks must the company sell to break even? How do you know?

8. APPLICATION Winning times for men and women in the 1500-meter Olympic speed skating event are given below, in minutes and seconds.

1500-Meter Olympic Speed Skating

Year	1972	1976	1980	1984	1988	1992	1994	1998	2002	2006
Men	2:02.96	1:59.38	1:55.44	1:58.36	1:52.06	1:54.81	1:51.29	1:47.87	1:43.95	1:45.97
Women	2:20.85	2:16.58	2:10.95	2:03.42	2:00.68	2:05.87	2:02.19	1:57.58	1:54.02	1:55.27

(*The World Almanac and Book of Facts 2007*)

a. Analyze the data and predict when the winning times for men and women will be the same if the current trends continue.

b. How reasonable do you think your prediction is? Explain.

c. Predict the winning times for the 2010 Winter Olympics. ⓐ

d. Is it appropriate to use a linear model for these data? Why?

Derek Parra of the United States won the 2002 Olympic gold medal for the men's 1500-meter speed skating event.

9. APPLICATION Suppose the long-distance phone companies in Example A calculate their charges so that a call of exactly 3 min will cost the same as a call of 3.25 min or 3.9 min, and there is no increase in cost until you have been connected for 4 min. Increases are calculated after each additional minute. A function that models this situation is the **greatest integer function,** $y = [x]$, which outputs the greatest integer less than or equal to x. Because the graph of this function looks like a staircase, it is called a **step function.** [▶ See **Calculator Note 3F** for instructions on using the greatest integer function on your calculator. ◀]

a. Use the greatest integer function to write cost equations for the two companies in Example A.

b. Graph the two new equations representing the Phrequent Phoner Plan and the Small Business Plan.

c. Now determine when each plan is more desirable. Explain your reasoning. ⓗ

10. APPLICATION An anthropologist can use the lengths of certain bones from skeletal remains to estimate the height of the living person. The humerus bone is the single large bone that extends from the elbow to the shoulder socket. The following formulas, attributed to the work of Mildred Trotter and G. C. Gleser, have been used to estimate a male's height, m, or a female's height, f, when the length, h, of the humerus bone is known: $m = 3.08h + 70.45$ and $f = 3.36h + 58.0$. All measurements are in centimeters.

a. Graph the two lines on the same set of axes.

b. If a humerus bone is found and it measures 42 cm, how tall would the person have been according to the model if the bone was determined to come from a male? From a female?

c. At what point do the two equations intersect? What does the point of intersection mean in this context?

Science CONNECTION

Physical anthropology is a science that deals with the biological evolution of human beings, the study of human ancestors and nonhuman primates, and the in-depth analysis of the human skeleton. By studying bones and bone fragments, physical anthropologists have developed methods that can provide a wealth of information on the age, sex, ancestry, height, and diet of a person who lived in ancient times, just by studying his or her skeleton.

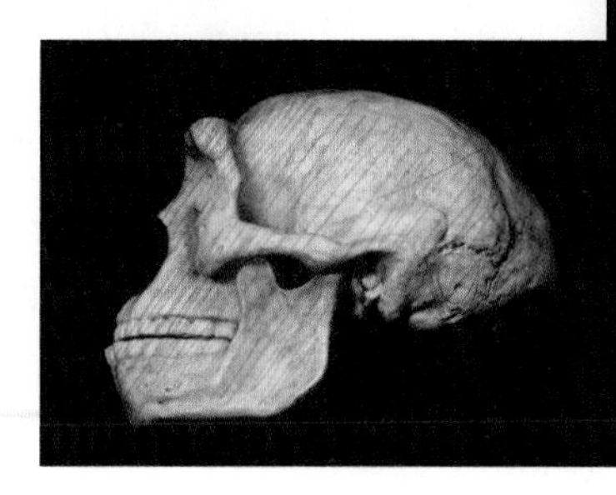

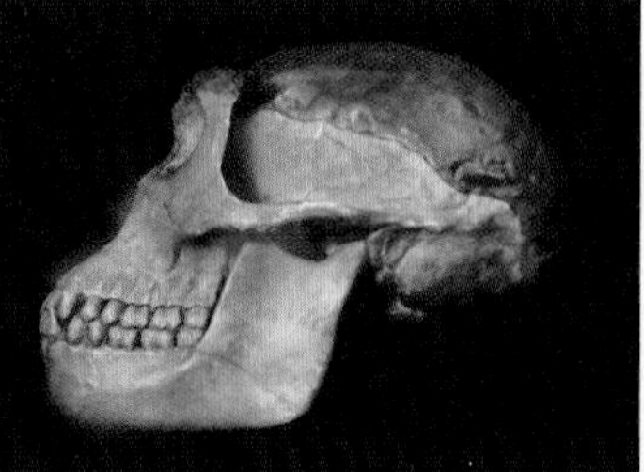

These skulls are from human ancestors—Peking man on the left and Java man on the right.

11. Write a system of equations to model each situation, and solve for the values of the appropriate variables.

a. The perimeter of a rectangle is 44 cm. Its length is 2 cm more than twice its width. ⓐ

b. The perimeter of an isosceles triangle is 40 cm. The base length is 2 cm less than the length of a leg of the triangle. ⓗ

c. The Fahrenheit reading on a dual thermometer is 0.4 degree less than three times the Celsius reading. (*Hint:* Your second equation needs to be a conversion formula between degrees Fahrenheit and degrees Celsius.)

Review

12. Use the model $\hat{y} = 373.1 + 1.6(x - 2002)$ that was found in the example in Lesson 3.3. Recall that x represents the year and y represents the concentration of CO_2 in parts per million (ppm) at Mauna Loa.

a. Predict the concentration of CO_2 in the year 2002.

b. Use the model to predict the concentration of CO_2 in the year 2010.

c. According to this model, when will the level of CO_2 be double the preindustrial level of 280 ppm?

13. The histogram shows the average annual cost of insuring a motor vehicle in the United States.

a. How many jurisdictions are included in the histogram?

b. South Carolina is the median jurisdiction. South Carolina is in what bin?

c. In what percentage of the jurisdictions is the average cost less than $700?

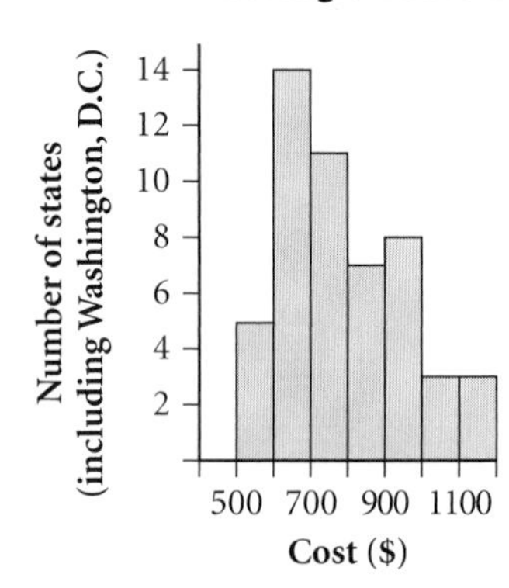

14. Solve these equations for y.

a. $3x - 8y = 12$ ⓐ

b. $5x + 2y = 12$

c. $-3x + 4y = 5$

IMPROVING YOUR REASONING SKILLS

Cartoon Watching Causes Small Feet

Lisa did a study for her health class about the effects of cartoon watching on foot size. Based on a graph of her data, she finds that there was an inverse relationship between foot size and hours spent watching cartoons per week. She concludes that "cartoon watching causes small feet." Is this true? Explain any flaws in Lisa's reasoning.

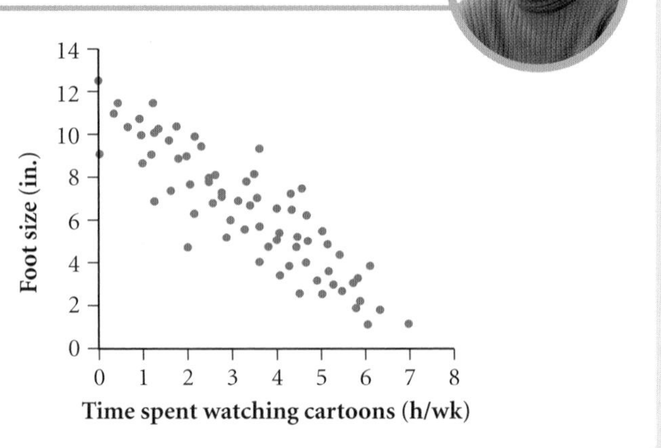

LESSON

3.7

Substitution and Elimination

A place for everything and everything in its place.

ENGLISH PROVERB

A solution to a system of equations in two variables is a pair of values that satisfies both equations and represents the intersection of their graphs. In Lesson 3.6, you reviewed solving a system of equations using substitution, when both equations are in intercept form. Suppose you want to solve a system and one or both of the equations are not in intercept form. You can rearrange them into intercept form, but sometimes there's an easier method.

If one equation is in intercept form, you can still use substitution.

In *Modern Warrior Series, Shirt #1* (1998), northern Cheyenne artist Bently Spang (b 1960) reflects on identity. He presents opposing influences as an equivalence of forms—the modern and the traditional, the spiritual and the mundane, the Cheyenne and the non-Cheyenne—all intersecting in an arrangement of family photographs.

EXAMPLE A

Solve this system for x and y.

$$\begin{cases} y = 15 + 8x \\ -10x - 5y = -30 \end{cases}$$

▶ Solution

You can solve the second equation for y so that both equations will be in intercept form and substitute the right side of one equation for y in the other equation. However, you may find it easier to substitute the right side of the first equation for y in the second equation.

$-10x - 5y = -30$ Original form of the second equation.

$-10x - 5(15 + 8x) = -30$ Substitute the right side of the first equation for y.

$-10x - 75 - 40x = -30$ Distribute -5.

$-50x = 45$ Add 75 to both sides and combine like terms.

$x = -0.9$ Divide both sides by -50.

Now that you know the value of x, you can substitute it into either equation to find the value of y.

$y = 15 + 8(-0.9)$ Substitute -0.9 for x in the first equation.

$y = 7.8$ Multiply and combine like terms.

Write your solution as an ordered pair. The solution to this system is $(-0.9, 7.8)$.

The substitution method relies on the substitution property, which says that if $a = b$, then a may be replaced by b in an algebraic expression. Substitution is a powerful mathematical tool that allows you to rewrite expressions and equations in forms that are easier to use and solve. Notice that substituting an expression for y, as you did in Example A, eliminates y from the equation, allowing you to solve a single equation for a single variable, x.

A third method for solving a system of equations is the **elimination** method. The elimination method uses the addition property of equality, which says that if $a = b$ and $c = d$, then $a + c = b + d$. In other words, if you add equal quantities to both sides of an equation, the resulting equation is still true. If necessary, you can also use the multiplication property of equality, which says that if $a = b$, then $ac = bc$, or if you multiply both sides of an equation by equal quantities, then the resulting equation is still true.

EXAMPLE B

Solve these systems for x and y.

a. $\begin{cases} 4x + 3y = 14 \\ 3x - 3y = 13 \end{cases}$ **b.** $\begin{cases} -3x + 5y = 6 \\ 2x + y = 6 \end{cases}$

▶ Solution

Because neither of these equations is in intercept form, it is probably easier to solve the systems using the elimination method.

a. You can solve the system without changing either equation to intercept form by adding the two equations.

$4x + 3y = 14$	Original equations.
$3x - 3y = 13$	Addition property of equality.
$7x \quad = 27$	The variable y is eliminated.
$x = \frac{27}{7}$	Multiplication property of equality.
$4\left(\frac{27}{7}\right) + 3y = 14$	Substitution property of equality.
$y = -\frac{10}{21}$	Addition and multiplication properties of equality.

The solution to this system is $\left(\frac{27}{7}, -\frac{10}{21}\right)$. You can substitute the coordinates back into both equations to check that the point is a solution for both.

$$4\left(\frac{27}{7}\right) + 3\left(-\frac{10}{21}\right) = \frac{108}{7} - \frac{30}{21} = \frac{294}{21} = 14$$

$$3\left(\frac{27}{7}\right) - 3\left(-\frac{10}{21}\right) = \frac{81}{7} + \frac{30}{21} = \frac{273}{21} = 13$$

b. Adding the equations as they are written will not eliminate either of the variables. You need to multiply one or both equations by some value so that if you add the equations, one of the variables will be eliminated.

The easiest choice is to multiply the second equation by -5 and then add it to the first equation.

$-3x + 5y = 6$	$\rightarrow -3x + 5y = 6$	Original form of the first equation.
$-5(2x + y) = -5(6)$	$\rightarrow -10x - 5y = -30$	Multiply both sides of the second equation by -5.
	$-13x \quad = -24$	Add the equations.

This eliminates the y-variable and gives $x = \frac{24}{13}$. Substituting this x-value back into either of the original equations gives the y-value. Or you can use the same process to eliminate the x-variable.

$$\begin{aligned} -3x + 5y = 6 \quad &\rightarrow \quad -6x + 10y = 12 && \text{Multiply both sides by 2.} \\ 2x + y = 6 \quad &\rightarrow \quad \underline{6x + 3y = 18} && \text{Multiply both sides by 3.} \\ & \qquad\qquad 13y = 30 \\ & \qquad\qquad y = \frac{30}{13} \end{aligned}$$

The solution to this system is $\left(\frac{24}{13}, \frac{30}{13}\right)$. You can use your calculator to verify the solution.

It would take a lot of effort to solve this last system using a table on your calculator. If you had used the substitution method to solve the systems in Example B, you would have had to work with fractions to get an accurate answer. To solve these systems, the easiest method to use is the elimination method.

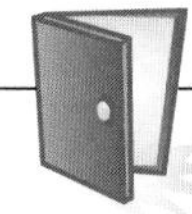

Investigation
What's Your System?

You will need

- graph paper

In this investigation you will discover different classifications of systems and their properties. You can divide up the work among group members, but make sure each problem is solved by one person and checked by another.

Step 1 Use the method of elimination to solve each system. (Don't be surprised if it doesn't always work.)

a. $\begin{cases} 2x + 5y = 6 \\ 2x - 3y = 22 \end{cases}$ **b.** $\begin{cases} 3x + 2y = 12 \\ -6x - 4y = -24 \end{cases}$ **c.** $\begin{cases} 4x - 8y = 5 \\ -3x + 6y = 11 \end{cases}$

d. $\begin{cases} -2x + y = 5 \\ 6x - 3y = -15 \end{cases}$ **e.** $\begin{cases} x + 3y = 6 \\ 5x - 3y = 6 \end{cases}$ **f.** $\begin{cases} x + 3y = 8 \\ 3x + 9y = -4 \end{cases}$

Step 2 Graph each system in Step 1.

Step 3 A system that has a solution (a point or points of intersection) is called **consistent.** Which of the six systems in Step 1 are consistent?

Step 4 A system that has no solution (no point of intersection) is called **inconsistent.** Which of the systems in Step 1 are inconsistent?

Step 5 A system that has infinitely many solutions is called **dependent.** For linear systems, this means the equations are equivalent (though they may not look identical). A system that has a single solution is called **independent.** Which of the systems in Step 1 are dependent? Independent?

Step 6 Your graphs from Step 2 helped you classify each system as inconsistent or consistent and as dependent or independent. Now look at your solutions from Step 1. Make a conjecture about how the results of the elimination method can be used to classify a system of equations.

In the elimination method, you combine equations to eliminate one of the variables. Solving for the variable that remains gives you the x- or y-coordinate of the point of intersection, if there is a point of intersection.

EXERCISES

You will need

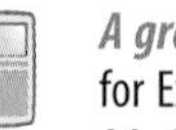

A graphing calculator for Exercises **2, 7, 11, 14, 15,** and **17.**

Practice Your Skills

1. Solve each equation for the specified variable.

a. $w - r = 11$, for w

b. $2p + 3h = 18$, for h (a)

c. $w - r = 11$, for r

d. $2p + 3h = 18$, for p

2. Graph each system and find an approximate solution. Then choose a method and find the exact solution. List each solution as an ordered pair.

a. $\begin{cases} y = 15 + 3x \\ y = 27 - 3x \end{cases}$ (a)

b. $\begin{cases} y = 4x - 5 \\ y = 2x + 1 \end{cases}$

c. $\begin{cases} y = 5 + \frac{2}{3}x \\ y = 21 - 2x \end{cases}$

d. $\begin{cases} -2x + y = 5 \\ x - 3y = -30 \end{cases}$

e. $\begin{cases} x + 3y = 6 \\ 5x - 3y = 6 \end{cases}$

3. Solve each system of equations.

a. $\begin{cases} 4s + 3t = 7 \\ 2s - t = 8 \end{cases}$ (a)

b. $\begin{cases} 5x + 2y = 12 \\ 6x - 4y = -7 \end{cases}$

c. $\begin{cases} 4x - 3y = 5 \\ -x + 6y = 11 \end{cases}$

d. $\begin{cases} \frac{1}{4}a - \frac{2}{5}b = 3 \\ \frac{3}{8}a + \frac{2}{5}b = 2 \end{cases}$

e. $\begin{cases} f = 3d + 5 \\ 10d - 4f = 16 \end{cases}$

4. Classify each system as consistent or inconsistent and as dependent or independent.

a. $\begin{cases} 3x + 4y = 7 \\ y = -\frac{3}{4}x + 2 \end{cases}$

b. $\begin{cases} y = 3x - 4 \\ 6x - 2y = 8 \end{cases}$

c. $\begin{cases} y = 3x - 5 \\ 6x + 2y = 10 \end{cases}$

Reason and Apply

5. Solve each problem.

a. If $4x + y = 6$, then what is $(4x + y - 3)^2$? (h)

b. If $4x + 3y = 14$ and $3x - 3y = 13$, what is $7x$?

6. The formula to convert from Fahrenheit to Celsius is $C = \frac{5}{9}(F - 32)$. What reading on the Fahrenheit scale is three times the equivalent temperature on the Celsius scale? ⓗ

7. APPLICATION Ellen must decide between two cameras. The first camera costs \$47.00 and uses two alkaline AA batteries. The second camera costs \$59.00 and uses one \$4.95 lithium battery. She plans to use the camera frequently enough that she probably would replace the AA batteries six times a year for a total cost of \$11.50 per year. The lithium battery, however, will last an entire year.

a. Let x represent the number of years, and let y represent the cost in dollars. Write an equation to represent the overall expense for each camera. ⓐ

b. Use a graph to estimate when the overall cost of the less expensive camera will equal that of the more expensive camera.

c. Use another method to find exactly when the overall cost of the less expensive camera will equal that of the more expensive camera.

d. Is it important to find an exact answer, as in 7c? Explain.

8. Write a system of two equations that has the solution $(-1.4, 3.6)$.

9. The two sequences below have one term that is the same. Determine which term this is and find its value.

$u_1 = 12$
$u_n = u_{n-1} + 0.3 \quad \text{where } n \geq 2$

$v_1 = 15$
$v_n = v_{n-1} + 0.2 \quad \text{where } n \geq 2$

10. Formulas play an important part in many fields of mathematics and science. You can create a new formula using substitution to combine formulas.

a. Using the formulas $A = s^2$ and $d = s\sqrt{2}$, write a formula for A in terms of d.

b. Using the formulas $P = IE$ and $E = IR$, write a formula for P in terms of I and R.

c. Using the formulas $A = \pi r^2$ and $C = 2\pi r$, write a formula for A in terms of C.

Science CONNECTION

In this simple circuit, a 9-volt battery lights a small light bulb. The power P supplied to the bulb in watts is a product of the voltage E of the battery times the electrical current I in the wire. The electrical current, in turn, can be calculated as the battery voltage E divided by the resistance R of the circuit, which depends on the length and gauge, or diameter, of the wire. It is important to know basic functions and relations when designing a circuit so that there will be enough power supplied to the components, but not excessive power, which would damage the components.

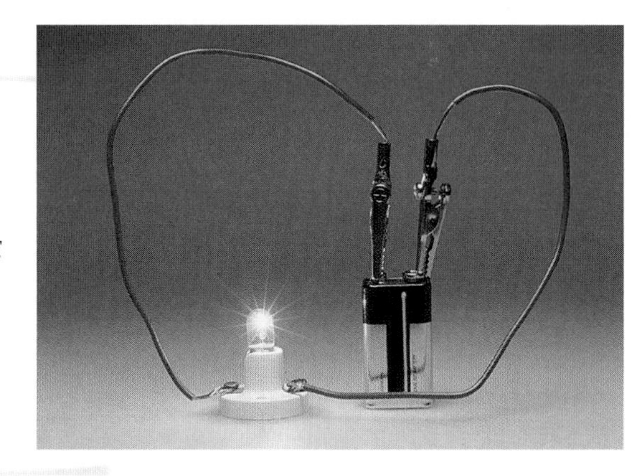

11. Consider the system

$$\begin{cases} 3x - 4y = 7 \\ 2x + 2y = 5 \end{cases}$$

a. Add the two equations.

b. Solve the two original equations and their sum for y. Sketch a graph of all three equations on the same axes.

c. What do you observe about the graph of the third equation?

d. Repeat 11a–c for the system

$$\begin{cases} 5x - 7y = 3 \\ -5x + 3y = 5 \end{cases}$$

e. Make a conjecture about the graphs of two linear equations and their sum.

12. APPLICATION A support bar will be in equilibrium (balanced) at the fulcrum, O, if $m_1x + m_2y = m_3z$, where m_1, m_2, and m_3 represent masses and x, y, and z represent the distance of the masses to the fulcrum. Draw a diagram for each question and calculate the answer.

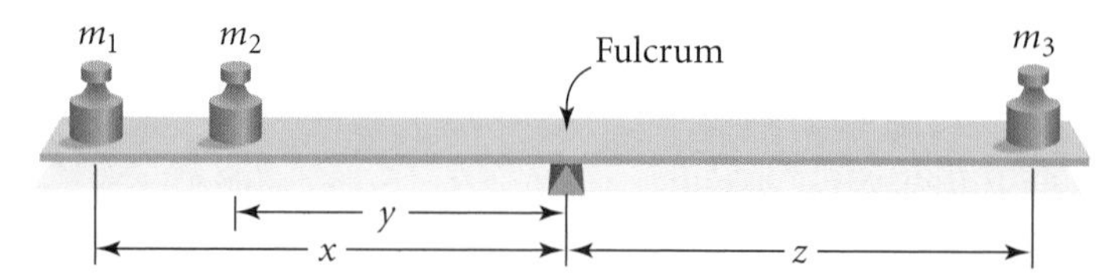

a. A 40 in. bar is in equilibrium when a weight of 6 lb is hung from one end and a weight of 9 lb is hung from the other end. Find the position of the fulcrum.

b. While in the park, Michael and his two sons, Justin and Alden, go on a 16 ft seesaw. Michael, who weighs 150 lb, sits at the edge of one end while Justin and Alden move to the other side and try to balance. The seesaw balances with Justin at the other edge and Alden 3 ft from him. After some additional experimentation, the seesaw balances once again with Alden at the edge and Justin 5.6 ft from the fulcrum. How much does each boy weigh?

Art CONNECTION

Alexander Calder (1898–1976) was an artistic pioneer who invented the mobile as an art form. After getting a degree in mechanical engineering, he went to art school, supporting himself through school by working as an illustrator. Once out of school, he began creating small three-dimensional sculptures made from wire, wood, and cloth that balanced perfectly, whether or not they were symmetrical. Eventually, he designed sculptures with painted elements that moved mechanically and then went on to produce pieces that moved with the air. These free-moving, hanging sculptures became known as "mobiles." He also designed "stabiles," essentially mobiles that balance on a fixed support.

Alexander Calder builds a mobile in his studio. At left is one of his stabiles, *Boomerang and Sickle Moon.*

Review

13. Classify each statement as true or false. If the statement is false, change the right side to make it true.

a. $x^2 + 8x + 15 = (x + 3)(x + 5)$

b. $x^2 - 16 = (x - 4)(x - 4)$

c. $(x + 5)^2 = x^2 + 25$

14. Consider the equation $3x + 2y - 7 = 0$.

a. Solve the equation for y. @

b. Graph this equation.

c. What is the slope?

d. What is the y-intercept?

e. Write an equation for a line perpendicular to this one and having the same y-intercept. Graph this equation.

15. APPLICATION This table shows the normal monthly precipitation in inches for Pittsburgh, Pennsylvania, and Portland, Oregon.

Month	J	F	M	A	M	J	J	A	S	O	N	D
Pittsburgh	2.7	2.4	3.2	3.0	3.8	4.1	4.0	3.4	3.2	2.3	3.0	2.9
Portland	5.1	4.2	3.7	2.6	2.4	1.6	0.7	0.9	1.7	2.9	5.6	5.7

(*The World Almanac and Book of Facts 2007*)

a. Display the data in two box plots on the same axis.

b. Give the five-number summary of each data set.

c. Describe the differences in living conditions with respect to precipitation.

d. Which city generally has more rain annually?

16. Consider these three sequences.

i. 243, −324, 432, −576, . . .

ii. 22, 26, 31, 37, 44, . . .

iii. 24, 25.75, 27.5, 29.25, 31, . . .

a. Find the next two terms in each sequence.

b. Identify each sequence as arithmetic, geometric, or other.

c. If a sequence is arithmetic or geometric, write a recursive routine to generate the sequence.

d. If a sequence is arithmetic, give an explicit formula that generates the sequence.

17. Melina and Angus are designing a fixed speed hot-rod car for a remote-control rally. They need to construct a car that travels at a constant 0.50 m/s. In order to qualify for the rally, they must show time trial results with a root mean square error of less than 0.05 m when using a direct variation equation with $k = 0.50$ m/s. This table shows their time trial results. Will Melina and Angus qualify, or do they need to go back to the drawing board?

Time (s)	1	2	3	4	5	6	7	8
Distance (m)	0.48	0.95	1.6	2.0	2.52	2.93	3.49	4.05

CHAPTER 3 REVIEW

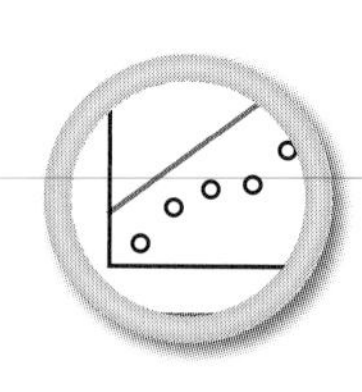

In this chapter you analyzed sets of two-variable data. A plot of two-variable data may be linear, or it may appear nearly linear over a short domain. If it is linear, you can find a **line of fit** to model the data. You can write a **linear equation** for this line and use it to **interpolate** or **extrapolate** points for which data are not available. To write the equation of a line, you can use its slope and y-intercept to write the equation in **intercept form,** or you can use the coordinates of two points to write the equation in **point-slope form.**

You learned two methods for finding a line of fit for a set of data: estimating by looking at the trend of the data and fitting a line based on certain criteria, and using the more systematic **median-median** method. Regardless of the method you use, you can examine the **residuals** to determine whether your model is a good fit. With a good model, the residuals should be randomly positive and negative, the sum of the residuals should be zero, and each residual should be as small as possible. You also learned about **root mean square error,** a measure of the accuracy of a model as a predictor.

You can solve a **system of equations** to find the intersection of two lines, using the methods of **substitution, elimination,** or using a graph or table of values. With two linear equations, a system will be **consistent and independent** (intersecting lines, with one solution), **consistent and dependent** (the same line, with infinitely many solutions), or **inconsistent** (parallel lines, with no solution).

EXERCISES

You will need

A graphing calculator for Exercises **8, 15,** and **17.**

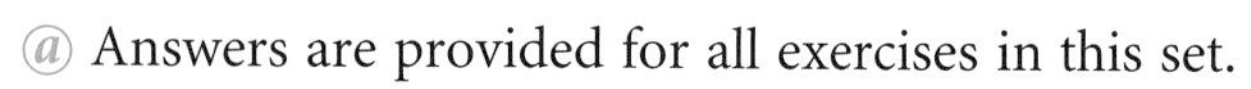

@ Answers are provided for all exercises in this set.

1. Find the slope of the line containing the points (16, 1300) and (−22, 3250).

2. Consider the line $y = -5.02 + 23.45x$.

a. What is the slope of this line?

b. Write an equation for a line that is parallel to this line.

c. Write an equation for a line that is perpendicular to this line.

3. Find the point on each line where y is equal to 740.0.

a. $y = 16.8x + 405$

b. $y = -7.4 + 4.3(x - 3.2)$

4. Write an equation in point-slope form for each line described.

a. slope −1 passing through (−2, 3)

b. parallel to $y = 5 + 2(x - 4)$ passing through (0, −8)

c. passing through (1999, 13.2) and (2009, 8.6)

d. horizontal line through (2, 7)

5. The graphs below show three different lines of fit for the same set of data. For each graph, decide whether the line is a good line of fit or not, and explain why.

a.

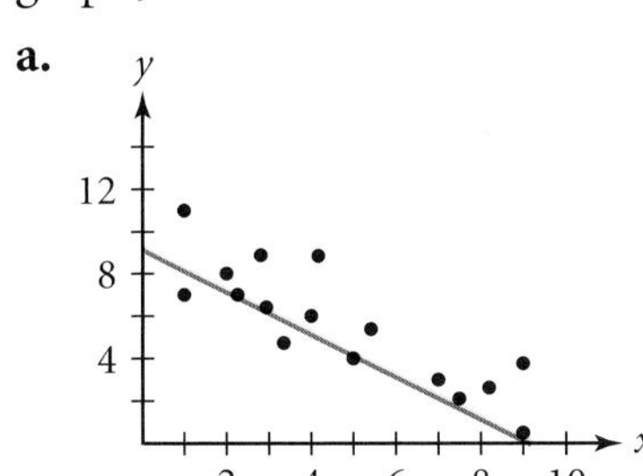

b.

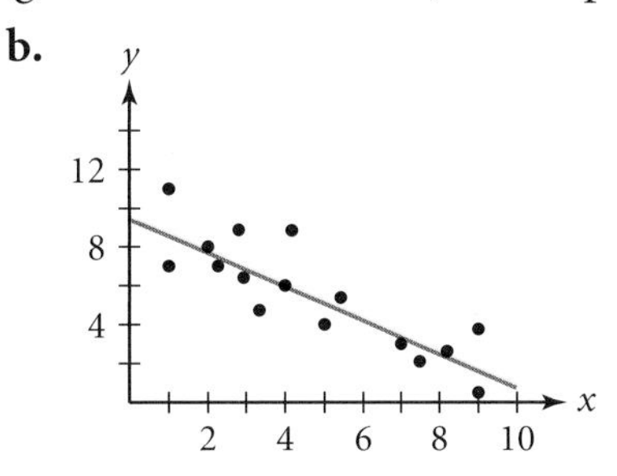

c.

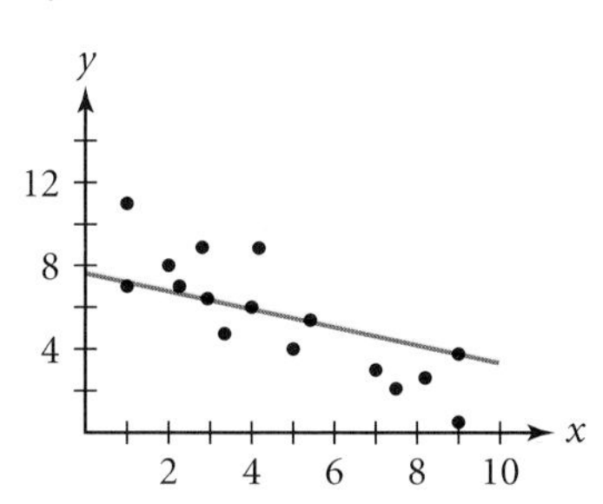

6. Solve each system (when possible) and classify each system as consistent or inconsistent and as independent or dependent.

a. $\begin{cases} y = 3.2x - 4 \\ y = 3.1x - 3 \end{cases}$

b. $\begin{cases} y = \frac{1}{4}(x - 8) + 5 \\ y = 0.25x + 3 \end{cases}$

c. $\begin{cases} 3x + 2y = 4 \\ -3x + 5y = 3 \end{cases}$

d. $\begin{cases} 5x - 4y = 5 \\ 2x + 10y = 2 \end{cases}$

e. $\begin{cases} y = \frac{3}{4}x - 1 \\ \frac{7}{10}x + \frac{2}{5}y = 8 \end{cases}$

f. $\begin{cases} \frac{3}{5}x - \frac{2}{5}y = 3 \\ 0.6x - 0.4y = -3 \end{cases}$

7. The ratio of the weight of an object on Mercury to its weight on Earth is 0.38.

a. Explain why you can use the direct variation equation $m = 0.38e$ to model the weight of an object on Mercury.

b. How much would a 160-pound student weigh on Mercury?

c. The ratios for the Moon and Jupiter are 0.17 and 2.54, respectively. The equations $y_1 = 0.38x$, $y_2 = 1x$, $y_3 = 0.17x$, and $y_4 = 2.54x$ are graphed at right. Match the Moon, Mercury, Earth, and Jupiter with its graph and equation.

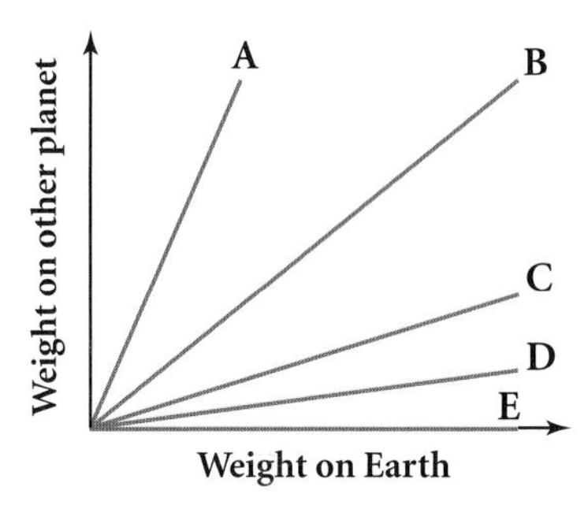

Science CONNECTION

Planning a trip to outer space? This table gives the ratio of an object's weight on each of the planets and the Moon to its weight on Earth.

You can calculate your weight on each of these celestial bodies. Some of these calculations will make you seem like a lightweight, and others will make you seem heavy! But, of course, your body will still be the same. Your *mass* won't change. Your *weight* on other planets depends on your mass, *m*, the planet's mass, *M*, and the distance, *r*, from the center of the planet.

Mercury	0.38
Venus	0.91
Earth	1
Moon	0.17
Mars	0.38
Jupiter	2.54
Saturn	0.93
Uranus	0.80
Neptune	1.19
Pluto	0.06

Architecture CONNECTION

In 1173 C.E., the Tower of Pisa was built on soft ground and, ever since, it has been leaning to one side as it sinks in the soil. This 8-story, 56-meter tower was built with only a 3-meter-deep foundation. The tower was completed in the mid-1300s even though it started to lean after the first 3 stories were completed. The tower's structure consists of a cylindrical body, arches, and columns. Rhombuses and rectangles decorate the surface.

The Tower of Pisa, in Pisa, Italy

8. Read the Architecture Connection. The table lists the amount of lean, measured in millimeters, for nine different years.

Tower of Pisa

Year	Lean
1910	5336
1920	5352
1930	5363
1940	5391
1950	5403

Year	Lean
1960	5414
1970	5428
1980	5454
1990	5467

a. Make a scatter plot of the data. Let x represent the year, and let y represent the amount of lean in millimeters.

b. Find a median-median line for the data.

c. What is the slope of the median-median line? Interpret the slope in the context of the problem.

d. Find the amount of lean predicted by your equation for 1992 (the year work was started to secure the foundation).

e. Find the root mean square error of the median-median line. What does this error tell you about your answer to 8d?

f. What are the domain and range for your linear model? Give an explanation for the numbers you chose.

9. The 4th term of an arithmetic sequence is 64. The 54th term is -61. Find the 23rd term.

MIXED REVIEW

10. State whether each recursive formula defines a sequence that is arithmetic, geometric, shifted geometric, or none of these. State whether a graph of the sequence would be linear or curved. Then list the first 5 terms of the sequence.

a. $u_1 = 4$ and $u_n = 3u_{n-1}$ where $n \geq 2$

b. $u_0 = 20$ and $u_n = 2u_{n-1} + 7$ where $n \geq 1$

11. APPLICATION You receive a $500 gift for high school graduation and deposit it into a savings account on June 15. The account has an annual interest rate of 5.9%, compounded annually.

a. Write a recursive formula that models this situation.

b. List the first 3 terms of the sequence.

c. What is the meaning of the value of u_3?

d. How much money will you have in your account when you retire, 35 years later?

e. If you deposit an additional $100 into your account each year on June 15, how much will you have in the account 35 years after graduation?

12. APPLICATION A website gives the current world population and projects the population for future years. Its projections for the number of people on Earth on January 1 in the years 2010–2014 are given in the table at right.

a. Find a recursive formula to model the population growth. What kind of sequence is this?

b. Predict the population on January 1, 2015.

c. In what year will the world's population exceed 10 billion?

d. Is this a realistic model that could predict world population in the next millennium? Explain.

Population Projections

Year	World population
2010	6,972,791,646
2011	7,074,193,720
2012	7,177,070,437
2013	7,281,731,271
2014	7,387,626,111

(*www.ibiblio.org*)

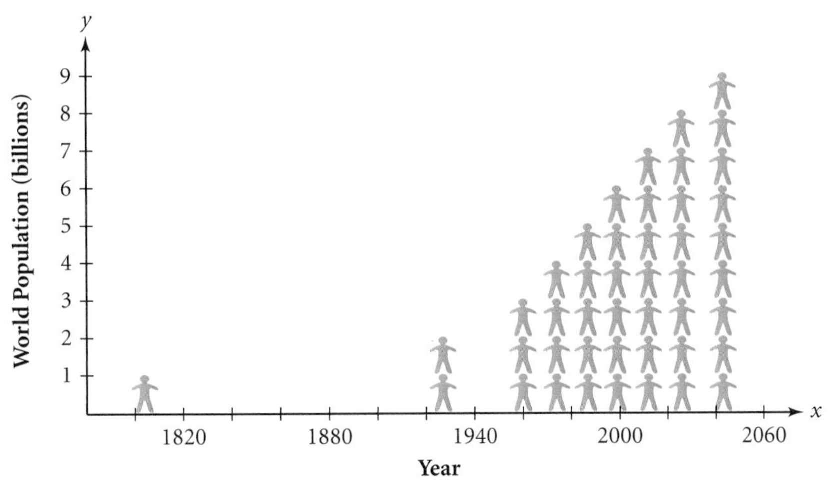

According to estimates from the United Nations, the world population will reach 8 billion around the year 2025. The graph depicts the years the world population reached or will reach multiples of 1 billion people, with each person representing 1 billion people.

13. APPLICATION Jonah must take an antibiotic every 12 hours. Each pill is 25 milligrams, and after every 12 hours, 50% of the drug remains in his body. What is the amount of antibiotic in his body over the first 2 days? What amount will there be in his body in the long run?

14. Create a box-and-whisker plot that has this five-number summary: 5, 7, 12, 13, 17.

a. Are the data skewed left, skewed right, or symmetric?

b. What is the median of the data?

c. What is the *IQR*?

d. What percentage of data values are above 12? Above 13? Below 5?

15. The table shows high school dropout rates reported by states in 2001–2002. Data are unavailable for some states.

High School Dropout Rates, 2001–2002

State	Rate (%)	State	Rate (%)	State	Rate (%)
Alabama	3.7	Maine	2.8	Oregon	4.9
Alaska	8.1	Maryland	3.9	Pennsylvania	3.3
Arizona	10.5	Minnesota	3.8	Rhode Island	4.3
Arkansas	5.3	Mississippi	3.9	South Carolina	3.3
Connecticut	2.6	Missouri	3.6	South Dakota	2.8
Delaware	6.2	Montana	3.9	Tennessee	3.8
Florida	3.7	Nebraska	4.2	Texas	3.8
Georgia	6.5	Nevada	6.4	Utah	3.7
Hawaii	5.1	New Hampshire	4.0	Vermont	4.0
Idaho	3.9	New Jersey	2.5	Virginia	2.9
Illinois	6.4	New Mexico	5.2	Washington	7.1
Indiana	2.3	New York	7.1	West Virginia	3.7
Iowa	2.4	North Carolina	5.7	Wisconsin	1.9
Kansas	3.1	North Dakota	2.0	Wyoming	5.8
Kentucky	4.0	Ohio	3.1		
Louisiana	7.0	Oklahoma	4.4		

(National Center for Education Statistics)

a. What are the mean, median, mode, and standard deviation of the data?

b. Do any states lie more than 2 standard deviations above or below the mean?

c. Draw a histogram of the data, using an appropriate bin width.

16. Use an appropriate method to solve each system of equations.

a. $\begin{cases} 2.1x - 3y = -11.43 \\ 5x + 3y = 44.8 \end{cases}$

b. $\begin{cases} y = 1 + \frac{1}{3}x \\ y = \frac{28}{3} - 3x \end{cases}$

c. $\begin{cases} 3x + 4y = 23 \\ 2x - 6y = 3.2 \end{cases}$

17. Consider these data on the estimated median age of U.S. women who married for the first time in these years between 1970 and 2006. Approximately 0.08% of Americans get married each year.

a. Create a scatter plot of the data. Do these data seem linear?

b. Find a median-median line for the data.

c. Use your median-median line to predict the median age of women at first marriage in 2010.

d. The equation for your age is $y = x - \textit{your birth year}$. Find the intersection of the median-median line and this line. What does this point tell you?

e. Calculate the residuals for each data point, and find the root mean square error for your equation. What does this tell you about your model?

U.S. Women's Median Age at First Marriage

Year	Age	Year	Age
1970	20.8	1990	23.9
1974	21.1	1994	24.5
1978	21.8	2002	25.3
1982	22.5	2006	25.5
1986	23.1		

(*www.census.gov*)

18. Consider the arithmetic sequence 6, 13, 20, 27, 34, ... Let u_1 represent the first term.

a. Write a recursive formula that describes this sequence.

b. Write an explicit formula for this sequence.

c. What is the slope of your equation in 18b? What relationship does this have to the arithmetic sequence?

d. Determine the value of the 32nd term. Is it easier to use your formula from 18a or 18b for this?

19. For an arithmetic sequence, $u_1 = 12$ and $u_{10} = 52.5$.

a. What is the common difference of the sequence?

b. Find the equation of the line through the points (1, 12) and (10, 52.5).

c. What is the relationship between 19a and 19b?

TAKE ANOTHER LOOK

1. You plan to borrow $18,000 to purchase a new car. The lender is advertising an interest rate of 9% compounded monthly. You want to determine how much your payment must be to pay off the loan in 5 years (60 months). Your first step is to guess a payment amount and use the recursive formula from Chapter 1 to determine the balance after exactly 60 payments. The balance may be negative if your guess is too high. In a table, record (*payment, final balance*).

Make two or three more guesses and record each guess and the resulting final balance in the table. Then plot the ordered pairs you have recorded. Fit a line to the data points and locate the x-intercept of the line. What does this point tell you? Check and see if it is correct using the recursive procedure. Find the slope and y-intercept of the line and interpret their meanings.

2. The data at right show the average price of a movie ticket for selected years. Find a median-median line for the years 1935–2001. Does your line seem to fit the data well? Which years are not predicted well by your equation? Consider whether or not two or more line segments would fit the data better. Sketch several connected line segments that fit the data.

Year x	Average ticket price ($) y	Year x	Average ticket price ($) y
1935	0.25	1974	1.89
1940	0.28	1978	2.33
1948	0.38	1982	2.93
1954	0.50	1986	3.70
1958	0.69	1990	4.21
1963	0.87	1994	4.10
1967	1.43	1998	4.68
1970	1.56	2001	5.65

(Motion Picture Association of America)

Which model, the single median-median line or the connected segments, do you think is more accurate for predicting the current price of a ticket? Is there another line or curve you might draw that you think would be better? Why do you think these data might not be best modeled by a single linear equation?

3. In this chapter you learned three methods for solving a system of linear equations—graphing, substitution, and elimination. These methods also can be applied to systems of nonlinear equations. Use all three methods to solve this system:

$$\begin{cases} y = x^2 - 4 \\ y = -2x^2 + 2 \end{cases}$$

Did you find the same solution(s) with all three methods? Describe how the process of solving this system was different from solving a system of linear equations. If you were given another system similar to this one, which method of solution would you choose? What special things would you look out for?

Assessing What You've Learned

In Chapters 0, 1, and 2, you were introduced to different ways to assess what you learned. Maybe you have tried all six ways—writing in your journal, giving a presentation, organizing your notebook, doing a performance assessment, keeping a portfolio, and writing test items. By now you should realize that assessment is more than just taking tests and more than your teacher giving you a grade.

In the working world, performance in only a few occupations can be measured with tests. All employees, however, must communicate and demonstrate to their employers, coworkers, clients, patients, or customers that they are skilled in their field. Assessing your own understanding and demonstrating your ability to apply what you've learned gives you practice in this important life skill. It also helps you develop good study habits, and that, in turn, will help you advance in school and give you the best possible opportunities in your life.

WRITE IN YOUR JOURNAL Use one of these prompts to write a paragraph in your journal.

- Find an exercise from this chapter that you could not fully solve. Write out the problem and as much of the solution as possible. Then clearly explain what is keeping you from solving the problem. Be as specific as you can.
- Compare and contrast arithmetic sequences and linear equations. How do you decide which to use?

PERFORMANCE ASSESSMENT Show a classmate, family member, or teacher different ways to find a line of fit for a data set. You may want to go back and use one of the data sets presented in an example or exercise, or you may want to research your own data. Discuss how well the line fits and whether you think a linear model is a good choice for the data.

CHAPTER

4 Functions, Relations, and Transformations

American artist Benjamin Edwards (b 1970) used a digital camera to collect images of commercial buildings for this painting, *Convergence.* He then projected all the images in succession on a 97-by-146-inch canvas, and filled in bits of each one. The result is that numerous buildings are transformed into one busy impression—much like the impression of seeing many things quickly out of the corner of your eye when driving through a city.

OBJECTIVES

In this chapter you will

- interpret graphs of functions and relations
- review function notation
- learn about the linear, quadratic, square root, absolute-value, and semicircle families of functions
- apply transformations—translations, reflections, and dilations—to the graphs of functions and relations
- transform functions to model real-world data

Solving Equations

When you evaluate an expression, you must follow the order of operations: parentheses, exponents, multiplication/division, addition/subtraction. When you solve equations, it is often helpful to think of reversing this order of operations in order to "undo" all that was done to the variable.

The **absolute value** of a number is its distance from zero on the number line. The equation $|x| = 5$ has two solutions, either $x = 5$ or $x = -5$, because both 5 and -5 are 5 units from zero on the number line.

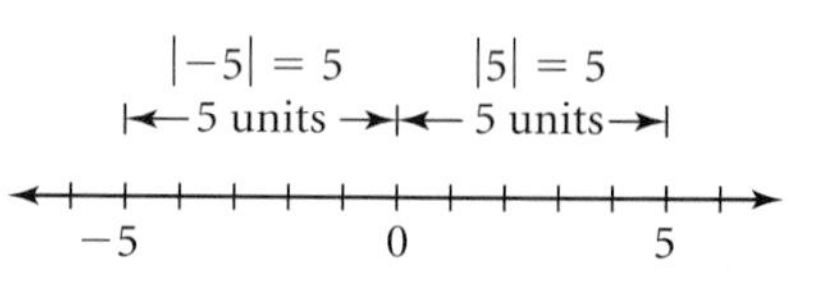

EXAMPLE A | Solve $5|a - 2| = 12$.

▶ ***Solution***

Consider the operations performed on a. First subtract 2 from a, then take the absolute value of the result, and finally, multiply by 5. To solve this equation, you can undo these steps in reverse order.

$$5|a-2| = 12$$ Original equation.

$$\frac{1}{5}\cdot 5|a-2| = \frac{1}{5}\cdot 12$$ Multiply by the reciprocal of 5 (to undo multiplying by 5).

$$|a-2| = \frac{12}{5} = 2.4$$ Multiply and change to decimal form.

To undo the absolute value, you'll need to consider two possibilities. The value $(a - 2)$ is 2.4 units from 0 on the number line, so $(a - 2)$ might equal 2.4 or $(a - 2)$ might equal -2.4.

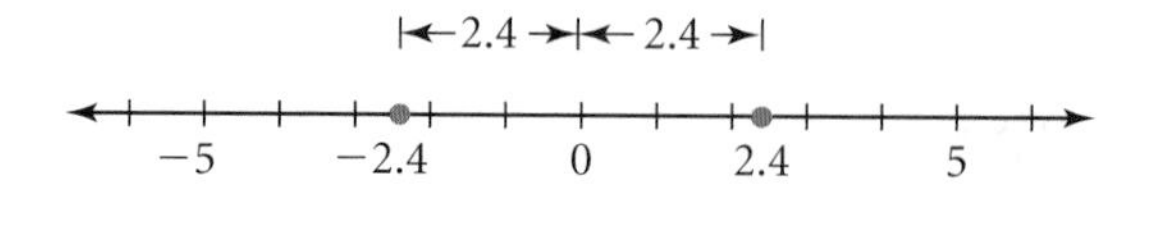

$a - 2 = 2.4$ or $a - 2 = -2.4$ Undo the absolute value.

$a = 4.4$ or $a = -0.4$ Add 2 to undo subtracting 2.

Check both answers to verify that they satisfy the original equation.

$$5|a-2| = 12$$

$5|4.4 - 2| \stackrel{?}{=} 12$ $\qquad$ $5|-0.4 - 2| \stackrel{?}{=} 12$

$5(2.4) \stackrel{?}{=} 12$ $\qquad$ $5(2.4) \stackrel{?}{=} 12$

$$12 = 12$$

Just as there are two solutions to the equation $|x| = 5$, there are two solutions to the equation $x^2 = 25$. You can take the square root of both sides of an equation if both sides are positive, but be careful! Note that for negative values of x, $\sqrt{x^2} \neq x$.

For example, $\sqrt{(-5)^2}$ equals 5, not -5. An equation that is true for all values of x is $\sqrt{x^2} = |x|$. Convince yourself of this by substituting some positive and negative values for x into $\sqrt{x^2}$.

If you use the absolute value in solving equations with x-squared, you won't forget to find both solutions.

EXAMPLE B | Solve $8 + 2(b - 6)^2 = 26$.

▶ **Solution**

Undo the operations performed on the variable b in reverse order.

$8 + 2(b - 6)^2 = 26$	Original equation.
$2(b - 6)^2 = 18$	Add -8 to each side to undo adding 8.
$(b - 6)^2 = 9$	Multiply by $\frac{1}{2}$ to undo multiplying by 2.
$\sqrt{(b - 6)^2} = \sqrt{9}$	Take the square root of each side to undo squaring.
$\lvert b - 6 \rvert = 3$	Use the relationship $\sqrt{x^2} = \lvert x \rvert$.
$b - 6 = 3$ or $b - 6 = -3$	Undo the absolute value.
$b = 9$ or $b = 3$	Add 6 to each side to undo subtracting 6.

Once again, you should check your answers in the original equation.

If you are solving an equation in which the variable is inside a square root, you can reverse the square root by squaring each side of the equation.

EXAMPLE C | Solve $\sqrt{c + 3} = 9$.

▶ **Solution**

To solve, undo the operations in reverse order.

$(\sqrt{c + 3})^2 = 9^2$	Square each side to undo the square root.
$d + 3 = 81$	Square.
$c = 78$	Add -3 to each side to undo adding 3.

You can check this answer mentally to see that it works in the original equation.

EXERCISES

1. Identify the first step in solving each of the equations for the variable. (It may be helpful to first identify the order of operations.)

a. $\frac{2}{3}x - 7 = 15$ @ **b.** $3|x + 8| = 21$ **c.** $2 + 5(x - 1)^2 = 82$ @

d. $\sqrt{y - 8} = 7$ **e.** $|x - 3| + 6 = 1$

2. Solve the equations in Exercise 1. @

3. Check the answers you found in Exercise 2. Did all of your answers check? Explain.

LESSON

4.1

Interpreting Graphs

A picture can be worth a thousand words, if you can interpret the picture. In this lesson you will investigate the relationship between real-world situations and graphs that represent them.

Wigs (portfolio) (1994), by American artist Lorna Simpson (b 1960), uses photos of African-American hairstyles through the decades, with minimal text, to critique deeper issues of race, gender, and assimilation.

Lorna Simpson, *Wigs (portfolio)*, 1994, waterless lithograph on felt, 72 x 162 in. overall installed. Collection Walker Art Center, Minneapolis/T. B. Walker Acquisition Fund, 1995

What is the real-world meaning of the graph at right, which shows the relationship between the number of customers getting haircuts each week and the price charged for each haircut?

The number of customers depends on the price of the haircut. So the price in dollars is the independent variable and the number of customers is the dependent variable. As the price increases, the number of customers decreases linearly. As you would expect, fewer people are willing to pay a high price; a lower price attracts more customers. The slope indicates the number of customers lost for each dollar increase. The x-intercept represents the haircut price that is too high for anyone. The y-intercept indicates the number of customers when haircuts are free.

EXAMPLE

Students at Central High School are complaining that the juice vending machine is frequently empty. Several student council members decide to study this problem. They record the number of cans in the machine at various times during a typical school day and make a graph.

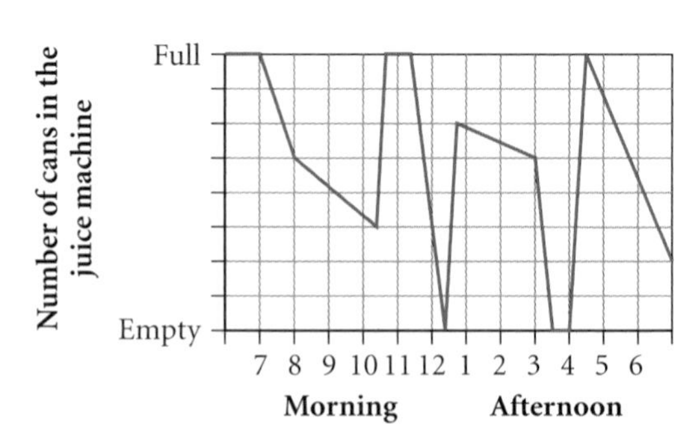

a. Based on the graph, at what times is juice consumed most rapidly?

b. When is the machine refilled? How can you tell?

c. When is the machine empty? How can you tell?

d. What do you think the student council will recommend to solve the problem?

▶ *Solution*

Each horizontal segment indicates a time interval when juice does not sell. Negative slopes represent when juice is consumed, and positive slopes show when the machine is refilled.

a. The most rapid consumption is pictured by the steep, negative slopes from 11:30 A.M. to 12:30 P.M. and from 3:00 to 3:30 P.M.

b. The machine is completely refilled overnight, again at 10:30 A.M., and again just after school lets out for the day. The machine is also refilled at 12:30 P.M., but only to 75% capacity.

c. The machine is empty from 3:30 to 4:00 P.M., and briefly at about 12:30 P.M.

d. The student council might recommend refilling the machine once more at about 2:00 or 3:00 P.M. in order to solve the problem of its frequently being empty. Refilling the machine completely at 12:30 P.M. may also solve the problem.

Health CONNECTION

Many school districts and several states have banned vending machines and the sale of soda pop and junk foods in their schools. Proponents say that schools have a responsibility to promote good health. The U.S. Department of Agriculture already bans the sale of foods with little nutritional value, such as soda, gum, and popsicles, in school cafeterias, but candy bars and potato chips don't fall under the ban because they contain some nutrients.

These recycled aluminum cans are waiting to be melted and made into new cans. Although 65% of the United States' aluminum is currently recycled, 1 million tons are still thrown away each year.

Although the student council members in the example are interested in solving a problem related to juice consumption, they could also use the graph to answer many other questions about Central High School: When do students arrive at school? What time do classes begin? When is lunch? When do classes let out for the day?

Both the graph of haircut customers and the graph in the example are shown as continuous graphs. In reality, the quantity of juice in the machine can take on only discrete values, because the number of cans must be a whole number. The graph might more accurately be drawn with a series of short horizontal segments, as shown at right. The price of a haircut and the number of customers can also take on only discrete values. This graph might be more accurately drawn with separate points. However, in both cases, a continuous "graph sketch" makes it easier to see the trends and patterns.

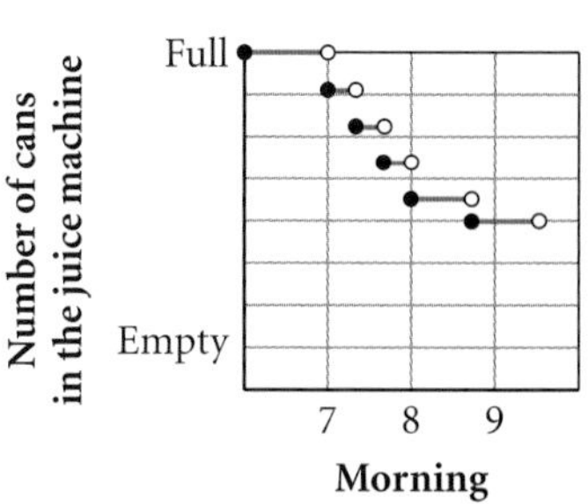

Investigation
Graph a Story

Every graph tells a story. Make a graph to go with the story in Part 1. Then invent your own story to go with the graph in Part 2.

Part 1

Sketch a graph that reflects all the information given in this story.

"It was a dark and stormy night. Before the torrents of rain came, the bucket was empty. The rain subsided at daybreak. The bucket remained untouched through the morning until Old Dog Trey arrived as thirsty as a dog. The sun shone brightly through the afternoon. Then Billy, the kid next door, arrived. He noticed two plugs in the side of the bucket. One of them was about a quarter of the way up, and the second one was near the bottom. As fast as you could blink an eye, he pulled out the plugs and ran away."

PEANUTS reprinted by permission of United Feature Syndicate, Inc.

Part 2

This graph tells a story. It could be a story about a lake, a bathtub, or whatever you imagine. Spend some time with your group discussing the information contained in the graph. Write a story that conveys all of this information, including when and how the rates of change increase or decrease.

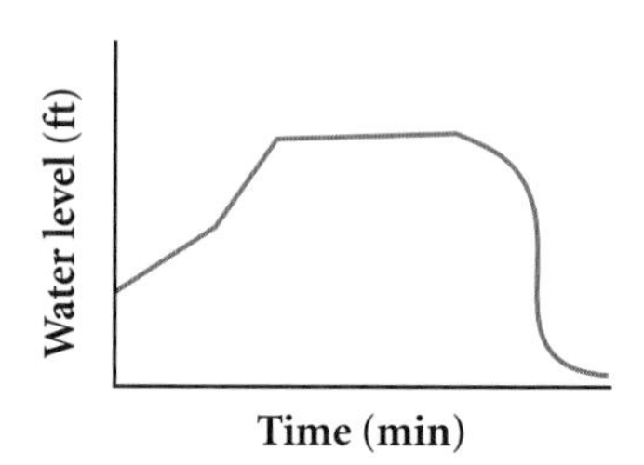

Science CONNECTION

Contour maps are a way to graphically represent altitude. Each line marks all of the points that are the same height in feet (or meters) above sea level. Using the distance between two contour lines, you can calculate the rate of change in altitude. These maps are used by hikers, forest fire fighters, and scientists.

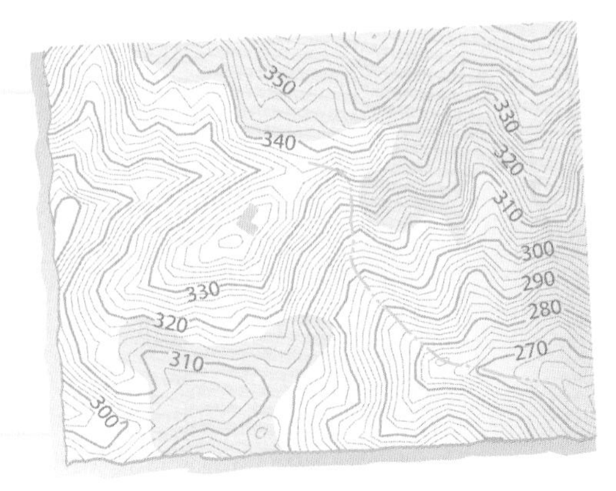

As you interpret data and graphs that show a relationship between two variables, you must always decide which is the independent variable and which is the dependent variable. You should also consider whether the variables are discrete or continuous.

EXERCISES

You will need

A graphing calculator for Exercise 12.

Practice Your Skills

1. Match a description to each graph.

a. 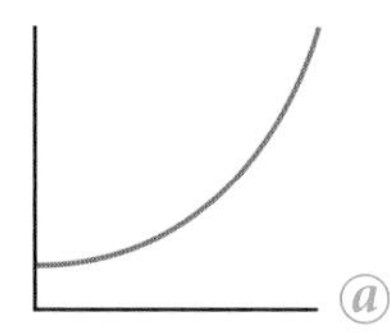

b. 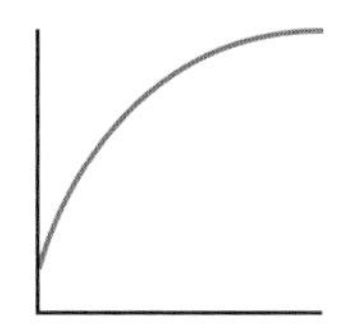

c. 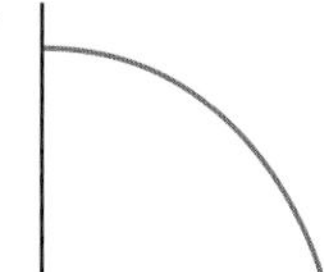

d. 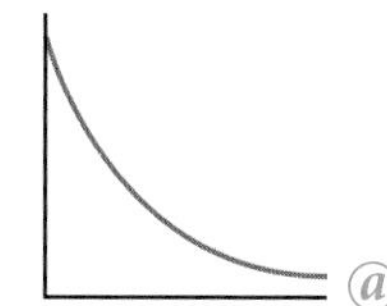

A. increasing more and more rapidly

B. decreasing more and more slowly

C. increasing more and more slowly

D. decreasing more and more rapidly

2. Sketch a graph to match each description.

a. increasing throughout, first slowly and then at a faster rate

b. decreasing slowly, then more and more rapidly, then suddenly becoming constant

c. alternately increasing and decreasing without any sudden changes in rate

3. For each graph, write a description like those in Exercise 2.

a. 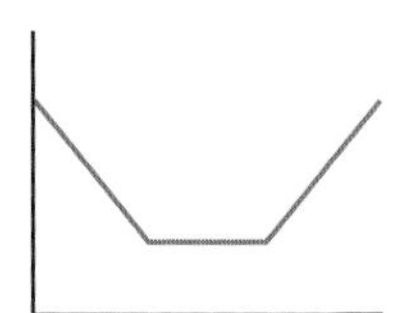

b. 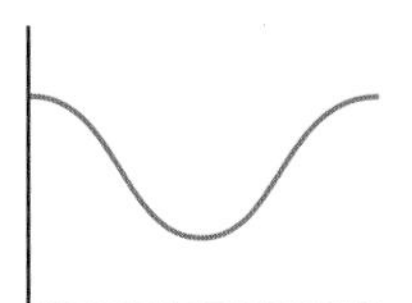

c. 

American minimalist painter and sculptor Ellsworth Kelly (b 1923) based many of his works on the shapes of shadows and spaces between objects.

Ellsworth Kelly *Blue Green Curve,* 1972, oil on canvas, 87-3/4 x 144-1/4 in. The Museum of Contemporary Art, Los Angeles, The Barry Lowen Collection

Reason and Apply

4. Harold's concentration often wanders from the game of golf to the mathematics involved in his game. His scorecard frequently contains mathematical doodles and graphs.

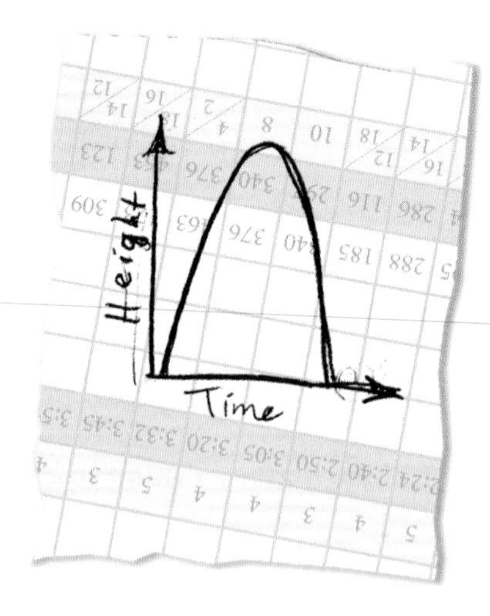

   a. What is a real-world meaning for this graph found on one of his recent scorecards?
   b. What units might he be using?
   c. Describe a realistic domain and range for this graph.
   d. Does this graph show how far the ball traveled? Explain.

5. Make up a story to go with the graph at right. Be sure to interpret the x- and y-intercepts. (h)

Depth (cm)

Time (s)

6. Sketch what you think is a reasonable graph for each relationship described. In each situation, identify the variables and label your axes appropriately.
   a. the height of a ball during a game of catch with a small child
   b. the distance it takes to brake a car to a full stop, compared to the car's speed when the brakes are first applied (@)
   c. the temperature of an iced drink as it sits on a table for a long period of time (@)
   d. the speed of a falling acorn after a squirrel drops it from the top of an oak tree (h)
   e. your height above the ground as you ride a Ferris wheel

7. Sketch what you think is a reasonable graph for each relationship described. In each situation, identify the variables and label your axes appropriately. In each situation, will the graph be continuous or will it be a collection of discrete points or pieces? Explain why.
   a. the amount of money you have in a savings account that is compounded annually, over a period of several years, assuming no additional deposits are made
   b. the same amount of money that you started with in 7a, hidden under your mattress over the same period of several years
   c. an adult's shoe size compared to the adult's foot length (@)
   d. the price of gasoline at the local station every day for a month
   e. the daily maximum temperature of a town for a month

8. Describe a relationship of your own and draw a graph to go with it.

9. Car A and Car B are at the starting line of a race. At the green light, they both accelerate to 60 mi/h in 1 min. The graph at right represents their velocities in relation to time.

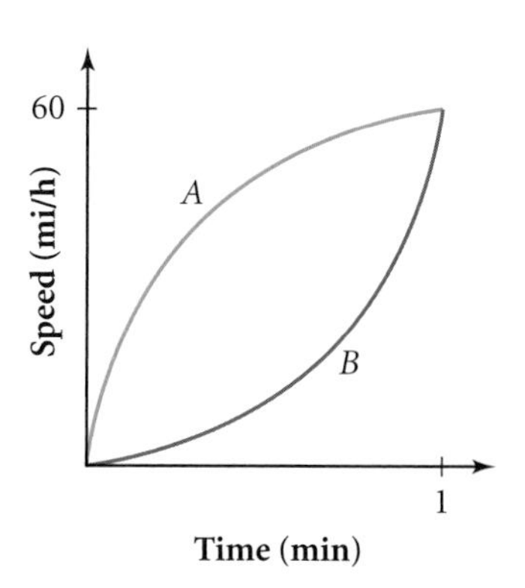

a. Describe the rate of change for each car. ⓗ

b. After 1 minute, which car will be in the lead? Explain your reasoning.

Review

10. Write an equation for the line that fits each situation.

a. The length of a rope is 1.70 m, and it decreases by 0.12 m for every knot that is tied in it.

b. When you join a CD club, you get the first 8 CDs for \$7.00. After that, your bill increases by \$9.50 for each additional CD you purchase.

11. APPLICATION Albert starts a business reproducing high-quality copies of pictures. It costs \$155 to prepare the picture and then \$15 to make each print. Albert plans to sell each print for \$27. ⓐ

a. Write a cost equation and graph it.

b. Write an income equation and graph it on the same set of axes.

c. How many pictures does Albert need to sell before he makes a profit?

d. What do the graphs tell you about the income and the cost for eight pictures?

American photographer Gordon Parks (1912–2006) holds a large, framed print of one of his photographs.

12. APPLICATION Suppose you have a \$200,000 home loan with an annual interest rate of 6.5%, compounded monthly.

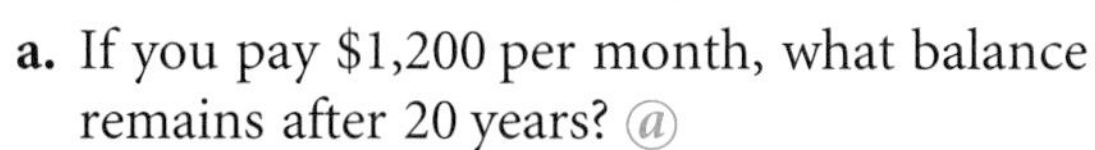

a. If you pay \$1,200 per month, what balance remains after 20 years? ⓐ

b. If you pay \$1,400 per month, what balance remains after 20 years?

c. If you pay \$1,500 per month, what balance remains after 20 years?

d. Make an observation about the answers to 12a–c.

13. Follow these steps to solve this system of three equations in three variables.

$$\begin{cases} 2x + 3y - 4z = -9 & \text{(Equation 1)} \\ x + 2y + 4z = 0 & \text{(Equation 2)} \\ 2x - 3y + 2z = 15 & \text{(Equation 3)} \end{cases}$$

a. Use the elimination method with Equation 1 and Equation 2 to eliminate z. The result will be an equation in two variables, x and y. ⓐ

b. Use the elimination method with Equation 1 and Equation 3 to eliminate z. ⓐ

c. Use your equations from 13a and b to solve for both x and y.

d. Substitute the values from 13c into one of the original equations and solve for z. What is the solution to the system?

LESSON

4.2 Function Notation

She had not understood mathematics until he had explained to her that it was the symbolic language of relationships. "And relationships," he had told her, "contained the essential meaning of life."

PEARL S. BUCK
THE GODDESS ABIDES, 1972

Rachel's parents keep track of her height as she gets older. They plot these values on a graph and connect the points with a smooth curve. For every age you choose on the x-axis, there is only one height that pairs with it on the y-axis. That is, Rachel is only one height at any specific time during her life.

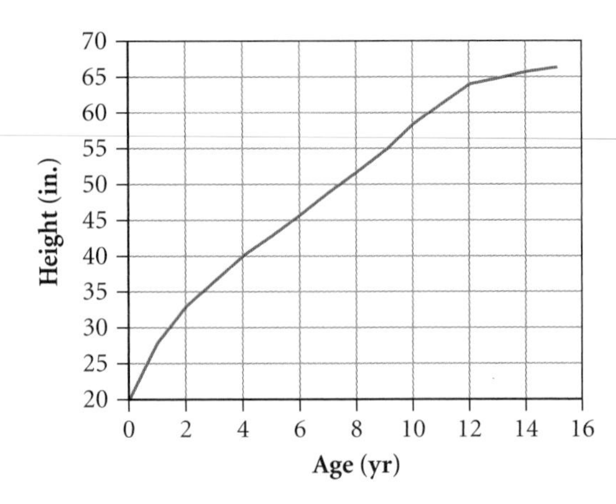

A **relation** is any relationship between two variables. A **function** is a special type of relation such that for every value of the independent variable, there is at most one value of the dependent variable. If x is your independent variable, a function pairs at most one y with each x. You can say that Rachel's height is a function of her age.

You may remember the vertical line test from previous mathematics classes. It helps you determine whether or not a graph represents a function. If no vertical line crosses the graph more than once, then the relation is a function. Take a minute to think about how you could apply this technique to the graph of Rachel's height and the graph in the next example.

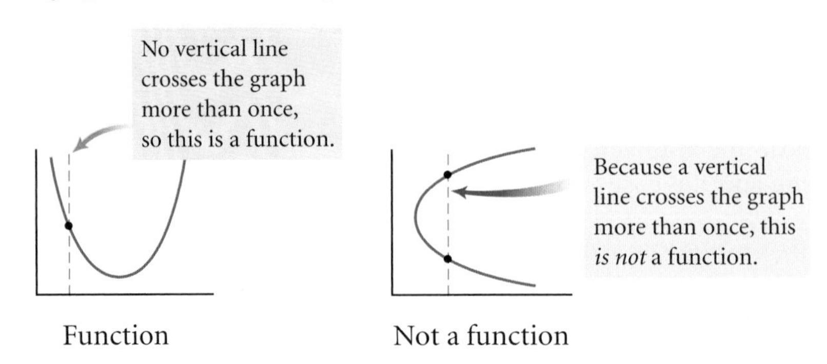

Function notation emphasizes the dependent relationship between the variables that are used in a function. The notation $y = f(x)$ indicates that values of the dependent variable, y, are explicitly defined in terms of the independent variable, x, by the function f. You read $y = f(x)$ as "y equals f of x."

Graphs of functions and relations can be continuous, such as the graph of Rachel's height, or they can be made up of discrete points, such as a graph of the maximum temperatures for each day of a month. Although real-world data often have an identifiable pattern, a function does not necessarily need to have a rule that connects the two variables.

Technology CONNECTION

A computer's desktop represents a function. Each icon, when clicked on, opens only one file, folder, or application.

This handwritten music manuscript by Norwegian composer Edvard Grieg (1843–1907) shows an example of functional relationships. Each of the four simultaneous voices for which this hymn is written can sing only one note at a time, so for each voice the pitch is a function of time.

EXAMPLE

Function f is defined by the equation $f(x) = \frac{2x+5}{x-3}$. Function g is defined by the graph at right.

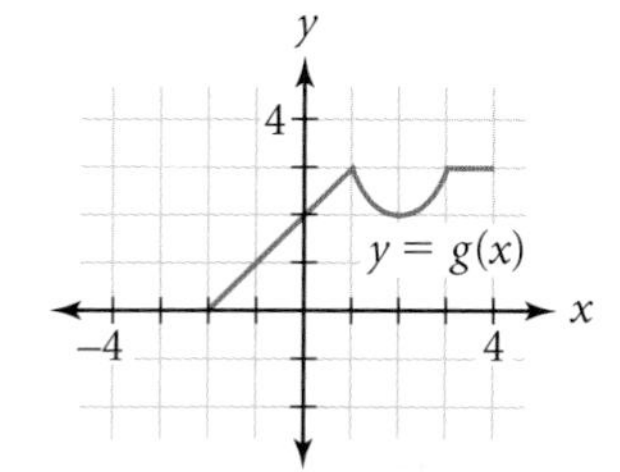

Find these values.

a. $f(8)$ **b.** $f(-7)$

c. $g(1)$ **d.** $g(-2)$

e. Find x when $g(x) = 0$.

▶ Solution

When a function is defined by an equation, you simply replace each x with the x-value and evaluate.

a. $f(x) = \dfrac{2x+5}{x-3}$

$$f(8) = \frac{2 \cdot 8 + 5}{8-3} = \frac{21}{5} = 4.2$$

b. $f(-7) = \dfrac{2 \cdot (-7) + 5}{-7-3} = \dfrac{-9}{-10} = 0.9$

You can check your work with your calculator. [▶ See **Calculator Note 4A** to learn about defining and evaluating functions. ◀]

Define $f(x) = \frac{2 \cdot x + 5}{x - 3}$ Done
$f(8)$ $\frac{21}{5}$
$f(-7)$ $\frac{9}{10}$

c. The notation $y = g(x)$ tells you that the values of y are explicitly defined, in terms of x, by the graph of the function g. To find $g(1)$, locate the value of y when x is 1. The point $(1, 3)$ on the graph means that $g(1) = 3$.

y
4
(1, 3)
y = g(x)
x
−4
(−2, 0)
4

d. The point $(-2, 0)$ on the graph means that $g(-2) = 0$.

e. To find x when $g(x) = 0$, locate points on the graph with a y-value of 0. There is only one, at $(-2, 0)$, so $x = -2$ when $g(x) = 0$.

Award-winning tap dancers Gregory Hines (1946–2003) and Savion Glover (b 1973) perform at the 2001 New York City Tap Festival.

At far right is Labanotation, one way of graphically representing movement, including dance. A single symbol shows you the direction, level, length of time, and part of the body performing a movement. This is a type of function notation because each part of the body can perform only one motion at any given time. For more information on dance notation, see the links at www.keymath.com/DAA .

In the investigation you will practice identifying functions and using function notation. As you do so, notice how you can identify functions in different forms.

Investigation
To Be or Not to Be (a Function)

Below are nine representations of relations.

a.
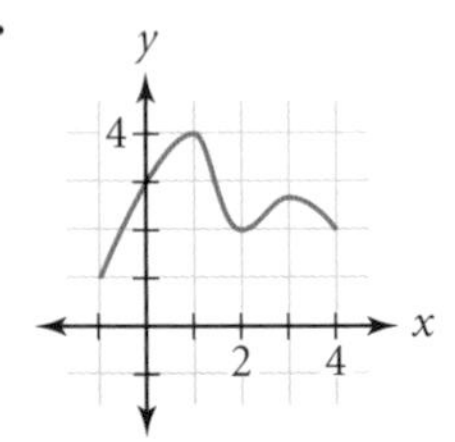

b.
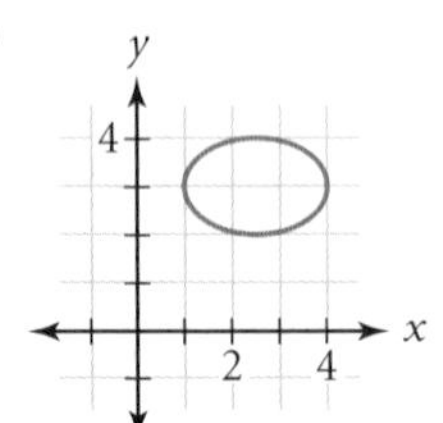

c.
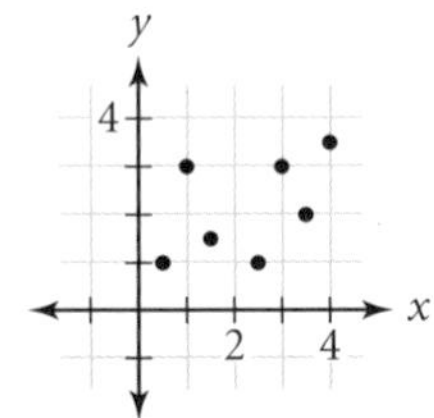

d. x → y
- 1 → 1
- 2 → 3
- 3 → 4
- 4 → 2

e. x → y
- 1 → 2
- 2 → 2
- 3 → 2
- 4 → 2

(y values 1, 2, 3, 4)

f. x → y
- 1 → 1
- 1 → 3
- 3 → 2
- 3 → 4

(x values 1, 2, 3, 4; y values 1, 2, 3, 4)

g. independent variable: the age of each student in your class
dependent variable: the height of each student

h. independent variable: an automobile in the state of Kentucky
dependent variable: that automobile's license plate number

i. independent variable: the day of the year
dependent variable: the time of sunset

Step 1 Identify each relation that is also a function. For each relation that is not a function, explain why not.

Step 2 For each graph or table that represents a function in parts a–f, find the y-value when $x = 2$, and find the x-value(s) when $y = 3$. Write each answer in function notation using the letter of the subpart as the function name. For example, if graph a represents a function, $a(2) = \underline{?}$ and $a(\underline{?}) = 3$.

When you use function notation to refer to a function, you can use any letter you like. For example, you might use $y = h(x)$ if the function represents height, or $y = p(x)$ if the function represents population. Often in describing real-world situations, you use a letter that makes sense. However, to avoid confusion, you should avoid using the independent variable as the function name, as in $y = x(x)$. Choose freely but choose wisely.

When looking at real-world data, it is often hard to decide whether or not there is a functional relationship. For example, if you measure the height of every student in your class and the weight of his or her backpack, you may collect a data set in which each student height is paired with only one backpack weight. But does that mean no two students of the same height could have backpacks of different weights? Does it mean you shouldn't try to model the situation with a function?

EXERCISES

You will need

A graphing calculator for Exercise **10**.

Practice Your Skills

1. Which of these graphs represent functions? Why or why not?

a.

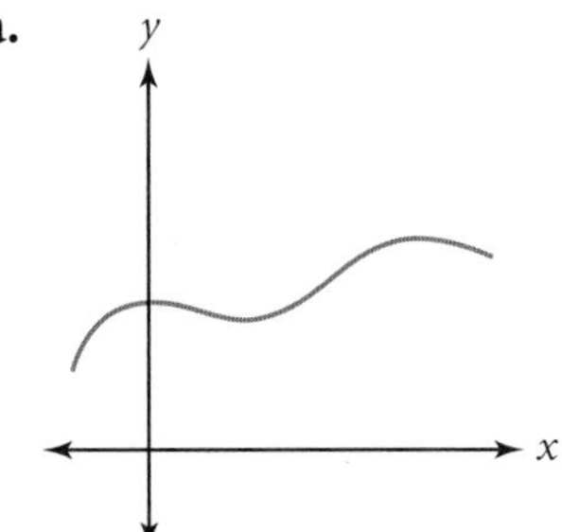

b.

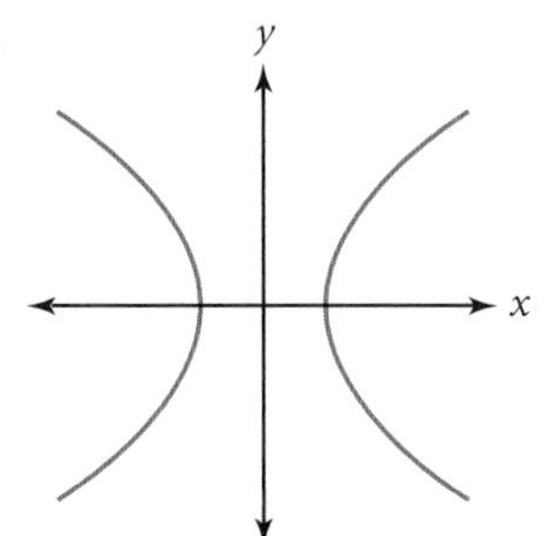

c. @

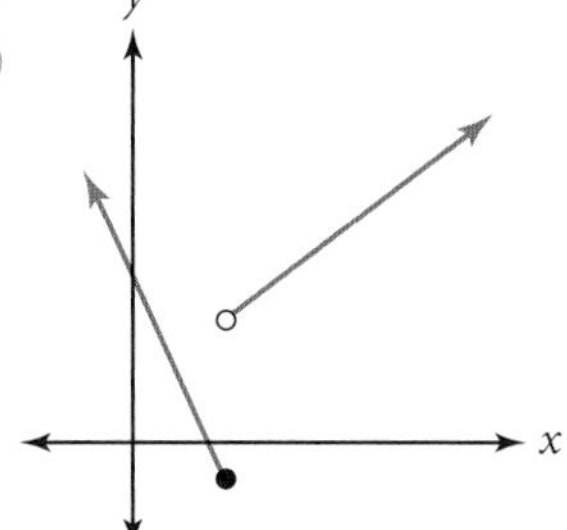

2. Use the functions $f(x) = 3x - 4$ and $g(x) = x^2 + 2$ to find these values.

a. $f(7)$ **b.** $g(5)$ **c.** $f(-5)$ **d.** $g(-3)$ @ **e.** x when $f(x) = 7$ @

3. Miguel works at an appliance store. He gets paid \$7.25 an hour and works 8 hours a day. In addition, he earns a 3% commission on all items he sells. Let x represent the total dollar value of the appliances that Miguel sells, and let the function m represent Miguel's daily earnings as a function of x. Which function describes how much Miguel earns in a day?

A. $m(x) = 7.25 + 0.03x$ **B.** $m(x) = 58 + 0.03x$

C. $m(x) = 7.25 + 3x$ **D.** $m(x) = 58 + 3x$

4. Use the graph at right to find each value. Each answer will be an integer from 1 to 26. Relate each answer to a letter of the alphabet (1 = A, 2 = B, and so on), and fill in the name of a famous mathematician.

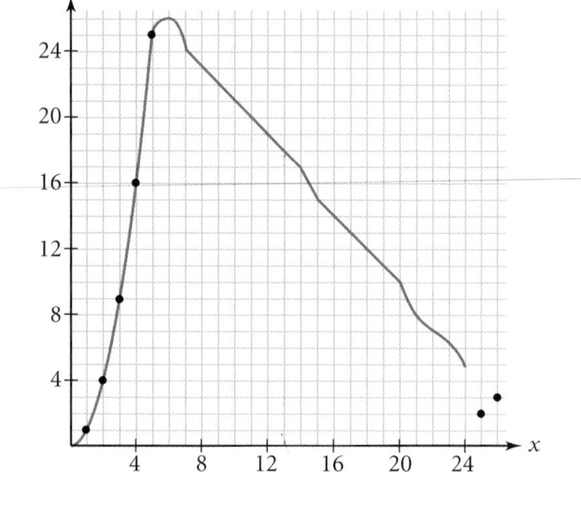

a. $f(13)$ 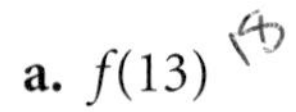

b. $f(25) + f(26)$

c. $2f(22)$

d. $\dfrac{f(3) + 11}{\sqrt{f(3+1)}}$ ⓐ

e. $\dfrac{f(1+4)}{f(1)+4} - \dfrac{1}{4}\left(\dfrac{4}{f(1)}\right)$

f. x when $f(x + 1) = 26$

g. $\sqrt[3]{f(21)} + f(14)$ ⓗ

h. x when $2f(x + 3) = 52$

i. x when $f(2x) = 4$

j. $f(f(2) + f(3))$ ⓐ

k. $f(9) - f(25)$

l. $f(f(5) - f(1))$

m. $f(4 \cdot 6) + f(4 \cdot 4)$

___	___	___	___		___	___	___	___	___	___	___	___	___
a	b	c	d		e	f	g	h	i	j	k	l	m

5. Identify the independent variable for each relation. Is the relation a function?

a. the price of a graphing calculator and the sales tax you pay

b. the amount of money in your savings account and the time it has been in the account

c. the amount your hair has grown since the time of your last haircut

d. the amount of gasoline in your car's fuel tank and how far you have driven since your last fill-up

6. Sketch a reasonable graph for each relation described in Exercise 5. In each situation, identify the variables and label your axes appropriately.

Reason and Apply

7. Suppose $f(x) = 25 - 0.6x$.

a. Draw a graph of this function.

b. What is $f(7)$?

c. Identify the point $(7, f(7))$ by marking it on your graph.

d. Find the value of x when $f(x) = 27.4$. Mark this point on your graph.

8. Identify the domain and range of the function g in the graph at right. ⓐ

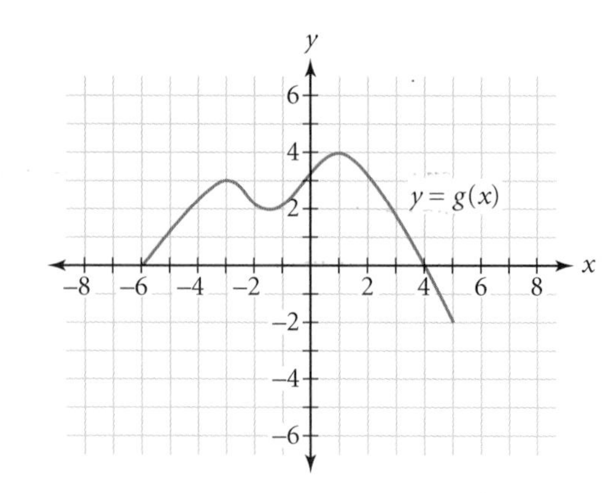

9. Sketch a graph for each function.

a. $y = f(x)$ has domain all real numbers and range $f(x) \leq 0$.

b. $y = g(x)$ has domain $x > 0$ and range all real numbers.

c. $y = h(x)$ has domain all real numbers and range $h(x) = 3$.

10. Consider the function $f(x) = 3(x + 1)^2 - 4$.

a. Find $f(5)$.

b. Find $f(n)$.

c. Find $f(x + 2)$.

d. Use your calculator to graph $y = f(x)$ and $y = f(x + 2)$ on the same axes. How do the graphs compare? ⓐ

11. Kendall walks toward and then away from a motion sensor. Is the (*time*, *distance*) graph of his motion a function? Why or why not?

12. APPLICATION The length of a pendulum in inches, L, is a function of its period, or the length of time it takes to swing back and forth, in seconds, t. The function is defined by the formula $L = 9.73t^2$.

a. Find the length of a pendulum if its period is 4 s.

b. The Foucault pendulum at the Panthéon in Paris has a 62-pound iron ball suspended on a 220-foot wire. What is its period?

Astronomer Jean Bernard Leon Foucault (1819–1868) displayed this pendulum for the first time in 1851. The floor underneath the swinging pendulum was covered in sand, and a pin attached to the ball traced the pendulum's path. While the ball swung back and forth in straight lines, it changed direction relative to the floor, proving that Earth was rotating underneath it.

13. The number of diagonals of a polygon, d, is a function of the number of sides of the polygon, n, and is given by the formula $d = \frac{n(n-3)}{2}$.

a. Find the number of diagonals in a dodecagon (a 12-sided polygon).

b. How many sides would a polygon have if it contained 170 diagonals? ⓗ

Language CONNECTION

You probably have noticed that some words, like biannual, triplex, and quadrant, have prefixes that indicate a number. Knowing the meaning of a prefix can help you determine the meaning of a word. The word "polygon" comes from the Greek *poly-* (many) and *-gon* (angle). Many mathematical words use the following Greek prefixes.

1 mono	6 hexa	
2 di	7 hepta	
3 tri	8 octa	
4 tetra	9 ennea	
5 penta	10 deca	20 icosa

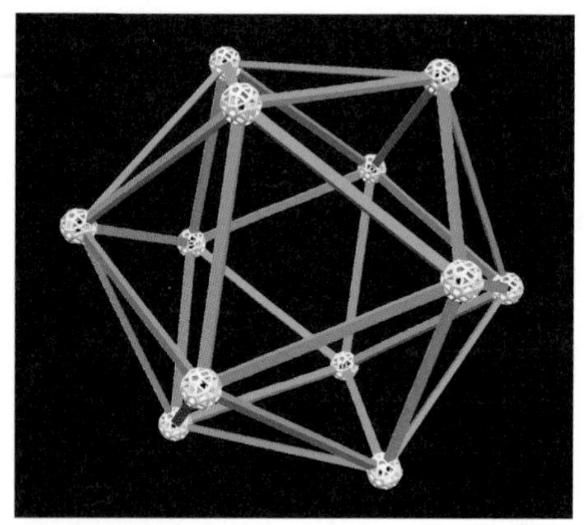

A polyhedron is a three-dimensional shape with many sides. Can you guess what the name of this shape is, using the prefixes given?

Review

14. Create graphs picturing the water height over time as each bottle is filled with water at a constant rate.

a. @

b.

c.

15. APPLICATION The five-number summary of this box plot is $2.10, $4.05, $4.95, $6.80, $11.50. The plot summarizes the amounts of money earned in a recycling fund drive by 32 members of the Oakley High School environmental club. Estimate the total amount of money raised. Explain your reasoning. @

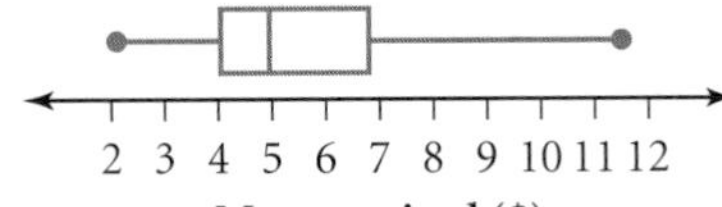

These photos show the breakdown of a biodegradable plastic during a one-hour period. Created by Australian scientists, the plastic is made of cornstarch and disintegrates rapidly when exposed to water. This technology could help eliminate the 24 million tons of plastic that end up in American landfills every year.

16. Given the graph at right, find the intersection of lines ℓ_1 and ℓ_2.

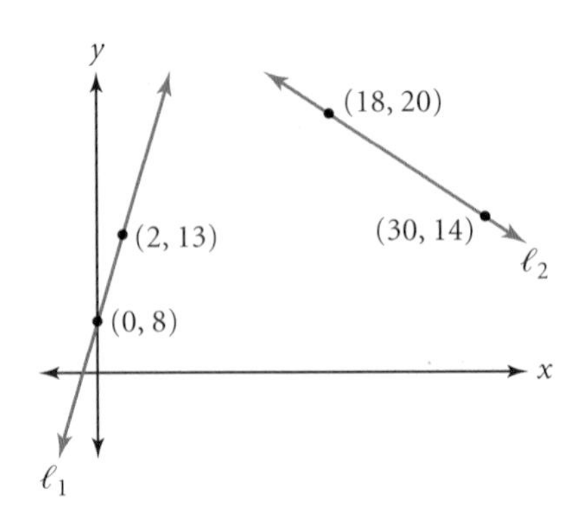

17. Sketch a graph for a function that has the following characteristics.

a. domain: $x \geq 0$
range: $f(x) \geq 0$
linear and increasing

b. domain: $-10 \leq x \leq 10$
range: $-3 < f(x) \leq 3$
nonlinear and increasing

c. domain: $x \geq 0$
range: $-2 < f(x) \leq 10$
increasing, then decreasing, then increasing, and then decreasing

18. You can use rectangle diagrams to represent algebraic expressions. For instance, this diagram demonstrates the equation $(x + 5)(2x + 1) = 2x^2 + 11x + 5$. Fill in the missing values on the edges or in the interior of each rectangle diagram.

	x	5
$2x$	$2x^2$	$10x$
1	x	5

a. (@)

	x	3
x		
7		

b.

	x^2	x
	$2x$	2

c.

	$2x^2$	$10x$
	$20x$	100

19. Alice and Carlos are each recording Bao's distance from where they stand. Initially Bao is between Alice and Carlos, standing 0.2 m from Alice and 4.2 m from Carlos. He walks at 0.5 m/s away from Alice and toward Carlos.

a. On the same axes, sketch graphs of Bao's distance from each student as a function of time.

b. Write an equation for each graph.

c. Find the intersection of the graphs and give the real-world meaning of that point.

Project

STEP FUNCTIONS

The graph at right represents a **step function.** The open circles mean that those points are not included in the graph. For example, the value of $f(3)$ is 5, not 2. The places where the graph "jumps" are called **discontinuities.**

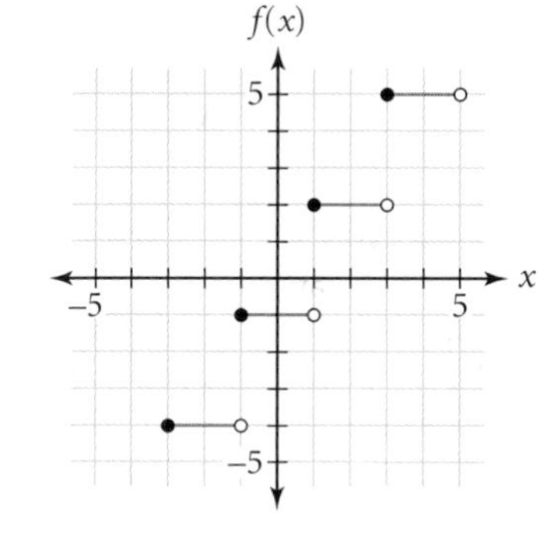

In Lesson 3.6, Exercise 9, you were introduced to an often-used step function—the **greatest integer function,** $f(x) = [x]$. Two related functions are the ceiling function, $f(x) = \lceil x \rceil$, and the floor function, $f(x) = \lfloor x \rfloor$.

Do further research on the greatest integer function, the ceiling function, and the floor function. Prepare a report or class presentation on the functions. Your project should include

- ▶ A graph of each function.
- ▶ A written or verbal description of how each function operates, including any relationships among the three functions. Be sure to explain how you would evaluate each function for different values of x.
- ▶ Examples of how each function might be applied in a real-world situation.

As you do your research, you might learn about other step functions that you'd like to include in your project.

LESSON

4.3 Lines in Motion

In Chapter 3, you worked with two forms of linear equations:

Intercept form $y = a + bx$

Point-slope form $y = y_1 + b(x - x_1)$

In this lesson you will see how these forms are related to each other graphically.

With the exception of vertical lines, lines are graphs of functions. That means you could write the forms above as $f(x) = a + bx$ and $f(x) = f(x_1) + b(x - x_1)$.

The investigation will help you see the effect that moving the graph of a line has on its equation. Moving a graph horizontally or vertically is called a **translation.** The discoveries you make about translations of lines will also apply to the graphs of other functions.

This skateboarding bowl, shown here at the Wexner Center for the Arts in Columbus, Ohio, is a functional sculpture designed by Simparch, an artists' collaborative in Chicago, Illinois. As former skateboarders, the makers of *Free Basin* (2002) wanted to create a piece formed like a kidney-shaped swimming pool, to pay tribute to the empty swimming pools that first inspired skateboarding on curved surfaces. The underside of the basin shows beams that lie on lines that are translations of each other.

Investigation
Movin' Around

You will need

- two motion sensors
- graph paper

In this investigation you will explore what happens to the equation of a linear function when you translate the graph of the line. You'll then use your discoveries to interpret data. Graph the lines in each step on the same set of axes and look for patterns.

Step 1 On graph paper, graph the line $y = 2x$ and then draw a line parallel to it, but 3 units higher. What is the equation of this new line? If $f(x) = 2x$, what is the equation of the new line in terms of $f(x)$?

Step 2 Draw a line parallel to the line $y = 2x$, but shifted down 4 units. What is the equation of this line? If $f(x) = 2x$, what is the equation of the new line in terms of $f(x)$?

Step 3 Mark the point where the line $y = 2x$ passes through the origin. Plot a point right 3 units from the origin. Draw a line parallel to the original line through this point. Use the point to write an equation in point-slope form for the new line. Then write an equation for the line in terms of $f(x)$.

Step 4 Plot a point left 1 unit and up 2 units from the origin. Draw a line parallel to the original line through this point and use the point to write an equation in point-slope form for the new line. Then write an equation for the line in terms of $f(x)$.

Step 5 If you move every point on the function $y = f(x)$ to a new point up k units and right h units, what is the equation of this translated function?

Your group will now use motion sensors to create a function and a translated copy of that function. [▶ See **Calculator Note 4C** for instructions on how to collect and retrieve data from two motion sensors. ◀]

Step 6 Arrange your group as in the photo to collect data.

Step 7 Person D coordinates the collection of data like this:

At 0 seconds:	C begins to walk slowly toward the motion sensors, and A begins to collect data.
About 2 seconds:	B begins to collect data.
About 5 seconds:	C begins to walk backward.
About 10 seconds:	A's sensor stops.
About 12 seconds:	B's sensor stops and C stops walking.

Step 8 After collecting the data, follow Calculator Note 4C to retrieve the data to two calculators and then transmit four lists of data to each group member's calculator. Be sure to keep track of which data each list contains.

Step 9 Graph both sets of data on the same screen. Record a sketch of what you see and answer these questions:

a. How are the two graphs related to each other?

b. If A's graph is $y = f(x)$, what equation describes B's graph? Describe how you determined this equation.

c. In general, if the graph of $y = f(x)$ is translated horizontally h units and vertically k units, what is the equation of this translated function?

If you know the effects of translations, you can write an equation that translates any function on a graph. No matter what the shape of a function $y = f(x)$ is, the graph of $y = f(x - 3) + 2$ will look just the same as $y = f(x)$, but it will be translated up 2 units and right 3 units. Understanding this relationship will enable you to graph functions and write equations for graphs more easily.

Translation of a Function

A **translation** moves a graph horizontally or vertically or both.

Given the graph of $y = f(x)$, the graph of

$$y = f(x - h) + k \text{ or, equivalently, of } y - k = f(x - h)$$

is a translation horizontally h units and vertically k units.

Pulitzer Prize–winning books *The Color Purple,* written in 1982 by Alice Walker (b 1944), and *The Grapes of Wrath,* written in 1939 by John Steinbeck (1902–1968), are shown here in Spanish translations.

Language CONNECTION

The word "translation" can refer to the act of converting between two languages. Similar to its usage in mathematics, *translation* of foreign languages is an attempt to keep meanings parallel. Direct substitution of words often destroys the subtleties of meaning of the original text. The complexity of the art and craft of translation has inspired the formation of Translation Studies programs in universities throughout the world.

In a translation, every point (x_1, y_1) is mapped to a new point, $(x_1 + h, y_1 + k)$. This new point is called an **image** of the original point. If you have difficulty remembering which way to move a function, recall the point-slope form of the equation of a line. In $y = y_1 + b(x - x_1)$, the point at $(0, 0)$ is translated to the new point at (x_1, y_1). In fact, every point is translated horizontally x_1 units and vertically y_1 units.

EXAMPLE A

Describe how the graph of $f(x) = 4 + 2(x - 3)$ is a translation of the graph of $f(x) = 2x$.

▶ Solution

The graph of $f(x) = 4 + 2(x - 3)$ passes through the point $(3, 4)$. Consider this point to be the translated image of $(0, 0)$ on $f(x) = 2x$. The point is translated right 3 units and up 4 units from its original location, so the graph of $f(x) = 4 + 2(x - 3)$ is the graph of $f(x) = 2x$ translated right 3 units and up 4 units.

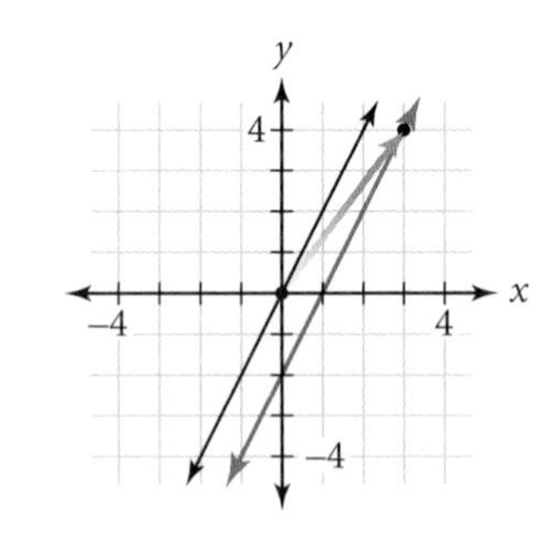

Note that you can distribute and combine like terms in $f(x) = 4 + 2(x - 3)$ to get $f(x) = -2 + 2x$. The fact that these two equations are equivalent means that translating the graph of $f(x) = 2x$ right 3 units and up 4 units is equivalent to translating the line down 2 units. In the graph in the example, this appears to be true.

Panamanian cuna (mola with geometric design on red background)

In the investigation and Example A, you translated a line that passed through the origin. If you are translating a graph of a function that does not pass through the origin, then you will need to identify points on the original function that will match up with points on the translated image.

EXAMPLE B The red graph is a translation of the graph of function f. Write an equation for the red function in terms of $f(x)$.

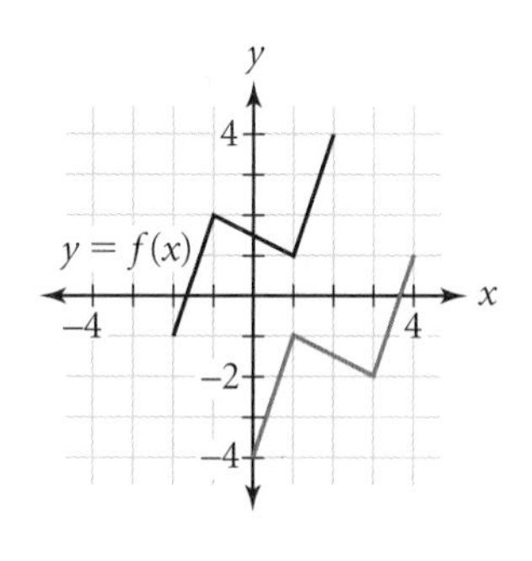

▶ Solution Any point on $f(x)$ can be matched with a point right 2 units and down 3 units on the red function. For example, the image of $(-1, 2)$ is $(1, -1)$. One notation to show this translation is $(x, y) \rightarrow (x + 2, y - 3)$. The equation of the red graph can be written $y - (-3) = f(x - 2)$, or $y + 3 = f(x - 2)$.

You can describe or graph a transformation of a function graph without knowing the equation of the function. But in the next few lessons, you will find that knowledge of equations for different families of functions can help you learn more about transformations.

EXERCISES

Practice Your Skills

1. The graph of the line $y = \frac{2}{3}x$ is translated right 5 units and down 3 units. Write an equation of the new line.

2. How does the graph of $y = f(x - 3)$ compare with the graph of $y = f(x)$? @

3. If $f(x) = -2x$, find

   a. $f(x + 3)$ @ b. $-3 + f(x - 2)$ @ c. $5 + f(x + 1)$

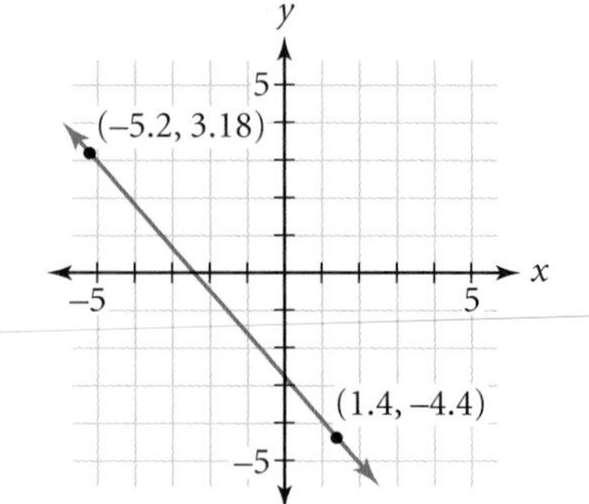

4. Consider the line that passes through the points $(-5.2, 3.18)$ and $(1.4, -4.4)$, as shown.

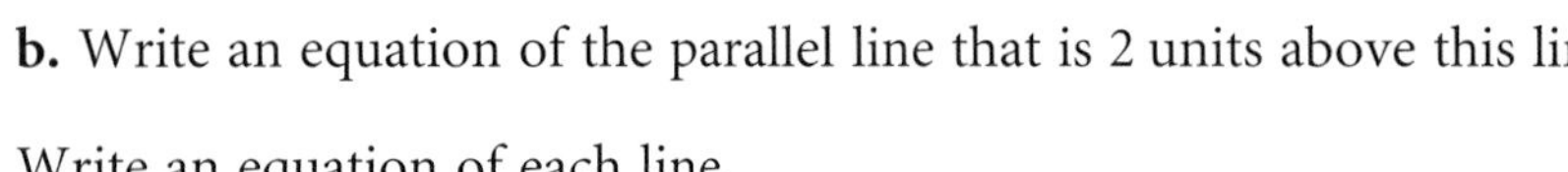

a. Find an equation of the line. ⓐ

b. Write an equation of the parallel line that is 2 units above this line.

5. Write an equation of each line.

a. the line $y = 4.7x$ translated down 3 units ⓐ

b. the line $y = -2.8x$ translated right 2 units

c. the line $y = -x$ translated up 4 units and left 1.5 units

Reason and Apply

6. The graph of $y = f(x)$ is shown in black. Write an equation for each of the red image graphs in terms of $f(x)$.

a. ⓐ

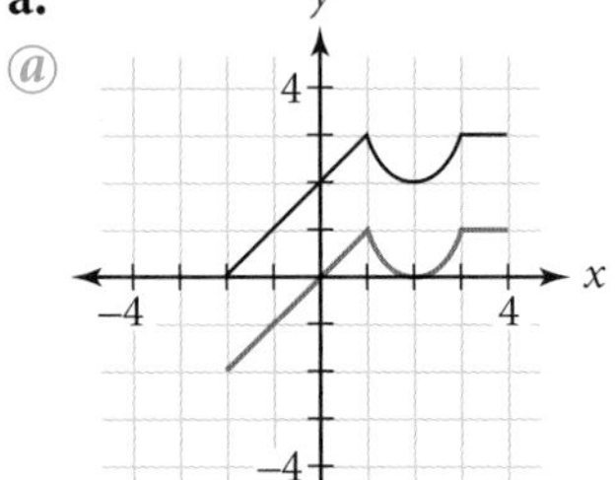

b.

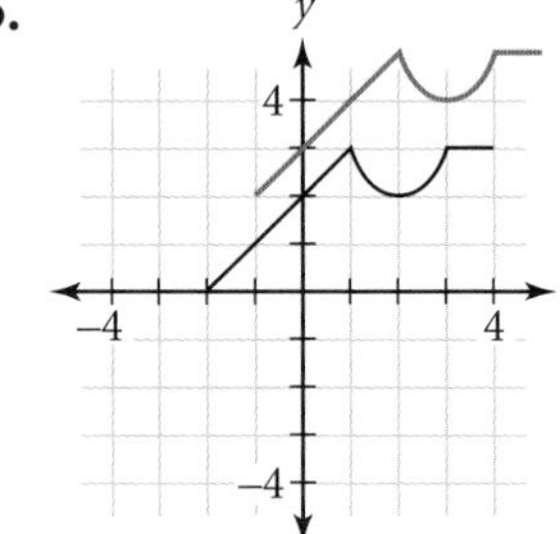

c.

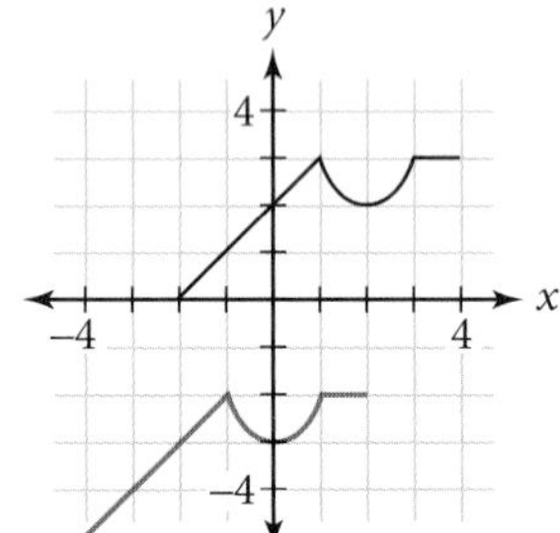

d.

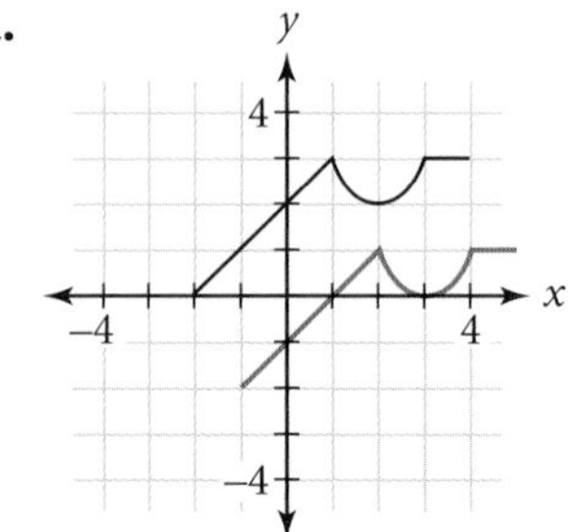

7. Jeannette and Keegan collect data about the length of a rope as knots are tied in it. The equation that fits their data is $y = 102 - 6.3x$, where x represents the number of knots and y represents the length of the rope in centimeters. Mitch had a piece of rope cut from the same source. Unfortunately he lost his data and can remember only that his rope was 47 cm long after he tied 3 knots. Write an equation that describes Mitch's rope.

8. Rachel, Pete, and Brian perform Steps 6–9 of the investigation in this lesson. Rachel walks while Pete and Brian hold the motion sensors. A graph of their results is shown at right.

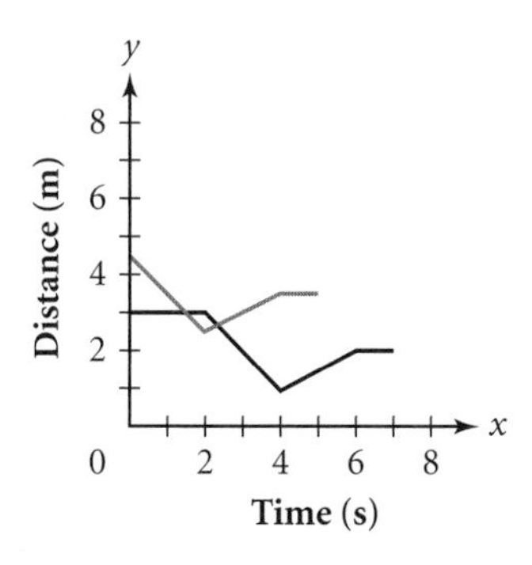

a. The black curve is made from the data collected by Pete's motion sensor. Where was Brian standing and when did he start his motion sensor to create the red curve? ⓗ

b. If Pete's curve is the graph of $y = f(x)$, what equation represents Brian's curve?

9. APPLICATION Kari's assignment in her computer programming course is to simulate the motion of an airplane by repeatedly translating it across the screen. The coordinate system in the software program is shown at right. In this program, coordinates to the right and down are positive.

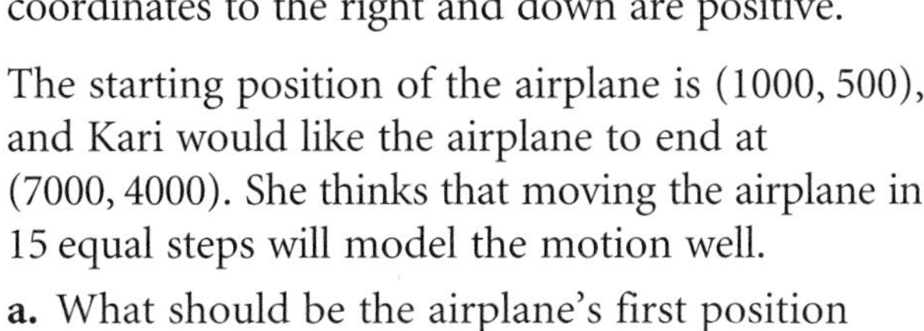

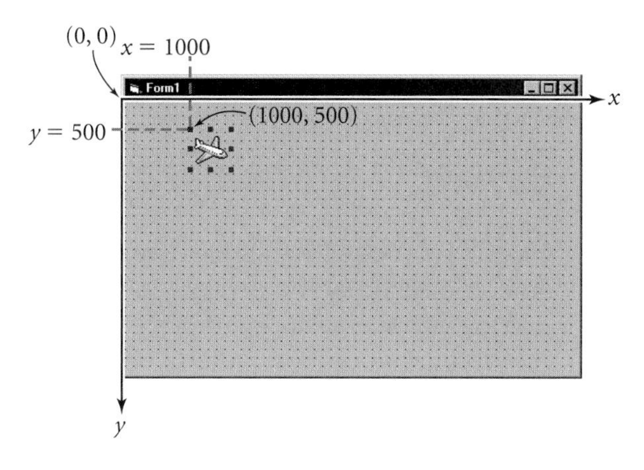

The starting position of the airplane is (1000, 500), and Kari would like the airplane to end at (7000, 4000). She thinks that moving the airplane in 15 equal steps will model the motion well.

a. What should be the airplane's first position after (1000, 500)?

b. If the airplane's position at any time is given by (x, y), what is the next position in terms of x and y?

c. If the plane moves down 175 units and right 300 units in each step, how many steps will it take to reach the final position of (7000, 4000)?

Art CONNECTION

Animation simulates movement. An old-fashioned way to animate is to make a book of closely related pictures and flip the pages. Flipbook technique is used in cartooning—a feature-length film might have more than 65,000 images. Today, hand drawing has been largely replaced by computer-generated special effects.

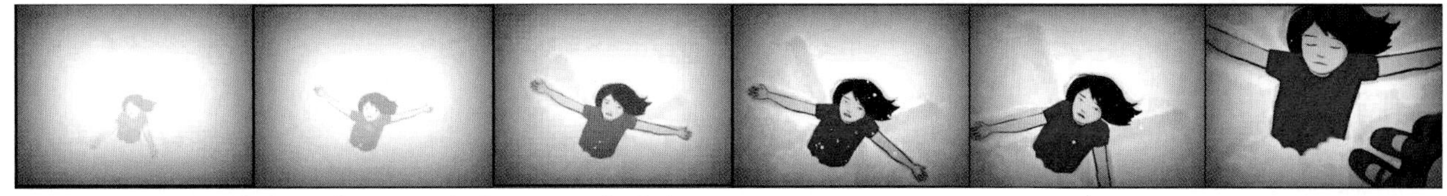

© 2002 Eun-Ha Paek. Stills from "L'Faux Episode 7" on www.MilkyElephant.com

10. *Mini-Investigation* Linear equations can also be written in standard form.

Standard form $\quad ax + by = c$

a. Identify the values of a, b, and c for each of these equations in standard form.

i. $4x + 3y = 12$ **ii.** $-x + y = 5$ **iii.** $7x - y = 1$

iv. $-2x + 4y = -2$ **v.** $2y = 10$ **vi.** $3x = -6$

b. Solve the standard form, $ax + by = c$, for y. The result should be an equivalent equation in intercept form. What is the y-intercept? What is the slope? ⓐ

c. Use what you've learned from 10b to find the y-intercept and slope of each of the equations in 10a.

d. The graph of $4x + 3y = 12$ is translated as described below. Write an equation in standard form for each of the translated graphs.

i. a translation right 2 units
ii. a translation left 5 units ⓐ
iii. a translation up 4 units
iv. a translation down 1 unit
v. a translation right 1 unit and down 3 units
vi. a translation up 2 units and left 2 units ⓐ

e. In general, if the graph of $ax + by = c$ is translated horizontally h units and vertically k units, what is the equation of the translated line in standard form?

Review

11. APPLICATION The Internal Revenue Service has approved ten-year linear depreciation as one method for determining the value of business property. This means that the value declines to zero over a ten-year period, and you can claim a tax exemption in the amount of the value lost each year. Suppose a piece of business equipment costs \$12,500 and is depreciated over a ten-year period. At right is a sketch of the linear function that represents this depreciation.

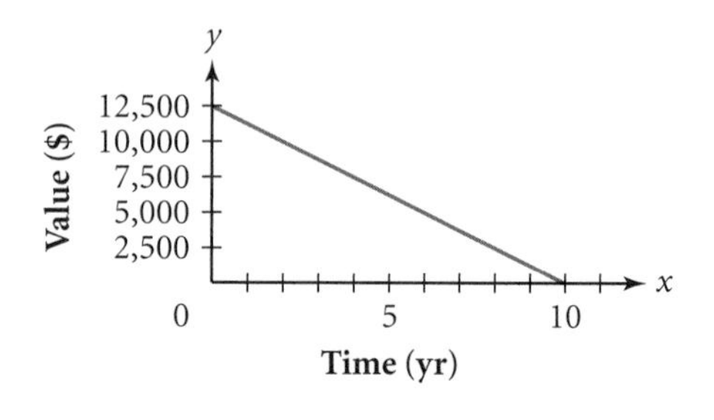

a. What is the y-intercept? Give the real-world meaning of this value.

b. What is the x-intercept? Give the real-world meaning of this value.

c. What is the slope? Give the real-world meaning of the slope.

d. Write an equation that describes the value of the equipment during the ten-year period.

e. When is the equipment worth \$6,500?

12. Suppose that your basketball team's scores in the first four games of the season were 86 points, 73 points, 76 points, and 90 points.

a. What will be your team's mean score if the fifth-game score is 79 points?

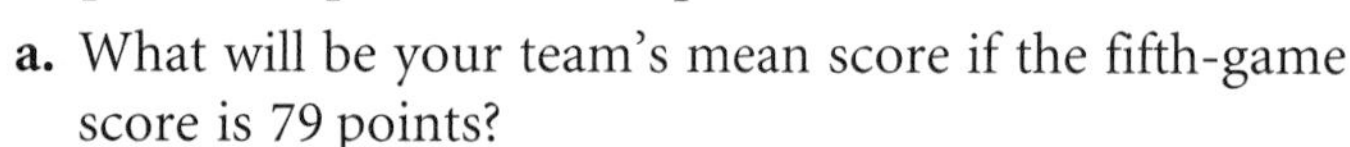

b. Write a function that gives the mean score in terms of the fifth-game score. ⓐ

c. What score will give a five-game average of 84 points? ⓗ

13. Solve.

a. $2(x + 4) = 38$

b. $7 + 0.5(x - 3) = 21$

c. $-2 + \frac{3}{4}(x + 1) = -17$

d. $4.7 + 2.8(x - 5.1) = 39.7$

14. The three summary points for a data set are $M_1(3, 11)$, $M_2(5, 5)$, and $M_3(9, 2)$. Find the median-median line.

LESSON

4.4 Translations and the Quadratic Family

I see music as the augmentation of a split second of time.

ERIN CLEARY

Music CONNECTION

When a song is in a key that is difficult to sing or play, it can be translated, or transposed, into an easier key. To transpose music means to change the pitch of each note without changing the relationships between the notes.

In the previous lesson, you looked at translations of the graphs of linear functions. Translations can occur in other settings as well. For instance, what will this histogram look like if the teacher decides to add five points to each of the scores?

What translation will map the black triangle on the left onto its red image on the right?

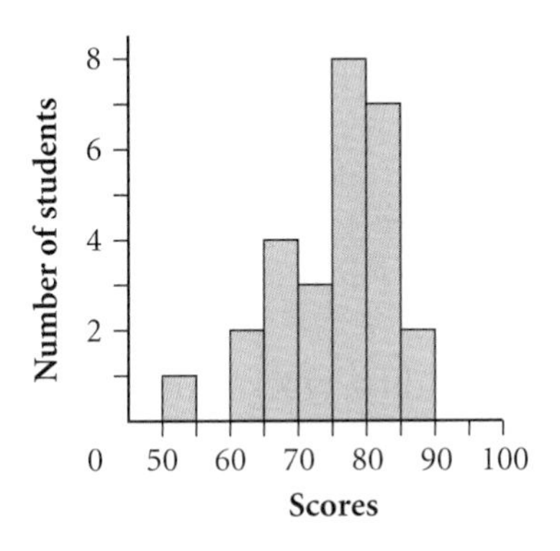

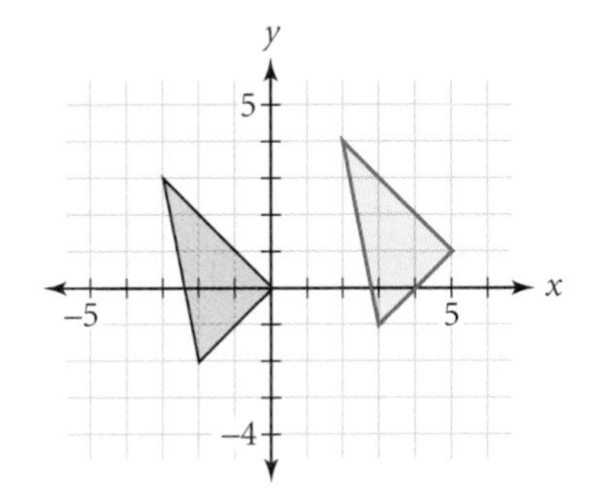

Translations are also a natural feature of the real world, including the world of art. Music can be transposed from one key to another. Melodies are often translated by a certain interval within a composition.

Jazz saxophonist Ornette Coleman (b 1930) grew up with strong interests in mathematics and science. Since the 1950s, he has developed award-winning musical theories, such as "free jazz," which strays from the set standards of harmony and melody.

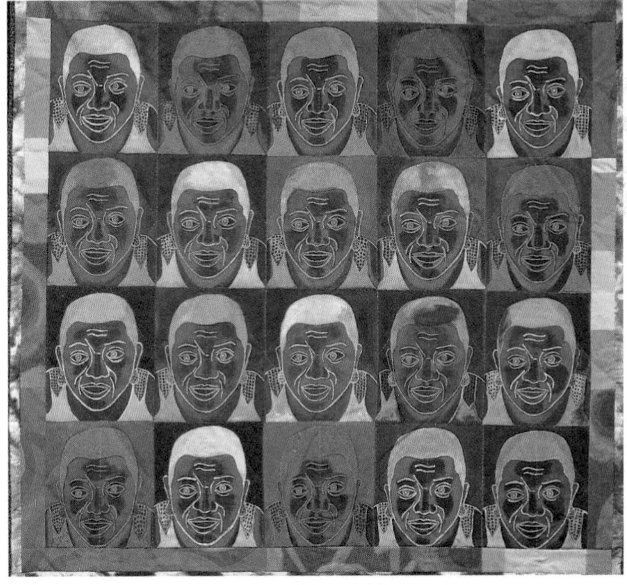

Bessie's Blues, by American artist Faith Ringgold (b 1930), shows 25 stenciled images of blues artist Bessie Smith. Was the stencil translated or reflected to make each image? How can you tell?

Bessie's Blues, by Faith Ringgold ©1997, acrylic on canvas, 76 × 79 in. Photo courtesy of the artist.

In mathematics, a change in the size or position of a figure or graph is called a **transformation.** Translations are one type of transformation. You may recall other types of transformations, such as reflections, dilations, stretches, shrinks, and rotations, from other mathematics classes.

In this lesson you will experiment with translations of the graph of the function $y = x^2$. The special shape of this graph is called a **parabola.** Parabolas always have a **line of symmetry** that passes through the parabola's **vertex.**

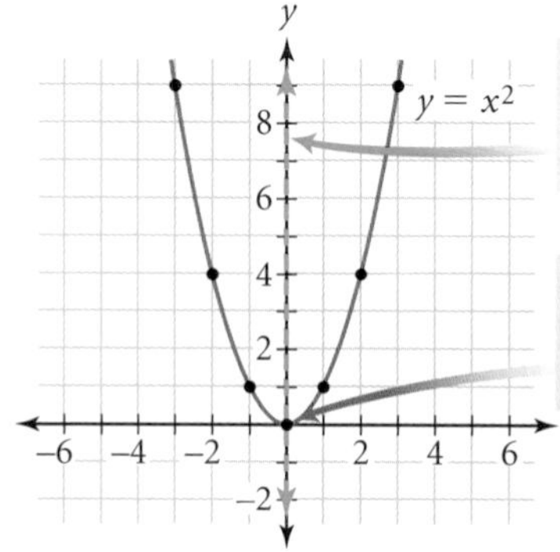

The line of symmetry divides the graph into mirror-image halves. The line of symmetry of $y = x^2$ is $x = 0$.

The vertex is the point where the graph changes direction. The vertex of $y = x^2$ is $(0, 0)$.

The function $y = x^2$ is a building-block function, or **parent function.** By transforming the graph of a parent function, you can create infinitely many new functions, or a **family of functions.** The function $y = x^2$ and all functions created from transformations of its graph are called **quadratic functions,** because the highest power of x is x-squared.

Quadratic functions are very useful, as you will discover throughout this book. You can use functions in the quadratic family to model the height of a projectile as a function of time, or the area of a square as a function of the length of its side.

The focus of this lesson is on writing the quadratic equation of a parabola after a translation and graphing a parabola given its equation. You will see that locating the vertex is fundamental to your success with understanding parabolas.

Engineering CONNECTION

Several types of bridge designs involve the use of curves modeled by nonlinear functions. Each main cable of a suspension bridge approximates a parabola. To learn more about the design and construction of bridges, see the links at www.keymath.com/DAA .

The Mackinac Bridge in Michigan was built in 1957.

Investigation
Make My Graph

Procedure Note

Different calculators have different resolutions. A good graphing window will help you make use of the resolution to better identify points. [► See **Calculator Note 4D** to find a good window setting for your calculator. ◄] Enter the parent function $y = x^2$ as the first equation. Enter the equation for the transformation as the second equation. Graph both equations to check your work.

Step 1 Each graph below shows the graph of the parent function $y = x^2$ in black. Find a quadratic equation that produces the congruent, red parabola. Apply what you learned about translations of the graphs of functions in Lesson 4.3.

a.

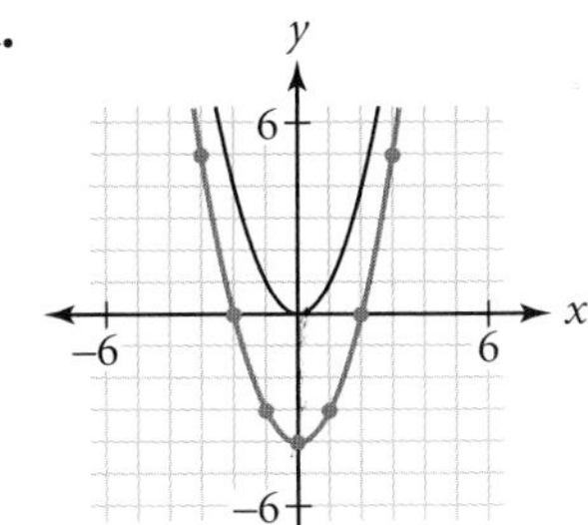

b.

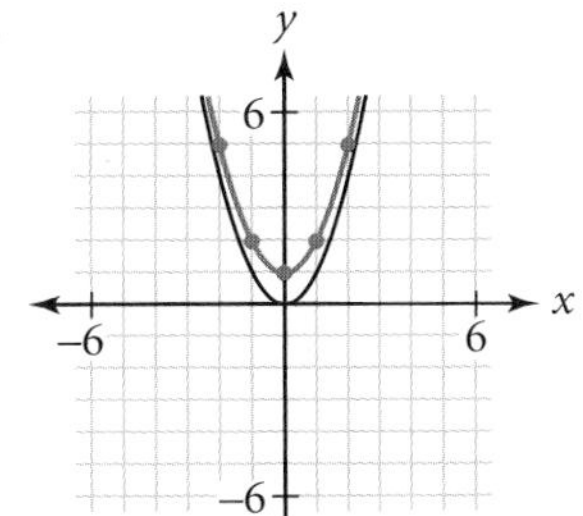

c.

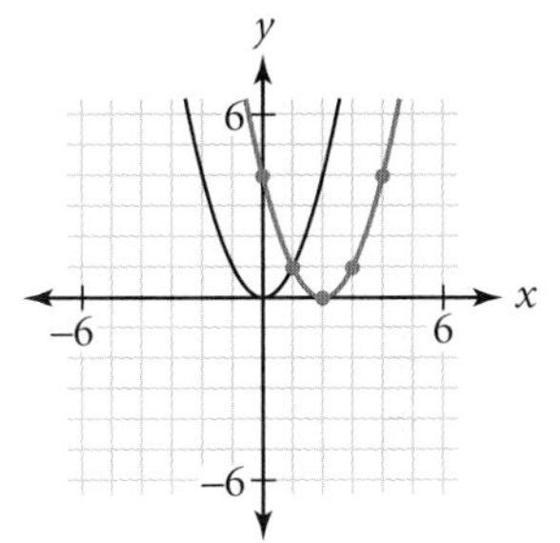

d.

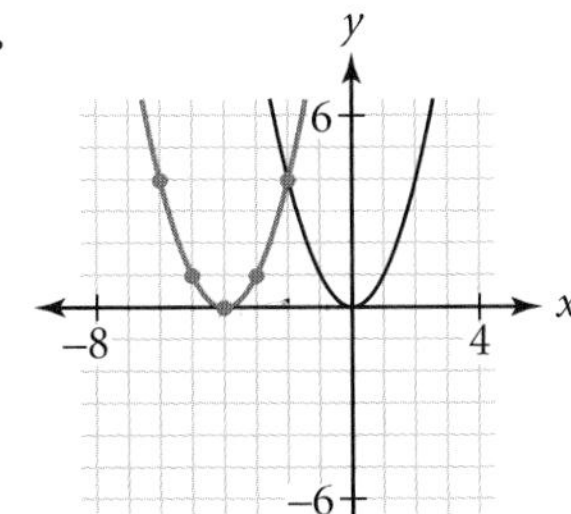

e.

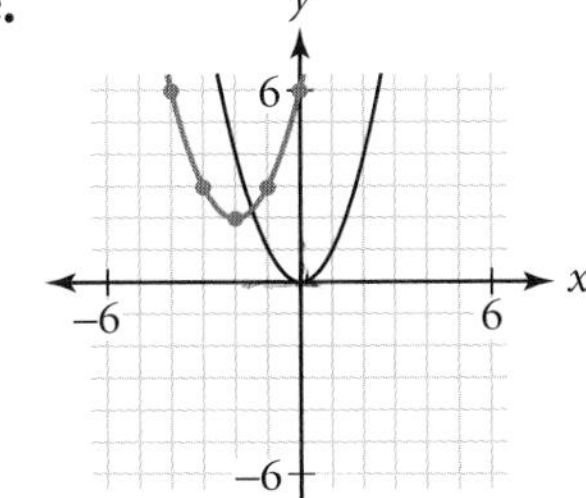

f.

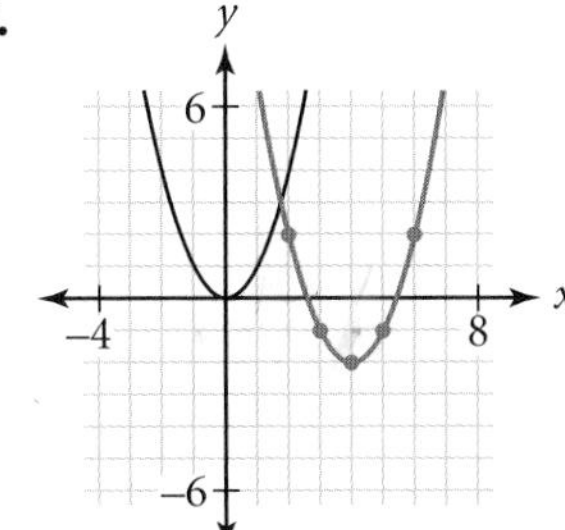

Step 2 Write a few sentences describing any connections you discovered between the graphs of the translated parabolas, the equation for the translated parabola, and the equation of the parent function $y = x^2$.

Step 3 In general, what is the equation of the parabola formed when the graph of $y = x^2$ is translated horizontally h units and vertically k units?

The following example shows one simple application involving parabolas and translations of parabolas. In later chapters you will discover many applications of this important mathematical curve.

EXAMPLE

This graph shows a portion of a parabola. It represents a diver's position (horizontal and vertical distance) from the edge of a pool as he dives from a 5 ft long board 25 ft above the water.

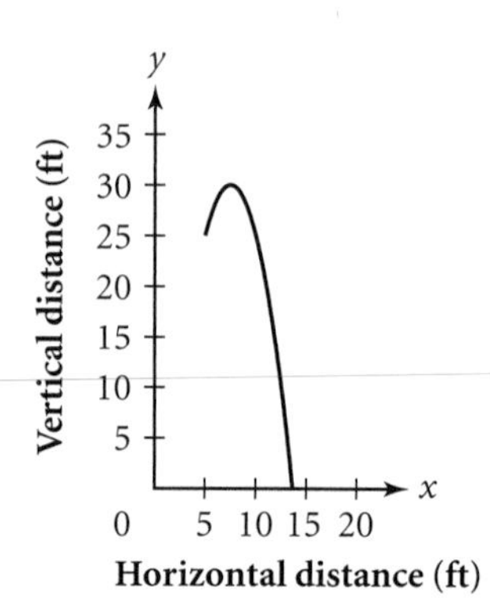

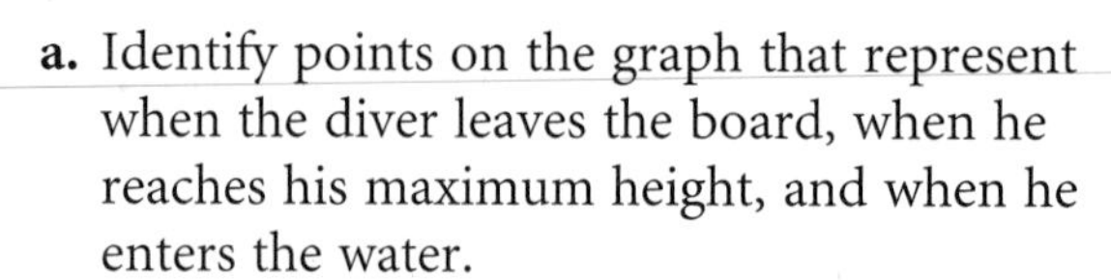

a. Identify points on the graph that represent when the diver leaves the board, when he reaches his maximum height, and when he enters the water.

b. Sketch a graph of the diver's position if he dives from a 10 ft long board 10 ft above the water. (Assume that he leaves the board at the same angle and with the same force.)

c. In the scenario described in part b, what is the diver's position when he reaches his maximum height?

▶ Solution

a. The point (5, 25) represents the moment when the diver leaves the board, which is 5 ft long and 25 ft high. The vertex, (7.5, 30), represents the position where the diver's height is at a maximum, or 30 ft; it is also the point where the diver's motion changes from upward to downward. The x-intercept, approximately (13.6, 0), indicates that the diver hits the water at approximately 13.6 ft from the edge of the pool.

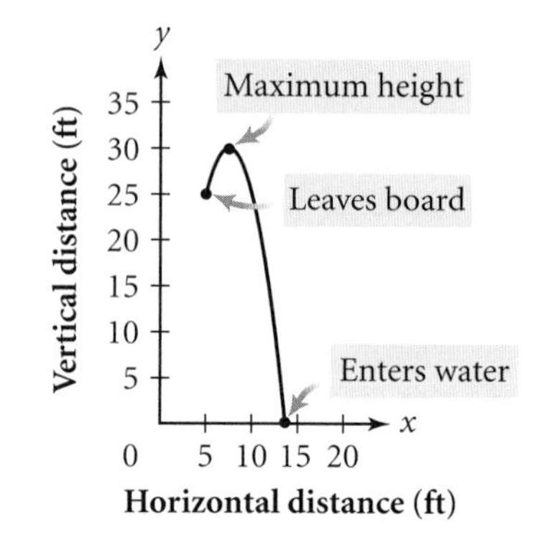

b. If the length of the board increases from 5 ft to 10 ft, then the parabola translates right 5 units. If the height of the board decreases from 25 ft to 10 ft, then the parabola translates down 15 units. If you define the original parabola as the graph of $y = f(x)$, then the function for the new graph is $y = f(x - 5) - 15$.

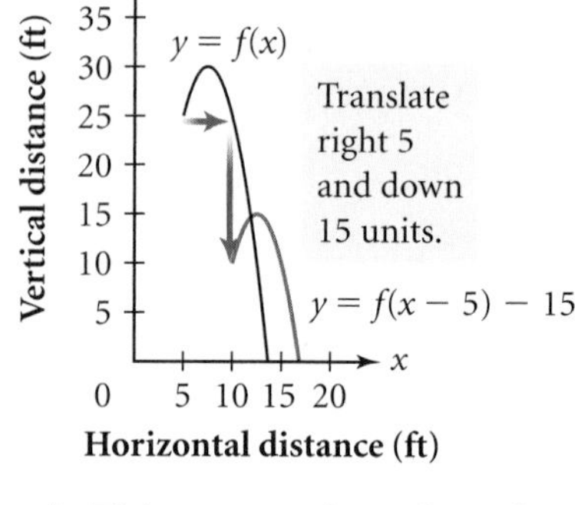

c. As with every point on the graph, the vertex translates right 5 units and down 15 units. The new vertex is $(7.5 + 5, 30 - 15)$, or $(12.5, 15)$. This means that when the diver's horizontal distance from the edge of the pool is 12.5 ft, he reaches his maximum height of 15 ft.

You can extend the ideas you've learned in translating linear and quadratic functions to functions in general. For a function $y = f(x)$, to translate the function horizontally h units, you can replace x in the equation with $(x - h)$. To translate the function vertically k units, replace y in the equation with $(y - k)$. If you translate the graph of $y = x^2$ horizontally h units and vertically k units, then the equation of the translated parabola is $y = (x - h)^2 + k$. You may also see this equation written as $y = k + (x - h)^2$ or $y - k = (x - h)^2$.

It is important to notice that the vertex of the translated parabola is (h, k). That's why finding the vertex is fundamental to determining translations of parabolas. In every function you study, there will be key points to locate. Finding the relationships between these points and the corresponding points in the parent function enables you to write equations more easily.

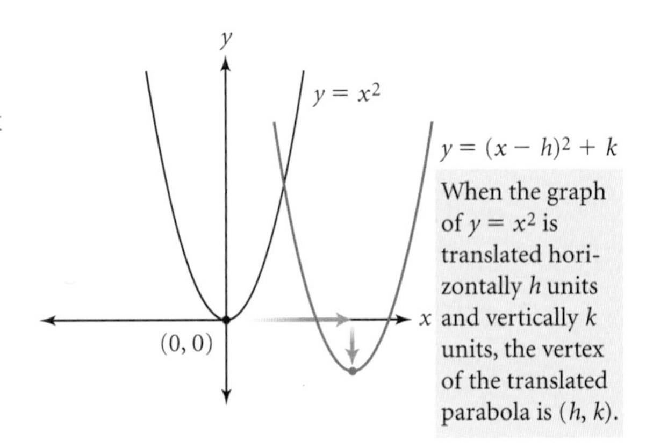

When the graph of $y = x^2$ is translated horizontally h units and vertically k units, the vertex of the translated parabola is (h, k).

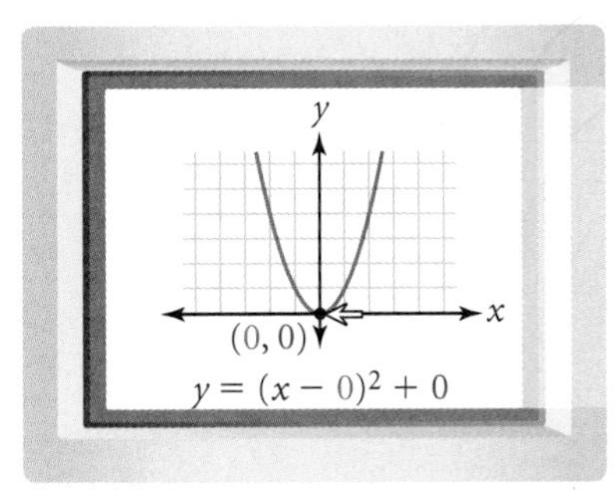

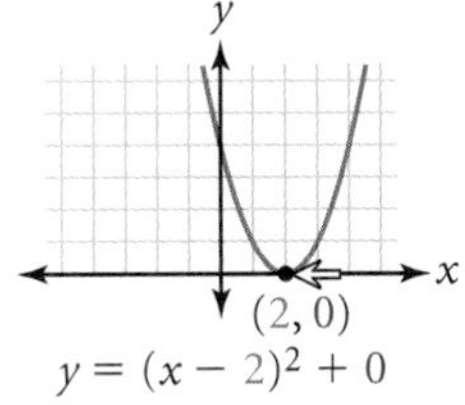

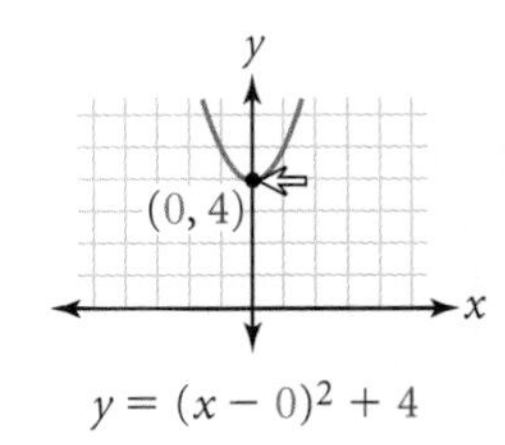

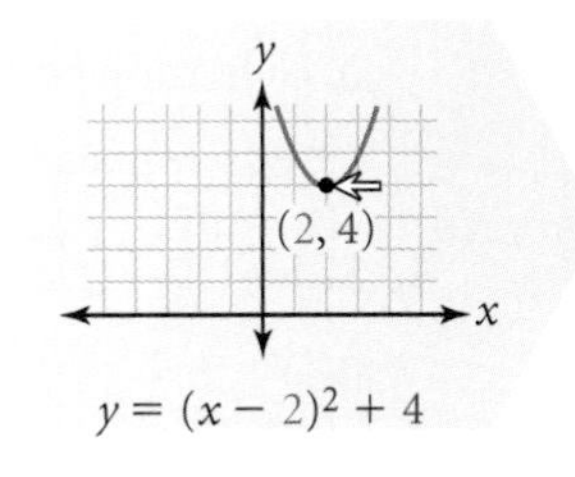

[▶ You can explore translations of parabolas using the **Dynamic Algebra Exploration** at www.keymath.com/DAA .◀]

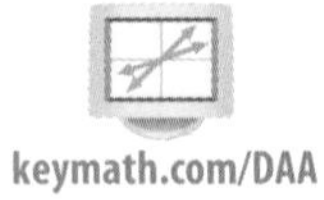
keymath.com/DAA

EXERCISES

You will need

A graphing calculator for Exercise **16.**

Practice Your Skills

1. Write an equation for each parabola. Each parabola is a translation of the graph of the parent function $y = x^2$.

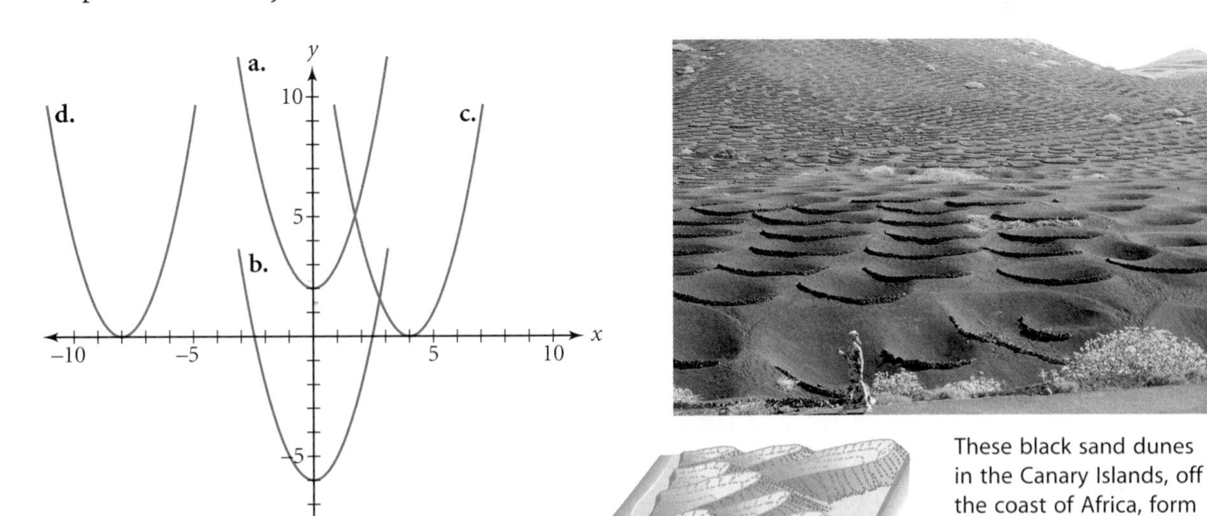

These black sand dunes in the Canary Islands, off the coast of Africa, form parabolic shapes called deflation hollows.

2. Each parabola described is congruent to the graph of $y = x^2$. Write an equation for each parabola and sketch its graph.

a. The parabola is translated vertically -5 units. ⓐ

b. The parabola is translated vertically 3 units.

c. The parabola is translated horizontally 3 units. ⓐ

d. The parabola is translated horizontally -4 units.

3. If $f(x) = x^2$, then the graph of each equation below is a parabola. Describe the location of the parabola relative to the graph of $f(x) = x^2$.

a. $y = f(x) - 3$

b. $y = f(x) + 4$

c. $y = f(x - 2)$

d. $y = f(x + 4)$ ⓐ

4. Describe what happens to the graph of $y = x^2$ in the following situations.

a. x is replaced with $(x - 3)$.

b. x is replaced with $(x + 3)$.

c. y is replaced with $(y - 2)$.

d. y is replaced with $(y + 2)$.

5. Solve. ⓗ

a. $x^2 = 4$

b. $x^2 + 3 = 19$ ⓐ

c. $(x - 2)^2 = 25$

Reason and Apply

6. Write an equation for each parabola at right.

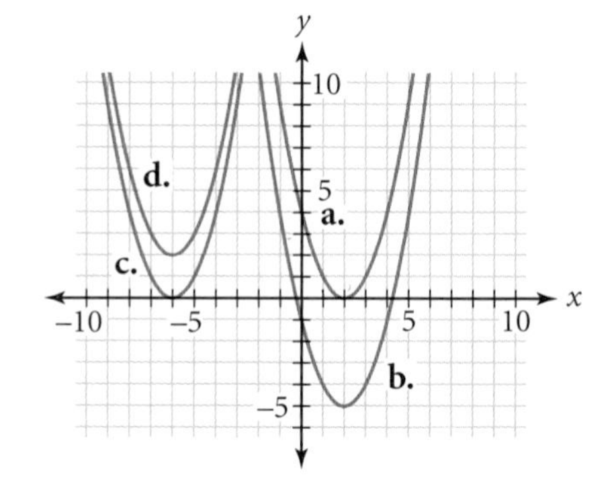

7. The red parabola below is the image of the graph of $y = x^2$ after a horizontal translation of 5 units and a vertical translation of -3 units.

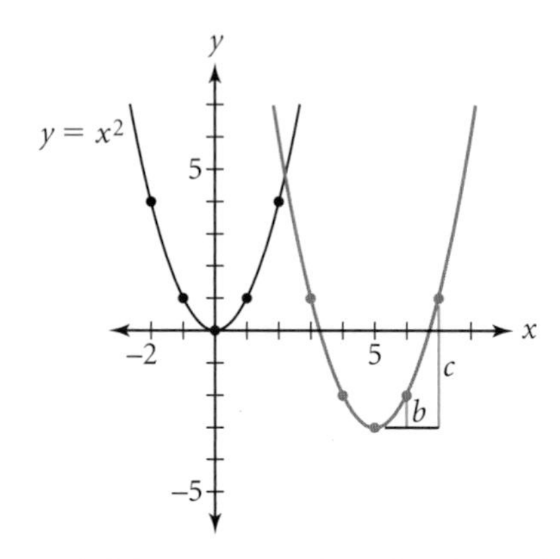

a. Write an equation for the red parabola.

b. Where is the vertex of the red parabola?

c. What are the coordinates of the other four points if they are 1 or 2 horizontal units from the vertex? How are the coordinates of each point on the black parabola related to the coordinates of the corresponding point on the red parabola? ⓐ

d. What is the length of blue segment b? Of green segment c? ⓐ

8. Given the graph of $y = f(x)$ at right, draw a graph of each of these related functions.

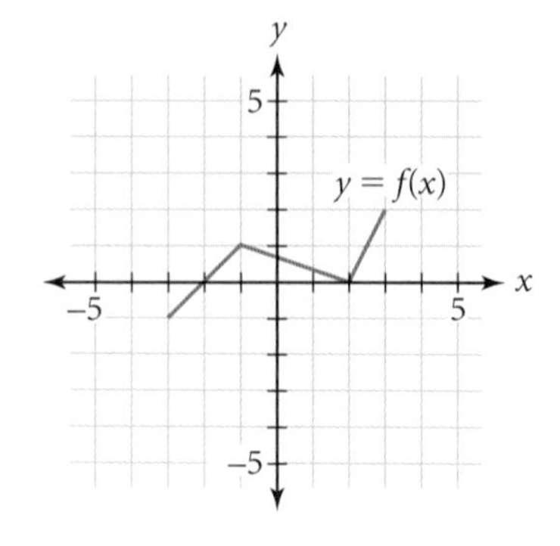

a. $y = f(x + 2)$

b. $y = f(x - 1) - 3$

9. APPLICATION This table of values compares the number of teams in a pee wee teeball league and the number of games required for each team to play every other team twice (once at home and once away from home). ⓗ

Number of teams (x)	1	2	3	...
Number of games (y)	0	2	6	...

a. Continue the table out to 10 teams.

b. Plot each point and describe the graph produced.

c. Write an explicit function for this graph.

d. Use your function to find how many games are required if there are 30 teams.

10. Solve.

a. $3 + (x - 5)^2 = 19$ ⓐ

b. $(x + 3)^2 = 49$

c. $5 - (x - 1)^2 = -22$

d. $-15 + (x + 6)^2 = -7$

11. This histogram shows the students' scores on a recent quiz in Ms. Noah's class. Describe what the histogram will look like if Ms. Noah

a. adds five points to everyone's score.

b. subtracts ten points from everyone's score.

Review

12. Match each recursive formula with the equation of the line that contains the sequence of points, (n, u_n), generated by the formula.

a. $u_0 = -8$
$u_n = u_{(n-1)} + 3$ where $n \geq 1$

b. $u_1 = 3$
$u_n = u_{(n-1)} - 8$ where $n \geq 2$

A. $y = 3x - 11$

B. $y = 3x - 8$

C. $y = 11 - 8x$

D. $y = -8x + 3$

13. APPLICATION You need to rent a car for one day. Mertz Rental charges \$32 per day plus \$0.10 per mile. Saver Rental charges \$24 per day plus \$0.18 per mile. Luxury Rental charges \$51 per day with unlimited mileage.

a. Write a cost equation for each rental agency.

b. Graph the three equations on the same axes.

c. Describe which rental agency is the cheapest alternative under various circumstances.

14. A car drives at a constant speed along the road pictured at right from point A to point X. Sketch a graph showing the straight line distance between the car and point X as it travels along the road. Mark points A, B, C, D, E, and X on your graph. @

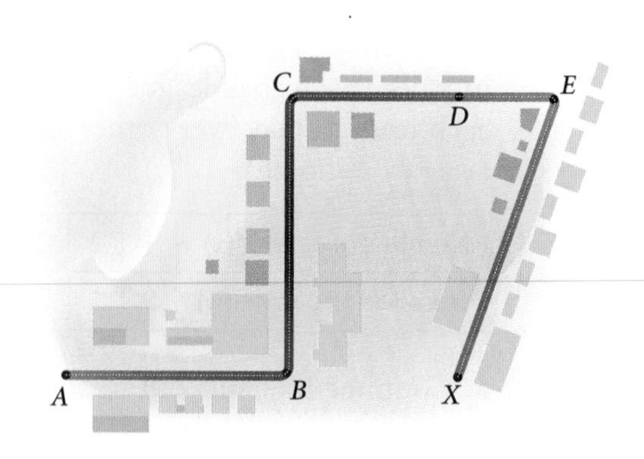

15. The distance between a walker and a stationary observer is shown at right.

a. Describe the actions of the walker.

b. What does the equation $3.8 - 0.84(x - 1.2) = 2$ mean in the context of the graph?

c. Solve the equation from 15b and interpret your solution.

16. Use a graphing calculator to investigate the form $y = ax + b$ of a linear function.

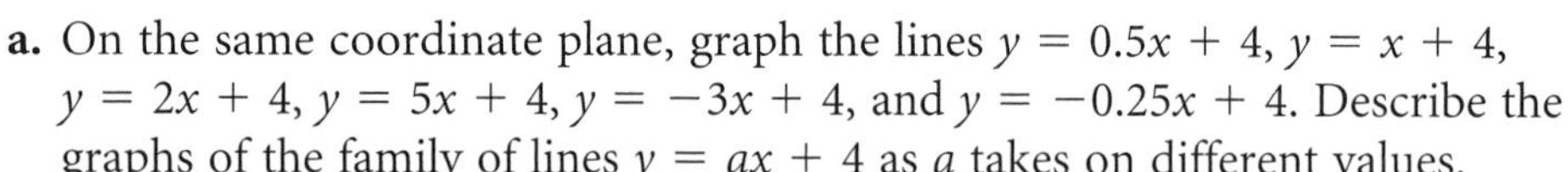

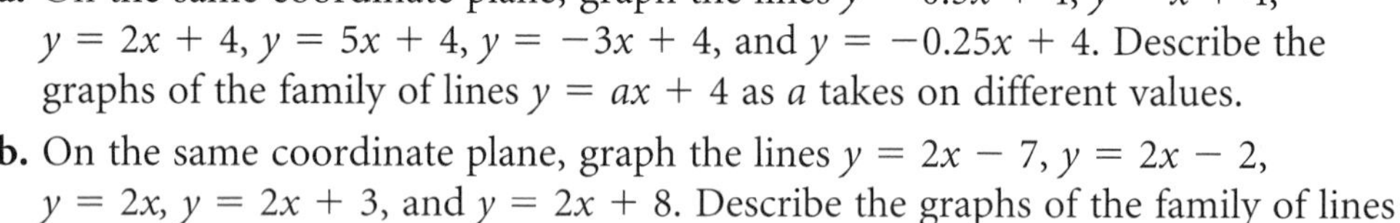

a. On the same coordinate plane, graph the lines $y = 0.5x + 4$, $y = x + 4$, $y = 2x + 4$, $y = 5x + 4$, $y = -3x + 4$, and $y = -0.25x + 4$. Describe the graphs of the family of lines $y = ax + 4$ as a takes on different values.

b. On the same coordinate plane, graph the lines $y = 2x - 7$, $y = 2x - 2$, $y = 2x$, $y = 2x + 3$, and $y = 2x + 8$. Describe the graphs of the family of lines $y = 2x + b$ as b takes on different values.

IMPROVING YOUR REASONING SKILLS

The Dipper

The group of stars known as the Big Dipper, which is part of the constellation Ursa Major, contains stars at various distances from Earth. Imagine translating the Big Dipper to a new position. Would all of the stars need to be moved the same distance? Why or why not?

Now imagine rotating the Big Dipper around the Earth. Do all the stars need to be moved the same distance? Why or why not?

LESSON

4.5

Reflections and the Square Root Family

Call it a clan, call it a network, call it a tribe, call it a family. Whatever you call it, whoever you are, you need one.

JANE HOWARD

The graph of the **square root function,** $y = \sqrt{x}$, is another parent function that you can use to illustrate transformations. From the graphs below, what are the domain and range of $f(x) = \sqrt{x}$? If you graph $y = \sqrt{x}$ on your calculator, you can show that $\sqrt{3}$ is approximately 1.732. What is the approximate value of $\sqrt{8}$? How would you use the graph to find $\sqrt{31}$? What happens when you try to find $f(x)$ for values of $x < 0$?

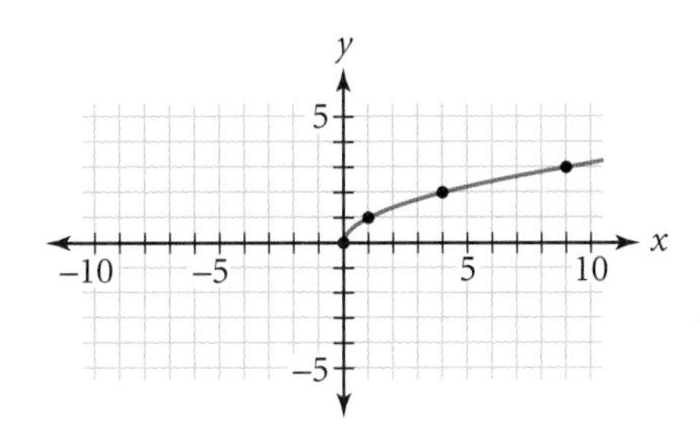

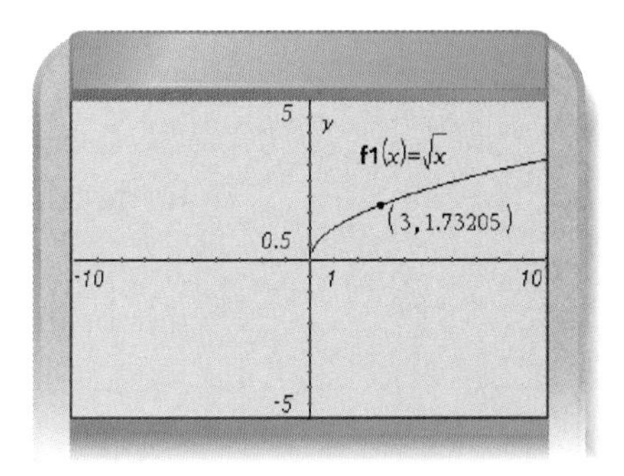

Investigation

Take a Moment to Reflect

In this investigation you first will work with linear functions to discover how to create a new transformation—a **reflection.** Then you will apply reflections to quadratic functions and square root functions.

Step 1 Graph $f_1(x) = 0.5x + 2$ on your calculator.

a. Predict what the graph of $-f_1(x)$ will look like. Then check your prediction by graphing $f_2(x) = -f_1(x)$.

b. Change f_1 to $f_1(x) = -2x - 4$, and repeat the instructions in Step 1a.

c. Change f_1 to $f_1(x) = x^2 + 1$ and repeat.

d. In general, how are the graphs of $y = f(x)$ and $y = -f(x)$ related?

Step 2 Graph $f_1(x) = 0.5x + 2$ on your calculator.

a. Predict what the graph of $f_1(-x)$ will look like. Then check your prediction by graphing $f_2(x) = f_1(-x)$.

b. Change f_1 to $f_1(x) = -2x - 4$, and repeat the instructions in Step 2a.

c. Change f_1 to $f_1(x) = x^2 + 1$ and repeat. Explain what happens.

d. Change f_1 to $f_1(x) = (x - 3)^2 + 2$ and repeat.

e. In general, how are the graphs of $y = f(x)$ and $y = f(-x)$ related?

Step 3 Graph $f_1(x) = \sqrt{x}$ on your calculator.

a. Predict what the graphs of $f_2 = -f_1(x)$ and $f_3 = f_1(-x)$ will look like. Use your calculator to verify your predictions. Write equations for both of these functions in terms of x.

b. Predict what the graph of $f_4 = -f_1(-x)$ will look like. Use your calculator to verify your prediction.

c. Notice that the graph of the square root function looks like half of a parabola, oriented horizontally. Why isn't it an entire parabola? What function would you graph to complete the bottom half of the parabola?

Reflections over the x- or y-axis are summarized below.

Reflection of a Function

A **reflection** is a transformation that flips a graph across a line, creating a mirror image.

Given the graph of $y = f(x)$,

the graph of $y = f(-x)$ is a horizontal reflection across the y-axis, and the graph of $-y = f(x)$, or $y = -f(x)$, is a vertical reflection across the x-axis.

Because the graph of the square root function looks like half a parabola, it's easy to see the effects of reflections. The square root family has many real-world applications, such as dating prehistoric artifacts, as discussed in the Science Connection below.

The next example shows how you can build a **piecewise function** by choosing particular domains for functions you have previously studied.

Science CONNECTION

Obsidian, a natural volcanic glass, was a popular material for tools and weapons in prehistoric times because it makes a very sharp edge. In 1960, scientists Irving Friedman and Robert L. Smith discovered that obsidian absorbs moisture at a slow, predictable rate and that measuring the thickness of the layer of moisture with a high-power microscope helps determine its age. Therefore, obsidian hydration dating can be used on obsidian artifacts, just as carbon dating can be used on organic remains. The age of prehistoric artifacts is predicted by a square root function similar to $d = \sqrt{5t}$, where t is time in thousands of years and d is the thickness of the layer of moisture in microns (millionths of a meter).

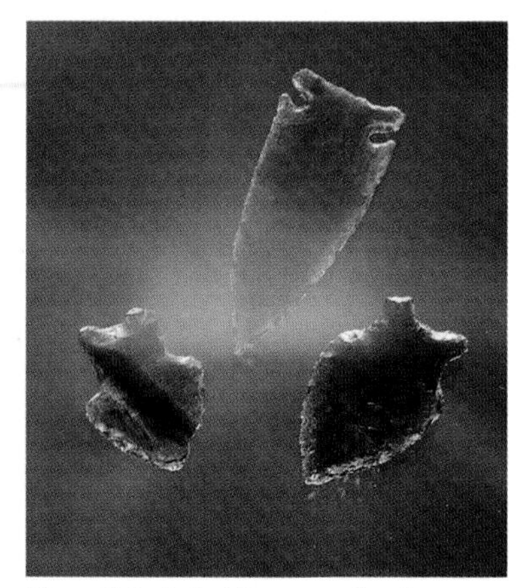

These flaked obsidian arrowheads—once used for cutting, carving, and hunting—were made by Native Americans near Jackson Lake, Wyoming, more than 8500 years ago.

EXAMPLE

A piecewise function is a function that consists of two or more ordinary functions defined on different domains.

a. Graph $f(x) = \begin{cases} 2x & -3 \leq x \leq 0 \\ \sqrt{x} & 0 < x \leq 4 \end{cases}$

b. Find an equation for the piecewise function pictured at right.

▶ Solution

a. The graph of the first part is a line with intercept 0 and slope 2. It is defined for x-values between -3 and 0, so sketch the line but keep only the segment from $(-3, -6)$ to $(0, 0)$.

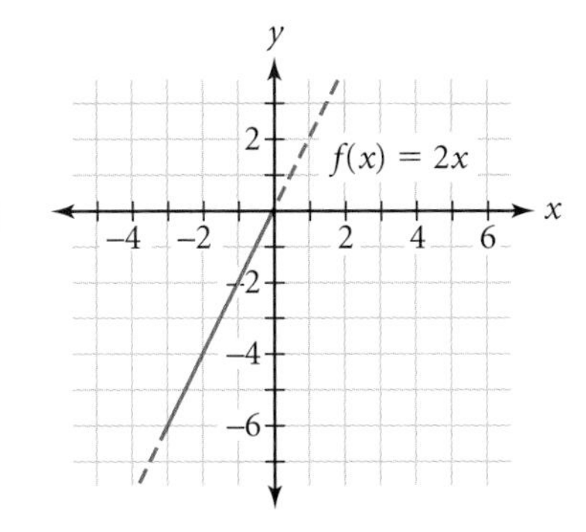

The second part of the function is a square root function. This part is defined for $0 < x \leq 4$. Graph the function over this domain. [▶ See **Calculator Note 4E** to learn about graphing piecewise equations on your calculator. ◀] This completes the graph of $f(x)$.

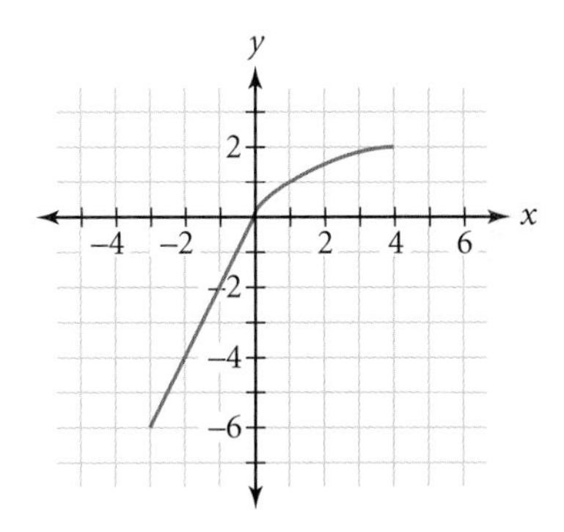

b. The graph has two pieces. The left piece appears to be a transformation of the square root parent function $f(x) = \sqrt{x}$. The parent function has been reflected horizontally across the y-axis and translated vertically 1 unit. Starting with $f(x) = \sqrt{x}$, a horizontal reflection of the function is $y = f(-x)$, or $y = \sqrt{-x}$. To translate this function vertically 1 unit, replace y with $y - 1$. This gives the equation $y - 1 = f(-x)$, or $y = 1 + \sqrt{-x}$. The domain for this piece of the function is $-4 \leq x \leq 0$.

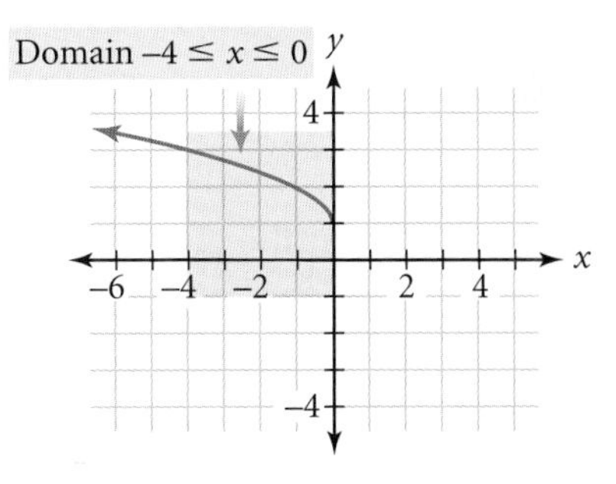

The right piece is a parabola that has been reflected vertically over the x-axis, and translated horizontally 1 unit and vertically 2 units. Applying these transformations to the parent function $g(x) = x^2$ gives the equation $y - 2 = -g(x - 1)$, or $y = -(x - 1)^2 + 2$. The domain is $0 < x \leq 3$. Combining the two pieces, you can represent the piecewise function as

$$y = \begin{cases} 1 + \sqrt{-x} & -4 \leq x \leq 0 \\ -(x - 1)^2 + 2 & 0 < x \leq 3 \end{cases}$$

Notice that even though the two pieces meet at $x = 0$, you include 0 in only one domain piece. It doesn't matter which piece, but it should not be included in both.

You've seen in previous lessons that you can transform complicated graphs without knowing their equations. However, writing the equations of piecewise graphs can give you practice working with transformations of the families of graphs you are studying in this chapter, as well as more practice working with domain and range.

EXERCISES

You will need

A graphing calculator for Exercises 7 and 13.

Practice Your Skills

1. Each graph at right is a transformation of the graph of the parent function $y = \sqrt{x}$. Write an equation for each graph. ⓐ

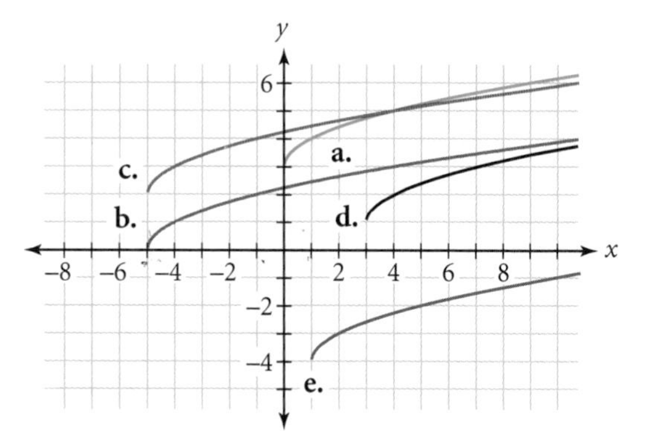

2. Describe what happens to the graph of $y = \sqrt{x}$ in the following situations.

a. x is replaced with $(x - 3)$. ⓐ

b. x is replaced with $(x + 3)$.

c. y is replaced with $(y - 2)$. ⓐ

d. y is replaced with $(y + 2)$.

3. Each graph at right is a transformation of the piecewise function $f(x)$. Match each equation to a graph.

a. $y = f(-x)$

b. $y = -f(x)$

c. $y = -f(-x)$

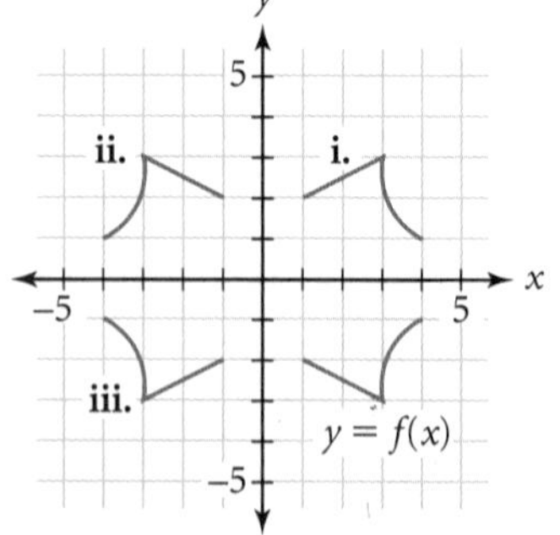

4. Given the graph of $y = f(x)$ below, draw a graph of each of these related functions.

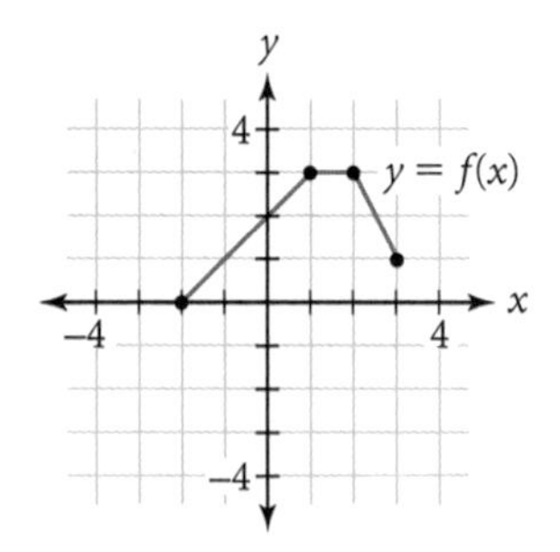

a. $y = f(-x)$ ⓐ **b.** $y = -f(x)$ **c.** $y = -f(-x)$

5. Each curve at right is a transformation of the graph of the parent function $y = \sqrt{x}$. Write an equation for each curve. ⓐ

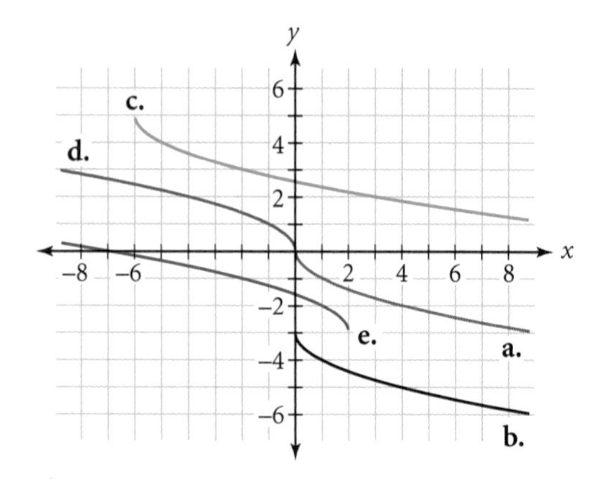

Reason and Apply

6. Consider the parent function $f(x) = \sqrt{x}$.

a. Name three pairs of integer coordinates that are on the graph of $y = f(x + 4) - 2$.

b. Write $y = f(x + 4) - 2$ using a **radical,** or square root symbol, and graph it.

c. Write $y = -f(x - 2) + 3$ using a radical, and graph it.

7. Consider the parabola at right.

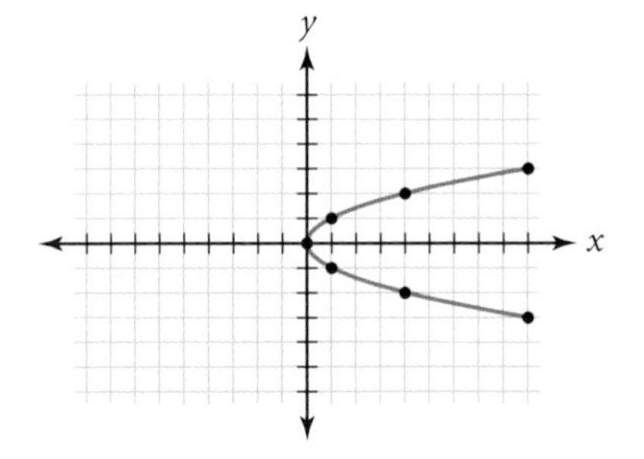

a. Graph the parabola on your calculator. What two functions did you use?

b. Combine both functions from 7a using $\pm$ notation to create a single relation. Square both sides of the relation. What is the resulting equation?

8. Refer to the two parabolas at right.

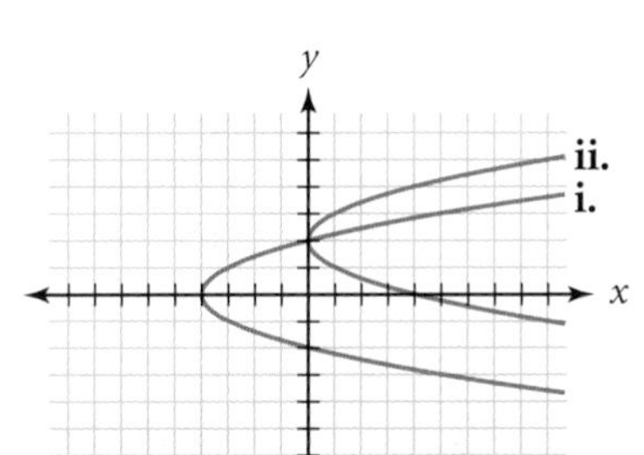

a. Explain why neither graph represents a function.

b. Write a single equation for each parabola using $\pm$ notation. ⓐ

c. Square both sides of each equation in 8b. What is the resulting equation of each parabola?

9. As Jake and Arthur travel together from Detroit to Chicago, each makes a graph relating time and distance. Jake, who lives in Detroit and keeps his watch on Detroit time, graphs his distance from Detroit. Arthur, who lives in Chicago and keeps his watch on Chicago time (1 hour earlier than Detroit), graphs his distance from Chicago. They both use the time shown on their watches for their x-axes. The distance between Detroit and Chicago is 250 miles.

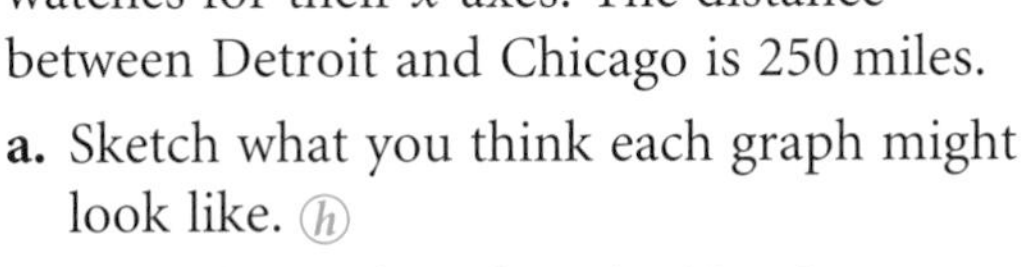

a. Sketch what you think each graph might look like. ⓗ

b. If Jake's graph is described by the function $y = f(x)$, what function describes Arthur's graph?

c. If Arthur's graph is described by the function $y = g(x)$, what function describes Jake's graph?

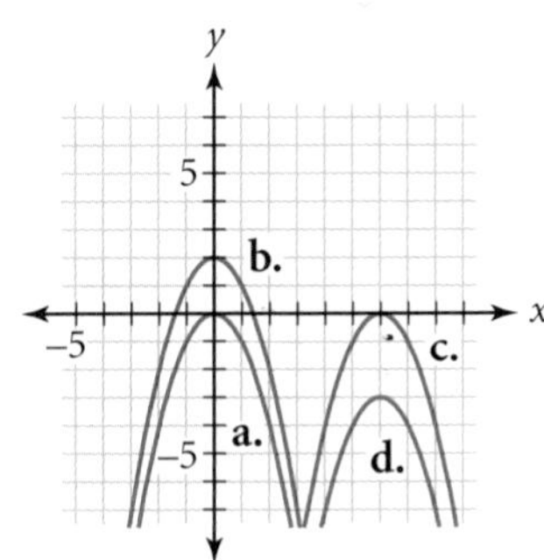

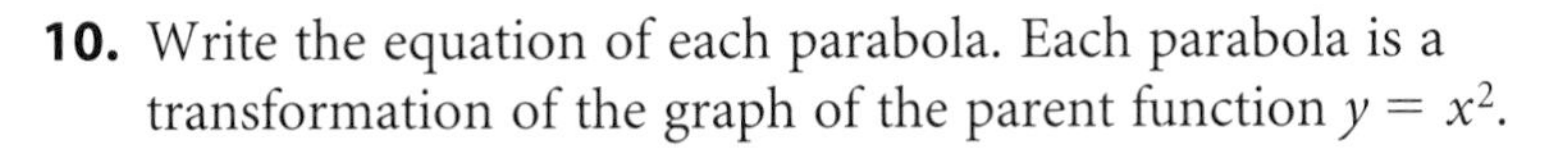

10. Write the equation of each parabola. Each parabola is a transformation of the graph of the parent function $y = x^2$.

11. Write the equation of a parabola that is congruent to the graph of $y = -(x + 3)^2 + 4$, but translated right 5 units and down 2 units.

12. Let $f(x)$ be defined as the piecewise function graphed at right, and let $g(x)$ be defined as

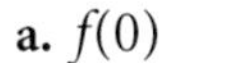
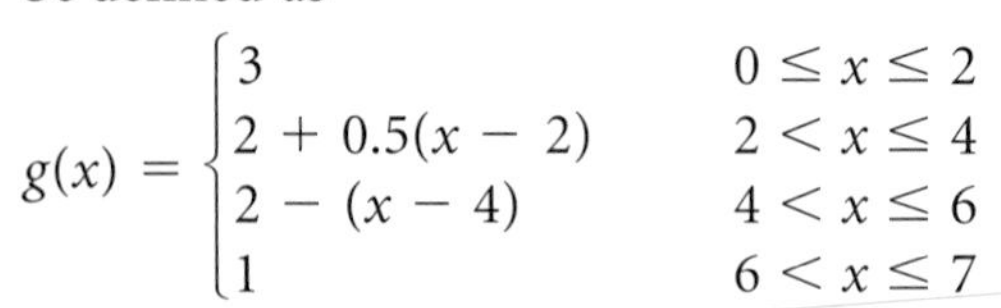

$$g(x) = \begin{cases} 3 & 0 \le x \le 2 \\ 2 + 0.5(x - 2) & 2 < x \le 4 \\ 2 - (x - 4) & 4 < x \le 6 \\ 1 & 6 < x \le 7 \end{cases}$$

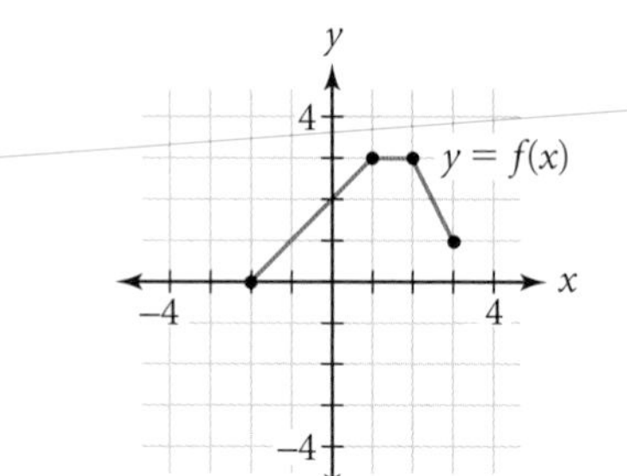

Find each value.

a. $f(0)$ b. x when $f(x) = 0$ c. x when $f(x) = 1$

d. $g(1.8)$ e. $g(2)$ ⓗ f. $g(4)$

g. $g(6.999)$

13. APPLICATION Police measure the lengths of skid marks to determine the initial speed of a vehicle before the brakes were applied. Many variables, such as the type of road surface and weather conditions, play an important role in determining the speed. The formula used to determine the initial speed is $S = 5.5\sqrt{D \cdot f}$, where S is the speed in miles per hour, D is the average length of the skid marks in feet, and f is a constant called the "drag factor." At a particular accident scene, assume it is known that the road surface has a drag factor of 0.7.

a. Write an equation that will determine the initial speed on this road as a function of the lengths of skid marks.

b. Sketch a graph of this function.

c. If the average length of the skid marks is 60 feet, estimate the initial speed of the car when the brakes were applied.

d. Solve your equation from 13a for D. What can you determine using this equation?

e. Graph your equation from 13d. What shape is it? ⓐ

f. If you traveled on this road at a speed of 65 miles per hour and suddenly slammed on your brakes, how long would your skid marks be?

Review

14. Identify each relation that is also a function. For each relation that is not a function, explain why not.

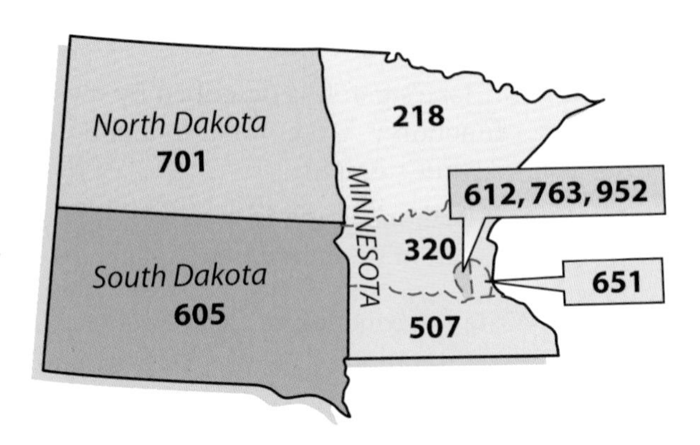

a. independent variable: state
dependent variable: area code

b. independent variable: any pair of whole numbers
dependent variable: their greatest common factor

c. independent variable: any pair of fractions
dependent variable: their common denominator

d. independent variable: the day of the year
dependent variable: the time of sunrise

15. Solve for x. Solving square root equations often results in **extraneous solutions,** or answers that don't work in the original equation, so be sure to check your work.

a. $3 + \sqrt{x - 4} = 20$ @

b. $\sqrt{x + 7} = -3$

c. $4 - (x - 2)^2 = -21$

d. $5 - \sqrt{-(x + 4)} = 2$ @

16. Find the equation of the parabola with vertex $(-6, 4)$, a vertical line of symmetry, and containing the point $(-5, 5)$.

17. The graph of the line ℓ_1 is shown at right.

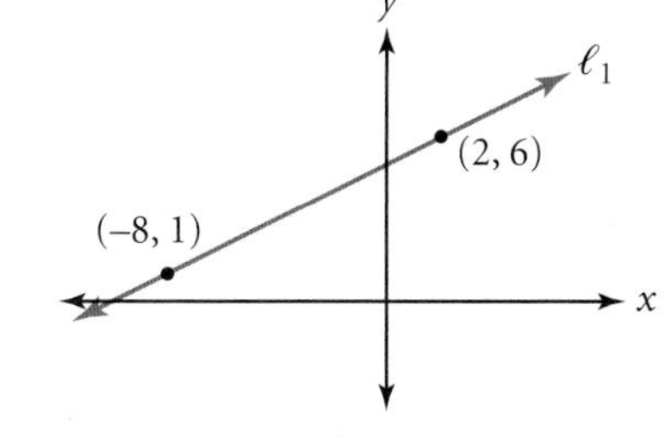

a. Write the equation of the line ℓ_1.

b. The line ℓ_2 is the image of the line ℓ_1 translated right 8 units. Sketch the line ℓ_2 and write its equation in a way that shows the horizontal translation. @

c. The line ℓ_2 also can be thought of as the image of the line ℓ_1 after a vertical translation. Write the equation of the line ℓ_2 in a way that shows the vertical translation.

d. Show that the equations in 17b and c are equivalent.

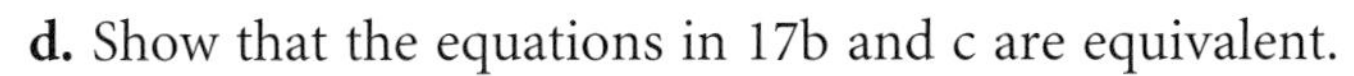

18. Consider this data set:

{37, 40, 36, 37, 37, 49, 39, 47, 40, 38, 35, 46, 43, 40, 47, 49, 70, 65, 50, 73} @

a. Give the five-number summary.

b. Display the data in a box plot.

c. Find the interquartile range.

d. Identify any outliers, based on the interquartile range.

19. Find the intersection of the lines $2x + y = 23$ and $3x - y = 17$.

IMPROVING YOUR GEOMETRY SKILLS

Lines in Motion Revisited

Imagine that a line is translated in a direction perpendicular to it, creating a parallel line. What vertical and horizontal translations would be equivalent to the translation along the perpendicular path? Find the slope of each line pictured. How does the ratio of the translations compare to the slope of the lines? Find answers both for the specific lines shown and, more generally, for any pair of parallel lines.

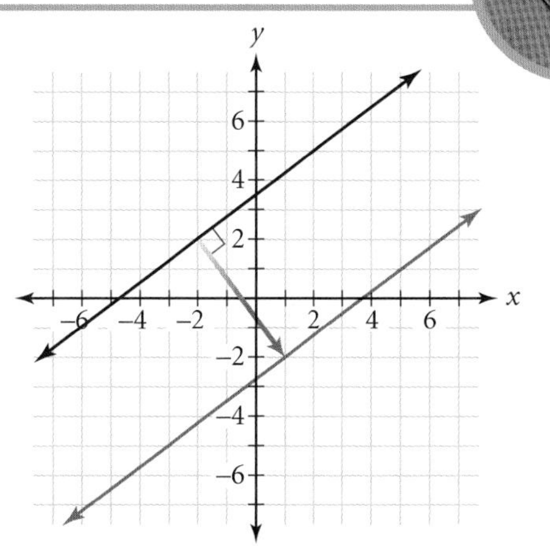

EXPLORATION

Rotation as a Composition of Transformations

keymath.com/DAA

You have learned rules that reflect and translate figures and functions on the coordinate plane. Is it possible to rotate figures on the coordinate plane using a rule? You will explore that question in this activity. When one transformation is followed by another, the resulting transformation is called a **composition** of the two transformations. In this activity you'll also explore how a composition of transformations can be equivalent to a single transformation.

Activity

Revolution

Step 1 Draw a figure using geometry software. Your figure should be nonsymmetric so that you can see the effects of various geometric transformations.

Step 2 Rotate your figure about the origin three times: once by 90° counterclockwise, once by 90° clockwise, and once by 180°. Change your original figure to a different color.

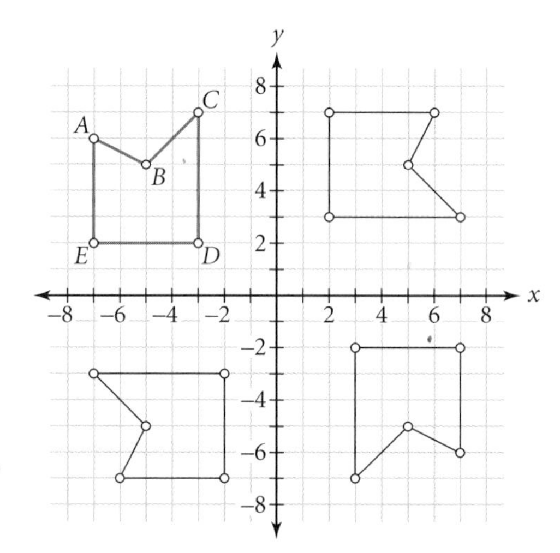

Step 3 Transform your original figure onto each of the three images using a composition of only reflections and translations. (You may use other lines of reflection besides the axes.) Keep track of the transformations you use. Find at least two different compositions of transformations that map the figure onto each of the three images.

Questions

1. Describe the effects of each rotation on the coordinates of the figure. Give a rule that describes the transformation of the x-coordinates and the y-coordinates for each of the three rotations. Do the rules change if your original figure is in a different quadrant?

2. Choose one of the compositions of transformations you found in Step 3. For each individual transformation you performed, explain the effect on the x- and y-coordinates. Show how the composition of these transformations confirms the rule you found by answering Question 1.

LESSON

4.6

Dilations and the Absolute-Value Family

A mind that is stretched by a new experience can never go back to its old dimensions.

OLIVER WENDELL HOLMES

Hao and Dayita ride the subway to school each day. They live on the same east-west subway route. Hao lives 7.4 miles west of the school, and Dayita lives 5.2 miles east of the school. This information is shown on the number line below.

H (Hao) *S* (School) *D* (Dayita)

West East

−7.4 mi 0 5.2 mi

The distance between two points is always positive. However, if you calculate Hao's distance from school, or *HS*, by subtracting his starting position from his ending position, you get a negative value:

$$-7.4 - 0 = -7.4$$

In order to make the distance positive, you use the absolute-value function, which gives the **magnitude** of a number, or its distance from zero on a number line. For example, the absolute value of -3 is 3, or $|-3| = 3$. For Hao's distance from school, you use the absolute-value function to calculate

$$HS = |-7.4 - 0| = |-7.4| = 7.4$$

What is the distance from *D* to *H*? What is the distance from *H* to *D*?

In this lesson you will explore transformations of the graph of the parent function $y = |x|$. [▶ See **Calculator Note 4F** to learn how to graph the absolute-value function. ◀] You will write and use equations in the form $\frac{y-k}{b} = \left|\frac{x-h}{a}\right|$. What you have learned about translating and reflecting other graphs will apply to these functions as well. You will also learn about transformations called **dilations** that stretch and shrink a graph.

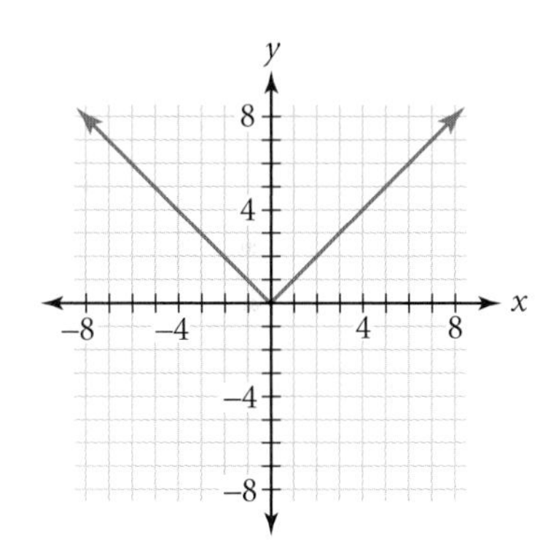

You may have learned about dilations of geometric figures in an earlier course. Now you will apply dilations to functions.

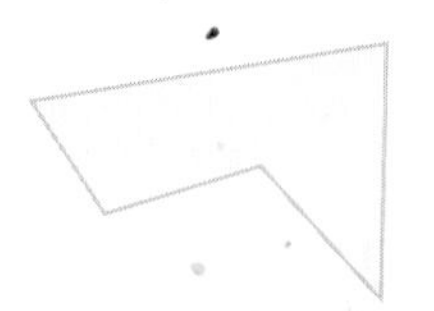

If you dilate a figure by the same **scale factor** both vertically and horizontally, then the image and the original figure will be similar and perhaps congruent.

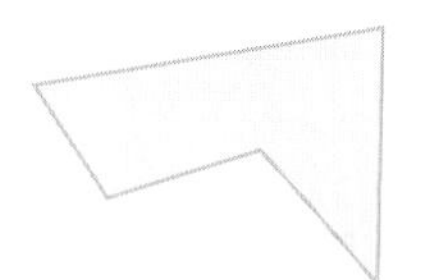

If you dilate by different vertical and horizontal scale factors, then the image and the original figure will not be similar.

EXAMPLE A

Graph the function $y = |x|$ with each of these functions. How does the graph of each function compare to the original graph?

a. $\frac{y}{2} = |x|$

b. $y = \left|\frac{x}{3}\right|$

c. $\frac{y}{2} = \left|\frac{x}{3}\right|$

▶ Solution

In the graph of each function, the vertex remains at the origin. Notice, however, how the points (1, 1) and (−2, 2) on the parent function are mapped to a new location.

a. Replacing y with $\frac{y}{2}$ pairs each x-value with twice the corresponding y-value in the parent function. The graph of $\frac{y}{2} = |x|$ is a vertical stretch, or a **vertical dilation,** of the graph of $y = |x|$ by a factor of 2.

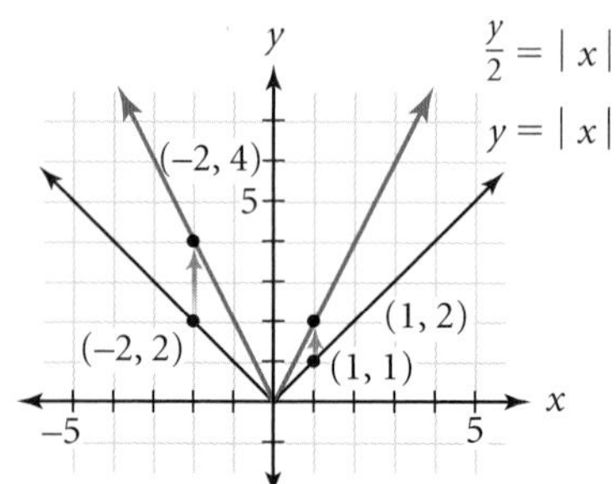

b. Replacing x with $\frac{x}{3}$ multiplies the x-coordinates by a factor of 3. The graph of $y = \left|\frac{x}{3}\right|$ is a horizontal stretch, or a **horizontal dilation,** of the graph of $y = |x|$ by a factor of 3.

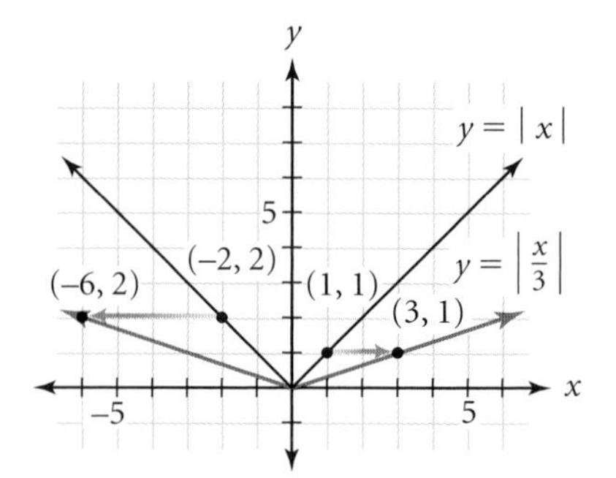

c. The combination of replacing y with $\frac{y}{2}$ and replacing x with $\frac{x}{3}$ results in a vertical dilation by a factor of 2 *and* a horizontal dilation by a factor of 3.

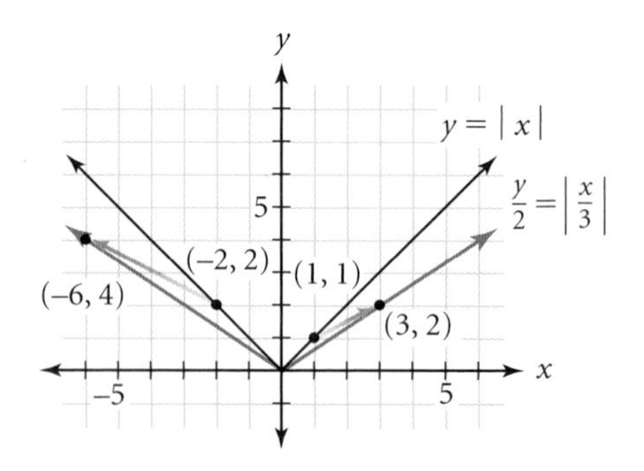

Translations and reflections are **rigid transformations**—they produce an image that is congruent to the original figure. Vertical and horizontal dilations are **nonrigid transformations**—the image is not congruent to the original figure (unless you use a factor of 1 or −1).

Using what you know about translations, reflections, and dilations, you can fit functions to data by locating only a few key points. For quadratic, square root, and absolute-value functions, first locate the vertex of the graph. Then use any other point to find the factors by which to dilate the image horizontally and/or vertically.

EXAMPLE B These data are from one bounce of a ball. Find an equation that fits the data over this domain.

Time (s) x	Height (m) y
0.54	0.05
0.58	0.18
0.62	0.29
0.66	0.39
0.70	0.46
0.74	0.52
0.78	0.57
0.82	0.59
0.86	0.60

Time (s) x	Height (m) y
0.90	0.59
0.94	0.57
0.98	0.52
1.02	0.46
1.06	0.39
1.10	0.29
1.14	0.18
1.18	0.05

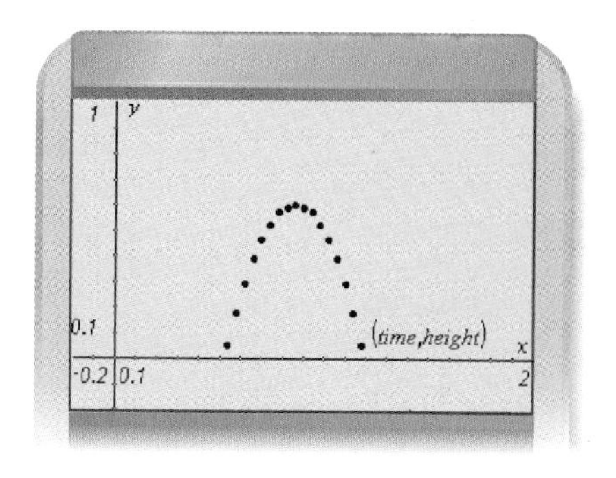

▶ Solution

The graph appears to be a parabola. However, the parent function $y = x^2$ has been reflected, translated, and dilated. Start by determining the translations. The vertex has been translated from (0, 0) to (0.86, 0.60). This is enough information for you to write the equation in the form $y = (x - h)^2 + k$, or $y = (x - 0.86)^2 + 0.60$. If you think of replacing x with $(x - 0.86)$ and replacing y with $(y - 0.60)$, you could also write the equivalent equation, $y - 0.6 = (x - 0.86)^2$.

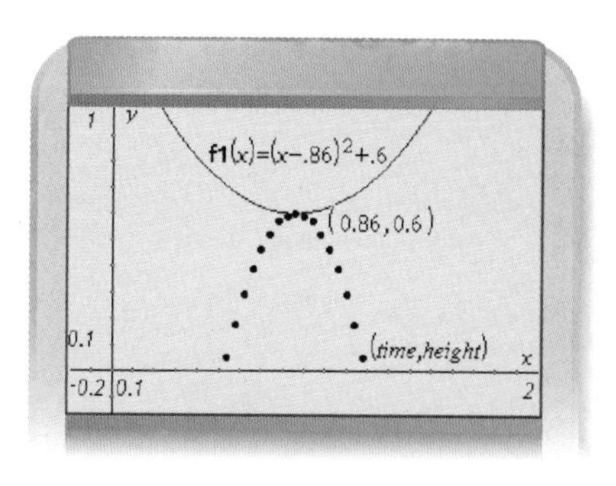

The graph still needs to be reflected and dilated. Select one other data point to determine the horizontal and vertical scale factors. You can use any point, but you will get a better fit if you choose one that is not too close to the vertex. For example, you can choose the data point (1.14, 0.18).

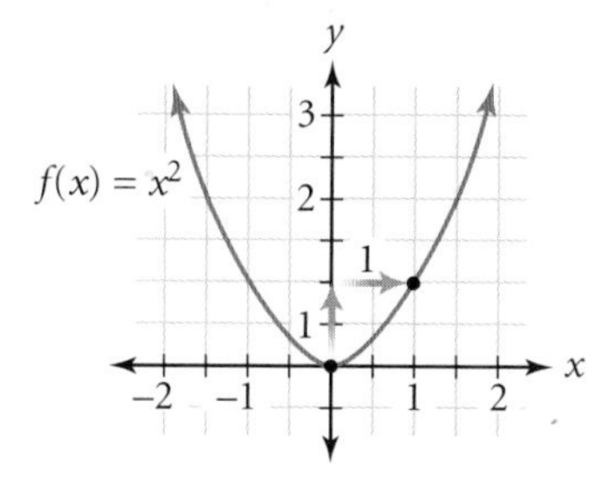

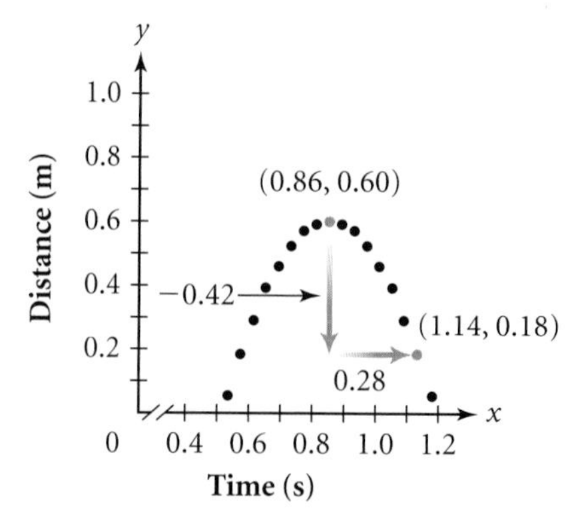

Assume this data point is the image of the point (1, 1) in the parent parabola $y = x^2$. In the graph of $y = x^2$, (1, 1) is 1 unit away from the vertex (0, 0) both horizontally and vertically. The data point we chose in this graph, (1.14, 0.18), is $1.14 - 0.86$, or 0.28, unit away from the x-coordinate of the vertex, and $0.18 - 0.60$, or -0.42, unit away from the y-coordinate of the vertex.

So the horizontal scale factor is 0.28, and the vertical scale factor is −0.42. The negative vertical scale factor also produces a vertical reflection.

Combine these scale factors with the translations to get the final equation

$$\frac{y-0.6}{-0.42}=\left(\frac{x-0.86}{0.28}\right)^2 \quad \text{or} \quad y=-0.42\left(\frac{x-0.86}{0.28}\right)^2+0.6$$

This model, graphed at right, fits the data nicely.

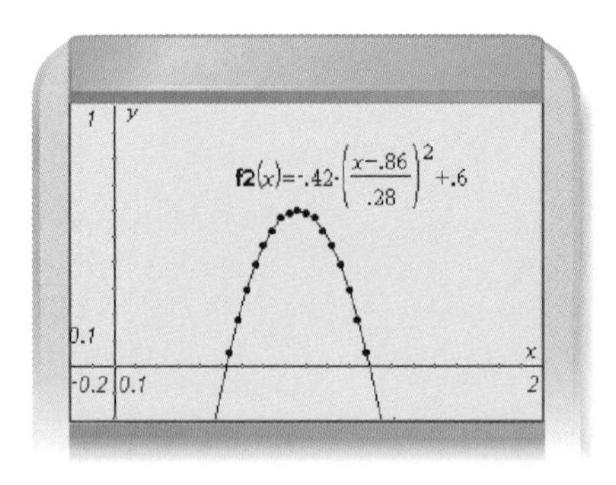

The same procedure works with the other functions you have studied so far. As you continue to add new functions to your mathematical knowledge, you will find that what you have learned about function transformations continues to apply.

keymath.com/DAA

[▶ You can explore vertical and horizontal dilations of absolute-value graphs, parabolas, and cubic graphs using the **Dynamic Algebra Exploration** at www.keymath.com/DAA .◀]

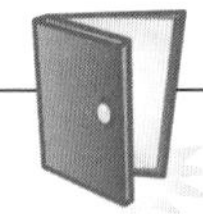

Investigation
The Pendulum

You will need

- string
- a small weight
- a stopwatch or a watch with a second hand

Italian mathematician and astronomer Galileo Galilei (1564–1642) made many contributions to our understanding of gravity, the physics of falling objects, and the orbits of the planets. One of his famous experiments involved the periodic motion of a pendulum. In this investigation you will carry out the same experiment and find a function to model the data.

This fresco, painted in 1841, shows Galileo at age 17, contemplating the motion of a swinging lamp in the Cathedral of Pisa. A swinging lamp is an example of a pendulum.

Step 1 Follow the Procedure Note to find the period of your pendulum. Repeat the experiment for several different string lengths and complete a table of values. Use a variety of short, medium, and long string lengths.

Step 2 Graph the data using *length* as the independent variable. What is the shape of the graph? What do you suppose is the parent function?

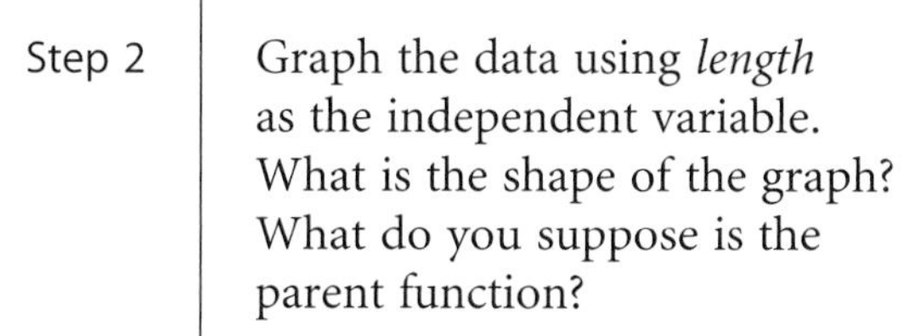

Step 3 The vertex is at the origin, (0, 0). Why do you suppose it is there?

Step 4 Have each member of your group choose a different data point and use that data point to find the horizontal and vertical dilations. Apply these transformations to find an equation to fit the data.

Step 5 Compare the collection of equations from your group. Which points are the best to use to fit the curve? Why do these points work better than others?

Procedure Note

1. Tie a weight at one end of a length of string to make a pendulum. Firmly hold the other end of the string, or tie it to something, so that the weight hangs freely.
2. Measure the length of the pendulum, from the center of the weight to the point where the string is held.
3. Pull the weight to one side and release it so that it swings back and forth in a short arc, about 10° to 20°. Time ten complete swings (forward and back is one swing).
4. The **period** of your pendulum is the time for one complete swing (forward and back). Find the period by dividing by 10.

In the exercises you will use techniques you discovered in this lesson. Remember that replacing y with $\frac{y}{b}$ dilates a graph by a factor of b vertically. Replacing x with $\frac{x}{a}$ dilates a graph by a factor of a horizontally. When graphing a function, you should do dilations before translations to avoid moving the vertex. When finding the equation for a graph, the process is reversed, so you estimate translations first and dilations second, as show in Example B.

Dilation of a Function

A **dilation** is a transformation that expands or compresses a graph either horizontally or vertically.

Given the graph of $y = f(x)$, the graph of

$$\frac{y}{b} = f(x) \quad \text{or} \quad y = bf(x)$$

is a vertical dilation by a factor of b. When $|b| > 1$, it is a stretch; when $0 < |b| < 1$, it is a shrink. When $b < 0$, a reflection across the x-axis also occurs.

Given the graph of $y = f(x)$, the graph of

$$y = f\left(\frac{x}{a}\right)$$

is a horizontal dilation by a factor of a. When $|a| > 1$, it is a stretch; when $0 < |a| < 1$, it is a shrink. When $a < 0$, a reflection across the y-axis also occurs.

EXERCISES

You will need

A graphing calculator for Exercises **5, 6, 12,** and **14.**

Practice Your Skills

1. Each graph is a transformation of the graph of one of the parent functions you've studied. Write an equation for each graph.

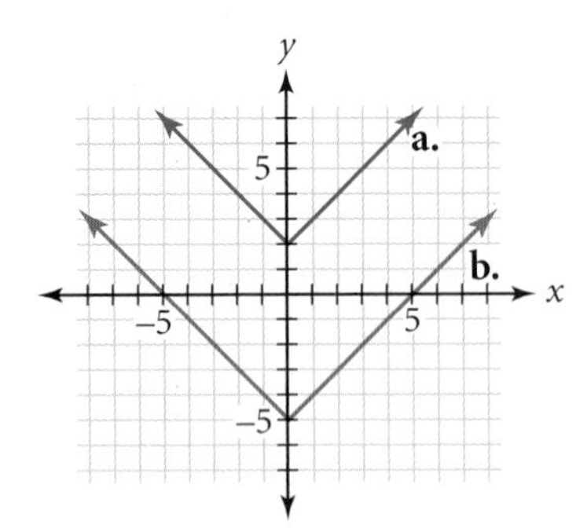

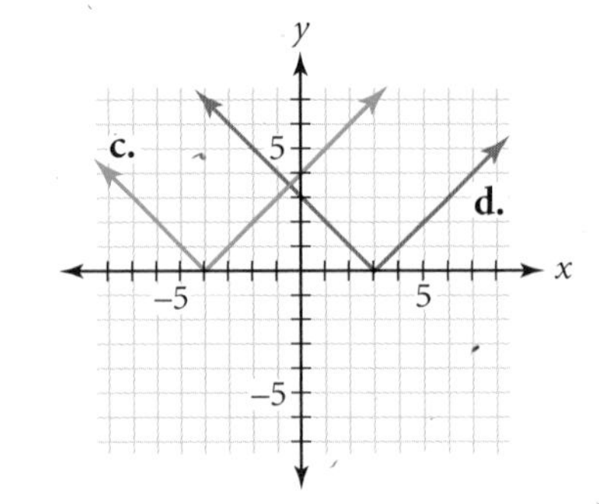

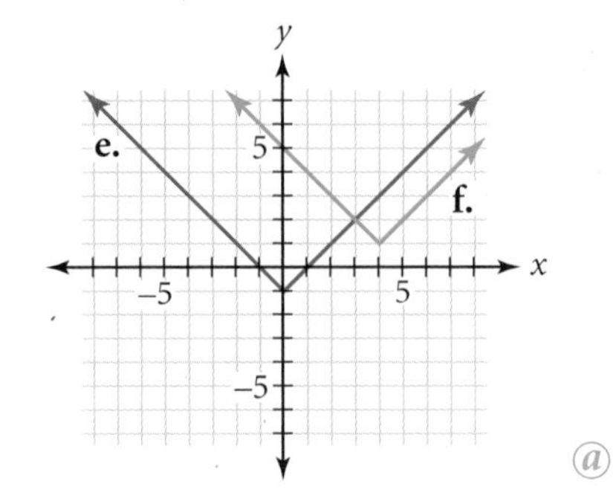

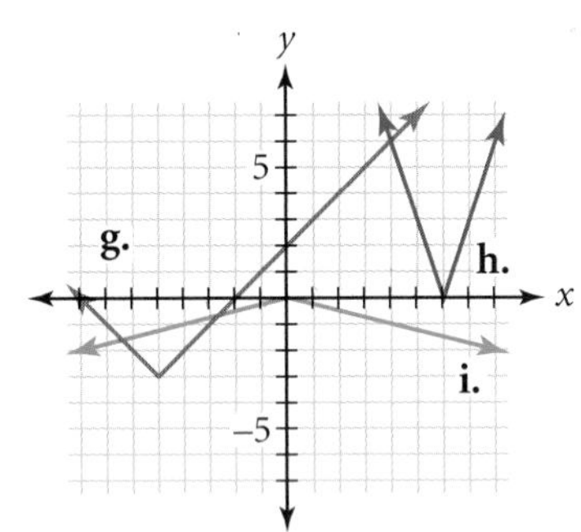

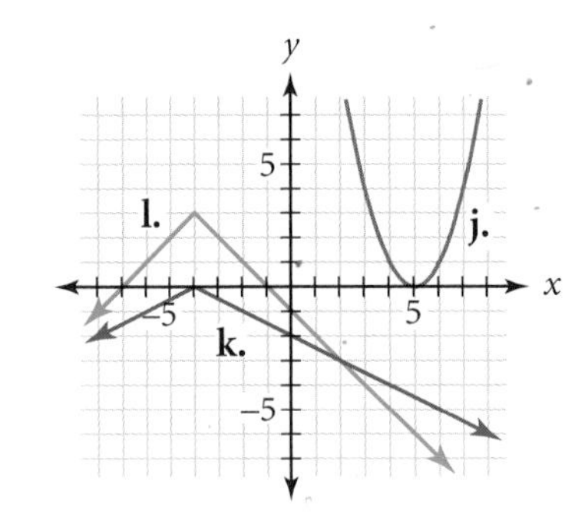

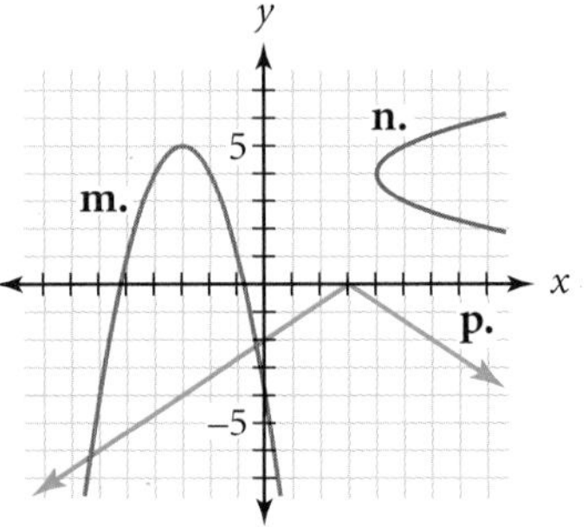

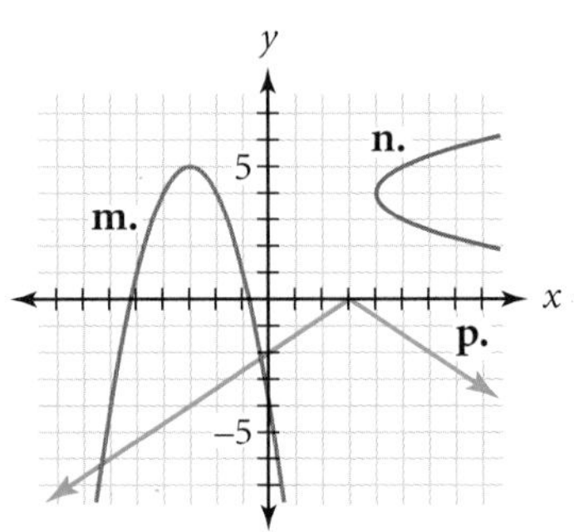

2. Describe what happens to the graph of $y = f(x)$ in these situations.

a. x is replaced with $\frac{x}{3}$.

b. x is replaced with $-x$.

c. x is replaced with $3x$.

d. y is replaced with $\frac{y}{2}$.

e. y is replaced with $-y$.

f. y is replaced with $2y$.

3. Solve each equation for y.

a. $\frac{y+3}{2} = (x-5)^2$

b. $\frac{y+5}{2} = \left|\frac{x+1}{3}\right|$

c. $\frac{y+7}{-2} = \sqrt{\frac{x-6}{-3}}$

Reason and Apply

4. Choose a few different values for b. What can you conclude about $y = b|x|$ and $y = |bx|$? Are they the same function? (h)

5. The graph at right shows how to solve the equation $|x - 4| = 3$ graphically. The equations $y = |x - 4|$ and $y = 3$ are graphed on the same coordinate axes.

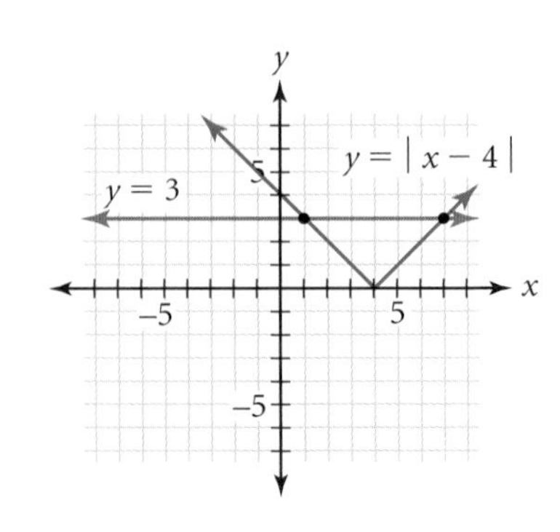

a. What is the x-coordinate of each point of intersection? What x-values are solutions of the equation $|x - 4| = 3$?

b. Solve the equation $|x + 3| = 5$ algebraically. Verify your solution with a graph.

6. **APPLICATION** You can use a single radio receiver to find the distance to a transmitter by measuring the strength of the signal. Suppose these approximate distances are measured with a receiver while you drive along a straight road. Find a model that fits the data. Where do you think the transmitter might be located? ⓐ

Miles traveled	0	4	8	12	16	20	24	28	32	36
Distance from transmitter (miles)	18.4	14.4	10.5	6.6	2.5	1.8	6.0	9.9	13.8	17.6

7. Assume that the parabola $y = x^2$ is translated so that its vertex is $(5, -4)$.

a. If the parabola is dilated vertically by a factor of 2, what are the coordinates of the point on the parabola 1 unit to the right of the vertex? ⓐ

b. If the parabola is dilated horizontally instead, by a factor of 3, what are the coordinates of the points on the parabola 1 unit above the vertex? ⓐ

c. If the parabola is dilated vertically by a factor of 2 *and* horizontally by a factor of 3, name two points on the new parabola that are symmetric with respect to the vertex.

8. Given the parent function $y = x^2$, describe the transformations represented by the function $\frac{y-2}{3} = \left(\frac{x+7}{4}\right)^2$. Sketch a graph of the transformed parabola.

9. A curve with parent function $f(x) = x^2$ has vertex $(7, 3)$ and passes through the point $(11, 11)$. ⓗ

a. What are the values of h and k in the equation of the curve?

b. Substitute the values for h and k from 9a into $y = k + a \cdot f(x - h)$. Substitute the coordinates of the other point into the equation as values for x and y.

c. Solve for a and write the complete equation of the curve. Confirm that the graph passes through both points.

d. Write the equation in the form $\frac{y-k}{b} = \left(\frac{x-h}{a}\right)^2$ by considering the horizontal and vertical dilations separately, as in Example B.

e. Use algebra to show that your answers from 9c and d are equivalent.

10. Sketch a graph of each of these equations.

a. $\frac{y-2}{3} = (x-1)^2$ **b.** $\left(\frac{y+1}{2}\right)^2 = \frac{x-2}{3}$ **c.** $\frac{y-2}{2} = \left|\frac{x+1}{3}\right|$

11. Given the graph of $y = f(x)$, draw graphs of these related functions.

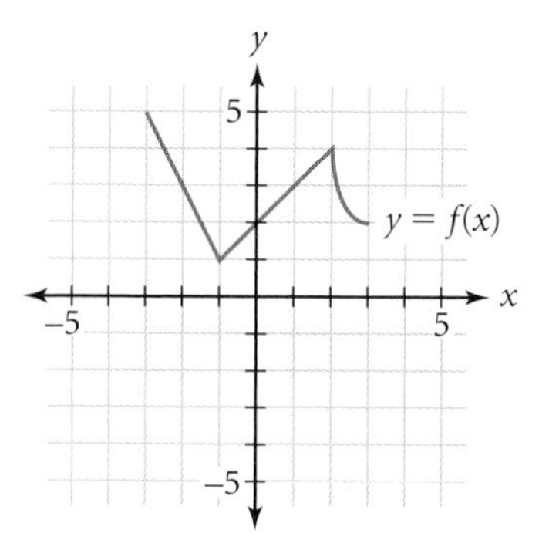

a. $\frac{y}{-2} = f(x)$ **b.** $y = f\left(\frac{x-3}{2}\right)$ **c.** $\frac{y+1}{\frac{1}{2}} = f(x+1)$ ⓐ

12. APPLICATION A chemistry class gathered these data on the conductivity of a base solution as acid is added to it. Graph the data and use transformations to find a model to fit the data.

Acid volume (mL) x	Conductivity (μS/cm³) y
0	4152.95
1	3140.97
2	2100.34
3	1126.55
4	162.299

Acid volume (mL) x	Conductivity (μS/cm³) y
5	1212.47
6	2358.11
7	3417.83
8	4429.81

Review

13. A panel of judges rate 20 science fair exhibits as shown. The judges decide that the top rating should be 100, so they add 6 points to each rating.

a. What are the mean and the standard deviation of the ratings before adding 6 points? ⓐ

b. What are the mean and the standard deviation of the ratings after adding 6 points?

c. What do you notice about the change in the mean? In the standard deviation?

Rank	Rating
1	94
2	92
3	92
4	92
5	90
6	89
7	89
8	88
9	86
10	85

Rank	Rating
11	84
12	83
13	83
14	81
15	79
16	79
17	77
18	73
19	71
20	68

14. APPLICATION This table shows the percentage of households with computers in the United States in various years.

Year	1995	1996	1997	1998	1999	2000
Households (%)	31.7	35.5	39.2	42.6	48.2	53.0

(*www.census.gov*)

a. Make a scatter plot of these data.

b. Find the median-median line.

c. Compare your model's prediction for 2003 with the actual census value of 61.8%.

d. Is a linear model for this situation good for long-term predictions? Explain your reasoning.

In 1946, inventors J. Presper Eckert and J. W. Mauchly created the first general-purpose electronic calculator, named ENIAC (Electronic Numerical Integrator and Computer). The calculator filled a large room and required a team of engineers and maintenance technicians to operate it.

LESSON

4.7 Transformations and the Circle Family

Many times the best way, in fact the only way, to learn is through mistakes. A fear of making mistakes can bring individuals to a standstill, to a dead center.

GEORGE BROWN

In this lesson you will investigate transformations of a relation that is not a function. A **unit circle** is centered at the origin with a radius of 1 unit. Suppose P is any point on a unit circle with center at the origin. Draw the slope triangle for the radius between the origin and point P.

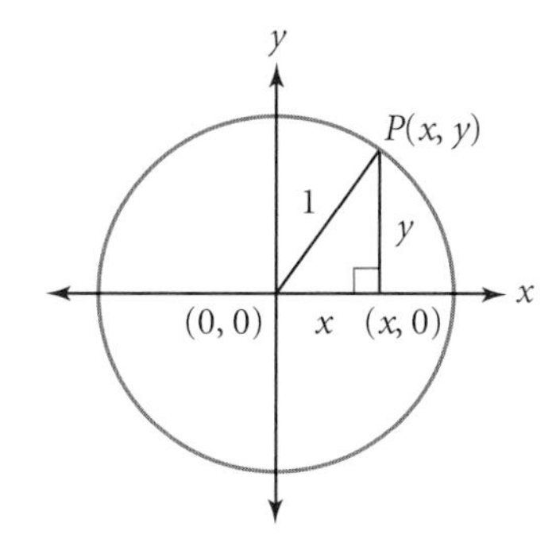

This photo shows circular housing developments in Denmark.

You can derive the equation of a unit circle from this diagram by using the Pythagorean Theorem. The legs of the right triangle have lengths x and y and the length of the hypotenuse is 1 unit, so its equation is $x^2 + y^2 = 1$. This is true for all points P on the unit circle.

What are the domain and the range of this relation? If a value, such as 0.5, is substituted for x, what are the output values of y? Why is the circle relation not a function?

In order to draw the graph of a circle on your calculator, you need to solve the equation $x^2 + y^2 = 1$ for y. When you do this, you get two equations, $y = +\sqrt{1 - x^2}$ and $y = -\sqrt{1 - x^2}$. Each of these is a function. You have to graph both of them to get the complete circle.

Equation of a Unit Circle

The equation of a **unit circle** is

$x^2 + y^2 = 1$ or, solved for y, $y = \pm\sqrt{1 - x^2}$

You can apply what you have learned about transformations of functions to find the equations of transformations of the unit circle.

EXAMPLE A

Find the equation for each graph.

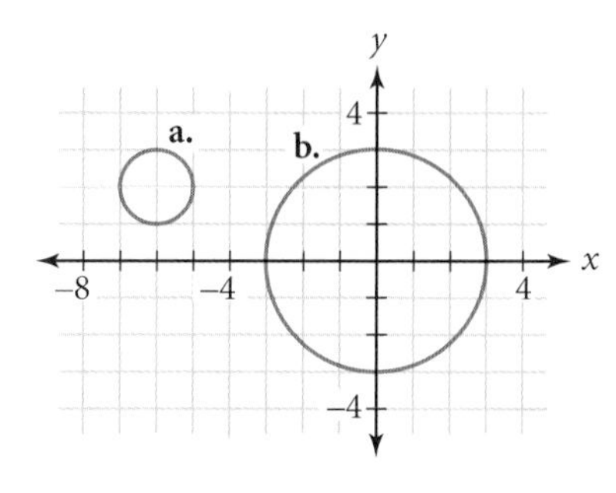

▶ Solution

a. Circle a is a translation of the unit circle horizontally -6 units and vertically 2 units. Replace x with $(x + 6)$ and y with $(y - 2)$ to get the equation $(x + 6)^2 + (y - 2)^2 = 1$. To check this result on your calculator, solve for y and graph:

$$(y - 2)^2 = 1 - (x + 6)^2$$
$$y - 2 = \pm\sqrt{1 - (x + 6)^2}$$
$$y = 2 \pm \sqrt{1 - (x + 6)^2}$$

You must enter two functions, $y = 2 + \sqrt{1 - (x + 6)^2}$ and $y = 2 - \sqrt{1 - (x + 6)^2}$ into your calculator.

b. Circle b is a dilation of the unit circle horizontally and vertically by the same scale factor of 3. Replacing x and y with $\frac{x}{3}$ and $\frac{y}{3}$, you find $\left(\frac{x}{3}\right)^2 + \left(\frac{y}{3}\right)^2 = 1$. This can also be written as $\frac{x^2}{9} + \frac{y^2}{9} = 1$ or $x^2 + y^2 = 9$.

You can transform a circle to get an **ellipse.** An ellipse is a circle where different horizontal and vertical scale factors have been used.

EXAMPLE B

What is the equation of this ellipse?

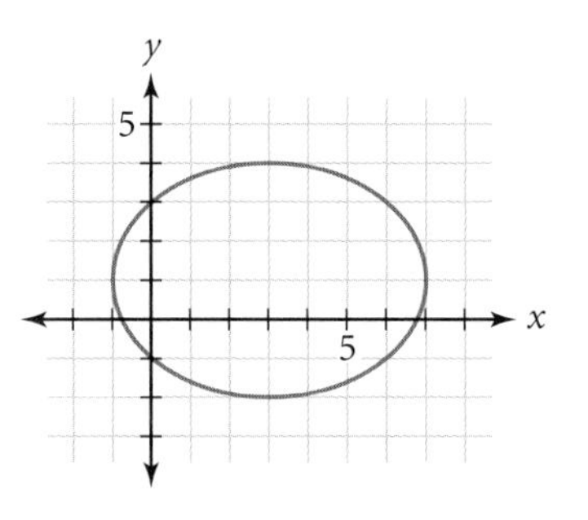

▶ Solution

The original unit circle has been translated and dilated. The new center is at (3, 1). In a unit circle, every radius measures 1 unit. In this ellipse, a horizontal segment from the center to the ellipse measures 4 units, so the horizontal scale factor is 4. Likewise, a vertical segment from the center to the ellipse measures 3 units, so the vertical scale factor is 3. The equation changes like this:

$x^2 + y^2 = 1$	Original unit circle.
$\left(\frac{x}{4}\right)^2 + y^2 = 1$	Dilate horizontally by a factor of 4. (Replace x with $\frac{x}{4}$.)
$\left(\frac{x}{4}\right)^2 + \left(\frac{y}{3}\right)^2 = 1$	Dilate vertically by a factor of 3. (Replace y with $\frac{y}{3}$.)
$\left(\frac{x-3}{4}\right)^2 + \left(\frac{y-1}{3}\right)^2 = 1$	Translate to new center at (3, 1). (Replace x with $x - 3$, and replace y with $y - 1$.)

To enter this equation into your calculator to check your answer, you need to solve for y. It takes two equations to graph this on your calculator. By graphing both of these equations, you can draw the complete ellipse and verify your answer.

$$y = 1 \pm 3\sqrt{1 - \left(\frac{x-3}{4}\right)^2}$$

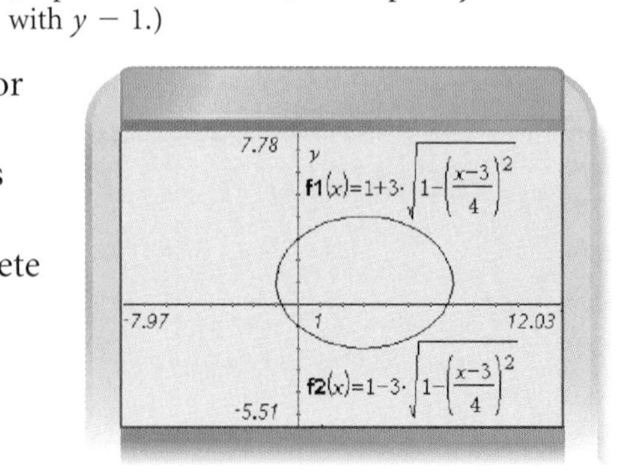

Investigation
When Is a Circle Not a Circle?

You will need

- the worksheet When Is a Circle Not a Circle?

If you look at a circle, like the top rim of a cup, from an angle, you don't see a circle; you see an ellipse. Choose one of the ellipses from the worksheet. Use your ruler carefully to place axes on the ellipse, and scale your axes in centimeters. Be sure to place the axes so that the longest dimension is parallel to one of the axes. Find the equation to model your ellipse. Graph your equation on your calculator and verify that it creates an ellipse with the same dimensions as on the worksheet.

The tops of these circular oil storage tanks look elliptical when viewed at an angle.

Equations for transformations of relations such as circles and ellipses are sometimes easier to work with in the general form before you solve them for y, but you need to solve for y to enter the equations into your calculator. If you start with a function such as the top half of the unit circle, $f(x) = \sqrt{1 - x^2}$, you can transform it in the same way you transformed any other function, but it may be a little messier to deal with.

EXAMPLE C

If $f(x) = \sqrt{1 - x^2}$, find $g(x) = 2f(3(x - 2)) + 1$. Sketch a graph of this new function.

► ***Solution***

In $g(x) = 2f(3(x - 2)) + 1$, note that $f(x)$ is the parent function, x has been replaced with $3(x - 2)$, and $f(3(x - 2))$ is then multiplied by 2 and 1 is added. You can rewrite the function g as

$$g(x) = 2\sqrt{1 - [3(x - 2)]^2} + 1 \quad \text{or} \quad g(x) = 2\sqrt{1 - \left(\frac{x - 2}{\frac{1}{3}}\right)^2} + 1$$

This indicates that the graph of $y = f(x)$, a semicircle, has been dilated horizontally by a factor of $\frac{1}{3}$, dilated vertically by a factor of 2, then translated right 2 units and up 1 unit. The transformed semicircle is graphed at right. What are the coordinates of the right endpoint of the graph? Describe how the original semicircle's right endpoint of $(1, 0)$ was mapped to this new location.

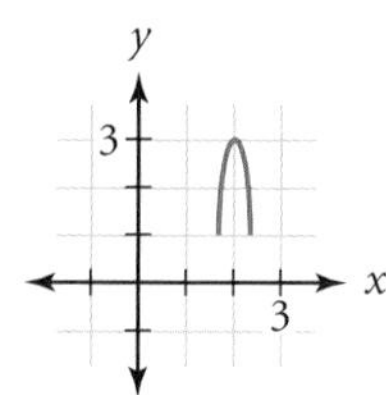

You have now learned to translate, reflect, and dilate functions and other relations. These transformations are the same for all equations.

Transformations of Functions and Other Relations

Translations

The graph of $y - k = f(x - h)$ translates the graph of $y = f(x)$ horizontally h units and vertically k units.

or

Replacing x with $(x - h)$ translates the graph horizontally h units. Replacing y with $(y - k)$ translates the graph vertically k units.

Reflections

The graph of $y = f(-x)$ is a reflection of the graph of $y = f(x)$ across the y-axis. The graph of $-y = f(x)$ is a reflection of the graph of $y = f(x)$ across the x-axis.

or

Replacing x with $-x$ reflects the graph across the y-axis. Replacing y with $-y$ reflects the graph across the x-axis.

Dilations

The graph of $\frac{y}{b} = f\left(\frac{x}{a}\right)$ is a dilation of the graph of $y = f(x)$ by a vertical scale factor of b and by a horizontal scale factor of a.

or

Replacing x with $\frac{x}{a}$ dilates the graph by a horizontal scale factor of a. Replacing y with $\frac{y}{b}$ dilates the graph by a vertical scale factor of b.

EXERCISES

You will need

A graphing calculator for Exercises **9–11, 14,** and **15.**

Practice Your Skills

1. Each equation represents a single transformation. Copy and complete this table.

Equation	Transformation (translation, reflection, dilation)	Direction	Amount or scale factor
$y + 3 = x^2$	Translation	Vertical	-3
$-y = \|x\|$			
$y = \sqrt{\frac{x}{4}}$			
$\frac{y}{0.4} = x^2$			
$y = \|x - 2\|$			
$y = \sqrt{-x}$			

2. The equation $y = \sqrt{1 - x^2}$ is the equation of the top half of the unit circle with center (0, 0) shown on the left. What is the equation of the top half of an ellipse shown on the right?

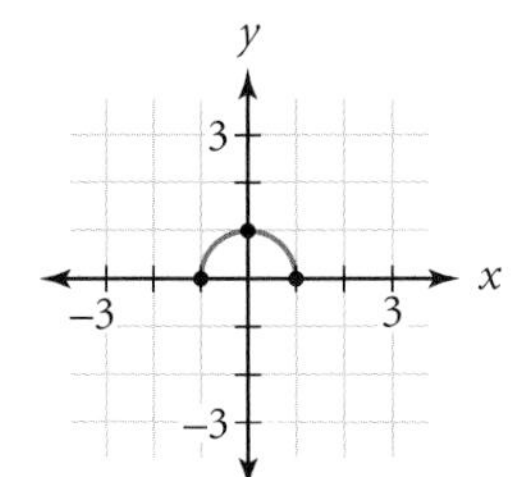

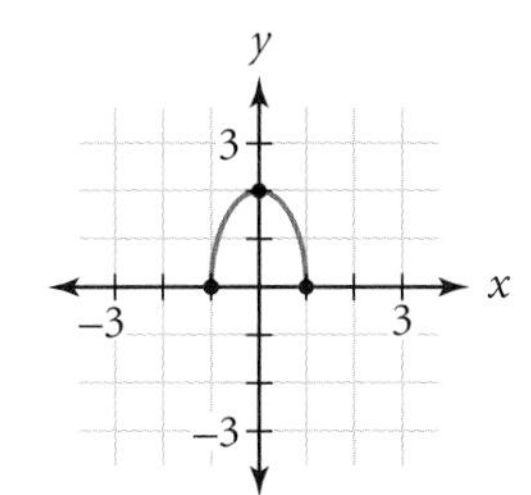

3. Use $f(x) = \sqrt{1 - x^2}$ to graph each of the transformations below.

a. $g(x) = -f(x)$ **b.** $h(x) = -2f(x)$ **c.** $j(x) = -3 + 2f(x)$ ⓐ

4. Each curve is a transformation of the graph of $y = \sqrt{1 - x^2}$. Write an equation for each curve.

a.

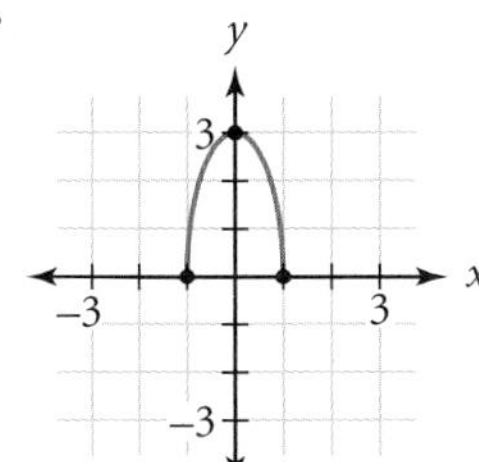

b.

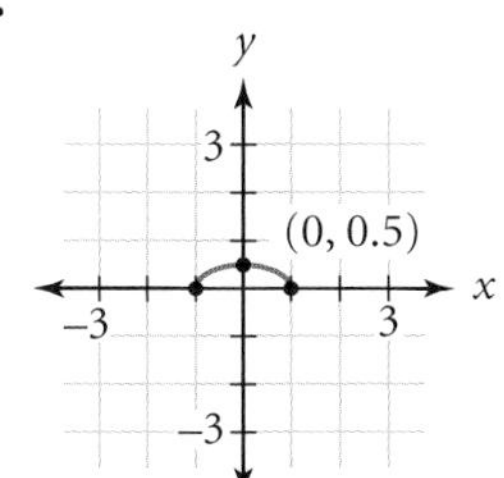

c.

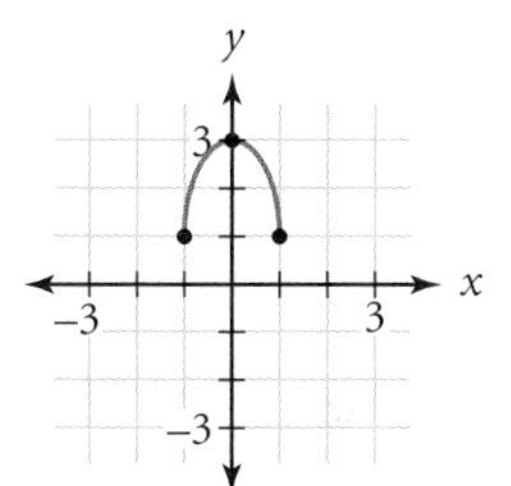

d. ⓐ

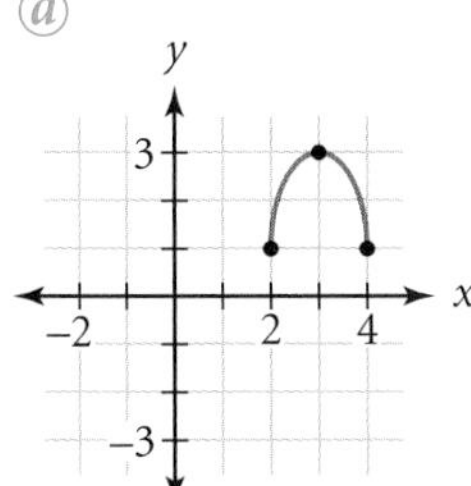

e. ⓐ

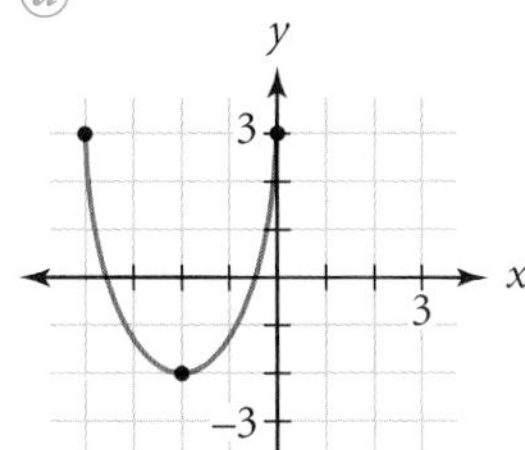

f.

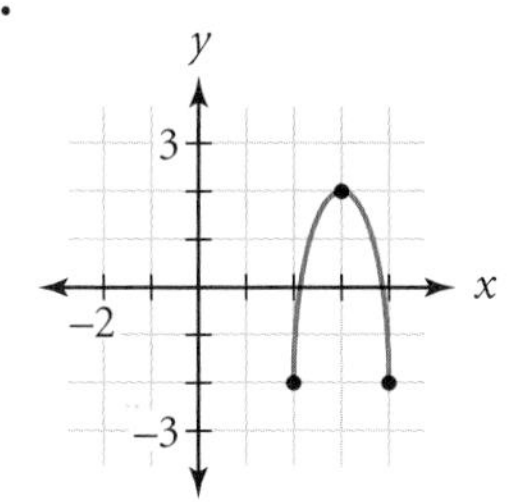

5. Write an equation and draw a graph for each transformation of the unit circle. Use the form $y = \pm\sqrt{1 - x^2}$.

a. Replace y with $(y - 2)$.

b. Replace x with $(x + 3)$.

c. Replace y with $\frac{y}{2}$. ⓐ

d. Replace x with $\frac{x}{2}$.

Reason and Apply

6. To create the ellipse at right, the x-coordinate of each point on a unit circle has been multiplied by a factor of 3.

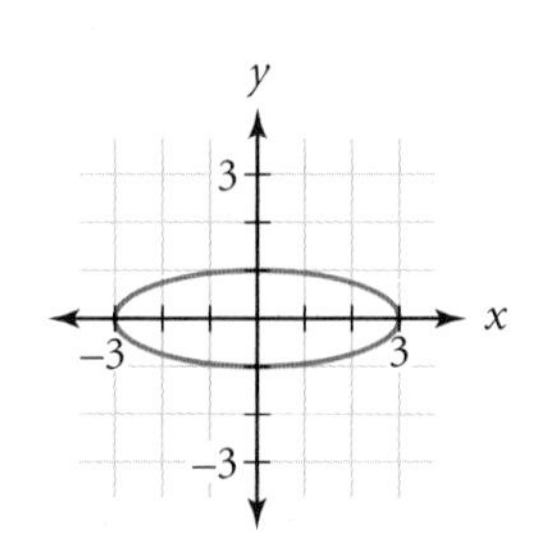

a. Write the equation of this ellipse.

b. What expression did you substitute for x in the parent equation? ⓐ

c. If $y = f(x)$ is the function for the top half of a unit circle, then what is the function for the top half of this ellipse, $y = g(x)$, in terms of f? ⓐ

7. Given the unit circle at right, write the equation that generates each transformation. Use the form $x^2 + y^2 = 1$.

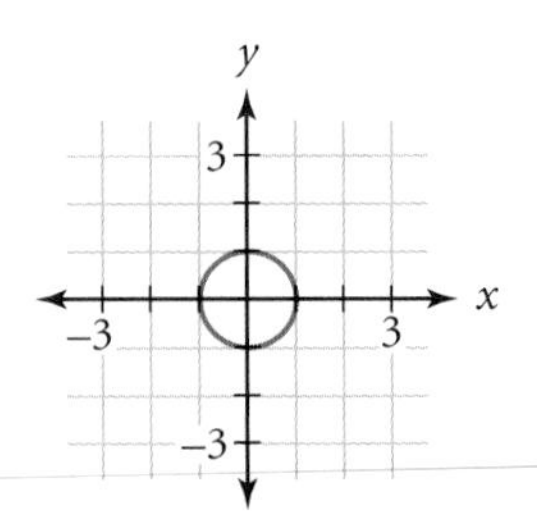

a. Each y-value is half the original y-value.

b. Each x-value is half the original x-value.

c. Each y-value is half the original y-value, and each x-value is twice the original x-value.

8. Consider the ellipse at right. ⓐ

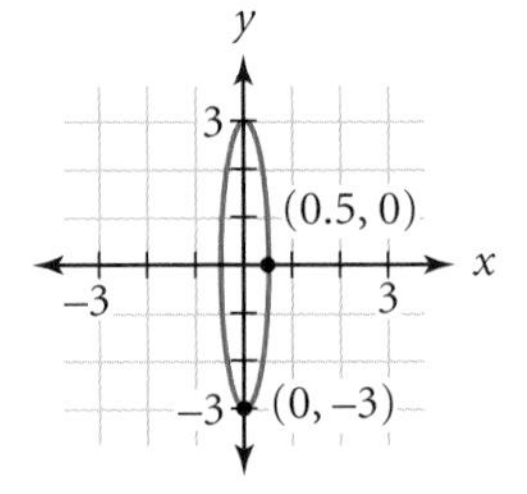

a. Write two functions that you could use to graph this ellipse.

b. Use ± to write one equation that combines the two equations in 8a.

c. Write another equation for the ellipse by squaring both sides of the equation in 8b.

9. ***Mini-Investigation*** Follow these steps to explore a relationship between linear, quadratic, square root, absolute-value, and semicircle functions. Use graphing windows of an appropriate size.

a. Graph these equations simultaneously on your calculator. The first four functions intersect in the same two points. What are the coordinates of these points? ⓐ

$y = x$ $\quad$ $y = x^2$ $\quad$ $y = \sqrt{x}$ $\quad$ $y = |x|$ $\quad$ $y = \sqrt{1 - x^2}$

b. Imagine using the intersection points that you found in 9a to draw a rectangle that just encloses the quarter-circle that is on the right half of the fifth function. How do the coordinates of the points relate to the dimensions of the rectangle? ⓗ

c. Solve these equations for y and graph them simultaneously on your calculator. Where do the first four functions intersect?

$\frac{y}{2} = \frac{x}{4}$ $\quad$ $\frac{y}{2} = \left(\frac{x}{4}\right)^2$ $\quad$ $\frac{y}{2} = \sqrt{\frac{x}{4}}$ $\quad$ $\frac{y}{2} = \left|\frac{x}{4}\right|$ $\quad$ $\frac{y}{2} = \sqrt{1 - \left(\frac{x}{4}\right)^2}$

d. Imagine using the intersection points that you found in 9c to draw a rectangle that just encloses the right half of the fifth function. How do the coordinates of the points relate to the dimensions of the rectangle?

10. Consider the parent function $y = \frac{1}{x}$ graphed at right. This function is not defined for $x = 0$. When the graph is translated, the center at (0, 0) is translated as well, so you can describe any translation of the figure by describing how the center is transformed.

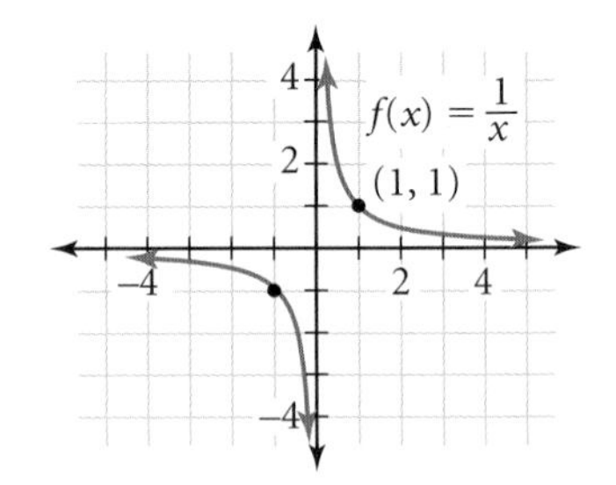

The parent function passes through the point (1, 1). You can describe any dilations of the function by describing how point (1, 1) is transformed. Use what you have learned about transformations to sketch each graph, then check your work with your graphing calculator.

a. $y = \frac{1}{x - 3}$

b. $y + 1 = \frac{1}{x + 4}$

c. $y = \frac{1}{3x}$

d. $\frac{y - 2}{-4} = \frac{1}{x}$

Science CONNECTION

Satellites are used to aid in navigation, communication, research, and military reconnaissance. The job the satellite is meant to do will determine the type of orbit it is placed in.

A satellite in a geosynchronous orbit revolves west to east above the diameter at the same speed Earth rotates, one revolution every 24 hours. To maintain this velocity, the satellite must have an altitude of about 22,000 miles. In order to stay above the same point on Earth, so that a satellite dish antenna can stay focused on it, the orbit of the satellite must be circular.

Another useful orbit is a north-south elliptical orbit that takes 12 hours to circle the planet. Satellites in these elliptical orbits cover areas of Earth that are not covered by geosynchronous satellites, and are therefore more useful for research and reconnaissance.

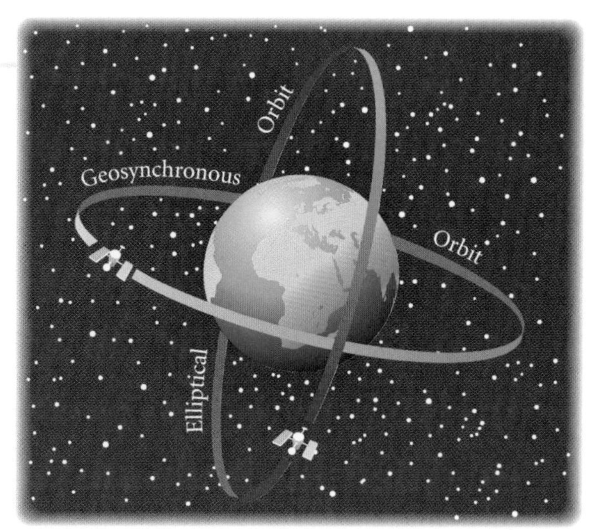

Satellites in a geosynchronous orbit follow a circular path above the equator. Another common orbit is an elliptical orbit in the north-south direction. For more information, see the links at www.keymath.com/DAA .

Review

11. Refer to Exercise 13 in Lesson 4.6. The original data are shown at right. Instead of adding the same number to each score, one of the judges suggests that perhaps they should *multiply* the original scores by a factor that makes the highest score equal 100. They decide to try this method.

Rank	Rating	Rank	Rating
1	94	11	84
2	92	12	83
3	92	13	83
4	92	14	81
5	90	15	79
6	89	16	79
7	89	17	77
8	88	18	73
9	86	19	71
10	85	20	68

a. By what factor should they multiply the highest score, 94, to get 100?

b. What are the mean and the standard deviation of the original ratings? Of the altered ratings?

c. Let x represent the exhibit number, and let y represent the rating. Plot the original and altered ratings on the same graph. Describe what happened to the ratings visually. How does this explain what happened to the mean and the standard deviation?

d. Which method do you think the judges should use? Explain your reasoning.

12. Find the next three terms in this sequence: 16, 40, 100, 250, ...

13. Solve. Give answers to the nearest 0.01.

a. $\sqrt{1-(a-3)^2}=0.5$

b. $1-|b+2|=-5$

c. $\sqrt{1-\left(\frac{c-2}{3}\right)^2}=0.8$ @

d. $3-5\left(\frac{d+1}{2}\right)^2=-7$

14. This table shows the distances needed to stop a car on dry pavement in a minimum length of time for various speeds. Reaction time is assumed to be 0.75 s.

Speed (mi/h) x	10	20	30	40	50	60	70
Stopping distance (ft) y	19	42	73	116	173	248	343

a. Construct a scatter plot of these data.

b. Find the equation of a parabola that fits the points and graph it.

c. Find the residuals for this equation and the root mean square error.

d. Predict the stopping distance for 56.5 mi/h.

e. How far off might your prediction in 14d be from the *actual* stopping distance?

15. This table shows passenger activity in the world's 30 busiest airports in 2005. ⓐ

a. Display the data in a histogram.

b. Estimate the total number of passengers who used the 30 airports. Explain any assumptions you make.

c. Estimate the mean usage among the 30 airports in 2005. Mark the mean on your histogram.

d. Sketch a box plot above your histogram. Estimate the five-number summary values. Explain any assumptions you make.

Number of passengers (in millions)	**Number of airports**
$30 \le p < 35$	10
$35 \le p < 40$	3
$40 \le p < 45$	9
$50 \le p < 55$	2
$55 \le p < 60$	1
$60 \le p < 65$	2
$65 \le p < 70$	1
$75 \le p < 80$	1
$85 \le p < 90$	1

(*The World Almanac and Book of Facts 2007*)

16. Consider the linear function $y = 3x + 1$.

a. Write an equation for the image of the graph of $y = 3x + 1$ after a reflection across the x-axis. Graph both lines on the same axes.

b. Write an equation for the image of the graph of $y = 3x + 1$ after a reflection across the y-axis. Graph both lines on the same axes.

c. Write an equation for the image of the graph of $y = 3x + 1$ after a reflection across the x-axis and then across the y-axis. Graph both lines on the same axes.

d. How does the image in 16c compare to the original line?

IMPROVING YOUR VISUAL THINKING SKILLS

4-in-1

Copy this trapezoid. Divide it into four congruent polygons.

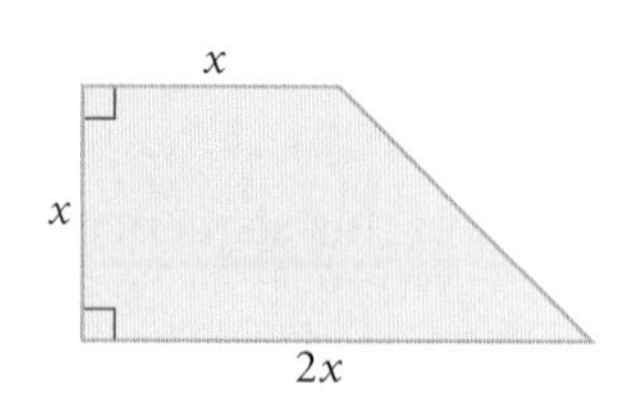

LESSON 4.8

Compositions of Functions

Sometimes you'll need two or more functions in order to answer a question or analyze a problem. Suppose an offshore oil well is leaking. Graph A shows the radius, r, of the spreading oil slick, growing as a function of time, t, so $r = f(t)$. Graph B shows the area, a, of the circular oil slick as a function of its radius, r, so $a = g(r)$. Time is measured in hours, the radius is measured in kilometers, and the area is measured in square kilometers.

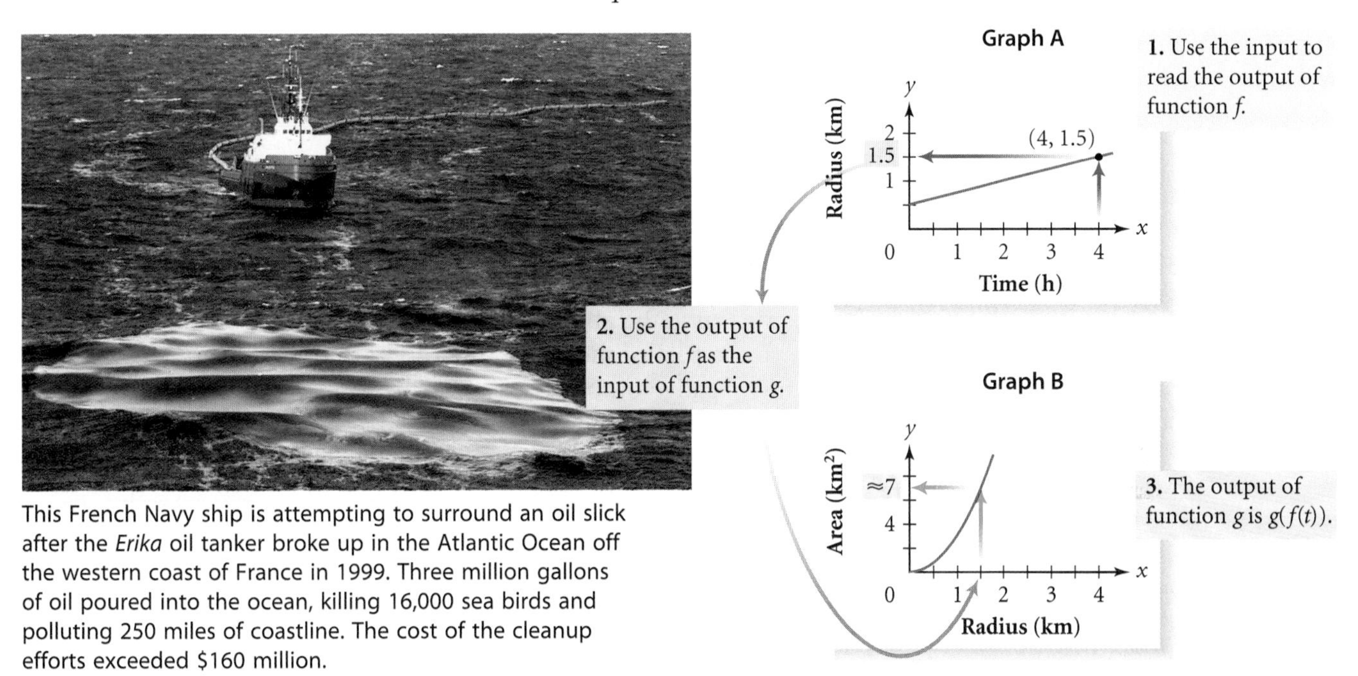

This French Navy ship is attempting to surround an oil slick after the *Erika* oil tanker broke up in the Atlantic Ocean off the western coast of France in 1999. Three million gallons of oil poured into the ocean, killing 16,000 sea birds and polluting 250 miles of coastline. The cost of the cleanup efforts exceeded $160 million.

Suppose you want to find the area of the oil slick after 4 hours. You can use function f on Graph A to find that when t equals 4, r equals 1.5. Next, using function g on Graph B, you find that when r equals 1.5, a is approximately 7. So, after 4 h, the radius of the oil slick is 1.5 km and its area is 7 km^2.

You used the graphs of two different functions, f and g, to find that after 4 h, the oil slick has area 7 km^2. You actually used the output from one function, f, as the input in the other function, g. This is an example of a **composition of functions** to form a new functional relationship between area and time, that is, $a = g(f(t))$. The symbol $g(f(t))$, read "g of f of t," is a composition of the two functions f and g. The composition $g(f(t))$ gives the final outcome when an x-value is substituted into the "inner" function, f, and its output value, $f(t)$, is then substituted as the input into the "outer" function, g.

EXAMPLE A

Consider these functions:

$$f(x) = \frac{3}{4}x - 3 \quad \text{and} \quad g(x) = |x|$$

What will the graph of $y = g(f(x))$ look like?

► Solution

Function f is the inner function, and function g is the outer function. Use equations and tables to identify the output of f and use it as the input of g.

Find several $f(x)$ output values.	Use the $f(x)$ output values as the input of $g(x)$.	Match the input of the inner function, f, with the output of the outer function, g, and plot the graph.

x	$f(x)$
−2	−4.5
0	−3
2	−1.5
4	0
6	1.5
8	3

$f(x)$	$g(f(x))$
−4.5	4.5
−3	3
−1.5	1.5
0	0
1.5	1.5
3	3

$f(x)$	$g(f(x))$
−2	4.5
0	3
2	1.5
4	0
6	1.5
8	3

The solution is the composition graph at right. All the function values of f, whether positive or negative, give positive output values under the rule of g, the absolute-value function. So, the part of the graph of function f showing negative output values is reflected across the x-axis in this composition.

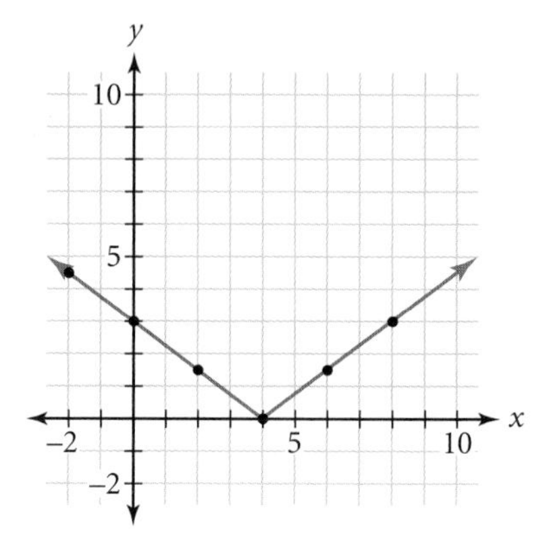

You can use what you know about transformations to get the specific equation for $y = g(f(x))$ in Example A. Use the parent function $y = |x|$, translate the vertex right 4 units, and then dilate horizontally by a factor of 4 and vertically by a factor of 3. This gives the equation $y = 3\left|\frac{x-4}{4}\right|$. You can algebraically manipulate this equation to get the equivalent equation $y = \left|\frac{3}{4}x - 3\right|$, which is the equation of f substituted for the input of g. You can always create equations of composed functions by substituting one equation into another.

Investigation
Looking Up

You will need

- a small mirror
- one or more tape measures or metersticks

Procedure Note

1. Place the mirror flat on the floor 0.5 m from a wall.
2. Use tape to attach tape measures or metersticks up the wall to a height of 1.5 to 2 m.

First, you'll establish a relationship between your distance from a mirror and what you can see in it.

Step 1 Set up the experiment as in the Procedure Note. Stand a short distance from the mirror, and look down into it. Move slightly left or right until you can see the tape measure on the wall reflected in the mirror.

Step 2 Have a group member slide his or her finger up the wall to help locate the highest height mark that is reflected in the mirror. Record the height in centimeters, h, and the distance from your toe to the center of the mirror in centimeters, d.

Step 3 Change your distance from the mirror and repeat Step 2. Make sure you keep your head in the same position. Collect several pairs of data in the form (d, h). Include some distances from the mirror that are small and some that are large.

Step 4 Find a function that fits your data by transforming the parent function $h = \frac{1}{d}$. Call this function f.

Now you'll combine your work from Steps 1–4 with the scenario of a timed walk toward and away from the mirror.

Step 5 Suppose this table gives your position at 1-second intervals:

Time (s) t	0	1	2	3	4	5	6	7
Distance to mirror (cm) d	163	112	74	47	33	31	40	62

Use one of the families of functions from this chapter to fit these data. Call this function g. It should give the distance from the mirror for seconds 0 to 7.

Step 6 Use your two functions to answer these questions:

a. How high up the wall can you see when you are 47 cm from the mirror?

b. Where are you at 1.3 seconds?

c. How high up the wall can you see at 3.4 seconds?

Step 7 Change each expression into words relating to the context of this investigation and find an answer. Show the steps you needed to evaluate each expression.

a. $f(60)$

b. $g(5.1)$

c. $f(g(2.8))$

Step 8 Find a single function, $H(t)$, that does the work of $f(g(t))$. Show that $H(2.8)$ gives the same answer as Step 7c above.

Don't confuse a composition of functions with the product of functions. In Example A, you saw that the composition of functions $f(x) = \frac{3}{4}x - 3$ and $g(x) = |x|$ is $g(f(x)) = \left|\frac{3}{4}x - 3\right|$. However, the product of the functions is $f(x) \cdot g(x) = \left(\frac{3}{4}x - 3\right) \cdot |x|$, or $\frac{3}{4}x\,|x| - 3|x|$. Multiplication of functions is commutative, so $f(x) \cdot g(x) = g(x) \cdot f(x)$.

Composing functions requires you to replace the independent variable in one function with the output value of the other function. This means that it is generally not commutative. That is, $f(g(x)) \neq g(f(x))$, except for certain functions.

To find the domain and range of a composite function, you must look closely at the domain and range of the original functions.

EXAMPLE B

Let $f(x)$ and $g(x)$ be the functions graphed below. What is the domain of $f(g(x))$?

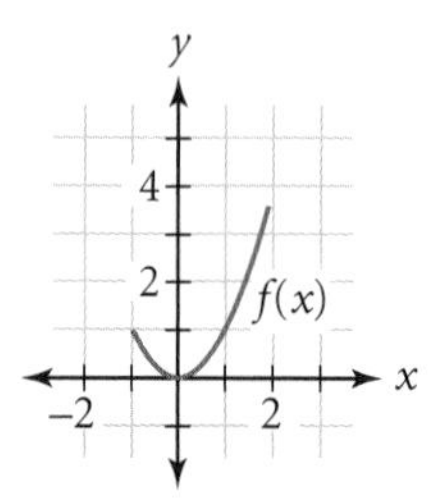

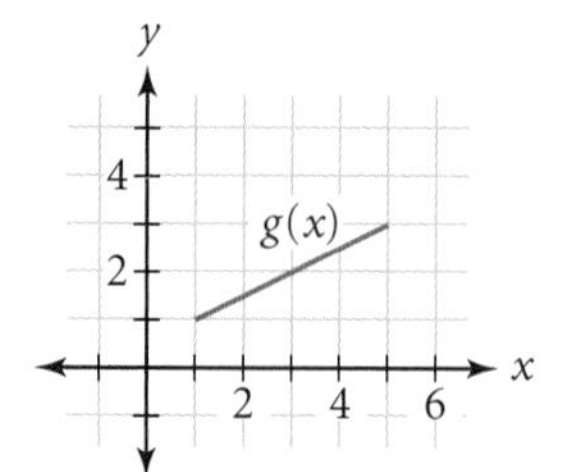

▶ Solution

Start by identifying the domain of the inner function, $g(x)$. This domain, as seen on the graph, is $1 \leq x \leq 5$. These values produce a range of $1 \leq g(x) \leq 3$. This is the input for the outer function, $f(x)$. However, notice that not all of these output values lie in the domain of $f(x)$. For example, there is no value for $f(2.5)$. Only the values $1 \leq g(x) \leq 2$ are in the domain of f. Now identify the x-values that produced this part of the range of $g(x)$. This is the domain of the composite function. The domain is $1 \leq x \leq 3$.

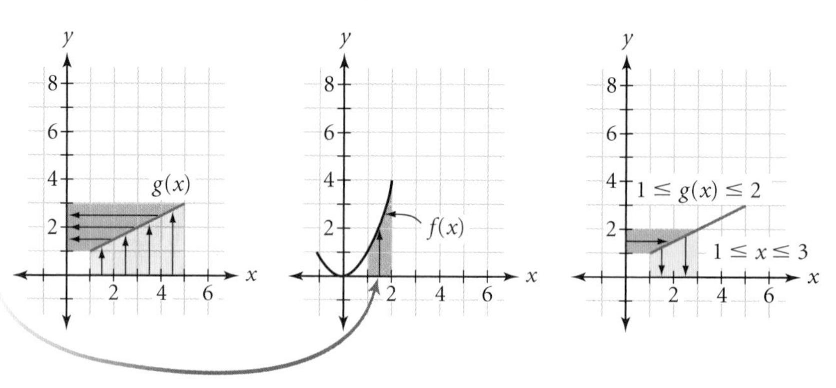

To find the domain of a composite function, first use the domain of the inner function to find its range. Then find the subset of the range that is within the domain of the outer function. The x-values that produce that subset of values are the domain of the composite function.

EXERCISES

You will need

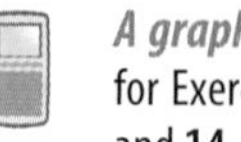

A graphing calculator for Exercises **8**, **10**, and **14**.

Practice Your Skills

1. Given the functions $f(x) = 3 + \sqrt{x + 5}$ and $g(x) = 2 + (x - 1)^2$, find these values.

a. $f(4)$ ⓐ **b.** $f(g(4))$ ⓐ **c.** $g(-1)$ **d.** $g(f(-1))$

2. The functions f and g are defined by these sets of input and output values.

$g = \{(1, 2), (-2, 4), (5, 5), (6, -2)\}$

$f = \{(0, -2), (4, 1), (3, 5), (5, 0)\}$

a. Find $g(f(4))$. **b.** Find $f(g(-2))$. @ **c.** Find $f(g(f(3)))$.

3. APPLICATION Graph A shows a swimmer's speed as a function of time. Graph B shows the swimmer's oxygen consumption as a function of her speed. Time is measured in seconds, speed in meters per second, and oxygen consumption in liters per minute. Use the graphs to estimate the values.

Graph A

Graph B

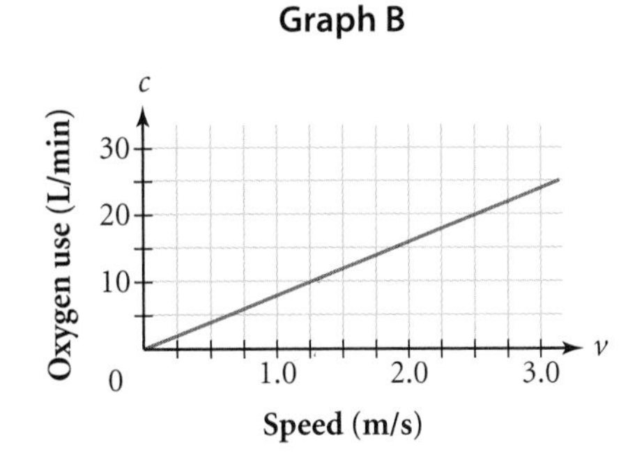

a. the swimmer's speed after 20 s of swimming

b. the swimmer's oxygen consumption at a swimming speed of 1.5 m/s

c. the swimmer's oxygen consumption after 40 s of swimming

4. Identify each equation as a composition of functions, a product of functions, or neither. If it is a composition or a product, then identify the two functions that combine to create the equation.

a. $y = 5\sqrt{3 + 2x}$

b. $y = 3 + (|x + 5| - 3)^2$ @

c. $y = (x - 5)^2(2 - \sqrt{x})$

Reason and Apply

5. Consider the graph at right.

a. Write an equation for this graph.

b. Write two functions, f and g, such that the figure is the graph of $y = f(g(x))$.

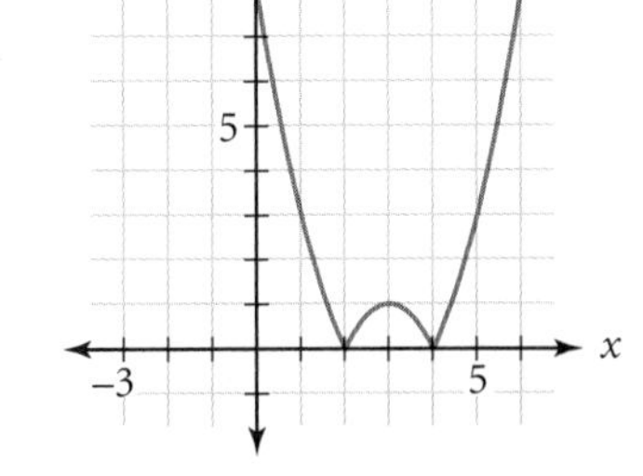

6. The functions f and g are defined by these sets of input and output values:

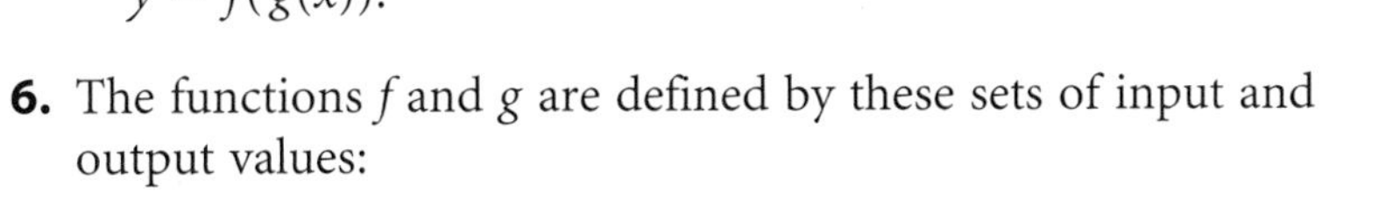

$g = \{(1, 2), (-2, 4), (5, 5), (6, -2)\}$

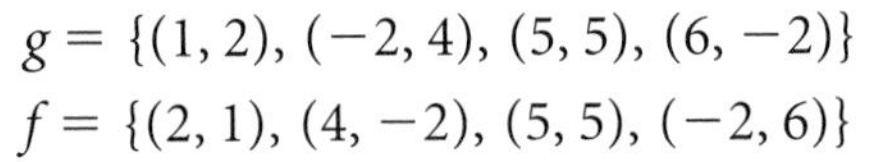

$f = \{(2, 1), (4, -2), (5, 5), (-2, 6)\}$

a. Find $g(f(2))$. @

b. Find $f(g(6))$.

c. Select any number from the domain of either g or f, and find $f(g(x))$ or $g(f(x))$, respectively. Describe what is happening.

7. A, B, and C are gauges with different linear measurement scales. When A measures 12, B measures 13, and when A measures 36, B measures 29. When B measures 20, C measures 57, and when B measures 32, C measures 84. (h)

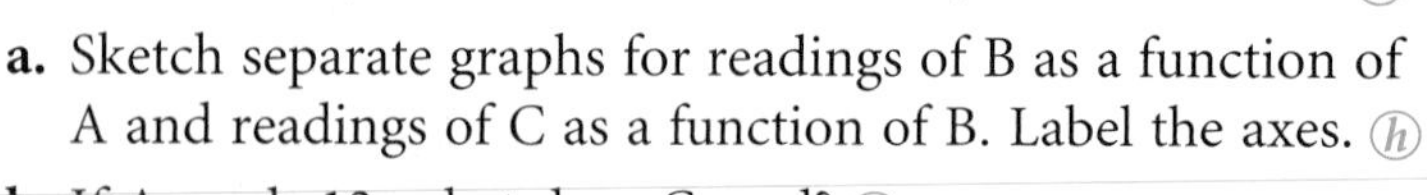

a. Sketch separate graphs for readings of B as a function of A and readings of C as a function of B. Label the axes. (h)

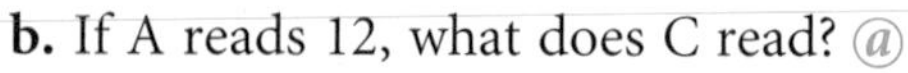

b. If A reads 12, what does C read? (a)

c. Write a function with the reading of B as the dependent variable and the reading of A as the independent variable. (a)

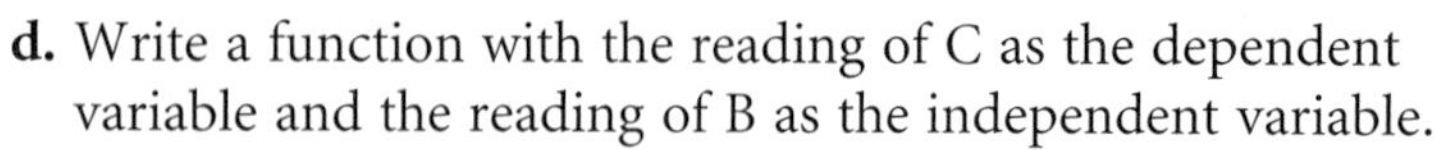

d. Write a function with the reading of C as the dependent variable and the reading of B as the independent variable.

e. Write a function with the reading of C as the dependent variable and the reading of A as the independent variable.

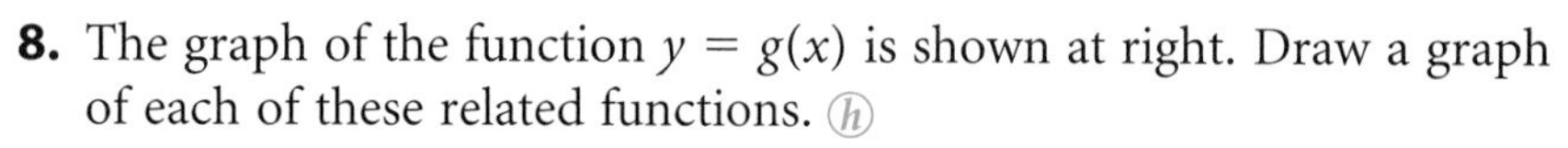

8. The graph of the function $y = g(x)$ is shown at right. Draw a graph of each of these related functions. (h)

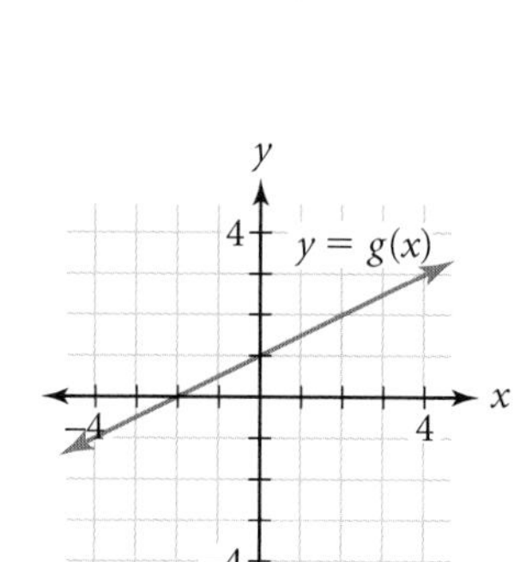

a. $y = \sqrt{g(x)}$

b. $y = |g(x)|$

c. $y = (g(x))^2$

d. What is the domain of each function in 8a–c?

9. The two lines pictured at right are $f(x) = 2x - 1$ and $g(x) = \frac{1}{2}x + \frac{1}{2}$. Solve each problem both graphically and numerically.

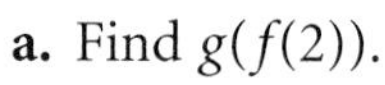

a. Find $g(f(2))$.

b. Find $f(g(-1))$.

c. Pick your own x-value in the domain of f, and find $g(f(x))$.

d. Pick your own x-value in the domain of g, and find $f(g(x))$.

e. Carefully describe what is happening in these compositions.

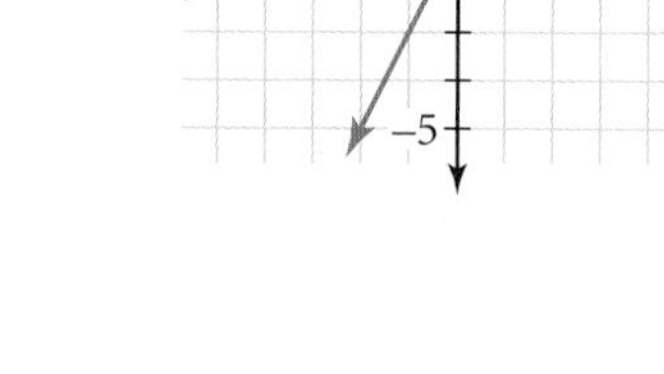

10. Given the functions $f(x) = -x^2 + 2x + 3$ and $g(x) = (x - 2)^2$, find these values.

a. $f(g(3))$ **b.** $f(g(2))$ **c.** $g(f(0.5))$ **d.** $g(f(1))$

e. $f(g(x))$. Simplify to remove all parentheses. (a)

f. $g(f(x))$. Simplify to remove all parentheses.

[▶ See **Calculator Note 4H** to learn how to use your calculator to check the answers to 10e and 10f. ◀]

11. Aaron and Davis need to write the equation that will produce the graph at right.

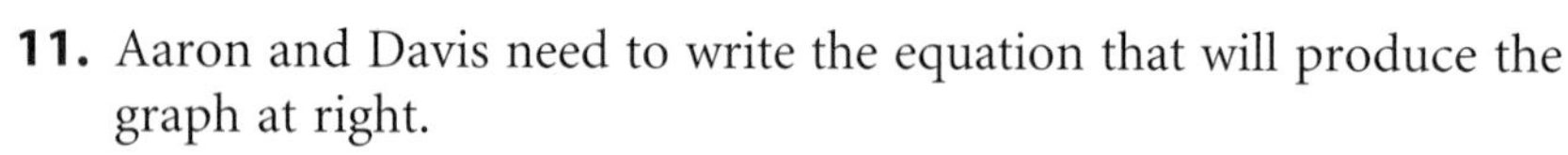

Aaron: "This is impossible! How are we supposed to know if the parent function is a parabola or a semicircle? If we don't know the parent function, there is no way to write the equation."

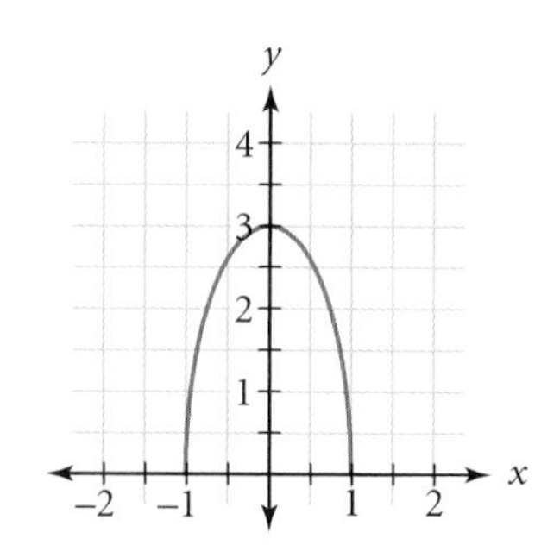

Davis: "Don't panic yet. I am sure we can determine its parent function if we study the graph carefully."

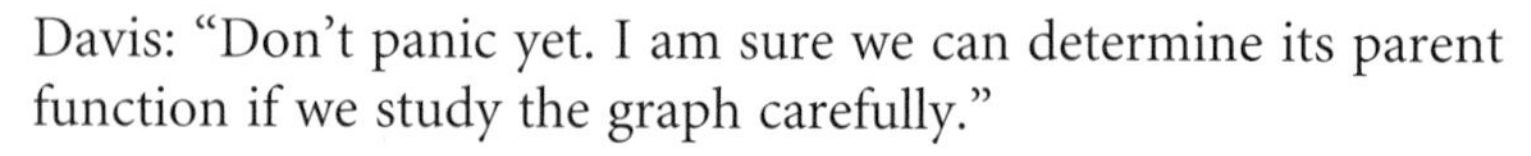

Do you agree with Davis? Explain completely and, if possible, write the equation of the graph. (h)

12. APPLICATION Jen and Priya decide to go out to the Hamburger Shack for lunch. They each have a 50-cent coupon from the Sunday newspaper for the Super-Duper-Deluxe $5.49 Value Meal. In addition, if they show their I.D. cards, they'll also get a 10% discount. Jen's server rang up the order as Value Meal, coupon, and then I.D. discount. Priya's server rang it up as Value Meal, I.D. discount, and then coupon.

a. How much did each girl pay?

b. Write a function, $C(x)$, that will deduct 50 cents from a price, x.

c. Write a function, $D(x)$, that will take 10% off a price, x.

d. Find $C(D(x))$.

e. Which server used $C(D(x))$ to calculate the price of the meal?

f. Is there a price for the Value Meal that would result in both girls paying the same price? If so, what is it?

Review

13. Solve. @

a. $\sqrt{|x-4|} = 3$

b. $(3-\sqrt{x+2})^2 = 4$

c. $|3-\sqrt{x}| = 5$

d. $3 + 5\sqrt{1+2x^2} = 13$

14. APPLICATION Bonnie and Mike are working on a physics project. They need to determine the ohm rating of a resistor. The ohm rating is found by measuring the potential difference in volts and dividing it by the electric current, measured in amperes (amps). In their project they set up the circuit at right. They vary the voltage and observe the corresponding readings of electrical current measured on the ammeter.

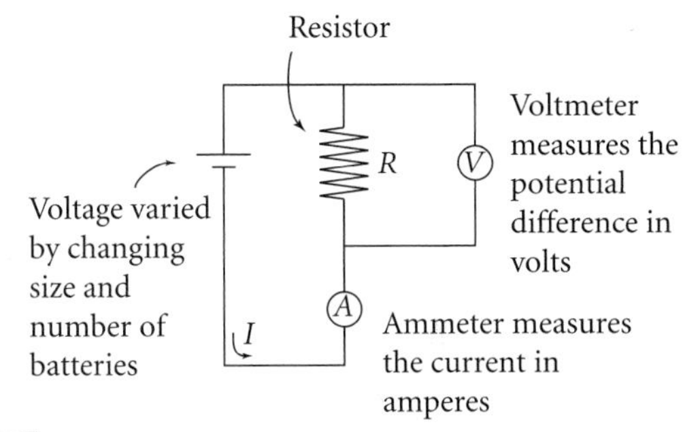

Potential difference (volts)	12	10	6	4	3	1
Current (amps)	2.8	2.1	1.4	1.0	0.6	0.2

a. Identify the independent and dependent variables.

b. Display the data on a graph.

c. Find the median-median line.

d. Bonnie and Mike reason that because 0 volts obviously yields 0 amps, the line they really want is the median-median line translated to go through (0, 0). What is the equation of the line through the origin that is parallel to the median-median line? @

e. How is the ohm rating Bonnie and Mike are trying to determine related to the line in 14d? @

f. What is their best guess of the ohm rating to the nearest tenth of an ohm?

15. Begin with the equation of the unit circle, $x^2 + y^2 = 1$.

a. Apply a horizontal dilation by a factor of 3 and a vertical dilation by a factor of 3, and write the equation that results.

b. Sketch the graph. Label the intercepts.

16. Imagine translating the graph of $f(x) = x^2$ left 3 units and up 5 units, and call the image $g(x)$.

a. Give the equation for $g(x)$.

b. What is the vertex of the graph of $y = g(x)$?

c. Give the coordinates of the image point on the parabola that is 2 units to the right of the vertex.

Project

PIECEWISE PICTURES

You can use piecewise functions to create designs and pictures. If you use several different functions together, you can create a picture that does not represent a function. [▶ See **Calculator Note 4E** to learn more about graphing piecewise functions. ◀]

You can use your calculator to draw this car by entering these functions:

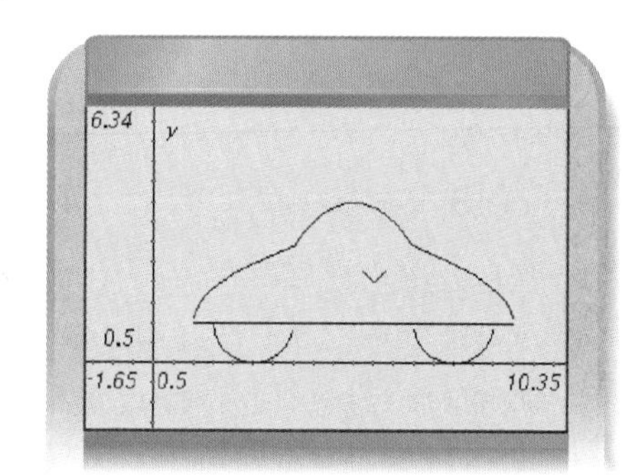

$$c(x) = \begin{cases} 1 + 1.2\sqrt{x-1},\ x \leq 3.5 \\ 4 - 0.5(x-5)^2,\ 3.5 \leq x \leq 6.5 \\ 1 + 1.2\sqrt{-(x-9)},\ 6.5 \leq x \end{cases}$$

$$d(x) = \{1,\ 1 \leq x \leq 9$$

$$e(x) = 1 - \sqrt{1-(x-2.5)^2}$$

$$g(x) = 1 - \sqrt{1-(x-7.5)^2}$$

$$h(x) = \{2 + |x - 5.5|,\ 5.2 \leq x \leq 5.8$$

Which function represents which part of the car? Explain why some of the functions do not have restricted domains.

Experiment with the given piecewise functions to see if you can modify the shape of the car or increase its size. Then write your own set of functions to draw a picture. Your project should include

- A screen capture or accurate graph of your drawing.
- The functions you used to create your drawing, including any restrictions on the domain.
- At least one piecewise function.

CHAPTER 4 REVIEW

This chapter introduced the concept of a **function** and reviewed **function notation.** You saw real-world situations represented by rules, sets, functions, graphs, and most importantly, equations. You learned to distinguish between functions and other **relations** by using either the definition of a function—at most one y-value per x-value—or the vertical line test.

This chapter also introduced several **transformations,** including **translations, reflections,** and vertical and horizontal **dilations.** You learned how to transform the graphs of **parent functions** to investigate several families of functions—linear, quadratic, square root, absolute value, and semicircle. For example, if you dilate the graph of the parent function $y = x^2$ vertically by a factor of 3 and horizontally by a factor of 2, and translate it right 1 unit and up 4 units, then you get the graph of the function $y = 3\left(\frac{x-1}{2}\right)^2 + 4$.

Finally, you looked at the **composition** of functions. Many times, solving a problem involves two or more related functions. You can find the value of a composition of functions by using algebraic or numeric methods or by graphing.

EXERCISES

You will need

A graphing calculator for Exercise **9.**

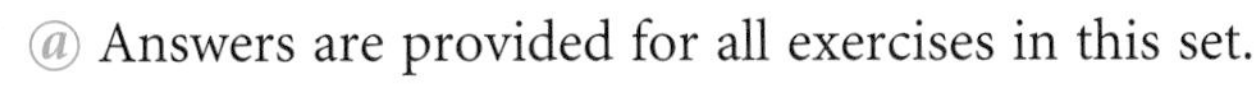

Ⓐ Answers are provided for all exercises in this set.

1. Sketch a graph that shows the relationship between the time in seconds after you start microwaving a bag of popcorn and the number of pops per second. Describe in words what your graph shows.

2. Use these three functions to find each value:

$f(x) = -2x + 7$
$g(x) = x^2 - 2$
$h(x) = (x + 1)^2$

a. $f(4)$ **b.** $g(-3)$ **c.** $h(x + 2) - 3$
d. $f(g(3))$ **e.** $g(h(-2))$ **f.** $h(f(-1))$
g. $f(g(a))$ **h.** $g(f(a))$ **i.** $h(f(a))$

3. The graph of $y = f(x)$ is shown at right. Sketch the graph of each of these functions:

a. $y = f(x) - 3$
b. $y = f(x - 3)$
c. $y = 3f(x)$
d. $y = f(-x)$

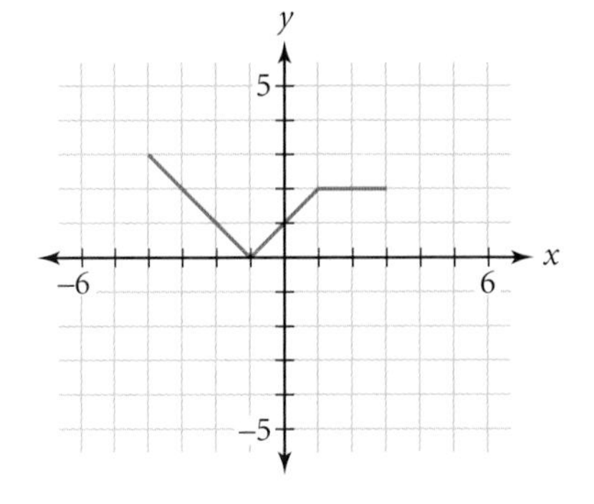

4. Assume you know the graph of $y = f(x)$. Describe the transformations, in order, that would give you the graph of these functions:

a. $y = f(x + 2) - 3$ **b.** $\frac{y - 1}{-1} = f\left(\frac{x}{2}\right) + 1$ **c.** $y = 2f\left(\frac{x - 1}{0.5}\right) + 3$

5. The graph of $y = f(x)$ is shown at right. Use what you know about transformations to sketch these related functions:

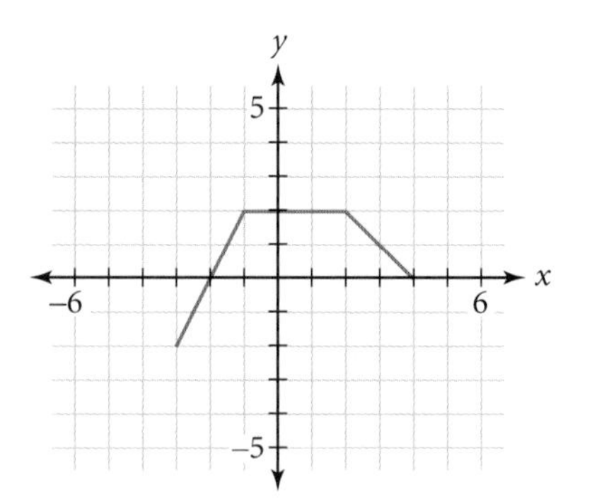

a. $y - 1 = f(x - 2)$ **b.** $\frac{y + 3}{2} = f(x + 1)$

c. $y = f(-x) + 1$ **d.** $y + 2 = f\left(\frac{x}{2}\right)$

e. $y = -f(x - 3) + 1$ **f.** $\frac{y + 2}{-2} = f\left(\frac{x - 1}{1.5}\right)$

6. For each graph, name the parent function and write an equation of the graph.

a.

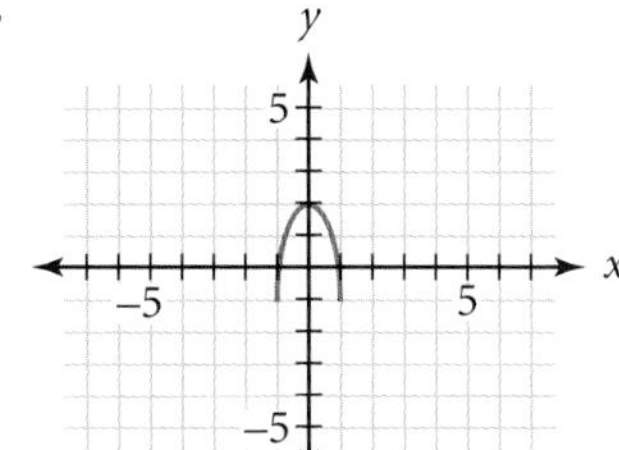

b.

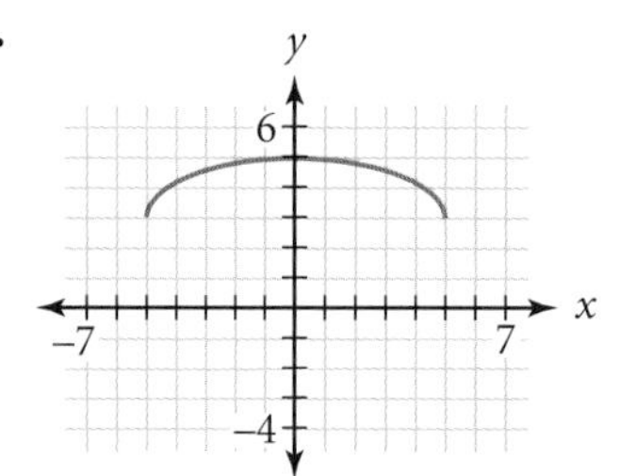

c.

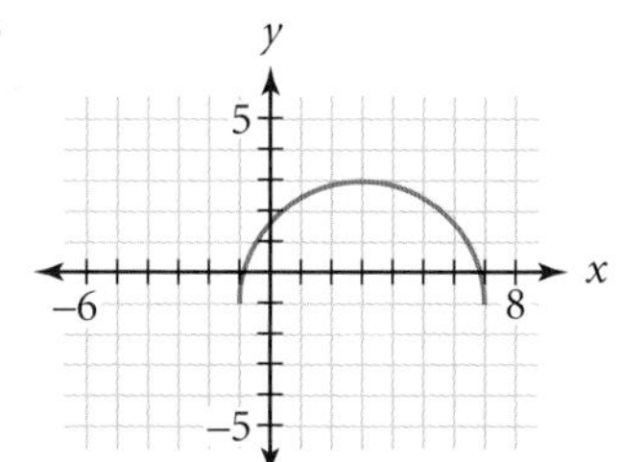

d.

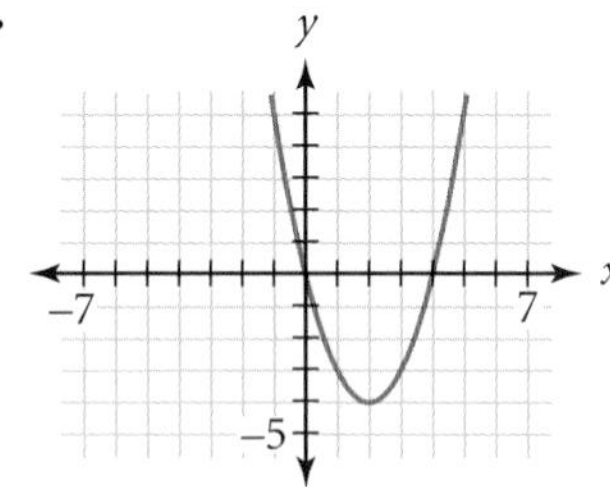

e.

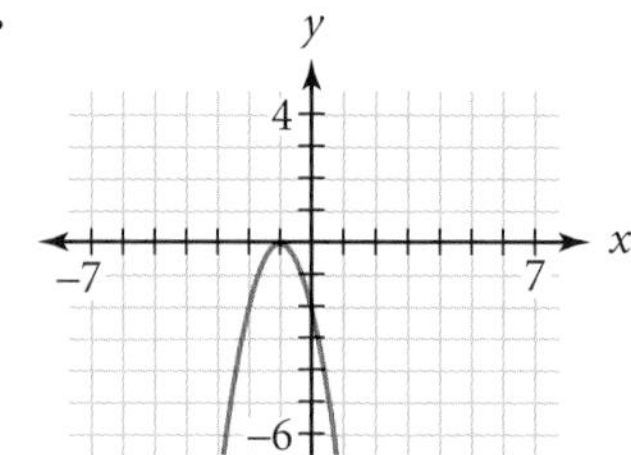

f.

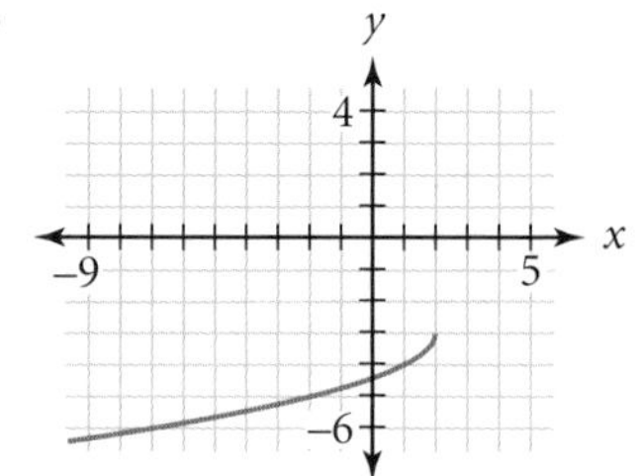

g.

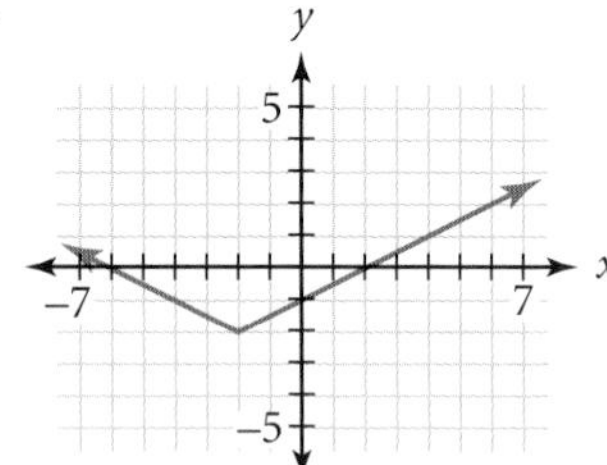

h.

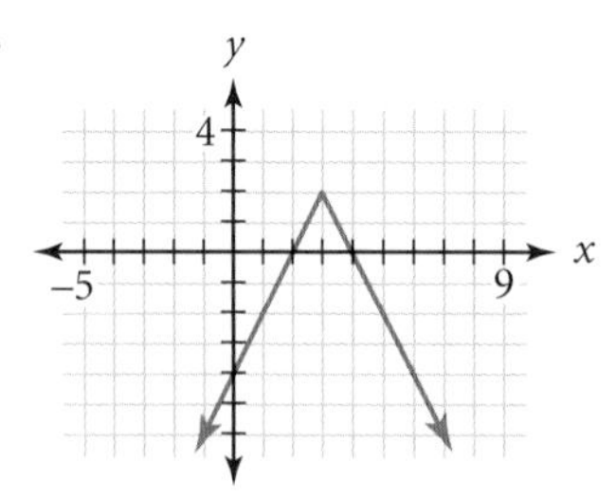

7. Solve for y.

a. $2x - 3y = 6$ **b.** $(y + 1)^2 - 3 = x$ **c.** $\sqrt{1 - y^2} + 2 = x$

8. Solve for x.

a. $4\sqrt{x - 2} = 10$ **b.** $\left(\frac{x}{-3}\right)^2 = 5$ **c.** $\left|\frac{x - 3}{2}\right| = 4$ **d.** $3\sqrt{1 + \left(\frac{x}{5}\right)^2} = 2$

9. The Acme Bus Company has a daily ridership of 18,000 passengers and charges \$1.00 per ride. The company wants to raise the fare yet keep its revenue as large as possible. (The revenue is found by multiplying the number of passengers by the fare charged.) From previous fare increases, the company estimates that for each increase of \$0.10 it will lose 1,000 riders.

a. Complete this table.

Fare (\$) x	1.00	1.10	1.20	1.30	1.40	1.50	1.60	1.70	1.80
Number of passengers	18,000								
Revenue (\$) y	18,000								

b. Make a graph of the revenue versus fare charged. You should recognize the graph as a parabola.

c. What are the coordinates of the vertex of the parabola? Explain the meaning of each coordinate of the vertex.

d. Find a quadratic function that models these data. Use your model to find

i. the revenue if the fare is \$2.00.

ii. the fare(s) that make no revenue (\$0).

TAKE ANOTHER LOOK

1. Some functions can be described as even or odd. An **even function** has the y-axis as a line of symmetry. If the function f is an even function, then $f(-x) = f(x)$ for all values of x in the domain. Which parent functions that you've seen are even functions? Now graph $y = x^3$, $y = \frac{1}{x}$, and $y = \sqrt[3]{x}$, all of which are **odd functions.** Describe the symmetry displayed by these odd functions. How would you define an odd function in terms of $f(x)$? If possible, give an example of a function that is neither even nor odd.

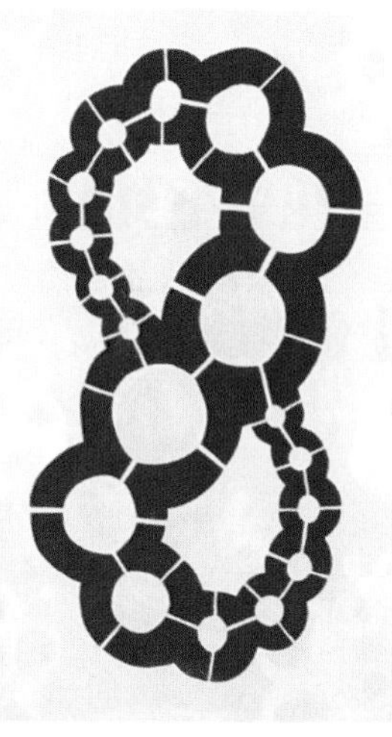

This painting by Laura Domela is titled *sense* (2002, oil on birch panel). The design on the left is similar to an even function, and the one on the right is similar to an odd function.

2. A line of reflection does not have to be the x- or y-axis. Draw the graph of a function and then draw its image when reflected across several different horizontal or vertical lines. Write the equation of each image. Try this with several different functions. In general, if the graph of $y = f(x)$ is reflected across the vertical line $x = a$, what is the equation of the image? If the graph of $y = f(x)$ is reflected across the horizontal line $y = b$, what is the equation of the image?

3. For the graph of the parent function $y = x^2$, you can think of any vertical dilation as an equivalent horizontal dilation. For example, the equations $y = 4x^2$ and $y = (2x)^2$ are equivalent, even though one represents a vertical dilation by a factor of 4 and the other represents a horizontal dilation by a factor of $\frac{1}{2}$. For the graph of any function or relation, is it possible to think of any vertical dilation as an equivalent horizontal dilation? If so, explain your reasoning. If not, give examples of functions and relations for which it is not possible.

4. Enter two linear functions into f_1 and f_2 on your calculator. Enter the compositions of the functions as $f_3 = f_1(f_2(x))$ and $f_4 = f_2(f_1(x))$. Graph f_3 and f_4 and look for any relationships between them. (It will help if you turn off the graphs of f_1 and f_2.) Make a conjecture about how the compositions of any two linear functions are related. Change the linear functions in f_1 and f_2 to test your conjecture. Can you algebraically prove your conjecture?

5. One way to visualize a composition of functions is to use a web graph. Here's how you evaluate $f(g(x))$ for any value of x, using a web graph:

Choose an x-value. Draw a vertical line from the x-axis to the function $g(x)$. Then draw a horizontal line from that point to the line $y = x$. Next, draw a vertical line from this point so that it intersects $f(x)$. Draw a horizontal line from the intersection point to the y-axis. The y-value at this point of intersection gives the value of $f(g(x))$.

Choose two functions $f(x)$ and $g(x)$. Use web graphs to find $f(g(x))$ for several values of x. Why does this method work?

Assessing What You've Learned

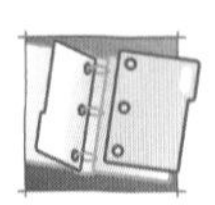

ORGANIZE YOUR NOTEBOOK Organize your notes on each type of parent function and each type of transformation you have learned about. Review how each transformation affects the graph of a function or relation and how the equation of the function or relation changes. You might want to create a large chart with rows for each type of transformation and columns for each type of parent function; don't forget to include a column for the general function, $y = f(x)$.

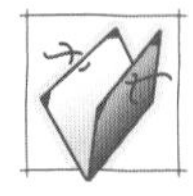

UPDATE YOUR PORTFOLIO Choose one piece of work that illustrates each transformation you have studied in this chapter. Try to select pieces that illustrate different parent functions. Add these to your portfolio. Describe each piece in a cover sheet, giving the objective, the result, and what you might have done differently.

WRITE TEST ITEMS Two important skills from this chapter are the ability to use transformations to write and graph equations. Write at least two test items that assess these skills. If you work with a group, identify other key ideas from this chapter and work together to write an entire test.

CHAPTER

5 Exponential, Power, and Logarithmic Functions

OBJECTIVES

In this chapter you will

- review the properties of square roots
- write explicit equations for geometric sequences
- use exponential functions to model real-world growth and decay scenarios
- review the properties of exponents and the meaning of rational exponents
- learn how to find the inverse of a function
- apply logarithms, the inverses of exponential functions

Art can take on living forms. To create *Tree Mountain,* conceptualized by artist Agnes Denes (b 1931), 11,000 people planted 11,000 trees on a human-made mountain in Finland, on a former gravel pit. The trees were planted in a mathematical pattern similar to a golden spiral, imitating the arrangement of seeds on a sunflower.

All living things grow and eventually decay. Both growth and decay can be modeled using exponential functions.

Tree Mountain–A Living Time Capsule–11,000 People, 11,000 Trees, 400 Years 1992–1996, Ylöjärvi, Finland, (420 × 270 × 28 meters) © Agnes Denes.

Working with Positive Square Roots

The square root of a number is the value that, when squared, gives back the original number. For example, $\sqrt{25}$ is 5 because $5^2 = 25$. You may recall that every positive number has both a positive and negative square root. In this chapter, though, we're considering only positive square roots, which are designated by the square root symbol.

Perfect squares are integers that have integer square roots. You can use your knowledge of perfect squares to estimate the square roots of other numbers.

EXAMPLE A Estimate the value of $\sqrt{85}$.

► **Solution**

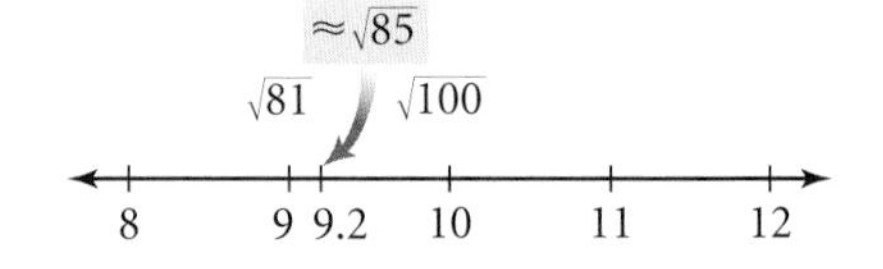

Find two perfect squares that are close to 85. If you square 9, you get 81. If you square 10, you get 100. Because 85 is between 81 and 100, $\sqrt{85}$ must be between 9 and 10. That approximation may be close enough. If you need to get closer, you can estimate pretty effectively. Because 85 is only a little more than 81, a good estimate of $\sqrt{85}$ is a little more than 9, maybe 9.2.

If you check your estimate by squaring 9.2, you find that $9.2^2 = 84.64$. So the estimate is a bit small, but not bad.

Square roots of some values can be written in different equivalent ways. For example, $\sqrt{72}$ can also be written as $2\sqrt{18}$ or $3\sqrt{8}$ or $6\sqrt{2}$. Which one is best? Actually, each of these forms can be useful at different times. In this text we will use either the original form, $\sqrt{72}$, or the form with the smallest possible value remaining inside the square root, $6\sqrt{2}$.

The next example shows how to change from one form to another. You will need to remember the property of square roots that states, "If a and b are nonnegative, then $\sqrt{a} \cdot \sqrt{b} = \sqrt{ab}$."

EXAMPLE B Rewrite each expression with the smallest possible value inside the square root.

a. $\sqrt{18}$ **b.** $\sqrt{500}$ **c.** $\sqrt{108}$

► **Solution**

a. The key is to find perfect-square factors of 18. Using the fact that $18 = 9 \cdot 2$, you can rewrite $\sqrt{18}$ as $\sqrt{9} \cdot \sqrt{2}$. Then you can evaluate $\sqrt{9}$ as 3 to get $\sqrt{18} = 3\sqrt{2}$.

b. Rewrite $\sqrt{500}$ as $\sqrt{100} \cdot \sqrt{5}$. Then $\sqrt{500} = \sqrt{100} \cdot \sqrt{5} = 10 \cdot \sqrt{5}$, so $\sqrt{500} = 10\sqrt{5}$.

c. You may not readily know factors of 108, so first try small numbers that you know are perfect squares. Start with 4. A quick check shows that $108 = 4 \cdot 27$. So, $\sqrt{108} = \sqrt{4} \cdot \sqrt{27} = 2\sqrt{27}$. However, 27 is divisible by 9, a perfect square, so keep going. Factoring again gives $\sqrt{108} = 2\sqrt{27} = 2\sqrt{9} \cdot \sqrt{3} = 2 \cdot 3\sqrt{3} = 6\sqrt{3}$.

You might realize that if you had factored 108 as $36 \cdot 3$, you could have arrived at the final result more quickly: $\sqrt{108} = \sqrt{36} \cdot \sqrt{3} = 6\sqrt{3}$.

In the last part of Example B, you rewrote $\sqrt{108}$ as $6\sqrt{3}$. Which form is easier for estimating the value? Because 108 is between the perfect squares 100 and 121, $\sqrt{108}$ must be between $\sqrt{100}$ and $\sqrt{121}$. So $\sqrt{108}$ is between 10 and 11, perhaps about 10.3. The other form, $6\sqrt{3}$, is harder to estimate. You would first need to estimate the value of $\sqrt{3}$ and then multiply by 6.

However, there are times when it is convenient to see the value written in this form. The important thing is to be able to work with both forms and convert one to the other.

EXAMPLE C Rewrite $4\sqrt{5}$ as a square root without a coefficient and estimate its value.

▶ ***Solution*** You can reverse the process from Example B to rewrite $4\sqrt{5}$ as $\sqrt{16}\sqrt{5}$. Then multiply to get $\sqrt{16}\sqrt{5} = \sqrt{16 \cdot 5} = \sqrt{80}$. In this form you can estimate the value to be almost 9, perhaps about 8.9.

EXERCISES

1. Graph each value on a number line without using your calculator.

a. $\sqrt{11}$ @ **b.** $\sqrt{47}$ **c.** $\sqrt{55}$ @ **d.** $\sqrt{67}$

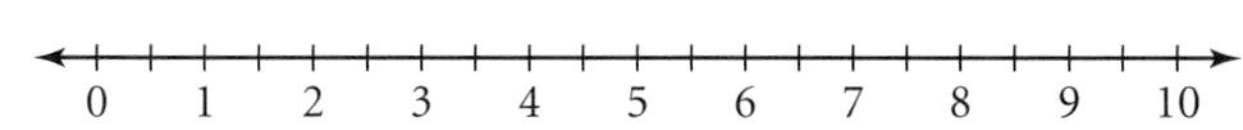

2. Rewrite each expression with the smallest possible value inside the square root.

a. $\sqrt{24}$ @ **b.** $\sqrt{75}$ **c.** $\sqrt{45}$ @ **d.** $\sqrt{40}$ **e.** $\sqrt{300}$

3. Match the equivalent values and estimates.

a. $\sqrt{12}$ @	**i.** $\sqrt{18}$	**A.** a number between 2 and 3
b. $3\sqrt{2}$	**ii.** $\sqrt{8}$	**B.** a number between 3 and 4
c. $\sqrt{28}$ @	**iii.** $2\sqrt{3}$	**C.** a number between 4 and 5
d. $2\sqrt{2}$	**iv.** $2\sqrt{7}$	**D.** a number between 5 and 6

4. Use estimation to determine which side of the expression is larger. Fill in the blank with > or < to make the statement true.

a. $\sqrt{17} + \sqrt{27}$ ___?___ $\sqrt{17 + 27}$ @ **b.** $8\sqrt{5}$ ___?___ $5\sqrt{8}$

c. $\sqrt{30} - \sqrt{10}$ ___?___ $\sqrt{30 - 10}$ @ **d.** $\sqrt{6^2 + 11^2}$ ___?___ $\sqrt{6^2} + \sqrt{11^2}$

LESSON

5.1

Exponential Functions

Life shrinks or expands in proportion to one's courage.

ANAÏS NIN

In Chapter 1, you used sequences and recursive rules to model geometric growth or decay of money, populations, and other quantities. Recursive formulas generate only discrete values, such as the amount of money after one year or two years, or the population in a certain year. Usually growth and decay happen continuously. In this lesson you will focus on finding explicit formulas for these patterns, which will allow you to model situations involving continuous growth and decay, or to find discrete points without using recursion.

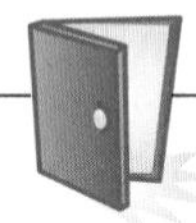

Investigation
Radioactive Decay

You will need

- one die per person

This investigation is a simulation of radioactive decay. Each person will need a standard six-sided die. [▶ See **Calculator Note 1L** to simulate this with your calculator instead. ◀] Each standing person represents a radioactive atom in a sample. The people who sit down at each stage represent the atoms that underwent radioactive decay.

Procedure Note

1. All members of the class should stand up, except for the recorder. The recorder counts and records the number standing at each stage.
2. Each standing person rolls a die, and anyone who gets a 1 sits down.
3. Wait for the recorder to count and record the number of people standing.
4. Repeat Steps 2 and 3 until fewer than three students are standing.

Step 1 Follow the Procedure Note to collect data in the form (*stage, number standing*).

Step 2 Graph your data. The graph should remind you of the sequence graphs you studied in Chapter 1. What type of sequence does this resemble?

Step 3 Identify u_0 and the common ratio, r, for your sequence. Complete the table below. Use the values of u_0 and r to help you write an explicit formula for your data.

n	u_n	u_n in terms of u_0 and r	u_n in terms of u_0 and r using exponents
0	u_0		
1	u_1	$u_0 \cdot r$	
2	u_2	$u_0 \cdot r \cdot r$	
3	u_3	$u_0 \cdot r \cdot r \cdot r$	

Step 4 Graph your explicit formula along with your data. Notice where the value of u_0 appears in your equation. Your graph should pass through the original data point $(0, u_0)$. Modify your equation so that it passes through $(1, u_1)$, the second data point. (Think about translating the graph horizontally and also changing the starting value.)

Step 5 Experiment with changing your equation to pass through other data points. Decide on an equation that you think is the best fit for your data. Write a sentence or two explaining why you chose this equation.

Step 6 What equation with ratio r would you write that contains the point $(6, u_6)$?

You probably recognized the geometric decay model in the investigation. As you learned in Chapter 1, geometric decay is nonlinear. At each step the previous term is multiplied by a common ratio. Because the common ratio appears as a factor more and more times, you can use exponential functions to model the geometric growth. An **exponential function** is a continuous function with a variable in the exponent, and it is used to model growth or decay.

Science CONNECTION

Certain atoms are unstable—their nuclei can split apart, emitting radiation and resulting in a more stable atom. This process is called radioactive decay. The time it takes for half the atoms in a radioactive sample to decay is called **half-life,** and the half-life is specific to the element. For instance, the half-life of carbon-14 is 5730 years, whereas the half-life of uranium-238 is 4.5 billion years.

Submerged in a storage tank in La Hague, France, this radioactive waste glows blue. The blue light is known as the "Cherenkov glow."

EXAMPLE A

Most automobiles depreciate as they get older. Suppose an automobile that originally costs \$14,000 depreciates by one-fifth of its value every year.

a. What is the value of this automobile after $2\frac{1}{2}$ years?

b. When is this automobile worth half of its initial value?

▶ Solution

a. The recursive formula gives automobile values only after one year, two years, and so on. The value decreases by $\frac{1}{5}$, or 0.2, each year, so to find the next term you multiply by another $(1 - 0.2)$. You can use this fact to write an explicit formula.

```
14000                 14000
Ans·(1-0.20)

1400·(1-.2)          11200.
11200.·(1-.2)         8960.
8960.·(1-.2)          7168.
```

$14{,}000 \cdot (1 - 0.2)$	Value after 1 year.
$14{,}000 \cdot (1 - 0.2) \cdot (1 - 0.2) = 14{,}000(1 - 0.2)^2$	Value after 2 years.
$14{,}000 \cdot (1 - 0.2) \cdot (1 - 0.2) \cdot (1 - 0.2) = 14{,}000(1 - 0.2)^3$	Value after 3 years.
$14{,}000 \cdot (1 - 0.2)^n$	Value after n years.

So the explicit formula for automobile value is $u_n = 14{,}000(1 - 0.2)^n$. The equation of the continuous function through the points of this sequence is

$$y = 14{,}000(1 - 0.2)^x$$

You can use the continuous function to find the value of the car at any point. To find the value after $2\frac{1}{2}$ years, substitute 2.5 for x.

$$y = 14{,}000(1 - 0.2)^{2.5} \approx \$8{,}014.07$$

It makes sense that the automobile's value after $2\frac{1}{2}$ years should be between the values of u_2 and u_3, \$8,960 and \$7,168. However, the value of the car after $2\frac{1}{2}$ years is not exactly halfway between those values because the function describing the value is not linear.

b. To find when the automobile is worth half of its initial value, substitute 7,000 for y and find x.

$$7{,}000 = 14{,}000(1 - 0.2)^x \quad \text{Substitute 7,000 for } y.$$
$$0.5 = (1 - 0.2)^x \quad \text{Divide both sides by 14,000.}$$
$$0.5 = (0.8)^x \quad \text{Combine like terms.}$$

You don't yet know how to solve for x when x is an exponent, but you can experiment to find an exponent that produces a value close to 0.5. The value of $(0.8)^{3.106}$ is very close to 0.5. This means that the value of the car is about \$7,000, or half of its original value, after 3.106 years (about 3 years 39 days). This is the **half-life** of the value of the automobile, or the amount of time needed for the value to decrease to half of the original amount.

$b < 1$

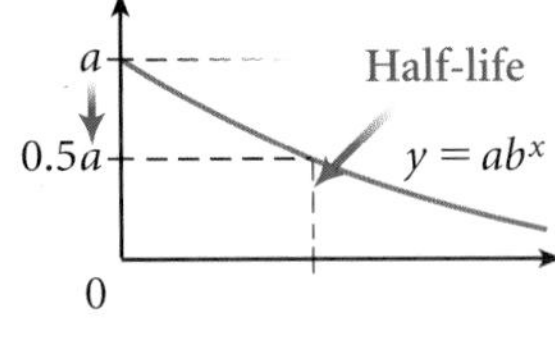

Exponential Function

The general form, or intercept form, of an exponential function is

$$y = ab^x$$

where the coefficient a is the y-intercept and the base b is the growth rate.

$b > 1$

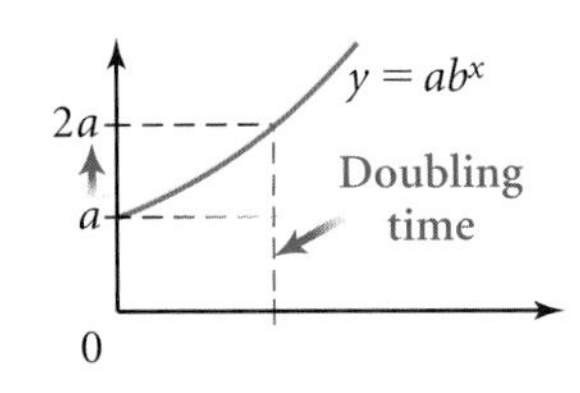

Exponential growth and decay are both modeled with the general form $y = ab^x$. Growth is modeled by a base that is greater than 1, and decay is modeled by a base that is less than 1. In general, a larger base models faster growth, and a base closer to 0 models faster decay.

All exponential growth curves have a **doubling time,** just as decay has a half-life. This time depends only on the ratio. For example, if the ratio is constant, it takes just as long to double \$1,000 to \$2,000 as it takes to double \$5,000 to \$10,000.

EXAMPLE B

The functions $g(x) = \frac{1}{2}(2)^x$ and $h(x) = 2^{x-1}$ are transformations of the parent function $f(x) = 2^x$. Describe the transformations and sketch the graphs.

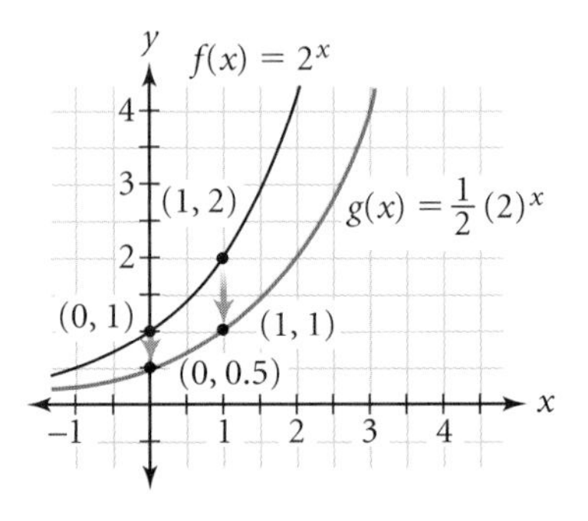

▸ Solution

The graph of $g(x)$ is a vertical dilation of the graph of $f(x)$ by a factor of $\frac{1}{2}$. The marked points on the red graph show how the y-values of the corresponding points on the black graph have been multiplied by $\frac{1}{2}$.

The graph of $h(x)$ is a horizontal translation of the graph of $f(x)$. The marked points on the red graph show how the corresponding points on the black graph have been moved right 1 unit.

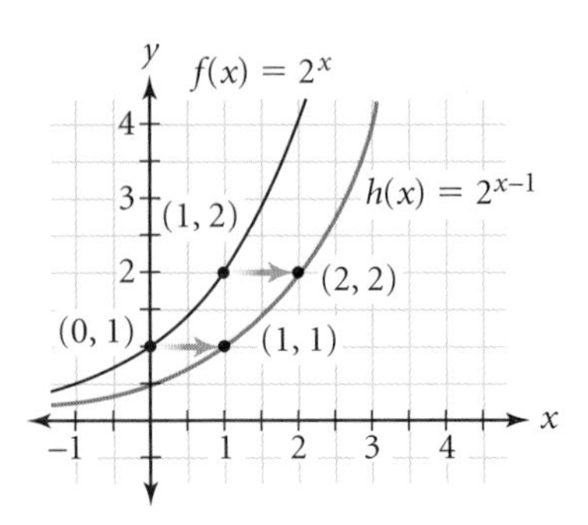

Do you notice anything special about the two red graphs? They appear to be the same. With exponential functions, as with linear functions, it is possible to find different transformations that produce the same graphs.

EXERCISES

You will need

A graphing calculator for Exercises **6–12** and **15.**

Practice Your Skills

1. Evaluate each function at the given value.

a. $f(x) = 4.753(0.9421)^x$, $x = 5$ ⓐ

b. $g(h) = 238(1.37)^h$, $h = 14$

c. $h(t) = 47.3(0.835)^t + 22.3$, $t = 24$ ⓐ

d. $j(x) = 225(1.0825)^{x-3}$, $x = 37$

2. Record three terms of the sequence and then write an explicit function for the sequence.

a. $u_0 = 16$
$u_n = 0.75u_{n-1}$ where $n \geq 1$ ⓐ

b. $u_0 = 24$
$u_n = 1.5u_{n-1}$ where $n \geq 1$

3. Evaluate each function at $x = 0$, $x = 1$, and $x = 2$ and then write a recursive formula for the pattern.

a. $f(x) = 125(0.6)^x$ ⓐ

b. $f(x) = 3(2)^x$

4. Calculate the ratio of the second term to the first term, and express the answer as a decimal value. State the percent increase or decrease.

a. 48, 36 **b.** 54, 72 **c.** 50, 47 ⓐ **d.** 47, 50

Reason and Apply

5. In 1995, the population of the People's Republic of China was approximately 1.211 billion, with an annual growth rate of 1.5%. (*www.chinability.com*) ⓐ

a. Write a recursive formula that models this growth. Let u_0 represent the population in 1995.

b. Complete a table recording the population for the years 1995 to 2002.

c. Define the variables and write an exponential equation that models this growth. Choose two data points from the table and show that your equation works.

d. One estimate for the population of China in 2006 was 1.314 billion. How does this compare with the value predicted by your equation? What does this tell you?

6. Jack planted a mysterious bean just outside his kitchen window. It immediately sprouted 2.56 cm above the ground. Jack kept a careful log of the plant's growth. He measured the height of the plant each day at 8:00 A.M. and recorded these data:

Day	0	1	2	3	4
Height (cm)	2.56	6.4	16	40	100

a. Define variables and write an exponential equation for this pattern. If the pattern continues, what will be the heights on the fifth and sixth days?

b. Jack's younger brother measured the plant at 8:00 P.M. on the evening of the third day and found it to be about 63.25 cm tall. Show how to find this value mathematically. (You may need to experiment with your calculator.) ⓗ

Kudzu, shown here in Oxford, Mississippi, is a fast-growing weed that was brought from Japan and once promoted for its erosion control. Today kudzu covers more than 7 million U.S. acres.

c. Find the height of the sprout at 12:00 noon on the sixth day.

d. Find the doubling time for this plant.

e. Experiment with the equation to find the day and time (to the nearest hour) when the plant reaches a height of 1 km.

7. ***Mini-Investigation*** For 7a–d, graph the equations on your calculator.

a. $y = 1.5^x$ b. $y = 2^x$ c. $y = 3^x$ d. $y = 4^x$

e. How do the graphs compare? What points (if any) do they have in common?

f. Predict what the graph of $y = 6^x$ will look like. Verify your prediction by using your calculator.

8. Each of the red curves is a transformation of the graph of $y = 2^x$, shown in black. Focus on how the two marked points on the black curve are transformed to become the corresponding points on the red curve. Write an equation for each red curve. (h)

a. (@)

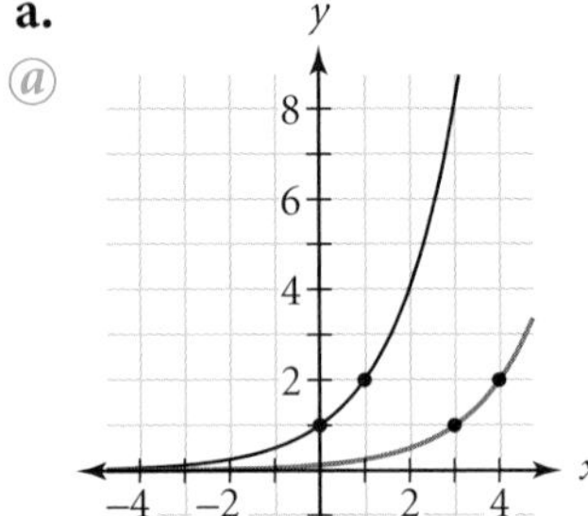

b.

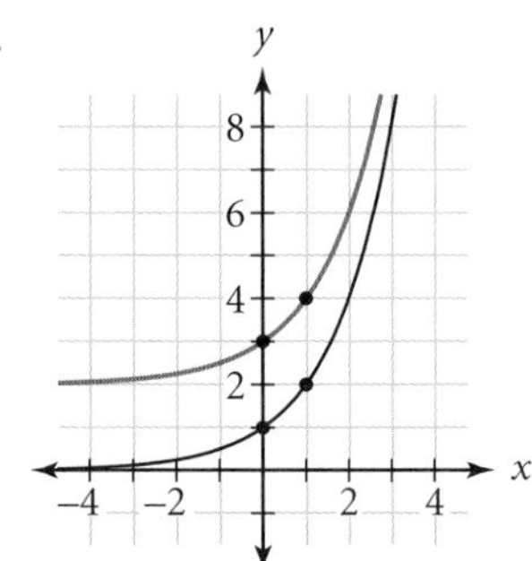

c. (@)

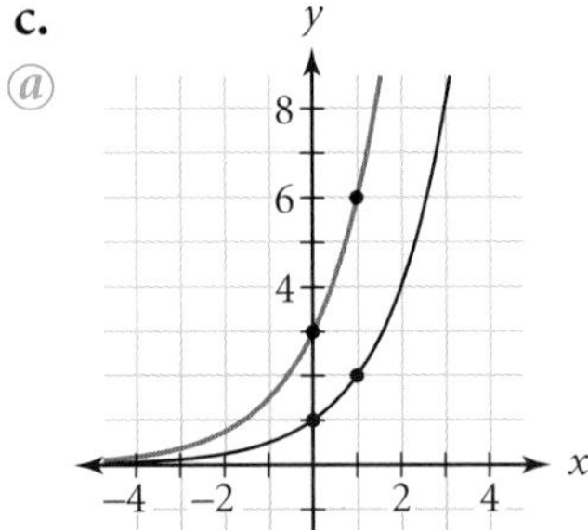

d.

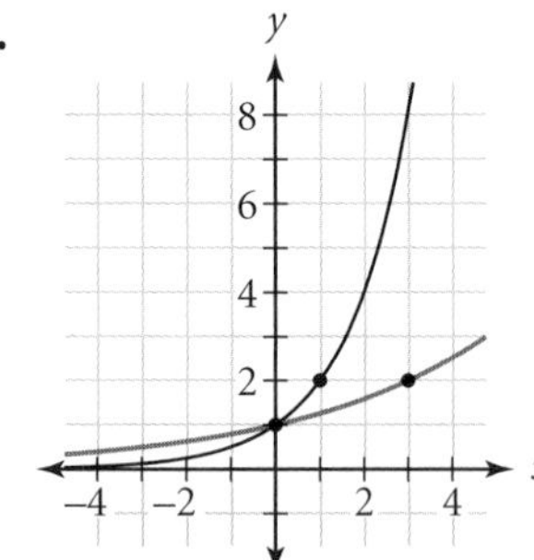

9. ***Mini-Investigation*** For 9a–d, graph the equations on your calculator.

a. $y = 0.2^x$ b. $y = 0.3^x$ c. $y = 0.5^x$ d. $y = 0.8^x$

e. How do the graphs compare? What points (if any) do they have in common?

f. Predict what the graph of $y = 0.1^x$ will look like. Verify your prediction by using your calculator.

10. Each of the red curves is a transformation of the graph of $y = 0.5^x$, shown in black. Focus on how the two marked points on the black curve are transformed to become the corresponding points on the red curve. Write an equation for each red curve. (h)

a. (@)

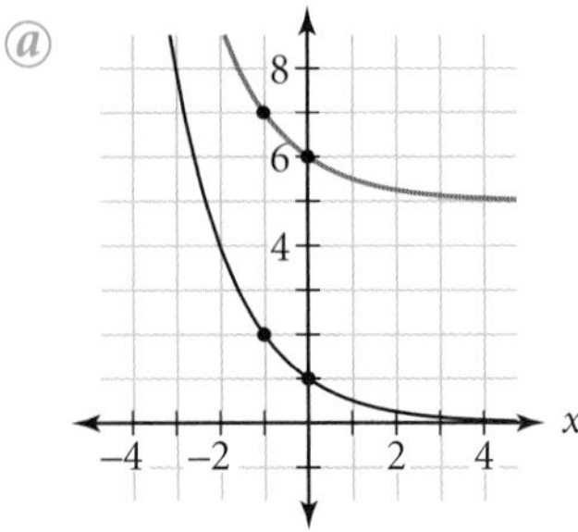

b.

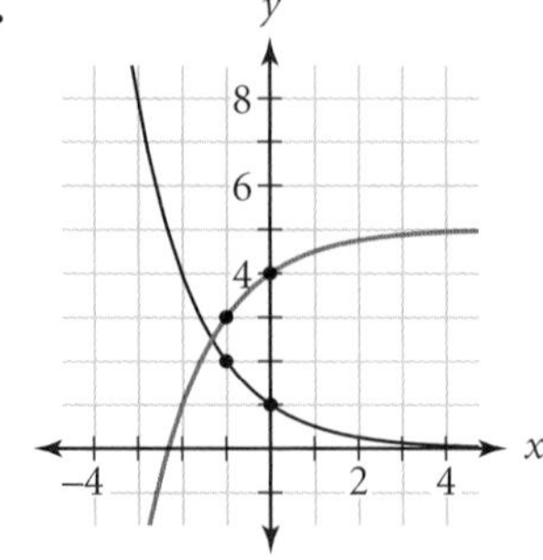

c. (@)

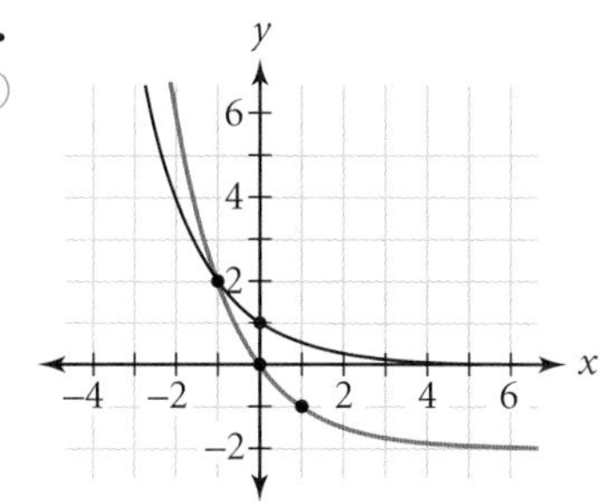

d.

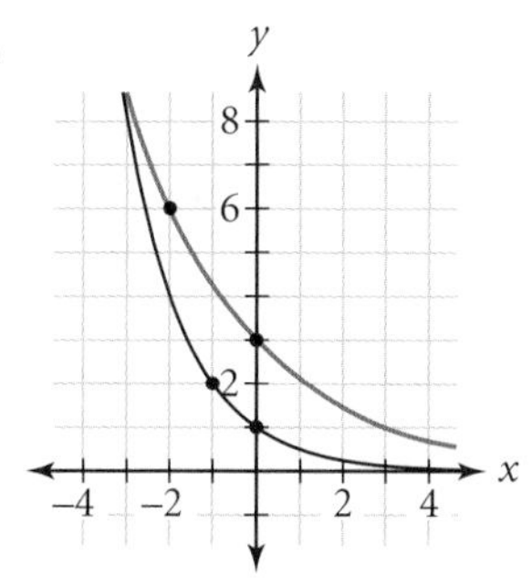

11. The general form of an exponential equation, $y = ab^x$, is convenient when you know the y-intercept. Start with $f(0) = 30$ and $f(1) = 27$.

a. Find the common ratio. (@)

b. Write the function $f(x)$ that passes through the two data points. (@)

c. Graph $f(x)$ and $g(x) = f(x - 4)$ on the same axes.

d. What is the value of $g(4)$?

e. Write an equation for $g(x)$ that does not use its y-intercept. (h)

f. Explain in your own words why $y = y_1 \cdot b^{x-x_1}$ might be called the point-ratio form. (h)

12. Write two equations in the form $y = y_1 \cdot b^{x-x_1}$ that contain the points listed in the table. (@)

x	0	1	2	3	4
y	4	7.2	12.96	23.328	41.9904

This painting, *Inspiration Point* (2000), by contemporary American artist Nina Bovasso (b 1965), implies an explosion of exponential growth.

Review

13. Janell starts 10 m from a motion sensor and walks at 2 m/s toward the sensor. When she is 3 m from the sensor, she instantly turns around and walks at the same speed back to the starting point.

a. Sketch a graph of the function that models Janell's walk.

b. Give the domain and range of the function.

c. Write an equation of the function.

14. The graph shows a line and the graph of $y = f(x)$.

a. Fill in the missing values to make a true statement.

$f(\underline{?}) = \underline{?}$.

b. Find the equation of the pictured line.

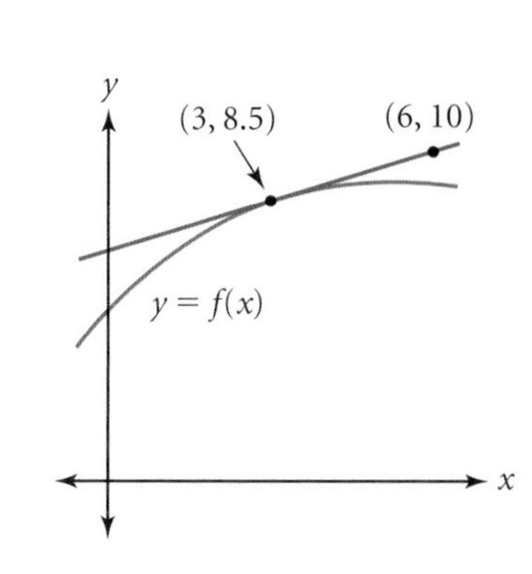

15. Austin deposits \$5,000 into an account that pays 3.5% annual interest. Sami deposits \$5,200 into an account that pays 3.2% annual interest.

a. Write an expression for the amount of money Austin will have in his account after 5 years if he doesn't deposit any more money.

b. Write an expression for the amount of money Sami will have in her account after 5 years if she doesn't deposit any more money.

c. How long will it take until Austin has more money than Sami?

16. You can use different techniques to find the product of two binomials, such as $(x - 4)(x + 6)$.

a. Use a rectangle diagram to find the product.

b. You can use the distributive property to rewrite the expression $(x - 4)(x + 6)$ as $x(x + 6) - 4(x + 6)$. Use the distributive property again to find all the terms. Combine like terms.

c. Compare your answers to 16a and b. Are they the same?

d. Compare the methods in 16a and b. How are they alike?

Project

THE COST OF LIVING

You may have heard someone say something like, "I remember when a hamburger cost five cents!" Did you know that the cost of living tends to increase exponentially? Talk to a relative or an acquaintance to see whether he or she remembers the cost of a specific item in a certain year. The item could be a meal, a movie ticket, a house in your neighborhood, or anything else the person recalls.

Research the current cost of that same item. How much has it increased? Use your two data points to write an exponential equation for the cost of the item. When can you expect the cost of the item to be double what it is now?

Research the cost of the item in a different year. How close is this third data point to the value predicted by your model?

In 1944, about 10,000 fans formed a line stretching several blocks outside of Manhattan's Paramount Theater to see Frank Sinatra. Although the cost of seeing a live concert has changed since 1944, the enthusiasm of fans has not.

Your project should include

- Your data and sources.
- Your equation and why you chose that model.
- The doubling time for the cost of your item.
- An analysis of how well your model predicted the third data point.
- An analysis of how accurate you think your model is.

LESSON

5.2

Properties of Exponents and Power Functions

Frequently, you will need to rewrite a mathematical expression in a different form to make the expression easier to understand or an equation easier to solve. Recall that in an exponential expression, such as 4^3, the number 4 is called the **base** and the number 3 is called the **exponent.** You say that 4 is **raised to the power** of 3. If the exponent is a positive integer, you can write the expression in **expanded form,** for example, $4^3 = 4 \cdot 4 \cdot 4$. Because 4^3 equals 64, you say that 64 is a power of 4.

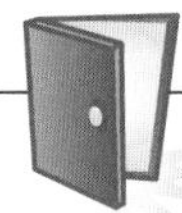

Investigation

Properties of Exponents

Use expanded form to review and generalize the properties of exponents.

Step 1 Write each product in expanded form, and then rewrite it in exponential form.

a. $2^3 \cdot 2^4$ **b.** $x^5 \cdot x^{12}$ **c.** $10^2 \cdot 10^5$

Step 2 Generalize your results from Step 1. $a^m \cdot a^n = \underline{?}$

Step 3 Write the numerator and denominator of each quotient in expanded form. Reduce by eliminating common factors, and then rewrite the factors that remain in exponential form.

a. $\frac{4^5}{4^2}$ **b.** $\frac{x^8}{x^6}$ **c.** $\frac{(0.94)^{15}}{(0.94)^5}$

Step 4 Generalize your results from Step 3. $\frac{a^m}{a^n} = \underline{?}$

Step 5 Write each quotient in expanded form, reduce, and rewrite in exponential form.

a. $\frac{2^3}{2^4}$ **b.** $\frac{4^5}{4^7}$ **c.** $\frac{x^3}{x^8}$

Step 6 Rewrite each quotient in Step 5 using the property you discovered in Step 4.

Step 7 Generalize your results from Steps 5 and 6. $\frac{1}{a^n} = \underline{?}$

Step 8 Write several expressions in the form $(a^n)^m$. Expand each expression, and then rewrite it in exponential form. Generalize your results.

Step 9 Write several expressions in the form $(a \cdot b)^n$. Don't multiply a times b. Expand each expression, and then rewrite it in exponential form. Generalize your results.

Step 10 Show that $a^0 = 1$, using the properties you have discovered. Write at least two exponential expressions to support your explanation.

Here's a summary of the properties of exponents. You discovered some of these in the investigation. Try to write an example of each property.

For $a > 0$, $b > 0$, and all values of m and n, these properties are true:

Product Property of Exponents

$$a^m \cdot a^n = a^{m+n}$$

Quotient Property of Exponents

$$\frac{a^m}{a^n} = a^{m-n}$$

Definition of Negative Exponents

$$a^{-n} = \frac{1}{a^n} \quad \text{or} \quad \left(\frac{a}{b}\right)^{-n} = \left(\frac{b}{a}\right)^n$$

Zero Exponents

$$a^0 = 1$$

Power of a Power Property

$$(a^m)^n = a^{mn}$$

Power of a Product Property

$$(ab)^m = a^m b^m$$

Power of a Quotient Property

$$\left(\frac{a}{b}\right)^n = \frac{a^n}{b^n}$$

Power Property of Equality

If $a = b$, then $a^n = b^n$.

Common Base Property of Equality

If $a^n = a^m$, and $a \neq 1$, then $n = m$.

In Lesson 5.1, you learned to solve equations that have a variable in the exponent by using a calculator to try various values for x. The properties of exponents allow you to solve these types of equations algebraically. One special case is when you can rewrite both sides of the equation with a common base. This strategy is fundamental to solving some of the equations you'll see later in this chapter.

EXAMPLE A

Solve.

a. $8^x = 4$ **b.** $27^x = \frac{1}{81}$ **c.** $\left(\frac{49}{9}\right)^x = \left(\frac{3}{7}\right)^{3/2}$

▶ Solution

If you use the power of a power property to convert each side of the equation to a common base, then you can solve without a calculator.

a.

$8^x = 4$	Original equation.
$(2^3)^x = 2^2$	$8 = 2^3$ and $4 = 2^2$.
$2^{3x} = 2^2$	Use the power of a power property to rewrite $(2^3)^x$ as 2^{3x}.
$3x = 2$	Use the common base property of equality.
$x = \frac{2}{3}$	Divide.

b. $27^x = \frac{1}{81}$ Original equation.

$(3^3)^x = \frac{1}{3^4}$ $27 = 3^3$ and $81 = 3^4$.

$3^{3x} = 3^{-4}$ Use the power of a power property and the definition of negative exponents.

$3x = -4$ Use the common base property of equality.

$x = -\frac{4}{3}$ Divide.

c. $\left(\frac{49}{9}\right)^x = \left(\frac{3}{7}\right)^{3/2}$ Original equation.

$\left(\frac{7^2}{3^2}\right)^x = \left(\frac{3}{7}\right)^{3/2}$ $49 = 7^2$ and $9 = 3^2$.

$\left[\left(\frac{7}{3}\right)^2\right]^x = \left[\left(\frac{7}{3}\right)^{-1}\right]^{3/2}$ Use the power of a quotient property and the definition of negative exponents.

$\left(\frac{7}{3}\right)^{2x} = \left(\frac{7}{3}\right)^{-3/2}$ Use the power of a power property.

$2x = -\frac{3}{2}$ Use the common base property of equality.

$x = -\frac{3}{4}$ Divide.

Remember, it's always a good idea to check your answer with a calculator. [▶ See **Calculator Note 5A** to find out how to calculate roots and powers. ◀]

An exponential function has a variable in the exponent. In the exercises you'll use exponent properties to verify the equivalence of exponential equations.

Exponential Function

The general form of an exponential function is

$y = ab^x$

where a and b are constants and $b > 0$.

In a **power function,** the variable is in the base. You must learn to distinguish between power functions and exponential functions.

Power Function

The general form of a power function is

$y = ax^n$

where a and n are constants.

You use different methods to solve power equations than to solve exponential equations.

EXAMPLE B Solve for positive values of x.

a. $x^4 = 3000$

b. $6x^{2.5} = 90$

▶ Solution To solve a power equation, use the power of a power property and choose an exponent that will undo the exponent on x.

a.

$x^4 = 3000$	Original equation.
$(x^4)^{1/4} = 3000^{1/4}$	Use the power of a power property. Raising both sides to the power of $\frac{1}{4}$, the reciprocal of 4, "undoes" the power of 4 on x.
$x \approx 7.40$	Use your calculator to approximate the value of $3000^{1/4}$.

b.

$6x^{2.5} = 90$	Original equation.
$x^{2.5} = 15$	Divide both sides by 6.
$(x^{2.5})^{1/2.5} = 15^{1/2.5}$	Use the power of a power property and choose the exponent $\frac{1}{2.5}$.
$x \approx 2.95$	Approximate the value of $15^{1/2.5}$.

In this book the properties of exponents are defined only for positive bases. So while there may be negative values that are solutions to power equations, in this context you will be asked to find only the positive solutions.

EXERCISES

You will need

A graphing calculator for Exercises **9**, **10**, **12**, and **17**.

Practice Your Skills

1. Rewrite each expression as a fraction without exponents or as an integer. Verify that your answer is equivalent to the original expression by evaluating each on your calculator.

a. 5^{-3} ⓐ

b. -6^2

c. -3^{-4} ⓐ

d. $(-12)^{-2}$

e. $\left(\frac{3}{4}\right)^{-2}$ ⓐ

f. $\left(\frac{2}{7}\right)^{-1}$

One of the authors of this book, Ellen Kamischke, works with two students in Interlochen, Michigan.

2. Rewrite each expression in the form a^n.

a. $a^8 \cdot a^{-3}$ ⓐ

b. $\frac{b^6}{b^2}$

c. $(c^4)^5$

d. $\frac{d^0}{e^{-3}}$ ⓐ

3. State whether each equation is true or false. If it is false, explain why.

a. $3^5 \cdot 4^2 = 12^7$

b. $100(1.06)^x = (106)^x$

c. $\frac{4^x}{4} = 1^x$

d. $\frac{6.6 \cdot 10^{12}}{8.8 \cdot 10^{-4}} = 7.5 \cdot 10^{15}$

4. Solve.

a. $3^x = \frac{1}{9}$ ⓐ

b. $\left(\frac{5}{3}\right)^x = \frac{27}{125}$

c. $\left(\frac{1}{3}\right)^x = 243$ ⓐ

d. $5 \cdot 3^x = 5$

5. Solve each equation for positive values of x. If answers are not exact, approximate to two decimal places.

a. $x^7 = 4000$ ⓐ

b. $x^{0.5} = 28$

c. $x^{-3} = 247$

d. $5x^{1/4} + 6 = 10.2$ ⓗ

e. $3x^{-2} = 2x^4$

f. $-3x^{1/2} + (4x)^{1/2} = -1$ ⓐ

Reason and Apply

6. Rewrite each expression in the form ax^n.

a. $x^6 \cdot x^6$

b. $4x^6 \cdot 2x^6$

c. $(-5x^3) \cdot (-2x^4)$ ⓐ

d. $\frac{72x^7}{6x^2}$

e. $\left(\frac{6x^5}{3x}\right)^3$

f. $\left(\frac{20x^7}{4x}\right)^{-2}$ ⓐ

7. *Mini-Investigation* You've seen that the power of a product property allows you to rewrite $(a \cdot b)^n$ as $a^n \cdot b^n$. Is there a power of a sum property that allows you to rewrite $(a + b)^n$ as $a^n + b^n$? Write some numerical expressions in the form $(a + b)^n$ and evaluate them. Are your answers equivalent to $a^n + b^n$ always, sometimes, or never? Write a short paragraph that summarizes your findings. ⓗ

8. Consider this sequence:

$7^2, 7^{2.25}, 7^{2.5}, 7^{2.75}, 7^3$

a. Use your calculator to evaluate each term in the sequence. If answers are not exact, approximate to four decimal places.

b. Find the differences between the consecutive terms of the sequence. What do these differences tell you?

c. Find the ratios of the consecutive terms in 8a. What do these values tell you?

d. What observation can you make about these decimal powers?

9. *Mini-Investigation* For 9a–d, graph the equations on your calculator.

a. $y = x^2$

b. $y = x^3$

c. $y = x^4$

d. $y = x^5$

e. How do the graphs compare? How do they contrast? What points (if any) do they have in common?

f. Predict what the graphs of $y = x^6$ and x^7 will look like. Verify your predictions by using your calculator.

g. Each of the equations you have graphed is a power function. How do the graphs of these functions differ from the graphs of exponential functions in Lesson 5.1?

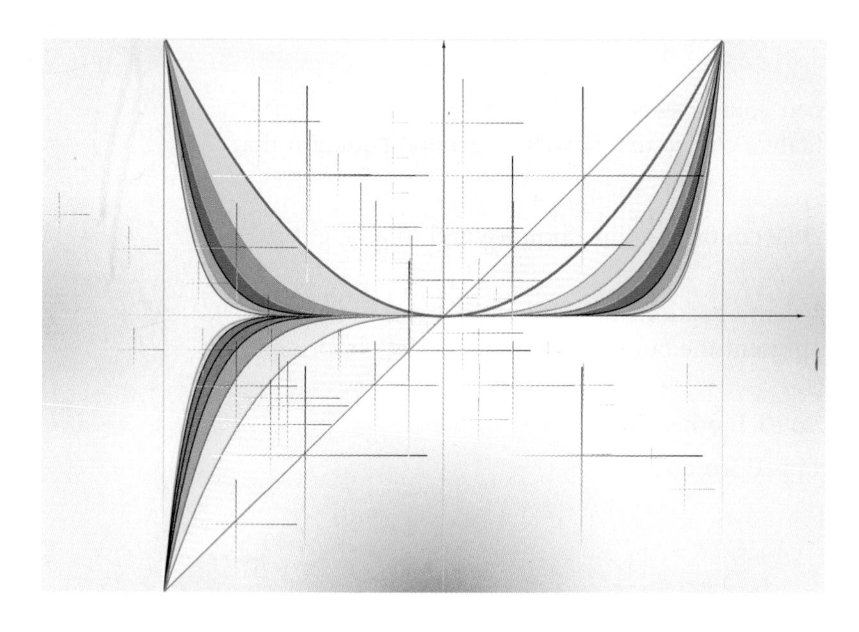

10. Each of the red curves is a transformation of the graph of $y = x^3$, shown in black. Write an equation for each red curve.

a.

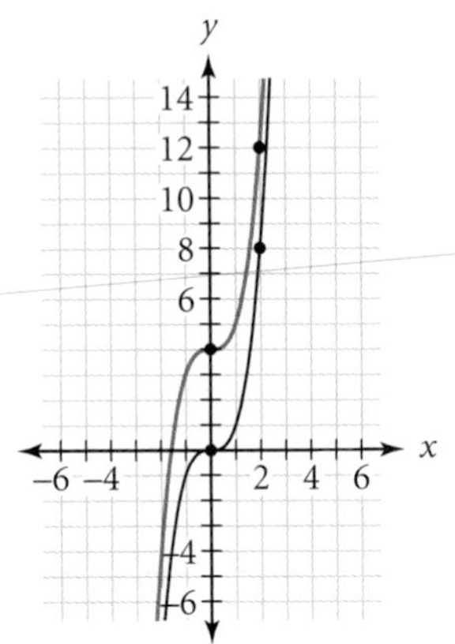

b.

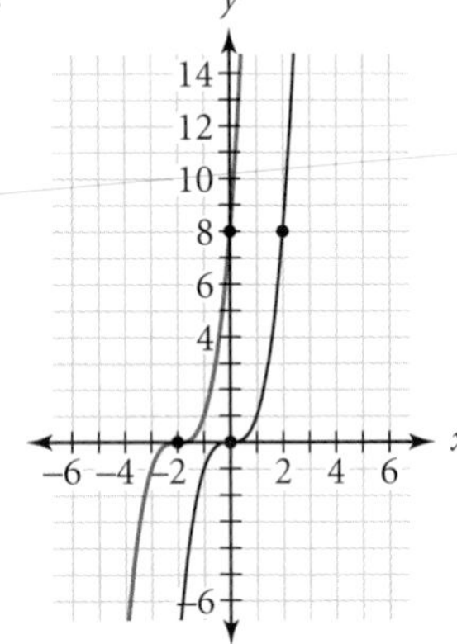

c. ⓐ

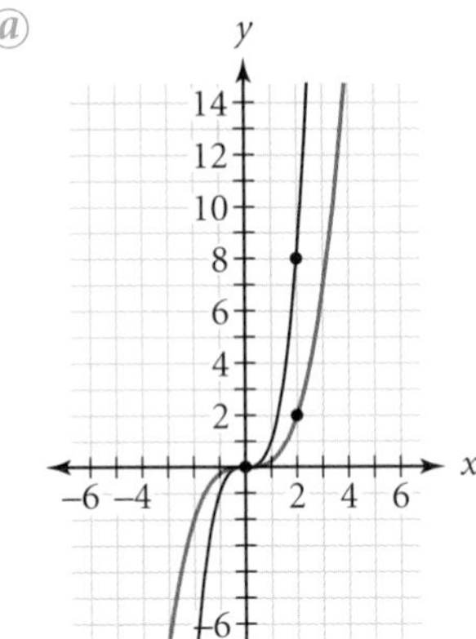

d.

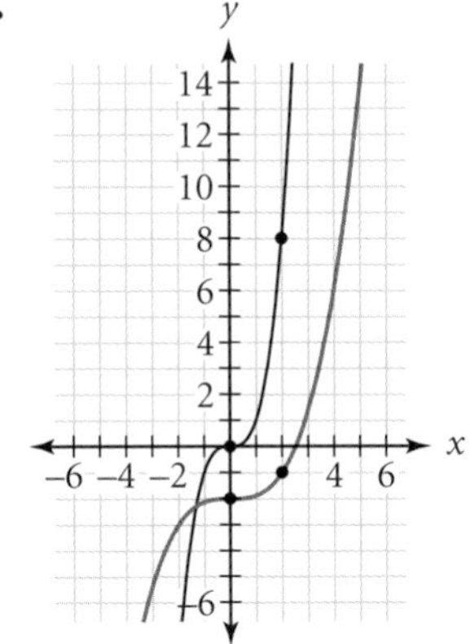

11. Consider the exponential function $f(x) = 47(0.9)^x$. Several points satisfying the function are shown in the calculator table. Notice that when $x = 0$, $f(x) = 47$.

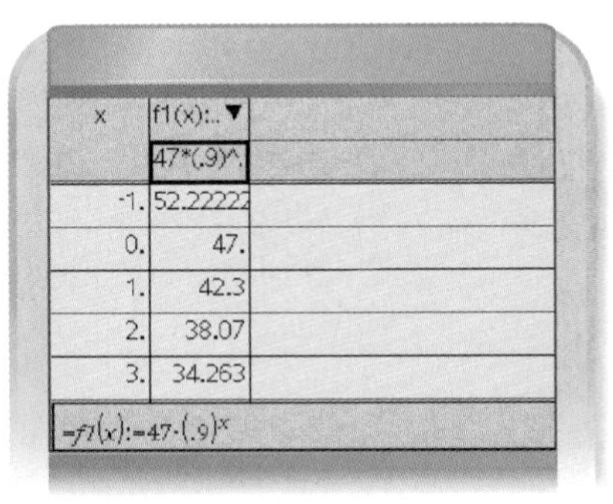

x	f1(x):= 47*(.9)^
-1.	52.22222
0.	47.
1.	42.3
2.	38.07
3.	34.263

f1(x):=47·(.9)^x

a. The expression $47(0.9)^x$ could be rewritten as $47(0.9)(0.9)^{x-1}$. Explain why this is true. Rewrite $47(0.9)(0.9)^{x-1}$ in the form $a \cdot b^{x-1}$.

b. The expression $47(0.9)^x$ could also be rewritten as $47(0.9)(0.9)(0.9)^{x-2}$. Rewrite $47(0.9)(0.9)(0.9)^{x-2}$ in the form $a \cdot b^{x-2}$.

c. Look for a connection between your answers to 11a and b and the values in the table. State a conjecture or write a general equation that summarizes your findings.

12. A ball rebounds to a height of 30.0 cm on the third bounce and to a height of 5.2 cm on the sixth bounce.

a. Write two different yet equivalent equations in point-ratio form, $y = y_1 \cdot b^{x-x_1}$, using r for the ratio. Let x represent the bounce number, and let y represent the rebound height in centimeters. ⓗ

b. Set the two equations equal to each other. Solve for r. ⓐ

c. What height was the ball dropped from? ⓗ

13. Solve.

a. $(x - 3)^3 = 64$

b. $256^x = \frac{1}{16}$

c. $\frac{(x+5)^3}{(x+5)} = x^2 + 25$ ⓗ

14. APPLICATION A radioactive sample was created in 1980. In 2002, a technician measures the radioactivity at 42.0 rads. One year later, the radioactivity is 39.8 rads.

a. Find the ratio of radioactivity between 2002 and 2003. Approximate your answer to four decimal places.

b. Let x represent the year, and let y represent the radioactivity in rads. Write an equation in point-ratio form, $y = y_1 \cdot b^{x-x_1}$, using the point $(x_1, y_1) = (2002, 42)$.

c. Write an equation in point-ratio form using the point (2003, 39.8).

d. Calculate the radioactivity in 1980 using both equations. @

e. Calculate the radioactivity in 2010 using both equations.

f. Use the properties of exponents to show that the equations in 14b and c are equivalent.

Review

15. Name the x-value that makes each equation true.

a. $37{,}000{,}000 = 3.7 \cdot 10^x$

b. $0.000801 = 8.01 \cdot 10^x$ @

c. $47{,}500 = 4.75 \cdot 10^x$ @

d. $0.0461 = x \cdot 10^{-2}$

16. Solve this equation for y. Then carefully graph it on your paper.

$$\frac{y+3}{2} = (x+4)^2$$

17. Paul collects these time-distance data for a remote-controlled car.

Time (s)	5	8	8	10	15	18	22	24	31	32
Distance (m)	0.8	1.7	1.6	1.9	3.3	3.4	4.1	4.6	6.4	6.2

a. Define variables and make a scatter plot of these data.

b. Use the median-median line to estimate the car's speed. (*Note:* Don't do more work than necessary.)

IMPROVING YOUR REASONING SKILLS

Breakfast Is Served

Mr. Higgins told his wife, a mathematics professor, that he would make her breakfast. She handed him this message:

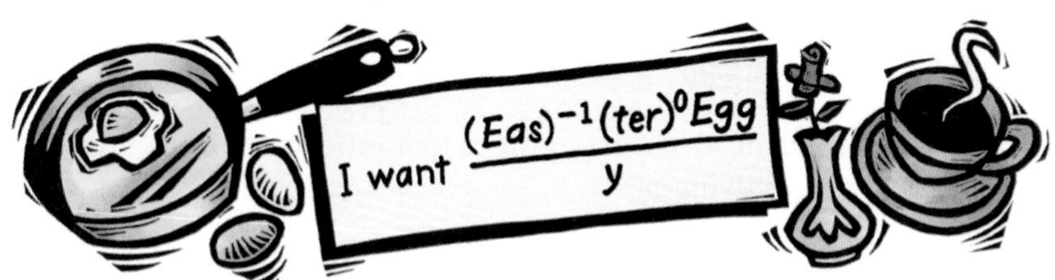

What should Mr. Higgins fix his wife for breakfast?

LESSON

5.3

Rational Exponents and Roots

In previous lessons you worked mostly with integer exponents and their properties. In Example B of Lesson 5.2, you saw how the reciprocal of an exponent can be used to undo the exponent. In this lesson you will investigate fractional and other **rational** exponents. Keep in mind that all of the properties you learned in Lesson 5.2 apply to this larger class of exponents as well.

The volume and surface area of a cube, such as this fountain in Osaka, Japan, are related by rational exponents.

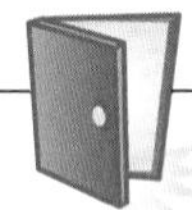

Investigation
Getting to the Root

In this investigation you'll explore the relationship between x and $x^{1/2}$ and learn how to find the values of some expressions with rational exponents.

Step 1 Use your calculator to create a table for $y = x^{1/2}$ at integer values of x. When is $x^{1/2}$ a positive integer? Describe the relationship between x and $x^{1/2}$.

Step 2 Graph $y = x^{1/2}$ in a graphing window with x- and y-values less than 10. This graph should look familiar to you. Make a conjecture about what other function is equivalent to $y = x^{1/2}$, enter your guess as a second equation, and verify that the equations give the same y-value at each x-value.

Step 3 State what you have discovered about raising a number to a power of $\frac{1}{2}$. Include an example with your statement.

Step 4 Clear the previous functions, and make a table for $y = 25^x$ with x incrementing by $\frac{1}{2}$.

Step 5 Study your table and explain any relationships you see. How could you find the value of $49^{3/2}$ without a calculator? Check your answer using a calculator.

Step 6 How could you find the value of $27^{2/3}$ without a calculator? Verify your response and then test your strategy on $8^{5/3}$. Check your answer.

Step 7 Describe what it means to raise a number to a rational exponent, and generalize a procedure for simplifying $a^{m/n}$.

Rational exponents with numerator 1 indicate positive roots. For example, $x^{1/5}$ is the same as $\sqrt[5]{x}$, or the "fifth root of x," and $x^{1/n}$ is the same as $\sqrt[n]{x}$, or the "nth root of x." The fifth root of x is the number that, raised to the power 5, gives x.

For rational exponents with numerators other than 1, such as $9^{3/2}$, the numerator is interpreted as the exponent to which to raise the root. That is, $9^{3/2}$ is the same as $(9^{1/2})^3$, or $(\sqrt{9})^3$, or 27.

Definition of Rational Exponents

The power of a power property shows that $a^{m/n} = (a^{1/n})^m$ and $a^{m/n} = (a^m)^{1/n}$, so

$$a^{m/n} = (\sqrt[n]{a})^m \text{ or } \sqrt[n]{a^m} \text{ for } a > 0$$

Properties of rational exponents are useful in solving equations with exponents.

EXAMPLE A

Rewrite with rational exponents, and find the positive solution.

a. $\sqrt[4]{a} = 14$

b. $\sqrt[9]{b^5} = 26$

c. $(\sqrt[3]{c})^8 = 47$

▶ Solution

Rewrite each expression with a rational exponent, then use properties of exponents to find the positive solution.

a.

$\sqrt[4]{a} = 14$	Original equation.
$a^{1/4} = 14$	Rewrite $\sqrt[4]{a}$ as $a^{1/4}$.
$(a^{1/4})^4 = 14^4$	Raise both sides to the power of 4.
$a = 38{,}416$	Evaluate 14^4.

b.

$\sqrt[9]{b^5} = 26$	Original equation.
$b^{5/9} = 26$	Rewrite $\sqrt[9]{b^5}$ as $b^{5/9}$.
$(b^{5/9})^{9/5} = 26^{9/5}$	Raise both sides to the power of $\frac{9}{5}$.
$b \approx 352.33$	Approximate the value of $26^{9/5}$.

c.

$(\sqrt[3]{c})^8 = 47$	Original equation.
$c^{8/3} = 47$	Rewrite $(\sqrt[3]{c})^8$ as $c^{8/3}$.
$(c^{8/3})^{3/8} = 47^{3/8}$	Raise both sides to the power of $\frac{3}{8}$.
$c \approx 4.237$	Approximate the value of $47^{3/8}$.

Recall that properties of exponents give only one solution to an equation, because they are defined only for positive bases. Will negative values of *a*, *b*, or *c* satisfy any of the equations in Example A?

In the previous lesson, you learned that functions in the general form $y = ax^n$ are power functions. A rational function, such as $y = \sqrt[9]{x^5}$, is considered to be a power function because it can be rewritten as $y = x^{5/9}$. All the transformations you discovered for parabolas and square root curves also apply to any function that can be written in the general form $y = ax^n$.

Recall that the equation of a line can be written using the point-slope form if you know a point on the line and the slope between points. Similarly, the equation for an exponential curve can be written using **point-ratio form** if you know a point on the curve and the common ratio between points that are 1 horizontal unit apart.

Point-Ratio Form

If an exponential curve passes through the point (x_1, y_1) and the function values have ratio b for values of x that differ by 1, the point-ratio form of the equation is

$$y = y_1 \cdot b^{x-x_1}$$

You have seen that if $x = 0$, then $y = a$ in the general exponential equation $y = a \cdot b^x$. This means that a is the initial value of the function at time 0 (the y-intercept) and b is the growth or decay ratio. This is consistent with the point-ratio form because when you substitute the point $(0, a)$ into the equation, you get $y = a \cdot b^{x-0}$, or $y = a \cdot b^x$.

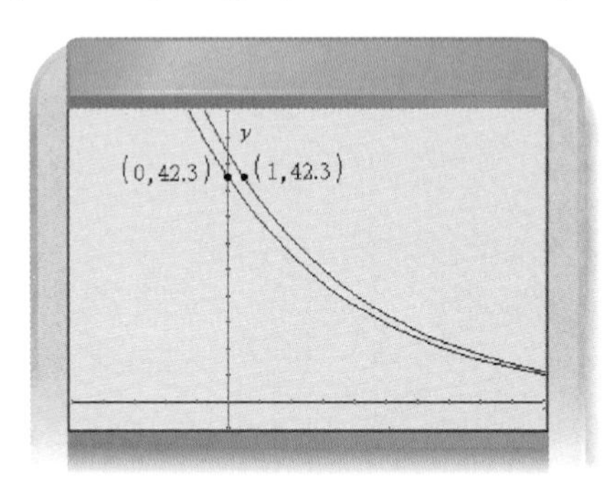

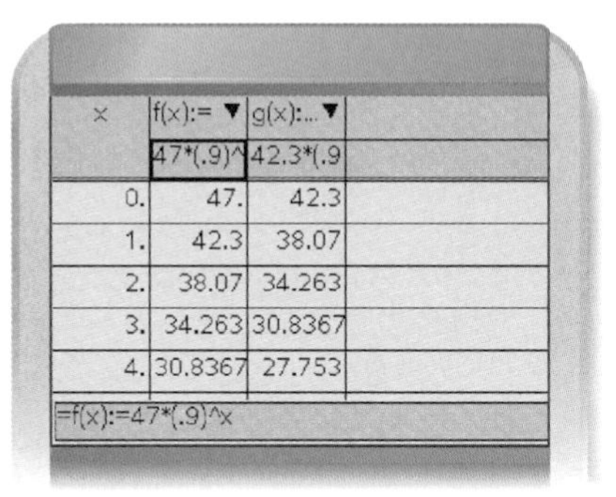

x	f(x):= 47*(.9)^	g(x):... 42.3*(.9
0.	47.	42.3
1.	42.3	38.07
2.	38.07	34.263
3.	34.263	30.8367
4.	30.8367	27.753

=f(x):=47*(.9)^x

The graph and table above show the exponential functions $f(x) = 47(0.9)^x$ and $g(x) = 42.3(0.9)^x$. Both the graph and table indicate that if the graph of function g is translated right 1 unit, it becomes the same as the graph of function f. So $f(x) = g(x - 1)$, or $f(x) = 42.3(0.9)^{(x-1)} = 47(0.9)^x$. This shows that using the point $(1, 42.3)$ in the point-ratio form gives you an equation equivalent to $y = a \cdot b^x$.

Try substituting another point, (x_1, y_1), along with the ratio $b = 0.9$ into the point-ratio form to convince yourself that any point (x_1, y_1) on the curve can be used to write an equation, $y = y_1 \cdot b^{x-x_1}$, that is equivalent to $y = a \cdot b^x$. You may want to use your graphing calculator or algebraic techniques.

EXAMPLE B

Casey hit the bell in the school clock tower. Her pressure reader, held nearby, measured the sound intensity, or loudness, at 40 lb/in^2 after 4 s had elapsed and at 4.7 lb/in^2 after 7.2 s had elapsed. She remembers from her science class that sound decays exponentially.

a. Name two points that the exponential curve must pass through.

b. Find an exponential equation that models these data.

c. How loud was the bell when it was struck (at 0 s)?

The bell tower at Oglethorpe University in Atlanta, Georgia.

▶ Solution

a. Time is the independent variable, x, and loudness is the dependent variable, y, so the two points are $(4, 40)$ and $(7.2, 4.7)$.

b. Start by substituting the coordinates of each of the two points into the point-ratio form, $y = y_1 \cdot b^{x-x_1}$.

$$y = 40b^{x-4} \quad \text{and} \quad y = 4.7b^{x-7.2}$$

Note that you don't yet know what b is. If you were given y-values for two consecutive integer points, you could divide to find the ratio. In this case, however, there are 3.2 horizontal units between the two points you are given, so you'll need to solve for b.

$40b^{x-4} = 4.7b^{x-7.2}$ Use substitution to combine the two equations.

$b^{x-4} = \dfrac{4.7b^{x-7.2}}{40}$ Divide both sides by 40.

$\dfrac{b^{x-4}}{b^{x-7.2}} = \dfrac{4.7}{40}$ Divide both sides by $b^{x-7.2}$.

$b^{(x-4)-(x-7.2)} = 0.1175$ Use the quotient property of exponents.

$b^{3.2} = 0.1175$ Combine like terms in the exponent.

$(b^{3.2})^{1/3.2} = (0.1175)^{1/3.2}$ Raise both sides to the power $\frac{1}{3.2}$.

$b \approx 0.5121$ Approximate the value of $0.1175^{1/3.2}$.

$y \approx 40(0.5121)^{x-4}$ Substitute 0.5121 for b in either of the two original equations.

The exponential equation that passes through the points $(4, 40)$ and $(7.2, 4.7)$ is $y \approx 40(0.5121)^{x-4}$.

c. To find the loudness at 0 s, substitute $x = 0$.

$$y \approx 40(0.5121)^{0-4} \approx 581$$

The sound intensity was approximately 581 lb/in^2 when the bell was struck.

Science CONNECTION

Sound is usually measured in bels, named after American inventor and educator Alexander Graham Bell (1847–1922). A decibel (dB) is one-tenth of a bel. The decibel scale measures loudness in terms of what an average human can hear. On the decibel scale, 0 dB is inaudible and 130 dB is the threshold of pain. However, sound can also be measured in terms of the pressure that the sound waves exert on a drum. In the metric system, pressure is measured in Pascals (Pa), a unit of force per square meter.

Prolonged or repeated exposure to loud noise can damage the inner ear and lead to noise-induced hearing loss. If you work in a noisy environment or attend loud concerts, you can reduce your risk of hearing damage by wearing protective ear coverings or earplugs.

In Example B, part b, note that the base, 0.5121, was an approximation for b found by dividing 4.7 by 40, then raising that quotient to the power $\frac{1}{3.2}$. You could use $\left(\frac{4.7}{40}\right)^{1/3.2}$ as an exact value of b in the exponential equation.

$$y = 40\left[\left(\frac{4.7}{40}\right)^{1/3.2}\right]^{x-4}$$

Using the power of a power property, you can rewrite this as

$$y = 40\left(\frac{4.7}{40}\right)^{(x-4)/3.2}$$

This equation indicates that the curve passes through the point $(4, 40)$ and that it has a ratio of $\frac{4.7}{40}$ spread over 3.2 units (from $x = 4$ to $x = 7.2$) rather than over 1 unit. Dividing the exponent by 3.2 stretches the graph horizontally by 3.2 units. Using this method, you can write an equation for an exponential curve in only one step.

EXERCISES

You will need

A graphing calculator for Exercises **6**, **7**, **10**, and **15**.

Practice Your Skills

1. Match all expressions that are equivalent.

a. $\sqrt[5]{x^2}$ ⓐ b. $x^{2.5}$ c. $\sqrt[3]{x}$ d. $x^{5/2}$ e. $x^{0.4}$

f. $\left(\frac{1}{x}\right)^{-3}$ g. $(\sqrt{x})^5$ h. x^3 i. $x^{1/3}$ j. $x^{2/5}$

2. Identify each function as a power function, an exponential function, or neither of these. (It may be translated, stretched, or reflected.) Give a brief reason for your choice.

a. $f(x) = 17x^5$ b. $f(t) = t^3 + 5$ c. $g(v) = 200(1.03)^v$ d. $h(x) = 2x - 7$ ⓐ

e. $g(y) = 3\sqrt{y - 2}$ f. $f(t) = t^2 + 4t + 3$ g. $h(t) = \frac{12}{3^t}$ ⓐ h. $g(w) = \frac{28}{w - 5}$

i. $f(y) = \frac{8}{y^4} + 1$ j. $g(x) = \frac{x^3 + 2}{1 - x}$ ⓐ k. $h(w) = \sqrt[5]{4w^3}$ l. $p(x) = 5(0.8)^{(x-4)/2}$

3. Rewrite each expression in the form b^n in which n is a rational exponent.

a. $\sqrt[6]{a}$ b. $\sqrt[10]{b^8}$ ⓐ c. $\frac{1}{\sqrt{c}}$ ⓐ d. $(\sqrt[5]{d})^7$

4. Solve each equation and show or explain your step(s).

a. $\sqrt[6]{a} = 4.2$ b. $\sqrt[10]{b^8} = 14.3$ ⓐ c. $\frac{1}{\sqrt{c}} = 0.55$ ⓐ d. $(\sqrt[5]{d})^7 = 23$

Reason and Apply

5. APPLICATION Dan placed three colored gels over the main spotlight in the theater so that the intensity of the light on stage was 900 watts per square centimeter (W/cm^2). After he added two gels, making a total of five over the spotlight, the intensity on stage dropped to 600 W/cm^2. What will be the intensity of the light on stage with six gels over the spotlight if you know that the intensity of light decays exponentially with the thickness of material covering it? ⓗ

Colorful stage lights surround Freddie Mercury (1946–1991) of the band Queen in 1978.

6. *Mini-Investigation* For 6a–d, graph the equations on your calculator.

a. $y = x^{1/2}$ b. $y = x^{1/3}$

c. $y = x^{1/4}$ d. $y = x^{1/5}$

e. How do the graphs compare? What points (if any) do they have in common?

f. Predict what the graph of $y = x^{1/7}$ will look like. Verify your prediction by using your calculator.

g. What is the domain of each function? Explain why.

7. *Mini-Investigation* For 7a–d, graph the equations on your calculator.

a. $y = x^{1/4}$

b. $y = x^{2/4}$

c. $y = x^{3/4}$

d. $y = x^{4/4}$

e. How do the graphs compare? What points (if any) do they have in common?

f. Predict what the graph of $y = x^{5/4}$ will look like. Verify your prediction by using your calculator.

8. Compare your observations of the power functions in Exercises 6 and 7 to your previous work with exponential functions and power functions with positive integer exponents. How do the shapes of the curves compare? How do they contrast?

9. Identify each graph as an exponential function, a power function, or neither of these. ⓐ

a.

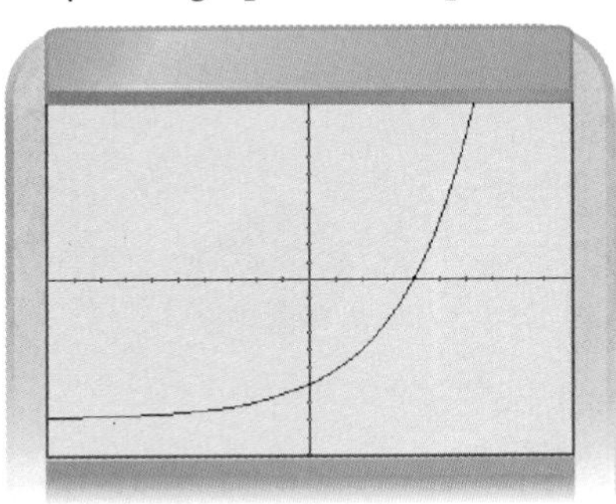

b.

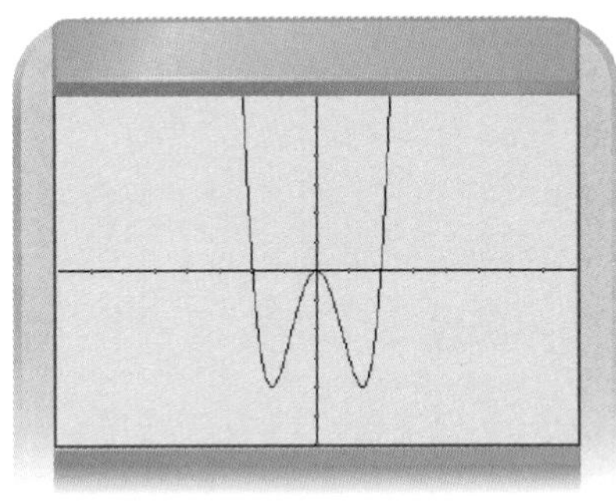

c.

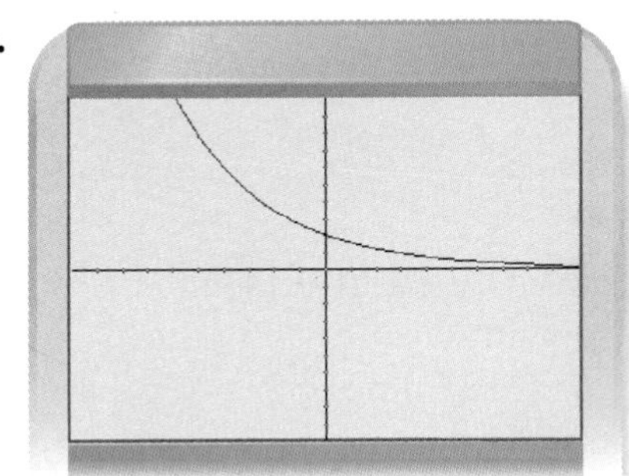

d.

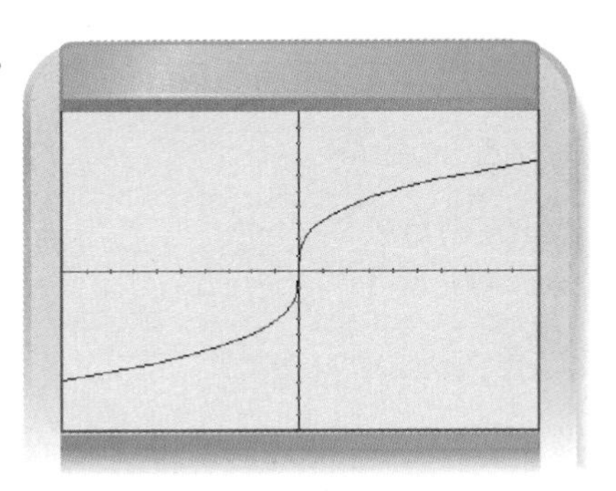

10. Each of the red curves is a transformation of the graph of the power function $y = x^{3/4}$, shown in black. Write an equation for each red curve. ⓗ

a. ⓐ

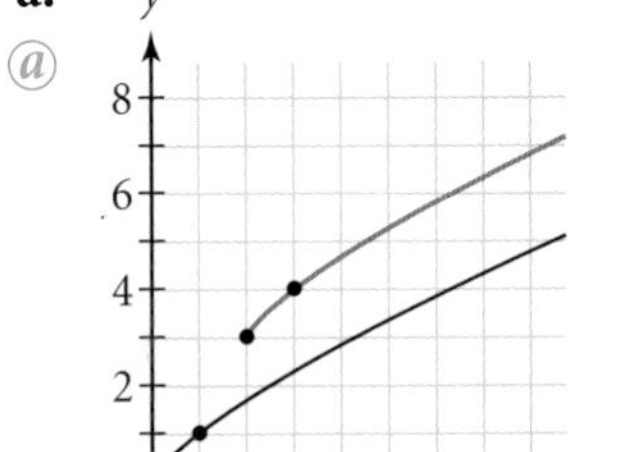

b. ⓐ

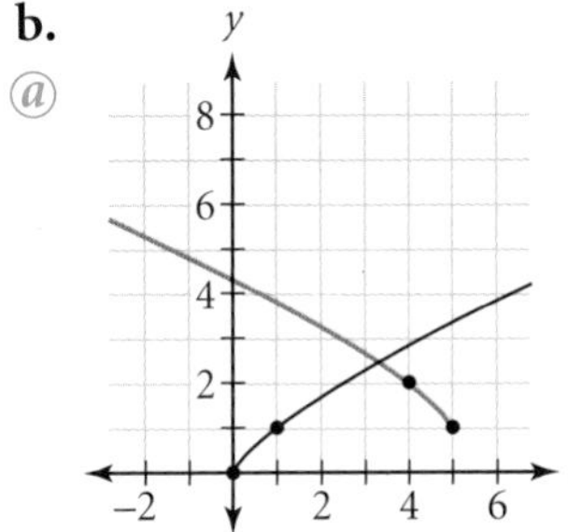

c.

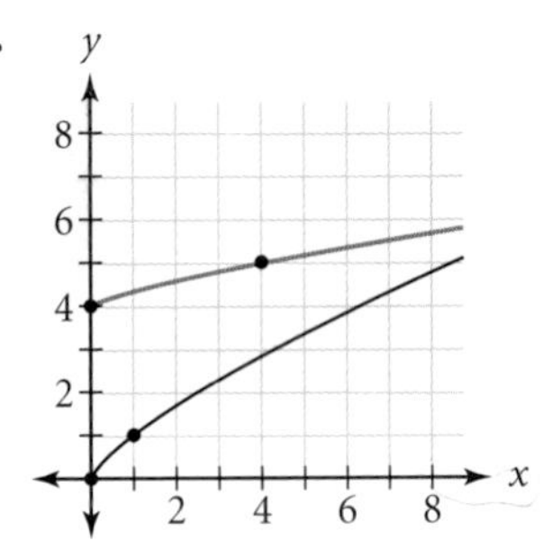

d.

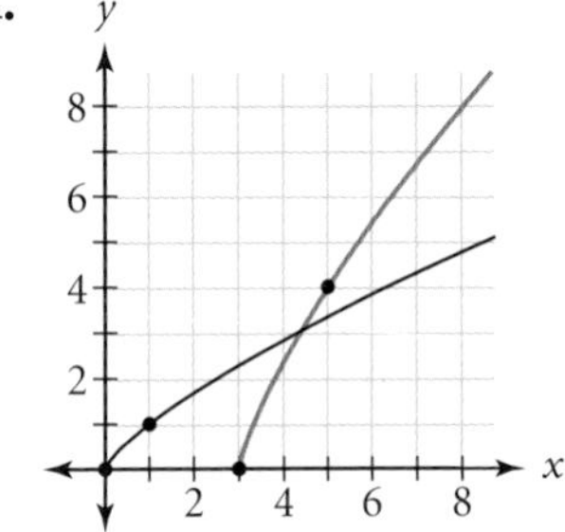

11. Solve. Approximate answers to the nearest hundredth.

a. $9\sqrt[5]{x} + 4 = 17$

b. $\sqrt{5x^4} = 30$ ⓐ

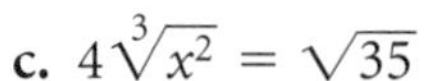

c. $4\sqrt[3]{x^2} = \sqrt{35}$

12. APPLICATION German astronomer Johannes Kepler (1571–1630) discovered in 1619 that the mean orbital radius of a planet, measured in astronomical units (AU), is equal to the time of one complete orbit around the sun, measured in years, raised to a power of $\frac{2}{3}$.

a. Venus has an orbital time of 0.615. What is its radius?

b. Saturn has a radius of 9.542 AU. How long is its orbital time?

c. Complete this table.

Planet	Mercury	Venus	Earth	Mars
Orbital radius (AU)	0.387			1.523
Orbital time (yr)		0.615	1.00	

Planet	Jupiter	Saturn	Uranus	Neptune
Orbital radius (AU)		9.542		30.086
Orbital time (yr)	11.861		84.008	

Clockwise from top left, this montage of separate planetary photos includes Mercury, Venus, Earth (and Moon), Mars, Jupiter, Saturn, Uranus, and Neptune.

13. APPLICATION Discovered by Irish chemist Robert Boyle (1627–1691) in 1662, Boyle's law gives the relationship between pressure and volume of gas if temperature and amount remain constant. If the volume in liters, V, of a container is increased, the pressure in millimeters of mercury (mm Hg), P, decreases. If the volume of a container is decreased, the pressure increases. One way to write this rule mathematically is $P = kV^{-1}$, where k is a constant.

a. Show that this formula is equivalent to $PV = k$. ⓗ

b. If a gas occupies 12.3 L at a pressure of 40.0 mm Hg, find the constant, k.

c. What is the volume of the gas in 13b when the pressure is increased to 60.0 mm Hg?

d. If the volume of the gas in 13b is 15 L, what would the pressure be?

Science CONNECTION

Scuba divers are trained in the effects of Boyle's law. As divers ascend, water pressure decreases, and so the air in the lungs expands. It is relatively safe to make an emergency ascent from a depth of 60 ft, but you must exhale as you do so. If you were to hold your breath while ascending, the expanding air in your lungs would cause your air sacs to rupture and your lungs to bleed.

A scuba diver swims below a coral reef in the Red Sea.

Review

14. Use properties of exponents to find an equivalent expression in the form ax^n.

a. $(3x^3)^3$ @

b. $(2x^3)(2x^2)^3$

c. $\frac{6x^4}{30x^5}$

d. $(4x^2)(3x^2)^3$ @

e. $\frac{-72x^5y^5}{-4x^3y}$ (Find an equivalent expression in the form ax^ny^m.)

15. For graphs a–h, write the equation of each graph as a transformation of $y = x^2$ or $y = \sqrt{x}$.

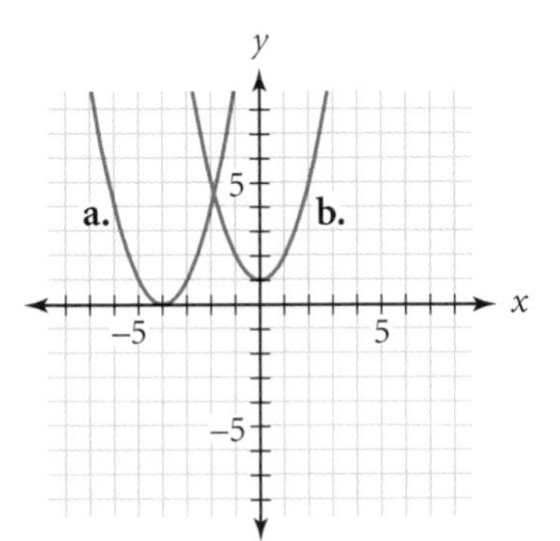

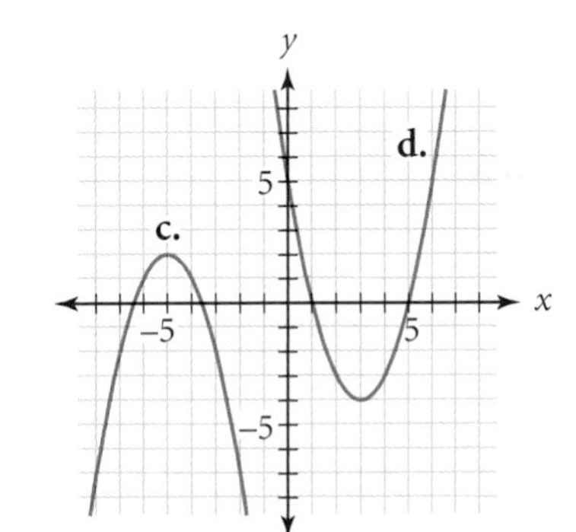

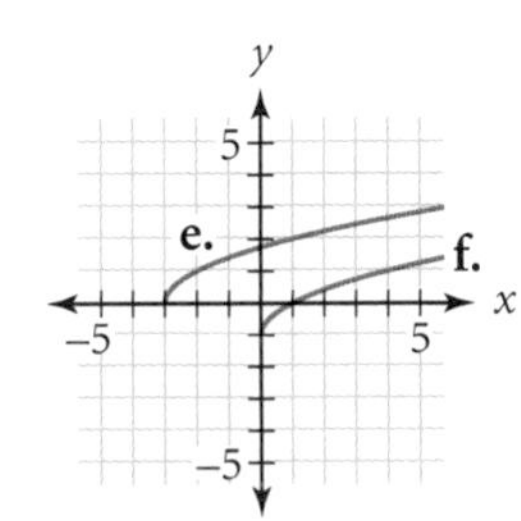

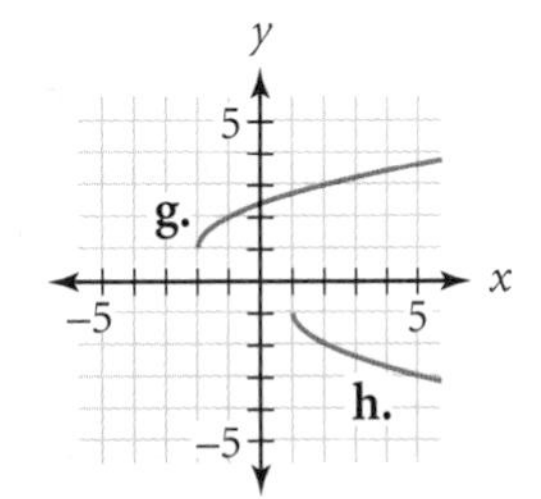

16. In order to qualify for the state dart championships, you must be in at least the 98th percentile of all registered dart players in the state. There are about 42,000 registered dart players. How many qualify for the championships?

17. The town of Hamlin has a growing rat population. Eight summers ago, there were 20 rat sightings, and the numbers have been increasing by about 20% each year.

a. Give a recursive formula that models the increasing rat population. Use the number of rats in the first year as u_1.

b. About how many rat sightings do you predict for this year?

c. Define variables and write an equation that models the continuous growth of the rat population.

German artist Katharina Fritsch (b 1956) designed *Rattenkönig (Rat-King)*, giant rat sculptures with tails knotted together, based on true accounts of this rare rat pack phenomenon.

Project

POWERS OF 10

How much longer does it take 1 billion seconds to go by than 1 million seconds? How much taller are you than an ant? Are there more grains of sand on a beach than stars in the sky? You've studied how quantities change exponentially, but just how different are 10^9 and 10^{10}?

In this project you'll identify and compare objects whose *orders of magnitude* in size or number are various powers of 10. When scientists describe a quantity as having a certain order of magnitude, they look only at the power of 10, not the decimal multiplier, when the quantity is expressed in *scientific notation.* For example, 9.2×10^3 is on the order of 10^3 even though it is very close to 10^4.

Decide what you're going to measure: length, area, volume, speed, or any other quantity. Then try to find at least one object with a measurement on the order of each power of 10. Your objects can be related in some way, but they don't have to be. For instance, what is the area of your kitchen? Your house? The state you live in? The land surface of Earth? You'll probably find some powers of 10 more easily than others.

Your project should include

- A list of the object or objects you found for each power of 10, and a source or calculation for each measurement. Try to include at least 15 powers of 10.
- An explanation of any powers of 10 you couldn't find, and the largest and smallest values you found, if there are any. Don't forget negative powers.
- A visual aid or written explanation showing the different scales of your objects. If your objects are related, include an explanation of how they're related.

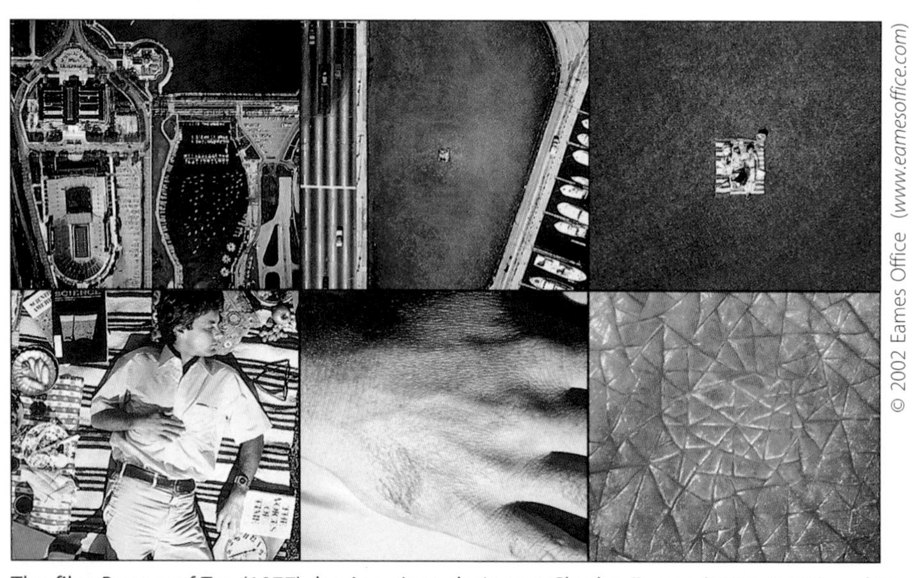

The film *Powers of Ten* (1977), by American designers Charles Eames (1907–1978) and Ray Eames (1912–1988), explores the vastness of the universe using the powers of 10. The film begins with a 1-meter-square image of a man in a Chicago park, which represents 10^0. Then the camera moves 10 times farther away each 10 seconds until it reaches the edge of the universe, representing 10^{25}. Then the camera zooms in so that the view is ultimately an atom inside the man, representing 10^{-18}. These stills from the film show images representing 10^3 to 10^{-2}.

LESSON

5.4 Applications of Exponential and Power Equations

You have seen that many equations can be solved by undoing the order of operations. In Lesson 5.2, you applied this strategy to some simple power equations. The strategy also applies for more complex power equations that arise in real-world problems.

EXAMPLE A

Rita wants to deposit \$500 into a savings account so that its doubling time will be 8 years. What annual percentage rate is necessary for this to happen? (Assume the interest on the account is compounded annually.)

▶ **Solution**

If the doubling time is 8 yr, the initial deposit of \$500 will double to \$1,000. The interest rate, r, is unknown. Write an equation and solve for r.

$1000 = 500(1 + r)^8$	Original equation.
$2 = (1 + r)^8$	Undo the multiplication by 500 by dividing both sides by 500.
$2^{1/8} = \left[(1 + r)^8\right]^{1/8}$	Undo the power of 8 by raising both sides to the power of $\frac{1}{8}$.
$2^{1/8} = 1 + r$	Use the properties of exponents.
$2^{1/8} - 1 = r$	Undo the addition of 1 by subtracting 1 from both sides.
$0.0905 \approx r$	Use a calculator to evaluate $2^{1/8} - 1$.

Rita will need to find an account with an annual percentage rate of approximately 9.05%.

You have seen how you can use the point-ratio form of an exponential equation in real-world applications. When an exponential graph models decay, the graph approaches the horizontal axis as x gets very large. When the context is growth, the graph approaches the horizontal axis as x gets increasingly negative. A horizontal line through any long-run value is called a **horizontal asymptote.** The graph approaches the asymptote but never intersects it. In the graphs below, the asymptote is the x-axis.

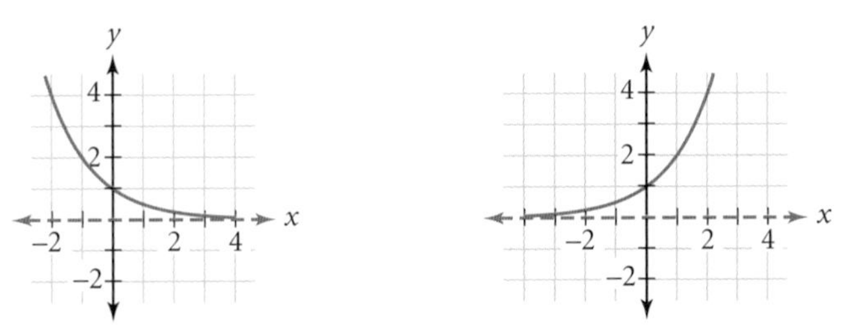

Exponential equations can also be used to model contexts that have long-run values that are not zero. You can still use the point-ratio form in these applications, although it is a bit more complex.

EXAMPLE B

A motion sensor is used to measure the distance between itself and a swinging pendulum. A table records the greatest distance for every tenth swing. At rest, the pendulum hangs 1.25 m from the motion sensor. Find an equation that models the data in the table below.

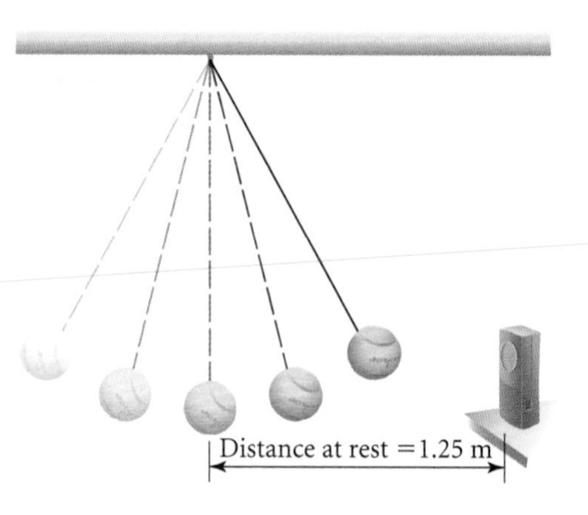

Swing number	0	10	20	30	40	50	60
Greatest distance (m)	2.35	1.97	1.70	1.53	1.42	1.36	1.32

▸ Solution

Plot these data. The graph shows a curved shape, so the data are not linear. As the pendulum slows, the greatest distance approaches a long-run value of 1.25 m. The graph appears to have a horizontal asymptote at $y = 1.25$. The pattern looks like a shifted decreasing geometric sequence, so an exponential decay equation is a good choice for the best model.

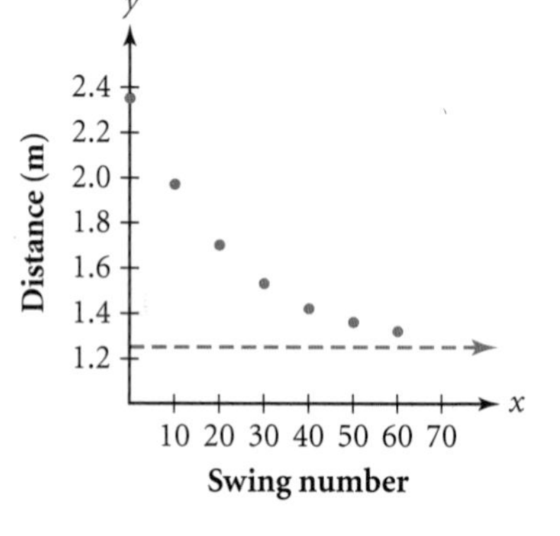

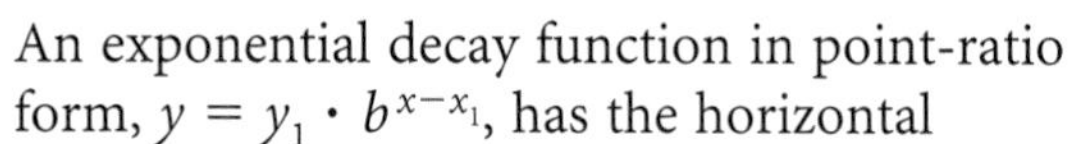

An exponential decay function in point-ratio form, $y = y_1 \cdot b^{x-x_1}$, has the horizontal asymptote $y = 0$. Because these data approach a long-run value of 1.25, the exponential function must be translated up 1.25 units. To do so, replace y with $y - 1.25$. The coefficient, y_1, is also a y-value, so you must also replace y_1 with $y_1 - 1.25$ in order to account for the translation.

The point-ratio equation is now $y - 1.25 = (y_1 - 1.25) \cdot b^{x-x_1}$. To find the value of b, select one point for (x_1, y_1) and a second point (not too close) for (x, y). If you choose (10, 1.97) and (50, 1.36) you will find

$$1.36 - 1.25 = (1.97 - 1.25) \cdot b^{50-10}$$ Substitute (10, 1.97) for (x_1, y_1) and (50, 1.36) for (x, y).

$$0.11 = 0.72b^{40}$$ Subtract.

$$b^{40} = \frac{0.11}{0.72}$$ Divide both sides by 0.72.

$$b = \left(\frac{0.11}{0.72}\right)^{1/40} \approx 0.9541$$ Raise both sides to the power $\frac{1}{40}$.

So a model for this data is $y = 1.25 + 0.72(0.9451)^{x-10}$. Using different pairs of points will generate different values for b, but all of the values should be similar. Graph the model to check whether the points chosen here resulted in a good-fitting model.

EXERCISES

You will need

A graphing calculator for Exercises **5, 6,** and **12.**

Practice Your Skills

1. Solve.

a. $x^5 = 50$ ⓐ **b.** $\sqrt[3]{x} = 3.1$ **c.** $x^2 = -121$ ⓐ

2. Solve.

a. $x^{1/4} - 2 = 3$ ⓐ **b.** $4x^7 - 6 = -2$ **c.** $3(x^{2/3} + 5) = 207$

d. $1450 = 800\left(1 + \frac{x}{12}\right)^{7.8}$ ⓐ **e.** $14.2 = 222.1 \cdot x^{3.5}$

3. Rewrite each expression in the form ax^n.

a. $(27x^6)^{2/3}$ ⓐ **b.** $(16x^8)^{3/4}$ **c.** $(36x^{-12})^{3/2}$

Reason and Apply

4. APPLICATION A sheet of translucent glass 1 mm thick is designed to reduce the intensity of light. If six sheets are placed together, then the outgoing light intensity is 50% of the incoming light intensity.

a. Let r represent the fraction of light that passes through one sheet of glass and let 100 represent the original amount of light. Write an expression for the amount of light passing through six sheets of glass.

b. If 50% of the light passes through six sheets of glass, write an equation that expresses this fact. Solve your equation for r.

c. What is the reduction rate of one sheet of glass? Write your answer as a percentage.

5. Natalie performs a decay simulation using small colored candies with a letter printed on one side. She starts with 200 candies and pours them onto a plate. She removes all the candies with the letter facing up, counts the remaining candies, and then repeats the experiment using the remaining candies. Here are her data for each stage: ⓗ

Stage number x	0	1	2	3	4	5	6
Candies remaining y	200	105	57	31	18	14	12

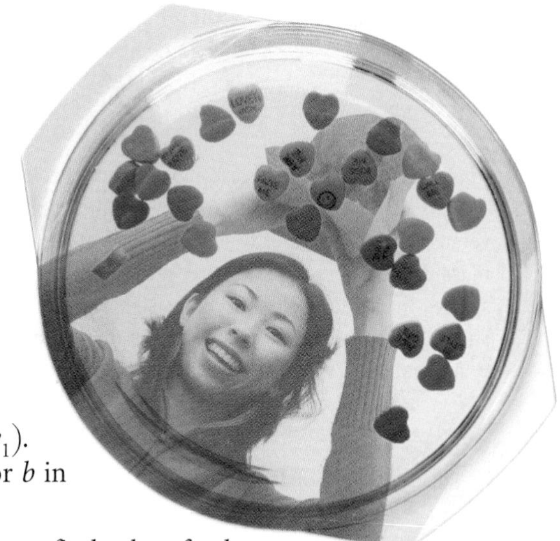

After stage 6, she checked the remaining candies and found that seven did not have a letter on either side.

a. Natalie wants to use the point-ratio equation, $y = y_1 \cdot b^{x-x_1}$, to model her data. What must she do to account for the seven unmarked candies? Write the equation. ⓐ

b. Natalie uses the second data point, (1, 105), as (x_1, y_1). Write her equation with this point and then solve for b in terms of x and y.

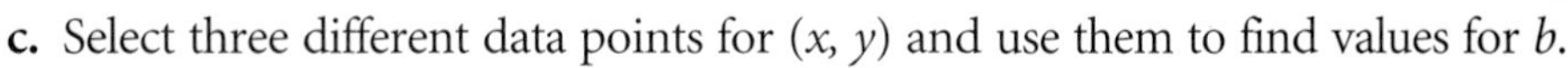

c. Select three different data points for (x, y) and use them to find values for b.

d. How should Natalie choose a value for b? What is her model for the data? Graph the equation with the data, and verify that the model fits reasonably well.

6. **APPLICATION** There is a power relationship between the radius of an orbit, x, and the time of one orbit, y, for the moons of Saturn. (The table at right lists 11 of Saturn's many moons.)

a. Make a scatter plot of these data.

b. Experiment with different values of a and b in the power equation $y = ax^b$ to find a good fit for the data. Work with a and b one at a time, first adjusting one and then the other until you have a good fit. Write a statement describing how well $y = ax^b$ fits the data. ⓗ

c. Use your model to find the orbital radius of Titan, which has an orbit time of 15.945 days.

d. Find the orbital time for Phoebe, which has an orbit radius of 12,952,000 km.

Moons of Saturn

Moon	Radius (100,000 km)	Orbital time (d)
Atlas	1.3767	0.602
Prometheus	1.3935	0.613
Pandora	1.4170	0.629
Epimetheus	1.5142	0.694
Janus	1.5147	0.695
Mimas	1.8552	0.942
Enceladus	2.3802	1.370
Tethys	2.9466	1.888
Dione	3.7740	2.737
Helene	3.7740	2.737
Rhea	5.2704	4.518

(*www.solarsystem.nasa.gov*)

These images of Saturn's system are a compilation of photos taken by the *Voyager I* spacecraft in 1980.

7. **APPLICATION** The relationship between the weight in tons, W, and the length in feet, L, of a sperm whale is given by the formula $W = 0.000137L^{3.18}$.

a. A male sperm whale is 52 ft long. What is its weight?

b. How long would a sperm whale be if it weighed 45 tons? ⓐ

8. **APPLICATION** In order to estimate the height of an *Ailanthus altissima* tree, botanists have developed the formula $h = \frac{5}{3}d^{0.8}$, where h is the height in meters and d is the diameter in centimeters.

a. If the height of an *Ailanthus altissima* tree is 18 m, find the diameter.

b. If the circumference of an *Ailanthus altissima* tree is 87 cm, estimate its height.

Science CONNECTION

Allometry is the study of size relationships between different features of an organism as a consequence of growth. Such relationships might involve weight versus length, height of tree versus diameter, or amount of fat versus body mass. Many characteristics vary greatly among different species, but within a species there may be a fairly consistent relationship or growth pattern. The study of these relationships produces mathematical models that scientists use to estimate one measurement of an organism based on another.

9. APPLICATION Fat reserves in birds are related to body mass by the formula $F = 0.033 \cdot M^{1.5}$, where F represents the mass in grams of the fat reserves and M represents the total body mass in grams.

a. How many grams of fat reserves would you expect in a 15 g warbler?

b. What percentage of this warbler's body mass is fat?

10. According to the consumer price index, the average cost of a gallon of whole milk was \$3.74 in July 2007. If the July 2007 rate of inflation continues, a gallon of whole milk will cost \$5.48 in July 2017. What was the rate of inflation in July 2007?

Review

11. APPLICATION A sample of radioactive material has been decaying for 5 years. Three years ago, there were 6.0 g of material left. Now 5.2 g are left.

a. What is the rate of decay? @

b. How much radioactive material was initially in the sample? @

c. Find an equation to model the decay.

d. How much radioactive material will be left after 50 years (45 years from now)?

e. What is the half-life of this radioactive material?

12. In his geography class, Juan makes a conjecture that more people live in U.S. cities that are warm (above 50°F) in the winter than live in U.S. cities that are cold (below 32°F). In order to test his conjecture, he collects the mean temperatures for January of the 25 largest U.S. cities.

a. Construct a box plot of these data.

b. List the five-number summary.

c. What are the range and the interquartile range for these data?

d. Do the data support Juan's conjecture? Explain your reasoning.

31.5°, 56.8°, 21.0°, 50.4°, 30.4°, 53.6°, 50.3°, 57.4°, 43.4°, 49.4°, 22.9°, 25.5°, 52.4°, 48.7°, 26.4°, 54.2°, 39.7°, 31.8°, 43.4°, 39.3°, 42.8°, 18.9°, 40.1°, 28.6°, 29.7°

(*www.allcountries.org, www.earthday.net*)

13. You have solved many systems of two equations with two variables. Use the same techniques to solve this system of three equations with three variables. @

$$\begin{cases} 2x + y + 4z = 4 \\ x + y + z = \frac{1}{4} \\ -3x - 7y + 2z = 5 \end{cases}$$

IMPROVING YOUR REASONING SKILLS

Cryptic Clue

Lieutenant Bolombo found this cryptic message containing a clue about where the stolen money was hidden: $\sqrt[\frac{1}{2}]{\text{cin}}$ nati.

Where should the lieutenant look?

LESSON

5.5

Building Inverses of Functions

Success is more a function of consistent common sense than it is of genius.

AN WANG

Gloria and Keith are sharing their graphs for the same set of data.

"I know my graph is right!" exclaims Gloria. "I've checked and rechecked it. Yours must be wrong, Keith."

Keith disagrees. "I've entered these data into my calculator too, and I made sure I entered the correct numbers."

The graphs are pictured below. Can you explain what is happening?

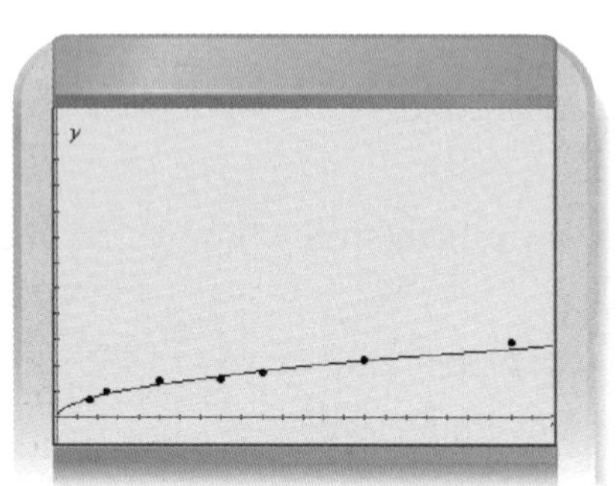

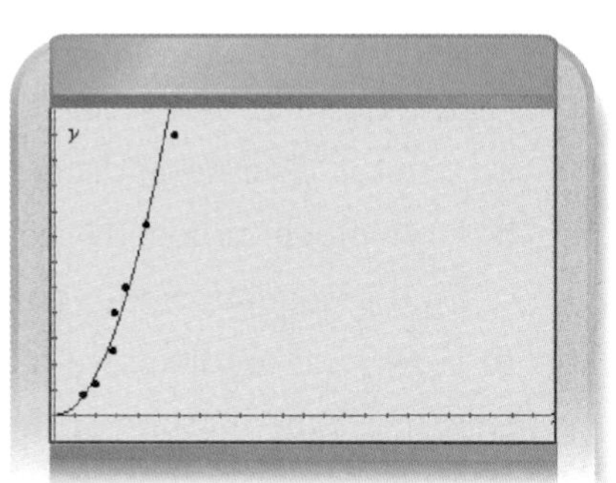

This lesson is about the **inverse** of a function—where the independent variable is exchanged with the dependent variable. Look again at Gloria's and Keith's graphs. If Gloria and Keith labeled the axes, they might see that the only difference is their choice of independent variables. In some real-world situations, it makes equal sense for either of two related variables to be used as the independent variable. In the investigation you will find equations for some inverses and then discover how the inverse equations relate to the original functions.

Investigation
The Inverse

In this investigation you will use graphs, tables, and equations to explore the inverses of several functions.

Step 1 Graph the equation $f(x) = 6 + 3x$ on your calculator. Complete the table for this function.

x	-1	0	1	2	3
y					

Step 2 Because the inverse is obtained by switching the independent and dependent variables, you can find five points on the inverse of function f by swapping the x- and y-coordinates in the table. Complete the table for the inverse.

x					
y	-1	0	1	2	3

Step 3 Graph the five points you found in Step 2 by creating a scatter plot. Describe the graph and write an equation for it. Graph your equation and verify that it passes through the points in the table from Step 2.

Step 4 Repeat Steps 1–3 for each of these functions. You may need to write more than one equation to describe the inverse.

i. $g(x) = \sqrt{x+1} - 3$ **ii.** $h(x) = (x-2)^2 - 5$

Step 5 Study the graphs of functions and their inverses that you made. What observations can you make about the graphs of a function and its inverse?

Step 6 You create the inverse by switching the x- and y-values of the points. How can you apply this idea to find the equation of the inverse from the original function? Verify that your method works by using it to find the equations for the inverses of functions f, g, and h.

EXAMPLE A

A 589 mi flight from Washington, D.C., to Chicago took 118 min. A flight of 1452 mi from Washington, D.C., to Denver took 222 min. Model this relationship both as (*time, distance*) data and as (*distance, time*) data. If a flight from Washington, D.C., to Seattle takes 323 min, what is the distance traveled? If the distance between Washington, D.C., and Miami is 910 mi, how long will it take to fly from one of these two cities to the other?

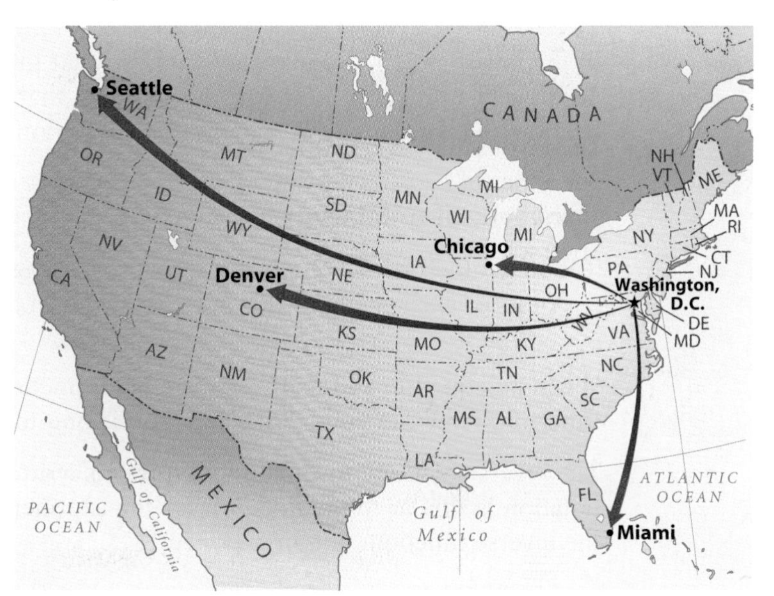

▸ Solution

If you know the time traveled and want to find the distance, then time is the independent variable, and the points known are (118, 589) and (222, 1452). The slope is $\frac{1452 - 589}{222 - 118} = \frac{863}{104}$, or approximately 8.3 mi/min. Using the first point to write an equation in point-slope form, you get $d = 589 + \frac{863}{104}(t - 118)$.

To find the distance between Washington, D.C., and Seattle, substitute 323 for the time, t:

$$d = 589 + \frac{863}{104}(323 - 118) \approx 2290.106$$

The distance is approximately 2290 mi.

If you know distance and want to find time, then distance is the independent variable. The two points then are (589, 118) and (1452, 222). This makes the slope $\frac{222 - 118}{1452 - 589} = \frac{104}{863}$, or approximately 0.12 min/mi. Using the first point again, the equation for time is $t = 118 + \frac{104}{863}(d - 589)$.

To find the time of a flight from Washington, D.C. to Miami, substitute 910 for the distance, d.

$$t = 118 + \frac{104}{863}(910 - 589) \approx 156.684$$

The flight will take approximately 157 min.

You can also use the first equation for distance and solve for t to get the second equation, for time.

$$d = 589 + \frac{863}{104}(t - 118) \quad \text{First equation.}$$

$$d - 589 = \frac{863}{104}(t - 118) \quad \text{Subtract 589 from both sides.}$$

$$\frac{104}{863}(d - 589) = (t - 118) \quad \text{Multiply both sides by } \tfrac{104}{863}.$$

$$118 + \frac{104}{863}(d - 589) = t \quad \text{Add 118 to both sides.}$$

These two equations are inverses of each other. That is, the independent and dependent variables have been switched. Graph the two equations on your calculator. What do you notice?

In the investigation you may have noticed that the inverse of a function is not necessarily a function. Recall from Chapter 4 that any set of points is called a relation. A relation may or may not be a function.

Inverse of a Relation

You get the **inverse** of a relation by exchanging the x- and y-coordinates of all points or exchanging the x- and y-variables in an equation.

When an equation and its inverse are *both* functions, it is called a **one-to-one function.** How can you tell if a function is one-to-one?

The inverse of a one-to-one function $f(x)$ is written as $f^{-1}(x)$. Note that this notation is similar to the notation for the exponent -1, but $f^{-1}(x)$ refers to the inverse function, not an exponent.

EXAMPLE B

Find the composition of this function with its inverse.

$$f(x) = 4 - 3x$$

▶ Solution

The first step is to find the inverse. Exchange the independent and dependent variables. Then, solve for the new dependent variable.

$$x = 4 - 3y \quad \text{Exchange } x \text{ and } y.$$

$$x - 4 = -3y \quad \text{Subtract 4 from both sides.}$$

$$\frac{x - 4}{-3} = y \quad \text{Divide by } -3.$$

$$f^{-1}(x) = \frac{x - 4}{-3} \quad \text{Write in function notation.}$$

The next step is to form the composition of the two functions.

$$f(f^{-1}(x)) = 4 - 3\left(\frac{x-4}{-3}\right) \qquad \text{Substitute } f^{-1}(x) \text{ for } x \text{ in } f(x).$$

Let's see what happens when you distribute and remove some parentheses.

$$f(f^{-1}(x)) = 4 + (x - 4)$$
$$f(f^{-1}(x)) = x$$

What if you had found $f^{-1}(f(x))$ instead of $f(f^{-1}(x))$?

$$f^{-1}(f(x)) = \frac{(4 - 3x) - 4}{-3} \qquad \text{Substitute } f(x) \text{ for } x \text{ in } f^{-1}(x).$$
$$f^{-1}(f(x)) = \frac{-3x}{-3} \qquad \text{Combine like terms in the numerator.}$$
$$f^{-1}(f(x)) = x \qquad \text{Divide.}$$

When you take the composition of a function and its inverse, you get x. How does the graph of $y = x$ relate to the graphs of a function and its inverse? Look carefully at the graphs below to see the relationship between a function and its inverse.

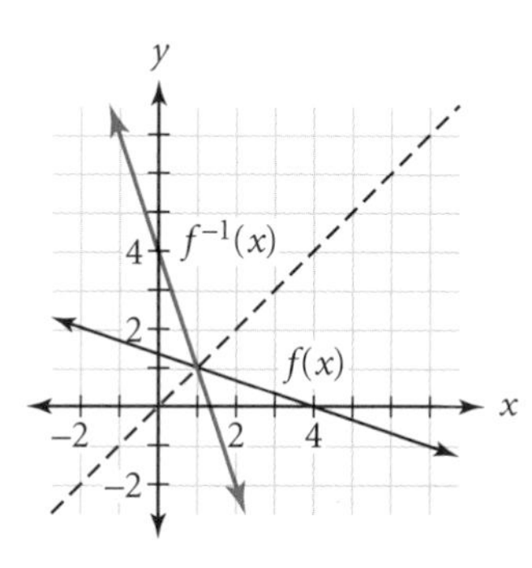

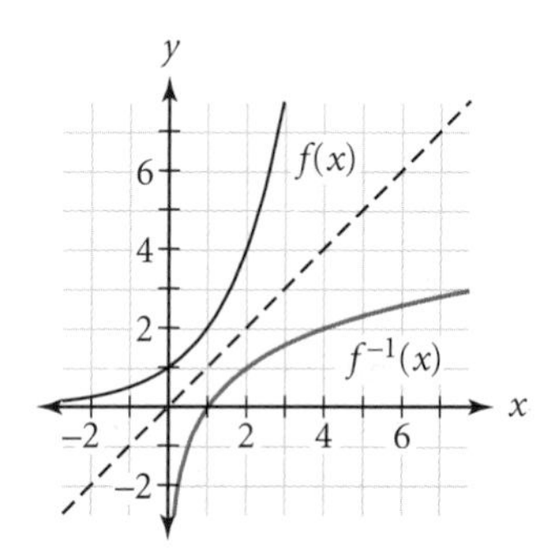

keymath.com/DAA

A turntable DJ, like DJ Spooky shown here, applies special effects and mixing techniques to alter an original source of music. If you consider the original record to be one function and the effects or a second record to be another function, the music that the DJ creates is a composition of functions.

EXERCISES

You will need

A graphing calculator for Exercise **10.**

Practice Your Skills

1. A function $f(x)$ contains the points $(-2, -3)$, $(0, -1)$, $(2, 2)$, and $(4, 6)$. Give the points known to be in the inverse of $f(x)$.

2. Given $g(t) = 5 + 2t$, find each value.

a. $g(2)$ **b.** $g^{-1}(9)$ ⓐ **c.** $g^{-1}(20)$ ⓗ

3. Which graph below represents the inverse of the relation shown in the graph at right? Explain how you know.

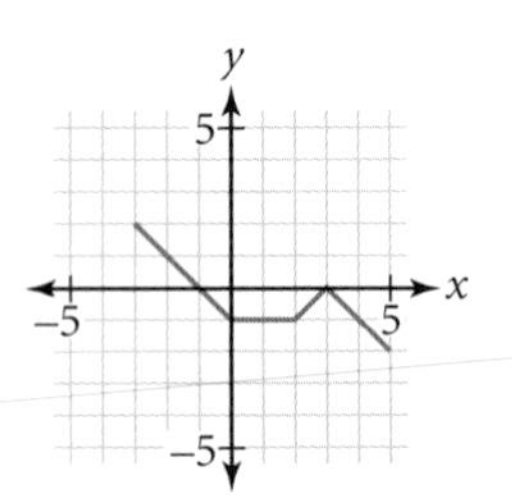

a.

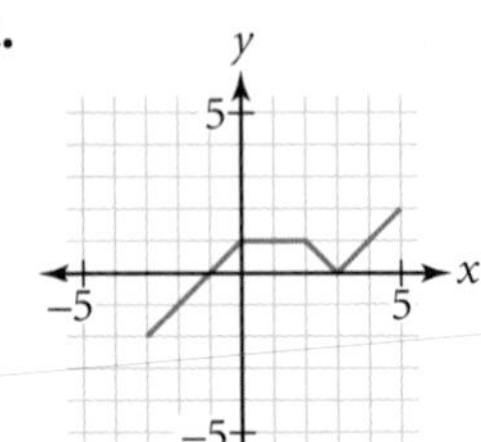

b.

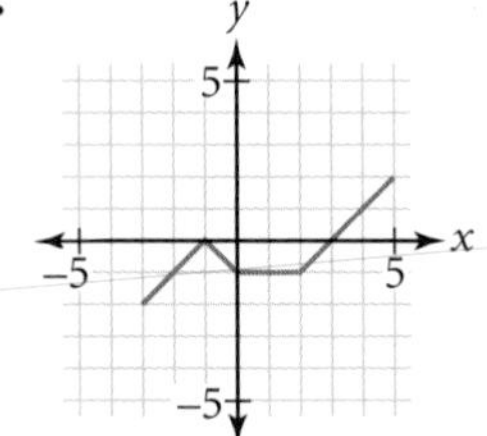

c.

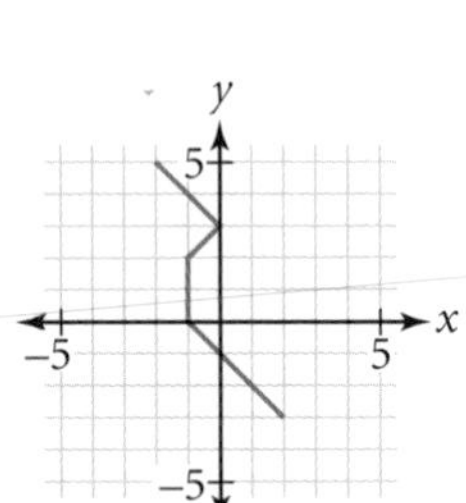

4. Match each function with its inverse. ⓗ

a. $y = 6 - 2x$ **b.** $y = 2 - \frac{6}{x}$ ⓐ **c.** $y = -6(x - 2)$ ⓐ **d.** $y = \frac{-6}{x - 2}$

e. $y = \frac{-1}{2}(x - 6)$ **f.** $y = \frac{2}{x - 6}$ **g.** $y = 2 - \frac{1}{6}x$ **h.** $y = 6 + \frac{2}{x}$

Reason and Apply

5. Given the functions $f(x) = -4 + 0.5(x - 3)^2$ and $g(x) = 3 + \sqrt{2(x + 4)}$:

a. Find $f(7)$ and $g(4)$. ⓐ

b. What do your answers to 5a imply? ⓐ

c. Find $f(1)$ and $g(-2)$.

d. What do your answers to 5c imply?

e. Over what domain are f and g inverse functions?

6. Given $f(x) = 4 + (x - 2)^{3/5}$:

a. Solve for x when $f(x) = 12$.

b. Find $f^{-1}(x)$ symbolically. ⓗ

c. How are solving for x and finding an inverse alike? How are they different?

7. Consider the graph of the piecewise function f shown at right.

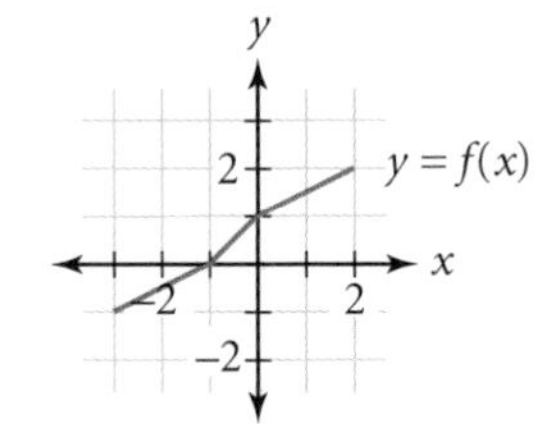

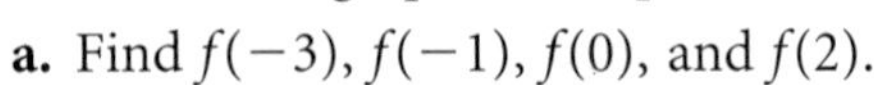

a. Find $f(-3)$, $f(-1)$, $f(0)$, and $f(2)$.

b. Name four points on the graph of the inverse of $f(x)$.

c. Draw the graph of the inverse. Is the inverse a function? Explain.

8. Write each function using $f(x)$ notation, then find its inverse. If the inverse is a function, write it using $f^{-1}(x)$ notation.

a. $y = 2x - 3$ ⓐ **b.** $3x + 2y = 4$ **c.** $x^2 + 2y = 3$ ⓐ

9. For each function in 9a and b, find the value of the expressions in i to iv.

a. $f(x) = 6.34x - 140$

b. $f(x) = 1.8x + 32$ ⓐ

i. $f^{-1}(x)$

ii. $f(f^{-1}(15.75))$

iii. $f^{-1}(f(15.75))$

iv. $f(f^{-1}(x))$ and $f^{-1}(f(x))$

Note that the equation in 9b will convert temperatures in °C to temperatures in °F. You will use either this function or its inverse in Exercises 10 and 11.

10. The data in the table describe the relationship between altitude and air temperature.

Feet	Meters	°F	°C
1,000	300	56	13
5,000	1,500	41	5
10,000	3,000	23	−5
15,000	4,500	5	−15
20,000	6,000	−15	−26
30,000	9,000	−47	−44
36,087	10,826	−69	−56

A caribou stands in front of Mount McKinley in Denali National Park, Alaska. Mount McKinley reaches an altitude of 6194 m above sea level.

a. Write a best-fit equation for $f(x)$ that describes the relationship (*altitude in meters, temperature in* °C). Use at least three decimal places in your answer. ⓐ

b. Use your results from 10a to write the equation for $f^{-1}(x)$.

c. Write a best-fit equation for $g(x)$, describing (*altitude in feet, temperature in* °F). ⓐ

d. Use your results from 10c to write an equation for $g^{-1}(x)$.

e. What would the temperature in °F be at the summit of Mount McKinley, which is 6194 m high? You will need to use the equation from Exercise 9b. ⓗ

11. APPLICATION On Celsius's original scale, freezing corresponded to 100° and boiling corresponded to 0°.

a. Write a formula that converts a temperature given by today's Celsius scale to a temperature on the scale that Celsius invented.

b. Write a formula that converts a temperature in degrees Fahrenheit to a temperature on the original scale that Celsius invented. You will need to use $f(x)$, or its inverse, from Exercise 9b.

History CONNECTION

Anders Celsius (1701–1744) was a Swedish astronomer. He created a thermometric scale using the freezing and boiling temperatures of water as reference points, on which freezing corresponded to 100° and boiling to 0°. His colleagues at the Uppsala Observatory reversed his scale five years later, giving us the current version. Thermometers with this scale were known as "Swedish thermometers" until the 1800s when people began referring to them as "Celsius thermometers."

12. Here is a paper your friend turned in for a recent quiz in her mathematics class:

If it is a four-point quiz, what is your friend's score? For each incorrect answer, provide the correct answer and explain it so that next time your friend will get it right! ⓐ

QUIZ

1. Rewrite x^{-1}.
   Answer: $\frac{1}{x}$

2. What does $f^{-1}(x)$ mean?
   Answer: $\frac{1}{f(x)}$

3. Rewrite $9^{-1/5}$.
   Answer: $\frac{1}{9^5}$

4. What number is 0^0 equal to?
   Answer: 0

13. Match each function with all the statements that apply to it.

a. $f(x) = 2x + 1$

b. $f(x) = x^2 + 1$

c. $f(x) = \sqrt{x - 1}$

d. $f(x) = \dfrac{x - 1}{2}$

i. one-to-one

ii. domain is all real numbers

iii. range is all real numbers

iv. domain is $x \geq 1$

v. range is $y \geq 1$

14. In looking over his water utility bills for the past year, Mr. Aviles saw that he was charged a basic monthly fee of \$7.18, and \$3.98 per thousand gallons (gal) used.

a. Write the monthly cost function in terms of the number of thousands of gallons used.

b. What is his monthly bill if he uses 8000 gal of water?

c. Write a function for the number of thousands of gallons used in terms of the cost.

d. If his monthly bill is \$54.94, how many gallons of water did he use?

e. Show that the functions from 14a and c are inverses. (h)

f. Mr. Aviles decides to fix his leaky faucets. He calculates that he is wasting 50 gal/d. About how much money will he save on his monthly bill? (h)

g. A gallon is 231 cubic inches. Find the dimensions of a rectangular container that will hold the contents of the water Mr. Aviles saves in a month by fixing his leaky faucets. (h)

Consumer **CONNECTION**

Leaks account for nearly 12% of the average household's annual water consumption. About one in five toilets leaks at any given time, and that can waste more than 50 gallons each day. Dripping sinks add up fast too. A faucet that leaks one drop per second can waste 30 gallons each day.

Review

15. Rewrite the expression $125^{2/3}$ in as many different ways as you can.

16. Find an exponential function that contains the points (2, 12.6) and (5, 42.525). (a)

17. Solve by rewriting with the same base.

a. $4^x = 8^3$

b. $3^{4x+1} = 9^x$

c. $2^{x-3} = \left(\frac{1}{4}\right)^x$

18. Give the equations of two different parabolas with vertex (3, 2) passing through the point (4, 5). (h)

19. Solve this system of equations.

$$\begin{cases} -x + 3y - z = 4 \\ 2z = x + y \\ 2.2y + 2.2z = 2.2 \end{cases}$$

LESSON

5.6

Logarithmic Functions

If all art aspires to the condition of music, all the sciences aspire to the condition of mathematics.

GEORGE SANTAYANA

You have used several methods to solve for x when it is contained in an exponent. You've learned that in special cases, it is possible to solve by finding a common base. For example, finding the value of x that makes each of these equations true is straightforward because of your experience with the properties of exponents.

$$10^x = 1000 \qquad 3^x = 81 \qquad 4^x = \frac{1}{16}$$

Solving the equation $10^x = 47$ isn't as straightforward because you may not know how to write 47 as a power of 10. You can, however, solve this equation by graphing $y = 10^x$ and $y = 47$ and finding the intersection—the solution to the system and the solution to $10^x = 47$. Take a minute to verify that $10^{1.672} \approx 47$ is true.

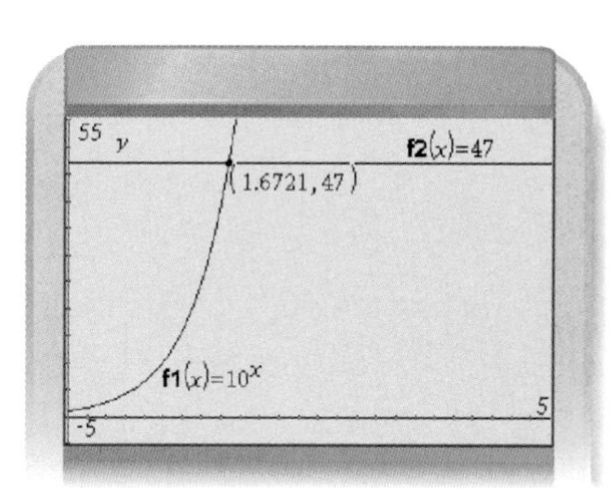

In the investigation you will discover a function called a **logarithm,** abbreviated log, used to solve for x in an exponential equation.

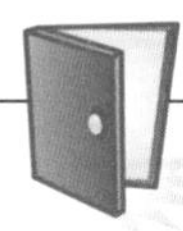

Investigation
Exponents and Logarithms

In this investigation you'll discover the connection between exponents on the base 10 and logarithms.

Step 1 Graph the function $f(x) = 10^x$ for $-1.5 \le x \le 1.5$ on your calculator. Sketch the graph and complete the table of information.

Domain	
Range	
***x*-intercept**	
***y*-intercept**	
Equation of asymptote	

Step 2 Complete the table of values for $f(x) = 10^x$ and its inverse.

x	-1.5	-1	-0.5	0	0.5	1	1.5
$f(x)$							

x							
$f^{-1}(x)$	-1.5	-1	-0.5	0	0.5	1	1.5

Step 3 Enter the points for the inverse of $f(x)$ into your calculator and plot them. You will need to adjust the graphing window in order to see these points. Sketch the graph of the inverse function, and complete the table of information about the inverse.

Domain	
Range	
***x*-intercept**	
***y*-intercept**	
Equation of asymptote	

Step 4 This inverse function is called the logarithm of x, or $\log(x)$. Enter the equation $y = \log(x)$ into your calculator. Trace your graphs or use tables to find the following values. [▶ See **Calculator Note 5C** to find out how to work with logarithms on your calculator. ◀]

a. $10^{1.5}$ **b.** $\log(10^{1.5})$ **c.** $\log 0.32$ **d.** $10^{\log 0.32}$

e. $10^{1.2}$ **f.** $\log(10^{1.2})$ **g.** $10^{\log 25}$ **h.** $\log 10^{2.8}$

Step 5 Based on your results from Step 4, what is $\log 10^x$? Explain.

Step 6 Based on your results from Step 4, what is $10^{\log x}$? Explain.

Step 7 Using your results from Steps 4–6, complete the following statements:

a. If $100 = 10^2$, then $\log 100 =$ _?_.

b. If $400 \approx 10^{2.6021}$, then $\log$ _?_ $\approx$ _?_.

c. If _?_ $\approx 10^?$, then $\log 500 \approx$ _?_.

Step 8 Complete the following statement: If $y = 10^x$, then $\log$ _?_ $=$ _?_.

The expression $\log x$ is another way of expressing x as a power of 10. Ten is the commonly used base for logarithms, so $\log x$ is called a **common logarithm** and is shorthand for writing $\log_{10} x$. You read this as "the logarithm base 10 of x." $\log x$ is the exponent you put on 10 to get x.

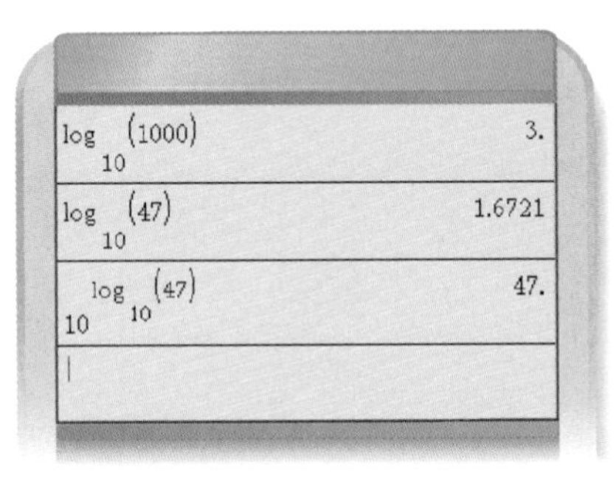

In the investigation you discovered that logarithms and exponents are inverse functions. This means you can use logarithms to "undo" exponents when solving equations.

EXAMPLE A

Solve $4 \cdot 10^x = 4650$.

▶ Solution

$4 \cdot 10^x = 4650$	Original equation.
$10^x = 1162.5$	Divide both sides by 4.
$x = \log_{10} 1162.5$	The logarithm base 10 of 1162.5 is the exponent you place on 10 to get 1162.5.
$x \approx 3.0654$	Use the log key on your calculator to evaluate.

The general **logarithmic function** is an exponent-producing function. Often you must solve for an exponent that is not on base 10. When bases other than 10 are used, you must specify the base by using a subscript. For example, $\log_2 x$ is an exponent on base 2. Regardless of the base, the logarithm is the exponent you put on the base to get x.

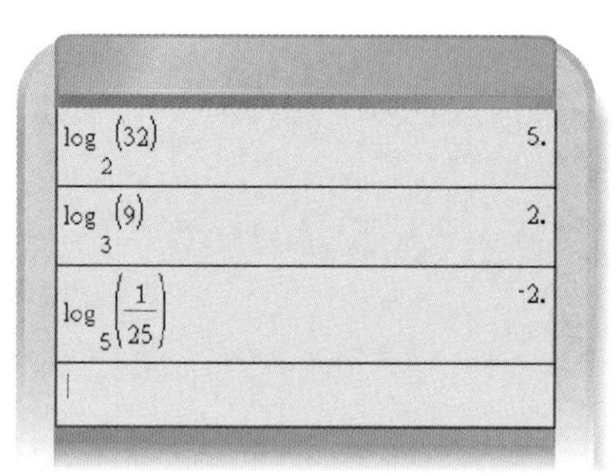

Definition of Logarithm

For $a > 0$ and $b > 0$, $\log_b a = x$ is equivalent to $a = b^x$.

Recall that composing a function with its inverse produces an output value that is the same as the input value. That is, $f(f^{-1}(x)) = x$. Because logarithms and exponents are inverse functions, $10^{\log x} = x$ and $b^{\log_b x} = x$. This enables you to solve equations involving any base.

EXAMPLE B Solve $4^x = 128$.

▶ Solution

You know that $4^3 = 64$ and $4^4 = 256$, so if $4^x = 128$, x must be between 3 and 4. You can rewrite the equation as $x = \log_4 128$ by the definition of a logarithm. But some calculators cannot calculate logarithms base 4. Have we hit a dead end?

One way to solve this equation is to use the inverse functions $\log(x)$ and 10^x to rewrite each side of the equation $4^x = 128$ as a power with a base of 10.

$4^x = 128$	Original equation.
$(10^{\log 4})^x = 10^{\log 128}$	Using inverse functions, $4 = 10^{\log 4}$ and $128 = 10^{\log 128}$.
$10^{x \log 4} = 10^{\log 128}$	Use the power of a power property of exponents.
$x \log 4 = \log 128$	Common base property of equality.
$x = \dfrac{\log 128}{\log 4} = 3.5$	Division property of equality.

You saw that you can rewrite the original equation as $\log_4 128 = x$. This value for x must be equivalent to the one found above. Because $x = \frac{\log 128}{\log 4}$ and $x = \log_4 128$, this means $\log_4 128 = \frac{\log 128}{\log 4}$.

The relationship above is called the logarithm change-of-base property. It enables you to solve problems involving logarithms with bases other than 10.

Logarithm Change-of-Base Property

$\log_b a = \frac{\log a}{\log b}$ where $a > 0$ and $b > 0$

This relationship allows you to rewrite a logarithmic expression of any base with base 10. You could also choose to rewrite a logarithmic expression with any other base, so $\log_b a = \frac{\log_c a}{\log_c b}$.

EXAMPLE C An initial deposit of $500 is invested at 8.5% interest, compounded annually. How long will it take until the balance grows to $800?

▶ Solution Let x represent the number of years the investment is held. Use the general formula for exponential growth, $y = a(1 + r)^x$.

$500(1 + 0.085)^x = 800$	Growth formula for compounding interest.
$(1.085)^x = 1.6$	Divide both sides by 500.
$x = \log_{1.085} 1.6$	Use the definition of logarithm.
$x = \dfrac{\log 1.6}{\log 1.085}$	Use the logarithm change-of-base property.
$x \approx 5.7613$	Evaluate.

It will take about 6 years for the balance to grow to at least $800.

EXERCISES

You will need

A graphing calculator for Exercises **4, 11, 12,** and **15.**

Practice Your Skills

1. Rewrite each logarithmic equation in exponential form using the definition of a logarithm.

a. $\log 1000 = x$ @ **b.** $\log_5 625 = x$ **c.** $\log_7 x = \frac{1}{2}$ @

d. $\log_x 8 = 3$ **e.** $\log_5 x = -2$ **f.** $\log_6 1 = x$

2. Solve each equation in Exercise 1 for x. @

3. Rewrite each exponential equation in logarithmic form using the definition of a logarithm. Then solve for x. (Give your answers rounded to four decimal places.)

a. $10^x = 0.001$ @ **b.** $5^x = 100$ **c.** $35^x = 8$ @

d. $0.4^x = 5$ **e.** $0.8^x = 0.03$ **f.** $17^x = 0.5$

4. Graph each equation. Write a sentence explaining how the graph compares to the graph of either $y = 10^x$ or $y = \log x$.

a. $y = \log(x + 2)$ **b.** $y = 3 \log x$ **c.** $y = -\log x - 2$

d. $y = 10^{x+2}$ **e.** $y = 3(10^x)$ **f.** $y = -(10^x) - 2$

Reason and Apply

5. Classify each statement as true or false. If false, change the second part to make it true.

a. If $6^x = 12$, then $x = \log_{12} 6$. **b.** If $\log_2 5 = x$, then $5^x = 2$.

c. If $2 \cdot 3^x = 11$, then $x = \dfrac{\log 11}{2 \log 3}$. **d.** If $x = \dfrac{\log 7}{\log 3}$, then $x = \log_7 3$.

6. The function $g(x) = 23(0.94)^x$ gives the temperature in degrees Celsius of a bowl of water x minutes after a large quantity of ice is added. After how many minutes will the water reach 5°C?

7. Assume the United States' national debt can be estimated with the model $y = 2.07596(1.083415)^x$, where x represents the number of years since 1900 and y represents the debt in billions of dollars.

a. According to the model, when did the debt pass $1 trillion ($1,000 billion)? ⓐ

b. According to the model, what is the annual growth rate of the national debt?

c. What is the doubling time for this growth model?

8. APPLICATION Carbon-14 is an isotope of carbon that is formed when cosmic rays strike nitrogen in the atmosphere. Trees, which get their carbon dioxide from the air, contain small amounts of carbon-14. Once a tree is cut down, it doesn't absorb any more carbon-14, and the amount that is present begins to decay slowly. The half-life of the carbon-14 isotope is 5730 yr.

a. Find an equation that models the percentage of carbon-14 in a sample of wood. (Consider that at time 0 there is 100% and that at time 5730 yr there is 50%.) ⓐ

b. A piece of wood contains 48.37% of its original carbon-14. According to this information, approximately how long ago did the tree that it came from die? What assumptions are you making, and why is this answer approximate?

Fossilized wood can be found in Petrified Forest National Park, Arizona. Some of the fossils are more than 200 million years old.

9. APPLICATION Crystal looks at an old radio dial and notices that the numbers are not evenly spaced. She hypothesizes that there is an exponential relationship involved. She tunes the radio to 88.7 FM. After six "clicks" of the tuning knob, she is listening to 92.9 FM. ⓗ

a. Write an exponential model in point-ratio form. Let x represent the number of clicks past 88.7 FM, and let y represent the station number. ⓐ

b. Use the equation you have found to determine how many clicks Crystal should turn to get from 88.7 FM to 106.3 FM.

Review

10. Solve.

a. $(x - 2)^{2/3} = 49$ **b.** $3x^{2.4} - 5 = 16$

11. The number of airline passengers has been increasing in the United States. The table shows the number of passengers on flights that start or end in the United States.

a. Plot the passenger data as a function of years since 1900 and find the median-median line. ⓐ

b. Calculate the residuals.

c. What is the root mean square error for this model? Explain what it means in this context. ⓐ

d. If the trend continues, what is a good estimate of the number of airline passengers in 2016?

Year	Passengers (millions)	Year	Passengers (millions)
1991	433.0	1999	596.4
1992	452.1	2000	621.7
1993	462.3	2001	579.4
1994	503.4	2002	571.2
1995	517.7	2003	603.4
1996	546.6	2004	650.4
1997	562.7	2005	676.0
1998	573.8	2006	676.1

(www.transtats.bts.gov)

12. APPLICATION The C notes on a piano $(C_1–C_8)$ are one octave apart. Their relative frequencies double from one C note to the next.

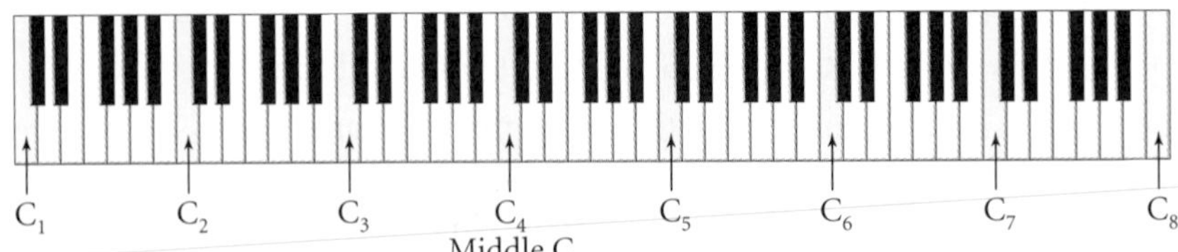

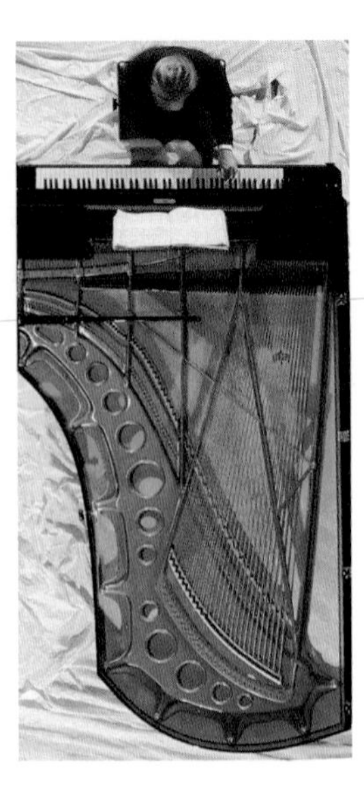

a. If the frequency of middle (C or C_4) is 261.6 cycles per second, and the frequency of C_5 is 523.2 cycles per second, find the frequencies of the other C notes.

b. Even though the frequencies of the C notes form a discrete function, you can model it using a continuous explicit function. Write a function model for these notes. ⓐ

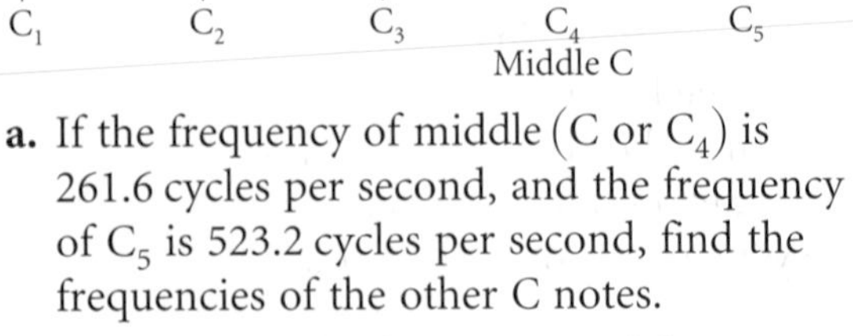

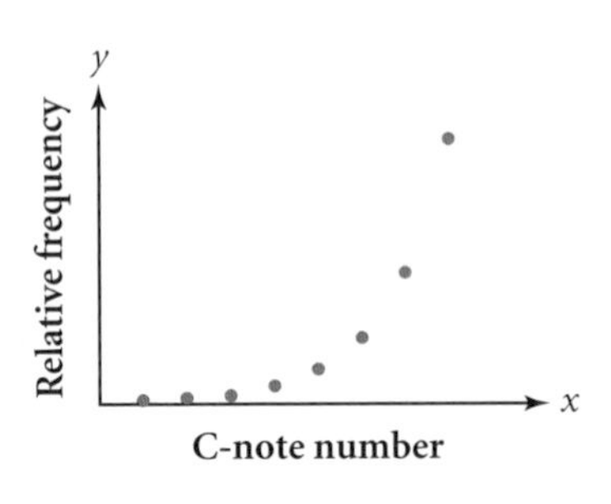

13. In each case below, use the graph and equation of the parent function to write an equation of the transformed image.

a.

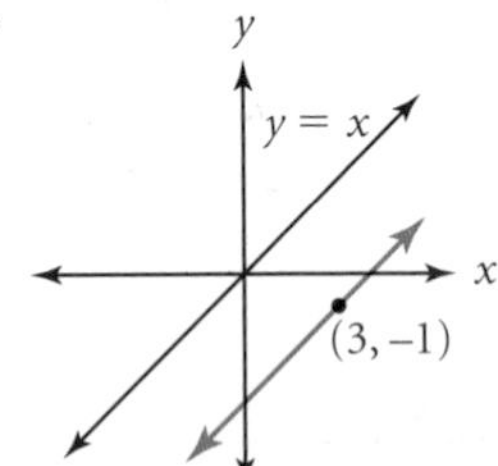

b.

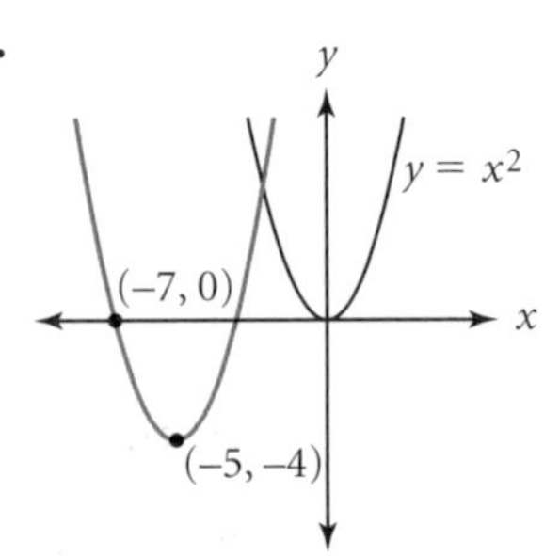

c.

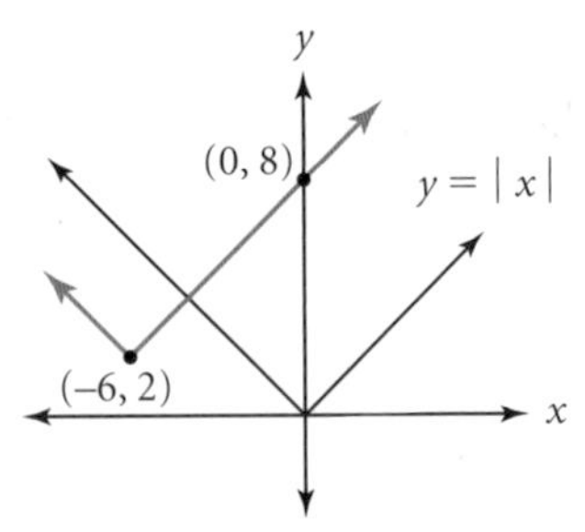

d.

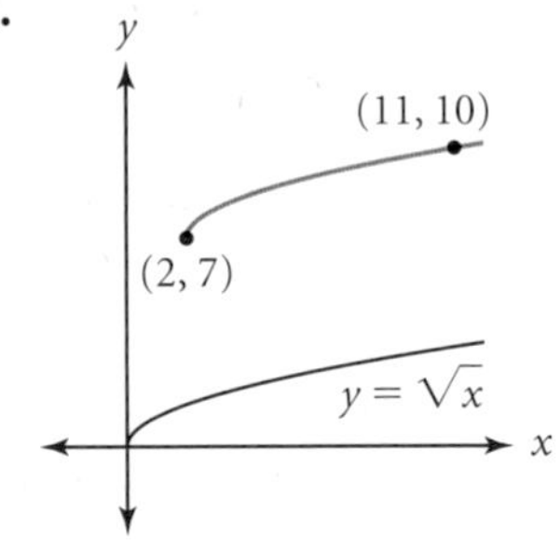

14. A rectangle has perimeter 155 inches. Its length is 7 inches more than twice its width.

a. Write a system of equations using the information given.

b. Solve the system and find the rectangle's dimensions.

15. ℓ_1: $2x - 3y = 9$ $\qquad$ ℓ_2: $2x - 3y = -1$

a. Graph ℓ_1 and ℓ_2. What is the relationship between these two lines?

b. Give the coordinates of any point A on ℓ_1, and any two points P and Q on ℓ_2.

c. Describe the transformation that maps ℓ_1 onto ℓ_2 and A onto P. Write the equation of the image of ℓ_1 showing the transformation. ⓐ

d. Describe the transformation that maps ℓ_1 onto ℓ_2 and A onto Q. Write the equation of the image of ℓ_1 showing the transformation.

e. Algebraically show that the equations in 15c and d are equivalent to ℓ_2.

LESSON

5.7

Properties of Logarithms

Each problem that I solved became a rule which served afterwards to solve other problems.

RENÉ DESCARTES

Before machines and electronics were adapted to do multiplication, division, and raising a number to a power, scientists spent long hours doing computations by hand. Early in the 17th century, Scottish mathematician John Napier (1550–1617) discovered a method that greatly reduced the time and difficulty of these calculations, using a table of numbers that he named logarithms. As you learned in Lesson 5.6, a common logarithm is an exponent—the power of 10 that equals a number—and you already know how to use the multiplication, division, and power properties of exponents. In the next example you will discover some shortcuts and simplifications.

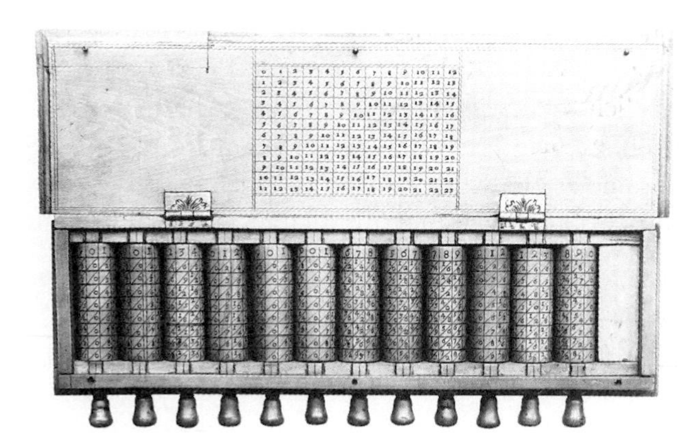

After inventing logarithms, John Napier designed a device for calculating with logarithms in 1617. Later called "Napier's bones," the device used multiplication tables carved on strips of wood or bone. The calculator at left has an entire set of Napier's bones carved on each spindle. You can learn more about Napier's bones and early calculating devices at www.keymath.com/DAA.

EXAMPLE Convert numbers to logarithms to solve these problems.

a. Multiply 183.47 by 19.628 without using the multiplication key on your calculator.

b. Divide 183.47 by 19.628 without using the division key on your calculator.

c. Evaluate $4.70^{2.8}$ without the exponentiation key on your calculator. (You may use the 10^x key.)

▶ Solution You can do parts a and b by hand. Or you can convert to logarithms and use alternative functions.

a. $183.47 = 10^{\log 183.47}$ and $19.628 = 10^{\log 19.628}$

$$183.47 \cdot 19.628 = 10^{\log 183.47} \cdot 10^{\log 19.628} = 10^{\log 183.47 + \log 19.628} \approx 10^{3.556441} \approx 3601.149$$

b. $$\frac{183.47}{19.628} = \frac{10^{\log 183.47}}{10^{\log 19.628}} = 10^{\log 183.47 - \log 19.628} \approx 10^{0.970689} \approx 9.34736$$

c. $4.70 = 10^{\log 4.70}$

$$4.70^{2.8} = (10^{\log 4.70})^{2.8} = 10^{2.8 \log 4.70} \approx 10^{1.8819} \approx 76.2$$

People did these calculations with a table of base-10 logarithms before there were calculators. For example, they looked up log 183.47 and log 19.628 in a table and added them. Then they worked backward in their table to find the **antilog,** or antilogarithm, of that sum.

	0	1	2	3	4
4.0	.6021	.6031	.6042	.6053	.6064
4.1	.6128	.6138	.6149	.6160	.6170
4.2	.6232	.6243	.6253	.6263	.6274
4.3	.6335	.6345	.6355	.6365	.6375
4.4	.6435	.6444	.6454	.6464	.6474
4.5	.6532	.6542	.6551	.6561	.6571
4.6	.6628	.6637	.6646	.6656	.6665
4.7	.6721	.6730	.6739	.6749	.6758
4.8	.6812	.6821	.6830	.6839	.6848
4.9	.6902	.6911	.6920	.6928	.6937

$$\log 4.70 \approx 0.6721$$

	0	1	2	3	4
7.0	.8451	.8457	.8463	.8470	.8476
7.1	.8513	.8519	.8525	.8531	.8537
7.2	.8573	.8579	.8585	.8591	.8597
7.3	.8633	.8639	.8645	.8651	.8657
7.4	.8692	.8698	.8704	.8710	.8716
7.5	.8751	.8756	.8762	.8768	.8774
7.6	.8808	.8814	.8820	.8825	.8831
7.7	.8865	.8871	.8876	.8882	.8887
7.8	.8921	.8927	.8932	.8938	.8943
7.9	.8976	.8982	.8987	.8993	.8998

$$\text{antilog}\, 0.8819 \approx 7.62$$
$$\text{antilog}\, 1 = 10$$
$$\text{antilog}\, 1.8819 \approx 7.62 \cdot 10 \approx 76.2$$

Can you see why "10 to the power" came to be called the antilog? The antilog of 3 is the same as 10^3, which equals 1000. Later, slide rules were invented to shorten this process, although logarithm tables were still used for more precise calculations.

Because logarithms are exponents, they must have properties similar to the properties of exponents. In the following investigation you will use your calculator to discover these properties.

Investigation
Properties of Logarithms

Step 1 Use your calculator to complete the table. Record the values to three decimal places.

Log form	Decimal form
Log 2	0.301
Log 3	
Log 5	
Log 6	
Log 8	
Log 9	
Log 10	
Log 12	
Log 15	
Log 16	
Log 25	
Log 27	

Step 2 Look closely at the values for the logarithms in the table. Look for pairs of values that add up to a third value in the table. For example, add log 2 and log 3. Where can you find that sum in the table?

Record the equations that you find in the form $\log 2 + \log 3 = \log$ _?_. (*Hint*: You should find at least six equations.)

Step 3 Write a conjecture based on your results from Step 2.

Step 4 Use your conjecture to write log 90 as the sum of two logs. Do the same for log 30 and log 72. Then use the table and your calculator to test your conjecture.

Complete the following statement: $\log a + \log b = \log$ _?_.

Step 5 Now find pairs of values in the table that subtract to equal another value in the table. Record your results in the form $\log 9 - \log 3 = \log$ _?_. Describe any patterns you see.

Complete the following statement: $\log a - \log b = \log$ _?_.

Step 6 Now find values in the table that can be multiplied by a small integer to give another value in the table, such as $3 \cdot \log 2 = \log$ _?_. Describe any patterns you see. You may want to think about different ways to express numbers such as 25 or 27 using exponents.

Complete the following statement: $b \cdot \log a = \log$ _?_.

Step 7 How do the properties you recorded in Steps 4–6 relate to the properties of exponents?

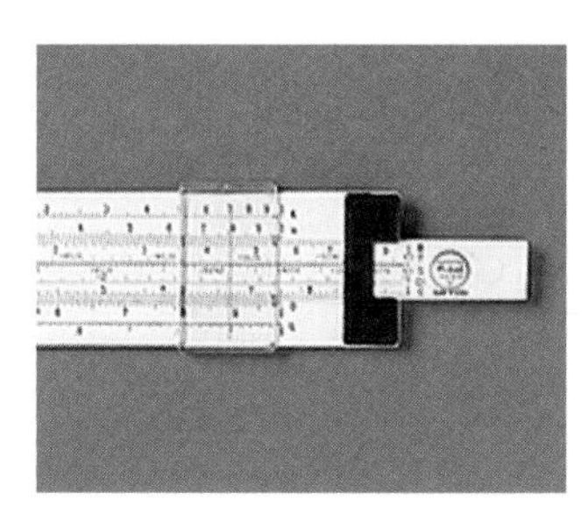

Technology CONNECTION

A few years after Napier's discovery, English mathematician William Oughtred (1574–1660) realized that sliding two logarithmic scales next to each other makes calculations easier, and he invented the slide rule. Over the next three centuries, many people made improvements to the slide rule, making it an indispensable tool for engineers and scientists, until computers and calculators became widely available in the 1970s. For more on the history of computational machines, see the links at www.keymath.com/DAA .

In this chapter you have learned the properties of exponents and logarithms, summarized on the following page. You can use these properties to solve equations involving exponents. Remember to look carefully at the order of operations and then work step by step to undo each operation.

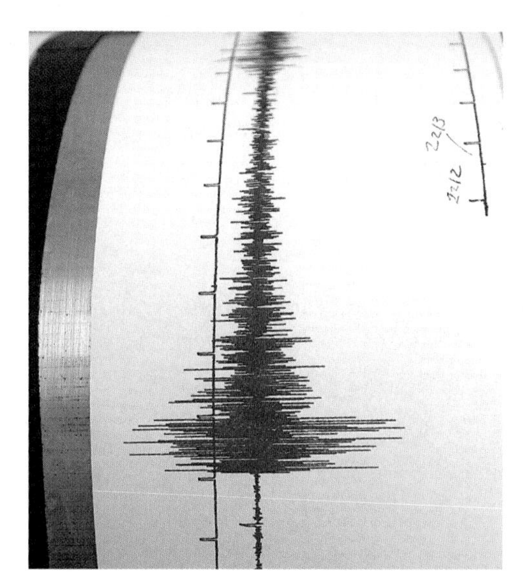

Developed in 1935 by American scientist Charles F. Richter (1900–1985), the Richter scale measures the magnitude of an earthquake by taking the logarithm of the amplitude of waves recorded by a seismograph, shown at left. Because it is a logarithmic scale, each whole-number increase in magnitude represents an increase in amplitude by a power of 10.

Properties of Exponents and Logarithms

For $a > 0$, $b > 0$ and all values of m and n, these properties are true:

Definition of Logarithm

If $x = a^m$, then $\log_a x = m$.

Product Property

$a^m \cdot a^n = a^{m+n}$ or $\log_a xy = \log_a x + \log_a y$

Quotient Property

$\frac{a^m}{a^n} = a^{m-n}$ or $\log_a \frac{x}{y} = \log_a x - \log_a y$

Power Property

$\log_a x^n = n \log_a x$

Power of a Power Property

$(a^m)^n = a^{mn}$

Power of a Product Property

$(ab)^m = a^m b^m$

Power of a Quotient Property

$\left(\frac{a}{b}\right)^n = \frac{a^n}{b^n}$

Change-of-Base Property

$\log_a x = \frac{\log_b x}{\log_b a}$

Definition of Rational Exponents

$a^{m/n} = \left(\sqrt[n]{a}\right)^m$ or $\sqrt[n]{a^m}$

Definition of Negative Exponents

$a^{-n} = \frac{1}{a^n}$ or $\left(\frac{a}{b}\right)^{-n} = \left(\frac{b}{a}\right)^n$

EXERCISES

You will need

A graphing calculator for Exercises **9** and **13**.

Practice Your Skills

1. Use the properties of logarithms to rewrite each expression as a single logarithm.

a. $\log 5 + \log 11$ ⓐ **b.** $3 \log 2$ **c.** $\log 28 - \log 7$ ⓐ

d. $-2 \cdot \log 6$ **e.** $\log 7 + 2 \cdot \log 3$

2. Rewrite each expression as a sum or difference of logarithms by using the properties of logarithms.

a. Write $\log 22$ as a sum of two logs. ⓐ

b. Write $\log 13$ as the difference of two logs. ⓐ

c. Write $\log 39$ as a sum of two logs.

d. Write $\log 7$ as the difference of two logs.

3. Use the power property of logarithms to rewrite each expression.

a. $\log 5^x$ @ b. $\log x^2$ c. $\log \sqrt{3}$ @ d. $2 \log 7^x$

4. Determine whether each equation is true or false. If false, rewrite one side of the equation to make it true. Check your answer on your calculator.

a. $\log 3 + \log 7 = \log 21$ @ b. $\log 5 + \log 3 = \log 8$ c. $\log 16 = 4 \log 2$ @

d. $\log 5 - \log 2 = \log 2.5$ e. $\log 9 - \log 3 = \log 6$ f. $\log \sqrt{7} = \log \frac{7}{2}$

g. $\log 35 = 5 \log 7$ h. $\log \frac{1}{4} = -\log 4$ i. $\frac{\log 3}{\log 4} = \log \frac{3}{4}$

j. $\log 64 = 1.5 \log 16$

5. Change the form of each expression below using properties of logarithms or exponents, without looking back in the book. Name each property or definition you use.

a. g^{h+k} b. $\log s + \log t$ c. $\frac{f^w}{f^v}$ d. $\log \frac{h}{k}$ e. $(j^s)^t$

f. $\log b^g$ g. $\sqrt[n]{k^m}$ h. $\frac{\log_s t}{\log_s u}$ i. $w^t w^s$ j. p^{-h}

Reason and Apply

6. **APPLICATION** The half-life of carbon-14, which is used in dating archaeological finds, is 5730 yr.

a. Assume that 100% of the carbon-14 is present at time 0 yr, or $x = 0$. Write the equation that expresses the percentage of carbon-14 remaining as a function of time. (This should be the same equation you found in Lesson 5.6, Exercise 8a.) @

b. Suppose some bone fragments have 25% of their carbon-14 remaining. What is the approximate age of the bones?

c. In the movie *Raiders of the Lost Ark* (1981), a piece of the Ark of the Covenant found by Indiana Jones contained 62.45% of its carbon-14. What year would this indicate that the ark was constructed in?

d. Coal is formed from trees that lived about 100 million years ago. Could carbon-14 dating be used to determine the age of a lump of coal? Explain your answer.

The spiral shape of this computer-generated shell was created by a logarithmic function.

7. APPLICATION This table lists the consecutive notes from middle C to the next C note. This scale is called a chromatic scale and it increases in 12 steps, called half-tones. The frequencies measured in cycles per second, or hertz (Hz), associated with the consecutive notes form a geometric sequence, in which the frequency of the last C note is double the frequency of the first C note.

a. Find a function that will generate the frequencies.

b. Fill in the missing table values.

	Note	Frequency (Hz)
Do	C_4	261.6
	C#	
Re	D	
	D#	
Mi	E	
Fa	F	
	F#	
Sol	G	
	G#	
La	A	
	A#	
Ti	B	
Do	C_5	523.2

Music CONNECTION

If an instrument is tuned to the mathematically simple intervals that make one key sound in tune, it will sound out of tune in a different key. With some adjustments, it will be a well-tempered scale—a scale that is approximately in tune for any key. However, not all music is based on an 8- or 12-note scale. Indian musical compositions are based on a *raga,* a structure of 5 or more notes. There are 72 *melas,* or parent scales, on which all ragas are based.

Anoushka Shankar (b 1981) plays the sitar in the tradition of classical Indian music.

8. Use the properties of logarithms and exponents to solve these equations.

a. $5.1^x = 247$

b. $17 + 1.25^x = 30$

c. $27(0.93^x) = 12$ ⓐ

d. $23 + 45(1.024^x) = 147$ ⓐ

9. APPLICATION The altitude of an airplane is calculated by measuring atmospheric pressure on the surface of the airplane. This pressure is exponentially related to the plane's height above Earth's surface. At ground level, the pressure is 14.7 pounds per square inch (abbreviated lb/in^2, or psi). At an altitude of 2 mi, the pressure is reduced to 9.46 lb/in^2.

a. Write an exponential equation for altitude in miles as a function of air pressure. ⓗ

b. Sketch the graph of air pressure as a function of altitude. Sketch the graph of altitude as a function of air pressure. Graph your equation from 9a and its inverse to check your sketches.

c. What is the pressure at an altitude of 12,000 ft? (1 mi = 5280 ft) ⓐ

d. What is the altitude of an airplane if the atmospheric pressure is 3.65 lb/in^2?

Science CONNECTION

Air pressure is the weight of the atmosphere pushing down on objects within the atmosphere, including Earth itself. Air pressure decreases with increasing altitude because there is less air above you as you ascend. A barometer is an instrument that measures air pressure, usually in millibars or inches of mercury, both of which can be converted to lb/in^2, which is the weight of air pressing down on each square inch of surface.

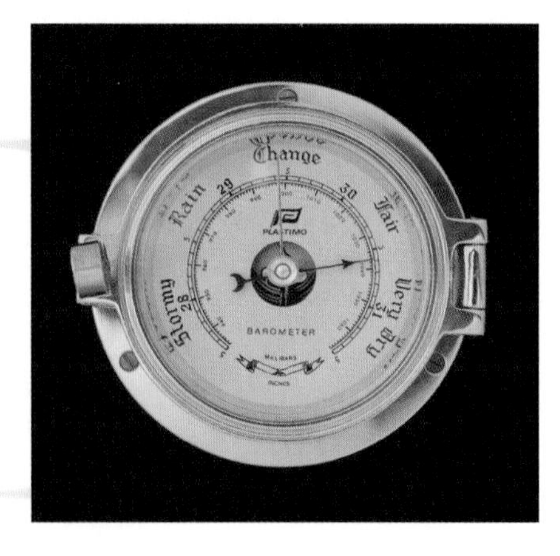

10. APPLICATION Carbon-11 decays at a rate of 3.5% per minute. Assume that 100% is present at time 0 min.

a. What percentage remains after 1 min?

b. Write the equation that expresses the percentage of carbon-11 remaining as a function of time. ⓐ

c. What is the half-life of carbon-11?

d. Explain why carbon-11 is not used for dating archaeological finds.

Review

11. Draw the graph of a function whose inverse is not a function. Carefully describe what must be true about the graph of a function if its inverse is not a function. ⓗ

12. Find an equation to fit each set of data.

a.

x	y
1	8
4	17
6	23
7	26

b.

x	y
0	2
3	54
4	162
6	1458

13. Describe how each function has been transformed from the parent function $y = 2^x$ or $y = \log x$. Then graph the function.

a. $y = -4 + 3(2)^{x-1}$

b. $y = 2 - \log\left(\frac{x}{3}\right)$

14. Answer true or false. If the statement is false, explain why or give a counterexample.

a. A grade of 86% is always better than being in the 86th percentile. ⓐ

b. A mean is always greater than a median.

c. If the range of a set of data is 28, then the difference between the maximum and the mean must be 14.

d. The mean for a box plot that is skewed left is to the left of the median. ⓐ

15. A driver charges \$14 per hour plus \$20 for chauffeuring if a client books directly with her. If a client books her through an agency, the agency charges 115% of what the driver charges plus \$25.

a. Write a function to model the cost of hiring the driver directly. Identify the domain and range.

b. Write a function to model what the agency charges. Identify the domain and range.

c. Give a single function that you can use to calculate the cost of using an agency to hire the driver for h hours.

LESSON

5.8 Applications of Logarithms

Drowning problems in an ocean of information is not the same as solving them.

RAY E. BROWN

In this lesson you will explore applications of the techniques and properties you discovered in the previous lesson. You can use logarithms to rewrite and solve problems involving exponential and power functions that relate to the natural world as well as to life decisions. You will be better able to interpret information about investing money, borrowing money, disposing of nuclear and other toxic waste, interpreting chemical reaction rates, and managing natural resources if you have a good understanding of these functions and problem-solving techniques.

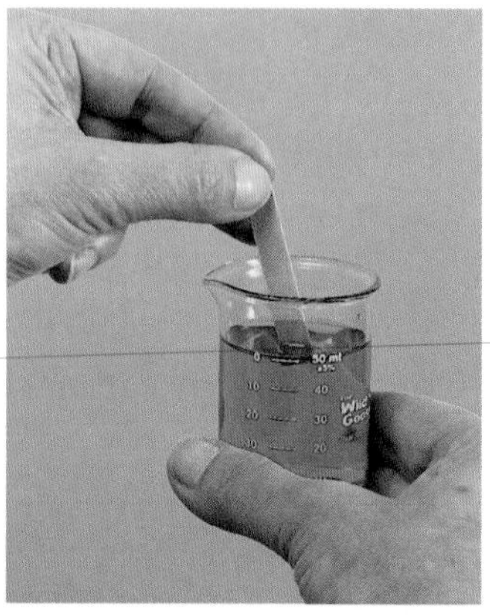

The pH scale is a logarithmic scale. A pH of 7 is neutral. A pH reading below 7 indicates an acid, and each whole-number decrease increases acidity by a power of 10. A pH above 7 indicates an alkaline, or base, and each whole-number increase increases alkalinity by a power of 10.

EXAMPLE A

Recall the pendulum example from Lesson 5.4. The equation $y = 1.25 + 0.72(0.954)^{x-10}$ gave the greatest distance from a motion sensor for each swing of the pendulum based on the number of the swing. Use this equation to find the swing number when the greatest distance will be closest to 1.47 m. Explain each step.

▸ Solution

$y = 1.25 + 0.72(0.954)^{x-10}$	Original equation.
$1.25 + 0.72(0.954)^{x-10} = 1.47$	Substitute 1.47 for y.
$0.72(0.954)^{x-10} = 0.22$	Subtract 1.25 from both sides.
$(0.954)^{x-10} = \frac{0.22}{0.72} \approx 0.3056$	Divide both sides by 0.72.
$\log\left[(0.954)^{x-10}\right] \approx \log(0.3056)$	Take the logarithm of both sides.
$(x - 10) \cdot \log(0.954) \approx \log(0.3056)$	Use the power property of logarithms.
$-0.02045(x - 10) \approx -0.51485$	Evaluate the logarithms.
$x - 10 \approx \frac{-0.51485}{-0.02045} \approx 25.18$	Divide both sides by -0.02045.
$x \approx 35.18$	Add 10 to both sides.

On the 35th swing, the pendulum will be closest to 1.47 m from the motion sensor.

As you can do with other operations on equations, you can take the logarithm of both sides, as long as the value of each side is known to be positive. In Example A, you knew that both sides were equal to the positive number 0.3056 before you took the logarithm of each side.

Solving with Logarithms

If $a^x = b$, and a and b are both positive, then $\log a^x = \log b$.

It is sometimes difficult to determine whether a relationship is logarithmic, exponential, or neither. You can use a technique called **curve straightening** to help you decide. After completing steps to straighten a curve, all you have to decide is whether or not the new graph is linear.

EXAMPLE B

Eva convinced the mill workers near her home to treat their wastewater before returning it to the lake. She then began to sample the lake water for toxin levels (measured in parts per million, or ppm) once every five weeks. Here are the data she collected.

Week	0	5	10	15	20	25	30
Toxin level (ppm)	349.0	130.2	75.4	58.1	54.2	52.7	52.1

Eva hoped that the level would be much closer to zero after this much time. Does she have evidence that the toxin is still getting into the lake? Find an equation that models these data that she can present to the mill to prove her conclusion.

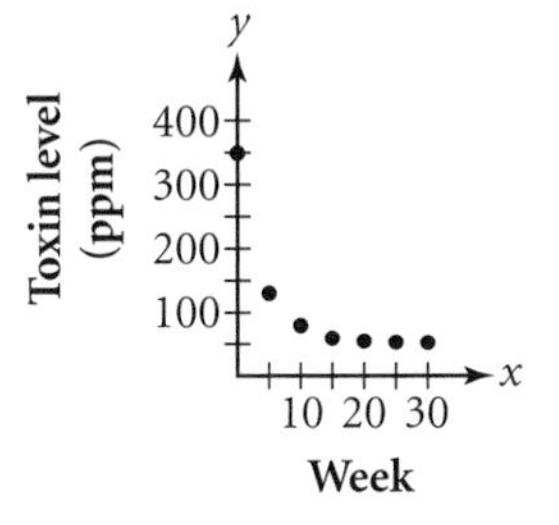

Two scientists measure toxin levels at a lake clean-up project.

▸ Solution

The scatter plot of the data shows exponential decay. So the model she must fit this data to is $y = k + ab^x$, where k is the toxin level the lake is dropping toward. If k is 0, then the lake will eventually be clean. If not, some toxins are still being released into the lake. If k is 0, then the general equation becomes $y = ab^x$. Take the logarithm of both sides of this equation.

$\log y = \log(ab^x)$	Take the logarithm of both sides.
$\log y = \log a + \log b^x$	Use the product property of logarithms.
$\log y = \log a + x \log b$	Use the power property of logarithms.
$\log y = c + dx$	Because $\log a$ and $\log b$ are numbers, replace them with the letters c and d for simplicity.

In this equation, c and d are the y-intercept and slope of a line, respectively. So the graph of the logarithm of the toxin level over time should be linear. Let w represent the week and T represent the toxin level.

The graph of $(w, \log T)$ shown is not linear. This tells Eva that if the relationship is exponential decay, k is not 0. From the table, it appears that the toxin level may be leveling off at 52 ppm. Subtract 52 from each toxin-level measurement to test again whether $(w, \log T)$ will be linear.

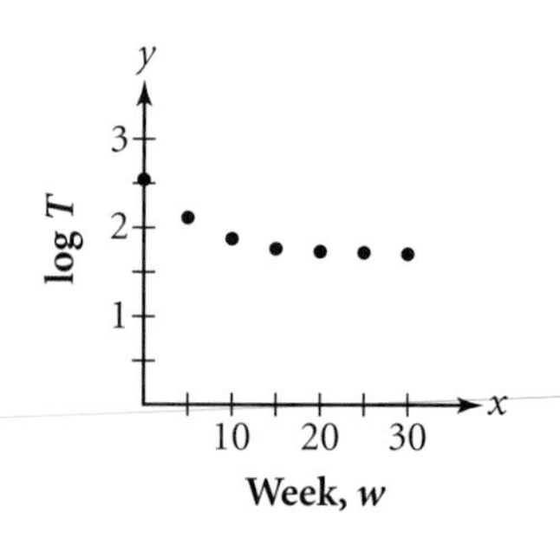

w	0	5	10	15	20	25	30
$T - 52$	297	78.2	23.4	6.1	2.2	0.7	0.1
$\log(T - 52)$	2.47	1.89	1.37	0.79	0.34	−0.15	−1.0

The graph of $(w, \log(T - 52))$ does appear linear, so Eva can be sure that the relationship is one of exponential decay with a vertical translation of approximately 52. She now knows that the general form of this relationship is $y = 52 + ab^x$.

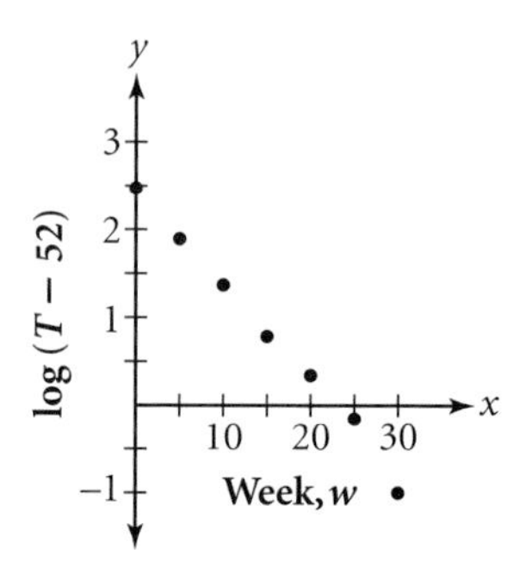

You could now use the same process as in Lesson 5.4 to solve for a and b, but the work you've done so far allows you to use an alternate method. Start by finding the median-median line for the linear data $(w, \log(T - 52))$. [▶ See **Calculator Note 3D** to review how to find a median-median line on your calculator. ◀]

$y = 2.453 - 0.110x$	Find the median-median line.
$\log(T - 52) = 2.453 - 0.110w$	Substitute $\log(T - 52)$ for y and w for x.
$T - 52 = 10^{2.453-0.110w}$	Use the definition of logarithm.
$T = 10^{2.453-0.110w} + 52$	Add 52 to both sides.

This is not yet in the form $y = k + ab^x$, so continue to simplify.

$T = 10^{2.453} \cdot 10^{-0.110w} + 52$	Use the product property of exponents.
$T = 283.93 \cdot 10^{-0.110w} + 52$	Evaluate $10^{2.453}$.
$T = 283.93 \cdot \left(10^{-0.110}\right)^w + 52$	Use the power property of exponents.
$T = 283.93 \cdot (0.776)^w + 52$	Evaluate $10^{-0.110}$.
$T = 52 + 283.93(0.776)^w$	Reorder in the form $y = k + ab^x$.

The equation that models the amount of toxin, T, in the lake after w weeks is $T = 52 + 283.93(0.776)^w$.

If you graph this equation with the original data, you see that it fits quite well.

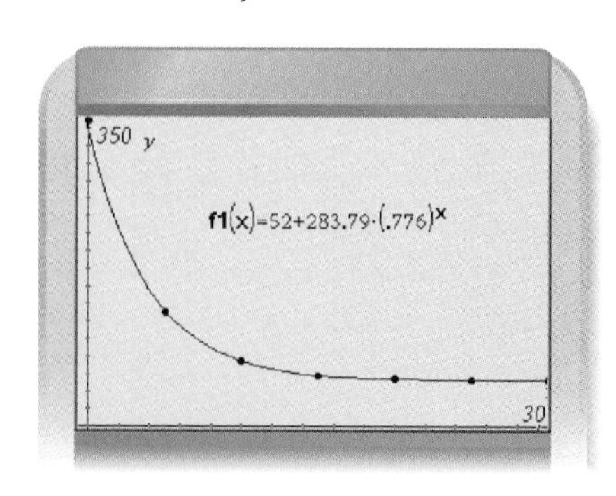

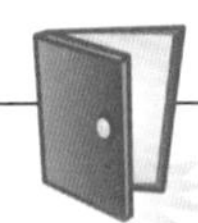

Investigation
Cooling

You will need

- a cup of hot water (optional)
- a temperature probe
- a data-collection device (optional)
- a second temperature probe (optional)

In this investigation you will find a relationship between temperature of a cooling object and time.

Step 1 Connect a temperature probe to your calculator or a data collector and set it up to collect 18 data points over 180 seconds, or 1 data point every 10 seconds. Heat the end of the probe by placing it in hot water or holding it tightly in the palm of your hand. When it is hot, set the probe on a table so that the tip is not touching anything and begin data collection. [▶ See **Calculator Note 5D.** ◀]

Step 2 Let t be the time in seconds, and let p be the temperature of the probe. While you are collecting the data, draw a sketch of what you expect the graph of (t, p) data to look like as the temperature probe cools. Label the axes and mark the scale on your graph. Did everyone in your group draw the same graph? Discuss any differences of opinion.

Step 3 Plot the data in the form (t, p) on an appropriately scaled graph. Your graph should appear to be an exponential function. Study the graph and the data, and guess the temperature limit, L. You could also use a second temperature probe to measure the room temperature, L.

Step 4 Subtract this limit from your temperatures and find the logarithm of this new list. Plot data in the form $(t, \log(p - L))$. If the data are not linear, then try a different limit.

Step 5 Find the equation that models the data in Step 4, and use this to find an equation that models the (t, p) data in Step 3. Give the real-world meanings of the values in the final equation.

EXERCISES

You will need

A graphing calculator for Exercises **4, 5, 7–10,** and **12.**

Practice Your Skills

1. Solve each equation. Check your solution by substituting your answer for x.

a. $800 = 10^x$ ⓐ **b.** $2048 = 2^x$ **c.** $16 = 0.5^x$

d. $478 = 18.5(10^x)$ ⓐ **e.** $155 = 24.0(1.89^x)$ **f.** $0.0047 = 19.1(0.21^x)$

2. Prove that these statements of equality are true. Take the logarithm of both sides, then use the properties of logarithms to re-express each side until you have two identical expressions.

a. $10^{n+p} = (10^n)(10^p)$ ⓐ **b.** $\frac{10^d}{10^e} = 10^{d-e}$

3. Suppose you invest \$3,000 at 6.75% annual interest, compounded monthly. How long will it take to triple your money? ⓗ

Reason and Apply

4. APPLICATION The length of time that milk (and many other perishable substances) will stay fresh depends on the storage temperature. Suppose that milk will stay fresh for 146 hours in a refrigerator at 4°C. Milk that is left out of the refrigerator at 22°C will stay fresh for only 42 hours. Because bacteria grow exponentially, you can assume that freshness decays exponentially.

a. Write an equation that expresses the number of hours, h, that milk will keep in terms of the temperature, T. ⓐ

b. Use your equation to predict how long milk will keep at 30°C and at 16°C.

c. If a container of milk soured after 147 hours, what was the temperature at which it was stored?

d. Graph the relationship between hours and temperature, using your equation from 4a and the five data points you have found.

e. What is a realistic domain for this relationship? Why?

Science CONNECTION

In 1860, French chemist Louis Pasteur (1822–1895) developed a method of killing bacteria in fluids. Today the process, called pasteurization, is routinely used on milk. It involves heating raw milk not quite to its boiling point, which would affect its taste and nutritional value, but to 63°C (145°F) for 30 min or 72°C (161°F) for 15 s. This kills most, but not all, harmful bacteria. Refrigerating milk slows the growth of the remaining bacteria, but eventually the milk will spoil, when there are too many bacteria for it to be healthful. The bacteria in milk change the lactose to lactic acid, which smells and tastes bad to humans.

Louis Pasteur

5. The equation $f(x) = \frac{12{,}000}{1 + 499(1.09)^{-x}}$ gives the total sales x days after the release of a new video game. Find each value and give a real-world meaning.

a. $f(20)$

b. $f(80)$

c. x when $f(x) = 6000$ ⓐ

d. Show the steps to solve 5c symbolically.

e. Graph the equation in a window large enough for you to see the overall behavior of the curve. Use your graph to describe how the number of games sold each day changes. Does this model seem reasonable?

6. APPLICATION The loudness of sound, D, measured in decibels (dB) is given by the formula

$$D = 10\log\left(\frac{I}{10^{-16}}\right)$$

where I is the intensity measured in watts per square centimeter (W/cm^2) and 10^{-16} W/cm^2 is the approximate intensity of the least sound audible to the human ear.

a. Find the loudness of a 10^{-13} W/cm^2 whisper in decibels. ⓗ

b. Find the loudness of a normal conversation at $3.16 \cdot 10^{-10}$ W/cm^2.

c. Find the intensity of the sound (in W/cm^2) experienced by the orchestra members seated in front of the brass section, measured at 107 dB. ⓐ

d. How many times more intensity does a sound of 47 dB have than a sound of 42 dB?

7. APPLICATION The table gives the loudness of spoken words, measured at the source, and the maximum distance at which another person can recognize the speech. Find an equation that expresses the maximum distance as a function of loudness.

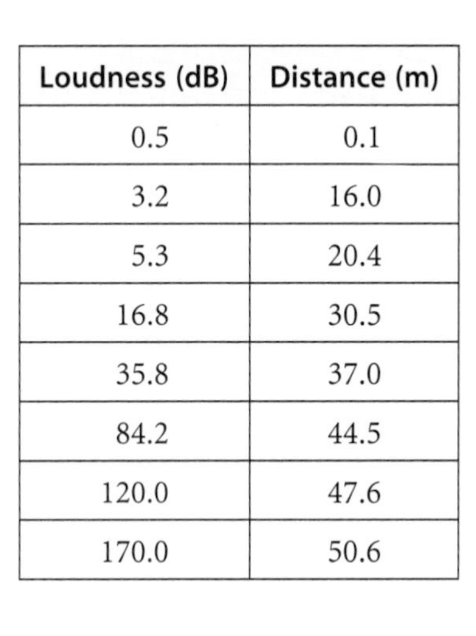

Loudness (dB)	Distance (m)
0.5	0.1
3.2	16.0
5.3	20.4
16.8	30.5
35.8	37.0
84.2	44.5
120.0	47.6
170.0	50.6

a. Plot the distance data as a function of the loudness.

b. Experiment to find the relationship between x and y by plotting different combinations of x, y, $\log x$, and $\log y$ until you have found the graph that best linearizes the data. Sketch this graph on your paper and label the axes with x, y, $\log x$, or $\log y$, as appropriate. ⓐ

c. Find the equation of a line that fits the plot you chose in 7b. Remember that your axes did not represent x and y, so substitute $(\log x)$ or $(\log y)$ into your equation as appropriate.

d. Graph this new equation with the original data. Does it seem to be a good model?

8. A container of juice is left in a room at a temperature of 74°F. After 8 minutes, the temperature is recorded at regular intervals.

Time (min)	8	10	12	14	16	18	20	22	24	26	28	30
Temp. (°F)	35	40	45	49	52	55	57	60	61	63	64	66

a. Plot the temperature data as a function of time using an appropriate window.

b. Find an exponential model for temperature as a function of time. ⓗ

9. APPLICATION In clear weather, the distance you can see from a window on a plane depends on your height above Earth, as shown in the table at right.

a. Using x for height and y for viewing distance, graph various combinations of x, y, $\log x$, and $\log y$ until you find a combination that linearizes the data. ⓐ

b. Use your results from 9a to find a best-fit equation for data in the form (*height, view*) using the data in this table.

Height (m)	Viewing distance (km)
305	62
610	88
914	108
1,524	139
3,048	197
4,572	241
6,096	278
7,620	311
9,144	340
10,668	368
12,192	393

10. Quinn starts treating her pool for the season with a shock treatment of 4 gal of chlorine. Every 24 h, 15% of the chlorine evaporates. The next morning, she adds 1 qt $\left(\frac{1}{4}\text{gal}\right)$ of chlorine to the pool, and she continues to do so each morning.

a. How much chlorine is there in the pool after one day (after she adds the first daily quart of chlorine)? After two days? Write a recursive formula for this pattern. ⓐ

b. Use the formula from 10a to make a table of values and sketch a graph of 20 terms. Find an explicit model that fits the data.

Review

11. Find these functions.

a. Find an exponential function that passes through the points (4, 18) and (10, 144).

b. Find a logarithmic function that passes through the points (18, 4) and (144, 10).

12. The Highland Fish Company is starting a new line of frozen fish sticks. It will cost \$19,000 to set up the production line and \$1.75 per pound to buy and process the fish. Highland Fish will sell the final product at a wholesale cost of \$1.92 per pound.

a. Write a cost function and an income function for HFC's new venture.

b. Graph both functions on the same axes over the domain $0 \le x \le 1{,}000{,}000$.

c. How many pounds of fish sticks will HFC have to produce before it starts making a profit on the new venture?

d. How much profit can HFC expect to make on the first 500,000 pounds of fish?

13. Sketch the graph of $[4(x+5)]^2 + \left(\frac{y-8}{2}\right)^2 = 1$. Give coordinates of a few points that define the shape.

14. Solve each equation. Round to the nearest hundredth.

a. $x^5 = 3418$

b. $(x - 5.1)^4 = 256$

c. $7.3x^6 + 14.4 = 69.4$

Project

INCOME BY GENDER

The median annual incomes of year-round full-time workers in the United States are listed in this table. Examine different relationships, such as data in the form (*time, men*), (*time, women*), (*women, men*), (*time, men − women*), (*time, men/women*), and so on. Find best-fit models for those relationships that seem meaningful. Write an article in which you interpret some of your models and make predictions about the future. Research some recent data to see if your predictions are accurate so far.

Year	Men	Women
1975	\$12,934	\$7,719
1978	\$16,062	\$9,641
1981	\$20,692	\$12,457
1984	\$24,004	\$15,422
1987	\$26,681	\$17,564
1990	\$28,979	\$20,591
1993	\$31,077	\$22,469
1996	\$33,538	\$24,935
1999	\$37,450	\$27,366
2002	\$40,507	\$30,970
2005	\$42,188	\$33,256

(*www.census.gov*)

Your project should include

- Your article, including relevant graphs, models, and predictions.
- More recent data (remember to cite your source).
- An analysis of how well the recent data fit your predictions.

EXPLORATION

The Number *e*

You've solved problems exploring the amount of interest earned in a savings account when interest is compounded yearly, monthly, or daily. But what if interest is compounded continuously? This means that at every instant, your interest is redeposited into your account and the new interest is calculated on the new balance. This type of continuous growth is related to a number, ***e***. This number has a value of approximately 2.718, and like π, it is a **transcendental number**—a special kind of irrational number. The logarithm function with base e, $\log_e x$, is also written as $\ln x$, and is called the **natural logarithm** function.

Activity

Continuous Growth

The Swiss mathematician Jacob Bernoulli (1654–1705) explored the following problem in 1683. Suppose you put $1 into an account that earns 100% interest per year. If the interest is compounded only once a year, at the end of the year you will have earned $1 in interest and your balance will be $2. What if interest is compounded more frequently? Follow the steps below to analyze this situation.

Step 1 If interest is compounded 10 times annually, how much money will be in your account at the end of one year? Remember that if 100% interest is compounded ten times, you earn 10% each time. Check your answer with another group to be sure you did this correctly.

Step 2 Predict what will happen if your money earns interest compounded continuously.

Step 3 What will be the balance of your account at the end of one year if interest is compounded 100 times? 1,000 times? 10,000 times? 1,000,000 times? How do your answers compare to your prediction in Step 2?

Step 4 Write an equation that would tell you the balance if the interest were compounded x times annually, and graph it on your calculator. Does this equation seem to be approaching one particular long-run value, or limit? If so, what is it? If not, what happens in the long run?

Step 5 The number e is called the natural exponential base. Look for e on your calculator, and find its value to six decimal places. What is the relationship between this number and your answer to Step 4?

Step 6 When interest is compounded continuously, the formula $y = P\left(1 + \frac{r}{n}\right)^{nt}$ becomes $y = Pe^{rt}$, where P is the principal (the initial value of your investment), r is the interest rate, and t is the amount of time the investment is left in the account. Use this formula to calculate the value of $1 deposited into an account earning 100% annual interest for one year. Calculate the value if the principal and the interest are left in the same account for ten years.

Questions

1. The formula $y = Pe^{rt}$ is often an accurate model for growth or decay problems, because things usually grow or decay continuously, not at specific time intervals. For example, a bacteria population may double every four hours, but it is increasing throughout those four hours, not suddenly doubling in value when exactly four hours have passed. Return to any problem involving growth or decay from this chapter, and use this new formula to solve the problem. How does your answer compare with your initial answer?

2. The intensity, I, of a beam of light after passing through t cm of liquid is given by $I = Ae^{-kt}$, where A is the intensity of the light when it enters the liquid, e is the natural exponential base, and k is a constant for a particular liquid. On a field trip, a marine biology class took light readings at Deep Lake. At a depth of 50 cm, the light intensity was 80% of the light intensity at the surface.
   a. Find the value of k for Deep Lake.
   b. If the light intensity is measured as 1% of the intensity at the surface, at what depth was the reading taken?

Science CONNECTION

Phytoplankton is a generic name for a great variety of microorganisms, including algae, that live in lakes and oceans and provide the lowest step of the food chain in some ecosystems. Phytoplankton require light for photosynthesis. In water, light intensity decreases with depth. As a result, the phytoplankton production rate, which is determined by the local light intensity, decreases with depth. Light levels determine the maximum depth at which these organisms can grow. Limnologists, scientists who study inland waters, estimate this depth to be the point at which the amount of light available is reduced to between 0.5% and 1% of the amount of light available at the lake surface.

Algae grow in abundance on this pond that holds waste water for the nearby milk house. At right is a microscopic image of freshwater phytoplankton.

Project

ALL ABOUT *e*

Mathematicians have been exploring e and calculating more digits of the decimal approximation of e since the 1600s. There are a number of procedures that can be used to calculate digits of e. Do some research at the library or on the Internet, and find at least two. Write a report explaining the formulas you find and how they are used to calculate e. Include any interesting historical facts you find on e as well.

CHAPTER 5 REVIEW

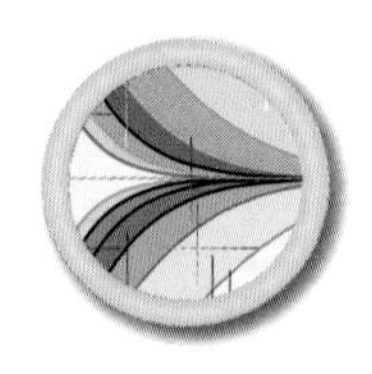

Exponential functions provide explicit and continuous equations to model geometric sequences. They can be used to model the growth of populations, the decay of radioactive substances, and other phenomena. The general form of an exponential function is $y = ab^x$, where a is the initial amount and b is the base that represents the rate of growth or decay. Because the exponent can take on all real number values, including negative numbers and fractions, it is important that you understand the meaning of these exponents. You also used the **point-ratio form** of this equation, $y = y_1 \cdot b_1^{x-x_1}$.

Until you read this chapter, you had no way to solve an exponential equation, other than guess-and-check. Once you defined the **inverse** of the exponential function—the **logarithmic function**—you were able to solve exponential functions symbolically. The inverse of a function is the relation you get when all values of the independent variable are exchanged with the values of the dependent variable. The graphs of a function and its inverse are reflected across the line $y = x$.

The definition of the logarithmic function is that $\log_b y = x$ means that $y = b^x$. You learned that the properties of logarithms parallel those of exponents: the logarithm of a product is the sum of the logarithms, the logarithm of a quotient is the difference of the logarithms, and the logarithm of a number raised to a power is the product of the logarithm and that number. By looking at the logarithms of the x-value, or the y-value, or both values in a set of data, you can determine what type of equation will best model the data by finding which of these creates the most linear graph.

EXERCISES

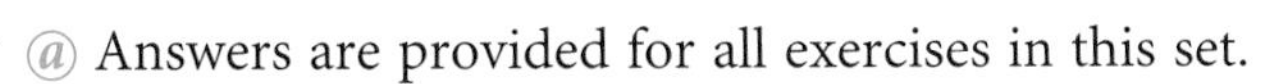

You will need

A graphing calculator for Exercises **12** and **14**.

@ Answers are provided for all exercises in this set.

1. Evaluate each expression without using a calculator. Then check your work with a calculator.

a. 4^{-2} **b.** $(-3)^{-1}$ **c.** $\left(\frac{1}{5}\right)^{-3}$ **d.** $49^{1/2}$ **e.** $64^{-1/3}$

f. $\left(\frac{9}{16}\right)^{3/2}$ **g.** -7^0 **h.** $(3)(2)^2$ **i.** $\left(0.6^{-2}\right)^{-1/2}$

2. Rewrite each equation in logarithm form so that the x is no longer an exponent.

a. $3^x = 7$ **b.** $5 = 4^x$ **c.** $10^x = 7^5$

3. Rewrite each equation in exponential form.

a. $\log x = 1.72$ **b.** $2 \log x = 4.8$ **c.** $\log_5 x = -1.47$ **d.** $3 \log_2 x = 15$

4. Rewrite each expression in another form.

a. $\log x + \log y$ **b.** $\log \frac{z}{v}$ **c.** $(7x^{2.1})(0.3x^{4.7})$

d. $\log w^k$ **e.** $\sqrt[5]{x}$ **f.** $\log_5 t$

5. Use the properties of exponents and logarithms to solve each equation. Confirm your answers by substituting them for x.

a. $4.7^x = 28$ **b.** $4.7^{x^2} = 2209$ **c.** $\log_x 2.9 = 1.25$ **d.** $\log_{3.1} x = 47$

e. $7x^{2.4} = 101$ **f.** $9000 = 500(1.065)^x$ **g.** $\log x = 3.771$ **h.** $\sqrt[5]{x^3} = 47$

6. Solve for x. Round your answers to the nearest thousandth.

a. $\sqrt[8]{2432} = 2x + 1$ **b.** $4x^{2.7} = 456$ **c.** $734 = 11.2(1.56)^x$

d. $f(f^{-1}(x)) = 20.2$ **e.** $147 = 12.1(1 + x)^{2.3}$ **f.** $2\sqrt{x - 3} + 4.5 = 16$

7. Once a certain medicine is in the bloodstream, its half-life is 16 h. How long (to the nearest 0.1 h) will it be before an initial 45 cm^3 of the medicine has been reduced to 8 cm^3?

8. Given $f(x) = (4x - 2)^{1/3} - 1$, find:

a. $f(2.5)$ **b.** $f^{-1}(x)$ **c.** $f^{-1}(-1)$ **d.** $f(f^{-1}(12))$

9. Find the equation of an exponential curve through the points (1, 5) and (7, 32).

10. Draw the inverse of $f(x)$, shown below.

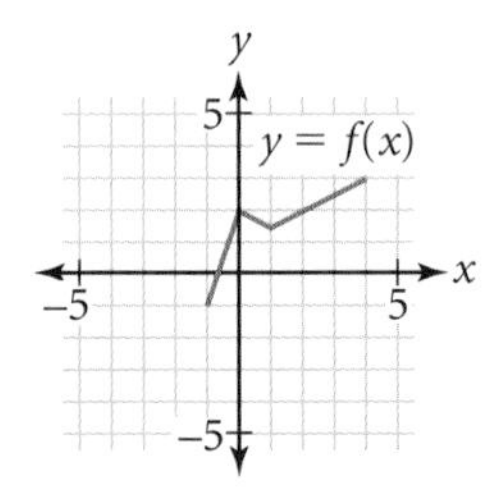

11. A new incentive plan for the Talk Alot long-distance phone company varies the cost of a call according to the formula $cost = a + b \log t$, where t represents time in minutes. When calling long distance, the cost for the first minute is \$0.50. The cost for 15 min is \$3.44.

a. Find the a-value in the equation.

b. Find the b-value in the equation.

c. What is the x-intercept of the graph of the equation? What is the real-world meaning of the x-intercept?

d. Use your equation to predict the cost of a 30-minute call.

e. If you decide you can afford only to make a \$2 call, how long can you talk?

12. Your head circumference gets larger as you grow. Most of the growth comes in the first few years of life, and there is very little additional growth after you reach adolescence. The estimated percentage of adult size for your head is given by the formula $y = 100 - 80(0.75)^x$, where x is your age in years and y is the percentage of the average adult size.

a. Graph this function.

b. What are the reasonable domain and range of this function?

c. Describe the transformations of the graph of $y = (0.75)^x$ that produce the graph in 12a.

d. A 2-year-old child's head is what percentage of the average adult-size head?

e. About how old is a person if his or her head circumference is 75% that of an average adult?

13. APPLICATION A "learning curve" describes the rate at which a task can be learned. Suppose the equation

$$t = -144 \log\left(1 - \frac{N}{90}\right)$$

predicts the time t (in number of short daily sessions) it will take to achieve a goal of typing N words per minute (wpm).

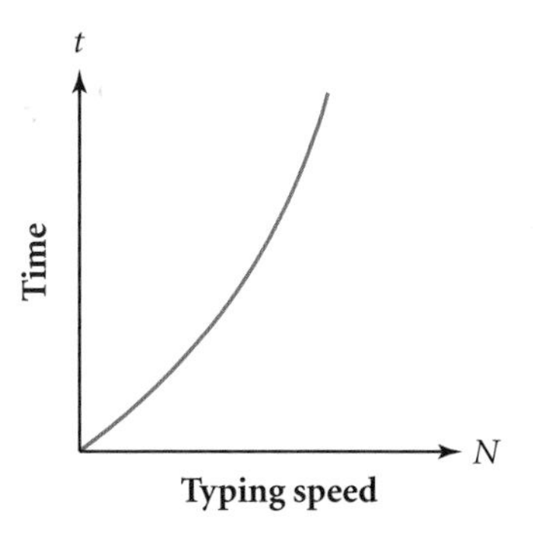

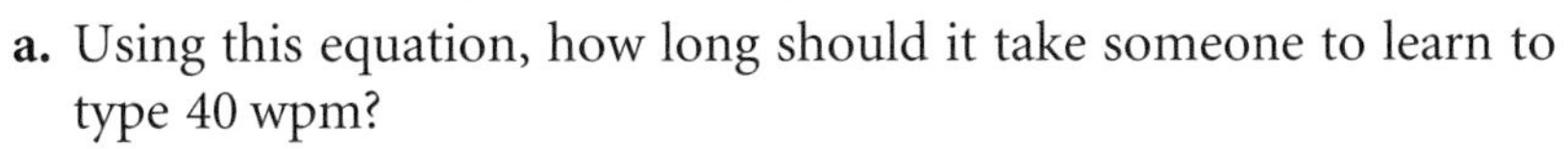

a. Using this equation, how long should it take someone to learn to type 40 wpm?

b. If the typical person had 47 lessons, then what speed would you expect him or her to have achieved?

c. Interpret the shape of the graph as it relates to learning time. What domain is realistic for this problem?

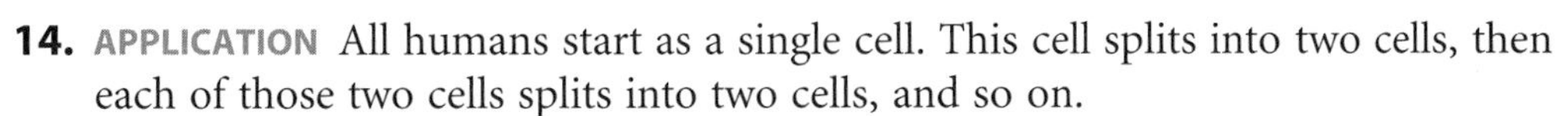

14. APPLICATION All humans start as a single cell. This cell splits into two cells, then each of those two cells splits into two cells, and so on.

a. Write a recursive formula for cell division starting with a single cell.

b. Write an explicit formula for cell division.

c. Sketch a graph to model the formulas in 14a and b.

d. Describe some of the features of the graph.

e. After how many divisions were there more than 1 million cells?

f. If there are about 1 billion cells after 30 divisions, after how many divisions were there about 500 million cells?

Science CONNECTION

Embryology is the branch of biology that deals with the formation, early growth, and development of living organisms. In humans, the growth of an embryo takes about 9 months. During this time a single cell will grow into many different cell types with different shapes and functions in the body.

A similar process occurs in the embryo of any animal. A chicken embryo develops and hatches in 20–21 days. Cutting a window in the eggshell allows direct observation for the study of embryonic growth.

These X-rays show the 21-day growth of a chicken embryo to a newborn chick.

TAKE ANOTHER LOOK

1. Is $(x^{1/m})^n$ always equivalent to $(x^n)^{1/m}$? Try graphing $f_1(x) = (x^{1/m})^n$ and $f_2(x) = (x^n)^{1/m}$ for various integer values of m and n. Make sure you try positive and negative values for m and n, as well as different combinations of odd and even numbers. Check to see if the expressions are equal by inspecting the graphs and looking at table values for positive and negative values of x. Make observations about when output values are different and when output values do not exist. Make conjectures about the reasons for the occurrence of different values or no values.

2. You have learned to use the point-ratio form to find an exponential curve that fits two data points. You could also use the general exponential equation $y = ab^x$ and your knowledge of solving systems of equations to find an appropriate exponential curve. For example, you can use the general form to write two equations for the exponential curve through (4, 40) and (7, 4.7):

$$40 = ab^4$$

$$4.7 = ab^7$$

Which constant will be easier to solve for in each equation, a or b? Solve each equation for the constant you have chosen. Use substitution and the properties you have learned in this chapter to solve this system of equations. Substitute the values you find for a and b into the general exponential form to write a general equation for this function.

Find an exercise or example in this chapter that you solved using the point-ratio form, and solve it again using this new method. Did you find this method easier or more difficult to use than the point-ratio form? Are there situations in which one method might be preferable to the other?

Assessing What You've Learned

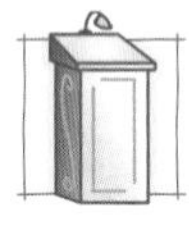

GIVE A PRESENTATION Give a presentation about how to fit an exponential curve to data or one of the Take Another Look activities. Prepare a poster or visual aid to explain your topic. Work with a group, or give a presentation on your own.

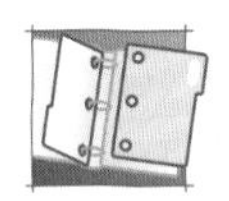

ORGANIZE YOUR NOTEBOOK Review your notebook to be sure it's complete and well organized. Make sure your notes include all of the properties of exponents and logarithms, including the meanings of negative and fractional exponents. Write a one-page chapter summary based on your notes.

CHAPTER

6

Matrices and Linear Systems

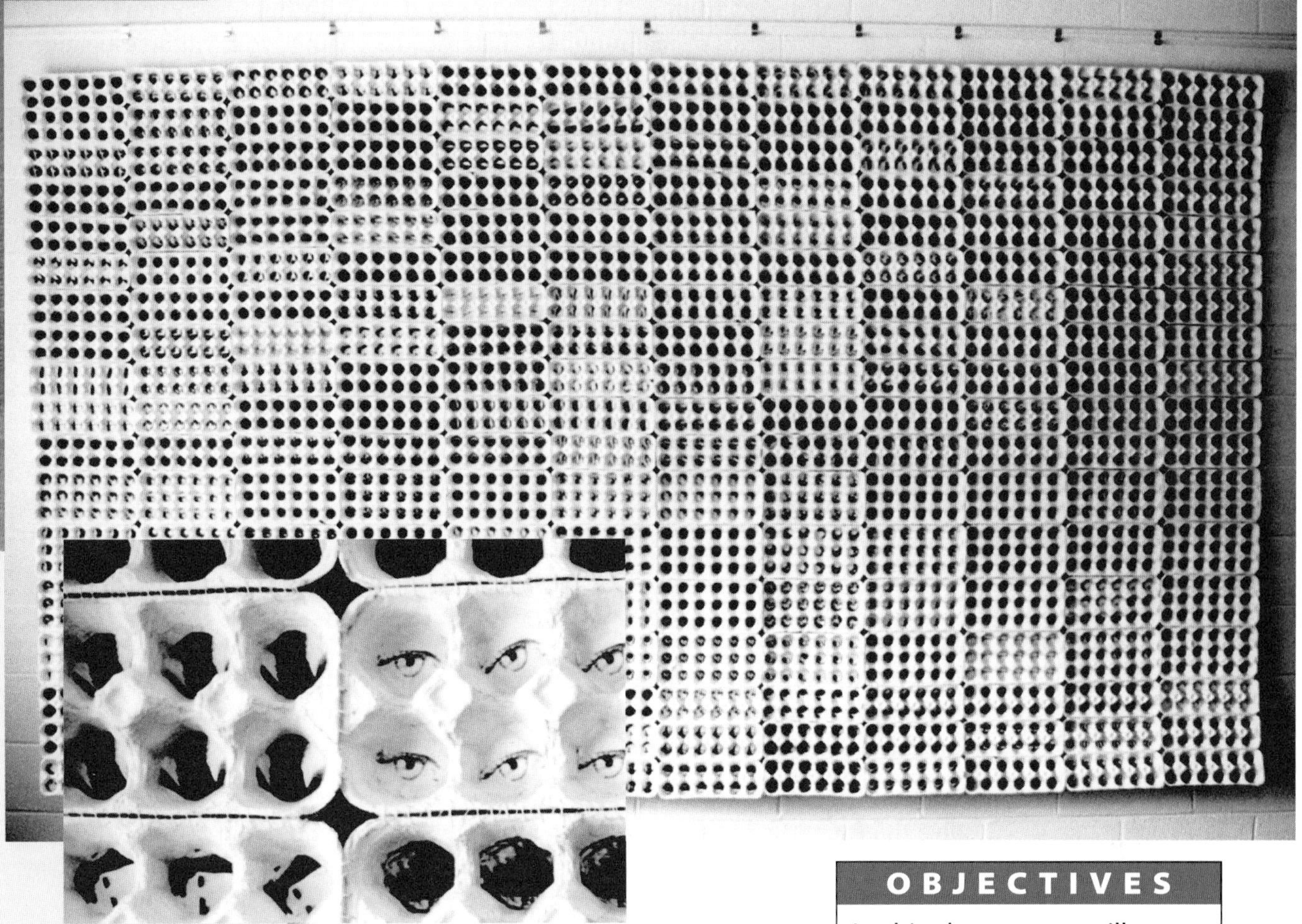

American installation artist Amy Stacey Curtis (b 1970) created this sculpture. The rectangular arrangement of egg cartons is used to organize an even larger arrangement of photocopied images. The egg cartons and their compartments divide the piece into rows and columns, while the small images—some darker or lighter than others—help certain elements of the piece to stand out more prominently.

Fragile and detail from *Fragile* by Amy Stacey Curtis
Egg cartons, acrylic, dye, thread, beads, photocopies

OBJECTIVES

In this chapter you will

- review properties of real numbers
- use matrices to organize information
- add, subtract, and multiply matrices
- solve systems of linear equations with matrices
- graph inequalities on a coordinate plane and solve systems of inequalities
- write and graph inequalities that represent conditions that must be met simultaneously

Properties of Real Numbers

In Chapter 6, you will use **matrices,** rectangular arrangements of numbers, to organize information and solve problems. Just as you can add, subtract, and multiply real numbers, you can perform these operations on matrices. However, matrices are different from real numbers in some ways. In this lesson you'll review some of the properties of real numbers. Later in the chapter you will investigate which of these properties matrices have.

A designer works on a computer animation in Hangzhou, China. In the software used for designing three-dimensional models, matrices are used to organize data and make renderings more lifelike and complex.

One property of real numbers is that multiplication is **commutative.** That means you can multiply two numbers in either order and get the same result. In symbols, $a \cdot b = b \cdot a$ (where a and b are any real numbers). On the other hand, division of real numbers is not commutative, because $3 \div 4 \neq 4 \div 3$.

Real numbers also have inverse and identity properties. Consider how you would solve the equation $3x = 47$. To undo multiplying by 3, you could multiply both sides by $\frac{1}{3}$. This approach is successful because the product of 3 and $\frac{1}{3}$ is 1. We say that $\frac{1}{3}$ is the **multiplicative inverse** of 3. To emphasize that it's the inverse of 3, you might write $\frac{1}{3}$ as 3^{-1}.

The number 1 is special because multiplying by 1 doesn't change a number's value. For example, $5 \cdot 1 = 5$ and $x \cdot 1 = x$. The number 1 is called the **multiplication identity element.** The multiplicative inverse and the multiplication identity element are closely related, because multiplying a number by its inverse gives the identity element.

EXAMPLE A

Test several cases as needed to answer each question.

a. Is addition a commutative operation? Is subtraction a commutative operation?

b. Is there an identity element for addition?

c. Does 3 have an additive inverse?

▶ Solution

Remember to check negative values as well as positive when testing cases.

a. Addition is commutative; that is $a + b = b + a$ for all real values of a and b. Subtraction is not commutative; $4 - 3 \neq 3 - 4$.

b. Zero is the identity element for addition. For all real values of a, $a + 0 = a$.

c. The additive inverse of 3 is the number that gives the identity element when added to 3. The additive inverse of 3 is -3 because $3 + -3 = 0$.

Another property of real-number multiplication is that it's **associative.** Consider the product $3 \cdot 4 \cdot 5$. You can first multiply $3 \cdot 4$, and then multiply by 5 to get $3 \cdot 4 \cdot 5 = 12 \cdot 5 = 60$. If you multiply $4 \cdot 5$ and then multiply by 3, the result is the same: $3 \cdot 4 \cdot 5 = 3 \cdot 20 = 60$. Parentheses help to indicate the order of multiplication: $(3 \cdot 4) \cdot 5 = 3 \cdot (4 \cdot 5)$. In general, $(a \cdot b) \cdot c = a \cdot (b \cdot c)$ for all real numbers a, b, and c.

EXAMPLE B

Test several cases to answer each question.

a. Are addition and subtraction of real numbers associative operations?

b. Multiplication is **distributive** over addition: $a \cdot (b + c) = ab + ac$. Is multiplication distributive over subtraction? Is addition distributive over multiplication?

▶ Solution

a. Testing some cases shows that addition is associative: $(a + b) + c = a + (b + c)$ for all real numbers a, b, and c.

Subtraction is not associative. For example, $(10 - 4) - 3 \neq 10 - (4 - 3)$ because $3 \neq 9$.

b. Multiplication is distributive over subtraction: $a \cdot (b - c) = ab - ac$ for all real numbers a, b, and c. You can verify this by rewriting the subtraction as the addition of the opposite: $a \cdot (b - c) = a \cdot (b + -c) = ab + a(-c) = ab - ac$.

Addition is not distributive over multiplication. For example, $3 + (4 \cdot 5) \neq (3 + 4) \cdot (3 + 5)$ because $23 \neq 56$.

Above we showed several times that a property *doesn't* hold by citing a single case where it fails. That case is called a **counterexample.** But to show that a property *does* hold, you must show that it is true for every number.

Other important properties of real numbers involve their order. Given any two real numbers, the first number is less than, greater than, or equal to the second number. If you represent real numbers on a number line, larger numbers lie to the right of smaller numbers.

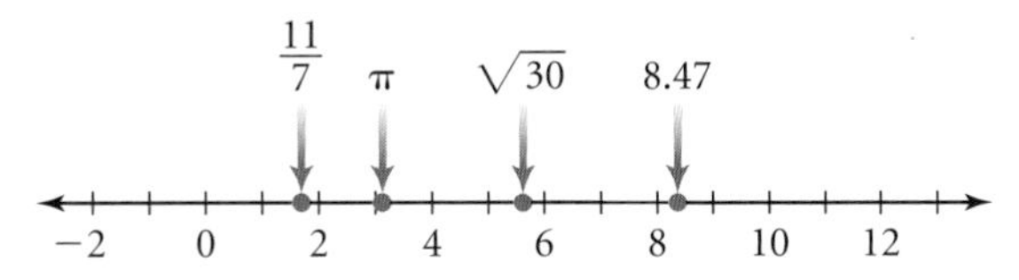

Because adding or subtracting shifts a number along the number line, adding the same value to two numbers (or expressions) does not change their order. In symbols, if a is larger than b (written $a > b$) then $a + c > b + c$, no matter what c is. Multiplication or division by a positive number also doesn't change the order. But multiplication or division by a negative value reverses the order of the inequality. For example, $2 > -3$, but if you multiply both sides by -5, you're comparing -10 with 15. Because $-10 < 15$, the inequality reverses direction. It's important to remember this property when solving inequalities.

EXAMPLE C | Graph the values of x that satisfy the inequality $2(5 - x) \leq 6$.

▶ **Solution**

$\frac{1}{2} \cdot 2(5 - x) \leq \frac{1}{2} \cdot 6$	Multiply both sides by $\frac{1}{2}$ to undo multiplying by 2.
$5 - x \leq 3$	Multiply.
$-5 + 5 - x \leq -5 + 3$	Add -5 to both sides to undo adding 5.
$-x \leq -2$	Add.
$-1 \cdot -x \geq -1 \cdot -2$	Multiply both sides by -1, and reverse the inequality symbol.
$x \geq 2$	Multiply.

−2 0 2 4 6 8 10 12

You can check your solution using any of the values in the shaded part of the number line. For example, you might substitute 3 or 5 for x in the original inequality to verify your answer.

The filled-in circle in the graph indicates that the number 2 is part of the solution. If the solution had been $x > 2$, you would have used an open (unfilled) circle to show that the endpoint is not included in the solution.

For all real values of a, b, and c, these properties are true:

Distributive Property

$a(b + c) = a(b) + a(c)$

Commutative Property

$a + b = b + a$

$ab = ba$

Associative Property

$a + (b + c) = (a + b) + c$

$a(bc) = (ab)c$

Additive Identity Property

$a + 0 = a$

Additive Inverse Property

$a + -a = 0$

Multiplicative Identity Property

$a \cdot 1 = a$

Multiplicative Inverse Property

$a \cdot \frac{1}{a} = 1$ for $a \neq 0$.

EXERCISES

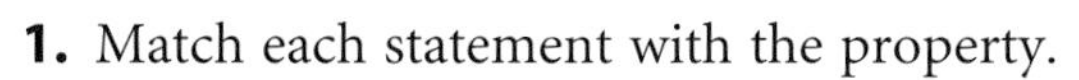

1. Match each statement with the property.

a. associative property
b. commutative property
c. distributive property
d. identity element property
e. inverse element property

A. $8 + 0 = 8$ @
B. $7 \cdot 5 = 5 \cdot 7$
C. $(12 + x) + 3x = 12 + (x + 3x)$ @
D. $11 \cdot 11^{-1} = 1$
E. $5x + 9x = (5 + 9)x$

2. Identify each statement as always true, sometimes true, or never true. Then state the name of the property that is being tested and tell whether it is a property for all real numbers.

a. $5 - x = x - 5$ @
b. $7x = 9x - 2x$
c. $(60 \div x) \div 2 = 60 \div (x \div 2)$ @
d. $x^{3+4} = x^3 + x^4$

3. Graph the solution to each inequality.

a. $4 + 2x \leq 8 + 4x$ @
b. $7 - 3(x - 5) > 16$

IMPROVING YOUR REASONING SKILLS

One and the Same?

Lee's math teacher put the following proof up on the board and said, "I have proved that two equals one!" Can you find the flaw in the teacher's proof?

$a = b$	Let a and b be two equal numbers.
$a^2 = a \cdot b$	Multiply each side by equal values.
$a^2 - b^2 = a \cdot b - b^2$	Subtract the same value from each side.
$(a + b)(a - b) = b(a - b)$	Factor.
$(a + b) = b$	Divide each side by the same value.
$a + a = a$	Substitute a for b in the previous equation.
$2a = a$	Add.
$2 = 1$	Divide each side by the same value.

LESSON

6.1

Matrix Representations

All dimensions are critical dimensions, otherwise why are they there?

RUSS ZANDBERGEN

On Saturday, Karina surveyed visitors to Snow Mountain with weekend passes and found that 75% of skiers planned to ski again the next day and 25% planned to snowboard. Of the snowboarders, 95% planned to snowboard the next day and 5% planned to ski. In order to display the information, she made this diagram.

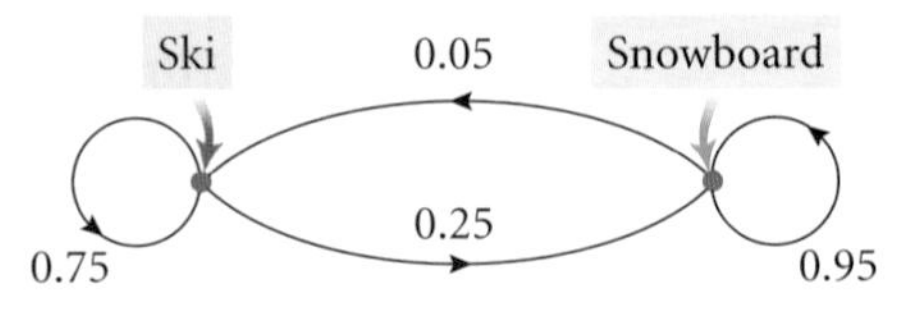

The arrows and labels show the patterns of the visitors' next-day activities. For instance, the circular arrow labeled 0.75 indicates that 75% of the visitors skiing one day plan to ski again the next day. The arrow labeled 0.25 indicates that 25% of the visitors who ski one day plan to snowboard the next day.

Diagrams like these are called **transition diagrams** because they show how something changes from one time to the next. The same information is sometimes represented in a **transition matrix.** A **matrix** is a rectangular arrangement of numbers. For the Snow Mountain information, the transition matrix looks like this:

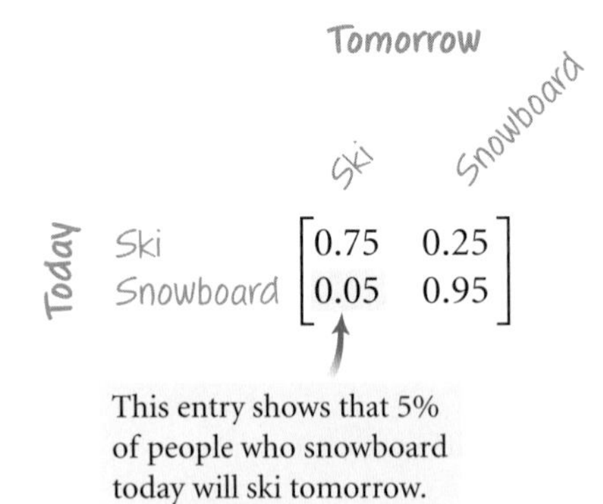

In the investigation you will create a transition diagram and matrix for another situation. You will also use the information to determine how the numbers of people in two different categories change over a period of time.

Investigation
Chilly Choices

The school cafeteria offers a choice of ice cream or frozen yogurt for dessert once a week. During the first week of school, 220 students choose ice cream but only 20 choose frozen yogurt. During each of the following weeks, 10% of the frozen-yogurt eaters switch to ice cream and 5% of the ice-cream eaters switch to frozen yogurt.

Step 1 Complete a transition diagram that displays this information.

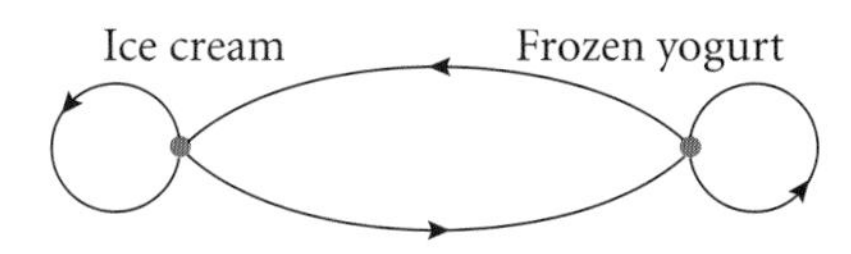

Step 2 Complete a transition matrix that represents this information. The rows should indicate the present condition, and the columns should indicate the next condition after the transition.

Next week

This week	Ice cream	Yogurt
Ice cream	?	?
Yogurt	?	?

Step 3 In the second week, how many students choose ice cream and how many students choose frozen yogurt?

Step 4 How many students will choose each option in the third week?

Step 5 Write a recursive routine to take any week's values and give the next week's values.

Step 6 What do you think will happen to the long-run values of the number of students who choose ice cream and the number who choose frozen yogurt?

You can use matrices to organize many kinds of information. For example, the matrix below can be used to represent the number of sports drinks, fruit juices, and waters sold this week from the vending machines at the main entrance and the back entrance of the school. The rows represent sports drinks, juices, and waters, from top to bottom. The columns represent the main and back entrances, from left to right.

The **dimensions** of the matrix give the numbers of rows and columns, in this case, 3×2 (read "three by two"). Each number in the matrix is called an **entry,** or **element,** and is identified as a_{ij}, where i and j are the row number and column number, respectively. In matrix $[A]$ at right, $a_{31} = 98$ because 98 is the entry in row 3, column 1.

$$[A] = \begin{bmatrix} 83 & 33 \\ 65 & 20 \\ 98 & 50 \end{bmatrix}$$

This element is the number of bottles of water sold at the main entrance.

Example A shows how to use matrices to represent coordinates of geometric figures.

EXAMPLE A Represent quadrilateral $ABCD$ as a matrix, $[M]$.

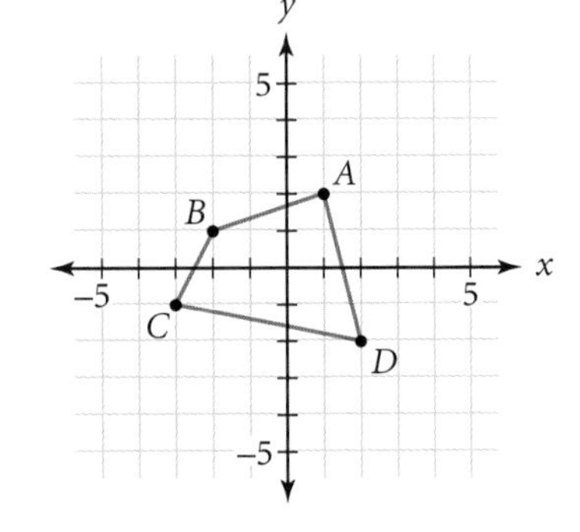

▶ Solution

You can use a matrix to organize the coordinates of the consecutive vertices of a geometric figure. Because each vertex has 2 coordinates and there are 4 vertices, use a 2×4 matrix with each column containing the x- and y-coordinates of a vertex. Row 1 contains consecutive x-coordinates and row 2 contains the corresponding y-coordinates.

$$[M] = \begin{bmatrix} 1 & -2 & -3 & 2 \\ 2 & 1 & -1 & -2 \end{bmatrix}$$

Example B shows how a transition matrix can be used to organize data and predictions. In Lesson 6.2, you'll learn how to do computations with matrices.

EXAMPLE B In Karina's survey from the beginning of this lesson, she interviewed 260 skiers and 40 snowboarders. How many people will do each activity the next day if her transition predictions are correct?

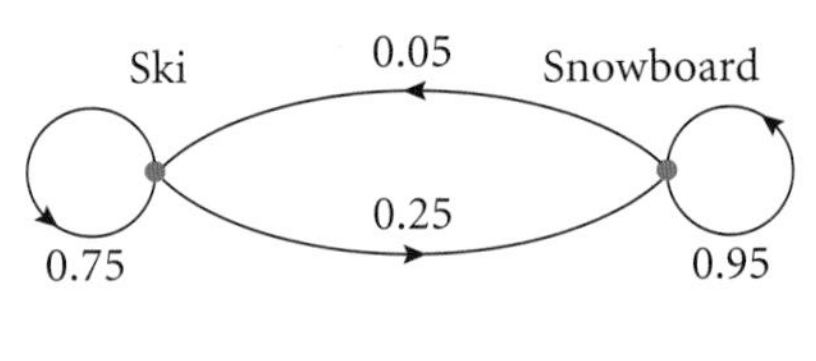

$$\begin{bmatrix} 0.75 & 0.25 \\ 0.05 & 0.95 \end{bmatrix}$$

▶ Solution

The next day, 75% of the 260 skiers will ski again and 5% of the 40 snowboarders will switch to skiing.

Skiers: $260(0.75) + 40(0.05) = 197$

So, 197 people will ski the next day.

The next day, 25% of the 260 skiers will switch to snowboarding and 95% of the 40 snowboarders will snowboard again.

Snowboarders: $260(0.25) + 40(0.95) = 103$

So, 103 people will snowboard the next day.

You can organize the information for the first day and second day as matrices in the form [*number of skiers*   *number of snowboarders*].

$[260 \quad 40]$

$[197 \quad 103]$

You can use transition diagrams and transition matrices to show changes in a closed system. (A closed system is one in which items may change, but nothing is added or removed.) The diagram, though very informative for simple problems, is difficult to use when you have 5 or more starting conditions, as this would create 25 or more arrows, or paths. The transition matrix is just as easy to read for any number of starting conditions as it is for two. It grows in size, but each entry shows what percentage changes from one condition to another.

EXERCISES

You will need

A graphing calculator for Exercise **15**.

Practice Your Skills

1. Russell collected data similar to Karina's at Powder Hill Resort. He found that 86% of the skiers planned to ski the next day and 92% of the snowboarders planned to snowboard the next day. ⓐ

a. Draw a transition diagram for Russell's information.

b. Write a transition matrix for the same information. Remember that rows indicate the present condition and columns indicate the next condition. List skiers first and snowboarders second.

2. Complete this transition diagram. ⓐ

Ski   Snowboard

0.60   0.47

3. Write a transition matrix for the diagram in Exercise 2. Order your information as in Exercise 1b.

4. Matrix $[M]$ represents the vertices of $\triangle ABC$.

$$[M] = \begin{bmatrix} -3 & 1 & 2 \\ 2 & 3 & -2 \end{bmatrix}$$

a. Name the coordinates of the vertices and graph the triangle. ⓐ

b. What matrix represents the image of $\triangle ABC$ after a translation down 4 units?

c. What matrix represents the image of $\triangle ABC$ after a translation right 4 units?

5. During a recent softball tournament, information about which side players bat from was recorded in a matrix. Row 1 represents girls and row 2 represents boys. Column 1 represents left-handed batters, column 2 represents right-handed batters, and column 3 represents those who can bat with either hand.

$$[A] = \begin{bmatrix} 5 & 13 & 2 \\ 4 & 18 & 3 \end{bmatrix}$$

a. How many girls and how many boys participated in the tournament?

b. How many boys always batted right-handed? ⓐ

c. What is the meaning of the value of a_{12}?

Reason and Apply

6. A mixture of 40 g of NO and 200 g of N_2O_2 is heated. During each second at the new temperature, 10% of the NO changes to N_2O_2 and 5% of the N_2O_2 changes to NO.

a. Draw a transition diagram that displays this information. ⓐ

b. Write a transition matrix that represents the same information. List NO first and N_2O_2 second. ⓗ

c. If the total mass remains 240 g and the transition percentages stay the same, what are the amounts in grams of NO and N_2O_2 after 1 s? After 2 s? Write your answers as matrices in the form $[\text{NO} \quad N_2O_2]$.

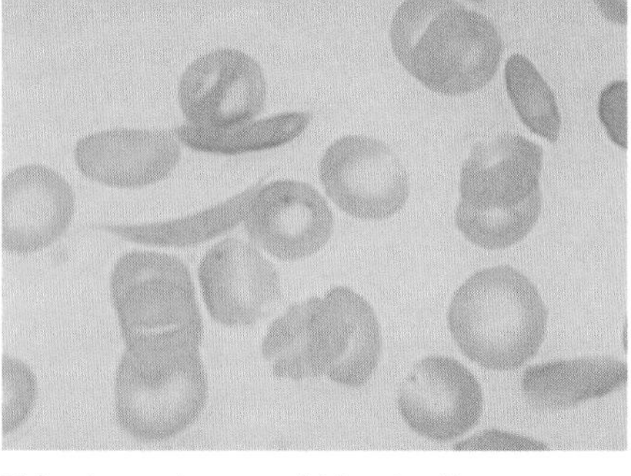

This photo shows red blood cells, some deformed by sickle-cell anemia. Researchers have found that nitric oxide (NO) counteracts the effects of sickle-cell anemia.

7. In many countries, more people move into urban areas than out of them. Suppose this diagram shows the movement of people between urban and rural areas in a country with a stable population.

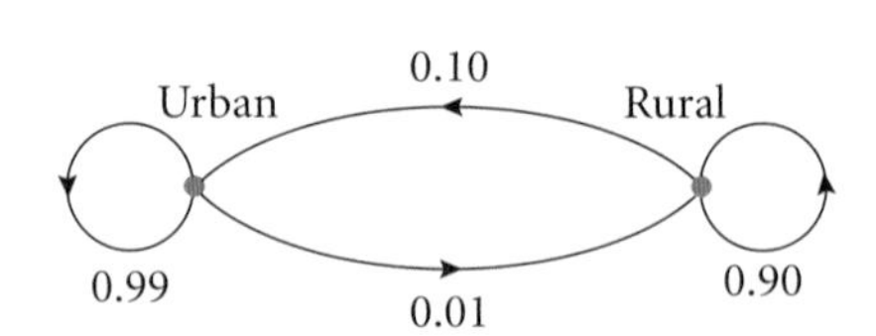

a. Explain what each number in the diagram means in this context.

b. Write a transition matrix that represents the same information. List people who live in urban areas first and those who live in rural areas second. ⓐ

c. If 16 million of the country's 25 million people live in the city initially, what are the urban and rural populations in millions after 1 yr? After 2 yr? Write your answers as matrices in the form $[urban \quad rural]$.

8. Recall the matrix $[A]$ on page 320 that represents the number of sports drinks, juices, and waters sold at the main and back entrances this week.

$$[A] = \begin{bmatrix} 83 & 33 \\ 65 & 20 \\ 98 & 50 \end{bmatrix}$$

a. Explain the meaning of the value of a_{32}. ⓐ

b. Explain the meaning of the value of a_{21}.

c. Matrix $[B]$ represents last week's sales. Compare this week's sales of sports drinks with last week's sales.

$$[B] = \begin{bmatrix} 80 & 25 \\ 65 & 15 \\ 105 & 55 \end{bmatrix}$$

d. Write a matrix that represents the total sales during last week and this week.

9. The three largest categories of motor vehicles sold in Springfield are sedan, SUV, and minivan. Suppose that of the buyers in Springfield who now own a minivan, 18% will change to an SUV and 20% will change to a sedan. Of the buyers who now own a sedan, 35% will change to a minivan and 20% will change to an SUV, and of those who now own an SUV, 12% will buy a minivan and 32% will buy a sedan.

a. Draw a transition diagram that displays these changes.

b. Write a transition matrix that represents this scenario. List the rows and columns in the order minivan, sedan, SUV.

c. What is the sum of the entries in row 1? Row 2? Row 3? Why does this sum make sense?

10. Lisa Crawford is getting into the moving-truck rental business in three nearby counties. She has the funds to buy about 100 trucks. She records where the trucks rented in Bay, Sage, and Thyme counties are returned. The results are reflected in this transition diagram.

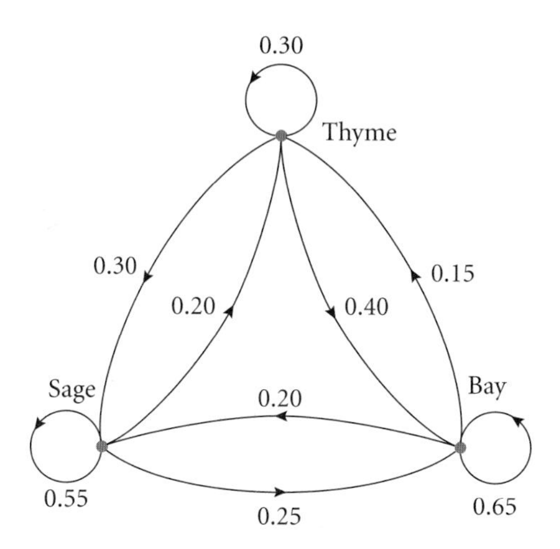

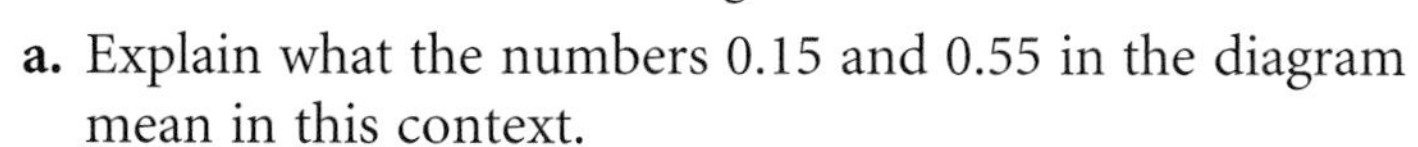

a. Explain what the numbers 0.15 and 0.55 in the diagram mean in this context.

b. Write a transition matrix that represents this scenario. List your rows and columns in the order Bay, Sage, Thyme.

c. What is the sum of the entries in row 1? Row 2? Row 3? Why does this sum make sense?

d. If Lisa starts with 45 trucks in Bay County, 30 trucks in Sage County, and 25 trucks in Thyme County and all trucks are rented one Saturday, how many trucks will she expect to be in each county the next morning?

11. APPLICATION Fly-Right Airways operates routes out of five cities as shown in the route map below. The route map is a **vertex-edge graph.** Each segment connecting two cities represents a round-trip flight between them. Matrix $[M]$ displays the information from the map in matrix form with the cities, A, B, C, D, and E, listed in order in the rows and columns. The rows represent starting conditions (departure cities), and the columns represent next conditions (arrival cities). This matrix is called an adjacency matrix. For instance, the value of the entry in row 1, column 5 shows that there are two round-trip flights between City A and City E.

$$[M] = \begin{bmatrix} 0 & 1 & 0 & 1 & 2 \\ 1 & 0 & 1 & 0 & 1 \\ 0 & 1 & 0 & 0 & 0 \\ 1 & 0 & 0 & 0 & 0 \\ 2 & 1 & 0 & 0 & 0 \end{bmatrix}$$

a. What are the dimensions of this matrix?

b. What is the value of m_{32}? What does this entry represent? ⓐ

c. Which city has the most flights? Explain how you can tell using the route map and using the matrix. ⓗ

d. Matrix $[N]$ represents Americana Airways' routes connecting four cities, J, K, L, and M. Sketch a possible route map.

$$[N] = \begin{bmatrix} 0 & 1 & 2 & 1 \\ 1 & 0 & 2 & 0 \\ 2 & 2 & 0 & 1 \\ 1 & 0 & 1 & 0 \end{bmatrix}$$

Mathematics CONNECTION

Graph theory is a branch of mathematics that deals with connections between items. In Exercise 11, a paragraph description of the flight routes could have been made, but a vertex-edge graph of the routes allows you to show the material quickly and clearly. You could also use a graph to diagram a natural gas pipeline, the chemical structure of a molecule, a family tree, or a computer network. The data in a graph can be represented, manipulated mathematically, and further investigated using matrices.

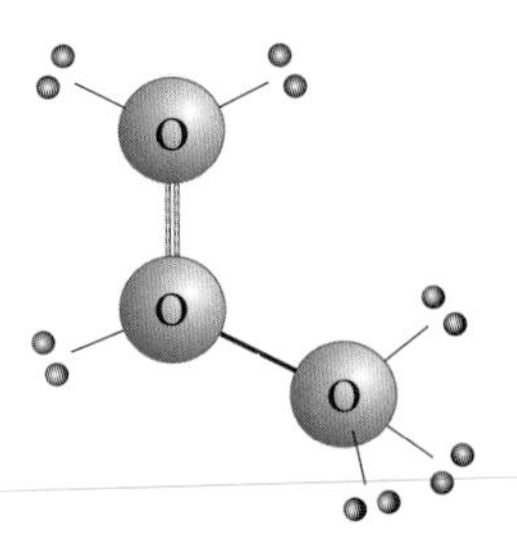

Ozone (O_3)

Review

12. Solve this system using either substitution or elimination.

$$\begin{cases} 5x - 4y = 25 \\ x + y = 3 \end{cases}$$

13. Each slice of pepperoni pizza has approximately 7.4 pieces of pepperoni on it, and each slice of supreme pizza has approximately 4.7 pieces of pepperoni on it. Write an equation that shows that p slices of pepperoni pizza and s slices of supreme pizza would have a total of 100 pieces of pepperoni. @

14. Solve the equation $2x + 3y = 12$ for y and then graph it.

15. APPLICATION The table at right shows the number of cellular telephone subscribers in the United States in selected years from 1985 to 2005.

a. Create a scatter plot of the data.

b. Find an exponential function to model the data.

c. Use your model to predict the number of subscribers in 2007. Do you think this is a realistic prediction? Why or why not? Do you think an exponential model is appropriate? Why or why not?

Cellular Phone Subscribers

Year	Number of subscribers	Year	Number of subscribers
1985	340,000	1997	55,312,000
1987	1,231,000	1999	86,047,000
1989	3,509,000	2001	128,375,000
1991	7,557,000	2003	158,722,000
1993	16,009,000	2004	182,140,000
1995	33,786,000	2005	207,900,000

(*The World Almanac and Book of Facts 2007*)

16. APPLICATION The equation $y = 20\log\left(\frac{x}{0.00002}\right)$ measures the intensity of a sound as a function of the pressure it creates on the eardrum. The intensity, y, is measured in decibels (dB), and the pressure, x, is measured in Pascals (Pa).

a. What is the intensity of the sound of a humming refrigerator, if it causes 0.00356 Pa of pressure on the eardrum? @

b. A noise that causes 20 Pa of pressure on the eardrum brings severe pain to most people. What is the intensity of this noise?

c. Write the inverse function that measures pressure on the eardrum as a function of intensity of a sound.

d. How much pressure on the eardrum is caused by a 90 dB sound?

LESSON

6.2

Matrix Operations

It is not once nor twice but times without number that the same ideas make their appearance in the world.

ARISTOTLE

You've seen that a matrix is a compact way of organizing data, similar to a table. But unlike tables, matrices can be added and multiplied to help you solve problems.

Consider this problem from Lesson 6.1. Matrix $[A]$ represents sports drinks, fruit juices, and waters sold this week from the vending machines at the main entrance and the back entrance of the school. Matrix $[B]$ contains the same information for last week. What are the total sales, by category and location, for both weeks?

$$[A] = \begin{bmatrix} 83 & 33 \\ 65 & 20 \\ 98 & 50 \end{bmatrix} \qquad [B] = \begin{bmatrix} 80 & 25 \\ 65 & 15 \\ 105 & 55 \end{bmatrix}$$

To solve this problem, you add matrices $[A]$ and $[B]$.

$$83 + 80 = 163$$

$$\begin{bmatrix} 83 & 33 \\ 65 & 20 \\ 98 & 50 \end{bmatrix} + \begin{bmatrix} 80 & 25 \\ 65 & 15 \\ 105 & 55 \end{bmatrix} = \begin{bmatrix} 163 & 58 \\ 130 & 35 \\ 203 & 105 \end{bmatrix}$$

If 83 sports drinks were sold at the main entrance this week and 80 sports drinks were sold at the main entrance last week, a total of 163 sports drinks were sold at the main entrance for both weeks.

To add two matrices, you simply add corresponding entries. So, in order for you to add (or subtract) two matrices, they both must have the same dimensions. The corresponding rows and columns should also have similar interpretations if the results are to make sense. [▶ See **Calculator Note 6A** to learn how to enter matrices into your calculator. **Calculator Note 6B** shows how to perform operations with matrices. ◀]

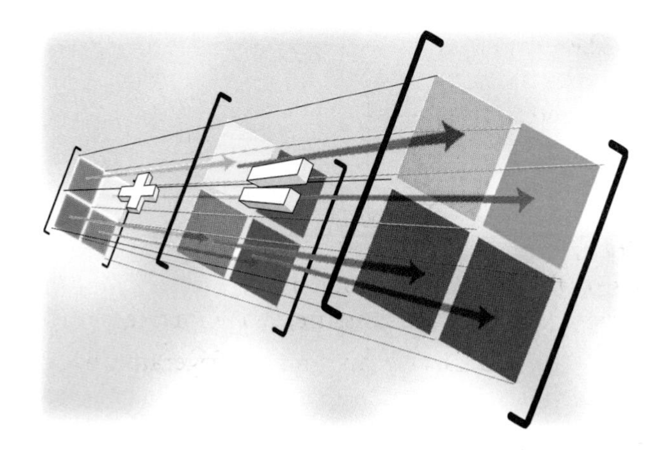

When you add matrices, you add corresponding entries. This illustration uses the addition of color to show how the addition carries through to the matrix representing the sum.

In Lesson 6.1, you used a matrix to organize the coordinates of the vertices of a triangle. You can use matrix operations to transform a figure such as a triangle just as you transformed the graph of a function.

EXAMPLE A

This matrix represents a triangle.

$$\begin{bmatrix} -3 & 1 & 2 \\ 2 & 3 & -2 \end{bmatrix}$$

a. Graph the triangle and its image after a translation left 3 units. Write a matrix equation to represent the transformation.

b. Describe the transformation represented by this matrix expression.

$$\begin{bmatrix} -3 & 1 & 2 \\ 2 & 3 & -2 \end{bmatrix} + \begin{bmatrix} -4 & -4 & -4 \\ -3 & -3 & -3 \end{bmatrix}$$

c. Describe the transformation represented by this matrix expression.

$$2 \cdot \begin{bmatrix} -3 & 1 & 2 \\ 2 & 3 & -2 \end{bmatrix}$$

▶ *Solution*

The original matrix represents a triangle with vertices $(-3, 2)$, $(1, 3)$, and $(2, -2)$.

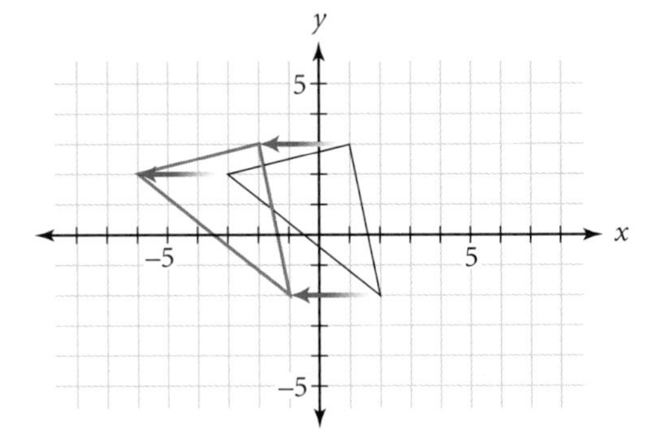

a. After a translation left 3 units, the x-coordinates of the image are reduced by 3. There is no change to the y-coordinates. You can represent this transformation as a subtraction of two matrices.

$$\begin{bmatrix} -3 & 1 & 2 \\ 2 & 3 & -2 \end{bmatrix} - \begin{bmatrix} 3 & 3 & 3 \\ 0 & 0 & 0 \end{bmatrix} = \begin{bmatrix} -6 & -2 & -1 \\ 2 & 3 & -2 \end{bmatrix}$$

b. $\begin{bmatrix} -3 & 1 & 2 \\ 2 & 3 & -2 \end{bmatrix} + \begin{bmatrix} -4 & -4 & -4 \\ -3 & -3 & -3 \end{bmatrix} = \begin{bmatrix} -7 & -3 & -2 \\ -1 & 0 & -5 \end{bmatrix}$

This matrix addition represents a translation left 4 units and down 3 units.

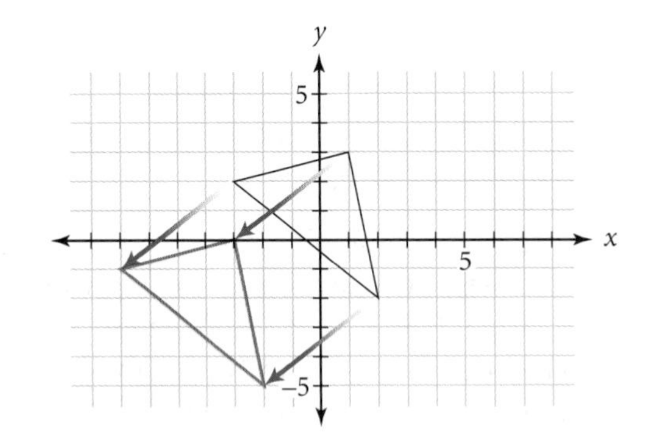

c. $2 \cdot \begin{bmatrix} -3 & 1 & 2 \\ 2 & 3 & -2 \end{bmatrix} = \begin{bmatrix} -3 & 1 & 2 \\ 2 & 3 & -2 \end{bmatrix} + \begin{bmatrix} -3 & 1 & 2 \\ 2 & 3 & -2 \end{bmatrix} = \begin{bmatrix} -6 & 2 & 4 \\ 4 & 6 & -4 \end{bmatrix}$

or

$$2 \cdot \begin{bmatrix} -3 & 1 & 2 \\ 2 & 3 & -2 \end{bmatrix} = \begin{bmatrix} 2 \cdot (-3) & 2 \cdot 1 & 2 \cdot 2 \\ 2 \cdot 2 & 2 \cdot 3 & 2 \cdot (-2) \end{bmatrix} = \begin{bmatrix} -6 & 2 & 4 \\ 4 & 6 & -4 \end{bmatrix}$$

Multiplying a matrix by a number is called **scalar multiplication.** Each entry in the matrix is simply multiplied by the **scalar,** which is 2 in this case.

The resulting matrix represents dilations, both horizontally and vertically, by the scale factor 2. When the term *dilation* is used without specifying whether it is a horizontal or vertical dilation, you can assume that the horizontal and vertical scale factors are the same. In this case the original triangle was dilated by the scale factor 2.

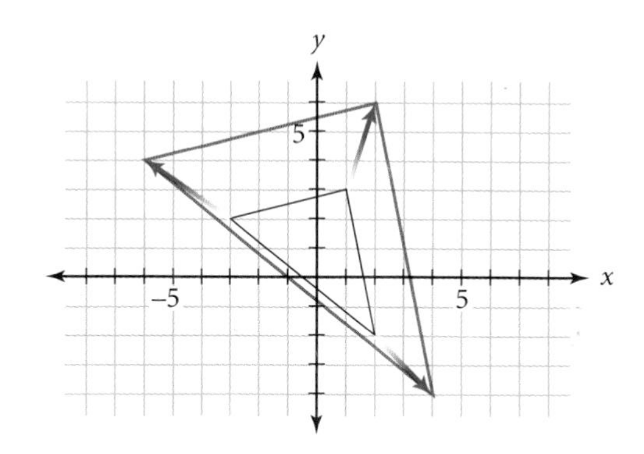

[▶ See **Calculator Note 6C** to learn how to graph polygons with matrices. ◀]

Addition and scalar multiplication operate on one entry at a time. The multiplication of two matrices is more involved and uses several entries to find one entry of the answer matrix. In the next example, you'll revisit the situation from the investigation in Lesson 6.1.

EXAMPLE B

The school cafeteria offers a choice of ice cream or frozen yogurt for dessert once a week. During the first week of school, 220 students choose ice cream and 20 choose frozen yogurt. During each of the following weeks, 10% of the frozen-yogurt eaters switch to ice cream and 5% of the ice-cream eaters switch to frozen yogurt. How many students will choose each dessert in the second week? In the third week?

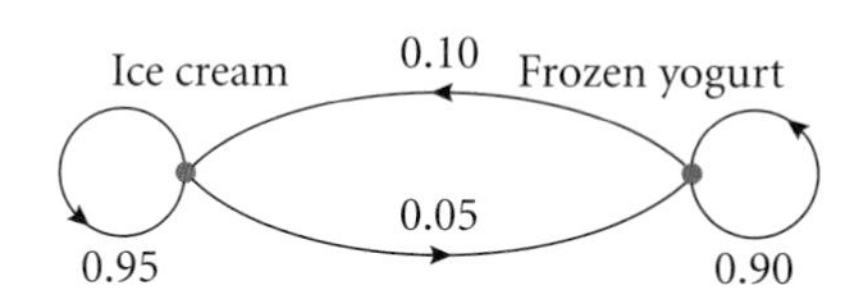

▶ Solution

You can use this matrix equation to find the answer for the second week:

$$[220 \quad 20]\begin{bmatrix} 0.95 & 0.05 \\ 0.10 & 0.90 \end{bmatrix} = [\textit{ice cream} \quad \textit{frozen yogurt}]$$

The initial matrix, $[A] = [220 \quad 20]$, represents the original numbers of ice-cream eaters and frozen-yogurt eaters.

In the transition matrix $[B] = \begin{bmatrix} 0.95 & 0.05 \\ 0.10 & 0.90 \end{bmatrix}$, the top row represents the transitions in the number of ice-cream eaters, and the bottom row represents the transitions in the number of frozen-yogurt eaters.

You can define matrix multiplication by looking at how you calculate the numbers for the second week. The second week's number of ice-cream eaters will be 220(0.95) + 20(0.10), or 211 students, because 95% of the 220 original ice-cream eaters don't switch and 10% of the 20 original frozen-yogurt eaters switch to ice cream. In effect, you multiply the two entries in row 1 of $[A]$ by the two entries in column 1 of $[B]$ and add the products. The result, 211, is entry c_{11} in the answer matrix, $[C]$.

Initial matrix		**Transition matrix**		**Answer matrix**
$[A]$	$\cdot$	$[B]$	$=$	$[C]$

$$[220 \quad 20]\begin{bmatrix} 0.95 & 0.05 \\ 0.10 & 0.90 \end{bmatrix} = [211 \quad \textit{frozen yogurt}]$$

$$220(0.95) + 20(0.10) = 211$$

Likewise, the second week's number of frozen-yogurt eaters will be 220(0.05) + 20(0.90), or 29 students, because 5% of the ice-cream eaters switch to frozen yogurt and 90% of the frozen-yogurt eaters don't switch. The number of frozen-yogurt eaters in the second week is the sum of the products of the entries in row 1 of $[A]$ and column 2 of $[B]$. The answer, 29, is entry c_{12} in the answer matrix, $[C]$.

$$\begin{bmatrix} 220 & 20 \end{bmatrix} \begin{bmatrix} 0.95 & 0.05 \\ 0.10 & 0.90 \end{bmatrix} = \begin{bmatrix} 211 & 29 \end{bmatrix}$$

To get the numbers for the third week, multiply the result of your previous calculations by the transition matrix again.

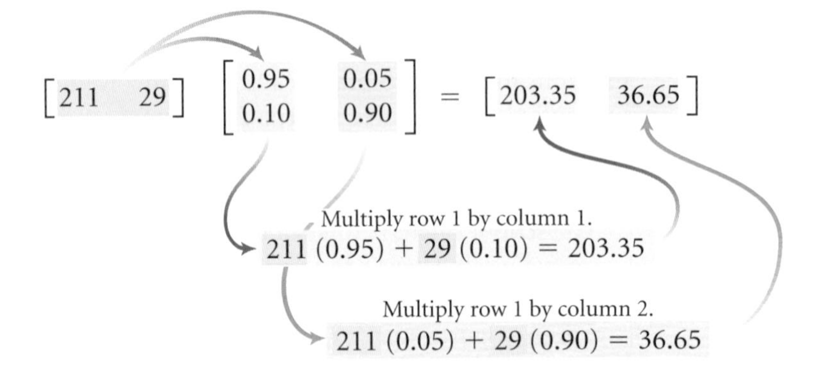

Approximately 203 students will choose ice cream and 37 will choose frozen yogurt in the third week.

You can continue multiplying to find the numbers in the fourth week, the fifth week, and so on. [▶ Revisit **Calculator Note 6B** to learn how to multiply matrices on your calculator. ◀]

In the investigation you will model a real-world situation with matrices. You'll also practice multiplying matrices.

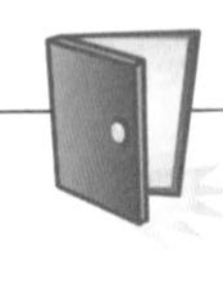

Investigation
Find Your Place

In this investigation you will simulate the weekly movement of rental cars between cities and analyze the results.

Each person represents a rental car starting at City A, City B, or City C.

Step 1 Follow the Procedure Note to simulate the movement of cars. At the beginning of the simulation and after each move, a class recorder should make note of the number of cars at each city.

Procedure Note

Rental Car Simulation

1. All students should position themselves in one of three designated locations in the room, City A, City B, or City C. Report the starting number of cars at each city to the recorder.
2. Use your calculator to generate a random number, x, between 0 and 1. [► See **Calculator Note 1L** to learn how to generate random numbers. ◄] Determine your location for next week as follows:

Starting city	City A		City B		City C	
Next location	$x \leq 0.2$	go to **B**	$x \leq 0.5$	go to **A**	$x \leq 0.1$	go to **B**
	$0.2 < x \leq 0.7$	go to **C**	$x > 0.5$	stay at **B**	$0.1 < x \leq 0.3$	go to **A**
	$x > 0.7$	stay at **A**			$x > 0.3$	stay at **C**

3. On the teacher's signal, you move (or stay) as indicated. Count the number of cars at each city and report this value to the recorder. Repeat the simulation five times. Each time, record the number of cars at each city.

Step 2 Work with your group to make a transition diagram and a transition matrix that represent the rules of the simulation.

Step 3 Write an initial condition matrix for the starting quantities at each city. Then, show how to multiply the initial condition matrix and the transition matrix for the first transition. How do these theoretical results for week 1 compare with the experimental data from your simulation?

Step 4 Use your calculator to find the theoretical number of cars at each city for the next four weeks. Find the theoretical long-run values of the number of cars at each city.

Step 5 Compare these results with the experimental values in your table. If they are not similar, explain why.

Just as only some matrices can be added (those with the same dimensions), only some matrices can be multiplied. Example C and Exercise 3 will help you explore the kinds of matrices that can be multiplied.

American artist Robert Silvers (b 1968) combined thousands of worldwide money images in a matrix-like arrangement to create this piece titled *Washington.*

EXAMPLE C

Consider this product.

$$\begin{bmatrix} -1 & 0 \\ 0 & 1 \end{bmatrix}\begin{bmatrix} -3 & 1 & 2 \\ 2 & 3 & -2 \end{bmatrix}$$

a. Determine the dimensions of this product.

b. Describe how to calculate entries in the product.

▶ Solution

a. You can multiply a 2×2 matrix by a 2×3 matrix because the inside dimensions are the same—the 2 row entries match up with the 2 column entries.

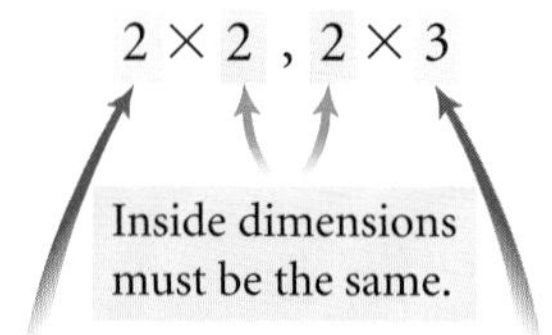

The outside dimensions tell you the dimensions of the product.

The product has dimensions 2×3.

$$\begin{bmatrix} -1 & 0 \\ 0 & 1 \end{bmatrix}\begin{bmatrix} -3 & 1 & 2 \\ 2 & 3 & -2 \end{bmatrix} = \begin{bmatrix} c_{11} & c_{12} & c_{13} \\ c_{21} & c_{22} & c_{23} \end{bmatrix}$$

b. To find the values of entries in the first row of your solution matrix, you add the products of the entries in the first row of the first matrix and the entries in the columns of the second matrix.

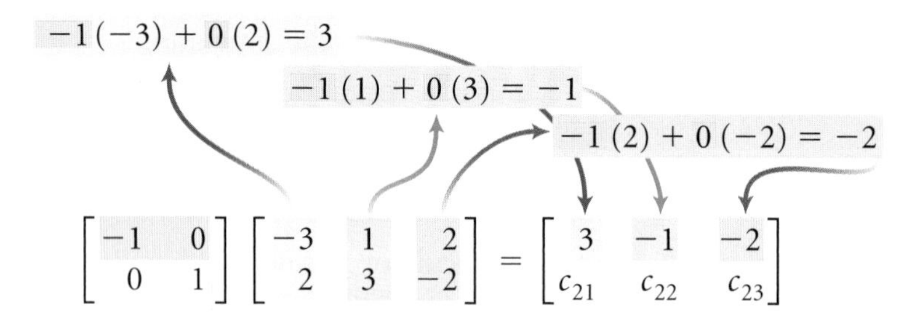

$$\begin{bmatrix} -1 & 0 \\ 0 & 1 \end{bmatrix}\begin{bmatrix} -3 & 1 & 2 \\ 2 & 3 & -2 \end{bmatrix} = \begin{bmatrix} 3 & -1 & -2 \\ c_{21} & c_{22} & c_{23} \end{bmatrix}$$

To find the values of entries in the second row of your solution matrix, you add the products of the entries in the second row of the first matrix and the entries in the columns of the second matrix.

$$\begin{bmatrix} -1 & 0 \\ 0 & 1 \end{bmatrix}\begin{bmatrix} -3 & 1 & 2 \\ 2 & 3 & -2 \end{bmatrix} = \begin{bmatrix} 3 & -1 & -2 \\ 2 & 3 & -2 \end{bmatrix}$$

$0(-3) + 1(2) = 2$

$0(1) + 1(3) = 3$

$0(2) + 1(-2) = -2$

The product is

$$\begin{bmatrix} 3 & -1 & -2 \\ 2 & 3 & -2 \end{bmatrix}$$

The following definitions review the matrix operations you've learned in this lesson.

Matrix Operations

Matrix Addition

To add matrices, you add corresponding entries.

$$\begin{bmatrix} 2 & 3 \\ -1 & 0 \end{bmatrix} + \begin{bmatrix} -1 & 2 \\ 1 & 4 \end{bmatrix} = \begin{bmatrix} 1 & 5 \\ 0 & 4 \end{bmatrix}$$

$$-1 + 1 = 0$$

You can add only matrices that have the same dimension.

Scalar Multiplication

To multiply a scalar by a matrix, you multiply the scalar by each value in a matrix.

$$3 \cdot \begin{bmatrix} 2 & -1 \\ 1 & 3 \\ 0 & -2 \end{bmatrix} = \begin{bmatrix} 6 & -3 \\ 3 & 9 \\ 0 & -6 \end{bmatrix}$$

$$3(-2) = -6$$

Matrix Multiplication

To multiply two matrices, $[A]$ and $[B]$, you multiply each entry in a row of matrix $[A]$ by corresponding entries in a column of matrix $[B]$.

$$\begin{bmatrix} 0 & -1 \\ 2 & 1 \end{bmatrix} \begin{bmatrix} 3 & -2 & 4 \\ 0 & -3 & 5 \end{bmatrix} = \begin{bmatrix} 0 & 3 & -5 \\ 6 & -7 & 13 \end{bmatrix}$$

$$2(-2) + 1(-3) = -7$$

Entry c_{ij} in the answer matrix, $[C]$, represents the sum of the products of each entry in row i of the first matrix and the entry in the corresponding position in column j of the second matrix. The number of entries in a row of matrix $[A]$ must equal the number of entries in a column of matrix $[B]$. That is, the inside dimensions must be equal. The answer matrix will have the same number of rows as matrix $[A]$ and the same number of columns as matrix $[B]$, or the outside dimensions.

EXERCISES

You will need

A graphing calculator for Exercises **8**, **10**, **11**, and **13**.

Practice Your Skills

1. Look back at the calculations in Example B. Calculate how many students will choose each dessert in the fourth week by multiplying these matrices:

$$[203.35 \quad 36.65]\begin{bmatrix} 0.95 & 0.05 \\ 0.10 & 0.90 \end{bmatrix} = [\textit{ice cream} \quad \textit{frozen yogurt}]$$

2. Find the missing values.

a. $[13 \quad 23] + [-6 \quad 31] = [x \quad y]$ @

b. $\begin{bmatrix} 0.90 & 0.10 \\ 0.05 & 0.95 \end{bmatrix}\begin{bmatrix} 0.90 & 0.10 \\ 0.05 & 0.95 \end{bmatrix} = \begin{bmatrix} c_{11} & c_{12} \\ c_{21} & c_{22} \end{bmatrix}$ @

c. $\begin{bmatrix} 18 & -23 \\ 5.4 & 32.2 \end{bmatrix} + \begin{bmatrix} -2.4 & 12.2 \\ 5.3 & 10 \end{bmatrix} = \begin{bmatrix} a & b \\ c & d \end{bmatrix}$

d. $10 \cdot \begin{bmatrix} 18 & -23 \\ 5.4 & 32.2 \end{bmatrix} = \begin{bmatrix} m_{11} & m_{12} \\ m_{21} & m_{22} \end{bmatrix}$

e. $\begin{bmatrix} 7 & -4 \\ 18 & 28 \end{bmatrix} + 5 \cdot \begin{bmatrix} -2.4 & 12.2 \\ 5.3 & 10 \end{bmatrix} = \begin{bmatrix} a & b \\ c & d \end{bmatrix}$

American painter Chuck Close (b 1940) creates photo-realistic portraits by painting a matrix-like grid of rectangular cells. Close is a quadriplegic and paints with a mouth brush. This portrait is from 1992.

Janet by Chuck Close, oil on canvas, 102 × 84 in.

3. Perform matrix arithmetic. If a particular operation is impossible, explain why.

a. $\begin{bmatrix} 1 & 2 \\ 3 & -2 \\ 0 & 1 \end{bmatrix}\begin{bmatrix} -3 & -1 & 2 \\ 5 & 2 & -1 \end{bmatrix}$ @

b. $\begin{bmatrix} 1 & -2 \\ 6 & 3 \end{bmatrix} + \begin{bmatrix} -3 & 7 \\ 2 & 4 \end{bmatrix}$

c. $[5 \quad -2 \quad 7]\begin{bmatrix} -2 & 3 \\ -1 & 0 \\ 3 & 2 \end{bmatrix}$ @

d. $\begin{bmatrix} 3 & -8 & 10 & 2 \\ -1 & 2 & 3 & 4 \end{bmatrix}\begin{bmatrix} 2 & -5 & 3 & 12 \\ 8 & -4 & 0 & 2 \end{bmatrix}$ @

e. $\begin{bmatrix} 3 & 6 \\ -4 & 1 \end{bmatrix} - \begin{bmatrix} -1 & 7 \\ -8 & 3 \end{bmatrix}$

f. $\begin{bmatrix} 4 & 11 \\ 7 & 3 \\ 4 & 2 \end{bmatrix} + \begin{bmatrix} 3 & -2 & 7 \\ 5 & 0 & 2 \end{bmatrix}$

4. Find matrix $[B]$ such that

$$\begin{bmatrix} 8 & -5 & 4.5 \\ -6 & 9.5 & 5 \end{bmatrix} - [B] = \begin{bmatrix} 5 & -1 & 2 \\ -4 & 3.5 & 1 \end{bmatrix}$$

5. This matrix represents a triangle:

$$\begin{bmatrix} -3 & 1 & 2 \\ 2 & 3 & -2 \end{bmatrix}$$

a. Graph the triangle. @

b. Find the result of this matrix multiplication:

$$\begin{bmatrix} -1 & 0 \\ 0 & 1 \end{bmatrix}\begin{bmatrix} -3 & 1 & 2 \\ 2 & 3 & -2 \end{bmatrix}$$ @

c. Graph the image represented by the matrix in 5b.

d. Describe the transformation.

Reason and Apply

6. Matrix $[T]$ represents a triangle.

$$[T] = \begin{bmatrix} -3 & 1 & 2 \\ 2 & 3 & -2 \end{bmatrix}$$

Find matrix $[A]$ and matrix $[C]$ such that the triangle represented by matrix $[T]$ is reflected across the x-axis.

$$\begin{bmatrix} a_{11} & a_{12} \\ a_{21} & a_{22} \end{bmatrix} \begin{bmatrix} -3 & 1 & 2 \\ 2 & 3 & -2 \end{bmatrix} = \begin{bmatrix} c_{11} & c_{12} & c_{13} \\ c_{21} & c_{22} & c_{23} \end{bmatrix}$$

7. Of two-car families in a small city, 88% remain two-car families in the following year and 12% become one-car families in the following year. Of one-car families, 72% remain one-car families and 28% become two-car families. Suppose these trends continue for a few years. At present, 4800 families have one car and 4200 have two cars.

a. Draw a transition diagram that displays this information.

b. What matrix represents the present situation? Let a_1 represent one-car families. ⓐ

c. Write a transition matrix that represents the same information as your transition diagram.

d. Write a matrix equation to find the numbers of one-car and two-car families one year from now.

e. Find the numbers of one-car and two-car families two years from now.

8. *Mini-Investigation* Enter these matrices into your calculator.

$$[A] = \begin{bmatrix} 2 & 3 \\ -1 & 1 \end{bmatrix} \quad [B] = \begin{bmatrix} 3 & 4 \\ 0 & -2 \end{bmatrix} \quad [C] = \begin{bmatrix} -2 & 3 & 0 \\ -1 & 5 & 4 \end{bmatrix} \quad [D] = \begin{bmatrix} 1 & 0 \\ 0 & 1 \end{bmatrix}$$

a. Find $[A][B]$ and $[B][A]$. Are they the same?

b. Find $[A][C]$ and $[C][A]$. Are they the same? What do you notice?

c. Find $[A][D]$ and $[D][A]$. Are they the same? What do you notice?

d. Is matrix multiplication commutative? That is, does order matter?

9. Find the missing values.

a. $\begin{bmatrix} 2 & a \\ b & -1 \end{bmatrix} \begin{bmatrix} 5 \\ 3 \end{bmatrix} = \begin{bmatrix} 19 \\ 17 \end{bmatrix}$ ⓐ

b. $\begin{bmatrix} a & -2 \\ 3 & 1 \end{bmatrix} \begin{bmatrix} -3 \\ b \end{bmatrix} = \begin{bmatrix} -29 \\ -5 \end{bmatrix}$

10. A spider is in a building with three rooms. The spider moves from room to room by choosing a door at random. If the spider starts in room 1 initially, then there is a 100% chance it is in room 1 and 0% chance it is in either room 2 or room 3, so the initial probability matrix is $[1 \;\; 0 \;\; 0]$. What is the probability matrix after four room changes? What are the long-run probabilities? ⓗ

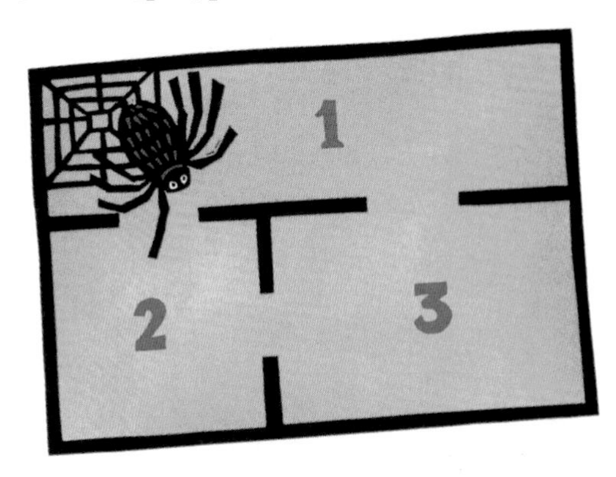

11. Recall the ice cream and frozen yogurt problem from Example B. Enter these matrices into your calculator, and use them to find the long-run values for the number of students who choose ice cream and the number of those who choose frozen yogurt. Explain why your answer makes sense.

$$[A] = [220 \quad 20] \qquad [B] = \begin{bmatrix} 0.95 & 0.05 \\ 0.10 & 0.90 \end{bmatrix}$$

12. APPLICATION A researcher studies the birth weights of women and their daughters. The weights are split into three categories: low (below 6 lb), average (between 6 and 8 lb), and high (above 8 lb). This transition diagram shows how birth weights changed from mother to daughter.

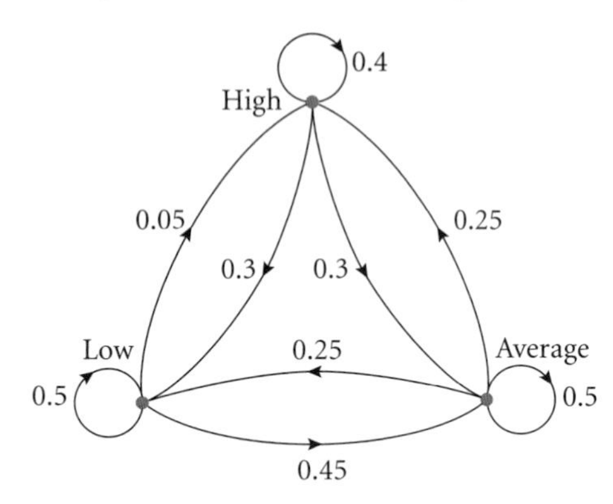

a. Write a transition matrix that represents the same information as the diagram. Put the rows and columns in the order low, average, high. @

b. Assume the changes in birth weights can be applied to any generation. If, in the initial generation of women, 25% had birth weights in the low category, 60% in the average category, and 15% in the high category, what were the percentages in each category after one generation? After two generations? After three generations? In the long run?

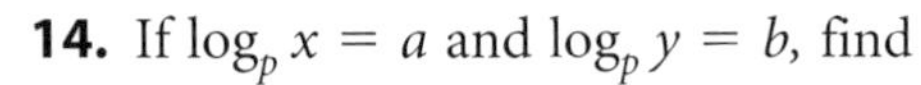

Review

13. For each segment shown in the figure at right, write an equation in point-slope form for the line that contains the segment. Check your equations by graphing them on your calculator.

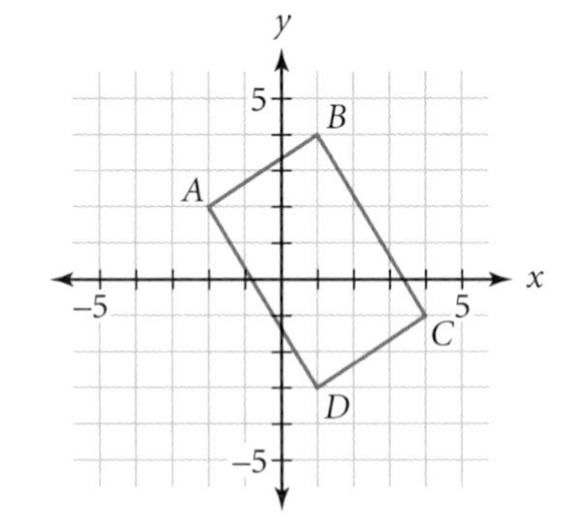

14. If $\log_p x = a$ and $\log_p y = b$, find

a. $\log_p xy$ @

b. $\log_p x^3$ @

c. $\log_p \frac{y^2}{x}$

d. $\log_{p^2} y$

e. $\log_p \sqrt{x}$

f. $\log_m xy$

15. Solve this system of equations for x, y, and z.

$$\begin{cases} x + 2y + z = 0 \\ 3x - 4y + 5z = -11 \\ -2x - 8y - 3z = 1 \end{cases}$$

LESSON

6.3

Solving Systems with Inverse Matrices

Things that oppose each other also complement each other.

CHINESE SAYING

Consider the following scenario: Three friends, Duane, Marsha, and Parker, decide to take their younger siblings to the movies. Before the movie, they buy some snacks at the concession stand. Duane buys two candy bars, a small drink, and two boxes of chocolate-covered peanuts for a total of \$11.85. Marsha spends \$9.00 on a candy bar, two small drinks, and one box of chocolate-covered peanuts. Parker spends \$12.35 on two small drinks and three boxes of chocolate-covered peanuts, but doesn't buy any candy bars. If all the prices include tax, what is the price of each item?

Set up a system of equations using the prices of the items as the unknowns. Let c represent the price of a candy bar in dollars, let d represent the price of a small drink in dollars, and let p represent the price of a box of chocolate-covered peanuts in dollars. This system represents the three friends' purchases:

$$\begin{cases} 2c + 1d + 2p = 11.85 \\ 1c + 2d + 1p = 9.00 \\ 0c + 2d + 3p = 12.35 \end{cases}$$

Solving this system using the techniques of Lesson 3.7 would be tedious. Instead, you'll learn a technique for solving it using matrices. First, translate these equations into a matrix equation in the form $[A][X] = [B]$.

$$\begin{bmatrix} 2 & 1 & 2 \\ 1 & 2 & 1 \\ 0 & 2 & 3 \end{bmatrix} \begin{bmatrix} c \\ d \\ p \end{bmatrix} = \begin{bmatrix} 11.85 \\ 9.00 \\ 12.35 \end{bmatrix}$$

Solving this equation is similar to solving the equation $ax = b$. To solve for x, you multiply both sides by the multiplicative inverse of a, which is written as $\frac{1}{a}$ or a^{-1}. You know this is the inverse of a because $a \cdot a^{-1} = 1$. Recall that 1 is the multiplication identity element, or **multiplicative identity,** because any number multiplied by 1 remains unchanged.

Similarly, to solve a system by using matrices, you can use an **inverse matrix.** If an inverse matrix exists, then when you multiply it by the system matrix you will get the matrix equivalent of 1, which is called the **identity matrix.** Any square matrix multiplied on either side by the identity matrix of the same dimensions remains unchanged, just as any number multiplied by 1 remains unchanged. In Example A, you will first use this multiplicative identity to find a 2×2 identity matrix. You'll solve the problem posed above in Example C.

EXAMPLE A Find an identity matrix for $\begin{bmatrix} 2 & 1 \\ 4 & 3 \end{bmatrix}$.

▶ Solution You want to find a matrix, $\begin{bmatrix} a & b \\ c & d \end{bmatrix}$, that satisfies the definition of the identity matrix.

$$\begin{bmatrix} 2 & 1 \\ 4 & 3 \end{bmatrix}\begin{bmatrix} a & b \\ c & d \end{bmatrix} = \begin{bmatrix} 2 & 1 \\ 4 & 3 \end{bmatrix}$$ Multiplying by an identity matrix leaves the matrix unchanged.

$$\begin{bmatrix} 2a + c & 2b + d \\ 4a + 3c & 4b + 3d \end{bmatrix} = \begin{bmatrix} 2 & 1 \\ 4 & 3 \end{bmatrix}$$ Multiply the left side.

Because the two matrices are equal, their entries must be equal. Setting corresponding entries equal produces these equations:

$$2a + c = 2 \qquad 2b + d = 1$$
$$4a + 3c = 4 \qquad 4b + 3d = 3$$

You can treat these as two systems of equations. Use substitution or elimination to solve each system.

You should find that the solution to the first system is $a = 1$ and $c = 0$, and the solution to the second system is $b = 0$ and $d = 1$. Substitute these values back into the identity matrix $\begin{bmatrix} a & b \\ c & d \end{bmatrix}$. The 2×2 identity matrix is

$$\begin{bmatrix} 1 & 0 \\ 0 & 1 \end{bmatrix}$$

Can you see why multiplying this matrix by any 2×2 matrix results in the same 2×2 matrix?

The identity matrix in Example A is the identity matrix for all 2×2 matrices. Take a minute to multiply $[I][A]$ and $[A][I]$, with $[A]$ being any 2×2 matrix. There are corresponding identity matrices for larger square matrices.

Identity Matrix

An **identity matrix,** symbolized by $[I]$, is the square matrix that does not alter the entries of a square matrix $[A]$ under multiplication.

$$[A][I] = [A] \text{ and } [I][A] = [A]$$

Matrix $[I]$ must have the same dimensions as matrix $[A]$, and it has entries of 1's along the main diagonal (from top left to bottom right) and 0's in all other entries.

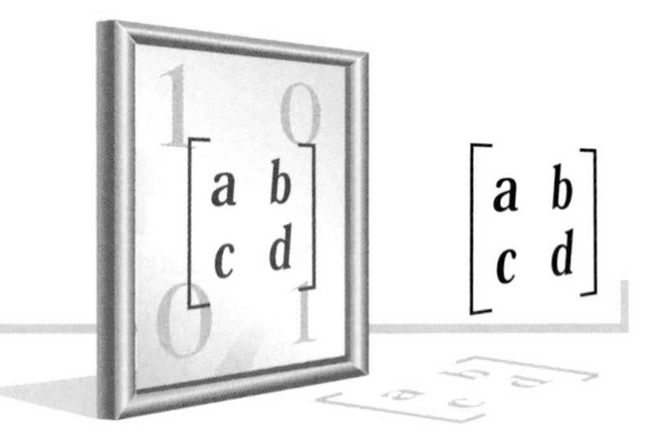

Now that you know the identity matrix for a 2×2 matrix, you can look for a way to find the inverse of a 2×2 matrix.

Inverse Matrix

The **inverse matrix** of $[A]$, symbolized by $[A]^{-1}$, is the matrix that will produce an identity matrix when multiplied by $[A]$.

$$[A][A]^{-1} = [I] \text{ and } [A]^{-1}[A] = [I]$$

Investigation
The Inverse Matrix

In this investigation you will learn ways to find the inverse of a 2 × 2 matrix.

Step 1 Use the definition of an inverse matrix to set up a matrix equation. Use these matrices and the 2 × 2 identity matrix for $[I]$.

$$[A] = \begin{bmatrix} 2 & 1 \\ 4 & 3 \end{bmatrix} \qquad [A]^{-1} = \begin{bmatrix} a & b \\ c & d \end{bmatrix}$$

Step 2 Use matrix multiplication to find the product $[A][A]^{-1}$. Set that product equal to matrix $[I]$.

Step 3 Use the matrix equation from Step 2 to write equations that you can solve to find values for a, b, c, and d. Solve the systems to find the values in the inverse matrix.

Step 4 Use your calculator to find $[A]^{-1}$. If this answer does not match your answer to Step 3, check your work for mistakes. [▶ See **Calculator Note 6D** to learn how to find the inverse on your calculator. ◀]

Step 5 Find the products of $[A][A]^{-1}$ and $[A]^{-1}[A]$. Do they both give you $\begin{bmatrix} 1 & 0 \\ 0 & 1 \end{bmatrix}$? Is matrix multiplication always commutative?

Step 6 Not every square matrix has an inverse. Try to find the inverse of each of these matrices. Make a conjecture about what types of 2 × 2 square matrices do not have inverses.

a. $\begin{bmatrix} 2 & 1 \\ 4 & 2 \end{bmatrix}$ **b.** $\begin{bmatrix} 50 & -75 \\ 10 & -15 \end{bmatrix}$ **c.** $\begin{bmatrix} 10.5 & 1 \\ 31.5 & 3 \end{bmatrix}$ **d.** $\begin{bmatrix} 2 & 1 \\ 2 & 1 \end{bmatrix}$

Step 7 Can a nonsquare matrix have an inverse? Why or why not?

You now know how to find the inverse of a square matrix, both by hand and on your calculator. If an inverse matrix exists, you can use it to solve a system of equations.

Solving a System Using the Inverse Matrix

A system of equations in standard form can be written in matrix form as $[A][X] = [B]$, where $[A]$ is the coefficient matrix, $[X]$ is the variable matrix, and $[B]$ is the constant matrix. Multiplying both sides by the inverse matrix, $[A]^{-1}$, with the inverse on the left, gives the values of variables in matrix $[X]$, which is the solution to the system.

$[A][X] = [B]$	The system in matrix form.
$[A]^{-1}[A][X] = [A]^{-1}[B]$	Left-multiply both sides by the inverse.
$[I][X] = [A]^{-1}[B]$	By the definition of inverse, $[A]^{-1}[A] = [I]$.
$[X] = [A]^{-1}[B]$	By the definition of identity, $[I][X] = [X]$.

EXAMPLE B

Solve this system using an inverse matrix.

$$\begin{cases} 2x + 3y = 7 \\ x = 6 - 4y \end{cases}$$

▶ Solution

First, rewrite the second equation in standard form.

$$\begin{cases} 2x + 3y = 7 \\ x + 4y = 6 \end{cases}$$

The matrix equation for this system is $\begin{bmatrix} 2 & 3 \\ 1 & 4 \end{bmatrix}\begin{bmatrix} x \\ y \end{bmatrix} = \begin{bmatrix} 7 \\ 6 \end{bmatrix}$.

In the matrix equation $\begin{bmatrix} 2 & 3 \\ 1 & 4 \end{bmatrix}\begin{bmatrix} x \\ y \end{bmatrix} = \begin{bmatrix} 7 \\ 6 \end{bmatrix}$, the variable matrix, $[X]$, is $\begin{bmatrix} x \\ y \end{bmatrix}$.

The coefficient matrix, $[A]$, is $\begin{bmatrix} 2 & 3 \\ 1 & 4 \end{bmatrix}$, and the constant matrix, $[B]$, is $\begin{bmatrix} 7 \\ 6 \end{bmatrix}$.

Use your calculator to find the inverse of $[A]$.

$$[A]^{-1} = \begin{bmatrix} 0.8 & -0.6 \\ -0.2 & 0.4 \end{bmatrix}$$

Multiply both sides of the equation by this inverse, with the inverse on the left, to find the solution to the system of equations.

$\begin{bmatrix} 2 & 3 \\ 1 & 4 \end{bmatrix}\begin{bmatrix} x \\ y \end{bmatrix} = \begin{bmatrix} 7 \\ 6 \end{bmatrix}$	$[A][X] = [B]$	The system in matrix form.
$\begin{bmatrix} 0.8 & -0.6 \\ -0.2 & 0.4 \end{bmatrix}\begin{bmatrix} 2 & 3 \\ 1 & 4 \end{bmatrix}\begin{bmatrix} x \\ y \end{bmatrix} = \begin{bmatrix} 0.8 & -0.6 \\ -0.2 & 0.4 \end{bmatrix}\begin{bmatrix} 7 \\ 6 \end{bmatrix}$	$[A]^{-1}[A][X] = [A]^{-1}[B]$	Left-multiply both sides by the inverse.
$\begin{bmatrix} 1 & 0 \\ 0 & 1 \end{bmatrix}\begin{bmatrix} x \\ y \end{bmatrix} = \begin{bmatrix} 0.8 & -0.6 \\ -0.2 & 0.4 \end{bmatrix}\begin{bmatrix} 7 \\ 6 \end{bmatrix}$	$[I][X] = [A]^{-1}[B]$	By the definition of inverse, $[A]^{-1}[A] = [I]$.
$\begin{bmatrix} x \\ y \end{bmatrix} = \begin{bmatrix} 0.8 & -0.6 \\ -0.2 & 0.4 \end{bmatrix}\begin{bmatrix} 7 \\ 6 \end{bmatrix}$	$[X] = [A]^{-1}[B]$	By the definition of identity, $[I][X] = [X]$.
$\begin{bmatrix} x \\ y \end{bmatrix} = \begin{bmatrix} 2 \\ 1 \end{bmatrix}$		Multiply the right side.

The solution to the system is (2, 1). Substitute the values into the original equations to check the solution.

$2x + 3y = 7$	$x = 6 - 4y$
$2(2) + 3(1) \stackrel{?}{=} 7$	$2 \stackrel{?}{=} 6 - 4(1)$
$4 + 3 \stackrel{?}{=} 7$	$2 \stackrel{?}{=} 6 - 4$
$7 = 7$	$2 = 2$

The solution checks.

You can also solve larger systems of equations using an inverse matrix. First, decide what quantities are unknown and write equations using the information given in the problem. Then rewrite the system of equations as a matrix equation, and use an inverse matrix to solve the system. Even systems of equations with many unknowns can be solved quickly this way.

EXAMPLE C | Use an inverse matrix to solve the problem posed at the beginning of the lesson. What is the cost of each snack item?

▶ Solution | Recall that the matrix below represents the cost of the items purchased by Duane, Marsha, and Parker, with c representing the price of a candy bar, d representing the price of a small drink, and p representing the price of a bag of chocolate-covered peanuts.

$$\begin{bmatrix} 2 & 1 & 2 \\ 1 & 2 & 1 \\ 0 & 2 & 3 \end{bmatrix} \begin{bmatrix} c \\ d \\ p \end{bmatrix} = \begin{bmatrix} 11.85 \\ 9.00 \\ 12.35 \end{bmatrix}$$

The solution to the system is simply the product $[A]^{-1}[B]$.

$$[X] = [A]^{-1}[B] = \begin{bmatrix} 2.15 \\ 2.05 \\ 2.75 \end{bmatrix}$$

A candy bar costs \$2.15, a small drink costs \$2.05, and a bag of chocolate-covered peanuts costs \$2.75.

2.15→c	2.15
2.05→d	2.05
2.75→p	2.75
2·c+d+2·p	11.85
c+2·d+p	9.
2·d+3·p	12.35

Substituting these answers into the original system shows that they are correct. You can use your calculator to evaluate the expressions quickly and accurately.

You have probably noticed that in order to solve systems of equations with two variables, you must have two equations. To solve a system of equations with three variables, you must have at least three equations. In general, you must have as many equations as variables. Otherwise, there will not be enough information to solve the problem. For matrix equations, this means that the coefficient matrix must be square.

If there are more equations than variables, often one equation is equivalent to another equation and therefore just repeats the same information. Or the extra information may contradict the other equations, and thus there is no solution that will satisfy all the equations.

EXERCISES

You will need

A graphing calculator for Exercises **2** and **4–14**.

▶ Practice Your Skills

1. Rewrite each system of equations in matrix form.

a. $\begin{cases} 3x + 4y = 11 \\ 2x - 5y = -8 \end{cases}$ @

b. $\begin{cases} x + 2y + z = 0 \\ 3x - 4y + 5z = -11 \\ -2x - 8y - 3z = 1 \end{cases}$ @

c. $\begin{cases} 5.2x + 3.6y = 7 \\ -5.2x + 2y = 8.2 \end{cases}$

d. $\begin{cases} \frac{1}{4}x - \frac{2}{5}y = 3 \\ \frac{3}{8}x + \frac{2}{5}y = 2 \end{cases}$

2. Multiply each pair of matrices. If multiplication is not possible, explain why.

a. $\begin{bmatrix} 5 & 2 \\ 7 & 3 \end{bmatrix}\begin{bmatrix} 1 & -3 \\ 5 & -2 \end{bmatrix}$ @

b. $\begin{bmatrix} 4 & -1 \\ 3 & 6 \\ 2 & -3 \end{bmatrix}\begin{bmatrix} 2 & -5 & 0 \\ 1 & -2 & 7 \end{bmatrix}$

c. $\begin{bmatrix} 9 & -3 \end{bmatrix}\begin{bmatrix} 4 & -6 \\ 0 & -2 \\ -1 & 3 \end{bmatrix}$

3. Use matrix multiplication to expand each system. Then solve for each variable by using substitution or elimination.

a. $\begin{bmatrix} 1 & 5 \\ 6 & 2 \end{bmatrix}\begin{bmatrix} a & b \\ c & d \end{bmatrix} = \begin{bmatrix} -7 & 33 \\ 14 & -26 \end{bmatrix}$ @

b. $\begin{bmatrix} 1 & 5 \\ 6 & 2 \end{bmatrix}\begin{bmatrix} a & b \\ c & d \end{bmatrix} = \begin{bmatrix} 1 & 0 \\ 0 & 1 \end{bmatrix}$

4. Multiply each pair of matrices. Are the matrices inverses of each other?

a. $\begin{bmatrix} 5 & 2 \\ 7 & 3 \end{bmatrix}\begin{bmatrix} 3 & -2 \\ -7 & 5 \end{bmatrix}$ @

b. $\begin{bmatrix} 1 & 5 & 4 \\ 6 & 2 & -2 \\ 0 & 3 & 1 \end{bmatrix}\begin{bmatrix} 0.16 & 0.14 & -0.36 \\ -0.12 & 0.02 & 0.52 \\ 0.36 & -0.06 & -0.56 \end{bmatrix}$

5. Find the inverse of each matrix by solving the matrix equation $[A][A]^{-1} = [I]$. Then find the inverse matrix on your calculator to check your answer.

a. $\begin{bmatrix} 4 & 3 \\ 5 & 4 \end{bmatrix}$ @

b. $\begin{bmatrix} 6 & 4 & -2 \\ 3 & 1 & -1 \\ 0 & 0 & 3 \end{bmatrix}$ h

c. $\begin{bmatrix} 5 & 3 \\ 10 & 7 \end{bmatrix}$

d. $\begin{bmatrix} 1 & 2 \\ 2 & 4 \end{bmatrix}$

Reason and Apply

6. Rewrite each system in matrix form and solve by using the inverse matrix. Check your solutions.

a. $\begin{cases} 8x + 3y = 41 \\ 6x + 5y = 39 \end{cases}$ @

b. $\begin{cases} 11x - 5y = -38 \\ 9x + 2y = -25 \end{cases}$

c. $\begin{cases} 2x + y - 2z = 1 \\ 6x + 2y - 4z = 3 \\ 4x - y + 3z = 5 \end{cases}$

d. $\begin{cases} 4w + x + 2y - 3z = -16 \\ -3w + 3x - y + 4z = 20 \\ 5w + 4x + 3y - z = -10 \\ -w + 2x + 5y + z = -4 \end{cases}$

7. At the High Flying Amusement Park there are three kinds of rides: Jolly rides, Adventure rides, and Thrill rides. Admission is free when you buy a book of tickets, which includes ten tickets for each type of ride. Or you can pay \$5.00 for admission and then buy individual tickets for the rides. Noah, Rita, and Carey decide to pay the admission price and buy individual tickets. Noah pays \$19.55 for 7 Jolly rides, 3 Adventure rides, and 9 Thrill rides. Rita pays \$13.00 for 9 Jolly rides, 10 Adventure rides, and no Thrill rides. Carey pays \$24.95 for 8 Jolly rides, 7 Adventure rides, and 10 Thrill rides. (The prices above do not include the admission price.)

a. How much does each type of ride cost?

b. What is the total cost of a 30-ride book of tickets? @

c. Would Noah, Rita, or Carey have been better off purchasing a ticket book?

8. A family invested a portion of \$5,000 in an account at 6% annual interest and the rest in an account at 7.5% annual interest. The total interest they earned in the first year was \$340.50. How much did they invest in each account? (h)

9. Find a, b, and c such that the graph of $y = ax^2 + bx + c$ passes through the points $(1, 3)$, $(4, 24)$, and $(-2, 18)$. (h)

10. Being able to solve a system of equations is definitely not "new" mathematics. Mahāvīra, the best-known Indian mathematician of the 9th century, worked the following problem. See if you can solve it.

The mixed price of 9 citrons and 7 fragrant wood apples is 107; again, the mixed price of 7 citrons and 9 fragrant wood apples is 101. O you arithmetician, tell me quickly the price of a citron and of a wood apple here, having distinctly separated those prices well.

History CONNECTION

Mahāvīra (ca. 800–870 C.E.) wrote *Ganita Sara Samgraha,* the first Indian text exclusively about mathematics. Writing in his home of Mysore, India, he considered this book to be a collection of insights from other Indian mathematicians, such as Āryabata I, Bhāskara I, and Brahmagupta, who wrote their findings in astronomy texts.

11. APPLICATION The circuit here is made of two batteries (6 volt and 9 volt) and three resistors (47 ohms, 470 ohms, and 280 ohms). The batteries create an electric current in the circuit. Let x, y, and z represent the current in amps flowing through each resistor.

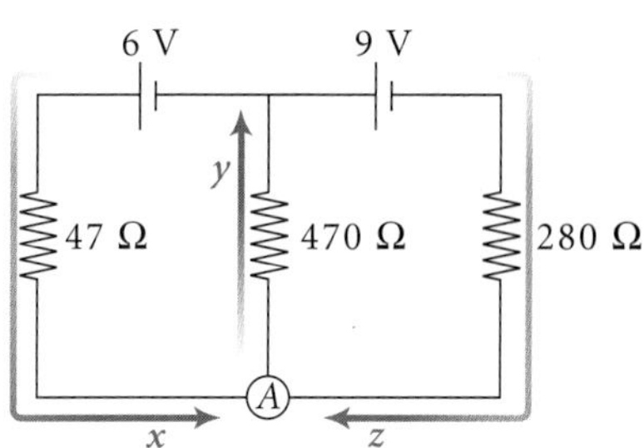

The voltage across each resistor is current times resistance ($V = IR$). This gives two equations for the two loops of the circuit:

$47x + 470y = 6$ $\qquad$ $280z + 470y = 9$

The electric current flowing into any point along the circuit must flow out. So, for instance, at junction A, $x + z - y = 0$. Find the current flowing through each resistor.

12. When you use your calculator to find the inverse of the coefficient matrix for this system, you get an error message. What does this mean about the system?

$$\begin{cases} 3.2x + 2.4y = 9.6 \\ 2x + 1.5y = 6 \end{cases}$$

13. Consider the system $7x + 2y = 35$, $4x - 3y = -9$, $2x + 3y = 7$. Because there are three equations and only two unknowns, there may not be a unique solution to the system. Choose any two equations and solve the system. Then choose a different pair and solve that system. If the two solutions match, there is a unique solution to the original system. Is there a unique solution to this system? What do your results tell you about a graph of the system?

14. APPLICATION An important application in the study of economics is the study of the relationship between industrial production and consumer demand. In creating an economic model, Russian-American economist Wassily Leontief (1906–1999) noted that the total output less the internal consumption equals consumer demand. Mathematically his input-output model looks like $[X] - [A][X] = [D]$, where $[X]$ is the total output matrix, $[A]$ is the input-output matrix, and $[D]$ is the matrix representing consumer demand.

Here is an input-output matrix, $[A]$, for a simple three-sector economy:

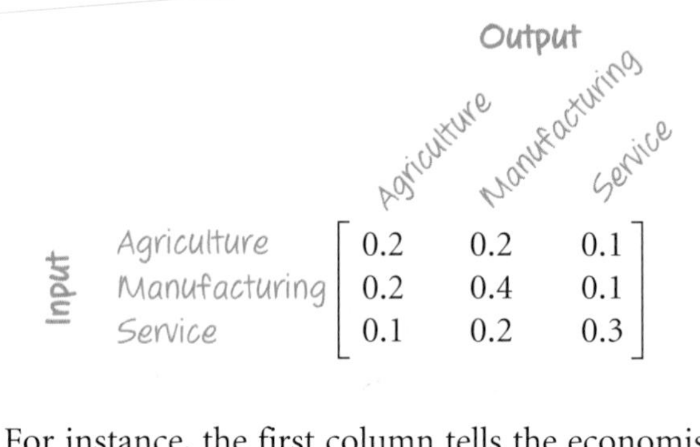

$$\begin{bmatrix} 0.2 & 0.2 & 0.1 \\ 0.2 & 0.4 & 0.1 \\ 0.1 & 0.2 & 0.3 \end{bmatrix}$$

For instance, the first column tells the economist that to produce an output of 1 unit of agricultural products requires the consumption (input) of 0.2 unit of agricultural products, 0.2 unit of manufacturing products, and 0.1 unit of service products.

The demand matrix, $[D]$, represents millions of dollars. Use the equation $[X] - [A][X] = [D]$ to find the output matrix, $[X]$.

$$[D] = \begin{bmatrix} 100 \\ 80 \\ 50 \end{bmatrix}$$

Economics
CONNECTION

During World War II, Wassily Leontief's method became a critical part of planning for wartime production in the United States. As a consultant to the U.S. Labor Department, he developed an input-output table for more than 90 economic sectors. During the early 1960s, Leontief and economist Marvin Hoffenberg used input-output analysis to forecast the economic effects of reduction or elimination of militaries. In 1973, Leontief was awarded a Nobel Prize in Economics for his contributions to the field.

Wassily Leontief

Review

15. For each equation, write a second linear equation that would create a consistent and dependent system.

a. $y = 2x + 4$ ⓐ **b.** $y = -\frac{1}{3}x - 3$ **c.** $2x + 5y = 10$ **d.** $x - 2y = -6$

16. For each equation, write a second linear equation that would create an inconsistent system.

a. $y = 2x + 4$ @ **b.** $y = -\frac{1}{3}x - 3$ **c.** $2x + 5y = 10$ **d.** $x - 2y = -6$

17. Four towns, Lenox, Murray, Davis, and Terre, are connected by a series of roads.

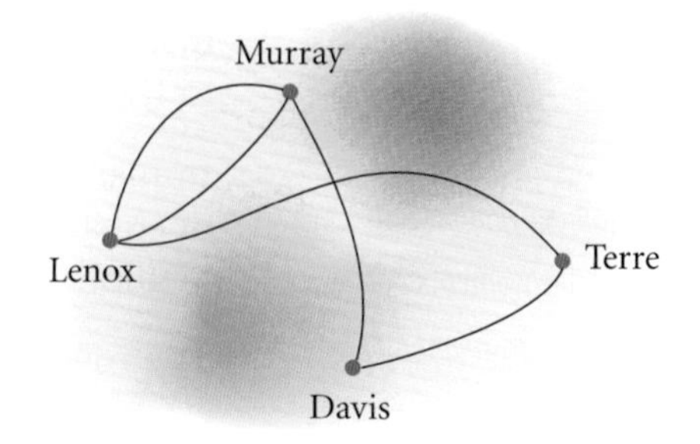

a. Represent the number of direct road connections between the towns in a matrix, $[A]$. List the towns in the order Lenox, Murray, Davis, and Terre.

b. Explain the meaning of the value of a_{22}.

c. Describe the symmetry of your matrix.

d. How many roads are there? What is the sum of the entries in the matrix? Explain the relationship between these two answers.

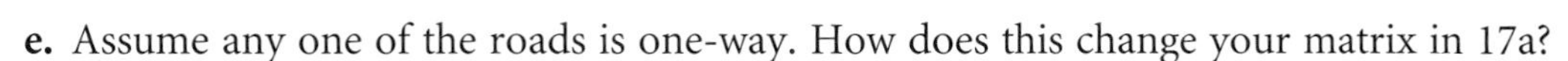

e. Assume any one of the roads is one-way. How does this change your matrix in 17a?

18. The third term of an arithmetic sequence is 28. The seventh term is 80. What is the first term?

IMPROVING YOUR REASONING SKILLS

Secret Survey

Eric is doing a survey. He has a deck of cards and two questions written on a sheet of paper. He says, “Pick a card from the deck. Don’t show it to me. If it is a red card, answer Question 1. If it is a black card, answer Question 2.”

Question 1 (red card): Does your phone number end in an even number?

Question 2 (black card): Do you own a stuffed animal?

You pick a card and look at the paper, and you respond, “Yes.” Eric records your answer, shuffles the cards, and goes on to the next person.

At the end of the survey, Eric has gathered 37 yeses and 23 noes. He estimates that 73% of the respondents own a stuffed animal.

Explain how Eric was able to find this result without knowing which question each person was answering.

LESSON

6.4

Row Reduction Method

Rather than denying problems, focus inventively, intentionally on what solutions might look or feel like . . .

MARSHA SINETAR

In Lesson 6.3, you used an inverse matrix to solve systems represented by matrices. This is a powerful strategy to use when the number of variables is equal to the number of equations.

In many applications, however, the number of variables is not equal to the number of equations. Moreover, you might have hundreds of equations! So you need a method that can be programmed into a computer to solve general systems.

The row reduction method is such a method. It requires only one matrix, and it extends the method of elimination in a systematic way.

In the previous lesson you saw how systems of equations can be converted into matrix equations.

$$\begin{cases} 2x + y = 5 \\ 5x + 3y = 13 \end{cases} \rightarrow \begin{bmatrix} 2 & 1 \\ 5 & 3 \end{bmatrix} \begin{bmatrix} x \\ y \end{bmatrix} = \begin{bmatrix} 5 \\ 13 \end{bmatrix}$$

You can also write the system as an **augmented matrix,** which is a single matrix that contains columns for the coefficients of each variable and a final column for the constant terms.

$$\begin{cases} 2x + y = 5 \\ 5x + 3y = 13 \end{cases} \rightarrow \left[\begin{array}{cc|c} 2 & 1 & 5 \\ 5 & 3 & 13 \end{array}\right]$$

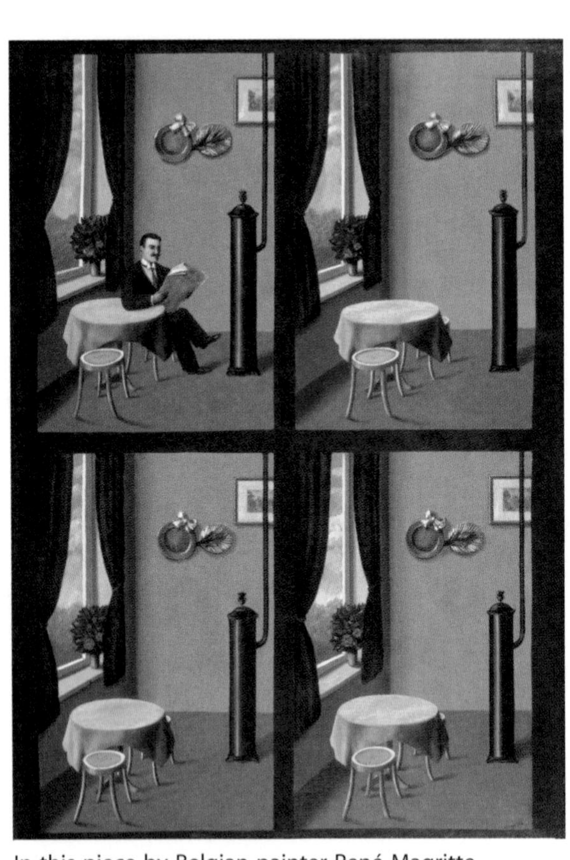

In this piece by Belgian painter René Magritte (1898–1967), a man appears in one frame, but he is "eliminated" from the others.

Man with a Newspaper (1928) by René Magritte, oil on canvas

You can use the augmented matrix to carry out a process similar to elimination.

The **row reduction method** transforms an augmented matrix into a solution matrix. Instead of combining equations and multiples of equations until you are left with an equation in one variable, you add multiples of rows to other rows until you obtain the solution matrix. The solution matrix for a consistent and independent system contains the solutions in the last column. The rest of the matrix consists of 1's along the main diagonal and 0's above and below it.

For example, the augmented matrix below represents the system

$$\begin{cases} 1x + 0y = a \\ 0x + 1y = b \end{cases}, \text{ or } x = a \text{ and } y = b.$$

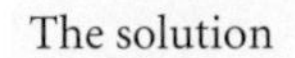

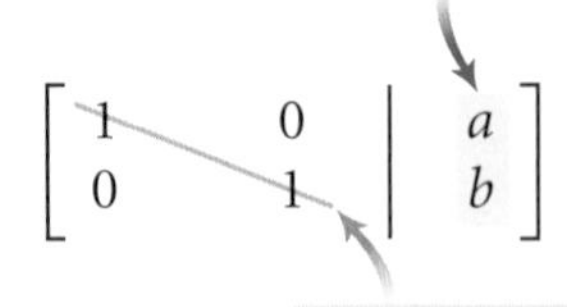

This solution matrix is in **reduced row-echelon form** because each row is reduced to a 1 and a solution, and the rest of the matrix entries are 0's. The 1's are in echelon, or step, form. The ordered pair (a, b) is the solution to the system.

An augmented matrix represents a system of equations, so the same rules apply to row operations in a matrix as to equations in a system of equations.

Row Operations in a Matrix

- You can multiply (or divide) all numbers in a row by a nonzero number.
- You can add all numbers in a row to corresponding numbers in another row.
- You can add a multiple of the numbers in one row to the corresponding numbers in another row.
- You can exchange two rows.

EXAMPLE

Solve this system of equations.

$$\begin{cases} 2x + y = 5 \\ 5x + 3y = 13 \end{cases}$$

Solution

You can solve the system using matrices or equations. Let's compare the row reduction method using matrices with the elimination method using equations.

Because the equations are in standard form, you can copy the coefficients and constants from each equation into corresponding rows of the augmented matrix.

$$\begin{cases} 2x + y = 5 \\ 5x + 3y = 13 \end{cases} \rightarrow \left[\begin{array}{cc|c} 2 & 1 & 5 \\ 5 & 3 & 13 \end{array}\right]$$

Let's call this augmented matrix $[M]$. Using only the elementary row operations, you can transform this matrix into the solution matrix. You need both m_{21} and m_{12} to be 0, and you need both m_{11} and m_{22} to be 1.

Add -2.5 times row 1 to row 2 to get 0 for m_{21}.

$$\left[\begin{array}{cc|c} 2 & 1 & 5 \\ 0 & 0.5 & 0.5 \end{array}\right]$$

Multiply equation 1 by -2.5 and add to equation 2 to eliminate x.

$$\begin{array}{rcr} -5x - 2.5y &=& -12.5 \\ 5x + 3y &=& 13 \\ \hline 0.5y &=& 0.5 \end{array}$$

Multiply row 2 by 2 to change m_{22} to 1.

$$\left[\begin{array}{cc|c} 2 & 1 & 5 \\ 0 & 1 & 1 \end{array}\right]$$

Multiply the equation by 2 to find y.

$$y = 1$$

Add -1 times row 2 to row 1 to get 0 for m_{12}.

$$\left[\begin{array}{cc|c} 2 & 0 & 4 \\ 0 & 1 & 1 \end{array}\right]$$

Multiply this new equation by -1, and add the result to equation 1 to eliminate y.

$$\begin{array}{rcr} 2x + y &=& 5 \\ -y &=& -1 \\ \hline 2x &=& 4 \end{array}$$

Multiply row 1 by 0.5.

$$\left[\begin{array}{cc|c} 1 & 0 & 2 \\ 0 & 1 & 1 \end{array}\right]$$

Multiply the equation by 0.5 to find x.

$$x = 2$$

The last column of the solution matrix indicates that the solution to the system is (2, 1).

You can represent row operations symbolically. For example, you can use R_1 and R_2 to represent the two rows of a matrix, as in Example A, and show the steps this way:

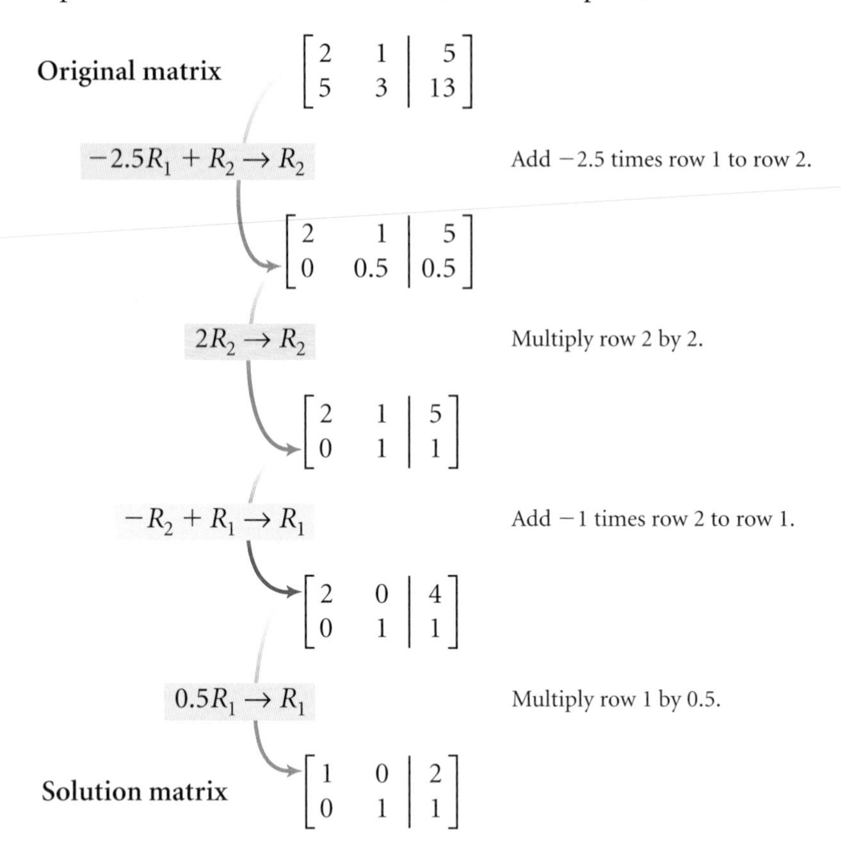

You have learned how to interpret a typical solution matrix. For example,

$$\left[\begin{array}{ccc|c} 1 & 0 & 0 & 12 \\ 0 & 1 & 0 & 30 \\ 0 & 0 & 1 & 8 \end{array}\right]$$

means $x = 12$, $y = 30$, and $z = 8$.

But in some cases the final matrix may be more difficult to interpret. Consider the solution matrix

$$\left[\begin{array}{ccc|c} 1 & 0 & 0 & 12 \\ 0 & 1 & 3 & 54 \\ 0 & 0 & 0 & 0 \end{array}\right].$$

This matrix represents the equations $x = 12$ and $y + 3z = 54$. These equations indicate that there was not enough information to find a single solution. Therefore, the system has infinitely many solutions. Can you find several ordered triples that satisfy the equations?

Or the final matrix may look like

$$\left[\begin{array}{ccc|c} 1 & 0 & 0 & 12 \\ 0 & 1 & 3 & 54 \\ 0 & 0 & 0 & 5 \end{array}\right].$$

The last row claims that 0 equals 5, which is not true. Therefore, this matrix shows that the system of equations has no solution.

EXERCISES

You will need

A graphing calculator for Exercises **5–8, 12,** and **13.**

Practice Your Skills

1. Write a system of equations for each augmented matrix.

a. $\left[\begin{array}{cc|c} 2 & 5 & 8 \\ 4 & -1 & 6 \end{array}\right]$ ⓐ

b. $\left[\begin{array}{ccc|c} 1 & -1 & 2 & 3 \\ 1 & 2 & -3 & 1 \\ 2 & 1 & -1 & 2 \end{array}\right]$

2. Write an augmented matrix for each system.

a. $\begin{cases} x + 2y - z = 1 \\ 2x - y + 3z = 2 \\ 2x + y + z = -1 \end{cases}$ ⓐ

b. $\begin{cases} 2x + y - z = 12 \\ 2x + z = 4 \\ 2x - y + 3z = -4 \end{cases}$

3. Perform each row operation on this matrix.

$$\left[\begin{array}{ccc|c} 1 & -1 & 2 & 3 \\ 1 & 2 & -3 & 1 \\ 2 & 1 & -1 & 2 \end{array}\right]$$

a. $-R_1 + R_2 \rightarrow R_2$ ⓐ

b. $-2R_1 + R_3 \rightarrow R_3$

4. Give the missing row operation or matrix in the table.

	Description	Matrix	
a.	The original system. $\begin{cases} 2x + 5y = 8 \\ 4x - y = 6 \end{cases}$	$\left[\begin{array}{cc|c} & & \\ & & \end{array}\right]$	ⓐ
b.	$-2R_1 + R_2 \rightarrow R_2$	$\left[\begin{array}{cc|c} & & \\ & & \end{array}\right]$	
c.		$\left[\begin{array}{cc|c} 2 & 5 & 8 \\ 0 & 1 & \frac{10}{11} \end{array}\right]$	ⓐ
d.		$\left[\begin{array}{cc|c} 2 & 0 & \frac{38}{11} \\ 0 & 1 & \frac{10}{11} \end{array}\right]$	
e.		$\left[\begin{array}{cc|c} 1 & 0 & \frac{19}{11} \\ 0 & 1 & \frac{10}{11} \end{array}\right]$	

Reason and Apply

5. Rewrite each system of equations as an augmented matrix. If possible, transform the matrix into its reduced row-echelon form using row operations on your calculator.

a. $\begin{cases} x + 2y + 3z = 5 \\ 2x + 3y + 2z = 2 \\ -x - 2y - 4z = -1 \end{cases}$ ⓐ

b. $\begin{cases} -x + 3y - z = 4 \\ 2z = x + y \\ 2.2y + 2.2z = 2.2 \end{cases}$

c. $\begin{cases} 3x - y + z = 7 \\ x - 2y + 5z = 1 \\ 6x - 2y + 2z = 14 \end{cases}$ ⓐ

d. $\begin{cases} 3x - y + z = 5 \\ x - 2y + 5z = 1 \\ 6x - 2y + 2z = 14 \end{cases}$

6. A farmer raises only goats and chickens on his farm. All together he has 47 animals, and they have a total of 118 legs.

a. Write a system of equations and an augmented matrix. How many of each animal does he have? [▶ See **Calculator Note 6F** to learn how to transform a matrix to reduced row-echelon form on your calculator. ◀]

b. The farmer's neighbor also has goats and chickens. She reports having 118 animals with a total of 47 legs. Write a system of equations and an augmented matrix. How many of each animal does she have?

7. The largest angle of a triangle is 4° more than twice the smallest angle. The smallest angle is 24° less than the midsize angle. What are the measures of the three angles? ⓗ

8. The yearbook staff sells ads in three sizes. The full-page ads sell for \$200, the half-page ads sell for \$125, and the business-card-size ads sell for \$20. All together they earned \$1,715 from selling 22 ads. There were four times as many business-card-size ads sold as full-page ads. How many of each ad type did they sell? ⓗ

9. One way to find an inverse of a matrix $\begin{bmatrix} a & b \\ c & d \end{bmatrix}$, if it exists, is to perform row operations on the augmented matrix $\left[\begin{array}{cc|cc} a & b & 1 & 0 \\ c & d & 0 & 1 \end{array}\right]$ to change it to the form $\left[\begin{array}{cc|cc} 1 & 0 & e & f \\ 0 & 1 & g & h \end{array}\right]$. The matrix $\begin{bmatrix} e & f \\ g & h \end{bmatrix}$ is the required inverse. Use this strategy to find the inverse of each matrix.

a. $\begin{bmatrix} 4 & 3 \\ 5 & 4 \end{bmatrix}$

b. $\begin{bmatrix} 6 & 4 & -2 \\ 3 & 1 & -1 \\ 0 & 7 & 3 \end{bmatrix}$

10. The junior class treasurer is totaling the sales and receipts from the last book sale. She has 50 receipts for sales of three different titles of books priced at \$14.00, \$18.50, and \$23.25. She has a total of \$909.00 and knows that 22 more of the \$18.50 books were sold than the \$23.25 books. How many of each book were sold?

Review

11. APPLICATION The Life is a Dance troupe has two choices in how it will be paid for its next series of performances. The first plan is to receive \$12,500 for the series plus 5% of all ticket sales. The second plan is to receive \$6,800 for the series plus 15% of ticket sales. The troupe will perform three consecutive nights in a hall that seats 2,200 people. All tickets will cost \$12.

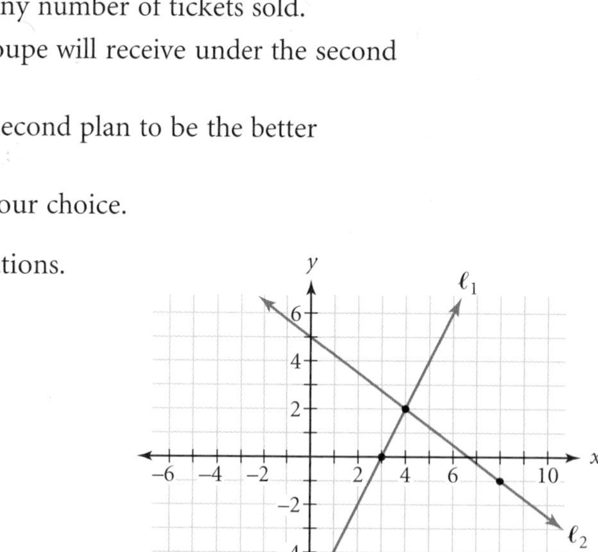

Honduran-American dancer Homer Avila (1955–2004) performed a solo work titled *Not/Without Words* in February 2002, one year after losing his leg and part of his hip to cancer.

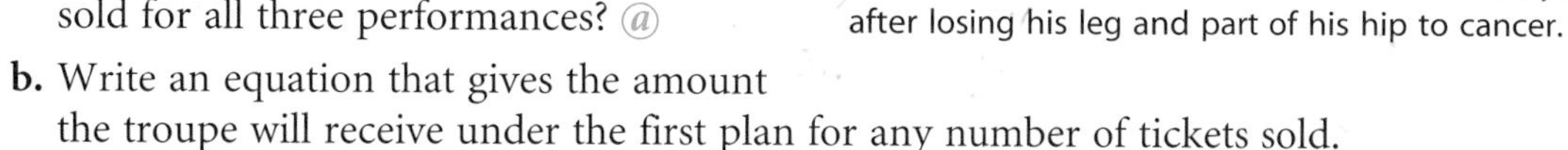

a. How much will the troupe receive under each plan if a total of 3,500 tickets are sold for all three performances? @

b. Write an equation that gives the amount the troupe will receive under the first plan for any number of tickets sold.

c. Write an equation that gives the amount the troupe will receive under the second plan for any number of tickets sold.

d. How many tickets must the troupe sell for the second plan to be the better choice?

e. Which plan should the troupe choose? Justify your choice.

12. Consider this graph of a system of two linear equations.

a. What is the solution to this system?

b. Write equations for the two lines.

13. Recall that equation $|x| = 5$ has two solutions, $x = 5$ or $x = -5$, because both values are 5 units from 0 on the number line. Now consider $|x| < 5$. This inequality represents all values of x that are less than 5 units from 0 on the number line.

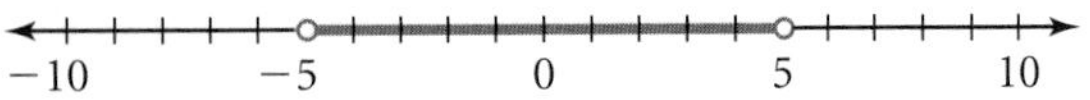

To be a solution, x must satisfy both inequalities $x < 5$ and $x > -5$. These inequalities can be combined into one statement: $-5 < x < 5$.

What about $|x| > 5$? This inequality represents all values of x that are greater than 5 units from 0 on the number line.

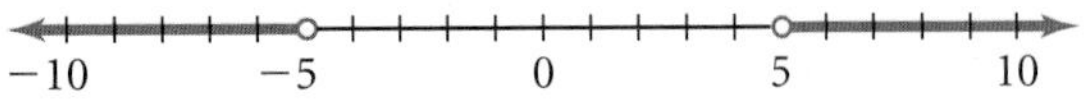

To be a solution, x must satisfy one of the inequalities $x < -5$ or $x > 5$. Because x can't satisfy both inequalities, it's not appropriate to combine the two inequalities into one statement.

Match each inequality to its graph and write the solution inequalities.

a. $|x - 3| > 2$

A. 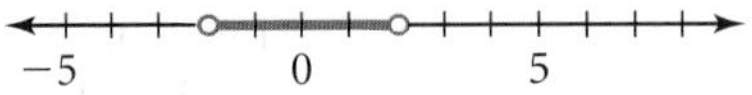

b. $2|x| < 4$

B. 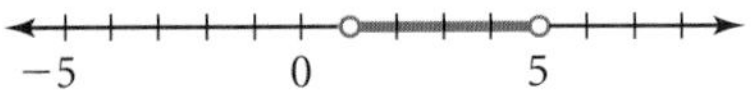

c. $2|x - 3| < 4$

C. 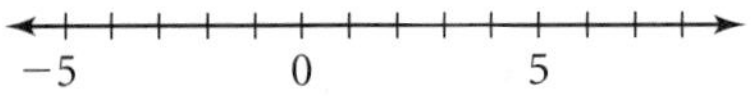

d. $|2x| < -4$

D. 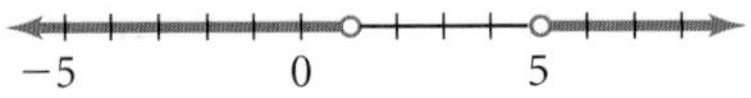

14. For each segment shown in the pentagon at right, write an equation in point-slope form for the line that contains the segment. Check your equations by graphing them on your calculator. @

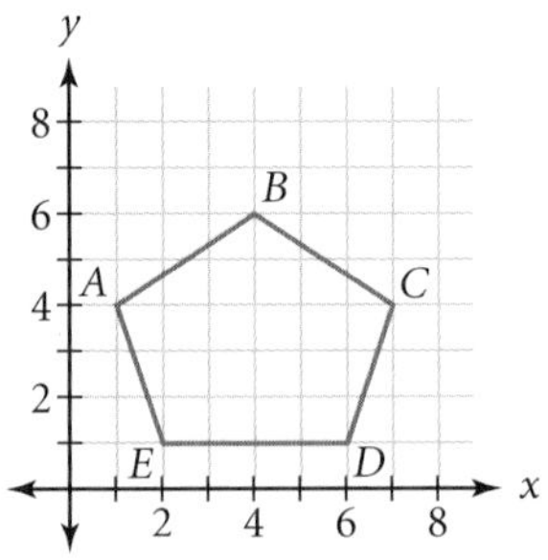

15. Consider the graph of $\triangle ABC$.

a. Represent $\triangle ABC$ with a matrix $[M]$.

b. Find each product and graph the image of the triangle represented by the result.

i. $\begin{bmatrix} 0 & 1 \\ 1 & 0 \end{bmatrix}[M]$

ii. $\begin{bmatrix} 0 & -1 \\ 1 & 0 \end{bmatrix}[M]$

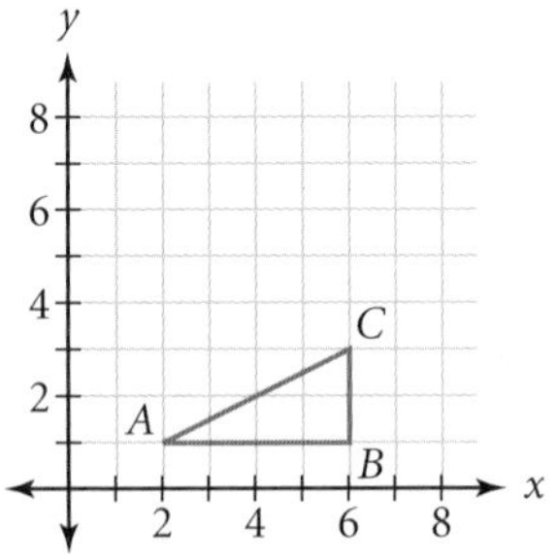

IMPROVING YOUR REASONING SKILLS

Coding and Decoding

One of the uses of matrices is in the mathematical field of cryptography, the science of enciphering and deciphering encoded messages. Here's one way that a matrix can be used to make a secret code:

Imagine encoding the word "CODE." First you convert each letter to a numerical value, based on its location in the alphabet. For example, C = 3 because it is the third letter in the alphabet. So, CODE = 3, 15, 4, 5. Arrange these numbers in a 2 × 2 matrix: $\begin{bmatrix} 3 & 15 \\ 4 & 5 \end{bmatrix}$.

Now multiply by an encoding matrix.

Let's use $[E] = \begin{bmatrix} 1 & -1 \\ -1 & 2 \end{bmatrix}$, so $\begin{bmatrix} 3 & 15 \\ 4 & 5 \end{bmatrix}[E] = \begin{bmatrix} -12 & 27 \\ -1 & 6 \end{bmatrix}$.

Now you have to convert back to letters. Notice that three of these numbers are outside of the numbers 1–26 that represent A–Z. To convert other numbers, simply add or subtract 26 to make them fall within the range of 1 to 26, like this:

$$\begin{bmatrix} -12 & 27 \\ -1 & 6 \end{bmatrix} \rightarrow \begin{bmatrix} 14 & 1 \\ 25 & 6 \end{bmatrix} \rightarrow \begin{bmatrix} N & A \\ Y & F \end{bmatrix}$$

Now, to *decode* the encoded message "NAYF," you convert back to numbers and multiply by the inverse of the coding matrix, and then you convert each number back to the corresponding letter in the alphabet:

$$\begin{bmatrix} 14 & 1 \\ 25 & 6 \end{bmatrix}[E]^{-1} = \begin{bmatrix} 29 & 15 \\ 56 & 31 \end{bmatrix} = \begin{bmatrix} C & O \\ D & E \end{bmatrix}$$

See if you can decode this Navajo saying. Begin by taking out the spaces and breaking the letters into groups of four letters each.

CC FSLG GTQN YP OPIIY UCB DKIC BYF BEQQ WW URQLPRE.

Language CONNECTION

The Navajo language was used in coding messages by the U.S. Marines in World War II, because it is a complex unwritten language that is unintelligible to anyone without extensive training. Navajo code talkers could encode, transmit, and decode three-line English messages in 20 seconds, whereas machines of the time required 30 minutes to perform the same job. Navajo recruits developed the code, including a dictionary and numerous words for military terms. Learn more about Navajo code talkers by using the links at www.keymath.com/DAA .

Navajo recruits, like these two shown in the Pacific island of Bougainville in 1943, shared their complex language with U.S. Marines to communicate in code during World War II.

LESSON 6.5

Systems of Inequalities

Frequently, real-world situations involve a range of possible values. Algebraic statements of these situations are called **inequalities.**

Situation	Inequality
Write an essay between two and five pages in length.	$2 \leq E \leq 5$
Practice more than an hour each day.	$P > 1$
The post office is open from nine o'clock until noon.	$9 \leq H \leq 12$
Do not spend more than \$10 on candy and popcorn.	$0 \leq c + p \leq 10$
A college fund has \$40,000 to invest in stocks and bonds.	$0 \leq s + b \leq 40{,}000$

Recall that you can perform operations on inequalities very much like you do on equations. You can add or subtract the same quantity on both sides, multiply by the same number or expression on both sides, and so on. The one exception to remember is that when you multiply or divide by a negative quantity or expression, the inequality symbol reverses.

In this lesson you will learn how to graphically show solutions to inequalities with two variables, such as the last two statements in the table above.

The cost of a college education continues to rise. The good news is that public and private institutions provide billions of dollars of financial aid to students every year. At four-year public colleges, for example, over 65% of students receive some form of financial aid. Financial aid makes college affordable for many students, despite increasing costs.

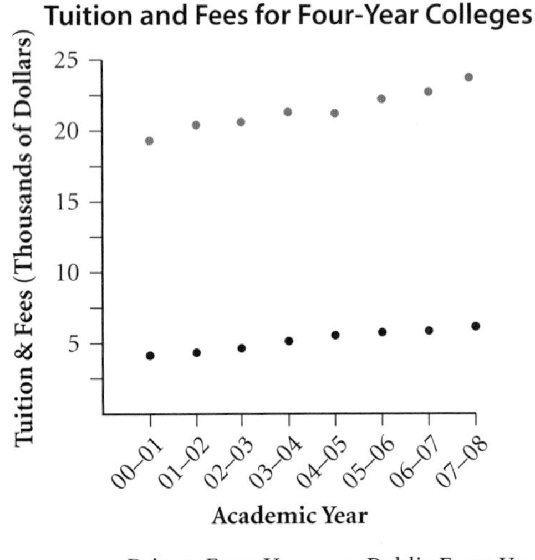

(www.collegeboard.com)

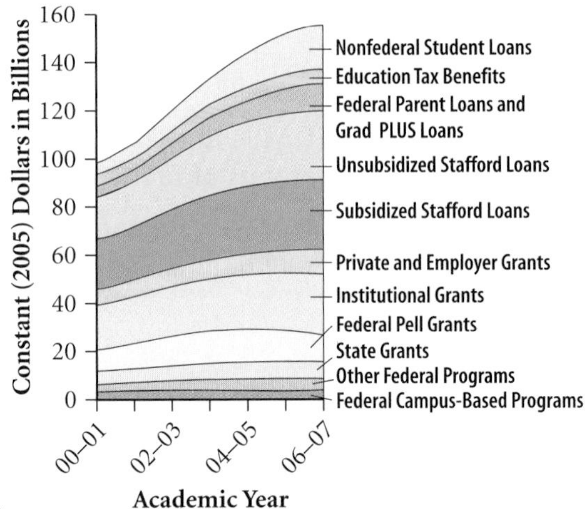

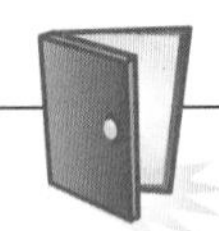

Investigation
Paying for College

A total of \$40,000 has been donated to a college scholarship fund. The administrators of the fund are considering how much to invest in stocks and how much to invest in bonds. Stocks usually pay more but are often a riskier investment, whereas bonds pay less but are usually safer.

Step 1 Let x represent the amount in dollars invested in stocks, and let y represent the amount in dollars invested in bonds. Graph the equation $x + y = 40{,}000$.

Step 2 Name at least five pairs of x- and y-values that satisfy the inequality $x + y < 40{,}000$ and plot them on your graph. In this problem, why can $x + y$ be less than \$40,000?

Step 3 Describe where all possible solutions to the inequality $x + y < 40{,}000$ are located. Shade this region on your graph.

Step 4 Describe some points that fit the condition $x + y \leq 40{,}000$ but do not make sense for the situation.

Assume that each option—stocks or bonds—requires a minimum investment of \$5,000, and that the fund administrators want to purchase some stocks and some bonds. Based on the advice of their financial advisor, they decide that the amount invested in bonds should be at least twice the amount invested in stocks.

Step 5 Translate all of the limitations, or **constraints,** into a system of inequalities. A table might help you to organize this information.

Step 6 Graph all of the inequalities and determine the region of your graph that will satisfy all the constraints. Find each corner, or **vertex,** of this region.

When there are one or two variables in an inequality, you can represent the solution as a set of ordered pairs by shading the region of the coordinate plane that contains those points.

The solid boundary line indicates that the region *includes* the line.

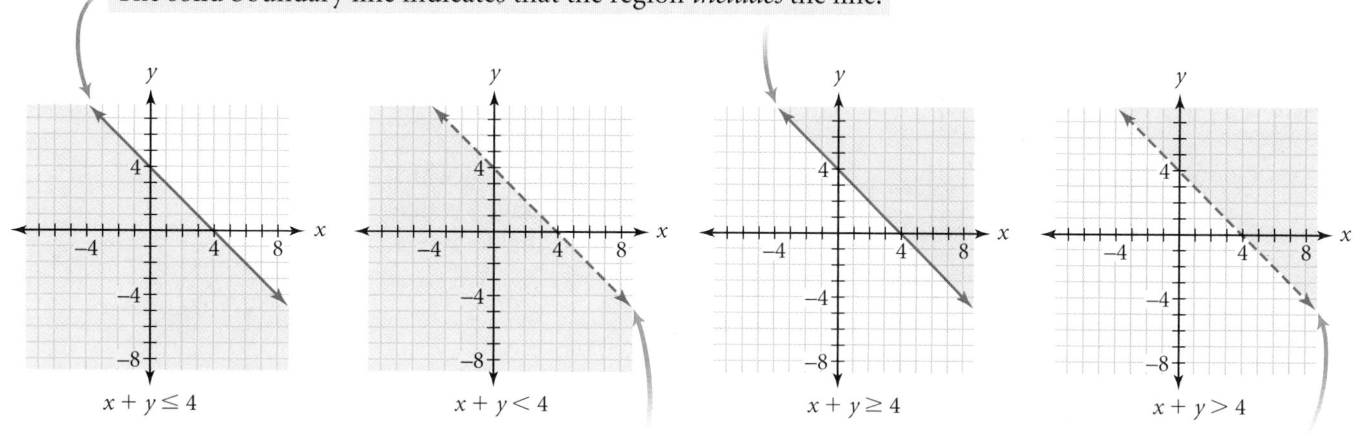

The dashed boundary line indicates that the region does *not* include the line.

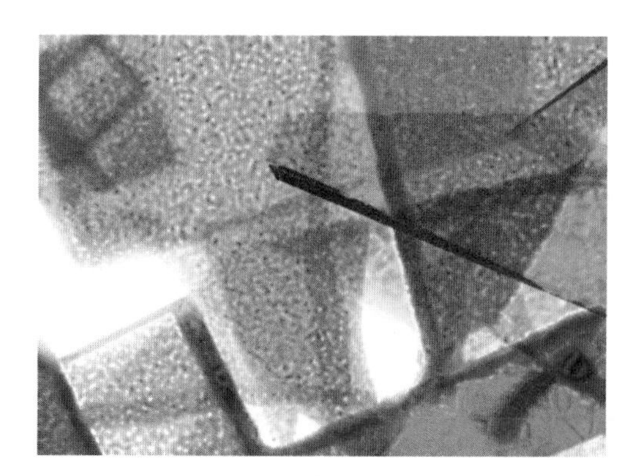

When you have several inequalities that must be satisfied simultaneously, you have a system. The solution to a system of inequalities with two variables will be a set of points. This set of points is called a **feasible region.** The feasible region can be shown graphically as part of a plane, or sometimes it can be described as a geometric shape with its vertices given.

EXAMPLE A

Rachel has 3 hours to work on her homework tonight. She wants to spend more time working on math than on chemistry, and she must spend at least a half hour working on chemistry.

a. Let x represent time in hours spent on math, and let y represent time in hours spent on chemistry. Write inequalities to represent the three constraints of the system.

b. Graph your inequalities and shade the feasible region.

c. Find the coordinates of the vertices of the feasible region.

d. Name two points that are solutions to the system, and describe what they mean in the context of the problem.

▶ Solution

First, read the given information carefully to make sure you understand the constraints.

a. Convert each constraint into an algebraic inequality.

$x + y \leq 3$, or $y \leq -x + 3$	Rachel has 3 h to work on homework.
$x > y$, or $y < x$	She wants to spend more time working on math than on chemistry.
$y \geq 0.5$	She must spend at least a half hour working on chemistry.

b. To graph the inequality $y \leq -x + 3$, first graph $y = -x + 3$. This line is part of the solution. Then shade the region that contains the points that satisfy the inequality $y < -x + 3$.

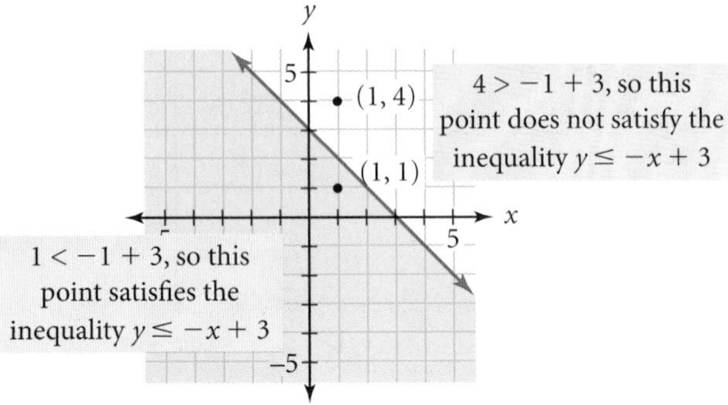

Shade the region in which all points (x, y) have y-values that are less than $-x + 3$.

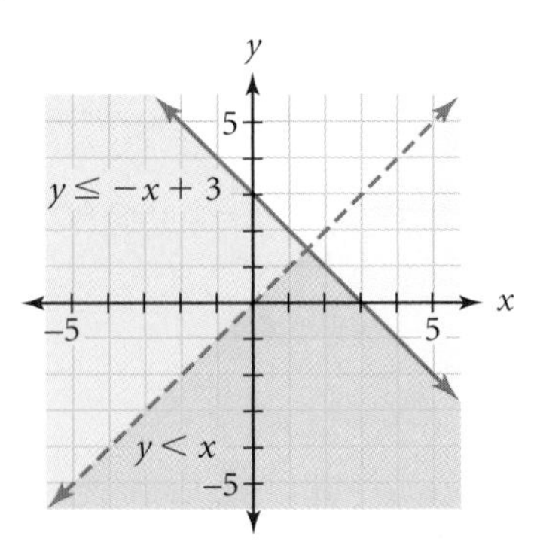

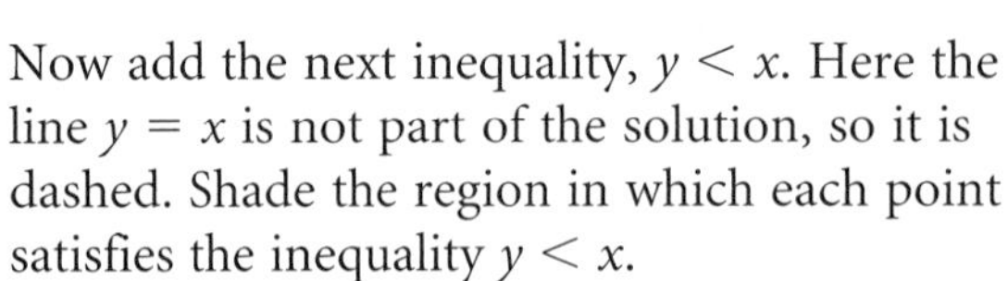

Now add the next inequality, $y < x$. Here the line $y = x$ is not part of the solution, so it is dashed. Shade the region in which each point satisfies the inequality $y < x$.

Then add the last inequality, $y \geq 0.5$. The solution includes $y = 0.5$ and all the points where y is greater than 0.5. [▶ See **Calculator Note 6G** to learn how to graph systems of inequalities on your calculator. ◀]

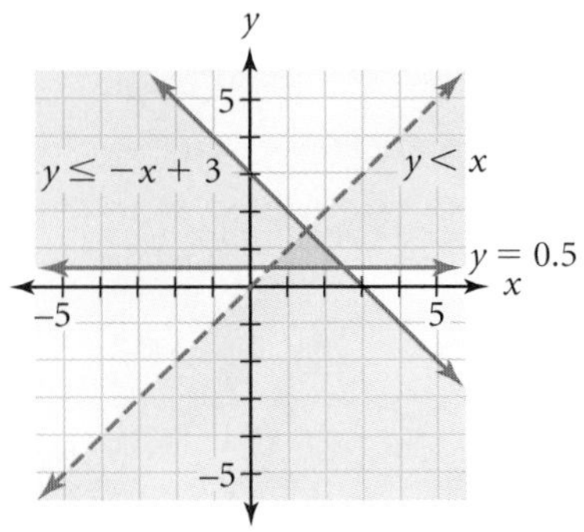

The solution to the system, the feasible region, is the set of points common to all three regions.

c. The solution may be identified by giving the coordinates of the vertices of the feasible region. You can find those coordinates by solving systems of equations representing adjacent boundary lines.

Equations	**Intersections**
$x + y = 3$ and $x = y$	$(1.5, 1.5)$
$x + y = 3$ and $y = 0.5$	$(2.5, 0.5)$
$x = y$ and $y = 0.5$	$(0.5, 0.5)$

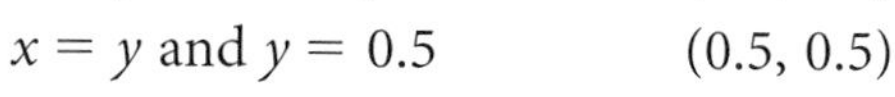

Only one of these vertices, (2.5, 0.5), is part of the solution. The other two vertices of the feasible region, (1.5, 1.5) and (0.5, 0.5), are on the dashed line, showing that they do not satisfy the constraint $y < x$.

d. The points (1.5, 1) and (2.5, 0.5) are two solutions to the system. Every point in the feasible region represents a way that Rachel could divide her time. The solution point (1.5, 1) means she could spend 1.5 h on mathematics and 1 h on chemistry and still meet all her constraints. The point (2.5, 0.5) means that Rachel could spend 2.5 h on math and 0.5 h on chemistry. This point represents the boundaries of two constraints: She can't spend less than 0.5 h on chemistry or more than 3 h total on homework.

Sometimes you will need to write nonlinear inequalities to represent constraints.

EXAMPLE B

Anna throws a ball straight up next to a building. The ball's height in feet after t seconds is given by $-16t^2 + 51t + 3$. Tom rides a glass elevator down the outside of the same building. His height from the time Anna throws the ball can be expressed as $43 - 5t$. As Tom is riding down, he sees a bird fly by above the elevator but below the ball. When did Tom see the bird? Give a range of possible times.

▶ Solution

At the time Tom sees the bird, the bird's height, h, must satisfy $43 - 5t < h < -16t^2 + 51t + 3$. You can graph these two inequalities separately. You might find it easier to graph the boundaries first and then shade the feasible region.

The feasible region shows times between about 1 and 2.5 seconds and heights between about 44 and 30 feet. The times at the intersection points can be found by substitution.

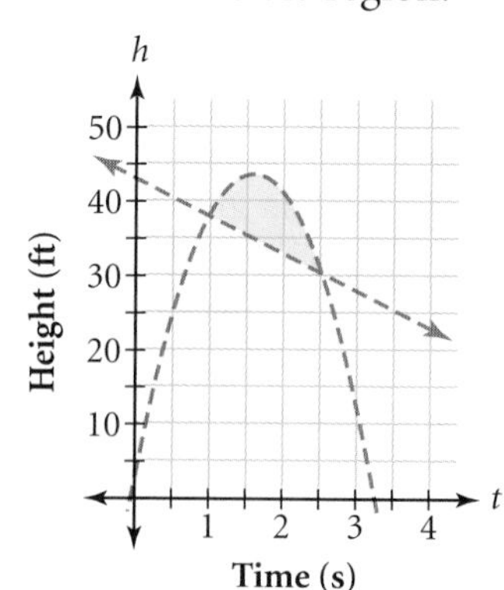

$$
\begin{aligned}
43 - 5t &= -16t^2 + 51t + 3 \\
16t^2 - 56t + 40 &= 0 \\
2t^2 - 7t + 5 &= 0 \\
(2t - 5)(1t - 1) &= 0 \\
t = 2.5 \text{ or } t &= 1
\end{aligned}
$$

Tom saw the bird between 1 and 2.5 seconds after Anna threw the ball.

When you are solving a system of equations based on real-world constraints, it is important to note that sometimes there are constraints that are not specifically stated in the problem. In Example A, negative values for x and y would not make sense, because you can't study for a negative number of hours. You could have added the commonsense constraints $x \geq 0$ and $y \geq 0$, although in this case it would not affect the feasible region.

EXERCISES

You will need

A graphing calculator for Exercise **9**.

Practice Your Skills

1. Solve each inequality for y.

a. $2x - 5y > 10$ @ **b.** $4(2 - 3y) + 2x > 14$

2. Graph each linear inequality on the coordinate plane.

a. $y \leq -2x + 5$ @ **b.** $2y + 2x > 5$ **c.** $x > 5$

3. For 3a–d, write the equation of each graph.

keymath.com/DAA

a. @

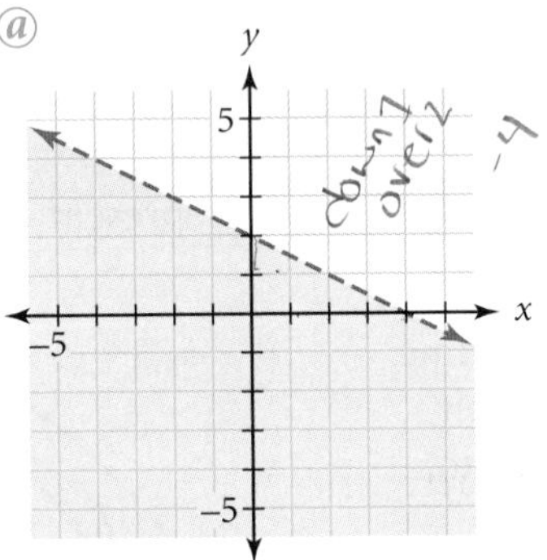

b.

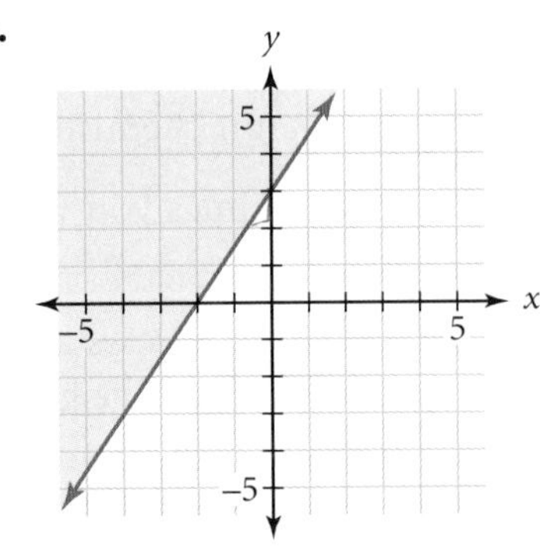

c. @

d.

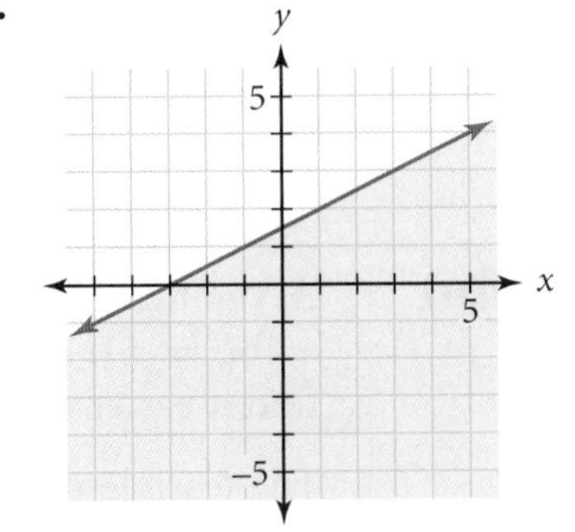

4. The graphs of $y = 2.4x + 2$ and $y = -x^2 - 2x + 6.4$ serve as the boundaries of this feasible region. What two inequalities identify this region?

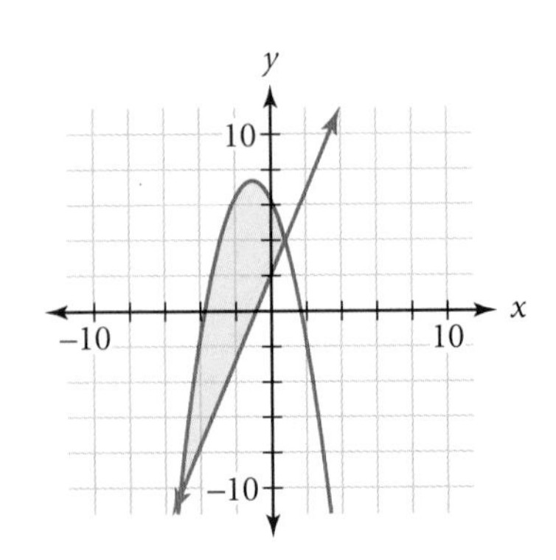

Reason and Apply

For Exercises 5–8, sketch the feasible region of each system of inequalities. Find the coordinates of each vertex.

5. $\begin{cases} y \leq -0.51x + 5 \\ y \leq -1.6x + 8 \\ y \geq 0.1x + 2 \\ y \geq 0 \\ x \geq 0 \end{cases}$ ⓐ

6. $\begin{cases} y \geq 1.6x - 3 \\ y \leq -(x - 2)^2 + 4 \\ y \geq 1 - x \\ y \geq 0 \\ x \geq 0 \end{cases}$

7. $\begin{cases} 4x + 3y \leq 12 \\ 1.6x + 2y \leq 8 \\ 2x + y \geq 2 \\ y \geq 0 \\ x \geq 0 \end{cases}$ ⓐ

8. $\begin{cases} y \geq |x - 1| \\ y \leq \sqrt{9 - x^2} \\ y \leq 2.5 \\ y \geq 0 \end{cases}$

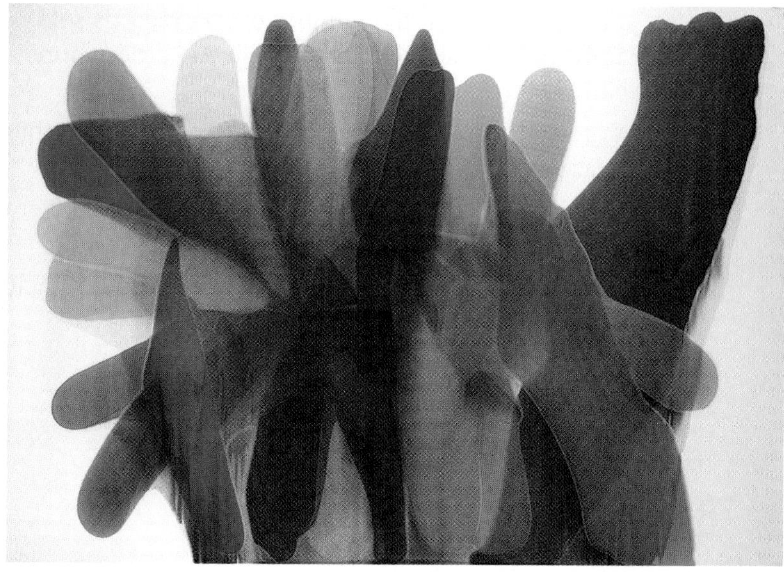

This color-stain painting by American artist Morris Louis (1912–1962) shows overlapping regions similar to the graph of a system of inequalities.

Floral (1959) by Morris Louis

9. In the Lux Art Gallery, rectangular paintings must have an area between 200 in^2 and 300 in^2 and a perimeter between 66 in. and 80 in. ⓗ

a. Write four inequalities involving length and width that represent these constraints. ⓐ

b. Graph this system of inequalities to identify the feasible region.

c. Will the gallery accept a painting that measures

i. 12.4 in. by 16.3 in.?

ii. 16 in. by 17.5 in.?

iii. 14.3 in. by 17.5 in.?

American artist Nina Bovasso (b 1965) carries one of her paintings to an exhibition called Vir Heroicus Snoopicus. Would this piece, titled *Black and White Crowd,* be accepted at the Lux Art Gallery in Exercise 9?

10. Al just got rid of 40 of his dad's old records. He sold each classical record for \$5 and each jazz record for \$2. He arranged to donate any records that he couldn't sell to a thrift shop. Al knows that he sold fewer than 10 jazz records and that he earned more than \$100.

a. Let x represent the number of classical records sold, and let y represent the number of jazz records sold. Write an inequality expressing that Al earned more than \$100.

b. Write an inequality expressing that he sold fewer than 10 jazz records.

c. Write an inequality expressing that the total number of records sold was no more than 40.

d. Graph the solution to the system of inequalities, including any commonsense constraints.

e. Name all the vertices of the feasible region.

Review

11. APPLICATION As the altitude of a spacecraft increases, an astronaut's weight decreases. The weight of a 180 lb astronaut, w, at a given altitude in kilometers above Earth's sea level, x, is given by the formula

$$w = 180 \cdot \frac{6400^2}{(6400 + x)^2}$$

a. At what altitudes will the astronaut weigh less than 20 lb?

b. At an altitude of 400 km, how much will the astronaut weigh?

c. Astronauts get the feeling of weightlessness because they are in free fall as they orbit Earth. Could they ever be truly weightless? Why or why not?

Science CONNECTION

A typical space shuttle orbits at an altitude of about 400 km. At this height, an astronaut still weighs about 89% of her weight on Earth. You have probably seen pictures in which astronauts in orbit on a space shuttle or space station appear to be weightless. This is actually not due to the absence of gravity but, rather, to an effect called microgravity. In orbit, astronauts and their craft are being pulled toward Earth by gravity, but their speed is such that they are in free fall around Earth, rather than toward Earth. Because the astronauts and their spacecraft are falling through space at the same rate, the astronauts appear to be floating inside the craft. This is similar to the fact that a car's driver can appear to be sitting still, although he is actually traveling at a speed of 60 mi/h.

Astronauts floating in space during the 1994 testing of rescue system hardware appear to be weightless. One astronaut floats without being tethered to the spacecraft by using a small control unit.

12. A parabola with an equation in the form $y = ax^2 + bx + c$ passes through the points $(-2, -32)$, $(1, 7)$, and $(3, 63)$.

a. Set up systems and use matrices to find the values of a, b, and c for this parabola.

b. Write the equation of this parabola.

c. Describe how to verify that your answer is correct.

13. These data were collected from a bouncing-ball experiment. Recall that the height in centimeters, y, is exponentially related to the number of the bounce, x. Find the values of a and b for an exponential model in the form $y = ab^x$.

Bounce number	3	7
Height (cm)	34.3	8.2

14. Complete the reduction of this augmented matrix to row-echelon form. Give each row operation and find each missing matrix entry.

$$\left[\begin{array}{cc|c} 3 & -1 & 5 \\ -4 & 2 & 1 \end{array}\right] \xrightarrow{\ ?\ } \left[\begin{array}{cc|c} 1 & -\frac{1}{3} & ? \\ -4 & 2 & 1 \end{array}\right] \xrightarrow{\ ?\ } \left[\begin{array}{cc|c} 1 & ? & ? \\ 0 & ? & ? \end{array}\right] \xrightarrow{\ ?\ } \left[\begin{array}{cc|c} 1 & ? & ? \\ 0 & 1 & ? \end{array}\right] \xrightarrow{\ ?\ } \left[\begin{array}{cc|c} 1 & 0 & ? \\ 0 & 1 & ? \end{array}\right]$$

15. APPLICATION The population of water fungus is modeled by the function $y = f(x) = 2.68(3.84)^x$, where x represents the number of hours elapsed and y represents the number of fungi spores.

a. How many spores are there initially?

b. How many spores are there after 10 h?

c. Find an inverse function that uses y as the dependent variable and x as the independent variable.

d. Use your answer to 15c to find how long it will take until the number of spores exceeds 1 billion.

IMPROVING YOUR VISUAL THINKING SKILLS

Intersection of Planes

Graphically, a system of two linear equations in two variables can be represented by two lines. If the lines intersect, the point of intersection is the solution and the system is called consistent and independent. If the lines are parallel, they never intersect, there is no solution, and the system is called inconsistent. If the lines are the same, there are infinitely many solutions, and the system is called consistent and dependent.

An equation such as $3x + 2y + 6z = 12$ is also called a linear equation because the highest power of any variable is 1. But, because there are three variables, the graph of this equation is a plane.

Graphically, a system of three linear equations in three variables can be represented by three planes. Sketch all the possible outcomes for the graphs of three planes. Classify each outcome as consistent, inconsistent, dependent, and/or independent.

LESSON

6.6

Linear Programming

Love the moment and the energy of that moment will spread beyond all boundaries.

CORITA KENT

Industrial managers often investigate more economical ways of doing business. They must consider physical limitations, standards of quality, customer demand, availability of materials, and manufacturing expenses as restrictions, or constraints, that determine how much of an item they can produce. Then they determine the optimum, or best, amount of goods to produce—usually to minimize production costs or maximize profit. The process of finding a feasible region and determining the point that gives the maximum or minimum value to a specific expression is called **linear programming.**

Polish artist Roman Cieslewicz (1930–1996) formed this 1988 work by gluing strips of paper and fragments of photographs on cardboard.

Les dieux ont soif (1988) by Roman Cieslewicz

Problems that can be modeled with linear programming may involve anywhere from two variables to hundreds of variables. Computerized modeling programs that analyze up to 200 constraints and 400 variables are regularly used to help businesses choose their best plan of action. In this lesson you will look at problems that involve two variables because you will be using two-dimensional graphs to help you find the feasible region.

In this investigation you'll explore a linear programming problem and make conjectures about how to find the optimum value in the most efficient way.

Investigation
Maximizing Profit

The Elite Pottery Shoppe makes two kinds of birdbaths: a fancy glazed and a simple unglazed. An unglazed birdbath requires 0.5 h to make using a pottery wheel and 3 h in the kiln. A glazed birdbath takes 1 h on the wheel and 18 h in the kiln. The company's one pottery wheel is available for at most 8 hours per day (h/d). The three kilns can be used a total of at most 60 h/d, and each kiln can hold only one birdbath. The company has a standing order for 6 unglazed birdbaths per day, so it must produce at least that many. The pottery shop's profit on each unglazed birdbath is \$10, and the profit on each glazed birdbath is \$40. How many of each kind of birdbath should the company produce each day in order to maximize profit?

Step 1 Organize the information into a table like this one:

	Amount per unglazed birdbath	Amount per glazed birdbath	Constraining value
Wheel hours			
Kiln hours			
Profit			Maximize

Step 2 Use your table to help you write inequalities that reflect the constraints given, and be sure to include any commonsense constraints. Let x represent the number of unglazed birdbaths, and let y represent the number of glazed birdbaths. Graph the feasible region to show the combinations of unglazed and glazed birdbaths the shop could produce, and label the coordinates of the vertices. (*Note:* Profit is not a constraint; it is what you are trying to maximize.)

Step 3 It will make sense to produce only whole numbers of birdbaths. List the coordinates of all integer points within the feasible region. (There should be 23.) Remember that the feasible region may include points on the boundary lines.

Step 4 Write the equation that will determine profit based on the number of unglazed and glazed birdbaths produced. Calculate the profit that the company would earn at each of the feasible points you found in Step 3. You may want to divide this task among the members of your group.

Step 5 What number of each kind of birdbath should the Elite Pottery Shoppe produce to maximize profit? What is the maximum profit possible? Plot this point on your feasible region graph. What do you notice about this point?

Step 6 Suppose that you want profit to be exactly \$100. What equation would express this? Carefully graph this line on your feasible region graph.

Step 7 Suppose that you want profit to be exactly \$140. What equation would express this? Carefully add this line to your graph.

Step 8 Suppose that you want profit to be exactly \$170. What equation would express this? Carefully add this line to your graph.

Step 9 How do your results from Steps 6–8 show you that (14, 1) must be the point that maximizes profit? Generalize your observations to describe a method that you can use with other problems to find the optimum value. What would you do if this vertex point did not have integer coordinates? What if you wanted to *minimize* profit?

Linear programming is a very useful real-world application of systems of inequalities. Its value is not limited to business settings, as the following example shows.

EXAMPLE

Marco is planning to provide a snack of graham crackers and blueberry yogurt at his school's track practice. He wants to make sure that the snack contains no more than 700 calories and no more than 20 g of fat. He also wants at least 17 g of protein and at least 30% of the daily recommended value of iron. The nutritional content of each food is given below. Each serving of yogurt costs $0.30 and each graham cracker costs $0.06. What combination of servings of graham crackers and blueberry yogurt should Marco provide to minimize cost?

	Serving	Calories	Fat	Protein	Iron (percent of daily recommended value)
Graham crackers	1 cracker	60	2 g	2 g	6%
Blueberry yogurt	4.5 oz	130	2 g	5 g	1%

▶ Solution

First organize the constraint information into a table, then write inequalities that reflect the constraints. Be sure to include any commonsense constraints. Let x represent the number of servings of graham crackers, and let y represent the number of servings of yogurt.

	Amount per graham cracker	Amount per serving of yogurt	Limiting value
Calories	60	130	≤ 700
Fat	2 g	2 g	≤ 20 g
Protein	2 g	5 g	≥ 17 g
Iron	6%	1%	$\geq 30\%$
Cost	$0.06	$0.30	Minimize

$$\begin{cases} 60x + 130y \leq 700 & \text{Calories} \\ 2x + 2y \leq 20 & \text{Fat} \\ 2x + 5y \geq 17 & \text{Protein} \\ 6x + 1y \geq 30 & \text{Iron} \\ x \geq 0 & \text{Common sense} \\ y \geq 0 & \text{Common sense} \end{cases}$$

Now graph the feasible region and find the vertices.

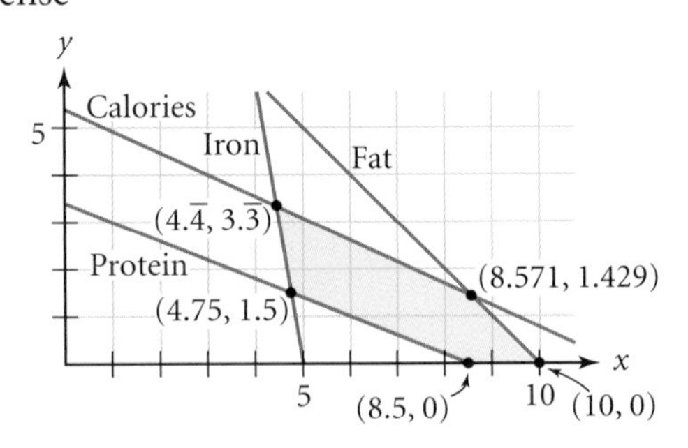

Next, write an equation that will determine the cost of a snack based on the number of servings of graham crackers and yogurt.

$$Cost = 0.06x + 0.30y$$

You could try any possible combination of graham crackers and yogurt that is in the feasible region, but recall that in the investigation it appeared that optimum values will occur at vertices. Calculate the cost at each of the vertices to see which vertex provides a minimum value.

The least expensive combination would be 8.5 crackers and no yogurt. What if Marco wants to serve only whole numbers of servings? The points (8, 1), (9, 1), and (9, 0) are the integer points within the feasible region closest to (8.5, 0), so test which point has a lower cost. The point (8, 1) gives a cost of $0.78, (9, 1) gives a cost of $0.84, and (9, 0) will cost $0.54. Therefore, if Marco wants to serve only whole numbers of servings, he should serve 9 graham crackers and no yogurt.

x	y	Cost
4.444	3.333	$1.27
4.75	1.5	$0.74
8.571	1.429	$0.94
10	0	$0.60
8.5	0	$0.51

The following box summarizes the steps of solving a linear programming problem. Refer to these steps as you do the exercises.

Solving a Linear Programming Problem

1. Define your variables, and write constraints using the information given in the problem. Don't forget commonsense constraints.
2. Graph the feasible region, and find the coordinates of all vertices.
3. Write the equation of the function you want to optimize, and decide whether you need to maximize or minimize it.
4. Evaluate your optimization function at each of the vertices of your feasible region, and decide which vertex provides the optimum value.
5. If your possible solutions need to be limited to whole-number values, and your optimum vertex does not contain integers, test the whole-number values within the feasible region that are closest to this vertex.

EXERCISES

You will need

A graphing calculator for Exercises **3**, **4**, and **6–11**.

Practice Your Skills

1. Carefully graph this system of inequalities and label the vertices. ⓐ

$$\begin{cases} x + y \leq 10 \\ 5x + 2y \geq 20 \\ -x + 2y \geq 0 \end{cases}$$

2. For the system in Exercise 1, find the vertex that optimizes these expressions:

a. maximize: $5x + 2y$ ⓐ

b. minimize: $x + 3y$

c. maximize: $x + 4y$

d. minimize: $5x + y$

e. What generalizations can you make about which vertex provides a maximum or minimum value?

3. Graph this system of inequalities, label the vertices of the feasible region, and name the integer coordinates that maximize the function $P = 0.08x + 0.10y$. What is this maximum value of P?

$$\begin{cases} x \geq 5{,}500 \\ y \geq 5{,}000 \\ y \leq 3x \\ x + y \leq 40{,}000 \end{cases}$$

4. APPLICATION During nesting season, two different bird species inhabit a region with area 180,000 m². Dr. Chan estimates that this ecological region can provide 72,000 kg of food during the season. Each nesting pair of species X needs 39.6 kg of food during a specified time period and 120 m² of land. Each nesting pair of species Y needs 69.6 kg of food and 90 m² of land. Let x represent the number of pairs of species X, and let y represent the number of pairs of species Y.

a. Describe the meaning of the constraints $x \geq 0$ and $y \geq 0$.

b. Describe the meaning of the constraint $120x + 90y \leq 180{,}000$.

c. Describe the meaning of the constraint $39.6x + 69.6y \leq 72{,}000$.

d. Graph the system of inequalities, and identify each vertex of the feasible region.

e. Maximize the total number of nesting pairs, N, by considering the function $N = x + y$.

Reason and Apply

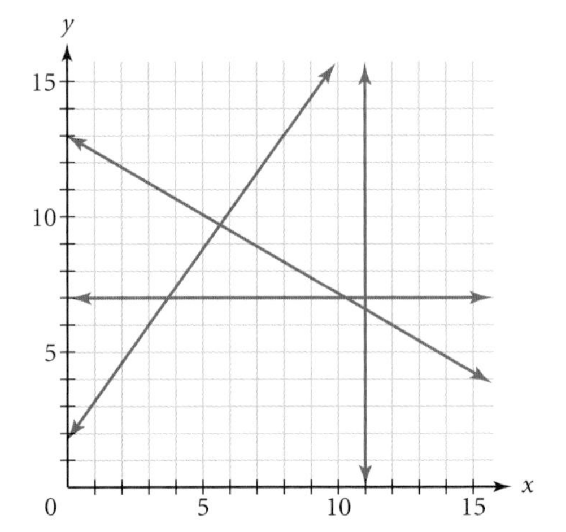

5. Use a combination of the four lines shown on the graph along with the axes to create a system of inequalities whose graph satisfies each description.

a. The feasible region is a triangle. ⓐ

b. The feasible region is a quadrilateral with one side on the y-axis.

c. The feasible region is a pentagon with sides on both the x-axis and the y-axis.

6. APPLICATION The International Canine Academy raises and trains Siberian sled dogs and dancing French poodles. Breeders can supply the academy with at most 20 poodles and 15 Siberian sled dogs each year. Each poodle eats 2 pounds per day (lb/d) of food and each sled dog eats 6 lb/d. Food supplies are restricted to at most 100 lb/d. A poodle requires 1,000 h/yr of training, whereas a sled dog requires 250 h/yr. The academy cannot provide more than 15,000 h/yr of training time. If each poodle sells for a profit of \$200 and each sled dog sells for a profit of \$80, how many of each kind of dog should the academy raise in order to maximize profits? ⓗ

7. APPLICATION The Elite Pottery Shoppe budgets a maximum of \$1,000 per month for newspaper and radio advertising. The newspaper charges \$50 per ad and requires at least four ads per month. The radio station charges \$100 per minute and requires a minimum of 5 minutes of advertising per month. It is estimated that each newspaper ad reaches 8,000 people and that each minute of radio advertising reaches 15,000 people. What combination of newspaper and radio advertising should the pottery shop use in order to reach the maximum number of people? What assumptions did you make in solving this problem? How realistic do you think they are? ⓐ

8. APPLICATION A small electric generating plant must decide how much low-sulfur (2%) and high-sulfur (6%) oil to buy. The final mixture must have a sulfur content of no more than 4%. At least 1200 barrels of oil are needed. Low-sulfur oil costs \$18.50 per barrel and high-sulfur oil costs \$14.70 per barrel. How much of each type of oil should the plant use to keep the cost at a minimum? What is the minimum cost? ⓗ

9. APPLICATION A fair-trade farmers' cooperative in Chiapas, Mexico, is deciding how much coffee and cocoa to recommend to their members to plant. Their 1,000 member families have 7,500 total acres to farm. Because of the geography of the region, 2,450 acres are suitable only for growing coffee and 1,230 acres are suitable only for growing cocoa. A coffee crop produces 30 lb/acre and a crop of cocoa produces 40 lb/acre. The cooperative has the resources to ship a total of 270,000 lb of product to the United States. Fair-trade organizations mandate a minimum price of $1.51 per pound for organic coffee and $0.88 per pound for organic cocoa (note that price is per pound, not per acre). How many acres of each crop should the cooperative recommend planting in order to maximize income?

These coffee beans are growing in the Mexican state of Chiapas.

Consumer CONNECTION

Many small coffee and cocoa farmers receive prices for their crop that are less than the costs of production, causing them to live in poverty and debt. Fair-trade certification has been developed to show consumers which products are produced with the welfare of farming communities in mind. To become fair-trade certified, an importer must meet stringent international criteria: paying a minimum price per pound, providing credit to farmers, and providing technical assistance in farming upgrades. Fair-trade prices allow farmers to make enough money to provide their families with food, education, and health care.

10. APPLICATION Teo sells a set of videotapes on an online auction. The postal service he prefers puts these restrictions on the size of a package:

Up to 150 lb
Up to 130 in. in length and girth combined
Up to 108 in. in length

Length is defined as the longest side of a package or object. Girth is the distance all the way around the package or object at its widest point perpendicular to the length. Teo is not concerned with the weight because the videotapes weigh only 15 lb.

a. Write a system of inequalities that represents the constraints on the package size. @

b. Graph the feasible region for the dimensions of the package.

c. Teo packages the videotapes in a box whose dimensions are 20 in. by 14 in. by 8 in. Does this box satisfy the restrictions?

Consumer CONNECTION

In the packaging industry, two sets of dimensions are used. Inside dimensions are used to ensure proper fit around a product to prevent damage. Outside dimensions are used in shipping classifications and determining how to stack boxes on pallets. In addition, the type of packaging material and its strength are important concerns. Corrugated cardboard is a particularly strong, yet economical, packaging material.

Review

11. Solve each of these systems in at least two different ways.

a. $\begin{cases} 8x + 3y = 41 \\ 9x + 2y = 25 \end{cases}$

b. $\begin{cases} 2x + y - 2z = 5 \\ 6x + 2y - 4z = 3 \\ 4x - y + 3z = 5 \end{cases}$

12. Sketch a graph of the feasible region described by this system of inequalities:

$$\begin{cases} y \geq (x-3)^2 + 5 \\ y \leq -|x-2| + 10 \end{cases}$$

13. Give the system of inequalities whose solution set is the polygon at right. ⓗ

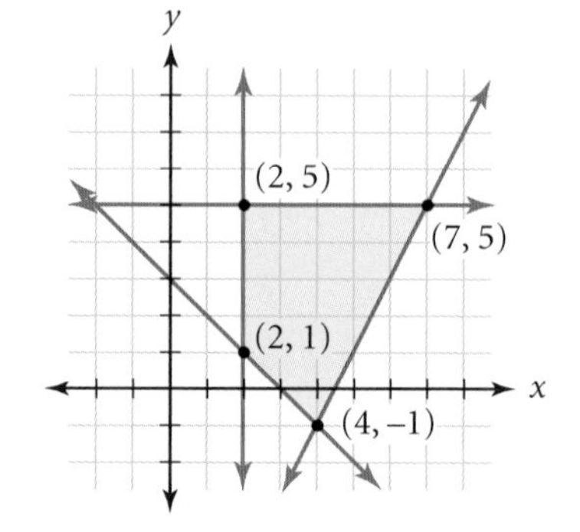

14. Consider this system of equations:

$$\begin{cases} 3x - y = 5 \\ -4x + 2y = 1 \end{cases}$$

a. Write a matrix equation for this system.

b. Solve the system and write the solution as an ordered pair of coordinates.

c. Check your solution values for x and y by substituting them into the original equations.

15. Find the equation of the parabola at right.

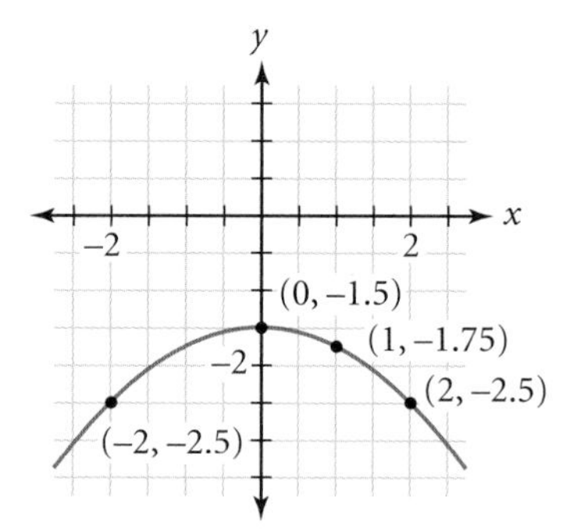

Project

NUTRITIONAL ELEMENTS

Write and solve your own linear programming problem similar to the cracker-and-yogurt snack combination problem in this lesson. Choose two food items from your home or a grocery store, and decide on your constraints and what you wish to minimize or maximize. Record the necessary information. Then write inequalities, and find the feasible region and optimum value for your problem. Your project should include

- Your linear programming problem
- A complete solution, including a graph of the feasible region
- An explanation of your process

CHAPTER 6 REVIEW

Matrices have a variety of uses. They provide ways to organize data about such things as inventory or the coordinates of vertices of a polygon. A **transition matrix** represents repeated changes that happen to a system. You can use matrix arithmetic to combine data or transform polygons on a coordinate graph. You can also use **matrix multiplication** to determine quantities at various stages of a transition simulation.

Another important application of matrices is to solve systems of equations. In Chapter 3, you solved systems of linear equations by looking for a point of intersection on a graph or by using substitution or elimination. Matrix methods are generally the simplest way to solve systems that involve more than three equations and three variables. You can use an **inverse matrix** and matrix multiplication to solve a system, or you can use an **augmented matrix** and the **row reduction** method.

When a system is made up of inequalities, the solution usually consists of many points that can be represented by a region in the plane. One important use of systems of linear inequalities is in **linear programming.** In linear programming problems, an equation for a quantity that is to be optimized (maximized or minimized) is evaluated at the vertices of the **feasible region.**

EXERCISES

You will need

A graphing calculator for Exercises **2, 7, 14, 15, 17,** and **21.**

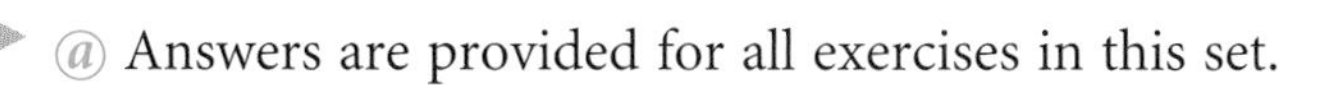

@ Answers are provided for all exercises in this set.

1. Use these matrices to do the arithmetic problems. If a particular operation is impossible, explain why.

$[A] = \begin{bmatrix} 1 & -2 \end{bmatrix}$ $\quad [B] = \begin{bmatrix} -3 & 7 \\ 6 & 4 \end{bmatrix}$ $\quad [C] = \begin{bmatrix} 1 & 0 \\ 5 & 2 \end{bmatrix}$ $\quad [D] = \begin{bmatrix} -3 & 1 & 2 \\ 2 & 3 & -2 \end{bmatrix}$

a. $[A] + [B]$ **b.** $[B] - [C]$ **c.** $4 \cdot [D]$ **d.** $[C][D]$ **e.** $[D][C]$ **f.** $[A][D]$

2. Find the inverse, if it exists, of each matrix.

a. $\begin{bmatrix} 2 & -3 \\ 1 & -4 \end{bmatrix}$ **b.** $\begin{bmatrix} 5 & 2 & -2 \\ 6 & 1 & 0 \\ -2 & 5 & 3 \end{bmatrix}$ **c.** $\begin{bmatrix} -2 & 3 \\ 8 & -12 \end{bmatrix}$ **d.** $\begin{bmatrix} 5 & 2 & -3 \\ 4 & 3 & -1 \\ 7 & -2 & -1 \end{bmatrix}$

3. Solve each system by any method.

a. $\begin{cases} 8x - 5y = -15 \\ 6x + 4y = 43 \end{cases}$

b. $\begin{cases} 5x + 3y - 7z = 3 \\ 10x - 4y + 6z = 5 \\ 15x + y - 8z = -2 \end{cases}$

4. Solve each system by using an inverse matrix.

a. $\begin{cases} 8x - 5y = -15 \\ 6x + 4y = 43 \end{cases}$

b. $\begin{cases} 5x + 3y - 7z = 3 \\ 10x - 4y + 6z = 5 \\ 15x + y - 8z = -2 \end{cases}$

5. Create a system of inequalities that has the feasible region graphed here.

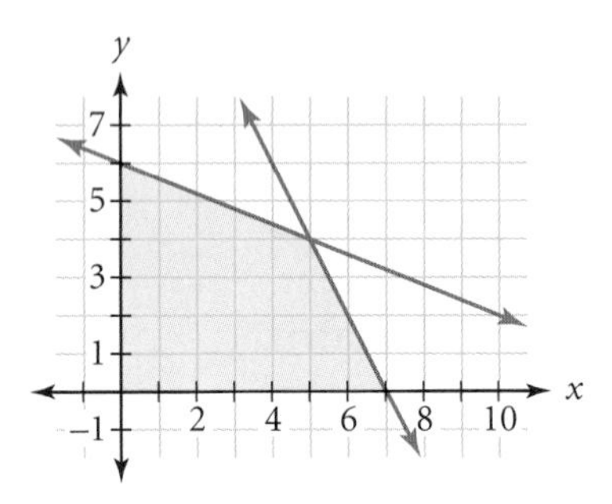

6. Graph the feasible region of each system of inequalities. Find the coordinates of each vertex. Then identify the point that maximizes the given expression.

a. $\begin{cases} 2x + 3y \leq 12 \\ 6x + y \leq 18 \\ x + 2y \geq 4 \\ x \geq 0 \\ y \geq 0 \end{cases}$

maximize: $1.65x + 5.2y$

b. $\begin{cases} x + y \leq 50 \\ 10x + 5y \leq 440 \\ 40x + 60y \leq 2400 \\ x \geq 0 \\ y \geq 0 \end{cases}$

maximize: $6x + 7y$

American artist Joseph Cornell (1903–1972) assembled sculptures from collections of objects, frequently drugstore items. This piece places medicine bottles in a matrix-like arrangement.

Untitled (Grand Hotel Pharmacy) (1947), Joseph Cornell

7. APPLICATION Heather's water heater needs repair. The plumber says it will cost \$300 to fix the unit, which currently costs \$75 per year to operate. Or Heather could buy a new energy-saving water heater for \$500, including installation, and the new heater would save 60% on annual operating costs. How long would it take for the new unit to pay for itself?

8. Interlochen Arts Academy 9th and 10th graders are housed in three dormitories: Picasso, Hemingway, and Mozart. Mozart is an all-female dorm, Hemingway is an all-male dorm, and Picasso is coed. In September, school started with 80 students in Mozart, 60 in Picasso, and 70 in Hemingway. Students are permitted to move from one dorm to another on the first Sunday of each month. This transition graph shows the movements this past year.

a. Write a transition matrix for this situation. List the dorms in the order Mozart, Picasso, Hemingway.

b. What were the populations of the dorms in

i. October?

ii. November?

iii. May?

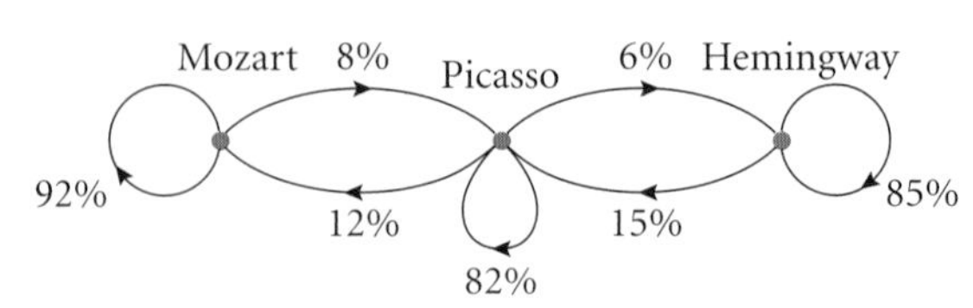

9. Find the values of a, b, and c such that the equation $y = ax^2 + bx + c$ will pass through the points $(2, -15)$, $(-1, 27)$, and $(5, 15)$.

10. Solve each inequality and graph the solution.

a. $|2x - 1| \leq 3$

b. $|x + 2| > 3$

c. $-2|x| < -4$

11. APPLICATION Yolanda, Myriam, and Xavier have a small business producing handmade shawls and blankets. They spin the yarn, dye it, and weave it. A shawl requires 1 h of spinning, 1 h of dyeing, and 1 h of weaving. A blanket needs 2 h of spinning, 1 h of dyeing, and 4 h of weaving. They make a \$16 profit per shawl and a \$20 profit per blanket.

Each person has a full-time job, so they have limited time to spend on the business. Xavier does the spinning on Monday, when he can spend at most 8 h spinning. Yolanda dyes the yarn on Wednesday, when she has at most 6 h. Myriam does all the weaving on Friday and Saturday, when she has at most 14 h available. How many of each item should they make each week to maximize their profit?

MIXED REVIEW

12. Graphs 12a–c were produced by a recursive formula in the form

$$u_0 = a$$
$$u_n = u_{n-1}(1 + p) + d \quad \text{where } n \geq 1$$

For each case, tell if a, p, and d are greater than zero (positive), equal to zero, or less than zero (negative).

a.

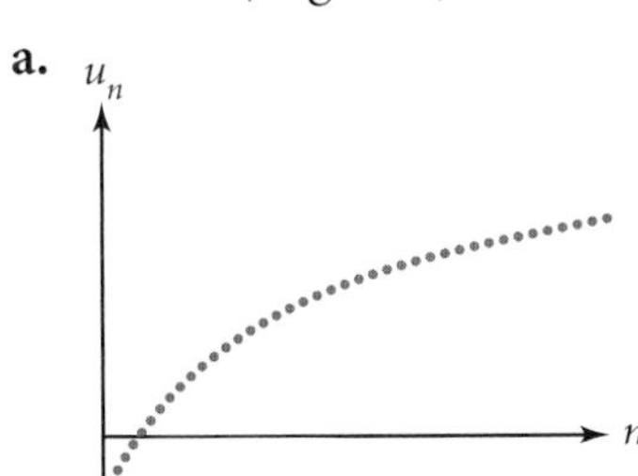

b.

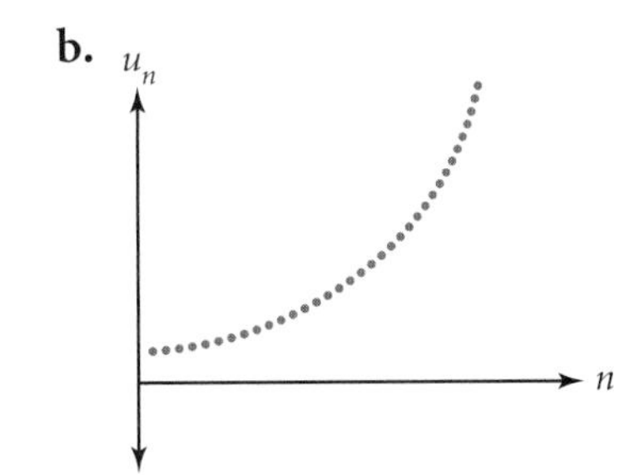

c.

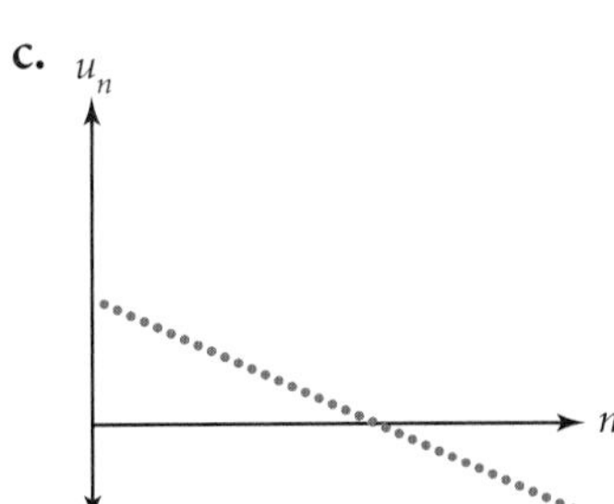

13. The graph of $y = f(x)$ is shown at right.

a. Find $f(-3)$.

b. Find x such that $f(x) = 1$.

c. How can you use the graph to tell whether or not f is a function?

d. What is the domain of f?

e. What is the range of f?

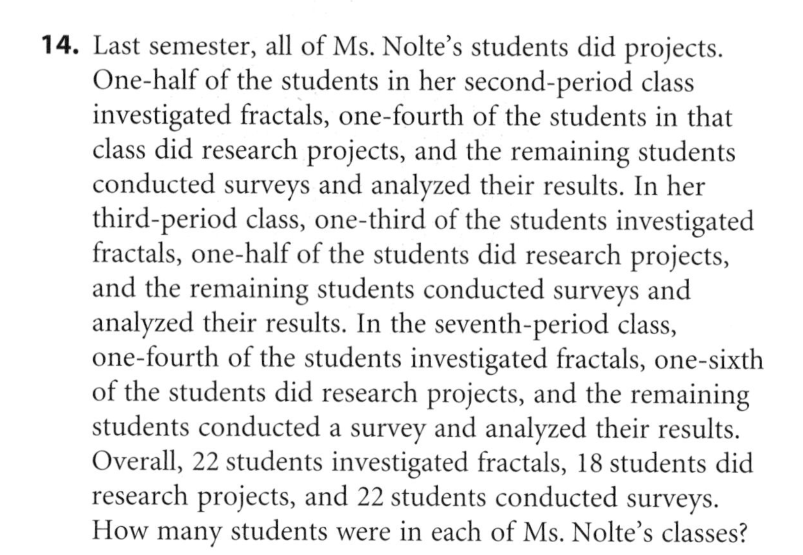

14. Last semester, all of Ms. Nolte's students did projects. One-half of the students in her second-period class investigated fractals, one-fourth of the students in that class did research projects, and the remaining students conducted surveys and analyzed their results. In her third-period class, one-third of the students investigated fractals, one-half of the students did research projects, and the remaining students conducted surveys and analyzed their results. In the seventh-period class, one-fourth of the students investigated fractals, one-sixth of the students did research projects, and the remaining students conducted a survey and analyzed their results. Overall, 22 students investigated fractals, 18 students did research projects, and 22 students conducted surveys. How many students were in each of Ms. Nolte's classes?

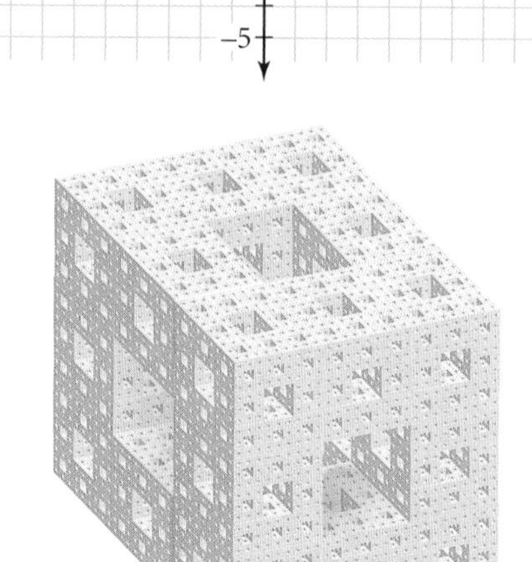

15. APPLICATION Canada's oil production has increased over the last half century. This table gives the production of oil per day for various years.

Canada's Oil Production

Year	1960	1970	1980	1990	1995	2000	2005	2006
Barrels per day (millions)	0.52	1.26	1.44	1.55	1.80	1.98	2.37	2.49

(*The New York Times Almanac 2007*)

a. Define variables and make a scatter plot of the data.

b. Find M_1, M_2, and M_3, and write the equation of the median-median line.

c. Use the median-median line to predict Canada's oil production in 2010.

This oil-drilling ship was frozen in six feet of ice on the Beaufort Sea, Northwest Territories, Canada, in 1980.

16. Solve.

a. $\log 35 + \log 7 = \log x$

b. $\log 500 - \log 25 = \log x$

c. $\log \sqrt{\frac{1}{8}} = x \log 8$

d. $15(9.4)^x = 37{,}000$

e. $\sqrt[3]{x + 6} + 18.6 = 21.6$

f. $\log_6 342 = 2x$

17. APPLICATION Suppose the signal strength in a fiber-optic cable diminishes by 15% every 10 mi.

a. What percentage of the original signal strength is left after a stretch of 10 mi?

b. Create a table of the percentage of signal strength remaining in 10 mi intervals, and make a graph of the sequence.

c. If a phone company plans to boost the signal just before it falls to 1%, how far apart should the company place its booster stations?

Technology CONNECTION

Fiber-optic technology uses light pulses to transmit information from one transmitter to another down fiber lines made of silica (glass). Fiber-optic strands are used in telephone wires, cable television lines, power systems, and other communications. These strands operate on the principle of total internal reflection, which means that the light pulses cannot escape out of the glass tube and instead bounce information from transmitter to transmitter.

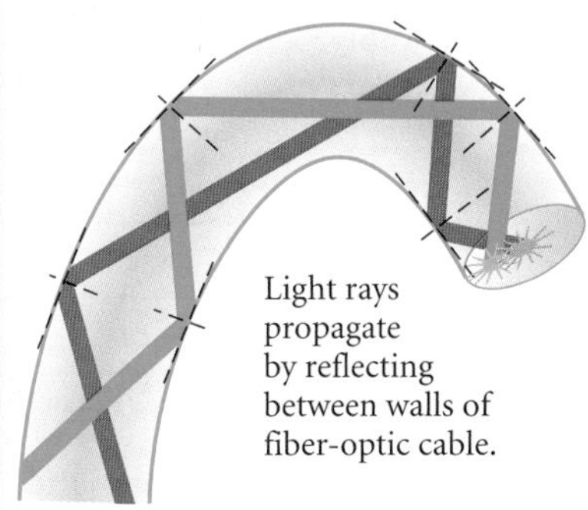

The photo on the left shows strands of fiber-optic cable. The illustration on the right shows how light is reflected along a strand of fiber-optic cable, creating total internal reflection.

18. The graph of an exponential function passes through the points (4, 50) and (6, 25.92).

a. Find an equation for the exponential function.

b. Does the equation model growth or decay? What is the rate of growth or decay?

c. What is the y-intercept?

d. What is the long-run value?

19. This data set gives the weights in kilograms of the crew members participating in the 2007 Boat Race between Oxford University and Cambridge University. (*www.theboatrace.org*)

{90, 100.5, 96, 98.5, 88, 101.5, 100.5, 111, 50, 95, 96, 89, 97, 92, 100.5, 86.5, 95, 55}

a. Make a histogram of these data.

b. Are the data skewed left, skewed right, or symmetric?

c. Identify any outliers.

d. What is the percentile rank of the crew member who weighs 98.5 kg?

Photographed on London's Thames River in 1949, this television crew films the Oxford-Cambridge Boat Race, which was the British Broadcasting Corporation's first broadcast with equipment in motion.

20. Identify each system as consistent and independent (has one solution), inconsistent (has no solution), or consistent and dependent (has infinitely many solutions).

a. $\begin{cases} y = -1.5x + 7 \\ y = -3x + 14 \end{cases}$

b. $\begin{cases} y = \frac{1}{4}(x - 8) + 5 \\ y = 0.25x + 3 \end{cases}$

c. $\begin{cases} 2x + 3y = 4 \\ 1.2x + 1.8y = 2.6 \end{cases}$

d. $\begin{cases} \frac{3}{5}x - \frac{2}{5}y = 3 \\ 0.6x - 0.4y = -3 \end{cases}$

21. APPLICATION This table gives the mean population per U.S. household for various years from 1890 to 2005.

Mean Household Population

Year	1890	1930	1940	1950	1960	1970	1980	1990	2000	2005
Mean population	4.93	4.11	3.67	3.37	3.35	3.14	2.76	2.63	2.62	2.57

(*The New York Times Almanac 2007*)

a. Define variables and find a linear equation that fits these data.

b. Write and solve a problem that you could solve by interpolation with your line of fit.

c. Write and solve a problem that you could solve by extrapolation with your line of fit.

A family poses before their sod house in Nebraska in 1887. Made from strips of earth and plant roots collected with a plow, sod was plastered with clay and ashes in a brick-like pattern. Sod houses were common in the western U.S. plains because timber was scarce.

22. Consider the graph of $y = x^2$. Let matrix $[P]$, which organizes the coordinates of the points shown, represent five points on the parabola.

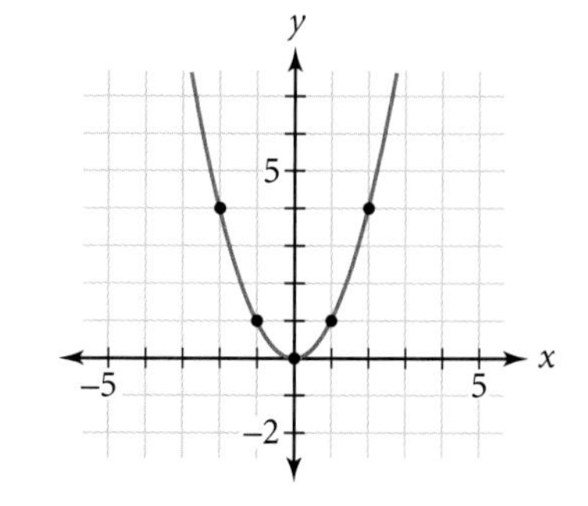

$$[P] = \begin{bmatrix} -2 & -1 & 0 & 1 & 2 \\ 4 & 1 & 0 & 1 & 4 \end{bmatrix}$$

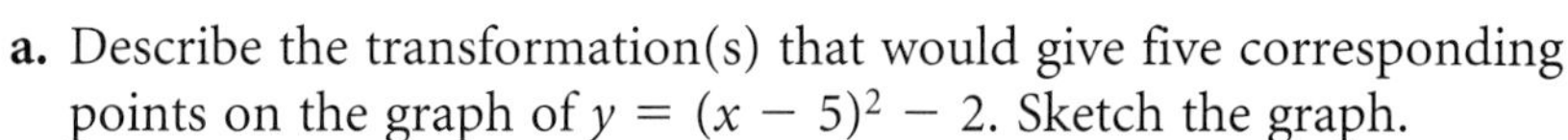

a. Describe the transformation(s) that would give five corresponding points on the graph of $y = (x - 5)^2 - 2$. Sketch the graph.

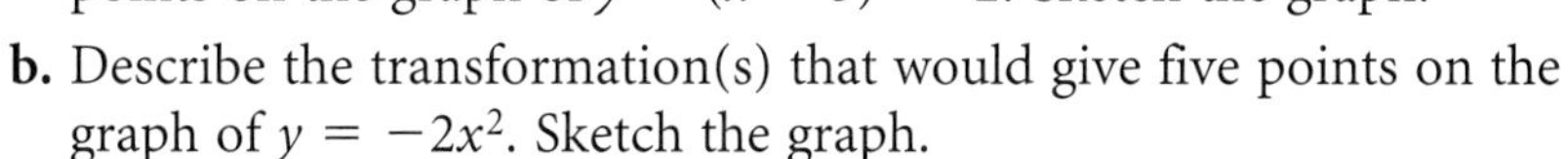

b. Describe the transformation(s) that would give five points on the graph of $y = -2x^2$. Sketch the graph.

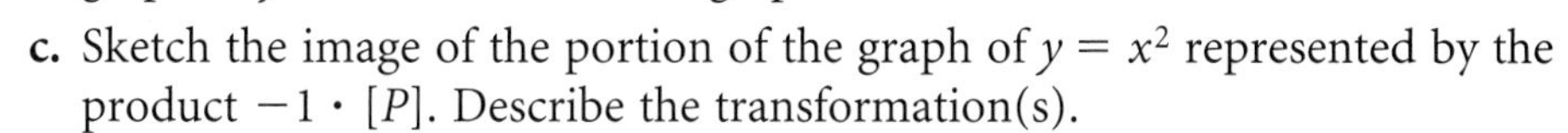

c. Sketch the image of the portion of the graph of $y = x^2$ represented by the product $-1 \cdot [P]$. Describe the transformation(s).

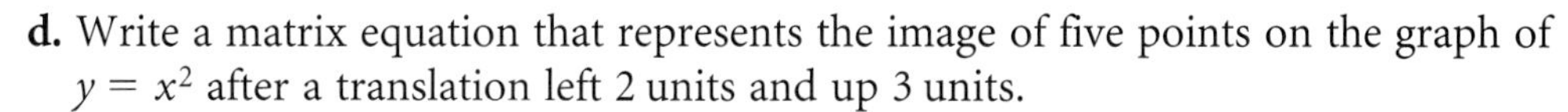

d. Write a matrix equation that represents the image of five points on the graph of $y = x^2$ after a translation left 2 units and up 3 units.

TAKE ANOTHER LOOK

1. You have probably noticed that when a matrix has no inverse, one of the rows is a multiple of another row. For a 2×2 matrix, this also means that the products of the diagonals are equal, or that the difference of these products is 0.

$$\begin{bmatrix} 2 & -3 \\ -6 & 9 \end{bmatrix} \quad 18 \quad 18 \qquad 18 - 18 = 0$$

This difference of the products of the diagonals is called the **determinant** of the matrix. For any 2×2 matrix $\begin{bmatrix} a & b \\ c & d \end{bmatrix}$, the determinant is $ad - bc$.

Make up some 2×2 matrices that have a determinant with value 1. Find the inverses of these matrices. Describe the relationship between the entries of each matrix and its inverse matrix.

Make up some 2×2 matrices that have a determinant with value 2. Find the inverses of these matrices. Describe the relationship between the entries of each matrix and its inverse matrix.

Write a conjecture about the inverse of a matrix and how it relates to the determinant. Test your conjecture with several other 2×2 matrices. Does your conjecture hold true regardless of the value of the determinant?

2. In Lesson 6.4, Exercise 13, you multiplied $\begin{bmatrix} 0 & -1 \\ 1 & 0 \end{bmatrix}$ by a matrix that represented a triangle. The result was a rotation. Try multiplying this matrix by matrices representing other types of polygons. Does it have the same effect? Find a different rotation matrix—one that will rotate a polygon by a different amount or in a different direction. Show that it works for at least three different polygons. Explain why your matrix works.

3. You have learned how to do linear programming problems, but how would you do *nonlinear* programming? Carefully graph this system of inequalities and label all the vertices. Then find the point within the feasible region that maximizes the value of P in the equation $P = (x + 2)^2 + (y - 2)^2$. Explain your solution method and how you know that your answer is correct.

$$\begin{cases} y \leq -|x + 2| + 10 \\ 10^{y+2} \geq x + 8 \\ 3x + 8y \leq 50 \\ -3y + x^2 \geq 9 \end{cases}$$

4. In Lesson 6.4, Exercise 9, you learned how to find the inverse of a square matrix using an augmented matrix. Use this method to find the inverse of the general 2×2 matrix, $\begin{bmatrix} a & b \\ c & d \end{bmatrix}$. Compare your results with your work on Take Another Look 1, where you found the inverse using the determinant. Did you get the same results?

Assessing What You've Learned

PERFORMANCE ASSESSMENT While a classmate, a friend, a family member, or a teacher observes, show how to solve a system of equations using both an inverse matrix and the row reduction method. Explain all of your steps and why each method works.

WRITE IN YOUR JOURNAL Choose one or more of the following questions to answer in your journal.

- What kinds of matrices can be added to or subtracted from one another? What kinds of matrices can you multiply? Is matrix multiplication commutative? Why or why not? What are the identity matrix and the inverse matrix, and what are they used for?
- You have learned five methods to find a solution to a system of linear equations: graphing, substitution, elimination, multiplication by an inverse matrix, and matrix row reduction. Which method do you prefer? Which one is the most challenging to you? What are the advantages and disadvantages of each method?
- You have now studied half the chapters of this book. What mathematical skills in the previous chapters were most crucial to your success in this chapter? Which concepts are your strengths and weaknesses?

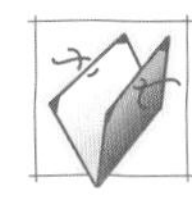

UPDATE YOUR PORTFOLIO Pick a linear programming problem for which you are especially proud of your work, and add it to your portfolio. Describe the steps you followed and how your graph helped you to solve the problem.

CHAPTER

7 Quadratic and Other Polynomial Functions

American artist Sarah Sze (b 1969) creates flowing sculptures, such as this one created for an exhibit at the San Francisco Museum of Modern Art. This piece features a fractured sport utility vehicle whose pieces have been replaced with disposable household items, including foam packing peanuts. Some of the curves in this artwork's cascade resemble the graphs of polynomial functions.

Things Fall Apart: 2001, mixed media installation with vehicle; variable dimensions/San Francisco Museum of Modern Art, Accessions Committee Fund purchase © Sarah Sze

OBJECTIVES

In this chapter you will

- add, subtract, multiply, and divide polynomial expressions
- find polynomial functions that fit a set of data
- study quadratic functions in general form, vertex form, and factored form
- find roots of a quadratic equation from a graph, by factoring, and by using the quadratic formula
- define complex numbers and operations with them
- identify features of the graph of a polynomial function
- use division and other strategies to find roots of higher-degree polynomials

Polynomial Expressions

In Chapter 7, you will learn about polynomial functions and their graphs. In this lesson you'll review some of the terms and properties of polynomial expressions.

Expressions such as $-3.2x$, $\frac{3}{4}x^2$, $4x^3$, and $2x^0$ are **monomials.** More generally, the expression ax^n is a monomial when a is a real number and n is a nonnegative integer. A sum of monomials, like $4x^3 + \frac{3}{4}x^2 - 3.2x + 2$, is a **polynomial.** You can add, subtract, multiply, and divide monomials and polynomials just as you can combine numbers.

EXAMPLE A | Find the sum and difference of $(2x^3 + 6x^2 - 3x + 9)$ and $(4x^3 - 2x^2 - x + 2)$.

▶ ***Solution***

Polynomials can be added by adding similar or **like terms** (monomials with the same base and the same exponent).

$$\begin{aligned}&(2x^3 + 6x^2 - 3x + 9) + (4x^3 - 2x^2 - x + 2)\\ &\quad= (2x^3 + 4x^3) + (6x^2 - 2x^2) + (-3x - x) + (9 + 2)\\ &\quad= 6x^3 + 4x^2 - 4x + 11\end{aligned}$$

You subtract two polynomials by adding the opposite of the polynomial to be subtracted.

$$\begin{aligned}&(2x^3 + 6x^2 - 3x + 9) - (4x^3 - 2x^2 - x + 2)\\ &\quad= (2x^3 + 6x^2 - 3x + 9) + (-4x^3 + 2x^2 + x - 2)\\ &\quad= (2x^3 + -4x^3) + (6x^2 + 2x^2) + (-3x + x) + (9 + -2)\\ &\quad= -2x^3 + 8x^2 - 2x + 7\end{aligned}$$

You may find it convenient to line up the like terms vertically and then add or subtract.

$$\begin{array}{r} 2x^3 + 6x^2 - 3x + 9\\ +\ \underline{4x^3 - 2x^2 - x + 2}\\ 6x^3 + 4x^2 - 4x + 11\end{array} \qquad \begin{array}{r} 2x^3 + 6x^2 - 3x + 9\\ +\ \underline{-4x^3 + 2x^2 + x - 2}\\ -2x^3 + 8x^2 - 2x + 7\end{array}$$

To multiply polynomials, it often helps to think of areas of rectangles. In calculating the area of a rectangle, you multiply length times width. If the sides of the rectangle are polynomial expressions, the area will also be a polynomial expression. The area can be written as the product of the length and width, or as the sum of the areas of the interior regions.

	2	3
1	2	3
2	4	6

Area $= (2 + 3)(1 + 2) = 15$ square units
Area $= 2 + 3 + 4 + 6 = 15$ square units

	x	3
x	x^2	$3x$
2	$2x$	6

Area $= (x + 3)(x + 2)$ square units
Area $= x^2 + 3x + 2x + 6 = x^2 + 5x + 6$ square units

EXAMPLE B

Even though lengths and areas are not negative, you can use rectangle diagrams to represent individual terms and products.

a. Copy each rectangle diagram and fill in the missing values to show the products and quotients of two polynomials.

i.

	$2x$	-3
x		$-3x$
4		

ii.

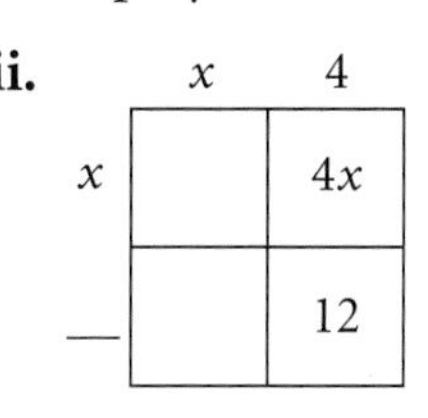

iii.

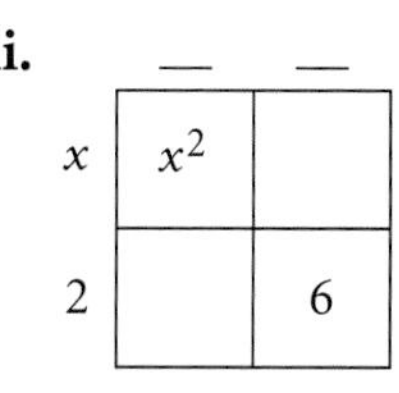

b. Write the two factors and the product for each diagram from part a.

▸ *Solution*

a. Multiply to find missing terms inside the rectangles, and divide to find missing factors.

i.

	$2x$	-3
x	$2x^2$	$-3x$
4	$8x$	-12

ii.

	x	4
x	x^2	$4x$
3	$3x$	12

iii.

	x	3
x	x^2	$3x$
2	$2x$	6

b. The factors may be written in either order, because multiplication of polynomials is commutative.

i. $(2x - 3)(x + 4) = 2x^2 + 5x - 12$ **ii.** $(x + 3)(x + 4) = x^2 + 7x + 12$

iii. $(x + 3)(x + 2) = x^2 + 5x + 6$

EXERCISES

1. Find each sum or difference.

a. $(2x + 7) + (6x - 1)$ @

b. $(4x^3 + 7x^2 - x + 3) + (3x^2 + 2.8x - 0.5)$

c. $(3a^2 - 11a + 4) - (6a^2 - 2a - 3)$ @

d. Subtract $(4x^2 + 5x - 6)$ from $(4x^2 - 3x + 2)$

2. Use a rectangle diagram to find the product of each pair of factors.

a. $(x + 3)$ and $(x + 3)$ @

b. $(4x + 7)$ and $(-x + 4)$

c. $(x + 4)$ and $(x - 4)$

d. $(x + 2)$ and $(2x^2 + 3x - 1)$

3. Copy and complete the rectangle diagrams to find the missing factors.

a.

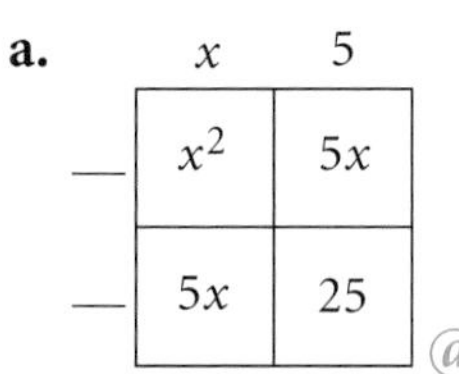

@

b.

	__	__
$2x$	$2x^2$	$8x$
-3	$-3x$	-12

c.

	__	__
x	x^2	$-6x$
6	$6x$	-36

4. Write the product and the factors for each diagram in Exercise 3. Which product might be described as a difference of two squares? Explain. @

LESSON

7.1 Polynomial Degree and Finite Differences

Differences challenge assumptions.

ANNE WILSON SCHAEF

In Chapter 1, you studied arithmetic sequences, which have a common difference between consecutive terms. This common difference is the slope of a line through the graph of the points. So, if you choose x-values along the line that form an arithmetic sequence, the corresponding y-values will also form an arithmetic sequence.

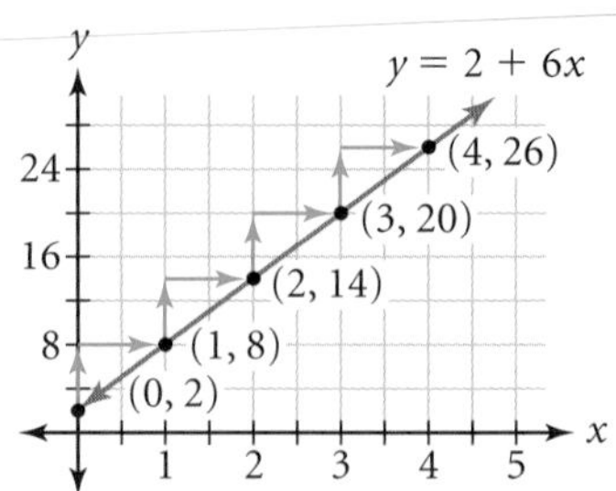

You have also studied several kinds of nonlinear sequences and functions, which do not have a common difference or a constant slope. In this lesson you will discover that even nonlinear sequences sometimes have a special pattern in their differences. These patterns are often described by polynomials.

Definition of a Polynomial

A **polynomial** in one variable is any expression that can be written in the form

$$a_nx^n + a_{n-1}x^{n-1} + \cdots + a_1x^1 + a_0$$

where x is a variable, the exponents are nonnegative integers, the coefficients are real numbers, and $a_n \neq 0$. Each monomial being added to make the polynomial is a **term.**

When a polynomial is set equal to a second variable, such as y, it defines a **polynomial function.** The **degree** of a polynomial or polynomial function is the power of the term that has the greatest exponent. For example, linear functions are 1st-degree polynomial functions because the largest power of x is 1. The polynomial function below has degree 3. If the degrees of the terms of a polynomial decrease from left to right, the polynomial is in **general form.**

Coefficients (if a term is written without a coefficient, the coefficient is 1)

Polynomial Function: $y = 1x^3 + 9x^2 + 26x + 24$

Highest-degree term

Constant term

Polynomial

A polynomial that has only one term is called a **monomial.** A polynomial with two terms is a **binomial,** and a polynomial with three terms is a **trinomial.** Polynomials with more than three terms are usually just called "polynomials."

In modeling linear functions, you have already discovered that for x-values that are evenly spaced, the differences between the corresponding y-values must be the same. With 2nd- and 3rd-degree polynomial functions, the differences between the corresponding y-values are not the same. However, finding the differences between those differences produces an interesting pattern.

1st degree
$y = 3x + 4$

x	y	D_1
2	10	
		3
3	13	
		3
4	16	
		3
5	19	
		3
6	22	
		3
7	25	

2nd degree
$y = 2x^2 - 5x - 7$

x	y	D_1	D_2
3.7	1.88		
		1	
3.8	2.88		0.04
		1.04	
3.9	3.92		0.04
		1.08	
4.0	5.00		0.04
		1.12	
4.1	6.12		0.04
		1.16	
4.2	7.28		

3rd degree
$y = 0.1x^3 - x^2 + 3x - 5$

x	y	D_1	D_2	D_3
−5	−57.5			
		52.5		
0	−5		−50	
		2.5		75
5	−2.5		25	
		27.5		75
10	25		100	
		127.5		75
15	152.5		175	
		302.5		
20	455			

Note that in each case the x-values are spaced equally. You find the first set of differences, D_1, by subtracting each y-value from the one after it. You find the second set of differences, D_2, by finding the differences of consecutive D_1 values in the same way.

For the 2nd-degree polynomial function, the D_2 values are constant, and for the 3rd-degree polynomial function, the D_3 values are constant. What do you think will happen with a 4th- or 5th-degree polynomial function?

Analyzing differences to find a polynomial's degree is called the **finite differences method.** You can use this method to help find the equations of polynomial functions modeling certain sets of data.

Similar types of foods are grouped together at this market in Camden Lock, London, England. Polynomials are often grouped into similar types as well.

EXAMPLE Find a polynomial function that models the relationship between the number of sides and the number of diagonals of a convex polygon. Use the function to find the number of diagonals of a dodecagon (a 12-sided polygon).

▶ Solution You need to create a table of values with evenly spaced x-values. Sketch polygons with increasing numbers of sides. Then draw all of their diagonals.

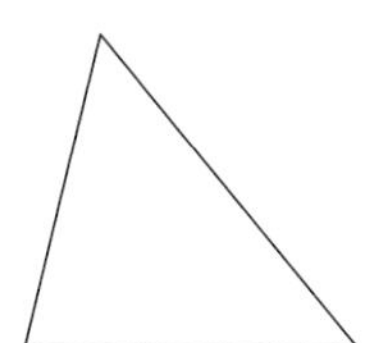 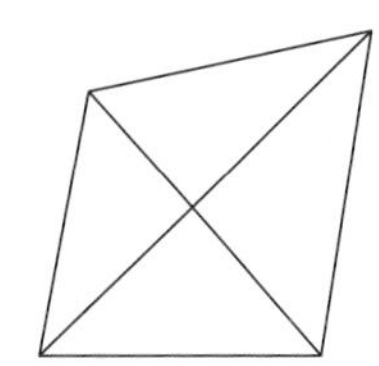 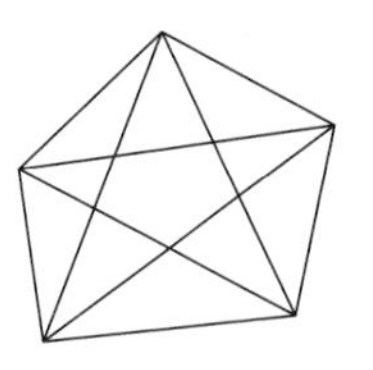 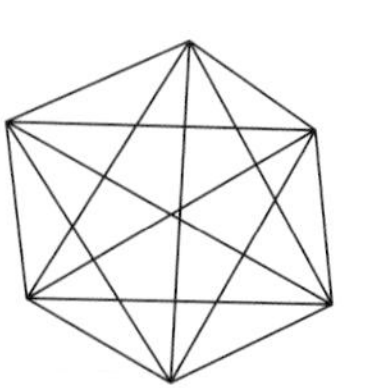

Let x be the number of sides and y be the number of diagonals. You may notice a pattern in the number of diagonals that will help you extend your table beyond the sketches you make. Calculate the finite differences to determine the degree of the polynomial function. (Remember that your x-values must be spaced equally in order to use finite differences.)

Number of sides x	Number of diagonals y	D_1	D_2
3	0		
4	2	2	
5	5	3	1
6	9	4	1
7	14	5	1
8	20	6	1

You can stop finding differences when the values of a set of differences are constant. Because the values of D_2 are constant, you can model the data with a 2nd-degree polynomial function like $y = ax^2 + bx + c$.

To find the values of a, b, and c, you need a system of three equations. Choose three of the points from your table, say (4, 2), (6, 9), and (8, 20), and substitute the coordinates into $y = ax^2 + bx + c$ to create a system of three equations in three variables. Can you see how these three equations were created?

$$\begin{cases} 16a + 4b + c = 2 \\ 36a + 6b + c = 9 \\ 64a + 8b + c = 20 \end{cases}$$

Solve the system to find $a = 0.5$, $b = -1.5$, and $c = 0$. Use these values to write the function $y = 0.5x^2 - 1.5x$. This equation gives the number of diagonals of any polygon as a function of the number of sides.

Now substitute 12 for x to find that a dodecagon has 54 diagonals.

$$y = 0.5x^2 - 1.5x$$
$$y = 0.5(12)^2 - 1.5(12)$$
$$y = 54$$

With exact function values, you can expect the differences to be equal when you find the right degree. But with experimental or statistical data, as in the investigation, you may have to settle for differences that are nearly constant and that do not show an increasing or decreasing pattern when graphed.

Science CONNECTION

Italian mathematician, physicist, and astronomer Galileo Galilei (1564–1642) performed experiments with free-falling objects. He discovered that the speed of a falling object at any moment is proportional to the amount of time it has been falling. In other words, the longer an object falls, the faster it falls. To learn more about Galileo's experiments and discoveries, see the links at www.keymath.com/DAA .

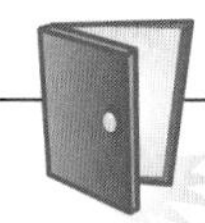

Investigation
Free Fall

You will need

- a motion sensor
- a small pillow or other soft object

What function models the height of an object falling due to the force of gravity? Use a motion sensor to collect data, and analyze the data to find a function.

Procedure Note

1. Set the sensor to collect distance data approximately every 0.05 s for 2 to 5 s. [▶ See **Calculator Note 7A** to learn how to set up your calculator. ◀]
2. Place the sensor on the floor. Hold a small pillow at a height of about 2 m, directly above the sensor.
3. Start the sensor and drop the pillow.

Step 1 Follow the Procedure Note to collect data for a falling object. Let x represent time in seconds, and let y represent height in meters. Select about 10 points from the free-fall portion of your data, with x-values forming an arithmetic sequence. Record this information in a table. Round all heights to the nearest 0.001.

Step 2 Use the finite differences method to find the degree of the polynomial function that models your data. Stop when the differences are nearly constant.

Step 3 Create scatter plots of the original data (*time, height*), then a scatter plot of (*time, first difference*), and finally a scatter plot of (*time, second difference*). [▶ See **Calculator Note 7B** to learn how to calculate finite differences and how to graph them. ◀]

Step 4 Write a description of each graph from Step 3 and what these graphs tell you about the data.

Step 5 Based on your results from using finite differences, what is the degree of the polynomial function that models free fall? Write the general form of this polynomial function.

Step 6 Follow the example on page 380 to write a system of three equations in three variables for your data. Solve your system to find an equation to model the position of a free-falling object dropped from a height of 2 m.

When using experimental data, you must choose your points carefully. When you collect data, as you did in the investigation, your equation will most likely not fit all of the data points exactly due to some errors in measurement and rounding. To minimize the effects of these errors, choose representative points that are not close together, just as you did when fitting a line to data.

History CONNECTION

The method of finite differences was used by the Chinese astronomer Li Shun-Fêng in the 7th century to find a quadratic equation to model the Sun's apparent motion across the sky as a function of time. The Persian astronomer Jamshid Masud al-Kashi, who worked at the Samarkand Observatory in the 15th century, also used the finite differences method when calculating the celestial longitudes of planets.

Scientists continued to develop and use the finite differences method in 17th- and 18th-century Europe. In the 19th century, early calculating machines were programmed to calculate differences and were called difference engines.

English mathematician and inventor Charles Babbage (1792–1871) designed the first difference engine in the early 1820s, and completed his Difference Engine No. 1, shown here, in 1832.

Note that some functions, such as logarithmic and exponential functions, cannot be expressed as polynomials. The finite differences method will not produce a set of constant differences for functions other than polynomial functions.

EXERCISES

You will need

A graphing calculator for Exercises **6** and **8**.

Practice Your Skills

1. Identify the degree of each polynomial.

a. $x^3 + 9x^2 + 26x + 24$

b. $7x^2 - 5x$

c. $x^7 + 3x^6 - 5x^5 + 24x^4 + 17x^3 - 6x^2 + 2x + 40$

d. $16 - 5x^2 + 9x^5 + 36x^3 + 44x$ ⓐ

2. Determine which of these expressions are polynomials. For each polynomial, state its degree and write it in general form. If it is not a polynomial, explain why not.

a. $-3 + 4x - 3.5x^2 + \frac{5}{9}x^3$ ⓐ

b. $5p^4 + 3.5p - \frac{4}{p^2} + 16$

c. $4\sqrt{x^3} + 12$ ⓐ

d. $x^2\sqrt{15} - x - 4^{-2}$

3. For each data set, decide whether the last column of differences shows constant values. If it does not, calculate the next set of finite differences.

a.

x	y
2	4.4
3	6.6
4	9.2
5	11.0
6	10.8
7	7.4

b. ⓐ

x	y	D_1
3.7	−8.449	
3.8	−8.706	−0.257
3.9	−8.956	−0.250
4.0	−9.200	−0.244
4.1	−9.436	−0.236
4.2	−9.662	−0.226

c.

x	y	D_1	D_2
−5	−101		
0	−6	95	
5	−11	−5	−100
10	34	45	50
15	279	245	200
20	874	595	350

4. Find the degree of the polynomial function that models these data.

x	0	2	4	6	8	10	12
y	12	−4	−164	−612	−1492	−2948	−5124

Reason and Apply

5. Consider the data at right.

n	1	2	3	4	5	6
s	1	3	6	10	15	21

a. Calculate finite differences to find the degree of the polynomial function that models these data.

b. Describe how the degree of this polynomial function is related to the finite differences you calculated.

c. What is the minimum number of data points required to determine the degree of this polynomial function? Why? ⓐ

d. Find the polynomial function that models these data and use it to find s when n is 12.

e. The values in the s row are called triangular numbers. Why do you think they are called triangular? ⓗ

6. You can use blocks to build pyramids such as these. All of the pyramids are solid with no empty space inside.

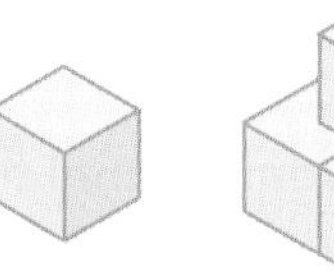

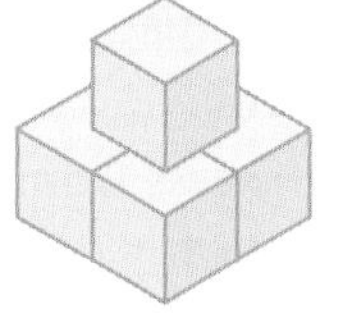

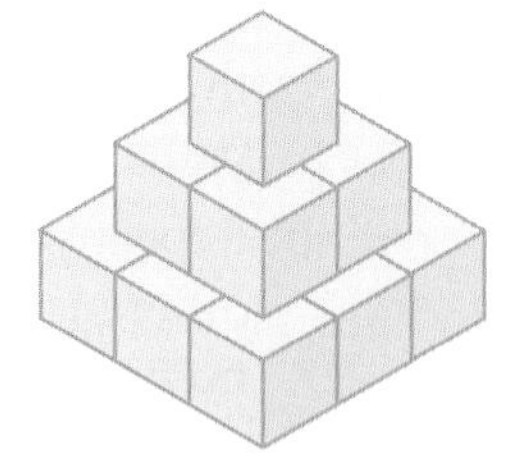

a. Create a table to record the number of layers, x, in each pyramid and the total number of blocks, y, needed to build it. You may need to build or sketch a few more pyramids, or look for patterns in the table.

b. Use finite differences to find a polynomial function that models these data.

c. Find the number of blocks needed to build a pyramid with eight layers.

d. Find the number of layers in a pyramid built with 650 blocks.

7. The data in these tables represent the heights of two objects at different times during free fall.

i.

Time (s) t	0	1	2	3	4	5	6
Height (m) h	80	95.1	100.4	95.9	81.6	57.5	23.6

ii.

Time (s) t	0	1	2	3	4	5	6
Height (m) h	4	63.1	112.4	151.9	181.6	201.5	211.6

a. Calculate the finite differences for each table.

b. What is the degree of the polynomial function that you would use to model each set of data?

c. Write a polynomial function to model each set of data. Check your answer by substituting one of the data points into your function.

8. Andy has measured his height every three months since he was $9\frac{1}{2}$ years old. Below are his measurements in meters.

Age (yr)	Height (m)	Age (yr)	Height (m)
9.5	1.14	11.5	1.35
9.75	1.21	11.75	1.35
10	1.27	12	1.36
10.25	1.31	12.25	1.37
10.5	1.33	12.5	1.39
10.75	1.34	12.75	1.42
11	1.35	13	1.47
11.25	1.35	13.25	1.54

a. Find the first differences for Andy's heights and make a scatter plot of the differences as a function of age. Remember to shorten the list of ages to match D_1. Describe the pattern you see.

b. Repeat your process from 8a until the differences are nearly constant and show no pattern. ⓗ

c. What type of function would you use to model these data? Why?

d. Define variables and find a model. For what domain values do you think this model is reasonable?

9. APPLICATION In an atom, electrons spin rapidly around a nucleus. An electron can occupy only specific energy levels, and each energy level can hold only a certain number of electrons. This table gives the greatest number of electrons that can theoretically be in any one level (although there are no known atoms that actually have more than 32 electrons in any one level).

Energy level	1	2	3	4	5	6	7
Maximum number of electrons	2	8	18	32	50	72	98

Is it possible to find a polynomial function that expresses the relationship between the energy level and the maximum number of electrons? If so, find the function. If not, explain why not. ⓐ

Science CONNECTION

The electrons in an atom exist in various energy levels. When an electron moves from a lower energy level to a higher energy level, the atom absorbs energy. When an electron moves from a higher to a lower energy level, energy is released (often as light). This is the principle behind neon lights. The electricity running through a tube of neon gas makes the electrons in the neon atoms jump to higher energy levels. When they drop back to their original level, they give off light.

A visitor admires a neon art display at the Yerba Buena Center for the Arts in San Francisco, California.

Review

10. Sketch a graph of each function without using your calculator.

a. $y = (x - 2)^2$

b. $y = x^2 - 4$

c. $y = (x + 4)^2 + 1$

11. Solve.

a. $12x - 17 = 13$

b. $2(x - 1)^2 + 3 = 11$

c. $3(5^x) = 48$ ⓐ

12. The rectangle diagram at right represents the product $(x + 2)(x + 3)$, which you can write as the trinomial $x^2 + 5x + 6$.

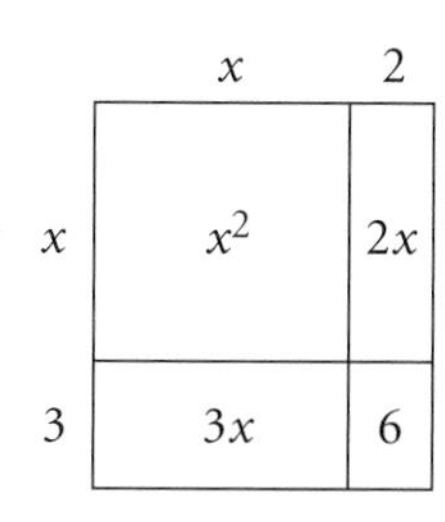

a. Draw a rectangle diagram that represents the product $(2x + 3)(3x + 1)$.

b. Express the area in 12a as a polynomial in general form.

c. Draw a rectangle diagram whose area represents the polynomial $x^2 + 8x + 15$. ⓗ

d. Express the area in 12c as a product of two binomials.

13. Write a system of inequalities that describes the feasible region graphed at right. ⓐ

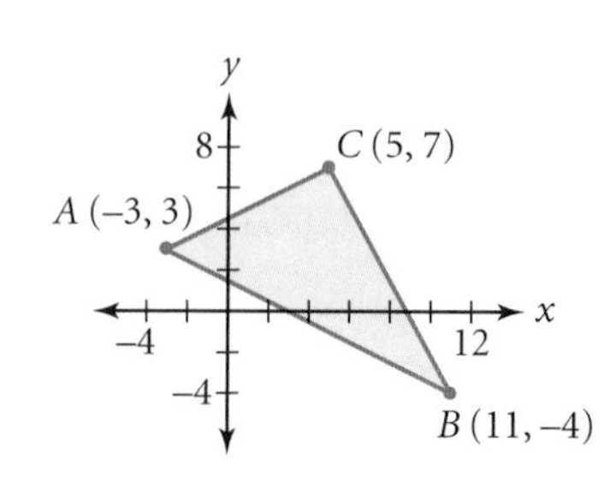

14. Find the product $(x + 3)(x + 4)(x + 2)$.

LESSON

7.2 Equivalent Quadratic Forms

I'm very well acquainted, too,
with matters mathematical.
I understand equations both the
simple and quadratical.

WILLIAM S. GILBERT AND ARTHUR SULLIVAN

In Lesson 7.1, you were introduced to polynomial functions, including 2nd-degree polynomial functions, or **quadratic functions.** The **general form** of a quadratic function is $y = ax^2 + bx + c$. In this lesson you will work with two additional, equivalent forms of quadratic functions.

This fountain near the Centre Pompidou in Paris, France, contains 16 animated surreal sculptures inspired by the music of Russian-American composer Igor Stravinsky (1882–1971). It was designed by artists Jean Tinguely (1925–1991) and Niki de Saint-Phalle (1930–2002). The arc formed by spouting water can be described with a quadratic equation.

Recall from Chapter 4 that every quadratic function can be considered as a transformation of the graph of the parent function $y = x^2$. A quadratic function in the form $\frac{y-k}{b} = \left(\frac{x-h}{a}\right)^2$ or $y = b\left(\frac{x-h}{a}\right)^2 + k$ identifies the location of the vertex, (h, k), and the horizontal and vertical scale factors, a and b.

EXAMPLE A

Find the horizontal and vertical scale factors of the parabola at right with vertex $(4, -2)$ and write its equation. Then rewrite the equation with a single scale factor.

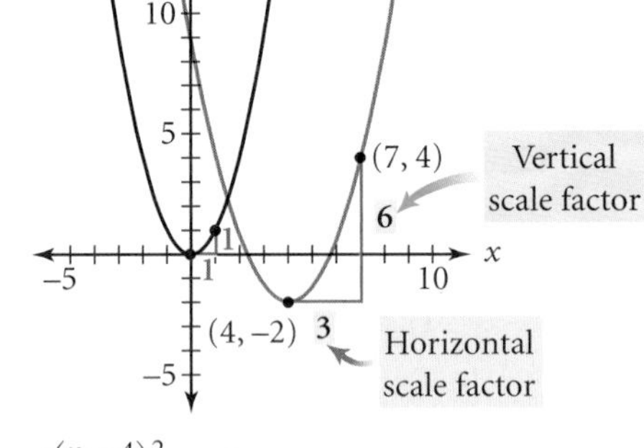

▶ Solution

If you consider the point $(7, 4)$ to be the image of the point $(1, 1)$ on the graph of $y = x^2$, the horizontal scale factor is 3 and the vertical scale factor is 6. So the quadratic function is $\frac{y+2}{6} = \left(\frac{x-4}{3}\right)^2$, or $y = 6\left(\frac{x-4}{3}\right)^2 - 2$.

To rewrite the equation with a single scale factor, first move the denominator out of the parentheses, which gives the equation $y = \frac{6}{3^2}(x-4)^2 - 2$. This equation is equivalent to $y = \frac{6}{9}(x-4)^2 - 2$, or $y = \frac{2}{3}(x-4)^2 - 2$. This form combines the original horizontal and vertical scale factors into a single vertical scale factor, $\frac{2}{3}$.

In general, the combined scale factor is $\frac{b}{a^2}$, where a and b are the horizontal and vertical scale factors. The coefficient a is often used to represent the combined scale factor. This new form, $y = a(x-h)^2 + k$, is called the **vertex form** of a quadratic function because it identifies the vertex, (h, k), and a single vertical scale factor, a. If you know the vertex of a parabola and one other point, then you can write the quadratic function in vertex form.

Now consider these parabolas. The x-intercepts are marked.

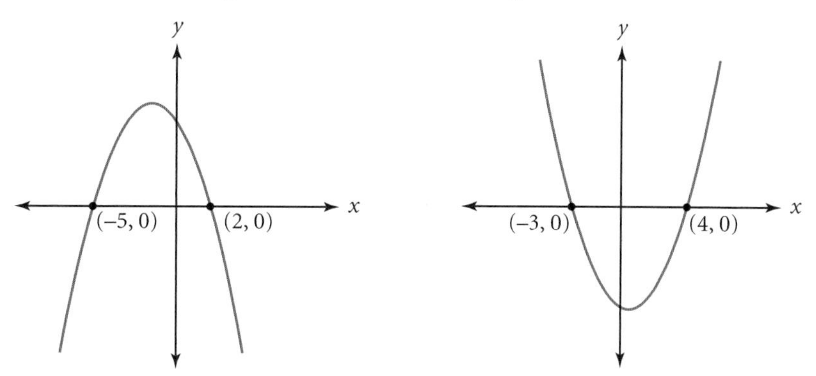

The y-coordinate of any point along the x-axis is 0, so the y-coordinate is 0 at each x-intercept. For this reason, the x-intercepts of the graph of a function are called the **zeros** of the function. You will use this information and the **zero-product property** to find the zeros of a function without graphing.

Zero-Product Property

For all real numbers a and b, if $ab = 0$, then $a = 0$, or $b = 0$, or $a = 0$ and $b = 0$.

To understand the zero-product property, think of numbers whose product is zero. Whatever numbers you think of will have this characteristic: *At least one of the factors must be zero.* Before moving on, think about numbers that satisfy each equation below.

$\underline{?} \cdot 16.2 = 0$

$3(\underline{?} - 4)(\underline{?} - 9) = 0$

These Mayan representations of zero were used as placeholders, as in "100," rather than to symbolize "nothingness."

EXAMPLE B | Find the zeros of the function $y = -1.4(x - 5.6)(x + 3.1)$.

▶ Solution

The zeros will be the x-values that make y equal 0. First, set the function equal to zero.

$$0 = -1.4(x - 5.6)(x + 3.1)$$

Because the product of three factors equals zero, the zero-product property tells you that at least one of the factors must equal zero.

$-1.4 = 0$	or	$x - 5.6 = 0$	or	$x + 3.1 = 0$
not possible		$x = 5.6$		$x = -3.1$

So the solutions, or **roots**, of the equation $0 = -1.4(x - 5.6)(x + 3.1)$ are $x = 5.6$ or $x = -3.1$. That means the zeros of the function $y = -1.4(x - 5.6)(x + 3.1)$ are $x = 5.6$ and $x = -3.1$.

Use your graphing calculator to check your work. You should find that the x-intercepts of the graph of $y = -1.4(x - 5.6)(x + 3.1)$ are 5.6 and -3.1.

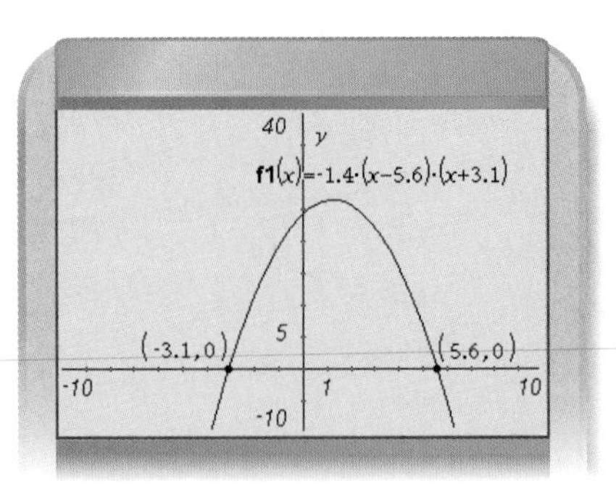

If you know the x-intercepts of a parabola, then you can write the quadratic function in **factored form**, $y = a(x - r_1)(x - r_2)$. This form identifies the locations of the x-intercepts, r_1 and r_2, and a vertical scale factor, a.

EXAMPLE C

Consider the parabola at right.

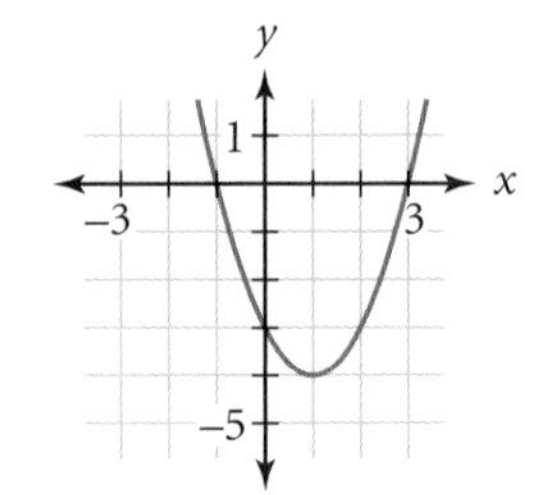

a. Write an equation of the parabola in vertex form.

b. Write an equation of the parabola in factored form.

c. Show that both equations are equivalent by converting them to general form.

▶ Solution

The vertex is $(1, -4)$. If you consider the point $(2, -3)$ to be the image of the point $(1, 1)$ on the graph of $y = x^2$, then the vertical and horizontal scale factors are both 1. So the single vertical scale factor is $a = \frac{1}{1^2} = 1$.

a. The vertex form is $y = (x - 1)^2 - 4$.

b. The x-intercepts are -1 and 3, so the factored form is $y = a(x + 1)(x - 3)$. To verify that the scale factor you found earlier is correct, you can solve for it analytically. Substitute the coordinates of another point on the parabola such as $(2, -3)$ into the equation for x and y.

$$-3 = a(2 + 1)(2 - 3)$$

$$a = \frac{-3}{(3)(-1)} = 1$$

The value of a is 1, so the factored form is $y = (x + 1)(x - 3)$.

c. To convert to general form, multiply the binomials and then combine like terms. The use of rectangle diagrams may help you multiply the binomials.

$$y = (x - 1)^2 - 4$$

$$y = (x - 1)(x - 1) - 4$$

	x	-1
x	x^2	$-x$
-1	$-x$	1

$$y = (x^2 - x - x + 1) - 4$$

$$y = x^2 - 2x - 3$$

$$y = (x + 1)(x - 3)$$

	x	1
x	x^2	x
-3	$-3x$	-3

$$y = x^2 + x - 3x - 3$$

$$y = x^2 - 2x - 3$$

The vertex form and the factored form are equivalent because they are both equivalent to the same general form.

You now know three different forms of a quadratic function.

Three Forms of a Quadratic Function

General form	$y = ax^2 + bx + c$
Vertex form	$y = a(x - h)^2 + k$
Factored form	$y = a(x - r_1)(x - r_2)$

The investigation will give you practice in using the three forms with real data. You'll find that the form you use guides which features of the data you focus on. Conversely, if you know only a few features of the data, you may need to focus on a particular form of the function.

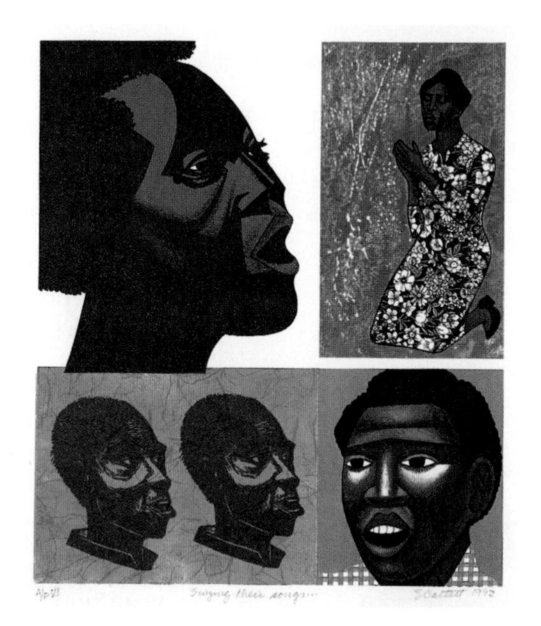

This painting by Elizabeth Catlett, which depicts people singing songs in different forms, was inspired by a poem titled "For My People" by American writer Margaret Walker Alexander (1915–1998).

Elizabeth Catlett (American, b 1915), *Singing Their Songs* (1992). Lithograph on paper (a.p.#6) 15-3/4 x 13-3/4 in. / National Museum of Women in the Arts, purchased with funds donated in memory of Florence Davis by her family, friends, and the NMWA Women's Committee

Investigation
Rolling Along

You will need

- a motion sensor
- an empty coffee can
- a long table

Procedure Note

Prop up one end of the table slightly. Place the motion sensor at the low end of the table and aim it toward the high end. With tape or chalk, mark a starting line 0.5 m from the sensor on the table.

Step 1 Practice rolling the can up the table directly in front of the motion sensor. Start the can behind the starting line. Give the can a gentle push so that it rolls up the table on its own momentum, stops near the end of the table, and then rolls back. Stop the can after it crosses the line and before it hits the motion sensor.

Step 2	Set up your calculator to collect data for 6 seconds. [▶ See **Calculator Note 7C.** ◀] When the sensor begins, roll the can up the table.
Step 3	The data collected by the sensor will have the form (*time, distance*). Adjust for the position of the starting line by subtracting 0.5 from each value in the distance list.
Step 4	Let x represent time in seconds, and let y represent distance from the line in meters. Draw a graph of your data. What shape is the graph of the data points? What type of function would model the data? Use finite differences to justify your answer.
Step 5	Mark the vertex and another point on your graph. Approximate the coordinates of these points and use them to write the equation of a quadratic model in vertex form.
Step 6	From your data, find the distance of the can at 1, 3, and 5 seconds. Use these three data points to find a quadratic model in general form.
Step 7	Mark the x-intercepts on your graph. Approximate the values of these x-intercepts. Use the zeros and the value of a from Step 5 to find a quadratic model in factored form.
Step 8	Verify by graphing that the three equations in Steps 5, 6, and 7 are equivalent, or nearly so. Write a few sentences explaining when you would use each of the three forms to find a quadratic model to fit parabolic data.

keymath.com/DAA

You can find a model for data in different ways, depending on the information you have. Conversely, different forms of the same equation give you different kinds of information. Being able to convert one form into another allows you to compare equations written in different forms. In the exercises you will convert both the vertex form and the factored form into the general form. In later lessons you will learn other conversions.

EXERCISES

You will need

A graphing calculator for Exercises **2–6** and **10–12**.

Practice Your Skills

For the exercises in this lesson, you may find it helpful to show a background grid on your calculator.

1. Identify each quadratic function as being in general form, vertex form, factored form, or none of these forms.

a. $y = -4.9(x + 4.5)^2 + 2$
b. $y = -4.9(x - 4.5)(x - 0.5)$
c. $y = -3.2(x + 4.5)^2$ ⓐ
d. $y = 2.5(x + 1.25)(x - 1.25) + 4$ ⓐ
e. $y = 2x(3 + x)$ ⓐ
f. $y = 2x^2 - 4.2x - 10$

2. Each quadratic function below is written in vertex form. What are the coordinates of each vertex? Graph each equation to check your answers.

a. $y = (x - 2)^2 + 3$
b. $y = 0.5(x + 4)^2 - 2$ ⓐ
c. $y = 4 - 2(x - 5)^2$

3. Each quadratic function below is written in factored form. What are the zeros of each function? Graph each equation to check your answers.

a. $y = (x + 1)(x - 2)$ @ **b.** $y = 0.5(x - 2)(x + 3)$ **c.** $y = -2(x - 2)(x - 5)$

4. Convert each function to general form. Graph both forms to check that the equations are equivalent.

a. $y = (x - 2)^2 + 3$ **b.** $y = 0.5(x + 4)^2 - 2$ **c.** $y = 4 - 2(x - 5)^2$ @

5. Convert each function to general form. Graph both forms to check that the equations are equivalent.

a. $y = (x + 1)(x - 2)$ **b.** $y = 0.5(x - 2)(x + 3)$ **c.** $y = -2(x - 2)(x - 5)$

When you see an image in a different form, your attention is drawn to different features.

Reason and Apply

6. As you learned in Chapter 4, the graphs of all quadratic functions have a line of symmetry that contains the vertex and divides the parabola into mirror-image halves. Consider this table of values generated by a quadratic function.

x	y
1.5	−8
2.5	7
3.5	16
4.5	19
5.5	16
6.5	7
7.5	−8

a. What is the line of symmetry for the graph of this quadratic function?

b. The vertex of a parabola represents either the **maximum** or **minimum** value of the quadratic function. Name the vertex of this function and determine whether it is a maximum or minimum.

c. Use the table of values to write the quadratic function in vertex form. @

7. Write each function in general form.

a. $y = 4 - 0.5(x + h)^2$ **b.** $y = a(x - 4)^2$ @

c. $y = a(x - h)^2 + k$ **d.** $y = -0.5(x + r)(x + 4)$ @

e. $y = a(x - 4)(x + 2)$ **f.** $y = a(x - r)(x - s)$ @

8. At right is the graph of the quadratic function that passes through $(-2.4, 0)$, $(0.8, 0)$, and $(1.2, -2.592)$.

a. Use the x-intercepts to write the quadratic function in factored form. For now, leave the scale factor as a.

b. Substitute the coordinates of $(1.2, -2.592)$ into your function from 8a, and solve for a. Write the complete quadratic function in factored form.

c. The line of symmetry for the graph of this quadratic function passes through the vertex and the point on the x-axis halfway between the two x-intercepts. What is the x-coordinate of the vertex? What is the y-coordinate?

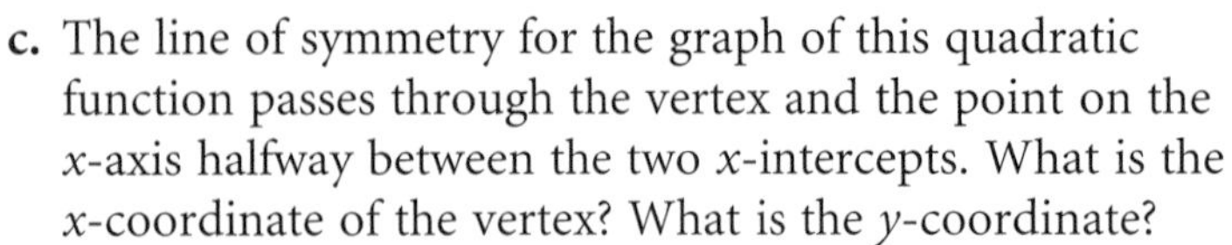

d. Write this quadratic function in vertex form.

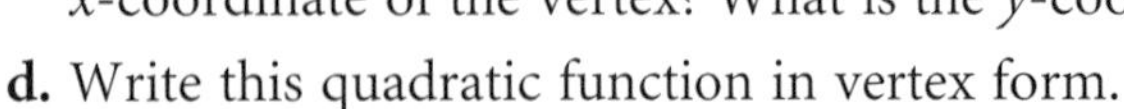

9. Write the factored form for each polynomial function. (h)

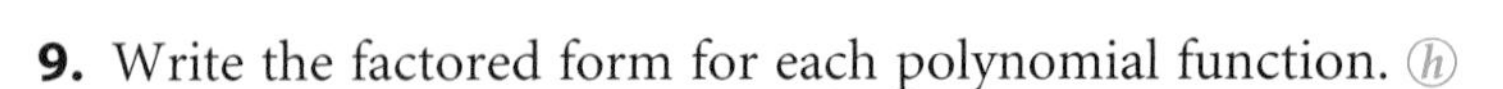

a.

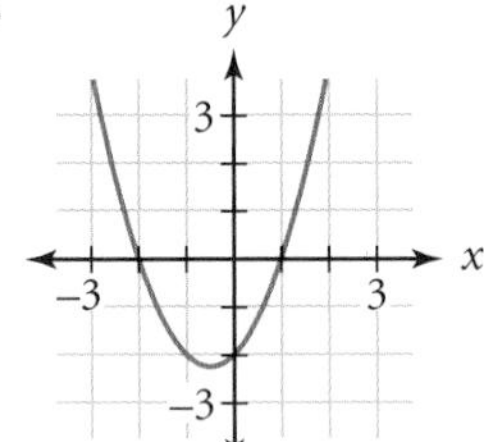

b. (a)

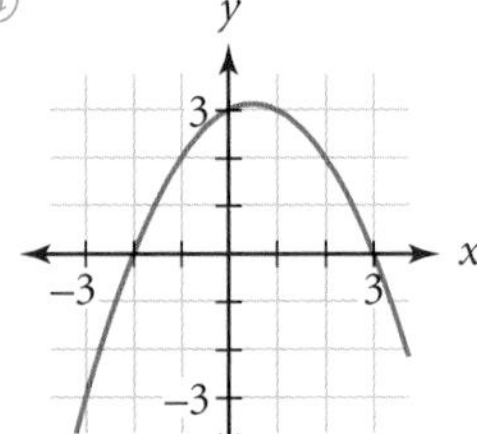

c.

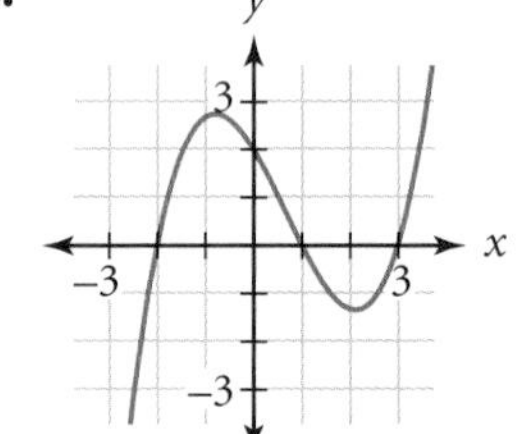

10. APPLICATION A local outlet store charges \$2.00 for a pack of four AA batteries. On an average day, 200 packs are sold. A survey indicates that the number sold will decrease by 5 packs per day for each \$0.10 increase in price.

Selling price (\$)	2.00	2.10	2.20	2.30	2.40
Number sold	200	?	?	?	?
Revenue (\$)	400	?	?	?	?

a. Complete the table based on the results of the survey.

b. Calculate the first and second differences for the revenue. (a)

c. Let x represent the selling price in dollars, and let y represent the revenue in dollars. Write a function that describes the relationship between the revenue and the selling price. (h)

d. Graph your function and find the maximum revenue. What selling price provides maximum revenue?

11. APPLICATION Delores has 80 m of fence to surround an area where she is going to plant a vegetable garden. She wants to enclose the largest possible rectangular area.

a. Copy and complete this table.

Width (m)	5	10	15	20	25
Length (m)	?	?	?	?	?
Area (m^2)	?	?	?	?	?

b. Let x represent the width in meters, and let y represent the area in square meters. Write a function that describes the relationship between the area and the width of the garden.

c. Which width provides the largest possible area? What is that area?

d. Which widths result in an area of 0 m^2?

12. APPLICATION Photosynthesis is the process by which plants use energy from the sun, together with CO_2 (carbon dioxide) and water, to make their own food and produce oxygen. Various factors affect the rate of photosynthesis, such as light intensity, light wavelength, CO_2 concentration, and temperature. Below is a graph of how temperature relates to the rate of photosynthesis for a particular plant. (All other factors are assumed to be held constant.)

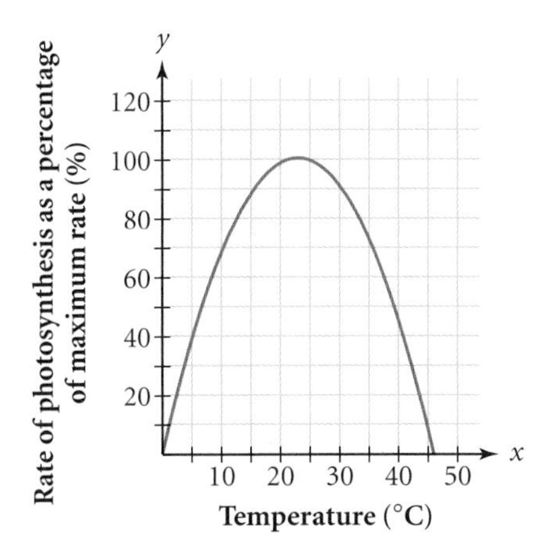

When chlorophyll (a green pigment in plants) breaks down in the winter, leaves change color. When they appear red or purple, it is because glucose trapped in the leaves produces a pigment when exposed to sunlight and cool nights.

a. Describe the general shape of the graph. What does the shape of the graph mean in the context of photosynthesis?

b. Approximate the optimum temperature for photosynthesis in this plant and the corresponding rate of photosynthesis. @

c. Temperature has to be kept within a certain range for photosynthesis to occur. If it gets too hot, then the enzymes in chlorophyll are killed and photosynthesis stops. If the temperature is too cold, then the enzymes stop working. At approximately what temperatures does the rate of photosynthesis fall to zero? @

d. Write a function in at least two forms that will produce this graph.

Review

13. Use a rectangle diagram to find each product.

a. $3x(4x - 5)$ b. $(x + 3)(x - 5)$ c. $(x + 7)(x - 7)$ d. $(3x - 1)^2$ @

14. Recall that the distributive property allows you to distribute a factor through parentheses. The factor that is distributed doesn't have to be a monomial. Here's an example with a binomial.

$$(x+1)(x-3)$$
$$x(x-3)+1(x-3)$$
$$x^2-3x+x-3$$
$$x^2-2x-3$$

Use the distributive property to find each product in Exercise 13. (h)

15. You can also use a rectangle diagram to help you factor some trinomials, such as $x^2 - 7x + 12$.

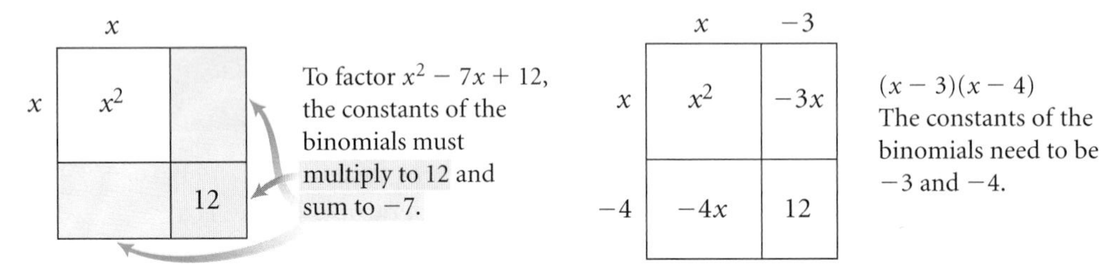

Use rectangle diagrams to help you factor these trinomials.

a. $x^2 + 3x - 10$ **b.** $x^2 + 8x + 16$ **c.** $x^2 - 25$

d. The polynomial $x^2 + 8x + 16$ is called a **perfect-square trinomial.** Use the factored form to help explain why. Give two more examples of perfect-square trinomials. (h)

e. The polynomial $x^2 - 25$ is called a **difference of two squares.** Give two more examples of differences of two squares.

16. Use the function $f(x) = 3x^3 - 5x^2 + x - 6$ to find these values.

a. $f(2)$ **b.** $f(-1)$ **c.** $f(0)$ **d.** $f\left(\frac{1}{2}\right)$ **e.** $f\left(-\frac{4}{3}\right)$

17. Tell whether each statement is always, sometimes, or never true. Give a counterexample if the statement is never true. If the statement is sometimes true, give an example of a true case and an example of a false case.

a. $x^{-3} = \frac{1}{x^3}$ **b.** $x^3 > x^2$ **c.** $2^x < 2^{x+1}$

IMPROVING YOUR REASONING SKILLS

Sums and Differences

Use the method of finite differences to find a formula for the sum of the first n terms in the arithmetic sequence 1, 2, 3, 4, ... In other words, find a formula for the sum

$$1 + 2 + 3 + \cdots + u_n$$

Then find a formula for the first n terms of the sequence 1, 3, 5, 7, ... How about 1, 4, 7, 10, . . . ? Look for patterns, and see if you can write a formula that will determine the sum of the first n terms of any arithmetic sequence with first term 1 and common difference d.

LESSON

7.3

Completing the Square

Everything should be made as simple as possible, but not simpler.

ALBERT EINSTEIN

The graph of $y = -4.9(x - 0.86)^2 + 0.6$ at right models one bounce of a ball, where x is time in seconds and y is height in meters. The maximum height of this ball occurs at the vertex (0.86, 0.6), which means that after 0.86 s the ball reaches its maximum height of 0.6 m.

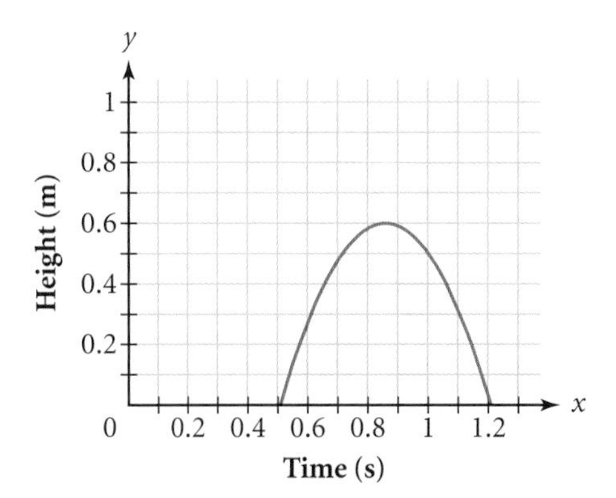

Finding the **maximum** or **minimum** value of a quadratic function is often necessary to answer questions about data. Finding the vertex is straightforward when you are given an equation in vertex form and sometimes when you are using a graph. However, you often have to estimate values on a graph. In this lesson you will learn a procedure called **completing the square** to convert a quadratic equation from general form to vertex form accurately.

An object that rises and falls under the influence of gravity is called a projectile. You can use quadratic functions to model **projectile motion,** or the height of the object as a function of time.

The height of a projectile depends on three things: the height from which it is thrown, the upward velocity with which it is thrown, and the effect of gravity pulling downward on the object. So the polynomial function that describes projectile motion has three terms. The leading coefficient of the polynomial is based on the acceleration due to gravity, g. On Earth, g has an approximate numerical value of 9.8 m/s^2 when height is measured in meters and 32 ft/s^2 when height is measured in feet. The leading coefficient of a projectile motion function is always $-\frac{1}{2}g$.

Projectile Motion Function

The height of an object rising or falling under the influence of gravity is modeled by the function

$$y = ax^2 + v_0x + s_0$$

where x represents time in seconds, y represents the object's height from the ground in meters or feet, a is half the downward acceleration due to gravity (on Earth, a is -4.9 m/s^2 or -16 ft/s^2), v_0 is the initial upward velocity of the object in meters per second or feet per second, and s_0 is the initial height of the object in meters or feet.

The water erupting from these geysers in Black Rock Desert, Nevada, follows a path that can be described as projectile motion.

EXAMPLE A A stopwatch records that when Julie jumps in the air, she leaves the ground at 0.25 s and lands at 0.83 s. How high did she jump, in feet?

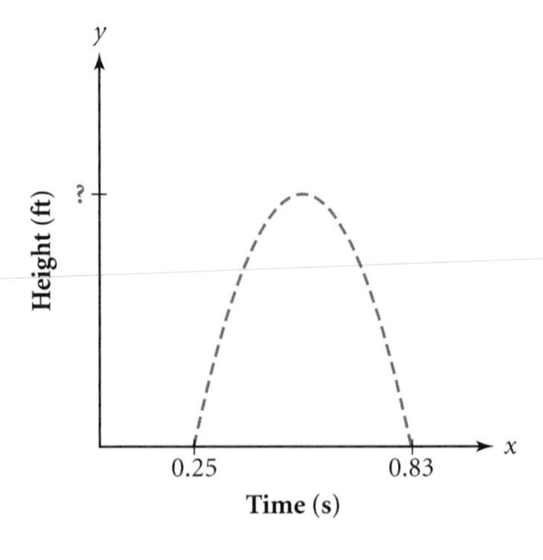

▸ ***Solution*** You don't know the initial velocity, so you can't yet use the projectile motion function. But you do know that height is modeled by a quadratic function and that the leading coefficient must be approximately -16 when using units of feet. Use this information along with 0.25 and 0.83 as the x-intercepts (when Julie's jump height is 0) to write the function

$$y = -16(x - 0.25)(x - 0.83)$$

The vertex of the graph of this equation represents Julie's maximum jump height. The x-coordinate of the vertex will be midway between the two x-intercepts, 0.25 and 0.83. The mean of 0.25 and 0.83 is $\frac{0.25 + 0.83}{2}$, or 0.54.

$y = -16(x - 0.25)(x - 0.83)$	Original function.
$y = -16(0.54 - 0.25)(0.54 - 0.83)$	Substitute the x-coordinate of the vertex.
$y \approx 1.35$	Evaluate.

Julie jumped 1.35 ft.

In Example A, you learned how to average the x-intercepts to find the vertex of a parabola if its equation is in factored form. In the investigation you will convert quadratic functions from general form to the equivalent vertex form by completing the square.

Investigation
Complete the Square

You can use rectangle diagrams to help convert quadratic functions to other equivalent forms.

Step 1

a. Complete a rectangle diagram to find the product $(x + 5)(x + 5)$, which can be written $(x + 5)^2$. Write out the four-term polynomial, and then combine any like terms you see and express your answer as a trinomial.

b. What binomial expression is being squared, and what is the perfect-square trinomial represented in the rectangle diagram at right?

x^2	$-8x$
$-8x$	64

c. Use a rectangle diagram to show the binomial factors for the perfect-square trinomial $x^2 + 24x + 144$.

d. Find the perfect-square trinomial equivalent to $(a + b)^2 = \underline{?} + \underline{?} + \underline{?}$. Describe how you can find the first, second, and third terms of the perfect-square trinomial (written in general form) when squaring a binomial.

Step 2 Consider the expression $x^2 + 6x$.

	x	3
x	x^2	$3x$
3	$3x$	?

a. What could you add to the expression to make it a perfect square? That is, what must you add to complete this rectangle diagram?

b. If you add a number to an expression, then you must also subtract the same number in order to preserve the value of the original expression. Fill in the blanks to rewrite $x^2 + 6x$ as the difference between a perfect square and a number.

$$x^2 + 6x = x^2 + 6x + \underline{\ ?\ } - \underline{\ ?\ } = (x+3)^2 - \underline{\ ?\ }$$

c. Use a graph or table to verify that your expression in the form $(x - h)^2 + k$ is equivalent to the original expression, $x^2 + 6x$.

Step 3 Consider the expression $x^2 + 6x - 4$.

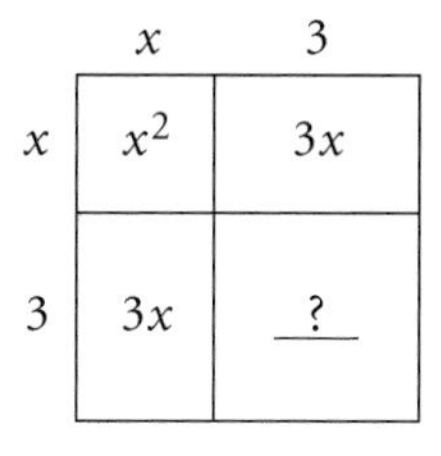

a. Focus on the 2nd- and 1st-degree terms of the expression, $x^2 + 6x$. What must be added to and subtracted from these terms to complete a perfect square yet preserve the value of the expression?

b. Rewrite the expression $x^2 + 6x - 4$ in the form $(x - h)^2 + k$.

c. Use a graph or table to verify that your expression is equivalent to the original expression, $x^2 + 6x - 4$.

Step 4 Rewrite each expression in the form $(x - h)^2 + k$. If you use a rectangle diagram, focus on the 2nd- and 1st-degree terms first. Verify that your expression is equivalent to the original expression.

a. $x^2 - 14x + 3$ **b.** $x^2 - bx + 10$

When the 2nd-degree term has a coefficient, you can first factor it out of the 2nd- and 1st-degree terms. For example, $3x^2 + 24x + 5$ can be written $3(x^2 + 8x) + 5$. Completing a diagram for $x^2 + 8x$ can help you rewrite the expression in the form $a(x - h)^2 + k$.

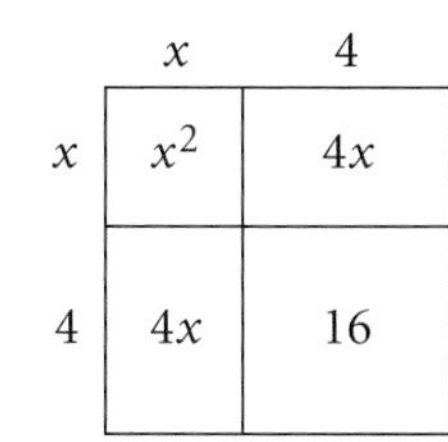

$3x^2 + 24x + 5$	Original expression.
$3(x^2 + 8x) + 5$	Factor the 2nd- and 1st-degree terms.
$3(x^2 + 8x + 16) - 3(16) + 5$	Complete the square. You add $3 \cdot 16$, so you must subtract $3 \cdot 16$.
$3(x + 4)^2 - 43$	An equivalent expression in the form $a(x - h)^2 + k$.

Step 5 Rewrite each expression in the form $a(x - h)^2 + k$. For Step 5a, use a graph or table to verify that your expression is equivalent to the original expression.

a. $2x^2 - 6x + 1$ **b.** $ax^2 + 10x + 7$

Step 6 If you graph the quadratic function $y = ax^2 + bx + c$, what will be the x-coordinate of the vertex in terms of a, b, and c? How can you use this value and the equation to find the y-coordinate?

In the investigation you saw how to convert a quadratic expression from the form $ax^2 + bx + c$ to the form $a(x - h)^2 + k$. This process is called **completing the square.**

Composition with Blue and Yellow (1931) was created by Dutch abstract painter Piet Mondrian (1872–1944).

EXAMPLE B

Convert each quadratic function to vertex form. Identify the vertex.

a. $y = x^2 - 18x + 100$ **b.** $y = ax^2 + bx + c$

▶ Solution

To convert to vertex form, you must complete the square.

a. First separate the constant from the first two terms.

$y = (x^2 - 18x) + 100$ Original equation.

To find what number you must add to complete the square, use a rectangle diagram.

	x	-9
x	x^2	$-9x$
-9	$-9x$	81

$x^2 - 18x + 100 \rightarrow (x^2 - 18x + ?) + 100 - ? \rightarrow$
$(x^2 - 18x + 81) + 100 - 81 \rightarrow (x - 9)^2 + 19$

When you add 81 to complete the square, you must also subtract 81 to keep the expression equivalent.

$y = (x^2 - 18x + 81) + 100 - 81$ Add 81 to complete the square, and subtract 81 to keep the equation equivalent.

$y = (x - 9)^2 + 19$ Factor into a perfect square, and add the constant terms.

The vertex form is $y = (x - 9)^2 + 19$, and the vertex is $(9, 19)$.

b. This solution will provide a formula for vertex (h, k) and an alternative to completing the square.

$y = ax^2 + bx + c$ Original equation.

$= a\left(x^2 + \frac{b}{a}x + \underline{\ ?\ }\right) + c$ Factor a from the first two terms.

$= a\left[x^2 + \frac{b}{a}x + \left(\frac{b}{2a}\right)^2\right] + c - a \cdot \left(\frac{b}{2a}\right)^2$ Add and subtract the missing term to complete the square.

$$y = a\left(x + \frac{b}{2a}\right)^2 + c - \frac{b^2}{4a}$$

The final equation is in vertex form, $y = a(x-h)^2 + k$. This means $h = -\frac{b}{2a}$ and $k = c - \frac{b^2}{4a}$.

You may find that using the formulas for h and k is often simpler than completing the square. Both methods will allow you to find the vertex of a quadratic equation and write the equation in vertex form. However, the method of completing the square will be used again in your work with ellipses and other geometric shapes, so you should become comfortable using it.

EXAMPLE C

Nora hits a softball straight up at a speed of 120 ft/s. If her bat contacts the ball at a height of 3 ft above the ground, how high does the ball travel? When does the ball reach its maximum height?

U.S. Olympic Softball Team player Natasha Watley at bat

▶ Solution

Using the projectile motion function, you know that the height of the object at time x is represented by the equation $y = ax^2 + v_0x + s_0$. The initial velocity, v_0, is 120 ft/s, and the initial height, s_0, is 3 ft. Because the distance is measured in feet, the approximate leading coefficient is -16. Thus, the function is $y = -16x^2 + 120x + 3$. To find the maximum height, locate the vertex.

$y = -16x^2 + 120x + 3$ — Original equation. Identify the coefficients, $a = -16$, $b = 120$, and $c = 3$.

$h = -\frac{b}{2a} = -\frac{120}{2(-16)} = 3.75$ — Use the formula for h to find the x-coordinate of the vertex.

$y = -16(3.75)^2 + 120(3.75) + 3 = 228$ — Substitute 3.75 for x into the equation to find the y-coordinate of the vertex.

The softball reaches a maximum height of 228 ft at 3.75 s.

You now have several strategies for finding the vertex of a quadratic function. You can convert from general form to vertex form by completing the square or by using the formula for h and substituting to find k.

EXERCISES

You will need

A graphing calculator for Exercises **6**, **10**, and **15**.

Practice Your Skills

1. Factor each quadratic expression.

a. $x^2 - 10x + 25$ **b.** $x^2 + 5x + \frac{25}{4}$ ⓐ **c.** $4x^2 - 12x + 9$ **d.** $x^2 - 2xy + y^2$ ⓐ

2. What value is required to complete the square?

a. $x^2 + 20x +$ _?_ **b.** $x^2 - 7x +$ _?_ ⓐ **c.** $4x^2 - 16x +$ _?_ **d.** $-3x^2 - 6x +$ _?_ ⓐ

3. Convert each quadratic function to vertex form by completing the square.

a. $y = x^2 + 20x + 94$ **b.** $y = x^2 - 7x + 16$ ⓐ **c.** $y = 6x^2 - 24x + 147$ **d.** $y = 5x^2 + 8x$ ⓐ

4. Rewrite each expression in the form $ax^2 + bx + c$, and then identify the coefficients, a, b, and c.

a. $3x^2 + 2x - 5$ **b.** $14 + 2x^2$ **c.** $-3 + 4x^2 - 2x + 8x$ ⓐ **d.** $3x - x^2$

Reason and Apply

5. What is the vertex of the graph of the quadratic function $y = -2x^2 - 16x - 20$?

6. Convert the function $y = 7.51x^2 - 47.32x + 129.47$ to vertex form. Use a graph or table to verify that the functions are equivalent. ⓗ

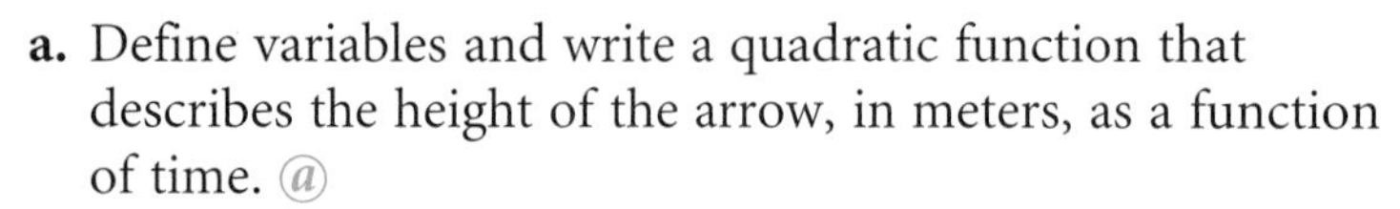

7. Imagine that an arrow is shot from the bottom of a well. It passes ground level at 1.1 s and lands on the ground at 4.7 s. ⓗ

a. Define variables and write a quadratic function that describes the height of the arrow, in meters, as a function of time. ⓐ

b. What was the initial velocity of the arrow in meters per second?

c. How deep was the well in meters?

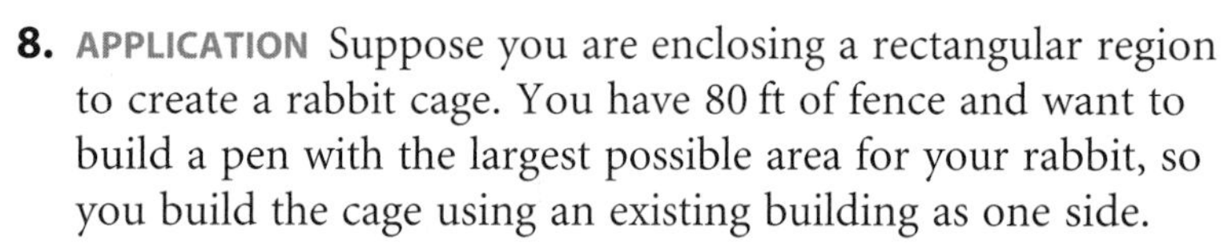

8. APPLICATION Suppose you are enclosing a rectangular region to create a rabbit cage. You have 80 ft of fence and want to build a pen with the largest possible area for your rabbit, so you build the cage using an existing building as one side.

a. Make a table showing the areas for some selected values of x, and write a function that gives the area, y, as a function of the width, x.

b. What width maximizes the area? What is the maximum area?

9. A rock is thrown upward from the edge of a 50 m cliff overlooking Lake Superior, with an initial velocity of 17.2 m/s. Define variables and write an equation that models the height of the rock.

10. An object is projected upward, and these data are collected. ⓐ

Time (s) t	1	2	3	4	5	6
Height (m) h	120.1	205.4	280.9	346.6	402.5	448.6

a. Write a function that relates time and height for this object.

b. What was the initial height? The initial velocity?

c. When does the object reach its maximum height? What is the maximum height?

11. APPLICATION The members of the Young Entrepreneurs Club decide to sell T-shirts in their school colors for Spirit Week. In a marketing survey, the members ask students whether or not they would buy a T-shirt for a specific price. In analyzing the data, club members find that at a price of \$20 they would sell 60 T-shirts. For each \$5 increase in price, they would sell 10 fewer T-shirts.

a. Find a linear function that relates the price in dollars, p, and the number of T-shirts sold, n. ⓗ

b. Write a function that gives revenue as a function of price. (Use your function in 11a as a substitute for the number of T-shirts sold.) ⓐ

c. Convert the revenue function to vertex form. What is the real-world meaning of the vertex?

d. If the club members want to receive at least \$1,050 in revenue, what price should they charge for the T-shirts?

Business CONNECTION

Entrepreneurs are individuals who take a risk to create a new product or a new business. Some famous American entrepreneurs include Sarah Breedlove ("Madam C. J. Walker") (1867–1919), who was a pioneer in the cosmetics industry for African-Americans; Clarence Birdseye (1886–1956), who made frozen food available; Ray Kroc (1902–1984), who developed a way to provide fast service and created McDonald's; and Jeff Bezos (b 1964), who founded Amazon.com and popularized a way to sell merchandise without the traditional retail stores.

Sarah Breedlove Walker expanded her cosmetics company across the United States, the Caribbean, and Europe.
The Granger Collection, New York City

Around the time of World War II (1939–1945), Clarence Birdseye, shown here dehydrating chopped carrots, created foods that were fast and easy to prepare.

Review

12. Multiply.

a. $(x - 3)(2x + 4)$

b. $(x^2 + 1)(x + 2)$

13. Solve $(x - 2)(x + 3)(2x - 1) = 0$. @

14. Consider a graph of the unit circle, $x^2 + y^2 = 1$. Dilate it vertically by a scale factor of 3, and translate it left 5 units and up 7 units.

a. Write the equation of this new shape. What is it called?

b. Sketch a graph of this shape. Label the center and at least four points.

15. APPLICATION This table shows the number of endangered species in the United States for selected years from 1980 to 2000. @

Endangered Species

Year	1980	1985	1990	1995	1996	1997	1998	1999	2000
Number of endangered species	224	300	442	756	837	896	924	939	961

(*The New York Times Almanac 2007*)

a. Define variables and create a scatter plot of these data.

b. Find the equation of a line of fit for these data.

c. Use your equation from 15b to predict the number of endangered species in 2005 and in 2050.

d. The actual number of endangered species in 2005 was 988. How does this value compare with your prediction from 15c? How might this information change your prediction for 2050?

e. Add the point (2005, 988) to your scatter plot. What type of function might be a good model for the more recent data? Explain.

Environmental CONNECTION

Most species become endangered when humans damage their ecosystems through pollution, habitat destruction, and introduction of nonnative species. Over-hunting and over-collecting also threaten animal and plant populations. The current global extinction rate is about 20,000 species a year. For information about what is being done to protect endangered species, see the links at www.keymath.com/DAA .

The Karner blue butterfly, native to the Great Lakes region, has become endangered as the availability of its primary food, blue lupine, has become scarce due to development and fire suppression.

The aye-aye is a nocturnal primate native to Madagascar. Their numbers have declined due to habitat destruction and also because, as they are considered to be a bad omen, they are often killed.

LESSON

7.4

The Quadratic Formula

Although you can always use a graph of a quadratic function to approximate the x-intercepts, you are often not able to find exact solutions. This lesson will develop a procedure to find the exact roots of a quadratic equation. Consider again this situation from Example C in the previous lesson.

EXAMPLE A

Nora hits a softball straight up at a speed of 120 ft/s. Her bat contacts the ball at a height of 3 ft above the ground. Recall that the equation relating height in feet, y, and time in seconds, x, is $y = -16x^2 + 120x + 3$. How long will it be until the ball hits the ground?

▶ Solution

The height will be zero when the ball hits the ground, so you want to find the solutions to the equation $-16x^2 + 120x + 3 = 0$. You can approximate the x-intercepts by graphing, but you may not be able to find the exact x-intercept.

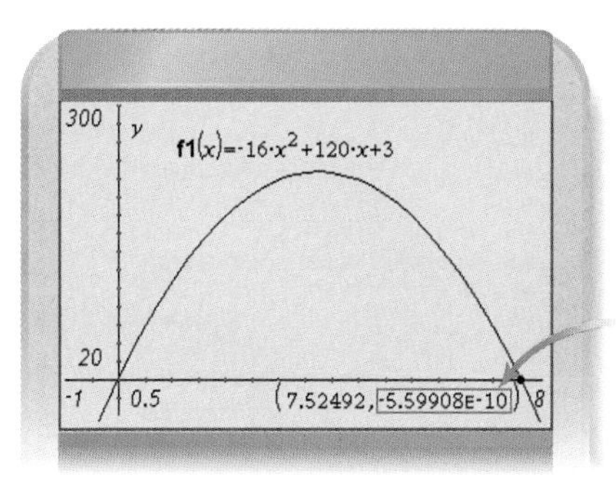

This value represents -5.59908×10^{-10}, or $-0.000000000559908...$, which is very close to zero.

You will not be able to factor this equation using a rectangle diagram, so you can't use the zero-product property. Instead, to solve this equation symbolically, first write the equation in the form $a(x - h)^2 + k = 0$.

$-16x^2 + 120x + 3 = 0$	Original equation.
$-16x^2 + 120x = -3$	Subtract the constant from both sides.
$-16(x^2 - 7.5x + \underline{\ ?\ }) = -3$	Factor to get the leading coefficient 1.
$-16[x^2 - 7.5x + (-\underline{3.75})^2] = -3 + -16(-\underline{3.75})^2$	Complete the square.
$-16(x - 3.75)^2 = -228$	Factor and combine like terms.
$(x - 3.75)^2 = 14.25$	Divide by -16.
$x - 3.75 = \pm\sqrt{14.25}$	Take the square root of both sides.
$x = 3.75 \pm \sqrt{14.25}$	Add 3.75 to both sides.
$x = 3.75 + \sqrt{14.25}$ or $x = 3.75 - \sqrt{14.25}$	Write the two exact solutions to the equation.
$x \approx 7.525$ or $x \approx -0.025$	Approximate the values of x.

The zeros of the function are $x \approx 7.525$ and $x \approx -0.025$. The negative time, -0.025 s, does not make sense in this situation, so the ball hits the ground after approximately 7.525 s.

If you follow the same steps with a general quadratic equation, then you can develop the **quadratic formula.** This formula provides solutions to $ax^2 + bx + c = 0$ in terms of a, b, and c.

$ax^2 + bx + c = 0$	Original equation.
$ax^2 + bx = -c$	Subtract c from both sides.
$a\left(x^2 + \frac{b}{a}x + \underline{\ ?\ }\right) = -c$	Factor to get the leading coefficient 1.
$a\left[x + \frac{b}{a}x + \left(\frac{b}{2a}\right)^2\right] = a\left(\frac{b}{2a}\right)^2 - c$	Complete the square.
$a\left(x + \frac{b}{2a}\right)^2 = \frac{b^2}{4a} - c$	Factor the perfect-square trinomial on the left side and multiply on the right side.
$a\left(x + \frac{b}{2a}\right)^2 = \frac{b^2}{4a} - \frac{4ac}{4a}$	Rewrite the right side with a common denominator.
$a\left(x + \frac{b}{2a}\right)^2 = \frac{b^2 - 4ac}{4a}$	Add terms with a common denominator.
$\left(x + \frac{b}{2a}\right)^2 = \frac{b^2 - 4ac}{4a^2}$	Divide both sides by a.
$x + \frac{b}{2a} = \pm\sqrt{\frac{b^2 - 4ac}{4a^2}}$	Take the square root of both sides.
$x + \frac{b}{2a} = \pm\frac{\sqrt{b^2 - 4ac}}{2a}$	Use the power of a quotient property to take the square roots of the numerator and denominator.
$x = -\frac{b}{2a} \pm \frac{\sqrt{b^2 - 4ac}}{2a}$	Subtract $\frac{b}{2a}$ from both sides.
$x = \frac{-b \pm \sqrt{b^2 - 4ac}}{2a}$	Add terms with a common denominator.

The Quadratic Formula

Given a quadratic equation written in the form $ax^2 + bx + c = 0$, the solutions are

$$x = \frac{-b \pm \sqrt{b^2 - 4ac}}{2a}$$

To use the quadratic formula on the equation in Example A, $-16x^2 + 120x + 3 = 0$, first identify the coefficients as $a = -16$, $b = 120$, and $c = 3$. The solutions are

$$x = \frac{-120 \pm \sqrt{120^2 - 4(-16)(3)}}{2(-16)}$$

$$x = \frac{-120 + \sqrt{14592}}{-32} \quad \text{or} \quad x = \frac{-120 - \sqrt{14592}}{-32}$$

$$x \approx -0.025 \quad \text{or} \quad x \approx 7.525$$

The quadratic formula gives you a way to find the roots of any equation in the form $ax^2 + bx + c = 0$. The investigation will give you an opportunity to apply the quadratic formula in different situations.

Investigation
How High Can You Go?

Salvador hits a baseball at a height of 3 ft and with an initial upward velocity of 88 feet per second.

Step 1 Let x represent time in seconds after the ball is hit, and let y represent the height of the ball in feet. Write an equation that gives the height as a function of time.

Step 2 Write an equation to find the times when the ball is 24 ft above the ground.

Step 3 Rewrite your equation from Step 2 in the form $ax^2 + bx + c = 0$, then use the quadratic formula to solve. What is the real-world meaning of each of your solutions? Why are there two solutions?

Step 4 Find the y-coordinate of the vertex of this parabola. How many different x-values correspond to this y-value? Explain.

Step 5 Write an equation to find the time when the ball reaches its maximum height. Use the quadratic formula to solve the equation. At what point in the solution process does it become obvious that there is only one solution to this equation?

Step 6 Write an equation to find the time when the ball reaches a height of 200 ft. What happens when you try to solve this impossible situation with the quadratic formula?

It's important to note that a quadratic equation must be in the general form $ax^2 + bx + c = 0$ before you use the quadratic formula.

EXAMPLE B Solve $3x^2 = 5x + 8$.

► Solution

To use the quadratic formula, first write the equation in the form $ax^2 + bx + c = 0$ and identify the coefficients.

$$3x^2 - 5x - 8 = 0$$

$$a = 3,\ b = -5,\ c = -8$$

Substitute a, b, and c into the quadratic formula.

$$x = \frac{-b \pm \sqrt{b^2 - 4ac}}{2a}$$

$$= \frac{-(-5) \pm \sqrt{(-5)^2 - 4(3)(-8)}}{2(3)}$$

$$= \frac{5 \pm \sqrt{121}}{6}$$

$$= \frac{5 \pm 11}{6}$$

$$x = \frac{5 + 11}{6} = \frac{8}{3} \quad \text{or} \quad x = \frac{5 - 11}{6} = -1$$

The solutions are $x = \frac{8}{3}$ or $x = -1$.

To check your work, substitute these values into the original equation. Here's a way to use your calculator to check.

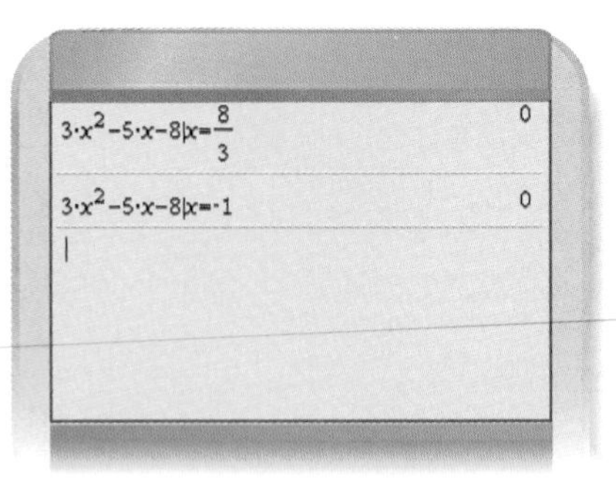

Remember, you can find exact solutions to some quadratic equations by factoring. However, most quadratic equations don't factor easily. The quadratic formula can be used to solve any quadratic equation.

EXERCISES

You will need

A graphing calculator for Exercises **8**, **11**, and **17**.

Practice Your Skills

1. Rewrite each equation in general form, $ax^2 + bx + c = 0$. Identify a, b, and c.

a. $3x^2 - 13x = 10$ **b.** $x^2 - 13 = 5x$ @ **c.** $3x^2 + 5x = -1$ @

d. $3x^2 - 2 - 3x = 0$ **e.** $14(x - 4) - (x + 2) = (x + 2)(x - 4)$

2. Evaluate each expression using your calculator. Round your answers to the nearest thousandth.

a. $\dfrac{-30 + \sqrt{30^2 - 4(5)(3)}}{2(5)}$ **b.** $\dfrac{-30 - \sqrt{30^2 - 4(5)(3)}}{2(5)}$ @

c. $\dfrac{8 - \sqrt{(-8)^2 - 4(1)(-2)}}{2(1)}$ **d.** $\dfrac{8 + \sqrt{(-8)^2 - 4(1)(-2)}}{2(1)}$ @

3. Solve by any method.

a. $x^2 - 6x + 5 = 0$ **b.** $x^2 - 7x - 18 = 0$ **c.** $5x^2 + 12x + 7 = 0$ @

4. Use the roots of the equations in Exercise 3 to write each of these functions in factored form, $y = a(x - r_1)(x - r_2)$.

a. $y = x^2 - 6x + 5$ **b.** $y = x^2 - 7x - 18$ **c.** $y = 5x^2 + 12x + 7$ @

5. Use the quadratic formula to find the zeros of each function.

a. $f(x) = 2x^2 + 7x - 4$ **b.** $f(x) = x^2 - 6x + 3$ **c.** $y - 6 = -2x^2$

d. $5x + 4 + 2x^2 = y$

Reason and Apply

6. Beth uses the quadratic formula to solve an equation and gets

$$x = \frac{-9 \pm \sqrt{9^2 - 4(1)(10)}}{2(1)}$$

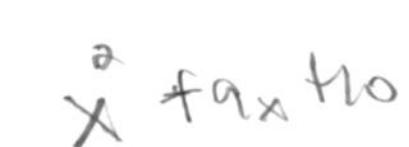

a. Write the quadratic equation Beth started with. @

b. Write the simplified forms of the exact answers.

c. What are the x-intercepts of the graph of this quadratic function?

7. Write a quadratic function whose graph has these x-intercepts.

a. 3 and -3 ⓐ **b.** 4 and $-\frac{2}{5}$ **c.** r_1 and r_2

8. Use the quadratic formula to find the zeros of $y = 2x^2 + 2x + 5$. Explain what happens. Graph $y = 2x^2 + 2x + 5$ to confirm your observation. How can you recognize this situation before using the quadratic formula? ⓗ

9. Write a quadratic function that has no x-intercepts. ⓗ

10. Show that the mean of the two solutions provided by the quadratic formula is $-\frac{b}{2a}$. Explain what this tells you about a graph.

11. These data give the amount of water in a draining bathtub and the amount of time after the plug was pulled. ⓗ

Time (min) x	1	1.5	2	2.5
Amount of water (L) y	38.4	30.0	19.6	7.2

a. Write a function that gives the amount of water as a function of time. ⓐ

b. How much water was in the tub when the plug was pulled?

c. How long did it take the tub to empty?

12. A **golden rectangle** is a rectangle that can be divided into a square and another smaller rectangle that is similar to the original rectangle. In the figure at right, $ABCD$ is a golden rectangle because it can be divided into square $ABFE$ and rectangle $FCDE$, and $FCDE$ is similar to $ABCD$. Setting up a proportion of the side lengths of the similar rectangles leads to $\frac{a}{a+b} = \frac{b}{a}$. Let $b = 1$ and solve this equation for a.

B a F b C
a
A E D

History
CONNECTION

Many people and cultures throughout history have felt that the golden rectangle is one of the most visually pleasing geometric shapes. It was used in the architectural designs of the Cathedral of Notre Dame in Paris, as well as in music and famous works of art. It is believed that the early Egyptians knew the value of the **golden ratio** (the ratio of the length of the golden rectangle to the width) to be $\frac{1+\sqrt{5}}{2}$ and that they used the ratio when building their pyramids, temples, and tombs.

The Fibonacci Fountain in Bowie, Maryland, was designed by mathematician Helaman Ferguson using Fibonacci numbers and the golden ratio. It has 14 water spouts arranged horizontally at intervals proportional to Fibonacci numbers.

Review

13. Complete each equation.

a. $x^2 + \underline{?} + 49 = (x + \underline{?})^2$ ⓐ **b.** $x^2 - 10x + \underline{?} = (\underline{?})^2$ ⓐ

c. $x^2 + 3x + \underline{?} = (\underline{?})^2$ **d.** $2x^2 + \underline{?} + 8 = 2(x^2 + \underline{?} + \underline{?}) = \underline{?}\,(x + \underline{?})^2$

14. Find the inverse of each function. (The inverse does not need to be a function.)

a. $y = (x + 1)^2$ **b.** $y = (x + 1)^2 + 4$ **c.** $y = x^2 + 2x - 5$ ⓐ

15. Convert these quadratic functions to general form.

a. $y = (x - 3)(2x + 5)$

b. $y = -2(x - 1)^2 + 4$

16. A 20 ft ladder leans against a building. Let x represent the distance between the building and the foot of the ladder, and let y represent the height the ladder reaches on the building.

a. Write an equation for y in terms of x. ⓐ

b. Find the height the ladder reaches on the building if the foot of the ladder is 10 ft from the building. ⓐ

c. Find the distance of the foot of the ladder from the building if the ladder must reach 18 ft up the wall.

17. APPLICATION The main cables of a suspension bridge typically hang in the shape of parallel parabolas on both sides of the roadway. The vertical support cables, labeled *a–k,* are equally spaced, and the center of the parabolic cable touches the roadway at *f.* If this bridge has a span of 160 ft between towers, and the towers reach a height of 75 ft above the road, what is the length of each support cable, *a–k*? What is the total length of vertical support cable needed for the portion of the bridge between the two towers?

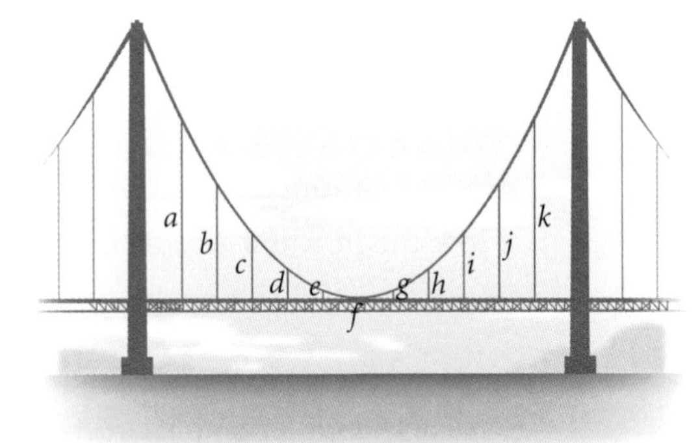

Engineering CONNECTION

The roadway of a suspension bridge is suspended, or hangs, from large steel support cables. By itself, a cable hangs in the shape of a *catenary* curve. However, with the weight of a roadway attached, the curvature changes, and the cable hangs in a parabolic curve. It is important for engineers to ensure that cables are the correct lengths to make a level roadway.

A chain hangs in the shape of a catenary curve.

Project

CALCULATOR PROGRAM FOR THE QUADRATIC FORMULA

Write a calculator program that uses the quadratic formula to solve equations. The program should calculate and display the two solutions for a quadratic equation in the form $ax^2 + bx + c$. Depending on the type of calculator you have, the user can give the a, b, and c values as parameters for the program (as shown) or the program can prompt the user for those values.

```
quad(1,-1,-6)
                3
               -2
             Done
```

Your project should include

- A written record of the steps your program uses.
- An explanation of how the program works.
- The results of solving at least two equations by hand and with your program to verify that your program works.

LESSON

7.5

Complex Numbers

Things don't turn up in the world until somebody turns them up.

JAMES A. GARFIELD

You have explored several ways to solve quadratic equations. You can find the x-intercepts on a graph, you can solve by completing the square, or you can use the quadratic formula. What happens if you try to use the quadratic formula on an equation whose graph has no x-intercepts?

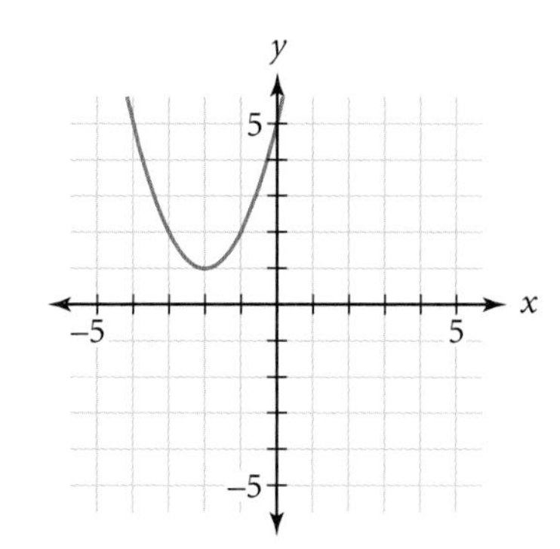

The graph of $y = x^2 + 4x + 5$ at right shows that this function has no x-intercepts. Using the quadratic formula to try to find x-intercepts, you get

$$x = \frac{-4 \pm \sqrt{16 - 4(1)(5)}}{2(1)} = \frac{-4 \pm \sqrt{-4}}{2}$$

How do you take the square root of a negative number? The two numbers $\frac{-4 + \sqrt{-4}}{2}$ and $\frac{-4 - \sqrt{-4}}{2}$ are unlike any of the numbers you have worked with this year—they are nonreal, but they are still numbers.

Throughout the development of mathematics, new sets of numbers have been defined in order to solve problems. Mathematicians have defined fractions and not just whole numbers, negative numbers and not just positive numbers, irrational numbers and not just fractions. For the same reasons, we also have square roots of negative numbers, not just square roots of positive numbers. Numbers that include the real numbers as well as the square roots of negative numbers are called **complex numbers.**

History CONNECTION

Since the 1500s, the square root of a negative number has been called an **imaginary number.** In the late 1700s, the Swiss mathematician Leonhard Euler (1707–1783) introduced the symbol i to represent $\sqrt{-1}$. He wrote:

> It is evident that we cannot rank the square root of a negative number amongst possible numbers, and we must therefore say that it is an impossible quantity. . . . But notwithstanding this these numbers present themselves to the mind; they exist in our imagination, and we still have a sufficient idea of them; since we know that by $\sqrt{-4}$ is meant a number which, multiplied by itself, produces -4; for this reason also, nothing prevents us from making use of these imaginary numbers, and employing them in calculation.

Leonhard Euler

Defining imaginary numbers made it possible to solve previously unsolvable problems.

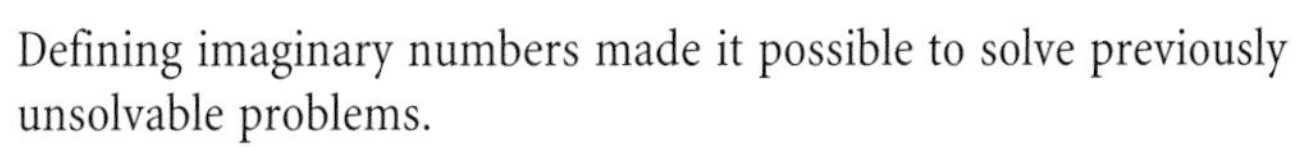

To express the square root of a negative number, we use an **imaginary unit** called i, defined by $i^2 = -1$ or $i = \sqrt{-1}$. You can rewrite $\sqrt{-4}$ as $\sqrt{4} \cdot \sqrt{-1}$, or $2i$. Therefore, you can write the two solutions to the quadratic equation above as the complex numbers $\frac{-4 + 2i}{2}$ and $\frac{-4 - 2i}{2}$, or $-2 + i$ and $-2 - i$. These two solutions are a **conjugate pair.** That is, one is $a + bi$ and the other is $a - bi$. The two numbers in a conjugate pair are **complex conjugates.** Why will nonreal solutions to the quadratic formula always give answers that are a conjugate pair?

Roots of polynomial equations can be real numbers or nonreal complex numbers, or there may be some of each. If the polynomial has real coefficients, any nonreal roots will come in conjugate pairs such as $2i$ and $-2i$ or $3 + 4i$ and $3 - 4i$.

> **Complex Numbers**
>
> A **complex number** is a number in the form $a + bi$, where a and b are real numbers and $i = \sqrt{-1}$.

For any complex number in the form $a + bi$, a is the real part and bi is the imaginary part. The set of complex numbers contains all real numbers and all imaginary numbers. This diagram shows the relationship between these numbers and some other sets you may be familiar with, as well as examples of numbers within each set.

Complex numbers
$3 + 4i, 2i, \frac{5}{4}, -3 - i\sqrt{2}$

Real numbers
$\frac{5}{4}, \sqrt{3}, -15, 2 + \sqrt{7}$

Imaginary numbers
$2i, -i, i\sqrt{2}, \frac{1}{5}i, \sqrt{-5}$

Rational numbers
$\frac{5}{4}, -15, \frac{6}{11}, 1.2$

Irrational numbers
$-2\sqrt{3}, 2 + \sqrt{7}, e, \pi$

Integers
$\ldots, -1, 0, 1, \ldots$

Natural numbers
$1, 2, 3, 4, \ldots$

EXAMPLE Solve $x^2 + 3 = 0$.

▶ Solution You can use the quadratic formula, or you can isolate x^2 and take the square root of both sides.

$$\begin{aligned} x^2 + 3 &= 0 \\ x^2 &= -3 \\ x &= \pm\sqrt{-3} \\ x &= \pm\sqrt{3} \cdot \sqrt{-1} \\ x &= \pm\sqrt{3} \cdot i \\ x &= \pm i\sqrt{3} \end{aligned}$$

To check the two solutions, substitute them into the original equation.

$$\begin{aligned} x^2 + 3 &= 0 \\ (i\sqrt{3})^2 + 3 &\stackrel{?}{=} 0 \\ i^2 \cdot 3 + 3 &\stackrel{?}{=} 0 \\ -1 \cdot 3 + 3 &\stackrel{?}{=} 0 \\ -3 + 3 &\stackrel{?}{=} 0 \\ 0 &= 0 \end{aligned} \qquad \begin{aligned} x^2 + 3 &= 0 \\ (-i\sqrt{3})^2 + 3 &\stackrel{?}{=} 0 \\ i^2 \cdot 3 + 3 &\stackrel{?}{=} 0 \\ -1 \cdot 3 + 3 &\stackrel{?}{=} 0 \\ -3 + 3 &\stackrel{?}{=} 0 \\ 0 &= 0 \end{aligned}$$

The two imaginary numbers $\pm i\sqrt{3}$ are solutions to the original equation, but because they are not real numbers, the graph of $y = x^2 + 3$ shows no x-intercepts.

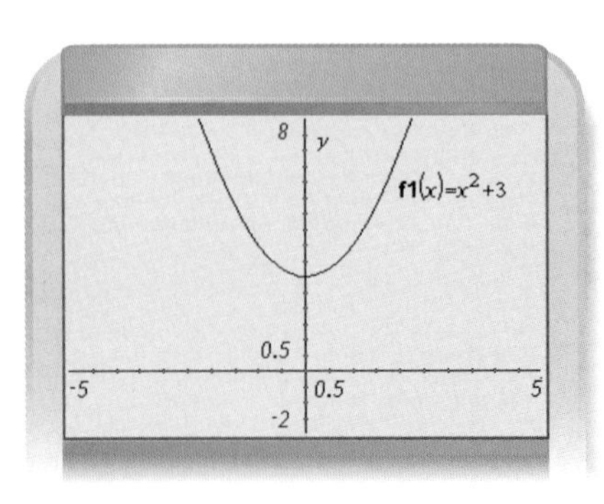

Complex numbers have many applications, particularly in science and engineering. To measure the strength of an electromagnetic field, a real number represents the amount of electricity and an imaginary number represents the amount of magnetism. The state of a component in an electronic circuit is also measured by a complex number, where the voltage is a real number and the current is an imaginary number.

The properties of calculations with complex numbers apply to these types of physical states more accurately than calculations with real numbers do. In the investigation you'll explore patterns in arithmetic with complex numbers.

The Heart Revealed: Portrait of Tita Thirifays (1936), by Belgian Surrealist painter René Magritte (1898–1967), is a portrait with an imaginary element.

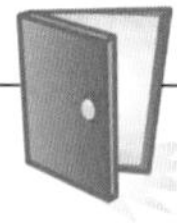

Investigation
Complex Arithmetic

When computing with complex numbers, there are conventional rules similar to those you use when working with real numbers. In this investigation you will discover these rules. You may use your calculator to check your work or to explore other examples. [▶ See **Calculator Note 7E** to learn how to enter complex numbers into your calculator. ◀]

Part 1: Addition and Subtraction

Addition and subtraction of complex numbers is similar to combining like terms. Use your calculator to add these complex numbers. Make a conjecture about how to add complex numbers without a calculator.

a. $(2 - 4i) + (3 + 5i)$

b. $(7 + 2i) + (-2 + i)$

c. $(2 - 4i) - (3 + 5i)$

d. $(4 - 4i) - (1 - 3i)$

Part 2: Multiplication

Use your knowledge of multiplying binomials to multiply these complex numbers. Express your products in the form $a + bi$. Recall that $i^2 = -1$.

a. $(2 - 4i)(3 + 5i)$

b. $(7 + 2i)(-2 + i)$

c. $(2 - 4i)^2$

d. $(4 - 4i)(1 - 3i)$

Part 3: The Complex Conjugates

Recall that every complex number $a + bi$ has a complex conjugate, $a - bi$. Complex conjugates have some special properties and uses.

Each expression below shows either the sum or product of a complex number and its conjugate. Simplify these expressions into the form $a + bi$, and generalize what happens.

a. $(2 - 4i) + (2 + 4i)$

b. $(7 + 2i) + (7 - 2i)$

c. $(2 - 4i)(2 + 4i)$

d. $(-4 + 4i)(-4 - 4i)$

Part 4: Division

Recall that you can create equivalent fractions by multiplying the numerator and denominator of a fraction by the same quantity. For example, $\frac{3}{4} = \frac{3}{4} \cdot \frac{k}{k} = \frac{3k}{4k}$, and $\frac{3}{\sqrt{2}} \cdot \frac{\sqrt{2}}{\sqrt{2}} = \frac{3\sqrt{2}}{2}$. In the second case, multiplying the fraction $\frac{3}{\sqrt{2}}$ by $\frac{\sqrt{2}}{\sqrt{2}}$ produced an equivalent fraction with an integer denominator instead of an irrational denominator.

You will use a similar technique to change the complex number in each denominator into a real number. Use your work from Part 3 to find a method for changing each denominator into a real number. (Your method should produce an equivalent fraction.) Once you have a real number in the denominator, divide to get an answer in the form $a + bi$.

a. $\dfrac{7 + 2i}{1 - i}$

b. $\dfrac{10 - 11i}{4 + 6i}$

c. $\dfrac{2 - i}{8 - 6i}$

d. $\dfrac{2 - 4i}{2 + 4i}$

You cannot graph a complex number, such as $3 + 4i$, on a real number line, but you can graph it on a **complex plane**, where the horizontal axis is the **real axis** and the vertical axis is the **imaginary axis.** In the graph, $3 + 4i$ is located at the point with coordinates $(3, 4)$. Any complex number $a + bi$ has (a, b) as its coordinates on a complex plane.

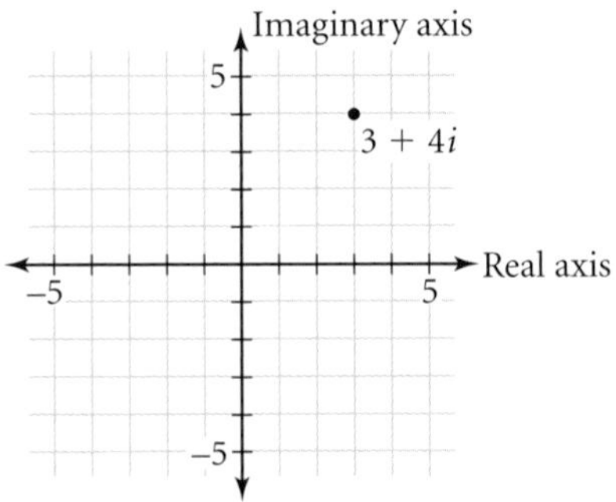

You'll observe some properties of a complex plane in the exercises.

EXERCISES

You will need

A graphing calculator for Exercise **16**.

Practice Your Skills

1. Add or subtract.

a. $(5 - 1i) + (3 + 5i)$

b. $(6 + 2i) - (-1 + 2i)$ ⓐ

c. $(2 + 3i) + (2 - 5i)$ ⓐ

d. $(2.35 + 2.71i) - (4.91 + 3.32i)$

2. Multiply.

a. $(5 - 1i)(3 + 5i)$ ⓐ

b. $6(-1 + 2i)$

c. $3i(2 - 5i)$ ⓐ

d. $(2.35 + 2.71i)(4.91 + 3.32i)$

3. Find the conjugate of each complex number.

a. $5 - i$ **b.** $-1 + 2i$

c. $2 + 3i$ **d.** $-2.35 - 2.71i$ ⓐ

4. Name the complex number associated with each point, A through G, on the complex plane at right. ⓐ

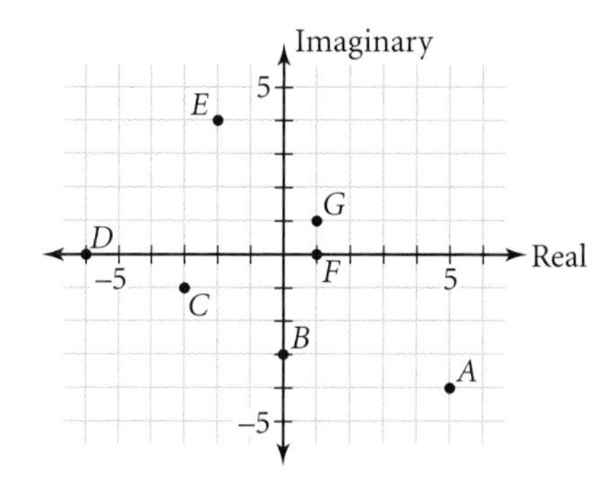

5. Draw **Venn diagrams** to show the relationships between these sets of numbers.

a. real numbers and complex numbers ⓐ

b. rational numbers and irrational numbers

c. imaginary numbers and complex numbers ⓐ

d. imaginary numbers and real numbers

e. complex numbers, real numbers, and imaginary numbers

History CONNECTION

In his book *Symbolic Logic*, English logician John Venn (1834–1923) proposed using diagrams as a method of representing logic relationships. For example, this diagram shows "if it's an eagle, then it's a bird" and "if it's a bird, then it's not a whale." It must also be true, therefore, that "if it's an eagle, then it's not a whale." A Venn diagram of this situation shows this conclusion clearly. Venn diagrams have become a tool for representing many kinds of relationships.

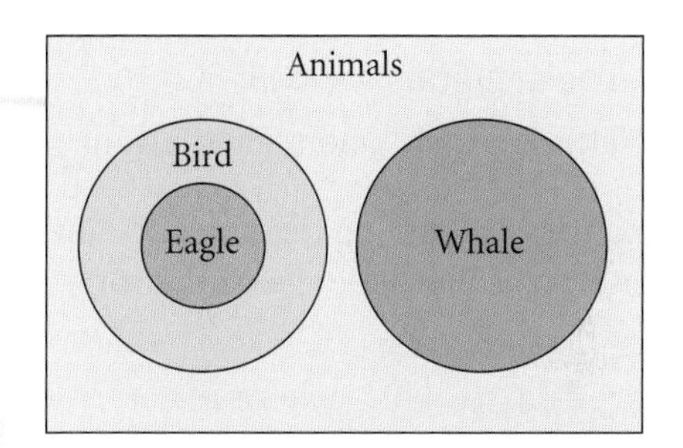

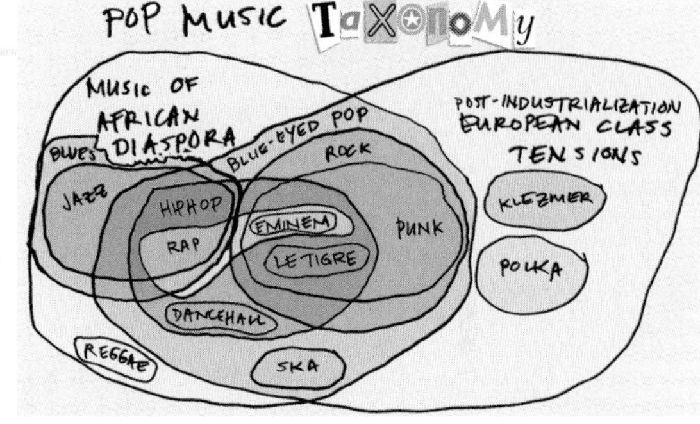

This Venn diagram by Jason Luz shows a taxonomy of music. Do you think speed metal should be included? The Beatles? Madonna? Make your own Venn diagram.

Reason and Apply

6. Rewrite this quadratic equation in general form.

$$[x - (2 + i)][x - (2 - i)] = 0$$

7. Use the definitions $i = \sqrt{-1}$ and $i^2 = -1$ to rewrite each power of i as 1, i, -1, or $-i$.

a. i^3 ⓐ **b.** i^4 ⓐ

c. i^5 **d.** i^{10}

8. *Mini-Investigation* Plot the numbers i, i^2, i^3, i^4, i^5, i^6, i^7, and i^8 on a complex plane. What pattern do you see? What do you expect the value of i^{17} to be? Verify your conjecture.

9. Rewrite the quotient $\frac{2+3i}{2-i}$ in the form $a + bi$. ⓐ

10. Solve each equation. Use substitution to check your solutions. Label each solution as real, imaginary, and/or complex.

a. $x^2 - 1 = 0$ **b.** $x^2 + 1 = 0$ **c.** $x^2 - 4x + 6 = 0$

d. $x^2 + x = -1$ **e.** $-2x^2 + 4x = 3$

11. Write a quadratic function in general form that has the given zeros, and leading coefficient of 1. ⓗ

a. $x = -3$ and $x = 5$

b. $x = -3.5$ (double root)

c. $x = 5i$ and $x = -5i$

d. $x = 2 + i$ and $x = 2 - i$

12. Write a quadratic function with real coefficients in general form that has a zero of $x = 4 + 3i$ and whose graph has y-intercept 50. ⓗ

13. Solve.

a. $x^2 - 10ix - 9i^2 = 0$

b. $x^2 - 3ix = 2$

c. Why don't the solutions to 13a and 13b come in conjugate pairs? ⓐ

14. The quadratic formula, $x = \frac{-b \pm \sqrt{b^2 - 4ac}}{2a}$, provides solutions to $ax^2 + bx + c = 0$. The quantity inside the square root, $b^2 - 4ac$, is called the **discriminant** of the quadratic equation. How can you use the value of the discriminant to determine each of these solution possibilities?

a. The solutions are nonreal.

b. The solutions are real.

c. There is only one real solution.

15. Use these recursive formulas to find the first six terms $(z_0$ to $z_5)$ of each sequence. Describe what happens in the long run for each sequence.

a. $z_0 = 0$
$z_n = z_{n-1}^2 + 0$ where $n \geq 1$

b. $z_0 = 0$
$z_n = z_{n-1}^2 + i$ where $n \geq 1$ ⓐ

c. $z_0 = 0$
$z_n = z_{n-1}^2 + 1 - i$ where $n \geq 1$

d. $z_0 = 0$
$z_n = z_{n-1}^2 + 0.2 + 0.2i$ where $n \geq 1$ ⓐ

Review

16. Consider the function $y = 2x^2 + 6x - 3$.

a. List the zeros in exact radical form and as approximations to the nearest hundredth.

b. Graph the function and label the exact coordinates of the vertex and points where the graph crosses the x-axis and the y-axis.

17. Consider two positive integers that meet these conditions:

- three times the first added to four times the second is less than 30
- twice the first is less than five more than the second

a. Define variables and write a system of linear inequalities that represents this situation.

b. Graph the feasible region.

c. List all integer pairs that satisfy the conditions listed above.

Project

THE MANDELBROT SET

You have seen geometric **fractals** such as the Sierpiński triangle, and you may have seen other fractals that look much more complicated. Fractal geometry can be used to describe natural objects such as clouds, coastlines, and trees.

The Mandelbrot set is a famous fractal that relies on repeated calculations with complex numbers. To create the Mandelbrot set, you use the recursive formula $z_0 = 0$ and $z_n = z_{n-1}^2 + c$, where $n \geq 1$. Depending on the complex number you choose for the constant, c, one of two things will happen: either the magnitude of the values of z will get increasingly large or they will not. (The magnitude of a complex number is defined as its distance from the origin of the complex plane, or $\sqrt{a^2 + b^2}$.)

This Mandelbrot set shows how fractal geometry creates order out of what seem like irregular patterns. Points that are not in the Mandelbrot set are colored based on how quickly they diverge. Benoit Mandelbrot (b 1924), left, was the first person to study and name fractal geometry.

You already explored a few values of c in Exercise 15. Try a few more. Which values of c make the magnitude of z get increasingly large? Which values of c make z converge to a single value or alternate between values?

Use your calculator to determine what happens if $z_0 = 0$ and $c = 0.25$. What happens if c or z is a complex number, for example, if $z_0 = 0$ and $c = -0.4 + 0.5i$?

The Mandelbrot set is all of the values of c that do not make the magnitude of z get increasingly large. If you plot these points on a complex plane, then you'll get a pattern that looks like this one. Your project is to choose a small region on the boundary of the black area of this graph and create a graph of that smaller region. [▶ Calculator Note 7F includes a program that analyzes every point in the window to determine whether it is in the Mandelbrot set. ◀] Look at this graph, select a window, and then run the program. It may take several hours.

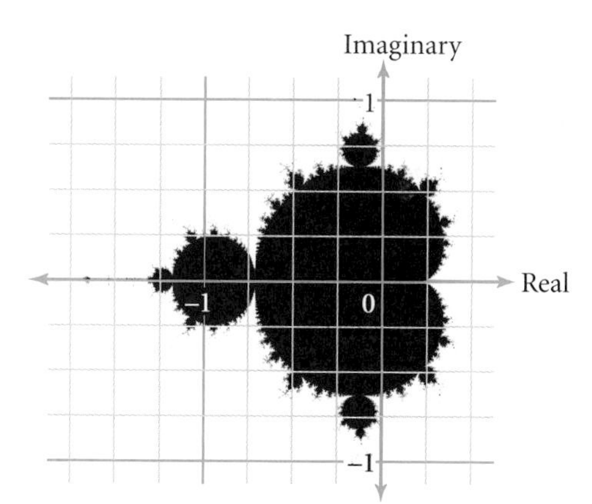

Your project should include

- A sketch of your graph.
- A report that describes any similarities between your portion of the Mandelbrot set and the complete graph shown above.
- Any additional research you do on the Mandelbrot set or on fractals in general.

You can learn more about the Mandelbrot set and other fractals by using the links at www.keymath.com/DAA .

LESSON

7.6

Factoring Polynomials

Ideas are the factors that lift civilization.

JOHN H. VINCENT

Imagine a cube with any side length. Imagine increasing the height by 2 cm, the width by 3 cm, and the length by 4 cm.

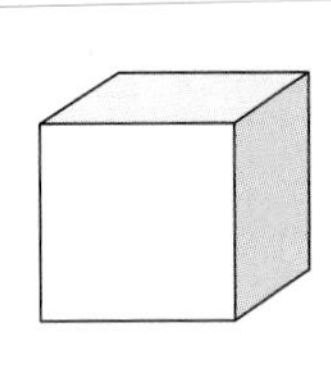

Imagine a cube.

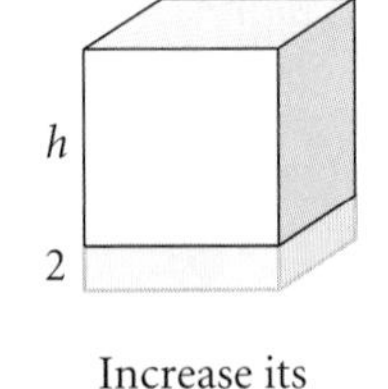

Increase its height 2 cm.

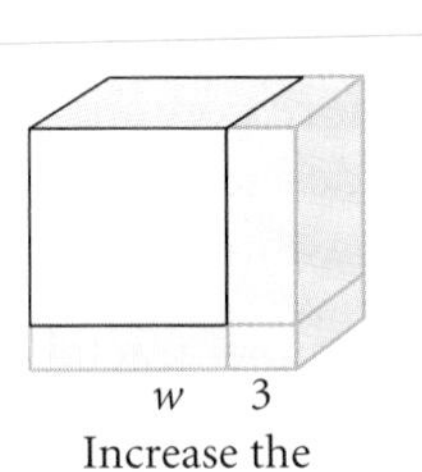

Increase the width 3 cm.

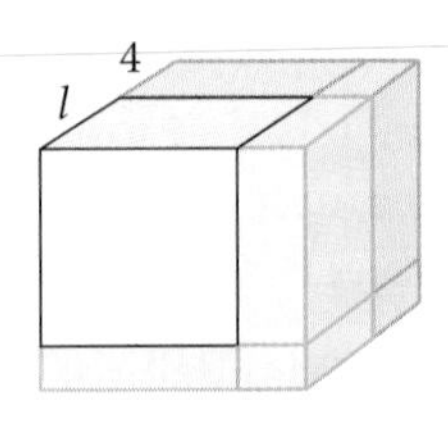

Increase the length 4 cm.

The starting figure is a cube, so you can let x be the length of each of its sides. So, $l = w = h = x$. The volume of the starting figure is x^3. To find the volume of the expanded box, you can see it as the sum of the volumes of eight different boxes. You find the volume of each piece by multiplying length by width by height.

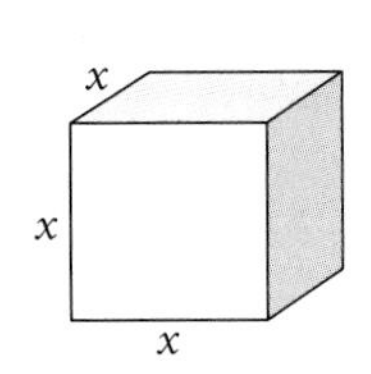

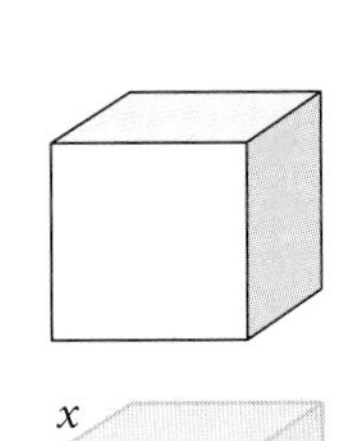

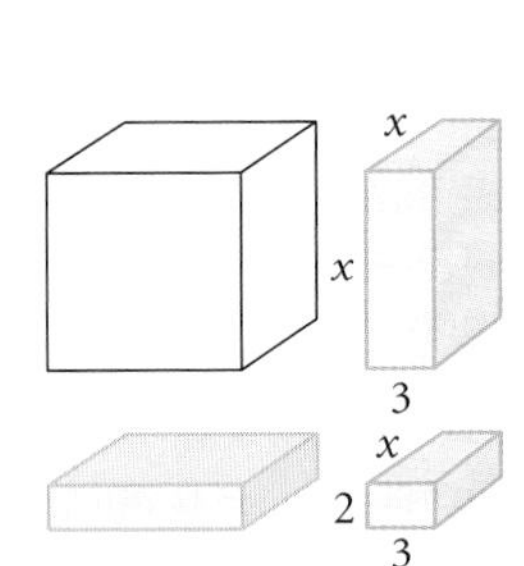

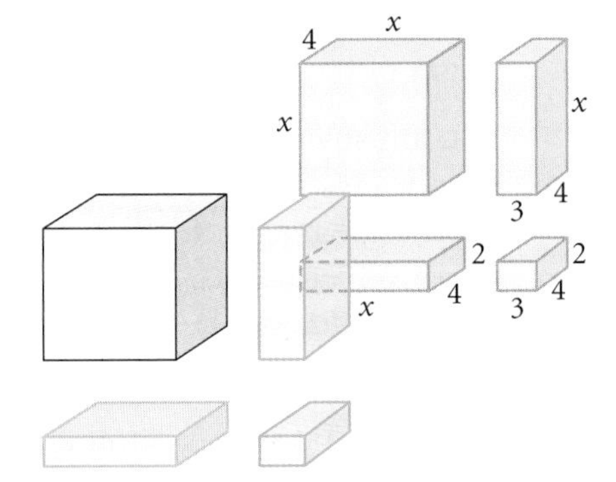

The total expanded volume is this sum:

$$V = x^3 + 2x^2 + 3x^2 + 6x + 4x^2 + 8x + 12x + 24 = x^3 + 9x^2 + 26x + 24$$

You can also think of the expanded volume as the product of the new height, width, and length.

$$V = (x + 2)(x + 3)(x + 4)$$

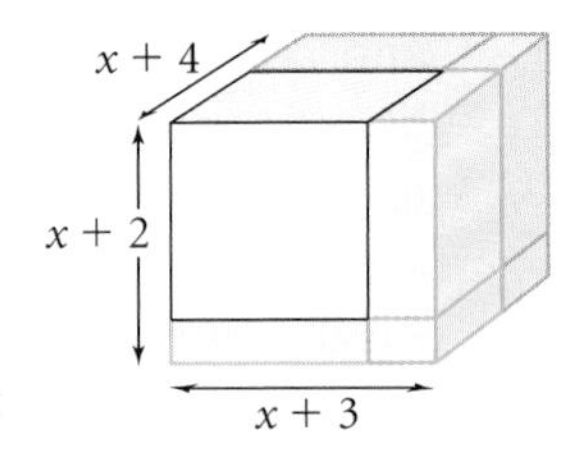

A 3rd-degree polynomial function is called a **cubic function.** This cubic function in factored form is equivalent to the polynomial function in general form. (Try graphing both functions on your calculator.)

You already know that there is a relationship between the factored form of a quadratic equation, and the roots and x-intercepts of that quadratic equation. In this lesson you will learn how to write higher-degree polynomial equations in factored form when you know the roots of the equation. You'll also discover useful techniques for converting a polynomial in general form to factored form.

EXAMPLE A Write cubic functions for the graphs below.

▶ Solution

Both graphs have the same x-intercepts: -2.5, 3.2, and 7.5. So both functions have the factored form $y = a(x + 2.5)(x - 7.5)(x - 3.2)$. But the vertical scale factor, a, is different for each function.

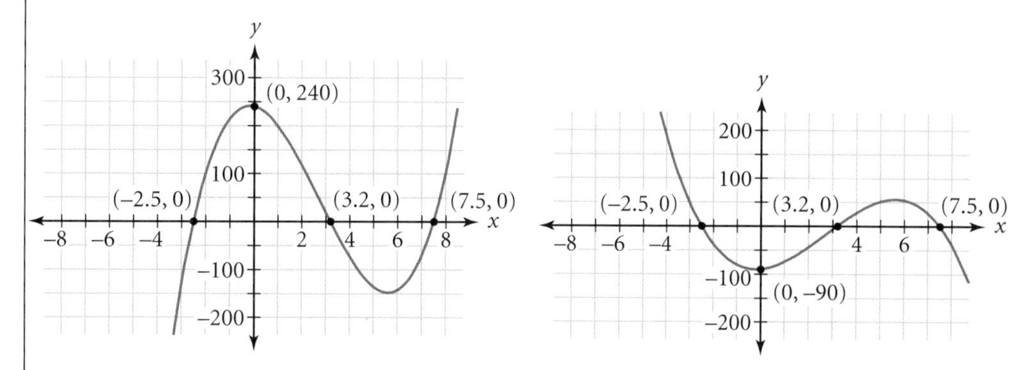

One way to find a is to substitute the coordinates of one other point, such as the y-intercept, into the function.

The curve on the left has y-intercept $(0, 240)$. Substituting this point into the equation gives $240 = a(2.5)(-7.5)(-3.2)$. Solving for a, you get $a = 4$. So the equation of the cubic function on the left is

$$y = 4(x + 2.5)(x - 7.5)(x - 3.2)$$

The curve on the right has y-intercept $(0, -90)$. Substituting this point into the equation gives $-90 = a(2.5)(-7.5)(-3.2)$. So $a = -1.5$, and the equation of the cubic function on the right is

$$y = -1.5(x + 2.5)(x - 7.5)(x - 3.2)$$

The factored form of a polynomial function tells you the zeros of the function and the x-intercepts of the graph of the function. Recall that zeros are solutions to the equation $f(x) = 0$. Factoring, if a polynomial can be factored, is one strategy for finding the real solutions of a polynomial equation. In the investigation you will practice writing a higher-degree polynomial function in factored form.

English sculptor Cornelia Parker (b 1956) creates art from damaged objects that have cultural or historical meaning. *Mass (Colder Darker Matter)* (1997) is made of the charred remains of a building struck by lightning—the building has been reduced to its charcoal factors. The sculpture looks flat when viewed from the front, but it is actually constructed in the shape of a cube.

Cornelia Parker *Mass (Colder Darker Matter)* (1997) charcoal, wire, and black string, Collection of Phoenix Art Museum, Gift of Jan and Howard Hendler 2002.

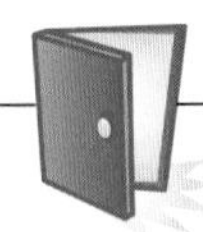

Investigation
The Box Factory

You will need

- graph paper
- scissors
- tape (optional)

What are the different ways to construct an open-top box from a 16-by-20–unit sheet of material? What is the maximum volume this box can have? What is the minimum volume? Your group will investigate this problem by constructing open-top boxes using several possible integer values for x.

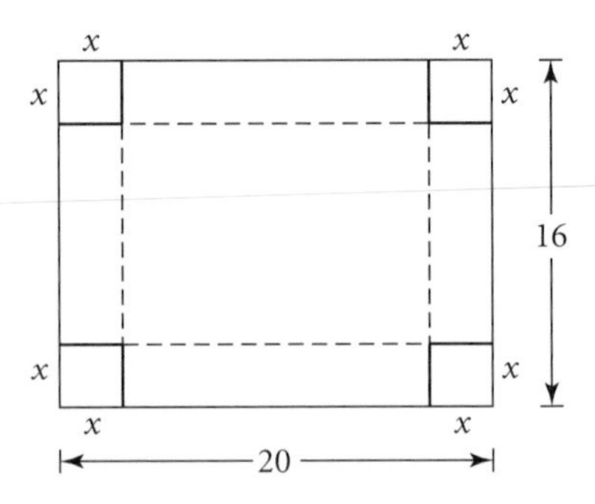

Procedure Note

1. Cut several 16-by-20–unit rectangles out of graph paper.
2. Choose several different values for x.
3. For each value of x, construct a box by cutting squares with side length x from each corner and folding up the sides.

Step 1 Follow the Procedure Note to construct several different-size boxes from 16-by-20–unit sheets of paper. Record the dimensions of each box and calculate its volume. Make a table to record the x-values and volumes of the boxes.

Step 2 For each box, what are the length, width, and height, in terms of x? Use these expressions to write a function that gives the volume of a box as a function of x.

Step 3 Graph your volume function from Step 2. Plot your data points on the same graph. How do the points relate to the function?

Step 4 What is the degree of this function? Give some reasons to support your answer.

Step 5 Locate the x-intercepts of your graph. (There should be three.) Call these three values r_1, r_2, and r_3. Use these values to write the function in the form $y = (x - r_1)(x - r_2)(x - r_3)$.

Step 6 Graph the function from Step 5 with your function from Step 2. What are the similarities and differences between the graphs? How can you alter the function from Step 5 to make both functions equivalent?

Step 7 What happens if you try to make boxes by using the values r_1, r_2, and r_3 as x? What domain of x-values makes sense in this context? What x-value maximizes the volume of the box?

The connection between the roots of a polynomial equation and the x-intercepts of a polynomial function helps you factor any polynomial that has real roots.

EXAMPLE B

Use the graph of each function to determine its factored form.

a. $y = x^2 - x - 2$

b. $y = 4x^3 + 8x^2 - 36x - 72$

▶ Solution

You can find the x-intercepts of each function by graphing. The x-intercepts tell you the real roots, which help you factor the function.

a. The graph shows that the x-intercepts are -1 and 2. Because the coefficient of the highest-degree term, x^2, is 1, the vertical scale factor is 1. The factored form is $y = (x + 1)(x - 2)$.

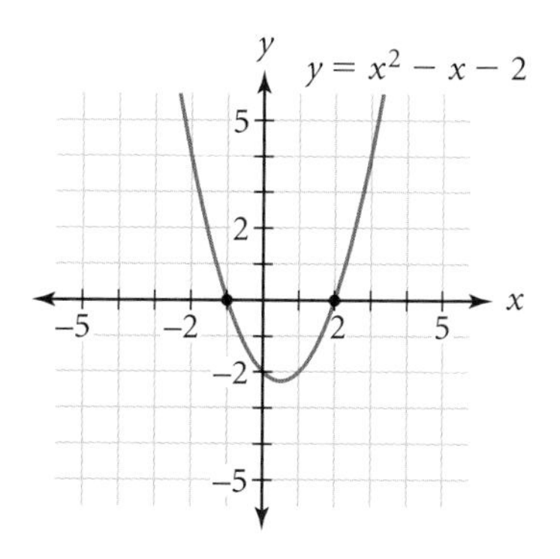

You can verify that the expressions $x^2 - x - 2$ and $(x + 1)(x - 2)$ are equivalent by graphing $y = x^2 - x - 2$ and $y = (x + 1)(x - 2)$. You can also check your work algebraically by finding the product $(x + 1)(x - 2)$. This rectangle diagram confirms that the product is $x^2 - x - 2$.

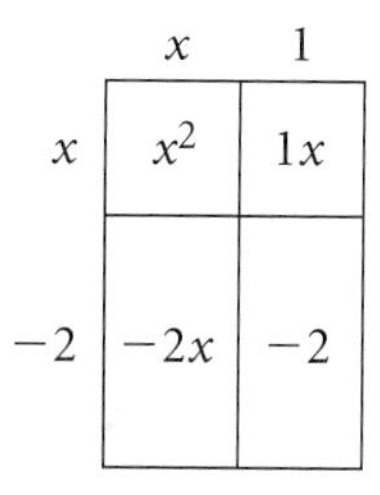

	x	1
x	x^2	$1x$
-2	$-2x$	-2

b. The x-intercepts are -3, -2, and 3. So, you can write the function as

$$y = a(x + 3)(x + 2)(x - 3)$$

Because the leading coefficient needs to be 4, the vertical scale factor is also 4.

$$y = 4(x + 3)(x + 2)(x - 3)$$

To check your answer, you can compare graphs or find the product of the factors algebraically.

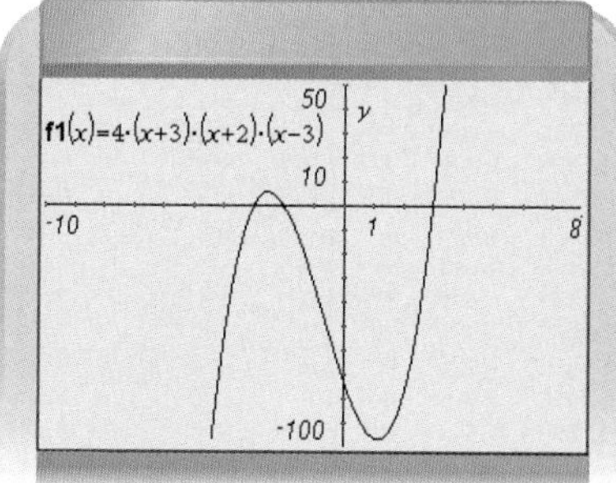

In Example B, you converted a function from general form to factored form by using a graph and looking for the x-intercepts. This method works especially well when the zeros are integer values. Once you know the zeros of a polynomial function, r_1, r_2, r_3, and so on, you can write the factored form,

$$y = a(x - r_1)(x - r_2)(x - r_3) \cdots$$

You can also write a polynomial function in factored form when the zeros are not integers, or even when they are nonreal.

Polynomials with real coefficients can be separated into three types: polynomials that can't be completely factored with real numbers; polynomials that can be factored with real numbers, but some of the roots are not "nice" integer or fractional values; and polynomials that can be factored and have all integer or fractional roots. For example, consider these cases of quadratic functions:

$y = x^2 - x + 1$

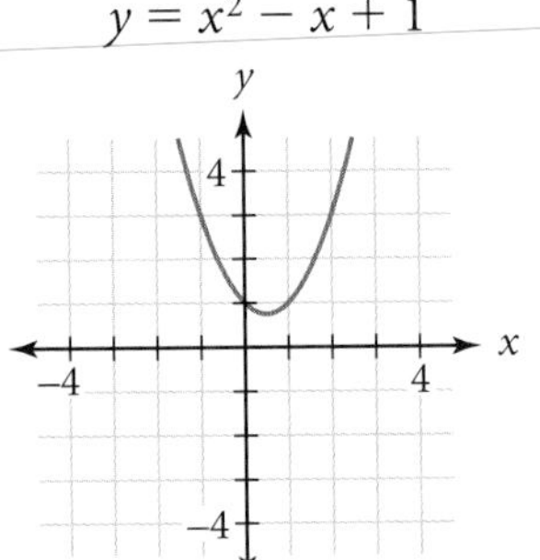

If the graph of a quadratic function does not intersect the x-axis, then you cannot factor the polynomial using real numbers. However, you can use the quadratic formula to find the complex zeros, which will be a conjugate pair.

$y = x^2 - x - 1$

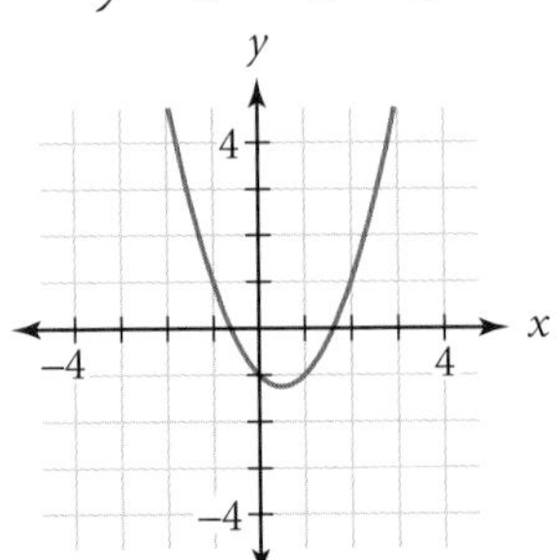

If the graph of a quadratic function intersects the x-axis, but not at integer or fractional values, then you can use the quadratic formula to find the real zeros.

$y = x^2 - x - 2$

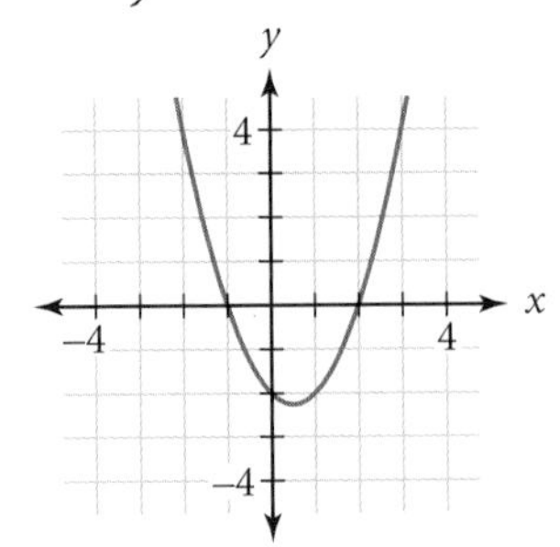

If the graph of a quadratic function intersects the x-axis at integer or fractional values, then you can use the x-intercepts to factor the polynomial. This is often quicker and easier than using the quadratic formula or a rectangle diagram.

What happens when the graph of a quadratic function has exactly one point of intersection with the x-axis?

EXERCISES

You will need

A graphing calculator for Exercises **1**, **4**, **5**, and **15**.

Practice Your Skills

1. Without graphing, find the x-intercepts and the y-intercept for the graph of each equation. Check each answer by graphing.

a. $y = -0.25(x + 1.5)(x + 6)$

b. $y = 3(x - 4)(x - 4)$

c. $y = -2(x - 3)(x + 2)(x + 5)$ @

d. $y = 5(x + 3)(x + 3)(x - 3)$

2. Write the factored form of the quadratic function for each graph. Don't forget the vertical scale factor.

a.

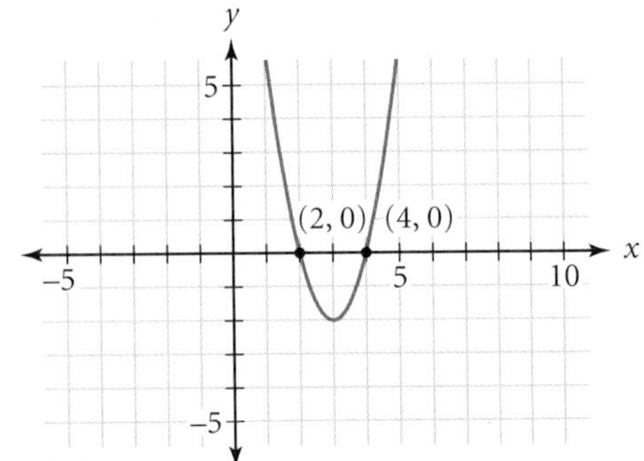

b.

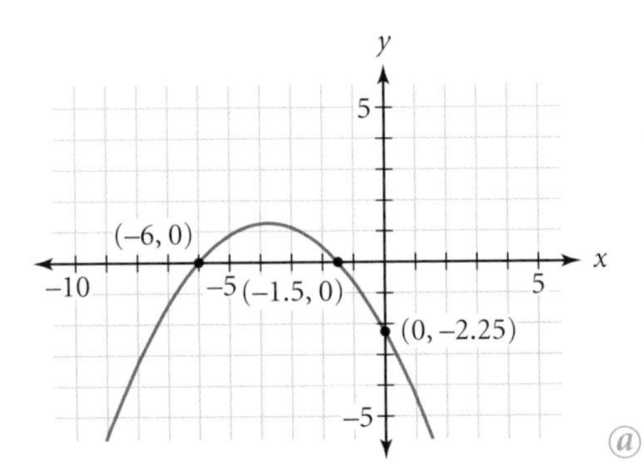

@

3. Convert each polynomial function to general form.

a. $y = (x - 4)(x - 6)$

b. $y = (x - 3)(x - 3)$

c. $y = x(x + 8)(x - 8)$ @

d. $y = 3(x + 2)(x - 2)(x + 5)$

4. Given the function $y = 2.5(x - 7.5)(x + 2.5)(x - 3.2)$:

a. Find the x-intercepts without graphing.

b. Find the y-intercept without graphing.

c. Write the function in general form.

d. Graph both the factored form and the general form of the function to check your work.

Reason and Apply

5. Use your work from the investigation to answer these questions.

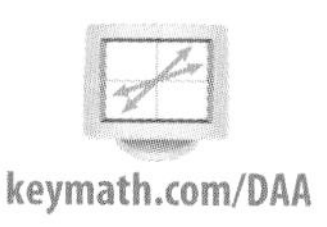

keymath.com/DAA

a. What x-value maximizes the volume of your box? What is the maximum volume possible?

b. What x-value or values give a volume of 300 cubic units?

c. The portion of the graph with domain $x > 10$ shows positive volume. What does this mean in the context of the problem?

d. Explain the meaning of the parts of the graph showing negative volume.

These boxes, on display during the 2002 Cultural Olympiad in Salt Lake City, Utah, through the organization Children Beyond Borders, were created by children with disabilities in countries worldwide. The children decorated identical 4-inch square boxes with their own creative visions, often in scenes from their countries and themes of love and unity.

1. The original cardboard box; **2.** © Diana Edna Cruz Yunes, *Corazón Mío, Mi Corazón*/Mexico; **3.** © Manal Deibes, *Are You Hungry?*/Jordan; **4.** © Leslie Hendricks, *Art Is What Makes the World Go 'Round*/USA; **5.** © Earl Hasith Vanabona, *Untitled*/Sri Lanka; **6.** © Renato Pinho, *Ocean*/Portugal. Participating artists of VSA arts (*www.vsarts.org*)

6. Write each polynomial as a product of factors. Some factors may include irrational numbers.

a. $4x^2 - 88x + 480$

b. $6x^2 - 7x - 5$ @

c. $x^3 + 5x^2 - 4x - 20$

d. $2x^3 + 16x^2 + 38x + 24$

e. $a^2 + 2ab + b^2$

f. $x^2 - 64$ @

g. $x^2 + 64$

h. $x^2 - 7$ @

i. $x^2 - 3x$

7. Sketch a graph for each situation if possible.

a. a quadratic function with only one real zero

b. a quadratic function with no real zeros

c. a quadratic function with three real zeros

d. a cubic function with only one real zero

e. a cubic function with two real zeros

f. a cubic function with no real zeros

8. Write the equation for a cubic function with x-intercepts -4.5, -1, and 2 that contains the point $(-4, -2.7)$.

9. Consider the function in this graph.

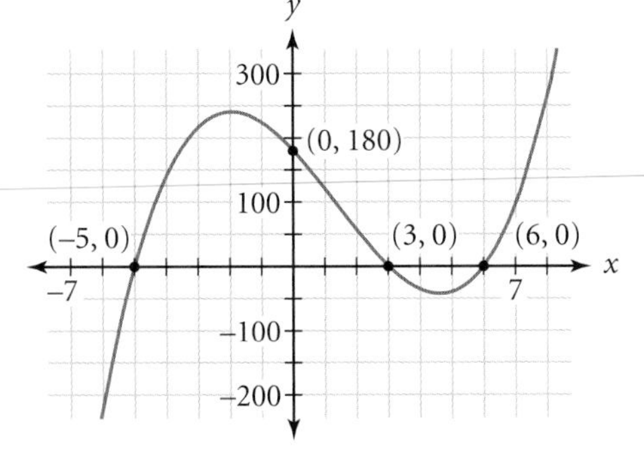

a. Write the equation of a polynomial function that has the x-intercepts shown. Use a for the scale factor.

b. Use the y-intercept to determine the vertical scale factor. Write the function from 9a, replacing a with the value of the scale factor. ⓐ

c. Imagine that this graph is translated up 100 units. Write the equation of the image. ⓐ

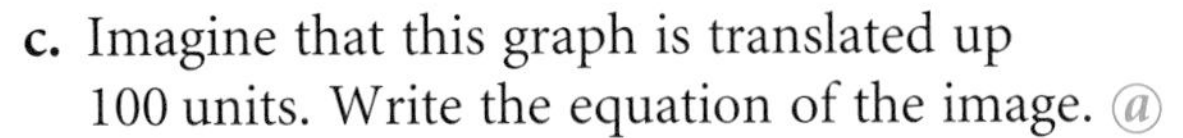

d. Imagine that this graph is translated left 4 units. Write the equation of the image.

10. APPLICATION The way you taste certain flavors is a genetic trait. For instance, the ability to taste the bitter compound phenyltheocarbamide (PTC) is inherited as a dominant trait in humans. In the United States, approximately 70% of the population can taste PTC, whereas 30% cannot.

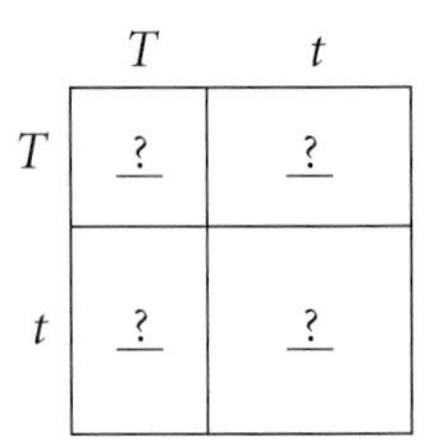

Every person inherits a pair of genes, one from each parent. A person who inherits the taster gene from *either* parent will be able to taste PTC. Let T represent the percentage of taster genes found in the population, and t represent the percentage of nontaster genes.

a. Complete the rectangle diagram of possible gene-pair combinations. What algebraic expression does this diagram represent? ⓐ

b. The sum of all possible gene-pair combinations must equal 1, or 100%, for the entire population. Write an equation to express this relationship. ⓐ

c. Use the fact that 70% of the U.S. population are tasters to write an equation that you can use to solve for t. ⓗ

d. Solve for t, the frequency of the recessive gene in the population.

e. What is the frequency of the dominant gene in the population?

f. What percentage of the U.S. population has two taster genes?

Review

11. Is it possible to find a quadratic function that contains the points $(-4, -2)$, $(-1, 7)$, and $(2, 16)$? Explain why or why not. ⓐ

12. Find the quadratic function whose graph has vertex $(-2, 3)$ and contains the point $(4, 12)$.

13. Find all real solutions.

a. $x^2 = 50.4$ ⓐ

b. $x^4 = 169$ ⓐ

c. $(x - 2.4)^2 = 40.2$

d. $x^3 = -64$

14. Find the inverse of each function algebraically. Then choose a value of x and check your answer.

a. $f(x) = \frac{2}{3}(x + 5)$

b. $g(x) = -6 + (x + 3)^{2/3}$

c. $h(x) = 7 - 2^x$

15. Use the finite differences method to find the function that generates this table of values. Explain your reasoning.

x	2.2	2.6	3.0	3.4
$f(x)$	-4.5	-5.5	-6.5	-7.5

LESSON

7.7

Higher-Degree Polynomials

It is good to have an end to journey towards, but it is the journey that matters in the end.

URSULA K. LEGUIN

Polynomials with degree 3 or higher are called higher-degree polynomials. Frequently, 3rd-degree polynomials are associated with volume measures, as you saw in Lesson 7.6.

If you create a box by removing small squares of side length x from each corner of a square piece of cardboard that is 30 inches on each side, the volume of the box in cubic inches is modeled by the function $y = x(30 - 2x)^2$, or $y = 4x^3 - 120x^2 + 900x$. The zero-product property tells you that the zeros are $x = 0$ or $x = 15$, the two values of x for which the volume is 0. The x-intercepts on the graph below confirm this.

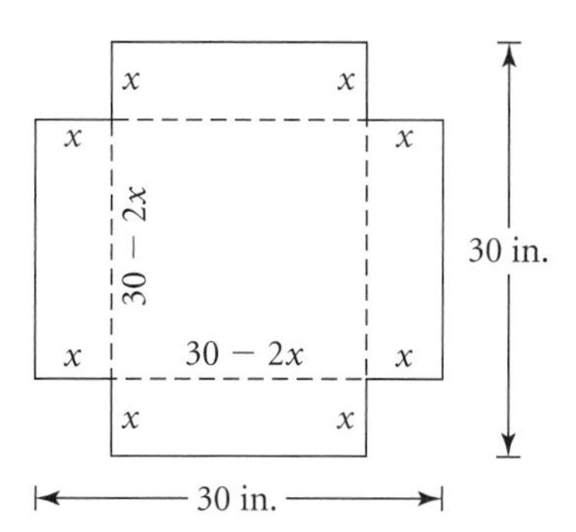

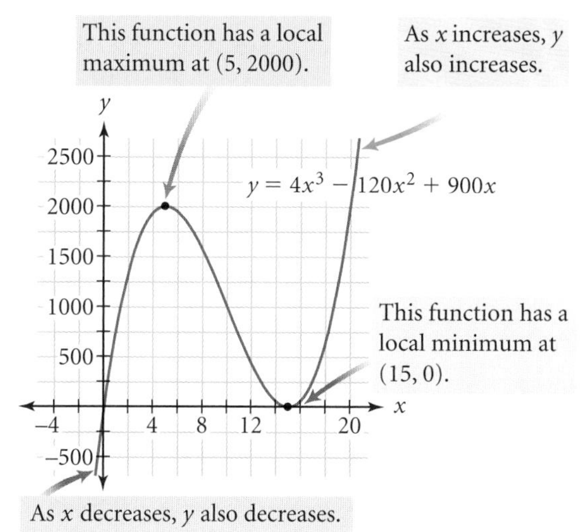

The shape of this graph is typical of the higher-degree polynomial graphs you will work with in this lesson. Note that this graph has one **local maximum** at (5, 2000) and one **local minimum** at (15, 0). These are the points that are higher or lower than all other points near them. You can also describe the **end behavior**—what happens to $f(x)$ as x takes on extreme positive and negative values. In the case of this cubic function, as x increases, y also increases. As x decreases, y also decreases.

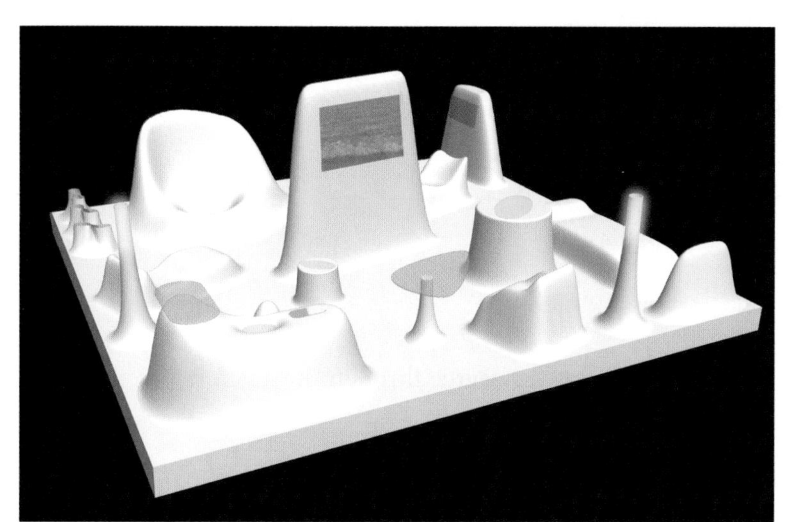

Graphs of all polynomials with real coefficients have a y-intercept and possibly one or more x-intercepts. You can also describe other features of polynomial graphs, such as local maximums or minimums and end behavior. Maximums and minimums, collectively, are called **extreme values.**

Egyptian-American artist Karim Rashid (b 1960) is a contemporary designer of fine art, as well as commercial products—from trash cans to sofas. This piece, *Softscape* (2001), is the artist's vision of a futuristic living room with chairs, tables, and a television melding together.

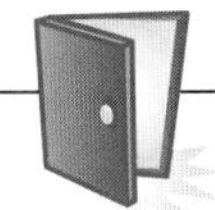

Investigation
The Largest Triangle

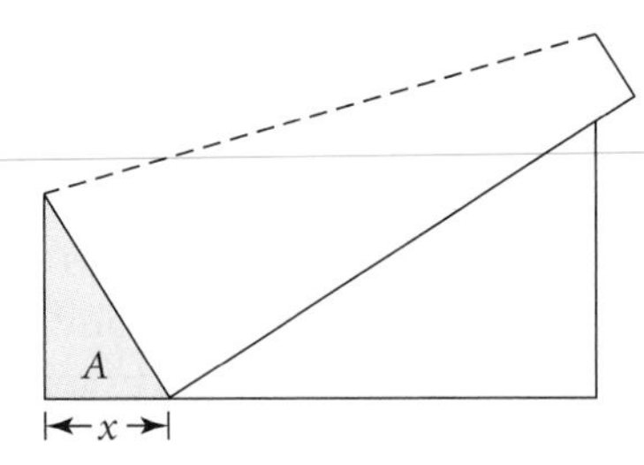

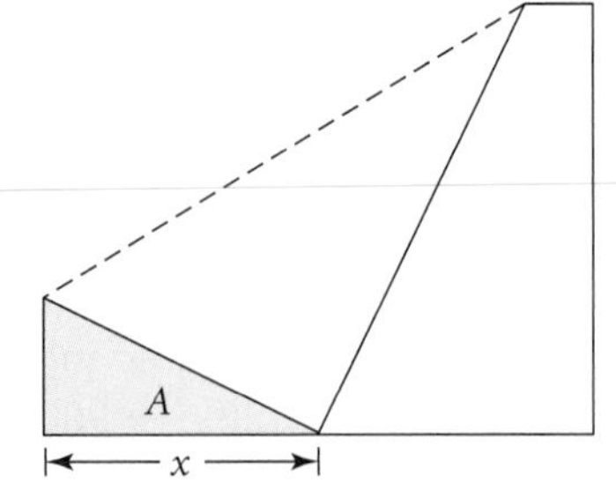

Take a sheet of notebook paper and orient it so that the longest edge is closest to you. Fold the upper-left corner so that it touches some point on the bottom edge. Find the area, A, of the triangle formed in the lower-left corner of the paper. What distance, x, along the bottom edge of the paper produces the triangle with the greatest area?

Work with your group to find a solution. You may want to use strategies you've learned in several lessons in this chapter. Write a report that explains your solution and your group's strategy for finding the largest triangle. Include any diagrams, tables, or graphs that you used.

In the remainder of this lesson you will explore the connections between a polynomial equation and its graph, which will allow you to predict when certain features will occur in the graph.

EXAMPLE A

Find a polynomial function whose graph has x-intercepts 3, 5, and -4, and y-intercept 180. Describe the features of its graph.

▶ Solution

A polynomial function with three x-intercepts has too many x-intercepts to be a quadratic function. It could be a 3rd-, 4th-, 5th-, or higher-degree polynomial function. Consider a 3rd-degree polynomial function, because that is the lowest degree that has three x-intercepts. Use the x-intercepts to write the equation $y = a(x - 3)(x - 5)(x + 4)$, where $a \neq 0$.

Substitute the coordinates of the y-intercept, $(0, 180)$, into this function to find the vertical scale factor.

$$180 = a(0 - 3)(0 - 5)(0 + 4)$$
$$180 = a(60)$$
$$a = 3$$

The polynomial function of the lowest degree through the given intercepts is

$$y = 3(x - 3)(x - 5)(x + 4)$$

Graph this function to confirm your answer and look for features.

This graph shows a local minimum at about $(4, -24)$ because that is the lowest point in its immediate neighborhood of x-values. There is also a local maximum at about $(-1.4, 220)$ because that is the highest point in its immediate neighborhood of x-values.

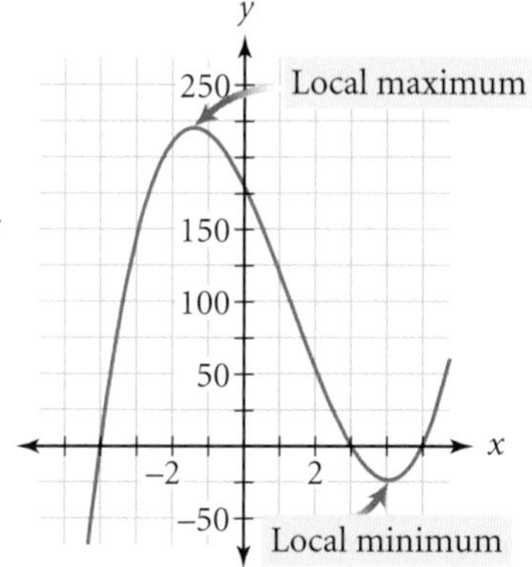

Even the small part of the domain shown in the graph suggests the end behavior. As x increases, y also increases. As x decreases, y also decreases. If you increase the domain of this graph to include more x-values at the right and left extremes of the x-axis, you'll see that the graph does continue this end behavior.

You can identify the degree of many polynomial functions by looking at the shapes of their graphs. Every 3rd-degree polynomial function has essentially one of the shapes shown below. Graph A shows the graph of $y = x^3$. It can be translated, stretched, or reflected. Graph B shows one possible transformation of Graph A. Graphs C and D show the graphs of general cubic functions in the form $y = ax^3 + bx^2 + cx + d$. In Graph C, a is positive, and in Graph D, a is negative.

Graph A

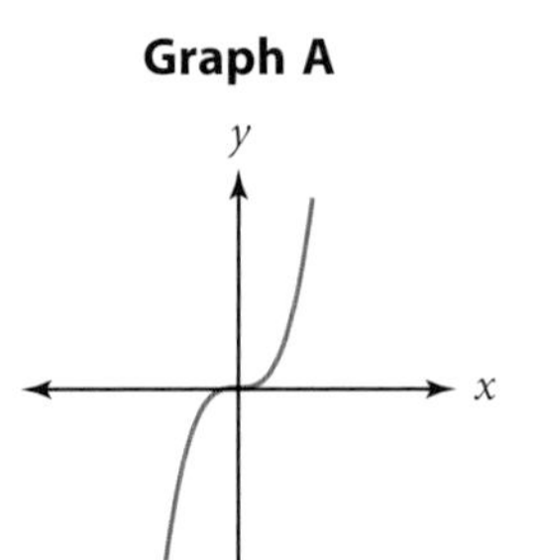

Graph B

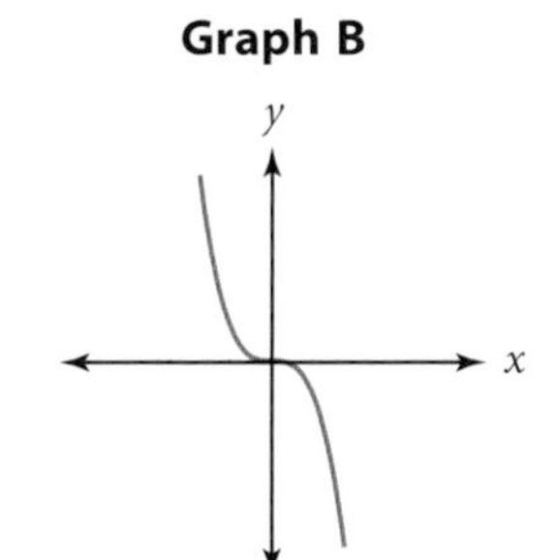

Graph C

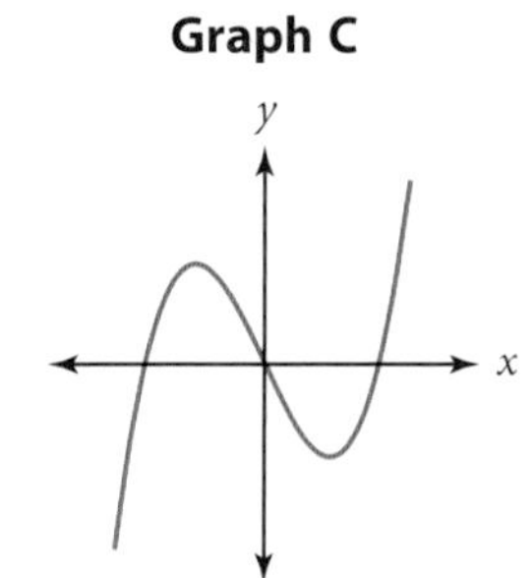

Graph D

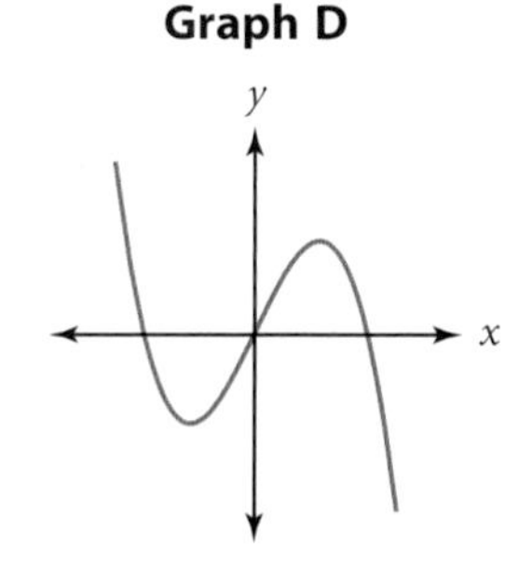

You'll explore the general shapes and characteristics of other higher-degree polynomials in the exercises.

EXAMPLE B

Write a polynomial function with real coefficients and zeros $x = 2$, $x = -5$, and $x = 3 + 4i$.

▸ Solution

For a polynomial function with real coefficients, complex zeros occur in conjugate pairs, so $x = 3 - 4i$ must also be a zero. In factored form the polynomial function of the lowest degree is

$$y = (x - 2)(x + 5)[x - (3 + 4i)][x - (3 - 4i)]$$

Multiplying the last two factors to eliminate complex numbers gives $y = (x - 2)(x + 5)\left(x^2 - 6x + 25\right)$. Multiplying all factors gives the polynomial function in general form:

$$y = x^4 - 3x^3 - 3x^2 + 135x - 250$$

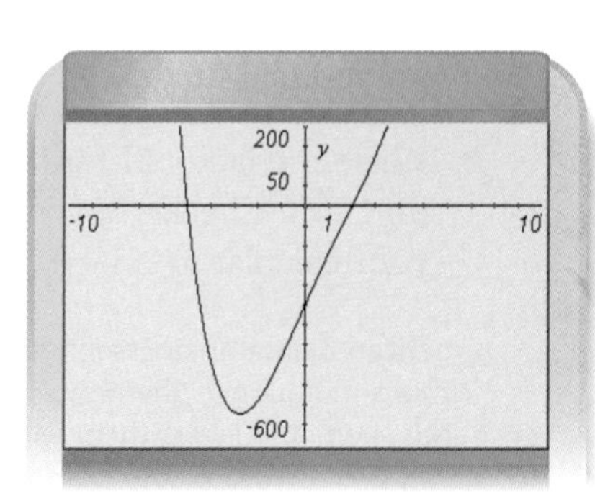

Graph this function to check your solution. You can't see the complex zeros, but you can see x-intercepts 2 and -5.

Note that in Example B, the solution was a 4th-degree polynomial function. It had four complex zeros, but the graph had only two x-intercepts, corresponding to the two real zeros. Any polynomial function of degree n always has n complex zeros (including repeated zeros) and at most n x-intercepts. Remember that complex zeros of polynomial functions with real coefficients always come in conjugate pairs.

EXERCISES

You will need

A graphing calculator for Exercises **4, 6, 10, 11,** and **17.**

Practice Your Skills

For Exercises 1–4, use these four graphs.

a.

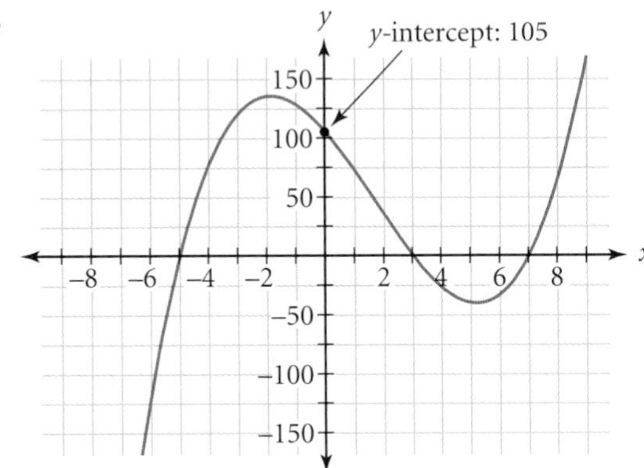

b. @

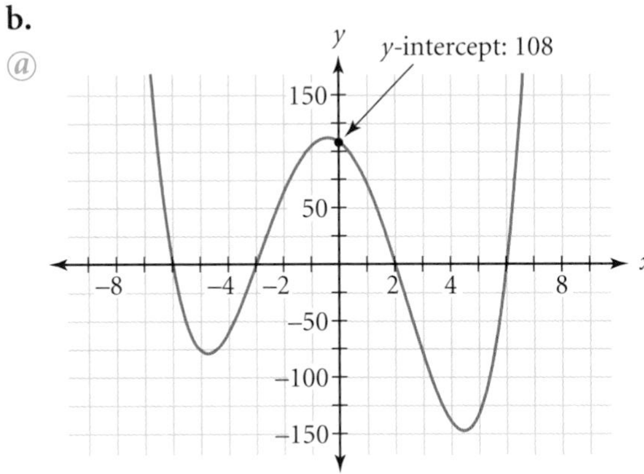

c.

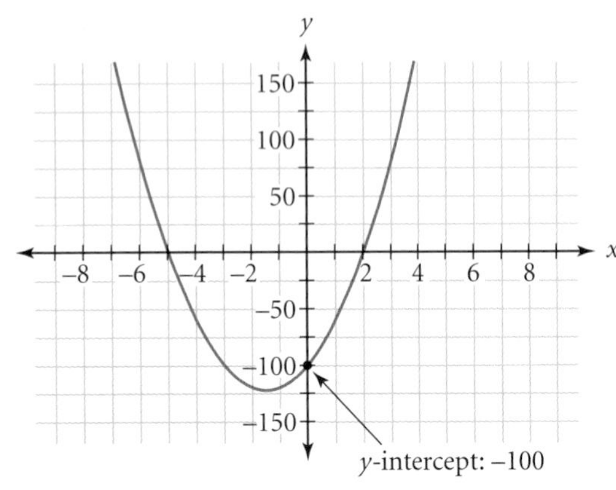

d.

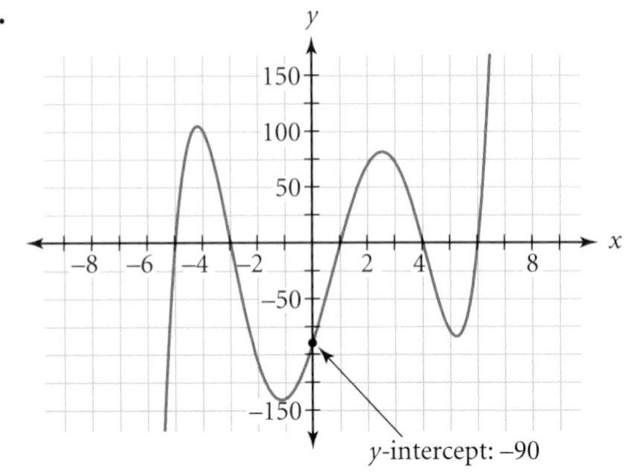

1. Identify the zeros of each function.

2. Give the coordinates of the y-intercept of each graph.

3. Identify the lowest possible degree of each polynomial function.

4. Write the factored form for each polynomial function. Check your work by graphing on your calculator.

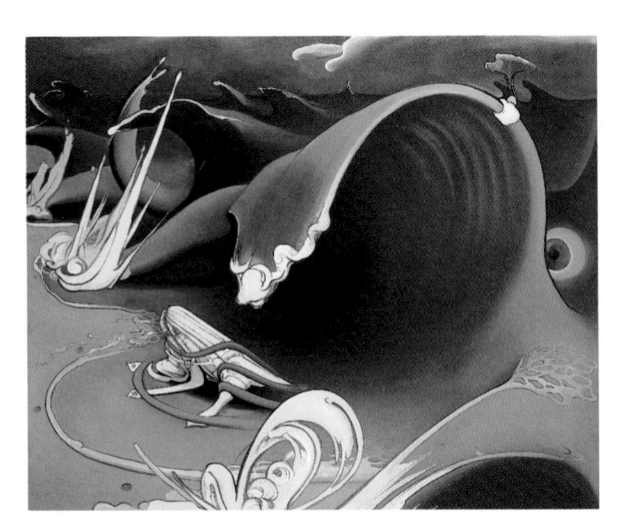

American painter Inka Essenhigh (b 1969) calls her works "cyborg mutations." She draws and paints images, and then sands them and layers them with enamel-based oil paint. This piece, *Green Wave* (2002), contains polynomial-like waves.

Reason and Apply

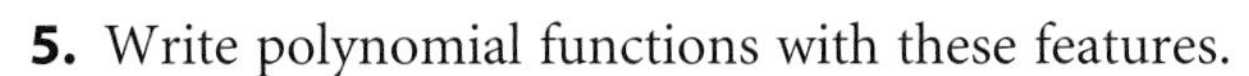

5. Write polynomial functions with these features.

a. A linear function whose graph has x-intercept 4.

b. A quadratic function whose graph has only one x-intercept, 4. ⓐ

c. A cubic function whose graph has only one x-intercept, 4.

6. **Mini-Investigation** The graph of $y = 2(x - 3)(x - 5)(x + 4)^2$ has x-intercepts 3, 5, and -4 because they are the only possible x-values that make $y = 0$. This is a 4th-degree polynomial, but it has only three x-intercepts. The root $x = -4$ is called a **double root** because the factor $(x + 4)$ occurs twice.

a. Complete the missing table entries for polynomial functions ii–vi.

Polynomial function	Degree	Roots
i. $y = 2(x - 3)(x - 5)(x + 4)^2$	4	one at 3, one at 5, two at -4
ii. $y = 2(x - 3)^2(x - 5)(x + 4)$		
iii. $y = 2(x - 3)(x - 5)^2(x + 4)$		
iv.		two at 3, one at 5, two at -4
v. $y = 2(x - 3)(x - 5)(x + 4)^3$		
vi. $y = 2(x - 3)(x - 5)^2(x + 4)^3$		

b. Make a complete graph of each of the functions i–vi. Each graph should display all relevant features, including local extreme values. ⓗ

c. Based on your graphs from 6b and the table from 6a, describe a connection between the power of a factor and what happens at that x-intercept. ⓗ

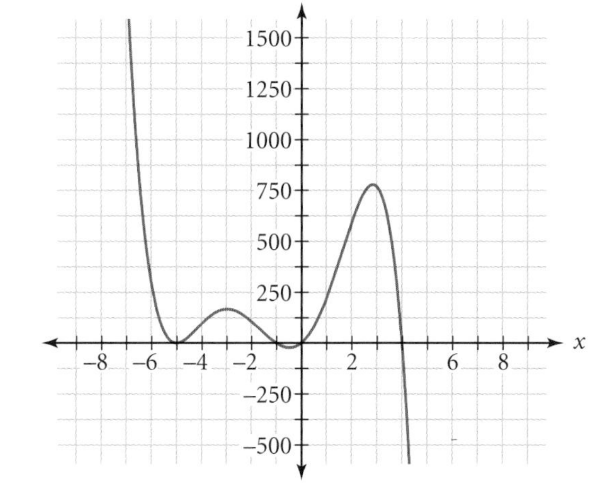

7. The graph at right is a complete graph of a polynomial function. ⓐ

a. How many x-intercepts are there?

b. What is the lowest possible degree of this polynomial function?

c. Write the factored form of this function if the graph includes the points $(0, 0)$, $(-5, 0)$, $(4, 0)$, $(-1, 0)$, and $(1, 216)$.

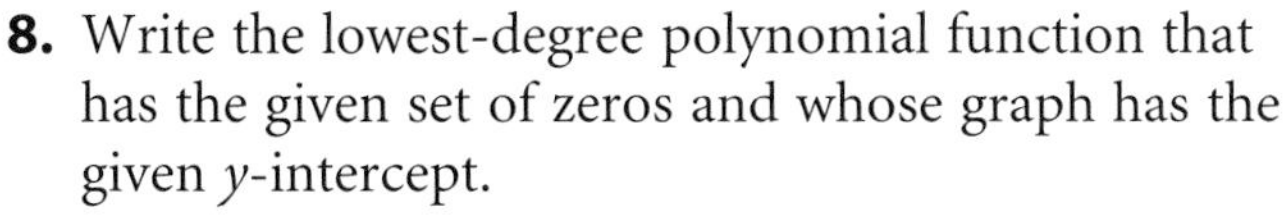

8. Write the lowest-degree polynomial function that has the given set of zeros and whose graph has the given y-intercept.

a. zeros: $x = -4$, $x = 5$, $x = -2$ (double root); y-intercept: -80

b. zeros: $x = -4$, $x = 5$, $x = -2$ (double root); y-intercept: 160 ⓐ

c. zeros: $x = \frac{1}{3}$, $x = -\frac{2}{5}$, $x = 0$; y-intercept: 0

d. zeros: $x = -5i$, $x = -1$ (triple root), $x = 4$; y-intercept: -100 ⓗ

9. Look back at Exercises 1–4. Find the products of the zeros in Exercise 1. How does the value of the leading coefficient, a, relate to the y-intercept, the product of the zeros, and the degree of the function? ⓐ

10. A 4th-degree polynomial function has the general form $y = ax^4 + bx^3 + cx^2 + dx + e$ for real values of a, b, c, d, and e, where $a \neq 0$. Graph several 4th-degree polynomial functions by trying different values for each coefficient. Be sure to include positive, negative, and zero values. Make a sketch of each different type of curve you get. Concentrate on the shape of the curve. You do not need to include axes in your sketches. Compare your graphs with the graphs of your classmates, and come up with six or more different shapes that describe all 4th-degree polynomial functions.

11. Each of these is the graph of a polynomial function with leading coefficient $a = 1$ or $a = -1$.

i.

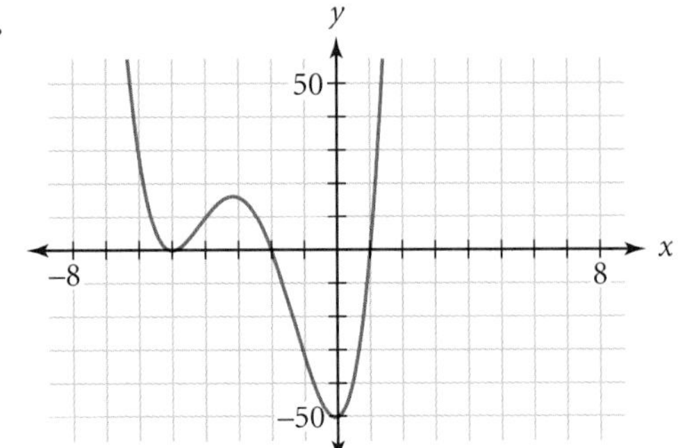

ii. @

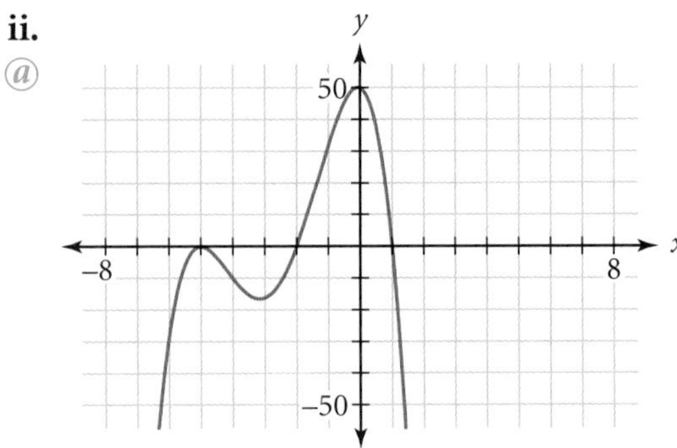

iii.

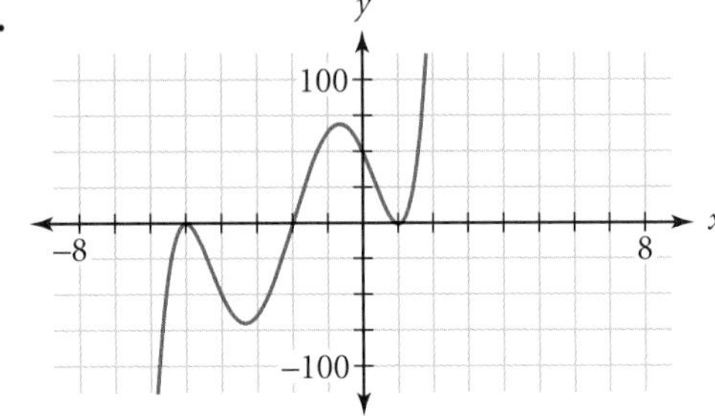

iv.

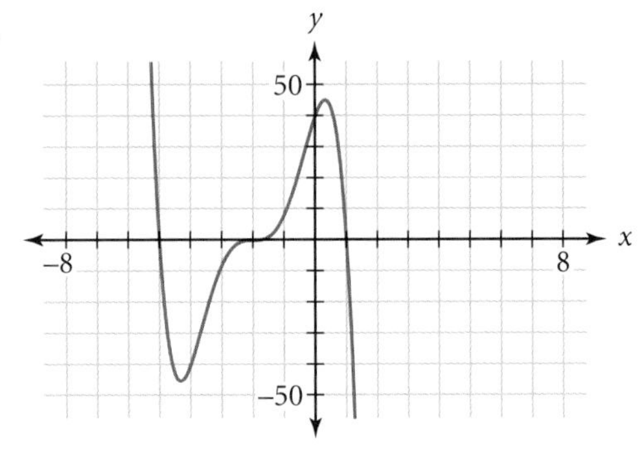

a. Write a function in factored form that will produce each graph.

b. Name the zeros of each polynomial function in 11a. If a factor is raised to the power of n, list the zero n times.

12. Consider the polynomial functions in Exercise 11.

a. What is the degree of each polynomial function?

b. How many extreme values does each graph have?

c. What is the relationship between the degree of the polynomial function and the number of extreme values?

d. Complete these statements:

i. The graph of a polynomial curve of degree n has at most _?_ x-intercepts.
ii. A polynomial function of degree n has at most _?_ real zeros.
iii. A polynomial function of degree n has _?_ complex zeros.
iv. The graph of a polynomial function of degree n has at most _?_ extreme values.

13. In the lesson you saw various possible appearances of the graph of a 3rd-degree polynomial function, and in Exercise 10 you explored possible appearances of the graph of a 4th-degree polynomial function. In Exercises 11 and 12, you found a relationship between the degree of a polynomial function and the number of zeros and extreme values. Use all the patterns you have noticed in these problems to sketch one possible graph of (h)

a. A 5th-degree function.

b. A 6th-degree function.

c. A 7th-degree function.

Review

14. Find the roots of these quadratic equations. Express them as fractions.

a. $0 = 3x^2 - 13x - 10$

b. $0 = 6x^2 - 11x + 3$

c. List all the factors of the constant term, c, and the leading coefficient, a, for 14a and b. What do you notice about the relationship between the factors of a and c, and the roots of the functions?

15. If $3 + 5\sqrt{2}$ is a solution of a quadratic equation with rational coefficients, then what other number must also be a solution? Write an equation in general form that has these solutions. (a)

16. Given the function $Q(x) = x^2 + 2x + 10$, find these values.

a. $Q(-3)$ **b.** $Q\left(-\frac{1}{5}\right)$ **c.** $Q(2 - 3\sqrt{2})$ (a) **d.** $Q(-1 + 3i)$

17. Solve this system using each method specified.

$$\begin{cases} 4x + 9y = 4 \\ 2x = 7 + 3y \end{cases}$$

a. Use an inverse matrix and matrix multiplication.

b. Write an augmented matrix, and reduce it to reduced row-echelon form.

18. **APPLICATION** According to Froude's Law, the speed at which an aquatic animal can swim is proportional to the square root of its length. (*On Growth and Form*, Sir D'Arcy Thompson, Cambridge University Press, 1961.) If a 75-foot blue whale can swim at a maximum speed of 20 knots, write a function that relates its speed to its length. How fast would a similar 60-foot-long blue whale be able to swim? (a)

A blue whale surfaces for air in the Gulf of St. Lawrence near Les Escoumins, Québéc, Canada. The blue whale is the world's largest mammal.

LESSON

7.8 More About Finding Solutions

It isn't that they can't see the solution. It is that they can't see the problem.

G. K. CHESTERTON

You can find zeros of a quadratic function by factoring or by using the quadratic formula. How can you find the zeros of a higher-degree polynomial?

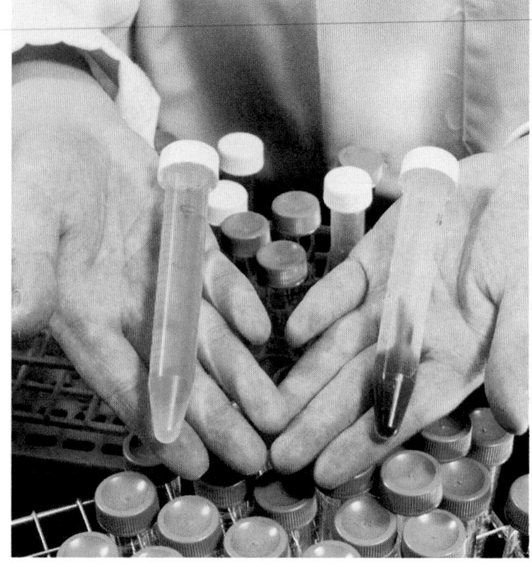

Blood is often separated into two of its factors, plasma and red blood cells.

Sometimes a graph will show you zeros in the form of x-intercepts, but only if they are real. And this method is often accurate only if the zeros have integer values. Fortunately, there is a method for finding exact zeros of many higher-degree polynomial functions. It is based on the procedure of long division.

You first need to find one or more zeros. Then you divide your polynomial function by the factors associated with those zeros. Repeat this process until you have a polynomial function that you can find the zeros of by factoring or using the quadratic formula. Let's start with an example in which we already know several zeros. Then you'll learn a technique for finding some zeros when they're not obvious.

EXAMPLE A What are the zeros of $P(x) = x^5 - 6x^4 + 20x^3 - 60x^2 + 99x - 54$?

► Solution The graph appears to have x-intercepts at 1, 2, and 3. You can confirm that these values are zeros of the function by substituting them into $P(x)$.

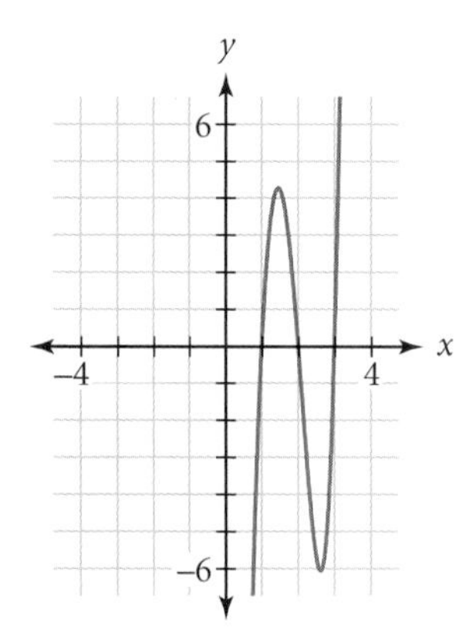

$$P(1) = (1)^5 - 6(1)^4 + 20(1)^3 - 60(1)^2 + 99(1) - 54 = 0$$
$$P(2) = (2)^5 - 6(2)^4 + 20(2)^3 - 60(2)^2 + 99(2) - 54 = 0$$
$$P(3) = (3)^5 - 6(3)^4 + 20(3)^3 - 60(3)^2 + 99(3) - 54 = 0$$

For all three values, you get $P(x) = 0$, which shows that $x = 1$, $x = 2$, and $x = 3$ are zeros. This also means that $(x - 1)$, $(x - 2)$, and $(x - 3)$ are factors of $x^5 - 6x^4 + 20x^3 - 60x^2 + 99x - 54$. None of the x-intercepts has the appearance of a repeated root, and you know that a 5th-degree polynomial function has five complex zeros, so this function must have two additional nonreal zeros.

You know that $(x - 1)$, $(x - 2)$, and $(x - 3)$ are all factors of the polynomial, so the product of these three factors, $x^3 - 6x^2 + 11x - 6$, must be a factor also. Your task is to find another factor such that

$$\left(x^3 - 6x^2 + 11x - 6\right)(\textit{factor}) = x^5 - 6x^4 + 20x^3 - 60x^2 + 99x - 54$$

You can find this factor by using long division.

First, divide x^5 by x^3 to get x^2.

$$\begin{array}{r} x^2 \\ x^3 - 6x^2 + 11x - 6 \overline{)\,x^5 - 6x^4 + 20x^3 - 60x^2 + 99x - 54} \\ \underline{x^5 - 6x^4 + 11x^3 - 6x^2 } \\ 9x^3 - 54x^2 + 99x - 54 \end{array}$$

Then, multiply x^2 by the divisor.

Subtract.

Now, divide $9x^3$ by x^3 to get 9.

$$\begin{array}{r} x^2 + 9 \\ x^3 - 6x^2 + 11x - 6 \overline{)\,x^5 - 6x^4 + 20x^3 - 60x^2 + 99x - 54} \\ \underline{x^5 - 6x^4 + 11x^3 - 6x^2 } \\ 9x^3 - 54x^2 + 99x - 54 \\ \underline{9x^3 - 54x^2 + 99x - 54} \\ 0 \end{array}$$

Then, multiply 9 by the divisor.

Subtract.

The remainder is zero, so the division is finished, resulting in two factors.

Now you can rewrite the original polynomial as a product of factors:

$$\begin{aligned} x^5 - 6x^4 + 20x^3 - 60x^2 + 99x - 54 &= \left(x^3 - 6x^2 + 11x - 6\right)\left(x^2 + 9\right) \\ &= (x - 1)(x - 2)(x - 3)\left(x^2 + 9\right) \end{aligned}$$

Now that the polynomial is in factored form, you can find the zeros. You knew three of them from the graph. The two additional zeros are contained in the factor $x^2 + 9$. What values of x make $x^2 + 9$ equal zero? If you solve the equation $x^2 + 9 = 0$, you get $x = \pm 3i$.

Therefore, the five zeros are $x = 1$, $x = 2$, $x = 3$, $x = 3i$, and $x = -3i$.

To confirm that a number is a zero, you can use the **Factor Theorem.**

Factor Theorem

$(x - r)$ is a factor of a polynomial function $P(x)$ if and only if $P(r) = 0$.

In the example you showed that $P(1)$, $P(2)$, and $P(3)$ equal zero. You can check that $P(3i)$ and $P(-3i)$ will also equal zero.

Division of polynomials is similar to the long-division process that you may have learned in elementary school. Both the original polynomial and the divisor are written in descending order of the powers of x. If any degree is missing, insert a term with coefficient 0 as a placeholder. For example, you can write the polynomial $x^4 + 3x^2 - 5x + 8$ as

$$x^4 + 0x^3 + 3x^2 - 5x + 8$$

Insert a zero placeholder because the polynomial did not have a 3rd-degree term.

Often you won't be able to find any zeros for certain by looking at a graph. However, there is a pattern to rational numbers that might be zeros. The **Rational Root Theorem** helps you narrow down the values that might be zeros of a polynomial function.

Rational Root Theorem

If the polynomial equation $P(x) = 0$ has rational roots, they are in the form $\frac{p}{q}$, where p is a factor of the constant term and q is a factor of the leading coefficient.

Notice that this theorem will identify only possible *rational* roots. It won't find roots that are irrational or contain imaginary numbers.

EXAMPLE B

Find the roots of this polynomial equation:

$$3x^3 + 5x^2 - 15x - 25 = 0$$

▶ Solution

First, graph the function $y = 3x^3 + 5x^2 - 15x - 25$ to see if there are any identifiable integer x-intercepts.

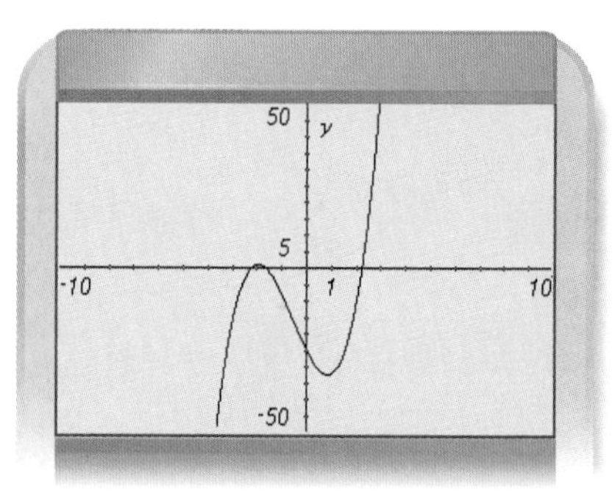

There are no integer x-intercepts, but the graph shows x-intercepts between -3 and -2, -2 and -1, and 2 and 3. Any rational root of this polynomial will be a factor of -25, the constant term, divided by a factor of 3, the leading coefficient. The factors of -25 are ±1, ±5, and ±25, and the factors of 3 are ±1 and ±3, so the possible rational roots are ±1, ±5, ±25, $\pm\frac{1}{3}$, $\pm\frac{5}{3}$, or $\pm\frac{25}{3}$. The only one of these that looks like a possibility on the graph is $-\frac{5}{3}$. Try substituting $-\frac{5}{3}$ into the original polynomial.

$$3\left(-\frac{5}{3}\right)^3 + 5\left(-\frac{5}{3}\right)^2 - 15\left(-\frac{5}{3}\right) - 25 = 0$$

Because the result is 0, you know that $-\frac{5}{3}$ is a root of the equation. If $-\frac{5}{3}$ is a root of the equation, then $\left(x+\frac{5}{3}\right)$ is a factor. Use long division to divide out this factor.

$$\begin{array}{r} 3x^2 - 15 \\ x+\frac{5}{3}\,\overline{)\,3x^3 + 5x^2 - 15x - 25} \\ \underline{3x^3 + 5x^2 } \\ 0 - 15x - 25 \\ \underline{-15x - 25} \\ 0 \end{array}$$

So $3x^3 + 5x^2 - 15x - 25 = 0$ is equivalent to $\left(x+\frac{5}{3}\right)(3x^2 - 15) = 0$. You already knew $-\frac{5}{3}$ was a root. Now solve $3x^2 - 15 = 0$.

$$3x^2 = 15$$
$$x^2 = 5$$
$$x = \pm\sqrt{5}$$

The three roots are $x = -\frac{5}{3}$, $x = \sqrt{5}$, and $x = -\sqrt{5}$. As decimal approximations, $-\frac{5}{3}$ is about -1.7, $\sqrt{5}$ is about 2.2, and $-\sqrt{5}$ is about -2.2. These values appear to be correct based on the graph.

Now that you know the roots, you could write the equation in factored form as $3\left(x+\frac{5}{3}\right)(x-\sqrt{5})(x+\sqrt{5}) = 0$, or $(3x+5)(x-\sqrt{5})(x+\sqrt{5}) = 0$. You need the coefficient of 3 to make sure you have the correct leading coefficient in general form.

When you divide a polynomial by a linear factor, such as $\left(x+\frac{5}{3}\right)$, you can use a shortcut method called **synthetic division.** Synthetic division is simply an abbreviated form of long division.

Consider this division of a cubic polynomial by a linear factor:

$$\frac{6x^3 + 11x^2 - 17x - 30}{x+2}$$

Here are the procedures using both long division and synthetic division:

Long Division

$$\begin{array}{r} 6x^2 - 1x - 15 \\ x+2\,\overline{)\,6x^3 + 11x^2 - 17x - 30} \\ (-)\ \underline{6x^3 + 12x^2 } \\ -1x^2 - 17x \\ (-)\ \underline{-1x^2 - 2x } \\ -15x - 30 \\ (-)\ \underline{-15x - 30} \\ 0 \end{array}$$

Synthetic Division

$$\begin{array}{r|rrrr} -2 & 6 & 11 & -17 & -30 \\ & & -12 & 2 & 30 \\ \hline & 6 & -1 & -15 & \boxed{0} \end{array}$$

$$6x^2 - 1x - 15$$

Both methods give a quotient of $6x^2 - 1x - 15$, but synthetic division certainly looks faster. The corresponding numbers in each process are shaded. Notice that synthetic division contains all of the same information, but in a condensed form.

Here's how to do synthetic division:

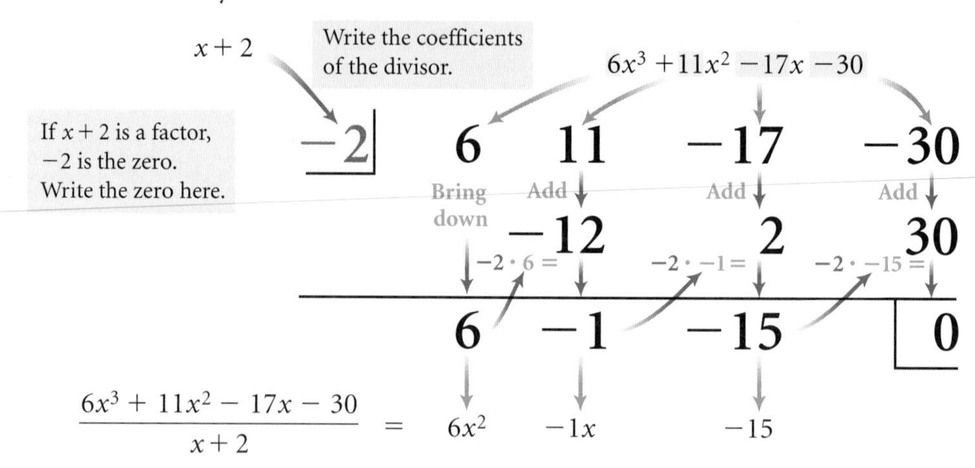

The number farthest to the right in the last row of a synthetic division problem is the remainder, which in this case is 0. When the remainder in a division problem is 0, you know that the divisor is a factor. This means -2 is a zero and the polynomial $6x^3 + 11x^2 - 17x - 30$ factors into the product of the divisor and the quotient, or $(x + 2)(6x^2 - 1x - 15)$. You could now use any of the methods you've learned—simple factoring, the quadratic formula, synthetic division, or perhaps graphing—to factor the quotient even further.

EXERCISES

You will need

A graphing calculator for Exercises **7**, **8**, **11**, and **15**.

Practice Your Skills

1. Find the missing polynomial in each long-division problem.

a. @

$$\begin{array}{r} a \\ x + 5\,\overline{)\,3x^3 + 22x^2 + 38x + 15} \\ \underline{3x^3 + 15x^2} \\ 7x^2 + 38x \\ \underline{7x^2 + 35x} \\ 3x + 15 \\ \underline{3x + 15} \\ 0 \end{array}$$

b.

$$\begin{array}{r} 2x^2 + 5x - 3 \\ 3x - 2\,\overline{)\,6x^3 + 11x^2 - 19x + 6} \\ \underline{b} \\ 15x^2 - 19x \\ \underline{15x^2 - 10x} \\ -9x + 6 \\ \underline{-9x + 6} \\ 0 \end{array}$$

2. Use the dividend, divisor, and quotient to rewrite each long-division problem in Exercise 1 as a factored product in the form $P(x) = D(x) \cdot Q(x)$. For example, $x^3 + 2x^2 + 3x - 6 = (x - 1)(x^2 + 3x + 6)$. @

3. Find the missing value in each synthetic-division problem.

a. $\begin{array}{r|rrrr} 4 & 3 & -11 & 7 & -44 \\ & & a & 4 & 44 \\ \hline & 3 & 1 & 11 & 0 \end{array}$ @

b. $\begin{array}{r|rrrr} -3 & 1 & 5 & -1 & -21 \\ & & -3 & -6 & 21 \\ \hline & 1 & b & -7 & 0 \end{array}$

c. $\begin{array}{r|rrrr} 1.5 & 4 & -8 & c & -6 \\ & & 6 & -3 & 6 \\ \hline & 4 & -2 & 4 & 0 \end{array}$ @

d. $\begin{array}{r|rrrr} d & 1 & 7 & 11 & -4 \\ & & -4 & -12 & 4 \\ \hline & 1 & 3 & -1 & 0 \end{array}$

4. Use the dividend, divisor, and quotient to rewrite each synthetic-division problem in Exercise 3 as a factored product in the form $P(x) = D(x) \cdot Q(x)$. @

5. Make a list of the possible rational roots of $0 = 2x^3 + 3x^2 - 32x + 15$.

Reason and Apply

6. Division often results in a remainder. In each of these problems, use the polynomial that defines P as the dividend and the polynomial that defines D as the divisor. Write the result of the division in the form $P(x) = D(x) \cdot Q(x) + R$, where R is an integer remainder. For example, $x^3 + 2x^2 + 3x - 4 = (x - 1)(x^2 + 3x + 6) + 2$. ⓗ

a. $P(x) = 47$, $D(x) = 11$ ⓐ

b. $P(x) = 6x^4 - 5x^3 + 7x^2 - 12x + 15$, $D(x) = x - 1$ ⓐ

c. $P(x) = x^3 - x^2 - 10x + 16$, $D(x) = x - 2$

7. Consider the function $P(x) = 2x^3 - x^2 + 18x - 9$.

a. Verify that $3i$ is a zero.

b. Find the remaining zeros of the function P.

8. Consider the function $y = x^4 + 3x^3 - 11x^2 - 3x + 10$. ⓐ

a. How many zeros does this function have?

b. Name the zeros.

c. Write the polynomial function in factored form.

9. Factor to find the zeros of these polynomial functions.

a. $y = x^3 - 5x^2 - 14x$

b. $y = x^3 + 3x^2 - 28x - 60$ if one zero is 5

10. Use your list of possible rational roots from Exercise 5 to write the function $y = 2x^3 + 3x^2 - 32x + 15$ in factored form.

11. When you trace the graph of a function on your calculator to find the value of an x-intercept, you often see the y-value jump from positive to negative when you pass over the zero. By using smaller windows, you can find increasingly accurate approximations for x. Use your calculator to find good approximations for the zeros of these functions. Then use synthetic or long division to find any nonreal zeros. [▸ See **Calculator Note 7H** for help finding zeros on your calculator. ◂]

a. $y = x^5 - x^4 - 16x + 16$ ⓐ

b. $y = 2x^3 + 15x^2 + 6x - 6$ ⓐ

c. $y = 0.2(x - 12)^5 - 6(x - 12)^3 - (x - 12)^2 + 1$

d. $y = 2x^4 + 2x^3 - 14x^2 - 9x - 12$

12. *Mini-Investigation* The difference of two squares, $a^2 - b^2$, can be factored over the set of real numbers, $a^2 - b^2 = (a + b)(a - b)$. Note that you could use polynomial division to find one factor if you know the other factor:

$$a + b\,\overset{\displaystyle a \qquad -b}{\overline{)\,a^2 + 0ab - b^2}}$$

a. Complete this polynomial division to find the missing factor for the difference of two cubes, $a^3 - b^3 = (a - b)(\underline{\ ?\ })$.

$$a - b\,\overline{)\,a^3 + 0a^2b + 0ab^2 - b^3}$$

b. Use polynomial division to find the missing factor for the sum of two cubes, $a^3 + b^3 = (a + b)(\underline{\ ?\ })$.

c. Use what you have learned to factor these polynomials.

i. $x^3 - 8$ **ii.** $x^3 + 8y^3$ **iii.** $a^2 - b^2i^2$ **iv.** $27x^3 - 8y^3$

d. Summarize how to factor the difference and sum of two cubes using

i. $x^3 - y^3$ **ii.** $x^3 + y^3$

13. You know how to use rectangle diagrams to multiply polynomials. You can also work "backward" to use a rectangle diagram to do division with polynomials.

a. Complete this rectangle diagram to find the quotient $\frac{2x^3 + 14x^2 + 17x - 15}{x + 5}$. Number your steps so another student could follow your work. (h)

x	$2x^3$		
5			

b. What are the factors of $2x^3 + 14x^2 + 17x - 15$?

Review

14. APPLICATION The relationship between the height and the diameter of a tree is approximately determined by the equation $f(x) = kx^{3/2}$, where x is the height in feet, $f(x)$ is the diameter in inches, and k is a constant that depends on the kind of tree you are measuring. (@)

a. A 221 ft British Columbian pine is about 21 in. in diameter. Find the value of k, and use it to express diameter as a function of height.

b. Give the inverse function.

c. Find the diameter of a 300 ft British Columbian pine.

d. What would be the height of a similar pine that is 15 in. in diameter?

15. Find a polynomial function of lowest possible degree whose graph passes through the points $(-2, -8.2)$, $(-1, 6.8)$, $(0, 5)$, $(1, -1)$, $(2, 1.4)$, and $(3, 24.8)$.

16. APPLICATION Sam and Beth have started a hat business in their basement. They make baseball caps and sun hats. Let b represent the number of baseball caps, and let s represent the number of sun hats. Manufacturing demands and machinery constraints confine the production per day to the feasible region defined by

$$\begin{cases} b \geq 0 \\ s \geq 0 \\ 4s - b \leq 20 \\ 2s + b \leq 22 \\ 7b - 8s \leq 77 \end{cases}$$

They make a profit of \$2 per baseball cap and \$1 per sun hat. (@)

a. Graph the feasible region. Give the coordinates of the vertices.

b. How many of each type of hat should they produce per day for maximum profit? (*Note:* At the end of each day, all partially made hats are recycled at no profit.) What is the maximum daily profit?

17. Write each quadratic function in general form and in factored form. Identify the vertex, y-intercept, and x-intercepts of each parabola.

a. $y = (x - 2)^2 - 16$ (@)

b. $y = 3(x + 1)^2 - 27$ (@)

c. $y = -\frac{1}{2}(x - 5)^2 + \frac{49}{2}$

d. $y = 2(x - 3)^2 + 3$

18. Solve.

a. $6x + x^2 + 5 = -4 + 4(x + 3)$

b. $7 = x(x + 3)$

c. $2x^2 - 3x + 1 = x^2 - x - 4$

CHAPTER 7 REVIEW

Polynomials can be used to represent the motion of projectiles, the areas of regions, and the volumes of boxes. When examining a set of data whose x-values form an arithmetic sequence, you can calculate the **finite differences** to find the **degree** of a polynomial that will fit the data. When you know the degree of the polynomial, you can define a system of equations to solve for the coefficients. Polynomial equations can be written in several forms. The form $a_nx^n + a_{n-1}x^{n-1} + \cdots + a_1x^1 + a_0$ is called **general form.** Quadratic equations, which are 2nd-degree polynomial equations, can also be written in **vertex form** or **factored form,** with each factor corresponding to a **root** of the equation. In a quadratic equation, you can find the roots by using the **quadratic formula.**

There are the same number of roots as the degree of the polynomial. In some cases these roots may include **imaginary numbers** or **complex numbers.** If the coefficients of a polynomial are real, then any nonreal roots come in **conjugate pairs.** The degree of a polynomial function and the values of its coefficients determine the shape of its graph. The graphs may have hills and valleys where you will find **local minimums** and **maximums.** The **Factor Theorem** helps you identify zeros, and the **Rational Root Theorem** helps you identify possible rational roots. You can use **long division** or **synthetic division** to find factors of complicated polynomial expressions.

EXERCISES

You will need

A graphing calculator for Exercises **3**, **5**, and **7**.

@ Answers are provided for all exercises in this set.

1. Factor each expression completely.

a. $2x^2 - 10x + 12$ **b.** $2x^2 + 7x + 3$ **c.** $x^3 - 10x^2 - 24x$

2. Solve each equation by setting it equal to zero and factoring.

a. $x^2 - 8x = 9$ **b.** $x^4 + 2x^3 = 15x^2$

3. Using three noncollinear points as vertices, how many different triangles can you draw? Given a choice of four points, no three of which are collinear, how many different triangles can you draw? Given a choice of five points? n points?

4. Tell whether each equation is written in general form, vertex form, or factored form. Write each equation in the other two forms, if possible.

a. $y = 2(x - 2)^2 - 16$ **b.** $y = -3(x - 5)(x + 1)$

c. $y = x^2 + 3x + 2$ **d.** $y = (x + 1)(x - 3)(x + 4)$

e. $y = 2x^2 + 5x - 6$ **f.** $y = -2 - (x + 7)^2$

5. Sketch a graph of each function. Label all zeros and the coordinates of all maximum and minimum points. (Each coordinate should be accurate to the nearest hundredth.)

a. $y = 2(x - 2)^2 - 16$ **b.** $y = -3(x - 5)(x + 1)$

c. $y = x^2 - 3x + 2$ **d.** $y = (x + 1)(x - 3)(x + 4)$

e. $y = x^3 + 2x^2 - 19x + 20$ **f.** $y = 2x^5 - 3x^4 - 11x^3 + 14x^2 + 12x - 8$

6. Write the equation of each graph in factored form.

a.

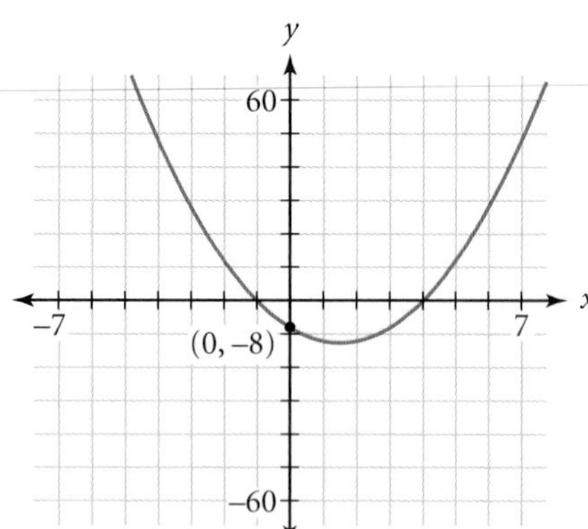

b.

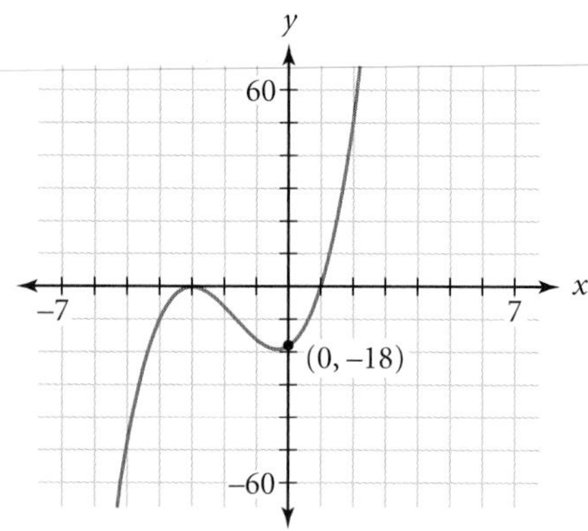

c.

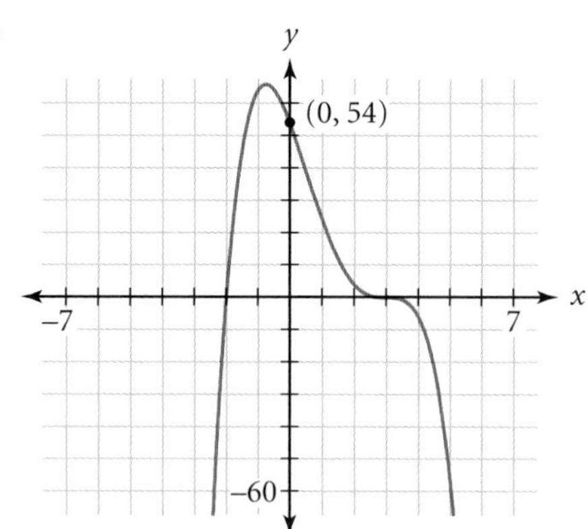

d.

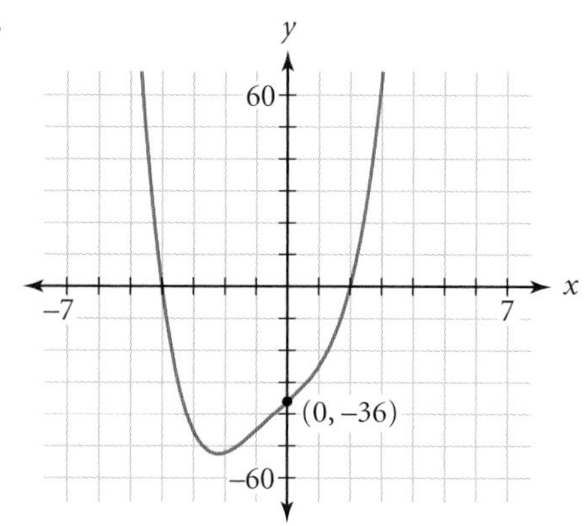

(*Hint:* One of the zeros occurs at $x = 3i$.)

7. APPLICATION By postal regulations, the maximum combined girth and length of a rectangular package sent by Priority Mail is 108 in. The length is the longest dimension, and the girth is the perimeter of the cross section. Find the dimensions of the package with maximum volume that can be sent through the mail. (Assume the cross section is always a square with side length x.) Making a table might be helpful.

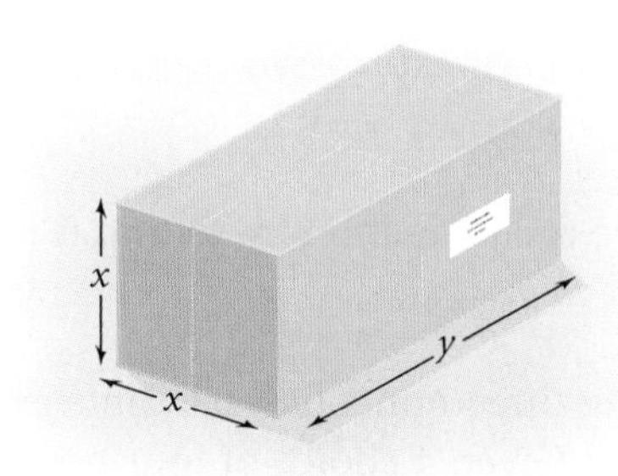

8. An object is dropped from the top of a building into a pool of water at ground level. There is a splash 6.8 s after the object is dropped. How high is the building in meters? In feet?

9. Consider this puzzle:

a. Write a formula relating the greatest number of pieces of a circle, y, you can obtain from x cuts.

b. Use the formula to find the maximum number of pieces with five cuts and with ten cuts.

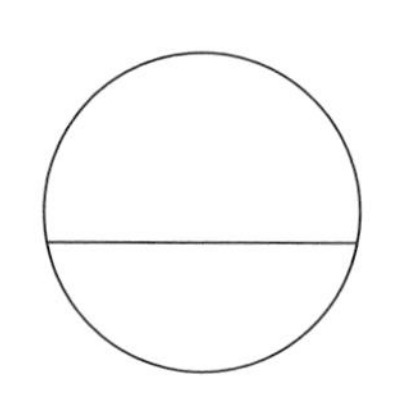

One cut

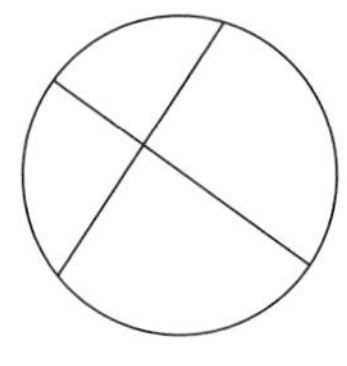

Two cuts

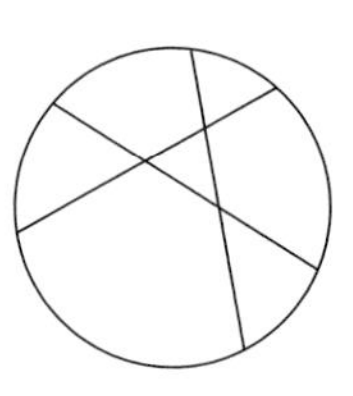

Three cuts

10. This 26-by-21 cm rectangle has been divided into two regions. The width of the unshaded region is x cm, as shown.

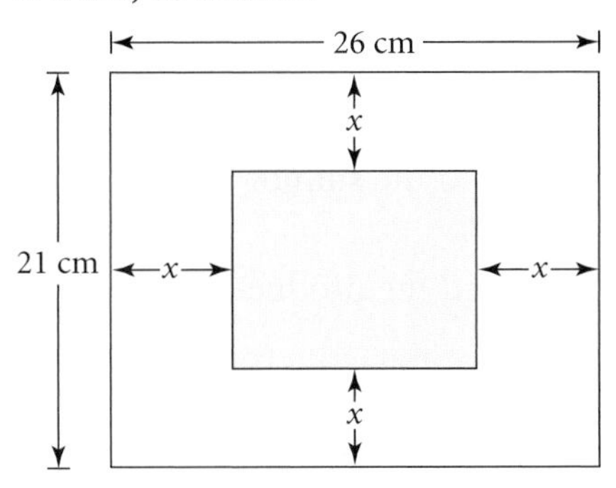

a. Express the area of the shaded part as a function of x, and graph it.

b. Find the domain and range for this function.

c. Find the x-value that makes the two regions (shaded and unshaded) equal in area.

11. Consider the polynomial equation $0 = 3x^4 - 20x^3 + 68x^2 - 92x - 39$.

a. List all possible rational roots.

b. Find the four roots of the equation.

The Grande Arche building at La Defense in Paris, France, is in the form of a hollowed-out cube and functions as an office building, conference center, and exhibition gallery.

12. Write each expression in the form $a + bi$.

a. $(4 - 2i)(-3 + 6i)$

b. $(-3 + 4i) - (3 + 13i)$

c. $\dfrac{2 - i}{3 - 4i}$

13. Divide.

$$\frac{6x^3 + 8x^2 + x - 6}{3x - 2}$$

TAKE ANOTHER LOOK

1. Use the method of finite differences to find the degree of a polynomial function that fits the data at right. What do you notice? Plot the points. What type of curve do you think might best fit the data?

x	y
0	60
1	42
2	28
3	20
4	14
5	10
6	7

2. How many points do you need to determine a line? How many points do you need to determine a parabola? Cubic curve? Quartic curve? Justify your conjectures using the method you learned in this chapter for finding the equations of polynomial curves.

3. You've already discovered several facts about quadratic polynomials in the form $f(x) = ax^2 + bx + c$. You know the vertex has coordinates $\left(\frac{-b}{2a}, c - \frac{b^2}{4a}\right)$, or $\left(\frac{-b}{2a}, \frac{4ac - b^2}{4a}\right)$, the line of symmetry is $x = -\frac{b}{2a}$, and the discriminant is $b^2 - 4ac$. You also know the quadratic formula for finding roots of the equation $ax^2 + bx + c = 0$: $x = \frac{-b \pm \sqrt{b^2 - 4ac}}{2a}$. Use the quadratic formula to complete 3a and b.

a. Find a relationship between the coefficients of $ax^2 + bx + c$ and the sum of the two roots.

b. Find a relationship between the coefficients of $ax^2 + bx + c$ and the product of the two roots.

c. Use your results for 3a and b to find the sum and product of the roots of these equations.

i. $2x^2 + x - 3 = 0$ **ii.** $14x^2 - 29x - 15 = 0$ **iii.** $4(3.5 - x)(x + 5) = 0$

d. Check your answers for 3c by finding the roots.

4. Choose any two complex numbers and plot them on a complex plane. Now add them and plot the resulting point. Try this with a few combinations of points. Do you see a geometric relationship between the third point and the first two? What if you subtract two complex numbers? Repeat the process: Choose two complex numbers, subtract, and plot the resulting point. Make a conjecture about the geometric relationship among the points.

5. Multiply each complex number represented by the vertices of $\triangle ABC$ by i. Plot the numbers associated with the results. Make a conjecture about the geometric meaning of $i(a + bi)$. Confirm your conjecture using other points and figures. Explore the geometric meaning of $i^2(a + bi)$. One way to do this is to multiply each complex number associated with the vertices of $\triangle ABC$ by i^2. Plot the points resulting from each multiplication. Make a conjecture for $i^n(a + bi)$. Confirm your conjecture.

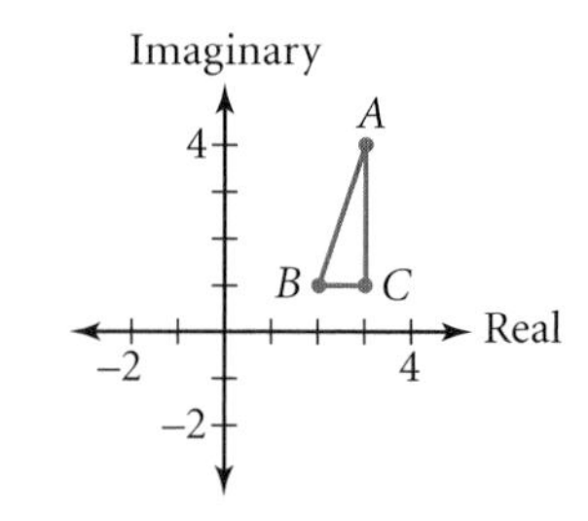

Assessing What You've Learned

WRITE TEST ITEMS In this chapter you learned what complex numbers are, how to do computations with them, and how they relate to polynomial functions. Write at least two test items that assess understanding of complex numbers. Be sure to include complete solutions.

PERFORMANCE ASSESSMENT While a classmate, a friend, a family member, or a teacher observes, show how you would find all zeros of a polynomial equation given in general form, or how you would find an equation in general form given the zeros. Explain the relationship between the zeros and the graph of a function, including what happens when a particular zero occurs multiple times.

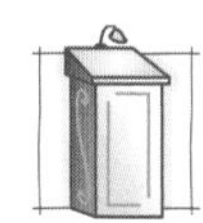

GIVE A PRESENTATION Give a presentation on how to do long division or synthetic division. Explain the advantages and the limitations of the method you have chosen, and describe a problem that could be solved using this method. If you like, solve the same problem using both methods, and show how they compare.

CHAPTER

8 Conic Sections and Rational Functions

Timetable (2000), designed by American architect and sculptor Maya Lin (b 1959), is a rotating circular clock/fountain located at Stanford University in Palo Alto, California. Made of black granite, steel, stone, and water, with the clock rings, motorized discs, and rotating parts submerged, the geometric sculpture gives the time in Pacific standard time, Pacific daylight saving time, and Universal time. Maya Lin's message, "Although we tend to think of time as an absolute, it is relative to location," is inscribed on a panel near the fountain.

OBJECTIVES

In this chapter you will

- use the distance formula to find the distance between two points on a plane and to solve distance and rate problems
- learn about conic sections—circles, ellipses, parabolas, and hyperbolas—which are created by intersecting a plane and a cone
- investigate the properties of conic sections
- write the equations of conic sections
- study rational functions and learn special properties of their graphs
- add, subtract, multiply, and divide rational expressions

The Distance Formula

In some parts of the United States, land is surveyed and divided into 1-mile squares, called *sections*. For instance, this portion of a map shows such a network of roads in northern Michigan. If you know there is 1 mile between roads, you can find the driving distance between two points by counting sections. However, if you could travel off-road and go directly from one point to another, you would need another way to determine the distance.

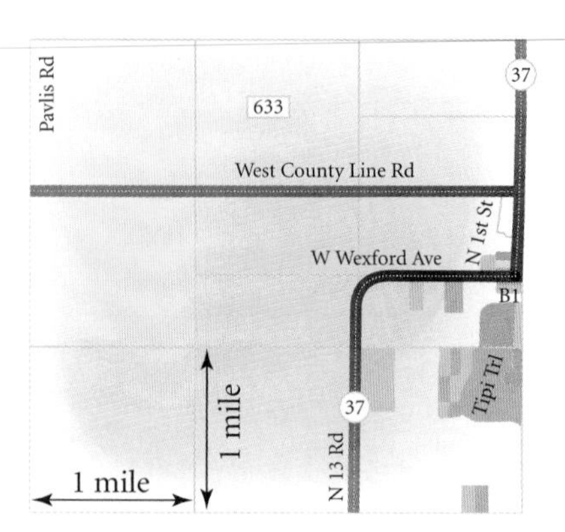

In fact, when looking for the distance between two points on the coordinate plane, you usually want the length of the direct path between the points. The Pythagorean Theorem provides a way to find that distance.

EXAMPLE A | Find the distance between the points $(-2, 1)$ and $(3, 4)$.

▶ Solution

Plot the points on a graph and draw a right triangle. Find the horizontal distance by finding the positive difference of the x-coordinates, and the vertical distance by finding the positive difference of the y-coordinates. Then you can find the length of the hypotenuse.

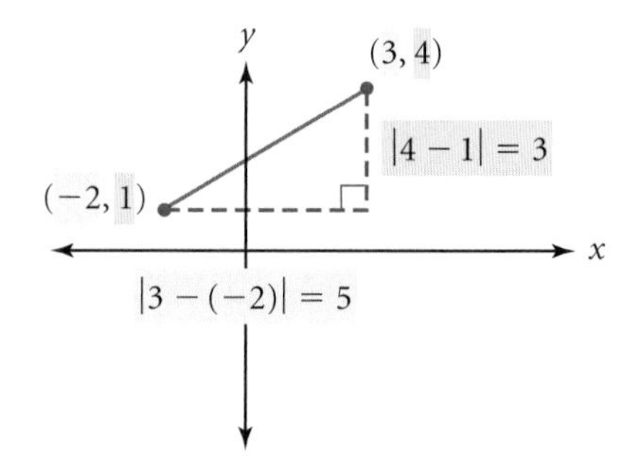

$$5^2 + 3^2 = d^2$$
$$34 = d^2$$
$$d = \sqrt{34}$$

The distance is $\sqrt{34}$, or about 5.83 units.

In general, suppose that d is the distance between two points (x_1, y_1) and (x_2, y_2). The Pythagorean Theorem gives $(x_2 - x_1)^2 + (y_2 - y_1)^2 = d^2$. Taking the square root of both sides gives you a formula for distance on a coordinate plane.

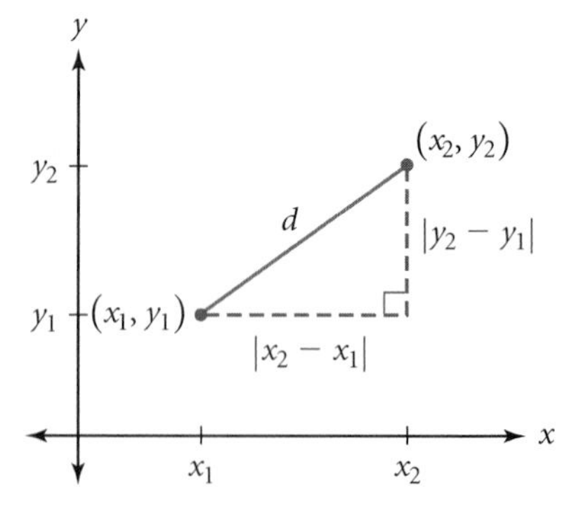

Distance Formula

The distance, d, between two points on a coordinate plane, $P_1(x_1, y_1)$ and $P_2(x_2, y_2)$, is given by the formula

$$d = \sqrt{(x_2 - x_1)^2 + (y_2 - y_1)^2}$$

EXAMPLE B

Antonio and Vikash are kicking a ball on a field. Their friend Maya is walking along the southern edge of the field and decides to join in. The three friends are positioned as shown in the diagram. Use x to represent Maya's distance from the southwest corner of the field and write an expression to answer each question.

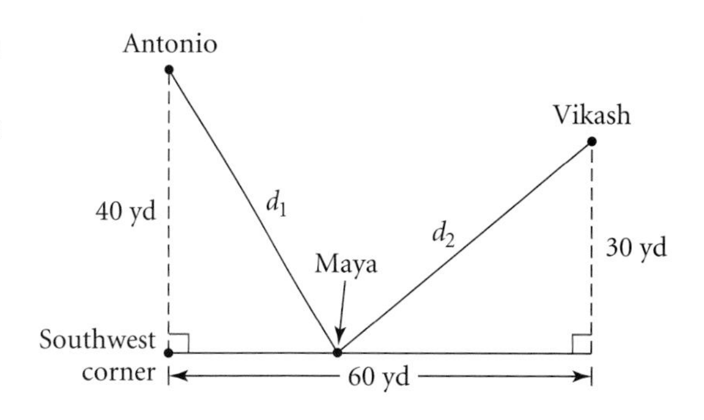

a. How far east of Maya is Vikash?

b. What is the distance from Antonio to Maya? From Vikash to Maya?

c. Write an equation that represents Antonio and Vikash being the same distance from Maya.

▶ ***Solution***

Sketch the situation recording all of the information you know.

a. You can express this distance without using the distance formula. Because Vikash is 60 yd east of the southwest corner, and Maya is x yd east of it, the rest of the distance must be the difference between 60 and x, or $60 - x$.

b. The direct path from Antonio to Maya is the hypotenuse of a right triangle with legs x and 40. Use the Pythagorean Theorem to find the length of the path: $x^2 + 40^2 = d_1^2$, so $d_1 = \sqrt{x^2 + 1600}$.

The direct path from Vikash to Maya is the hypotenuse of a right triangle with legs $60 - x$ and 30. Use the Pythagorean Theorem: $(60 - x)^2 + 30^2 = d_2^2$, so $d_2 = \sqrt{(60 - x)^2 + 900}$.

c. To represent Antonio and Vikash being the same distance from Maya, set the expressions from part b equal to each other: $\sqrt{x^2 + 1600} = \sqrt{(60 - x)^2 + 900}$.

EXERCISES

1. Find the distance between each pair of points.

a. (2, 7), (5, 11) ⓐ **b.** (−3, −1), (2, −5) **c.** (a, b), (c, d)

2. Make a sketch to represent each situation and label all distances.

a. Smallville and Bigcity are 147 miles apart. Middletown is x miles from Smallville on the way to Bigcity. ⓐ

b. A 13-foot ladder is leaning against the side of a house. The base of the ladder is x feet from the house.

3. A triangle has vertices $A(1, 2)$, $B(3, -1)$, and $C(5, 3)$.

a. Which side is the longest? ⓐ

b. What is the perimeter of the triangle? Find the exact value and the value to the nearest tenth of a unit.

LESSON

8.1

Using the Distance Formula

It is impossible to be a mathematician without being a poet in soul.

SOPHIA KOVALEVSKAYA

Imagine a race in which you carry an empty bucket from the starting line to the edge of a pool, fill the bucket with water, and then carry the bucket to the finish line. Luck, physical fitness, common sense, a calm attitude, and a little mathematics will make a difference in how you perform. Your performance depends on your speed and on the distance you travel. As you'll see in the investigation, mathematics can help you find the shortest path.

This illustration of "Jack and Jill" was created by the English illustrator Walter Crane (1845–1915).

Investigation
Bucket Race

You will need

- centimeter graph paper
- a ruler

The starting line of a bucket race is 5 m from one end of a pool, the pool is 20 m long, and the finish line is 7 m from the opposite end of the pool, as shown. In this investigation you will find the shortest path from point A to a point C on the edge of the pool to point B. That is, you will find the value of x, the distance in meters from the end of the pool to point C, such that $AC + CB$ is the shortest path possible.

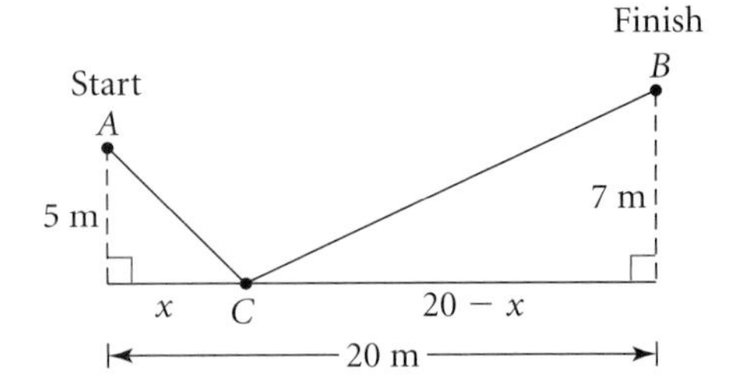

Step 1 Make a scale drawing of the situation on graph paper.

Step 2 Plot several different locations for point C. For each, measure the distance x and find the total length $AC + CB$. Record your data.

Step 3 What is the best location for C such that the length $AC + BC$ is minimized? What is the distance traveled? Is there more than one best location? Describe at least two different methods for finding the best location for C.

Step 4 Make a scale drawing of your solution.

Imagine that the amount of water you empty out at point B is an important factor in winning the race. This means you must move carefully so as not to spill water, and you'll be able to move faster with the empty bucket than you can with the bucket full of water. Assume that you can carry an empty bucket at a rate of 1.2 m/s and that you can carry a full bucket, without spilling, at a rate of 0.4 m/s.

Step 5 | Go back to the data collected in Step 2 and find the time needed for each x-value.

Step 6 | Now find the best location for point C so that you minimize the time from point A to the pool edge, then to point B. What is your minimum time? What is the distance traveled? How does this compare with your answer in Step 3? Describe your solution process.

In Refreshing Your Skills for Chapter 8, you found the distance between points and you wrote equations involving distances. You can use expressions created from the distance formula to solve many problems.

History CONNECTION

In 1790, the U.S. Coast Guard was formed to prevent smuggling and maintain customs laws. Combined with the Life Saving Service in 1915, it now oversees rescue missions, environmental protection, navigation, safety during weather hazards, port security, boat safety, and oil tanker transfers. The U.S. Coast Guard has both military and volunteer divisions.

U.S. Coast Guard ship *Acushnet*

EXAMPLE A

An injured worker must be rushed from an oil rig 15 mi offshore to a hospital in the nearest town 98 mi down the coast from the oil rig.

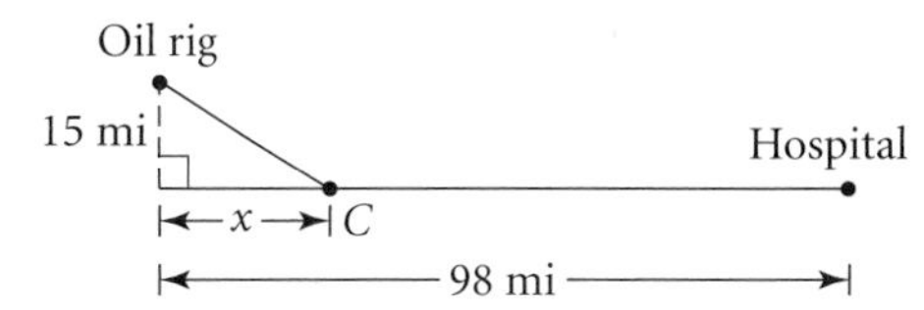

a. Let x represent the distance in miles from the point on the shore closest to the oil rig and another point, C, on the shore.

How far does the injured worker travel, in terms of x, if a boat takes him to C and then an ambulance takes him to the hospital?

b. Assume the boat travels at an average rate of 23 mi/h and the ambulance travels at an average rate of 70 mi/h. What value of x makes the trip 3 h?

▶ Solution

Use the distance formula, $d = \sqrt{(x_2 - x_1)^2 + (y_2 - y_1)^2}$.

a. The boat must travel $\sqrt{15^2 + x^2}$, and the ambulance must travel $98 - x$. The total distance in miles is $\sqrt{15^2 + x^2} + 98 - x$.

b. Distance equals rate times time, or $d = rt$. Solving for time, $t = \frac{d}{r}$. So the boat's time is $\frac{\sqrt{15^2 + x^2}}{23}$, and the ambulance's time is $\frac{98 - x}{70}$. The total time in hours, y, is represented by

$$y = \frac{\sqrt{15^2 + x^2}}{23} + \frac{98 - x}{70}$$

One way to find the value of x that gives a trip of 3 h is to graph the total time equation and $y = 3$, and use the graph to approximate the intersection. The graphs intersect when x is approximately 51.6. For the trip to be 3 h, the boat and the ambulance should meet at the point on the shore 51.6 mi from the point closest to the oil rig.

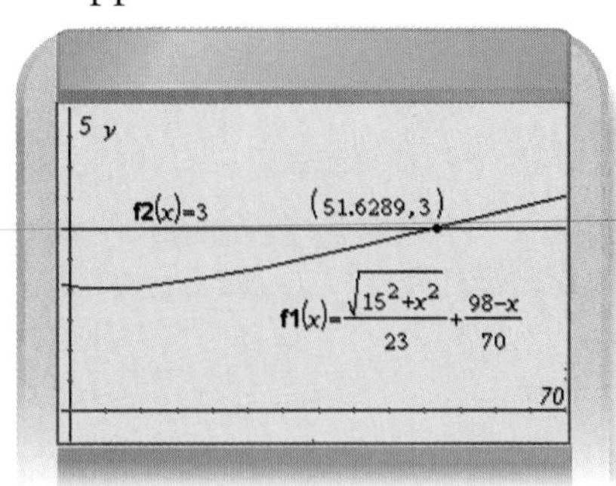

You can use the distance formula to describe the points on a circle. If a circle is centered at the origin and the points on the circle are all r units away from the center, then the equation describing the location of those points is $\sqrt{(x-0)^2+(y-0)^2}=r$, or $\sqrt{x^2+y^2}=r$, or $x^2+y^2=r^2$.

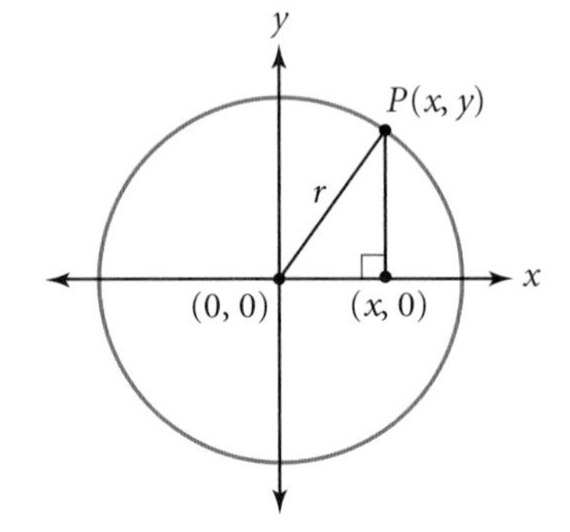

An equation that describes a set of points that meet a certain condition is called a **locus.** For example, the locus of points that are 3 units from the point (0, 0) is the circle with equation $x^2 + y^2 = 9$. The locus of points midway between $(-4, 0)$ and $(4, 0)$ is the line $x = 0$, the y-axis. In this chapter you will explore equations describing a variety of different **loci** (the plural of locus).

EXAMPLE B

Find the equation of the locus of points that are equidistant from the points (1, 3) and (5, 6).

▶ Solution

First make a sketch of the situation. Plot the two given points and then estimate the location of a few points that are the same distance from the given points. Do they appear to be in any sort of pattern? In this case they appear to lie on a line. Therefore, you should expect to find the equation of a line as your final answer. Let d_1 represent the distance between (1, 3) and any point, (x, y), on the locus. By the distance formula,

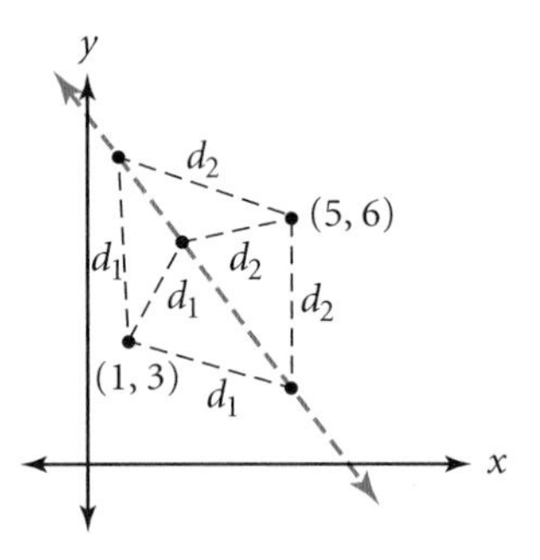

$$d_1 = \sqrt{(x-1)^2 + (y-3)^2}$$

Let d_2 represent the distance between (5, 6) and the same point on the locus, so

$$d_2 = \sqrt{(x-5)^2 + (y-6)^2}$$

The locus of points contains all points whose coordinates satisfy the equation $d_1 = d_2$, or

$$\sqrt{(x-1)^2 + (y-3)^2} = \sqrt{(x-5)^2 + (y-6)^2}$$

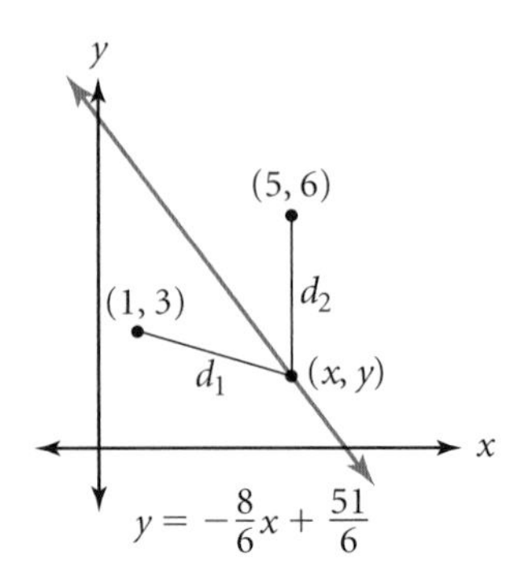

Use algebra to transform this equation into something more familiar.

$\sqrt{(x-1)^2+(y-3)^2} = \sqrt{(x-5)^2+(y-6)^2}$	Original equation.
$(x-1)^2+(y-3)^2 = (x-5)^2+(y-6)^2$	Square both sides.
$x^2-2x+1+y^2-6y+9 = x^2-10x+25+y^2-12y+36$	Expand the binomials.
$-2x+1-6y+9 = -10x+25-12y+36$	Subtract x^2 and y^2 from both sides.
$8x+6y=51$	Combine like terms and rewrite in the form $ax + by = c$.

As you predicted, the locus is a line. It has equation $8x + 6y = 51$, or $y = -\frac{8}{6}x + \frac{51}{6}$. How could you show that this locus is the perpendicular bisector of the segment joining the two points?

Just as the Pythagorean Theorem and the distance formula are useful for finding the equation of a locus of points, they are helpful for solving real-world problems.

EXERCISES

You will need

A graphing calculator for Exercises **8–11**.

Practice Your Skills

1. Find the distance between each pair of points.

a. (2, 5) and (8, 13)

b. (0, 3) and (5, 10)

c. (−4, 6) and (−2, −3)

d. $(3, d)$ and $(-6, 3d)$

2. Sketch the locus of points for each situation.

a. A point is 3 units from the point (1, −1).

b. A point is 5 units from the point (−2, 0).

c. A point is the same distance from the point (1, 2) as it is from the point (5, 7).

3. Write an equation for each locus in Exercise 2.

4. The distance between the points (2, 7) and $(5, y)$ is 5 units. Make a sketch of the situation and find the possible value(s) of y. @

5. The distance between the points (−1, 5) and $(x, -2)$ is 47. Make a sketch of the situation and find the possible value(s) of x.

Reason and Apply

6. A point on the line $y = 1$ has coordinates $(x, 1)$. Two other points have coordinates (2, 5) and (4, 9).

a. Write an expression for the distance between the point on the line and the point (2, 5). @

b. Write an expression for the distance between the point on the line and the point (4, 9).

7. In a set of points, each point is twice as far from the point (2, 0) as it is from (5, 0).

a. Sketch this locus of points. (h)

b. Write an equation for the locus and solve it for y.

8. If you are too close to a radio tower, you will be unable to pick up its signal. Let the center of a town be represented by the origin of a coordinate plane. Suppose a radio tower is located 2 mi east and 3 mi north of the center of town, or at the point (2, 3). A highway runs north-south 2.5 mi east of the center of town, along the line $x = 2.5$. Where on the highway will you be less than 1 mi from the tower and therefore unable to pick up the signal?

9. Josh is riding his mountain bike when he realizes that he needs to get home quickly for dinner. He is 2 mi from the road, and home is 3 mi down that road. He can ride 9 mi/h through the field separating him from the road and can ride 22 mi/h on the road.

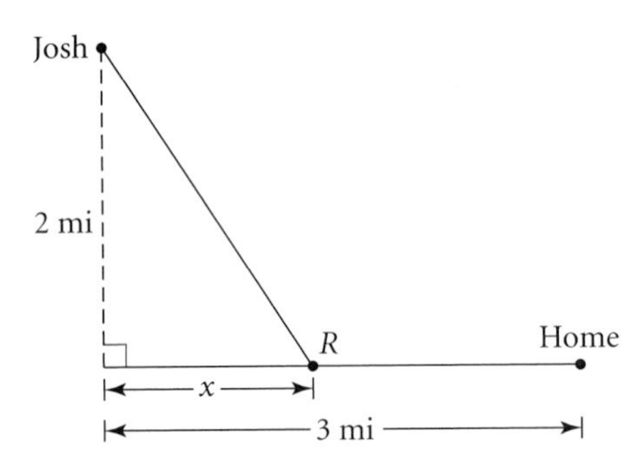

a. If Josh rides through the field to a point R on the road, and then home along the road, how far will he ride through the field? How far on the road? Let x represent the distance in miles between point R and the point on the road that is closest to his current location. (a)

b. How much time will Josh spend riding through the field? How much time on the road? (h)

c. What value of x gets Josh home the fastest? What is the minimum time?

10. A 10 m pole and a 13 m pole are 20 m apart at their bases. A wire connects the top of each pole with a point on the ground between them.

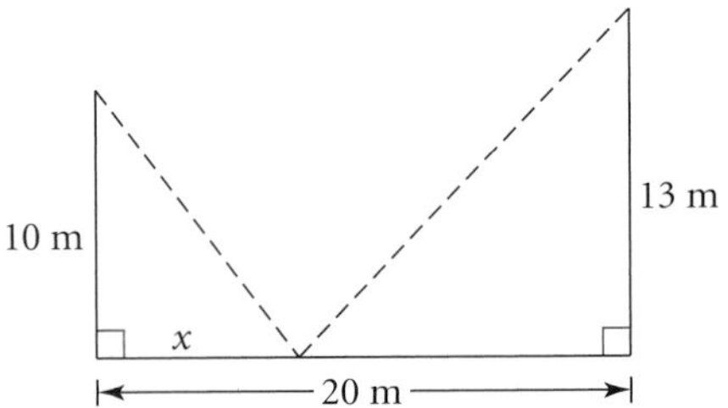

a. Let y represent the total length of the wire. Write an equation that relates x and y. (a)

b. What domain and range make sense in this situation?

c. Where should the wire be fastened to the ground so that the length of wire is minimized? What is the minimum length?

This cable-stay bridge, the Fred Hartman Bridge, connects Baytown, Texas, and La Porte, Texas. Taut cables stretch from the tops of two towers to support the roadway. For more information on cable-stay bridges, see the links at www.keymath.com/DAA .

11. A 24 ft ladder is placed upright against a wall. Then the top of the ladder slides down the wall while the foot of the ladder is pulled outward along the ground at a steady rate of 2 ft/s.

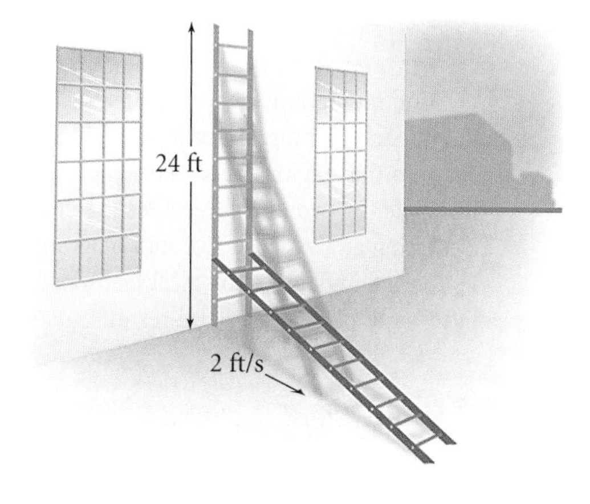

a. Find the heights that the ladder reaches at 1 s intervals while the ladder slides down the wall. @

b. How long will it take before the ladder is lying on the ground?

c. Does the top of the ladder also slide down the wall at a steady rate of 2 ft/s? Explain your reasoning.

12. Let d represent the distance between the point $(5, -3)$ and any point, (x, y), on the parabola $y = 0.5x^2 + 1$.

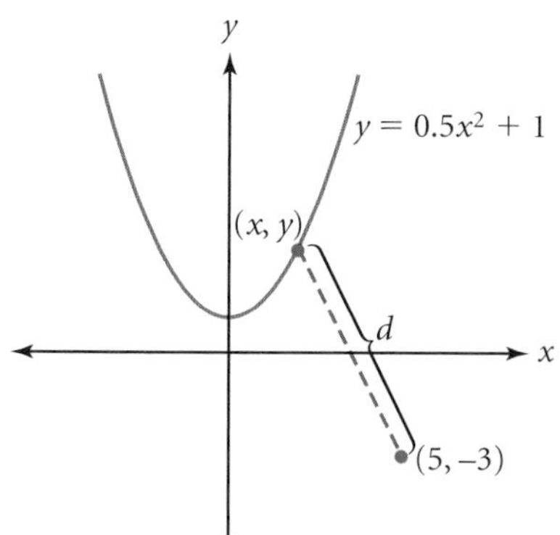

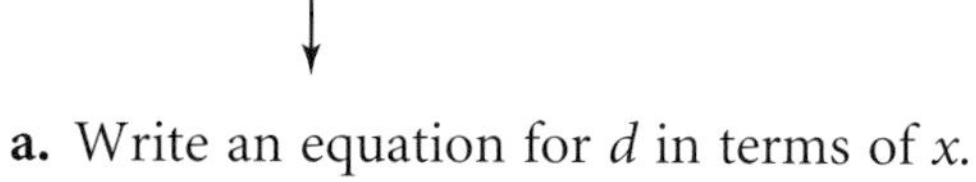

a. Write an equation for d in terms of x.

b. What is the minimum distance? What are the coordinates of the point on the parabola that is closest to the point $(5, -3)$?

13. Two towns are 6 mi apart. Chanda often bicycles from one town to the other. However, she prefers not to bicycle down the straight highway but rather to explore the region between the towns. She decides that she can bicycle at most 10 mi.

a. Make a sketch of the situation, locating the two towns at the points $(3, 0)$ and $(-3, 0)$. Estimate the locations of several points such that the straight line distances from each point to the two towns sum to 10 units on your graph.

b. Suppose Chanda is at the point (x, y). What is her distance from the point $(3, 0)$? What is her distance from the point $(-3, 0)$?

c. Write an equation expressing the fact that the sum of these distances is 10 mi.

14. APPLICATION The city councils of three neighboring towns—Ashton, Bradburg, and Carlville—decide to pool their resources and build a recreation center. To be fair, they decide to locate the recreation center equidistant from all three towns.

a. When a coordinate plane is placed on a map of the towns, Ashton is at $(0, 4)$, Bradburg is at $(3, 0)$, and Carlville is at $(12, 8)$. At what point on the map should the recreation center be located?

b. If the three towns were collinear (along a line), could the recreation center be located equidistant from all three towns? Explain your reasoning.

c. What other factors might the three city councils consider while making their decision as to where to locate the recreation center?

Career CONNECTION

A Voronoi diagram shows regions formed by a set of points such that any point inside one of the regions is closer to that region's site than to any other site. Voronoi diagrams for two and three sites are shown below. Marketing analysts use Voronoi diagrams as well as demographic data, topological features, and traffic patterns to place stores and restaurants strategically.

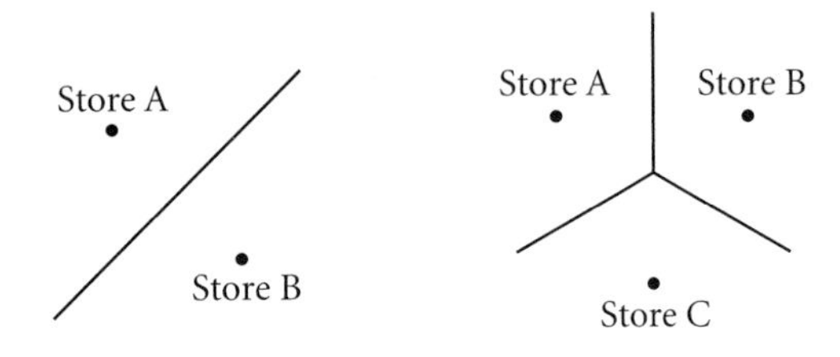

Boundary functions, by Scott Snibbe, is an interactive art piece that divides space into a Voronoi diagram as people move on a section of floor. This piece is a commentary on personal space and the separation of people.

Review

15. Complete the square in each equation so that the left side represents a perfect square or a sum of perfect squares.

a. $x^2 + 6x = 5$ ⓐ

b. $y^2 - 4y = -1$

c. $x^2 + 6x + y^2 - 4y = 4$ ⓐ

16. Triangle ABC has vertices $A(8, -2)$, $B(1, 5)$, and $C(4, -5)$.

a. Find the midpoint of each side.

b. Write the equations of the three medians of the triangle. (A median of a triangle is a segment connecting one vertex to the midpoint of the opposite side.) ⓐ

c. Locate the point where the medians meet.

17. Perform the indicated operation and reduce your answer to lowest terms.

a. $\frac{10}{x} + \frac{5}{3x}$ ⓐ

b. $\frac{3x}{2} - \frac{5}{4}$

c. $\frac{7+x}{2} + \frac{5-3x}{6}$

18. Factor each polynomial.

a. $x^2 + 7x + 6$

b. $x^2 - 8x + 12$

c. $x^2 + 4x - 21$

19. Give the domain and range of the function $f(x) = x^2 + 6x + 7$.

20. Find w and x.

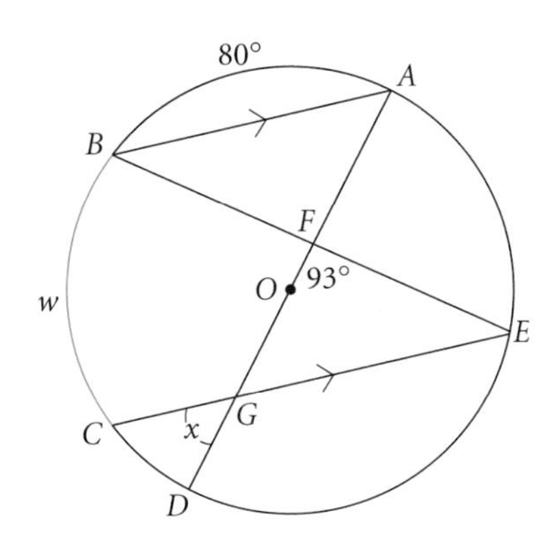

Andrea Champlin's paintings are often called "cyber" or "digital" landscapes. Can you identify curves in this painting that resemble functions you've studied in previous chapters?

Wandee Love (2001), Andrea Champlin, oil on canvas, 70 in. × 46 in. Courtesy of the artist and Clifford-Smith Gallery, Boston; photo courtesy of the artist.

IMPROVING YOUR VISUAL THINKING SKILLS

Miniature Golf

Shannon and Lori are playing miniature golf. The third hole is in a 22-by-14-foot walled rectangular playing area with a large rock in the center. Each player's ball comes to rest as shown. The rock makes a direct shot into the hole impossible.

At what point on the south wall should Shannon aim in order to have the ball bounce off and head directly for the hole? Recall from geometry that the angle of incidence is equal to the angle of reflection. Lori cannot aim at the south wall. Where should she aim?

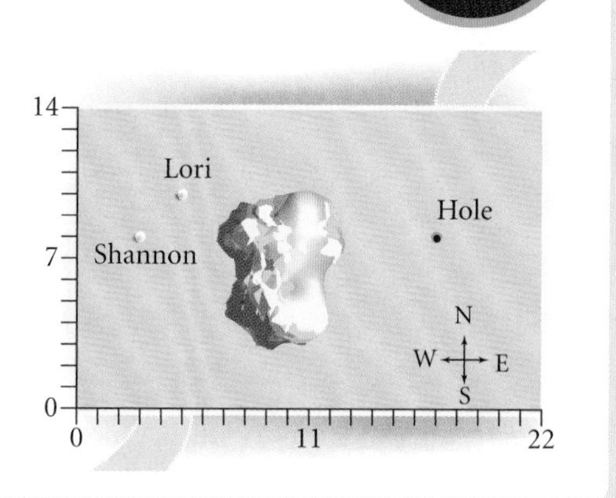

LESSON

8.2

Circles and Ellipses

Before you can be eccentric you must know where the circle is.

ELLEN TERRY

The orbital paths of the planets around the Sun are not exactly circular. These paths are examples of an important mathematical curve—the ellipse. The curved path of a stream of water from a water fountain, the path of a football kicked into the air, and the pattern of cables hanging between the towers of the Golden Gate Bridge are all examples of parabolas. The design of nuclear cooling towers, transmission gears, and the long-range navigational system known as LORAN all depend on hyperbolas.

You may be surprised to learn that all of these curves belong to the same family. These curves are all called **conic sections** because they can be created by slicing a double cone. In the next few lessons you will learn about the major conic sections: circles, ellipses, parabolas, and hyperbolas.

Sonia Delaunay's (1885–1979) oil-on-canvas painting, *Rhythm and Colors* (1939), contains many circles.

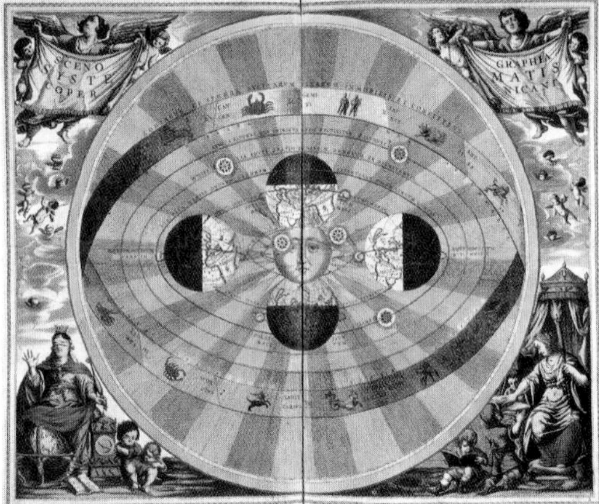

Dutch-German mathematician and cosmographer Andreas Cellarius (ca. 1595–1665) showed elliptical orbits in this celestial atlas titled *Harmonia Macrocosmica* (1660).

The Swann Memorial Fountain at Logan Circle in Philadelphia, Pennsylvania, was designed by American sculptor Alexander Stirling Calder (1898–1976). The jets of water form parabolas.

The McDonnell Planetarium, built in 1963 in St. Louis, Missouri, is a **hyperboloid,** a three-dimensional shape created by revolving a hyperbola about an axis.

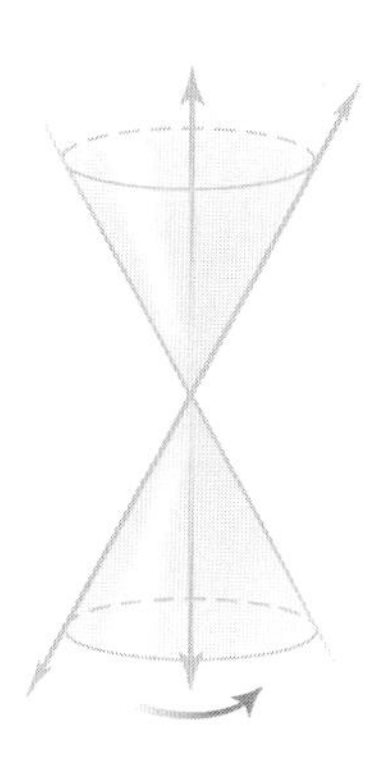

When two lines meet at an acute angle, revolving one of the lines about the other creates a double cone. These cones do not have bases. They continue infinitely, like the original lines.

Slicing these cones with a plane at different angles produces different conic sections.

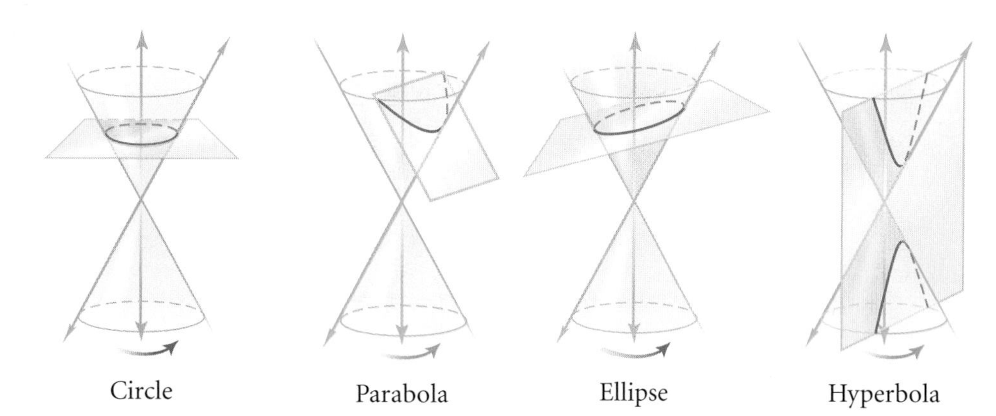

History CONNECTION

Students of mathematics have studied conic sections for more than 2000 years. Greek mathematician Menaechmus (ca. 380–320 B.C.E.) thought of conic sections as coming from different kinds of cones. Parabolas came from right-angled cones, ellipses from acute-angled cones, and hyperbolas from obtuse-angled cones. Apollonius of Perga (ca. 262–190 B.C.E.) later showed how all three kinds of curves could be obtained from the same cone. Known as the "Great Geometer," he wrote the eight-book *Treatise on Conic Sections* and named the ellipse, hyperbola, and parabola.

Conic sections have some interesting properties. Each of the shapes can also be defined as a locus of points. For example, you can describe a circle as a locus of points that are a fixed distance from a fixed point.

Definition of a Circle

A **circle** is a locus of points P in a plane, that are a constant distance, r, from a fixed point, C. Symbolically, $PC = r$. The fixed point is called the **center** and the constant distance is called the **radius.**

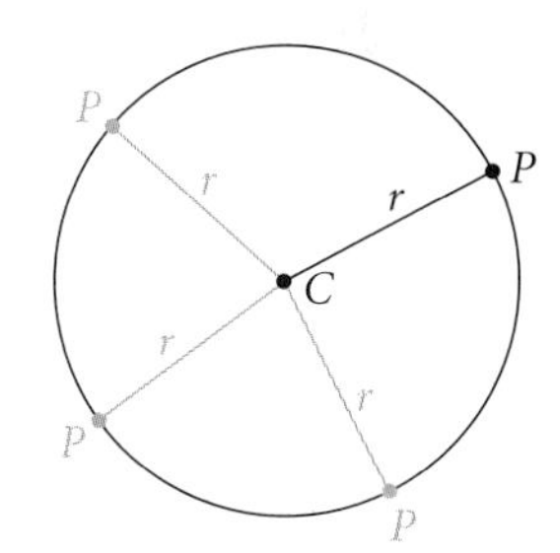

You can use the locus definition to write an equation that describes all the points on a circle.

EXAMPLE A Write an equation for the locus of points (x, y) that are 4 units from the point $(0, 0)$.

▶ Solution The locus describes a circle with radius 4 and center $(0, 0)$.

$\sqrt{(x-0)^2 + (y-0)^2} = 4$ The distance formula with points (x, y) and $(0, 0)$.

$x^2 + y^2 = 16$ Square both sides.

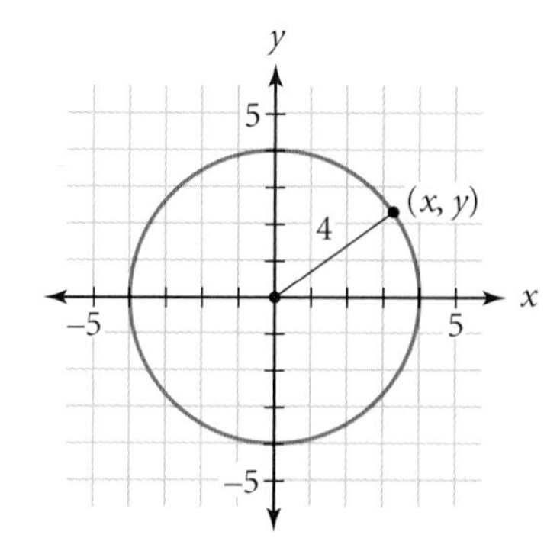

In general, the equation of a circle with center (0, 0) is $x^2 + y^2 = r^2$. As you learned in Lesson 4.7, if the circle is translated horizontally and/or vertically, you can modify the equations by replacing x with $(x - h)$ and replacing y with $(y - k)$.

Equation of a Circle

The **standard form** of the equation of a circle with center (h, k) and radius r is

$$(x - h)^2 + (y - k)^2 = r^2$$

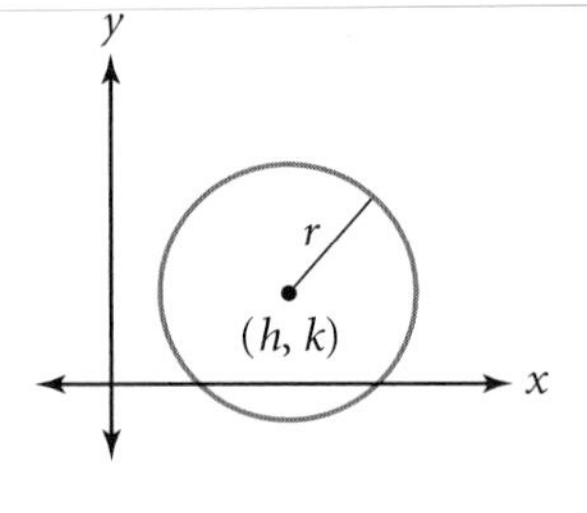

EXAMPLE B A circle has center (3, −2) and is tangent to the line $y = 2x + 1$. Write the equation of the circle.

▶ Solution

To write the equation of a circle, you need to know the center and the radius. You know the center, (3, −2), but you need to find the radius. The radius is the distance between the point of tangency and the center. To find this distance, you'll need the coordinates of the point of tangency.

Recall from geometry that a line tangent to a circle intersects the circle at only one point and is perpendicular to a diameter of the circle at the point of tangency. The tangent line has slope 2. So, the line containing the diameter through the point of tangency will have slope $-\frac{1}{2}$, and will pass through the center of the circle, (3, −2).

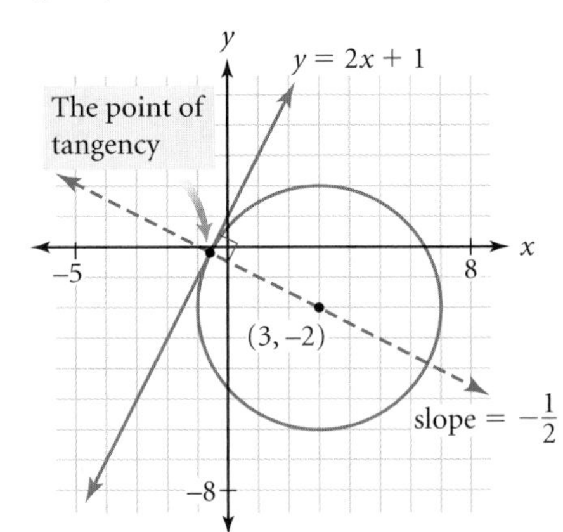

Use this information to write the equation of the line that contains the diameter.

$$y = -2 - \frac{1}{2}(x - 3)$$

Now find the point of intersection of this line with the tangent line by solving the system of equations. You should find that the point of intersection, which is the point of tangency, is (−0.6, −0.2). You can now find the radius.

$$\sqrt{(3 + 0.6)^2 + (-2 + 0.2)^2} = \sqrt{16.2} \approx 4.025$$

The radius of the circle is $\sqrt{16.2}$ units. Therefore, the equation of this circle is

$$(x - 3)^2 + (y + 2)^2 = 16.2$$

In Chapter 4, you dilated a circle horizontally and vertically by different scale factors to create ellipses. You can think of the equation of an ellipse as the equation of a unit circle that has been translated and dilated.

Equation of an Ellipse

The standard form of the equation of an ellipse with center (h, k), horizontal scale factor a, and vertical scale factor b is

$$\left(\frac{x-h}{a}\right)^2 + \left(\frac{y-k}{b}\right)^2 = 1$$

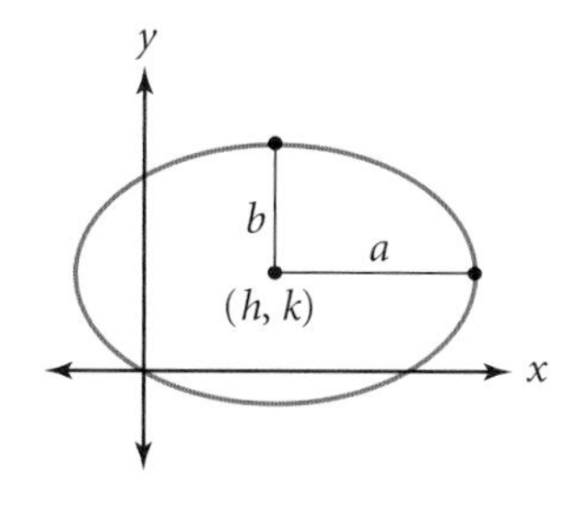

An ellipse is like a circle, except that it involves two fixed points called **foci** instead of just one point at the center. You can construct an ellipse by tying a string around two pins and tracing a set of points, as shown.

The sum of the distances, $d_1 + d_2$, is the same for any point on the ellipse.

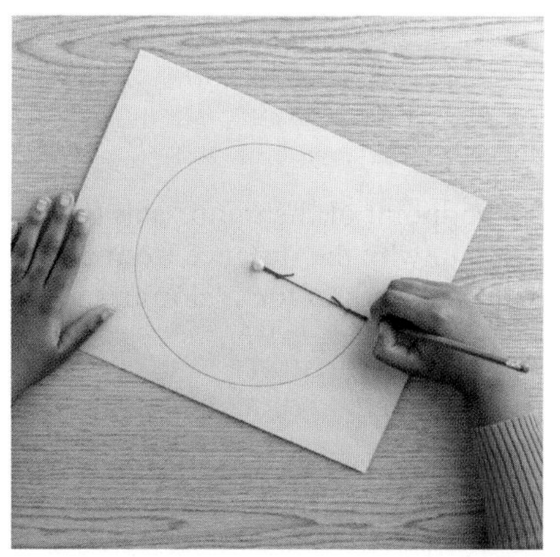

A piece of string attached to one pin helps you draw a circle. This is the same concept as using a compass to construct a circle.

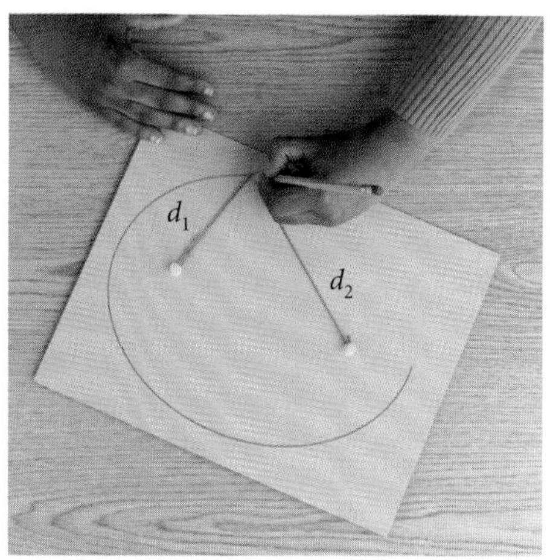

A piece of string attached to two pins helps you draw an ellipse. The string length is the sum $d_1 + d_2$.

Definition of an Ellipse

An **ellipse** is a locus of points in a plane the sum of whose distances from two fixed points is always a constant. In the diagram, the two fixed points, or **foci** (the plural of focus), are labeled F_1 and F_2. For all points on the ellipse, the distances d_1 and d_2 sum to the same value.

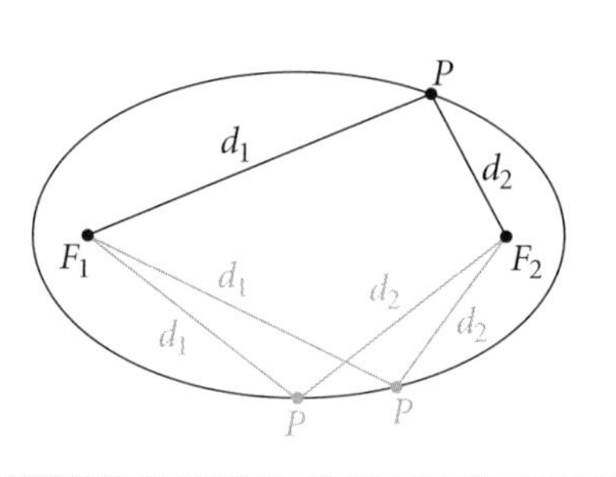

The length of half of the horizontal axis of an ellipse is the horizontal scale factor, a, and the length of half of the vertical axis is the vertical scale factor, b.

The segment that forms the longer dimension of an ellipse and contains the foci is the **major axis.** The segment along the shorter dimension is the **minor axis.**

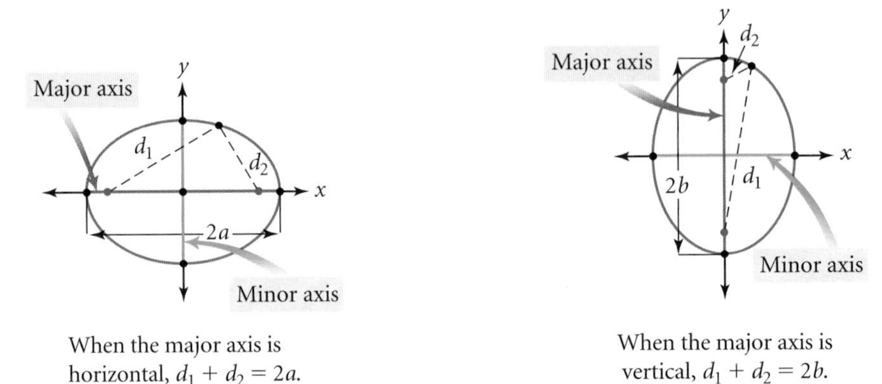

If you connect an endpoint of the minor axis to the foci, you form two congruent right triangles. Explain why the distance from a focus to an end of the minor axis is the same as half the length of the major axis.

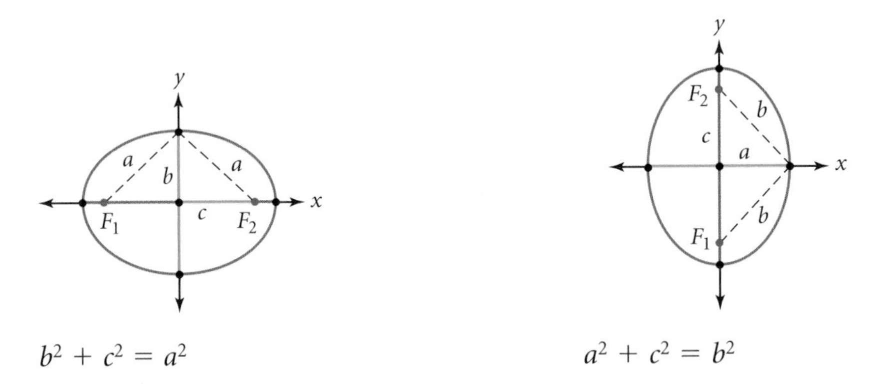

$b^2 + c^2 = a^2$ $a^2 + c^2 = b^2$

From these facts you can conclude that the distance, c, between the center and a focus is related to a and b through the Pythagorean Theorem.

To find the coordinates of the foci, add or subtract c from the appropriate coordinate of the center.

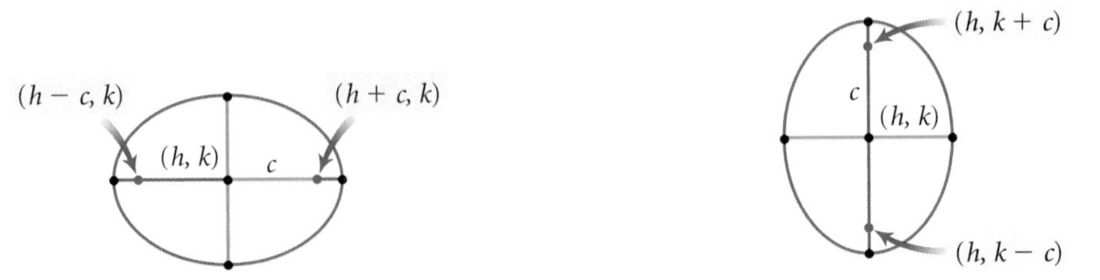

EXAMPLE C Graph an ellipse that is centered at the origin, with a vertical major axis of 6 units and a minor axis of 4 units. Where are the foci?

▶ ***Solution*** First you need to find the equation of the ellipse. Start with a unit circle, $x^2 + y^2 = 1$. The radius is 1 unit and the diameter is 2 units. You can dilate this circle vertically by a factor of 3 to make it 6 units tall. To make it 4 units wide, you must dilate it horizontally by a factor of 2.

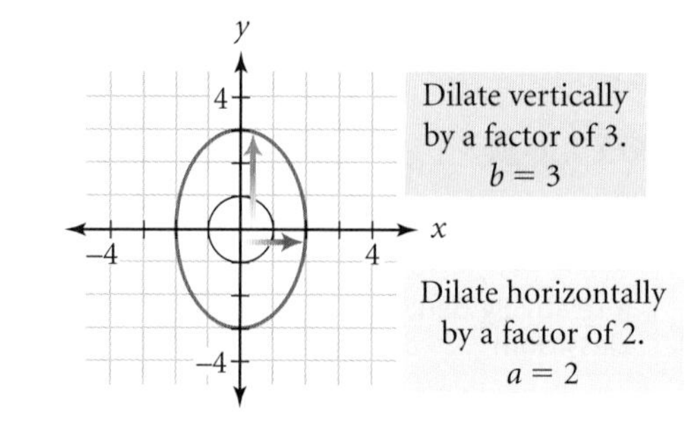

Replace y with $\frac{y}{3}$ and replace x with $\frac{x}{2}$. The equation is

$$\left(\frac{x}{2}\right)^2 + \left(\frac{y}{3}\right)^2 = 1 \quad \text{or} \quad \frac{x^2}{4} + \frac{y^2}{9} = 1$$

To sketch $\left(\frac{x}{2}\right)^2 + \left(\frac{y}{3}\right)^2 = 1$ by hand, plot the center, the endpoints of the major axis, and the endpoints of the minor axis. Then connect the endpoints with a smooth curve.

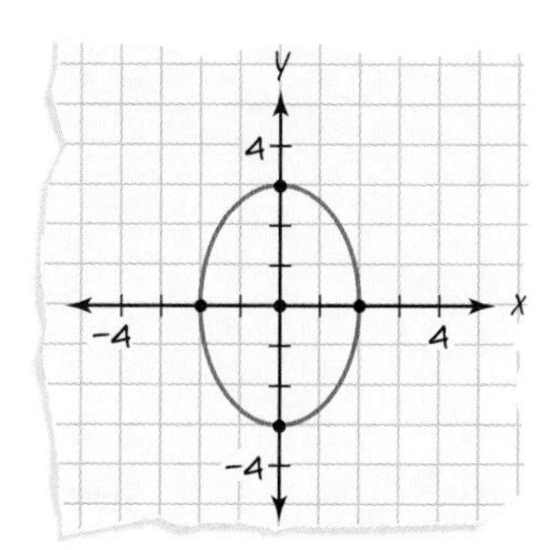

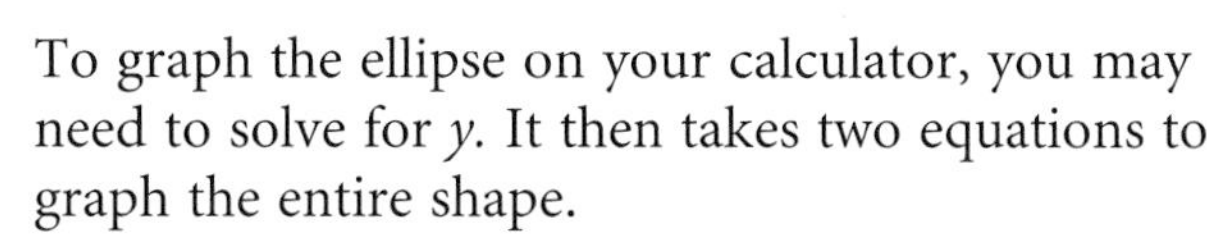

To graph the ellipse on your calculator, you may need to solve for y. It then takes two equations to graph the entire shape.

$$\left(\frac{x}{2}\right)^2 + \left(\frac{y}{3}\right)^2 = 1 \qquad \text{Equation in standard form.}$$

$$\left(\frac{y}{3}\right)^2 = 1 - \left(\frac{x}{2}\right)^2 \qquad \text{Subtract } \left(\tfrac{x}{2}\right)^2 \text{ from both sides.}$$

$$\frac{y}{3} = \pm\sqrt{1 - \left(\frac{x}{2}\right)^2} \qquad \text{Take the square root of both sides.}$$

$$y = \pm 3\sqrt{1 - \left(\frac{x}{2}\right)^2} \qquad \text{Multiply both sides by 3.}$$

Use your calculator to check the graph.

To locate the foci, apply the Pythagorean relationship between a, b, and c.

$$a^2 + c^2 = b^2$$
$$2^2 + c^2 = 3^2$$
$$c^2 = 5$$
$$c = \pm\sqrt{5}$$

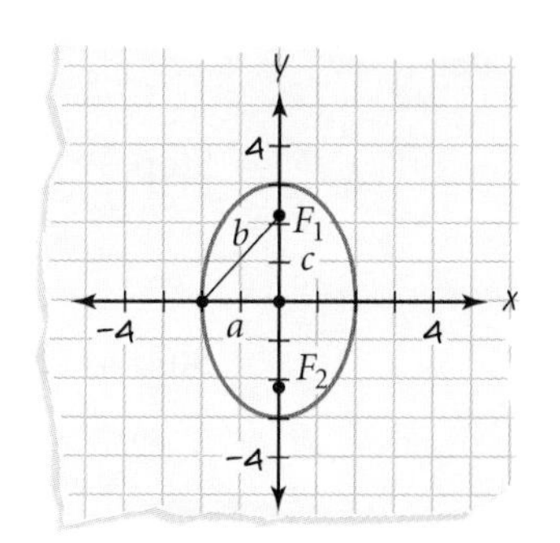

The foci are $\sqrt{5}$ units above and below the center, (0, 0).

The coordinates of the foci are $(0, \sqrt{5})$ and $(0, -\sqrt{5})$, or approximately (0, 2.24) and (0, −2.24).

In Example C, what is the result if you move the two foci closer until they merge at one point?

In the investigation you will find the equation of an ellipse that you create yourself.

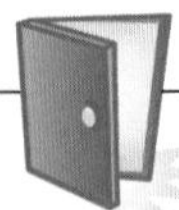

Investigation
A Slice of Light

You will need

- graph paper
- a flashlight
- a relatively dark classroom

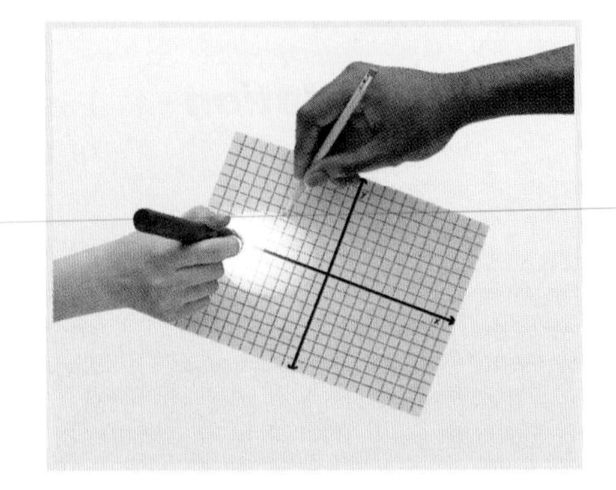

The beam of a flashlight is close to the shape of a cone. A sheet of paper held in front of the flashlight shows different slices, or sections, of the cone of light.

Work with a partner, then share results with your group.

Procedure Note

1. Shine a flashlight on the graph paper at an angle.
2. Align the major axis of the ellipse formed by the beam with one axis of the paper. You might start by placing four points on the paper to help the person holding the flashlight stay on target.
3. Carefully trace the edge of the beam as your partner holds the light steady.

Step 1 Draw a pair of coordinate axes at the center of your graph paper. Follow the Procedure Note and trace an ellipse.

Step 2 Write an equation that fits the data as closely as possible. Find the lengths of both the major and minor axes. Use the values in your equation to locate the foci. Finally, verify your equation by selecting any two pairs of points on the ellipse and checking that the sum of the distances to the foci is constant.

Eccentricity is a measure of how elongated an ellipse is. Eccentricity is defined as the ratio $\frac{c}{a}$, for an ellipse with a horizontal major axis, or $\frac{c}{b}$, for an ellipse with a vertical major axis. If the eccentricity is close to 0, then the ellipse looks almost like a circle. The higher the ratio, the more elongated the ellipse.

Step 3 Use your flashlight to make ellipses with different eccentricities. Trace three different ellipses. Calculate the eccentricity of each one and label it on your paper. What is the range of possible values for the eccentricity of an ellipse?

Step 4 Continue to tilt your flashlight until the eccentricity becomes too large and you no longer have an ellipse. What shape can you trace now?

Circles and ellipses are two of the conic sections. Besides their descriptions as intersections of cones, they have definitions involving the distance formula. In upcoming lessons you will learn similar definitions for the other conic sections.

EXERCISES

You will need

A graphing calculator for Exercise **9**.

Practice Your Skills

1. Sketch each circle on your paper, and label the center and the radius.

a. $x^2 + y^2 = 4$ @ **b.** $(x - 3)^2 + y^2 = 1$ **c.** $(x + 1)^2 + (y - 2)^2 = 9$ @

d. $x^2 + (y - 1.5)^2 = 0.25$ **e.** $(x - 1)^2 + (y - 2)^2 = 4$ **f.** $(x + 3)^2 + y^2 = 16$

2. Write an equation in standard form for each graph.

a. 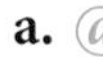

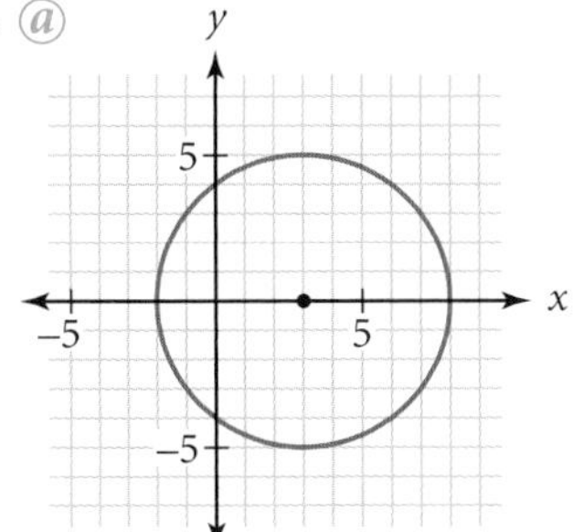

b. 

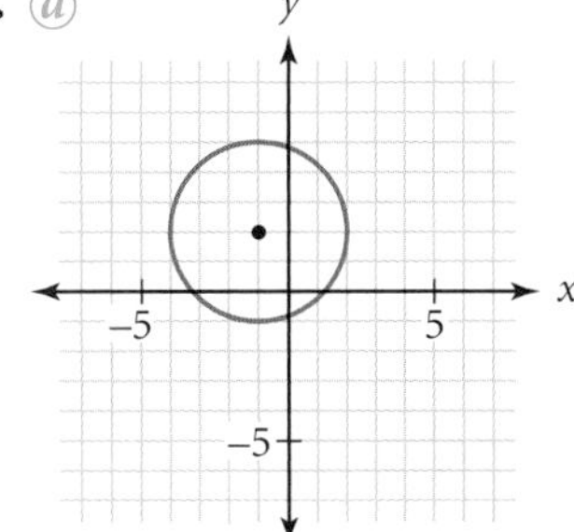

c. 

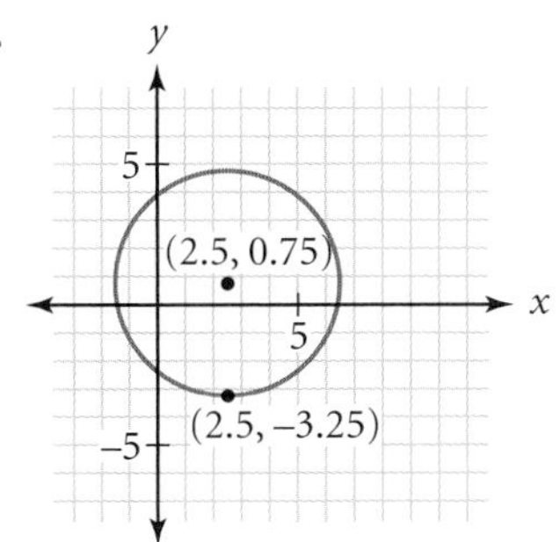

d.

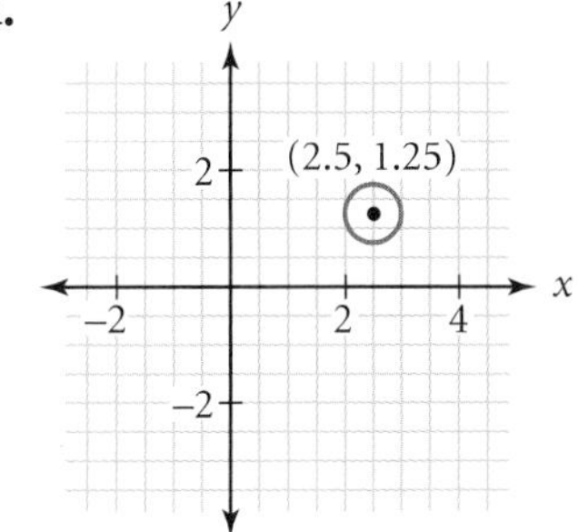

3. Write an equation for each circle described.

a. center $(2, -3)$, radius 5 @ **b.** center $(1, 0)$, radius 7 **c.** center $(-4, -2)$, radius 0.3

4. Sketch a graph of each equation. Label the coordinates of the endpoints of the major and minor axes.

a. $\left(\frac{x}{2}\right)^2 + \left(\frac{y}{4}\right)^2 = 1$ @

b. $\left(\frac{x-2}{3}\right)^2 + \left(\frac{y+2}{1}\right)^2 = 1$

c. $\left(\frac{x-4}{3}\right)^2 + \left(\frac{y-1}{3}\right)^2 = 1$ @

d. $y = \pm 2\sqrt{1 - \left(\frac{x+2}{3}\right)^2} - 1$

e. $\frac{(x+1)^2}{16} + \frac{(y-3)^2}{4} = 1$

f. $\frac{(x-3)^2}{9} + \frac{y^2}{25} = 1$

The Meeting Center (1973), nicknamed "The Egg," at the Governor Nelson A. Rockefeller Empire State Plaza in Albany, New York, contains two interior auditoriums. It appears to be the lower half of an **ellipsoid,** a three-dimensional shape created by revolving an ellipse about one of its axes.

5. Write an equation for each ellipse described.

a. center (1, 4), horizontal axis has length 14, vertical axis has length 6

b. center (−5, 0), horizontal axis has length 10, vertical axis has length 18

6. Write an equation in standard form for each graph.

a. (a)

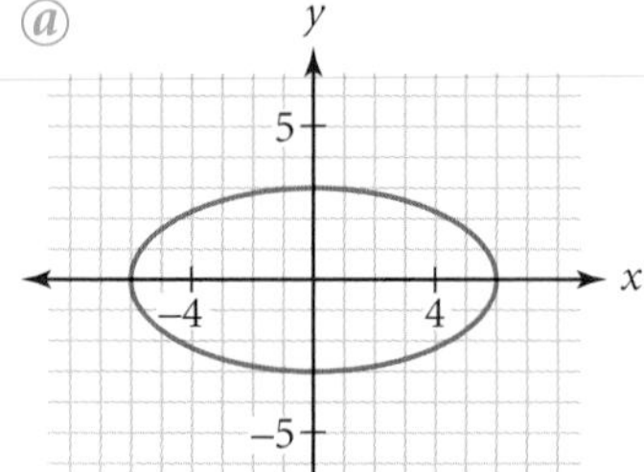

b.

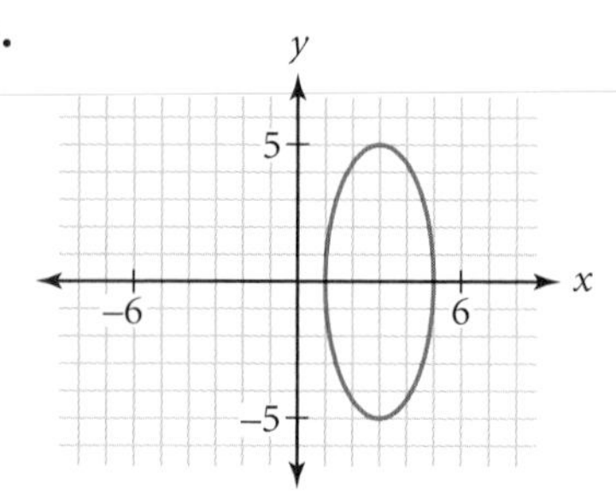

c. (a)

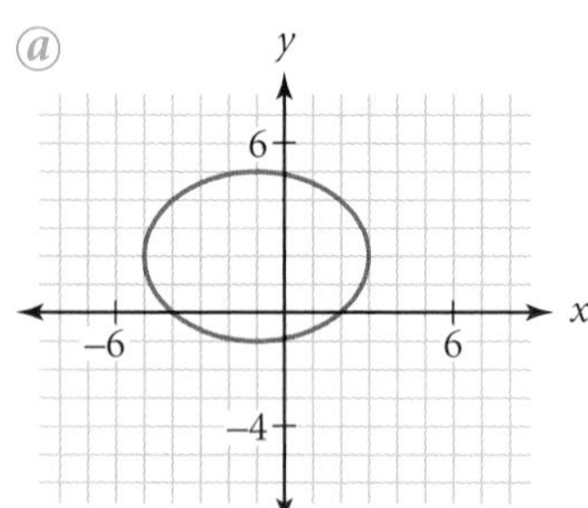

d.

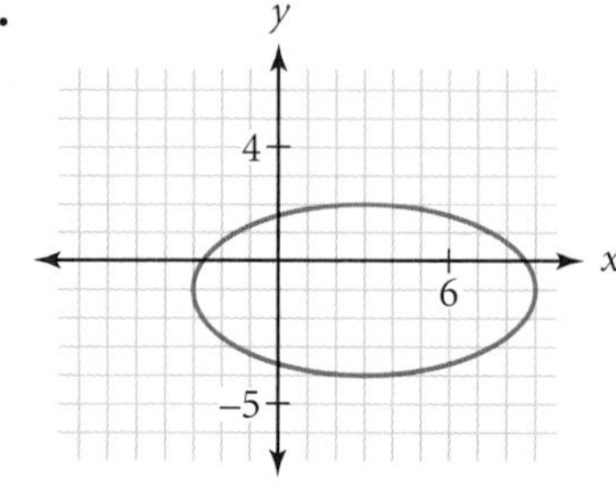

7. Find the exact coordinates of the foci for each ellipse in Exercise 6. (a)

Reason and Apply

8. A 10 cm string is tacked at (3, 0) and (−3, 0). What is an equation of the ellipse that can be created?

9. Suppose you placed a grid on the plane of a comet's orbit, with the origin at the Sun and the x-axis running through the longer axis of the orbit, as shown in the diagram below. The table gives the approximate coordinates of the comet as it orbits the Sun. Both x and y are measured in astronomical units (AU).

x	−2.1	12.9	62.6	244.5	579.3	778.1	900.1	982.4	923.4	663.0	450.0	141.6
y	5.5	16.3	31.5	54.6	62.0	51.6	36.1	10.9	−31.5	−59.2	−62.8	−44.5

a. Find an equation to fit these data. (h)

b. Find the y-coordinate when the x-coordinate is 493.0 AU.

c. What is the greatest distance of the comet from the Sun?

d. What are the coordinates of the foci?

10. APPLICATION The top of a doorway is designed to be half an ellipse. The width of the doorway is 1.6 m, and the height of the half-ellipse is designed to be 62.4 cm. The crew have nails and string available. They want to trace the half-ellipse with a pencil before they cut the plywood to go over the doorway.

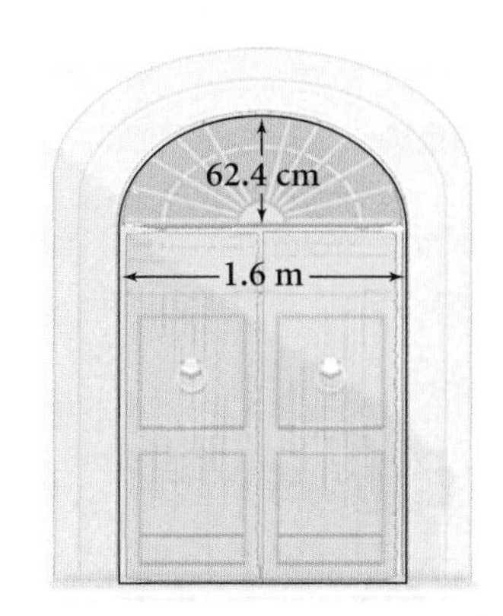

a. How far apart should they place the nails? ⓗ

b. How long should the string be?

11. The Moon's greatest distance from Earth is 252,710 mi, and its shortest distance is 221,643 mi.

a. Earth is at one focus of the Moon's elliptical orbit. Write an equation that describes the Moon's orbit around Earth.

b. What is the eccentricity of the Moon's orbit?

12. Read the Science Connection below about the reflection property of an ellipse. If a room is constructed in the shape of an ellipse and you stand at one focus and speak softly, a person standing at the other focus will hear you clearly. Such rooms are often called whispering chambers. Consider a whispering chamber that is 12 m long and 6 m wide.

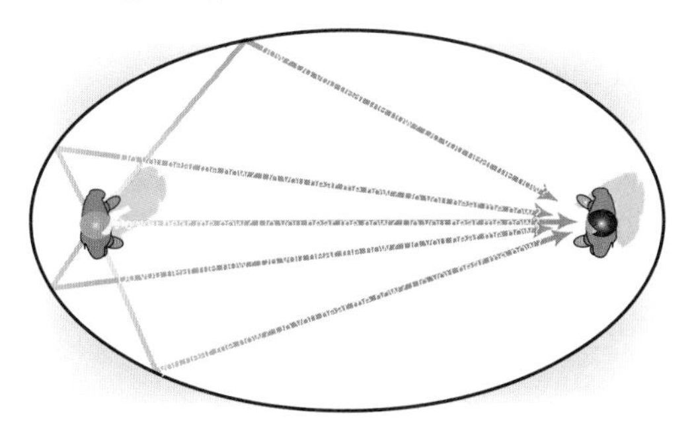

The Whispering Gallery (1937) at Chicago's Museum of Science and Industry is constructed in an ellipsoid shape with two parabolic dishes that reflect the quietest sounds in perfect clarity from one dish's focus to the other's.

a. Where should two whisperers stand to talk to each other? ⓐ

b. How far does the sound travel from one person to the other, bouncing off the wall in between?

Science CONNECTION

A signal from one focus of an ellipse will always bounce off the ellipse in such a way that it will travel to the other focus.

This is called the *reflection property of an ellipse.*

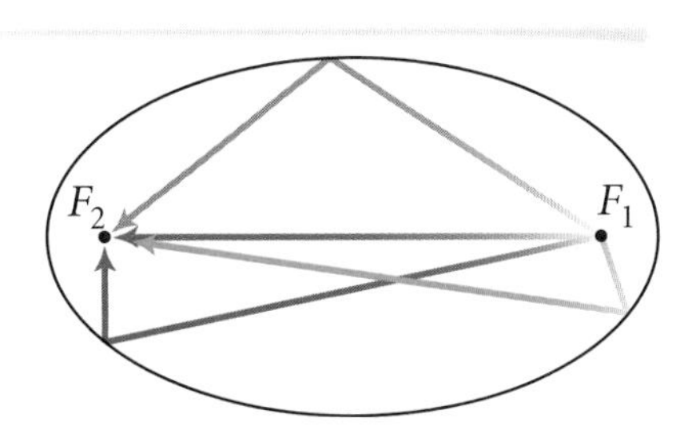

13. A contractor needs to install a vent pipe through the roof of a new house. The pipe has a diameter of 4 in. The roof has a 5:12 pitch, meaning a rise of 5 ft over a horizontal run of 12 ft. Because of the slope, the hole must be an ellipse.

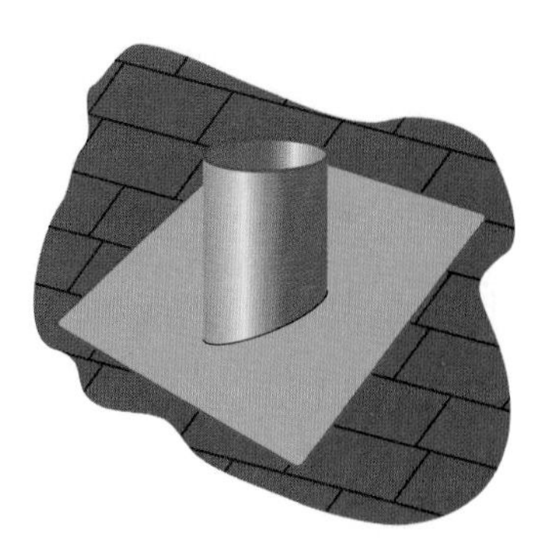

a. What is the length of the minor axis of this ellipse?

b. Use the Pythagorean Theorem to find the length of the major axis of this ellipse.

Review

14. Write the equation of a parabola congruent to $y = x^2$ that has been reflected across the x-axis and translated left 3 units and up 2 units. @

15. Solve for y.

$$\frac{y-4}{0.5} = \left(\frac{x+2}{3}\right)^2$$

16. Solve a system of equations and find the quadratic equation, $y = ax^2 + bx + c$, that fits these data points.

x	0	1	2	3	4	5
y	117	95	77	63	53	47

17. The vertices of $\triangle ABC$ are $A(-4, 2)$, $B(2, 5)$, and $C(6, -8)$. Find the perimeter of $\triangle ABC$.

18. A canal running east and west is 2 mi south of the center of town. The power plant is located 3 mi east and 1 mi north of the center of town.

a. Make a graph showing the location of the canal and the power plant. Assume the center of town is (0, 0).

b. Liann wants to build her new business at some point that is the same distance from the canal as it is from the power plant. Find at least three possible locations and add them to your graph.

c. Write an expression for the distance from Liann's business to the canal.

d. Write an expression for the distance from Liann's business to the power plant.

e. Write an equation expressing the fact that these two distances are equal. Verify that your points from 18b satisfy the equation.

IMPROVING YOUR REASONING SKILLS

Elliptical Pool

You could use the reflection property of an ellipse to design an unusual pool table. On an elliptical pool table, if you start with a ball at one focus and hit it in any direction, it will always rebound off the side and roll toward the other focus. Suppose a pool table is designed in the shape of an ellipse with a pocket at one focus. Describe how you will hit ball 1 to land in the pocket, even though ball 2 is in the way.

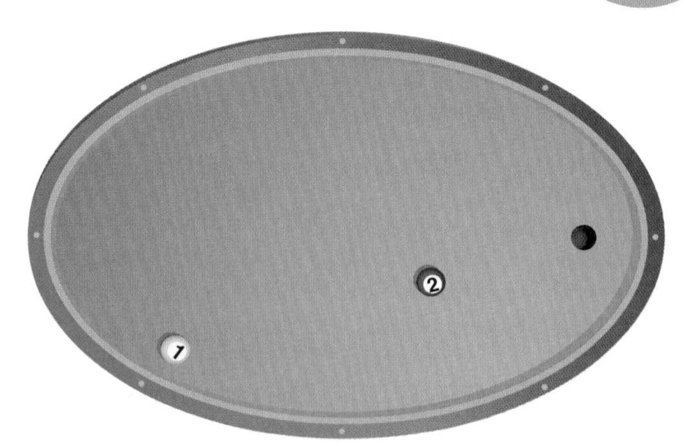

LESSON

8.3

Parabolas

There are two ways of spreading light: to be the candle or the mirror that reflects it.

EDITH WHARTON

You have studied parabolas in several different lessons, and you have transformed parabolic graphs to model a variety of situations. Now a locus definition will reveal properties of parabolas that you can use to solve other practical problems. One important property is described in the Technology and Science Connections below.

Built in 1969, one side of this building in Odeillo, France, is a large parabolic mirror that focuses sunlight on the building's solar-powered furnace.

Technology CONNECTION

A reflecting telescope is a type of optical telescope that uses a curved, mirrored lens to magnify objects. The most powerful reflecting telescope uses a parabolic or hyperbolic mirror and can bring the faintest light rays into clear view. Telescopes with larger mirrors can detect objects at greater distances. To avoid the expense and weight of producing one massive lens, today's most sophisticated telescopes have a tile-like combination of hexagonal mirrors that produce the same effect as one concave mirror.

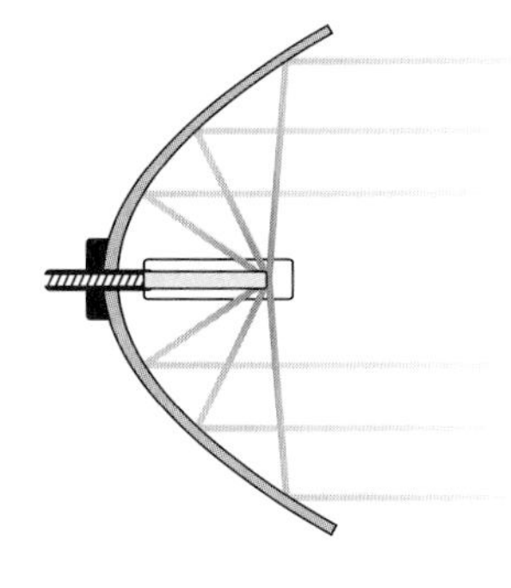

The designs of telescope lenses, spotlights, satellite dishes, and other parabolic reflecting surfaces are based on a remarkable property of parabolas: A ray that travels parallel to the axis of symmetry will strike the surface of the parabola or paraboloid and reflect toward the **focus.** Likewise, when a ray from the focus strikes the curve, it will reflect in a ray that is parallel to the axis of symmetry. A **paraboloid** is a three-dimensional parabola, formed when a parabola is rotated about its line of symmetry.

Science CONNECTION

Satellite dishes, used for television, radio, and other communications, are always parabolic. A satellite dish is set up to aim directly at a satellite. As the satellite transmits signals to a dish, the signals are reflected off the dish surface and toward the receiver, which is located at the focus of the paraboloid. In this way, every signal that hits a parabolic dish can be directed into the receiver.

A large satellite dish

This reflective property of parabolas can be proved based on the locus definition of a parabola. Compare this locus definition of a parabola to the locus definition of an ellipse.

Definition of a Parabola

A **parabola** is a locus of points in a plane whose distance from a fixed point called the **focus** is the same as the distance from a fixed line called the **directrix.** That is, $d_1 = d_2$. In the diagram, F is the focus and l is the directrix.

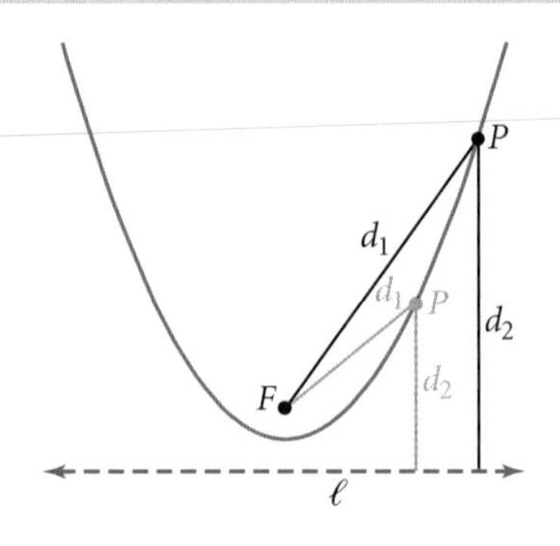

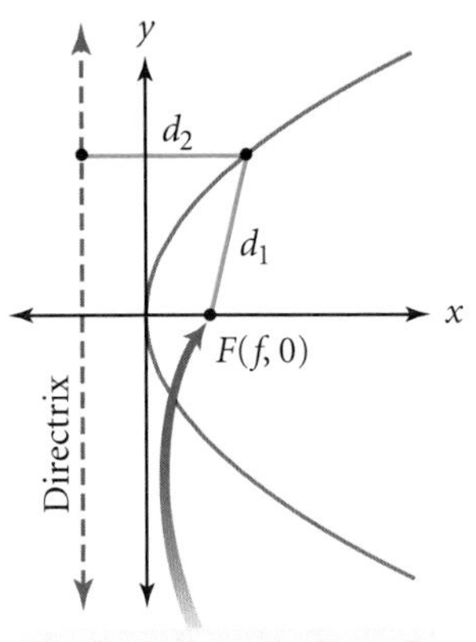

The focus is on the line of symmetry of the parabola.

A parabola is the set of points for which the distances d_1 and d_2 are equal. If the directrix is a horizontal line, the parabola is vertically oriented, like the one in the definition box above. If the directrix is a vertical line, the parabola is horizontally oriented, like the one at left. The directrix can also be neither horizontal nor vertical, creating a parabola that is rotated at an angle.

How can you locate the focus of a given parabola? Suppose the parabola is horizontally oriented, with vertex $(0, 0)$. It has a focus inside the curve at a point, $(f, 0)$, as shown at left. The vertex is the same distance from the focus as it is from the directrix, so the equation of the directrix is $x = -f$, as shown below at left. You can use this information and the distance formula to find the value of f when the vertex is at the origin, as shown below at right.

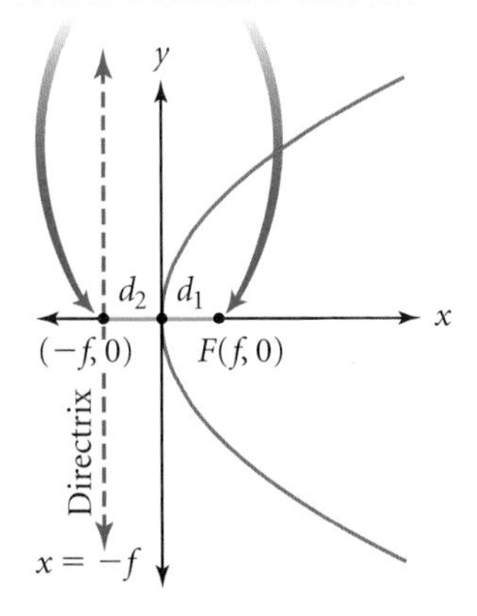

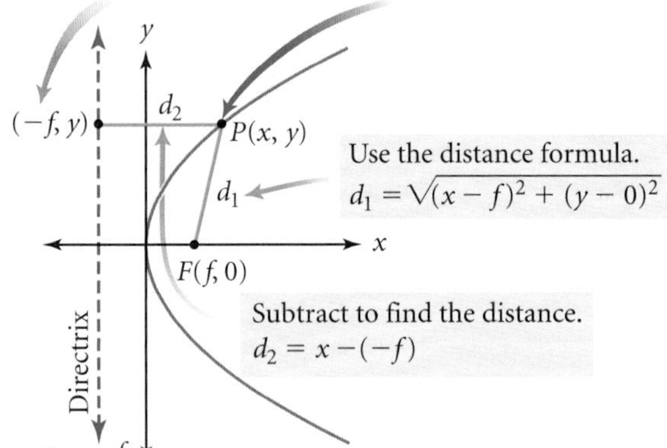

$\sqrt{(x - f)^2 + (y - 0)^2} = x - (-f)$	Definition of parabola states that $d_1 = d_2$.
$\sqrt{(x - f)^2 + y^2} = x + f$	Subtract.
$(x - f)^2 + y^2 = (x + f)^2$	Square both sides.
$x^2 - 2fx + f^2 + y^2 = x^2 + 2fx + f^2$	Expand.
$y^2 = 4fx$	Combine like terms.

This result means that the coefficient of the variable x is $4f$, where f is the distance from the vertex to the focus. What do you think it means if f is negative?

If the parabola is vertically oriented, the x- and y-coordinates are exchanged, for a final equation of $x^2 = 4fy$, or $y = \frac{1}{4f}x^2$.

Designed in 1960, the Theme Building at Los Angeles International Airport in California uses double parabolic arches.

EXAMPLE

Consider the parent equation of a horizontally oriented parabola, $y^2 = x$.

a. Write the equation of the image of this graph after the following transformations have been performed, in order: a vertical dilation by a factor of 3, a translation right 2 units, and then a translation down 1 unit. Graph the new equation.

b. Where is the focus of $y^2 = x$? Where is the directrix?

c. Where is the focus of the transformed parabola? Where is its directrix?

▶ *Solution*

Recall the transformations of functions that you studied in Chapter 4.

a. Begin with the parent equation, and perform the specified transformations.

$$y^2 = x$$ Original equation.

$$\left(\frac{y}{3}\right)^2 = x$$ Dilate vertically by a factor of 3.

$$\left(\frac{y+1}{3}\right)^2 = x - 2$$ Translate right 2 units and down 1 unit.

Graph the transformed parabola.

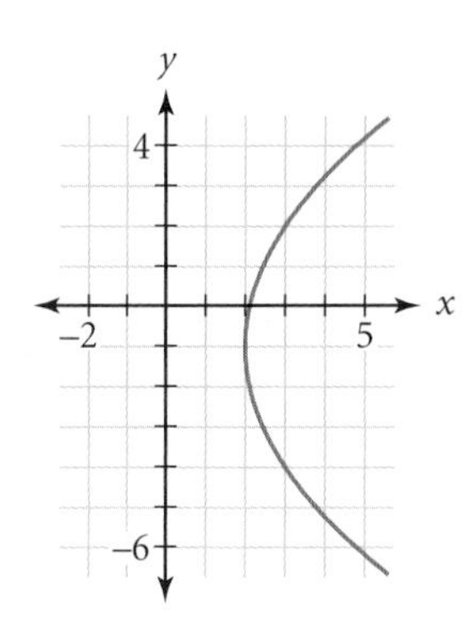

b. Use the standard form, $y^2 = 4fx$, to locate the focus and the directrix of $y^2 = x$. The coefficient of x is $4f$ in the general form, and 1 in the equation $y^2 = x$. So $4f = 1$, or $f = \frac{1}{4}$. Recall that f is the distance from the vertex to the focus and from the vertex to the directrix. The vertex is $(0, 0)$, so the focus is $\left(\frac{1}{4}, 0\right)$ and the directrix is the line $x = -\frac{1}{4}$.

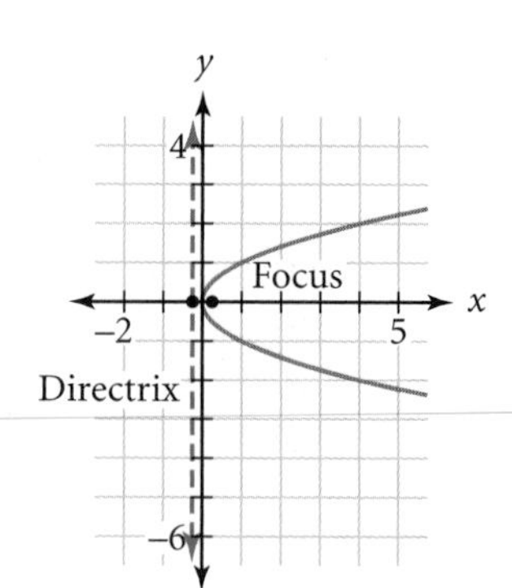

c. To locate the focus and the directrix of $\left(\frac{y+1}{3}\right)^2 = x - 2$, first rewrite the equation as $(y + 1)^2 = 9(x - 2)$. The coefficient of x in this equation is 9, so $4f = 9$, or $f = 2.25$. The focus and the directrix will both be 2.25 units from the vertex in the horizontal direction. The vertex is $(2, -1)$, so the focus is $(4.25, -1)$ and the directrix is the line $x = -0.25$.

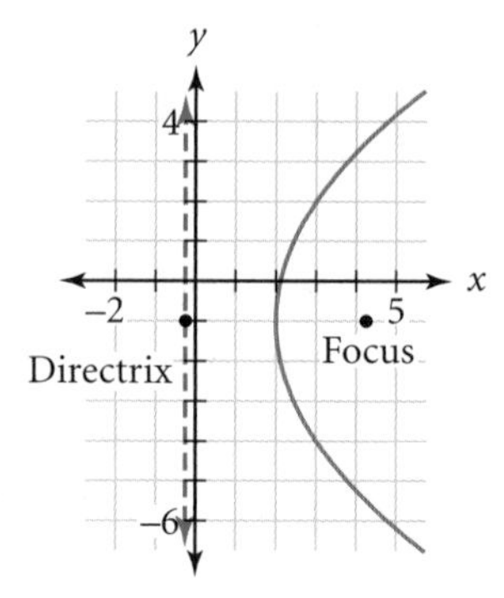

You can now extend what you know about parabolic equations to both horizontally and vertically oriented parabolas.

Equation of a Parabola

The standard form of the equation of a vertically oriented parabola with vertex (h, k), horizontal scale factor of a, and vertical scale factor of b is

$$\frac{y-k}{b} = \left(\frac{x-h}{a}\right)^2$$

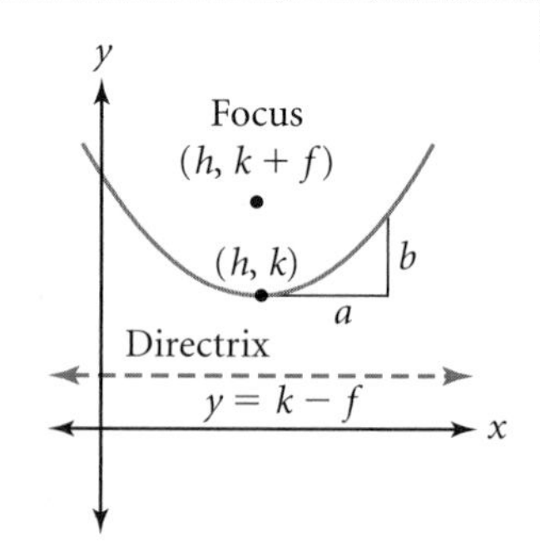

The focus of a vertically oriented parabola is $(h, k + f)$, where $\frac{a^2}{b} = 4f$, and the directrix is the line $y = k - f$.

The standard form of the equation of a horizontally oriented parabola with vertex (h, k), horizontal scale factor of a, and vertical scale factor of b is

$$\left(\frac{y-k}{b}\right)^2 = \frac{x-h}{a}$$

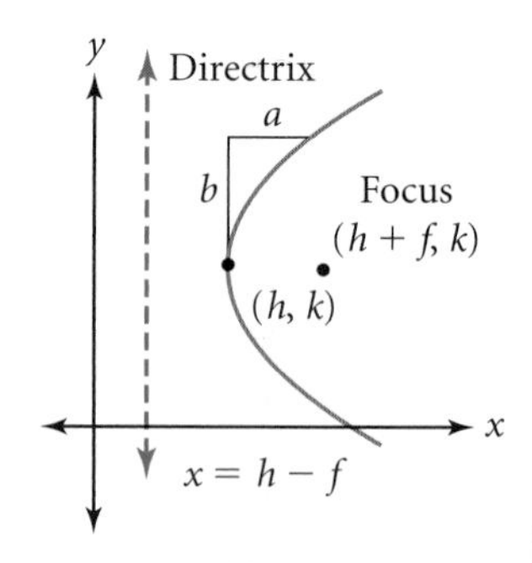

The focus of a horizontally oriented parabola is $(h + f, k)$, where $\frac{b^2}{a} = 4f$, and the directrix is the line $x = h - f$.

In the investigation you will construct a parabola. As you create your model, think about how your process relates to the locus definition of a parabola.

Investigation
Fold a Parabola

You will need

- patty paper
- graph paper

Fold the patty paper parallel to one edge to form the directrix for a parabola. Mark a point on the larger portion of the paper to serve as the focus for your parabola. Fold the paper so that the focus lies on the directrix. Unfold, and then fold again, so that the focus is at another point on the directrix. Repeat this many times. The creases from these folds should create a parabola. Lay the patty paper on top of a sheet of graph paper. Identify the coordinates of the focus and the equation of the directrix, and write an equation for your parabola.

EXERCISES

You will need

A graphing calculator for Exercises **13** and **16.**

Practice Your Skills

1. For each parabola described, use the information given to find the location of the missing feature. It may help to draw a sketch.

a. If the focus is $(1, 4)$, and the directrix is $y = -3$, where is the vertex? @

b. If the vertex is $(-2, 2)$, and the focus is $(-2, -4)$, what is the equation of the directrix? @

c. If the directrix is $x = 3$, and the vertex is $(6, 2)$, where is the focus?

2. Sketch each parabola, and label the vertex and line of symmetry.

a. $\left(\frac{x}{2}\right)^2 + 5 = y$ @

b. $(y + 2)^2 - 2 = x$ @

c. $-(x + 3)^2 + 1 = 2y$

d. $2y^2 = -x + 4$

e. $\frac{y-3}{2} = \left(\frac{x+1}{4}\right)^2$

f. $\frac{x-2}{3} = \left(\frac{y}{5}\right)^2$

3. Locate the focus and directrix for each graph in Exercise 2. @

4. Write an equation in standard form for each parabola.

a. @

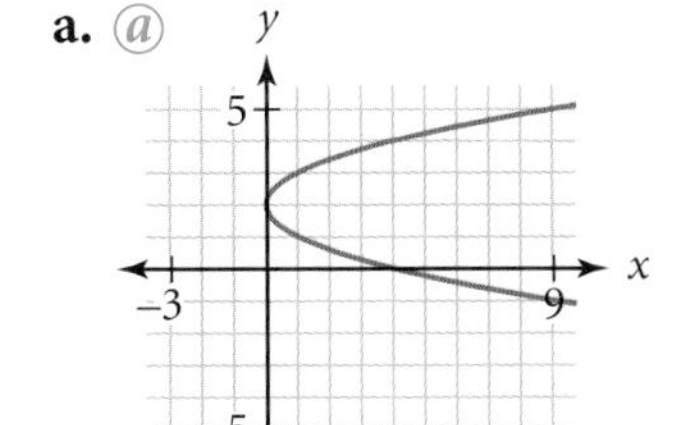

b. @

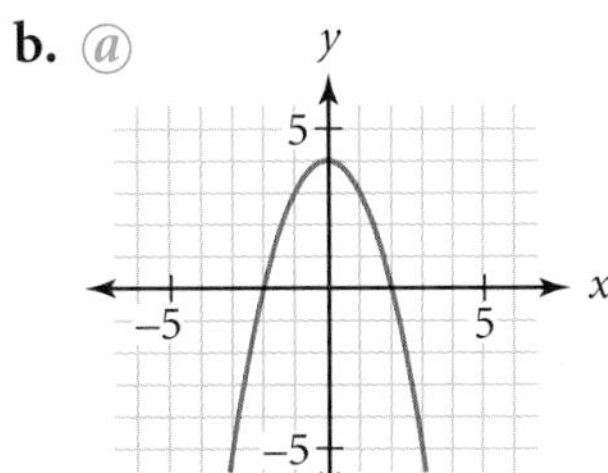

c.

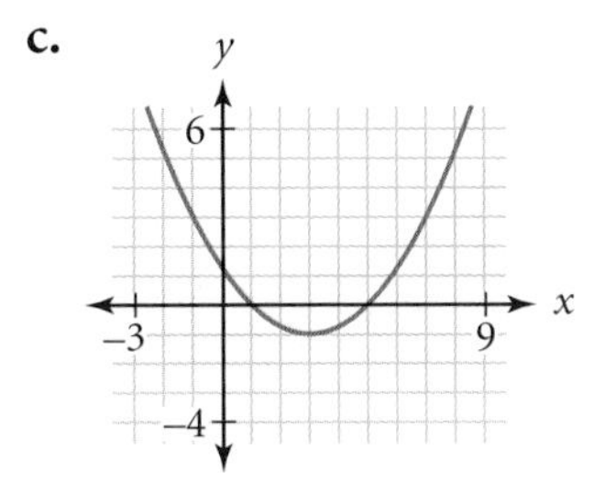

d.

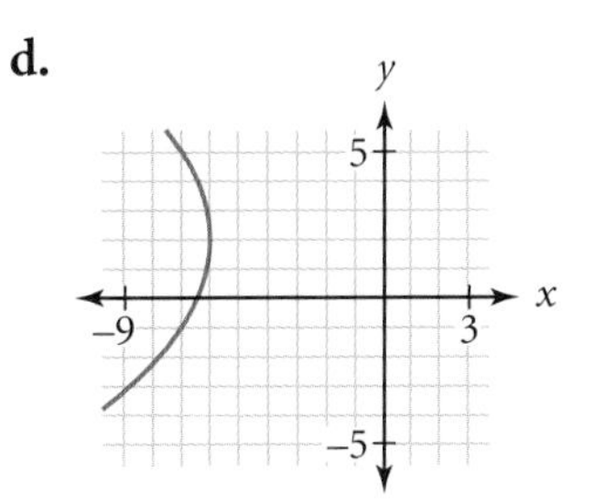

5. Locate the focus and directrix for each graph in Exercise 4. ⓐ

Reason and Apply

6. Find the equation of the parabola with directrix $x = 3$ and vertex $(0, 0)$. ⓗ

7. Write the equation of the parabola with focus $(1, 3)$ and directrix $y = -1$. ⓐ

8. Consider the graph at right.

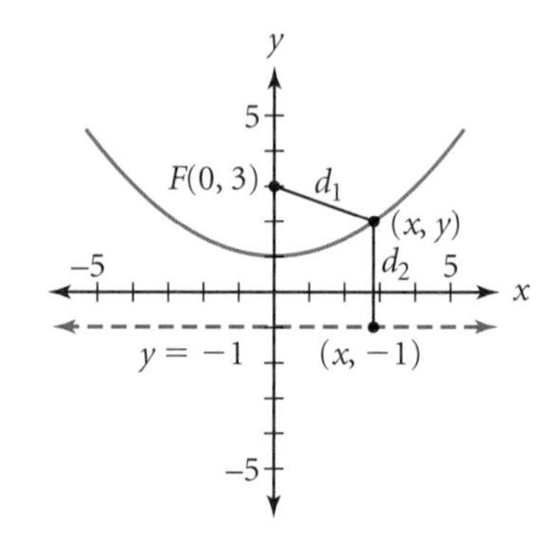

a. Because $d_1 = d_2$, you can write the equation

$$\sqrt{(x-0)^2 + (y-3)^2} = y - (-1)$$

Rewrite this equation by solving for y. ⓐ

b. Describe the graph represented by your equation from 8a.

9. The pilot of a small boat charts a course such that the boat will always be equidistant from an upcoming rock and the shoreline.

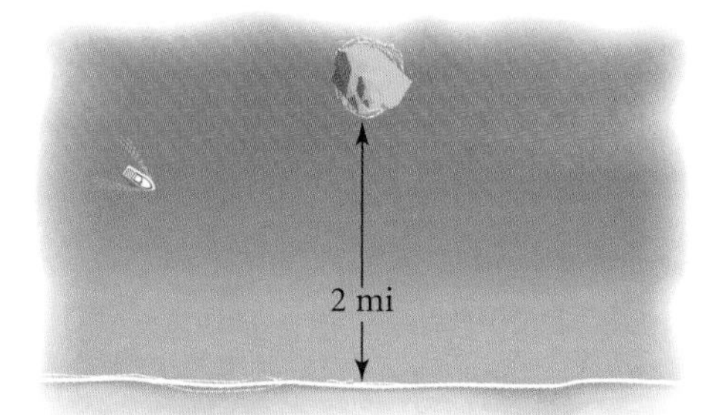

a. Make a sketch of the situation, showing the approximate path of the boat.

b. Describe the path of the boat.

c. If the rock is 2 mi offshore, write an equation for the path of the boat.

10. APPLICATION Sheila is designing a parabolic dish to use for cooking on a camping trip. She plans to make the dish 40 cm wide and 20 cm deep. Where should she locate the cooking grill so that all of the light that enters the parabolic dish will be reflected toward the food? ⓗ

Environmental CONNECTION

Solar cookers focus the heat of the Sun into a single spot, in order to boil water or cook food. A well-designed cooker can create heat up to 400°C. Solar cookers can be created with minimal materials and can help save natural resources, particularly firewood. Inexpensive solar cookers are now being designed and distributed for use in developing countries. For more information on solar cookers, see the links at www.keymath.com/DAA .

A solar cooker fries an omelette.

Review

11. Perform the indicated operation and reduce your answer to lowest terms.

a. $\frac{5}{3x} - \frac{7}{2x}$ @ b. $\frac{x}{x+2} + \frac{3x}{x+2}$ c. $\frac{3x+5}{2x} + \frac{5x-10}{4x}$

12. Find the equation that describes a parabola containing the points (3.6, 0.764), (5, 1.436), and (5.8, −2.404).

13. Find the minimum distance from the origin to the parabola $y = -x^2 + 1$. What point(s) on the parabola is closest to the origin?

14. Find the equation of the ellipse with foci (−6, 1) and (10, 1) that passes through the point (10, 13). @

15. Consider the polynomial function $f(x) = 2x^3 - 5x^2 + 22x - 10$.

a. What are the possible rational roots of $f(x)$? @

b. Find all rational roots.

c. Write the equation in factored form.

16. On a three-dimensional coordinate system with variables x, y, and z, the standard equation of a plane is in the form $ax + by + cz = d$. Find the intersection of the three planes described by $3x + y + 2z = -11$, $-4x + 3y + 3z = -2$, and $x - 2y - z = -3$.

17. Completely factor each expression.

a. $x^2 + 8x + 15$ b. $x^3 - 49x$ @ c. $x^2 - 3x - 28$

18. APPLICATION One possible gear ratio on Matthew's mountain bike is 4 to 1. This means that the front gear has four times as many teeth as the gear on the back wheel. So each revolution of the pedal causes the rear wheel to make four revolutions.

a. If Matthew is pedaling 60 revolutions per minute (r/min), how many revolutions per minute is the rear tire making?

b. If the diameter of the rear tire is 26 in., what speed in miles per hour will Matthew attain?

c. Matt downshifts to a front gear that has 22 teeth and a rear gear that has 30 teeth. If he keeps pedaling 60 r/min, what will his new speed be?

LESSON

8.4

Hyperbolas

The best paths usually lead to the most remote places.

SUSAN ALLEN TOTH

The fourth, and final, conic section is the hyperbola. It is like the ellipse in that it has two foci, but it differs from the other conics in that it consists of two separate pieces. The two light shadows on a wall next to a cylindrical lampshade form two branches of a hyperbola.

The definition of a hyperbola is similar to that of an ellipse. However, this time it is the difference in the distances to the two foci, which is constant.

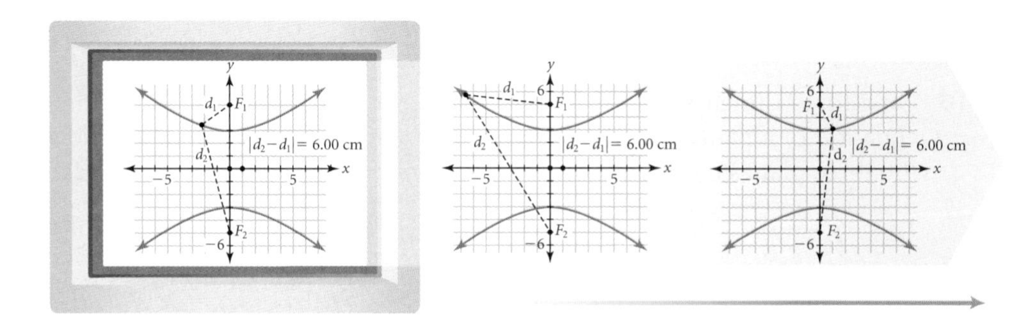

keymath.com/DAA

[► You can explore properties of hyperbolas using the **Dynamic Algebra Exploration** at www.keymath.com/DAA .◄]

Definition of a Hyperbola

A **hyperbola** is a locus of points in a plane the difference of whose distances from two fixed points, called **foci,** is always constant. In the diagram, F_1 and F_2 are the foci.

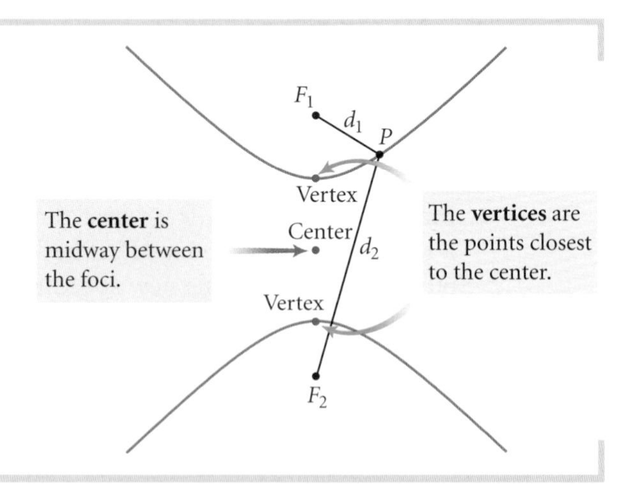

The **center** is midway between the foci.

The **vertices** are the points closest to the center.

Regardless of where a point is on a hyperbola, the difference in the distances from the point to the two foci is constant. Notice that this constant is equal to the distance between the two vertices of the hyperbola.

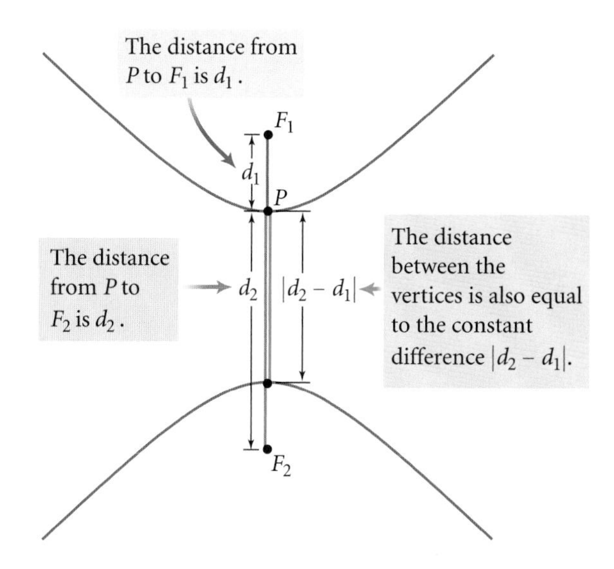

EXAMPLE A

Just as the parent equation of any circle is a unit circle, $x^2 + y^2 = 1$, the parent equation of a hyperbola is called a **unit hyperbola.** The horizontally oriented unit hyperbola has vertices $(1, 0)$ and $(-1, 0)$, and foci $(\sqrt{2}, 0)$ and $(-\sqrt{2}, 0)$. Find the equation of a unit hyperbola.

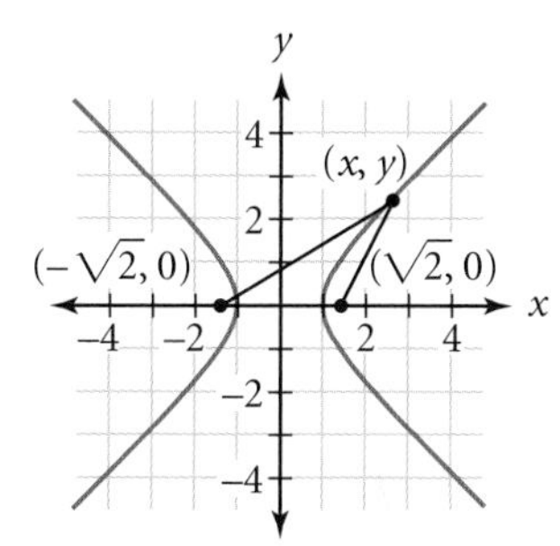

▸ **Solution**

The distance between the vertices is 2, so the difference in the distances from any point on the hyperbola to the two foci is 2. Label a point on the hyperbola (x, y). Then use the definition of a hyperbola.

$\left\lvert\sqrt{(x+\sqrt{2})^2+y^2}-\sqrt{(x-\sqrt{2})^2+y^2}\right\rvert = 2$	Definition of hyperbola states that $\lvert d_2 - d_1 \rvert$ is a constant, in this case 2.
$\sqrt{(x+\sqrt{2})^2+y^2}-\sqrt{(x-\sqrt{2})^2+y^2} = 2$	Consider the case $d_2 > d_1$.
$\sqrt{(x+\sqrt{2})^2+y^2} = \sqrt{(x-\sqrt{2})^2+y^2} + 2$	Add $\sqrt{(x-\sqrt{2})^2+y^2}$ to both sides.
$(x+\sqrt{2})^2+y^2 = (x-\sqrt{2})^2+y^2+4\sqrt{(x-\sqrt{2})^2+y^2}+4$	Square both sides.
$x^2+2x\sqrt{2}+2+y^2 = x^2-2x\sqrt{2}+2+y^2+4\sqrt{(x-\sqrt{2})^2+y^2}+4$	Expand.
$2x\sqrt{2} = -2x\sqrt{2}+4\sqrt{(x-\sqrt{2})^2+y^2}+4$	Subtract x^2, y^2, and 2 from both sides.
$4x\sqrt{2}-4 = 4\sqrt{(x-\sqrt{2})^2+y^2}$	Isolate the radical.
$x\sqrt{2}-1 = \sqrt{(x-\sqrt{2})^2+y^2}$	Divide by 4.
$2x^2-2x\sqrt{2}+1 = x^2-2x\sqrt{2}+2+y^2$	Square both sides and expand.
$x^2-y^2 = 1$	Combine like terms, and collect variables on one side of the equation.

If you consider the case $d_1 > d_2$, you find the same equation for the unit hyperbola.

Check your answer by graphing on a calculator. First you must solve for y.

$$x^2 - y^2 = 1$$
$$-y^2 = 1 - x^2$$
$$y^2 = x^2 - 1$$
$$y = \pm\sqrt{x^2 - 1}$$

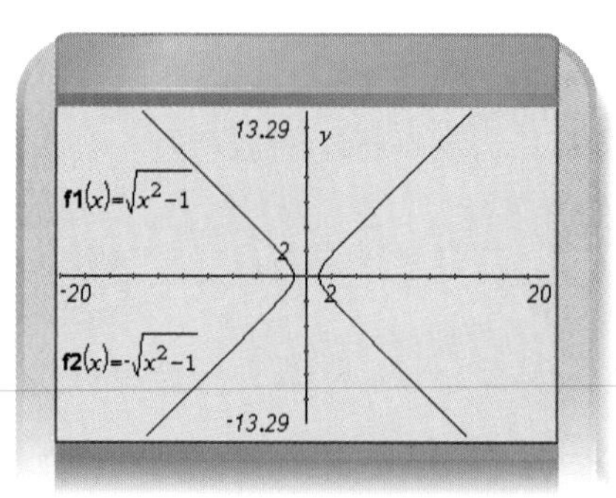

Like the exponential and logarithmic curves you studied in Chapter 5, hyperbolas are curves that approach asymptotes. If you zoom out on the graph of a hyperbola, eventually it looks just like an "X." The lines that form the apparent X are the asymptotes. For the unit hyperbola, the asymptotes are the lines $y = x$ and $y = -x$. These lines are the extended diagonals of the square with vertices $(1, 1)$, $(1, -1)$, $(-1, -1)$, and $(-1, 1)$.

The horizontally oriented unit hyperbola

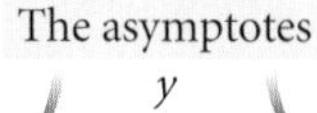

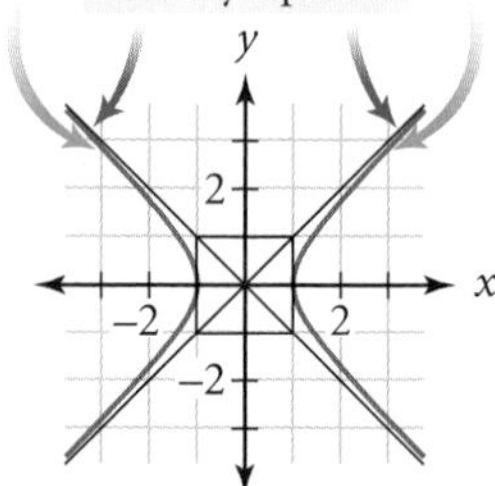

The curve approaches the asymptotes as it moves away from the vertices.

The equation $y^2 - x^2 = 1$ also describes a unit hyperbola. This hyperbola, shown below, is the vertically oriented unit hyperbola. If the hyperbola is centered at the origin and dilated, then the equation can be written in the form $\left(\frac{x}{a}\right)^2 - \left(\frac{y}{b}\right)^2 = 1$ or $\left(\frac{y}{b}\right)^2 - \left(\frac{x}{a}\right)^2 = 1$, where a is the horizontal scale factor and b is the vertical scale factor.

The equation of a hyperbola is similar to the equation of an ellipse, except that the terms are subtracted, rather than added. For example, the equation $\left(\frac{y}{4}\right)^2 - \left(\frac{x}{3}\right)^2 = 1$ describes a hyperbola, whereas $\left(\frac{y}{4}\right)^2 + \left(\frac{x}{3}\right)^2 = 1$ describes an ellipse.

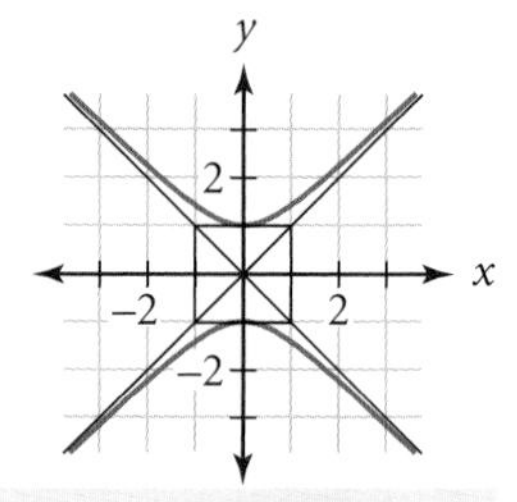

The vertically oriented unit hyperbola

EXAMPLE B

Graph $\left(\frac{y}{4}\right)^2 - \left(\frac{x}{3}\right)^2 = 1$.

▶ Solution

From the equation, you can tell that this is a vertically oriented hyperbola with a vertical scale factor of 4 and a horizontal scale factor of 3. The hyperbola is not translated, so its center is at the origin. To graph it on your calculator, you must solve for y.

$$\left(\frac{y}{4}\right)^2 - \left(\frac{x}{3}\right)^2 = 1$$
$$\left(\frac{y}{4}\right)^2 = 1 + \left(\frac{x}{3}\right)^2$$
$$\left(\frac{y}{4}\right) = \pm\sqrt{1 + \left(\frac{x}{3}\right)^2}$$
$$y = \pm 4\sqrt{1 + \left(\frac{x}{3}\right)^2}$$

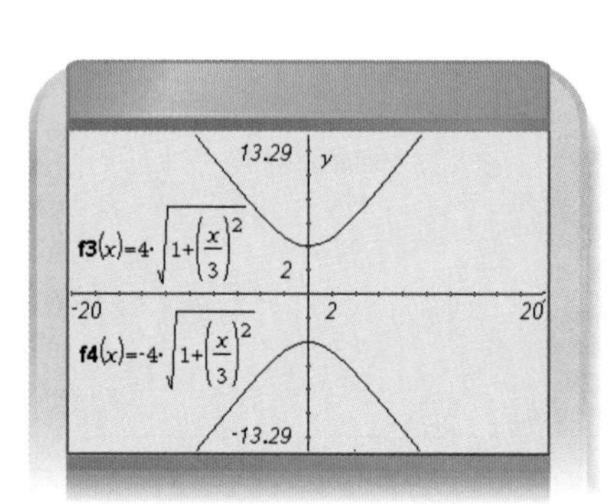

To sketch a hyperbola by hand, follow these steps:

1. Dilate the unit box by the horizontal and vertical scale factors.
2. Draw in the asymptotes (the diagonals of the box, extended).
3. Locate the vertices at the centers of the sides of the box. Because the y^2 term is positive, the vertices lie on the top and bottom sides of the box.
4. Draw the curve starting from the vertices and approaching the asymptotes.

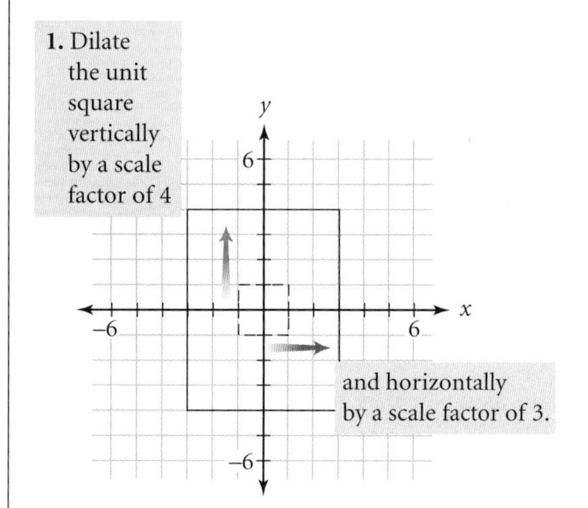

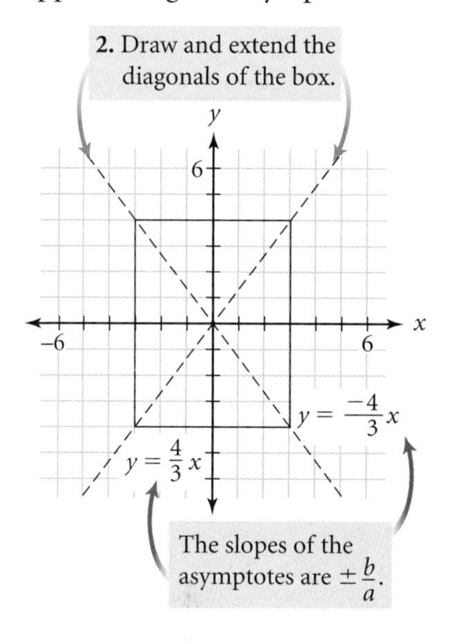

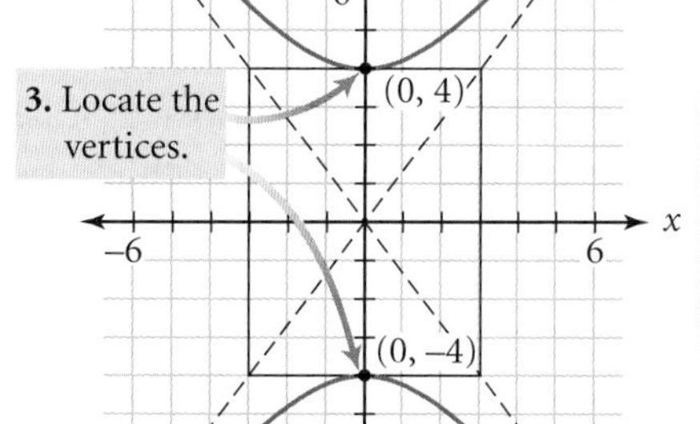

4. Sketch a curve through each vertex and extend the curves to approach the asymptotes.

You can graph the two asymptotes on your calculator to confirm that the hyperbola does approach them asymptotically.

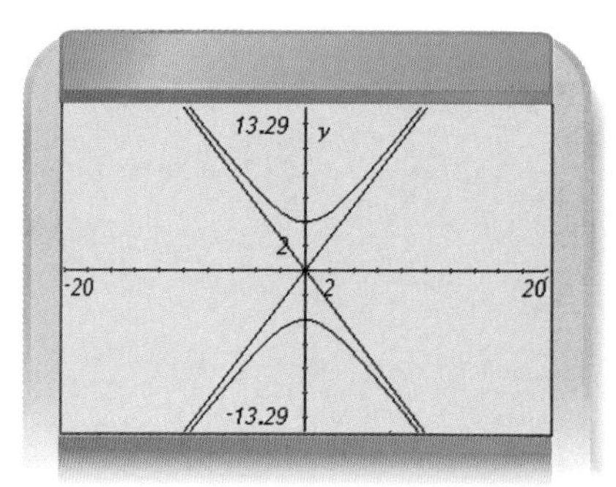

The location of foci in a hyperbola is related to a circle that can be drawn through the four corners of the asymptote rectangle. The distance from the center of the hyperbola to the foci is equal to the radius of the circle.

To locate the foci in a hyperbola, you can use the relationship $a^2 + b^2 = c^2$, where a and b are the horizontal and vertical scale factors. In the hyperbola from Example B, shown at right, $3^2 + 4^2 = c^2$, so $c = 5$, and the foci are 5 units above and below the center of the hyperbola at (0, 5) and (0, −5).

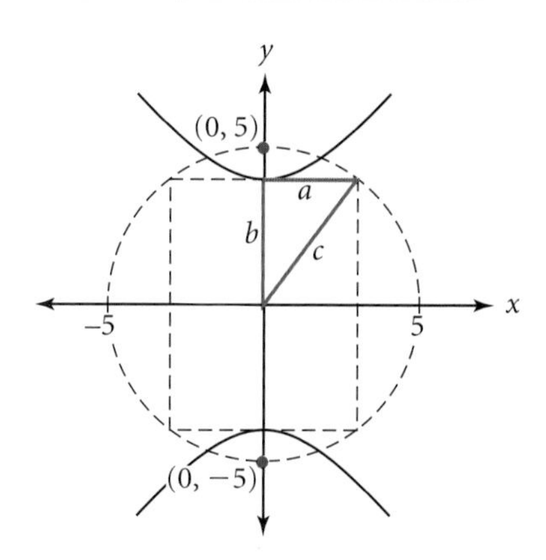

In the investigation you will explore a situation that produces hyperbolic data and find a curve to fit your data.

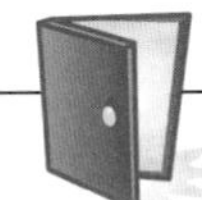

Investigation
Passing By

You will need

- a motion sensor

Procedure Note

1. One member of your group will use a motion sensor to measure the distance to the walker for 10 s. The motion sensor must be kept pointed at the walker.
2. The walker should start about 5 m to the left of the sensor holder. He or she should walk at a steady pace in a straight line, continuing past the sensor holder, and stop about 5 m to the right of the sensor holder.

Step 1 Collect data as described in the Procedure Note. Transfer these data from the motion sensor to each calculator in the group, and graph your data. They should form one branch of a hyperbola.

Step 2 Assume the sensor was held at the center of the hyperbola, and find an equation to fit your data. You may want to try to graph the asymptotes first.

Step 3 Transfer your graph to paper, and add the foci and the other branch of the hyperbola. To verify your equation, choose at least two points on the curve and measure their distances from the foci. Calculate the differences between the distances from each focus. What do you notice? Why?

Equation of a Hyperbola

The standard form of the equation of a horizontally oriented hyperbola with center (h, k), horizontal scale factor of a, and vertical scale factor of b is

$$\left(\frac{x-h}{a}\right)^2 - \left(\frac{y-k}{b}\right)^2 = 1$$

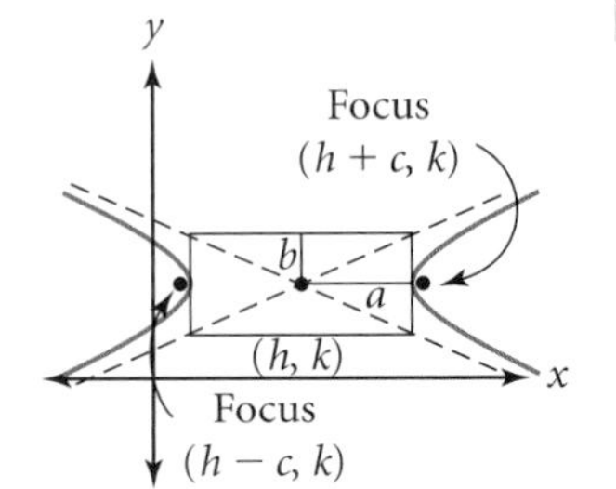

The equation of a vertically oriented hyperbola under the same conditions is

$$\left(\frac{y-k}{b}\right)^2 - \left(\frac{x-h}{a}\right)^2 = 1$$

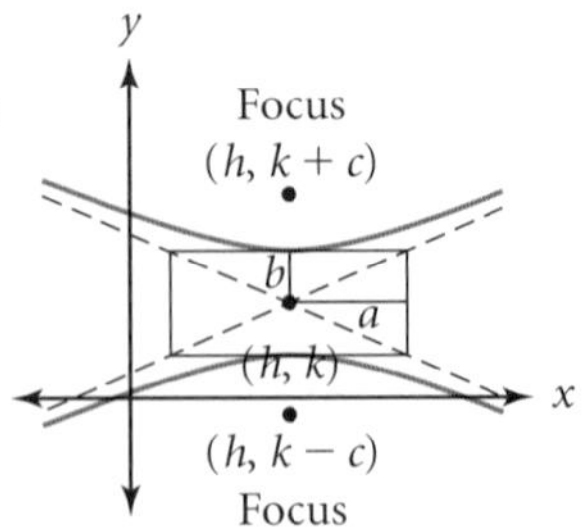

The foci are located c units from the center, where $a^2 + b^2 = c^2$. The asymptotes pass through the center and have slope $\pm\frac{b}{a}$.

EXAMPLE C Write the equation of this hyperbola in standard form, and find the foci.

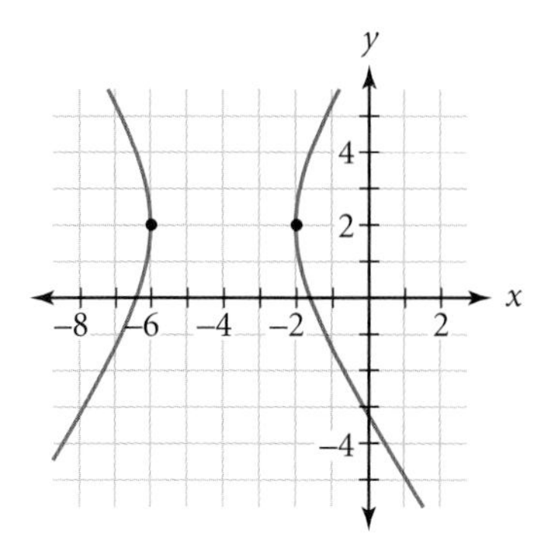

▶ Solution

The center is halfway between the vertices, at the point $(-4, 2)$. The horizontal distance from the center to the vertex, a, is 2. If you knew the location of the asymptotes, you could find the value of b using the fact that the slopes of asymptotes of a hyperbola are $\pm\frac{b}{a}$. In this case, to find the vertical scale factor you will need to estimate the asymptotes or, using a point from the graph, solve for the value of b. Write the equation, substituting the values you know.

$$\left(\frac{x+4}{2}\right)^2 - \left(\frac{y-2}{b}\right)^2 = 1$$

Estimating a point not too close to either vertex, such as $(0, -3.2)$, will allow you to approximate b.

$$\left(\frac{0+4}{2}\right)^2 - \left(\frac{-3.2-2}{b}\right)^2 = 1$$ Substitute 0 for x and -3.2 for y.

$$(2)^2 - \left(\frac{-5.2}{b}\right)^2 = 1$$ Add and divide.

$$4 - \frac{27.04}{b^2} = 1$$ Square.

$$-\frac{27.04}{b^2} = -3$$ Subtract 4 from both sides.

$$9.01\bar{3} = b^2$$ Multiply by b^2 and divide by -3 on both sides.

$$3.002 \approx b$$ Take the square root of both sides.

The value of b is approximately 3, so the equation of the hyperbola is close to

$$\left(\frac{x+4}{2}\right)^2 - \left(\frac{y-2}{3}\right)^2 = 1$$

You can find the distance to the foci by using the equation $a^2 + b^2 = c^2$.

$$2^2 + 3^2 = c^2$$
$$13 = c^2$$
$$\pm\sqrt{13} = c$$

So the foci are $\sqrt{13}$ to the right and left of the center, at $\left(-4+\sqrt{13}, 2\right)$ and $\left(-4-\sqrt{13}, 2\right)$, or approximately $(-0.39, 2)$ and $(-7.6, 2)$.

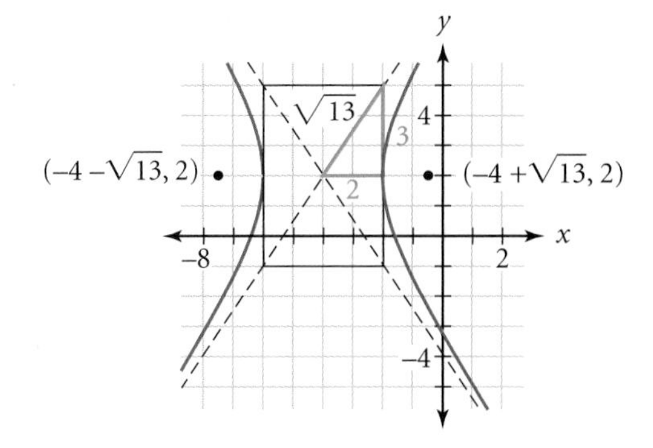

You will continue to explore the relationship between the equation and graph of a hyperbola in the exercises.

EXERCISES

You will need

A graphing calculator for Exercise **10.**

Practice Your Skills

1. Sketch each hyperbola on your paper. Write the coordinates of each vertex and the equation of each asymptote.

a. $\left(\frac{x}{2}\right)^2 - \left(\frac{y}{4}\right)^2 = 1$ @

b. $\left(\frac{y+2}{1}\right)^2 - \left(\frac{x-2}{3}\right)^2 = 1$ @

c. $\left(\frac{x-4}{3}\right)^2 - \left(\frac{y-1}{3}\right)^2 = 1$

d. $\left(\frac{y+1}{2}\right)^2 - \left(\frac{x+2}{3}\right)^2 = 1$

e. $\left(\frac{x+1}{4}\right)^2 - \left(\frac{y-3}{2}\right)^2 = 1$

f. $\left(\frac{y-3}{3}\right)^2 - \left(\frac{x}{5}\right)^2 = 1$

2. What are the coordinates of the foci of each hyperbola in Exercise 1? @

3. Write an equation in standard form for each graph.

a. @

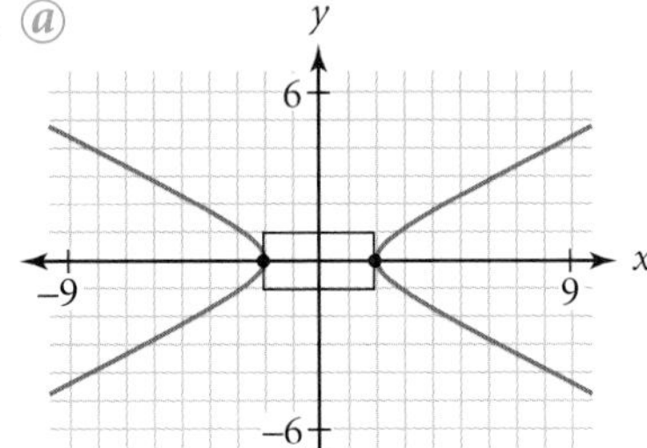

b. @

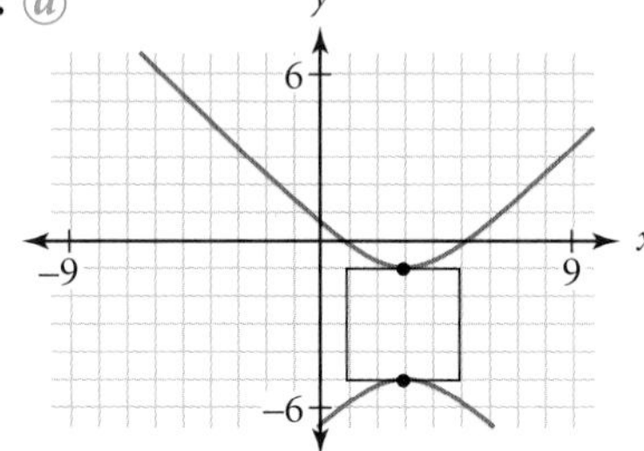

c.

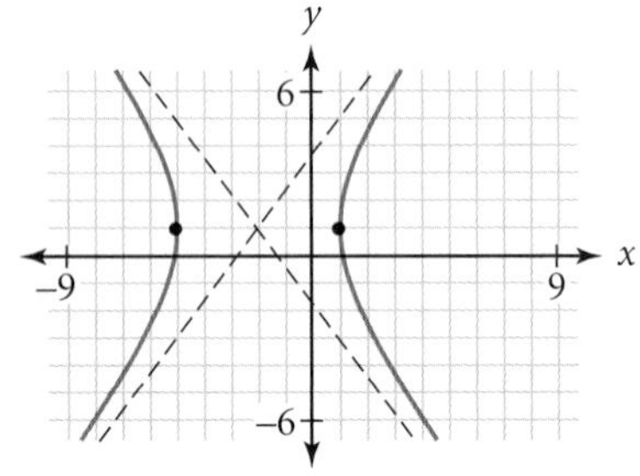

d.

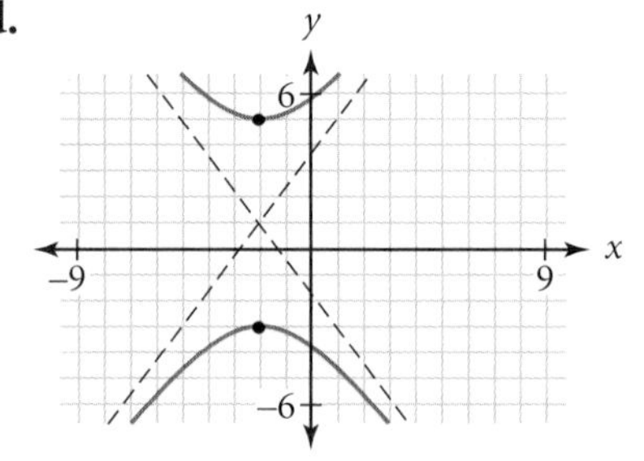

4. What are the coordinates of the foci of each hyperbola in Exercise 3? @

5. Write the equations of the asymptotes for each hyperbola in Exercise 3. @

Reason and Apply

6. A point moves in a plane so that the difference of its distances from $(-5, 1)$ and $(7, 1)$ is always 10 units. What is the equation of the path of this point? @

7. Graph and write the equation of a hyperbola that has an upper vertex at $(-2.35, 1.46)$ and has asymptote $y = 1.5x + 1.035$.

8. Approximate the equation of each hyperbola shown.

a. ⓐ

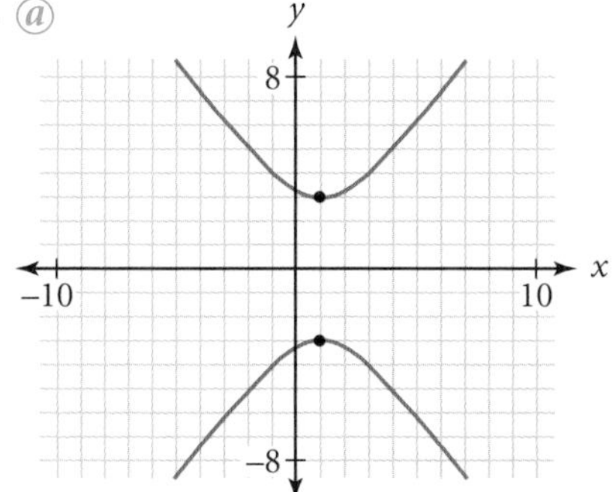

b.

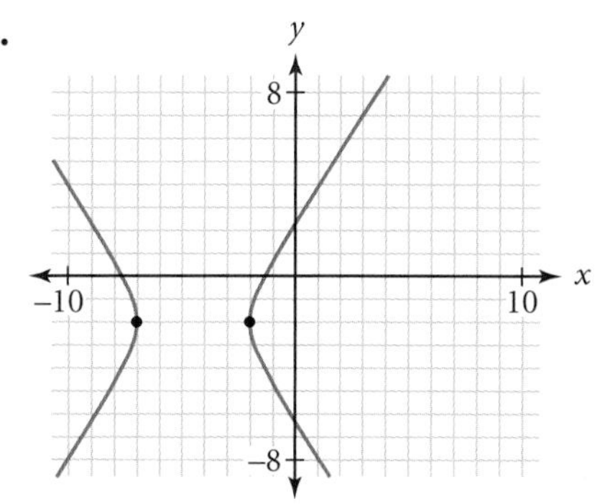

9. Another way to locate the foci of a hyperbola is by rotating the asymptote rectangle about its center so that opposite corners lie on the line of symmetry that contains the vertices of the hyperbola. From the diagram, you can see that the distance from the origin to a focus is one-half the length of the diagonal of the rectangle.

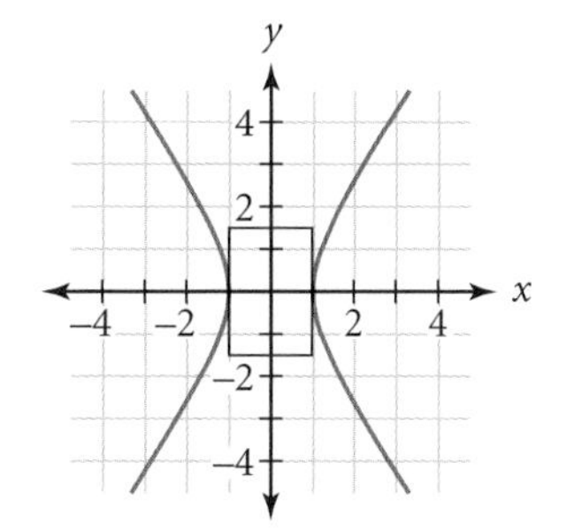

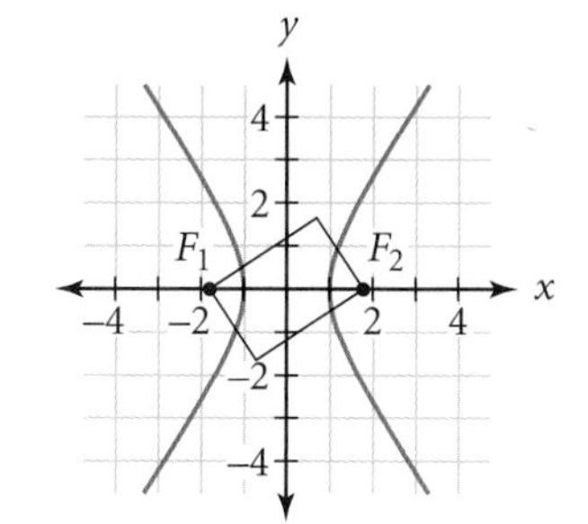

a. Show that this distance is $\sqrt{a^2 + b^2}$.

b. Find the coordinates of the foci for $\left(\frac{y+2}{1}\right)^2 - \left(\frac{x-2}{3}\right)^2 = 1$.

10. APPLICATION A receiver can determine the distance to a homing transmitter by its signal strength. These distances were determined using a receiver in a car traveling due north.

Distance traveled (mi)	0.0	2.0	4.0	6.0	8.0	10.0	12.0	14.0	16.0
Distance from transmitter (mi)	9.82	7.91	6.04	4.30	2.92	2.55	3.54	5.15	6.96

a. Find an equation for the hyperbola that best fits the data. ⓗ

b. Name the center of this hyperbola. What does this point tell you?

c. What are the possible locations of the homing transmitter?

11. Sketch the graphs of the conic sections in 11a–d.

a. $y = x^2$

b. $x^2 + y^2 = 9$

c. $\frac{x^2}{9} + \frac{y^2}{16} = 1$

d. $\frac{x^2}{9} - \frac{y^2}{16} = 1$

e. If each of the curves in 11a–d is rotated about the y-axis, describe the shape that is formed. Include a sketch.

Mathematics CONNECTION

One area in the study of calculus is the analysis of three-dimensional solids formed by revolving a curve about an axis. Revolving a hyperbola about the line through its foci or about the perpendicular bisector of the segment connecting the foci produces a hyperboloid. The hyperboloid is used in the design of cooling towers because the concrete shell can be relatively thin for its large size. Also, the structure of the hyperboloid allows cooling towers to use a natural draft design to bring air into the cooling process.

Cooling towers in Middletown, Pennsylvania

12. Find the vertical distance between a point on the hyperbola $\left(\frac{y+1}{2}\right)^2 - \left(\frac{x-2}{3}\right)^2 = 1$ and its nearest asymptote for each x-value shown at right. (h)

x-value	5	10	20	40
Distance				

Review

13. Solve the quadratic equation $0 = -x^2 + 6x - 5$ by completing the square.

14. Mercury's orbit is an ellipse with the Sun at one focus, eccentricity 0.206, and major axis approximately 1.158×10^8 km. If you consider Mercury's orbit with the Sun at the origin and the other focus on the positive x-axis, what equation models the orbit? (a)

15. The setter on a volleyball team makes contact with the ball at a height of 5 ft. The parabolic path of the ball reaches a maximum height of 17.5 ft when the ball is 10 ft from the setter.

a. Find an equation that models the ball's path. (h)

b. A hitter can spike the ball when it is 8.5 feet off the floor. How far from the setter is the hitter when she makes contact?

16. Sketch the graph of each parabola. Give the coordinates of each vertex and focus, and the equation of each directrix.

a. $y = -(x+1)^2 - 2$ **b.** $y = \frac{1}{2}x^2 - 3x + 5$ **c.** $x = \frac{1}{2}y^2 - 6$

17. The half-life of radium-226 is 1620 yr.

a. Write a function that relates the amount s of a sample of radium-226 remaining after t years.

b. After 1000 yr, how much of a 500 g sample of radium-226 will remain?

c. How long will it take for a 3 kg sample of radium-226 to decay so that only 10 g remain?

18. Factor each expression completely.

a. $x^2 + 10x + 25$ **b.** $x^2 - 12x + 36$ **c.** $y^2 + 2y + 1$ **d.** $y^2 - 4y + 4$

IMPROVING YOUR VISUAL THINKING SKILLS

Slicing a Cone

Describe how to slice a double cone to produce each of these geometric shapes: a circle, an ellipse, a parabola, a hyperbola, a point, one line, and two lines. Be sure to describe at what angle and where the plane must slice the cone. Sketch a diagram of each slicing. (*Hint:* Look back at the illustrations on page 453 for help.)

EXPLORATION

Constructing the Conic Sections

In geometry class you probably used a compass and straightedge to construct polygons, such as triangles and squares. You also know that a compass can easily construct a circle, one of the conic sections.

What about the other conic sections—the ellipse, the parabola, and the hyperbola? How could you possibly construct these complex curves? In this chapter you have learned locus definitions for these shapes, but they may seem impossible to construct geometrically.

Actually, there are many different ways to construct the conic sections. Geometry software, such as The Geometer's Sketchpad, makes these constructions even easier. In the activity you will learn one way to construct an ellipse. The follow-up questions then challenge you to construct a parabola and a hyperbola.

History CONNECTION

In 1646, Dutch mathematician Frans van Schooten (1615–1660) wrote *Sive de Organica Conicarum Sectionum in Plano Descriptione, Tractatus,* which translates to *A Treatise on Devices for Drawing Conic Sections.* This book describes several different ways to construct each of the conic sections. Some of the constructions used unique mechanical devices.

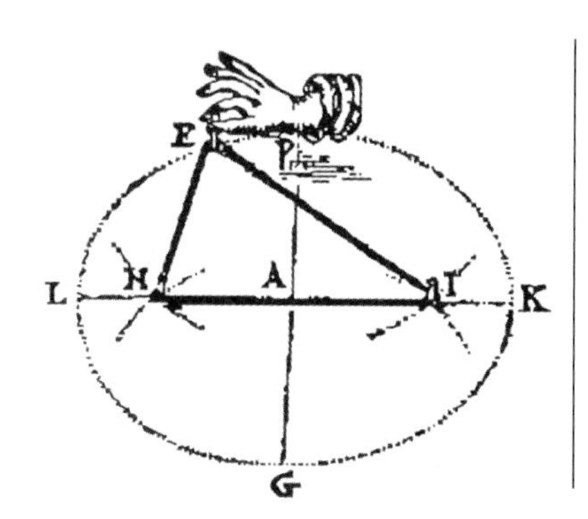

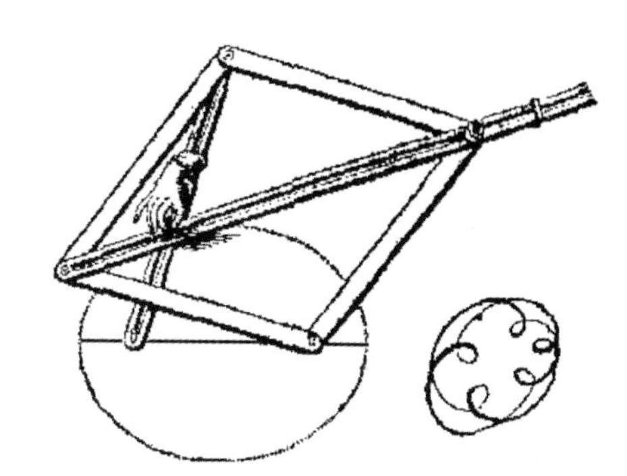

These illustrations from van Schooten's *Sive de Organica Conicarum Sectionum in Plano Descriptione, Tractatus* show two ways of constructing an ellipse.

Activity

From Circles to the Ellipse

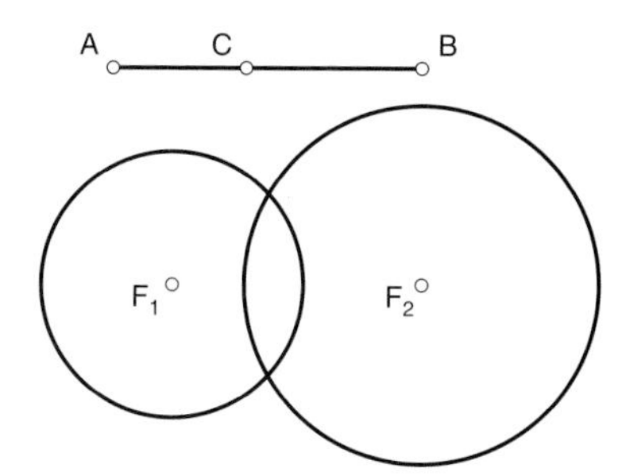

Step 1 In a new sketch, construct a segment and label the endpoints A and B.

Step 2 Construct a point on $\overline{AB}$. Label this point C.

Step 3 Construct segments AC and CB. What is true about $AC + CB$?

Step 4 Construct and label points F_1 and F_2, not on $\overline{AB}$. These will be the foci of your ellipse.

Step 5 Select F_1 and $\overline{AC}$, and choose **Circle By Center+Radius** from the Construct menu. Construct another circle with F_2 and $\overline{CB}$.

Step 6 If necessary, adjust your model so that the circles intersect. Select the two circles and choose **Intersections** from the Construct menu. What is the distance from either intersection to F_1? What is the distance from either intersection to F_2? What is true about the sum of these distances?

Step 7 Select the two intersections and choose **Trace Intersections** from the Display menu. Slowly drag point C back and forth along $\overline{AB}$. Describe the shape of the trace. Explain why this happens.

In this chapter you learned about loci. In Step 7 you traced a locus of points. Depending on how slowly you dragged point C, your ellipse may be smooth or it may have gaps. Sketchpad also allows you to see the locus as a smooth curve.

Step 8 Choose **Erase Traces** from the Display menu. Then select the two intersections and deselect **Trace Intersections** from the Display menu.

Step 9 Select point C and one of the intersections, and choose **Locus** from the Construct menu. You should see one half of your ellipse. Repeat this process with the other intersection to see both halves.

When you construct a locus, you can experiment by dragging objects and seeing how the locus changes.

keymath.com/DAA

Questions

1. Experiment by dragging one focus. How does the ellipse change? How far apart can the foci be before you no longer have an ellipse? What happens when both foci are at the same point?

2. Experiment by dragging point B. How does the ellipse change? Based on the locus definition of an ellipse, what is changing?

3. The illustration at right shows one method of constructing a parabola based on the geometric definition. Use Sketchpad to construct this locus. Explain how the construction satisfies the definition. (Note that $\overleftrightarrow{PA} \perp \overline{FB}$.)

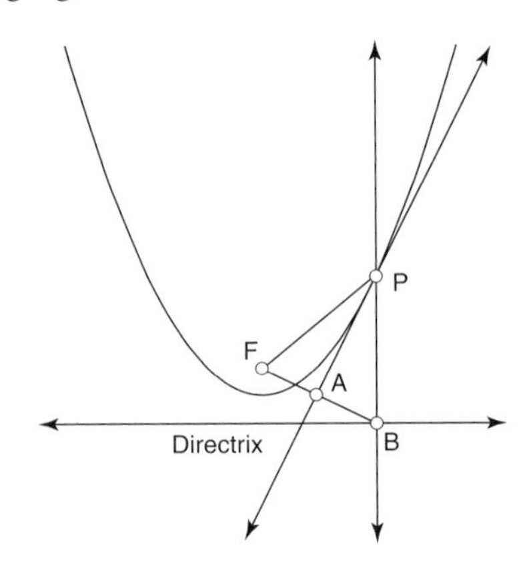

4. Find a way to construct a hyperbola with Sketchpad. (*Hint:* One possible construction is similar to the ellipse construction, except you want the difference $AC - BC$ to have a constant value.)

LESSON

8.5

The General Quadratic

A complex system that works is invariably found to have evolved from a simple system that works.

JOHN GAULE

You have seen that all of the conic sections come from slicing a double cone. Do their equations also have something in common?

Circle	$(x-h)^2+(y-k)^2=r^2$	
	Horizontal	Vertical
Parabola	$\left(\frac{y-k}{b}\right)^2=\frac{x-h}{a}$ with $\frac{b^2}{a}=4f$	$\frac{y-k}{b}=\left(\frac{x-h}{a}\right)^2$ with $\frac{a^2}{b}=4f$
Ellipse	$\left(\frac{x-h}{a}\right)^2+\left(\frac{y-k}{b}\right)^2=1$ with $a>b$	$\left(\frac{x-h}{a}\right)^2+\left(\frac{y-k}{b}\right)^2=1$ with $b>a$
Hyperbola	$\left(\frac{x-h}{a}\right)^2-\left(\frac{y-k}{b}\right)^2=1$	$\left(\frac{y-k}{b}\right)^2-\left(\frac{x-h}{a}\right)^2=1$

In fact, all of the equations can be converted into one standard equation.

General Quadratic Equation

The **general quadratic equation** in two variables, x and y, is

$$Ax^2+Bxy+Cy^2+Dx+Ey+F=0$$

where A, B, and C are not all zero.

In this lesson you will learn how to convert the general quadratic equation into standard form. You will also learn how to solve for y so that you can use your calculator to graph the curves.

In all the relationships you have seen so far in this chapter, B is equal to zero. When B does not equal zero, the graph of the equation will be a rotated conic section. You will explore these rotated curves in later courses.

In 1997, the comet Hale-Bopp passed by Earth and was one of the brightest comets seen in the 20th century. Hale-Bopp has an elliptical orbit.

Astronomy CONNECTION

When a comet passes through the solar system, its motion is influenced by the Sun's gravity. Some comets swing around the Sun and leave the solar system, never to return. The paths are described by one branch of a hyperbola with the Sun at one focus or by a parabola. However, if a comet is moving more slowly, it will be captured into an elliptical orbit with the Sun at one focus. The outcome depends on the speed of the comet and the angle at which it approaches the Sun. Some orbits appear parabolic, but are probably very long ellipses. A circular orbit is possible, but unlikely.

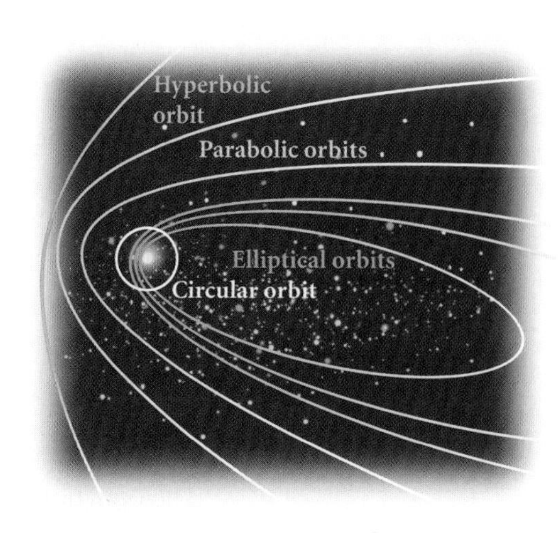

EXAMPLE A

Convert the equation $4x^2 - 9y^2 + 144 = 0$ to standard form. Then name the shape and graph it. Solve for y and graph on your calculator to confirm your answer.

▶ ***Solution***

Put the equation in standard form:

$$4x^2 - 9y^2 = -144$$ Subtract 144 from both sides.

$$\frac{4x^2 - 9y^2}{-144} = 1$$ Divide by -144.

$$-\frac{x^2}{36} + \frac{y^2}{16} = 1$$ Divide each term by -144 and reduce the fractions.

$$\left(\frac{y}{4}\right)^2 - \left(\frac{x}{6}\right)^2 = 1$$ Reorder and write the terms as squares to put in standard form.

Tractricious, a Robert Wilson–designed sculpture at the Fermi National Accelerator Laboratory, is a freestanding hyperboloid of stainless steel tubes.

This equation is the standard form of a vertically oriented hyperbola. It is centered at the origin and has values $a = 6$ and $b = 4$. Sketch the asymptote rectangle and the asymptotes, then plot the vertices and draw the curve.

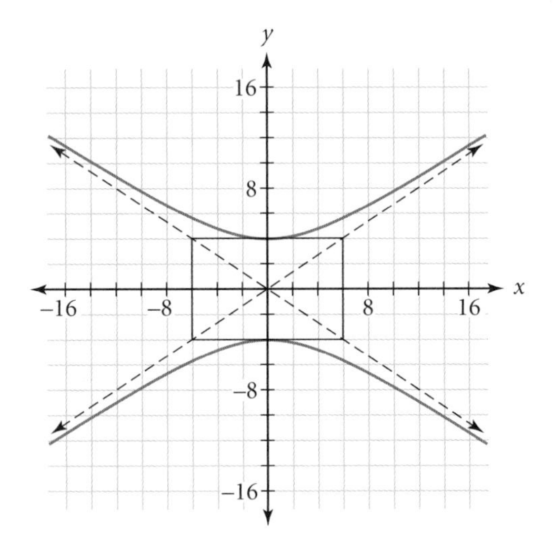

Now solve for y:

$$\left(\frac{y}{4}\right)^2 - \left(\frac{x}{6}\right)^2 = 1$$ Equation in standard form.

$$\frac{y^2}{16} = 1 + \frac{x^2}{36}$$ Square terms and add $\frac{x^2}{36}$ to both sides.

$$y^2 = 16\left(1 + \frac{x^2}{36}\right)$$ Multiply by 16.

$$y = \pm 4\sqrt{1 + \frac{x^2}{36}}$$ Take the square root of both sides.

Graph this equation on your calculator to confirm the sketch.

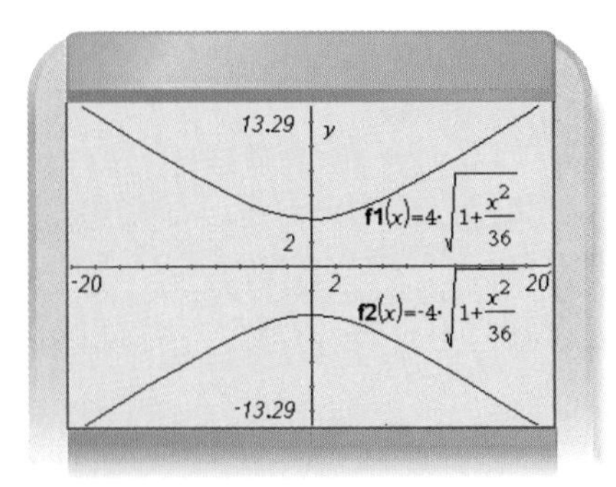

The equation in Example A was relatively easy to work with because the values of B, D, and E in the general equation were zero. When D or E is nonzero, you must use the process of completing the square to convert the equation to standard form.

EXAMPLE B | Describe the shape determined by the equation $x^2 + 4y^2 - 14x + 33 = 0$.

▶ Solution

Complete the square to convert from general form to standard form.

$$x^2 + 4y^2 - 14x + 33 = 0$$ Original equation.

$$\left(x^2 - 14x\right) + \left(4y^2\right) = -33$$ Group x terms and y terms and isolate constants on the other side.

$$\left(x^2 - 14x + 49\right) + \left(4y^2\right) = -33 + 49$$ To complete the square for x, add $\left(-\frac{14}{2}\right)^2$, or 49, to both sides.

$$(x - 7)^2 + 4y^2 = 16$$ Write the equation in perfect-square form.

$$\frac{(x-7)^2}{16} + \frac{y^2}{4} = 1$$ Divide by 16 and reduce.

$$\left(\frac{x-7}{4}\right)^2 + \left(\frac{y}{2}\right)^2 = 1$$ Write the equation in standard form.

In this form, it is clear that this is the equation of an ellipse. The center is (7, 0), the horizontal scale factor is 4, and the vertical scale factor is 2.

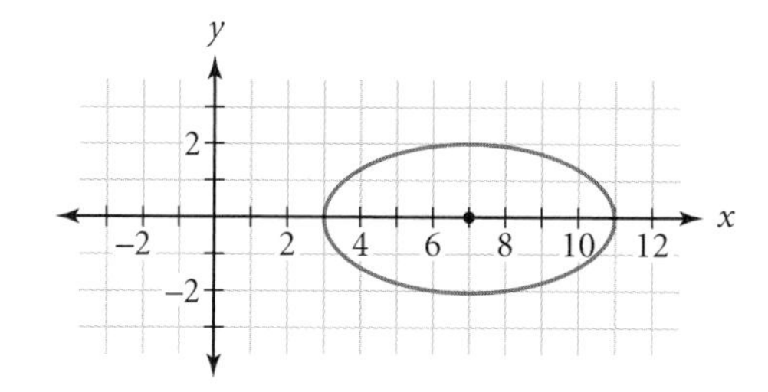

Look back at the original equation in Example B. How might you have known that this equation described an ellipse?

EXAMPLE C | Graph the equation $y^2 - 4x + 6y + 1 = 0$.

▶ Solution

This equation describes a parabola, because only one of the variables has an exponent of 2. You can now choose whether to convert this equation to standard form or solve for y. The first choice requires completing the square.

$$\left(y^2 + 6y\right) - (4x) = -1$$ Group x terms and y terms and isolate constants on the other side.

$$\left(y^2 + 6y + 9\right) - (4x) = -1 + 9$$ To complete the square for y, add 9 to both sides.

$$(y + 3)^2 - 4x = 8$$ Write in perfect-square form.

$$(y + 3)^2 = 4x + 8$$ Add $4x$ to both sides.

$$(y + 3)^2 = 4(x + 2)$$ Factor.

$$\left(\frac{y+3}{2}\right)^2 = (x + 2)$$ Write the equation in standard form.

This equation indicates a horizontally oriented parabola with vertex $(-2, -3)$ and a vertical scale factor of 2.

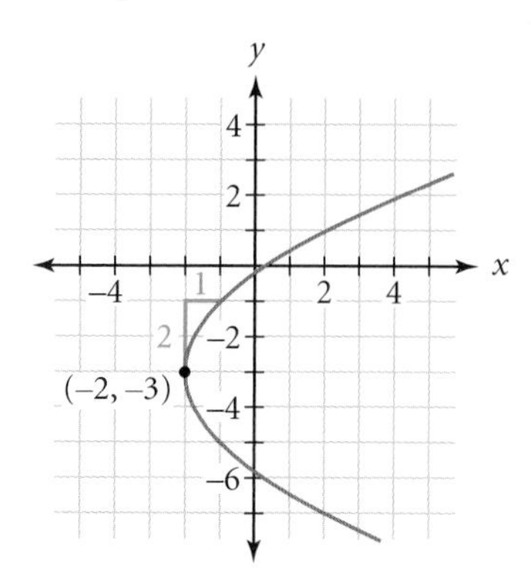

To follow the other approach, solving for y, you use the quadratic formula. In this case the variable y is the quadratic term, so rewrite the equation in the form $ay^2 + by + c = 0$.

$y^2 - 4x + 6y + 1 = 0$ Original equation.

$y^2 + 6y - 4x + 1 = 0$ Write the equation as $ay^2 + by + c = 0$.

So $a = 1$, $b = 6$, and $c = -4x + 1$. Substitute these values into the quadratic formula.

$$y = \frac{-b \pm \sqrt{b^2 - 4ac}}{2a}$$

$$= \frac{-6 \pm \sqrt{6^2 - 4(1)(-4x + 1)}}{(2)(1)}$$

$$= \frac{-6 \pm \sqrt{36 + 16x - 4}}{2}$$

$$= \frac{-6 \pm \sqrt{32 + 16x}}{2}$$

$$= \frac{-6 \pm \sqrt{16(2 + x)}}{2}$$

$$= \frac{-6 \pm 4\sqrt{2 + x}}{2}$$

$$y = -3 \pm 2\sqrt{2 + x}$$

The solution gives the equation $y = -3 \pm 2\sqrt{2 + x}$.

The standard form gives you information about the translations, dilations, foci, and directrix. However, if you wish to graph a conic section on your calculator, you can always use the quadratic formula to solve for y.

You can also use the quadratic formula to find out other information about conics. For example, how can you find the points of intersection of two conic sections? In this investigation you'll begin by exploring how many intersection points are possible for various combinations of curves.

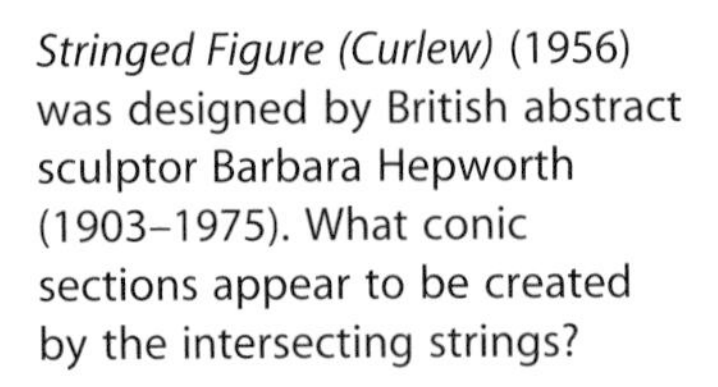

Stringed Figure (Curlew) (1956) was designed by British abstract sculptor Barbara Hepworth (1903–1975). What conic sections appear to be created by the intersecting strings?

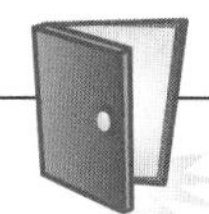

Investigation
Systems of Conic Equations

If you graph two conic sections on the same graph, in how many ways could they intersect?

There are four conic sections: circles, ellipses, parabolas, and hyperbolas. Among the members of your group, investigate the possible numbers of intersection points for all ten pairs of shapes. For example, an ellipse and a hyperbola could intersect in 0, 1, 2, 3, or 4 points, as shown below. For each pair of conic sections, list the possible numbers of intersection points.

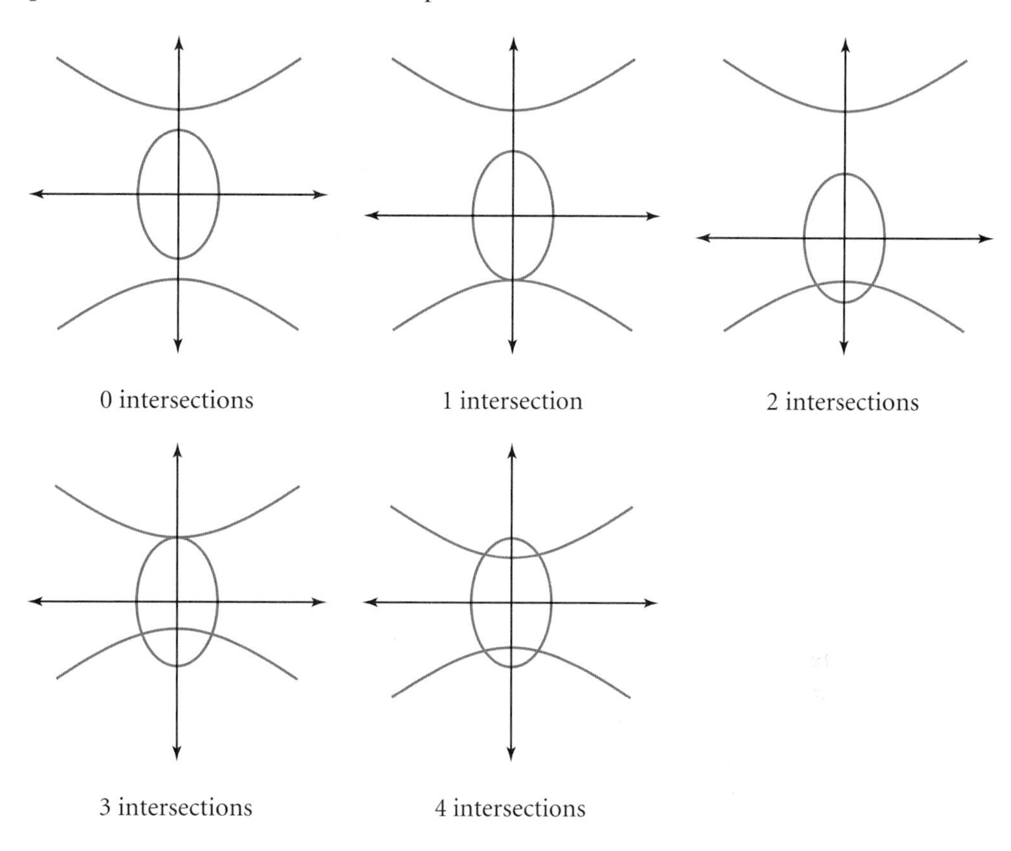

You can use algebra to identify the points of intersection.

EXAMPLE D Find the points of intersection of $\left(\frac{x-1}{4}\right)^2 + \left(\frac{y}{3}\right)^2 = 1$ and $\left(\frac{y}{3}\right)^2 = \frac{x+2}{3}$.

▶ Solution First make a sketch of the two curves. The first equation describes an ellipse with center (1, 0). It has a horizontal scale factor of 4 and a vertical scale factor of 3. The second equation describes a parabola with vertex (−2, 0) and both scale factors of 3. Looking at your sketch, you can expect to find two intersection points.

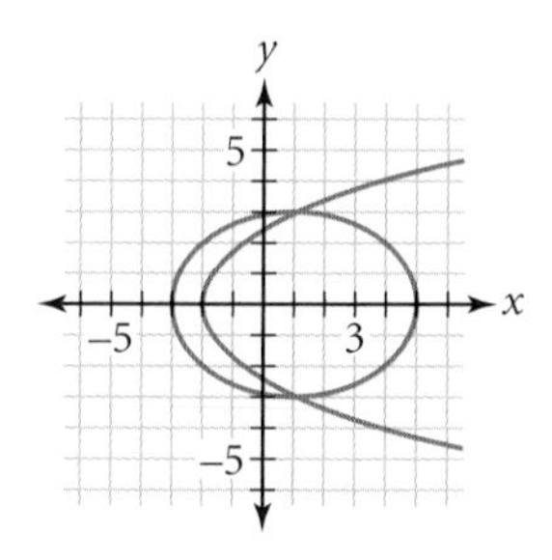

To solve for the points of intersection, notice that both equations contain the term $\left(\frac{y}{3}\right)^2$. So you can use substitution.

$\left(\frac{x-1}{4}\right)^2 + \left(\frac{y}{3}\right)^2 = 1$ and $\left(\frac{y}{3}\right)^2 = \frac{x+2}{3}$	Original equations.
$\frac{(x-1)^2}{16} + \frac{x+2}{3} = 1$	Substitute $\frac{x+2}{3}$ for $\left(\frac{y}{3}\right)^2$.
$3(x-1)^2 + 16(x+2) = 48$	Multiply both sides by 48 to eliminate the fractions.
$3x^2 - 6x + 3 + 16x + 32 = 48$	Expand.
$3x^2 + 10x - 13 = 0$	Combine like terms.
$(3x+13)(x-1) = 0$	Factor.
$x = -\frac{13}{3}$ or $x = 1$	Set each factor equal to 0 and solve.

The first value for x is beyond the end of the ellipse, so you can ignore this solution. The other solution, $x = 1$, looks as if it fits the sketch. Now you need to find the y-values for the intersection points.

$\left(\frac{y}{3}\right)^2 = \frac{1+2}{3} = 1$	Substitute 1 for x into the second equation.
$y^2 = 9$	Solve for y^2.
$y = \pm 3$	Take the square root.

So the intersection points are (1, 3) and (1, −3).

In Example D, you solved algebraically and found the exact points of intersection. Sometimes using graphs to find approximate points of intersection is sufficient.

EXAMPLE E Approximate the points of intersection of $\frac{(x-5)^2}{4} + y^2 = 1$ and $x = y^2 + 5$.

► **Solution**

First solve each equation for y.

$\frac{(x-5)^2}{4} + y^2 = 1$	First equation.
$y^2 = 1 - \frac{(x-5)^2}{4}$	Solve for y^2.
$y = \pm\sqrt{1 - \frac{(x-5)^2}{4}}$	Solve for y.
$x = y^2 + 5$	Second equation.
$y^2 = x - 5$	Solve for y^2.
$y = \pm\sqrt{x-5}$	Solve for y.

Then graph the two equations, and use the graph to approximate the points of intersection.

There are two points of intersection, approximately (5.8, 0.9) and (5.8, −0.9).

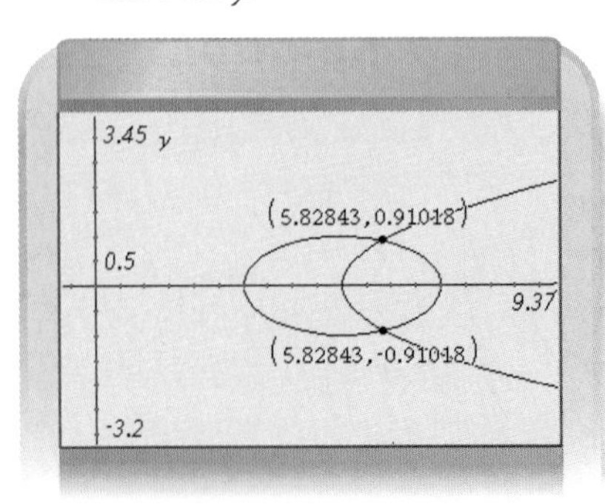

You can always use graphing to estimate real solutions. Graphing is valuable even if you are finding solutions algebraically, because a graph will tell you how many intersection points to look for. It can also help confirm whether your algebraic solutions are correct.

EXERCISES

You will need

A graphing calculator for Exercises **9** and **11**.

Practice Your Skills

1. Identify each equation as true or false. If it is false, correct it to make it true.

a. $y^2 + 11y + 121 = (y + 11)^2$ **b.** $x^2 - 18x + 81 = (x - 9)^2$

c. $5y^2 + 10y + 5 = 5(y + 1)^2$ **d.** $4x^2 + 24x + 36 = 4(x + 6)^2$

2. Rewrite each equation in the form $Ax^2 + Bxy + Cy^2 + Dx + Ey + F = 0$.

a. $(x + 7)^2 = 9(y - 11)$ @ **b.** $\frac{(x-7)^2}{9} + \frac{(y+11)^2}{1} = 1$ @

c. $(x - 1)^2 + (y + 3)^2 = 5$ **d.** $\frac{(x-2)^2}{4} - \frac{(y+3)^2}{9} = 1$

3. Find the values of a, b, c, and d as you follow these steps to complete the square for $3x^2 + 18x$.

$3x^2 + 18x$

$3(x^2 + ax)$

$3(x^2 + ax + b) - 3b$

$3(x^2 + ax + b) - c$

$3(x + d)^2 - c$

4. Convert each equation to the standard form of a conic section. Name the shape described by each equation.

a. $x^2 - y^2 + 8x + 10y + 2 = 0$ @ **b.** $2x^2 + y^2 - 12x - 16y + 10 = 0$ @

c. $3x^2 + 30x + 5y - 4 = 0$ **d.** $5x^2 + 5y^2 + 20x - 6 = 0$

Reason and Apply

5. Solve each equation for y. You may need to use the quadratic formula.

a. $25x^2 - 4y^2 + 100 = 0$ @ **b.** $4y^2 - 10x + 16y + 36 = 0$ @

c. $4x^2 + 4y^2 + 24x - 8y + 39 = 0$ **d.** $3x^2 + 5y^2 - 12x + 20y + 8 = 0$

6. Solve each system of equations algebraically, using the substitution or elimination method.

a. $\begin{cases} y = x^2 + 4 \\ y = (x-2)^2 + 3 \end{cases}$ **b.** $\begin{cases} 3x^2 + 9y^2 = 9 \\ 3x^2 + 5y^2 = 8 \end{cases}$ **c.** $\begin{cases} x^2 - \frac{y^2}{4} = 1 \\ x^2 + (y+4)^2 = 9 \end{cases}$

7. APPLICATION Two seismic monitoring stations recorded the vibrations of an earthquake. The second monitoring station is 50 mi due east of the first. The epicenter was determined to be 30 mi from the first station and 27 mi from the second station. Where could the epicenter of the earthquake be located?

8. Match each equation to one of the graphs.

A. $9x^2 + 4y^2 - 36 = 0$

B. $x^2 - 4y^2 - 8x = 0$

C. $3x^2 - 30x + 5y + 55 = 0$

D. $x^2 + y^2 + 2x - 6y - 15 = 0$

i.

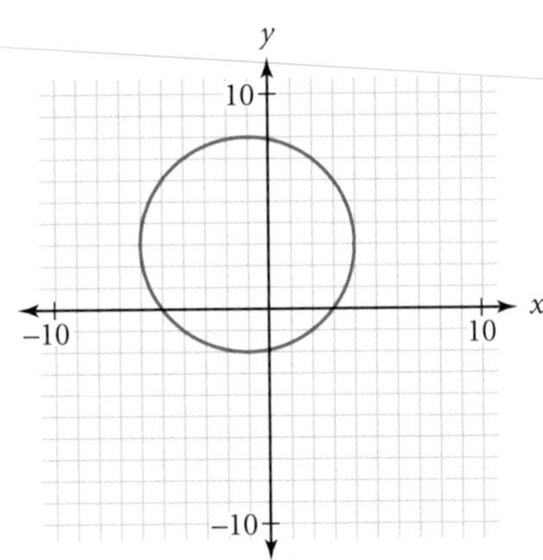

ii.

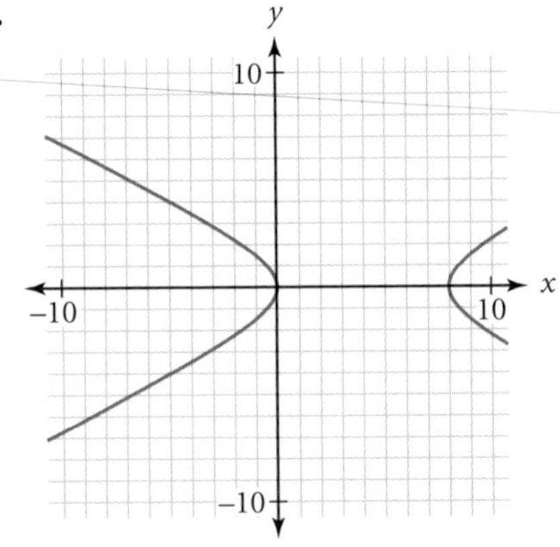

iii.

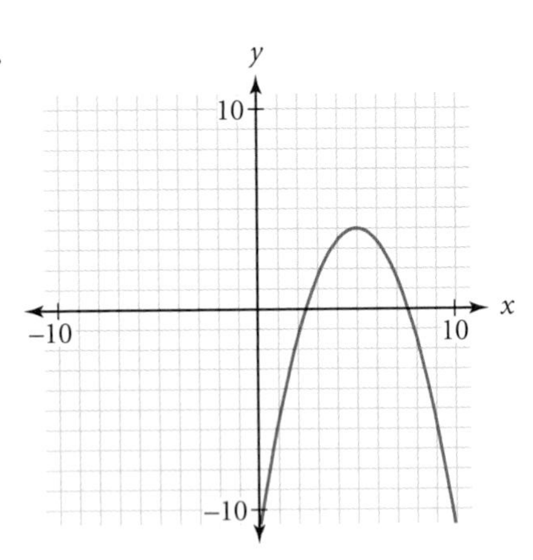

iv.

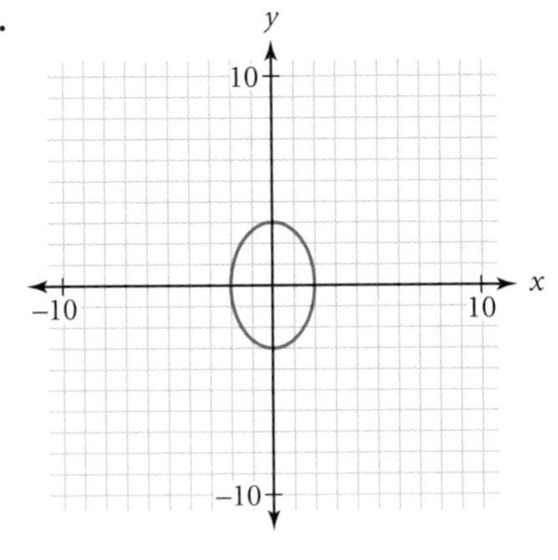

9. Graph each system of equations and approximate the solutions. ⓐ

a. $\begin{cases} x^2 + y^2 = 10 \\ \left(\frac{x+1}{4}\right)^2 + (y+2)^2 = 1 \end{cases}$

b. $\begin{cases} (x-2)^2 - \left(\frac{y-3}{4}\right)^2 = 1 \\ y = (x-1)^2 - 8 \end{cases}$

10. Find the equation of the circle that passes through the four intersection points of the ellipses $\frac{x^2}{16} + \frac{y^2}{9} = 1$ and $\frac{x^2}{9} + \frac{y^2}{16} = 1$. ⓗ

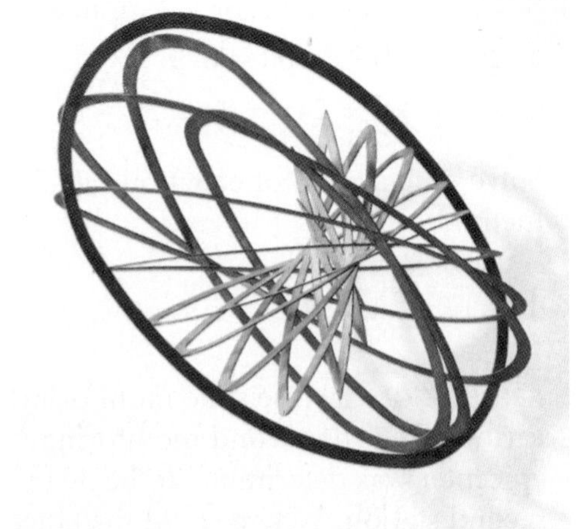

This plywood and wire elliptical sculpture by Russian artist Alexander Rodchenko (1891–1956) is titled *Oval Hanging Construction Number 12* (1920).

11. APPLICATION Three LORAN radio transmitters, *A*, *B*, and *C*, are located 200 miles apart along a straight coastline. They simultaneously transmit radio signals at regular intervals. The signals travel at a speed of 980 feet per microsecond. A ship, at *S*, first receives a signal from transmitter *B*. After 264 microseconds, the ship receives the signal from transmitter *C*, and then another 264 microseconds later it receives the signal from transmitter *A*. Use the diagram to answer these questions to find the location of the ship.

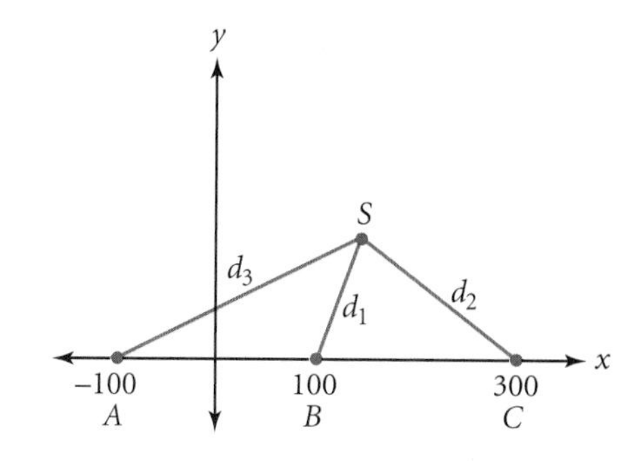

a. Find $d_2 - d_1$. Express your answer in miles (1 mi = 5280 ft).

b. Find $d_3 - d_1$. Express your answer in miles.

c. Use the fact that $d_2 - d_1$ is constant to write the equation of the hyperbola represented in 10a. Note that transmitters *B* and *C* are located at the foci. ⓗ

d. Write the equation of the hyperbola represented in 10b.

e. Graph the hyperbolas and find the coordinates of the location of the ship. How can you be sure which of the intersection points represents the ship?

History
CONNECTION

During World War II, LORAN, a long-range navigation system developed at MIT, used radio waves and the definition of a hyperbola to determine the exact location of ships. Today, LORAN-C is operated by the U.S. Coast Guard to monitor U.S. coastal waters. Civil and military air, land, and marine users are provided navigation, location, and timing services by the Coast Guard. Although today global positioning satellites can provide accurate locations, LORAN is a lower-cost alternative because it does not require the launch of satellites.

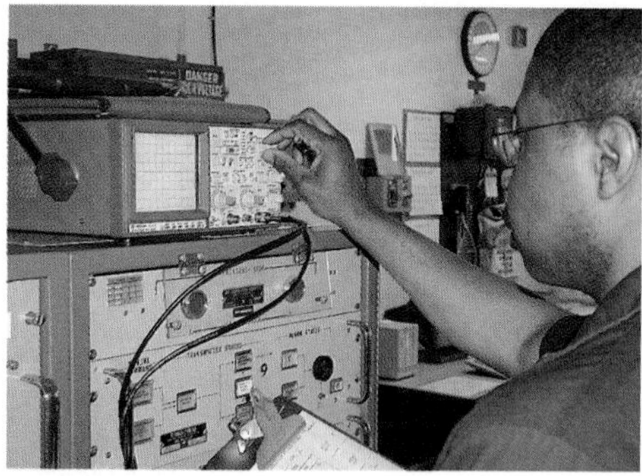
A technician performs a system check at the LORAN Station at Kodiak, Alaska.

Review

12. Find the equations of two parabolas that pass through the points (2, 5), (0, 9), and (−6, 7). Sketch each parabola.

13. Find the coordinates of the foci of each ellipse.

a. $\left(\frac{x+2}{4}\right)^2 + \left(\frac{y-5}{6}\right)^2 = 1$

b. $(x-1)^2 + 4(y+2)^2 = 1$

14. Find the equations of the asymptotes of the hyperbola with vertices (5, 8.5) and (5, 3.5), and foci (5, 12.5) and (5, −0.5).

15. If the vertices of a triangle are $A(10, 16)$, $B(4, 9)$, and $C(8, 1)$, find the perimeter of $\triangle ABC$.

16. Write the polynomial expression $x^4 - 3x^3 + 4x^2 - 6x + 4$ in factored form.

17. You have seen that a double cone can be intersected with a plane to form a circle, an ellipse, a parabola, and a hyperbola. What shapes can be formed by the intersection of a plane and a square-based pyramid? Draw a sketch of each possibility.

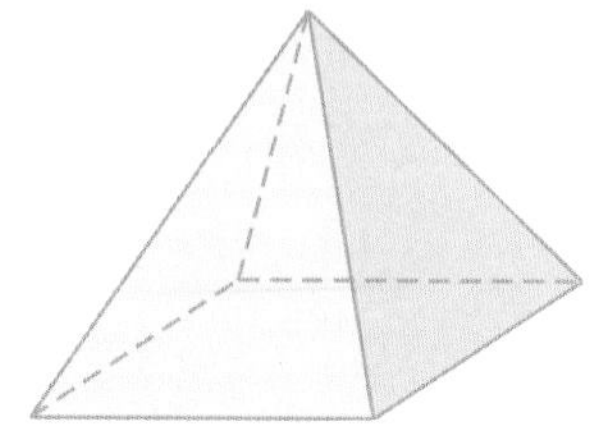

LESSON

8.6 Introduction to Rational Functions

You probably know that a lighter tree climber can crawl farther out on a branch than a heavier climber can, before the branch is in danger of breaking. What do you think the graph of (*length, mass*) data will look like when mass is added to a length of pole until it breaks? Is the relationship linear, like line A, or does it resemble one of the curves, B or C?

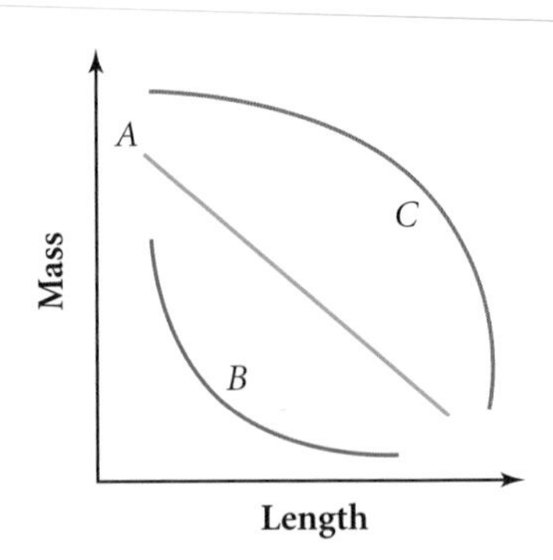

Engineers study problems like this because they need to know the weight that a beam can safely support. In the next investigation you will collect data and experiment with this relationship.

The Louise M. Davies Symphony Hall, built in 1980, is part of the Civic Center in San Francisco, California. The design of both the balcony and the covered entrance rely upon cantilevers—projecting beams that are supported at only one end.

Investigation
The Breaking Point

You will need

- several pieces of dry linguine
- a small film canister
- string
- some weights (pennies, beans, or other small units of mass)
- a ruler
- tape

Procedure Note

1. Lay a piece of linguine on a table so that its length is perpendicular to one side of the table and the end extends over the edge of the table.
2. Tie the string to the film canister so that you can hang it from the end of the linguine. (You may need to use tape to hold the string in place.)
3. Measure the length of the linguine between the edge of the table and the string. (See the photo on the next page.) Record this information in a table of (*length, mass*) data.
4. Place mass units into the container one at a time until the linguine breaks. Record the maximum number of weights that the length of linguine was able to support.

Step 1 Work with a partner. Follow the Procedure Note to record at least five data points, and then compile your results with those of other group members.

Step 2 Make a graph of your data with length as the independent variable, x, and mass as the dependent variable, y. Does the relationship appear to be linear? If not, describe the appearance of the graph.

The relationship between length and mass is an **inverse variation.** The parent function for an inverse variation curve, $f(x) = \frac{1}{x}$, is the simplest **rational function.**

Step 3 Your data should fit a dilated version of the parent function $f(x) = \frac{1}{x}$. Write an equation that is a good fit for the plotted data.

Rational Function

A **rational function** is one that can be written as a quotient, $f(x) = \frac{p(x)}{q(x)}$, where $p(x)$ and $q(x)$ are both polynomial expressions. The denominator polynomial cannot equal the constant 0.

This type of function can be transformed just like all the other functions you have previously studied. In the investigation you created a dilation of $f(x) = \frac{1}{x}$.

Graph the function $f(x) = \frac{1}{x}$ on your calculator and observe some of its special characteristics. The graph is made up of two branches. One part occurs where x is negative and the other where x is positive. There is no value for this function when $x = 0$. What happens when you try to evaluate $f(0)$?

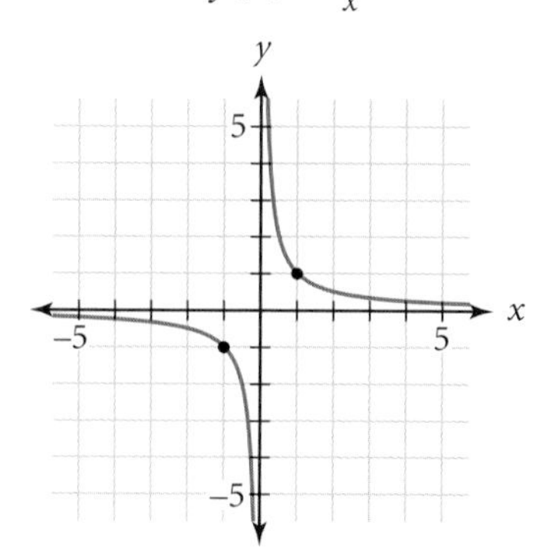

This graph is a hyperbola. It's like the hyperbolas you studied in Lesson 8.4, but it has been rotated 45°. It has vertices (1, 1) and (−1, 1), and its asymptotes are the x- and y-axes.

To understand the behavior of the graph close to the axes, make a table with values of x very close to zero and very far from zero and examine the corresponding y-values. Consider these values of the function $f(x) = \frac{1}{x}$.

x	-1	-0.1	-0.01	-0.001	0	0.001	0.01	0.1	1
y	-1	-10	-100	-1000	undefined	1000	100	10	1

As x approaches zero from the negative side, the y-values have an increasingly large absolute value.

So $x = 0$ is a vertical asymptote.

As x approaches zero from the positive side, the y-values have an increasingly large absolute value.

The behavior of the y-values as x gets closer to zero shows that the y-axis is a vertical asymptote for this function.

x	-10000	-1000	-100	-10	0	10	100	1000	10000
y	-0.0001	-0.001	-0.01	-0.1	undefined	0.1	0.01	0.001	0.0001

As x takes on negative values farther from zero, the y-values approach zero.

So $y = 0$ is a horizontal asymptote.

As x takes on larger positive values, the y-values approach zero.

As x approaches the extreme values at the left and right ends of the x-axis, the curve approaches the x-axis. The horizontal line $y = 0$, then, is a horizontal asymptote. This asymptote is called an end behavior model of the function. In general, the end behavior of a function is its behavior for x-values that are large in absolute value.

If you think of $y = \frac{1}{x}$ as a parent function, then $y = \frac{1}{x} + 1$, $y = \frac{1}{x-2}$, and $y = 3\left(\frac{1}{x}\right)$ are examples of transformed rational functions. What happens to a function when x is replaced with $(x - 2)$? The function $y = \frac{1}{x-2}$ is shown at right.

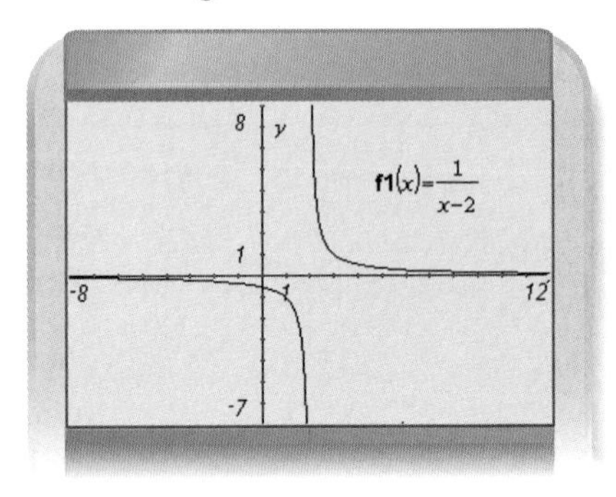

EXAMPLE A

Describe the function $f(x) = \frac{2x-5}{x-1}$ as a transformation of the parent function, $f(x) = \frac{1}{x}$. Then sketch a graph.

▶ Solution

You can change the form of the equation so that the transformations are more obvious. Because the numerator and denominator both have degree 1, you can use division to rewrite the expression.

$$\begin{array}{r} 2 \\ x-1\overline{)2x-5} \\ \underline{2x-2} \\ -3 \end{array}$$

The remainder, -3, means the function is the same as $f(x) = 2 + \frac{-3}{x-1}$, or $f(x) = 2 - \frac{3}{x-1}$. In this form, you can see that the parent function has been dilated vertically by a factor of -3, then translated right 1 unit and up 2 units.

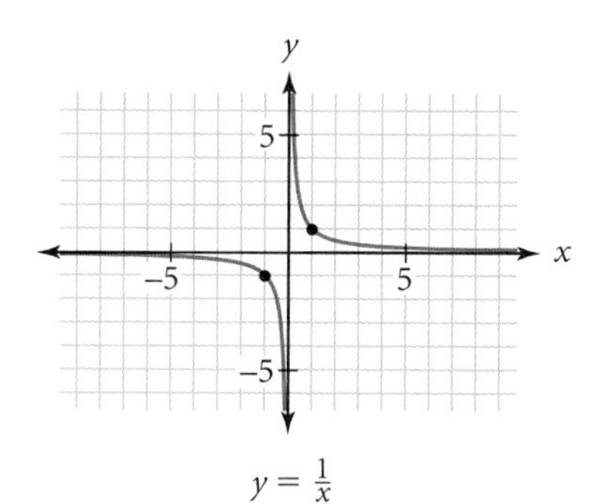

$y = \frac{1}{x}$

The parent rational function, $y = \frac{1}{x}$, has vertices $(1, 1)$ and $(-1, -1)$.

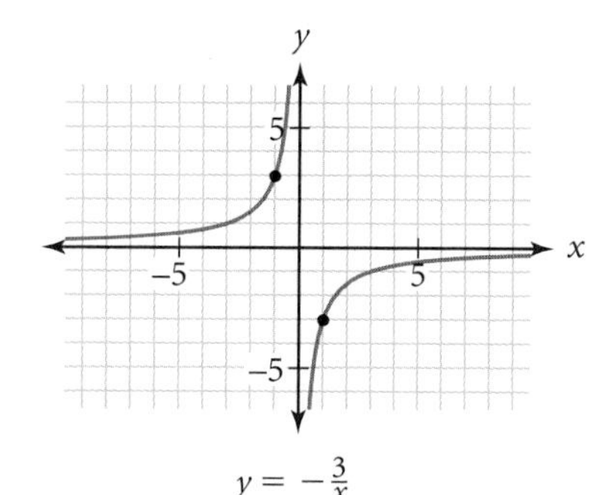

$y = -\frac{3}{x}$

A vertical dilation of –3 moves the vertices of the original hyperbola to $(1, -3)$ and $(-1, 3)$. The points $(3, -1)$ and $(-3, 1)$ are also on the curve. Notice that this graph looks more “spread out” than the graph of $y = \frac{1}{x}$.

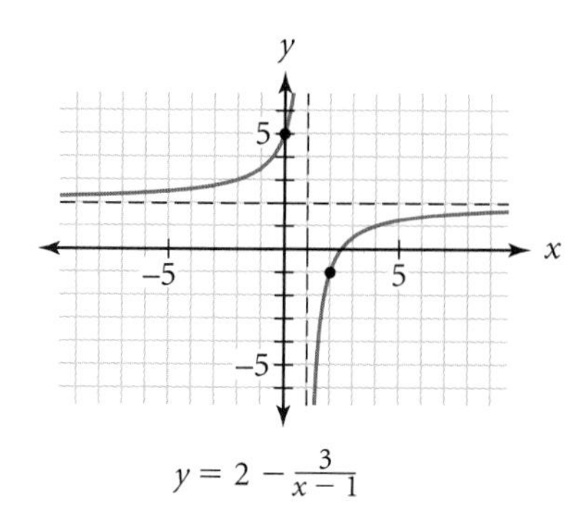

$y = 2 - \frac{3}{x-1}$

A translation right 1 unit and up 2 units moves the vertices of the original hyperbola to $(2, -1)$ and $(0, 5)$. The asymptotes are also translated to $x = 1$ and $y = 2$.

Notice that the asymptotes have been translated. How are the equations of the asymptotes related to your final equation above?

To identify an equation that will produce a given graph, reverse the procedure you used to graph Example A. You can identify translations by looking at the translations of the asymptotes. To identify scale factors, pick a point, such as a vertex, whose coordinates you would know after the translation of $f(x) = \frac{1}{x}$. Then find a point on the dilated graph that has the same x-coordinate. The ratio of the vertical distances from the horizontal asymptote to those two points is the vertical scale factor.

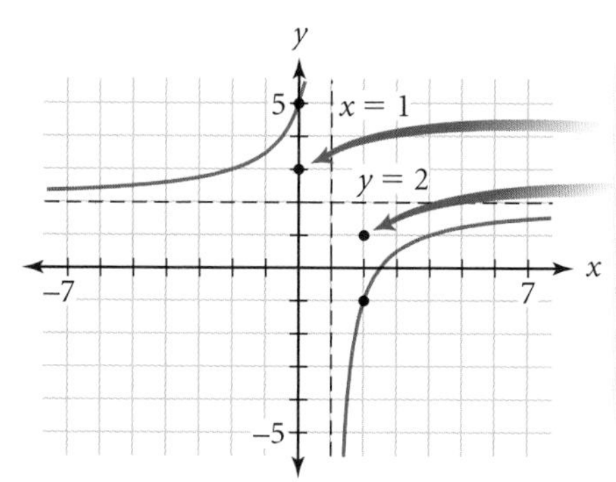

An undilated (but reflected) inverse variation function with these translations would have vertices $(2, 1)$ and $(0, 3)$, 1 horizontal unit and 1 vertical unit away from the center. Because the distance is now 3 vertical units from the center, include a vertical scale factor of 3 to get the equation $\frac{y-2}{3} = \frac{-1}{x-1}$, or $y = 2 - \frac{3}{x-1}$

Rational expressions are very useful in chemistry. Scientists use them to model many situations, including the concentration of a solution or mixture as it is diluted.

EXAMPLE B

Suppose you have 100 mL of a solution that is 30% acid and 70% water. How many mL of acid do you need to add to make a solution that is 60% acid? To make it a 90% acid solution? Can it ever be 100% acid?

▶ Solution

Of the 100 mL of solution, 30%, or 30 mL, is acid. The percentage, P, can be written as $P = \frac{30}{100}$. If x milliliters of acid are added, there will be more acid, but also more solution. The concentration of acid will be

$$P = \frac{30 + x}{100 + x}$$

To find when the solution is 60% acid, substitute 0.6 for P and solve the equation.

$$0.6 = \frac{30 + x}{100 + x}$$ Substitute 0.6 for P.

$$0.6(100 + x) = 30 + x$$ Multiply both sides by $(100 + x)$.

$$60 + 0.6x = 30 + x$$ Distribute.

$$30 = 0.4x$$ Collect like terms.

$$75 = x$$ Divide by 0.4.

Adding 75 mL of acid will make a 60% acid solution.

To find when the solution is 90% acid, solve the equation $0.9 = \frac{30 + x}{100 + x}$. You will find that 600 mL of acid must be added.

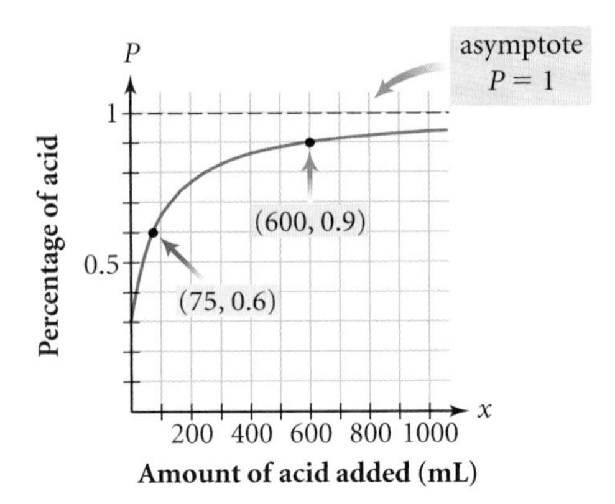

You can also find the solutions by graphing the function $P(x) = \frac{30 + x}{100 + x}$ and finding the x-values for the P-values 0.6 and 0.9, shown at right. The graph shows horizontal asymptote $P = 1$. No matter how many milliliters of acid you add, you will never have a mixture that is 100% acid. This is because the original 70 mL of water will remain, even though it is a smaller and smaller percentage of the entire solution as you continue to add acid.

EXERCISES

You will need

A graphing calculator for Exercises **1, 10, 12,** and **16.**

Practice Your Skills

1. Write an equation and graph each transformation of the parent function $f(x) = \frac{1}{x}$.

a. Translate the graph up 2 units. ⓐ

b. Translate the graph right 3 units.

c. Translate the graph down 1 unit and left 4 units.

d. Vertically dilate the graph by a scale factor of 2.

e. Horizontally dilate the graph by a factor of 3, and translate it up 1 unit. ⓐ

2. Describe how each function was transformed from the parent function $f(x) = \frac{1}{x}$.

a. $g(x) = \frac{1}{x + 3} - 2$ ⓐ

b. $h(x) = \frac{5}{x - 7}$

c. $j(x) = 3 - \frac{4}{x - 1}$

3. As the rational function $y = \frac{1}{x}$ is translated, its asymptotes are translated also. Write an equation for the translation of $y = \frac{1}{x}$ that has the asymptotes described.

a. horizontal asymptote $y = 2$ and vertical asymptote $x = 0$ ⓐ

b. horizontal asymptote $y = -4$ and vertical asymptote $x = 2$

c. horizontal asymptote $y = 3$ and vertical asymptote $x = -4$ ⓐ

4. What are the equations of the asymptotes for each hyperbola?

a. $y = \frac{2}{x} + 1$ ⓐ **b.** $y = \frac{3}{x-4}$ **c.** $y = \frac{4}{x+2} - 1$ ⓐ **d.** $y = \frac{-2}{x+3} - 4$

5. Solve.

a. $12 = \frac{x-8}{x+3}$ **b.** $21 = \frac{3x+8}{x-5}$ **c.** $3 = \frac{2x+5}{4x-7}$ **d.** $-4 = \frac{-6x+5}{2x+3}$

Reason and Apply

6. If a basketball team's present record is 42 wins and 36 losses, how many consecutive games must it win so that its winning record reaches 60%? ⓐ

7. Match the graphs of the rational functions with their equations.

A.

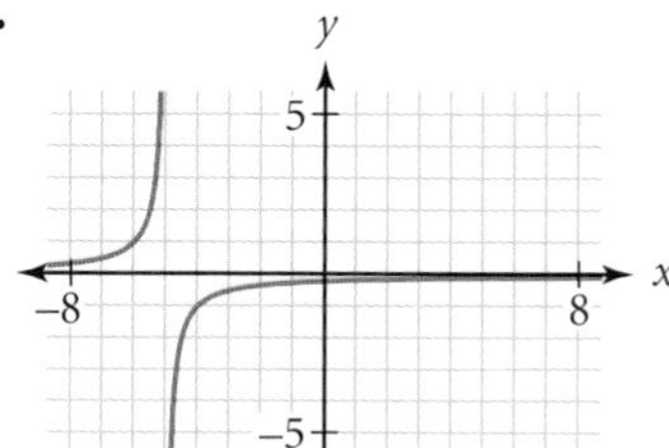

B.

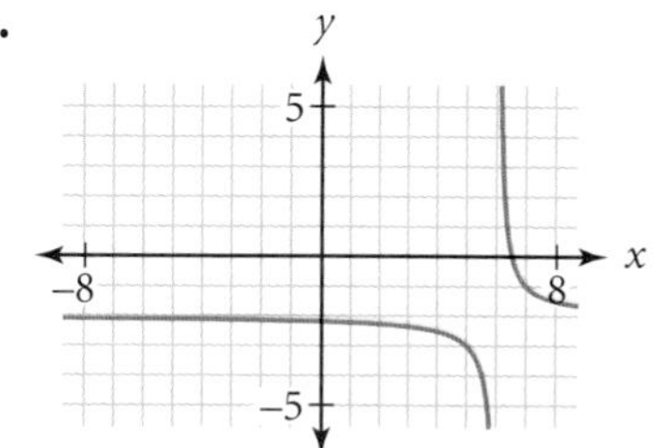

C.

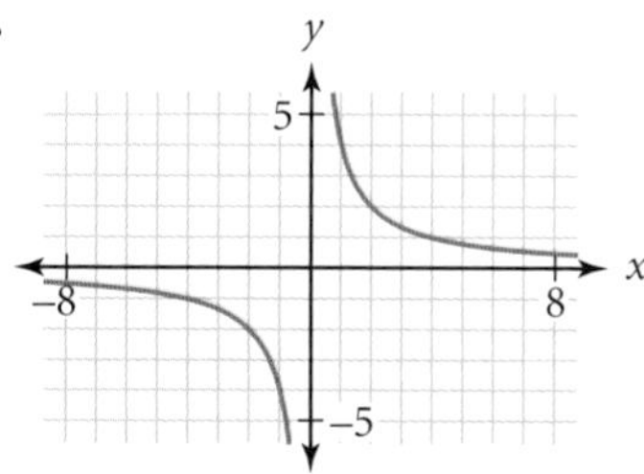

D.

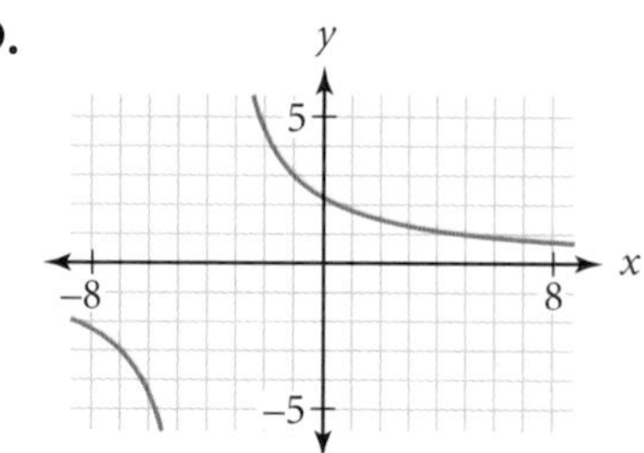

i. $f(x) = \frac{4}{x}$ **ii.** $f(x) = \frac{1}{x-6} - 2$ **iii.** $f(x) = \frac{9}{x+4}$ **iv.** $f(x) = \frac{-1}{x+5}$

8. APPLICATION The graph at right shows the concentration of acid in a solution as pure acid is added. The solution began as 55 mL of a 38% acid solution.

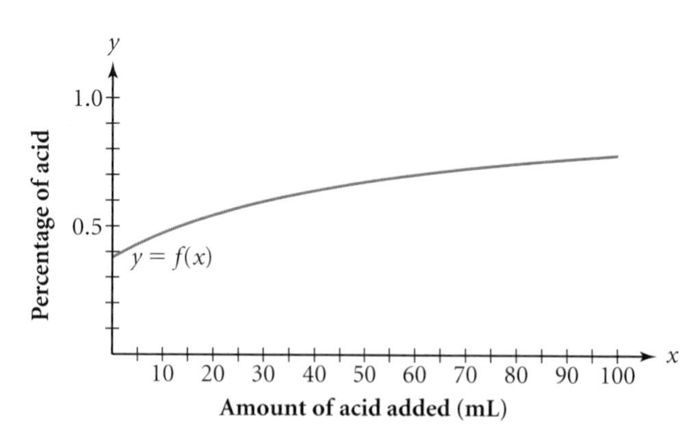

a. How many milliliters of pure acid were in the original solution? ⓐ

b. Write an equation for $f(x)$.

c. Find the amount of pure acid that must be added to create a solution that is 64% acid.

d. Describe the end behavior of $f(x)$.

9. APPLICATION In a container of 2% milk, 2% of the mixture is fat. How much of the liquid in a 1 gal container of 2% milk would need to be emptied and replaced with pure fat so that the container could be labeled as whole (3.25%) milk?

10. Consider these functions.

i. $y = \frac{2x - 13}{x - 5}$ **ii.** $y = \frac{3x + 11}{x + 3}$

a. Rewrite each rational function to show how it is a transformation of $y = \frac{1}{x}$. (h)

b. Describe the transformations of the graph of $y = \frac{1}{x}$ that will produce graphs of the equations in 10a.

c. Graph each equation on your calculator to confirm your answers to 10b.

11. Draw the graph of $y = \frac{1}{x}$.

a. Label the vertices of the hyperbola.

b. The x- and y-axes are the asymptotes for this hyperbola. Between the two branches of the hyperbola, draw the box whose diagonals lie on the asymptotes. The vertices should lie on the box.

c. The dimensions of the box are $2a$ and $2b$. Find the values of a and b.

d. The foci of the hyperbola lie on the line passing through the two vertices. In this case, that line is $y = x$. The foci are c units from the center of the hyperbola where $a^2 + b^2 = c^2$. Find the value of c and the coordinates of the two foci.

12. Recall that the general quadratic equation is $Ax^2 + Bxy + Cy^2 + Dx + Ey + F = 0$. Let $A = 0$, $B = 4$, $C = 0$, $D = 0$, $E = 0$, and $F = -1$.

a. Graph this equation. What type of conic section is formed? (h)

b. What is the relationship between the inverse variation function, $y = \frac{1}{x}$, and the conic sections?

c. Convert the rational function $y = \frac{1}{x - 2} + 3$ to general quadratic form. What are the values of A, B, C, D, E, and F in the general quadratic equation?

13. APPLICATION Ohm's law states that $I = \frac{V}{R}$. This law can be used to determine the amount of current I, in amps, flowing in the circuit when a voltage V, in volts, is applied to a resistance R, in ohms.

a. If a hairdryer set on high is using a maximum of 8.33 amps on a 120-volt line, what is the resistance in the heating coils? (a)

b. In the United Kingdom, power lines use 240 volts. If a traveler were to plug in a hairdryer, and the resistance in the hairdryer were the same as in 13a, what would be the flow of current?

c. The additional current flowing through the hairdryer would cause a meltdown of the coils and the motor wires. In order to reduce the current flow in 13b back to the value in 13a, how much resistance would be needed?

14. The cost of a square piece of glass varies jointly with the thickness of the glass and the square of the side length. A **joint variation** means that the cost varies directly with each of the other variables taken one at a time. Suppose a $\frac{1}{4}$ in. thick piece of glass that measures 15 in. on a side costs \$2.40. Find the cost of a $\frac{1}{2}$ in. thick piece of glass whose sides measure 18 in.

15. The equation $P = \frac{kT}{V}$ is an example of a **combined variation** because P varies directly with T and inversely with V. Suppose that when $P = 3$, $V = 8$ and $T = 300$.

a. Find the value of the constant of variation, k.

b. When $T = 350$ and $V = 14$, what is P?

16. Graph the function $g(x) = \frac{2}{x-3} + 1$ on your calculator.

a. Describe the transformations of the graph of $f(x) = \frac{1}{x}$ that will produce the graph of $g(x)$.

b. What are the domain and range of $f(x) = \frac{1}{x}$?

c. What are the domain and range of $g(x) = \frac{2}{x-3} + 1$?

d. How does identifying the transformations help you determine the domain and range?

Review

17. Factor each expression completely.

a. $x^2 - 7x + 10$

b. $x^3 - 9x$

18. Write the equation of the circle with center $(2, -3)$ and radius 4.

19. A 2 m rod and a 5 m rod are mounted vertically 10 m apart. One end of a 15 m wire is attached to the top of each rod. Suppose the wire is stretched taut and fastened to the ground between the two rods. How far from the base of the 2 m rod is the wire fastened?

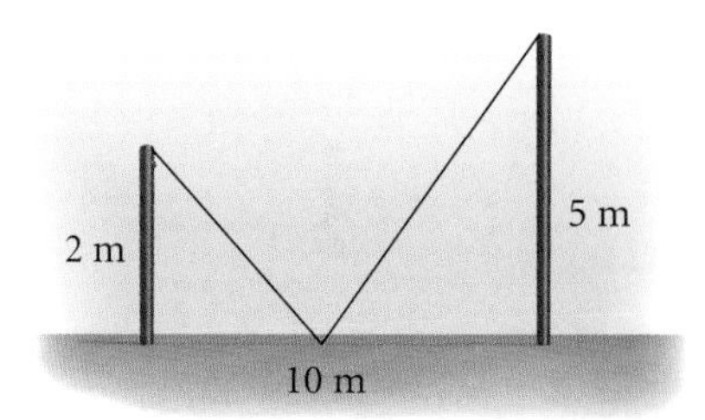

20. Write the general quadratic equations of two concentric circles with center $(6, -4)$ and radii 5 and 8. ⓗ

21. The radius of the circle is 1. Find the coordinates of the labeled point.

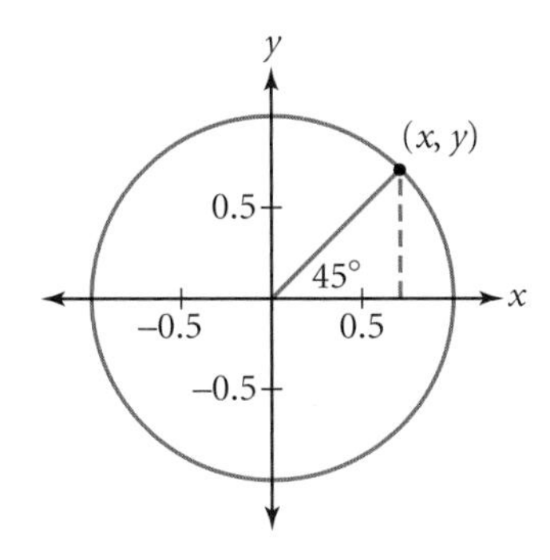

22. Find the length of each labeled segment.

a.

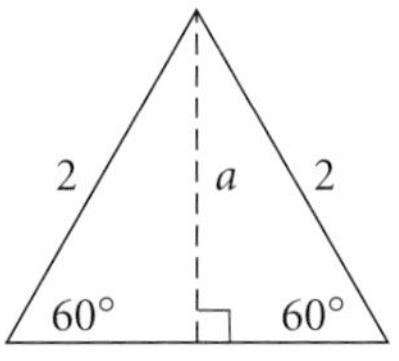

b.

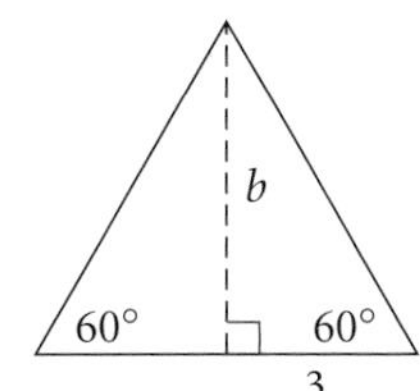

c.

6
60°
60°
c

LESSON

8.7

Graphs of Rational Functions

Besides learning to see, there is another art to be learned—to see what is not.

MARIA MITCHELL

Some rational functions create very different kinds of graphs from those you have studied previously. The graphs of these functions often have two or more parts. This is because the denominator, a polynomial function, may be equal to zero at some point, so the function will be undefined at that point.

Sometimes it's difficult to see the different parts of the graph because they may be separated by only one missing point, called a **hole.** At other times you will see two parts that look very similar—one part may look like a reflection or rotation of the other part. Or you might get multiple parts that look totally different from each other. Look for these features in the graphs below.

$$y = \frac{x^3 + 2x^2 - 5x - 6}{x - 2}$$

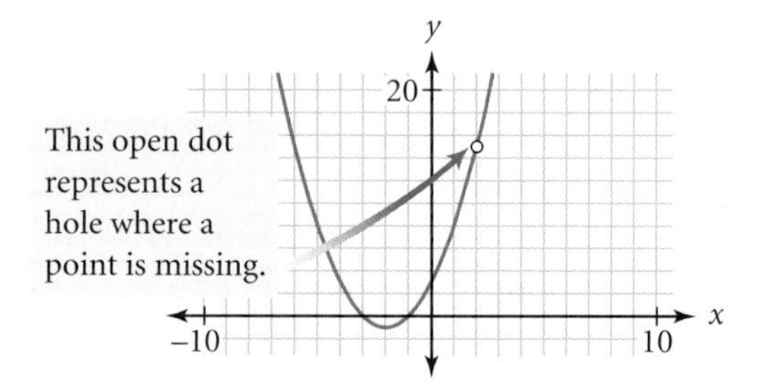

$$y = \frac{1}{x - 2}$$

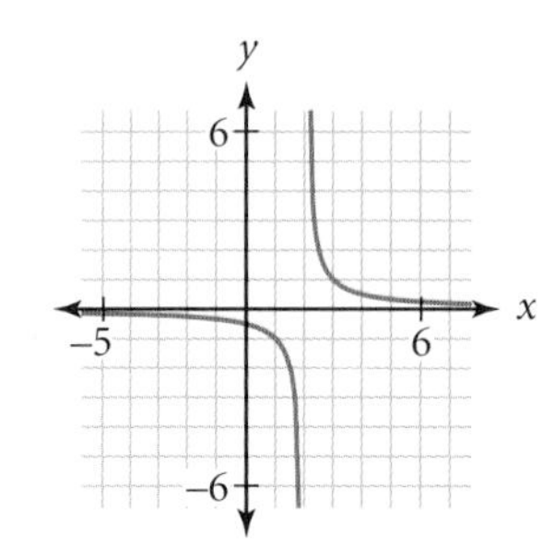

$$y = \frac{x^3 - x^2 - 8x + 12}{6x^2 + 6x - 12}$$

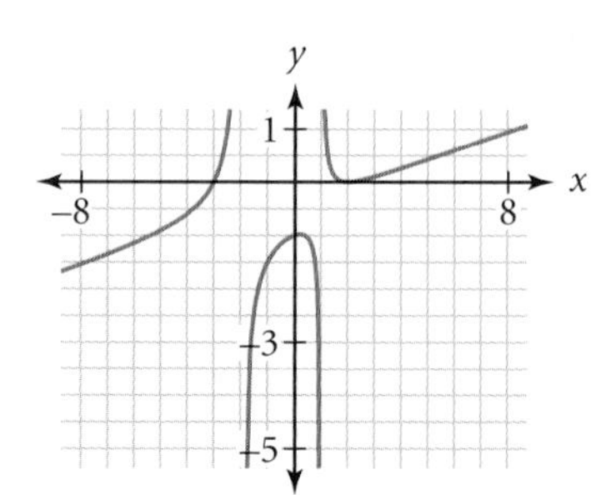

$$y = \frac{1}{x^2 + 1}$$

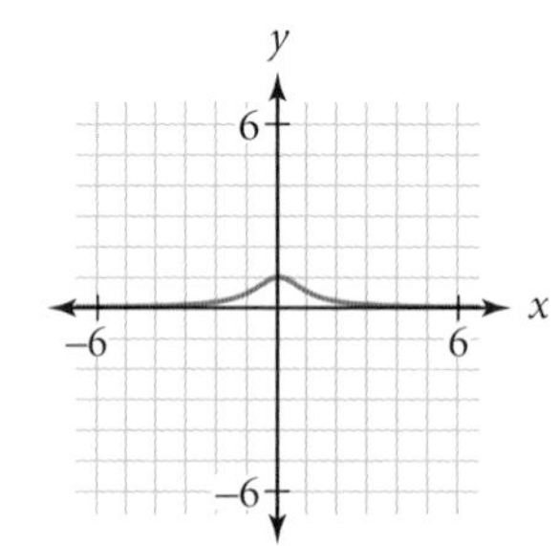

$$y = \frac{x - 1}{x^2 + 4}$$

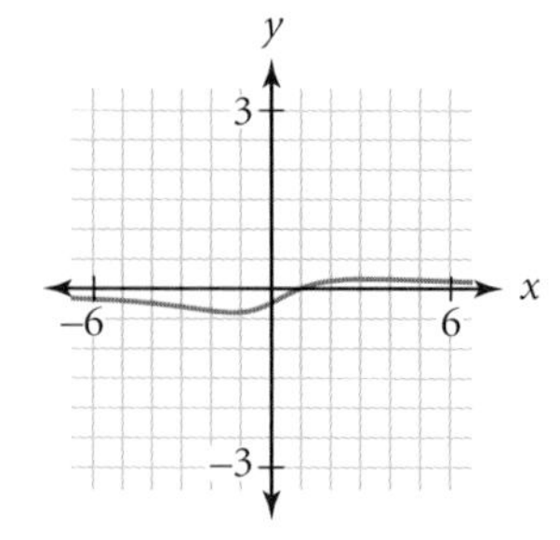

In this lesson you will explore local and end behavior of rational functions, and you will learn how to predict some of the features of a rational function's graph by studying its equation. When examining a rational function, you will often find it helpful to look at the equation in factored form.

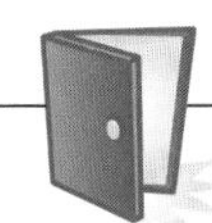

Investigation
Predicting Asymptotes and Holes

In this investigation you will consider the graphs of four rational functions.

Step 1 Match each rational function with a graph. Investigate each graph by dragging a point on the graph, tracing, or looking at a table of values. Describe the unusual occurrences at exactly $x = 2$ and other values nearby. Try to explain what features in the equation cause the different types of graph behavior. (You will not see an actual "hole," as pictured in graphs b and d.)

a.

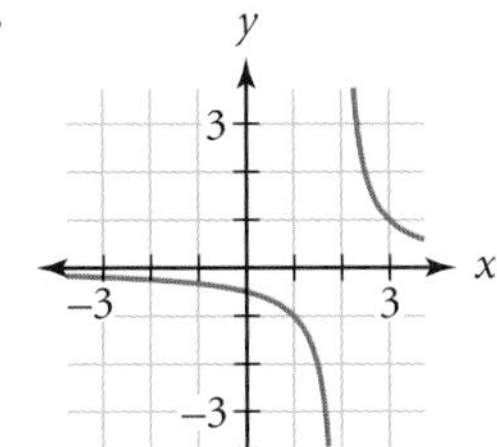

b.

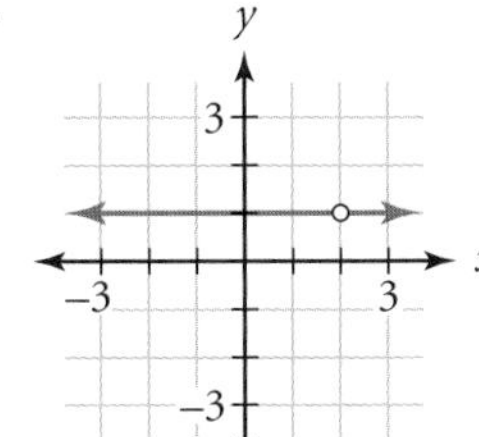

c.

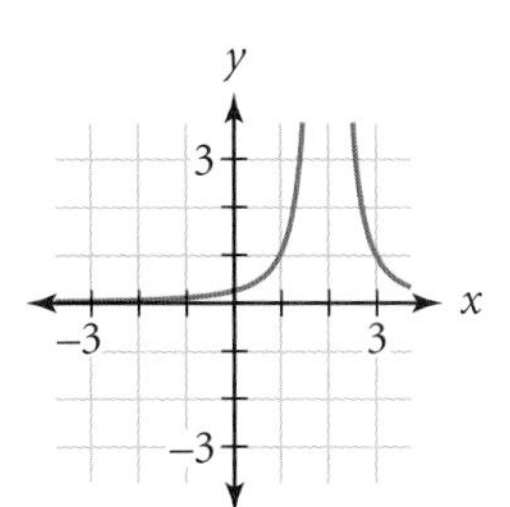

d.

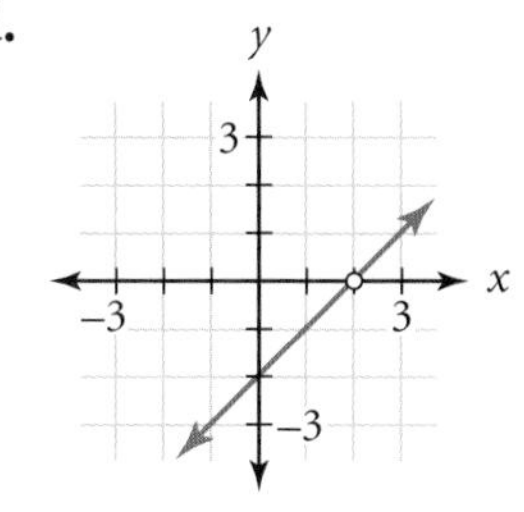

A. $y = \dfrac{1}{(x-2)^2}$ **B.** $y = \dfrac{1}{x-2}$ **C.** $y = \dfrac{(x-2)^2}{x-2}$ **D.** $y = \dfrac{x-2}{x-2}$

Step 2 Have each group member choose one of the graphs below. Find a rational function equation for your graph, and write a few sentences that explain the appearance of your graph. Share your answers with your group.

a.

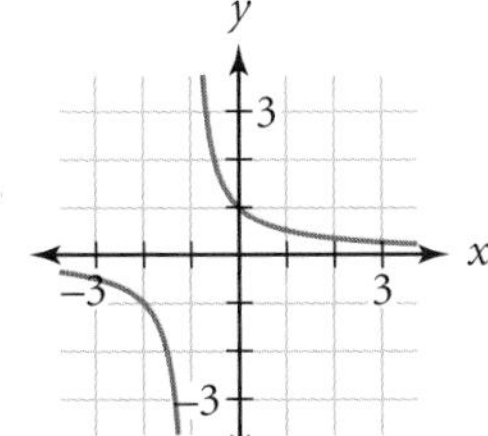

b.

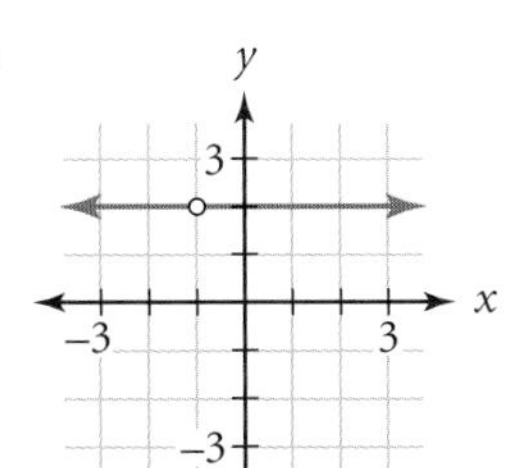

c.

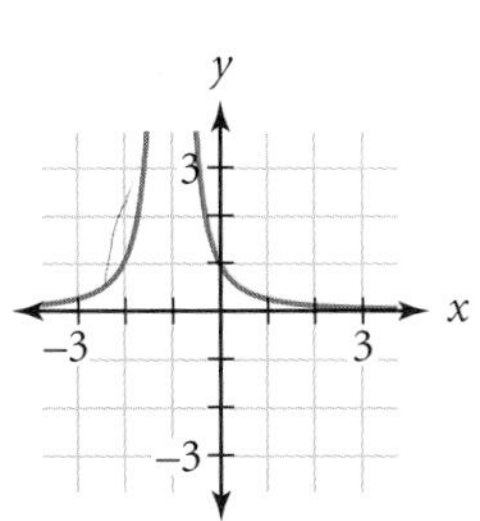

d.

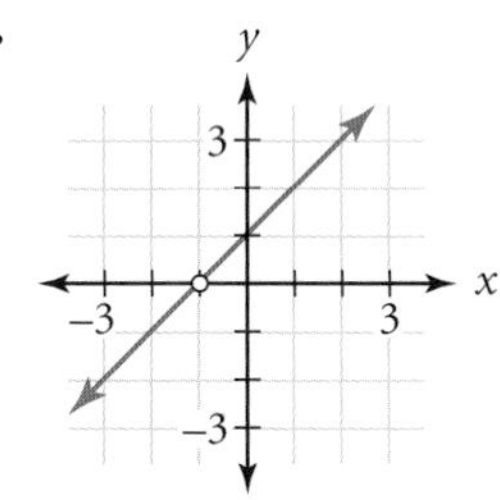

Step 3 Write a paragraph explaining how you can use an equation to predict where holes and asymptotes will occur, and how you can use these features in a graph to write an equation.

Step 4 Consider the graph of $y = \frac{x-2}{(x-2)^2}$. What features does it have? What can you generalize about the graph of a function that has a factor that occurs more times in the denominator than in the numerator?

In the investigation you saw that the graph of a rational function contains many clues about the equation. The next example shows how you can put that information to use.

EXAMPLE A

Write an equation for each graph.

a.

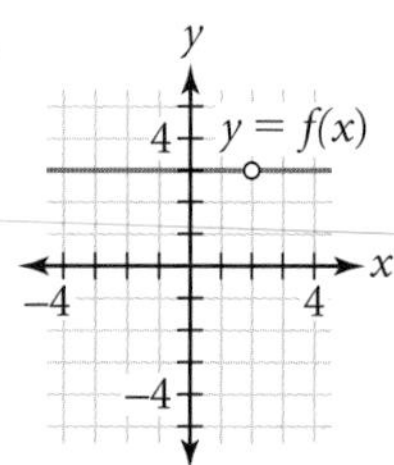

b.

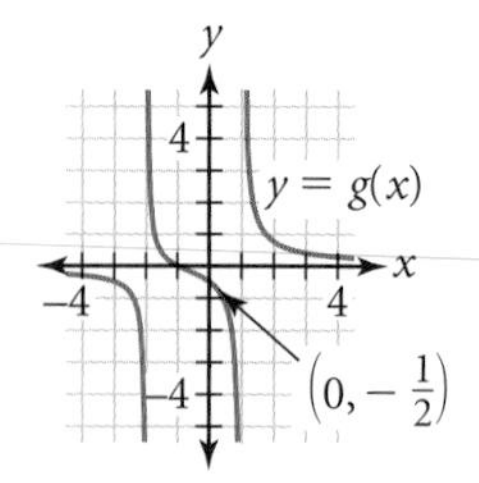

c.

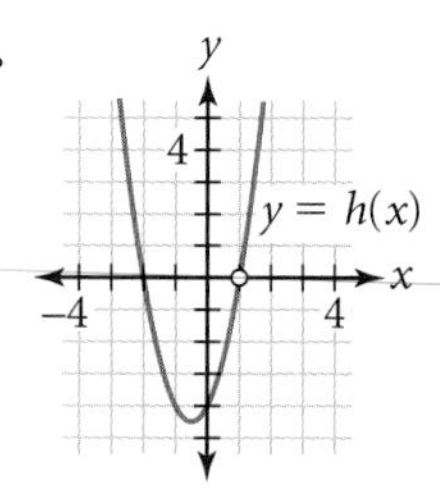

▶ Solution

a. This graph looks like the line $y = 3$, but there is a hole when $x = 2$. This means the function is rational with the factor $(x - 2)$ in both the numerator and denominator. So its equation is $f(x) = \frac{3(x-2)}{(x-2)}$.

b. This graph has vertical asymptotes $x = -2$ and $x = 1$. So the denominator contains the factors $(x + 2)$ and $(x - 1)$. Also, the graph has x-intercept $x = -1$, which means the numerator contains the factor $(x + 1)$. Write the equation $g(x) = \frac{a(x+1)}{(x+2)(x-1)}$. To find a, check the value of the y-intercept by evaluating $g(0)$. It is $-\frac{1}{2}a$, and the graph shows y-intercept $-\frac{1}{2}$. So $a = 1$ and the final equation is $g(x) = \frac{(x+1)}{(x+2)(x-1)}$.

c. This graph looks like a parabola with x-intercepts $x = -2$ and $x = 1$, so its basic equation is $h(x) = a(x + 2)(x - 1)$. But the graph has a hole at $x = 1$, so its equation is a rational function with the factor $(x - 1)$ in both the numerator and the denominator. Write the equation $h(x) = \frac{a(x+2)(x-1)^2}{(x-1)}$. To find a, check the value of the y-intercept. You find that $h(0) = -2a$. The graph has y-intercept -4, so the function needs to be dilated vertically by the scale factor $a = 2$. The final equation is $h(x) = \frac{2(x+2)(x-1)^2}{(x-1)}$.

Similar thinking can help you determine the features of a rational function graph before you actually graph it. Values that make the denominator or numerator equal to zero give you important clues about the appearance of the graph.

EXAMPLE B

Describe the features of the graph of $y = \frac{x^2 + 2x - 3}{x^2 - 2x - 8}$

▶ Solution

Features of rational functions are apparent when the numerator and denominator are factored.

$$y = \frac{x^2 + 2x - 3}{x^2 - 2x - 8} = \frac{(x+3)(x-1)}{(x-4)(x+2)}$$

This table summarizes how the equation can help you identify features in the graph.

Graph feature	Location	Equation feature
Holes	none	Common factors in numerator and denominator
x-intercepts	-3 and 1	Numerator (but not denominator) is 0 at these values
Vertical asymptotes	$x = 4$ and $x = -2$	Denominator (but not numerator) is 0 at these values
y-intercepts	0.375	y-value when $x = 0$
Horizontal asymptotes	$y = 1$	Long-run value of y (see table on next page)

To find any horizontal asymptotes, consider what happens to y-values as x-values get increasingly large in absolute value.

x	$-10{,}000$	$-1{,}000$	-100	100	1,000	10,000
y	0.9996001	0.9960129	0.9612441	1.0413603	1.0040131	1.0004001

A table shows that the y-values get closer and closer to 1 as x gets farther from 0. So, $y = 1$ is a horizontal asymptote. Note that for extreme positive values of x, y-values decrease to 1, whereas for extreme negative values of x, y-values increase to 1.

A graph of the function confirms these features.

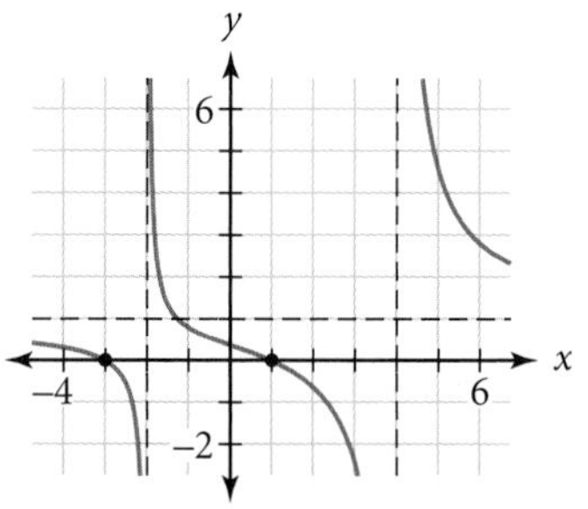

Rational functions can be written in different forms. The factored form is convenient for locating asymptotes and intercepts. And you saw in the previous lesson how rational functions can be written in a form that shows you clearly how the parent function has been transformed. You can use properties of arithmetic with fractions to convert from one form to another.

EXAMPLE C

Rewrite $f(x) = \frac{3}{x-2} - 4$ in fractional form.

▶ Solution

The original form shows that this function is related to the parent function, $f(x) = \frac{1}{x}$. It has been vertically dilated by a factor of 3 and translated right 2 units and down 4 units. To change to fractional form, you must add the two parts to form a single fraction.

$$f(x) = \frac{3}{x-2} - 4 \qquad \text{Original equation.}$$

$$= \frac{3}{x-2} - \frac{4}{1} \cdot \frac{x-2}{x-2} \qquad \text{Create a common denominator of } (x-2).$$

$$= \frac{3}{x-2} - \frac{4x-8}{x-2} \qquad \text{Rewrite second fraction.}$$

$$= \frac{3 - 4x + 8}{x-2} \qquad \text{Combine the two fractions.}$$

$$f(x) = \frac{11 - 4x}{x-2} \qquad \text{Add like terms.}$$

You can check this result by using long division. The expression is indeed equivalent to the original form, which shows that the horizontal asymptote is $y = -4$.

$$\begin{array}{r} -4 \\ x-2\,\overline{)\,-4x + 11} \\ \underline{-4x + 8} \\ 3 \end{array}$$

EXERCISES

You will need

A graphing calculator for Exercises **6**, **8–11**, and **13**.

Practice Your Skills

1. Rewrite each rational expression in factored form.

a. $\frac{x^2+7x+12}{x^2-4}$ ⓐ **b.** $\frac{x^3-5x^2-14x}{x^2+2x+1}$

2. Identify the vertical asymptotes for each equation.

a. $y=\frac{x^2+7x+12}{x^2-4}$ ⓐ **b.** $y=\frac{x^3-5x^2-14x}{x^2+2x+1}$

3. What is the domain for each equation in Exercise 2?

4. Add by finding a common denominator.

a. $3+\frac{2}{x-4}$ ⓐ **b.** $2x+\frac{x-1}{x+3}$ **c.** $5+2x-\frac{3}{x+1}$

5. Rewrite each expression in rational form (as the quotient of two polynomials). ⓗ

a. $3+\frac{4x-1}{x-2}$ ⓐ **b.** $\frac{3x+7}{2x-1}-5$

6. Graph each equation on your calculator, and make a sketch of the graph on your paper. Indicate any holes on your sketches.

a. $y=\frac{5-x}{x-5}$ **b.** $y=\frac{3x+6}{x+2}$ **c.** $y=\frac{(x+3)(x-4)}{x-4}$

d. What causes a hole to appear in the graph? ⓗ

Reason and Apply

7. Write an equation for each graph.

a. 

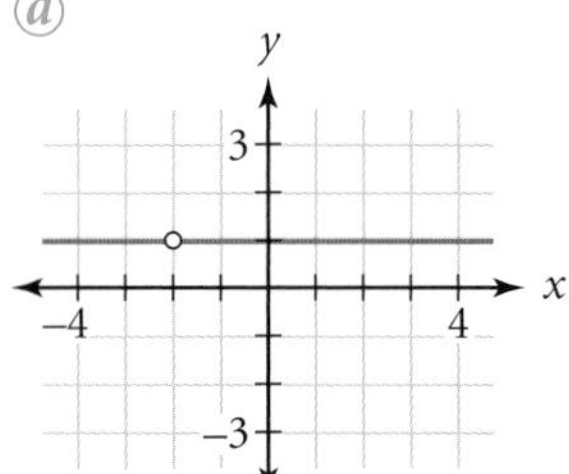

b.

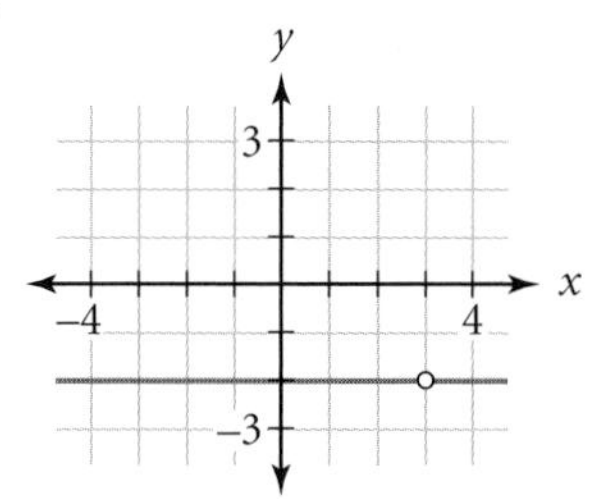

c.

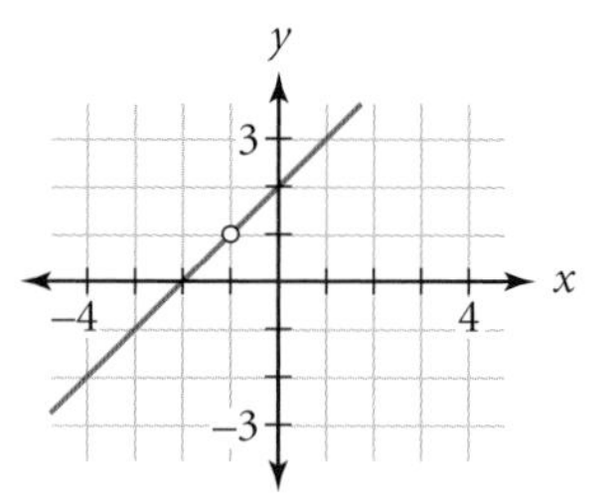

8. Graph each function on your calculator. List all holes and asymptotes.

a. $y=3+\frac{1}{x+2}-\frac{1}{x-1}$ **b.** $y=\frac{2}{x+3}+\frac{6-2x}{x-3}$ **c.** $y=\frac{x}{x+2}-\frac{x}{x-2}$

9. Write an equation for each graph.

a.

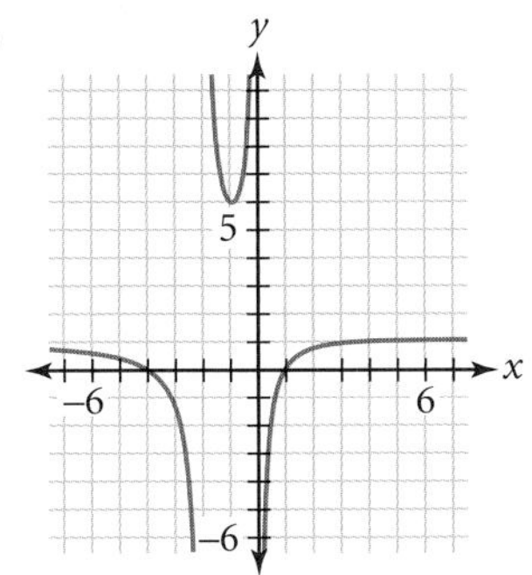

b.

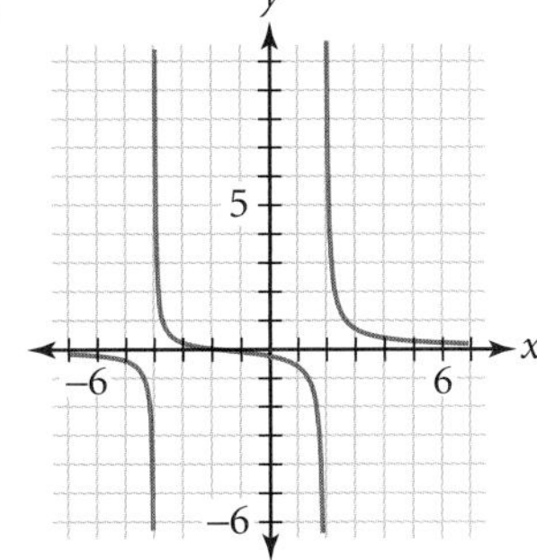

c.

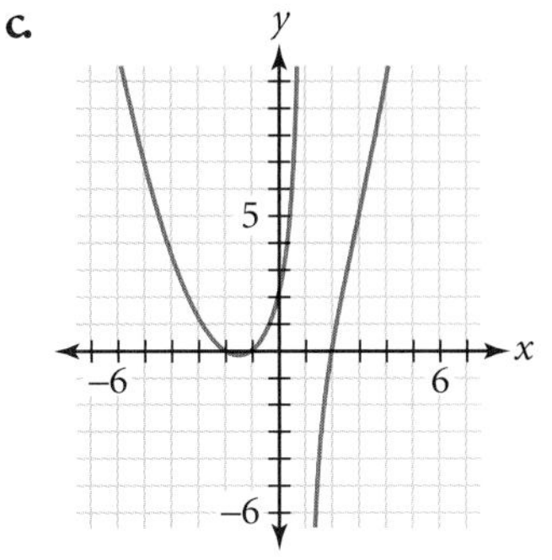

10. Write an equation that fits each description.

a. vertical asymptotes $x = 2$ and $x = 5$, x-intercept $x = -1$, y intercept $y = 1$

b. vertical asymptote $x = 4$, no x-intercept, horizontal asymptote $y = 2$, y-intercept $y = 3$

c. vertical asymptote $x = 1$, hole at $x = 3$, y-intercept $y = 4$

11. Consider the equation $y = \frac{(x-1)(x+4)}{(x-2)(x+3)}$.

a. Describe the features of the graph of this function.

b. Describe the end behavior of the graph.

c. Sketch the graph.

12. Solve. Give exact solutions.

a. $\frac{2}{x-1} + x = 5$

b. $\frac{2}{x-1} + x = 2$

13. APPLICATION The functional response curve given by the function $y = \frac{60x}{1 + 0.625x}$ models the number of moose attacked by wolves as the density of moose in an area increases. In this model, x represents the number of moose per 1000 km^2, and y represents the number of moose attacked every 100 days.

a. How many moose are attacked every 100 days if there is a herd containing 260 moose in a land preserve with area 1000 km^2?

b. Graph the function.

c. What are the asymptotes for this function?

d. Describe the significance of the asymptotes in this problem.

Environmental CONNECTION

Ecologists often look for a mathematical model to describe the interrelationship of organisms. C. S. Holling, a Canadian researcher, came up with an equation in the late 1950s for what he called a Type II functional response curve. The equation describes the relationship between the number of prey attacked by a predator and the density of the prey. For example, the wolf population increases through reproduction as moose density increases. Eventually, wolf populations stabilize at about 40 per 1000 km^2, which is the optimum size of their range based on defense of their territories.

The functional response curve applies to all species of animals. It could be larvae-eating insects and mosquito larvae, fishers and a particular species of fish, or panda and amount of bamboo in a forest.

A moose rises from the surface of a pond in Baxter State Park, Maine.

14. A machine drill removes a core from any cylinder. Suppose you want the amount of material left after the core is removed to remain constant. The table below compares the height and radius needed if the volume of the hollow cylinder is to remain constant.

Radius x	2.5	3.0	3.5	4.0	4.5	5.0	5.5	6.0	6.5
Height h	56.6	25.5	15.4	10.6	7.8	6.1	4.9	4.0	3.3

a. Plot the data points, (x, h), and draw a smooth curve through them.

b. Explain what happens to the height of the figure as the radius gets smaller. How small can x be?

c. Write a formula for the volume of the hollow cylinder, V, in terms of x and h.

d. Solve the formula in 14c for h to get a function that describes the height as a function of the radius.

e. What is the constant volume?

Review

15. Find the points of intersection, if any, of the circle with center (2, 1) and radius 5 and the line $x - 7y + 30 = 0$.

16. A 500 g jar of mixed nuts contains 30% cashews, 20% almonds, and 50% peanuts.

a. How many grams of cashews must you add to the mixture to increase the percentage of cashews to 40%? What is the new percentage of almonds and peanuts?

b. How many grams of almonds must you add to the original mixture to make the percentage of almonds the same as the percentage of cashews? Now what is the percentage of each type of nut?

17. Solve each quadratic equation.

a. $2x^2 - 5x - 3 = 0$ **b.** $x^2 + 4x - 4 = 0$ **c.** $x^2 + 4x + 1 = 0$

LESSON

8.8 Operations with Rational Expressions

Images/split the truth/
in fractions.
DENISE LEVERTOV

In this lesson you will learn to add, subtract, multiply, and divide rational expressions. In the previous lesson you combined a rational expression with a single constant or variable by finding a common denominator. That process was much like adding a fraction to a whole number. Likewise, all of the other arithmetic operations you will do with rational expressions have their counterparts in working with fractions. Keeping the operations with fractions in mind will help you understand the procedures.

Recall that to add $\frac{3}{10} + \frac{4}{15}$, you need a common denominator. The smallest number that has both 10 and 15 as factors is 30. So use 30 as the common denominator.

$\frac{3}{10} + \frac{4}{15}$ Original expression.

$\frac{3}{10} \cdot \frac{3}{3} + \frac{4}{15} \cdot \frac{2}{2}$ Multiply each fraction by an equivalent of 1 to get a denominator of 30.

$\frac{9}{30} + \frac{8}{30}$ Multiply.

$\frac{17}{30}$ Add.

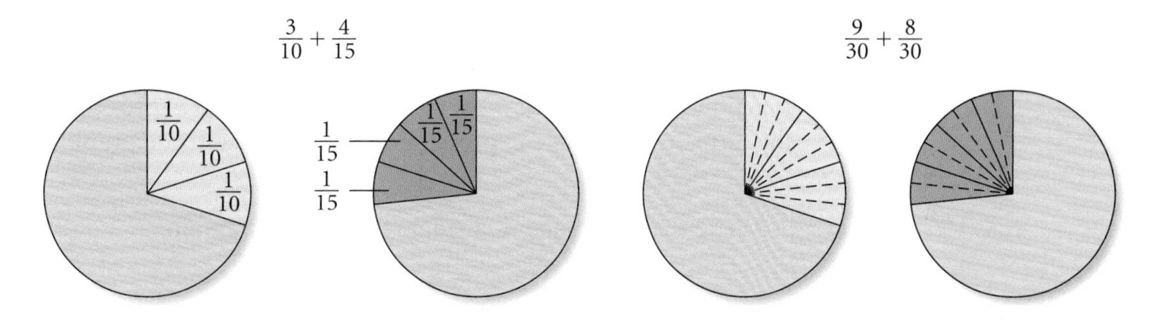

You could use other numbers, such as 150, as a common denominator, but using the least common denominator keeps the numbers as small as possible and eliminates some of the reducing afterward. Recall that you can find the least common denominator by factoring the denominators to see which factors they share and which factors are unique to each one. In the example above, 10 factors to $2 \cdot 5$ and 15 factors to $3 \cdot 5$. The least common denominator must include factors that multiply to give each denominator, with no extras. So in this case you need a 2, a 3, and a 5. You can use this same process to add two rational expressions.

EXAMPLE A

Add the rational expressions and write the sum as a single rational expression.

$$\frac{x-1}{(x-2)(x+1)} + \frac{x}{(x+1)(x-1)}$$

▸ Solution

The common denominator contains the factors of each denominator. Here the common denominator will contain $(x-2)$, $(x+1)$, and $(x-1)$, so it can be written $(x-2)(x+1)(x-1)$.

$\dfrac{x-1}{(x-2)(x+1)}+\dfrac{x}{(x+1)(x-1)}$	Original expression.
$\dfrac{(x-1)(x-1)}{(x-2)(x+1)(x-1)}+\dfrac{(x-2)x}{(x-2)(x+1)(x-1)}$	Multiply by an equivalent of 1 to get a common denominator.
$\dfrac{x^2-2x+1}{(x-2)(x+1)(x-1)}+\dfrac{x^2-2x}{(x-2)(x+1)(x-1)}$	Multiply factors in the numerators.
$\dfrac{2x^2-4x+1}{(x-2)(x+1)(x-1)}$	Add and combine like terms.

The numerator cannot be factored, so the expression cannot be factored further.

To verify your results, try substituting a value for x into the original expressions and the final expression. We'll try $x=4$.

Original expressions: $\dfrac{4-1}{(4-2)(4+1)}+\dfrac{4}{(4+1)(4-1)}=\dfrac{3}{2\cdot 5}+\dfrac{4}{5\cdot 3}=\dfrac{3}{10}+\dfrac{4}{15}$

Final expression: $\dfrac{2(4)^2-4(4)+1}{(4-2)(4+1)(4-1)}=\dfrac{32-16+1}{2\cdot 5\cdot 3}=\dfrac{17}{30}$

So this problem is a generalization of the problem with fractions that opened this lesson.

In Lesson 8.7, you learned what to look for to graph a single rational function. If a function equation contains more than one rational expression, you must combine the terms into a single expression in order to use those graphing clues you learned.

EXAMPLE B

Write the difference as a single rational expression.

$$\frac{x+2}{(x-3)(x+4)}-\frac{5}{x+1}$$

▸ Solution

Begin by finding a common denominator so that you can write the expression as a single rational expression in factored form. The common denominator is $(x-3)(x+4)(x+1)$.

$\dfrac{(x+2)}{(x-3)(x+4)}\cdot\dfrac{(x+1)}{(x+1)}-\dfrac{5}{(x+1)}\cdot\dfrac{(x-3)(x+4)}{(x-3)(x+4)}$	Multiply each fraction by an equivalent of 1 to get a common denominator.
$\dfrac{x^2+3x+2}{(x-3)(x+4)(x+1)}-\dfrac{5(x^2+x-12)}{(x+1)(x-3)(x+4)}$	Expand the numerators.
$\dfrac{x^2+3x+2-5(x^2+x-12)}{(x-3)(x+4)(x+1)}$	Write with a single denominator.
$\dfrac{x^2+3x+2-5x^2-5x+60}{(x-3)(x+4)(x+1)}$	Use the distributive property.
$\dfrac{-4x^2-2x+62}{(x-3)(x+4)(x+1)}$	Combine like terms.

Check to see whether the numerator factors. There is a common factor of -2, so you can rewrite the numerator.

$$\frac{-2(2x^2 + x - 31)}{(x-3)(x+4)(x+1)}$$ Factor the numerator.

No further factoring is possible.

To multiply and divide rational expressions, you don't need to find a common denominator. In this sense they're like fractions. The picture below of a 1-by-1 square shows the product of $\frac{3}{4}$ and $\frac{5}{12}$. The square is divided into $12 \cdot 4$, or 48 parts. The shaded area shows $3 \cdot 5$, or 15 of those parts. Thus, $\left(\frac{3}{4}\right)\left(\frac{5}{12}\right) = \frac{15}{48} = \frac{5}{16}$.

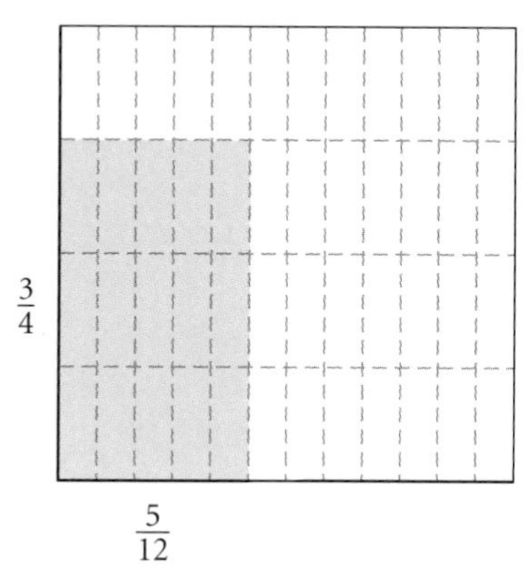

Here's a problem to remind you of how to divide fractions.

$$\frac{\frac{2}{5}}{\frac{6}{7}} = \frac{2}{5} \cdot \left(\frac{6}{7}\right)^{-1} = \frac{2}{5} \cdot \frac{7}{6} = \frac{\cancel{2} \cdot 7}{5 \cdot \cancel{2} \cdot 3} = \frac{7}{15}$$ To divide, multiply by the reciprocal of the denominator.

In multiplication and division problems with rational expressions, it is best to factor all expressions first. This will make it easy to reduce common factors and identify x-intercepts, holes, and vertical asymptotes.

EXAMPLE C Multiply $\frac{(x+2)}{(x-3)} \cdot \frac{(x+1)(x-3)}{(x-1)(x^2-4)}$.

▸ Solution

$$\frac{(x+2)}{(x-3)} \cdot \frac{(x+1)(x-3)}{(x-1)(x-2)(x+2)}$$ Factor any expressions that you can. $(x^2 - 4)$ factors to $(x-2)(x+2)$.

$$\frac{(x+2)(x+1)(x-3)}{(x-3)(x-1)(x-2)(x+2)}$$ Combine the two expressions.

$$\frac{\cancel{(x+2)}(x+1)\cancel{(x-3)}}{\cancel{(x-3)}(x-1)(x-2)\cancel{(x+2)}}$$ Reduce common factors.

$$\frac{(x+1)}{(x-1)(x-2)}$$ Rewrite.

Does the final expression equal the original expression for every value of x?

Multiplication and division with rational expressions will give you a good chance to practice your factoring skills.

EXAMPLE D Divide $\dfrac{\frac{x^2-1}{x^2+5x+6}}{\frac{x^2-3x+2}{x+3}}$.

▶ Solution

$\dfrac{x^2-1}{x^2+5x+6} \cdot \dfrac{x+3}{x^2-3x+2}$ Invert the fraction in the denominator and multiply.

$\dfrac{(x+1)(x-1)}{(x+2)(x+3)} \cdot \dfrac{(x+3)}{(x-2)(x-1)}$ Factor all expressions.

$\dfrac{(x+1)(x-1)(x+3)}{(x+2)(x+3)(x-2)(x-1)}$ Multiply.

$\dfrac{(x+1)}{(x+2)(x-2)}$ Reduce all common factors.

Rational expressions like those in Example D can look intimidating. However, the rules are the same as for regular fraction arithmetic. Work carefully and stay organized.

EXERCISES

You will need

A graphing calculator for Exercises **6–10**.

Practice Your Skills

1. Factor the numerator and denominator of each expression completely, and reduce common factors.

a. $\dfrac{x^2+2x}{x^2-4}$ @ b. $\dfrac{x^2-5x+4}{x^2-1}$ c. $\dfrac{3x^2-6x}{x^2-6x+8}$ @ d. $\dfrac{x^2+3x-10}{x^2-25}$

2. What is the least common denominator for each pair of rational expressions? Don't forget to factor the denominators completely first.

a. $\dfrac{x}{(x+3)(x-2)}, \dfrac{x-1}{(x-3)(x-2)}$ @

b. $\dfrac{x^2}{(2x+1)(x-4)}, \dfrac{x}{(x+1)(x-2)}$

c. $\dfrac{2}{x^2-4}, \dfrac{x}{(x+3)(x-2)}$ @

d. $\dfrac{x+1}{(x-3)(x+2)}, \dfrac{x-2}{x^2+5x+6}$

3. Add or subtract as indicated.

a. $\dfrac{x}{(x+3)(x-2)} + \dfrac{x-1}{(x-3)(x-2)}$ @

b. $\dfrac{2}{x^2-4} - \dfrac{x}{(x+3)(x-2)}$ @

c. $\dfrac{x+1}{(x-3)(x+2)} + \dfrac{x-2}{x^2+5x+6}$

d. $\dfrac{2x}{(x+1)(x-2)} - \dfrac{3}{x^2-1}$

Rational expressions can be combined, just like fractions, to express the sum of parts. *Circles* (1916–1923) by Johannes Itten (1888–1967) shows many circles divided into fractional parts.

4. Multiply or divide as indicated. Reduce any common factors to simplify.

a. $\frac{x+1}{(x+2)(x-3)} \cdot \frac{x^2-4}{x^2-x-2}$ ⓐ

b. $\frac{x^2-16}{x+5} \div \frac{x^2+8x+16}{x^2+3x-10}$ ⓐ

c. $\frac{x^2+7x+6}{x^2+5x-6} \cdot \frac{2x^2-2x}{x+1}$

d. $\dfrac{\frac{x+3}{x^2-8x+15}}{\frac{x^2-9}{x^2-4x-5}}$

Reason and Apply

5. Rewrite as a single rational expression.

a. $\dfrac{1-\frac{x}{x+2}}{\frac{x+1}{x^2-4}}$ ⓐ

b. $\dfrac{\frac{1}{x-1}+\frac{1}{x+1}}{\frac{x}{x-1}-\frac{x}{x+1}}$

6. Graph $y = \frac{x+1}{x^2-7x-8} - \frac{x}{2(x-8)}$ on your calculator.

a. List all asymptotes, holes, and intercepts based on your calculator's graph.

b. Rewrite the right side of the equation as a single rational expression.

c. Use your answer from 6b to verify your observations in 6a. Explain.

7. Consider the equation $y = \frac{x-3}{x^2-4}$.

a. Without graphing, identify the zeros and asymptotes of the graph of the equation. Explain your methods. ⓗ

b. Verify your answers by graphing the function.

8. Consider the equation $y = 1 + \frac{1}{x-1} + \frac{2}{x-2}$.

a. Without graphing, name the asymptotes of the function. ⓗ

b. Rewrite the equation as a single rational function. ⓐ

c. Sketch a graph of the function without using your calculator.

d. Confirm your work by graphing the function with your calculator.

9. *Mini-Investigation* In Lessons 8.6 and 8.7, you saw many examples of graphs with vertical and horizontal asymptotes, including the parent function $f(x) = \frac{1}{x}$. Are there other types of asymptotes?

a. Graph the function $y = x + \frac{1}{x}$ and describe any asymptotes you see.

b. The graph in 9a has a **slant asymptote.** What is its equation?

c. Graph the function $y = x - 1 + \frac{2}{x+1}$ and find the equation of its slant asymptote. (*Hint:* You may want to guess-and-check to find this equation.)

d. Graph the function $y = 2x + 1 - \frac{1}{x-1}$. Look back at your equations for the slant asymptotes in 9b and c, and compare them with the equations for the functions. Predict the equation for the slant asymptote in this new function and verify your prediction by graphing.

e. What is the equation of the slant asymptote for the equation $y = ax + b + \frac{c}{x-d}$?

10. APPLICATION How long should a traffic light stay yellow before turning red? One study suggests that for a car approaching a 40 ft wide intersection under normal driving conditions, the length of time, y, that a light should stay yellow, in seconds, is given by the equation $y = 1 + \frac{v}{25} + \frac{50}{v}$, where v is the velocity of an approaching car in feet per second.

a. Rewrite the equation in rational function form. ⓐ

b. Enter both the original equation and your simplified equation into your calculator. Check using the table feature that the values of both functions are the same. What does this tell you?

c. If the speed limit at a particular intersection is 45 mi/h, how long should the light stay yellow?

d. If cars typically travel at speeds ranging from 25 mi/h to 55 mi/h at that intersection, what is the possible range of times that a light should stay yellow?

Career
CONNECTION

Traffic engineers time traffic signals to minimize the "dilemma zone." The dilemma zone occurs when a driver has to decide to brake hard to stop or to accelerate to get through the intersection. Either decision can be risky. It is estimated that 22% of traffic accidents occur when a driver runs a red light.

Review

11. This graph is the image of $y = \frac{1}{x}$ after a transformation.

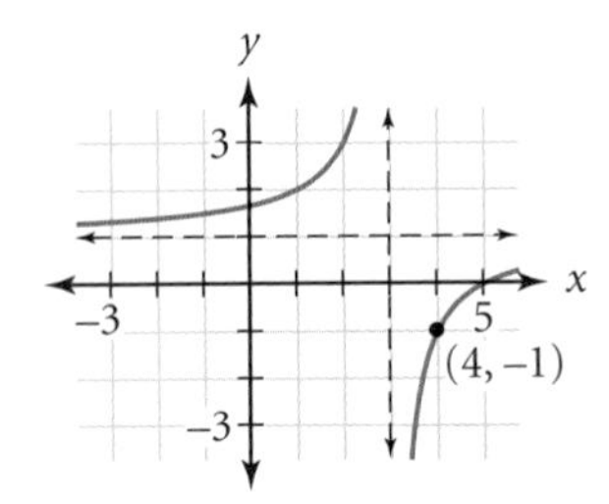

a. Write an equation for each asymptote.

b. What translations are involved in transforming $y = \frac{1}{x}$ into its image? ⓐ

c. The point $(4, -1)$ is on the image. What is the vertical scale factor in the transformation?

d. Write an equation of the image.

e. Name the intercepts.

12. Factor each expression completely.

a. $x^2 + 9x + 8$

b. $x^2 - 9$

c. $x^2 - 7x - 18$

d. $3x^3 - 15x^2 + 18x$

13. If you invest \$1,000 at 6.5% interest for 5 years, how much interest do you earn in each of these situations?

a. The interest is compounded annually. @

b. The interest is compounded monthly.

c. The interest is compounded weekly.

d. The interest is compounded daily.

Hungarian modern sculptor Màrton Vàrò (b 1943) titled this piece *15 Cubes* (1998). It is located at the Elektro building in Trondheim, Norway.

Project

CYCLIC HYPERBOLAS

Consider the rational function $f(x) = \frac{x-3}{x+1}$, whose graph is a hyperbola. You can use this function as a recursive formula. Choose any starting value for x and find the first six terms of the sequence. The values should repeat. Choose another value for x. Does the same thing happen?

Explore this question with several values of x. Then use function composition to prove that such repetition always happens for this function. Hyperbolas with this property are called *cyclic hyperbolas.*

Your project is to explore what can happen when you use a rational function whose graph is a hyperbola as a recursive rule and to look for other cyclic hyperbolas. You might also explore what happens when you use other functions you've studied as a recursive rule. Are any of them cyclic?

Your project should include

- Your calculations for $f(x) = \frac{x-3}{x+1}$, and other functions you explore.
- Any other cyclic hyperbolas you find, including a proof that they are cyclic.
- Any research you do about recursive rules based on hyperbolas or other functions.

CHAPTER 8 REVIEW

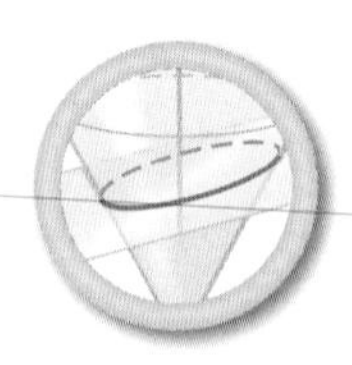

In this chapter you saw some special relations. Each relation was described as a set of points, or **locus,** that satisfied some criteria. These relations are called **conic sections** because the shapes can be formed by slicing a double cone at various angles. The simplest of the conic sections is the **circle,** the set of points a fixed distance from a fixed point called the **center.** Closely related to the circle is the **ellipse.** The ellipse can be defined as the set of points such that the sum of their distances from two fixed points, the **foci,** is a constant. The **parabola** is another conic section, which you studied in earlier chapters. It can be defined as the set of points that are equidistant from a fixed point called the focus and a fixed line called the **directrix.** The last of the conic sections is the **hyperbola.** The definition of a hyperbola is similar to the definition of an ellipse, except that the difference between the distances from the foci remains constant. The equations for these conic sections can be written in standard form or as a **general quadratic equation.** And each of these conic sections can be either vertically oriented or horizontally oriented. You learned how to convert between the general quadratic equation and standard form, and how to solve systems of quadratic equations to find the intersections of two conic sections.

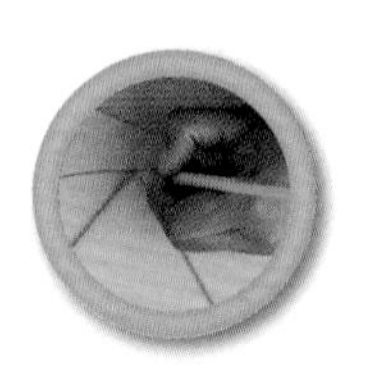

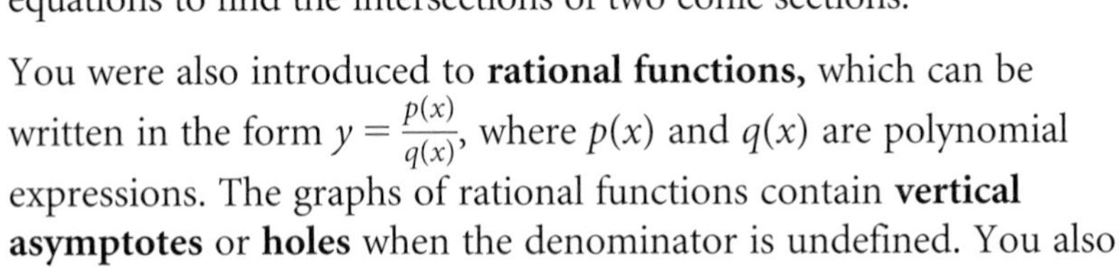

You were also introduced to **rational functions,** which can be written in the form $y = \frac{p(x)}{q(x)}$, where $p(x)$ and $q(x)$ are polynomial expressions. The graphs of rational functions contain **vertical asymptotes** or **holes** when the denominator is undefined. You also learned how to do arithmetic with rational expressions.

EXERCISES

You will need

A graphing calculator for Exercises **8, 9,** and **12.**

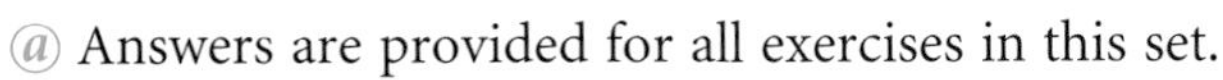

@ Answers are provided for all exercises in this set.

1. Sketch the graph of each equation. Label foci when appropriate.

a. $\frac{x-3}{-2} = (y-2)^2$

b. $x^2 + (y-2)^2 = 16$

c. $\left(\frac{y+2.5}{3}\right)^2 - \left(\frac{x-1}{4}\right)^2 = 1$

d. $\left(\frac{x}{5}\right)^2 + 3y^2 = 1$

2. Consider this ellipse.

a. Write the equation for the graph shown in standard form.

b. Name the coordinates of the center and foci.

c. Write the general quadratic form of the equation for this graph.

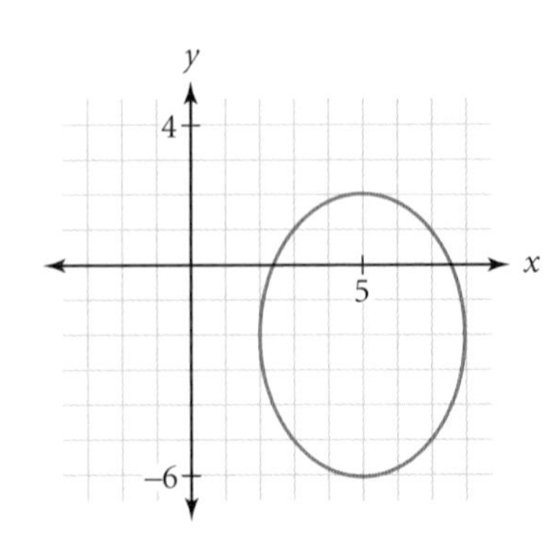

3. Consider the hyperbola graphed at right.

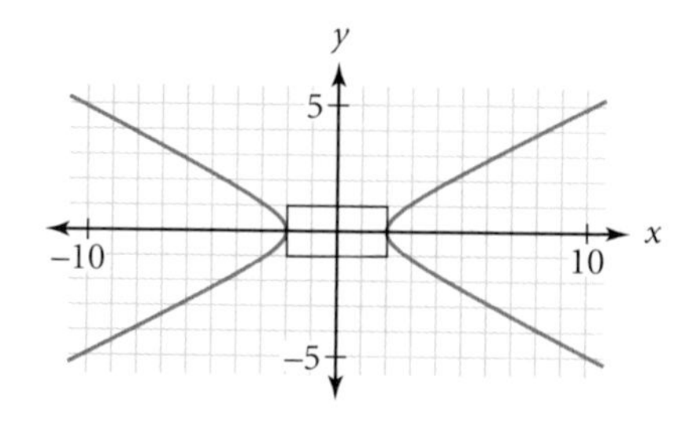

a. Write the equations of the asymptotes for this hyperbola.

b. Write the general quadratic equation for this hyperbola.

c. Write an equation that will give the vertical distance, d, between the asymptote with positive slope and a point on the upper portion of the right branch of the hyperbola as a function of the point's x-coordinate, x.

d. Use the function from 3c to fill in this table. What does this tell you about the relationship between the function and its asymptote?

x	2	10	20	100
d				

4. Write the general quadratic equation $x^2 + y^2 + 8x - 2y - 8 = 0$ in standard form. Identify the shape described by the equation and describe its features.

5. Write the general quadratic equation $y^2 - 8y - 4x + 28 = 0$ in standard form. Determine the vertex, focus, and directrix of the parabola defined by this equation. Sketch a graph.

6. APPLICATION Pure gold is too soft to be used for jewelry, so gold is always mixed with other metals. 18-karat gold is 75% gold and 25% other metals. How much pure gold must be mixed with 5 oz of 18-karat gold to make a 22-karat (91.7%) gold mixture?

7. Write an equation of each rational function described as a translation of the graph of $y = \frac{1}{x}$.

a. The rational function has asymptotes $x = -2$ and $y = 1$.

b. The rational function has asymptotes $x = 0$ and $y = -4$.

8. Graph $y = \frac{2x - 14}{x - 5}$. Write equations for the horizontal and vertical asymptotes.

9. How can you modify the equation $y = \frac{2x - 14}{x - 5}$ so that the graph of the new equation is the same as the original graph except for a hole at $x = -3$? Verify your new equation by graphing it on your calculator.

10. On her way to school, Ellen drives at a steady speed for the first 2 mi. After glancing at her watch, she drives 20 mi/h faster during the remaining 3.5 mi. How fast does she drive during the two portions of this trip if the total time of her trip is 10 min?

11. Rewrite each expression as a single rational expression in factored form.

a. $\frac{2x}{(x-2)(x+1)} + \frac{x+3}{x^2-4}$ **b.** $\frac{x^2}{x+1} \cdot \frac{3x-6}{x^2-2x}$ **c.** $\frac{x^2-5x-6}{x} \div \frac{x^2-8x+12}{x^2-1}$

12. Solve this system of equations algebraically, then confirm your answer graphically.

$$\begin{cases} x^2 + y^2 = 4 \\ (x+1)^2 - \frac{y^2}{3} = 1 \end{cases}$$

TAKE ANOTHER LOOK

1. The equation $\left(\frac{x}{a}\right)^2 + \left(\frac{y}{b}\right)^2 = 1$ is the standard form of an ellipse centered at the origin. What shape will the equations $\left(\frac{x}{a}\right)^3 + \left(\frac{y}{b}\right)^3 = 1$, $\left(\frac{x}{a}\right)^4 + \left(\frac{y}{b}\right)^4 = 1$, or generally $\left(\frac{x}{a}\right)^n + \left(\frac{y}{b}\right)^n = 1$ create? These equations are called Lamé curves, or superellipses, when $n > 2$. Investigate several Lamé curve graphs. Assume that $a > 0$ and $b > 0$. You already know what effect the values of a and b have on an ellipse. Do they have the same effect on a Lamé curve? For fixed values of a and b, try different n-values, including positive, negative, whole-number, and rational values. Explore the graph shapes and properties for different values of n. Summarize your discoveries.

2. How can you find the horizontal asymptote of a rational function without graphing or using a table? Use a calculator to explore these functions, and look for patterns. Some equations may not have horizontal asymptotes. Make a conjecture about how to determine the equation of a horizontal asymptote just by looking at the equation.

$y = \dfrac{3x^2 + 4x - 5}{2x^4 + 2}$ $\qquad y = \dfrac{4x^4 - 2x^3 - 2}{x^5 - 5x}$ $\qquad y = \dfrac{3x^2 + 4x - 5}{2x^2}$

$y = \dfrac{4x^5 - 2x^3 - 2}{x^5 - 5x}$ $\qquad y = \dfrac{-x^3}{x^2 - 2x + 5}$ $\qquad y = \dfrac{3x^4 + 2x}{5x^2 - 1}$

3. You have seen that for an ellipse, the sum of the distances from any point to the two foci is constant. For a hyperbola, the difference of the distances from any point to the two foci is constant. What shape is created if the product of the distances is constant? What if the ratio of the distances is constant? You may want to use geometry software to explore these patterns.

4. In Lesson 8.5, you explored the number of points of intersection of two conic sections. You saw, for example, that a hyperbola and an ellipse can intersect 0, 1, 2, 3, or 4 times. However, points of intersection include only the real solutions to a system of equations. If you include nonreal answers, how many solutions can a system of conic sections have? Consider each pair of the four conic sections. Explain how the number of solutions is related to the exponents in the original equations.

Assessing What You've Learned

GIVE A PRESENTATION By yourself or with a group, demonstrate how to factor the equation of a rational function. Then describe how to find asymptotes, holes, and intercepts, and graph the equation. Or present your solution of a project or Take Another Look activity from this chapter.

PERFORMANCE ASSESSMENT While a classmate, teacher, or family member observes, demonstrate how to convert the equation of a conic section in general quadratic form to standard form. Describe the features of the conic section, then draw a graph of the equation.

ORGANIZE YOUR NOTEBOOK Update your notebook to include the equations for the four conic sections. Include methods of graphing and how to find foci, vertices, centers, and asymptotes.

CHAPTER

9 Series

Korean artist Do-Ho Suh (b 1962) is perhaps best known for his sculptures that use numerous miniature items that contribute to a greater, larger mass. Shown here is a detail from his installation *Floor* (1997–2000), made of thousands of plastic figurines that support a 40 m^2 floor of glass.

OBJECTIVES

In this chapter you will

- learn about mathematical patterns called series, and distinguish between arithmetic and geometric series
- write recursive and explicit formulas for series
- find the sum of a finite number of terms of an arithmetic or geometric series
- determine when an infinite geometric series has a sum and find the sum if it exists

Arithmetic and Geometric Sequences

This lesson will help you recall earlier work with sequences, which are ordered lists. Remember that arithmetic sequences like 4, 7, 10, 13, . . . have a common difference between consecutive terms, and geometric sequences such as 4, 8, 16, 32, . . . have a common ratio between consecutive terms. Recursive definitions of sequences specify one or more initial terms and a recursive rule that defines the nth term relative to one or more previous terms.

EXAMPLE A

Create recursive formulas to define the sequences 4, 7, 10, 13, . . . and 4, 8, 16, 32, ... Then find the 75th term of each sequence.

▶ ***Solution***

The first term, $u_1 = 4$, and recursive rule $u_n = u_{n-1} + 3$ where $n \geq 2$ describe the arithmetic sequence. You can use technology to find later terms. For example, the spreadsheet shows terms 71 to 75 of each sequence. For the arithmetic sequence, the 75th term is 226.

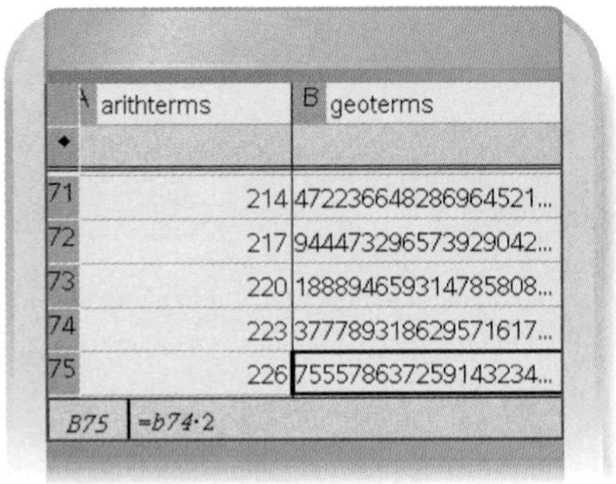

	A arithterms	B geoterms
◆		
71	214	472236648286964521...
72	217	944473296573929042...
73	220	188894659314785808...
74	223	377789318629571617...
75	226	755578637259143234...

B75 =b74·2

The geometric sequence can be defined with $t_1 = 4$ and $t_n = 2 \cdot t_{n-1}$ where $n \geq 2$. The 75th term is $4 \cdot 2^{74}$, which is huge!

In contrast to recursive formulas, explicit representations of sequences define each term as a function of n.

EXAMPLE B

Find the first 3 terms and the 20th term of the sequences a and u, where $a_n = 4.2 + 5n$ and $u_n = 40\left(1 + \frac{0.05}{12}\right)^{n-1}$.

▶ ***Solution***

Substituting 1, 2, 3, and 20 for n gives the requested terms, as shown in the table.

n	$a_n = 4.2 + 5n$	$u_n = 40\left(1 + \frac{0.05}{12}\right)^{n-1}$
1	9.2	40
2	14.2	40.1667
3	19.2	40.3340
. . .	. . .	. . .
20	104.2	43.2883

Discrete points (n, a_n) defined by the arithmetic sequence $a_n = 4.2 + 5n$ lie on the line $y = 4.2 + 5x$. The common difference 5 of the discrete sequence is called the *slope,* or *rate of change,* of the line. The discrete points (n, u_n) of the geometric sequence above are found on the exponential curve $y = 40\left(1 + \frac{0.05}{12}\right)^{x-1}$. This geometric sequence might represent the monthly balances of a \$40 deposit that earns 5%, compounded monthly.

EXAMPLE C Find the long-run value of the sequence with initial term 140 and successive terms decreased by 10% and then increased by 20. What will happen if the terms are increased by 14 rather than 20?

▶ ***Solution*** The original sequence can be described recursively as $u_1 = 140$ and $u_n = u_{n-1} \cdot (1 - 0.10) + 20$ where $n \geq 2$. Technology shows that the sequence will eventually level off at 200.

When the sequence is changed to $u_1 = 140$ and $u_n = u_{n-1} \cdot (1 - 0.10) + 14$ where $n \geq 2$, the long-run value is 140 because the amount decreased, 10% of 140, is the same as the amount of increase, 14.

	A terms	B	C	D	E
•	=seqn(u(n−1)*.9+20,(				
110	199.999				
111	199.999				
112	199.999				
113	200.000				
114	200.000				
A114	=a113·.9+20				

To review how to find the long-run values algebraically, refer to the example in Lesson 1.3.

EXERCISES

1. Identify each sequence as arithmetic, geometric, or neither. For arithmetic and geometric sequences, give the common difference or common ratio.
   a. 14, 7, 3.5, 1.75, . . . @ **b.** 47, 41, 35, 29, . . . **c.** 1, 1, 2, 3, 5, 8, . . . @

2. Write a recursive definition for each sequence in Exercise 1. @

3. Find the 10th term of each sequence.
   a. $u_1 = 140$ and $u_n = u_{n-1} \cdot (1 - 0.10) + 10$ @
   b. $a_n = 60(0.40)^n$
   c. $a_n = 305 + (n - 1) \cdot 0.8$

4. A flower farm starts with 10,000 bulbs.
   a. What would happen in the long run if each year the farm plans to sell 50% of its bulbs and plant 2,000 new bulbs? Explain. @
   b. How many bulbs would the farm need to plant to maintain an inventory of 12,000 bulbs?

5. Fill in the missing entries in the table. Describe how to generate the entries in the bottom row.

n	1	2	3	4	5	. . .	10
$2n + 1$	3					. . .	
	3	8	15	24		. . .	

LESSON

9.1 Arithmetic Series

According to the U.S. Environmental Protection Agency, each American produced an average of 978 lb of trash in 1960. This increased to 1336 lb in 1980. By 2000, trash production had risen to 1646 lb/yr per person. You have learned in previous chapters how to write a sequence to describe the amount of trash produced per person each year. If you wanted to find the total amount of trash a person produced in his or her lifetime, you would add the numbers in this sequence.

Environmental CONNECTION

Mount Everest, part of the Himalaya range of southern Asia, reaches an altitude of 29,035 ft and is the world's highest mountain above sea level. It has been nicknamed "the world's highest junkyard" because decades of litter—climbing gear, plastic, glass, and metal—have piled up along Mount Everest's trails and in its camps. Environmental agencies like the World Wildlife Fund have cleared garbage from the mountain's base camp, but removing waste from higher altitudes is more challenging. An estimated 50 tons of junk remain.

A 1998 cleanup of Mt. Everest

Series

The indicated sum of terms of a sequence is a **series.**

The sum of the first n terms in a sequence is represented by S_n. You can calculate S_n by adding the terms $u_1 + u_2 + u_3 + \cdots + u_n$.

History CONNECTION

Chu Shih-chieh (ca. 1280–1303) was a celebrated mathematician from Beijing, China, known for his theories on arithmetic series, geometric series, and finite differences. His two mathematical works, *Introduction to Mathematical Studies* and *Precious Mirror of the Four Elements,* were discovered in the 19th century.

Finding the value of a series is a problem that has intrigued mathematicians for centuries. Chinese mathematician Chu Shih-chieh called the sum $1 + 2 + 3 + \cdots + n$ a "pile of reeds" because it can be pictured like the diagram at right. The diagram shows S_9, the sum of the first nine terms of this sequence, $1 + 2 + 3 + \cdots + 9$. The sum of any **finite,** or limited, number of terms is called a **partial sum** of the series.

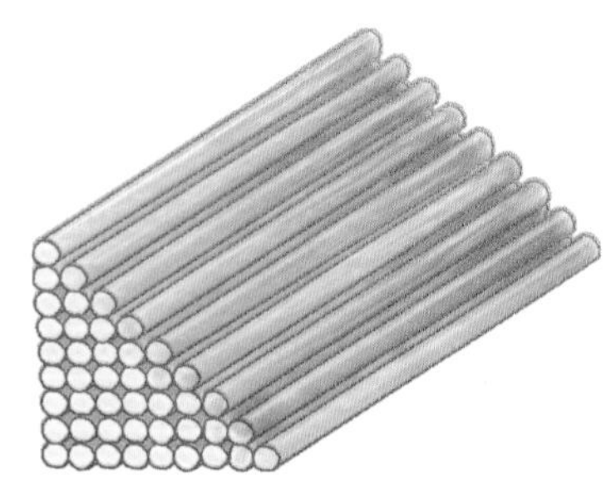

The expressions S_9 and $\sum_{n=1}^{9} u_n$ are shorthand ways of writing $u_1 + u_2 + u_3 + \cdots + u_9$.

For this series, $u_n = n$, so you can express the partial sum S_9 with sigma notation as $\sum_{n=1}^{9} n$.

The expression $\sum_{n=1}^{9} n$ tells you to substitute the integers 1 through 9 for n in the explicit formula $u_n = n$, and then sum the resulting nine values. You get $1 + 2 + 3 + \cdots + 9 = 45$.

How could you find the sum of the integers 1 through 100? The most obvious method is to add the terms, one by one. You can use technology and a recursive formula to do this quickly.

The sum of the first 100 terms is the sum of the first 99 terms plus the 100th term.

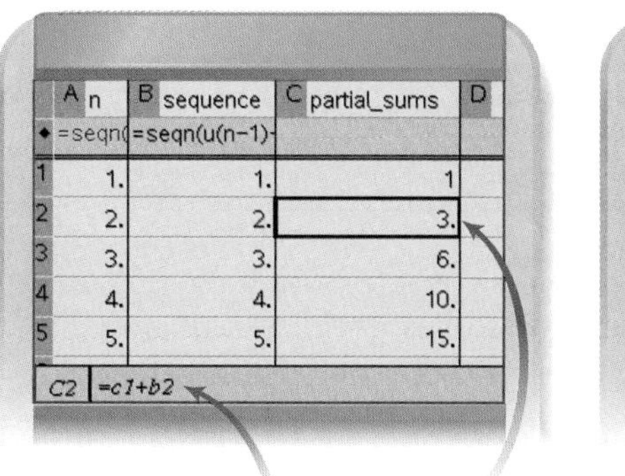

	A n	B sequence	C partial_sums	D
◆	=seqn(	=seqn(u(n-1)-		
1	1.	1.	1	
2	2.	2.	3.	
3	3.	3.	6.	
4	4.	4.	10.	
5	5.	5.	15.	

C2 =c1+b2

S_2 is the sum of S_1 and u_2.

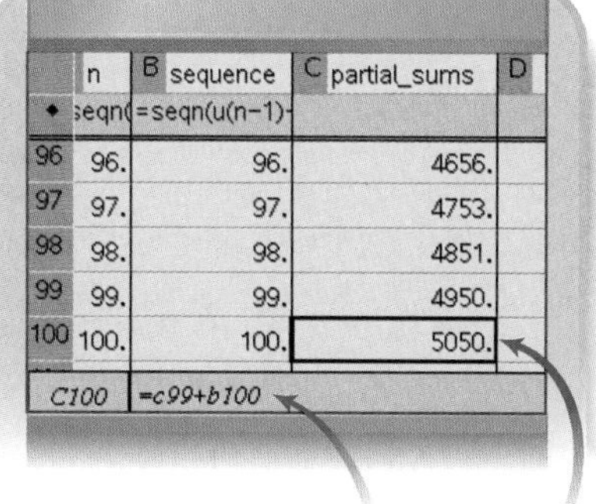

	n	B sequence	C partial_sums	D
◆	seqn(	=seqn(u(n-1)-		
96	96.	96.	4656.	
97	97.	97.	4753.	
98	98.	98.	4851.	
99	99.	99.	4950.	
100	100.	100.	5050.	

C100 =c99+b100

S_{100} is the sum of S_{99} and u_{100}.

The recursive definitions for the sequence and series described above are:

Sequence	**Series**
$u_1 = 1$	$S_1 = 1$
$u_n = u_{n-1} + 1 \quad$ where $n \geq 2$	$S_n = S_{n-1} + u_n \quad$ where $n \geq 2$

The graph of S_n appears to form a solid curve, but it is actually a discrete set of 100 points representing each partial sum from S_1 through S_{100}. Each point is in the form (n, S_n) for integer values of n, for $1 \leq n \leq 100$. You can trace to find that the sum of the first 100 terms, S_{100}, is 5050. [► See **Calculator Note 9A** for more information on graphing and calculating partial sums. ◄]

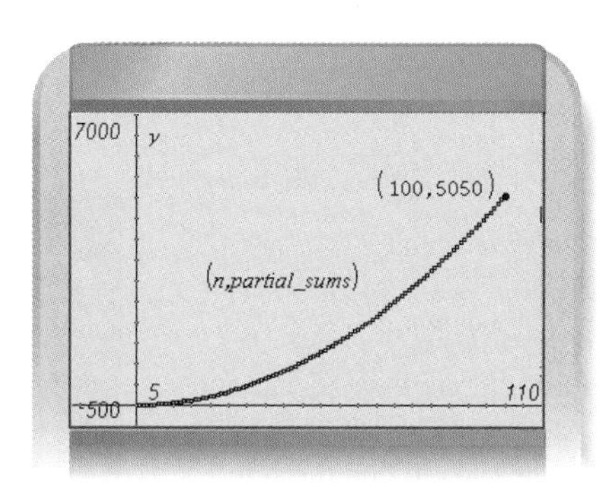

When you compute this sum recursively, you or the calculator must compute each of the individual terms. The investigation will give you an opportunity to discover at least one explicit formula for calculating the partial sum of an **arithmetic series** without finding all terms and adding.

Investigation
Arithmetic Series Formula

You will need

- graph paper
- scissors

In this investigation you will use a geometric model to help you develop an explicit formula for the partial sum of an arithmetic series.

Step 1 The lengths of the rows of this step-shaped figure represent terms of an arithmetic sequence. Write the sequence u_1, u_2, u_3, u_4, u_5, represented by the figure. What is the sum of the series?

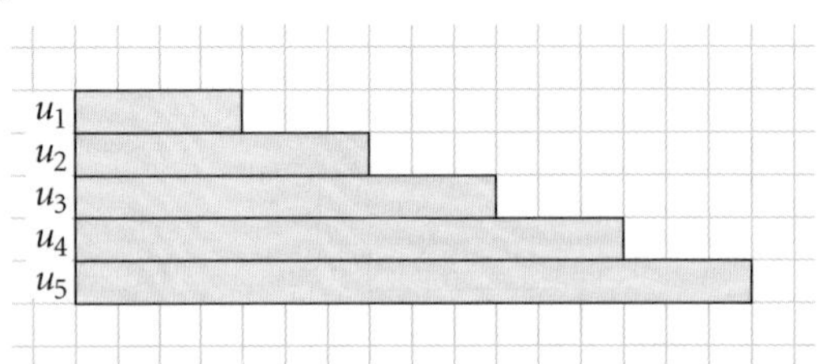

Step 2 If you cut out two copies of this figure and slid them together to make a rectangle, what would the dimensions of your rectangle be?

Step 3 Have each member of your group create a different arithmetic sequence. Each of you can choose a starting value and a common difference. (Use positive numbers less than 10 for each of these.) On graph paper, draw two copies of a step-shaped figure representing your sequence.

Step 4 Cut out both copies of the step-shaped figure. Slide the two congruent shapes together to form a rectangle, and then calculate the dimensions and area of the rectangle. Now express this area in terms of the number of rows, n, and the first and last terms of the sequence.

Step 5 Based on what you have discovered, what is a formula for the partial sum, S_n, of an arithmetic series? Describe the relationship between your formula and the dimensions of your rectangle.

In the investigation you found a formula for a partial sum of an arithmetic series. Use your formula to check that when you add $1 + 2 + 3 + \cdots + 100$ you get 5050.

History CONNECTION

According to legend, when German mathematician and astronomer Carl Friedrich Gauss (1777–1855) was 9 years old, his teacher asked the class to find the sum of the integers 1 through 100. The teacher was hoping to keep his students busy, but Gauss quickly wrote the correct answer, 5050. The example shows Gauss's solution method.

This stamp of Gauss was issued by Nicaragua in 1994 as part of a series about astronomers.

EXAMPLE Find the sum of the integers 1 through 100, without using a calculator.

▶ **Solution**

Carl Friedrich Gauss solved this problem by adding the terms in pairs. Consider the series written in ascending and descending order, as shown.

$$
\begin{array}{r}
1 + 2 + 3 + \cdots + 98 + 99 + 100 = S_{100} \\
100 + 99 + 98 + \cdots + 3 + 2 + 1 = S_{100} \\
\hline
101 + 101 + 101 + \cdots + 101 + 101 + 101 = 2S_{100}
\end{array}
$$

The sum of every column is 101, and there are 100 columns. Thus, the sum of the integers 1 through 100 is

$$\frac{100(101)}{2} = 5050$$

You must divide the product 100(101) by 2 because the series was added twice.

You can extend the method in the example to any arithmetic series. Before continuing, take a moment to consider why the sum of the reeds in the original pile can be calculated using the expression

$$\frac{9(1+9)}{2}$$

What do the 9, 1, 9, and 2 represent in this context?

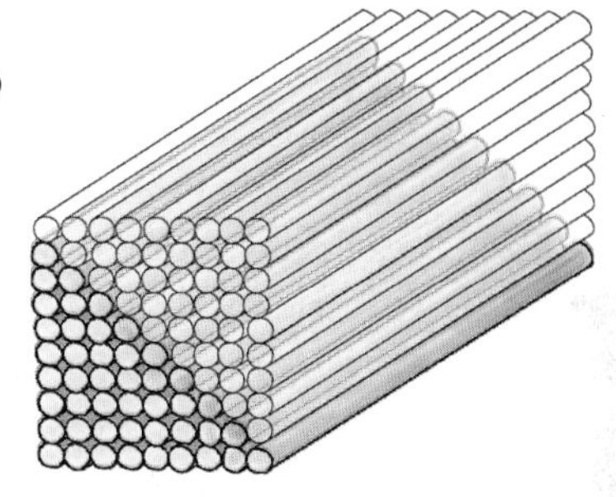

Partial Sum of an Arithmetic Series

The partial sum of an arithmetic series is given by the explicit formula

$$S_n = \frac{n(u_1 + u_n)}{2}$$

where n is the number of terms, u_1 is the first term, and u_n is the last term.

In the exercises you will use the formula for partial sums to find the sum of consecutive terms of an arithmetic sequence.

EXERCISES

You will need

A graphing calculator for Exercise **16.**

Practice Your Skills

1. List the first five terms of this sequence. Identify the first term and the common difference.

$u_1 = -3$
$u_n = u_{n-1} + 1.5 \quad \text{where } n \geq 2$

2. Find S_1, S_2, S_3, S_4, and S_5 for this sequence: 2, 6, 10, 14, 18. ⓐ

3. Write each expression as a sum of terms, then calculate the sum.

a. $\sum_{n=1}^{4}(n+2)$

b. $\sum_{n=1}^{3}(n^2-3)$ ⓐ

4. Find the sum of the first 50 multiples of 6: $\{6, 12, 18, \ldots, u_{50}\}$. ⓗ

5. Find the sum of the first 75 even numbers, starting with 2. ⓐ

Reason and Apply

6. Find these values.

a. Find u_{75} if $u_n = 2n - 1$.

b. Find $\sum_{n=1}^{75}(2n - 1)$. ⓐ

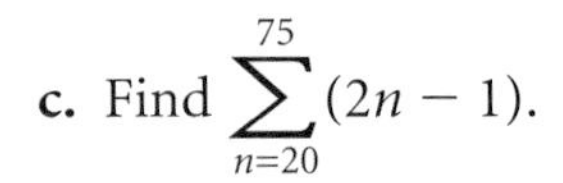

c. Find $\sum_{n=20}^{75}(2n - 1)$.

7. Consider the graph of the arithmetic sequence shown at right.

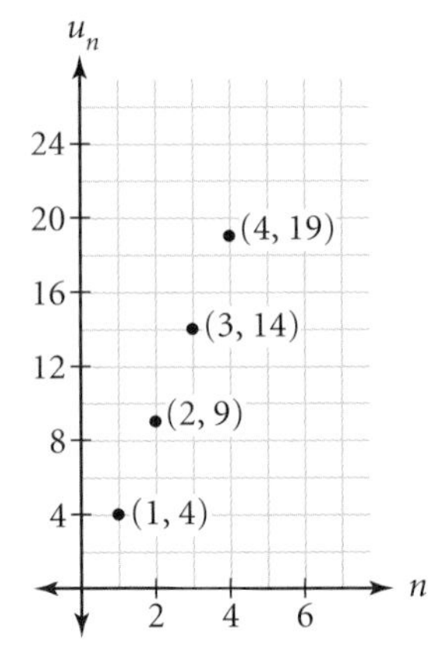

a. What is the 46th term? ⓐ

b. Write a formula for u_n.

c. Find the sum of the heights from the horizontal axis of the first 46 points of the sequence's graph.

8. Suppose you practice the piano 45 min on the first day of the semester and increase your practice time by 5 min each day. How much total time will you devote to practicing during

a. The first 15 days of the semester?

b. The first 35 days of the semester?

American pianist Van Cliburn (b 1934) caused a sensation by winning the first International Tchaikovsky Competition in Moscow in 1958, during the height of the Cold War. When asked how many hours he practiced per day, he replied, "Mother always told me that if you knew how long you had practiced, you hadn't done anything. She believed you had to become engrossed . . . or it just didn't count for anything. Nothing happens if you watch the clock."

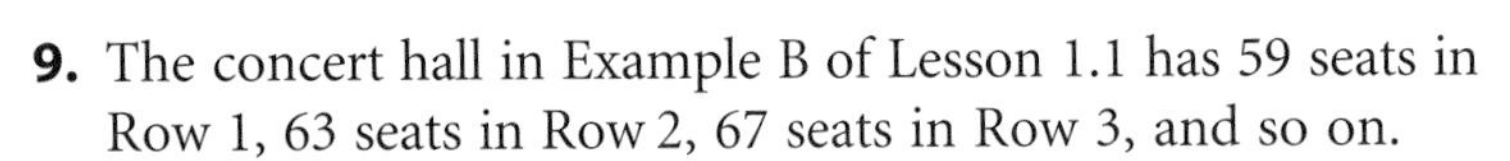

9. The concert hall in Example B of Lesson 1.1 has 59 seats in Row 1, 63 seats in Row 2, 67 seats in Row 3, and so on.

a. How many seats are there in all 35 rows of the concert hall?

b. Suppose that Seat 1 is at the left end of Row 1 and that Seat 60 is at the left end of Row 2. Describe the location of Seat 970.

10. Jessica arranges a display of soup cans as shown.

a. List the number of cans in the top layer, the second layer, the third layer, and so on, down to the tenth layer. ⓐ

b. Write a recursive formula for the terms of the sequence in 10a. ⓐ

c. If the cans are to be stacked 47 layers high, how many cans will it take to build the display?

d. If Jessica uses six cases, with 48 cans in each case, how tall can she make the display? ⓗ

11. Find each value.

a. Find the sum of the first 1000 positive integers (the numbers 1 through 1000). ⓗ

b. Find the sum of the second 1000 positive integers (1001 through 2000).

c. Guess the sum of the third 1000 positive integers (2001 through 3000).

d. Now calculate the sum for 11c.

e. Describe a way to find the sum of the third 1000 positive integers if you know the sum of the first 1000 positive integers. ⓐ

12. Create a formula for the sum S_n of an arithmetic series that is based on the number of terms n, the first term u_1, and the common difference d.

13. It takes 5 toothpicks to build the top trapezoid shown at right. You need 9 toothpicks to build 2 connected trapezoids and 13 toothpicks for 3 trapezoids.

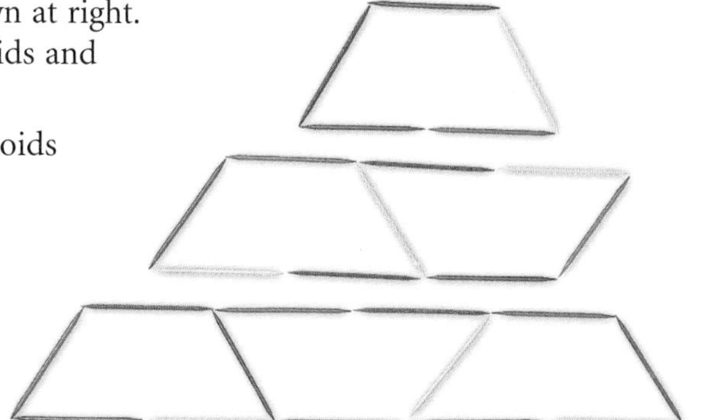

a. If 1000 toothpicks are available, how many trapezoids will be in the last complete row?

b. How many complete rows will there be?

c. How many toothpicks will you use to construct these rows?

d. Use the context of this exercise to explain the difference between a sequence and a series.

14. APPLICATION If an object falls from rest, then the distance it falls during the first second is about 4.9 m. In each subsequent second, the object falls 9.8 m farther than in the preceding second.

a. Write a recursive formula to describe the distance the object falls during each second of free fall.

b. Find an explicit formula for 14a. ⓐ

c. How far will the object fall during the 10th second?

d. How far does the object fall during the first 10 seconds? ⓐ

e. Find an explicit formula for the distance an object falls in n seconds.

f. Suppose you drop a quarter from the Royal Gorge Bridge. How long will it take to reach the Arkansas River 321 m below?

The Royal Gorge Bridge (built 1929) near Cañon City, Colorado, is the world's highest suspension bridge, with length 384 m (1260 ft) and width 5 m (18 ft).

15. Suppose $y = 65 + 2(x - 1)$ is an explicit representation of an arithmetic sequence, for integer values $x \geq 1$. Express the partial sum of the arithmetic series as a quadratic expression, with x representing the term number. @

16. Consider these two geometric sequences:

    i. 2, 4, 8, 16, 32, . . .

    ii. $2, 1, \frac{1}{2}, \frac{1}{4}, \frac{1}{8}, \ldots$

    a. What is the long-run value of each sequence?

    b. What is the common ratio of each sequence?

    c. What will happen if you try to sum all of the terms of each sequence?

17. APPLICATION There are 650,000 people in a city. Every 15 minutes, the local radio and television stations broadcast a tornado warning. During each 15-minute time period, 42% of the people who had not yet heard the warning become aware of the approaching tornado. How many people have heard the news

    a. After 1 hour? @

    b. After 2 hours?

In the central United States, an average of 800 to 1000 tornadoes occur each year. Tornado watches, forecasts, and warnings are announced to the public by the National Weather Service.

Review

18. Suppose you invest \$500 in a bank that pays 5.5% annual interest compounded quarterly.

    a. How much money will you have after five years?

    b. Suppose you also deposit an additional \$150 at the end of every three months. How much will you have after five years?

19. Consider the explicit formula $u_n = 81\left(\frac{1}{3}\right)^{n-1}$. @

    a. List the first six terms, u_1 to u_6.

    b. Write a recursive formula for the sequence.

20. Consider the recursive formula

    $u_1 = 0.39$
    $u_n = 0.01 \cdot u_{n-1}$ where $n \geq 2$

    a. List the first six terms.

    b. Write an explicit formula for the sequence.

21. Consider the rational equation $y = \frac{4x + 3}{2x - 1}$. @

    a. Rewrite the equation as a transformation of the parent function $y = \frac{1}{x}$.

    b. What are the asymptotes of $y = \frac{4x + 3}{2x - 1}$?

    c. The point (1, 1) is on the graph of the parent function. What is its image on the transformed function?

LESSON

9.2

Infinite Geometric Series

Beauty itself is but the sensible image of the infinite.

GEORGE BANCROFT

If you start adding the terms of the arithmetic sequence 1, 2, 3, . . . , you get larger and larger values. Even if the terms of an arithmetic sequence are small, as in 0.001, 0.002, 0.003, . . . , the partial sums eventually get large. As the number of terms increases, the magnitude of the partial sum increases.

These nested photographs represent an infinite sequence.

But consider the geometric sequence

0.4, 0.04, 0.004, 0.0004, ...

It has common ratio $\frac{1}{10}$, so the terms get smaller. The partial sums seem to follow a pattern.

$$S_3 = 0.4 + 0.04 + 0.004 = 0.444$$
$$S_4 = 0.4 + 0.04 + 0.004 + 0.0004 = 0.4444$$
$$S_5 = 0.4 + 0.04 + 0.004 + 0.0004 + 0.00004 = 0.44444$$

If you sum infinitely many terms of this sequence, would the result be infinitely large?

It appears that the partial sums will not get infinitely large; they are all less than 0.5. The indicated sum of a geometric sequence is a **geometric series.** An **infinite geometric series** is a geometric series with infinitely many terms. In this lesson you will specifically look at **convergent series,** for which the sequence of partial sums approaches a long-run value as the number of terms increases.

EXAMPLE A

Jack baked a pie and promptly ate one-half of it. Determined to make the pie last, he then decided to eat only one-half of the pie that remained each day.

a. Record the amount of pie eaten each day for the first seven days.

b. For each of the seven days, record the total amount of pie eaten since it was baked.

c. If Jack lives forever, then how much of this pie will he eat?

▶ Solution

The amount of pie eaten each day is a geometric sequence with first term $\frac{1}{2}$ and common ratio $\frac{1}{2}$.

a. The first seven terms of this sequence are

$$\frac{1}{2}, \frac{1}{4}, \frac{1}{8}, \frac{1}{16}, \frac{1}{32}, \frac{1}{64}, \frac{1}{128}$$

b. Find the partial sums, S_1 through S_7, of the terms in part a.

$$\frac{1}{2}, \frac{3}{4}, \frac{7}{8}, \frac{15}{16}, \frac{31}{32}, \frac{63}{64}, \frac{127}{128}$$

c. It may seem that eating pie "forever" would result in eating a lot of pie. However, if you look at the pattern of the partial sums, it seems as though for any finite number of days Jack's total is slightly less than 1. This leads to the conclusion that Jack would eat exactly one pie in the long run. This is a convergent infinite geometric series with long-run value 1.

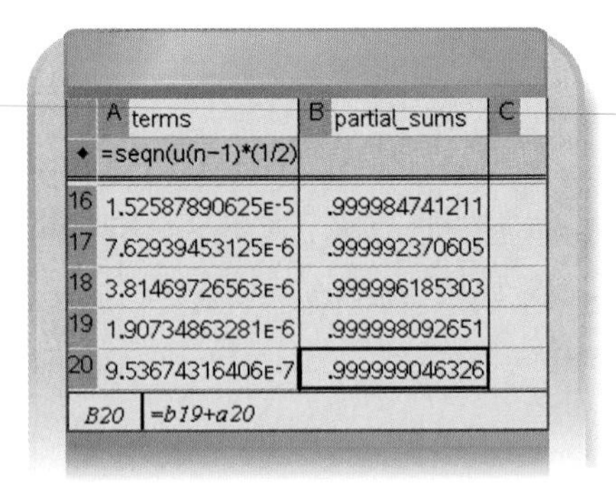

	A terms	B partial_sums	C
◆	=seqn(u(n−1)*(1/2))		
16	1.52587890625E-5	.999984741211	
17	7.62939453125E-6	.999992370605	
18	3.81469726563E-6	.999996185303	
19	1.90734863281E-6	.999998092651	
20	9.53674316406E-7	.999999046326	

B20 =b19+a20

Recall that a geometric sequence can be represented with an explicit formula in the form $u_n = u_1 \cdot r^{n-1}$ or $u_n = u_0 \cdot r^n$, where r represents the common ratio between the terms. The investigation will help you create an explicit formula for the sum of a convergent infinite geometric series.

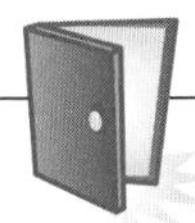

Investigation
Infinite Geometric Series Formula

In algebra you may have learned a method for writing a repeating decimal as a fraction. This method can also help you find the value of an infinite geometric series.

For example, consider the repeating decimal 0.44444. . . . It can be thought of as the sum of the infinite geometric series $0.4 + 0.04 + 0.004 + \cdots$. If $S = 0.44444\ldots$, then $10S = 4.44444\ldots$. Subtract $10S - S$ to eliminate the decimal portion:

$$\begin{aligned} 10S &= 4.44444\ldots \\ -S &= 0.44444\ldots \\ \hline 9S &= 4 \end{aligned}$$

Solving for S gives $S = \frac{4}{9}$, so the series $0.4 + 0.04 + 0.004 + \cdots$ has sum $\frac{4}{9}$.

Step 1 Consider the sequence 0.4, 0.04, 0.004, . . . underlying the series S. Identify the first term, u_1, and the common ratio, r, of the sequence. How is the multiplier 10 (in the expression $10S$) derived from r? What other multipliers could be used to eliminate the repeating decimal portion?

Step 2 Use the common ratio, r, as the multiplier instead of 10 and solve for S again. Is your answer equivalent to $\frac{4}{9}$?

Step 3 Consider the sequence 0.9, 0.09, 0.009, ... Identify the first term, u_1, and the common ratio, r. Now use the method from Step 2 to find the sum S of the series $0.9 + 0.09 + 0.009 + \cdots$.

Step 4 Repeat Step 3 for the sequence 0.27, 0.0027, 0.000027, $\cdots$. Remember to use r as the multiplier.

Step 5 Repeat Step 3 for the series $u_1 + r \cdot u_1 + r^2 \cdot u_1 + r^3 \cdot u_1 + \cdots$, assuming that it has sum S. Create a new series with sum $r \cdot S$. Then subtract to find a formula for S based on u_1 and r.

Step 6 Use a variety of r-values, including both positive and negative numbers, to create several geometric sequences. Look at the partial sums of each sequence as n gets very large. Use your formula from Step 5 to help you describe when the partial sums of a geometric sequence will converge to a unique number S. Use your examples to justify your answer.

Infinity by American sculptor José de Rivera (1904–1985)

In the investigation you used partial sums with large values of n to determine the long-run value of the sum of infinitely many terms. A table or graph of the sequence of partial sums is another tool you can use.

EXAMPLE B

Consider an ideal (frictionless) ball bouncing after it is dropped. The distances in inches that the ball falls on each bounce are represented by 200, 200(0.8), $200(0.8)^2$, $200(0.8)^3$, and so on. Summing these distances creates a series. Find the total distance the ball falls during infinitely many bounces.

▸ Solution

The sequence of partial sums is represented by the recursive formula

$$S_1 = 200$$
$$S_n = S_{n-1} + u_n \quad \text{where } n \geq 2$$

The explicit formula for the sequence of terms is $u_n = 200(0.8)^{n-1}$. So the recursive formula for the series is equivalent to

$$S_1 = 200$$
$$S_n = S_{n-1} + 200(0.8)^{n-1} \quad \text{where } n \geq 2$$

The graph of S_n levels off as the number of bounces increases. This means the total sum of all the distances continues to grow, but seems to approach a long-run value for larger values of n.

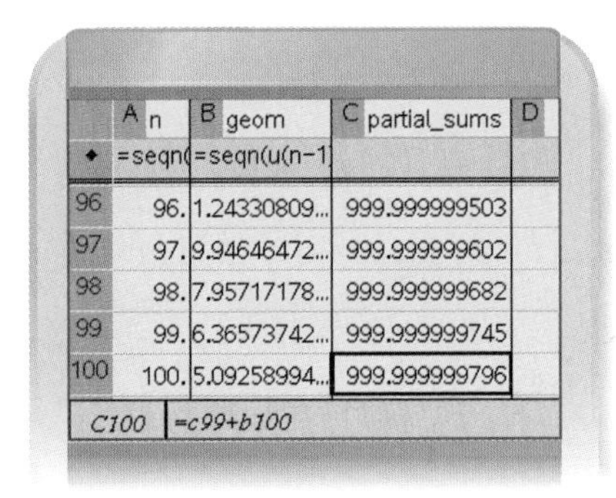

	A n	B geom	C partial_sums	D
◆	=seqn(	=seqn(u(n−1		
96	96.	1.24330809...	999.999999503	
97	97.	9.94646472...	999.999999602	
98	98.	7.95717178...	999.999999682	
99	99.	6.36573742...	999.999999745	
100	100.	5.09258994...	999.999999796	
C100	=c99+b100			

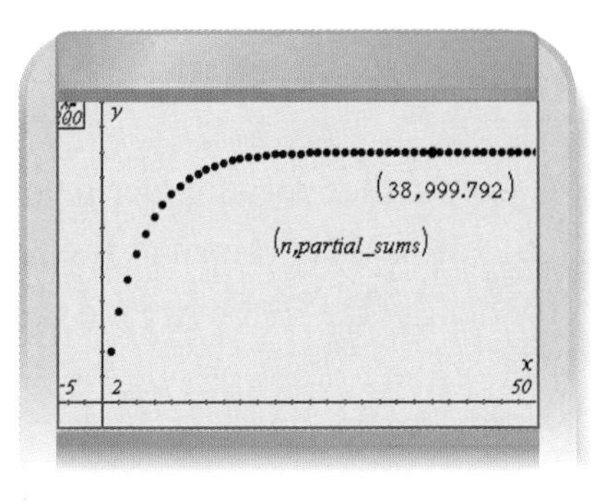

By looking at larger and larger values of n, you'll find the sum of this series is 1000 inches. So the sum of the distances the ball falls is 1000 inches. Use the formula that you found in the investigation to verify this answer.

You now have several ways to determine the long-run value of an infinite geometric series. If the series is convergent, the formula that you found in the investigation gives you the sum of infinitely many terms.

Convergent Infinite Geometric Series

An infinite geometric series is a convergent series if the absolute value of the common ratio is less than 1, $|r| < 1$. The sum of infinitely many terms, S, of a convergent infinite geometric series is given by the explicit formula

$$S = \frac{u_1}{1 - r}$$

where u_1 is the first term and r is the common ratio ($|r| < 1$).

In mathematics the symbol ∞ is used to represent infinity, or a quantity without bound. You can use ∞ and sigma notation to represent infinite series.

EXAMPLE C

Consider the infinite series

$$\sum_{n=1}^{\infty} 0.3(0.1)^{n-1}$$

a. Express this sum of infinitely many terms as a decimal.

b. Identify the first term, u_1, and the common ratio, r.

c. Express the sum as a ratio of integers.

▶ Solution

When you substitute $n = \{1, 2, 3, \ldots\}$ into the expression $0.3(0.1)^{n-1}$, you get

$$0.3 + 0.03 + 0.003 + \cdots$$

a. The sum is the repeating decimal 0.333..., or $0.\overline{3}$.

b. $u_1 = 0.3$ and $r = 0.1$

c. Use the formula for the infinite sum and reduce to a ratio of integers:

$$S = \frac{0.3}{1 - 0.1} = \frac{0.3}{0.9} = \frac{1}{3}$$

You'll work further with infinite geometric series and their sums in the exercises.

EXERCISES

You will need

A graphing calculator for Exercise **7**.

Practice Your Skills

1. Consider the repeating decimal 0.111..., or $0.\overline{1}$.

a. Express this decimal as the sum of terms of an infinite geometric series.

b. Identify the first term and the ratio.

c. Use the formula you learned in this lesson to express the sum as a ratio of integers.

2. Repeat Exercise 1 with the repeating decimal 0.474747..., or $0.\overline{47}$. ⓐ

3. Repeat Exercise 1 with the repeating decimal 0.123123123..., or $0.\overline{123}$.

4. An infinite geometric sequence has first term 20 and sum 400. What are the first five terms? ⓗ

Reason and Apply

5. An infinite geometric sequence contains the consecutive terms 128, 32, 8, and 2. The sum of the series is $43{,}690.\overline{6}$. What is the first term? ⓐ

6. Consider the sequence $u_1 = 47$ and $u_n = 0.8u_{n-1}$ where $n \geq 2$. Find

a. $\sum_{n=1}^{10} u_n$ ⓐ **b.** $\sum_{n=1}^{20} u_n$ **c.** $\sum_{n=1}^{30} u_n$ **d.** $\sum_{n=1}^{\infty} u_n$

7. Consider the sequence $u_n = 96(0.25)^{n-1}$.

a. List the first ten terms, u_1 to u_{10}. ⓐ

b. Find the sum $\sum_{n=1}^{10} 96(0.25)^{n-1}$.

c. Make a graph of partial sums for $1 \leq n \leq 10$.

d. Find the sum $\sum_{n=1}^{\infty} 96(0.25)^{n-1}$.

This digitally manipulated photo, depicting a never-ending roller coaster, is titled *Infinite Fun.*

8. A ball is dropped from an initial height of 100 cm. The rebound heights to the nearest centimeter are 80, 64, 51, 41, and so on. What is the total distance the ball will travel, both up and down? ⓗ

9. APPLICATION A sporting event is to be held at a stadium. Suppose 50,000 visitors attend the event and spend \$500 each. In the month after the event, the local businesses spend 60% of the income from the visitors. The next month, 60% is spent again, and so on.

a. What is the initial amount the visitors spent?

b. In the long run, how much money does this sporting event seem to add to the local economy? ⓐ

c. The ratio of the long-run amount to the initial amount is called the *economic multiplier.* What is the economic multiplier in this example?

d. If the initial amount spent by visitors is \$10,000,000 and the economic multiplier is 1.8, what percentage of the initial amount is spent again and again in the local economy?

10. A flea jumps $\frac{1}{2}$ ft right, then $\frac{1}{4}$ ft left, then $\frac{1}{8}$ ft right, and so on. To what point is the flea zooming in? ⓗ

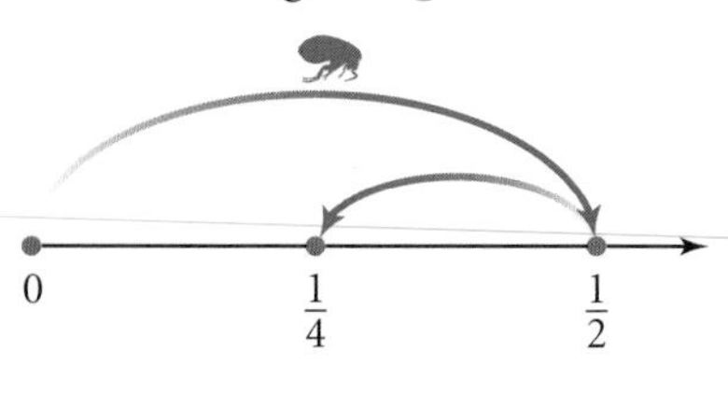

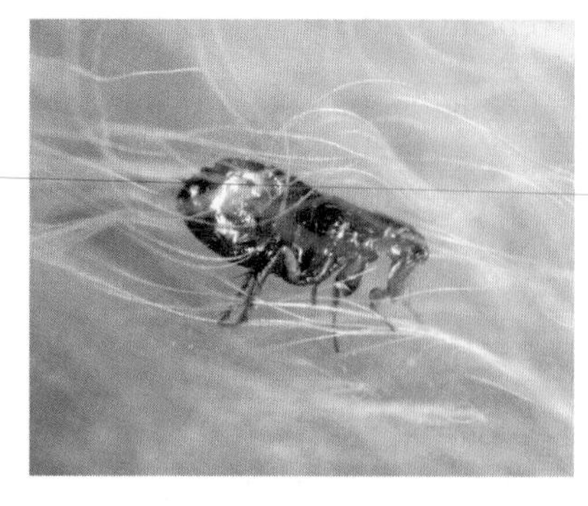

Magnified view of a flea in a dog's fur

11. Suppose square $ABCD$ with side length 8 in. is cut out of paper. Another square, $EFGH$, is placed with its corners at the midpoints of $ABCD$. A third square is placed with its corners at midpoints of $EFGH$, and so on.

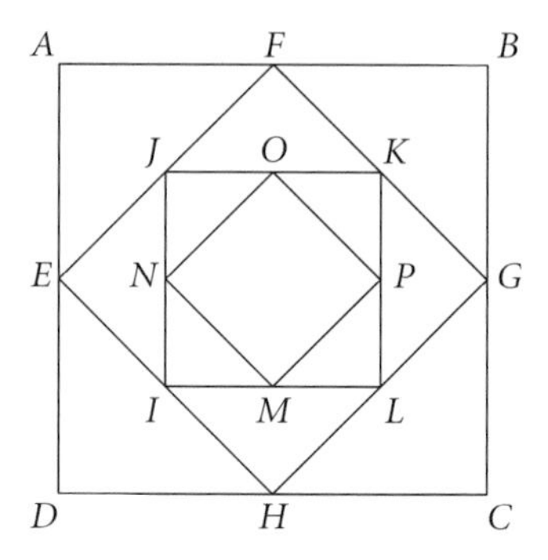

a. What is the perimeter of the tenth square? ⓐ

b. What is the area of the tenth square? ⓐ

c. If the pattern were repeated infinitely many times, what would be the sum of the perimeters of the squares?

d. What would be the sum of the areas?

12. The fractal known as the Sierpiński triangle begins as an equilateral triangle with side length 1 unit and area $\frac{\sqrt{3}}{4}$ square units. The fractal is created recursively by replacing the triangle with three smaller congruent equilateral triangles such that each smaller triangle shares a vertex with the larger triangle. This removes the area from the middle of the original triangle.

Stage 0

Stage 1

Stage 2

. . .

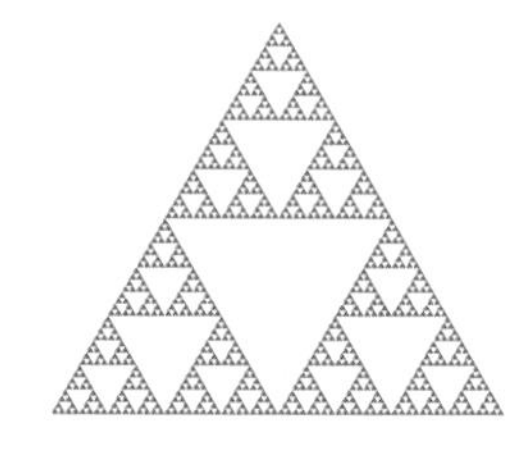

Stage 6

In the long run, what happens to

a. The perimeter of each of the smaller shaded triangles? ⓗ

b. The area of each of the smaller shaded triangles? ⓗ

c. The sum of the perimeters of the smaller shaded triangles? (*Hint:* You can't use the sum formula from this lesson.)

d. The sum of the areas of the smaller shaded triangles?

These Swedish stamps show the Koch snowflake, another fractal design that begins with an equilateral triangle. Can you determine the recursive procedure that creates the snowflake?

Review

13. Oil is being pumped at a rate of 4.2 gal/min from a large barrel of oil. If the barrel contains 12.4 gal of oil 18 min after the pumping starts, how many gallons of oil were in the barrel initially?

14. Match each equation to a graph.

A. $y = 10(0.8)^x$ ⓐ

B. $y = 10 - 10(0.75)^x$

C. $y = 3 + 7(0.7)^x$

D. $y = 10 - 7(0.65)^x$

i.

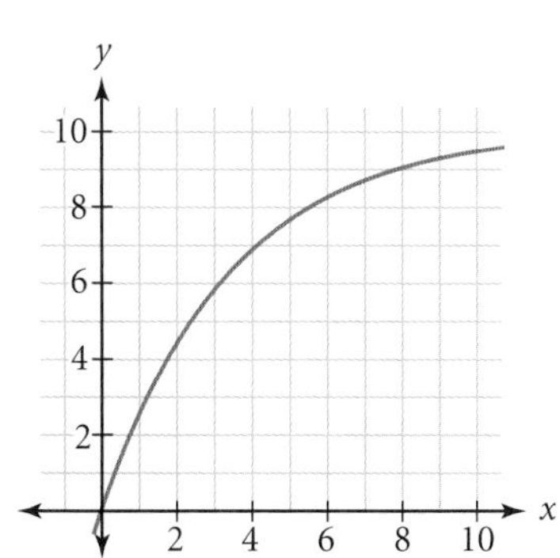

ii.

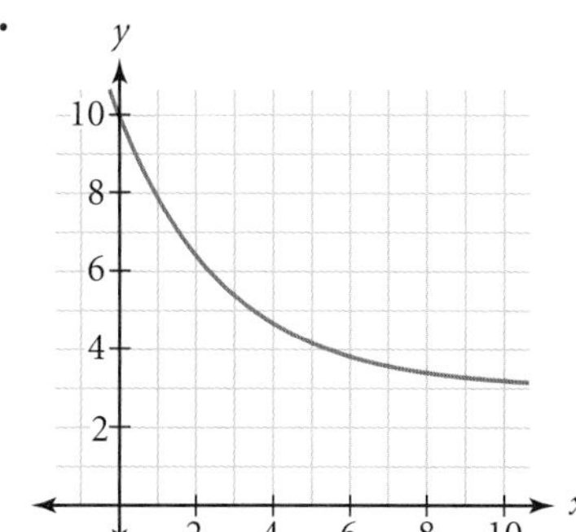

iii.

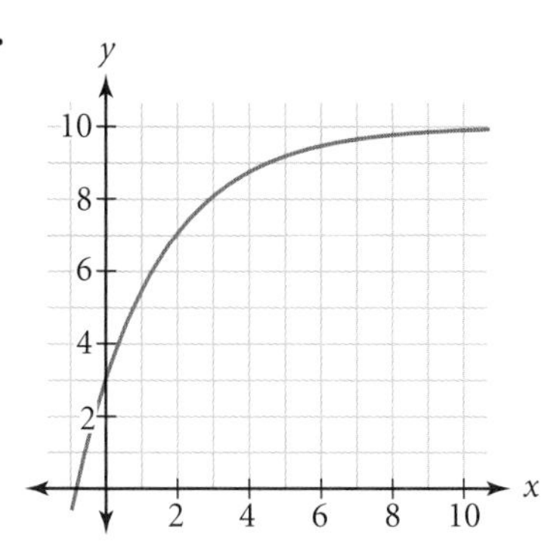

iv.

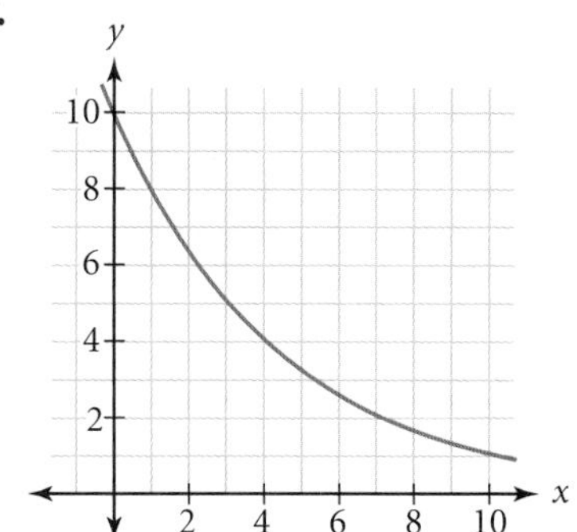

15. APPLICATION A computer software company decides to budget \$100,000 to develop a new video game. It estimates that development will cost \$955 the first week and that expenses will increase by \$65 each week.

a. After 25 weeks, how much of the development budget will be left? ⓐ

b. How long can the company keep the development phase going before the budget will not support another week of expenses?

16. Hans sees a dog. The dog has four puppies. Four cats follow each puppy. Each cat has four kittens. Four mice follow each kitten. How many legs does Hans see, if he sees all the animals? Express your answer using sigma notation.

IMPROVING YOUR VISUAL THINKING SKILLS

Toothpicks

You have eight toothpicks—four are short, and four are long. Each of the short toothpicks is one-half the length of any of the long toothpicks. Arrange the toothpicks to make exactly three congruent squares.

LESSON

9.3

Partial Sums of Geometric Series

If a pair of calculators can be linked and a file transferred from one calculator to the other in 20 s, how long will it be before everyone in a lecture hall of 250 students has the file? During the first time period, the file is transferred to one calculator; during the second time period, to two calculators; during the third time period, to four more calculators; and so on. The number of students who have the file doubles every 20 s.

To solve this problem, you must determine the minimum value of n for which S_n exceeds 250. The solution uses a partial sum of a geometric series. It requires the sum of a finite number of terms of the geometric sequence 1, 2, 4, 8, ...

EXAMPLE A

Consider the sequence 2, 6, 18, 54, ...

a. Find u_{15}.

b. Graph the partial sums S_1 through S_{15}, and find the partial sum S_{15}.

▶ ***Solution***

The sequence is geometric with $u_1 = 2$ and $r = 3$.

a. A recursive formula for the sequence is $u_1 = 2$ and $u_n = 3u_{n-1}$ where $n \geq 2$. The sequence can also be defined explicitly as $u_n = 2(3)^{n-1}$. Substituting 15 for n into either equation gives $u_{15} = 9{,}565{,}938$.

b. Use your calculator to graph the partial sums. Use a table or trace the graph to find that S_{15} is 14,348,906.

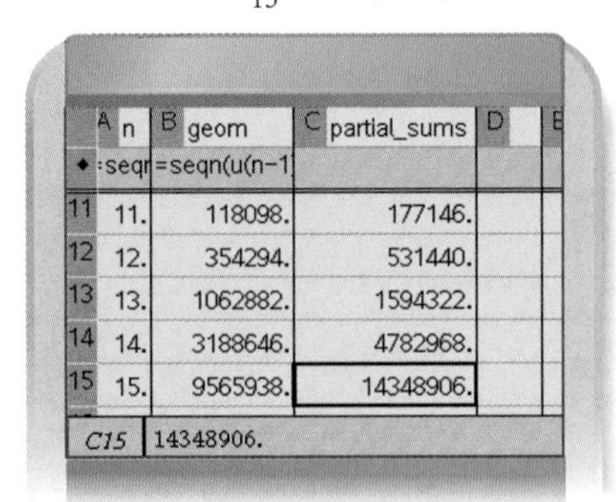

	A n	B geom	C partial_sums	D
◆	:seq	=seqn(u(n−1		
11	11.	118098.	177146.	
12	12.	354294.	531440.	
13	13.	1062882.	1594322.	
14	14.	3188646.	4782968.	
15	15.	9565938.	14348906.	

C15 14348906.

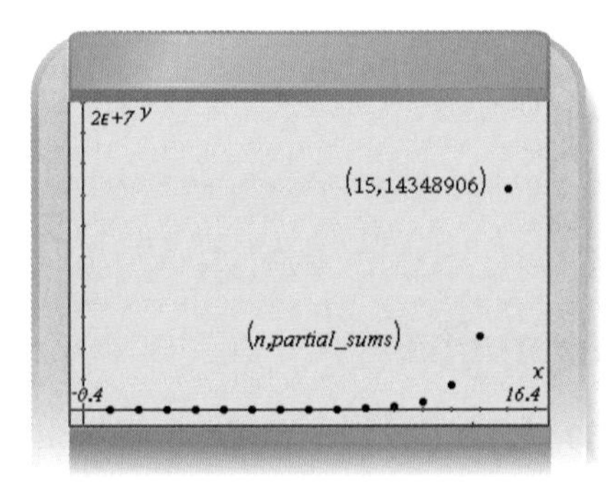

In the example you used a recursive method to find the partial sum of a geometric series. For some partial sums, especially those involving a large number of terms, it can be faster and easier to use an explicit formula. The investigation will help you develop an explicit formula.

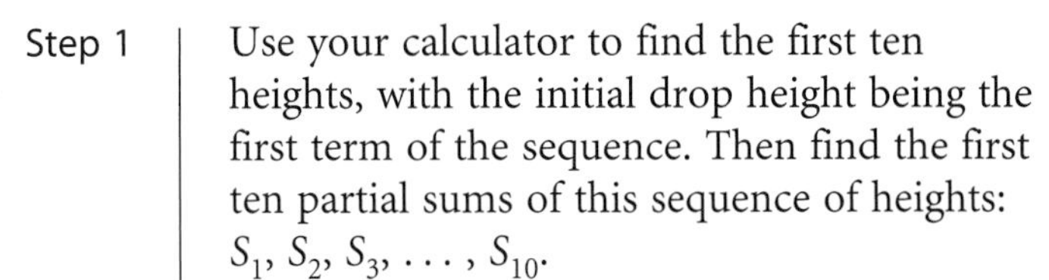

Investigation
Geometric Series Formula

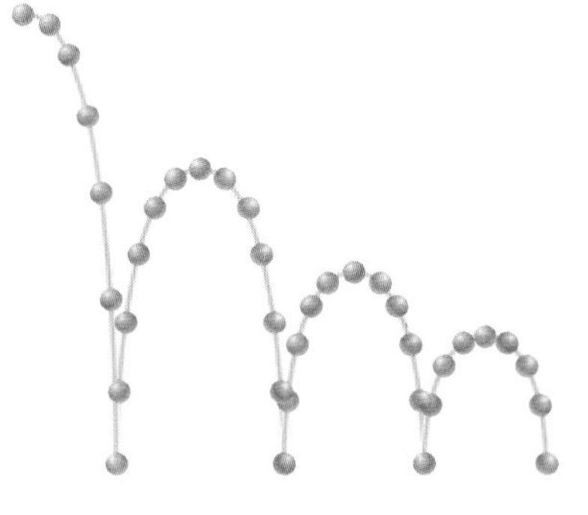

A ball is dropped from 180 cm above the floor. With each bounce, it rebounds to 65% of its previous height.

Step 1 Use your calculator to find the first ten heights, with the initial drop height being the first term of the sequence. Then find the first ten partial sums of this sequence of heights: $S_1, S_2, S_3, \ldots, S_{10}$.

Step 2 Create a scatter plot of points (n, S_n) and find a translated exponential equation to fit the data. The equation will be in the form $S_n = L - a \cdot b^n$. (*Hint:* L is the long-run value of the partial sums.)

Step 3 Rewrite your equation from Step 2 in terms of n, a or u_1, and r. Use algebraic techniques to write your explicit formula as a single rational expression.

Step 4 Here is a way to arrive at an equivalent formula. Complete the derivation by filling in the blanks.

$$S_n = u_1 + u_1 \cdot r + u_1 \cdot r^2 + \cdots + u_1 \cdot r^{n-1}$$

$$rS_n = \underline{\quad ? \quad}$$

$$S_n - rS_n = \underline{\quad ? \quad}$$

$$S_n = \underline{\quad ? \quad}$$

Use algebraic techniques to verify that the formula is equivalent to your formula from Step 3.

Step 5 Use your formula from Step 4 to find S_{10} for the bouncing ball. Then use it to find S_{10} for the geometric sequence 2, 6, 18, 54, ...

In the investigation you found an explicit formula for a partial sum of a geometric series that uses only three pieces of information—the first term, the common ratio, and the number of terms. Now you do not need to write out the terms to find a sum.

EXAMPLE B Find S_{10} for the series $16 + 24 + 36 + \cdots$.

▶ Solution The first term, u_1, is 16. The common ratio, r, is 1.5. The number of terms, n, is 10. Use the formula you developed in the investigation to calculate S_{10}.

$$S_{10} = \frac{16(1 - 1.5^{10})}{1 - 1.5} = 1813.28125$$

EXAMPLE C Each day, the imaginary caterpillarsaurus eats 25% more leaves than it did the day before. If a 30-day-old caterpillarsaurus has eaten 151,677 leaves in its brief lifetime, how many will it eat the next day?

▶ ***Solution*** To solve this problem, you must find u_{31}. The information in the problem tells you that r is $(1 + 0.25)$, or 1.25, and when n equals 30, S_n equals 151,677. Substitute these values into the formula for S_n and solve for the unknown value, u_1.

$$\frac{u_1(1 - 1.25^{30})}{1 - 1.25} = 151{,}677$$

$$u_1 = 151{,}677 \cdot \frac{1 - 1.25}{1 - 1.25^{30}}$$

$$u_1 \approx 47$$

Now you can write an explicit formula for the terms of the geometric sequence, $u_n = 47(1.25)^{n-1}$. Substitute 31 for n to find that on the 31st day, the caterpillarsaurus will eat 37,966 leaves.

The explicit formula for the sum of a geometric series can be written in several ways, but they are all equivalent. You probably found these two ways during the investigation:

Partial Sum of a Geometric Series

A partial sum of a geometric series is given by the explicit formula

$$S_n = \left(\frac{u_1}{1 - r}\right) - \left(\frac{u_1}{1 - r}\right) r^n \quad \text{or} \quad S_n = \frac{u_1(1 - r^n)}{1 - r}$$

where n is the number of terms, u_1 is the first term, and r is the common ratio $(r \neq 1)$.

EXERCISES

You will need

A graphing calculator for Exercises **4**, **5**, **12**, and **16**.

Practice Your Skills

1. For each partial sum equation, identify the first term, the ratio, and the number of terms.

a. $\frac{12}{1 - 0.4} - \frac{12}{1 - 0.4}(0.4)^8 \approx 19.9869$ ⓐ

b. $\frac{75(1 - 1.2^{15})}{1 - 1.2} \approx 5402.633$ ⓐ

c. $\frac{40 - 0.46117}{1 - 0.8} \approx 197.69$

d. $-40 + 40(2.5)^6 = 9725.625$

2. Consider the geometric sequence

256, 192, 144, 108, . . .

a. What is the eighth term?

b. Which term is the first one smaller than 20? @

c. Find u_7.

d. Find S_7. @

3. Find each partial sum of this sequence.

$u_1 = 40$
$u_n = 0.6u_{n-1}$ where $n \geq 2$

a. S_5

b. S_{15}

c. S_{25}

4. Identify the first term and the common ratio or common difference of each series. Then find the partial sum.

a. $3.2 + 4.25 + 5.3 + 6.35 + 7.4$

b. $3.2 + 4.8 + 7.2 + \cdots + 36.45$ @

c. $\sum_{n=1}^{27} (3.2 + 2.5n)$ @

d. $\sum_{n=1}^{10} 3.2(4)^{n-1}$

Reason and Apply

5. Find the missing value in each set of numbers.

a. $u_1 = 3$, $r = 2$, $S_{10} = \underline{?}$ @

b. $u_1 = 4$, $r = 0.6$, $S_{\underline{?}} \approx 9.999868378$ @

c. $u_1 = \underline{?}$, $r = 1.4$, $S_{15} \approx 1081.976669$

d. $u_1 = 5.5$, $r = \underline{?}$, $S_{18} \approx 66.30642497$

6. Find the nearest integer value for n if $\frac{3.2(1 - 0.8^n)}{1 - 0.8}$ is approximately 15.

7. Consider the sequence $u_1 = 8$ and $u_n = 0.5u_{n-1}$ where $n \geq 2$. Find

a. $\sum_{n=1}^{10} u_n$ @

b. $\sum_{n=1}^{20} u_n$

c. $\sum_{n=1}^{30} u_n$

d. Explain what is happening to these partial sums as you add more terms.

8. Suppose you begin a job with an annual salary of \$17,500. Each year, you can expect a 4.2% raise.

a. What is your salary in the tenth year after you start the job? @

b. What is the total amount you earn in ten years?

c. How long must you work at this job before your total earnings exceed \$1 million?

9. Consider the geometric series

$5 + 10 + 20 + 40 + \cdots$

a. Find the first seven partial sums, $S_1, S_2, S_3, \ldots, S_7$.

b. Do the partial sums create a geometric sequence?

c. If u_1 is 5, find value(s) of r such that the partial sums form a geometric sequence.

10. List terms to find

a. $\sum_{n=1}^{7} n^2$

b. $\sum_{n=3}^{7} n^2$

11. An Indian folktale, recounted by Arab historian and geographer Ahmad al-Yaqubi in the 9th century, begins, "It is related by the wise men of India that when Husiya, the daughter of Balhait, was queen . . . ," and goes on to tell how the game of chess was invented. The queen was so delighted with the game that she told the inventor, "Ask what you will." The inventor asked for one grain of wheat on the first square of the chessboard, two grains on the second, four grains on the third, and so on, so that each square contained twice the number of grains as on the square before. (There are 64 squares on a chessboard.)

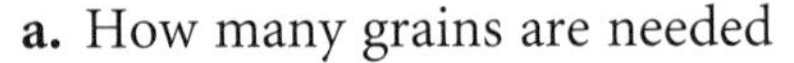

Sonfonisba Anguissola's (ca. 1531–1625) painting, titled *The Chess Game* (1555), includes a self-portrait of the Italian artist (far left).

a. How many grains are needed

i. For the 8th square?

ii. For the 64th square?

iii. For the first row?

iv. To fill the board?

b. In sigma notation, write the series you used to fill the board.

12. As a contest winner, you are given the choice of two prizes. The first choice awards \$1,000 the first hour, \$2,000 the second hour, \$3,000 the third hour, and so on. For one entire year, you will be given \$1,000 more each hour than you were given during the previous hour. The second choice awards 1¢ the first week, 2¢ the second week, 4¢ the third week, and so on. For one entire year, you will be given double the amount you received during the previous week. Which of the two prizes will be more profitable, and by how much? ⓗ

13. Consider the series

$$\sum_{n=1}^{8} \frac{1}{n} = \frac{1}{1} + \frac{1}{2} + \frac{1}{3} + \frac{1}{4} + \cdots + \frac{1}{8}$$ ⓐ

a. Is this series arithmetic, geometric, or neither?

b. Find the sum of this series.

14. ***Mini-Investigation*** The diagram shows a sequence of layered blue and white squares.

a. Explain how the areas of the squares are related to the series $1 + 3 + 5 + \cdots$.

b. Show two different ways to find the area of the largest square. (One should involve a series.)

c. The nth odd number can be represented by the expression $2n - 1$ for values of $n \geq 1$. Use the formula $1 + 3 + 5 + \cdots + (2n - 1) = n^2$ to find the sum of the first 100 odd integers.

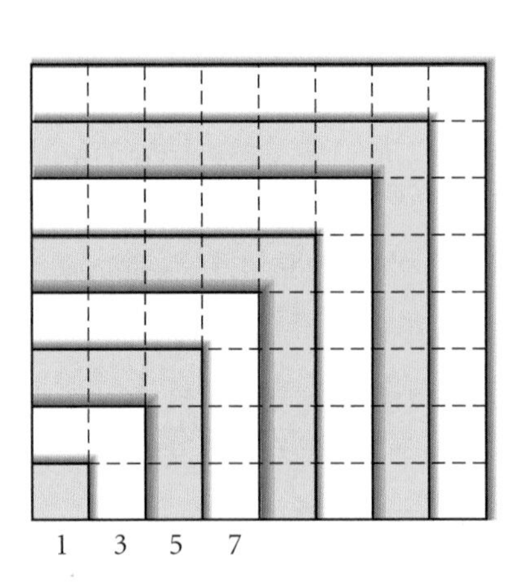

d. Is the formula from 14c true for *all* $n \geq 1$? You can use a technique called **mathematical induction** to prove that it is true.

To prove a statement is true for all natural numbers n,

i. Show the statement is true for $n = 1$.

ii. Show that if the statement is true for some natural number $n = k$, then it must also be true for the next natural number, $n = k + 1$.

Now apply these steps to the formula you found for the sum of odd integers.

i. Show that the formula, $1 + 3 + 5 + \cdots + (2n - 1) = n^2$, is true for $n = 1$.

ii. Assume the formula is true for all natural numbers up to k, so that $1 + 3 + 5 + \cdots + (2k - 1) = k^2$. Now consider the case $n = k + 1$. What is the next odd integer after $2k - 1$? Add it to both sides of the previous equation. Explain why the right side of your equation is equivalent to $(k + 1)^2$.

15. The 32 members of the Greeley High chess team are going to have a tournament. They need to decide whether to have a round-robin tournament or an elimination tournament. (Read the Recreation Connection.)

a. If the tournament is round-robin, how many games need to be scheduled? (h)

b. If it is an elimination tournament, how many games need to be scheduled?

Recreation **CONNECTION**

In setting up tournaments, organizers have to decide the type of play. Most intramural sports programs are set up in "round-robin" format, in which every player or team plays every other player or team. Scheduling is different for odd and even numbers of teams, and it can be tricky if there is to be a minimum number of rounds. Another method is the elimination tournament, in which teams or players are paired and only the winners progress to the next round. For this format, scheduling difficulties arise when the initial number of teams is not a power of 2.

Review

16. What monthly payment is required to pay off an \$80,000 mortgage at 8.9% interest in 30 years? (a)

17. The Magic Garden Seed Catalog advertises a bean with unlimited growth. It guarantees that with proper watering, the bean will grow 6 in. the first week and the height increase each subsequent week will be three-fourths of the previous week's height increase. "Pretty soon," the catalog claims, "Your beanstalk will touch the clouds!" Is this misleading advertising?

18. Write the polynomial equation of least degree that has integer coefficients and zeros $-3 + 2i$ and $\frac{2}{3}$. (a)

EXPLORATION

Seeing the Sum of a Series

The sum of an infinite geometric sequence is sometimes hard to visualize. Some sums clearly converge, whereas other sums do not. In this exploration you will use The Geometer's Sketchpad to simulate the sum of an infinite geometric series.

Activity

A Geometric Series

Step 1 Start Sketchpad. In a new sketch, construct horizontal segment $\overline{AB}$. Follow the Procedure Note to extend this segment by 60%.

Procedure Note

1. Mark center B.
2. Dilate point A by $\frac{-6}{10}$ about center B. Label the new point A'. Construct $\overline{BA'}$.
3. Construct lines perpendicular to $\overline{AA'}$ through point B and A'. These will help you see how the series develops.

Step 2 Measure the distance between A and B.

Step 3 Construct point C, not on $\overline{AA'}$. Measure the distance between C and B.

Step 4 Select point A, then point B. Choose **Iterate** from the Transform menu. Map point A to point B, and point B to point A'. When you iterate, you'll get a table of values for AB and CB. What do the values for AB represent? If you move point C to coincide with point A, what do the values for CB represent?

AB = 3.00 cm
CB = 3.00 cm

n	AB	CB
0	3.00 cm	3.00 cm
1	1.80 cm	4.80 cm
2	1.08 cm	5.88 cm
3	0.65 cm	6.53 cm

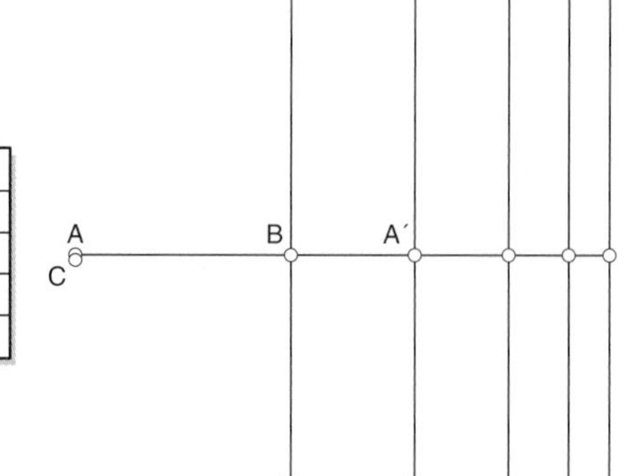

Step 5 Choose **Select All** from the Edit menu. Use the "+" key on your keyboard to add iterations to your sketch. Do approximately 25 iterations. Describe what happens as you add iterations.

Questions

1. What geometric series does your model represent? Write your answer in sigma notation.
2. According to the table values, what is the sum of your infinite series? Use the explicit formula $S = \frac{u_1}{1 - r}$ to confirm this sum.
3. Vary the length of $\overline{AB}$ by dragging point B. What conclusions can you draw? Does the sum always converge for this common ratio?
4. Repeat the activity, but this time extend the segment by 120% (dilate point A by a factor of $\frac{-12}{10}$ about B). What happens with this series model?
5. Consider any geometric series in the form $\sum_{n=1}^{\infty} ar^{n-1}$. Explain how to model the series with Sketchpad, using the values a and r.

keymath.com/DAA

[▸ You can explore a different geometric model of an infinite geometric series using the **Dynamic Algebra Exploration** at www.keymath.com/DAA .◂]

IMPROVING YOUR REASONING SKILLS

Planning for the Future

Pamela and Candice are identical 20-year-old twins with identical jobs and identical salaries, and they receive identical bonuses of $2,000 yearly.

Pamela is immediately concerned with saving money for her retirement. She invests her $2,000 bonus each year at an interest rate of 9% compounded annually. At age 30, when she receives her tenth bonus, she decides it is time to see the world, and from that point on she spends her annual bonus on a trip.

Candice is immediately concerned with enjoying her income while she is young. She spends her $2,000 bonus every year until she reaches 30. On her 30th birthday, when she receives her tenth bonus, she starts to worry about what will happen when she retires, so she starts saving her bonus money at an interest rate of 9% compounded annually.

How much will each twin have in her retirement account when she is 30 years old? Compare the value of each investment account when Pamela and Candice are 65 years old.

CHAPTER 9 REVIEW

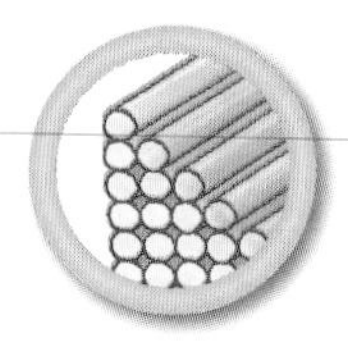

A **series** is the indicated sum of terms of a sequence, defined recursively with the rule $S_1 = u_1$ and $S_n = S_{n-1} + u_n$ where $n \geq 2$. Series can also be defined explicitly. With an explicit formula, you can find any **partial sum** of a sequence without knowing the preceding term(s). An explicit formula for **arithmetic series** is

$$S_n = \frac{n(u_1 + u_n)}{2}$$

Explicit formulas for **geometric series** are

$$S_n = \left(\frac{u_1}{1-r}\right) - \left(\frac{u_1}{1-r}\right)r^n \quad \text{and} \quad S_n = \frac{u_1(1-r^n)}{1-r}$$

where $r \neq 1$. If a geometric series is **convergent,** then you can calculate the sum of infinitely many terms with the formula $S = \frac{u_1}{1-r}$, where $|r| < 1$.

EXERCISES

You will need

A graphing calculator for Exercises **6, 7,** and **9.**

@ Answers are provided for all exercises in this set.

1. Consider the arithmetic sequence: 3, 7, 11, 15, . . .

a. What is the 128th term?

b. Which term has the value 159?

c. Find u_{20}.

d. Find S_{20}.

2. Consider the geometric sequence: 100, 84, 70.56, . . .

a. Which term is the first one smaller than 20?

b. Find the sum of all the terms that are greater than 20.

c. Find the value of $\sum_{n=1}^{20} u_n$.

d. What happens to S_n as n gets very large?

3. The population of a particular bug species can be modeled with a geometric sequence. When given plenty of food and water, each female hatches 24 young at age 5 days. Assume that half the newborn bugs are male and half female, and are ready to reproduce in five days. Also assume each female bug can reproduce only once. Initially, there are 12 bugs, half male and half female.

a. How many bugs are born on the 5th day? On the 10th day? On the 15th day? On the 35th day?

b. Write a recursive formula for the sequence in 3a.

c. Write an explicit formula for the sequence in 3a.

d. Find the total number of bugs after 60 days.

A swarm of ladybugs

4. Consider the series

$125.3 + 118.5 + 111.7 + 104.9 + \cdots$

a. Find S_{67}.

b. Write an expression for S_{67} using sigma notation.

5. Emma's golf ball lies 12 ft from the last hole on the golf course. She putts and, unfortunately, the ball rolls to the other side of the hole, $\frac{2}{3}$ as far away as it was before. On her next putt, the same thing happens.

a. If this pattern continues, how far will her ball travel in seven putts?

b. How far will the ball travel in the long run?

6. A flea jumps $\frac{1}{2}$ ft, then $\frac{1}{4}$ ft, then $\frac{1}{8}$ ft, and so on. It always jumps to the right.

a. Do the jump lengths form an arithmetic or geometric sequence? What is the common difference or common ratio?

b. How long is the flea's eighth jump, and how far is the flea from its starting point?

c. How long is the flea's 20th jump? Where is it after 20 jumps?

d. Write explicit formulas for jump length and the flea's location for any jump.

e. To what point is the flea zooming in?

7. For 7a–c, use $u_1 = 4$. Round your answers to the nearest thousandth.

a. For a geometric series with $r = 0.7$, find S_{10} and S_{40}.

b. For $r = 1.3$, find S_{10} and S_{40}.

c. For $r = 1$, find S_{10} and S_{40}.

d. Graph the partial sums of the series in 7a–c.

e. For which value of r (0.7, 1.3, or 1) do you have a convergent series?

8. Consider the series

$$0.8 + 0.08 + 0.008 + \cdots$$

a. Find S_{10}.

b. Find S_{15}.

c. Express the sum of infinitely many terms as a ratio of integers.

MIXED REVIEW

9. D'Andre surveyed a randomly chosen group of 15 teachers at his school and asked them how many students were enrolled in their third-period classes. Here is the data set he collected.

{27, 29, 18, 34, 42, 38, 34, 33, 25, 28, 45, 35, 32, 19, 36}

a. List the mean, median, and mode.

b. Make a box plot of the data.

c. Calculate the standard deviation. What does this tell you about the data? If the standard deviation were smaller, what would it tell you about the data?

10. Write an equation of the image of the absolute-value function, $y = |x|$, after performing each of the following transformations in order. Sketch a graph of your final equation.

a. Dilate vertically by a factor of 2.

b. Then translate horizontally 4 units.

c. Then translate vertically -3 units.

11. Earth's orbit is an ellipse with the Sun at one of the foci. Perihelion is the point at which Earth is closest to the Sun, and aphelion is the point at which it is farthest from the Sun. The distances from perihelion to the Sun and from the Sun to aphelion are in an approximate ratio of 59:61. If the total distance from aphelion to perihelion along the major axis is about 186 million miles, approximate

a. The distance from perihelion to the Sun.

b. The distance from aphelion to the Sun.

c. The distance from aphelion to the center.

d. The distance from the center to the Sun.

e. In an ellipse aligned on the axes, as at right, the distance from a focus (the Sun) to point P equals the distance from the intersection of the ellipse with the x-axis to the center. Using this information, find the distance from the center to P.

f. Write an equation that models the orbit of Earth around the Sun.

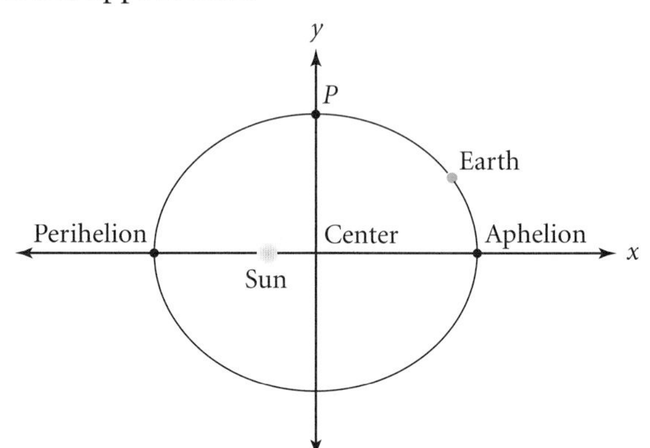

These four circular snow sculptures by British environmental sculptor Andy Goldsworthy (b 1956) were made from bricks of packed snow at the Arctic Circle.

12. Perform the following matrix computations. If a computation is not possible, explain why.

$$[A] = \begin{bmatrix} -2 & 0 \\ 1 & -3 \end{bmatrix} \qquad [B] = \begin{bmatrix} 0 & -2 \\ 5 & 0 \\ 3 & -1 \end{bmatrix} \qquad [C] = \begin{bmatrix} 1 & -1 \\ 0 & 2 \end{bmatrix}$$

a. $[A][B]$ **b.** $[A] + [B]$ **c.** $[A] - [C]$ **d.** $[B][A]$ **e.** $3[C] - [A]$

13. Rewrite each equation in standard form, and identify the type of curve.

a. $25x^2 - 4y^2 + 100 = 0$
b. $4y^2 - 10x + 16y + 36 = 0$
c. $4x^2 + 4y^2 + 24x - 8y + 39 = 0$
d. $3x^2 + 5y^2 - 12x + 20y + 8 = 0$

14. The towers of a parabolic suspension bridge are 400 m apart and reach 50 m above the suspended roadway. The cable is 4 m above the roadway at the halfway point. Write an equation that models the shape of the cable. Assume the origin, (0, 0), is located at the halfway point of the roadway.

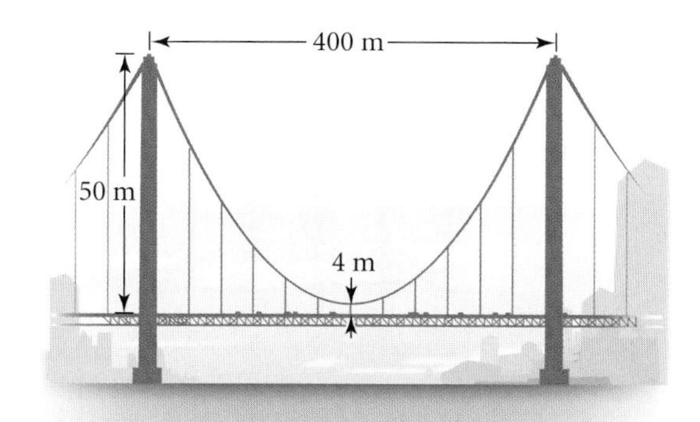

15. Solve algebraically. Round answers to the nearest hundredth.

a. $4 + 5^x = 18$
b. $12(0.5)^{2x} = 30$
c. $\log_3 15 = \dfrac{\log x}{\log 3}$
d. $\log_6 100 = x$
e. $2\log x = 2.5$
f. $\log_5 5^3 = x$
g. $4\log x = \log 16$
h. $\log(5 + x) - \log 5 = 2$
i. $x\log 5^x = 12$

16. Use $P = 1 + 3i$, $Q = -2 + i$, and $R = 3 - 5i$ to evaluate each expression. Give answers in the form $a + bi$.

a. $P + Q - R$
b. PQ
c. Q^2
d. $P \div Q$

TAKE ANOTHER LOOK

1. You know how to write the equation of a continuous function that passes through the discrete points of a sequence, (n, u_n). For example, the function $y = 200(0.8)^{x-1}$ passes through the sequence of points representing the distance in inches that a ball falls on each bounce.

Write a continuous function that passes through the points representing the *total* distance the ball has fallen on each bounce, (n, S_n). How can you use the function to find any partial sum or the sum of infinitely many terms? In general, what continuous function passes through the points of a geometric series? Through the points of an arithmetic series?

2. The explicit formula for a partial sum of a geometric series is $S_n = \frac{u_1(1 - r^n)}{1 - r}$. To find the sum of an infinite geometric series, you can imagine substituting ∞ for n. Explain what happens to the expression $\frac{u_1(1 - r^n)}{1 - r}$ when you do this substitution.

3. Since Chapter 1, you have solved problems about monthly payments, such as auto loans and home mortgages. You've learned how to find the monthly payment, P, required to pay off an initial amount, A_0, over n months with a monthly percentage rate, r. With series, you can find an explicit formula to calculate P. The recursive rule $A_n = A_{n-1}(1 + r) - P$ creates a sequence of the unpaid balances after the nth payment. The expanded equations for the unpaid balances are

$$A_1 = A_0(1 + r) - P$$
$$A_2 = A_0(1 + r)^2 - P(1 + r) - P$$
$$A_3 = A_0(1 + r)^3 - P(1 + r)^2 - P(1 + r) - P$$

and so on. Find the expanded equation for the last unpaid balance, A_n. Look at this equation for a partial sum of a geometric series, and use the explicit formula, $S_n = \frac{u_1(1 - r^n)}{1 - r}$, to simplify the equation for A_n. Then, substitute 0 for A_n (because after the last payment, the loan balance should be zero) and solve for P. This gives you an explicit formula for P in terms of A_0, n, and r. Test your explicit formula by solving these problems.

a. What monthly payment is required for a 60-month auto loan of \$11,000 at an annual interest rate of 4.9% compounded monthly? (Answer: \$207.08)

b. What is the maximum home mortgage for which Tina Fetzer can qualify if she can only afford a monthly payment of \$620? Assume the annual interest rate is fixed at 7.5%, compounded monthly, and that the loan term is 30 years. (Answer: \$88,670.93)

Assessing What You've Learned

WRITE TEST ITEMS Write a few test questions that explore series. You may use sequences that are arithmetic, geometric, or perhaps neither. You may want to include problems that use sigma notation. Be sure to include detailed solution methods and answers.

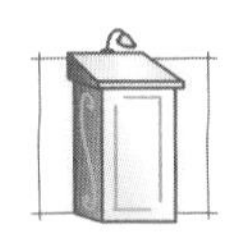

GIVE A PRESENTATION By yourself or with a partner, do a presentation showing how to find the partial sum of an arithmetic or geometric series using both a recursive formula and an explicit formula. Discuss the advantages and disadvantages of each method. Are there series that can be summed using one method, but not the other?

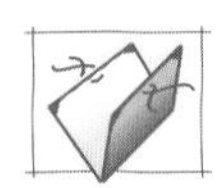

UPDATE YOUR PORTFOLIO Pick one of the three investigations from this chapter to include in your portfolio. Explain in detail the methods that you explored, and how you derived a formula for the partial sum of an arithmetic series, the partial sum of a geometric series, or the sum of an infinite convergent geometric series.

CHAPTER

10 Probability

American artist Carmen Lomas Garza (b 1948) created this color etching, *Lotería–Primera Tabla* (1972), which shows one game card in the traditional Mexican game *lotería.* As in bingo, a caller randomly selects one image that may appear on the game cards and players try to cover an entire row, column, diagonal, or all four corners of their game cards. The chance of winning many games, including *lotería,* can be calculated using probability.

OBJECTIVES

In this chapter you will

- learn about randomness and the definition of probability
- count numbers of possibilities to determine probabilities
- determine expected values of random variables
- discover how numbers of combinations relate to binomial probabilities

Basic Probability

If you flip a coin, the probability that it lands with heads up is $\frac{1}{2}$. If you roll a standard die, the probability that it lands with either the 1 or the 3 up is $\frac{2}{6}$. In each fraction, the denominator is the number of possible **outcomes,** or results, that are equally likely. The numerator is the number of those outcomes that are desirable. In this lesson you will review basic probability to prepare for an in-depth look at probability in the rest of the chapter.

EXAMPLE A If you roll a pair of standard dice, what is the probability that they land with a sum of 5 showing?

▶ Solution You need to determine the number of possible equally likely outcomes and the number of desirable outcomes for this experiment. You might think that there are 11 possible sums of two dice (from 2 through 12), so the probability of a 5 would be $\frac{1}{11}$. However, the 11 sums are not equally likely. One way to see this is to list all of the possible rolls in an organized fashion. In the table you can see there are 36 possible equally likely outcomes. The four pairs that sum to 5 are highlighted. So the probability that the sum of the two dice is 5 is $\frac{4}{36}$, or $\frac{1}{9}$.

Second Die (columns); First Die (rows)

Roll	1	2	3	4	5	6
1	(1, 1)	(1, 2)	(1, 3)	**(1, 4)**	(1, 5)	(1, 6)
2	(2, 1)	(2, 2)	**(2, 3)**	(2, 4)	(2, 5)	(2, 6)
3	(3, 1)	**(3, 2)**	(3, 3)	(3, 4)	(3, 5)	(3, 6)
4	**(4, 1)**	(4, 2)	(4, 3)	(4, 4)	(4, 5)	(4, 6)
5	(5, 1)	(5, 2)	(5, 3)	(5, 4)	(5, 5)	(5, 6)
6	(6, 1)	(6, 2)	(6, 3)	(6, 4)	(6, 5)	(6, 6)

Sometimes there are too many possibilities to list, so you need another way to determine the numbers of outcomes.

EXAMPLE B If you roll three dice, what is the probability that they sum to 5?

▶ Solution In the table for Example A, each of the six numbers on the first die is paired with the six numbers on the second die. So the total number of possible outcomes is $6 \cdot 6$, or 36. If you include a third die, then each of these 36 two-dice outcomes can be matched with each of the six numbers on the third die. So the total number of possible outcomes is $6 \cdot 6 \cdot 6$, or 216. How many of the outcomes sum to 5? Make an organized list:

(1, 1, 3), (1, 2, 2), (1, 3, 1)	The possibilities with a 1 on the first die.
(2, 1, 2), (2, 2, 1)	The possibilities with a 2 on the first die.
(3, 1, 1)	The only possibility with a 3 on the first die.

There are six desirable outcomes, so the probability of three dice summing to 5 is $\frac{6}{216}$, or $\frac{1}{36}$.

Playing cards are often used to demonstrate probability ideas.

EXAMPLE C

If you choose one card at random from a standard deck, what is the probability that the card you choose is from the hearts suit?

▶ Solution

A standard deck of cards contains 52 cards. There are four different suits: clubs, diamonds, hearts, and spades. Each suit contains 13 cards labeled ace, 2, 3, 4, 5, 6, 7, 8, 9, 10, jack, queen, and king. Two suits are red: diamonds and hearts. The other two suits are black: clubs and spades. If you want to get a heart, you're interested in 13 outcomes out of the 52 possible outcomes. So the probability that your card is a heart is $\frac{13}{52}$, or $\frac{1}{4}$.

EXERCISES

1. The spinner shown at right has ten equal-size sectors. Determine the probabilities of these outcomes.

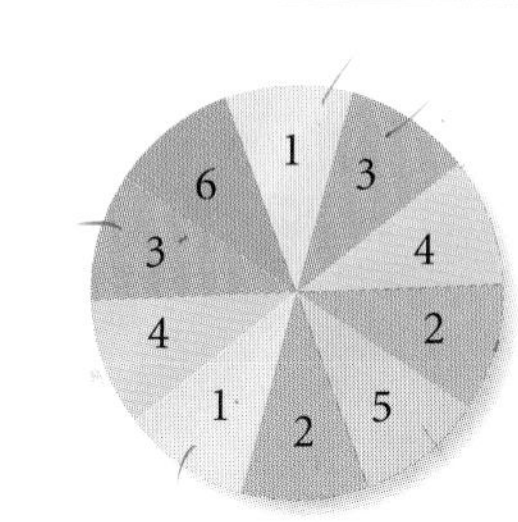

a. You spin an odd number. ⓐ

b. You spin a number less than 4.

c. You spin a number that is a perfect square. ⓐ

2. Consider rolling two fair dice.

a. What is the probability that the dice sum to more than 8? ⓐ

b. What is the most likely sum? What is its probability? ⓐ

c. What is the probability of rolling an even sum?

3. If you draw a card at random from a standard deck of playing cards, what are the probabilities of these outcomes?

a. You draw a red card. ⓐ

b. You draw a 5 or a king. ⓐ

c. You draw a black 7. ⓗ

4. A gumball machine contains five colors of gumballs as indicated in the table. When you insert a quarter, you get one gumball.

Color	Red	Yellow	Blue	Green	White
Number of gumballs	120	115	155	100	110

a. What is the probability that you get a blue gumball?

b. What is the probability that you get a blue or red gumball?

c. What is the probability that you don't get a green gumball?

LESSON 10.1

Randomness and Probability

It's choice—not chance—that determines your destiny.

JEAN NIDETCH

"It isn't fair," complains Noah. "My car insurance rates are much higher than yours." Rita replies, "Well, Noah, that's because insurance companies know the chances are good that it will cost them less to insure me."

How much do you think Demolition Derby drivers pay for auto insurance?

Insurance companies can't know for sure what kind of driving record you will have. So they use the driving records for people of your age group, gender, and prior driving experience to determine your chances of an accident and, therefore, your insurance rates. This is just one example of how probability theory and the concept of randomness affect your life.

Career CONNECTION

An actuary uses mathematics to solve financial problems. Actuaries often work for insurance, consulting, and investment companies. They use probability, statistics, and risk theory to decide the cost of a company's employee benefit plan, the cost of a welfare plan, and how much funding an insurance company will need to pay for expected claims.

Probability theory was developed in the 17th century as a means of determining the fairness of games, and it is still used to make sure that casino customers lose more money than they win. Probability is also important in the study of sociological and natural phenomena.

At the heart of probability theory is randomness. Rolling a die, flipping a coin, drawing a card, and spinning a game-board spinner are examples of **random processes.** In a random process, no individual outcome is predictable, even though the long-range pattern of many individual outcomes often is predictable.

Investigation
Flip a Coin

You will need

- a coin

In this investigation you will explore the predictability of random outcomes. You will use a familiar random process, the flip of a coin.

Step 1 Imagine you are flipping a **fair** coin, one that is equally likely to land heads or tails. Without flipping a coin, record a random arrangement of H's and T's, as though you were flipping a coin ten times. Label this Sequence A.

Step 2 Now flip a coin ten times and record the results on a second line. Label this Sequence B.

Step 3 How is Sequence A different from the result of your coin flips? Make at least two observations.

Step 4 Find the longest string of consecutive H's or T's in Sequence A. Do the same for Sequence B. Then find the second-longest string. Record these lengths for each person in the class as tally marks in a table like the one below. How do the lengths of the longest strings in Sequence A compare with the lengths of the longest strings in Sequence B?

Longest string	Sequence A	Sequence B	2nd-longest string	Sequence A	Sequence B
1			1		
2			2		
3			3		
4			4		
5 or more			5 or more		

Step 5 Count the number of H's in each set. Record the results of the entire class in a table like the one at right. What do you notice about the numbers of H's in Sequence A compared with Sequence B?

Step 6 If you were asked to write a new random sequence of H's and T's, how would it be different from what you recorded in Sequence A?

Number of H's	Sequence A	Sequence B
0		
1		
2		
3		
4		
5		
6		
7		
8		
9		
10		

History CONNECTION

Many games are based on random outcomes, or chance. Paintings and excavated material from Egyptian tombs show that games using *astragali* were established by the time of the First Dynasty, around 3500 B.C.E. An *astragalus* is a small bone in the foot and was used in games resembling modern dice games. In the Ptolemaic dynasty (323–30 B.C.E.), games with 6-sided dice became common in Egypt. People in ancient Greece made icosahedral (20-sided) and other polyhedral dice. The Romans were such enthusiastic dice players that laws were passed forbidding gambling except in certain seasons.

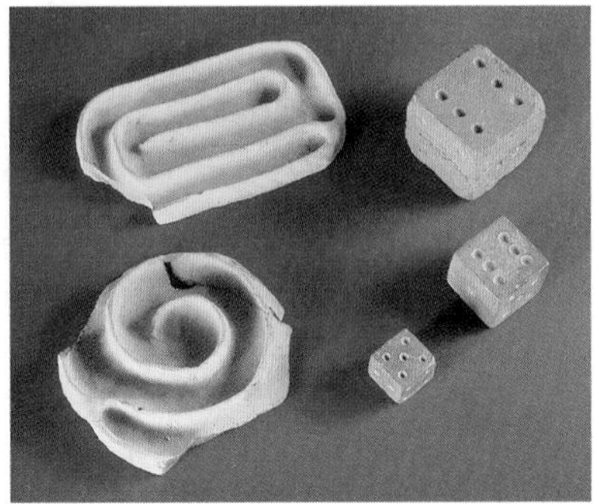

Playing dice, or *tesserae,* was a popular game during the Roman Empire (ca. 1st century B.C.E.–5th century C.E.). The mosaic on the left depicts three men playing *tesserae.* The photo on the right shows ancient Gallo-Roman counters and dice, dating from the second half of the 1st century B.C.E.

You often use a random process to generate **random numbers.** Over the long run, each number is equally likely to occur, and there is no pattern in any sequence of numbers generated. [▶ See **Calculator Note 1L** to learn how to generate random numbers. ◀]

EXAMPLE A

Use a random-number generator to find the probability of rolling a sum of 6 with a pair of dice.

▶ Solution

As you study this solution, follow along on your own calculator. Your results will be slightly different. [▶ To learn how to simulate rolling two dice, see **Calculator Note 10A.** ◀]

To find the probability of the event "the sum is 6," also written as $P(\text{sum is } 6)$, simulate a large number of rolls of a pair of dice. You can use three spreadsheet columns or calculator lists, one for the first die, one for the second die, and one for the sum. Create a list of 300 random integers from 1 to 6 to simulate 300 tosses of the first die. Store the results in the first column or list. Then create a second list of 300 outcomes and store it in the second column. Add the two columns to get a list of 300 random sums of two dice, and store the results in the third column.

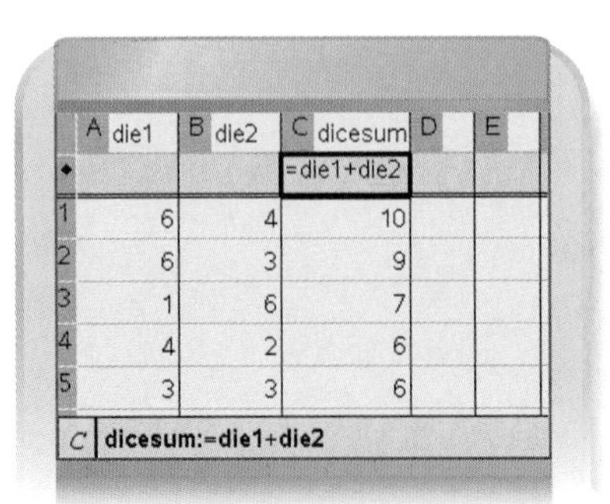

	A die1	B die2	C dicesum	D	E
•			=die1+die2		
1	6	4	10		
2	6	3	9		
3	1	6	7		
4	4	2	6		
5	3	3	6		

C dicesum:=die1+die2

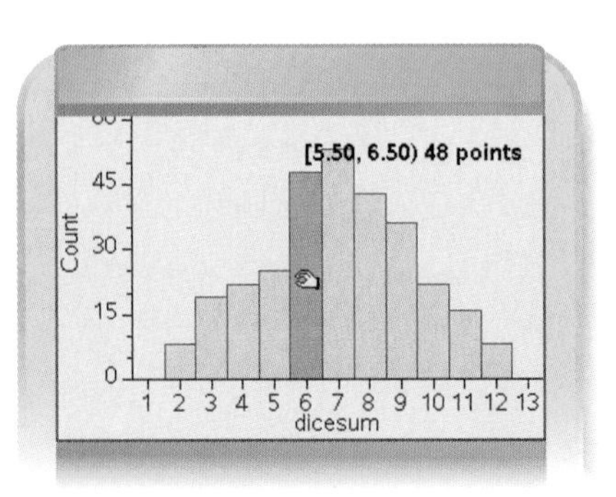

Create a histogram of the 300 entries in the sum list. The calculator screen on the previous page shows the number of each of the sums from 2 to 12. (Your lists and histogram will show different entries.) The bin height of the "6" bin is 48. So out of 300 simulated rolls, $P(\text{sum is } 6) = \frac{48}{300} = 0.16$.

Repeating the entire process five times gives slightly different results: $\frac{249}{1800} \approx 0.138$.

Rather than actually rolling a pair of dice 1800 times in Example A, you performed a **simulation,** representing the random process electronically. You can use dice, coins, spinners, or random-number generators to simulate trials and explore the probabilities of different **outcomes,** or results.

An **event** consists of one or more outcomes. A **simple event** consists of only one outcome. Events that aren't simple are called **compound.** You might recall that the probability of an event, such as "the sum of two dice is 6," must be a number between 0 and 1. The probability of an event that is certain to happen is 1. The probability of an impossible event is 0. The solution for Example A showed that $P(\text{sum is } 6)$ is approximately 0.14, or 14%.

Probabilities that are based on trials and observations like this are called **experimental probabilities.** A pattern often does not become clear until you observe a large number of trials. Find your own results for 300 or 1800 simulations of a sum of two dice. How do they compare with the outcomes in Example A?

Sometimes you can determine the **theoretical probability** of an event without conducting an experiment. To find a theoretical probability, you count the number of ways a desired event can happen and compare this number to the total number of equally likely possible outcomes. Outcomes that are "equally likely" have the same chance of occurring. For example, you are equally likely to flip a head or a tail with a fair coin.

Experimental Probability

If $P(E)$ represents the probability of an event, then

$$P(E) = \frac{\textit{number of occurrences of an event}}{\textit{total number of trials}}$$

Theoretical Probability

If $P(E)$ represents the probability of an event, then

$$P(E) = \frac{\textit{number of different ways the event can occur}}{\textit{total number of equally likely outcomes possible}}$$

In Refreshing Your Skills for Chapter 10, you saw one method for determining the theoretical probability of rolling a specific sum with two dice, organizing the possible outcomes in a table. Another way to find the theoretical probability is to use a graph.

EXAMPLE B Find the theoretical probability of rolling a sum of 6 with a pair of dice.

▶ Solution The possible equally likely outcomes, or sums, when you roll two dice are represented by the 36 grid points in this diagram. The point in the upper-left corner represents a roll of 1 on the green die and 6 on the white die, for a total of 7.

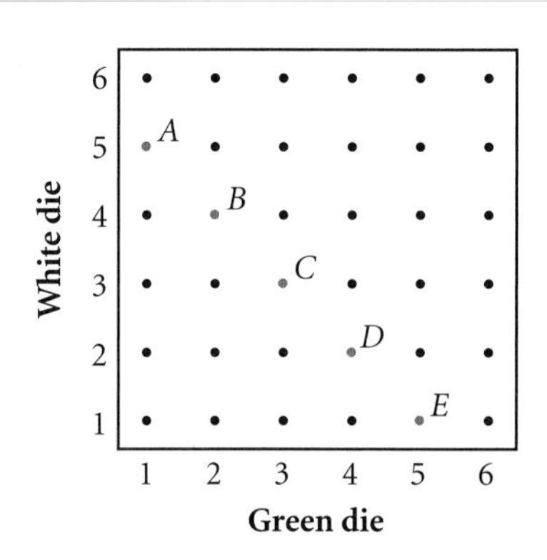

The five possible outcomes with a sum of 6 are labeled *A–E* in the diagram. Point *D*, for example, represents an outcome of 4 on the green die and 2 on the white die. What outcome does point *A* represent?

The theoretical probability is the number of ways the event can occur, divided by the number of equally likely events possible. So $P(\text{sum is } 6) = \frac{5}{36} = 0.13\bar{8}$, or about 13.89%.

Before moving on, compare the experimental and theoretical results of this event. Do you think the experimental probability of an event can vary? How about its theoretical probability?

When you roll dice, the outcomes are whole numbers. What if outcomes could be other kinds of numbers? In those cases, you can often use an area model to find probabilities.

EXAMPLE C What is the probability that any two real numbers you select at random between 0 and 6 have a sum that is less than or equal to 5?

▶ Solution Because the two values are no longer limited to integers, counting would be impossible. The possible outcomes are represented by all points within a 6-by-6 square.

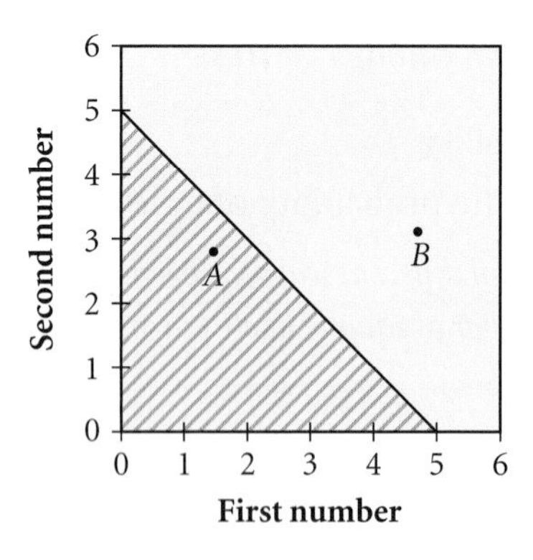

In the diagram, point *A* represents the outcome $1.47 + 2.8 = 4.27$, and point *B* represents $4.7 + 3.11 = 7.81$. The points in the triangular shaded region are all those with a sum less than or equal to 5. They satisfy the inequality $n_1 + n_2 \leq 5$, where n_1 is the first number and n_2 is the second number. The area of this triangle is $(0.5)(5)(5) = 12.5$. The area of all possible outcomes is $(6)(6) = 36$. The probability is therefore $\frac{12.5}{36} \approx 0.347$, or 34.7%.

A probability that is found by calculating a ratio of lengths or areas is called a **geometric probability.**

Experimental probabilities can help you estimate a trend if you have enough cases. But obtaining enough data to observe what happens in the long run is not always feasible. Calculating theoretical probabilities can help you predict these trends. In the rest of this chapter, you'll explore different ways to calculate numbers of outcomes in order to find theoretical probabilities.

In skee ball the probability of getting different point values is based on the geometric area of the regions and their distance from the player.

EXERCISES

You will need

A graphing calculator for Exercises **6, 8,** and **21.**

Practice Your Skills

1. Nina has observed that her coach does not coordinate the color of his socks to anything else that he wears. Guessing that the color is a random selection, she records these data during three weeks of observation: ⓐ

black, white, black, white, black, white, black, red, white, red, white, white, white, black, black

a. What is the probability that he will wear black socks the next day? White socks? Red socks?

b. Are the probabilities you found in 1a experimental or theoretical?

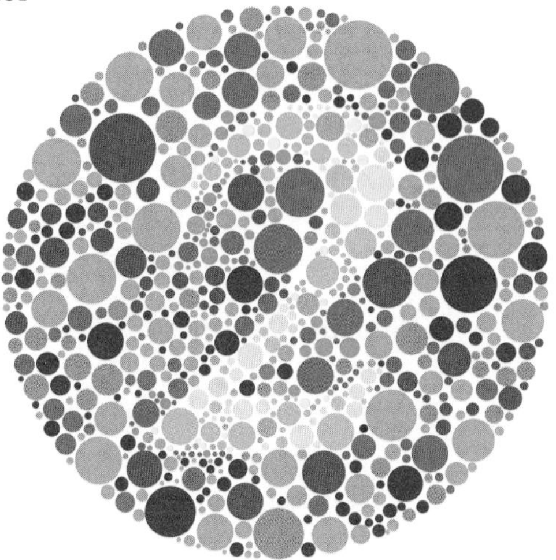

This is a visual test for color blindness, a condition that affects about 8% of men and 0.5% of women. A person with red-green color blindness (the most common type) will see the number 2 more clearly than the number 5 in this image. Dr. Shinobu Ishihara developed this test in 1917.

2. This table shows numbers of students in several categories at Ridgeway High. Find the probabilities described below. Express each answer to the nearest thousandth.

	Male	Female	Total
10th grade	263	249	512
11th grade	235	242	477
12th grade	228	207	435
Total	726	698	1424

a. What is the probability that a randomly chosen student is female?

b. What is the probability that a randomly chosen student is an 11th grader?

c. What is the probability that a randomly chosen 12th grader is male? ⓐ

d. What is the probability that a randomly chosen male is a 10th grader?

e. Are the probabilities in 2a–d experimental or theoretical?

3. Consider number pairs (x, y) selected from the shaded region of the graph at right. Use the graph and basic area formulas to answer each question. Express each answer to the nearest thousandth.

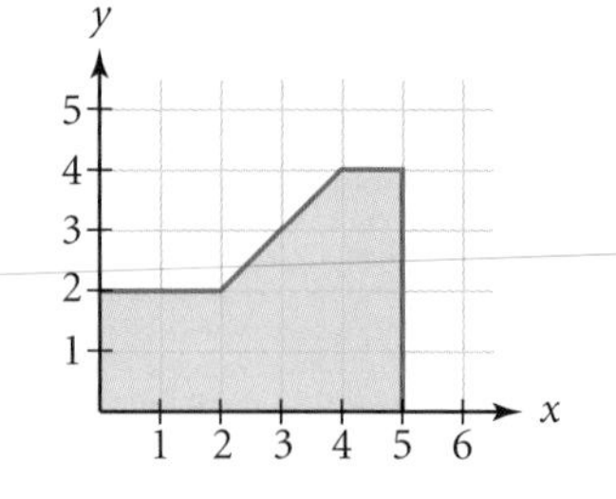

a. What is the probability that x is between 0 and 2? ⓐ

b. What is the probability that y is between 0 and 2?

c. What is the probability that x is greater than 3?

d. What is the probability that y is greater than 3?

e. What is the probability that $x + y$ is less than 2? ⓗ

4. Find each probability.

a. Each day, your teacher randomly calls on 5 students in your class of 30. What is the probability you will be called on today?

b. If 2.5% of the items produced by a particular machine are defective, then what is the probability that a randomly selected item will *not* be defective?

c. What is the probability that the sum of two tossed dice will not be 6? ⓗ

Reason and Apply

5. To prepare necklace-making kits, three camp counselors pull beads out of a box, one at a time. They discuss the probability that the next bead pulled out of the box will be red. Describe each probability as theoretical or experimental.

a. Claire said that $P(\text{red}) = \frac{1}{2}$, because 15 of the last 30 beads she pulled out were red.

b. Sydney said that $P(\text{red}) = \frac{1}{2}$, because the box label says that 1000 of the 2000 beads are red.

c. Mavis says $P(\text{red}) = \frac{1}{3}$, because 200 of the 600 beads the three of them have pulled out so far have been red.

6. Suppose you are playing a board game for which you need to roll a 6 on a die before you can start playing.

a. Predict the average number of rolls a player should expect to wait before starting to play.

b. Describe a simulation, using random numbers, that you could use to model this problem.

c. Do the simulation ten times and record the number of rolls you need to start playing in each game. (For example, the sequence of rolls 4, 3, 3, 1, 6 means you start playing on the fifth roll.)

d. Find the average number of rolls needed to start during these ten games.

e. Combine your results from 6d with those of three classmates, and approximate the average number of rolls a player should expect to wait.

This watercolor (ca. 1890) by an unknown Indian artist shows men playing dice.

7. Rank i–iii according to the method that will best produce a random integer from 0 to 9. Support your reasoning with complete statements. ⓐ

- **i.** the number of heads when you drop nine pennies
- **ii.** the length, to the nearest inch, of a standard 9 in. pencil belonging to the next person you meet who has a pencil
- **iii.** the last digit of the page number closest to you after you open a book to a random page and spin the book flat on the table so that a random corner is pointing to you

8. Simulate rolling a fair die 100 times with your calculator's random-number generator. Display the results in a histogram to see the number of 1's, 2's, 3's, and so on. [▶ To learn how to display random numbers in a histogram, see **Calculator Note 10A.** ◀] Do the simulation 12 times.

a. Make a table storing the results of each simulation. After each 100 rolls, calculate the experimental probability, so far, of rolling a 3.

Simulation number	1's	2's	3's	4's	5's	6's	Ratio of 3's	Cumulative ratio of 3's
1								$\frac{?}{100} = \underline{?}$
2								$\frac{?}{200} = \underline{?}$
3								$\frac{?}{300} = \underline{?}$

b. What do you think the long-run experimental probability will be?

c. Make a graph of the cumulative ratio of 3's versus the number of tosses. Plot the points (cumulative number of tosses, cumulative ratio of 3's). Then plot three more points as you extend the domain of the graph to 2400, 3600, and 4800 trials by adding the data from three classmates. Would it make any difference if you were considering 5's instead of 3's? Explain.

d. What is $P(3)$ for this experiment?

e. What do you think the theoretical probability, $P(3)$, should be? Explain.

9. Find the number of equally likely outcomes of each event described for a roll of two dice. Then write the probability of each event. ⓐ

a. The dice sum to 9.

b. The dice sum to 6.

c. The dice have a difference of 1.

d. The sum of the dice is 6, and their difference is 2.

e. The sum of the dice is at most 5.

10. A sportscaster makes the following statement about a table-tennis game between two rivals: "The odds are 3 to 2 that May will win the first game." A statement like this means that in general, if the rivals play five games, May will likely win three and lose two.

a. What is the probability that May will win the first game?

b. The probability that May will win a second game against a different player is $\frac{2}{7}$. What are her odds of winning the second game?

11. Consider the diagram at right.

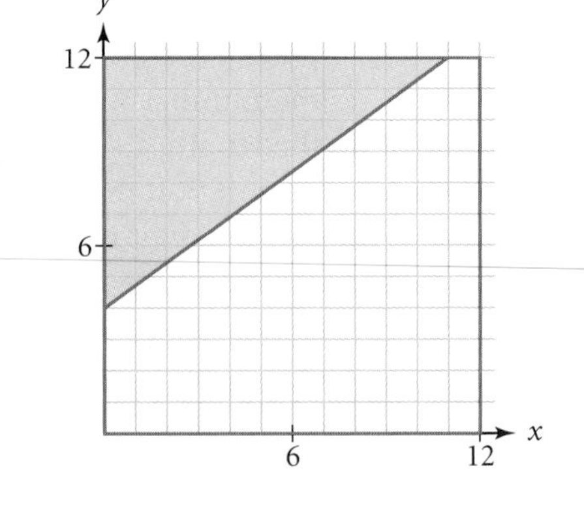

a. What is the total area of the square?

b. What is the area of the shaded region?

c. Suppose the horizontal and vertical coordinates are randomly chosen numbers between 0 and 12, inclusive. Over the long run, what ratio of these points will be in the shaded area? (@)

d. What is the probability that any randomly chosen point within the square will be in the shaded area?

e. What is the probability that the randomly chosen point will *not* land in the shaded area?

f. What is the probability that any point randomly selected within the square will land on a specific point? On a specific line? (h)

12. Suppose x and y are both randomly chosen numbers between 0 and 8. (The numbers are not necessarily integers.)

a. Write a symbolic statement describing the event that the sum of the two numbers is at most 6.

b. Draw a two-dimensional picture of all possible outcomes, and shade the region described in 12a.

c. Determine the probability of the event described in 12a.

13. Use the histogram at right for 13a–d.

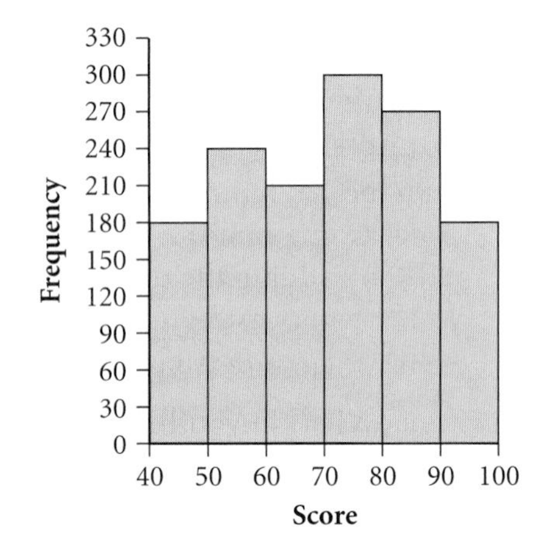

a. Approximate the frequency of scores between 80 and 90. (@)

b. Approximate the sum of all the frequencies.

c. Find P(a score between 80 and 90). (@)

d. Find P(a score that is not between 80 and 90).

14. A 6 in. cube painted on the outside is cut into 27 smaller congruent cubes. Find the probability that one of the smaller cubes, picked at random, will have the specified number of painted faces. (h)

a. exactly one

b. exactly two

c. exactly three

d. no painted face

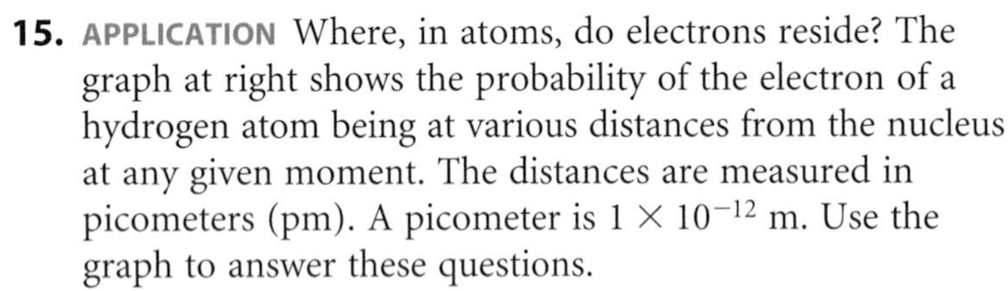

15. **APPLICATION** Where, in atoms, do electrons reside? The graph at right shows the probability of the electron of a hydrogen atom being at various distances from the nucleus at any given moment. The distances are measured in picometers (pm). A picometer is 1×10^{-12} m. Use the graph to answer these questions.

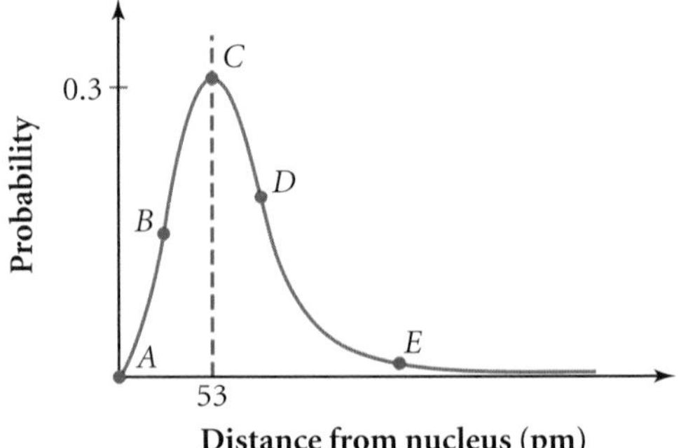

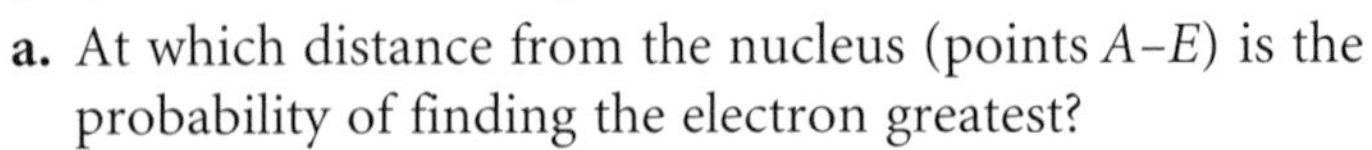

a. At which distance from the nucleus (points A–E) is the probability of finding the electron greatest?

b. At what distance is there zero probability of finding the electron?

c. As the distance from the nucleus increases, describe what happens to the probability of finding an electron.

Science CONNECTION

Electron position is important to scientists because it lays the foundation for understanding chemical changes at the atomic level and determining how likely it is that particular chemical changes will take place. In the early part of the 20th century, scientists thought that electrons orbit the nucleus of an atom much the way the planets orbit the Sun. Danish physicist Niels Bohr (1885–1962) theorized that the electron of a hydrogen atom always orbits at 53 pm from the nucleus. In the mid-1920s, Austrian physicist Erwin Schrödinger (1887–1961) first proposed using a probability model to describe the electron's position. Instead of a predefined orbit, Schrödinger's work led to the electron cloud model.

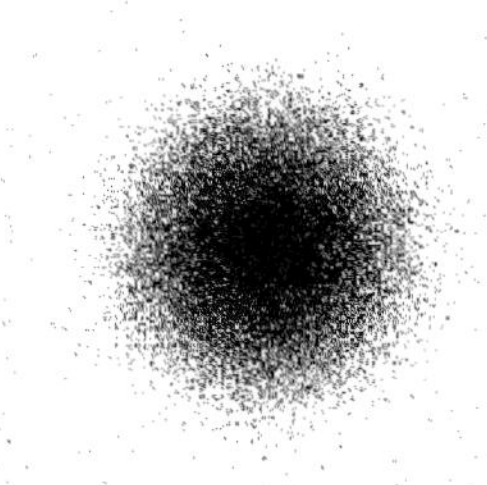

This electron cloud model shows possible locations of an electron in a hydrogen molecule. The density of points in a particular region indicates the probability that an electron will be located in that area.

Review

16. Expand $(x - y)^4$. @

17. Write $\log a - \log b + 2\log c$ as a single logarithmic expression.

18. Solve $\log 2 + \log x = 4$.

19. Consider this system of inequalities: @

$$\begin{cases} 3x + y \leq 15 \\ x + 6y \geq -12 \\ -5x + 4y \leq 26 \end{cases}$$

a. Graph the triangle defined by this system.

b. Give the coordinates of the vertices of the triangle in 19a.

c. Find the area of the triangle in 19a.

20. Describe the locus of points equidistant from the line $y = 6$ and the point $(3, 0)$. Then write the polynomial equation in general form. @

21. Consider these two sets of data. @

i. $\{5, 23, 36, 48, 63\}$ **ii.** $\{112, 115, 118, 119, 121\}$

a. Which set would you expect to have the larger standard deviation? Explain your reasoning.

b. Calculate the mean and the standard deviation of each set.

c. Predict how the mean and the standard deviation of each set will be affected if you multiply every data value by 10. Then do calculations to verify your answer. How do these measures compare with those you found in 21b?

d. Predict how the mean and the standard deviation of each set will be affected if you add 10 to every data value. Then do calculations to verify your answer. How do these measures compare with those you found in 21b?

EXPLORATION

Geometric Probability

French naturalist Georges Louis Leclerc, Comte de Buffon (1707–1788), posed one of the first geometric probability problems: If a coin is tossed randomly onto a floor of congruent tiles, what is the probability that it will land entirely within a tile, not touching any edges? The answer depends on the size of the coin and the size and shape of the tiles. You'll explore this problem through experimentation, then analyze your results.

Activity

The Coin Toss Problem

You will need

- a centimeter ruler
- a penny, a nickel, a dime, and a quarter
- grid paper in different sizes: 20 mm, 30 mm, 40 mm, and 40 mm with 5 mm borders

Work with your group to investigate Buffon's coin toss problem using different-size grids to simulate different-size tiles. For the first three grids, you can assume the borders around each tile have no thickness. Divide up the experimentation among your group members so that you can collect the data more quickly. Record your results in a table like the one below.

	Coin diameter (mm)	20 mm grid	30 mm grid	40 mm grid	40 mm grid with 5 mm borders
Penny					
Nickel					
Dime					
Quarter					

Step 1 Measure the diameter of your coin in millimeters and record your answer in your table. Toss your coin 100 times, and count the number of times the coin lands entirely within a square, not touching any lines. Record your number of successes, and the experimental probability of success. Collect data from your group members to complete the table.

Step 2 Where must the center of a coin fall in order to have a successful outcome? What is the area of this region for each combination of coin and grid paper?

Step 3 Use your answers from Step 2 to calculate the theoretical probability of success for each coin and grid-paper combination. How do these theoretical probabilities compare with your experimental probabilities? If they are significantly different, explain why.

Questions

1. Design a grid, different from any you used in the activity, that has a probability of success of 0.01 for a coin of your choice.

2. Determine a formula that will calculate the theoretical probability of success, given a coin with diameter d and a grid of squares with side length a and line thickness t.

3. What would be the theoretical probability of success if you tossed a coin with diameter 10 mm onto an infinite grid paper tiled with equilateral triangles with side length 40 mm, arranged as shown below?

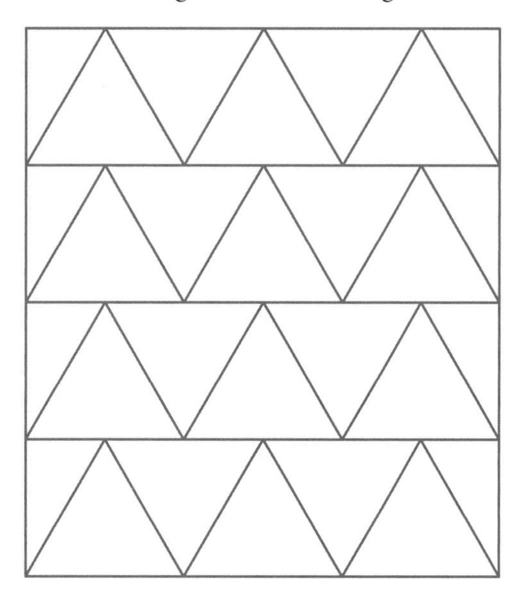

IMPROVING YOUR REASONING SKILLS

Beating the Odds

Mavis and Trish have two boxes. One box contains 50 blue marbles; the other box contains 50 red marbles. Mavis will blindfold Trish and place the two boxes in front of her. Trish will pick one marble from one of the boxes. If Trish picks a red marble, she wins. If she picks a blue marble, Mavis wins. Before being blindfolded, Trish requests that she be allowed to distribute the marbles between the boxes in any way she likes. Mavis thinks about the request and says, "Sure, as long as all one hundred marbles are there, what difference could it make?" How should Trish distribute the marbles to have the greatest chance of winning? What would be the probability that Trish will win?

LESSON

10.2 Counting Outcomes and Tree Diagrams

In Lesson 10.1, you determined some theoretical probabilities by finding the ratio of the number of desired outcomes to the number of possible equally likely outcomes. In some cases it can be difficult to count the number of possible or desired outcomes. You can make this easier by organizing information and counting outcomes using a **tree diagram.**

EXAMPLE A

A national advertisement says that every puffed-barley cereal box contains a toy and that the toys are distributed equally. Talya wants to collect a complete set of the different toys from cereal boxes.

a. If there are two different toys, what is the probability that she will find both of them in her first two boxes?

b. If there are three different toys, what is the probability that she will have them all after buying her first three boxes?

▶ Solution

Draw tree diagrams to organize the possible outcomes.

a. In this tree diagram, the first branching represents the possibilities for the first box and the second branching represents the possibilities for the second box. Thus, the four paths from left to right represent all outcomes for two boxes and two toys. Path 2 and Path 3 contain both toys. If the advertisement is accurate about equal distribution of toys, then the paths are equally likely. So the probability of getting both toys is $\frac{2}{4}$, or 0.5.

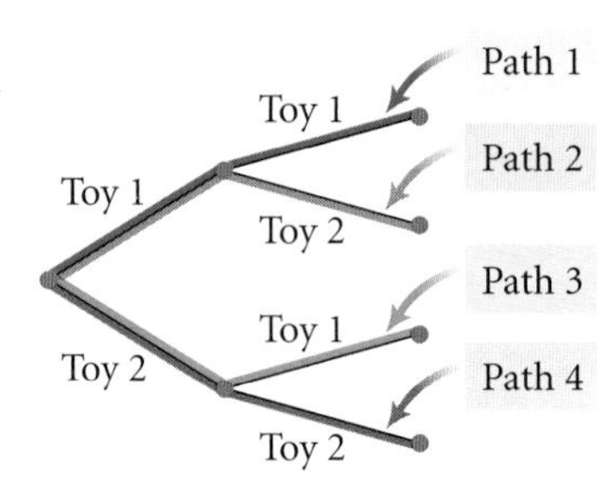

b. This tree diagram shows all the toy possibilities for three boxes. There are 27 possible paths. You can determine this quickly by counting the number of branches on the far right. Six of the 27 paths contain all three toys, as shown. Because the paths are equally likely, the probability of having all three toys is $\frac{6}{27}$, or $0.\overline{2}$.

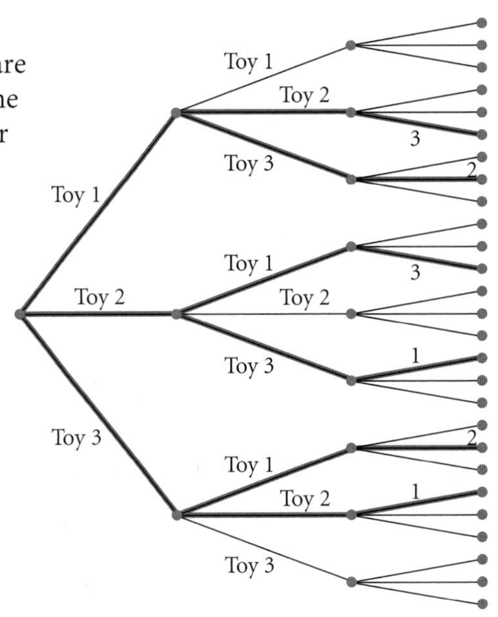

Any path from left to right through a tree diagram is an outcome, or simple event. Probabilities exist for each stage of the process as well as for each complete path. Tree diagrams can be helpful for organizing complicated situations.

Investigation
The Multiplication Rule

Step 1 On your paper, redraw the tree diagram for Example A, part a. This time, write the probability on each branch. Then find the probability of each path.

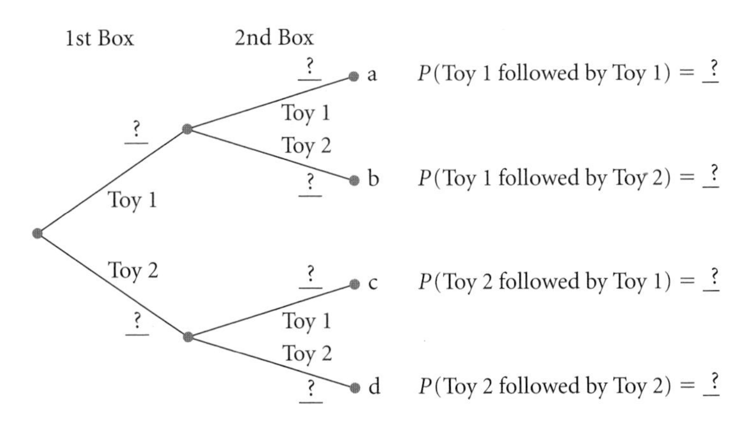

Step 2 Redraw the tree diagram for Example A, part b. Indicate the probability on each branch, and also write the probability of each path. What is the sum of the probabilities of all possible paths? What is the sum of the probabilities of the highlighted paths?

Step 3

Suppose the national advertisement mentioned in Example A listed four different toys distributed equally in a huge supply of boxes. Draw only as much of a tree diagram as you need to in order to answer these questions:

a. What would be *P*(Toy 2) in Talya's first box? Talya's second box? Third box? *P*(any particular toy in any particular box)?

b. In these situations, does the toy she finds in one box influence the probability of there being a particular toy in the next box?

c. One outcome that includes all four toys is Toy 3, followed by Toy 2, followed by Toy 4, followed by Toy 1. What is the probability of this outcome? Another outcome would be the same four toys in a different order. How many such outcomes are there? Are they all equally likely?

Step 4

Write a statement explaining how to use the probabilities of a path's branches to find the probability of the path.

Step 5

What is *P*(obtaining the complete set in the first four boxes)?

As you saw in Example A and the investigation, a tree diagram with many equally likely branches is a lot to draw. In some cases, as you'll see in Example B, a tree with branches of different probabilities may be practical.

EXAMPLE B

Mr. Roark teaches three classes. Each class has 20 students. His first class has 12 sophomores, his second class has 8 sophomores, and his third class has 10 sophomores. If he randomly chooses one student from each class to participate in a competition, what is the probability that he will select three sophomores?

▶ Solution

You could consider drawing a tree with 20 branches representing the students in the first class. These would split into 20 branches for the second class, and each of these paths would split into 20 branches for the third class. This would be a tree with 8000 paths!

Instead, you can draw two branches for each stage of the selection process. One branch represents a choice of a sophomore (S) and one represents a choice of a nonsophomore (NS). This tree shows all eight possible outcomes. However, the outcomes are not equally likely. For the first class, the probability of choosing a sophomore is $\frac{12}{20} = 0.6$, and the probability of choosing a nonsophomore is $1 - 0.6 = 0.4$. Calculate the probabilities for the second and third classes, and represent them on a tree diagram, as shown.

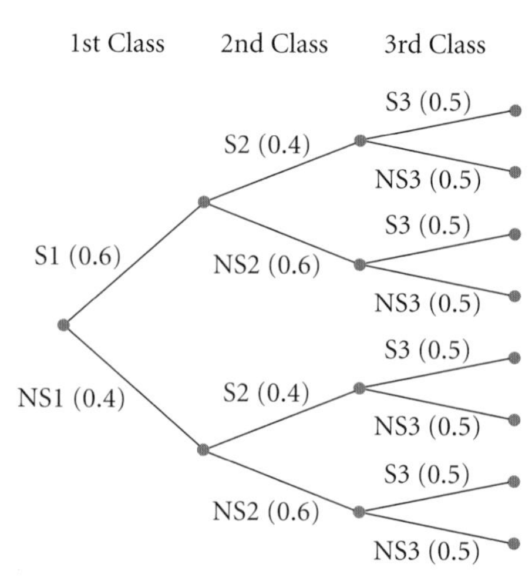

The uppermost path represents a sophomore being chosen from each class. In the investigation you learned to find the probability of a path by multiplying the probabilities of the branches. So the probability of choosing three sophomores is $(0.6)(0.4)(0.5)$, or 0.12.

In Example B, the probability of choosing a sophomore in the second class is the same, regardless of whether a sophomore was chosen in the first class. These events are called **independent.** Events are independent if the occurrence of one has no influence on the probability of the other.

Probability of a Path
(The Multiplication Rule for Independent Events)

If n_1, n_2, n_3, and so on, represent independent events, then the probability that this sequence of events will occur can be found by multiplying the probabilities of the events.

$$P(n_1 \text{ and } n_2 \text{ and } n_3 \text{ and} \ldots) = P(n_1) \cdot P(n_2) \cdot P(n_3) \cdot \cdots$$

When the probability of one branch does not depend on what occurred on the previous branches, you can multiply the probabilities of the branches. But what if the probability of the second branch *does* depend on which branch was taken first?

EXAMPLE C

Devon is going to draw three cards, one after the other, from a standard deck. What is the probability that she will draw exactly two hearts?

▶ Solution

The outcome of each draw can be represented by branches on a tree diagram. Rather than list all of the cards in the deck, classify the result of each draw as a heart (H) or nonheart (NH).

Study the tree diagram. There are 52 cards in a standard deck, with 13 cards in each suit. Notice that the denominator of each probability for the second draw is 51 because there are only 51 cards left to choose from.

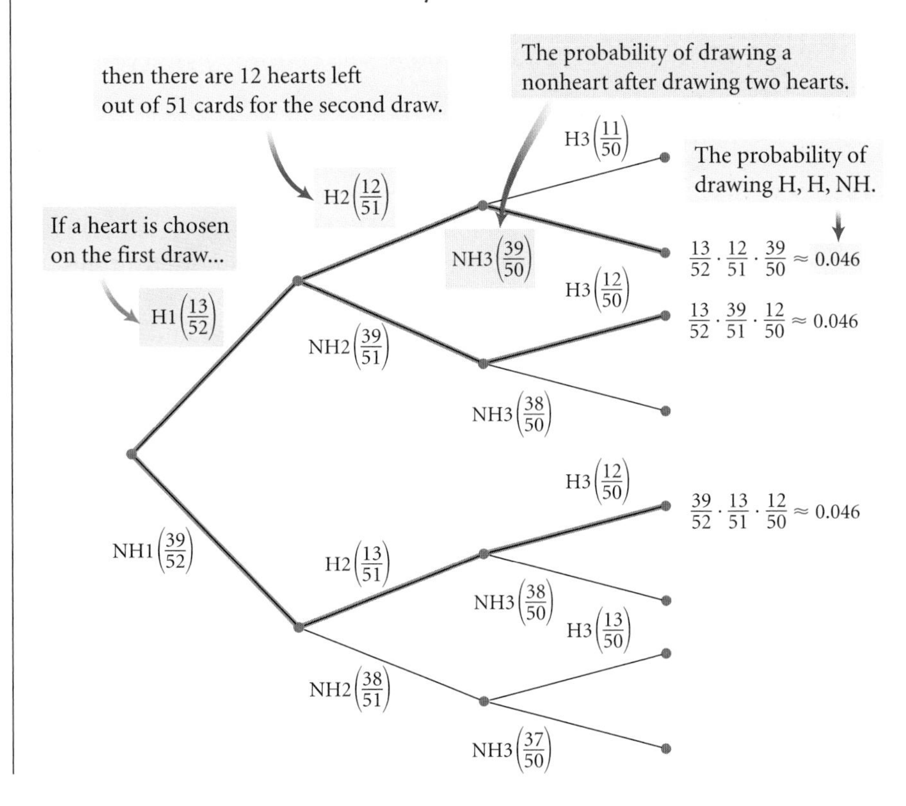

The three highlighted paths show the ways to get exactly two hearts. For example, the top highlighted path shows getting a heart on the first draw, a heart on the second draw, and a nonheart on the third draw.

The probability of any event that can occur along multiple paths is the sum of the probabilities of those paths. Thus, the probability that Devon gets exactly two hearts is the sum of the probabilities of the highlighted paths, which is about $0.046 + 0.046 + 0.046 = 0.138$.

In a tree diagram, the probability for each simple event along a path may depend on what occurred on the previous branches. In Example C, for instance, the probabilities for the second draw are dependent on the result of the first draw. That is, the events are **dependent events.**

You can use **conditional probability** notation to describe both independent and dependent events. The probability of event A given event B is denoted with a vertical line:

$$P(A \mid B)$$

For example, to denote the probability of drawing a heart on the second draw given that you have already drawn a nonheart, you write $P(\mathrm{H}_2 \mid \mathrm{NH}_1)$. And $P(\mathrm{H}_2 \mid \mathrm{H}_1)$ denotes the probability of drawing a heart on the second draw given that you have already drawn a heart. Using the tree diagram from Example C, you can see that $P(\mathrm{H}_2 \mid \mathrm{NH}_1) = \frac{13}{51}$ and $P(\mathrm{H}_2 \mid \mathrm{H}_1) = \frac{12}{51}$.

The Multiplication Rule (again)

If n_1, n_2, n_3, and so on, represent events, then the probability that this sequence of events will occur can be found by multiplying the conditional probabilities of the events.

$$P(n_1 \text{ and } n_2 \text{ and } n_3 \text{ and} \ldots) = P(n_1) \cdot P(n_2 \mid n_1) \cdot P(n_3 \mid (n_1 \text{ and } n_2)) \cdot \cdots$$

If events A and B are independent, then the probability of A is the same whether B happens or not. In this case, $P(A \mid B) = P(A)$. In Example B, the probability of choosing a sophomore in the second class (S2) is the same whether Mr. Roark chooses a sophomore or a nonsophomore from his first class. This means that $P(\mathrm{S2} \mid \mathrm{S1}) = P(\mathrm{S2} \mid \mathrm{NS1}) = P(\mathrm{S2})$. Can you explain the difference in the multiplication rule defined on page 563 and the rule defined above?

This tree could represent any two-stage event with two options at each stage. To find the probability of event C, you must add all paths, or outcomes, that contain C. $P(C) = P(A \text{ and } C) + P(B \text{ and } C) = P(A) \cdot P(C \mid A) + P(B) \cdot P(C \mid B)$.

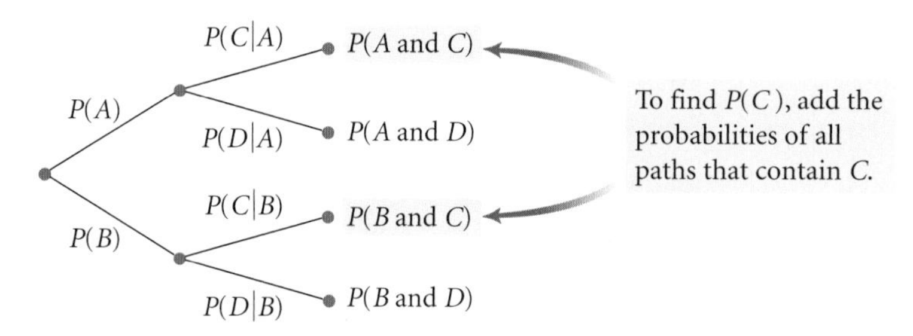

EXERCISES

Practice Your Skills

1. Create a tree diagram showing the different outcomes if the cafeteria has three main entrée choices, two vegetable choices, and two dessert choices. ⓐ

2. Find the probability of each path, a–d, in the tree diagram at right. What is the sum of the values of a, b, c, and d? ⓐ

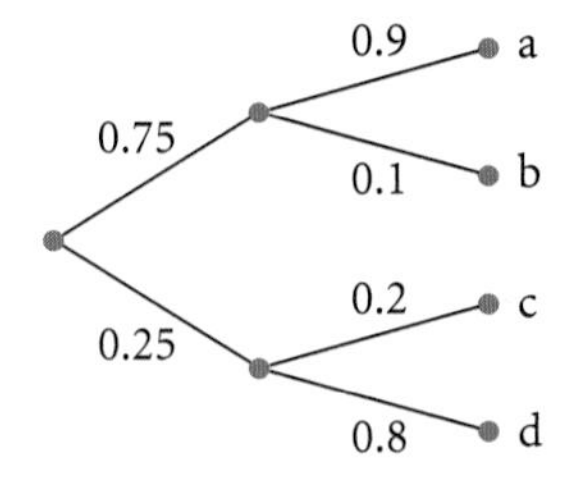

3. Find the probability of each branch or path, a–g, in the tree diagram below. ⓗ

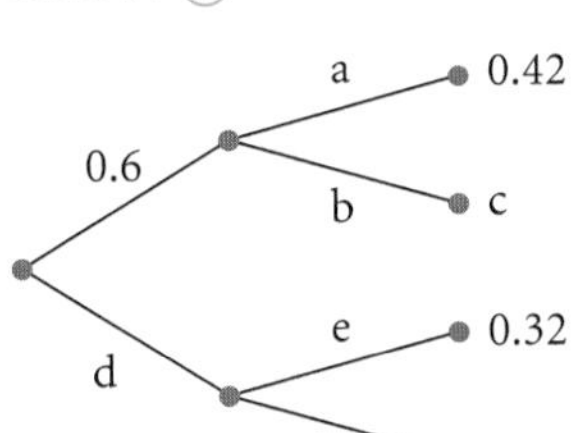

4. Three friends are auditioning for different parts in a comedy show. Each student has a 50% chance of success. Use the tree diagram at right to answer 4a–c.

a. Find the probability that all three students will be successful. ⓐ

b. Find the probability that exactly two students will be successful.

c. If you know that exactly two students have been successful, but do not know which pair, what is the probability that Celina was successful? ⓗ

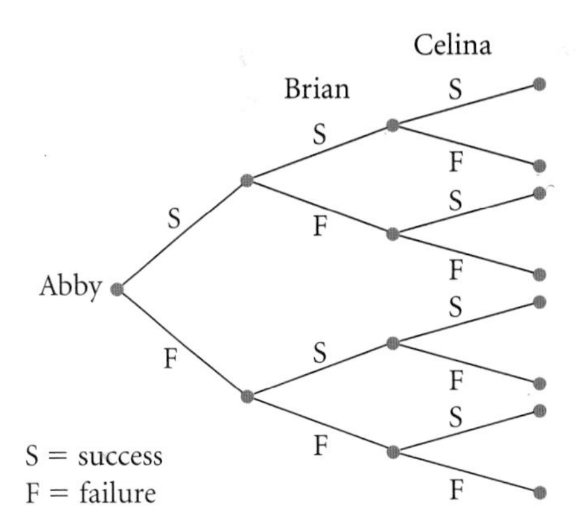

Reason and Apply

5. This tree diagram models the outcomes of selecting two different students from a class of 7 juniors and 14 sophomores.

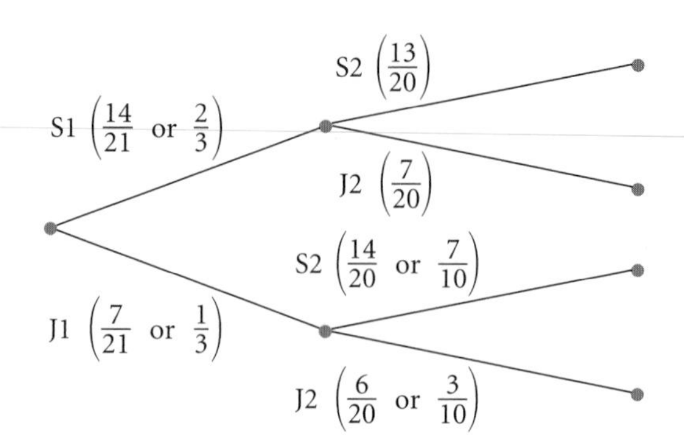

a. Locate the probability $\frac{7}{20}$ on the diagram and explain what it represents.

b. What is $P(\text{S2} \mid \text{S1})$?

c. What is $P(\text{S2} \mid \text{J1})$?

6. Use the diagram from Exercise 5 to answer each question. @

a. Use the multiplication rule to find the probability of each path.

b. Are the paths equally likely? Explain.

c. What is the sum of the four answers in 6a?

7. A recipe calls for four ingredients: flour, baking powder, shortening, and milk (F, B, S, M). But there are no directions stating the order in which they should be combined. Chris has never followed a recipe like this before and has no idea which order is best, so he chooses the order at random.

a. How many different possible orders are there? @

b. What is the probability that Chris will choose milk first?

c. What is the probability that Chris will choose flour first and shortening second?

d. What is the probability that Chris will choose FBSM?

e. What is the probability that Chris will not choose FBSM?

f. What is the probability that Chris will choose flour and milk next to each other? @

8. Draw a tree diagram that pictures all possible equally likely outcomes if a coin is flipped as specified.

a. two times **b.** three times **c.** four times

9. How many different equally likely outcomes are possible if a coin is flipped as specified?

a. two times **b.** three times **c.** four times @

d. five times **e.** ten times @ **f.** n times

10. Consider these questions about flipping a coin multiple times.

a. Janny flipped a coin and it came up heads. What is the probability it will come up heads if she flips it again?

b. Kevin has flipped a coin three times and it has come up heads each time. What is the probability it will come up heads the next time he flips it?

c. Are multiple flips of a coin independent or dependent events?

d. Jeremy flipped a coin five times and it came up heads each time. Sue says, "You're on a hot streak, it will be heads the next time you flip it." Dawn says, "It's come up heads so many times, you're due for tails next time." Evaluate each of their statements.

11. You are totally unprepared for a true-false quiz, so you decide to guess randomly at the answers. There are four questions. Find the probabilities described in 11a–e. ⓗ

a. P(none correct)

b. P(exactly one correct)

c. P(exactly two correct)

d. P(exactly three correct)

e. P(all four correct)

f. What should be the sum of the five probabilities in 11a–e?

g. If a passing grade means you get at least three correct answers, what is the probability that you passed the quiz? ⓐ

12. **APPLICATION** The ratios of the number of phones manufactured at three sites, M1, M2, and M3, are 20%, 35%, and 45%, respectively. The diagram at right shows some of the ratios of the numbers of defective (D) and good (G) phones manufactured at each site. The top branch indicates a 0.20 probability that a phone made by this manufacturer was manufactured at site M1. The ratio of these phones that are defective is 0.05. Therefore, 0.95 of these phones are good. The probability that a randomly selected phone is both from site M1 and defective is (0.20)(0.05), or 0.01.

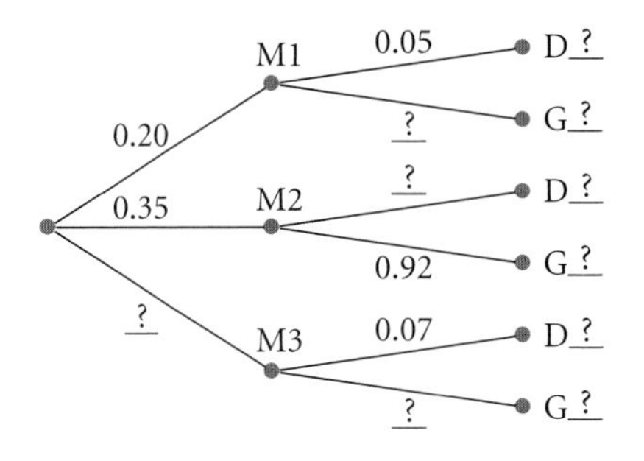

a. Copy the diagram and fill in the missing probabilities. ⓐ

b. Find P(a phone from site M2 is defective).

c. Find P(a randomly chosen phone is defective).

d. Find P(a phone was manufactured at site M2 if you already know it is defective). ⓐ

13. The Pistons and the Bulls are tied, and time has run out in the game. However, the Pistons have a player at the free throw line, and he has two shots to make. He generally makes 83% of the free throw shots he attempts. The shots are independent events, so each one has the same probability. Find these probabilities:

a. P(he misses both shots)

b. P(he makes at least one of the shots)

c. P(he makes both shots)

d. P(the Pistons win the game)

14. What is the probability that there are exactly two girls in a family with four children? Assume that girls and boys are equally likely. ⓗ

15. Consider one roll of a six-sided die. Let A represent the event "odd" and let B represent the event "3 or 5." Express each probability in words, and then give its value.

a. $P(A \text{ and } B)$

b. $P(B \mid A)$

c. $P(A \mid B)$

16. The table at right gives numbers of students in several categories. ⓐ

a. Find $P(\text{10th grade} \mid \text{female})$

b. Find P(10th grade).

c. Are the events "10th grade" and "female" dependent or independent? Explain your reasoning.

	Male	Female	Total
10th grade	163	349	512
11th grade	243	234	477
12th grade	220	215	435
Total	626	798	1424

17. Braille is a form of writing for sight-impaired people. Each Braille character consists of a cell containing six positions that can have a raised dot or not have one. How many different Braille characters are possible? @

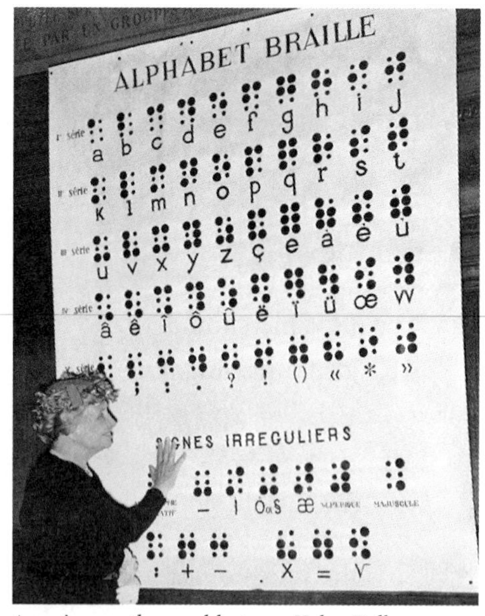

American author and lecturer Helen Keller (1880–1968) was both blind and deaf. This photo shows her with a Braille chart.

History CONNECTION

At age 12, French innovator and teacher Louis Braille (1809–1852) invented a code that enables sight-impaired people to read and write. He got the idea from a former soldier who used a code consisting of up to 12 raised dots that allowed soldiers to share information on a dark battlefield without having to speak. Braille modified the number of dots to 6, and added symbols for math and music. In 1829, he published the first book in Braille.

Review

18. Write each expression in the form $a + bi$. @

a. $(2 + 4i) - (5 + 2i)$

b. $(2 + 4i)(5 + 2i)$

c. 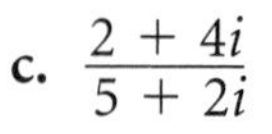 $\dfrac{2 + 4i}{5 + 2i}$

19. Refer to the diagram at right. What is the probability that a randomly selected point within the rectangle is in the orange region? The blue region?

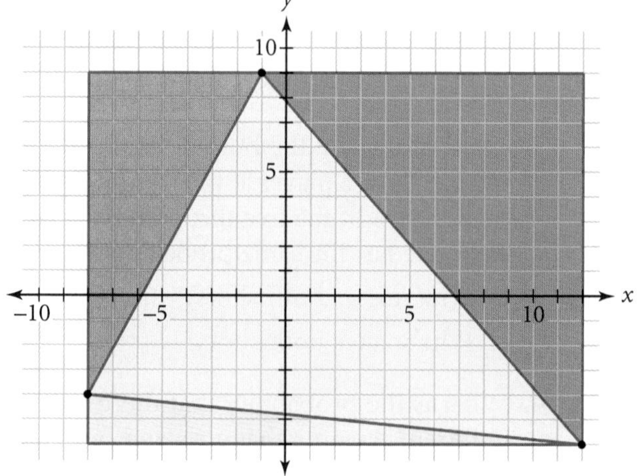

20. A sample of 230 students is categorized as shown.

	Male	Female
Junior	60	50
Senior	70	50

a. What is the probability that a junior is female?

b. What is the probability that a student is a senior?

21. The side of the largest square in the diagram at right is 4. Each new square has side length equal to half of the previous one. If the pattern continues infinitely, what is the long-run length of the spiral made by the diagonals?

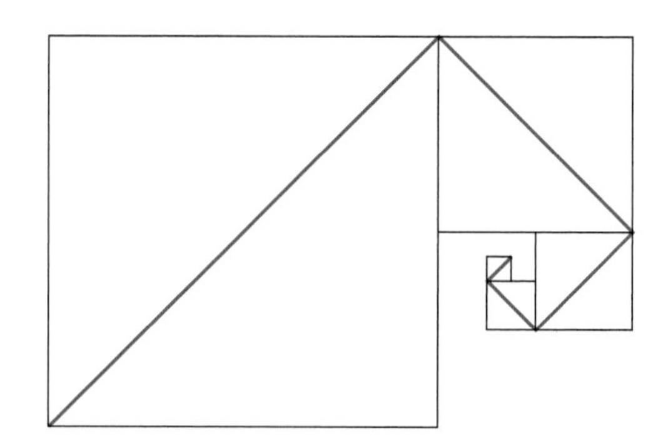

EXPLORATION

The Law of Large Numbers

You have calculated both experimental and theoretical probabilities of events. In some cases, these probabilities are close. But when might they be different? And how can you predict the outcome of a single random process? In the activity you will explore ways to generalize the probability of an event.

Activity

A Repeat Performance

In this activity you will use a random-number generator to simulate the roll of a die. You will then explore the effects of different numbers of trials.

Step 1 Open a new Fathom document. Insert a new case table. Enter the column heading, or attribute, Roll.

Step 2 Click on the attribute Roll. Choose **Edit Formula** from the Edit menu. To simulate the roll of a standard six-sided die, enter **randomPick(1, 2, 3, 4, 5, 6).** This command tells the computer to pick randomly any of the integers 1 through 6.

Step 3 From the Collection menu, choose **New Cases.** Enter 10 as the number of new cases. Your case table will fill with 10 randomly chosen integers between 1 and 6.

Predict the shape of the histogram of your 10 data items. Then make a histogram and compare it with your prediction. To make a histogram, open a new graph. Drag the attribute Roll to the x-axis of your graph. Then change the graph type to histogram.

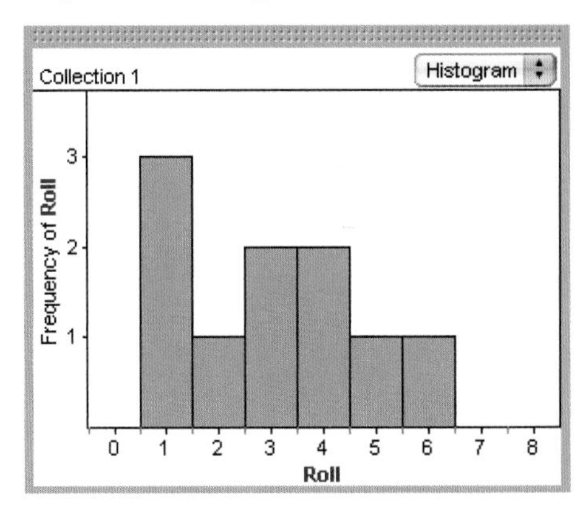

Step 4 What is the mode of your data? What are the maximum and minimum number of occurrences of outcomes? What is the difference between these values (the range)?

Step 5 Choose **Rerandomize** from the Collection menu to generate another set of cases. Does the random generator seem to favor one number over the others? Does the **randomPick** command seem to simulate a fair die? How can you check?

Step 6 Predict the shape of a histogram for 100 cases. How do you think it would compare with a histogram for 10 items?

Step 7 Choose **New Cases** from the Collection menu again and add 90 new cases. Your histogram should update automatically. Compare it with your prediction and with the histogram for 10 cases.

Step 8 What are the mode and range of possible outcomes? Rerandomize to generate another set of cases. Does the random generator seem to favor one number over the others?

Step 9 Predict the histogram for 1000 cases. Then make a histogram and repeat Step 8.

A die that is not fair is called "loaded." A loaded die is designed to roll certain numbers more often than other numbers.

Step 10 Create a loaded die by using **randomPick** and any selection of numbers between 1 and 6. Let a classmate guess how the die is loaded by observing only 10 cases at a time.

Step 11 Now repeat Step 10, but allow your classmate to use 1000 cases at a time.

Questions

1. How does increasing the number of cases affect the shape of the histogram?
2. How does increasing the number of cases affect the range of data?
3. What strategies did you use to judge whether a die was loaded? Suppose you suspect that a die is loaded. How can you show that it is?
4. The Law of Large Numbers states that if an experiment is repeated many times, the experimental probability will get closer and closer to the theoretical probability. How does the Law of Large Numbers apply to this activity?
5. How large is large? You were able to estimate the probability fairly accurately with 1000 cases. But if the event is rare, then *large* takes on a different meaning. For example, if an event had a probability (unknown to you) of 0.002, then how many trials would you have to observe to estimate the probability accurately?

IMPROVING YOUR REASONING SKILLS

The Fake Coin

Angelina's grandfather gives her a collection of rare coins. It includes a box with 81 identical-looking coins. He tells her that one coin is fake because it is slightly lighter than all the others, but he does not remember which one it is. Angelina has a pan balance. What is the smallest number of balancings that she needs to make to be sure she will find the fake coin?

LESSON

10.3

Mutually Exclusive Events and Venn Diagrams

Of course there is no formula for success except perhaps an unconditional acceptance of life and what it brings.

ARTHUR RUBENSTEIN

Two outcomes or events that cannot both happen are **mutually exclusive.** You already worked with theoretical probabilities and mutually exclusive events when you added probabilities of different paths.

For example, this tree diagram represents the possibilities of success in auditions by Abby, Bonita, and Chih-Lin. One path from left to right represents the outcome that Abby and Chih-Lin are successful but Bonita is not (outcome AC). Another outcome is success by all three (outcome ABC). These two outcomes cannot both take place, so they're mutually exclusive.

Abby Bonita Chih-Lin
S S S Outcome ABC
F
F S Outcome AC
F
F S S
F
F S
F

S = success
F = failure

Suppose that each student has a 0.5 probability of success. Then the probability of any single path in the tree is $(0.5)(0.5)(0.5) = 0.125$. So the probability that either AC or ABC occurs is the sum of the probabilities on two particular paths, $0.125 + 0.125$, or 0.25.

Just as tree diagrams allow breaking down sequences of dependent events into sequences of independent events, there's a tool for breaking down non–mutually exclusive events into mutually exclusive events. This tool is the Venn diagram, consisting of overlapping circles.

Untitled (1988) by British-American artist Judy Pfaff (b 1946) shows overlapping circles similar to Venn diagrams.

EXAMPLE A

Melissa has been keeping a record of probabilities of events involving

i. Her violin string breaking during orchestra rehearsal (Event B).

ii. A pop quiz in math (Event Q).

iii. Her team losing in gym class (Event L).

Although the three events are not mutually exclusive, they can be broken into eight mutually exclusive events. These events and their probabilities are shown in the Venn diagram.

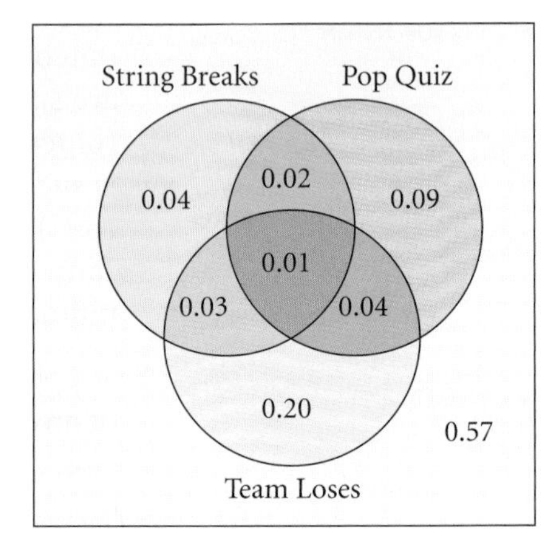

a. What is the meaning of the region labeled 0.01?

b. What is the meaning of the region labeled 0.03?

c. What is the probability of either a pop quiz or Melissa's team losing today?

d. Find the probability of a pretty good day, *P*(not B and not Q and not L). This means no string breaks and no quiz and no loss.

▶ Solution

The meaning of each region is determined by the circles that contain it.

a. The region labeled 0.01 represents the probability of a really bad day. In this **intersection,** or overlap, of all three circles, Melissa's string breaks, she gets a pop quiz, and her team loses. The word *and* is often associated with the intersection.

b. The region labeled 0.03 represents the probability that Melissa's string will break and her team will lose, but there will be no pop quiz in math.

c. You can find the probability of either a pop quiz or the team losing by adding the probabilities in regions inside the two circles: $0.02 + 0.09 + 0.01 + 0.04 + 0.03 + 0.20 = 0.39$. This is the same as finding the **union** of the two circles. The word *or* is often associated with the union.

d. The probability of a pretty good day, *P*(not B and not Q and not L), is pictured by the region outside the circles and is 0.57.

In general, the probability that one of a set of mutually exclusive events will occur is the sum of the probabilities of the individual events.

The Addition Rule for Mutually Exclusive Events

If n_1, n_2, n_3, and so on, represent mutually exclusive events, then the probability that any event in this collection of mutually exclusive events will occur is the sum of the probabilities of the individual events.

$$P(n_1 \text{ or } n_2 \text{ or } n_3 \text{ or} \ldots) = P(n_1) + P(n_2) + P(n_3) + \cdots$$

But what if you don't know all the probabilities that Melissa knew? In the investigation you'll discover one way to figure out probabilities of mutually exclusive events when you know probabilities of non-mutually exclusive events.

Investigation
Addition Rule

Of the 100 students in 12th grade, 70 are enrolled in mathematics, 50 are in science, 30 are in both subjects, and 10 are in neither subject.

Step 1 "A student takes mathematics" and "a student takes science" are two events. Are these events mutually exclusive? Explain.

Step 2 Complete a Venn diagram, similar to the one below, that shows enrollments in mathematics and science courses.

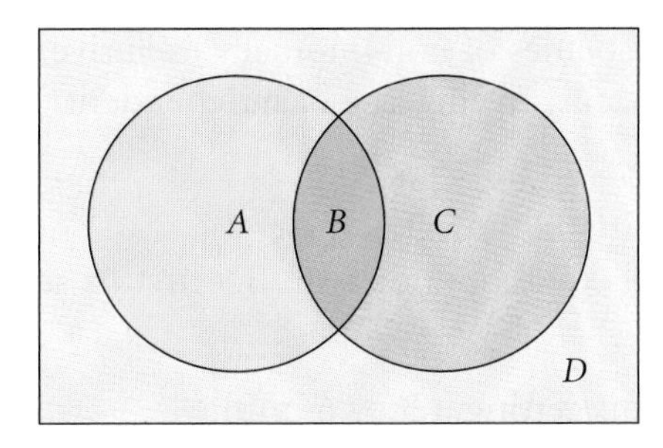

Step 3 Use the numbers of students in your Venn diagram to calculate probabilities.

Step 4 Explain why the probability that a randomly chosen student takes mathematics or science, $P(\text{M or S})$, does not equal $P(\text{M}) + P(\text{S})$.

Step 5 Create a formula for calculating $P(\text{M or S})$ that includes the expressions $P(\text{M})$, $P(\text{S})$, and $P(\text{M and S})$.

Step 6 Suppose two dice are tossed. Draw a Venn diagram to represent the events

A = "sum is 7"

B = "both dice > 2"

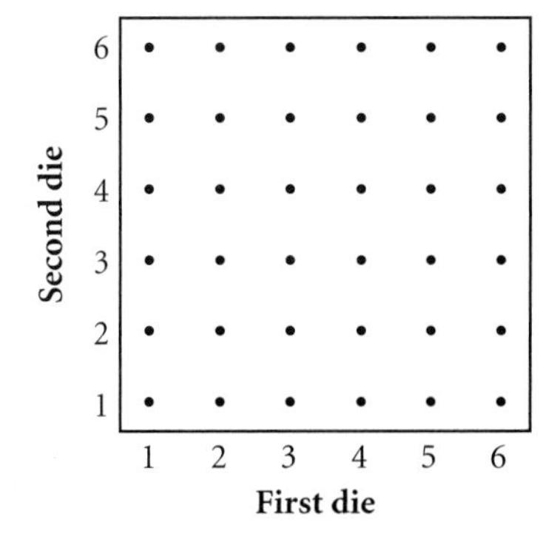

Find the probabilities in parts a–e by counting dots:

a. $P(\text{A})$

b. $P(\text{B})$

c. $P(\text{A and B})$

d. $P(\text{A or B})$

e. $P(\text{not A and not B})$

f. Find $P(\text{A or B})$ by using a rule or formula similar to your response in Step 5.

Step 7 Complete the statement: For any two events A and B, $P(\text{A or B}) = \underline{?}$.

A general form of the addition rule allows you to find the probability of an "or" statement even when two events are not mutually exclusive.

The General Addition Rule

If n_1 and n_2 represent event 1 and event 2, then the probability that at least one of the events will occur can be found by adding the probabilities of the events and subtracting the probability that both will occur.

$$P(n_1 \text{ or } n_2) = P(n_1) + P(n_2) - P(n_1 \text{ and } n_2)$$

You might wonder whether independent events and mutually exclusive events are the same. Independent events don't affect the probabilities of each other. Mutually exclusive events affect each other dramatically: If one occurs, the probability of the other occurring is 0.

But there is a connection between independent and mutually exclusive events. In calculating probabilities of non–mutually exclusive events, you use the probability that they both will occur. In the case of independent events, you know this probability.

EXAMPLE B

The probability that a rolled die comes up 3 or 6 is $\frac{1}{3}$. What's the probability that a die will come up 3 or 6 on the first or second roll?

► **Solution**

Let F represent getting a 3 or 6 on the first roll, regardless of what happens on the second roll. Let S represent getting a 3 or 6 on the second roll, regardless of what happens on the first roll.

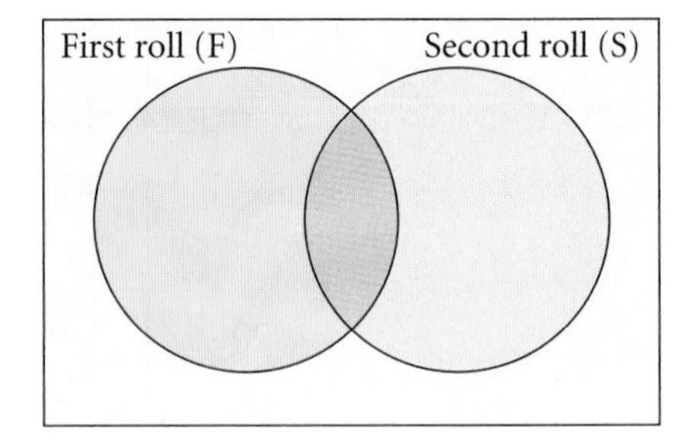

Notice that both F and S include the possibility of getting a 3 or 6 on both rolls, as shown by the overlap in the Venn diagram. To compensate for the overlap, the general addition rule subtracts $P(\text{F and S})$ once.

$P(\text{F or S}) = P(\text{F}) + P(\text{S}) - P(\text{F and S})$ — The general addition rule.

$P(\text{F or S}) = P(\text{F}) + P(\text{S}) - P(\text{F}) \cdot P(\text{S})$ — F and S are independent, so $P(\text{F and S}) = P(\text{F}) \cdot P(\text{S})$.

$P(\text{F or S}) = \frac{1}{3} + \frac{1}{3} - \left(\frac{1}{3}\right)\left(\frac{1}{3}\right)$ — Substitute the probability.

$P(\text{F or S}) = \frac{5}{9}$ — Multiply and add.

The probability of getting a 3 or 6 on the first or second roll is $\frac{5}{9}$.

When two events are mutually exclusive and make up all possible outcomes, they are referred to as **complements.** The complement of "a 1 or a 3 on the first roll" is "not a 1 and not a 3 on the first roll," an outcome that is represented by the regions outside the F circle. Because $P(\text{F})$ is $\frac{1}{3}$, the probability of the complement, $P(\text{not F})$, is $1 - \frac{1}{3}$, or $\frac{2}{3}$.

EXAMPLE C

Every student in the school music program is backstage, and no other students are present. Use O to represent the event that a student is in the orchestra, C to represent the event that a student is in the choir, and B to represent the event that a student is in the band. A reporter who approaches a student at random backstage knows these probabilities:

i. $P(\text{B or C}) = 0.8$

ii. $P(\text{not O}) = 0.6$

iii. $P(\text{C and not O and not B}) = 0.1$

iv. O and C are independent

v. O and B are mutually exclusive

a. Turn each of these statements into a statement about percentage in plain English.

b. Create a Venn diagram of probabilities describing this situation.

▸ Solution

a. Convert each probablility statement to a percentage statement.

i. $P(\text{B or C}) = 0.8$ means that 80% of the students are in band or in choir.

ii. $P(\text{not O}) = 0.6$ means that 60% of the students are not in orchestra.

iii. $P(\text{C and not O and not B}) = 0.1$ means that 10% of the students are in choir only.

iv. "O and C are independent" means that the percentage of orchestra students in choir is the same as the percentage of all students in choir. Being in orchestra does not make a student any more or less likely to be in choir, and vice versa.

v. "O and B are mutually exclusive" means that there are no students in both orchestra and band.

b. Start with a general Venn diagram showing the overlap of 3 events.

Because every student backstage is in the music program, the probability $P(\text{not C and not O and not B})$ is 0. So the region Z in the Venn diagram below has a probability of 0, and thus $Z = 0$.

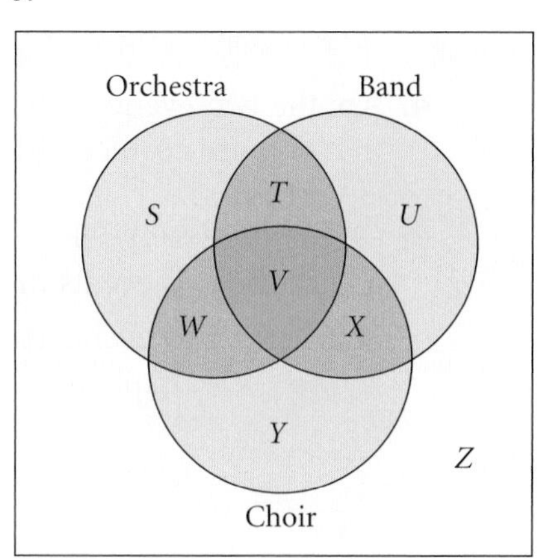

Because O and B are mutually exclusive, $T = 0$ and $V = 0$.

$P(\text{C and not O and not B}) = 0.1$ means that $Y = 0.1$.

$P(\text{B or C}) = 0.8$ means that $T + U + V + W + X + Y = 0.8$.

$P(\text{not O}) = 0.6$ means that $U + X + Y + Z = 0.6$.

The difference in the last two statements, $(T + U + V + W + X + Y) - (U + X + Y + Z)$, is $T + V + W - Z$. Because T, V, and Z all equal zero, $W = 0.8 - 0.6 = 0.2$.

$P(\text{not O}) = 0.6$ means $P(\text{O}) = 0.4$. Therefore, $S + W + T + V$ equals 0.4. Substituting the known values of W, T, and V gives $S + 0.2 = 0.4$, so $S = 0.2$.

Because O and C are independent,

$$P(\text{O}) \cdot P(\text{C}) = P(\text{O and C})$$
$$0.4(W + V + X + Y) = W + V$$
$$0.4(0.2 + 0 + X + 0.1) = 0.2 + 0$$
$$0.3 + X = \frac{0.2}{0.4} = 0.5$$
$$X = 0.2$$

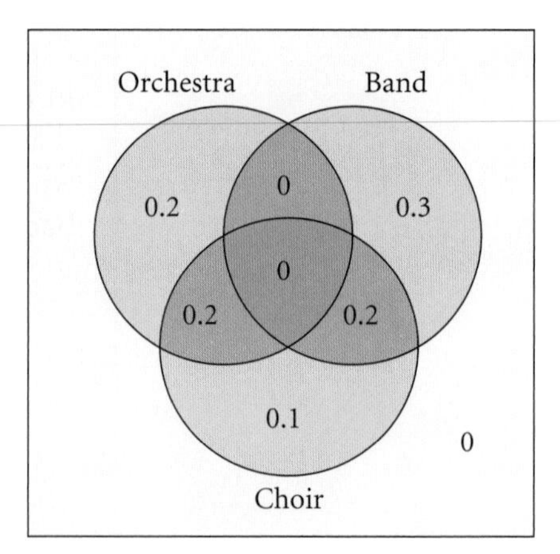

Finally returning to $U + X + Y + Z = 0.6$, substitute the values of X, Y, and Z to get $U = 0.3$.

Write the probabilities in your Venn diagram.

You should take the time to check that the sum of the probabilities in all the areas is 1. Solving this kind of puzzle exercises your understanding of probabilities and the properties of *and, or, independent,* and *mutually exclusive.*

EXERCISES

Practice Your Skills

Exercises 1–4 refer to the diagram at right. Let S represent the event that a student is a sophomore, and A represent the event that a student takes advanced algebra.

1. Describe the events represented by each of the four regions.

2. Find the probability of each event. ⓐ

a. $P(\text{S})$ **b.** $P(\text{A and not S})$

c. $P(\text{S} \mid \text{A})$ **d.** $P(\text{S or A})$

3. Suppose the diagram refers to a high school with 500 students. Change each probability into a frequency (number of students). ⓐ

4. Are the two events, S and A, independent? Show mathematically that you are correct. ⓗ

5. Events A and B are pictured in the Venn diagram at right. ⓐ

a. Are the two events mutually exclusive? Explain.

b. Are the two events independent? Assume $P(A) \neq 0$ and $P(B) \neq 0$. Explain.

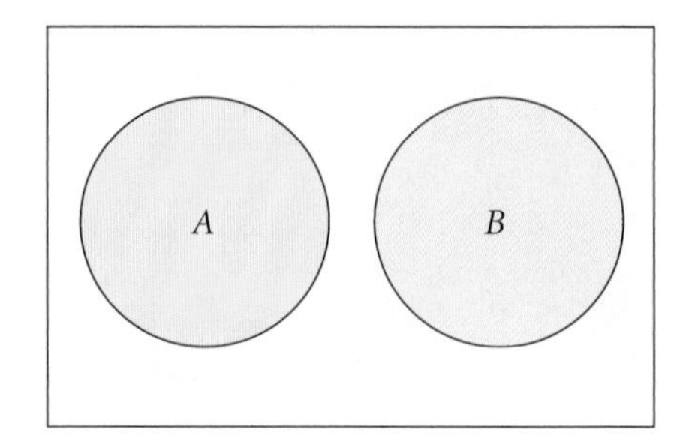

Reason and Apply

6. Of the 420 students at Middletown High School, 126 study French and 275 study music. Twenty percent of the music students take French. ⓐ

a. Create a Venn diagram of this situation.

b. What percentage of the students take both French and music?

c. How many students take neither French nor music?

7. Two events, A and B, have probabilities $P(A) = 0.2$, $P(B) = 0.4$, $P(A \mid B) = 0.2$.

a. Create a Venn diagram of this situation. ⓗ

b. Find the value of each probability indicated.

i. $P(A \text{ and } B)$
ii. $P(\text{not } B)$
iii. $P(\text{not } (A \text{ or } B))$

8. If $P(A) = 0.4$ and $P(B) = 0.5$, what range of values is possible for $P(A \text{ and } B)$? What range of values is possible for $P(A \text{ or } B)$? Use a Venn diagram to help explain how each range is possible. ⓐ

9. Assume that the diagram for Exercises 1–4 refers to a high school with 800 students. Draw a new diagram showing the probabilities if 20 sophomores moved from geometry to advanced algebra. ⓗ

10. At right are two color wheels. Figure A represents the mixing of light, and Figure B represents the mixing of pigments. ⓗ

Figure A Figure B

Using Figure A, what color is produced when equal amounts of

a. Red and green light are mixed? ⓐ

b. Blue and green light are mixed?

c. Red, green, and blue light are mixed?

Using Figure B, what color is produced when equal amounts of

d. Magenta and cyan pigments are mixed?

e. Yellow and cyan pigments are mixed?

f. Magenta, cyan, and yellow pigments are mixed?

Science CONNECTION

The three primary colors of light are red, green, and blue, each having its own range of wavelengths. When these waves reach our eyes, we see the color associated with the wave. These colors are also used to project images on TV screens and computer monitors and in lighting performances on stage. In the mixing of colors involving paint, ink, or dyes, the primary colors are cyan (greenish-blue), magenta (purplish-red), and yellow. Mixing other pigment colors cannot duplicate these three colors. Pigment color mixing is important in the textile industry as well as in art, design, and printing.

This magnified image shows how the color in a printed photograph appears on paper. The colors are combinations of cyan, magenta, yellow, and black. These four colors can be blended in varying amounts to create any color.

11. Kendra needs help on her math homework and decides to call one of her friends—Amber, Bob, or Carol. Kendra knows that Amber is on the phone 30% of the time, Bob is on the phone 20% of the time, and Carol is on the phone 25% of the time.

a. If the three friends' phone usage is independent, make a Venn diagram of the situation.

b. What is the probability that all three of her friends will be on the phone when she calls? ⓗ

c. What is the probability that none of her friends will be on the phone when she calls?

Review

12. The registered voters represented in the table have been interviewed and rated. Assume that this sample is representative of the voting public in a particular town. Find each probability. ⓐ

	Liberal	Conservative
Age under 30	210	145
Age 30–45	235	220
Age over 45	280	410

a. P(a randomly chosen voter is over 45 yr old and liberal)

b. P(a randomly chosen voter is conservative)

c. P(a randomly chosen voter is conservative if under 30 yr old)

d. P(a randomly chosen voter is under 30 yr old if conservative)

13. The most recent test scores in a chemistry class were

{74, 71, 87, 89, 73, 82, 55, 78, 80, 83, 72}

What was the average (mean) score?

14. Claire and Sydney toss a thumbtack and keep track of whether it lands point up (U) or point down (D). If $P(\text{U}) = \frac{2}{5}$ and $P(\text{D}) = \frac{3}{5}$, then what is the probability that six tosses come up with U, D, D, D, U, D in exactly this order?

15. Rewrite each expression in the form $a\sqrt{b}$, such that b contains no factors that are perfect squares.

a. $\sqrt{18}$

b. $\sqrt{54}$

c. $\sqrt{60x^3y^5}$ ⓐ

16. A bag contains two coins. One is a regular coin, with heads and tails. The second coin is a trick coin with two heads. If you choose one of the coins at random and flip it and it lands a head, what is the probability that the other side of the coin will also be a head?

17. Janie keeps track of her free throw percentage in basketball. Over the last year, she has made 50% of her free throw attempts. However, in the last three games, she made 5 out of 6 free throw attempts. Her teammate says, "You're on a hot streak, Janie!" What do you think? Explain your answer completely.

LESSON

10.4 Random Variables and Expected Value

The real measure of your wealth is how much you'd be worth if you lost all your money.

ANONYMOUS

Imagine that you are sitting near the rapids on the bank of a rushing river. Salmon are attempting to swim upstream. They must jump out of the water to pass the rapids. While sitting on the bank, you observe 100 salmon jumps and, of those, 35 are successful. Having no other information, you can estimate that the probability of success is 35% for each jump.

What is the probability that a salmon will make it on its second attempt? Note that this situation requires that two conditions be met: that the salmon fails on the first jump, and that it succeeds on the second jump. In the tree diagram, you see that this probability is $(0.65)(0.35) = 0.2275$, or about 23%.

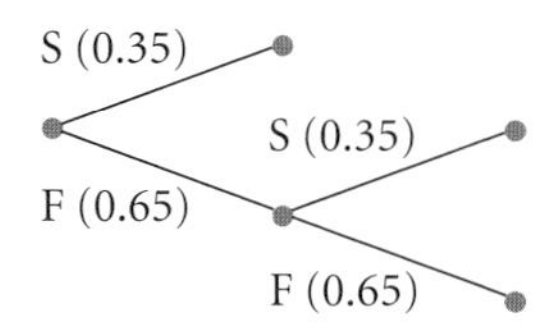

To determine the probability that the salmon makes it on the first or second jump, you sum the probabilities of the two mutually exclusive events: making it on the first jump and making it on the second jump. The sum is $0.35 + 0.2275 = 0.5775$, or about 58%.

Sockeye salmon swim upstream to spawn.

Probabilities of success such as these are often used to predict the number of independent trials required before the first success (or failure) is achieved. The salmon situation gave experimental probabilities. In the investigation you'll explore theoretical probabilities associated with dice.

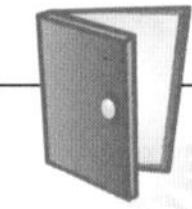

Investigation
"Dieing" for a Four

You will need

- a die

Each person will need a single die. [▸ See **Calculator Note 1L** to simulate rolling a die if you don't have one.◂] Imagine that you're about to play a board game in which you must roll a 4 on your die before taking your first turn.

Step 1 Record the number of rolls it takes for you to get a 4. Repeat this a total of ten times. Combine your results with those of your group and find the mean of all values. Then find the mean of all the group results in the class.

Step 2 Based on this experiment, how many rolls would you expect to make on average before a 4 comes up?

Step 3 To calculate the result theoretically, imagine a "perfect" sequence of rolls, with the results 1, 2, 3, 4, 5, 6, 1, 2, 3, 4, 5, 6, and so on. On average, how many rolls do you need after each 4 to get the next 4?

Step 4 Another theoretical approach uses the fact that the probability of success is $\frac{1}{6}$. Calculate the probability of rolling the first 4 on the first roll, the first 4 on the second roll, the first 4 on the third roll, and the first 4 on the fourth roll. (A tree diagram might help you do the calculations.)

Step 5 Find a formula for the probability of rolling the first 4 on the nth roll.

Step 6 Create a spreadsheet column or a calculator list with the numbers 1 through 100. Use your formula from Step 5 to make a second list of the probabilities of rolling the first 4 on the first roll, the second roll, the third roll, and so on, up to the 100th roll. Create a third list that is the product of these two lists. Calculate the sum of this third list. [▶ See **Calculator Note 2A.**◀]

Step 7 How close is the sum you found in Step 6 to your estimates in Steps 2 and 3?

The average value you found in Steps 2, 3, and 6 is called the **expected value.** It is also known as the long-run value, or mean value.

But it's the expected value of what? A numerical quantity whose value depends on the outcome of a chance experiment is called a **random variable.** In the investigation the random variable gave the number of rolls before getting a 4 on a die. You found the expected value of that random variable. The number of jumps a salmon makes before succeeding is another example of a random variable. Both of these variables are **discrete random variables** because their values are integers.

They're also called **geometric random variables** because the probabilities form a geometric sequence. Geometric random variables occur when counting the number of independent trials before something happens (success or failure).

The following example shows discrete random variables that are not geometric.

EXAMPLE A

When two fair dice are rolled, the sum of the results varies.

a. What are the possible values of the random variable, and what probabilities are associated with those values?

b. What is the expected value of this random variable?

▶ *Solution*

The random variable x has as its values all possible sums of the two dice.

a. The values of x are the integers such that $2 \leq x \leq 12$. The table shows each value and its probability, computed by counting numbers of outcomes.

x	2	3	4	5	6	7	8	9	10	11	12
Probability $P(x)$	$\frac{1}{36}$	$\frac{2}{36}$	$\frac{3}{36}$	$\frac{4}{36}$	$\frac{5}{36}$	$\frac{6}{36}$	$\frac{5}{36}$	$\frac{4}{36}$	$\frac{3}{36}$	$\frac{2}{36}$	$\frac{1}{36}$

b. The expected value of x is the theoretical average you'd expect to have after many rolls of the dice. Your intuition may tell you that the expected value is 7. One way of finding this weighted average is to imagine 36 "perfect" rolls, so every possible outcome occurs exactly once. The mean of the values is

$$\frac{2 + 3 + 3 + 4 + 4 + 4 + \cdots + 11 + 11 + 12}{36} = 7$$

If you distribute the denominator over the terms in the numerator, and group like outcomes, you get an equivalent expression that uses the probabilities:

$$\frac{1}{36} \cdot 2 + \frac{2}{36} \cdot 3 + \frac{3}{36} \cdot 4 + \frac{4}{36} \cdot 5 + \frac{5}{36} \cdot 6 + \frac{6}{36} \cdot 7 + \frac{5}{36} \cdot 8 +$$
$$\frac{4}{36} \cdot 9 + \frac{3}{36} \cdot 10 + \frac{2}{36} \cdot 11 + \frac{1}{36} \cdot 12 = 7$$

Note that each term in this expression is equivalent to the product of a value of x and the corresponding probability, $P(x)$, in the table on page 580.

Expected Value

The expected value of a random variable is an average value found by multiplying the value of each event by its probability and then summing all of the products. If $P(x)$ is the probability of an event with a value of x, then the expected value, $E(x)$, is given by the formula

$$E(x) = \sum xP(x)$$

Even if a discrete random variable has integer values, its expected value may not be an integer.

EXAMPLE B

When Nate goes to visit his grandfather, his grandfather always gives him a piece of advice, closes his eyes, and then takes out one bill and gives it to Nate. On this visit he sends Nate to get his wallet. Nate peeks inside and sees 8 bills: 2 one-dollar bills, 3 five-dollar bills, 2 ten-dollar bills, and 1 twenty-dollar bill. What is the expected value of the bill Nate will receive?

▶ Solution

The random variable x takes on 4 possible values, and each has a known probability.

Outcome x	\$1	\$5	\$10	\$20	
Probability $P(x)$	0.25	0.375	0.25	0.125	**Sum**
Product $x \cdot P(x)$	0.25	1.875	2.5	2.5	7.125

The expected value is \$7.125.

The other approach, using division, gives the same result:

$$\frac{1 + 1 + 5 + 5 + 5 + 10 + 10 + 20}{8} = \frac{57}{8} = 7.125$$

Nate doesn't actually expect Grandpa to pull out a \$7.125 bill. But if Grandpa did the activity over and over again, with the same bills, the value of the bill would average \$7.125. The expected value applies to a single trial, but it's based on an average over many imagined trials.

In Example B, suppose Nate always had the same choice of bills and that he had to pay his grandfather \$7 for the privilege of receiving a bill. Over the long run, he'd make \$0.125 per trial. On the other hand, if Nate had to pay \$8 each time, his grandfather would make \$0.875 on average per trial. This is the principle behind the way that raffles and casinos make money. Sometimes gamblers win, but on average gamblers lose.

Consumer CONNECTION

Casinos make money because on average gamblers bet more than they win. The amount by which the casino is favored is called the "house edge," calculated by finding the ratio of the casino's expected earnings to the player's initial wager. There are two ways that a casino creates an edge. In some games, like Blackjack, the probability is simply higher that the dealer will win. In a game like Roulette, the casino pays a winner at a rate less than the actual odds of winning. Either way, the casino always profits in the long run.

The name Roulette is derived from the Old French word *roelete,* meaning "little wheel."

You can use either the mean or the probability approach to find the expected value if the random variable takes on a finite number of values. But there's no theoretical limit to the number of times you might have to roll a die before getting a 4. If the random variable has infinitely many values, you must use the probability approach to find the expected value.

EXERCISES

You will need

A graphing calculator for Exercises **6** and **8.**

Practice Your Skills

1. Which of these numbers comes from a discrete random variable? Explain. @

a. the number of children who will be born to members of your class

b. the length of your pencil in inches

c. the number of pieces of mail in your mailbox today

2. Which of these numbers comes from a geometric random variable? Explain. *(h)*

a. the number of phone calls a telemarketer makes until she makes a sale

b. the number of cats in the home of a cat owner

c. the number of minutes until the radio plays your favorite song

3. Danielle conducts an experiment by tossing a fair coin three times.

a. Copy and complete the table showing the probabilities for the number of heads that can appear.

x	0	1	2	3
$P(x)$				

b. Calculate the expected number of heads for Danielle's experiment.

4. You have learned that 8% of the students in your school are left-handed. Suppose you stop students at random and ask whether they are left-handed. *(h)*

a. What is the probability that the first left-handed person you find is on your third try?

b. What is the probability that you will find exactly one left-handed person in three tries?

5. Suppose you are taking a multiple-choice test for fun. Each question has five choices (A–E). You roll a six-sided die and mark the answer according to the number on the die and leave the answer blank if the die is a 6. Each question is scored one point for a right answer, minus one-quarter point for a wrong answer, and no points for a question left blank. *(a)*

a. What is the expected value for each question?

b. What is the expected value for a 30-question test?

Reason and Apply

6. Sly asks Andy to play a game with him. They each roll a die. If the sum is greater than 7, Andy scores 5 points. If the sum is less than 8, Sly scores 4 points.

a. Find a friend and play the game ten times. [▸ See **Calculator Note 1L** if you have no dice. ◂] Record the final score.

b. What is the experimental probability that Andy will have a higher score after ten games?

c. Draw a tree diagram of one game showing the theoretical probabilities.

d. If you consider this game from Andy's point of view, his winning value is +5 and his losing value is −4. What is the expected value of the game from his point of view? *(a)*

e. Suggest a different distribution of points that would favor neither player.

7. In a concert hall, 16% of seats are in section A, 24% are in section B, 32% are in section C, and 28% are in section D. Section A seats sell for $35, section B for $30, section C for $25, and section D for $15. You see a ticket stuck high in a tree.

a. What is the expected value of the ticket? (h)

b. The markings on the ticket look like either section A or C. If this is true, then what is the probability that the ticket is for section C? (@)

c. If the ticket is for section A or C, then what is the expected value of the ticket?

8. Suppose each box of a certain kind of cereal contains a card with one letter from the word CHAMPION. The letters have been equally distributed in the boxes. You win a prize when you send in all eight letters. (@)

a. Predict the number of boxes you would expect to buy to get all eight letters.

b. Describe a method of modeling this problem using the random-number generator in your calculator.

c. Use your method to simulate winning the prize. Do this five times. Record your results.

d. What is the average number of boxes you will have to buy to win the prize?

e. Combine your results with those of several classmates. What seems to be the overall average number of boxes needed to win the prize?

9. APPLICATION Bonny and Sally are playing in a tennis tournament. On average, each of these players makes 80% of her shots. At right is a partial tree diagram showing possible sequences of hits for one point. In this point, Bonny serves. Note that once a shot is missed, that branch ends, and one of the players wins the point.

S
0.80
B
0.80
0.20
Sally
0.80
0.20
Bonny
0.20

a. What is the probability that Sally wins this point without hitting the ball? (@)

b. What is the probability that Bonny wins after she hits the ball twice (including the serve)?

c. Make a table for 0 to 5 hits showing the probability of exactly that many successful hits in one point. (@)

d. Give a formula for the probability of the point ending after exactly n successful hits.

e. What kind of sequence does the formula for 9d describe?

f. What is the probability that there will be more than 6 successful hits in a point?

Technology CONNECTION

In computer simulations of sporting events, software designers enter all the data and statistics they can find so that they can model the game as accurately as possible. In tennis, for instance, individual statistics on serves, backhand shots and forehand shots, and positions on the court are taken into account. When setting up a fantasy team or match, programmers base their formulas on actual probabilities, and even the best players or teams will lose some of the time.

American tennis players Venus and Serena Williams are shown here playing in the Women's Doubles at the 2003 Australian Open Tennis Championship in Melbourne, Australia.

10. The tree diagram at right represents a game played by two players.

a. Find a value of x that gives approximately the same expected value for each player.

b. Design a game that could be described with this tree diagram. (You may use coins, dice, spinners, or some other device.) Explain the rules of the game and the scoring of points.

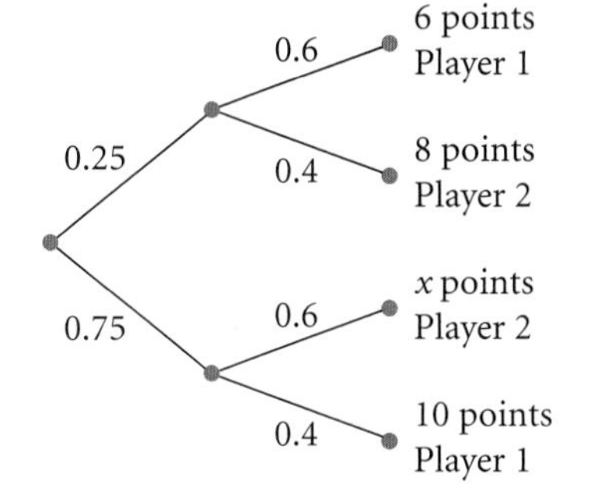

11. APPLICATION A quality-control engineer randomly selects five radios from an assembly line. Experience has shown that the probabilities of finding 0, 1, 2, 3, 4, or 5 defective radios are as indicated in the table.

a. What is the probability the engineer will find at least one defective radio in a random sample of five?

b. Make a copy of this table on your paper, and complete the entries for the column titled $x \cdot P(x)$.

c. Find the sum of the entries in the column titled $x \cdot P(x)$.

d. What is the real-world meaning of your answer in 11c?

Number of defective radios x	Probability $P(x)$	$x \cdot P(x)$
0	0.420	
1	0.312	
2	0.173	
3	0.064	
4	0.031	
5	0.000	

12. What is the probability that a 6 will not appear until the eighth roll of a fair die? (h)

Review

13. Find $P(E_1 \text{ or } E_2)$ if E_1 and E_2 are mutually exclusive and complementary. (a)

14. Two varieties of flu spread through a school one winter. The probability that a student gets both varieties is 0.18. The probability that a student gets neither variety is 0.42. Having one variety of flu does not make a student more or less likely to get the other variety. What is the probability that a student gets exactly one of the flu varieties?

15. This table gives counts of different types of paperclips in Maricela's paperclip holder.

a. Create a Venn diagram of the probabilities of picking each kind of clip if one is selected at random. Use metal, oval, and small as the categories for the three circles on your diagram.

b. Create a tree diagram of the outcomes and their probabilities. Use size on the first branch, shape on the second, and material on the third.

	Small	Large
Metal & oval	47	23
Plastic & oval	25	10
Metal & triangular	18	6
Plastic & triangular	10	5

16. Solve $\left(\frac{1}{2}\right)^n \leq 10^{-5}$.

17. Suppose the probability of winning a single game of chance is 0.9. How many wins in a row would it take before the likelihood of such a string of wins would be less than 1%? (a)

LESSON

10.5 Permutations and Probability

The numerator and denominator of a theoretical probability are numbers of possibilities. Sometimes those possibilities follow regular patterns that allow you to "count" them.

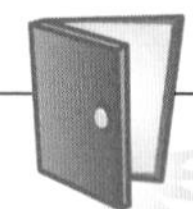

Investigation
Name That Tune

Suppose you want to create a random playlist from a library of songs on an MP3 player. If you do not repeat any songs, in how many different orders do you think the songs could be played? In this investigation you will discover a pattern allowing you to determine the number of possible orders without listing all of them.

Step 1 Start by investigating some simple cases. Consider libraries of up to five songs, and playlists of up to five of those songs. In the table, n represents the number of songs in the library ($1 \leq n \leq 5$) and r represents the length of the playlist ($1 \leq r \leq n$). For example, $n = 3$ and $r = 2$ represents the number of playlists you can make using two songs from a library of three songs. Let *A*, *B*, and *C* represent the three songs available. The two-song playlists that don't repeat songs are *AB*, *AC*, *BA*, *BC*, *CA*, and *CB*. So there are six playlists.

Copy and complete the table for different values of n and r.

Number of songs in playlist, r		Number of songs in library, n				
		$n = 1$	$n = 2$	$n = 3$	$n = 4$	$n = 5$
	$r = 1$					
	$r = 2$			6		
	$r = 3$					
	$r = 4$					
	$r = 5$					

Step 2 Compare your results with those of your group. Describe any patterns you found in either the rows or columns of the table.

Step 3 Use the patterns you found in the table to write an expression for the number of ways to arrange 10 songs in a playlist from a library of 150 songs.

There are many patterns you might have seen in the investigation. Consider each space for a song in the playlist as a slot that needs to be filled. Then one pattern can be described as a product of the numbers of outcomes that fill the slots. For example, for $n = 4$ and $r = 3$, you would make three slots because $r = 3$, and then fill the first slot with four choices and decrease by one choice for each slot as you go: $\underline{4}\ \underline{3}\ \underline{2}$. The product of those numbers is 24, the number in the third row, fourth column of the table you completed.

Why do you multiply the numbers in the slots? The problem might remind you of a familiar situation. How many different outfits—consisting of a sweater, pants, and shoes—could you wear if you were to select from four sweaters, six pairs of pants, and two pairs of shoes?

You can visualize a tree diagram with four choices of sweaters. For each of those sweaters, you can select six different pants. Each of those sweater and pants outfits can be matched with two pairs of shoes. (Actually drawing all of the paths would be difficult and messy.) Each different outfit is represented by a path of three segments representing a sweater *and* a pair of pants *and* a pair of shoes. The paths represent all the possible outcomes, or the different ways in which the entire sequence of choices can be made.

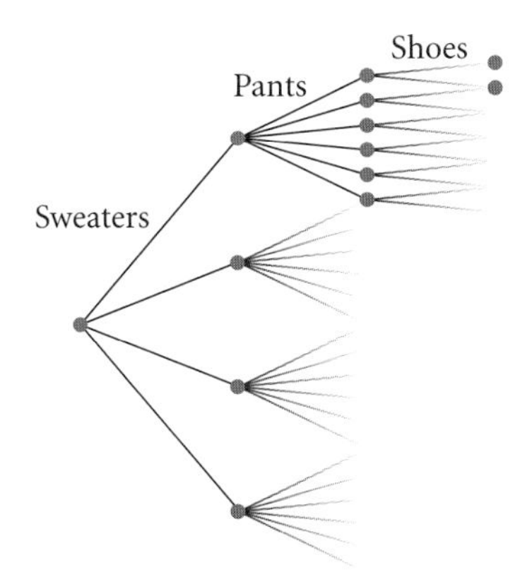

How many different outfits are there? From the tree you're imagining, you can see that the total number of different outfits with four choices, then six choices, and then two choices can be found by multiplying $4 \cdot 6 \cdot 2$. You are using the **counting principle.**

Counting Principle

Suppose there are n_1 ways to make a choice, and for each of these there are n_2 ways to make a second choice, and for each of these there are n_3 ways to make a third choice, and so on. The product $n_1 \cdot n_2 \cdot n_3 \cdot \cdots$ gives the number of possible outcomes.

The counting principle provides a quick method for calculating numbers of outcomes, using multiplication. Rather than memorizing the formula, you can look for patterns, and sketch or visualize a representative tree diagram.

EXAMPLE A

Suppose a set of license plates has any three letters from the alphabet, followed by any three digits.

a. How many different license plates are possible?

b. What is the probability that a license plate has no repeated letters or numbers?

▶ Solution

To better understand the problem, fill in some slots.

a. There are 26 possible letters for each of the first three slots and 10 possible digits for each of the last three slots. Remember that letters and digits can repeat. Using the counting principle, you can multiply the number of possibilities:

$$\underline{26} \cdot \underline{26} \cdot \underline{26} \cdot \underline{10} \cdot \underline{10} \cdot \underline{10} = 17{,}576{,}000$$

There are 17,576,000 possible different license plates.

b. Once the first letter is chosen, there are only 25 ways of choosing the second letter to avoid repetition. This pattern continues for the third letter and for the digits. Filling slots gives the product $\underline{26} \cdot \underline{25} \cdot \underline{24} \cdot \underline{10} \cdot \underline{9} \cdot \underline{8} = 11{,}232{,}000$. The probability that a license plate will be one of these arrangements is this number divided by the total number of possible outcomes:

$$\frac{11{,}232{,}000}{17{,}576{,}000} \approx 0.639$$

This means that about 63.9% of license plates would not have repeated letters or numbers.

This building is decorated with Colorado license plates.

When the objects cannot be used more than once, the number of possibilities decreases at each step. These are called "arrangements without replacement." In other words, once an item is chosen, that same item cannot be used again in the same arrangement. An arrangement of some or all of the objects from a set, without replacement, is called a **permutation.** The notation ${}_nP_r$ is read "the number of permutations of n things chosen r at a time." As in the investigation and part b of Example A, you can calculate ${}_nP_r$ by multiplying $n(n-1)(n-2)(n-3)\cdots(n-r+1)$. [▶ See **Calculator Note 10B** to learn how to compute ${}_nP_r$ on a calculator.◀]

EXAMPLE B

Seven flute players are performing in an ensemble.

a. The names of all seven players are listed in the program in random order. What is the probability that the names are in alphabetical order?

b. After the performance, the players are backstage. There is a bench with room for only four to sit. How many possible seating arrangements are there?

c. What is the probability that the group of four players is sitting in alphabetical order?

▶ Solution

You can find numbers of permutations by filling in slots.

a. There are seven choices for the first name on the list, six choices remaining for the second name, five for the third name, and so on.

$${}_7P_7 = \underline{7} \cdot \underline{6} \cdot \underline{5} \cdot \underline{4} \cdot \underline{3} \cdot \underline{2} \cdot \underline{1} = 5040$$

Only one of these arrangements is in alphabetical order. The probability is $\frac{1}{5040}$, or approximately 0.0002.

b. In this case there are only four slots to fill. There are seven choices for the first seat, six choices remaining for the second seat, five for the third seat, and four for the fourth seat.

$${}_7P_4 = \underline{7} \cdot \underline{6} \cdot \underline{5} \cdot \underline{4} = 840$$

c. With each arrangement of 4, there is only one correct order.

$${}_4P_4 = 4 \cdot 3 \cdot 2 \cdot 1 = 24$$

There are 24 ways to arrange each grouping of 4 players, so the probability is $\frac{1}{24}$, or approximately 0.04167.

Notice that the answer to part c does not depend on the answer to part b.

When $r = n$, as in parts a and c of Example B, you can see that ${}_nP_r$ equals the product of integers from n all the way down to 1. A product like this is called a **factorial** and is written with an exclamation point. For example, 7 factorial, or 7!, is $7 \cdot 6 \cdot 5 \cdot 4 \cdot 3 \cdot 2 \cdot 1$, or 5040. [▶ See **Calculator Note 10C** to learn how to evaluate factorials on a calculator. ◀]

Look again at Example B. You can write the solution to part a as

$${}_7P_7 = 7! = 5040$$

For part b, you need only the product of integers from 7 down to 4. Notice that

$$7 \cdot 6 \cdot 5 \cdot 4 = \frac{7 \cdot 6 \cdot 5 \cdot 4 \cdot \cancel{3} \cdot \cancel{2} \cdot \cancel{1}}{\cancel{3} \cdot \cancel{2} \cdot \cancel{1}} = \frac{7!}{3!} = \frac{7!}{(7-4)!}$$

You can use this idea to write a formula for any number of permutations, ${}_nP_r$. As you saw above, you can write the number as $\frac{n!}{(n-r)!}$.

You saw in Example B that ${}_7P_7 = 7! = 5040$. However, using the formula from the previous page,

$${}_7P_7 = \frac{7!}{(7-7)!} = \frac{7!}{0!}$$

To prevent inconsistency, we define 0! as equal to 1. So,

$${}_7P_7 = \frac{7!}{0!} = \frac{7!}{1} = 7!$$

Permutations

A **permutation** is an arrangement of some or all of the objects from a set, without replacement.

The number of permutations of n objects chosen r at a time ($r \leq n$) is

$${}_nP_r = n(n-1)(n-2)\cdots(n-r+1) = \frac{n!}{(n-r)!}$$

Verify that the formula above gives you the same values you found in the investigation.

In this lesson you used tree diagrams, the counting principle, your calculator, and perhaps other ways to count permutations. As you do the exercises, consider which strategy is best for each particular problem.

EXERCISES

You will need

A graphing calculator for Exercises **10** and **18.**

Practice Your Skills

1. Screamers Ice Cream Parlor sells a triple-scoop cone. Which of these situations are permutations? For those that are not, tell why. ⓐ

a. the different cones if all three scoops are different flavors and vanilla, then lemon, then mint is considered different from vanilla, then mint, then lemon

b. the different cones if all three scoops are different flavors and vanilla, then lemon, then mint is considered the same as vanilla, then mint, then lemon

c. the different cones if you can repeat a flavor two or three times and vanilla, then lemon, then lemon is different from lemon, then vanilla, then lemon

d. the different cones if you can repeat a flavor two or three times and vanilla, then lemon, then lemon is the same as lemon, then vanilla, then lemon

2. Evaluate the factorial expressions. (Some answers will be in terms of n.)

a. $\frac{12!}{11!}$ @

b. $\frac{7!}{6!}$

c. $\frac{(n+1)!}{n!}$

d. $\frac{n!}{(n-1)!}$

e. $\frac{120!}{118!}$ @

f. $\frac{n!}{(n-2)!}$ @

g. Find n if $\frac{(n+1)!}{n!} = 15$.

3. Evaluate each expression. (Some answers will be in terms of n.)

a. ${}_7P_3$ @

b. ${}_7P_6$

c. ${}_{n+2}P_n$ @

d. ${}_nP_{n-2}$

4. Consider making a four-digit I.D. number using the digits 3, 5, 8, and 0.

a. How many I.D. numbers can be formed using each digit once? @

b. How many can be formed using each digit once and not using 0 first?

c. How many can be formed if repetition is allowed and any digit can be first? @

d. How many can be formed if repetition is allowed but 0 is not used first?

5. APPLICATION A combination lock has four dials. On each dial are the digits 0 to 9.

a. Suppose you forget the correct combination to open the lock. How many combinations are possible? If it takes 10 s to enter each combination, how long will it take you to try every possibility? @

b. Suppose you replace your lock with one that has five dials, each with the digits 0 to 9. How many combinations are possible? If it still takes 10 s to enter each combination, how long will it take to try every possibility?

c. For a lock to be secure, it has to be difficult for someone to guess the correct combination. How many times as secure as a 4-dial lock is a 5-dial lock?

Consumer CONNECTION

Gaining access to personal bank, telephone, and e-mail accounts often requires using a password, usually a minimum of 4 digits and/or letters. Most businesses recommend that consumers change their passwords frequently and choose a password with more than six characters using a mix of digits and uppercase and lowercase letters to increase the safety of their accounts. Given a particular set of requirements for a password, you can use the counting principle to determine the number of different passwords that are possible.

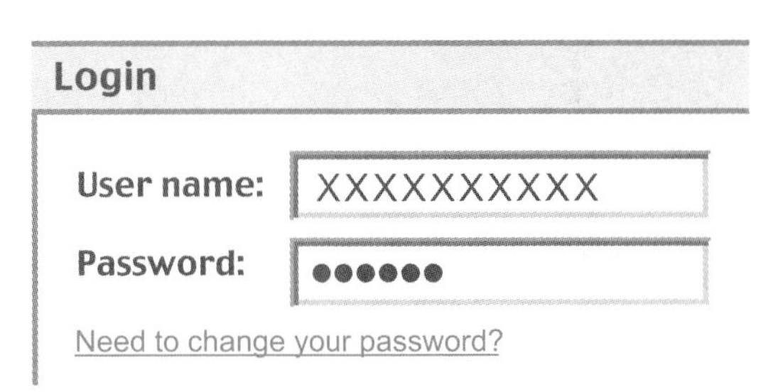

Reason and Apply

6. For what value(s) of n and r does ${}_nP_r = 720$? Is there more than one answer? (h)

7. How many factors are in the expression $n(n-1)(n-2)(n-3)\cdots(n-r+1)$?

8. Suppose each student in a school is assigned one locker. How many ways can three new students be assigned to five available lockers? @

9. An eight-volume set of reference books is kept on a shelf. The books are used frequently and put back in random order.

 a. How many ways can the eight books be arranged on the shelf? ⓐ

 b. How many ways can the books be arranged so that Volume 5 will be the rightmost book?

 c. Use the answers from 9a and b to find the probability that Volume 5 will be the rightmost book if the books are arranged at random.

 d. Explain how to compute the probability in 9c using another method.

 e. If the books are arranged randomly, what is the probability that the last book on the right is an even-numbered volume? Explain how you determined this probability.

 f. How many ways can the books be arranged so that they are in the correct order, with volume numbers increasing from left to right?

 g. How many ways can the books be arranged so that they are out of order?

 h. What is the probability that the books happen to be in the correct order?

10. Suppose a computer is programmed to list all permutations of N items. Use the values given in the table at right to figure out how long it will take for the computer to list all of the permutations for the remaining values of N listed. Use an appropriate time unit for each answer (minutes, hours, days, or years). ⓗ

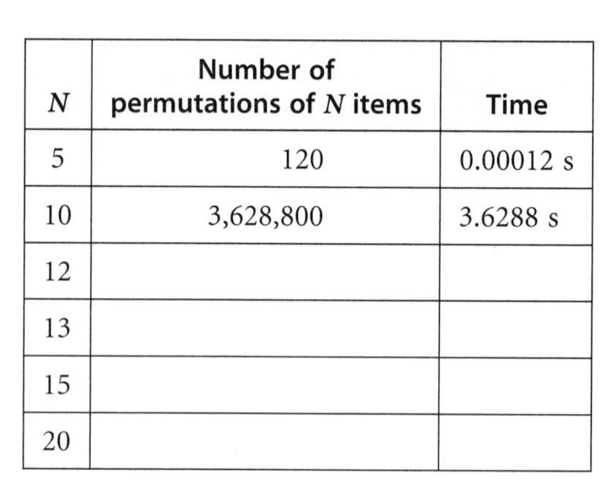

N	Number of permutations of N items	Time
5	120	0.00012 s
10	3,628,800	3.6288 s
12		
13		
15		
20		

11. **APPLICATION** In 1963, the U.S. Postal Service introduced the ZIP code to help process mail more efficiently.

 a. A ZIP code contains five digits, 0–9. How many possible ZIP codes are there?

 b. In 1983, the U.S. Postal Service introduced ZIP + 4. The extra four digits at the end of the ZIP code help pinpoint the destination of a parcel with greater accuracy and efficiency. How many possible ZIP + 4 codes are there?

The Canadian postal service uses a six-character mailing code of the form

letter, digit, letter, digit, letter, digit

 c. How many possible Canadian postal codes are there if no restrictions are placed on the letters and digits?

 d. In Canadian postal codes, the letters D, F, I, O, Q, and U are never used and the letters W and Z are not used as the first characters. How many possible postal codes are there now?

Postal workers sort through bulk mail.

12. Suppose you are creating a sound track of music for a video. You have 500 great songs in your library, but you need only 6 songs for the video.

  a. How many different sound tracks can you create with no repeated songs?

  b. If the songs average 4 min in length, about how long would it take to play all of the songs in your library?

  c. If the songs average 4 min in length, how long would it take to play all of the possible sound tracks for your video?

13. APPLICATION Biologists use Punnett squares to represent the ways that genes can be passed from parent to offspring. In the Punnett squares at right, B stands for brown eyes, a dominant trait, whereas b stands for blue eyes, a recessive trait. E represents brown hair, the dominant trait, whereas e represents the recessive trait, blonde hair. For any particular trait, if a dominant gene is present, a recessive trait will be hidden. The first row and column represent the parents' genes. The rest of the cells represent the possible pairs of genes the parents could pass on to their offspring.

Table 1

	B	b
B	BB	Bb
B	BB	Bb

Table 2

	B	b
B	BB	Bb
b	Bb	bb

Table 3

	b	b
B	Bb	Bb
b	bb	bb

Table 4

	b	b
b	bb	bb
b	bb	bb

Table 5

	BE	BE	bE	bE
Be	BBEe	BBEe	BbEe	BbEe
Be	BBEe	BBEe	BbEe	BbEe
be	BbEe	BbEe	bbEe	bbEe
be	BbEe	BbEe	bbEe	bbEe

  a. In Table 1, both parents are brown-eyed, but one parent has a Bb gene combination, whereas the other has a BB gene combination. What is the probability of a blue-eyed (bb) offspring?

  b. In Table 2, both parents are brown-eyed, each having the Bb gene combination. What is the probability of a blue-eyed (bb) offspring? @

  c. In Table 3, one parent is blue-eyed (bb), whereas the other is brown-eyed (Bb). What is the probability of a blue-eyed (bb) offspring?

  d. According to Table 4, is it possible for two blue-eyed parents to have a child with brown eyes? Explain.

  e. In Table 5, one parent is brown-eyed and brown-haired (BbEE). The other parent is brown-eyed and blonde-haired (Bbee). What is the probability of a blue-eyed, brown-haired child? @

Science CONNECTION

Genetics helps predict the likelihood of inheriting particular traits. Genetics can help animal breeders develop breeds with more desirable qualities than existing breeds have. And new strains of disease-resistant crops can be developed by crossbreeding existing plants. Genetic counselors can test the genes of human parents to see whether they are carriers of genetic diseases, such as sickle-cell anemia, Tay-Sachs disease, or hemophilia, which are passed on from generation to generation. To learn more about the science of genetics, visit the weblinks at www.keymath.com/DAA .

This photo shows an image of a karyotype, all 23 pairs of chromosomes found in every human cell. The genes in these chromosomes determine individual characteristics.

Review

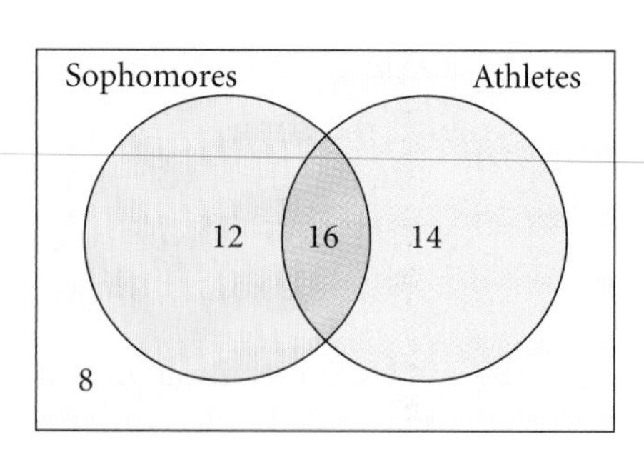

14. In a gym class of 50 students, 28 are sophomores and 30 are athletes. Refer to the Venn diagram at right.

a. What is the probability that a randomly selected member of the class is an athlete?

b. If you already know a randomly selected member is an athlete, what is the probability that this student is a sophomore?

15. A fair six-sided die is rolled. If the number showing is even, you lose a point for each dot showing. If the number showing is odd, you win a point for each dot showing.

a. Find $E(x)$ for one roll if x represents the number of points you win.

b. Find the expected winnings for ten rolls.

16. Assume that boys and girls are equally likely to be born.

a. If there are three consecutive births, what is the probability of two girls and one boy being born, in that order? ⓐ

b. What is the probability that two girls and one boy are born in any order?

c. Given that the first two babies born are girls, what is the probability that the third will be a boy?

17. Write the equation of the parabola at right in

a. Polynomial form. **b.** Vertex form. **c.** Factored form.

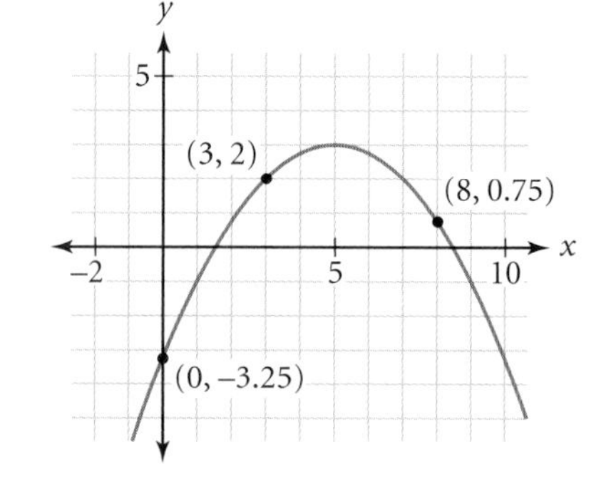

18. As an art project, Jesse is planning to make a set of nesting boxes (boxes that fit inside each other). Each box requires five squares that will fold into an open box with no lid. The largest box will measure 4 in. on each side. The side length of each successive box is 95% of the previous box. Jesse wants the smallest box to be no less than 0.5 in. on a side. ⓗ

a. How many boxes can Jesse make in one set?

b. What is the total amount of paper needed for one set of open boxes?

IMPROVING YOUR VISUAL THINKING SKILLS

A Perfect Arrangement

Craig has a summer landscape business planting trees and shrubs. His neighbor, Mr. Malone, contracted with him to plant 5 rows of fruit trees with 4 trees in each row. He said that he had all the trees ready for him to plant. When Craig arrived to do the planting, he found 10 trees. He was about to point out the mistake, but just completed the job instead. What did he do?

LESSON

10.6

Combinations and Probability

Math is like love—a simple idea but it can get complicated.

R. DRABEK

If three coins are flipped, a tree diagram and the counting principle both indicate that there are $\underline{2} \cdot \underline{2} \cdot \underline{2}$, or 8, equally likely outcomes: 2 choices, then 2 choices, then 2 choices. But if you are not concerned about the order in which the heads and tails occur, then you can describe paths 2, 3, and 5 as "2 heads and 1 tail" and paths 4, 6, and 7 as "1 head and 2 tails." So if you're not concerned about order, there are only 4 events, which are not equally likely:

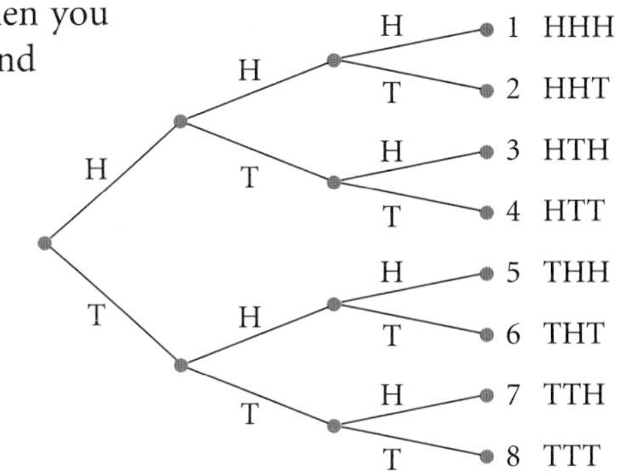

3 heads (one path)

2 heads and 1 tail (three paths)

1 head and 2 tails (three paths)

3 tails (one path)

In this lesson you will learn about counting outcomes and calculating theoretical probabilities when order doesn't matter. There are generally fewer possibilities if order doesn't matter than there are if order is important.

EXAMPLE A

At the first meeting of the International Club, the members get acquainted by introducing themselves and shaking hands. Each member shakes hands with every other member exactly once. How many handshakes are there in each of the situations listed below?

a. Three people meet.

b. Four people meet.

c. Five people meet.

d. Fifteen people meet.

▸ Solution

The points (vertices) pictured in the diagrams at right can represent the three, four, or five people in a room, and the lines (edges) can represent the handshakes. The diagrams show that there are 3 handshakes among 3 people, 6 handshakes among 4 people, and 10 handshakes among 5 people. You can find the number of handshakes by counting edges, but as you add more people to the group, it will become more difficult to draw and count. So look for patterns to determine the number of handshakes among 15 people.

3 people
3 handshakes

4 people
6 handshakes

5 people
10 handshakes

Each person shakes hands with everyone else, so if there are n people, each person shakes $n - 1$ hands. This might suggest that there are $n \cdot (n - 1)$ handshakes. However, this calculation counts each handshake twice, because you have counted A shaking hands with B and B shaking hands with A as two different handshakes. Therefore, n people will have $\frac{n(n-1)}{2}$ handshakes. So the number of handshakes for 15 people is $\frac{15 \cdot 14}{2}$, or 105 handshakes. You can verify that this formula is accurate for $n = 3$, 4, and 5.

You can think of each handshake as a pairing of two of the people in the room, or two of the vertices in the diagram. When you count collections of people or objects without regard to order, you are counting **combinations.**

The number of combinations of 4 people taken 2 at a time is symbolized by ${}_4C_2$. (The notation ${}_4C_2$ can be read as "four choose two.") Although there are ${}_4P_2$, or 12, permutations of 4 people taken 2 at a time, you have only half as many combinations:

In counting permutations, order matters, so AB and BA are both counted.

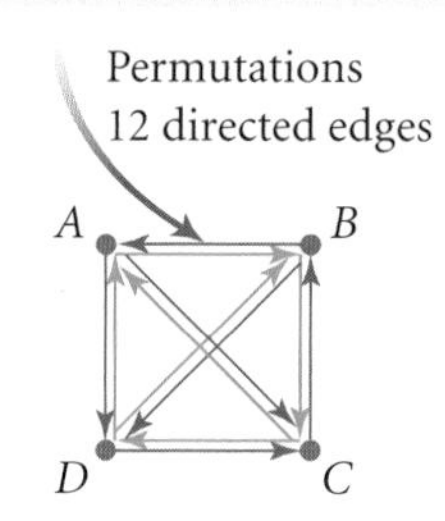

In counting combinations, order does not matter, so AB and BA count as a single arrangement.

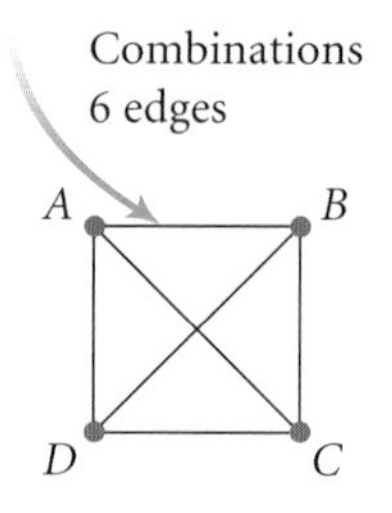

$${}_4C_2 = \frac{{}_4P_2}{2} = \frac{12}{2} = 6$$

You can calculate the number of handshakes in Example A in the same way:

$${}_{15}C_2 = \frac{{}_{15}P_2}{2} = \frac{15 \cdot 14}{2} = 15 \cdot 7 = 105$$

There are 105 handshakes.

EXAMPLE B

Anna, Ben, Chang, and Dena are members of the International Club, and they have volunteered to be on a committee that will arrange a reception for exchange students. Usually there are only three students on the committee. How many different three-member committees could be formed with these four students?

▸ Solution

Note that order isn't important in these committees. ABD and BDA are the same committee and shouldn't be counted more than once. The number of different committee combinations will be smaller than the number of permutations, ${}_4P_3$, or 24, listed on the next page.

ABC	**ABD**	**ACD**	**BCD**
ACB	ADB	ADC	BDC
BAC	BAD	CAD	CBD
BCA	BDA	CDA	CDB
CAB	DAB	DAC	DBC
CBA	DBA	DCA	DCB

Each of the four committees listed in the top row can represent all of the 3!, or 6, permutations listed in its column. Therefore, the number of combinations, ${}_4C_3$, is one-sixth the number of permutations. That is, ${}_4C_3 = \frac{{}_4P_3}{3!}$. You can evaluate ${}_4C_3$ using the factorial definition of ${}_nP_r$:

$${}_4C_3 = \frac{{}_4P_3}{3!} = \frac{4!}{3!(4-3)!} = \frac{4!}{3!\,1!} = \frac{4 \cdot \not{3} \cdot \not{2} \cdot \not{1}}{\not{3} \cdot \not{2} \cdot \not{1} \cdot 1} = 4$$

Combinations

A **combination** is a grouping of some or all of the objects from a set without regard to order.

The number of combinations of n objects chosen r at a time ($r \leq n$) is

$${}_nC_r = \frac{{}_nP_r}{r!} = \frac{n!}{r!(n-r)!}$$

keymath.com/DAA

Rather than simply memorizing the formula given above, try to understand how numbers of combinations relate to numbers of permutations and to a tree diagram. You may want to draw a representation of each problem you investigate.

You can count combinations to calculate theoretical probabilities.

EXAMPLE C

Suppose a coin is flipped 10 times.

a. What is the probability that it will land heads exactly five times?

b. What is the probability that it will land heads exactly five times, including on the third flip?

▶ *Solution*

The tree diagram for this problem has ten stages (one for each flip) and splits into two possibilities (heads or tails) at each point on the path. It's not necessary to draw the entire tree diagram. By the counting principle there are 2^{10}, or 1024, possibilities.

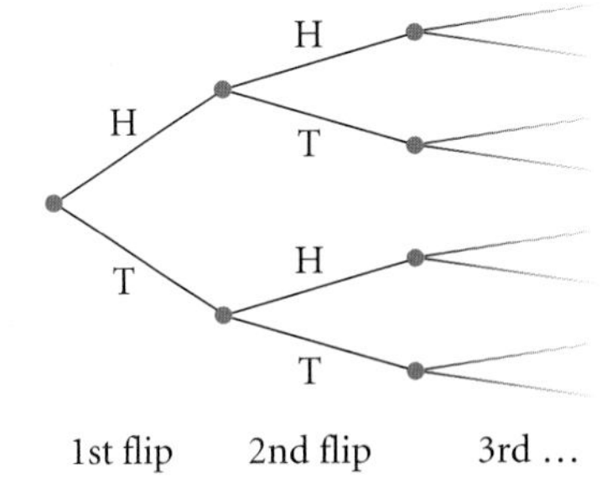

a. To find the numerator of the probability ratio, you must determine how many of the 1024 separate paths contain 5 heads. Because order is not important, you can find the number of paths that fit this description by counting combinations. There are ${}_{10}C_5 = \frac{10!}{5!\,5!} = 252$ ways of choosing 5 of the 10 flips to contain H's. [▶ See **Calculator Note 10D** to learn how to find numbers of combinations on a calculator. ◀] That is, 252 of the 1024 paths will contain exactly 5 heads.

Therefore, the probability that you will get exactly 5 heads and 5 tails is $\frac{252}{1024}$, or approximately 0.246.

b. If there's a head on the third flip, then the other four heads must be chosen from the other nine flips. There are ${}_9C_4 = \frac{9!}{5!\,4!} = 126$ ways of making that choice. Therefore, the probability of this event is $\frac{126}{1024}$, or approximately 0.123.

In the investigation you'll count combinations to discover mathematical reasons why playing the lottery is a losing proposition.

Investigation
Winning the Lottery

You will do Step 1 of this investigation with the whole class. Then, in Steps 2–8, you will work with your group to analyze the results.

Consider a state lottery called Lotto 47. Twice a week, players select six different numbers between 1 and 47. The state lottery commission also selects six numbers between 1 and 47. Selection order doesn't matter, but a player needs to match all six numbers to win Lotto 47.

Step 1 Follow these directions with your class to simulate playing Lotto 47.

a. For five minutes, write down as many sets of six different numbers as you can. Write only integers between 1 and 47.

b. After five minutes of writing, everyone stands up.

c. Your teacher will generate a random integer, 1–47. Cross out all of your sets of six numbers that do not contain the given number. If you cross out all of your sets, sit down.

d. Your teacher will generate a second number, 1–47. (If it's the same number as before, it will be skipped.) Again, cross out all of your sets that do not contain this number. If you cross out all of your sets, sit down.

e. Your teacher will continue generating different random numbers until no one is still standing or six numbers have been generated.

Work with your group to answer the questions in Steps 2–8.

Step 2 What is the probability that any one set of six numbers wins?

Step 3 At $1 for each set of six numbers, how much did each of your group members invest during the first five minutes? What was the total group investment?

Step 4 Estimate the total amount invested by the entire class during the first five minutes. Explain how you determined this estimate.

Step 5 Estimate the probability that someone in your class wins. Explain how you determined this estimate.

Step 6 Estimate the probability that someone in your school would win if everyone in the school participated in this activity. Explain how you determined this estimate.

Step 7 If each possible set of six numbers were written on a 1-inch chip and if all the chips were laid end to end, how long would the line of chips be? Convert your answer to an appropriate unit.

Step 8 Write a paragraph comparing Lotto 47 with some other event whose probability is approximately the same.

EXERCISES

Practice Your Skills

1. Evaluate each expression without using your calculator.

a. $\frac{10!}{3!\,7!}$ ⓐ **b.** $\frac{7!}{4!\,3!}$ **c.** $\frac{15!}{13!\,2!}$ **d.** $\frac{7!}{7!\,0!}$ ⓐ

2. Evaluate each expression.

a. ${}_{10}C_7$ **b.** ${}_7C_3$ ⓐ **c.** ${}_{15}C_2$ **d.** ${}_7C_0$

3. Consider each expression in the form ${}_nP_r$ and ${}_nC_r$.

a. What is the relationship between ${}_7P_2$ and ${}_7C_2$? ⓐ

b. What is the relationship between ${}_7P_3$ and ${}_7C_3$?

c. What is the relationship between ${}_7P_4$ and ${}_7C_4$?

d. What is the relationship between ${}_7P_7$ and ${}_7C_7$?

e. Describe how you can find ${}_nC_r$ if you know ${}_nP_r$.

4. Which is larger, ${}_{18}C_2$ or ${}_{18}C_{16}$? Explain.

Reason and Apply

5. For what value(s) of n and r does ${}_nC_r$ equal 35? ⓗ

6. Find a number r, $r \neq 4$, such that ${}_{10}C_r = {}_{10}C_4$. Explain why your answer makes sense.

7. Suppose you need to answer any four of seven essay questions on a history test and you can answer them in any order. ⓐ

a. How many different question combinations are possible?

b. What is the probability that you include Essay Question 5 if you randomly select your combination?

8. Does a "combination lock" really use combinations of numbers? Should it be called a "permutation lock"? Explain. ⓗ

9. Find the following sums. ⓐ

a. ${}_2C_0 + {}_2C_1 + {}_2C_2$

b. ${}_3C_0 + {}_3C_1 + {}_3C_2 + {}_3C_3$

c. ${}_4C_0 + {}_4C_1 + {}_4C_2 + {}_4C_3 + {}_4C_4$

d. Make a conjecture about the sum of ${}_nC_0 + {}_nC_1 + \cdots + {}_nC_n$. Test it by finding the sum for all possible combinations of five things.

10. Consider the Lotto 47 game, which you simulated in the Investigation Winning the Lottery.

a. If it takes someone 10 seconds to fill out a Lotto 47 ticket, how long would it take him or her to fill out all possible tickets?

b. If someone fills out 1000 tickets, what is his or her probability of winning Lotto 47? ⓐ

11. Draw a circle and place points on its circumference. Draw chords to connect all pairs of points. How many chords are there if you place ⓗ

a. 4 points? **b.** 5 points? **c.** 9 points? **d.** n points?

12. In most state and local courts, 12 jurors and 2 alternates are chosen from a pool of 30 prospective jurors. The order of the alternates is specified. If a juror is unable to serve, then the first alternate will replace that juror. The second alternate will be called on if another juror is dismissed.

a. In how many ways can 12 jurors and first and second alternates be chosen from 30 people? ⓐ

b. In federal court cases, 12 jurors and 4 alternates are usually selected from a pool of 64 prospective jurors. In how many ways can this be done?

Social Science CONNECTION

The U.S. Constitution guarantees a person accused of a crime the right to a speedy and public trial by an impartial jury. In the jury selection process, both the defense and the prosecution can "challenge" (request that the judge dismiss) potential jurors. Each side tries to seat a jury they feel will benefit their case. Often, jury consultants are brought in to evaluate prospective jurors and to investigate community and individual attitudes and biases.

A jury box

13. You have purchased four tickets to a charity raffle. Only 50 tickets were sold. Three tickets will be drawn, and first, second, and third prizes will be awarded.

a. What is the probability that you win the first prize (and no other prize)? ⓗ

b. What is the probability that you win both the first and second prizes, but not the third prize?

c. What is the probability that you win the second or third prize? ⓐ

d. If the first, second, and third prizes are gift certificates for \$25, \$10, and \$5, respectively, what is the expected value of your winnings?

Review

14. Expand each expression.

a. $(x + y)^2$ **b.** $(x + y)^3$ ⓗ **c.** $(x + y)^4$

15. How many speeds, or combinations of gears, does a bicycle have if it has three gears in front and five gears in the rear? ⓐ

16. Use the tree diagram at right. Find each probability and explain its meaning.

a. $P(\text{H and P})$ ⓐ **b.** $P(\text{P} \mid \text{H})$

c. $P(\text{P})$ **d.** $P(\text{H} \mid \text{P})$

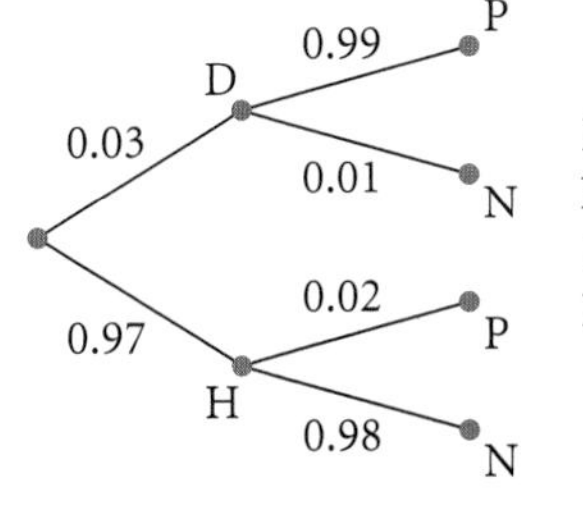

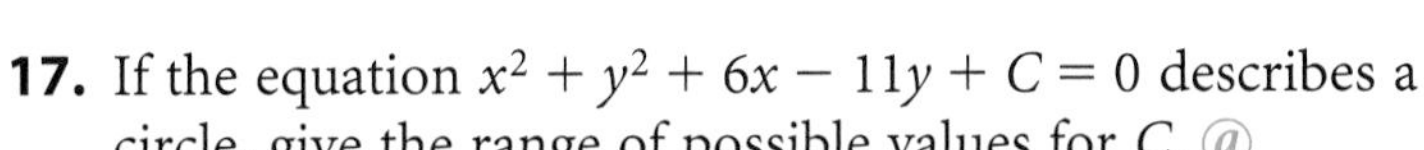

17. If the equation $x^2 + y^2 + 6x - 11y + C = 0$ describes a circle, give the range of possible values for C. ⓐ

18. Suppose the value of a certain building depreciates at a rate of 6% per year. When new, the building was worth $36,500.

a. How much is the building worth after 5 years and 3 months?

b. To the nearest month, when will the building be worth less than $10,000?

19. Find the coordinates of each labeled point.

a.

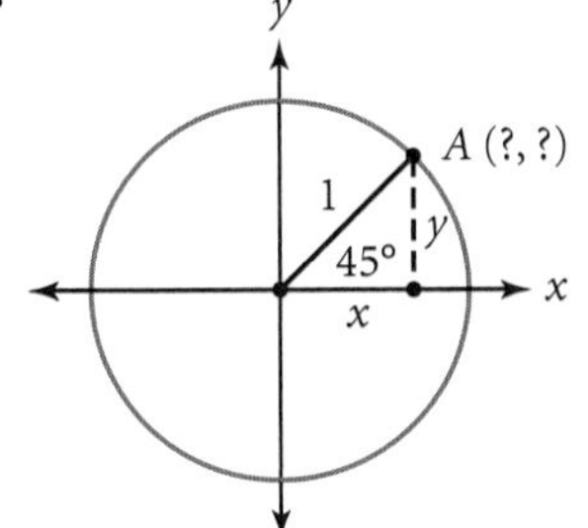

b.

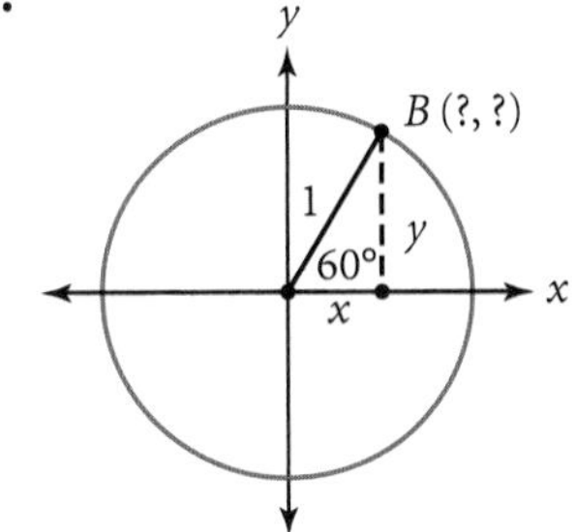

IMPROVING YOUR REASONING SKILLS

A Change of Plans

Angelo takes the train into the city for his weekly guitar lesson. After his lesson his mother picks him up at the train station knowing he will be on the 2:00 P.M. train. She always gets to the station just as the train arrives. This week when Angelo arrived for his lesson, he found out that his teacher was ill and had to cancel the lesson. Angelo went back to the train station and just managed to get the 1:00 P.M. train. He didn't have time to call his mother about the change. When he arrived at his station, he decided to start walking home until he met his mother en route to the station. When he saw her, he flagged her down and hopped into the car. They arrived home 20 min ahead of their usual time. How long did Angelo walk before he met his mother?

LESSON

10.7 The Binomial Theorem and Pascal's Triangle

Mathematics is the science of patterns.

LYNN ARTHUR STEEN

Probability is an area of mathematics that is rich with patterns. Many random processes, such as flipping coins, involve patterns in which there are two possible outcomes. In this lesson you will learn about using the binomial theorem and Pascal's triangle to find probabilities in those cases.

1

1 1

1 2 1

1 3 3 1

1 4 6 4 1

1 5 10 10 5 1

Shown at left are the first six rows of **Pascal's triangle.** It contains many different patterns that have been studied for centuries. The triangle begins with a 1, then there are two 1's below it. Each successive row is filled in with numbers formed by adding the two numbers above it. For example, each of the 4's is the sum of a 1 and a 3 in the previous row. Every row begins and ends with a 1.

In Lesson 10.6, you studied numbers of combinations. These numbers occur in the rows of Pascal's triangle. For example, the numbers 1, 5, 10, 10, 5, and 1 in the sixth row are the values of ${}_5C_r$:

$${}_5C_0 = 1 \quad {}_5C_1 = 5 \quad {}_5C_2 = 10 \quad {}_5C_3 = 10 \quad {}_5C_4 = 5 \quad {}_5C_5 = 1$$

Is this the case in all rows? If so, why? In the investigation you'll explore these questions.

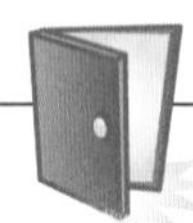

Investigation
Pascal's Triangle and Combination Numbers

A group of five students regularly eats lunch together, but each day only three of them can show up.

Step 1 How many groups of three students could there be? Express your answer in the form ${}_nC_r$ and as a numeral.

Step 2 If Leora is definitely at the table, how many other students are at the table? How many students are there to choose from? Find the number of combinations of students possible in this instance. Express your answer in the form ${}_nC_r$ and as a numeral.

Step 3 How many three-student combinations are there that don't include Leora? Consider how many students there are to select from and how many are to be chosen. Express your answer in the form ${}_nC_r$ and as a numeral.

Step 4 Repeat Steps 1–3 for groups of four of the five students.

Step 5 | What patterns do you notice in your answers to Steps 1–3 for groups of three students and four students? Write a general rule that expresses ${}_nC_r$ as a sum of other combination numbers.

Step 6 | How does this rule relate to Pascal's triangle?

Pascal's triangle can provide a shortcut for expanding binomials. For example, the **expansion** of $(x + y)^3$ is $1x^3 + 3x^2y + 3xy^2 + 1y^3$. Note that the coefficients of this expansion are the numbers in the fourth row of Pascal's triangle.

Why are the numbers in Pascal's triangle equal to the coefficients of a binomial expansion? Remember that the numbers in Pascal's triangle are evaluations of ${}_nC_r$. So the question could also be asked, why are the coefficients of a binomial expansion equal to values of ${}_nC_r$?

EXAMPLE A | Expand $(\text{H} + \text{T})^3$. Relate the coefficients in the expansion to combinations.

▸ Solution | You can write $(\text{H} + \text{T})^3$ as $(\text{H} + \text{T})(\text{H} + \text{T})(\text{H} + \text{T})$. To expand this product, multiply in steps. First you multiply the first binomial, $(\text{H} + \text{T})$, by the second binomial, $(\text{H} + \text{T})$:

$$(\text{H} + \text{T})(\text{H} + \text{T}) = (\text{HH} + \text{HT} + \text{TH} + \text{TT})$$

Then you multiply each term of that result with the H and the T in the third binomial:

$$(\text{HH} + \text{HT} + \text{TH} + \text{TT})(\text{H} + \text{T})$$
$$= \text{HHH} + \text{HHT} + \text{HTH} + \text{HTT} + \text{THH} + \text{THT} + \text{TTH} + \text{TTT}$$

Notice that if H represents flipping a head, and T represents flipping a tail, this result shows you all the possible outcomes of flipping a coin three times. If the order does not matter, you can also write this expression as

$$1\text{H}^3 + 3\text{H}^2\text{T} + 3\text{HT}^2 + 1\text{T}^3$$

This expression shows that there is one way of getting three heads, three ways of getting two heads and a tail, and so on. Considering combinations, the number of ways of choosing three heads from three coin tosses is ${}_3C_3$ and the number of ways of choosing exactly two heads from three coin tosses is ${}_3C_2$. Thus, the expansion can also be written ${}_3C_3\text{H}^3 + {}_3C_2\text{H}^2\text{T} + {}_3C_1\text{HT}^2 + {}_3C_0\text{T}^3$.

Here are binomial expansions of the first few powers of $(\text{H} + \text{T})$. Notice that the coefficients are the same as the rows in Pascal's triangle. Think about what these expansions tell you about the results of flipping a coin 0, 1, 2, and 3 times.

$(\text{H} + \text{T})^0 =$	$\mathbf{1}$	1st row
$(\text{H} + \text{T})^1 =$	$\mathbf{1}\text{H} + \mathbf{1}\text{T}$	2nd row
$(\text{H} + \text{T})^2 =$	$\mathbf{1}\text{H}^2 + \mathbf{2}\text{HT} + \mathbf{1}\text{T}^2$	3rd row
$(\text{H} + \text{T})^3 =$	$\mathbf{1}\text{H}^3 + \mathbf{3}\text{H}^2\text{T} + \mathbf{3}\text{HT}^2 + \mathbf{1}\text{T}^3$	4th row

Notice the pattern in the coefficients. You can use Pascal's triangle, or the values of ${}_nC_r$, to expand binomials without multiplying all the terms.

The Binomial Theorem

If a binomial $(p + q)$ is raised to a whole-number power, n, the coefficients of the expansion are the combination numbers ${}_nC_n$ to ${}_nC_0$, as shown:

$$(p+q)^n = {}_nC_n p^n q^0 + {}_nC_{(n-1)} p^{n-1} q^1 + {}_nC_{(n-2)} p^{n-2} q^2 + \cdots + {}_nC_0 p^0 q^n$$

or

$$(p+q)^n = \sum_{j=0}^{n} {}_nC_{(n-j)} \cdot p^{n-j} q^j$$

In a binomial event, it can be useful to designate one of the results as a success. For example, if you call the result of heads a success, then each outcome in Example A can be summarized by the number of heads. So the possible outcomes of three coin tosses are 3, 2, 1, or 0 heads.

When you flip a fair coin, the outcomes H and T are equally likely. Example B shows how you can use a binomial expansion to find probabilities of outcomes that are not equally likely.

EXAMPLE B

Suppose that a hatching yellow-bellied sapsucker has a 0.58 probability of surviving to adulthood. Assume the chance of survival for each egg in a nest is independent of the outcomes for the other eggs. Given a nest of six eggs, what is the probability that exactly four birds will survive to adulthood?

A yellow-bellied sapsucker

▶ Solution

You are given the probability of survival, $P(\text{S}) = 0.58$, so you know the probability of nonsurvival, $P(\text{N}) = 1 - 0.58 = 0.42$. If you represent these values as s and n respectively, then the probability that the first four birds survive and the last two do not is s^4n^2. But it may not be the first four that survive. The 4 surviving birds can be chosen from the 6 birds in ${}_6C_4$ ways, represented by SSNSNS, SNNSSS, and so on. So the probability that exactly 4 birds survive is ${}_6C_4 s^4 n^2 = 15(0.58)^4(0.42)^2 \approx 0.299$.

Can you extend this method to find the probabilities of survival for different numbers of birds? In Example B, $s + n$ is equal to 1, as is $(s + n)^6$. But if you expand the latter expression, you will find the probabilities of survival for 0, 1, 2, 3, 4, 5, and 6 birds.

$$\begin{aligned}(s+n)^6 &= {}_6C_6 s^6 n^0 + {}_6C_5 s^5 n^1 + {}_6C_4 s^4 n^2 + {}_6C_3 s^3 n^3 + {}_6C_2 s^2 n^4 \\ &\quad + {}_6C_1 s^1 n^5 + {}_6C_0 s^0 n^6 \\ &= 1(0.58)^6(0.42)^0 + 6(0.58)^5(0.42)^1 + 15(0.58)^4(0.42)^2 \\ &\quad + 20(0.58)^3(0.42)^3 + 15(0.58)^2(0.42)^4 + 6(0.58)^1(0.42)^5 \\ &\quad + 1(0.58)^0(0.42)^6 \\ &\approx 0.038 + 0.165 + 0.299 + 0.289 + 0.157 + 0.045 + 0.005\end{aligned}$$

You can put these numbers into a table to organize the information.

Number of birds that survive, x	6	5	4	3	2	1	0
Probability, $P(x)$	0.038	0.165	0.299	0.289	0.157	0.045	0.005

So it is most likely that 4 birds survive and very unlikely that 0 will survive.

You could show these probabilities with a histogram, as shown at right. This display allows you to compare the probabilities of different outcomes quickly. You'll learn more about probability diagrams like these in Chapter 11.

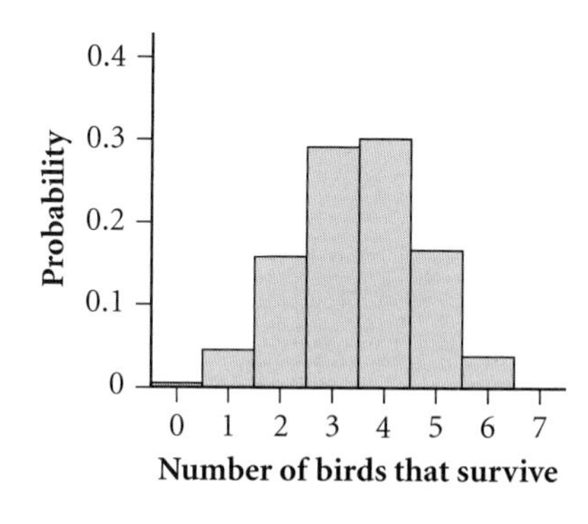

Suppose you want to know the probability that at least 4 birds survive or that at most 3 birds survive. The table and histogram can be extended to calculate probabilities such as these. How could you calculate the values in the table below? In Exercise 3 you will calculate the missing table values.

	6 birds	5 birds	4 birds	3 birds	2 birds	1 bird	0 birds
Exactly	0.038	0.165	0.299	0.289	0.157	0.045	0.005
At most	1	0.962			0.207	0.050	0.005
At least	0.038				0.949	0.995	1

Probability of Success in a Binomial Event

In a binomial event (independent trials with the same probability) the probability of x successes in n trials, where the probability is p for each trial, is

$$P(x) = {}_nC_x \cdot p^x \cdot (1 - p)^{n-x}$$

and the probability of x being between a and b is

$$P(a \leq x \leq b) = \sum_{x=a}^{b} {}_nC_x \cdot p^x \cdot (1 - p)^{n-x}.$$

You can think of the 6 birds as a **sample** taken from a **population** of many birds. You know the probability of success in the population, and you want to calculate various probabilities of success in the sample. You saw in Example B that a survival rate of 4 birds was most likely.

What if the situation is reversed? Suppose you don't know what the probability of success in the population is, but you do know that on average 4 out of 6 birds survive to adulthood. This ratio would indicate a population success probability of $\frac{4}{6}$, or 0.67; but you saw that this ratio is also highly likely with a population success probability of 0.58. So more than one population success probability yields the highest probability that 4 birds will survive in a sample of 6.

Reporters and political strategists use polls to try to predict election results. A pollster questions a sample of voters, not the entire population of voters. Statisticians then have to interpret the data from that sample to make predictions about the population.

EXAMPLE C A random sample of 32 voters is taken from a large population. In the sample, 20 voters favor passing a proposal. Is it likely that the proposal will pass?

▶ Solution You need to find out how likely it is that more than 50% of the voting population supports the proposal. Suppose f represents the number of survey respondents who favor the proposal. If exactly 50% of voters actually favor the proposal, then with a sample of 32 voters ($n = 32$), $P(f = 20)$ is ${}_{32}C_{20}(0.5)^{20}(0.5)^{12} \approx 0.05257$. So it's not very likely that exactly 20 of the 32 survey respondents would favor the proposal if half of the voters overall did not. And the probability that $f = 20$ would be even smaller if less than 50% of the voters actually favored the proposal.

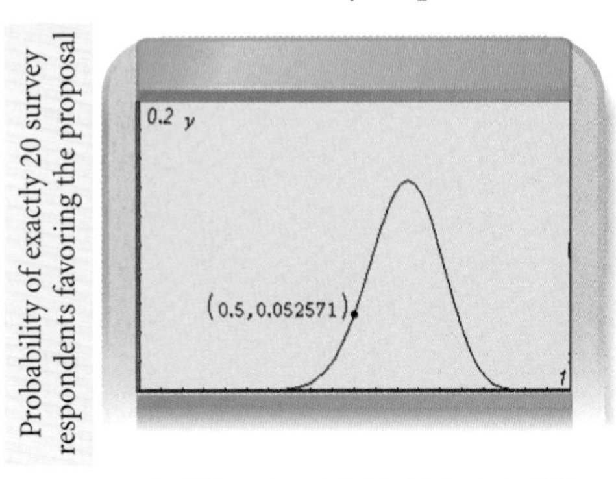

True ratio of voters favoring the proposal

But the probability of any single event, like exactly 20 respondents in favor of the proposal, is small. So instead, consider the probability of this particular event or a more extreme one. Thus, you look at the probability $P(f \geq 20) = \sum_{j=20}^{32} {}_{32}C_j \cdot x^j \cdot (1 - x)^{n-j}$, or the sum of all the probabilities that from 20 to 32 voters support the proposal. When $x = 0.5$, this probability is equal to 0.10766. If the actual percentage of supporters were 50% or less, then there would be only a 10.8% or less probability that you would have had this particular survey result (or a more extreme one). Because this value is small, the actual proportion supporting the proposal is most likely to be greater than 50%. Thus, the proposal is likely to pass.

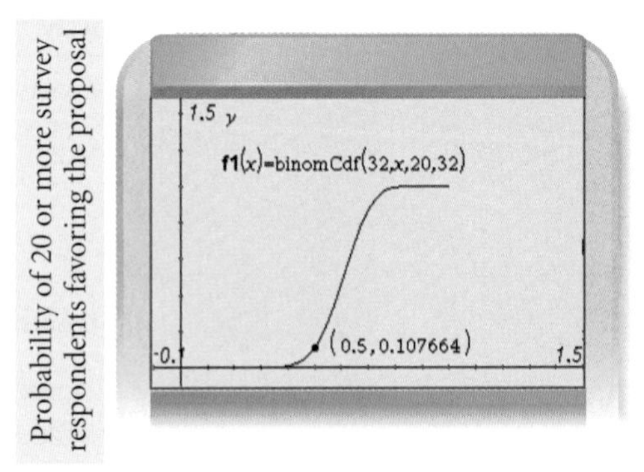

True ratio of voters favoring the proposal

You've seen how to use a population probability to make predictions about a sample, and how to use the results from a sample to make predictions about a population. You will use both types of thinking in the exercises.

EXERCISES

You will need

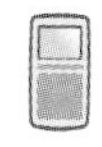
A graphing calculator for Exercises **9, 10, 15** and **19.**

Statistics software for Exercise **17**

Practice Your Skills

1. Given the expression $(x + y)^{47}$, find the

a. 1st term. @ **b.** 11th term. @

c. 41st term. **d.** 47th term.

2. If the probability of success for each trial is 0.25 and all trials are independent, then

a. What is the probability of failure for a single trial? @

b. What is the probability of two successes in two trials? @

c. What is the probability of n successes in n trials?

d. What is the probability that there will be some combination of two successes and three failures in five trials?

3. Return to the expanded table after Example B on page 605. @

a. Find the missing probability in the "exactly" row.

b. Find the two missing probabilities in the "at most" row. For example, to find the probability that *at most* 2 birds survive, you could find the sum of the probabilities of 0, 1, and 2 birds surviving.

c. Find the three missing probabilities in the "at least" row.

d. Why don't the "at most" and "at least" values for each number of birds sum to 1?

e. Make a statement about birds that incorporates the 20.3% entry in the "at least" row.

4. Suppose that the probability of success is 0.62. What probability question is answered by the expression ${}_{50}C_{35}(0.62)^{35}(0.38)^{15}$? @

5. Suppose that the probability of success is 0.62. What probability question is answered by the expression $\sum_{r=35}^{50} {}_{50}C_r(0.62)^r(0.38)^{50-r}$?

Reason and Apply

6. Answer each probability question. @

a. List the equally likely outcomes if a coin is tossed twice.

b. List the equally likely outcomes if two coins are tossed once.

c. Draw tree diagrams that illustrate the answers to 6a and b.

d. Describe the connection between the expressions ${}_2C_0 = 1$, ${}_2C_1 = 2$, and ${}_2C_2 = 1$ and the results of your answers to 6a–c.

e. Give a real-world meaning to the equation

$$(\text{H} + \text{T})^2 = 1\text{H}^2 + 2\text{HT} + 1\text{T}^2$$

7. Expand each binomial expression.

a. $(x + y)^4$ @ **b.** $(p + q)^5$ **c.** $(2x + 3)^3$ @ **d.** $(3x - 4)^4$

8. APPLICATION A survey of 50 people shows that only 10 support a new traffic circle.

a. Which term of $(p + q)^{50}$ would correspond to the results of this sample? ⓐ

b. The sum of which terms would correspond to the probability of *at most* 10 supporters?

c. What is the sum for 8b if 50% of the voters support the new traffic circle? [▸ See **Calculator Note 10E** to learn how to find the sum. ◂]

d. Based on your answer to 8c, do you think it is likely that 50% or more of the population supports the traffic circle?

9. Dr. Miller is using a method of treatment that is 97% effective. ⓗ

a. What is the probability that there will be no failure in 30 treatments?

b. What is the probability that there will be fewer than 3 failures in 30 treatments?

c. Let x represent the number of failures in 30 treatments. Write an equation that will provide a table of values representing the probability $P(x)$ for any value of x. ⓐ

d. Use the equation and table from 9c to find the probability that there will be fewer than 3 failures in 30 treatments.

10. A university medical research team has developed a new test that is 88% effective at detecting a disease in its early stages. What is the probability that there will be more than 20 incorrect readings in 100 applications of the test on subjects known to have the disease? [▸ See **Calculator Note 10E** to learn how to find the terms of a binomial expansion. ◂]

11. Suppose the probability is 0.12 that a randomly chosen penny was minted before 1975. What is the probability that you will find 25 or more such coins in

a. A roll of 100 pennies? ⓐ

b. Two rolls of 100 pennies each?

c. Three rolls of 100 pennies each?

12. Suppose that a blue-footed booby has a 47% chance of surviving from egg to adulthood. For a nest of four eggs, ⓗ

a. What is the probability that all four birds will hatch and survive to adulthood?

b. What is the probability that none of the four birds will hatch and survive to adulthood?

c. How many birds would you expect to survive?

A pair of blue-footed boobies

13. A fair coin is tossed five times and comes up heads four out of five times. In your opinion, is this event a rare occurrence? Defend your position.

14. Data collected over the last ten years show that in a particular town it will rain sometime during 30% of the days in the spring. How likely is it that there will be a week with

a. Exactly five rainy days? ⓐ

b. Exactly six rainy days?

c. Exactly seven rainy days? ⓐ

d. At least five rainy days?

15. Consider the function $y = f(x) = \left(1 + \frac{1}{x}\right)^x$. Note that the right side of the equation is a binomial raised to a power.

a. Fill in the table below using the Binomial Theorem or Pascal's triangle. Verify using your calculator. ⓐ

x	1	2	3	4
Sum of the first 2 terms				
Sum of all the terms				

b. Using your calculator, find $f(10)$, $f(100)$, $f(1000)$, and $f(10000)$.

c. Describe what happens to the values of $f(x)$ as you use larger and larger values of x.

Mathematics CONNECTION

The real number e is a mathematical constant whose value is 2.718... . When a quantity changes at a rate proportional to the quantity present, the growth can be modeled using an exponential function with base e. This kind of change occurs when money is continuously compounded, when a capacitor discharges, and when a radioactive compound decays. In statistics, e is part of the equation that gives us the normal distribution curve, and in engineering, e is part of the equation of a *catenary,* a hanging cable. In the study of calculus, e is used most often in the study of exponents and logarithms.

16. Mrs. Gutierrez has 25 students, and she sends 4 or 5 students at random to the board each day to solve a homework problem. You can calculate that there are ${}_{25}C_4$, or 12,650, ways that 4 students could be selected and ${}_{25}C_5$, or 53,130, ways that 5 students could be selected. Suppose the class has grown to 26 students. Without using a calculator, determine how many ways 5 students can be selected now. Explain your solution method. ⓗ

Review

17. *Technology* Use statistics software to simulate flipping a coin. (See the Exploration The Law of Large Numbers on page 569 for help creating a simulation.)

a. Simulate 10 flips of a coin and make a bar graph of the results. How do your experimental probabilities compare with the theoretical probabilities of getting heads or tails?

b. Simulate 1000 flips of a coin by adding 990 more trials to your simulation. How do the experimental probabilities now compare with the theoretical probabilities?

c. Explain how you could modify your simulation to model flipping a thumbtack that lands point up 60% of the time. Make the necessary modifications and test whether or not your simulation produces appropriate experimental results.

18. Suppose that 350 points are randomly selected within the rectangle at right and 156 of them fall within the closed curve. What is an estimate of the area within the curve?

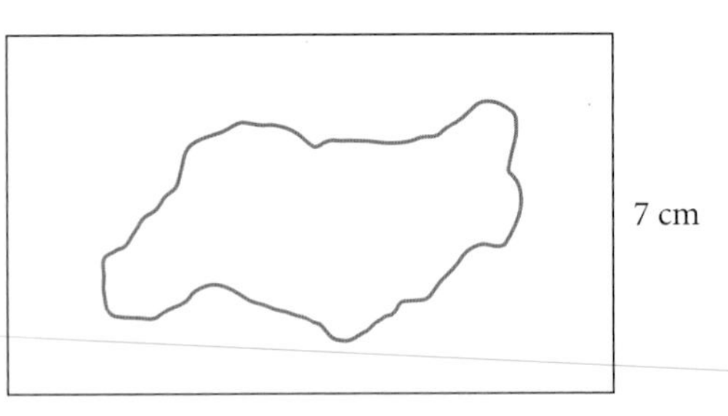

19. APPLICATION Kepler's third law of planetary motion states that the square of a planet's period of revolution is proportional to the cube of its average distance to the Sun. Planetary data are given in the tables.

a. Use logarithms and curve straightening to find an equation that fits data in the form (*distance, period*). Show all steps leading to your answer. ⓗ

b. Use your equation to verify the data for Uranus, Neptune, or Pluto. How can you explain any errors? ⓗ

Planet	Average distance from Sun (miles)	Period of revolution (days)
Mercury	35,900,000	88
Venus	67,200,000	224.7
Earth	92,960,000	365.26
Mars	141,600,000	687
Jupiter	483,600,000	4,332.6
Saturn	886,700,000	10,759.2

Planet	Average distance from Sun (miles)	Period of revolution (days)
Uranus	1,783,000,000	30,685.4
Neptune	2,794,000,000	60,189
Pluto	3,666,100,000	90,465

c. Rewrite your equation from 19a in the form $p^2 = ka^3$.

Science CONNECTION

In 1618, after working for more than ten years with the data, German astronomer Johannes Kepler (1571–1630) suddenly realized that "the proportion between the periodic times of any two planets is precisely one and a half times the proportion of the mean distances." English physicist and mathematician Sir Isaac Newton (1642–1727) later restated Kepler's third law to include the masses of the planets. Newton was then able to calculate the masses of the planets relative to the mass of Earth.

20. Write an equation to express each relationship.

a. The area of a circle varies directly with the square of its radius. The constant of variation is π.

b. The volume of a rectangular prism varies jointly with its length, width, and height.

c. The force in newtons (N) needed to open a door is inversely proportional to the distance from the hinge. Use k for the constant of variation.

CHAPTER 10 REVIEW

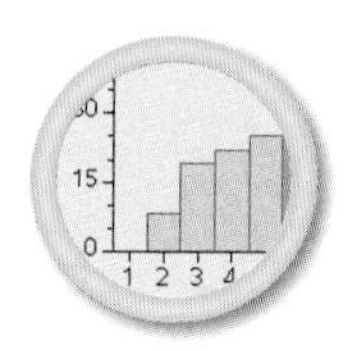

In this chapter you were introduced to the concept of randomness and you learned how to use technology to generate random numbers. Random numbers should all have an equal chance of occurring and should, in the long run, occur equally frequently. You can perform an experiment or use random-number procedures to simulate the experiment and determine the **experimental probability** of an **event.** You can determine **theoretical probability** by comparing the number of successful **outcomes** to the total number of possible outcomes. You represented situations involving probability with Venn diagrams and learned the meaning of events that are **dependent, independent,** and **mutually exclusive.** You used **tree diagrams** to help you count possibilities and learned that sometimes probability situations can be represented geometrically. Using what you learned about theoretical probability, you were able to calculate **expected value,** by multiplying the value of each event by its probability and then summing all the products.

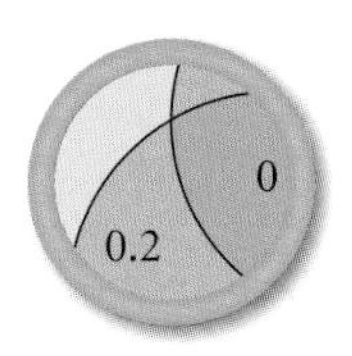

To help find theoretical probabilities, you were introduced to some formal counting techniques. The **counting principle** states that when there are n_1 ways to make the first choice, n_2 ways to make the second choice, n_3 ways to make the third choice, and so on, the product $n_1 \cdot n_2 \cdot n_3 \cdot \cdots$ gives the total number of different ways in which the entire sequence of choices can be made. These arrangements of choices, in which the order is important, are called **permutations.** The notation ${}_nP_r$ indicates the number of ways of arranging r things out of n possible choices. If the order is unimportant, then arrangements are called **combinations** and the notation ${}_nC_r$ represents the number of combinations of r things from a set of n choices. Your calculator can calculate permutation numbers and combination numbers, but before you use the calculator, be sure that you understand the situation and can visualize the possibilities. Combinations also appear in **Pascal's triangle** and as coefficients in **binomial expansions** that you can use to help calculate probabilities when there are two possible outcomes.

EXERCISES

@ Answers are provided for all exercises in this set.

1. Name two different ways to generate random numbers from 0 to 10.

2. Suppose you roll two octahedral (eight-sided) dice, numbered 1–8.

a. Draw a diagram that shows all possible outcomes of this experiment.

b. Indicate on your diagram all the possible outcomes for which the sum of the dice is less than 6.

c. What is the probability that the sum is less than 6?

d. What is the probability that the sum is more than 6?

3. Answer each geometric probability problem.

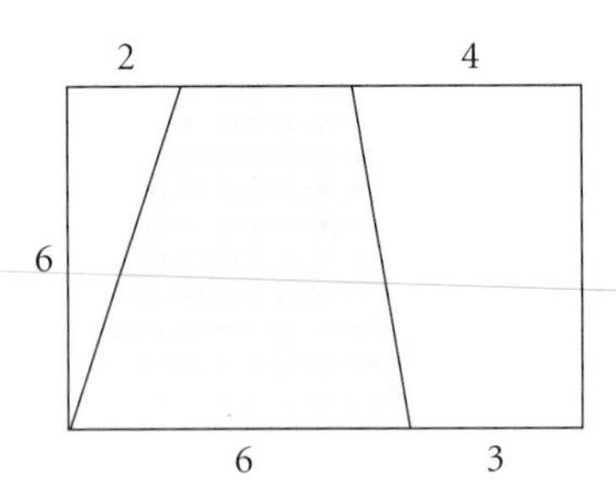

a. What is the probability that a randomly plotted point will land in the shaded region at right?

b. One thousand points are randomly plotted in the rectangular region shown below. Suppose that 374 of the points land in the shaded portion of the rectangle. What is an approximation of the area of the shaded portion?

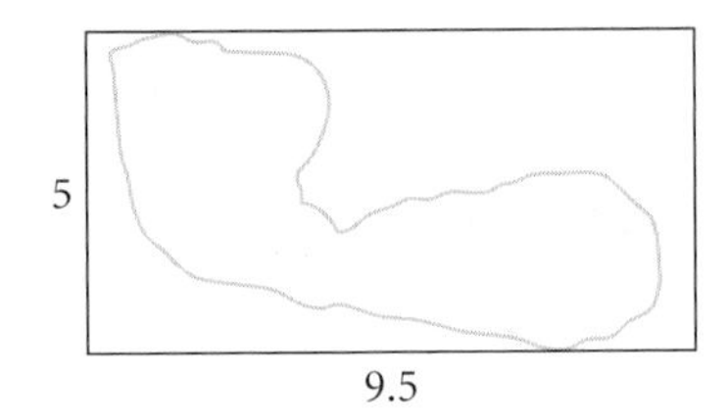

4. A true-false test has four questions.

a. Draw a tree diagram representing all of the possible results. (Assume all four questions are answered.)

b. How many possible ways are there of getting one false and three true answers?

c. How could you use combinations or permutations to answer 4b?

d. Suppose you are sure that the answers to the first two questions on the test are true, and you write these answers down. Then you guess the answers to the remaining two questions at random. What is the probability that there will be one false and three true answers on the test?

5. The local outlet of Frankfurter Franchise sells three types of hot dogs: plain, with chili, and veggie. The owners know that 47% of their sales are chili dogs, 36% are plain, and the rest are veggie. They also offer three types of buns: plain, rye, and multigrain. Sixty-two percent of their sales are plain buns, 27% are multigrain, and the rest are rye. Assume that the choice of bun is independent of the choice of type of hot dog.

a. Make a tree diagram showing this information.

b. What is the probability that the next customer will order a chili dog on rye?

c. What is the probability that the next customer will *not* order a veggie dog on a plain bun?

d. What is the probability that the next customer will order either a plain hot dog on a plain bun or a chili dog on a multigrain bun?

6. All students in a school were surveyed regarding their preference for whipped cream or ice cream to be served with chocolate cake. The results, tabulated by grade level, are reported in the table.

	9th grade	10th grade	11th grade	12th grade	Total
Ice cream	18	37	85	114	
Whipped cream	5	18	37	58	
Total					

a. Copy and complete the table.

b. What is the probability that a randomly chosen 10th grader will prefer ice cream?

c. What is the probability that a randomly chosen 11th grader will prefer whipped cream?

d. What is the probability that someone who prefers ice cream is a 9th grader?

e. What is the probability that a randomly chosen student will prefer whipped cream?

7. Rita is practicing darts. On this particular dartboard she can score 20 points for a bull's-eye and 10 points, 5 points, or 1 point for the other regions. Although Rita doesn't know exactly where her five darts will land, she has been a fairly consistent dart player over the years. She figures that she hits the bull's-eye 30% of the time, the 10-point circle 40% of the time, the 5-point circle 20% of the time, and the 1-point circle 5% of the time. What is the expected value of her score if she throws a dart ten times?

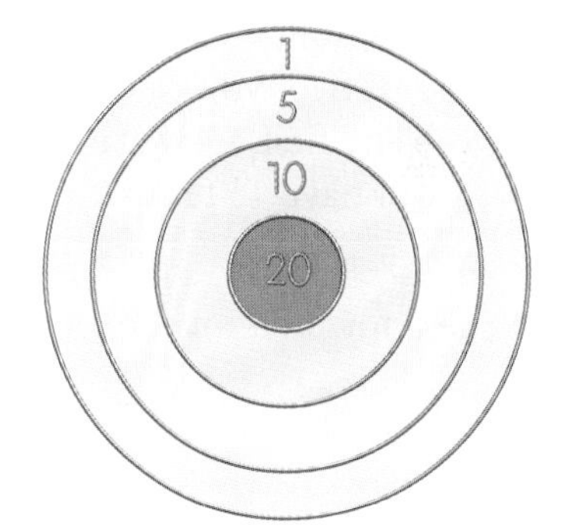

8. Misty polls residents of her neighborhood about the types of pets they have: cat, dog, other, or none. She determines these facts:

Ownership of cats and dogs is mutually exclusive.

32% of homes have dogs.

54% of homes have a dog or a cat.

16% of homes have only a cat.

42% of homes have no pets.

22% of homes have pets that are not cats or dogs.

Draw a Venn diagram of these data. Label each region with the probability of each outcome.

9. Elliott has to fill exactly 20 more pizza orders before closing. He has enough pepperoni for 16 more pizzas. On a typical night, 65% of orders are for pepperoni pizzas. What is the probability that Elliott will run out of pepperoni?

10. Find the term specified for each binomial expansion.

a. the first term of $\left(1 + \frac{x}{12}\right)^{99}$

b. the last term of $\left(1 + \frac{x}{12}\right)^{99}$

c. the tenth term of $(a + b)^{21}$

TAKE ANOTHER LOOK

1. Pascal's triangle is filled with patterns. Write the first ten rows of Pascal's triangle, and look for as many relationships as you can among the numbers. You may want to consider numerical patterns, sums of numbers in particular locations, and patterns of even or odd numbers, or numbers that share a factor.

2. In this chapter you have seen that a binomial expansion can help you find the probability of an outcome of a series of events when each event has only two possible results (such as success or failure). When there are three possible results instead of two, you can use trinomials in a similar way. Expand each trinomial given below. The first answer will have 6 terms, the second will have 10 terms, and the third will have 15 terms.

$(x + y + z)^2$ $\quad$ $(x + y + z)^3$ $\quad$ $(x + y + z)^4$

Write a formula for the expansion of $(x + y + z)^n$.

Assessing What You've Learned

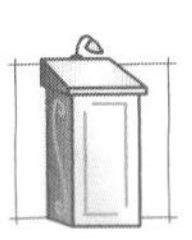

GIVE A PRESENTATION By yourself or with a group, write and present a probability exercise. Detail the outcome you want to find the probability of, and tell how to solve the problem. If possible, explain how to find both experimental and theoretical probabilities of the outcome occurring. You may even want to do a simulation with the class. Or you could present your work on a Take Another Look from this chapter.

WRITE IN YOUR JOURNAL You have seen how to represent situations involving probabilities using tree diagrams and Venn diagrams. Describe situations in which each of these approaches would be appropriate, and explain how each method can help you solve problems. Specify the kinds of questions that can be answered with each type of diagram.

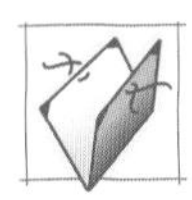

UPDATE YOUR PORTFOLIO Choose a couple of investigations or exercises from this chapter that you are particularly proud of. Write a paragraph about each piece of work. Describe the objective of the problem, how you demonstrated understanding in your solution, and anything you might have done differently.

CHAPTER

11 Applications of Statistics

This untitled painting (1990) by Japanese-born artist Naoki Okamoto creates a "crowd" of faces. In this chapter you'll explore how numerical data about a population can be represented with a few summary numbers. You'll also explore whether data about a small but random group of people can lead to valid generalizations about a whole population.

OBJECTIVES

In this chapter you will

- learn different ways to collect data as part of designing a study
- apply different methods for making predictions about a population based on a random sample
- discover the association between the statistics of a sample and the parameters of an entire population
- study population distributions, including normal distributions
- fit functions to data and make predictions using least squares lines and other regression equations

Statistics

You've seen in earlier chapters that statistics, such as mean and standard deviation, are numbers that describe a set of data. In Chapter 11, you'll learn how to use statistics describing a sample to make predictions about a population. To help with that new use of statistics, this lesson reviews some of the statistics you studied in Chapter 2.

EXAMPLE A

Ms. Silver asks her students to keep track of how many hours they spend on a semester project. The data below are from a simple random sample of the responses. Find the mean and the standard deviation of these data.

3.4 4.7 2.1 5.4 5.1 3.8 4.4 5.2 4.1 3.8

▶ Solution

The mean, $\bar{x}$, is the sum of all the data values divided by the number of data values. You can use sigma notation to express the calculation.

$$\bar{x} = \frac{1}{10}\sum_{i=1}^{10} x_i = \frac{3.4 + 4.7 + 2.1 + 5.4 + 5.1 + 3.8 + 4.4 + 5.2 + 4.1 + 3.8}{10}$$
$$= 4.2 \text{ h}$$

The mean is 4.2 h.

Recall that *deviations* are the signed differences between the data values and the mean. The standard deviation, s, is the sum of the squares of the deviations divided by one less than the number of data values.

$$s = \sqrt{\frac{1}{n-1}\sum_{i=1}^{n}(x_i - \bar{x})^2} = \sqrt{\frac{1}{9}\sum_{i=1}^{10}(x_i - 4.2)^2}$$

$$= \sqrt{\frac{(-0.8)^2+0.5^2+(-2.1)^2+(1.2)^2+0.9^2+(-0.4)^2+0.2^2+1^2+(-0.1)^2+(-0.4)^2}{9}}$$

$$\sqrt{\frac{8.92}{9}} \approx \sqrt{0.9911} \approx 0.9955 \text{ h}$$

The standard deviation is about 1 h.

Some statistics can be estimated from a histogram of the data.

EXAMPLE B

The histogram shows the heights of students in a sample of kindergartners. Calculate the percentile ranks for children of heights 105 cm and 115 cm.

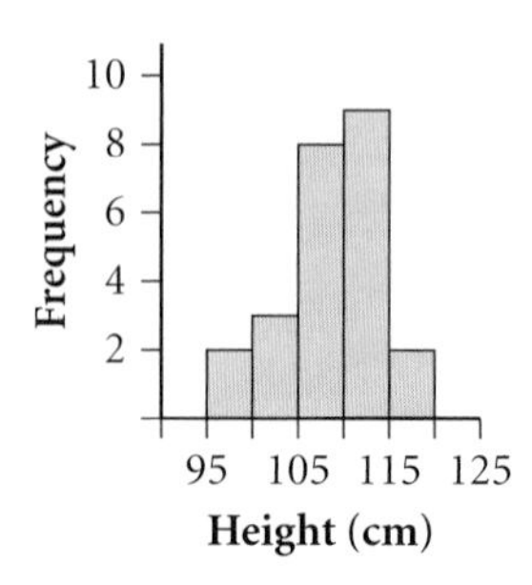

▶ ***Solution***

Recall from Chapter 2 that the percentile rank of a value is the percentage of values in a data set that are below it. First make a table of counts to find the number in the sample. Then use this total to determine the percentage in each bin.

Bin	95–100	100–105	105–110	110–115	115–120	Total
Frequency	2	3	8	9	2	**24**
Percent	$8.\overline{3}\%$	12.5%	$33.\overline{3}\%$	37.5%	$8.\overline{3}\%$	**100%**

To find the percentile of height 105 cm, add the percents in the bins below 105. That is, 105 is at the $8.3 + 12.5 \approx 20.8$th percentile. Height 115 cm is at the $8.3 + 12.5 + 33.3 + 37.5 \approx 91.7$th percentile.

EXERCISES

1. The high temperatures for a sample of eight summer days in Arizona are 112°, 92°, 107°, 118°, 91°, 93°, 104°, and 101°.

a. Find the mean, median, and interquartile range of the temperature data.

b. Find the standard deviation of the temperature data. @

2. The weights in kilograms of a sample of 12 dogs in a dog show are 16.8, 18.2, 22.4, 19.6, 21.7, 18.0, 16.2, 29.8, 25.4, 26.4, 19.9, and 20.1.

a. Find the mean, median, and interquartile range of the weight data. @

b. Find the standard deviation of the weight data.

3. Find the percentile ranks for heights 105 cm and 115 cm based on each histogram.

a. 

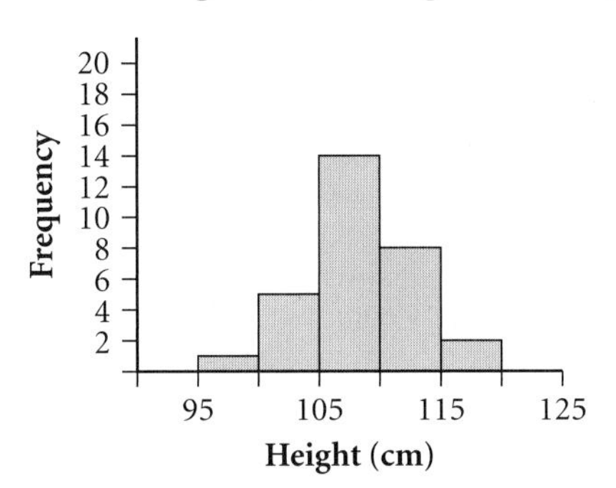

b. Heights of kindergartners

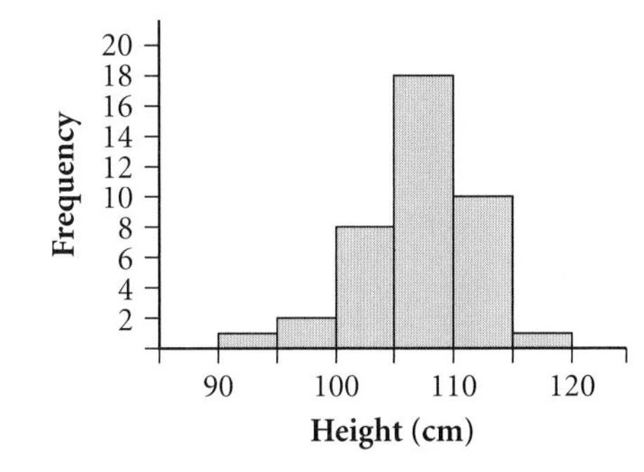

LESSON 11.1

Experimental Design

Rebecca and Tova have math class right after lunch. Rebecca always eats a hot lunch on days when she has an exam, because she has a theory that people score a little higher on math exams when they have eaten a hot meal. Tova is very skeptical of this theory. How can Rebecca and Tova test whether the theory is valid?

Rebecca and Tova write down as many of their own exam scores as they can remember, and they record whether they had a cold lunch or hot lunch on the days of those exams. They ask a few of their friends to give them the same information. Then they find the average of all the scores associated with a cold lunch and all the scores associated with a hot lunch.

This type of study, in which data are collected from a sample convenient to the investigators, is called **anecdotal.** Though it is not an accurate or scientific kind of study, this approach is used very frequently. Here are three better ways to collect sample data for a study.

Type of study	Choices, treatments, or conditions	Example
Experimental	Treatments (assignments) are applied to subjects as assigned by the researcher.	A number of students agree to help. The students (subjects) are randomly assigned to one of two groups. On the day of the exam, Group 1 eats a hot lunch and Group 2 eats a cold lunch.
Observational	Treatments are not applied by the researcher. Subjects may know they are being observed but don't know what is observed.	A number of subjects are observed at lunch and information is collected about the kind of lunch (treatment) they eat. The subjects are then classified into groups.
Survey	Information about treatments and/or results is collected from the subjects.	A questionnaire is given to members of the class asking them what kind of lunch (treatment) they had.

The diagram shows the stages of an experimental study.

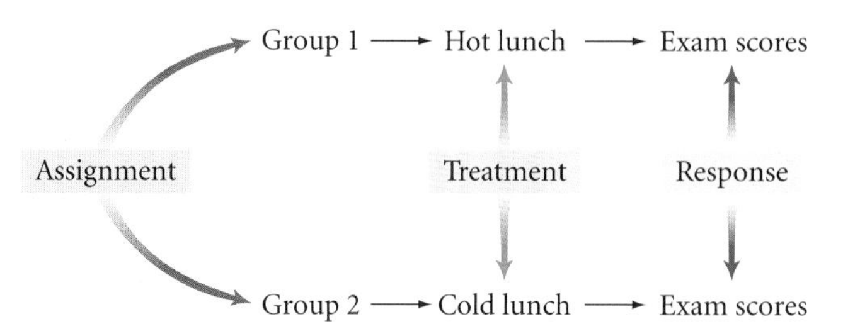

After using any of these methods to collect information about the type of lunch students ate, Rebecca and Tova would need to learn the exam scores for the students who participated in the study. For this study it would be best if they gave the names of the students in each lunch-type group to the teacher, who could provide the mean and standard deviation of the exam scores of each group. Each type of study has advantages and disadvantages. In the investigation you will use two of the methods to conduct a study.

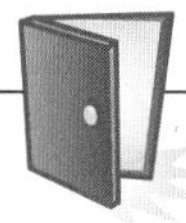

Investigation
Designing a Study

Consider this hypothesis: Listening to music while doing math homework will shorten the time it takes to complete the assignment.

Your group will design two studies that could test this hypothesis. Choose two of the types of study described on the previous page, and describe how you would collect sample data for the study. Provide enough detail so that another group could use your description to collect data in the same way. Keep in mind these points as you design your study:

- If you design an **experiment,** you must describe what treatments will be used and how subjects will be assigned to a treatment.
- To design an **observational study,** you must describe what is to be observed and measured and how you'll decide which subjects to observe.
- A **survey** plan must address who will be surveyed, the method of survey (interview, paper questionnaire, e-mail, and so on), and the exact questions to be asked.

When the assignment of treatments in an experiment is done so that the groups are not biased, then the conclusions of the experiment may indicate **causation.** If Rebecca implements a carefully designed experimental study, she might find that students having a hot lunch did much better on exams than those eating a cold lunch. If so, she would have evidence that eating the hot lunch "caused" the students to perform better on the exam. However, if Rebecca used either an observational study or a survey, she could conclude only that there's an **association** between the type of lunch and exam scores.

Careful experimental studies are the primary way to learn about causation. Only a huge observational study over a long period of time can indicate causation. On the other hand, observational studies and surveys are often more practical than experimental studies, and they do not require subjects to give up control of their own treatment.

In all types of studies, a major concern is to minimize bias that will invalidate the results. In experimental studies, the way subjects are assigned to a treatment may affect the data. Observational studies may omit consideration of some subjects. Surveys may be biased by the selection of the participants, by the truthfulness of the subjects, and by how the questions are phrased.

Health CONNECTION

Studies have found that insufficient sleep is associated with many chronic diseases, including diabetes, obesity, heart disease, and depression. Many motor vehicle accidents and industrial accidents are also linked to insufficient sleep. More than 25% of adults report not getting enough sleep on occasion, and almost 10% have chronic insomnia. The National Sleep Foundation recommends that teenagers get 8.5 to 9.5 hours of sleep a night.

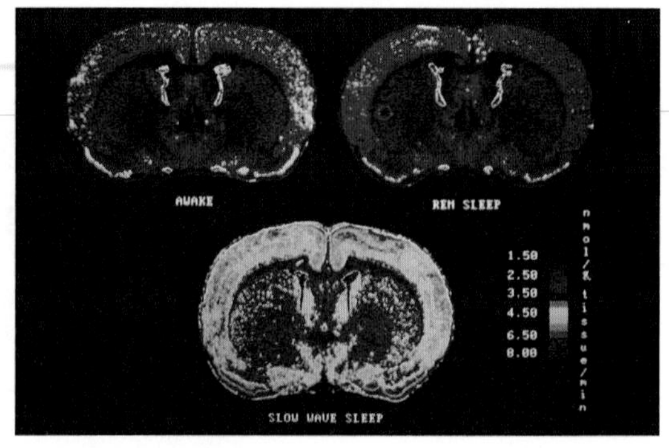

These brain images were used in a study of sleep at La Villette Science Project in Paris, France. During an extended period of sleep, we cycle through five phases: stages 1, 2, 3, and 4, and REM (rapid eye movement) sleep. Most dreams occur during REM sleep, and researchers believe that REM sleep stimulates areas of the brain involved in learning. *Slow wave* sleep describes the deep sleep of stages 3 and 4, during which there is no eye movement. (*www.ninds.nih.gov*)

EXAMPLE A

Brandon surveyed some students and found that the average amount of sleep for female students was much higher than that for male students. Later, a thorough follow-up study concluded that the average amount of sleep was the same for males and females. Give some possible reasons for the discrepancy in the results. How could Brandon design a better survey?

▶ Solution

The original survey may have been biased by how Brandon selected the students to be sampled. For example, the only males he sampled might have been friends who had similar sleeping patterns to his own. Or students who were reluctant to talk about their sleep habits may have declined to participate. Or the participants may have misunderstood the questions.

One way Brandon could minimize bias is to choose subjects randomly. In addition, Brandon should ask some other people to review the questions in his survey to make sure the phrasing of the questions does not influence the answers given or the likelihood that a person would choose not to respond.

No matter how the data are collected, a researcher must be concerned about over- or underrepresenting subgroups of the population. For instance, Brandon might have overrepresented his male friends.

As another example, suppose you want to design a study about the effectiveness of a new fertilizer. If all of your evidence comes from using the fertilizer on tomato plants, then your conclusion can be only about tomato plants, not about all plants. If your data come from several types of plants, but most of your data are from tomato plants, then the data will be biased because one segment of the population is overrepresented. A study based on biased data will not necessarily have incorrect conclusions, but the methods of data collection should be described so that those who read about the study know of the sampling bias.

Newspapers or other sources do not always include enough information about a study to determine whether the statistics are reliable. As an educated consumer, before accepting a conclusion from a study you need to know how the data were collected and whether any bias or other errors could be present.

EXAMPLE B

Evaluate each of these statistics.

a. Arthur says that 71.43% of students hate the new school dress code. He interviewed seven people in the parking lot after school and asked them, "Do you like the way the school is forcing us to dress?"

b. Betty claims that the typical student spends $19.93 on fast food every month. She passed out a survey to the 30 students in her algebra class, and 20 of the surveys were returned. The survey question read, "How much did you spend last week at fast-food restaurants?"

c. Carl states that the number of students with multiple suspensions has increased by 67% in the last 10 years. He found out from school records that last year there were 5 students with multiple suspensions and 11 years ago there were only 3 students.

▶ Solution

Although none of these students have lied about their results, all of them are using biased data or misrepresenting a statistic to make a point.

a. Arthur's conclusion does not match his question. A participant may object to being forced to dress in a certain way, but find nothing objectionable in the present dress code. Or subjects may have one small issue with the current code but overall think that it is fair.

Arthur's sample is biased because it didn't include students who weren't in the parking lot (such as those involved in after-school activities). Therefore, the students that he sampled may not share the same view about uniforms as the other students in the school.

b. Betty multiplied her one-week figures to obtain a monthly value. This assumes that spending is uniform every week. You do not know whether the value is a mean or a median value. You do not know why $\frac{1}{3}$ of the people did not return the survey. If participants had similar reasons for not returning the survey, it may be they had similar spending habits also. Because Betty chose to sample only students from algebra class, her coverage of students is biased.

c. Carl's claim may sound as if he has many years of data instead of only two years. Also, the population of the school needs to be considered. Suppose there are 600 students at the school. Although it is true that there was an increase of 67% in the actual number of students with multiple suspensions, the *ratio* of students with multiple suspensions in the student body increased by only $0.\bar{3}\%$ from $\frac{3}{600}$ to $\frac{5}{600}$. Also, if the population of the school has increased in the last ten years, the suspension rate may have actually decreased.

EXERCISES

You will need

A graphing calculator for Exercise **10.**

Practice Your Skills

For Exercises 1–3, use each of the four scenarios to answer the questions.

a. Nick stood outside the theater and asked many of the patrons if the movie was good. @

b. Elise gave each member of the class a cookie, some from recipe A and some from recipe B. Then she observed which students returned for a second cookie.

c. Steven recorded whether the first half of the students who arrived for class chose to sit in the front half of the room.

d. Jill searched the Internet to find comments from people who owned the type of cell phone she was thinking of buying.

1. Identify which type of data collection was used. Explain your reasoning.

2. Identify the treatments and what is being measured in each scenario.

3. Identify at least one source of bias in each study design in each scenario.

4. Erianna is designing a survey to determine how much sleep the students in her school get. Here are three versions of a question. Which is the best and why?

A. What time did you go to sleep last night?

B. How many hours of sleep did you get last night?

C. How many total hours of sleep did you get in the last week?

Reason and Apply

5. Darita's government class plans to conduct a poll to determine which political candidate has the most support in an upcoming election. They decide to place phone calls to the fourth name in the second column of every page of the local phone book. What problems, if any, do you see in their plan?

6. For each scenario, decide which of the three types of data collection you would use, give a reason for the choice, and give your design for the study. (Refer to the investigation for a description of each type of study.)

a. You wish to find which brand of microwave popcorn has the fewest unpopped kernels. @

b. You wish to find if there is a connection between sunny days and people's moods.

c. You wish to find out the proportion of students at your school who are planning to attend the school play two weekends from now.

Review

7. Find the area of each shape.

a. @

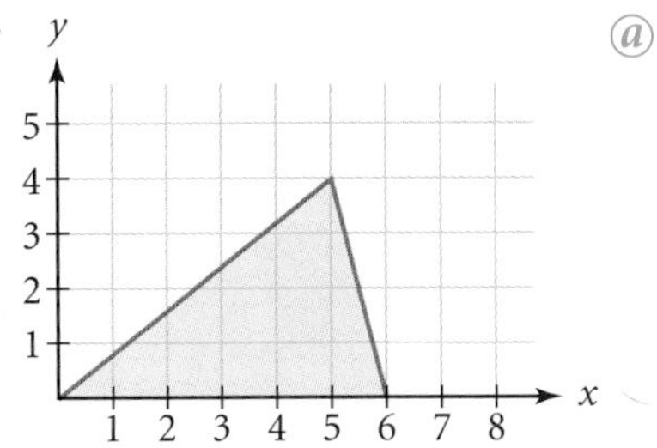

b.

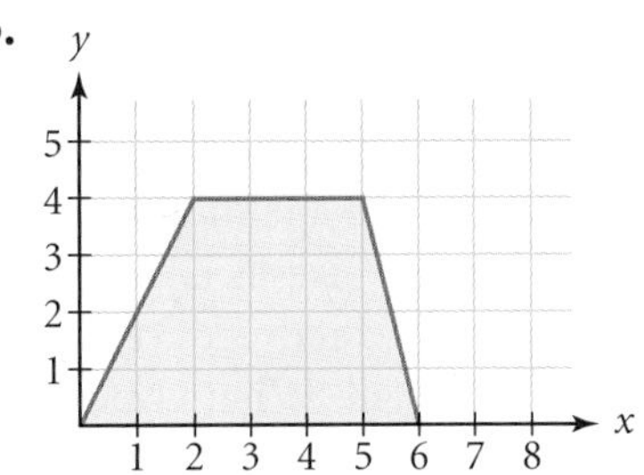

c.

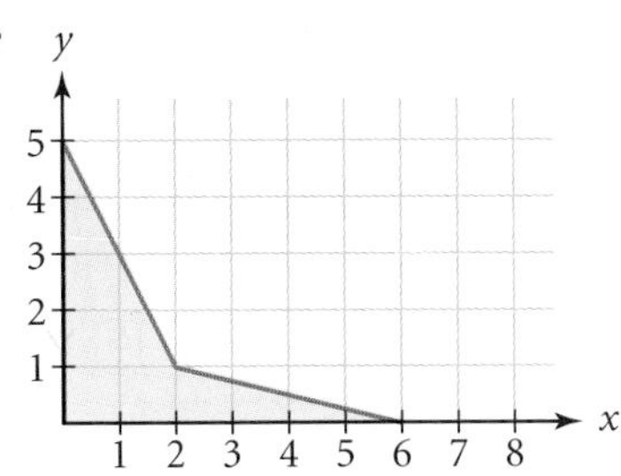

8. A poll shows 47% of voters favor Proposition B. What is the probability that exactly 11 of 20 voters end up voting in favor of the proposal? @

9. In your class, if three students are picked at random, what is the probability that

a. All were born on a Wednesday?

b. At least one was born on a Wednesday?

c. Each was born on a different day of the week?

10. At a maple tree nursery, a grower selects a random sample of 5-year-old trees. He measures their heights to the nearest inch.

Height (in.)	60	61	62	63	64	65	66	67	68	69	70	71	72	73	74	75
Frequency	1	1	4	6	8	10	9	8	13	10	5	6	6	3	6	3

a. Find the mean and the standard deviation of the heights.

b. Make a histogram of these data.

c. Find the percentile rank of 65 inches. @

History CONNECTION

Polls before the 1948 presidential election predicted that Republican Thomas Dewey would beat Democrat Harry Truman by 5 to 15 percentage points. The polls missed a late swing in favor of Truman, who won the election with a margin of more than 4 percentage points. Some analysts believe that reports of Dewey's lead may have encouraged more Democrats to come out and vote for Truman. In response to this embarrassing mistake, pollsters changed some of their methods. They extended polling until closer to the election day, and used probability sampling to poll voters chosen by chance throughout the country.

President Truman holds up the newspaper erroneously predicting his defeat in the 1948 presidential election.

LESSON

11.2 Probability Distributions

Statistical thinking will one day be as necessary for efficient citizenship as the ability to read and write.

H. G. WELLS

"I surveyed my chemistry class and found that, on average, the students are 16.31 years old," Sean says. "Therefore, the average age in all of the chemistry classes is 16.31 years." "Exactly 16.31 years?" asks Yiscah. What concerns do you think Yiscah might have about Sean's statement?

Sean is making a prediction about a population based on the mean of a sample. A **sample** is a part of a larger collection, called a **population,** that is selected to represent the entire population. Values that describe a population are called **parameters** instead of statistics. So Sean is predicting the value of a population parameter based on a sample statistic.

Sean is also considering ages. When you studied probabilities in Chapter 10, you considered random variables that had whole-number values, such as 5 heads, 3 tails, or 454 students. Those were *discrete* random variables. Ages have values between whole numbers. You might say that everyone in your class is 16 or 17; but actually no one is exactly 16, because a person is exactly 16 at only one instant. Ages are **continuous random variables,** and there are infinitely many possible ages.

This 14-meter installation, shown here in two views, is called *100 edition of 12* (1995). To create this piece, American artist Paul Ramirez Jonas (b 1965) chronologically arranged 100 photos of people aged 0–99.

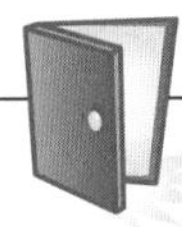

Investigation
Pencil Lengths

You will need

- centimeter rulers
- a few pencils
- graph paper

In this investigation you'll explore the difference between discrete and continuous random variables.

Begin by collecting all the pencils that your group has.

Step 1 Measure your pencils accurate to a tenth of a centimeter. Before you share data with other groups, predict the shape of a histogram of the class data.

Step 2 Share all measurements so that the class has one set of data. On graph paper, draw a histogram with bins representing 1 cm increments in pencil length.

Step 3 Divide the number of pencils in each bin by the total number of pencils. Make a new histogram, using these quotients as the values on the y-axis.

Step 4 Check that the area of your second histogram is 1. Why must this be true?

Step 5 Imagine that you collect more and more pencils and draw a histogram using the method described in Step 3. Sketch what this histogram of increasingly many pencil lengths would look like. Give reasons for your answer.

Step 6 Imagine doing a very complete and precise survey of all the pencils in the world. Assume that their distribution is about the same as the distribution of pencils in your sample. Also assume that you use infinitely many very narrow bins. What will this histogram look like?

To approximate this plot, sketch over the top of your histogram with a smooth curve, as shown at right. Make the area between the curve and the horizontal axis about the same as the area of the histogram. Make sure that the extra area enclosed by the curve above the histogram is about the same as the area cut off the corners of the bins as you smooth out the shape.

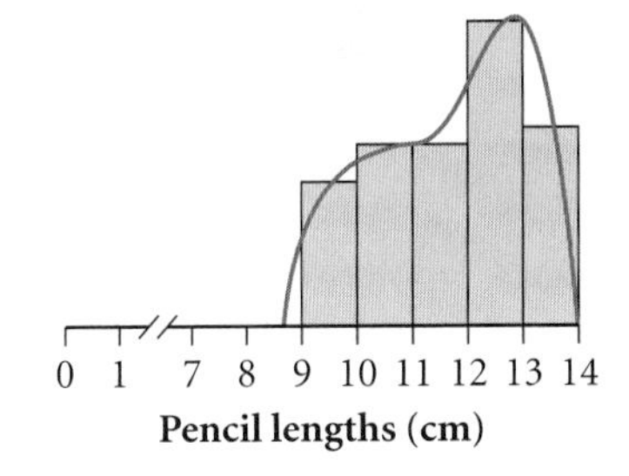

Step 7 Let x represent pencil length. Use your histogram from Step 6 to estimate the areas of various regions between the curve and the x-axis that satisfy these conditions:

a. $x < 10$

b. $11 < x < 12$

c. $x > 12.5$

d. $x = 11$

The histogram you made in Step 3 of the investigation, giving the proportions of pencils in the bins, is a **relative frequency histogram.** It shows what fraction of the time the value of a discrete random variable falls within each bin. The continuous curve you drew in Step 6 approximates a continuous random variable for the infinite set of measurements. This graph represents a function called a **probability distribution.**

The areas you estimated in Step 7 of the investigation give probabilities that a randomly chosen pencil length will satisfy a condition. As with geometric probabilities in Lesson 10.1, these probabilities are given by areas. If x represents the continuous random variable giving the pencil lengths in centimeters, then you can write these areas as

$$P(x < 10\text{ cm}),\ P(11\text{ cm} < x < 12\text{ cm}),\ P(x > 12.5\text{ cm}),\text{ and } P(x = 11\text{ cm})$$

In a continuous probability distribution, the probability of any single outcome, such as the probability that x is exactly 11 centimeters, is the area of a line segment. As you learned in geometry, a single point, line segment, or line has area 0. While it is possible for a pencil to be exactly 11 cm long, the probability of choosing any one value from infinitely many values is theoretically 0.

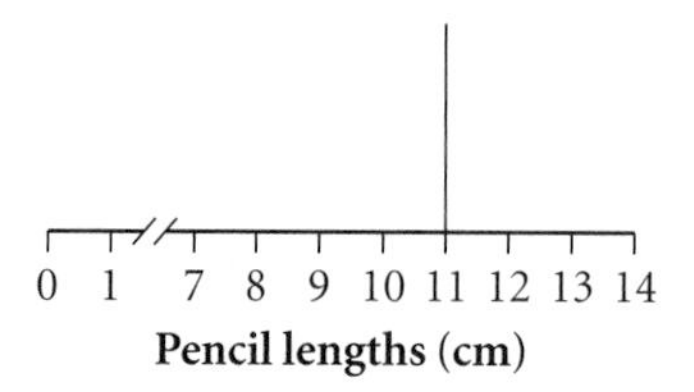

In the following example, you'll see how areas represent probabilities for a continuous random variable.

EXAMPLE A

A random-number generator selects a number between 0 and 6 according to the probability distribution at right. Because the random number can be any value of x with $0 \leq x \leq 6$, the graph is a continuous graph. Find the probability that a selected number is

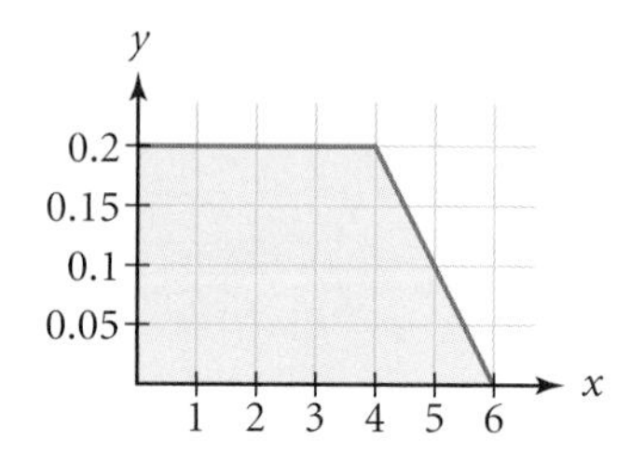

a. Less than 2.

b. Between 2.5 and 3.5.

c. More than 4.

▶ Solution

First, note that the region shaded for the entire distribution has area 1. To find the probability of a particular set of outcomes, find the area of the region that corresponds to it.

a. The region between 0 and 2 is a rectangle with width 2 and height 0.2. Its area is $2 \cdot 0.2$, or 0.4. So the probability is 0.4 that a randomly selected number is between 0 and 2.

b. The region between 2.5 and 3.5 is a rectangle with width 1 and height 0.2. The area of this rectangle is $1 \cdot 0.2$, or 0.2. So the probability is 0.2 that a number is between 2.5 and 3.5.

c. The region between 4 and 6 is a triangle with width 2 and height 0.2. The area of the triangle is $0.5 \cdot 2 \cdot 0.2 = 0.2$, so the probability is 0.2.

In Chapter 2, you learned three measures of center to describe a data set—mean, median, and mode. With a probability distribution you don't have a finite set of data. So these values must be defined and calculated in a slightly different way.

Measures of Center for Probability Distributions

Mode

The value or values of x at which the graph reaches its maximum value.

Median

The number d such that the line $x = d$ divides the area into two parts with equal areas.

Mean

For discrete distributions, the sum of each value of x multiplied by its probability. For continuous distributions, estimate the mean by breaking the area into vertical strips and treating it like a discrete distribution.

EXAMPLE B

A large number of people were asked to complete a puzzle. The time it took each person was recorded. The data are shown in the probability distribution graph at right, with times ranging between 0 and 8 minutes.

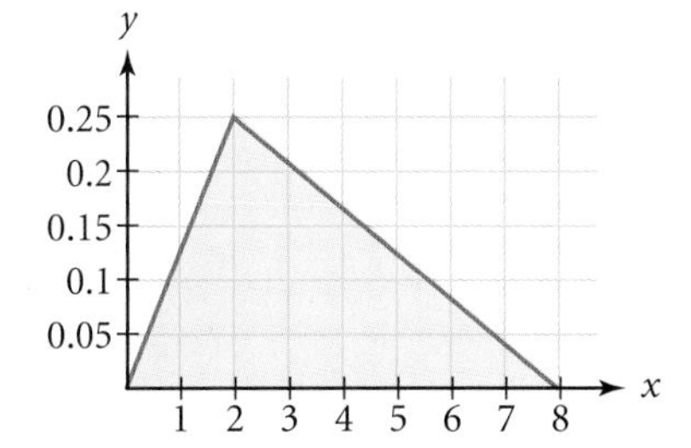

a. Find the mode.

b. Find the median.

c. Find the mean.

▶ *Solution*

Note that the shaded region has area 1.

a. The mode is the x-coordinate of the highest point, 2 minutes.

b. Find the vertical line that divides the triangle into two regions, each with area 0.5. Some trial and error shows the median is about 3 min. The smaller triangle has base 5 and height about 0.2.

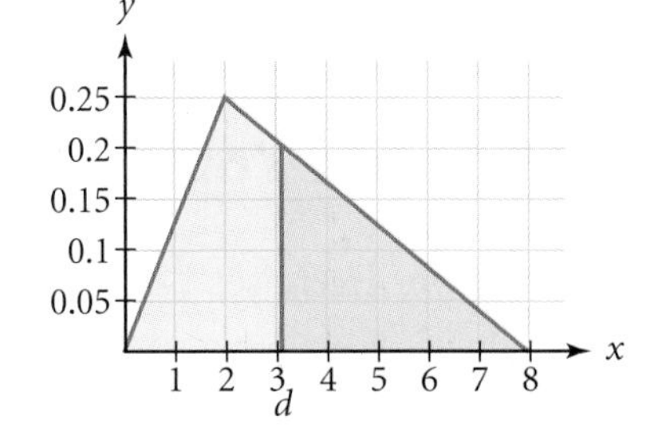

$$A \approx 0.5 \cdot (5)(0.2) \approx 0.5$$

So about half the area of the larger triangle falls after 3. To calculate the median exactly, you can use equations of the lines that form boundaries of the region. The equation of the line through $(2, 0.25)$ and $(8, 0)$ is $y = -\frac{1}{24}(x - 8)$. Use this equation to find the value of the median, d, such that the area of the triangle to the right of the median is 0.5.

$A = 0.5bh = 0.5(8 - d)\left[-\frac{1}{24}(d - 8)\right]$	The area of a triangle is half the product of the base $(8 - d)$ and height. The height is the y-value at $x = d$.
$0.5 = 0.5(8 - d)\left[-\frac{1}{24}(d - 8)\right]$	Solve for d when the area is 0.5.
$1 = -\frac{1}{24}(-d^2 + 16d - 64)$	Divide both sides by 0.5 and multiply the binomials.
$-24 = -d^2 + 16d - 64$	Multiply both sides by -24.
$0 = d^2 - 16d + 40$	Write the quadratic equation in general form.
$d \approx 3.101, 12.899$	Use the quadratic formula to solve for d.

The second value, 12.899, is outside of the domain, so the median of this distribution is about 3.101 min.

c. Slice the triangle into vertical strips and treat it like a discrete distribution. The mean is the sum of each value of x multiplied by its probability. A convenient number of strips for this triangle is eight, breaking at the integers. Take values of x to be the middle values between the integers. The probabilities are the areas of the strips. Calculate the area of each strip using the method from part b. For example, the probability $4 < x < 5$ can be found as the area of the strip between 4 and 5. Then the product of x times this probability is about $4.5 \cdot 0.1458 \approx 0.6563$.

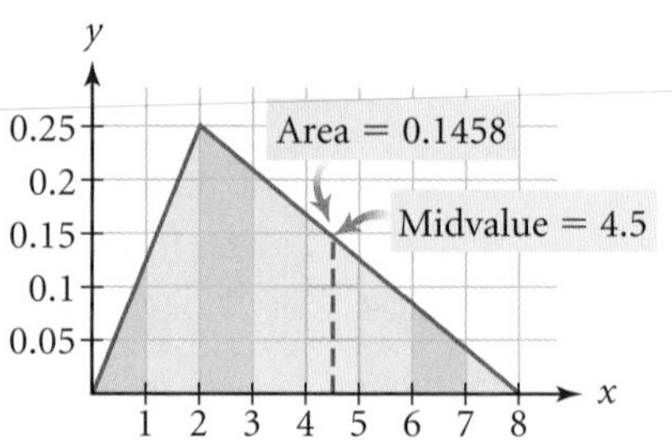

Interval	0–1	1–2	2–3	3–4	4–5	5–6	6–7	7–8
Midvalue	0.5	1.5	2.5	3.5	4.5	5.5	6.5	7.5
Probability	0.0625	0.1875	0.2292	0.1875	0.1458	0.1042	0.0625	0.0208
Product	0.0313	0.2813	0.5729	0.6563	0.6563	0.5729	0.4063	0.1563

The mean (sum of the products) is approximately 3.33 min.

A distribution is **symmetric** if its graph is symmetric about a vertical line. The mean of a symmetric distribution will always be the same as the median. Calculus can be used to find the measures of center of some nonsymmetric distributions. But for most probability distributions, finding the exact values of these parameters is almost impossible. They must be estimated using methods such as the vertical-strip approach.

EXERCISES

You will need

A graphing calculator for Exercises **9**, **12–14**, and **17**.

Practice Your Skills

Answer Exercises 1–4 on page 629 for each probability distribution below.

Distribution A

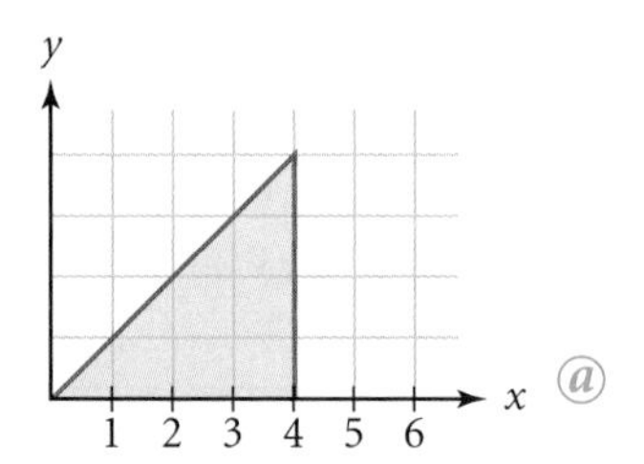

Distribution B

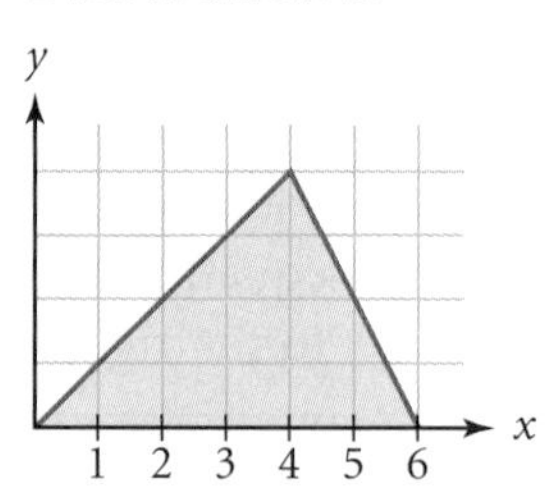

Distribution C

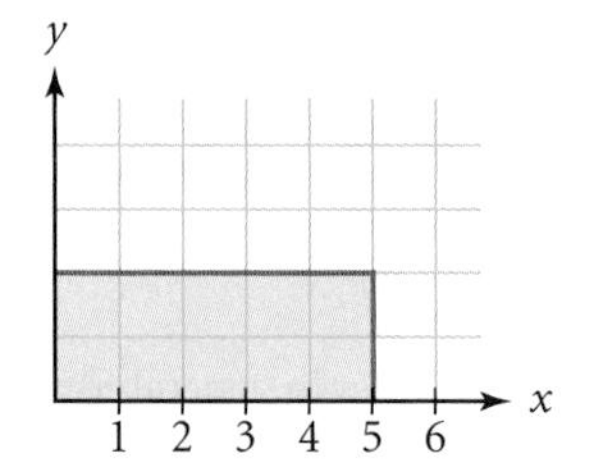

Distribution D

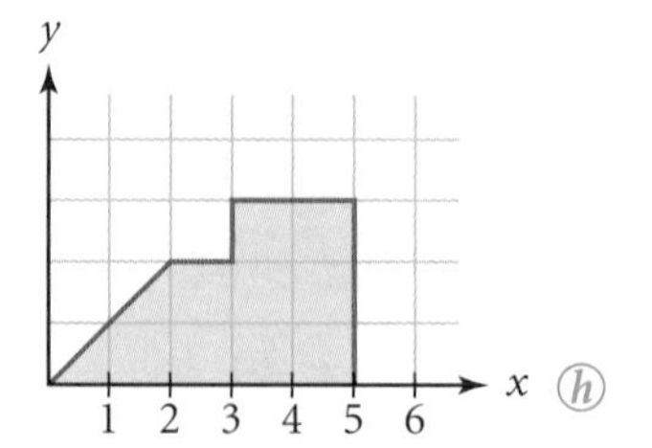

1. Find the height of one grid box on the y-scale so that the total area of the distribution is 1.
2. Find the probability that a randomly chosen value will be less than 3.
3. Estimate the median.
4. Estimate the mean.
5. Copy each histogram below and then draw a probability distribution. Try to keep the area under the curve the same as the area under the histogram.

   a. y

   x @

   1 2 3 4 5 6

   b.

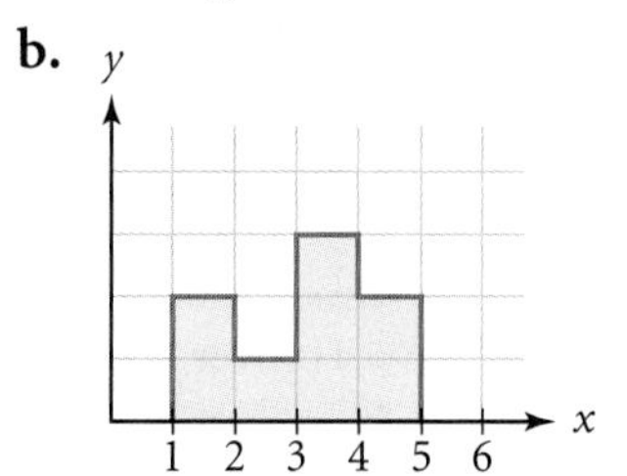

Reason and Apply

6. Suppose each person in your class selects a set of four numbers from 1 to 8 (repeats are allowed) and that each person calculates the mean of his or her own set.
   a. Sketch a possible histogram of these mean values. Explain the reasoning behind your histogram.
   b. Based on your histogram, estimate the mode and median of your distribution.
7. Classify each statement as true or false. If false, explain why.
   a. The y-value of the mode in a probability distribution can never be more than 1. @
   b. It is impossible to tell how many data values were used to create a probability distribution.
   c. The mean, median, and mode of a continuous distribution can never all be the same value.
8. Imagine finding many random numbers from 0 to 1 and substituting them into each expression below. Sketch what you think the relative frequency histograms or probability distributions of the results would look like, and explain your reasoning.

   a. $(random\ number)^2$ b. $(random\ number)^4$ c. $\sqrt{random\ number}$
9. ***Technology*** Use statistics software or your graphing calculator to investigate Exercise 8. (h)
10. Sketch a relative frequency histogram to fit each set of conditions. You may want to sketch these using the squares on graph paper to be certain you have a total area of 1. Label each axis with an appropriate scale.
    a. The data values are continuous from 0 to 10. The mean and median are the same value.
    b. The data values are continuous from 10 to 15. The mean is larger than the median.
11. Describe a data set that might produce each of these continuous distribution graphs. Indicate the range of values on the x-axis.

    a.

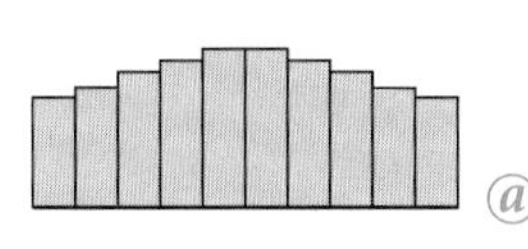

@

    b.

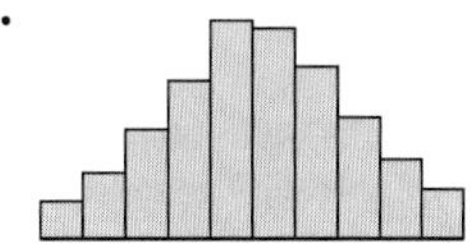

    c.

12. This table lists the ages of the presidents and vice presidents of the United States when they first took office through the 2004 election.

President	Age	President	Age
Washington	57	McKinley	54
J. Adams	61	T. Roosevelt	42
Jefferson	57	Taft	51
Madison	57	Wilson	56
Monroe	58	Harding	55
J. Q. Adams	57	Coolidge	51
Jackson	61	Hoover	54
Van Buren	54	F. D. Roosevelt	51
W. Harrison	68	Truman	60
Tyler	51	Eisenhower	62
Polk	49	Kennedy	43
Taylor	64	L. B. Johnson	55
Fillmore	50	Nixon	56
Pierce	48	Ford	61
Buchanan	65	Carter	52
Lincoln	52	Reagan	69
A. Johnson	56	G. H. W. Bush	64
Grant	46	W. Clinton	46
Hayes	54	G. W. Bush	54
Garfield	49		
Arthur	50		
Cleveland	47		
B. Harrison	55		

Vice president	Age	Vice president	Age
J. Adams	53	Hobart	52
Jefferson	53	T. Roosevelt	42
Burr	45	Fairbanks	52
G. H. Clinton	65	Sherman	53
Gerry	68	Marshall	58
Tompkins	42	Coolidge	48
Calhoun	42	Dawes	59
Van Buren	50	Curtis	69
R. M. Johnson	56	Garner	64
Tyler	50	Wallace	52
Dallas	52	Truman	60
Fillmore	49	Barkley	71
King	66	Nixon	40
Breckinridge	36	L. B. Johnson	52
Hamlin	51	Humphrey	53
A. Johnson	56	Agnew	50
Colfax	45	Ford	60
Wilson	61	Rockefeller	66
Wheeler	57	Mondale	49
Arthur	51	G. H. W. Bush	56
Hendricks	65	Quayle	41
Morton	64	Gore	44
Stevenson	57	Cheney	59

(*The World Almanac and Book of Facts 2007*)

a. What were the ages of the current vice president and president when they took office? Update the data sets to include these values.

b. Enter the two separate lists of data into your calculator and calculate the mean, $\bar{x}$, the standard deviation, s, the median, and the *IQR* for each list. Compare the data sets based on these statistics.

c. Graph a histogram for each data set. Use the same range and scale for each graph. Describe how the histograms reflect the statistics of each data set.

d. Calculate $\frac{x_i - \bar{x}}{s}$ for each entry, and create two new lists to convert the ages in each list to a standardized scale.

e. What is the range of values in each of the new distributions? Explain what the new distributions represent.

f. Graph a histogram for each of these standardized distributions. Use domain $-3.5 \leq x \leq 3.5$.

g. Compare and describe the graphs.

The U.S. Constitution specifies that presidents must be natural-born citizens, have lived in the United States for at least 14 years, and be at least 35 years old. The 26th U.S. president, Theodore Roosevelt (1858–1919), was the youngest president. Roosevelt became president when William McKinley (1843–1901) was assassinated. John F. Kennedy (1917–1963) was the youngest president to be elected.

13. APPLICATION In order to provide better service, a customer service call center investigated the hang-up rates of people who called in for help on a recent evening. These data were collected.

Hang-Up Rates

Length of call before hanging up (min)	0–3	3–6	6–9	9–12	12–15	15–18
Number of customers	1	3	4	6	13	9

Length of call before hanging up (min)	18–21	21–24	24–27	27–30	30–33
Number of customers	8	10	6	4	1

a. Make a table showing the probability distribution of the random variable x, where x represents the length of the call in three-minute intervals. @

b. Construct a relative frequency histogram for the probability distribution.

c. What is the median length of time a customer waited before hanging up?

d. Draw a smooth curve on your histogram so that the area under the curve is approximately the same as the area of the histogram.

Consumer CONNECTION

Customers often get frustrated when put on hold, and lost calls can mean lost business. Companies perform queue-time case studies to see how long a customer will wait on hold before hanging up. They can reschedule their staff to be available during peak calling periods.

Review

14. Consider the number of heads, x, when 15 fair coins are all tossed at once.

a. Use binomial expansion to find the probability distribution for $P(x)$. Then calculate the theoretical results for 500 trials of this experiment. Copy and complete the table below. Round off the frequencies in the bottom row to whole-number values.

Heads (x)	0	1	2	...	14	15	Total
$P(x)$	(0.5^{15})						1
Frequency	$500(0.5^{15}) = \underline{?}$			...			500

b. Draw a histogram showing the probability distribution for $P(x)$. ⓐ

c. Find the mean and standard deviation of the number of heads.

d. How many of the 500 trials are within one standard deviation of the mean?

e. What percentage of the data is within one standard deviation of the mean?

f. What percentage of the data is within two standard deviations of the mean?

g. What percentage of the data is within three standard deviations of the mean?

15. Suppose you roll a pair of standard six-sided dice five times. What is the probability of rolling a sum of 8 at least three times? ⓐ

16. The differences between scheduled and actual arrival times for Big Bird Airlines and Flying Bus Airlines are given in the table below. Negative values mean the flight was late and positive values mean the flight was early. The differences are given in minutes.

Big Bird	0	+2	−1	0	−3	+1	+1	0	−2	−2
Flying Bus	+5	−6	+12	0	−8	+2	+7	−1	−10	0

a. Give statistics to convince someone to fly with Big Bird. ⓐ

b. Now give statistics to convince someone to use Flying Bus.

17. Small data sets (fewer than 20 values) can have deceptive histograms. Using the times for Big Bird Airlines in Exercise 16, experiment with starting values and bin widths to create three misleading histograms. The histograms should give the impression that this data set is symmetric, skewed left, and skewed right. ⓗ

18. A spinner is divided into ten equal sectors numbered 1 through 10 in random order. If you get an even number, you add that number to your score. If you get an odd number, you subtract that number from your score. The game is over when your score reaches either +150 or −150. How many spins do you expect a typical game to last?

19. How is the area of a rectangle affected when its length is doubled and its width is halved? How is the area of a triangle affected when its base is doubled and its height is halved? How are the areas of a rectangle and triangle affected when their horizontal dimensions are multiplied by 3 and their vertical dimensions are multiplied by $\frac{1}{3}$? Do you think this relationship is true for any two-dimensional figure?

Project

SIMPSON'S PARADOX

This table shows the number of male and female applicants and the percentages admitted to the six largest graduate-school majors at the University of California, Berkeley, in Fall 1973.

	Men		Women		Total	
	Number of applicants	Percentage admitted	Number of applicants	Percentage admitted	Number of applicants	Percentage admitted
A	825	62	108	82	933	64
B	560	63	25	68	585	63
C	325	37	593	34	918	35
D	417	33	375	35	792	34
E	191	28	393	24	584	25
F	373	6	341	7	714	6

(D. Freedman, R. Pisani, R. Purves, and A. Adhikari (1991), *Statistics,* 2nd ed., New York: Norton, p. 17)

Compare the percentages for men, women, and total admitted. Does it seem as though there was a bias in favor of men or women in admissions? Why or why not?

Now calculate the total number of men and women admitted to these six majors. (You'll need to use the data given for number of applicants and admission rate for each major.) Then calculate the overall percentages of men and of women admitted. Does it appear that there was a bias in favor of men or women? What happened?

Your project should include

- Answers to the questions above.
- Any additional research you do on Simpson's paradox, including examples of other problems you find or one you make up yourself.

LESSON 11.3

Normal Distributions

Many continuous variables follow a distribution that is symmetric and mounded in a "bell" shaped distribution. Graphs of distributions for large populations, such as heights, clothing sizes, and test scores, often have this shape. The bell-shaped distribution is so common that it is called the **normal distribution,** and its graph is called the **normal curve.**

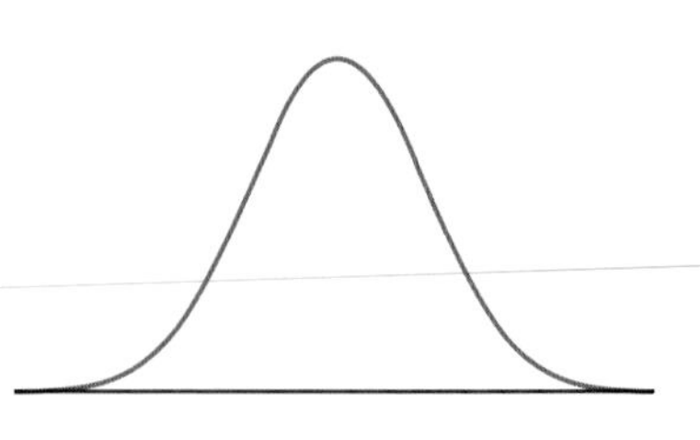

This 10-foot-tall machine drops balls through a grid of pins. The balls land in a bell-shaped curve—a visual representation of their probability distribution.

The formulas you have learned for the sample mean, $\bar{x}$, and sample standard deviation, s, are estimates for values in the population. When you find the mean and standard deviation for an entire population, they are called the **population mean,** μ, and **population standard deviation,** σ. These are the Greek letters mu (pronounced "mew") and sigma.

In this lesson you'll see some properties of normal distributions. The general equation for a normal distribution curve is in the form $y = ab^{-x^2}$. If you graph a function like $y = 3^{-x^2}$, you'll get a bell-shaped curve that is symmetric about the vertical axis.

To describe a particular distribution of data, you translate the curve horizontally to be centered at the mean of the data, and you dilate it horizontally to match the standard deviation of the data. Then you dilate it vertically so that the area is 1. These steps are shown graphically at the top of the next page. You'll want to begin with a parent function that has standard deviation 1.

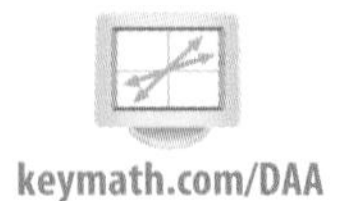
keymath.com/DAA

[▶ You can explore properties of the normal distribution using the **Dynamic Algebra Exploration** Normal Distributions at www.keymath.com/DAA .◀]

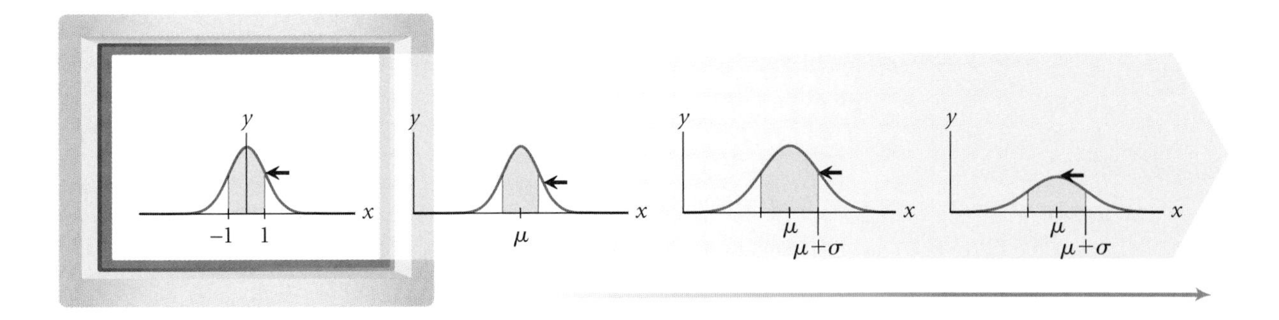

History CONNECTION

In 1733 French mathematician Abraham de Moivre (1667–1754) developed the normal distribution as a means to approximate the binomial distribution. Given a binomial sample of n trials with a probability of success of p for each trial, the probabilities can be estimated using a normal curve with mean $\mu = n \cdot p$ and standard deviation $\sigma = \sqrt{n \cdot p \cdot (1 - p)}$.

The parent function of a normal distribution has standard deviation 1, and is called the **standard normal distribution.** To meet the conditions for the standard normal distribution, statisticians have used advanced mathematics to determine the values of a and b in the equation $y = ab^{-x^2}$. The value of a is related to the number π.

$$a = \sqrt{\frac{1}{2\pi}} \approx 0.399$$

The value of b is related to another common mathematical constant, the transcendental number e.

$$b = \sqrt{e} \approx 1.649$$

Calculators allow you to work with these numbers fairly easily.

EXAMPLE

The general equation for a normal curve is in the form $y = ab^{-x^2}$.

a. Write the equation for a standard normal curve, using $a = \sqrt{\frac{1}{2\pi}}$ and $b = \sqrt{e}$. Find a good graphing window for this equation and describe the graph.
[▶ To learn how to enter the value of *e*, see **Calculator Note 11A.** ◀]

b. Write the equation for a normal distribution with mean μ and standard deviation σ.

c. How well will a normal curve fit a binomial distribution if there are 90 trials and the probability of success is 0.5?

▶ ***Solution***

Substitute the values given for the constants a and b into the general equation for a normal curve.

a. The equation for a standard normal curve is

$$y = \sqrt{\frac{1}{2\pi}}\left(\sqrt{e}\right)^{-x^2}$$

Note that $\sqrt{\frac{1}{2\pi}}$ is the same as $\frac{1}{\sqrt{2\pi}}$.

A good window for the graph is $-3.5 \leq x \leq 3.5$, $-0.1 \leq y \leq 0.5$. The graph is bell-shaped and symmetric about $x = 0$. So the mean, median, and mode are all 0. Almost all of the data are in the interval $-3 \leq x \leq 3$.

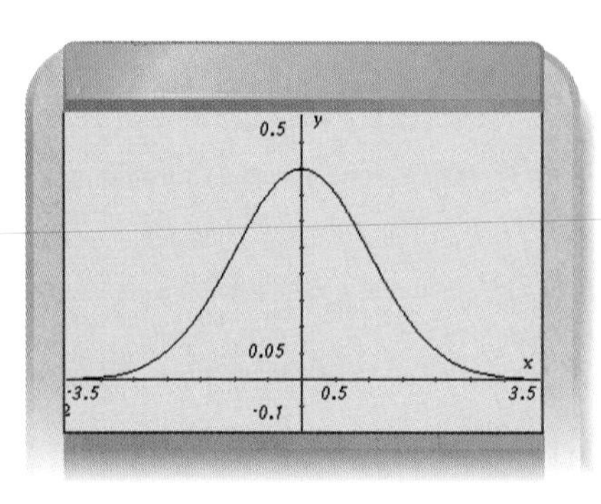

b. Translate the curve horizontally to shift the mean from 0 to μ. And dilate the curve horizontally to change the standard deviation from 1 to σ. The area under a probability distribution must be 1. The horizontal dilation will increase the area, so it must be accompanied by a vertical dilation.

$y = \sqrt{\frac{1}{2\pi}}(\sqrt{e})^{-x^2}$ Start with the parent function.

$y = \sqrt{\frac{1}{2\pi}}(\sqrt{e})^{-(x-\mu)^2}$ Substitute $(x - \mu)$ for x to translate the mean horizontally to μ.

$y = \sqrt{\frac{1}{2\pi}}(\sqrt{e})^{-[(x-\mu)/\sigma]^2}$ Divide $(x - \mu)$ by the horizontal scale factor, σ, so the curve reflects the correct standard deviation.

$y = \frac{1}{\sigma\sqrt{2\pi}}(\sqrt{e})^{-[(x-\mu)/\sigma]^2}$ Divide the right side of the equation by the vertical scale factor, σ, to keep the area under the curve equal to 1.

c. Use the equation $P(x) = {}_nC_x p^x(1 - p)^{n-x}$ to find the probabilities from the binomial distribution. When $n = 90$ and $p = 0.5$, then $P(x) = {}_{90}C_x \cdot 0.5^x(1 - 0.5)^{90-x} = {}_{90}C_x \cdot 0.5^{90}$. To write the equation of a normal curve, you must find the values of μ and σ. The mean is $\mu = 90(0.5) = 45$ and the standard deviation is $\sigma = \sqrt{90(0.5)(1 - 0.5)} \approx 4.7434$, so the equation is

$$y = \frac{1}{4.7434\sqrt{2\pi}}(\sqrt{e})^{-[(x-45)/4.7434]^2} \approx (0.084)(1.649)^{-[(x-45)/4.7434]^2}$$

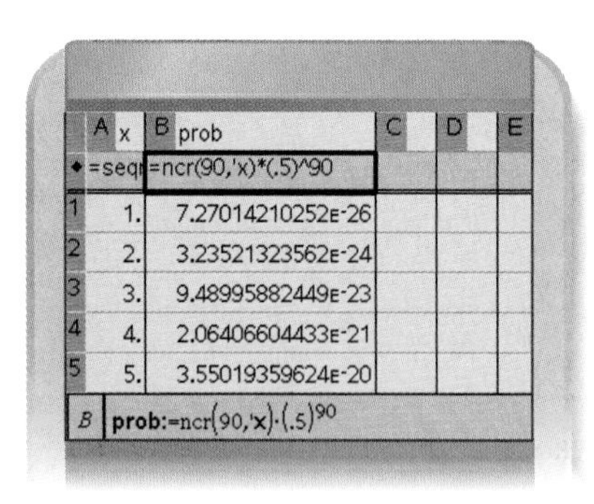

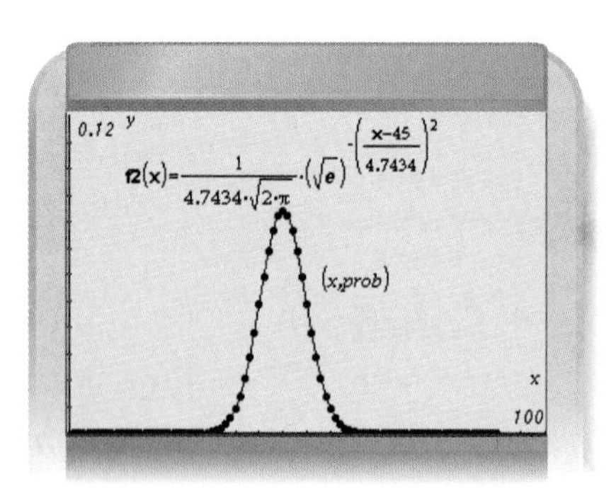

You can see that the graph of this normal curve fits the discrete points of the binomial distribution graph very closely.

The Normal Distribution

The equation for a **normal distribution** with mean μ and standard deviation σ is

$$y = \frac{1}{\sigma\sqrt{2\pi}}\left(\sqrt{e}\right)^{-[(x-\mu)/\sigma]^2}$$

You can view the graph of a normal distribution well in the window

$$\mu - 3\sigma \leq x \leq \mu + 3\sigma \quad \text{and} \quad 0 \leq y \leq \frac{0.4}{\sigma}$$

This window will show three standard deviations to the left and the right of the mean, and the minimum and maximum frequencies of the distribution.

In the equation for a normal distribution, the data values are represented by x and their relative frequencies by y. The area under a section of the curve gives the probability that a data value will fall in that interval.

Most graphing calculators provide the normal distribution equation as a built-in function, and you have to provide only the mean and standard deviation. In this chapter we will use the notation $n(x)$ to indicate a standard normal distribution function with mean 0 and standard deviation 1. Using this notation, a nonstandard normal distribution is written

n(x, mean, standard deviation)

For example, a normal distribution with mean 3.1 and standard deviation 0.14 is written $n(x, 3.1, 0.14)$. [▶ To learn how to graph a normal distribution on your calculator, see **Calculator Note 11B.** ◀] The area under a portion of a normal curve is written

N(lower, upper, mean, standard deviation)

This notation indicates the lower and upper endpoints of the interval, and the mean and standard deviation of the distribution. [▶ To learn how to find this value on your calculator, see **Calculator Note 11C.** ◀] This area determines the probability that a value in a normal distribution will fall within a particular range.

Investigation
The Normal Curve

Imagine that a group of students finds the weights of 500 U.S. pennies. Even if the actual weights of the pennies were exactly the same, there would be some variation in the measurements of those weights. Typically, the measurements made by a group of people of the same or very similar objects are normally distributed. You'll use this idea to explore areas under the normal curve.

Step 1 Each person in your group should use technology to simulate the weights of 500 pennies. Technically, this is a sample from a population of infinitely many weights. Assume that pennies are normally distributed about a mean of 2.5 g and have a standard deviation of 0.14 g. [▶ See **Calculator Note 11D.** ◀]

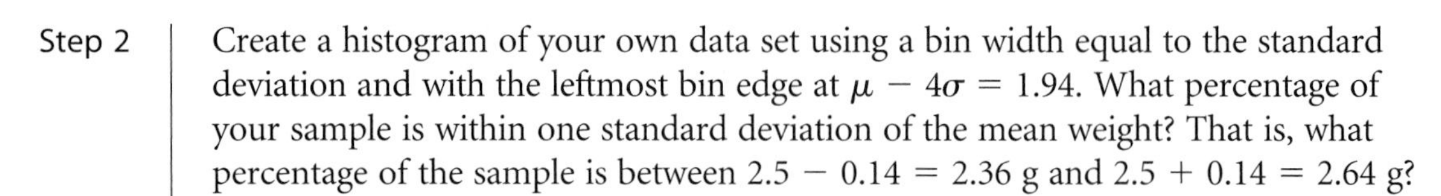

Step 2 Create a histogram of your own data set using a bin width equal to the standard deviation and with the leftmost bin edge at $\mu - 4\sigma = 1.94$. What percentage of your sample is within one standard deviation of the mean weight? That is, what percentage of the sample is between $2.5 - 0.14 = 2.36$ g and $2.5 + 0.14 = 2.64$ g?

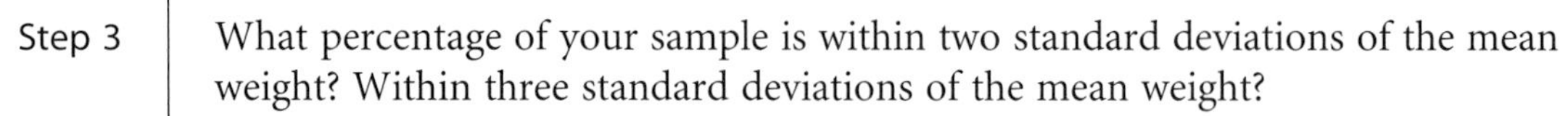

Step 3 What percentage of your sample is within two standard deviations of the mean weight? Within three standard deviations of the mean weight?

Step 4 There is a rule in statistics known as the "68-95-99.7 rule." Compare your results from Steps 2 and 3 with those of your group, and write a rule that might go by this name.

Step 5 Each person in your group should select a different mean and positive standard deviation for a normal distribution. Use your calculators to find the probability that a value lies within one, two, and three standard deviations of the mean. [► See **Calculator Note 11C** for help finding the area of a region under the normal curve. ◄]

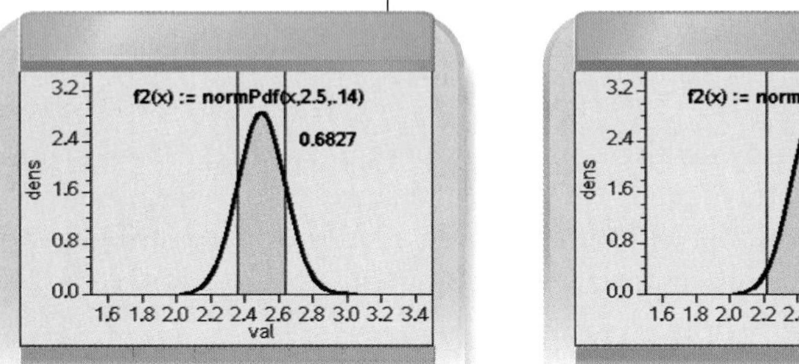

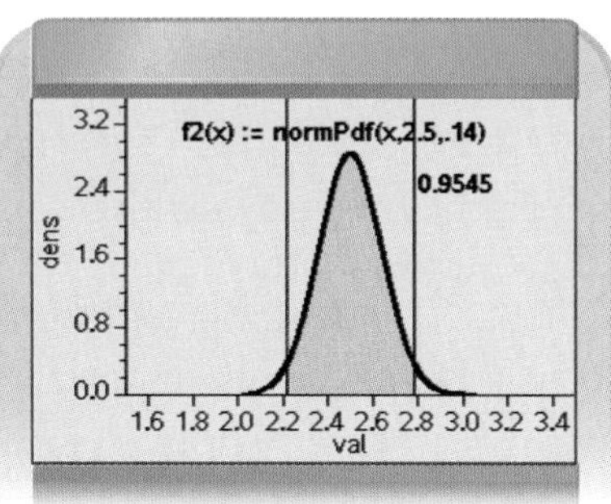

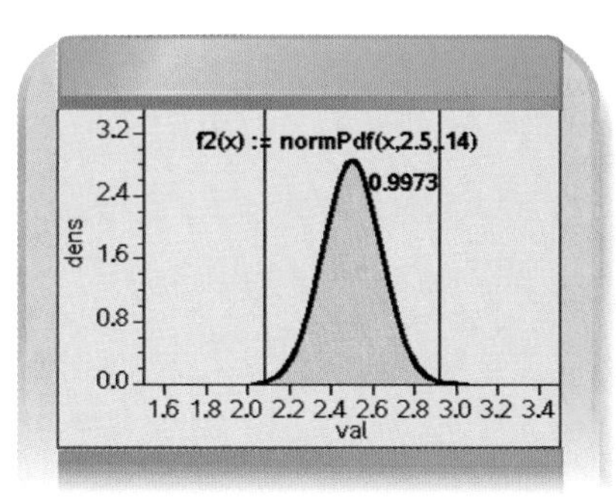

Step 6 Summarize the results of your group for Step 5. Do the probabilities follow the 68-95-99.7 rule?

Look at the curvature of a normal curve. At the points that are exactly one standard deviation from the mean, the curve changes between curving downward (the part of the curve with decreasing slope) and curving upward (the parts of the curve with increasing slope). These points are called **inflection points.** You can estimate the standard deviation of any normal distribution by locating the inflection points of its graph.

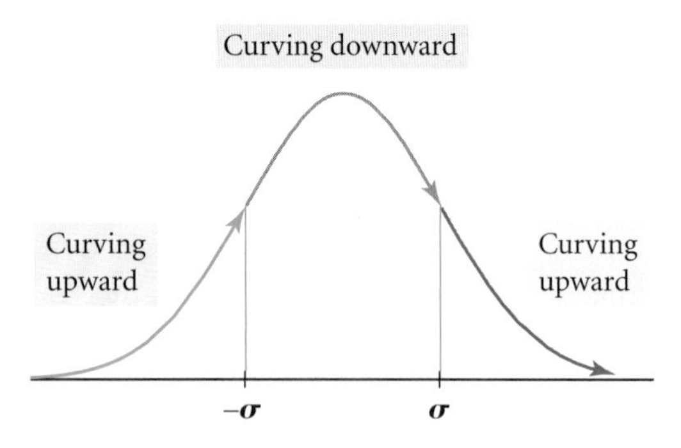

History CONNECTION

English nurse Florence Nightingale (1820–1910) contributed to the field of applied statistics by collecting and analyzing data during the Crimean War (1853–1856). While stationed in Turkey, she systematized data collection and record keeping at military hospitals and created a new type of graph, the polar-area diagram. Nightingale used statistics to show that improved sanitation in hospitals resulted in fewer deaths. For more information about Nightingale's contributions, see the links at www.keymath.com/DAA .

Florence Nightingale

EXERCISES

You will need

A graphing calculator for Exercises **1, 5–8,** and **10–12.**

Practice Your Skills

1. The standard normal distribution equation, $y = ab^{-x^2}$, where $a = \sqrt{\frac{1}{2\pi}}$ and $b = \sqrt{e}$, is equivalent to the calculator's built-in function $n(x, 0, 1)$.

a. Use a table or graph to verify that these functions are equivalent. ⓗ

b. Evaluate each function at $x = 1$. ⓐ

2. From each equation, estimate the mean and standard deviation.

a. $y = \frac{1}{5\sqrt{2\pi}}\left(\sqrt{e}\right)^{-[(x-47)/5]^2}$ ⓐ

b. $y = \frac{0.4}{23}(0.60653)^{[(x-250)/23]^2}$

c. $y = 1.29e^{-[(x-5.5)^2/0.1922]}$ ⓐ

d. $y = 0.054(0.99091)^{(x-83)^2}$

3. From each graph, estimate the mean and standard deviation.

a. ⓗ

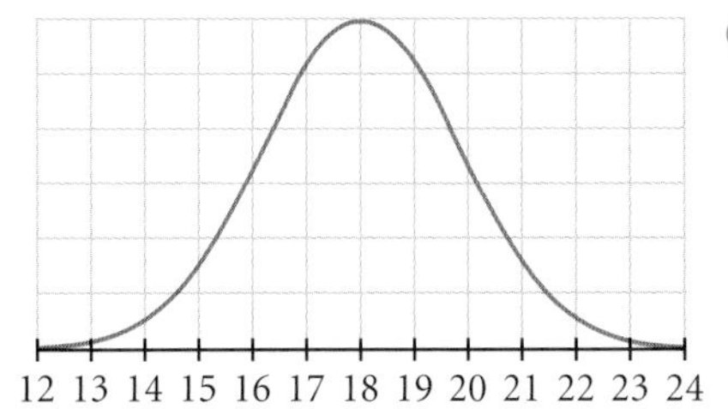

b.

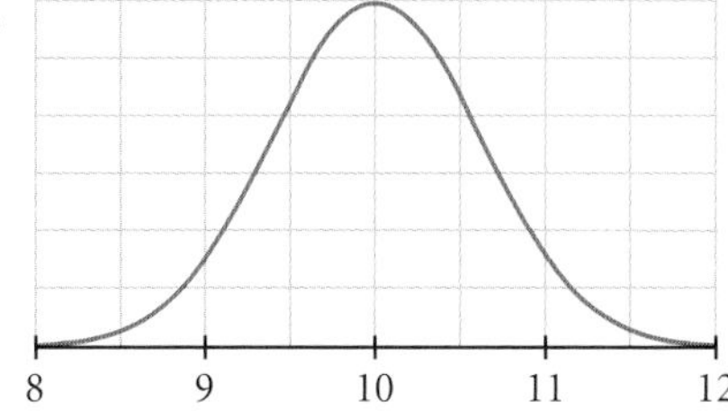

c.

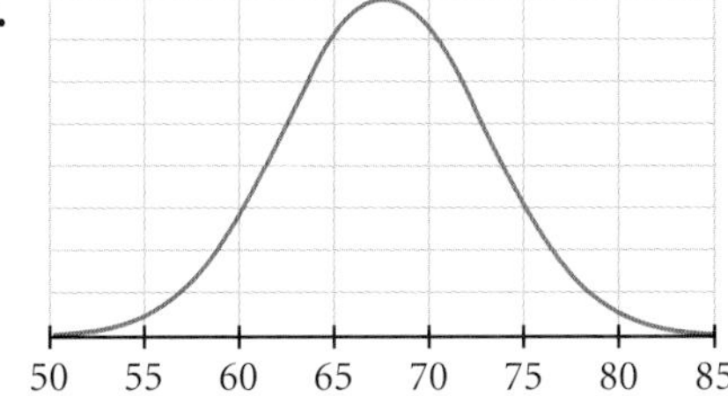

d.

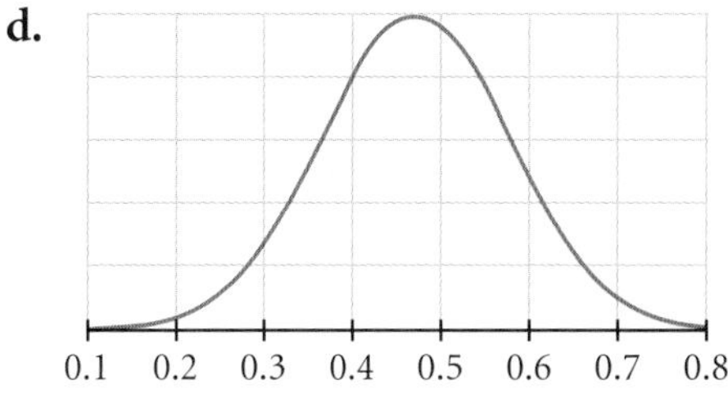

4. Estimate the equation of each graph in Exercise 3. ⓐ

Reason and Apply

5. The life spans of wild tribbles are normally distributed with a mean value of 1.8 years and a standard deviation of 0.8 year. Sketch the normal curve, and shade the portion of the graph showing tribble life spans of 1.0 to 1.8 years.

William Shatner (b 1931) played Captain James T. Kirk in the television series *Star Trek* (1966–1969). Captain Kirk and two fictional alien creatures called tribbles are shown here.

6. Assume that the mean height of an adult male gorilla is 5 ft 8 in., with a standard deviation of 7.2 in.

a. Sketch the graph of the normal distribution of gorilla heights. ⓐ

b. Sketch the graph if, instead, the standard deviation were 4.3 in.

c. Shade the portion of each graph representing heights greater than 6 ft. Compare your sketches and explain your reasoning.

7. Frosted Sugar Squishies are packaged in boxes labeled "Net weight: 16 oz." The filling machine is set to put 16.8 oz in the box, with a standard deviation of 0.7 oz.

a. Sketch a graph of the normal distribution of box weights. Shade the portion of the graph representing boxes that are below the labeled weight.

b. What percentage of boxes does the shading represent? Is this acceptable? Why or why not?

8. Makers of Sweet Sips 100% fruit drink have found that their filling machine will fill a bottle with a standard deviation of 0.75 oz. The control on the machine will change the mean value but will not affect the standard deviation.

a. Where should they set the mean so that 90% of the bottles have at least 12 oz of fruit drink in them? ⓗ

b. If a fruit drink bottle can hold 13.5 oz before overflowing, what percentage of the bottles will overflow at the setting suggested in 8a?

9. The pH scale measures the acidity or alkalinity of a solution. Water samples from different locations and depths of a lake usually have normally distributed pH values. The mean of those pH values, plus or minus one standard deviation, is defined to be the pH range of the lake.

Lake Fishbegon has a pH range of 5.8 to 7.2. Sketch the normal curve, and shade those portions that are outside the pH range of the lake. ⓐ

10. Data collected from 493 women are summarized in the table.

Height (cm)	148–50	150–52	152–54	154–56	156–58	158–60	160–62	162–64	164–66
Frequency	2	5	9	15	27	40	52	63	66

Height (cm)	166–68	168–70	170–72	172–74	174–76	176–78	178–80	180–82	182–84
Frequency	64	53	39	28	15	8	4	1	2

a. Find the mean and standard deviation of the heights, and sketch a histogram of the data. @

b. Write an equation based on the model $y = ab^{-x^2}$ that approximates the histogram.

c. Find the equation for a normal curve using the height data.

11. These data are the pulse rates of 50 people.

a. Find the mean and standard deviation of the data and sketch a histogram.

b. Sketch a distribution that approximates the histogram.

c. Find the equation of a normal curve using the pulse-rate data.

d. Are these pulse rates normally distributed? Why or why not?

66	75	83	73	87	94	79	93	87	64
80	72	84	82	80	73	74	80	83	68
86	70	73	62	77	90	82	85	84	80
80	79	81	82	76	95	76	82	79	91
82	66	78	73	72	77	71	79	82	88

12. Ridge counts in fingerprints are approximately normally distributed with a mean of about 150 and a standard deviation of about 50. Find the probability that a randomly chosen individual has a ridge count

a. Between 100 and 200.

b. Of more than 200.

c. Of less than 100.

Science CONNECTION

Dactyloscopy is the comparison of fingerprints for identification. Francis Galton (1822–1911), an English anthropologist, demonstrated that fingerprints do not change over the course of an individual's lifetime and that no two fingerprints are exactly the same. Even identical twins, triplets, and quadruplets have completely different prints. According to Galton's calculations, the odds of two individual fingerprints being the same are 1 in 64 billion. For more information on fingerprint identification, see the weblinks at www.keymath.com/DAA .

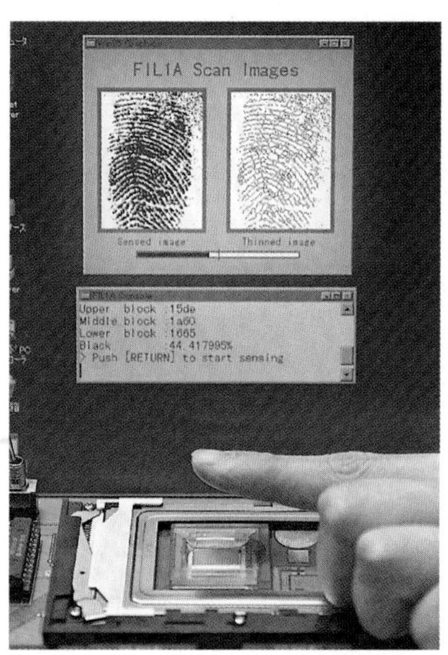

Developed by researchers in Atsugi, Japan, a microchip scans a finger to identify a fingerprint with 99% accuracy in half a second.

Review

13. Paul, Kenyatta, and Rosanna each took one national language exam. Paul took the French exam and scored 88. Kenyatta took the Spanish exam and scored 84. Rosanna took the Mandarin exam and scored 91. The national means and standard deviations for the tests are as follows:

French: $\mu = 72, \sigma = 8.5$

Spanish: $\mu = 72, \sigma = 5.8$

Mandarin: $\mu = 85, \sigma = 6.1$

a. Can you determine which test is most difficult? Why or why not? ⓗ

b. Which test had the widest range of scores nationally? Explain your reasoning.

c. Which of the three friends did best when compared with the national norms? Explain your reasoning. ⓗ

14. Mr. Hamilton gave his history class an exam in which a student must choose 3 out of 6 parts and complete 2 out of 4 questions in each part selected. How many different ways are there to complete the exam?

15. In the expansion of $(2x + y)^{12}$, what is the coefficient of the term containing y^7? ⓗ

16. Find the equation of a conic section that passes through the three points given at right if the conic section is ⓗ

a. A parabola. **b.** A circle.

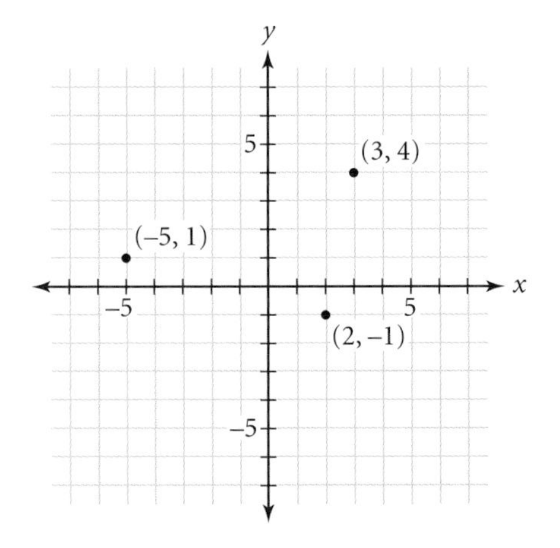

IMPROVING YOUR VISUAL THINKING SKILLS

Acorns

Acorns fall around the base of a particular tree in an approximately normal distribution with standard deviation 20 ft away from the base.

1. What is the probability that any 1 acorn will land more than 10 ft from the tree?

2. What is the probability that any 4 randomly chosen acorns will land an average of more than 10 ft from the tree?

3. What is the probability that any 16 randomly chosen acorns will land an average of more than 10 ft from the tree?

EXPLORATION

Fathom
Dynamic Data™ Software

Normally Distributed Data

What kinds of data are normally distributed? In this exploration you'll use census data and Fathom Dynamic Data Software to explore which attributes of the population of the United States are distributed normally.

Activity

Is This Normal?

Step 1 Start Fathom. From the File menu choose **Open Sample Document.** Open one of the census data files in the **Social Science** folder. You'll see a box of gold balls, called a collection, that holds data about a group of individuals.

Step 2 Click on the collection, and then choose **New | Case Table** from the Object menu. Scroll through the table. What numerical attributes are included? Which ones do you think might be normally distributed?

Step 3 A histogram can show whether a set of data is approximately normally distributed. To create a histogram, choose **New | Graph** from the Object menu. Drag and drop an attribute onto the horizontal axis, and then choose **Histogram** from the pull-down menu in the corner of the graph window.

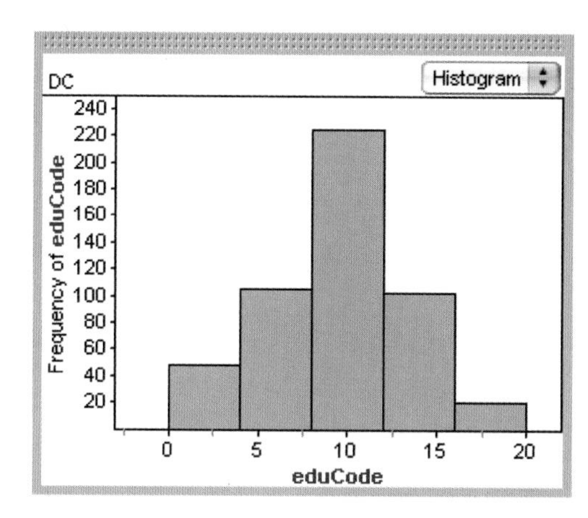

To experiment with different bin widths, double-click on any number on the horizontal axis. You will see a new window describing the bin width and axes scales. Click on any of the numbers to modify them. What bin width makes it easiest to see patterns for each of the numerical attributes?

Inspect Graph of DC

Cases Properties

Property	Value
binAlignm...	0
binWidth	4
xLower	-3
xUpper	22
yLower	0
yUpper	250

Step 4 Add a slider named *Scale*. Choose **Plot Function** from the Graph menu. You will graph the normal curve, using *Scale* as an adjustable scale factor. Enter the function as shown below the graph on the next page, using the attribute you chose. Move the slider until the function graph fits the histogram as well as possible.

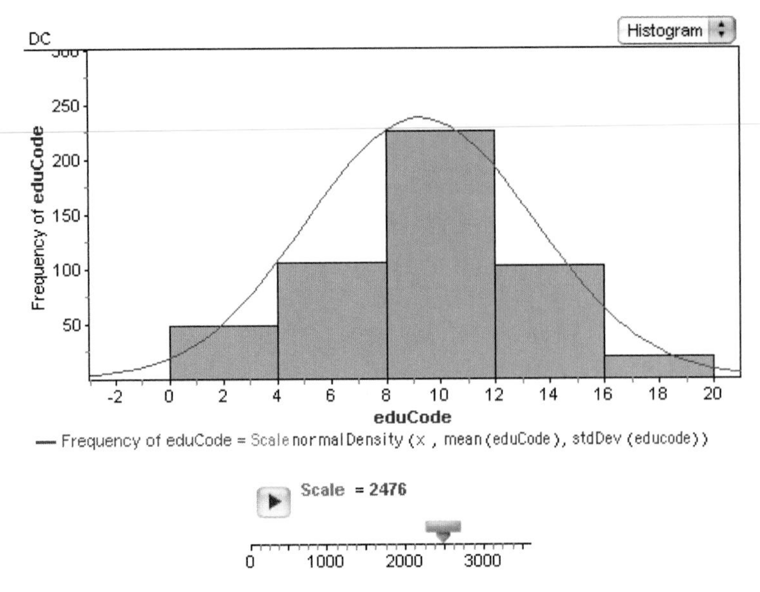

Step 5 Experiment with different attributes, bin widths, and regional data files to find data that are approximately normally distributed. You may want to filter your data. For example, income will not appear to be normally distributed because there are many people who have an income of $0. Many of these people are younger than 20 or older than 65. Select your collection and choose **Add Filter** from the Object menu. Type "(age > 20) and (age < 65)." Click **OK.** People younger than 20 and older than 65 will be removed from the case table and histogram. You may also want to separate histograms into categories, such as male and female. To do this, drag and drop a second attribute onto the vertical axis of the histogram.

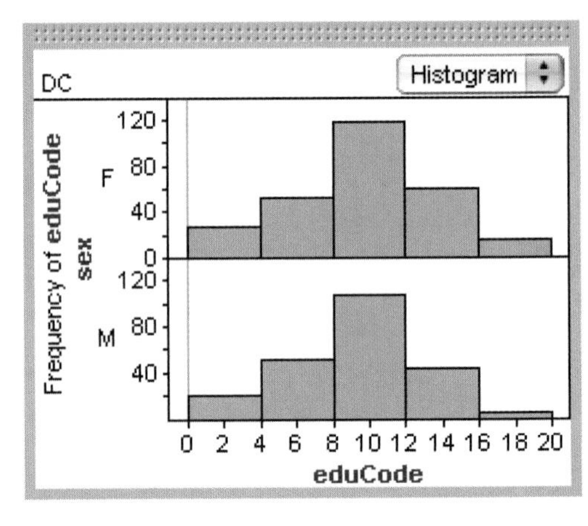

What normally distributed data do you find? What data are not normally distributed that you thought might be? What regional differences are there?

Questions

1. Make conjectures about why specific attributes in the census data that you explored are or are not normally distributed.

2. For census data that are not normally distributed, what histogram shape is most common?

LESSON

11.4 z-Values and Confidence Intervals

The way to do research is to attack the facts at the point of greatest astonishment.

CELIA GREEN

In Lesson 11.3, you learned about normally distributed populations and the probability that a randomly chosen item from such a population will fall in various intervals. That is, you knew something about the population and you saw how to find information about a sample.

In most real-life situations, however, you have statistics from one or more samples and want to estimate parameters of the population, which can be quite large. For example, suppose you know the mean height and standard deviation of 50 students that you survey, and you want to know the mean height and standard deviation of the entire population of students in your school. In this lesson you'll see how to describe some population parameters based on sample statistics.

First, it will be useful to learn a new method for describing a data value in a normal distribution. Knowing how a value relates to the mean value is important, but it does not tell you how typical the value is. For instance, to say that a penny's weight is 0.4 g less than the mean does not tell you whether this measurement is a rare event or a common event. In the investigation your group will devise a method for describing how typical or unusual a particular data value is.

Chinese-American artist Diana Ong (b 1940) titled this watercolor *So Very Crowded.*

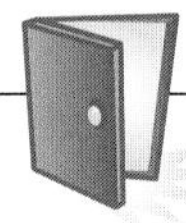

Investigation
Keeping Score

Andres and his cousin Imani have always been very competitive. This year they are both taking French at their respective schools. Andres boasted that he scored 86 on a recent exam. Imani had scored only 84 on her latest exam. But she countered by asking Andres about the mean and standard deviation on his test. The mean score on Andres's exam was 74 and the standard deviation was 9. The mean score on Imani's exam was 75 and the standard deviation was 6. Which of the two scores was actually better relative to the population of scores on the same exam?

Work with your group to decide who has the better score. Come up with a measure that uses the mean and standard deviation to support your conclusion. Be prepared to explain your measure and justify your conclusion.

In the investigation you may have found a measure that involved adding or subtracting some number of standard deviations to or from the mean. This measure could give you an idea of how rare an outcome is. For example, to say that a penny's weight is 0.4 g less than the mean doesn't indicate whether this is unusual. But to say that a penny's weight is 2.86 standard deviations from the mean tells you that this measurement is quite rare.

The number of standard deviations that a variable x is from the mean is called its **z-value,** or **z-score.** If the z-value of a measurement is between -1 and 1, then the measurement is within one standard deviation of the mean. As you saw in Lesson 11.3, if the variable is normally distributed, then the probability of a measurement with a z-value between -1 and 1 is about 68%. You also found that the probability of a z-value between -2 and 2 (within two standard deviations) is about 95%, and the probability of a z-value between -3 and 3 (within three standard deviations) is about 99.7%.

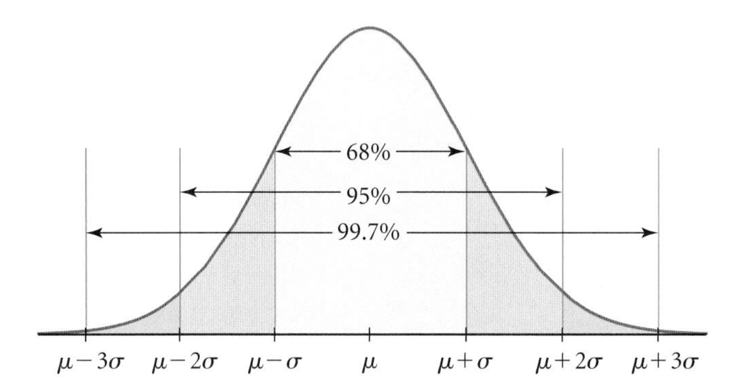

By the 68-95-99.7 rule, 68% of the area under a normal curve falls within one standard deviation of the mean, 95% within two standard deviations, and 99.7% within three standard deviations.

You can think of the z-value of x as the image of x under a transformation that translates and dilates the normal distribution to the standard normal distribution $n(x)$, with mean $\mu = 0$ and standard deviation $\sigma = 1$. This transformation from x-value to z-value is called **standardizing the variable** and can be calculated with the equation $z = \frac{x-\mu}{\sigma}$. The following example illustrates how to standardize values of a variable.

EXAMPLE A

The heights of a large group of men are distributed normally with mean 70 in. and standard deviation 2.5 in.

a. Find the z-values for 67.5 in. and 72.5 in.

b. What is the probability that a randomly chosen member of this group has height x between 65 and 75 in.?

c. Find an interval of x-values, symmetric about the mean, that contains 90% of the heights.

▸ Solution

For this population, $\mu = 70$ and $\sigma = 2.5$.

a. Use the formula $z = \frac{x-\mu}{\sigma}$ to standardize the variable.

$$z = \frac{67.5 - 70}{2.5} = -1 \quad \text{and} \quad z = \frac{72.5 - 70}{2.5} = 1$$

In this distribution, 67.5 in. corresponds to a z-value of -1, which means the value is one standard deviation below the mean. The height 72.5 in. corresponds to a z-value of 1, or one standard deviation above the mean.

b. The z-value of 65 in. is $\frac{65-70}{2.5}$, or -2, and the z-value of 75 in. is $\frac{75-70}{2.5}$, or 2. There is about a 95% probability that a randomly chosen value is within two standard deviations of the mean. A calculator graph confirms this prediction. [▸ See **Calculator Note 11C** to recall how to draw a normal curve with an area shaded. ◂]

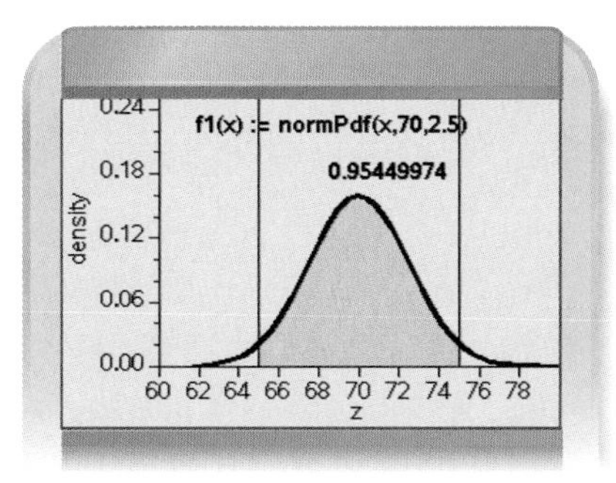

c. You already know that about 68% of the heights fall in the interval $-1 \leq z \leq 1$, which corresponds to $67.5 \leq x \leq 72.5$. You also know that about 95% of the heights fall in the interval $65 \leq x \leq 75$. So you can guess that an interval of about $66 \leq x \leq 74$ will contain 90% of the heights. Use your calculator and trial and error to obtain more precise endpoints. The calculator screen below shows that the interval $65.885 \leq x \leq 74.115$ contains about 90.0% of the data.

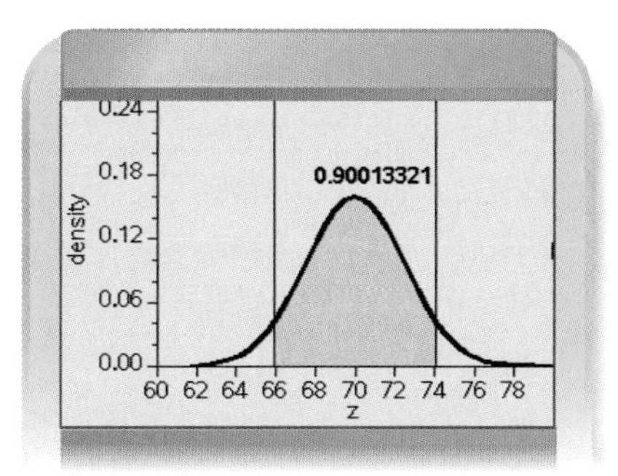

If you wanted to know for sure the average height of all the men in your state, you would need to measure every man's height and find the mean. That isn't a practical way to find the answer, so you must settle for taking a sample of the population and finding the mean height of the sample. The mean of the sample is not likely to equal the mean of the population, but it can provide a good estimate if the men in the sample are chosen randomly.

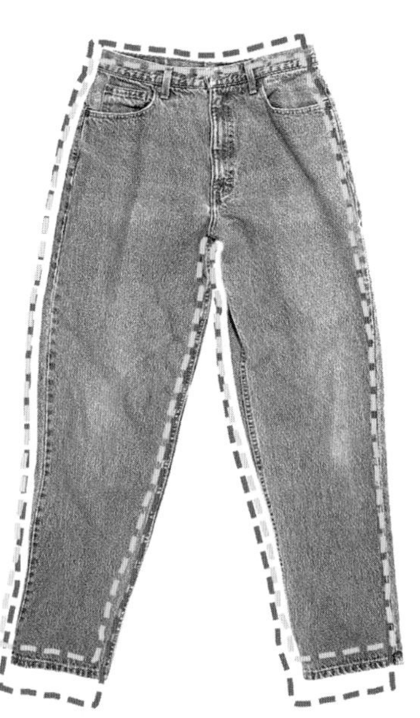
Clothing sizes are usually normally distributed.

The difference between the sample mean and the population mean is limited by the spread of the distribution (such as σ, the standard deviation) and the number of data points used to find the mean. As more values are added, the sample mean closes in on the population mean. The accuracy of the estimate can be given using a formula that involves these two measures. One measure of accuracy, the **margin of error,** is calculated by $\frac{z\sigma}{\sqrt{n}}$.

Confidence Interval

Suppose a sample from a normally distributed population has size n and mean $\bar{x}$, and the population standard deviation is σ. Then the $p\%$ **confidence interval** is an interval about $\bar{x}$ in which you can be $p\%$ confident that the population mean, μ, lies. If z is the number of standard deviations from the mean within which $p\%$ of normally distributed data lie, the $p\%$ confidence interval is

$$\bar{x} - \frac{z\sigma}{\sqrt{n}} < \mu < \bar{x} + \frac{z\sigma}{\sqrt{n}}$$

Suppose a science class is asked to estimate the mean pH of the water in a lake. The students take 30 samples of water. When they analyze their results, they find a mean pH of 5.7 and a standard deviation of 0.8. It is very unlikely that any sample will give the true mean pH of the lake water, so they express the estimate as an interval of possible values of the mean pH. The 68-95-99.7 rule tells the students that 95% of normally distributed values have a z-value between -2 and 2. Using this z-value, the class can predict with 95% confidence that the mean pH of the lake lies between $5.7 - \frac{2(0.8)}{\sqrt{30}}$ and $5.7 + \frac{2(0.8)}{\sqrt{30}}$, or between about 5.41 and 5.99. This interval can also be expressed as 5.7 ± 0.29, where 0.29 is the margin of error.

What if you want a confidence interval that is not 68%, 95%, or 99.7%? You will need a z-value other than 1, 2, or 3. You have used your calculator to find probabilities (areas) between z-values. You can use an inverse function to find the z-values for a given probability. A 97% confidence interval has 1.5% of the area below the interval and 1.5% above the interval.

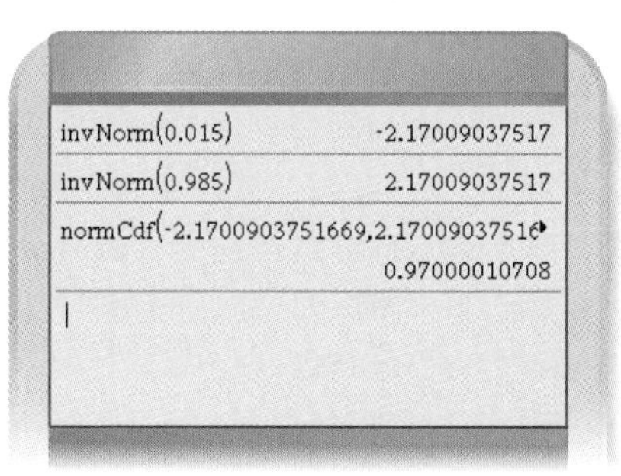

The calculator screen shows that a z-value of about 2.17 standard deviations corresponds to a 97% confidence interval.

EXAMPLE B

Jackson is training for the 100 m race. His coach timed his last run at 11.47 s. Experience in previous training sessions indicates that the standard deviation for timing this race is 0.28 s.

a. Find the 95% confidence interval.

b. What confidence interval corresponds to ± 2.3 standard deviations?

c. Find the 90% confidence interval.

Men race in the 100-meter final at the 2000 Olympic Games in Sydney, Australia.

▶ Solution

Because only one run time is known, the value for n is 1. All confidence intervals are given as $\left(11.47 - \frac{z(0.28)}{\sqrt{1}}, 11.47 + \frac{z(0.28)}{\sqrt{1}}\right)$, where the value of z is determined by the level of confidence. This is the interval with endpoints $11.47 \pm 0.28z$.

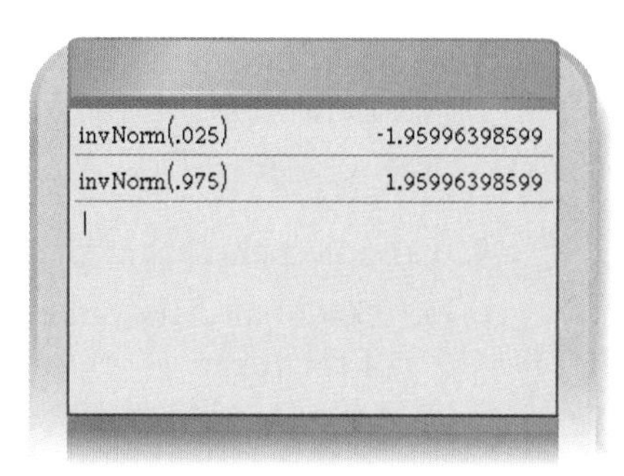

a. The center region under the normal curve is 95% of the total area, so the "tail" of the curve on each side represents 2.5% of the data. Use the calculator to find that $P(z < -1.96) \approx 0.025$. So the margin of error, with 95% confidence, corresponds to a z-score of approximately 1.96, and therefore is about $0.28(1.96) = 0.55$ s. The 95% confidence interval has endpoints at about 11.47 ± 0.55 s, or 10.92 s and 12.02 s.

Using the 68-95-99.7 rule to find the 95% confidence interval gives a result that differs only by the amount of round-off error. By the 68-95-99.7 rule, the z-value for 95% is about two standard deviations. The endpoints of the confidence interval, then, are $11.47 \pm 0.28(2)$. The coach is 95% confident that the actual time is between approximately 10.91 and 12.03 s.

b. The area of the region within 2.3 standard deviations of the mean is $P(-2.3 < z < 2.3) \approx 0.979$, so it represents a 98% confidence interval. Alternatively, because $11.47 \pm 0.28(2.3)$ gives the interval between 10.826 and 12.114, you can get the same confidence level by finding $P(-2.3 < x < 2.3 \mid \mu = 11.47, \sigma = 0.28)$.

c. If the center region under the normal curve has area 90%, then the tail on each side must have area 5%. Use the calculator to find that $P(z < -1.645) \approx 0.05$. The margin of error here, with 90% confidence, is $0.28(1.645) = 0.46$ s. Therefore, the time interval has endpoints 11.47 ± 0.46, so the coach can be 90% confident that the actual time lies between 11.01 s and 11.93 s.

In Example B, the coach had to rely on only one time measurement. If four people, instead of only one, had timed the run, and the mean of those times had been 11.47 s, then the 95% interval would have had endpoints $11.47 \pm \frac{2(0.28)}{\sqrt{4}}$, making it between 11.19 and 11.75 s. In general, the larger the sample size, the narrower the interval in which you can be confident that the population mean lies.

Exercises

You will need

A graphing calculator for Exercises **5–8**, **11**, and **15**.

Practice Your Skills

1. Trace the normal curve at right onto your paper. Add vertical lines demonstrating the 68-95-99.7 rule.

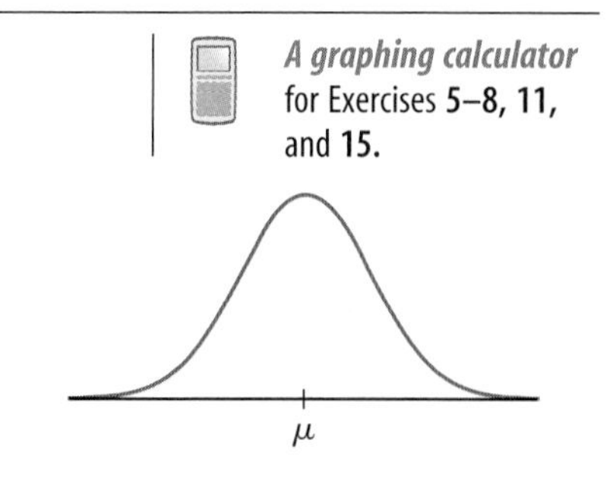

2. A set of normally distributed data has mean 63 and standard deviation 1.4. Find the z-value for each of these data values.

a. 64.4 ⓐ　　**b.** 58.8

c. 65.2 ⓐ　　**d.** 62

3. A set of normally distributed data has mean 125 and standard deviation 2.4. Find the data value for each of these z-values.

a. $z = -1$ @ **b.** $z = 2$ **c.** $z = 2.9$ @ **d.** $z = -0.5$

4. Complete these statements.

a. 95% of all data values in a normal distribution are within __?__ standard deviations of the mean.

b. 90% of all data values in a normal distribution are within __?__ standard deviations of the mean.

c. 99% of all data values in a normal distribution are within __?__ standard deviations of the mean.

Reason and Apply

5. The mean travel time between two bus stops is 58 min with standard deviation 4.5 min.

a. Find the z-value for a trip that takes 66.1 min. @

b. Find the z-value for a trip that takes 55 min.

c. Find the probability that the bus trip takes between 55 and 66.1 min.

Created by American sculptor Richard Beyer (b 1925), *Waiting for the Interurban* is a cluster of statues at a bus stop in Seattle, Washington. Local residents frequently change the statues' clothes. In this photo, they are wearing Batman costumes.

6. A set of normally distributed data has mean 47 s and standard deviation 0.6 s. Find the percentage of data within these intervals: (h)

a. Between 45 and 47 s

b. Greater than 1.5 s above or below the mean

7. A sample has mean 3.1 and standard deviation 0.14. Find each confidence interval, and round values to the nearest 0.001. Assume $n = 30$ in each case.

a. 90% confidence interval @ **b.** 95% confidence interval **c.** 99% confidence interval

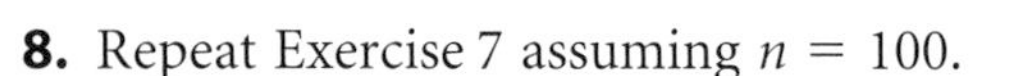

8. Repeat Exercise 7 assuming $n = 100$.

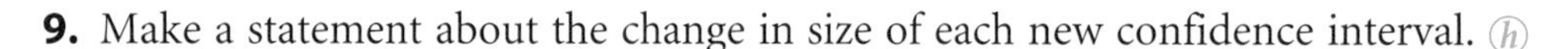

9. Make a statement about the change in size of each new confidence interval. (h)

a. If you increase the size of your sample, then the confidence interval will __?__.

b. If you increase your confidence from 90% to 99%, then the interval will __?__.

c. If your sample has a larger mean, then the interval will __?__.

d. If your sample has a larger standard deviation, then the interval will __?__.

10. APPLICATION Fifty recent tests of an automobile's mileage indicate it averages 31 mi/gal with standard deviation 2.6 mi/gal. Assuming the distribution is normal, find the 95% confidence interval.

11. APPLICATION A commercial airline finds that over the last 60 days a mean of 207.5 ticketed passengers actually show up for a particular 7:24 A.M. flight. The standard deviation of their data is 12 passengers.

a. Assuming this distribution is normal, find the 95% confidence interval.

b. If the plane seats 225 passengers, what is the probability the plane will be overbooked?

12. APPLICATION The BB Manufacturing Company mass-produces ball bearings. The optimum diameter of a bearing is 45 mm, but records show that the diameters follow a normal distribution with mean 45 mm and standard deviation 0.05 mm. A diameter between 44.95 and 45.05 mm is acceptable.

a. What percentage of the output is acceptable? ⓐ

b. What percentage of the output is unacceptable?

c. After retooling some of the equipment, the standard deviation is cut in half. What percentage of the output is now unacceptable?

d. If the engineers at the company want a 99.7% acceptability rate, what standard deviation should they aim for?

Review

13. A random-number generator selects a real number between 0 and 50, inclusive, according to the probability distribution at right. Find each value described.

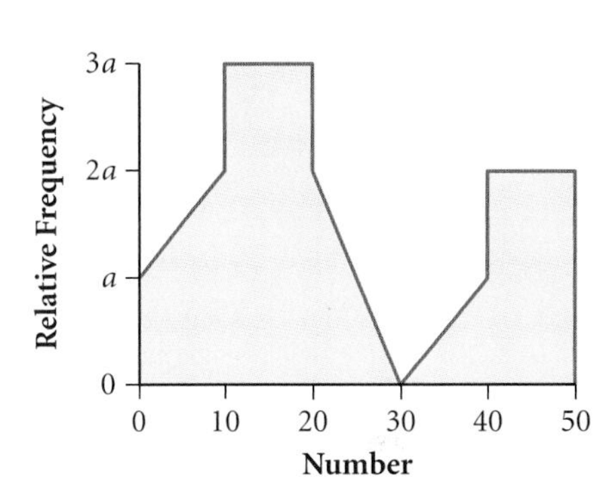

a. a ⓗ

b. P(a number is less than 30)

c. P(a number is between 20 and 40) ⓐ

d. P(a number is 30)

e. P(a number is 15)

f. the median value

14. Five hundred integer values, -3 through 3, are randomly selected (and replaced) from a hat containing tens of thousands of integers. The frequency table at right lists the results. Find these values:

Value	Frequency
-3	32
-2	60
-1	153
0	92
1	45
2	90
3	28

a. $P(-3)$

b. P(less than 0)

c. P(not 2)

d. the expected sum of the next ten selected values

15. You are about to sign a long-term rental agreement for an apartment. You are given two options:

Plan 1: Pay \$400 the first month with a \$4 increase each month.
Plan 2: Pay \$75 the first month with a 2.5% increase each month.

a. Write a function to model the accumulated total you will pay over time for each plan. ⓗ

b. Use your calculator to graph both functions on the same screen.

c. Which rental plan would you choose? Explain your reasoning.

EXPLORATION

Prediction Intervals

You have been hired by the Gone to the Dogs company, which is producing a new mega-size tartar-control dog biscuit. The company wants the biscuits to have a mean weight of 500 g with a standard deviation of 12 g. The production of the biscuits is automated, with no human involvement, and it is up to you to monitor the production and decide if the process needs to be adjusted. It will cost the company money to shut down the machine and adjust the process, but it will also cost the company money to keep producing biscuits that don't meet the weight specifications.

What biscuit weights would cause you to be concerned? Certainly biscuits weighing 400 g or 600 g would cause you to adjust the production process; but what if you found a biscuit that weighed 530 g?

Actually, you probably shouldn't make your decision based on a single biscuit. You should take means of, say, four biscuits in each production batch. Means are more reliable than single values for checking a process. One single bad value should not cause you to stop the process, but a bad mean may indicate that the process is truly in need of adjustment.

So you want to decide a **prediction interval** in which mean weights will fall if the process is satisfactory. A prediction interval is the reverse of a confidence interval. Confidence intervals, as you've seen, are based on sample means and are used to estimate the true mean. Prediction intervals are based on some target (true) mean and are used to predict sample means.

Now you'll use Fathom to simulate the biscuit production and experiment with prediction intervals for mean weights.

Consumer CONNECTION

Most manufacturers practice quality control. The strength of materials, sizes of parts, and reliability of the product are constantly monitored. In mass production and assembly lines, companies measure randomly chosen samples and compare the measurements with the desired mean and standard deviation. Using sampling, manufacturers can determine whether or not they are operating consistently, and they can determine when adjustments or improvements are necessary.

Quality control in plastics manufacturing

Activity
Quality Control

Step 1 Open a new Fathom document. Drag a new slider from the shelf. Label it *m* to represent the true mean and give it a value of 500 g. Drag a new collection from the shelf and name it Biscuits. Double-click to show its inspector. Create a new attribute *weight* with the formula randomNormal(*m*, 12g). Then choose **New Cases** from the Collection menu and add one new case.

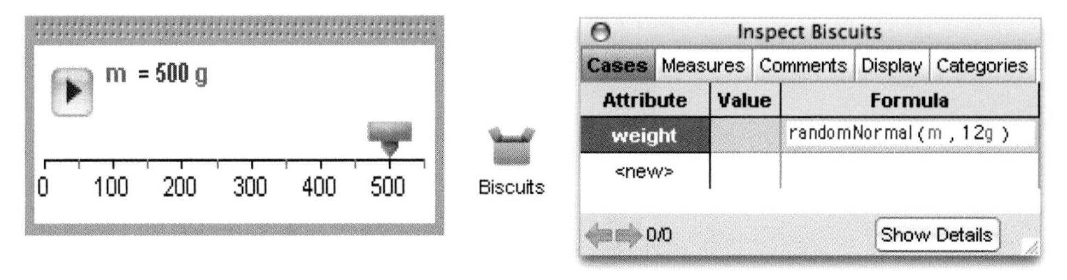

Step 2 Now you want to simulate selecting four biscuits in a production batch and finding the mean of their weights. Select the Biscuits collection and choose **Sample Cases** from the Collection menu. A sample collection, Sample of Biscuits, and its inspector will appear. Set the Sample panel to collect 4 cases. Then click on the Measures tab and create a new measure, *xbar,* with the formula mean(*weight*).

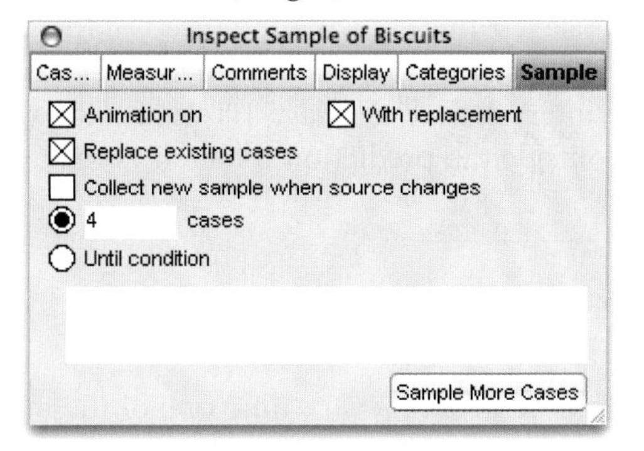

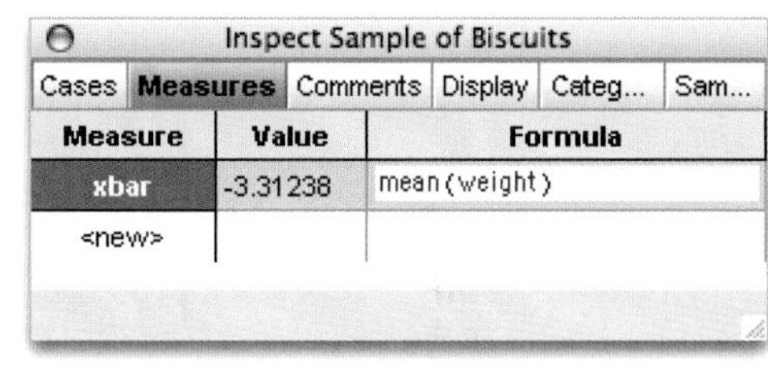

Step 3 Now you'll simulate measuring the mean weights of four biscuits from each of 100 production batches. Select the Sample of Biscuits collection and choose **Collect Measures** from the Collection menu. A new collection will be created that contains the measures. Change the settings of the new collection to match those shown below. To fill the measures collection, click **Collect More Measures.**

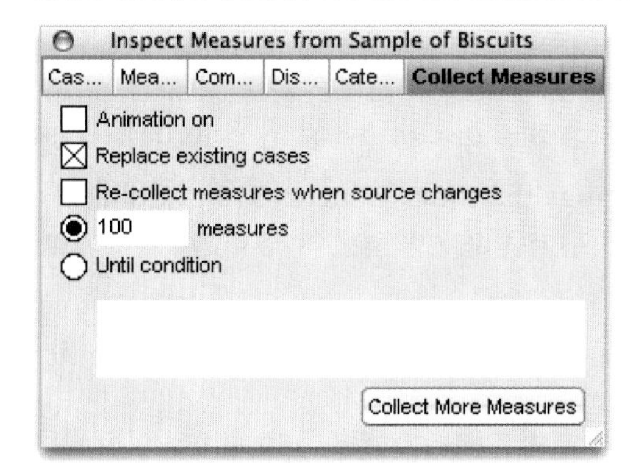

Step 4 Select the collection Measures from Sample of Biscuits and drag a new case table from the shelf. Drag a new graph down from the shelf and drag *xbar* from the case table to the vertical axis of the graph. Change the graph to a line plot. You know from Lesson 11.4, Example B, that a 90% confidence interval corresponds to a *z*-score of 1.645. The same will be true about a 90% prediction interval. So you want to consider the interval with endpoints $\mu \pm 1.645\frac{\sigma}{\sqrt{n}}$. In this situation those endpoints are $500 \pm 1.645\left(\frac{12}{\sqrt{4}}\right) = 500 \pm 9.87$, or about 490 g and 510 g. You can plot those endpoints on your Fathom graph. With the graph selected, choose **Plot Value** from the Graph menu and plot the value 490. Then plot the value 510.

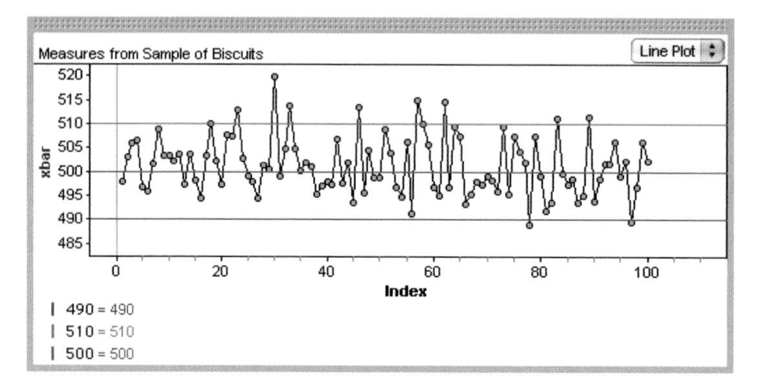

Step 5 This graph is an example of a *control chart.* A control chart shows data from a process, a center line showing the mean of the data, and upper and lower horizontal lines called *control limits.* Between those horizontal lines is a 90% prediction interval, and you have 100 values for *xbar*. How many values on your graph are outside the prediction interval? Is this what you would expect? Explain.

Questions

1. If you stop the process every time one of the means falls outside the limits, then you might waste a lot of time and money. Many producer of goods use a *z*-value of 3 (for a prediction interval spanning six standard deviations). This interval should produce very few "false alarms" (about 1 in 370). Reset your endpoints to $500 \pm 3\left(\frac{12}{\sqrt{4}}\right) = 500 \pm 18$. Now how many means are outside the prediction interval?

2. Suppose an adjustment slips and the process starts producing biscuits with a mean weight of 505 g. Change the slider value to 505. Show the inspector for the collection Measures from Samples of Biscuits, and collect more measures. How many batches of biscuits went by before you discovered this small problem?

3. Now change the slider value to 510 g. Collect more measures. How many batches of biscuits went by before you discovered this big problem?

LESSON

11.5 Bivariate Data and Correlation

It is a capital mistake to theorize before one has data. Insensibly one begins to twist facts to suit theories, instead of theories to suit facts.

SIR ARTHUR CONAN DOYLE

Dr. Aviles and Dr. Scott collected data on tree diameters and heights. Dr. Aviles thought that the height was closely associated with the diameter, but Dr. Scott claimed that the height was more closely associated with the square of the diameter. Which model is better? Each researcher plotted data and found a good line of fit. Their graphs are shown here.

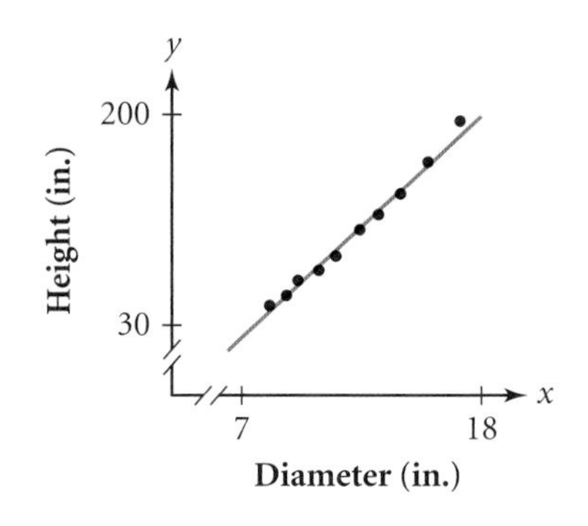

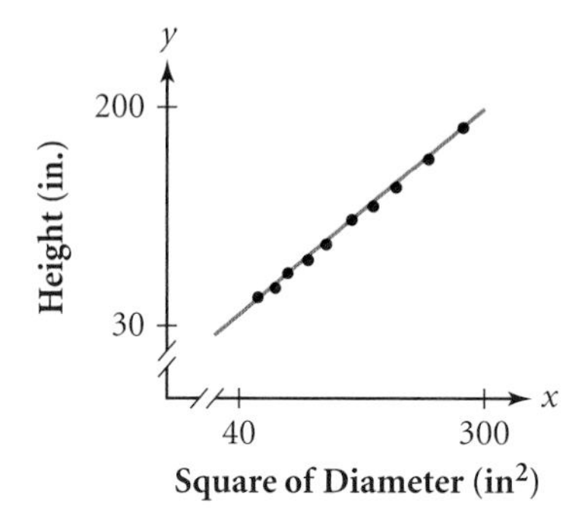

Does it appear that a line is the appropriate model? If so, is there a better linear relationship in Dr. Aviles's data or in Dr. Scott's data? How good is the fit for each line? In this lesson you'll learn how to answer questions like these.

In Lesson 11.4, you learned how to make predictions about population parameters from sample statistics. You can also predict associations between parameters, such as height and diameter or height and the squares of diameter, for a large population, such as all trees. You can even make predictions about populations that are infinitely large.

The process of collecting data on two possibly related variables is called **bivariate sampling.** How can you measure the strength of the association in the sample?

An association between two variables is called **correlation.** A commonly used statistical measure of linear association is called the **correlation coefficient.** In the investigation you'll use a calculator to explore properties of the correlation coefficient for a bivariate sample.

A giant Sequoia tree in Yosemite National Park, California.

Investigation
Looking for Connections

Step 1 Work with your group to create a survey with five questions that are all answered with a number, or use the sample survey here.

1. How many minutes of homework did you do last night?
2. How many minutes did you spend talking, calling, e-mailing, or writing to friends?
3. How many minutes did you spend just watching TV or listening to music?
4. At what time did you go to bed?
5. How many academic classes do you have?

Step 2 Conjecture with your group about the strengths of correlations between pairs of variables. For example, you may decide that the number of minutes of homework is strongly correlated with the number of academic classes. Consider each of the ten pairs of variables and identify which combinations you believe will have

i. A positive correlation (as one increases, the other tends to increase).
ii. A negative correlation (as one increases, the other tends to decrease).
iii. A weak correlation.

Step 3 Gather data from each student in your class. Then enter the data into five calculator lists. Plot points for each pair of lists, and find the correlation coefficients. [► See **Calculator Note 11E** to learn how to find the correlation coefficient. ◄] You may want to divide this work among members of your group. Describe the relationship between the appearance of the graph and the value of the correlation coefficient.

Step 4 Write a paragraph describing the correlations you discover. Include any pairs that are not correlated that you find surprising. You have collected a small and not very random sample; do you think these relationships would still be present if you collected answers from a random sample of your entire school population?

In 1896, English mathematician Karl Pearson (1857–1936) proposed the correlation coefficient, now abbreviated r. To compute the correlation coefficient, Pearson replaced each x- and y-value in a data set with its corresponding z-value. If a particular x- or y-value is larger than the mean value for that variable, then its z-value is positive. And if a particular x- or y-value is smaller than the mean value for that variable, then its z-value is negative.

Pearson then found the product of z_x and z_y for each data point and summed these products. In a data set that is generally increasing, the products of z_x and z_y are positive. This is because for every point, usually either both x and y are above the mean or both x and y are below the mean. In a data set that is generally decreasing, usually either z_x is positive and z_y is negative or z_x is negative and z_y is positive. Therefore, the products will be negative. After summing the products, Pearson divided by $(n - 1)$ to get a number between -1 and 1. So he defined the correlation coefficient as $\frac{\Sigma z_x z_y}{n-1}$. But what do values of this coefficient mean?

You may have noticed in the investigation that values of r can range from -1 to 1. A value of 1 means the x-values are positively correlated with the y-values in the strongest possible way. That is, as x-values increase, y-values increase proportionally. A value of -1 means the x-values are negatively correlated with the y-values in the strongest possible way. That is, as x-values increase, y-values decrease proportionally. A value of 0 means there's no linear correlation between the values of x and y.

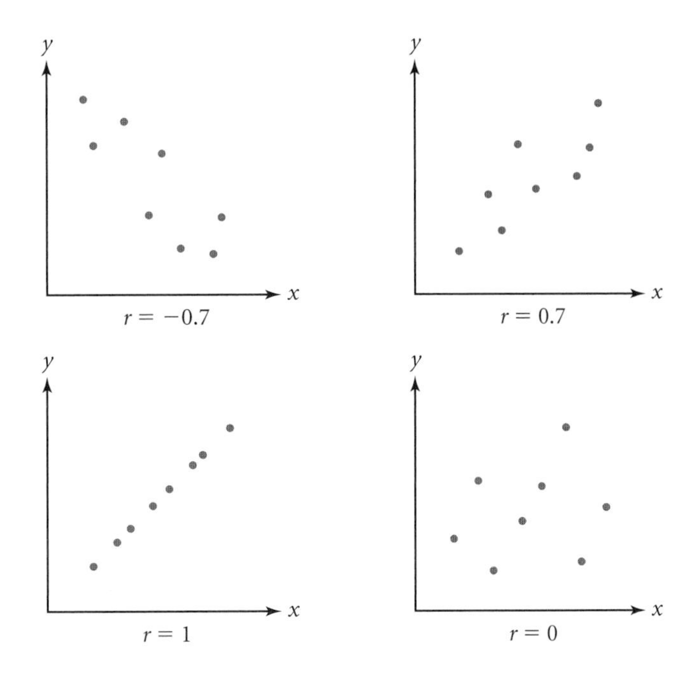

If data are highly correlated, a straight line will model the data points well.

For Dr. Aviles's data in the tree example at the beginning of this lesson, the correlation coefficient is 0.992. So his data are very close to linear. For Dr. Scott's data, the correlation coefficient is 0.999. That's even better. This means that for the trees measured, the squares of diameters are better predictors of the heights than are the diameters themselves.

The Correlation Coefficient

The correlation coefficient, r, can be calculated with the formula

$$r = \frac{\Sigma z_x z_y}{n-1} = \frac{\Sigma(x-\bar{x})(y-\bar{y})}{s_x s_y (n-1)}$$

A value of r close to ±1 indicates a strong correlation, whereas a value of r close to 0 indicates little or no correlation.

Note that the definition of the correlation coefficient includes no reference to any particular line, though it describes how linear a bivariate data set is. In contrast, the root mean square error you studied in Chapter 3 describes how well a *particular* line fits a data set.

Often, bivariate data are collected from a study or an experiment in which one variable represents some condition and the other represents measurements based on that condition, as shown in the next example. In statistics, the x- and y-variables are often called the **explanatory** and **response** variables instead of the independent and dependent variables.

EXAMPLE A

Kiane belongs to many committees and notices that different groups take different amounts of time to make decisions. She wonders if the time it takes to make a decision is linearly related to the size of the committee. So she collects some data. Find the correlation coefficient of this data set and interpret your result.

Size (people)	4	6	7	9	9	11	15	15	18	20	21	24	25
Time (min)	5.2	3.8	8.2	8.5	12.0	10.8	14.7	15.5	22.0	19.1	35.3	29.2	32.1

▸ **Solution**

In this instance, it makes sense to let the explanatory variable, x, represent the committee size and the response variable, y, represent the time it takes to make a decision. When plotted, the data show an approximately linear pattern.

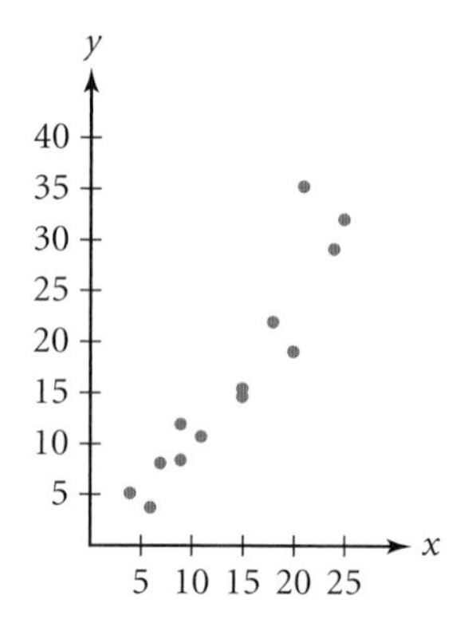

You can find the value of r directly using a calculator, but you should also try to use the formula—assisted by a calculator, of course. First, enter the data into lists and verify these statistics:

$$\bar{x} \approx 14.15 \text{ people}, \ s_x \approx 7.0456 \text{ people}$$

$$\bar{y} \approx 16.65 \text{ minutes}, \ s_y \approx 10.302 \text{ minutes}$$

Then use the lists in the formula

$$\frac{\Sigma(x - 14.15)(y - 16.65)}{(7.0456)(10.302)(13 - 1)} \approx 0.9380$$

A correlation coefficient $r \approx 0.9380$ means there is a strong positive correlation between the size of a committee and the time it takes to reach a decision. This means that as the size of a committee increases, the time it takes to reach a decision increases proportionately.

Be careful that you don't confuse the ideas of correlation and causation. A strong correlation may exist between two sets of data, but this does not necessarily imply a causal relationship. For instance, in Example A, Kiane found a strong correlation between committee size and decision-making time. But this does not necessarily mean that the size of the committee *caused* the decision to take longer. Whether it did or did not can be proved only by a carefully controlled experiment.

EXAMPLE B

The director of a summer camp has collected data for two weeks on both daily ice cream sales from the camp store and visits to the camp nurse for treatment of sunburn. What conclusions, if any, can you make?

Ice cream sales	$245.10	$45.25	$17.85	$205.00	$276.35	$428.25	$312.15
Visits to nurse	66	17	1	65	72	131	93

Ice cream sales	$288.25	$267.95	$74.10	$111.50	$371.55	$244.45	$115.75
Visits to nurse	81	99	2	84	113	78	79

▶ Solution

Graph the data and calculate the correlation coefficient. The graph of the data shows a clear upward trend. The correlation coefficient $r \approx 0.866$ indicates a fairly strong correlation.

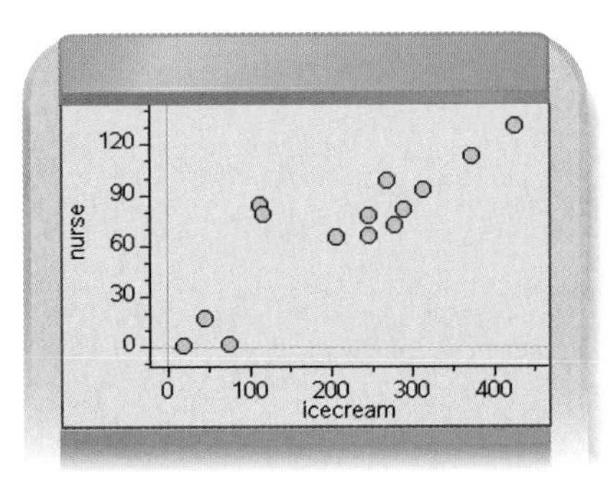

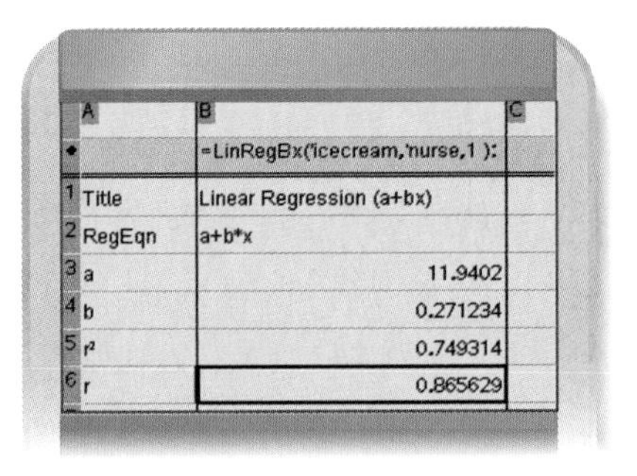

Can you conclude that buying ice cream causes a sunburn? Or does getting a sunburn cause ice cream buying? There is most likely another variable causing both of these effects. Perhaps the daily temperature might be a **lurking variable** behind both of these results.

So you can conclude that sunburn and ice cream sales are correlated, but not that one of these occurrences causes the other.

EXERCISES

You will need

A graphing calculator for Exercises **3**, **6–10**, **12**, and **14**.

Practice Your Skills

1. Approximate the correlation coefficient for each data set.

a.

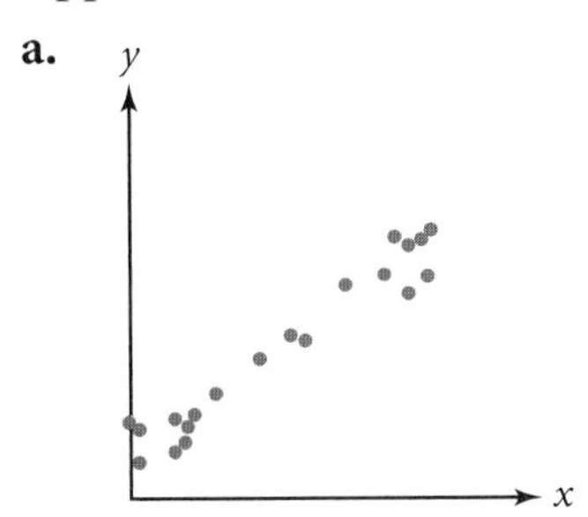

b. y x @

c.

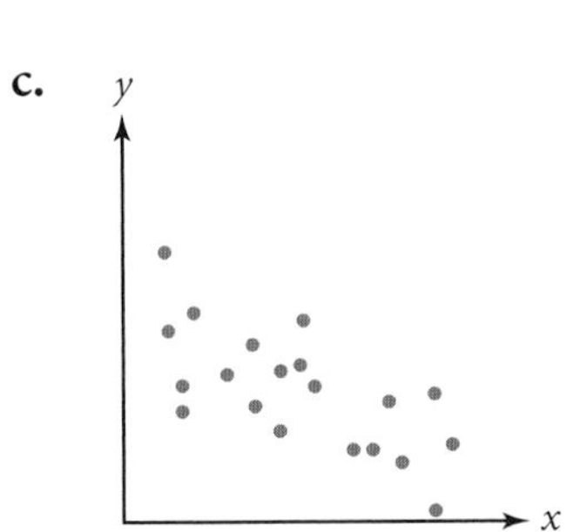

d.

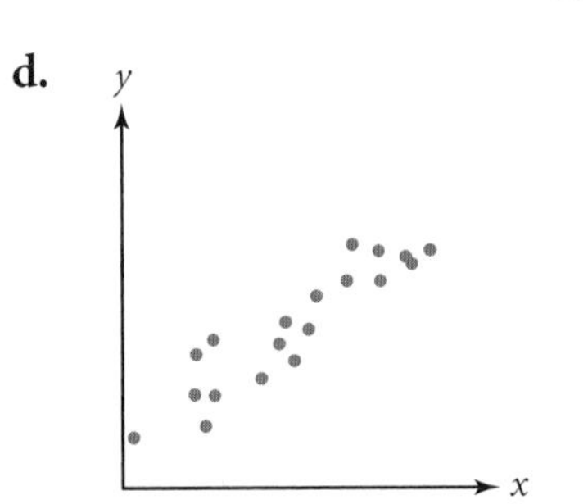

2. Sketch a graph of a data set with approximately these correlation coefficients.

a. $r = -0.8$ ⓐ

b. $r = -0.4$

c. $r = 0.4$

d. $r = 0.8$

3. Copy and complete the table at right. Then answer 3a–d to calculate the correlation coefficient.

x	y	$x - \bar{x}$	$y - \bar{y}$	$(x - \bar{x})(y - \bar{y})$
12	8			
14	8			
18	6			
19	5			
22	3			

a. What is the sum of the values for $(x - \bar{x})(y - \bar{y})$?

b. What are $\bar{x}$ and s_x?

c. What are $\bar{y}$ and s_y?

d. Calculate $r = \dfrac{\Sigma(x - \bar{x})(y - \bar{y})}{s_x s_y(n - 1)}$. ⓐ

e. What does this value of r tell you about the data?

f. Draw a scatter plot to confirm your conclusion in 3e. ⓗ

4. In each study described, identify the explanatory and response variables.

a. Doctors measured how well students learned finger-tapping patterns after various amounts of sleep. ⓗ

b. Scientists investigated the relationship between the weight of a mammal and the weight of its brain.

c. A university mathematics department collected data on the number of students enrolled each year in the school and the number of students who signed up for a basic algebra class.

5. For each research finding, decide whether there is evidence of causation, correlation, or both. If it is only a correlation, name a possible lurking variable that may be the cause of the results.

a. As the sales of television sets have increased, so has the number of overweight adults. Does owning a television cause weight gain? ⓗ

b. A study in an elementary school found that children with larger shoe sizes were better readers than those with smaller shoe sizes. Do big feet make children read better?

c. The more firefighters sent to a fire, the longer it takes to put out the fire. Does sending more firefighters cause a fire to burn longer?

Reason and Apply

6. An environmental science class conducted a research project to determine whether there was a relationship between the soil pH, x, and the percent dieback of new growth, y, for a particular type of tree. The table contains the data the class collected.

x	3.3	3.4	3.5	3.6	3.7	3.8	3.9	4.0	4.1	4.2	4.3	4.5	5.0	5.1
y	7.3	10.8	10.4	9.3	12.4	11.2	6.6	10.0	9.2	12.4	2.3	4.3	1.6	1.0

a. Make a scatter plot of the data.

b. Describe the relationship between the two variables.

c. Find the correlation coefficient. Does this confirm or refute your answer to 6b? How can you tell?

d. Can you conclude that higher soil pH causes less dieback of new growth? Explain.

Environmental CONNECTION

Trees take in carbon dioxide from the air, which helps offset carbon dioxide emissions caused by cars and industry. Destruction of forests has contributed to an increase in the amount of carbon dioxide in the atmosphere, leading to global warming and other problems. Although nature can eventually reforest most areas on its own, we can speed the process by improving soil quality and selecting the best species of trees for a particular area.

Deforestation from logging and pasture clearing have reduced the size of the Amazon rain forest in Brazil by more than 15% since 1970.

7. The table contains information from a selection of four-year colleges and universities for the 2005–2006 school year. Describe the correlation between the number of students and the number of faculty.

Four-Year Colleges

College	Number of students	Number of faculty
Alfred University	2,235	205
Brandeis University	5,189	472
Brown University	8,261	888
Bryn Mawr College	1,799	185
California College of the Arts	1,616	370
Carleton College	1,959	216
College of William & Mary	7,544	763
DePauw University	2,397	253
Drake University	5,277	388
Duquesne University	9,916	908
Gallaudet University	1,834	230
Hampshire College	1,376	137
Illinois Wesleyan University	2,146	222
Lehigh University	6,748	621
Maryland Institute, College of Art	1,717	267
Miami University of Ohio	16,338	1,198

College	Number of students	Number of faculty
Mills College	1,372	184
Morehouse College	3,029	223
Mt. Holyoke College	2,127	241
Princeton University	6,916	1,060
Rhode Island School of Design	2,258	494
Rhodes College	1,692	184
Saint John's University	1,196	176
St. Olaf College	3,058	332
Spelman College	2,318	245
Swarthmore College	1,479	195
Syracuse University	17,266	1,391
Tufts University	9,780	1,194
Tuskegee University	2,880	265
University of Tulsa	4,084	422
Wesleyan University	3,207	368
Wheaton College	1,568	161

(*The World Almanac and Book of Facts 2007*)

8. *Mini-Investigation* For each data set in 8a–c, draw a scatter plot and find the correlation coefficient, r. State what this value of r implies about the data, and note any surprising results you find.

a. {(0.5, 1), (0.6, 0.9), (0.7, 0.8), (0.8, 0.7), (0.9, 0.6)} ⓐ

b. {(0.5, 1), (0.6, 0.9), (0.7, 0.8), (0.8, 0.7), (0.9, 0.6), (1.9, 1.9)}

c. {(0.5, 1), (0.6, 0.9), (0.7, 0.8), (0.8, 0.7), (0.9, 0.6), (1.9, 0.9)}

d. Based on your answers to 8a–c, do you think the correlation coefficient is strongly affected by outliers?

9. *Mini-Investigation* For each data set in 9a and b, draw a scatter plot and find the correlation coefficient, r. State what this value of r implies about the data. ⓗ

a. {(0.3, 0.9), (0.4, 0.6), (0.6, 0.4), (1, 0.3), (1.4, 0.4), (1.6, 0.6), (1.7, 0.9), (0.3, 1.1), (0.4, 1.4), (0.6, 1.6), (1, 1.7), (1.4, 1.6), (1.6, 1.4), (1.7, 1.1)}

b. {(0.4, 1.4), (0.6, 1.1), (0.8, 0.8), (1, 0.5), (1.2, 0.8), (1.4, 1.1), (1.6, 1.4)}

c. For the data sets in 9a and b, does there appear to be a relationship between x- and y-values? Is this reflected in the values you found for r?

10. These data show numbers of country radio stations and numbers of oldies radio stations for several years.

Year	1994	1995	1996	1997	1998	1999	2001	2002	2003	2004	2005	2006
Country	2642	2613	2525	2491	2368	2306	2190	2131	2088	2047	2019	2097
Oldies	714	710	738	755	799	766	785	813	807	816	773	755

(*The World Almanac and Book of Facts 2007*)

a. What is the correlation coefficient for year and number of country stations, and for year and number of oldies stations? ⓐ

b. Is the number of country stations or the number of oldies stations more strongly correlated with the year? Explain your reasoning.

Singer Patsy Cline's (1932–1963) (top) recordings are considered some of the greatest in country music. Between 1998 and 2007, the Dixie Chicks (bottom) recorded five albums and released seven number-one singles.

Review

11. What is the slope of the line that passes through the point (4, 7) and is parallel to the line $y = 12(x - 5) + 21$? ⓐ

12. The data at right give the reaction times of ten people who were administered different dosages of a drug. Find the median-median line for these data and determine the root mean square error.

Dosage (mg)	Reaction time (s)
85	0.5
89	0.6
90	0.2
95	1.2
95	1.6
103	0.6
107	1.0
110	1.8
111	1.0
115	1.5

13. Find an equation of the line passing through (4, 0) and (6, −3).

14. Graph the equation $y = \log_5 x$.

15. On a car trip, your speed averages 50 km/h as you drive to your destination. You return by the same route and average 75 km/h. What's your average speed for the entire trip? ⓗ

16. David wants to send his nephew a new 5-foot fishing pole. David wraps up the pole and takes it to Bob's Courier Service. But Bob's has a policy of not accepting any parcels longer than 4 feet. David returns an hour later with the fishing pole wrapped and sends it with no problem. How did he do it? (Assume the pole is not broken into pieces.)

Project

CORRELATION VS. CAUSATION

Think of a relationship someone claims involves causation, but you think might involve only correlation. Your claim can be about anything—science, popular beliefs, sociology—but it must be something that can be tested. First, research data related to the claim and determine whether or not the data seem to show a correlation between the two variables. Then, think about whether or not one event really causes the other. What other factors might be involved? Might the data you found be misleading in some way? If you can, find the data for any other factors and see how these data are related to your claim. Write a report on your findings.

Your project should include

- The claim you researched and the data and analyses you found.
- Any graphs, tables, and equations that you used while analyzing the data.
- A summary of other factors that might be involved.
- Your own conclusion on the relationship of the data.

Fathom™
Dynamic Data™ Software

With Fathom Dynamic Data you can plot data related to different pairs of variables. You can also compare the fit of different equations through your data points.

LESSON

11.6

The Least Squares Line

In Chapter 3, you saw how to make predictions by fitting a line to a data set. For example, you learned how to find the median-median line. You now can see that you were making predictions about a population from the data points in a sample. In this lesson you'll see how to make predictions about a population from the most commonly used line of fit, the **least squares line,** which is related to correlation coefficients.

History CONNECTION

In the late 1800s, English anthropologist Francis Galton studied correlations among various measurements, including heights of fathers and sons. He found that sons' heights tended to be closer to the mean height for men than their fathers' heights are. This phenomenon is known today as "regression toward the mean." The term **regression analysis** now refers to finding a model with which to make predictions about one variable from another.

If this family follows the trend of regression toward the mean, then the height of the grandson after he is fully grown should be closer to the mean height for men than his grandfather's height is.

You find the least squares line by first standardizing both variables—the x-values and the y-values—which gives the bivariate data center $(0, 0)$ and standard deviation 1 in both the horizontal and vertical directions. Then fit a line that passes through the origin and has slope equal to the correlation coefficient, r. In terms of z-values for x and y, the equation of the least squares line is $z_y = rz_x$. In the extreme case where the data are all perfectly linear, the value of r is $+1$ or -1, so the equation is $z_y = z_x$ or $z_y = -z_x$. In the other extreme, when the data are very scattered, the value of r is 0 and the equation is $z_y = 0$, a horizontal line through the origin.

In practice, you don't want to standardize every piece of sample data. Instead, you can rewrite the equation using means and standard deviations. By the definition of z-values, the equation $z_y = rz_x$ is equivalent to $\frac{y-\bar{y}}{s_y} = r\left(\frac{x-\bar{x}}{s_x}\right)$, or, solving for y, $y = \bar{y} + r\left(\frac{s_y}{s_x}\right)(x - \bar{x})$. Notice that this equation represents a translation of the center from the origin to $(\bar{x}, \bar{y})$.

Finding a Least Squares Line

1. Find the values of r, s_x, s_y, $\bar{x}$, and $\bar{y}$ for the data set.
2. Calculate the slope, $b = r\left(\frac{s_y}{s_x}\right)$.
3. Substitute values of b, $\bar{x}$, and $\bar{y}$ to write the equation of the least squares line, $\hat{y} = \bar{y} + b(x - \bar{x})$.

EXAMPLE A

A photography studio offers several packages to students who pose for yearbook photos.

Number of pictures (x)	44	31	24	15
Total cost (y)	\$19.00	\$16.00	\$13.00	\$10.00

Find an equation of the least squares line. Use your line to decide how much a package of two photographs should cost.

▶ ***Solution***

Begin by finding the mean and standard deviation of both the x- and y-values.

$\bar{x} = 28.5 \qquad \bar{y} = 14.5 \qquad s_x = 12.23 \qquad s_y = 3.873$

Then create a table to calculate values of $(x - \bar{x})(y - \bar{y})$.

x	y	$x - \bar{x}$	$y - \bar{y}$	$(x - \bar{x})(y - \bar{y})$
44	19	15.5	4.5	69.75
31	16	2.5	1.5	3.75
24	13	−4.5	−1.5	6.75
15	10	−13.5	−4.5	60.75

So, $\Sigma(x - \bar{x})(y - \bar{y})$ is 141. Use the formula $r = \frac{\Sigma(x - \bar{x})(y - \bar{y})}{s_x s_y (n - 1)}$ to find $r \approx 0.992$.
The slope of the least squares line is $r\left(\frac{s_y}{s_x}\right) \approx \frac{0.992 \cdot 3.873}{12.23} \approx 0.314$ dollar per photo.
So the equation of the least squares line is $\hat{y} \approx 14.5 + 0.314(x - 28.5)$, or $\hat{y} \approx 5.55 + 0.314x$.

To find the cost of two photographs, substitute 2 for x. You get $y \approx 6.178$, so according to the model the package should cost \$6.18.

In the investigation you'll use the correlation coefficient and the least squares line to help you interpret the results of an experiment.

Investigation
Spin Time

You will need

- four different-size coins
- a stopwatch or watch with a second hand

You will conduct an experiment to determine if the spin time of a coin can be explained by some attribute of the coin.

Step 1 Two important characteristics of good experimental design are **randomization** and **replication.** You'll build in randomization by spinning the coins in a random order. To ensure replication, the same spinner and timer will perform the experiment four times on each coin. Follow the Procedure Note on the next page to collect data on four coins.

Procedure Note

1. Assign one student to be a spinner and one student to be a timer.
2. Randomize the order of the coins. You can shake them up or use a random-number generator to assign the order.
3. Spin and time each coin once. Record the amount of time the coin spins before it comes to a complete stop.
4. Repeat Steps 2 and 3 so that each coin is spun four times. The order in which the four coins are spun should not be the same in every round.

Step 2 What affects how long different coins spin? Is it the diameter of the coin? The weight? You'll test several characteristics to see whether they make a good predictor for spin time. For each coin, record the diameter, weight, and thickness, and calculate the area of the face and the volume of the coin. The table shows information about each type of coin.

Denomination	Cent	Nickel	Dime	Quarter	Half Dollar	Presidential $1	Golden Dollar
Weight	2.500 g	5.000 g	2.268 g	5.670 g	11.340 g	8.1 g	8.1 g
Diameter	0.750 in. 19.05 mm	0.835 in. 21.21 mm	0.705 in. 17.91 mm	0.955 in. 24.26 mm	1.205 in. 30.61 mm	1.043 in. 26.50 mm	1.043 in. 26.50 mm
Thickness	1.55 mm	1.95 mm	1.35 mm	1.75 mm	2.15 mm	2.00 mm	2.00 mm
Edge	Plain	Plain	Reeded	Reeded	Reeded	Edge-lettering	Plain

(*www.usmint.gov*)

Step 3 For each of the five possible predictors, calculate the coefficient of correlation and the equation of the least squares line for spin time versus that predictor. Which of these characteristics is the best linear predictor?

Step 4 A Golden Dollar weighs 8.1 g. It has a diameter of 26.50 mm and a thickness of 2.00 mm, so the area of its face is about 552 mm^2 and its volume is about 1100 mm^3. What would you estimate to be the average spin time of a Golden Dollar?

Step 5 With your group, discuss and answer these questions.

a. What is the purpose of replication? (Why spin the same coin more than one time?)

b. What is the purpose of randomization? (Why not spin the coins in a predetermined order?)

c. Why use only one spinner and one timer instead of giving everyone a turn at each job?

d. What did you consider when choosing the "best" linear predictor?

e. Suppose you did an observational study and saw that one student could spin a quarter longer than another student could spin a dime. What conclusion could you reach? What is the advantage of doing a well-designed experiment rather than an observational study?

You can measure the accuracy of the least squares line using the typical spread of the residuals, as calculated by the root mean square error.

EXAMPLE B

In Chapter 3, you estimated a line of fit for data on the concentration of CO_2 in the atmosphere around Mauna Loa in Hawaii as a function of time. Refer back to the table on page 139 to find

a. The least squares line for the data given.

b. The median-median line for the data given.

c. The root mean square error of both models.

▶ Solution

Enter the data into lists in your calculator and find each model. [▶ See **Calculator Note 11F** to learn how to find the equation of the least squares line.◀]

a. The equation for the least squares line is $\hat{y} \approx -2804.22 + 1.59x$. Notice that $r \approx 0.997$, so a linear model is a good fit for these data.

b. The equation of the median-median line is $\hat{y} \approx -2798.85 + 1.584x$.

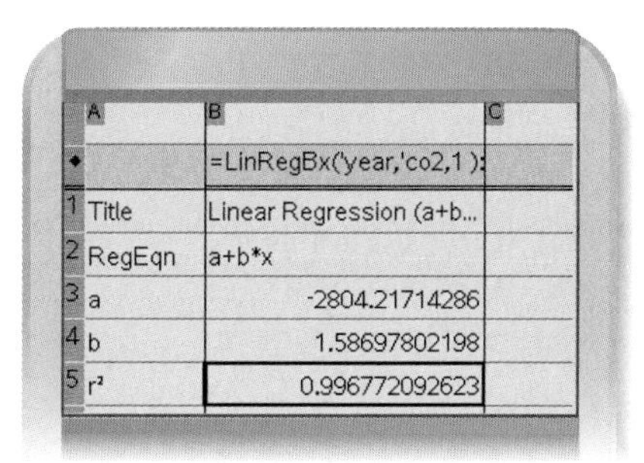

A	B	C
◆	=LinRegBx('year,'co2,1):	
1 Title	Linear Regression (a+b...	
2 RegEqn	a+b*x	
3 a	-2804.21714286	
4 b	1.58697802198	
5 r²	0.996772092623	

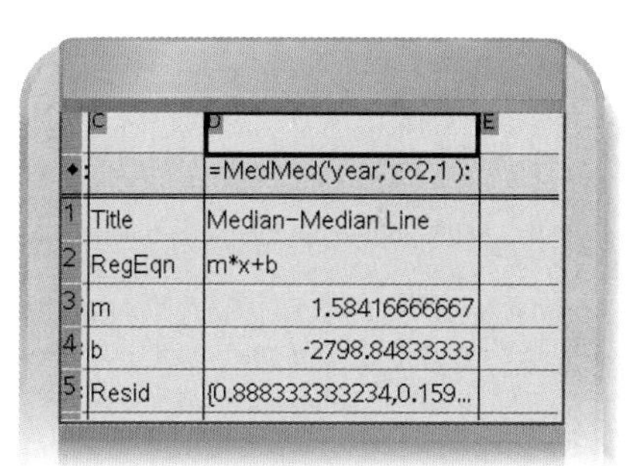

C	D	E
◆	=MedMed('year,'co2,1):	
1 Title	Median-Median Line	
2 RegEqn	m*x+b	
3 m	1.58416666667	
4 b	-2798.84833333	
5 Resid	{0.888333333234,0.159...	

c. To calculate the root mean square error, find the differences between the y-values in the data and the y-values predicted by the model. Then square the differences and sum the squares. Next, divide by two less than the number of data values, and finally, take the square root. Enter your equations for the least squares line and the median-median line into your calculator as f_1 and f_2. Then calculate the root mean square error as shown.

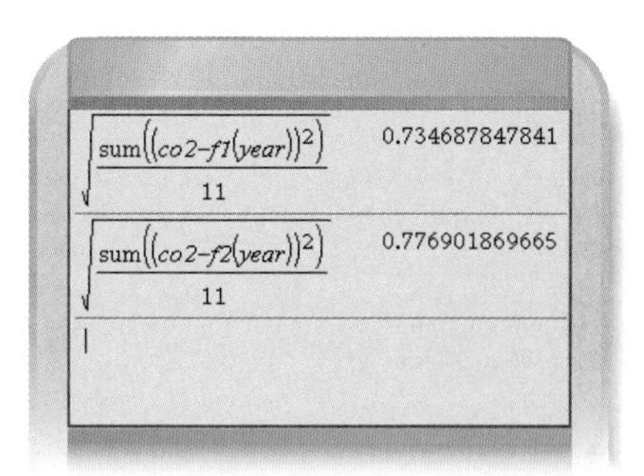

$\sqrt{\frac{\text{sum}((co2-f1(year))^2)}{11}}$ 0.734687847841

$\sqrt{\frac{\text{sum}((co2-f2(year))^2)}{11}}$ 0.776901869665

The root mean square error for the least squares line is about 0.735 ppm and for the median-median line is about 0.777 ppm. The root mean square error is smaller for the least squares line, so predictions you make with this model are likely to have smaller errors than predictions you make with the median-median line. In all cases, though, you should be careful not to predict too far in the future. Just because the trend has been quite linear in the past does not mean it will be so in the future.

keymath.com/DAA

[▶ You can explore an interactive model of a least squares lines using the **Dynamic Algebra Exploration** at www.keymath.com/DAA .◀]

There are many procedures to find lines of fit for linear data. Some, such as the median-median line, distill the data into a few points and are relatively unaffected by one or two outliers. Others, like the least squares line, place equal importance on each point. The least squares procedure produces the line that has the smallest sum of squares of errors between data points and predictions from the line. It is often called the "best-fit line" because of this property. However, when fitting a line to data, always check the line visually; you may sometimes find that a procedure that ignores outliers gives a line that models the overall trend better than the least squares line.

EXERCISES

You will need

A graphing calculator for Exercises **1**, **3**, **5–7**, and **11**.

Practice Your Skills

1. Use these data to find the values specified.

Year x	1950	1960	1970	1980	1990	2000
Percentage y	29.6	33.4	38.1	42.5	45.3	52.0

a. $\bar{x}$ @ **b.** $\bar{y}$ **c.** s_x @ **d.** s_y **e.** r

2. Use the following sets of statistics to calculate the least squares line for each data set described.

a. $\bar{x} = 18$, $s_x = 2$, $\bar{y} = 54$, $s_y = 5$, $r = 0.8$

b. $\bar{x} = 0.31$, $s_x = 0.04$, $\bar{y} = 5$, $s_y = 1.2$, $r = -0.75$ @

c. $\bar{x} = 88$, $s_x = 5$, $\bar{y} = 6$, $s_y = 2$, $r = -0.9$ @

d. $\bar{x} = 1975$, $s_x = 18.7$, $\bar{y} = 40$, $s_y = 7.88$, $r = 0.9975$

3. Use the data from Exercise 1 and the equation $\hat{y} = -818.13 + 0.434571x$ to calculate

a. The residuals. @

b. The sum of the residuals.

c. The squares of the residuals. @

d. The sum of the squares of the residuals.

e. The root mean square error.

4. Use the equation $\hat{y} = -818.13 + 0.434571x$ to predict a y-value for each x-value.

a. $x = 1954$ @ **b.** $x = 1978$ **c.** $x = 1989$ @ **d.** $x = 2004$

Reason and Apply

5. APPLICATION Carbon tetrachloride is an ozone-depleting chemical found in the atmosphere. The table shows the concentration of the chemical in parts per trillion (ppt) measured in the European Union.

Year x	1978	1980	1982	1984	1986	1988
Carbon tetrachloride (ppt) y	88.3	89.82	92.78	95.09	98.23	100.63

a. Make a scatter plot of the data, and find the least squares line to fit the data. ⓐ

b. Use your model to predict the amount of carbon tetrachloride present in 2010.

c. In 1987, 22 countries and the European Economic Community agreed on the Montreal Protocol to reduce ozone-depleting chemicals in the atmosphere. In 1989, the protocol went into effect. The data below represent the levels of carbon tetrachloride in even years from 1990 to 2006. Make a scatter plot of the data, and find the least squares line to fit the data.

Year x	1990	1992	1994	1996	1998	2000	2002	2004	2006
Carbon tetrachloride (ppt) y	102.34	102.15	100.91	98.96	97.07	95.12	93.2	91.76	90.05

(*The World Almanac and Book of Facts 2007*)

d. Use the model you found in 5c to predict the amount of carbon tetrachloride in 2010. How does this compare with your answer in 5b?

Environmental CONNECTION

Ozone depletion is a worldwide environmental concern that has been addressed by several international agreements. The ozone in the stratosphere (from 11 to 50 km above Earth's surface) protects us by blocking the Sun's ultraviolet (UV) radiation. Exposure to too much UV radiation has been linked to skin cancer, eye problems, and immune-system suppression. When ozone is depleted in the stratosphere, it can build up closer to Earth's surface, where it acts as a pollutant and contributes to lung damage. For more information on ozone depletion, see the links at www.keymath.com/DAA .

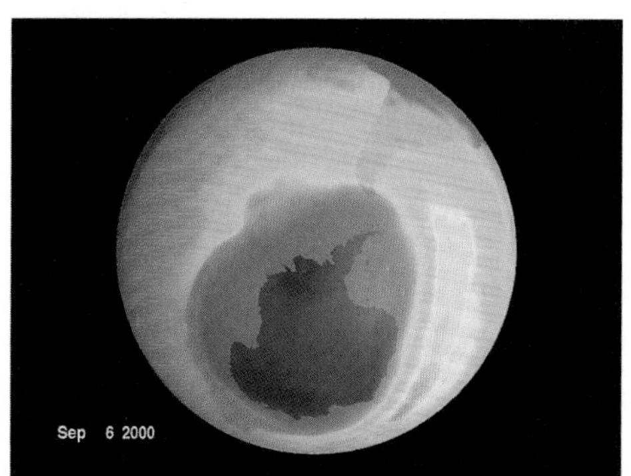

This color-coded image of Earth shows a hole in the ozone layer above Antarctica. Blue and purple indicate low ozone levels, and yellow and orange indicate higher ozone levels.

6. In the late 1990s, an upward trend was noticed in mean SAT math scores of college-bound seniors.

Year	1997	1998	1999	2000	2002	2003	2004	2005	2006
Score	511	512	511	514	516	519	518	520	518

(*The World Almanac and Book of Facts 2007*)

a. Find the equation of the least squares line. Let x represent the year, and let y represent the mean SAT score.

b. Is a linear model a good fit for these data? Justify your answer. ⓗ

c. Verify that the least squares line passes through the mean x-value and the mean y-value.

d. What does the equation predict for the year 2001? How does this compare with the actual mean score of 514?

e. What does the equation predict for the year 2010? How reasonable is this prediction?

7. This table shows the percentage of females in the U.S. labor force at various times.

Year	1950	1960	1970	1980	1990	2000
Percentage	29.6	33.4	38.1	42.5	45.3	52.0

a. Find the least squares line for these data. Let x represent the year, and let y represent the percentage. (Use 1900 as the reference year.) ⓐ

b. What is the real-world meaning of the slope? Of the y-intercept?

c. According to your model, what percentage of the current labor force is female? Check an almanac or the Internet to see how accurate your prediction is.

A firefighter manages a controlled fire in New Jersey.

8. Name at least two major differences between the median-median method and the least squares method for finding a line of fit. ⓗ

9. Explain how the root mean square error is related to the sum of the squares that is minimized by the least squares procedure. If the root mean square error is minimized, is the sum of the squares minimized as well?

Review

10. Solve each equation for y. ⓐ

a. $\log y = 3$

b. $\log x + 2\log y = 4$

11. This table shows the number of daily newspapers in the United States, the daily circulation, and the total U.S. population, for selected years.

Daily Newspapers in the United States

Year	Number of daily papers	Daily circulation (millions)	Total population (millions)
1900	2226	15.1	76.2
1920	2042	27.8	106.0
1940	1878	41.1	132.1
1960	1763	58.9	179.3
1980	1745	62.2	226.5
2000	1480	55.8	281.4

(*The New York Times Almanac 2007*)

a. What is the correlation coefficient between the population and the number of daily newspapers? What does this mean?

b. What is the correlation between the total population and the daily circulation?

c. For each year, what percentage of the population reads a daily paper? What does this mean? What are the trends? What are the implications of the data?

12. Write an equation that will produce the graph shown below, with intercepts $(-5, 0)$, $(-2, 0)$, $(1, 0)$, and $(0, 50)$. ⓗ

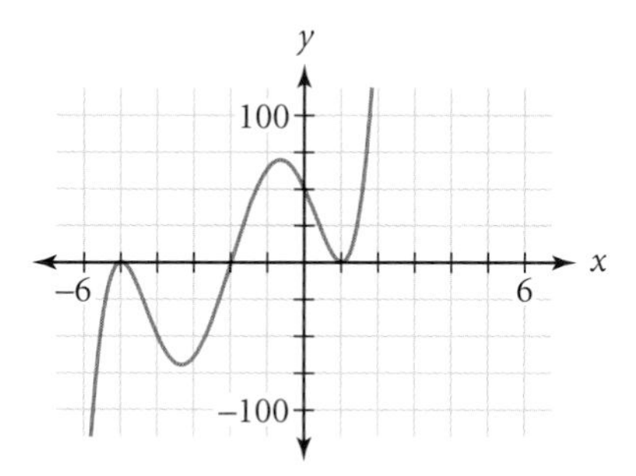

13. If you spin this spinner ten times, what is your expected score?

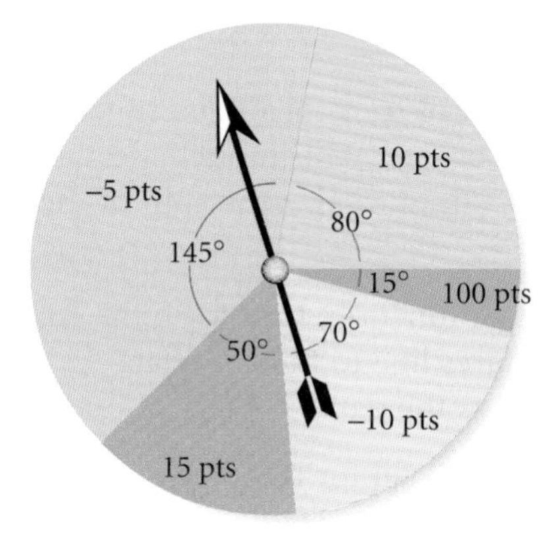

14. The Koch curve is a famous fractal introduced in 1906 by Swedish mathematician Niels Fabian Helge von Koch (1870–1924). To create a Koch curve, follow these steps:

- **i.** Draw a segment and divide it into thirds.
- **ii.** Make a bottomless triangle on the middle portion, with each side the length of the missing bottom.
- **iii.** Repeat steps i and ii with each shorter segment.

If the original segment is 18 cm and the process continues through infinitely many steps, how long will the Koch curve become?

IMPROVING YOUR REASONING SKILLS

A Set of Weights

You have a 40-ounce bag of sand, a balance scale, and many blocks of wood with unknown weights. How can you divide the sand into four smaller bags so that you can then use the smaller bags to weigh any block of wood with a whole-number weight between 1 and 40 ounces?

EXPLORATION

Nonlinear Regression

Science CONNECTION

These data, from the eastern part of Asia (China/Korea/Japan), are in thousands of metric tons of carbon released in the air from burning solid fuels. This part of the world has become the greatest producer of goods for the planet. But what is the cost to the global environment?

(Carbon Dioxide Information Analysis Center)

In many chapters of this book, you used transformations of parent functions to fit data "by eye" in order to determine the relationships between variables. The principles of the least squares sum of errors can also be applied to nonlinear models. In this exploration you will compare two methods for fitting nonlinear models to data.

(1) Transformation: Use the familiar method of transforming a function to fit data, and then compute a list of residuals between your model and the data. By making small changes in your model, you can minimize the sum of the squares of the residuals.

(2) Linearization: Rather than transforming a function, transform the data until they become linear. Then find the least squares line to fit those data. Lastly, reverse the transformations to find the model that fits the data.

A dynamic data program like Fathom is very useful for both of these methods of nonlinear regression.

Activity

How Does It Fit?

Step 1 Open a new Fathom document. Drag a new case table from the shelf and enter the data shown at right.

Step 2 Create a graph with *Year* on the horizontal axis and *CO2emissions* on the vertical axis. You can see that the data are not linear. To start the modeling process, you have to make a guess as to what kind of function might model the data.

Step 3 You will try to fit an exponential curve to the data in the form $y = ab^{(x-h)}$. Create two sliders with the names a and b. Then choose **Plot Function** from the Graph menu and plot the function $a \cdot b$ ^ $(year - 1949)$.

C02 Emissions

	Year	C02emissions
1	1949	19142
2	1954	23353
3	1959	31048
4	1964	42999
5	1969	47790
6	1974	55557
7	1979	81688
8	1984	119861
9	1989	178214
10	1994	239880
11	1999	310811
12	2004	402221

Adjust the sliders until the graph seems to fit the data.

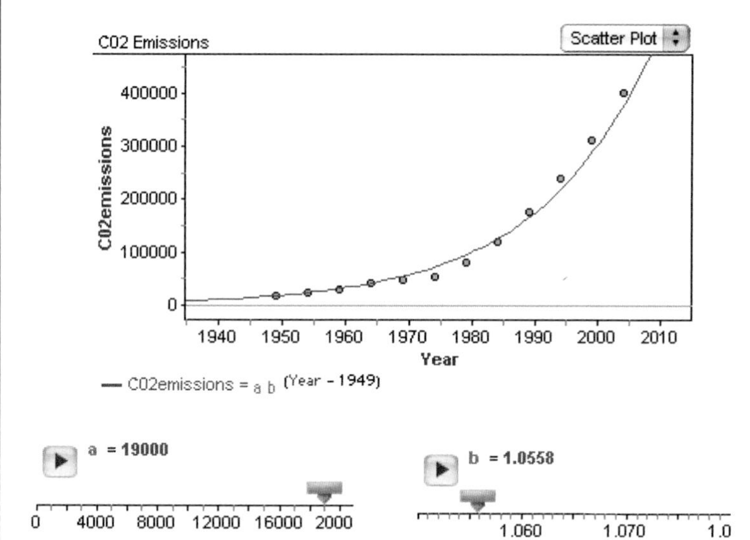

Step 4 Select the graph and choose **Show Squares** from the Graph menu. Below the equation you will see the sum of the squares of the residuals. Continue to make changes with the sliders until you have minimized this sum. (Don't spend more than a few minutes on this.) This is your least squares exponential model.

Step 5 If $y = ab^{x-1949}$ is the model, then $\log(y) = \log\left(ab^{x-1949}\right) = \log(a) + (x - 1949)\log(b)$. Create a new attribute, *yrs,* in the case table to represent the years since 1949. Choose **Edit Formula** from the Edit menu, and give it the formula *Year* − 1949. Then add another attribute, *log_y,* to the case table and give it the formula log(*CO2emissions*).

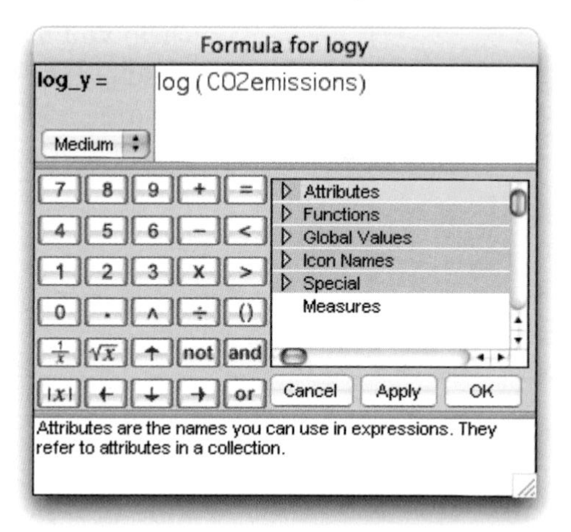

Questions

1. Create a new graph with *yrs* on the horizontal axis and *log_y* on the vertical axis. Find the least squares model for these data.

2. Use the least squares model for the linearized data to write an exponential equation for the original data. Return to the first graph and plot the equation $y = ab^{x-1949}$.

3. Each of the modeling procedures has certain desirable qualities. Identify one good feature for each method of regression.

CHAPTER 11 REVIEW

In this chapter you learned about three different methods for collecting data: **experimental studies, observational studies,** and **surveys.** You saw some statistical tools for estimating **parameters** of a very large (perhaps infinite) population from **statistics** of samples taken from that population. Many large populations can be described by **probability distributions** of **continuous random variables.** The area under the graph of a probability distribution is always 1. When using any probability distribution, you do not find the probability of a single exact value; you find the probability of an interval of values.

The probability distributions of many sets of data are **normal.** Their graphs, called **normal curves,** are bell-shaped. To write an equation of a normal distribution curve, you need to know the mean and standard deviation of the data set. To make predictions about a population based on a sample, you can make a nonstandard normal distribution standard by using ***z*-values,** and you can predict things about the population mean with a **confidence interval.**

You can also make predictions by collecting **bivariate data** and analyzing the relationship between two variables. The **correlation coefficient,** r, tells to what extent the variables have a linear relationship. If two variables are linearly related, the **least squares line** of fit, which passes through $(\bar{x}, \bar{y})$ and has slope $r\left(\frac{s_y}{s_x}\right)$, can help you make predictions about the population. You can also find curves to fit some nonlinear data by linearizing those data and finding a least squares line or by using statistics software or your calculator to transform a function to fit the data.

EXERCISES

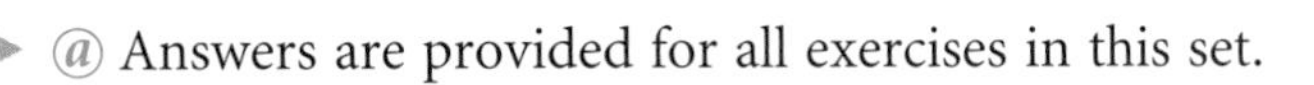

@ Answers are provided for all exercises in this set.

You will need

A graphing calculator for Exercises **5**, **6**, and **8**.

1. Classify each method for collecting data. Then describe a different method for collecting the data.

a. A random group of dogs was allowed to choose one of two types of dog food. Once the dogs made the choice, they were fed only that food for a week. Data were recorded on the number of minutes each dog spent actively playing each day for a week.

b. A number of students were asked if they would return money to a cashier if they were given back the incorrect amount of change.

2. A graph of a probability distribution consists of two segments. The first segment connects the points (0, 0) and (5, 0.1), and the second segment connects the points (5, 0.1) and (20, 0), as shown.

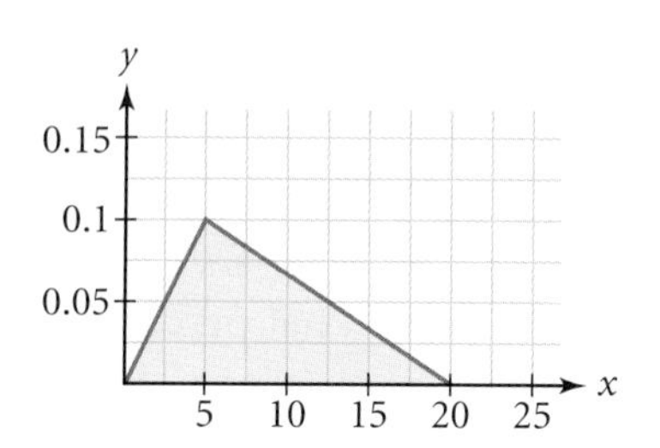

a. Verify that the area under the segments is equal to 1.

b. Find the median.

c. What is the probability that a data value will be less than 3?

d. What is the probability that a data value will be between 3 and 6?

3. Suppose a probability distribution has the shape of a semicircle. What is the radius of the semicircle?

4. The weights of house cats in northern Michigan can be modeled by a normal distribution, with the middle 68% of cats weighing between 8.4 and 12.7 lb.

a. What are the population mean and standard deviation for the weight of a house cat in northern Michigan?

b. What interval of weights would include the middle 95% of house cats in northern Michigan?

5. Recently, Adam put larger wheels on his skateboard and noticed it would coast farther. He decided to test this relationship. He collected some skateboards, attached wheels of various sizes to them, and rolled them to see how far they would roll on their own. He collected the data shown in the table.

Wheel diameter (in.) x	Rolling distance (in.) y
1	17
2	23
3.5	32
5	30
5.5	36
7	52
8.5	57
9.5	55
10	70

Dirtboards use wheels up to 10 inches in diameter.

a. Make a scatter plot of these data.

b. Does there appear to be a linear relationship between the variables? Find the correlation coefficient, and use this value to justify your answer.

c. Find the equation of the least squares line that models these data.

d. Describe the real-world meaning of the slope and the y-intercept of your line.

e. Use your model to determine the size of wheel of a skateboard that rolls 50.5 in.

6. Suppose the weights of all male baseball players who are 6 ft tall and between the ages of 18 and 24 are normally distributed. The mean is 175 lb, and the standard deviation is 14 lb.

a. What percentage of these males weigh between 180 and 200 lb?

b. What percentage of these males weigh less than 160 lb?

c. Find an interval that should include the weights of the middle 90% of these males.

d. Find the equation of a normal curve that provides a probability distribution for this information.

7. The length of a book is measured by many people. The measurements have mean 284 mm and standard deviation 1.3 mm. If four students measure the book, what is the probability that the mean of their measurements will be less than 283 mm?

8. Consider the monthly rainfall totals for three cities in northern Florida.

a. Which two of the cities show the strongest correlation?

b. Find the least squares line for the two cities from 8a. What does the slope of the line indicate?

c. Find the least squares line for the two cities that have the weakest correlation. What does the y-intercept of the line indicate?

	Rain per month (in.), 2006		
Month	**Crestview**	**Chipley**	**Panama City**
Jan	3.01	3.12	3.69
Feb	5.09	5.36	4.31
Mar	0.29	0.50	0.21
Apr	2.42	2.54	1.56
May	3.66	5.18	4.03
Jun	1.51	2.57	4.15
Jul	3.77	2.17	3.19
Aug	4.35	6.72	6.71
Sep	3.42	2.94	2.95
Oct	3.71	4.40	4.98
Nov	4.91	3.43	2.64
Dec	2.42	3.63	4.12

(Northwest Florida Water Management)

Library: Homage to Marcel Proust is a mixed media installation box constructed by French artist Charles Matton (b 1933). The miniature work (15.8 in. wide) evokes the world of French writer Marcel Proust (1871–1922).

TAKE ANOTHER LOOK

1. The least squares method minimizes the sum of the squares of the residuals. Why is this important? Try to think of another method you could use to find a line of fit. Explain what advantage or disadvantage the method of minimizing squares of residuals has that this other method does not have.

2. Consider a normal distribution with mean 0. Use your graphing calculator or geometry or statistics software to explore how the standard deviation, σ, affects the graph of the normal curve,

$$y = \frac{1}{\sigma\sqrt{2\pi}}\left(\sqrt{e}\right)^{-(x/\sigma)^2}$$

Summarize how the normal curve changes as the value of σ changes.

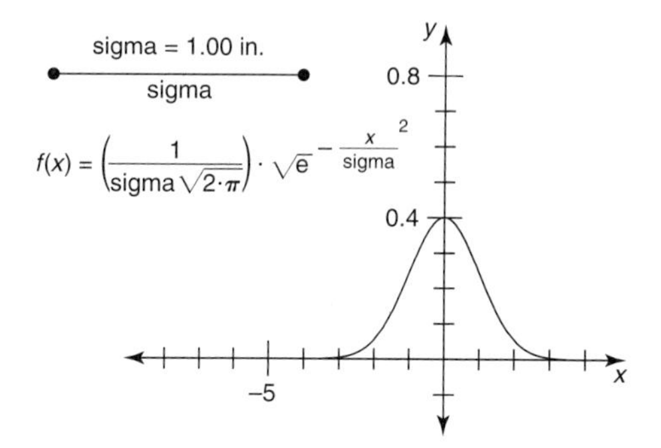

3. Consider a normal distribution with mean 0. You've already seen that the equation of the normal curve is

$$y = \frac{1}{\sigma\sqrt{2\pi}}\left(\sqrt{e}\right)^{-(x/\sigma)^2}$$

To avoid using *e*, you can use another equation that approximates the normal curve:

$$y = \frac{1}{\sigma\sqrt{2\pi}}\left(1 - \frac{1}{2\sigma^2}\right)^{x^2}$$

Use your graphing calculator or geometry or statistics software to explore how the graphs of these two equations compare for different values of σ. For which values of σ is the second equation a good approximation, a poor approximation, or even undefined?

Project

MAKING IT FIT

Find any bivariate data that you think might be related. You might look in an almanac, or search the library or the Internet. Then use techniques from this chapter to find a function that fits the data well.

Your project should include

- Your data and its source.
- A graph of your data with the equation you found to model it.
- A description of your process and an analysis of how well your curve fits the data.

Assessing What You've Learned

WRITING TEST QUESTIONS Write a few test questions that reflect the topics of this chapter. You may want to include questions on experimental design, probability distributions, confidence intervals, and bivariate data and correlation. Include detailed solutions.

ORGANIZE YOUR NOTEBOOK Make sure that your notebook has complete notes on all of the statistical tools and formulas that you have learned. Specify which statistical measures apply to populations and which apply to samples, and explain which tools allow you to make conclusions about a population based on a sample and vice versa. Be sure you know when to use the various statistical measures and exactly what each one allows you to predict.

PERFORMANCE ASSESSMENT As a friend, family member, or teacher watches, solve a problem from this chapter that deals with fitting a line to data and analyzing how well the function fits. Explain the various tools for analyzing how well a function fits data.

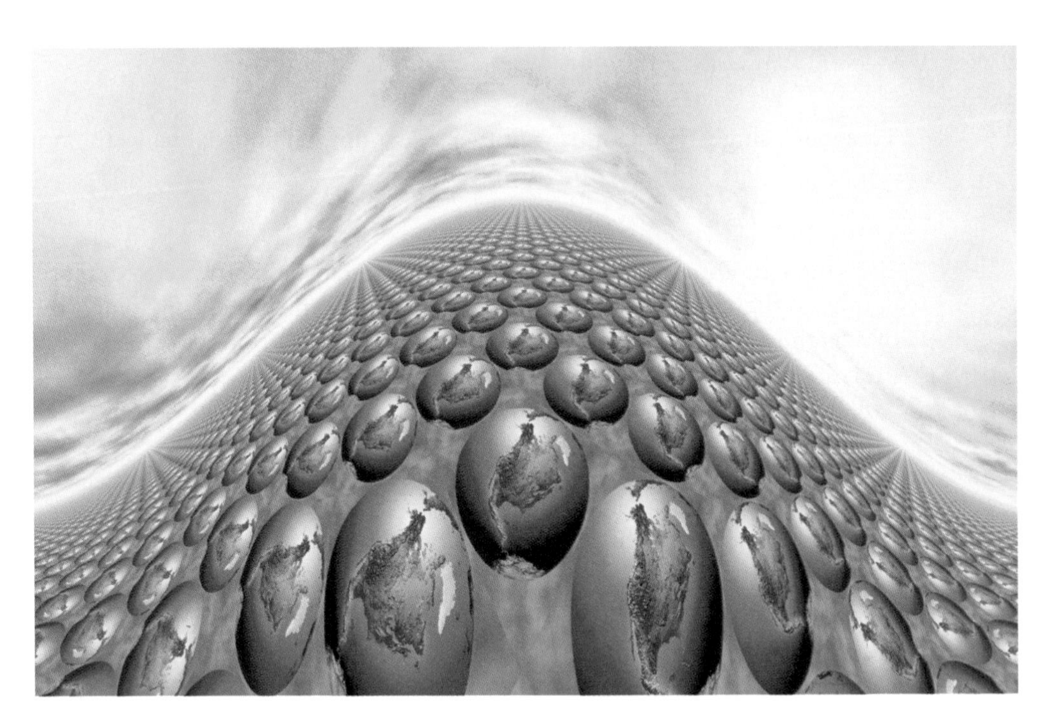

CHAPTER 12 Trigonometry

Project Bandaloop is a group of contemporary dancers who combine dance, rock climbing, and rapelling. Their choreography relies upon a blend of aerial, horizontal, and vertical movement. By performing in a variety of venues, ranging from skyscrapers to granite cliffs, the group hopes to stimulate the viewer's awareness of his or her natural and human-made environment.

OBJECTIVES

In this chapter you will

- review properties of similar triangles and special right triangles
- review the trigonometric ratios—sine, cosine, and tangent
- use the trigonometric ratios and their inverses to solve application problems
- solve problems using the Law of Sines and the Law of Cosines
- explore the properties of vectors
- use a third variable to write parametric equations that separately define x and y

Special Right Triangles

When one figure is a dilation of another figure by the same horizontal and vertical scale factors, the two figures are **similar.** Saying that two triangles are similar is equivalent to saying that their corresponding angles are **congruent** (equal in measure) or that their corresponding side lengths are *proportional* (having the same ratio). These properties allow you to calculate the lengths of unknown sides of a triangle when you know the side lengths of a similar triangle.

If you split an equilateral triangle in half along an altitude, you create two right triangles with angles of 30°, 60°, and 90°. If the original triangle has sides of length 2, you can use the Pythagorean Theorem to find the length of the altitude, which is the longer leg of each new triangle. You can use the side length relationships of this 30°-60°-90° triangle to find the side lengths of similar triangles.

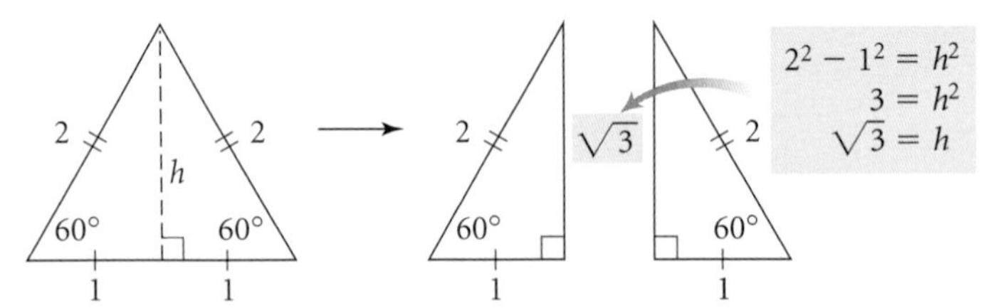

EXAMPLE A Find the lengths of the labeled sides.

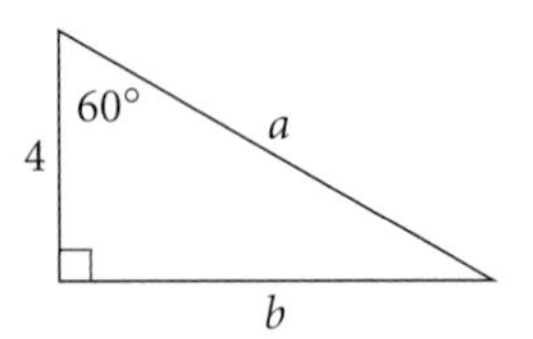

▶ Solution The unmarked angle of the triangle measures 30° because the sum of the angles of any triangle is 180°. All 30°-60°-90° triangles are similar, so you can use the comparison triangle and write the following proportions:

$$\frac{a}{4} = \frac{2}{1} \quad \text{and} \quad \frac{b}{4} = \frac{\sqrt{3}}{1}$$

$$a = 8 \quad \text{and} \quad b = 4\sqrt{3}$$

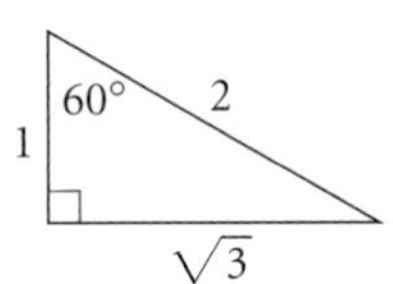

30°-60°-90° Triangle Properties

In a 30°-60°-90° triangle, the side lengths have the ratio $1:\sqrt{3}:2$. If the shorter leg has length a, then the longer leg has length $a\sqrt{3}$, and the hypotenuse has length $2a$.

You can make another comparison triangle by cutting a square in half along the diagonal to form two isosceles right triangles. You can use the Pythagorean Theorem to find the length of the hypotenuse.

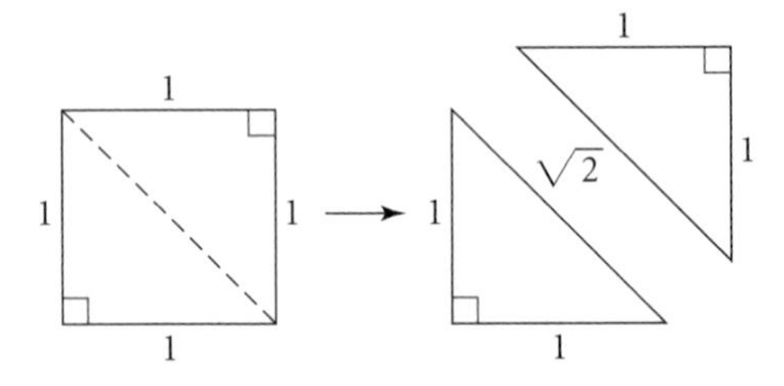

EXAMPLE B Find the lengths of the labeled sides.

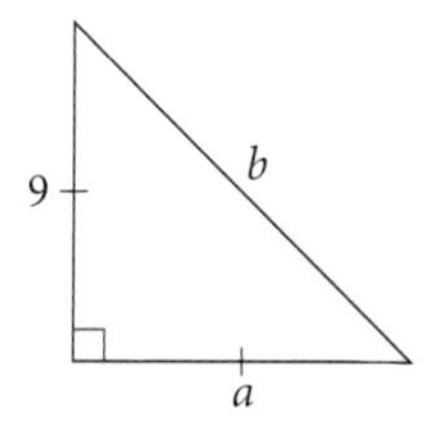

▶ ***Solution*** The hash marks indicate that the legs are equal in length, so $a = 9$. You can then use the isosceles right comparison triangle to write the proportion $\frac{b}{9} = \frac{\sqrt{2}}{1}$. Therefore, $b = 9\sqrt{2}$.

All isosceles right triangles are similar, so you can write the following general relationships:

Isosceles Right Triangle Properties

In an isosceles right triangle, also known as a 45°-45°-90° triangle, the side lengths have the ratio $1:1:\sqrt{2}$. If each leg has length l, then the hypotenuse has length $l\sqrt{2}$.

These two special types of triangles, the isosceles right triangle and the 30°-60°-90° triangle, occur frequently in mathematics so it's important that you become confident in working with them.

EXERCISES

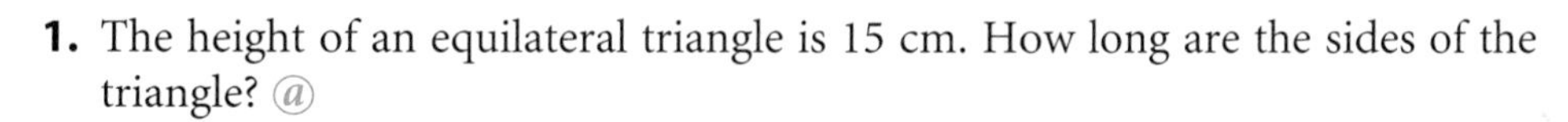

1. The height of an equilateral triangle is 15 cm. How long are the sides of the triangle? ⓐ

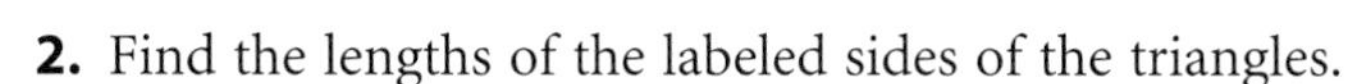

2. Find the lengths of the labeled sides of the triangles.

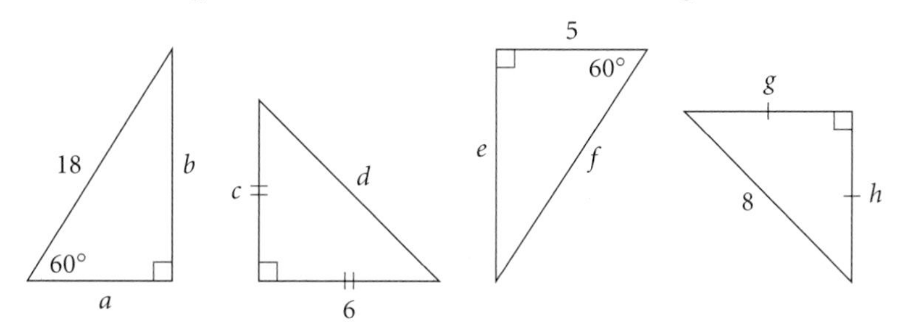

3. Identify each right triangle described as isosceles right, 30°-60°-90°, or neither.

a. legs with lengths $5\sqrt{3}$ and 15 ⓐ

b. hypotenuse with length 4, one leg with length $2\sqrt{2}$

c. hypotenuse with length $6\sqrt{2}$, one leg with length $3\sqrt{3}$ ⓐ

d. hypotenuse with length $4\sqrt{3}$, one leg with length 6

e. hypotenuse with length $7\sqrt{3}$, one leg with length 7

LESSON

12.1 Right Triangle Trigonometry

Happiness is not a destination.
It is a method of life.

BURTON HILLIS

The steepest paved road in the world is Baldwin Street in Dunedin, New Zealand. It runs up a hill for a quarter of a mile at an angle of 20.8° with the valley floor. Can you imagine riding a bicycle up this hill?

By aligning a camera with the surface of the road, this flat-roofed house along Baldwin Street in Dunedin, New Zealand appears to be sinking into the ground in a lopsided fashion.

How much higher are you at the top of the hill than at the bottom? You can make a diagram of this situation using a right triangle. But this isn't one of the special right triangles.

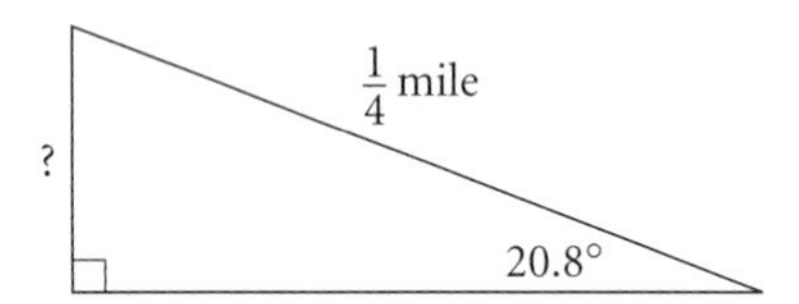

How can you find the height of this triangle? One way is to use the fact that corresponding side lengths of similar triangles have the same ratio. For example, in the right triangles shown below, all corresponding angles are congruent, so the triangles are similar. The ratio of the length of the shorter leg to the length of the longer leg is always 0.75, and the ratios of the lengths of other pairs of corresponding sides are also equal.

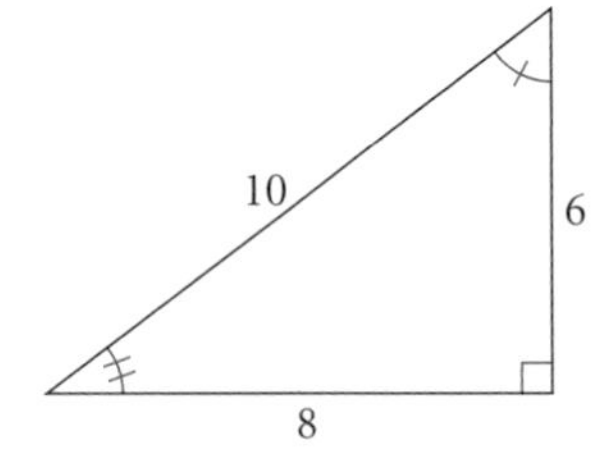

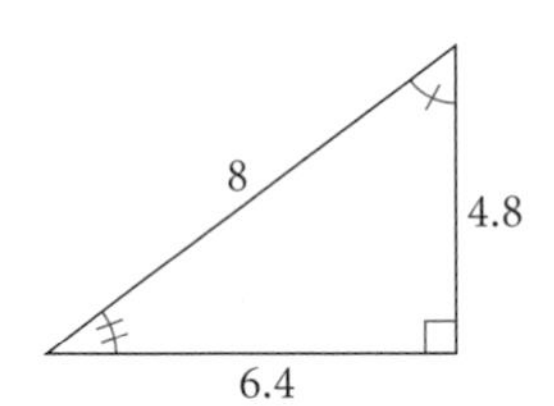

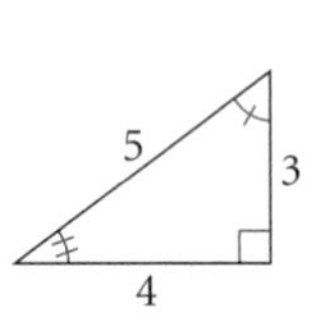

keymath.com/DAA

These ratios are called **trigonometric ratios.** The word **trigonometry** comes from the Greek words for "triangle" and "measure." [► You can deepen your understanding of trigometric ratios using the **Dynamic Algebra Exploration** at www.keymath.com/DAA .◄]

For the New Zealand road problem, the ratio of the length of the leg opposite the 20.8° to the length of the hypotenuse will be the same in every similar triangle. If you know this ratio, you can solve for the height.

Trigonometric Ratios

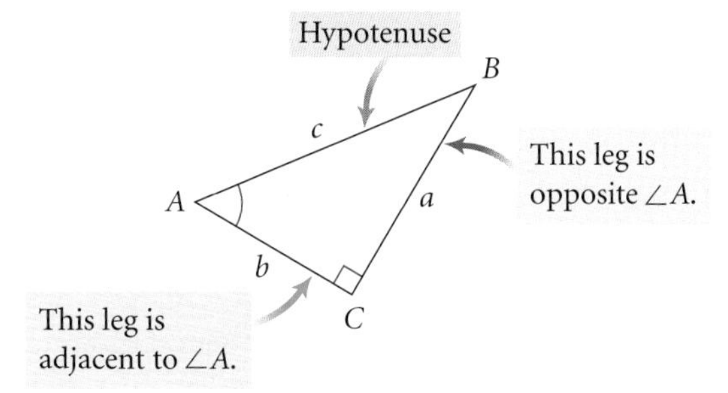

For any acute angle A in a right triangle, the **sine** of $\angle A$ is the ratio of the length of the leg opposite $\angle A$ to the length of the hypotenuse.

$$\sin A = \frac{\text{opposite leg}}{\text{hypotenuse}} = \frac{a}{c}$$

The **cosine** of $\angle A$ is the ratio of the length of the leg adjacent to $\angle A$ to the length of the hypotenuse.

$$\cos A = \frac{\text{adjacent leg}}{\text{hypotenuse}} = \frac{b}{c}$$

The **tangent** of $\angle A$ is the ratio of the length of the opposite leg to the length of the adjacent leg.

$$\tan A = \frac{\text{opposite leg}}{\text{adjacent leg}} = \frac{a}{b}$$

Language CONNECTION

The word *sine* has a curious history. The Sanskrit term for sine was *jya-ardha* ("half-chord"), later abbreviated to *jya.* Islamic scholars, who learned about sine from India, called it *jiba.* In Segovia, around 1140, Robert of Chester read *jiba* as *jaib* when he was translating al-Khwārizmī's book, *Kitāb al-jabr wa'al-muqābalah,* from Arabic into Latin. One meaning of *jaib* is "indentation" or "gulf." So *jiba* was translated into Latin as *sinus,* meaning "fold" or "indentation," and from that we get the word *sine.*

You can use the trigonometric ratios to find unknown side lengths of a right triangle when you know the measure of one acute angle and the length of one of the sides. [► See **Calculator Note 12B** to learn about calculating the trigonometric ratios on your calculator. ◄]

EXAMPLE A

Find the unknown length, x.

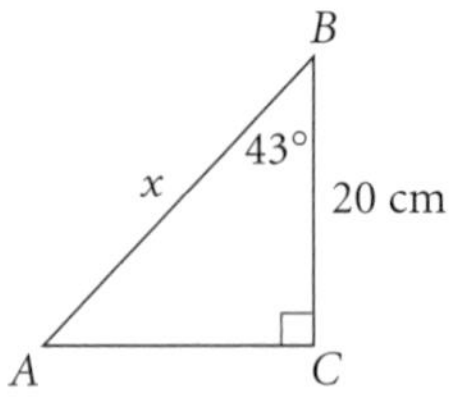

▸ Solution

In this problem, you know the length of one leg and the measure of one acute angle. You want to find the length of side $\overline{AB}$, which is the hypotenuse. The known leg is adjacent to the known angle. The cosine ratio relates the lengths of the adjacent leg and hypotenuse, so you can use it to find the length of the hypotenuse.

$\cos 43° = \dfrac{20}{x}$	Write the cosine ratio and substitute the known values.
$x \cos 43° = 20$	Multiply both sides by x.
$x = \dfrac{20}{\cos 43°}$	Divide both sides by $\cos 43°$.
$x \approx 27.35$	Find the cosine of 43° on a calculator and divide.

The length, x, of hypotenuse $\overline{AB}$ is approximately 27.35 cm.

Note that the trigonometric ratios apply to both acute angles in a right triangle. In the triangle below, b is the leg adjacent to A and the leg opposite B, whereas a is the leg opposite A and the leg adjacent to B. The hypotenuse, labeled c in this triangle, is always the side opposite the right angle.

$\sin A = \dfrac{a}{c}$ $\quad \sin B = \dfrac{b}{c}$

$\cos A = \dfrac{b}{c}$ $\quad \cos B = \dfrac{a}{c}$

$\tan A = \dfrac{a}{b}$ $\quad \tan B = \dfrac{b}{a}$

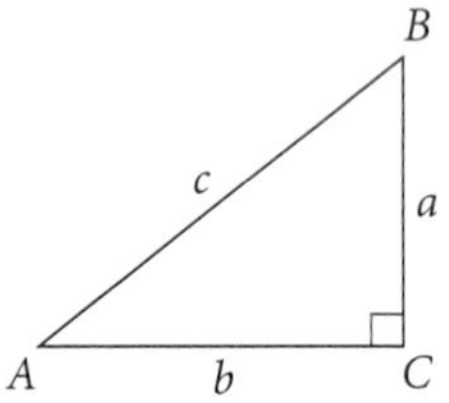

Each trigonometric ratio is a function of an acute angle measure because each angle measure has a unique ratio associated with it. The inverse of each trigonometric function gives the measure of the angle. For example, $\sin^{-1}\left(\frac{3}{5}\right) \approx 36.87°$. (*Note:* This can be read as "the angle whose sine is $\frac{3}{5}$ measures 36.87 degrees.")

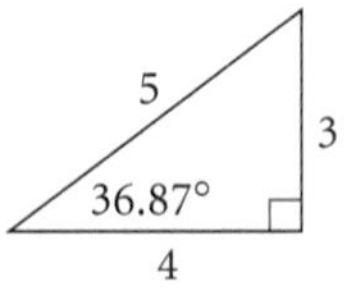

EXAMPLE B

Two hikers leave their campsite. Emily walks east 2.85 km and Savannah walks south 6.03 km.

a. After Savannah gets to her destination, she looks directly toward Emily's destination. What is the measure of the angle between the path Savannah walked and her line of sight to Emily's destination?

b. How far apart are Emily and Savannah?

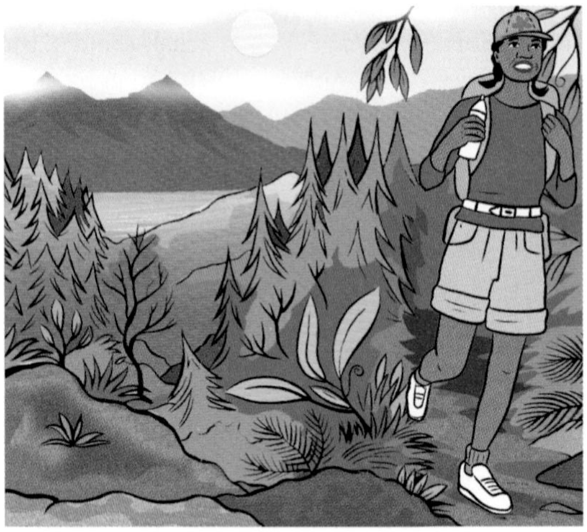

► ***Solution***

Draw a diagram using the information given.

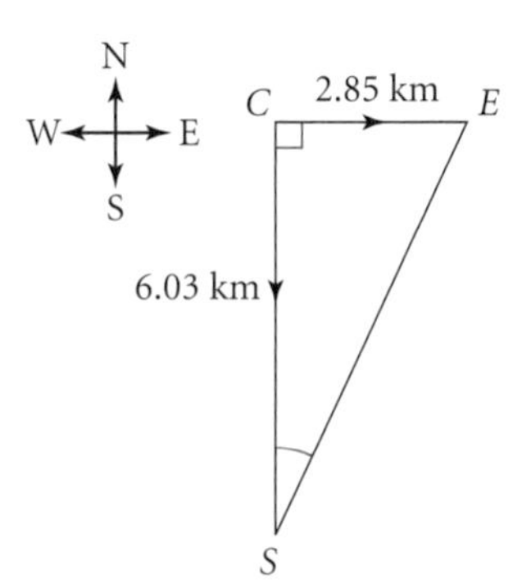

a. Angle S is the angle between Savannah's path and her line of sight to Emily's destination. Triangle SCE is a right triangle and you know the lengths of both legs, so you can use the tangent ratio to find S.

$$\tan S = \frac{2.85}{6.03}$$

Take the inverse tangent of both sides to find the angle measure. Because you are composing tangent with its inverse, $\tan^{-1}(\tan S)$ is equivalent to S. [► Revisit **Calculator Note 12B** to learn about calculating the inverse trigonometric functions. ◄]

$$S = \tan^{-1}\left(\frac{2.85}{6.03}\right) \approx 25°$$

The angle between Savannah's path and her line of sight to Emily's destination measures about 25°.

b. You can use the Pythagorean Theorem to find the distance from S to E.

$$6.03^2 + 2.85^2 = (SE)^2$$
$$SE = \sqrt{6.03^2 + 2.85^2}$$
$$SE \approx 6.67$$

The distance between the two hikers is approximately 6.67 km.

Recreation CONNECTION

A compass is a useful tool to have if you're hiking, camping, or sailing in an unfamiliar setting. The circular edge of a compass has 360 marks representing degrees of direction. North is 0° or 360°, east is 90°, south is 180°, and west is 270°. Each degree on a compass shows the direction of travel, or **bearing.** The needle inside a compass always points north. Magnetic poles on Earth guide the direction of the needle and allow you to navigate along your desired course. Learn how to use a compass with the links at www.keymath.com/DAA .

Your calculator will display each trigonometric ratio to many digits, so your final answer could be displayed to several decimal places. However, it is usually appropriate to round your final answer to the nearest degree or 0.1 unit of length, as in the solution to part a in Example B. If the problem gives more precise measurements, you can use that same amount of precision in your answer. In the example, distances were given to the nearest 0.01 km, so the answer in part b was rounded to the nearest 0.01 also.

Investigation
Steep Steps

Have you ever noticed that some sets of steps are steeper than others? Building codes and regulations place restrictions on how steep steps can be. Over time these codes change, so stairs built in different locations and at different times may vary quite a bit in their steepness.

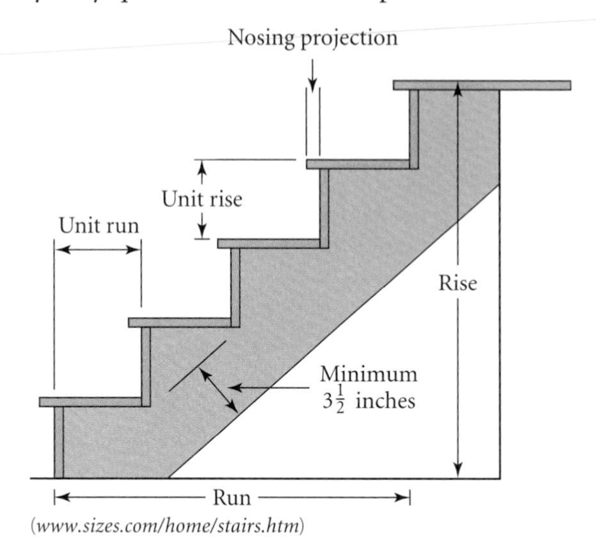

(*www.sizes.com/home/stairs.htm*)

Step 1 Refer to the diagram of stairs. According to the 1996 Council of American Building Officials and the 2000 International Code Council, the unit run should be not less than 10 inches, and the unit rise should not be more than 7.75 inches. With these limiting dimensions, what is the angle of inclination for the stairs?

Step 2 A rule of thumb for designing stairs is that the sum of the unit rise and unit run should be about 17.5 inches. Design three different sets of stairs that meet this condition. Make two of your designs within the approved building code given in Step 1. The third design should not meet the building code. Find the angle of inclination for each set of stairs.

Step 3 Consider designing steps to be built alongside Baldwin Street, Dunedin, at an angle of 20.8°.

a. How many designs are possible? Do all possible designs meet the code given in Step 1?

b. Create a design for the steps that meets the code. Does your design meet the rule of thumb in Step 2? If not, create a new design in which the sum of the rise and run is 17.5 in.

Step 4 Wheelchair ramps are supposed to have a slope between $\frac{1}{16}$ and $\frac{1}{20}$. For each of these slopes, design a ramp to get up to a door 24 inches above the surrounding ground. What is the angle of each ramp? (*www.mobility-advisor.com/wheelchair-ramp-specs.html*)

As you've seen in this lesson, trigonometry can be used to solve problems relating to roads, navigation, and construction. You'll get more practice using the trigonometric functions and their inverses in the exercises.

EXERCISES

You will need

A graphing calculator for Exercise **16.**

Practice Your Skills

1. Write all the trigonometric formulas (including inverses) relating the sides and angles in this triangle. There should be a total of 12. ⓐ

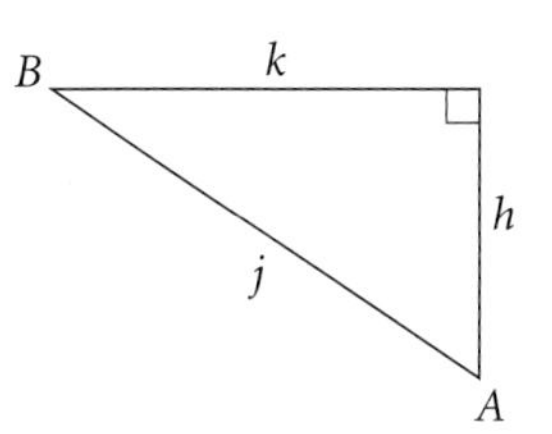

2. Draw a right triangle for each problem. Label the sides and angle, then solve to find the unknown measure.

a. $\sin 20° = \frac{a}{12}$ ⓐ **b.** $\cos 80° = \frac{25}{b}$ **c.** $\tan 55° = \frac{c+4}{c}$ ⓐ **d.** $\sin^{-1}\left(\frac{17}{30}\right) = D$

Reason and Apply

3. For each triangle, find the length of the labeled side.

a. ⓐ

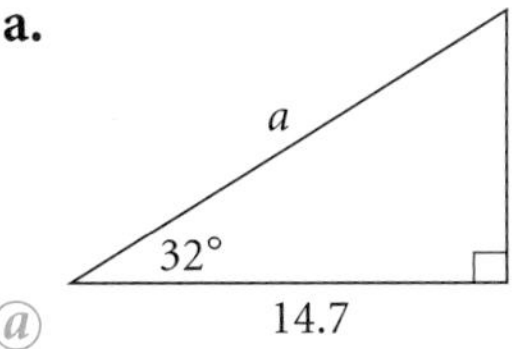

b. ⓐ

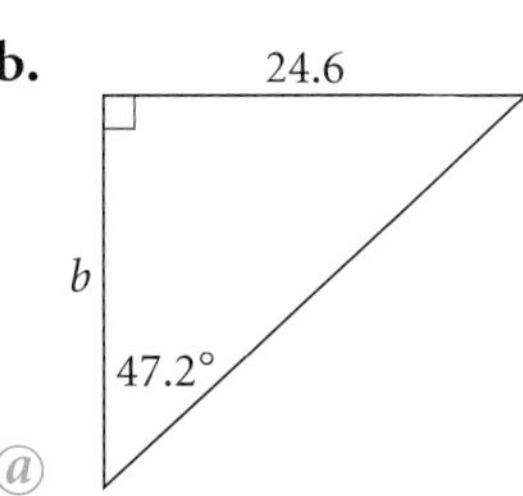

c.

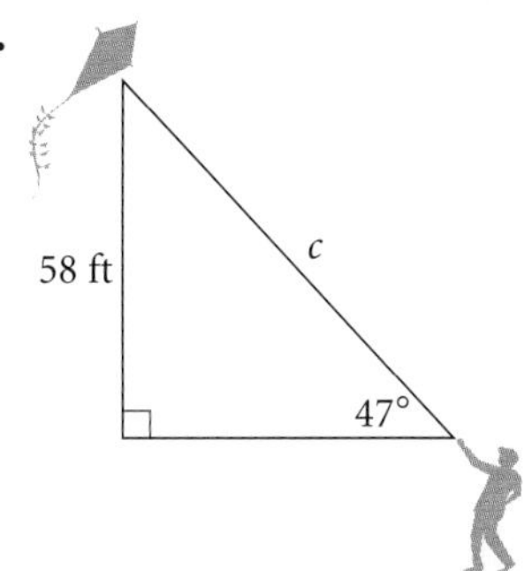

4. For each triangle, find the measure of the labeled angle.

a. ⓐ

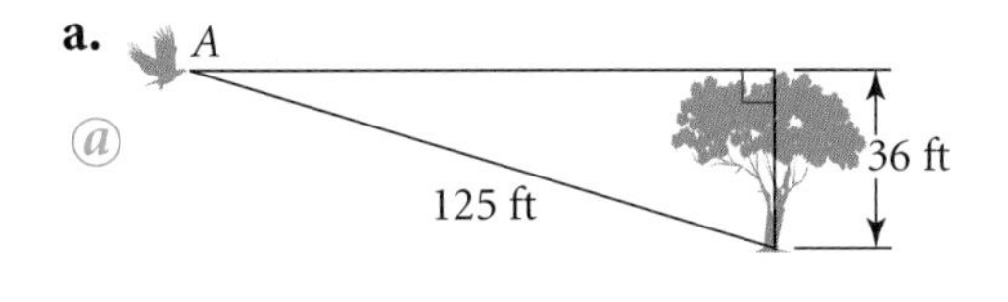

b.

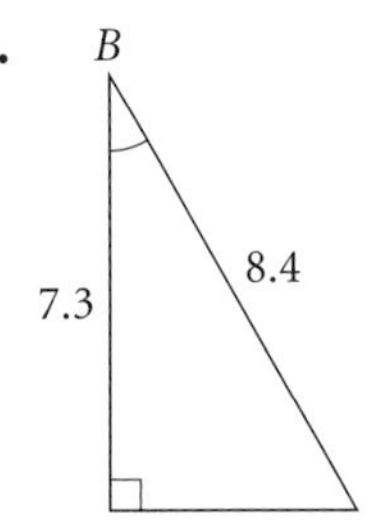

c.

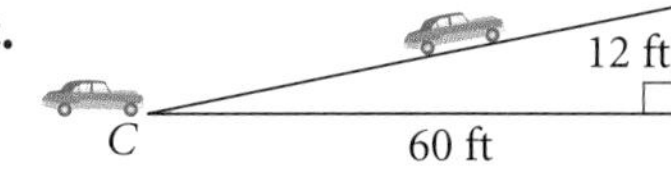

5. The steepest paved road in the world runs for a quarter of a mile at an angle of 20.8° to the floor of the valley. (*Note:* 1 mile = 5280 feet.)

a. How many feet higher are you at the top of the road than at the bottom?

b. If stair steps average 7.75 inches of rise for each step, how many steps would you have to climb to go up the same height as the hill? ⓐ

6. Use the properties of the special right triangles to complete the table using exact values. ⓗ

Angle	Sine	Cosine	Tangent
30°			
45°			
60°			

7. In the figure below, $AB = 75$ cm, $AC = 90$ cm, and $\angle A$ measures 20°.

a. Find BD, the height of $\triangle ABC$. ⓐ

b. Find the area of $\triangle ABC$.

c. Find AD and DC. ⓐ

d. Find the measure of $\angle C$.

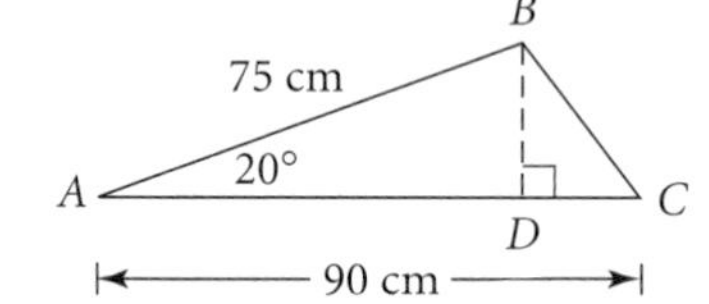

8. Woody owns a triangular shaped piece of land with a pond on it. He makes the measurements shown below. What are the area and perimeter of his property? ⓗ

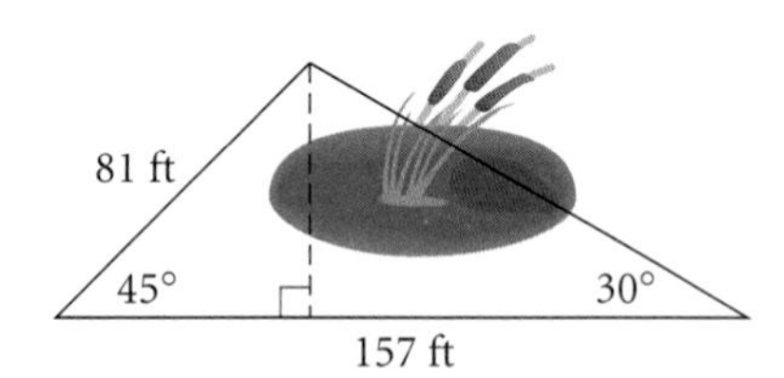

9. The term *cosine* is a shortened form of *complement's sine,* because the cosine of an angle is the sine of its complement. That is, $\cos A = \sin(90° - A)$. Explain why this is true, using a diagram and equations.

10. Shane says, "If you double the size of an acute angle that is less than 45° in a right triangle and keep the hypotenuse the same length, then the length of the leg opposite that angle also doubles." Is he correct? Use specific examples either to refute or to justify Shane's statement.

11. Graph the line $y = \frac{3}{4}x$, and plot a point on the line. What is the angle between the line and the x-axis? ⓗ

12. APPLICATION Civil engineers generally bank, or angle, a curve on a road so that a car going around the curve at the recommended speed does not skid off the road. Engineers use this formula to calculate the proper banking angle, θ, where v represents the velocity in meters per second, r represents the radius of the curve in meters, and g represents the gravitational constant, 9.8 m/s².

$$\tan\theta = \frac{v^2}{rg}$$

a. If the radius of an exit ramp is 60 m and the recommended speed is 40 km/h, at what angle should the curve be banked? ⓗ

b. A curve on a racetrack is banked at 36°. The radius of the curve is about 1.7 km. What speed is this curve designed for? Express your answer in km/h. ⓐ

Science CONNECTION

When a car rounds a curve, the driver must rely on the friction between the car's tires and the road surface to stay on the road. Unfortunately, this does not always work—especially if the road surface is wet!

In car racing, where cars travel at high speeds, tracks banked steeply allow cars to go faster, especially around the corners. Banking on NASCAR tracks ranges from 36° in the corners to just a slight degree of banking in the straighter portions.

Review

13. All of the right triangles in the figure below are 30°-60°-90° triangles. What is the length labeled x?

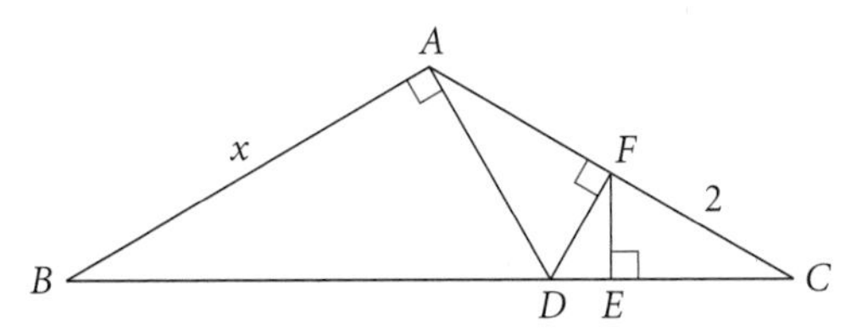

14. Solve this system. @

$$\begin{cases} 3x + 5y = 6 \\ 4y = x - 19 \end{cases}$$

15. Find the exact roots of this equation without using a calculator.

$$0 = x^3 - 4x^2 + 2x + 4$$

16. Consider this sequence.

$-6, -4, 3, 15, 32, \ldots$

a. Find the nth term using finite differences. Assume -6 is u_1.

b. Find u_{20}.

17. A sealed 10 cm tall cone resting on its base is filled to half its height with liquid. It is then turned upside down. To what height, to the nearest hundredth of a centimeter, does the liquid reach?

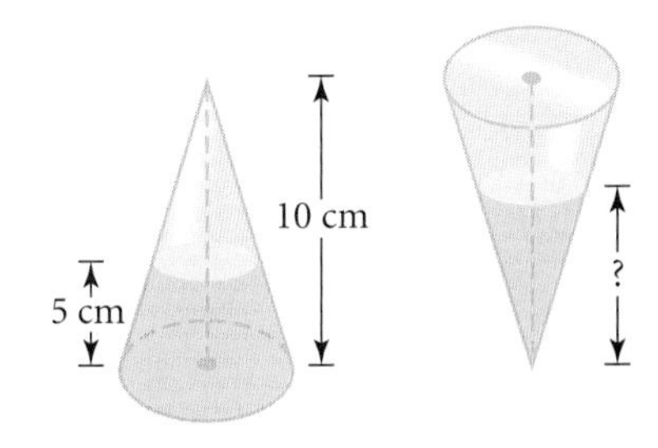

LESSON

12.2

The Law of Sines

Two airplanes pass over Chicago at the same time. Plane A is flying on a heading of 105° clockwise from north. Plane B is flying on a heading of 260° clockwise from north. After 2 hours, Plane A is 800 miles from Chicago and Plane B is 900 miles from Chicago. How far apart are the planes at this time?

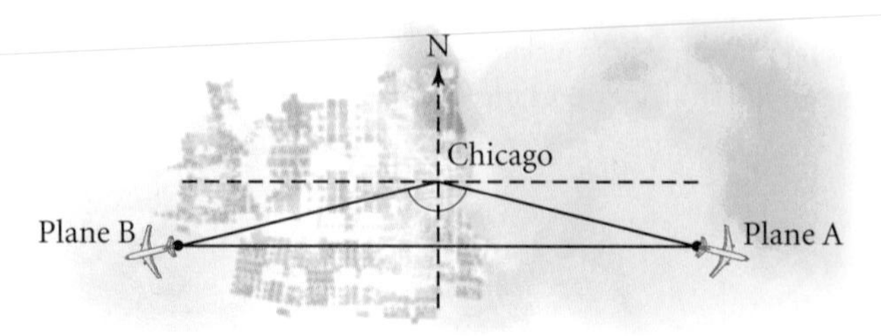

The problem above is a familiar scenario, but the triangle formed by the planes' paths is not a right triangle. In the next two lessons, you will discover useful relationships involving the sides and angles of nonright, or **oblique,** triangles and apply those relationships to situations like the distance problem presented above.

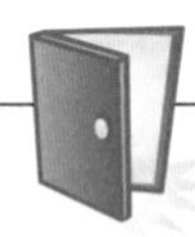

Investigation
Oblique Triangles

You will need

- a ruler
- a protractor

In this investigation you'll explore a special relationship between the sines of the angle measures of an oblique triangle and the lengths of the sides.

Step 1 Have each group member draw a different acute triangle ABC. Label the length of the side opposite $\angle A$ as a, the length of the side opposite $\angle B$ as b, and the length of the side opposite $\angle C$ as c. Draw the altitude from $\angle A$ to $\overline{BC}$. Label the height h.

Step 2 The altitude divides the original triangle into two right triangles, one containing $\angle B$ and the other containing $\angle C$. Use your knowledge of right triangle trigonometry to write an expression involving $\sin B$ and h, and an expression with $\sin C$ and h. Combine the two expressions by eliminating h. Write your new expression as a proportion in the form

$$\frac{\sin B}{?} = \frac{\sin C}{?}$$

Step 3 Now draw the altitude from $\angle B$ to $\overline{AC}$ and label the height j. Repeat Step 2 using expressions involving j, $\sin C$, and $\sin A$. What proportion do you get when you eliminate j?

Step 4 Compare the proportions that you wrote in Steps 2 and 3. Use the transitive property of equality to combine them into an extended proportion:

$$\frac{?}{?} = \frac{?}{?} = \frac{?}{?}$$

Step 5 | Share your results with the members of your group. Did everyone get the same proportion in Step 4?

Sine, cosine, and tangent are defined for all real angle measures. Therefore, you can find the sine of obtuse angles as well as the sines of acute angles and right angles. Does your work from Steps 1–5 hold true for obtuse triangles as well? (In Lesson 12.4, you will learn how these definitions are extended beyond right triangles.)

Step 6 | Have each group member draw a different obtuse triangle. Measure the angles and the sides of your triangle. Substitute the measurements and evaluate to verify that the proportion from Step 4 holds true for your obtuse triangles as well.

The relationships that you discovered in the investigation allow you to solve many problems involving oblique triangles.

EXAMPLE A

Towers A, B, and C are located in a national forest. From Tower B, the angle between Towers A and C is 53.3°, and from Tower C the angle between Towers A and B is 46.7°. The distance between Towers A and B is 4084 m. A lake between Towers A and C makes it difficult to measure the distance between them directly. What is the distance between Towers A and C?

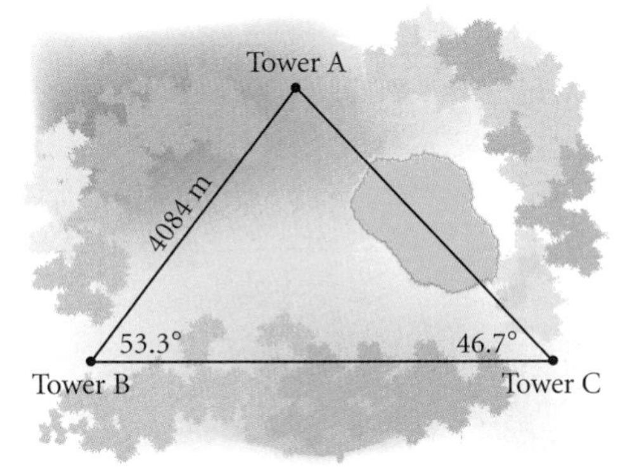

This forest fire observation tower is located in Great Smoky Mountains National Park, Tennessee.

▸ Solution

On your paper, sketch and label a diagram. Include the altitude from $\angle A$ to $\overline{BC}$. As in the investigation, two right triangles are formed. Set up sine ratios for each.

$$\sin 53.3^\circ = \frac{h}{4084} \quad \text{or} \quad h = 4084 \sin 53.3^\circ$$

and

$$\sin 46.7^\circ = \frac{h}{b} \quad \text{or} \quad h = b \sin 46.7^\circ$$

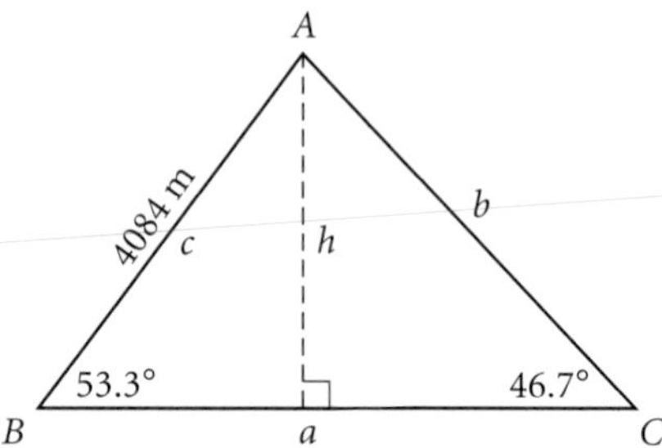

Substituting for h gives $4084 \sin 53.3^\circ = b \sin 46.7^\circ$. Solving for b gives

$$b = \frac{4084 \sin 53.3^\circ}{\sin 46.7^\circ}$$

$$b \approx 4499$$

The distance between Towers A and C is approximately 4499 m.

In Example A, notice that the height, h, was eliminated from the final calculations. You did not need a measurement for h! The equation $4084 \sin 53.3^\circ = b \sin 46.7^\circ$ can also be written as $\frac{\sin 53.3^\circ}{b} = \frac{\sin 46.7^\circ}{4084}$, or $\frac{\sin B}{b} = \frac{\sin C}{c}$, which is the relationship you found in the investigation.

For all three angles of a triangle, the ratio of the sine of an angle to the length of the opposite side is constant. This relationship is called the **Law of Sines.**

Law of Sines

For any triangle with angles A, B, and C, and sides of lengths a, b, and c (a is opposite $\angle A$, b is opposite $\angle B$, and c is opposite $\angle C$),

$$\frac{\sin A}{a} = \frac{\sin B}{b} = \frac{\sin C}{c}$$

You use the Law of Sines to find missing parts of triangles when you know the measures of two angles and the length of one side of a triangle.

EXAMPLE B

Find the length of $\overline{BC}$.

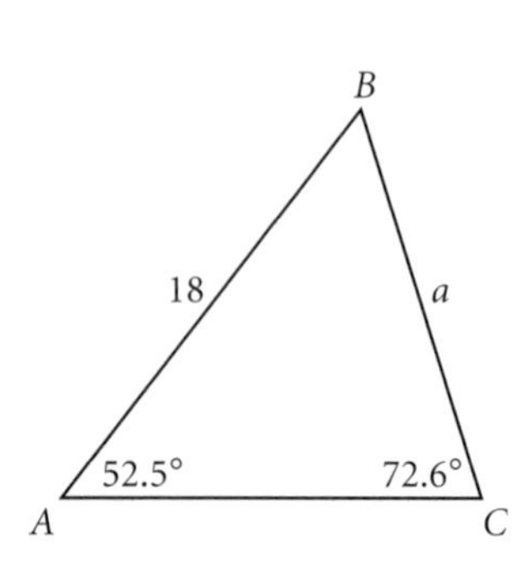

▶ Solution

Use the Law of Sines.

$\dfrac{\sin A}{a} = \dfrac{\sin C}{c}$	Select the proportion for the given angles and sides.
$\dfrac{\sin 52.5°}{a} = \dfrac{\sin 72.6°}{18}$	Substitute the angle measures and the known side length.
$18 \sin 52.5° = a \sin 72.6°$	Multiply both sides by $18a$ and reduce.
$\dfrac{18 \sin 52.5°}{\sin 72.6°} = a$	Divide both sides by sin 72.6° and reduce the right side.
$a \approx 15$	Evaluate.

The length of $\overline{BC}$ is approximately 15 units.

You may also use the Law of Sines when you know two side lengths and the measure of the angle opposite one of the sides. However, in this case you may find more than one possible solution. This is because two different angles—one acute and one obtuse—may share the same value of sine. Look at this diagram to see how this works.

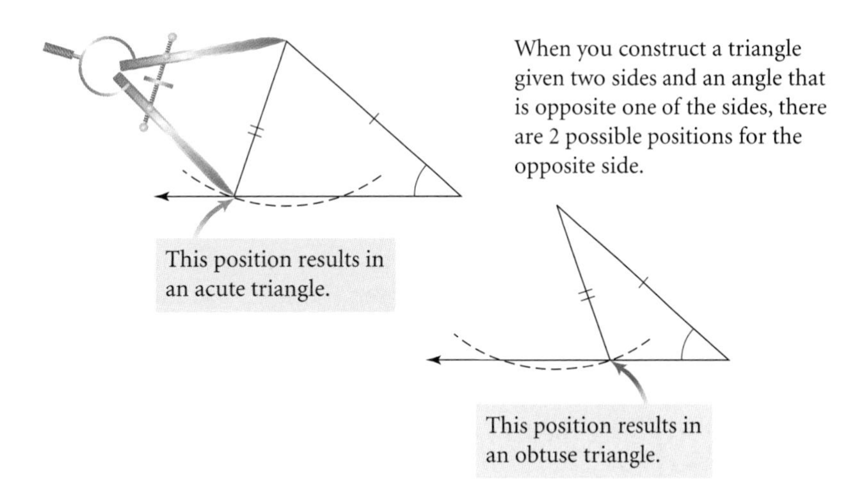

EXAMPLE C

Tara and Yacin find a map that they think will lead to buried treasure. The map instructs them to start at the 47° fork in the river. They need to follow the line along the southern branch for 200 m, then walk to a point on the northern branch that's 170 m away. Where along the northern branch should they dig for the treasure?

▶ Solution

As the map shows, there are two possible locations to dig. Consider the angles formed by the 170 m segment and the line along the northern branch as $\angle A$. Use the Law of Sines to find one possible measure of $\angle A$.

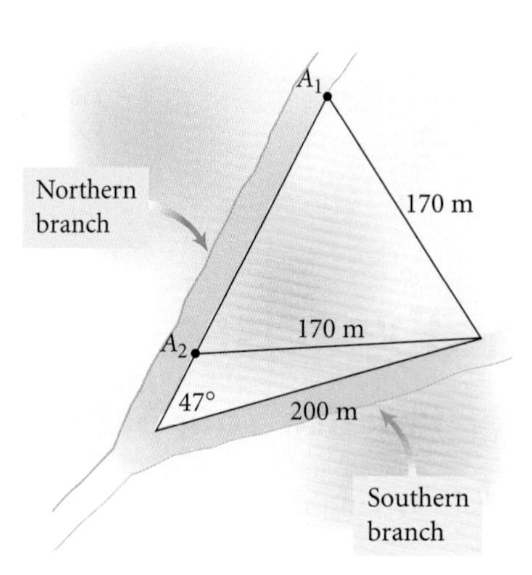

$$\frac{\sin 47°}{170} = \frac{\sin A}{200}$$ Set up a proportion using opposite sides and angles.

$$\sin A = \frac{200 \sin 47°}{170}$$ Solve for sin *A*.

$$A = \sin^{-1}\left(\frac{200 \sin 47°}{170}\right)$$ Take the inverse sine of both sides.

$$A \approx 59.4°$$ Evaluate.

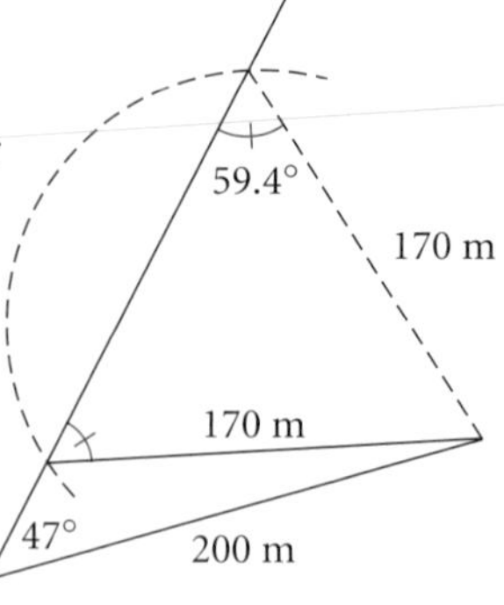

If $\angle A$ is acute, it measures approximately 59.4°. The other possibility for $\angle A$ is the obtuse supplement of 59.4°, or 120.6°. You can verify this with geometry, as shown in the diagrams. Use your calculator to check that sin 59.4° is equivalent to sin 120.6° and that both angle measures satisfy the Law of Sines equation $\frac{\sin 47°}{170} = \frac{\sin A}{200}$.

The two 170 m segments form an isosceles triangle, so the base angles are congruent.

If $\angle A$ is obtuse, it forms a linear pair with the 59.4° angle. It therefore measures 180° − 59.4°, or 120.6°.

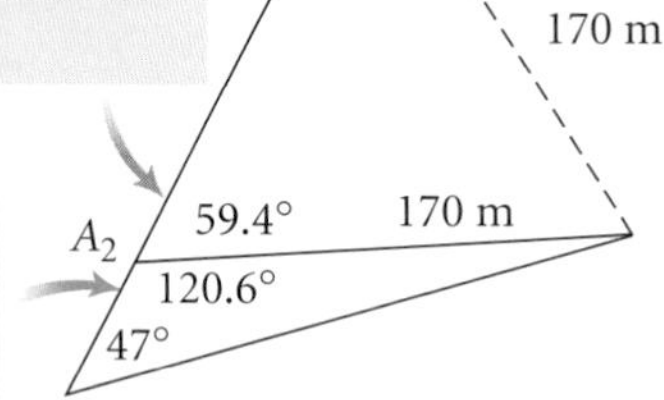

In order to find the distance along the northern branch, you need the measure of the third angle in the triangle. Use the known angle measure, 47°, and the approximations for the measure of $\angle A$.

$$180° - (47° + 59.4°) \approx 73.6° \quad \text{or} \quad 180° - (47° + 120.6°) \approx 12.4°$$

Use the Law of Sines to find the distance along the northern branch, *x*.

$$\frac{\sin 73.6°}{x} = \frac{\sin 47°}{170} \quad \text{or} \quad \frac{\sin 12.4°}{x} = \frac{\sin 47°}{170}$$

$$x = \frac{170 \sin 73.6°}{\sin 47°} \quad \text{or} \quad x = \frac{170 \sin 12.4°}{\sin 47°}$$

$$x \approx 223 \quad \text{or} \quad x \approx 50$$

Tara and Yacin should dig two holes along the northern branch, one 50 m and the other 223 m from the fork in the river.

A situation like that in Example C, where more than one possible solution exists, is called an **ambiguous case.** You can't tell which of the possibilities is correct unless there is more information in the problem, such as knowing whether the triangle is acute or obtuse. In general, you should report both solutions in cases like this.

EXERCISES

Practice Your Skills

1. Find the length of side $\overline{AC}$. @

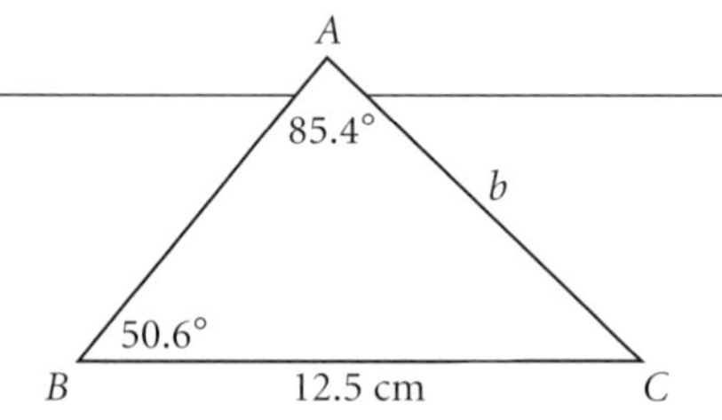

2. Assume $\triangle PQR$ is an acute triangle. Find the measure of $\angle P$. ⓐ

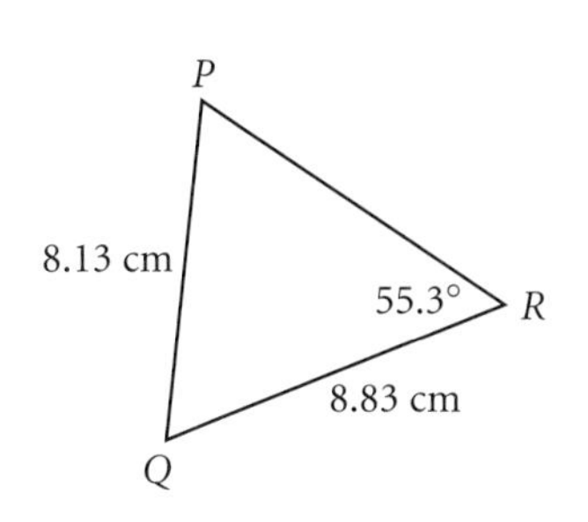

3. In $\triangle XYZ$, at right, $\angle Z$ is obtuse. Find the measures of $\angle X$ and $\angle Z$. ⓐ

4. Find the length of $\overline{XY}$ in Exercise 3. ⓐ

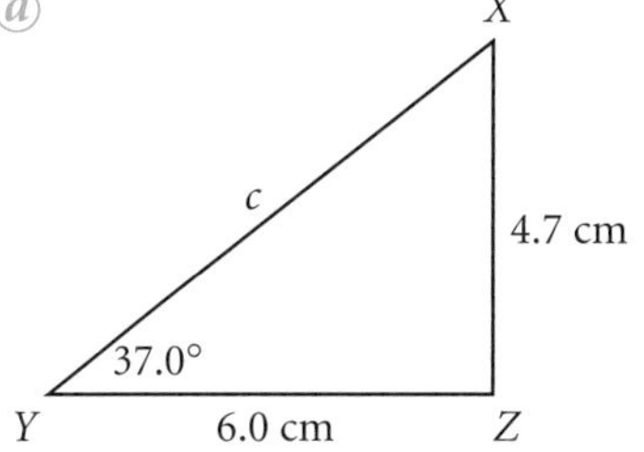

Reason and Apply

5. Find all of the unknown angle measures and side lengths.

a.

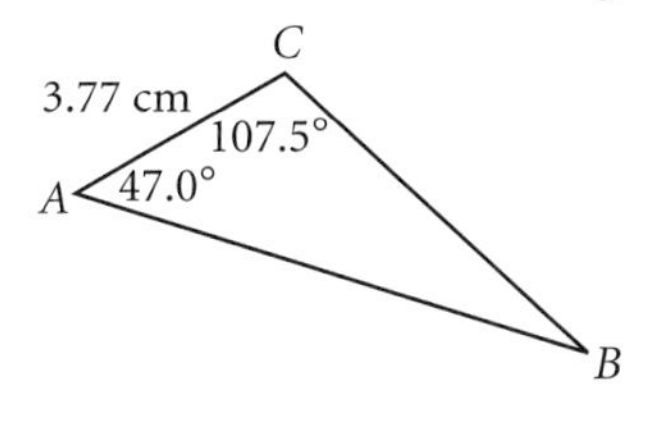

b.

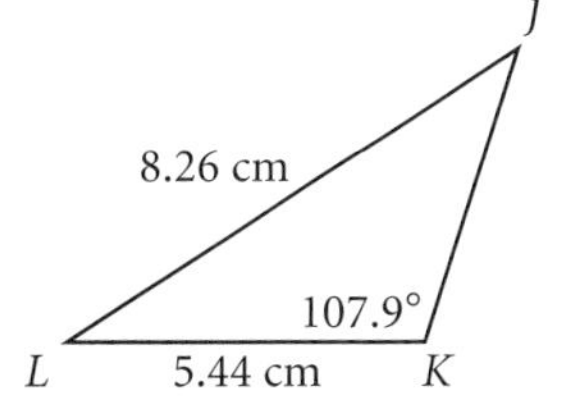

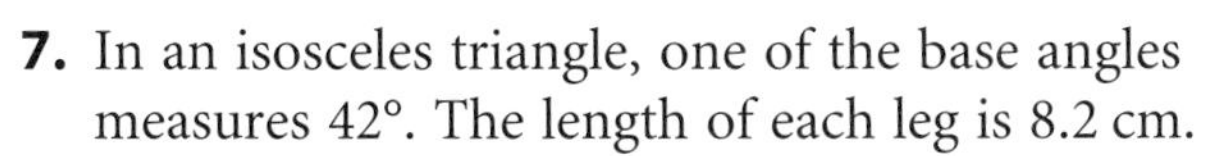

6. The Daredevil Cliffs rise vertically from the beach. The beach slopes gently down to the water at an angle of 3° from the horizontal. Scott lies at the water's edge, 50 ft from the base of the cliff, and determines that his line of sight to the top of the cliff makes a 70° angle with the line to the beach. How high is the cliff? ⓗ

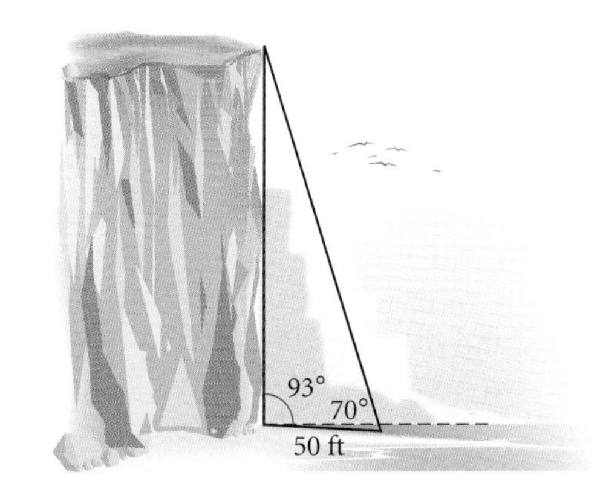

7. In an isosceles triangle, one of the base angles measures 42°. The length of each leg is 8.2 cm.

a. Find the length of the base. ⓐ

b. Even though you are given one angle and two sides not including the angle, this is not an ambiguous case. Why not?

8. **APPLICATION** Venus is 67 million mi from the Sun. Earth is 93 million mi from the Sun. Gayle measures the angle between the Sun and Venus as 14°. At that moment in time, how far is Venus from Earth? ⓗ

Chabot Observatory in Oakland, California

9. **APPLICATION** The SS *Minnow* is lost at sea in a deep fog. As the ship moves along a line at an angle of 107° clockwise from north, the skipper sees a light at an angle of 60° clockwise from north. The same light reappears through the fog after the skipper has sailed 1.5 km on his initial course. The second sighting of the light is at an angle of 34° clockwise from north. How far is the boat from the source of the light at the time of the second sighting? ⓗ

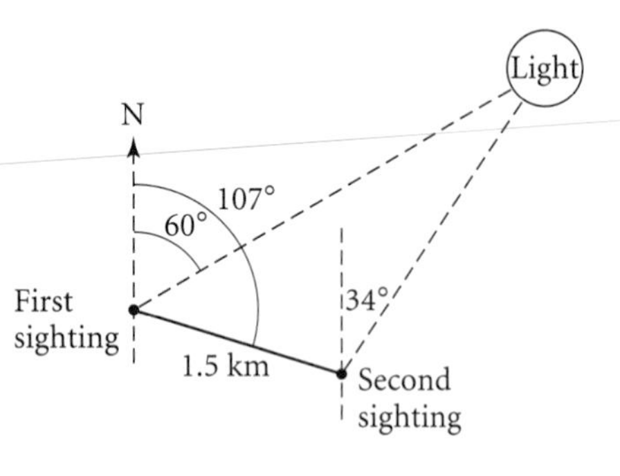

History CONNECTION

Lighthouses project light through darkness or poor weather to help guide ships ashore. The first lighthouses were actually bonfires—one was even mentioned in the *Iliad,* a Greek epic poem by Homer written sometime before 700 B.C.E. Today, most lighthouses hold powerful electric lights that flash automatically. Every lighthouse has a distinct sequence of flashes that allows ship captains to identify the nearby harbor.

10. **APPLICATION** One way to calculate the distance between Earth and a nearby star is to measure the angle between the star and the *ecliptic* (the plane of Earth's orbit) at 6-month intervals. A star is measured at a 42.13204° angle. Six months later, the angle is 42.13226°. The diameter of Earth's orbit is 3.13×10^{-5} light-years. What is the distance to the star at the time of each reading? Use this diagram to help you solve this problem. ⓐ

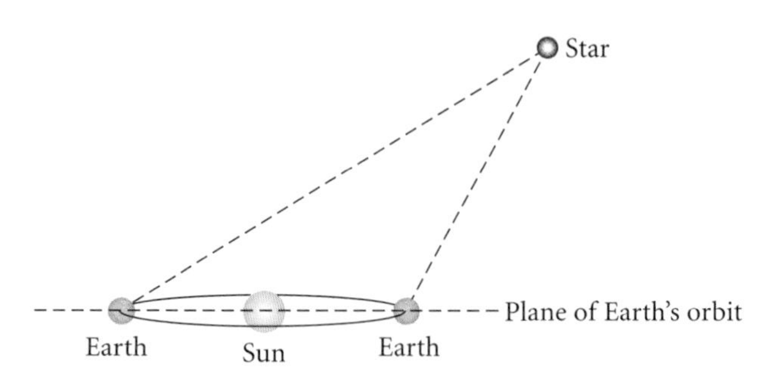

Science CONNECTION

A light-year is the distance that light can travel in one year. Light moves at a velocity of about 300,000 km/s, so one light-year is equal to about 9,461,000,000,000 km. Distances in space are so large that it's difficult to express them with relatively small units such as kilometers. For example, the distance to the next nearest big galaxy is 21×10^{18} km! Another unit of distance used by astronomers to measure distances within our solar system is the astronomical unit (AU). One AU is the average distance between Earth and the Sun, about 150 million km. Pluto averages about 40 AU from the Sun.

11. APPLICATION When light travels from one transparent medium into another, the rays bend, or refract. Snell's Law of Refraction states that

$$\frac{\sin\theta_1}{n_2} = \frac{\sin\theta_2}{n_1}$$

where θ_1 is the angle of incidence, θ_2 is the angle of refraction, and n_1 and n_2 are the indices of refraction for the two mediums, as shown in the diagram.

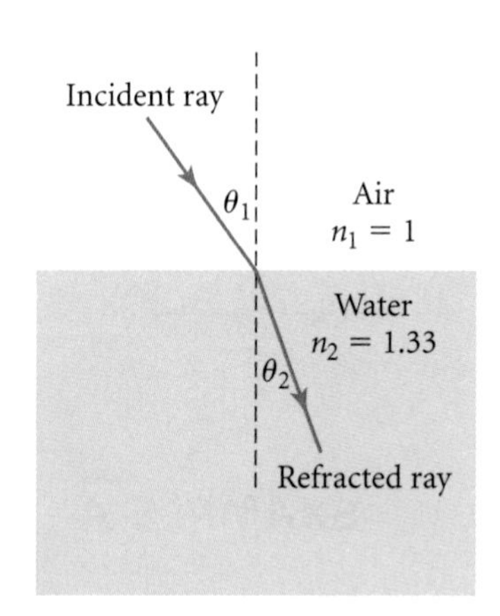

a. Find the angle of refraction in water if the angle of incidence from air is 60°. @

b. If the angle of refraction from air to water is 45°, at what angle did the ray enter the water?

c. If the angle of incidence is 0°, what is the angle of refraction? @

Review

12. Use the quadratic formula to solve each equation.

a. $2x^2 - 8x + 5 = 0$ @

b. $3x^2 + 4x - 2 = 7$

13. Use these three functions to find each value:

$f(x) = 2x + 7$

$g(x) = 2|x - 3| + 1$

$h(x) = x^2 - 5x$

a. $f(g(5))$

b. $g(h(2))$

c. $f(h(x))$

d. $h(f(x))$

14. The value of a building depreciates at a rate of 6% per year. When new, the building is worth $36,500.

a. How much is the building worth after 5 years 3 months?

b. To the nearest month, when will the building be worth less than $10,000?

15. Find the volume of the greenhouse at right. Round your answer to the nearest cubic foot.

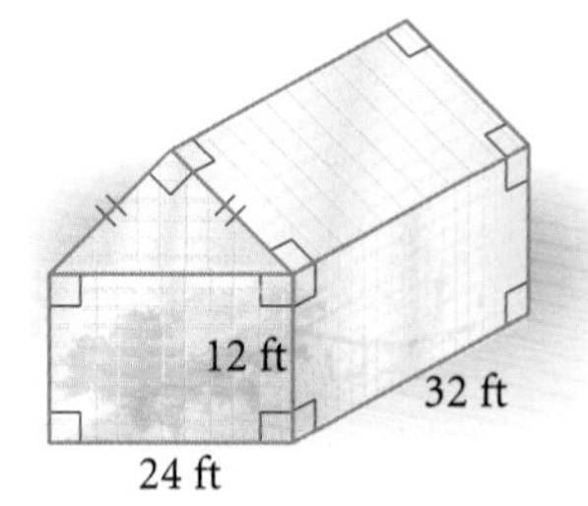

IMPROVING YOUR GEOMETRY SKILLS

A New Area Formula

You are given the lengths of two sides of a triangle, a and b, and the measure of the angle between them, C. How can you find the area of the triangle using only these three measurements?

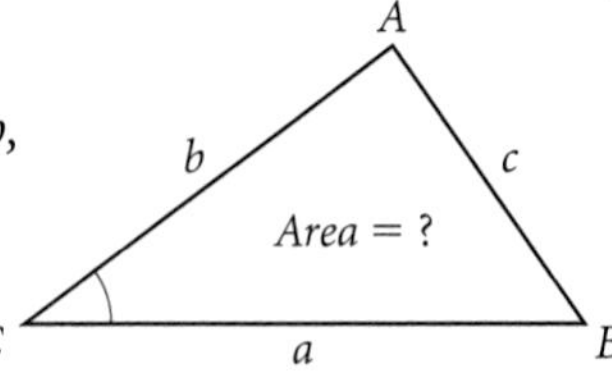

LESSON

12.3

The Law of Cosines

The Law of Sines enabled you to find the lengths of sides of a triangle or the measures of the angles in certain situations. To use the Law of Sines, you needed to know the measures of two angles and the length of any side, or the lengths of two sides and the measure of the angle opposite one of the sides. What if you know a different combination of sides and angles?

EXAMPLE A

Two hot-air balloons approach a landing field. One is 12 m from the landing point and the other is 17 m from the landing point. The angle between the balloons is 70°. How far apart are the two balloons?

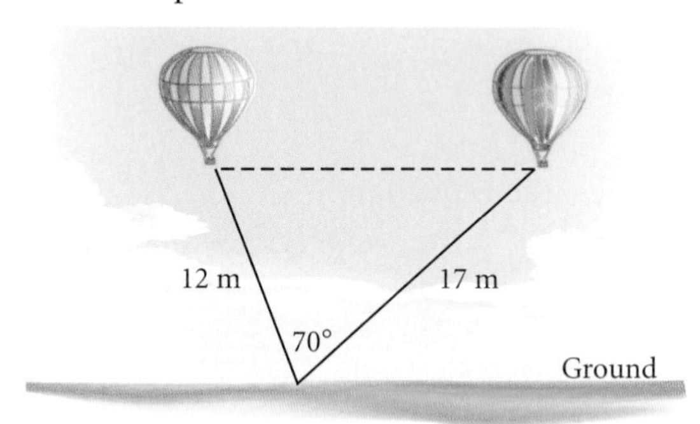

▸ Solution

In this case the Law of Sines does not help. The only angle you know is the included angle, or the angle between the two sides. If you try to set up an equation using the Law of Sines, you will always have more than one variable. So you must try something else.

Sketch one altitude to form two right triangles, so that one of the right triangles contains the 70° angle.

If you draw the altitude of the triangle from the balloon on the left, the 17 m side is split into two parts. Label one part p and the other part $17 - p$. Label the height h, and label the distance between the balloons d. You can now write two equations using the Pythagorean Theorem.

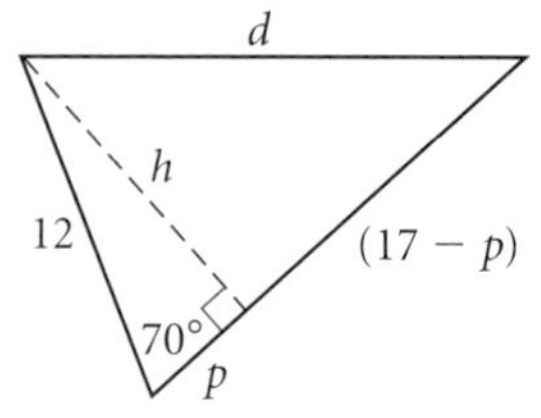

$$p^2 + h^2 = 12^2 \quad \text{and} \quad (17 - p)^2 + h^2 = d^2$$

You can multiply and expand the binomial in the second equation.

$$289 - 34p + p^2 + h^2 = d^2$$

Now you can rearrange the first equation and use substitution to solve the system of equations.

$h^2 = 144 - p^2$ — Solve $p^2 + h^2 = 144$ for h^2.

$289 - 34p + p^2 + (144 - p^2) = d^2$ — Substitute $144 - p^2$ for h^2 into the second equation.

$289 + 144 - 34p = d^2$ — Combine like terms.

Use the right triangle that contains the 70° angle to write

$$\cos 70° = \frac{p}{12}$$

Solve this equation for p to get $p = 12\cos 70°$, and substitute this value for p into the equation for d^2.

$$289 + 144 - 34(12\cos 70°) = d^2$$ Substitute $12\cos 70°$ for p.

$$\sqrt{289 + 144 - 34(12\cos 70°)} = d$$ Take the square root of both sides. You need only the positive root, because d represents a distance.

$$d \approx 17.1$$ Evaluate.

The distance between the two balloons is approximately 17.1 m.

You can repeat the procedure that is used in Example A any time you know the lengths of two sides of a triangle and the measure of the included angle and you need to find the length of the third side. Notice that you could also write the equation for d^2 as

$$d^2 = 17^2 + 12^2 - 2(17)(12)\cos 70°$$

which looks similar to the Pythagorean Theorem with an extra term that is twice the product of the length of the sides and the cosine of the angle between them. This modified Pythagorean relationship is called the **Law of Cosines.**

Law of Cosines

For any triangle with angles A, B, and C and sides of lengths a, b, and c (a is opposite $\angle A$, b is opposite $\angle B$, and c is opposite $\angle C$),

$$c^2 = a^2 + b^2 - 2ab\cos C$$

keymath.com/DAA

[► You can use the **Dynamic Algebra Exploration** at www.keymath.com/DAA to explore the Law of Cosines. ◄]

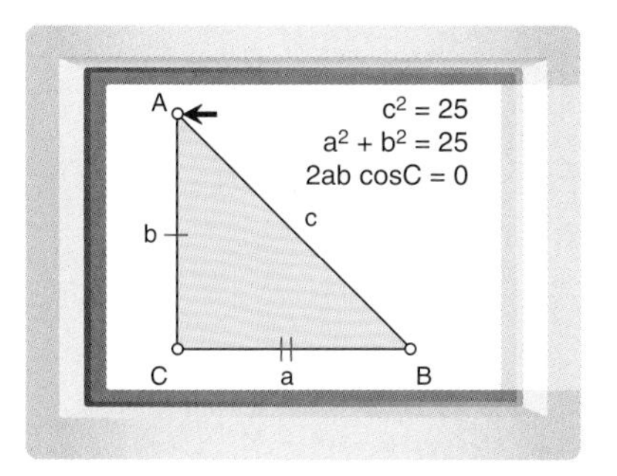

In this right triangle, c^2 is equivalent to $a^2 + b^2$.
$c^2 = a^2 + b^2$

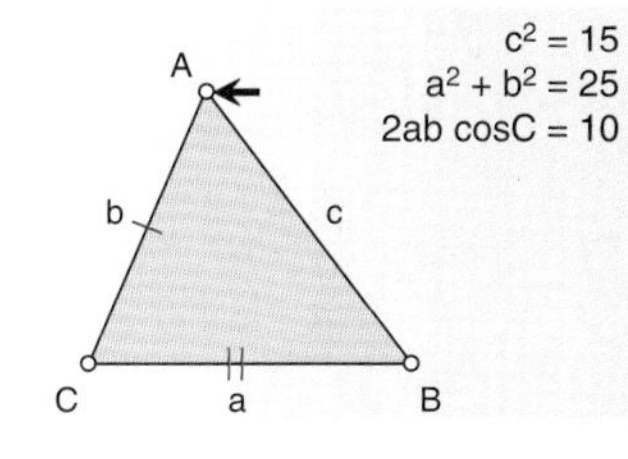

In this acute triangle, c^2 is less than $a^2 + b^2$. The difference is $2ab\cos C$.
$c^2 = a^2 + b^2 - 2ab\cos C$

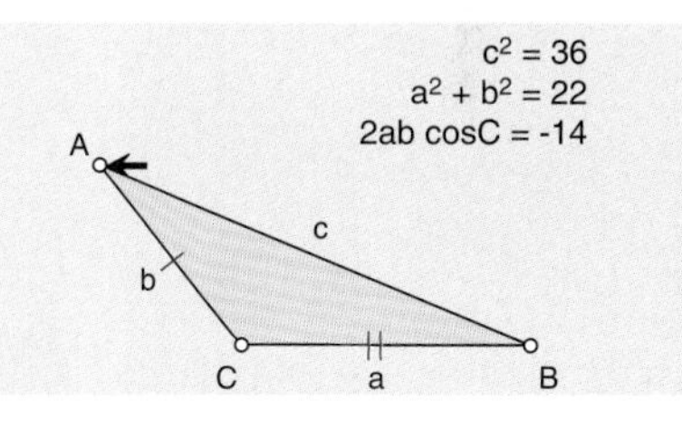

In this obtuse triangle, c^2 is more than $a^2 + b^2$. Again, the difference is $2ab\cos C$.
$c^2 = a^2 + b^2 - 2ab\cos C$

In the investigation you'll apply the Law of Cosines as you model a surveying problem.

Investigation
Around the Corner

You will need

- metersticks or a tape measure
- a protractor

The towns of Easton and Westville lie on opposite sides of a mountain. The townspeople wish to have a tunnel connecting the towns constructed through the mountain. A construction engineer positions herself so that she can see both towns. She plans to make some measurements and use trigonometry to determine the length of the proposed tunnel.

In this investigation you will simulate this situation. Position three members of your group so that two people are on opposite sides of a wall and the third person can see both of them. The first two group members represent the two towns, and the wall represents the mountain. The third member represents the engineer.

Find the distance between the two towns. Sketch an overhead view of the situation, show the measurements you make, and show your calculations.

You can use the Law of Cosines alone or in combination with other triangle properties.

EXAMPLE B

Find the unknown angle measures and side lengths.

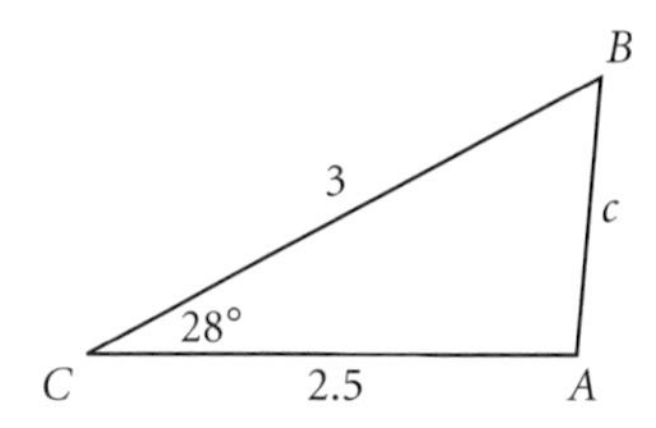

▶ Solution

First, use the Law of Cosines to find the length of $\overline{AB}$.

$c^2 = a^2 + b^2 - 2ab\cos C$	The Law of Cosines for finding c when a, b, and C are known.
$c^2 = 3^2 + 2.5^2 - 2(3)(2.5)\cos 28°$	Substitute 3 for a, 2.5 for b, and 28° for C.
$c^2 = 9 + 6.25 - 15\cos 28°$	Multiply.
$c = \sqrt{9 + 6.25 - 15\cos 28°}$	Take the positive square root of both sides.
$c \approx 1.42$	Evaluate.

The length of $\overline{AB}$ is approximately 1.42 units.

Now use the Law of Cosines to find the measure of $\angle A$.

$a^2 = b^2 + c^2 - 2bc\cos A$	The Law of Cosines for finding A when a, b, and c are known.
$3^2 \approx 2.5^2 + 1.42^2 - 2(2.5)(1.42)\cos A$	Substitute values for a, b, and c.
$9 \approx 8.2558 - 7.1\cos A$	Multiply.
$0.7442 \approx -7.1\cos A$	Subtract 8.2558 from both sides.
$\cos A \approx \frac{0.7442}{-7.1}$	Divide by -7.1.
$A \approx \cos^{-1}\left(\frac{0.7442}{-7.1}\right)$	Take the inverse cosine of both sides.
$A \approx 96°$	Evaluate.

Angle A measures approximately 96°.

To find the measure of the last angle, use the fact that the measures of the three angles in any triangle sum to 180°. The measure of $\angle B$ is approximately $180° - 28° - 96°$, or 56°.

During a calculation, it is best to use the entire previous answer for the next calculation. Rounding before the last step can reduce the accuracy of your answer. In Example B, you could find the measure of $\angle A$ with more accuracy by storing and using $\sqrt{9 + 6.25 - 15\cos 28°}$ for c instead of the approximation of 1.42. In all cases, you need to verify that the answers you get make sense in the context of the problem or in a sketch of the triangle.

In deciding whether to use the Law of Sines or the Law of Cosines, consider the triangle parts whose measurements you know and their relationships to each other.

Law of Sines	**Law of Cosines**
Side-Angle-Angle	Side-Angle-Side
Side-Side-Angle (ambiguous case)	Side-Side-Side

EXERCISES

You will need

A graphing calculator for Exercise **11**.

Practice Your Skills

1. Find the length of $\overline{AC}$. ⓐ

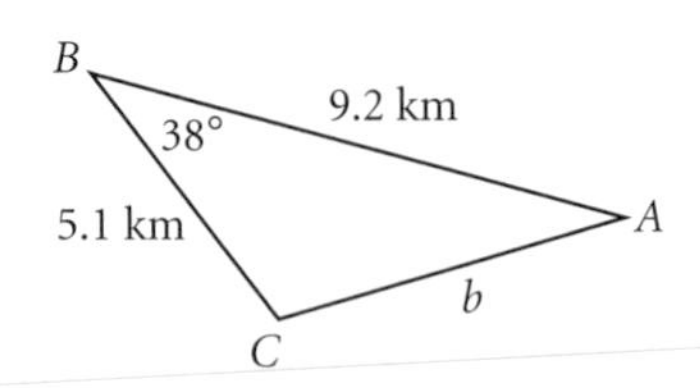

2. Find the measure of $\angle T$. ⓐ

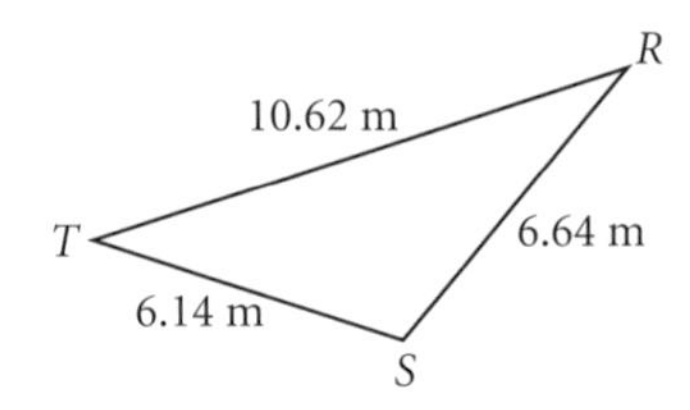

3. Solve for A and b. (Assume b is positive.)

a. $16 = 25 + 36 - 2(5)(6)\cos A$ ⓐ

b. $49 = b^2 + 9 - 2(3)(b)\cos 60°$

4. Find all of the unknown angle measures and side lengths.

a. ⓐ

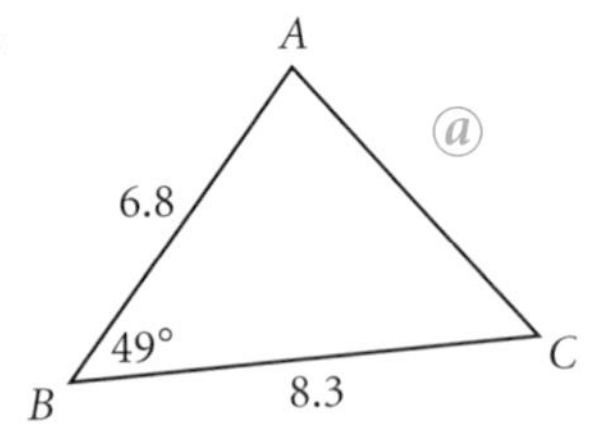

b. ⓐ

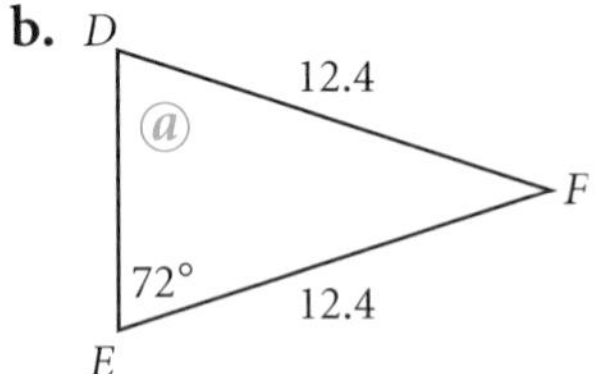

c.

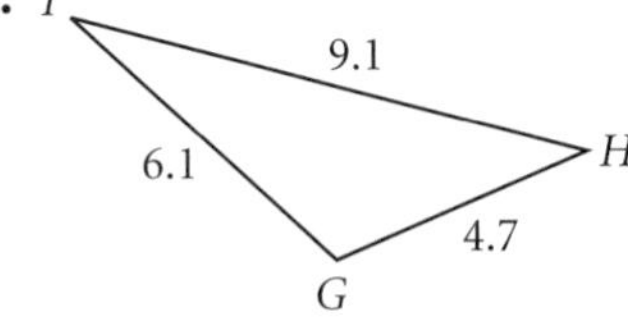

Reason and Apply

5. Two airplanes pass over the same point in Chicago, Illinois, at the same time. Plane A is flying on a heading of 105° clockwise from north. Plane B is flying on a heading of 260° clockwise from north. After 2 hours, Plane A is 800 miles from Chicago and Plane B is 900 miles from Chicago. How far apart are the planes at this time? ⓗ

Pilots use the control panels of this passenger aircraft to follow air traffic control commands, to adjust for wind currents, and to avoid the airspace of other planes.

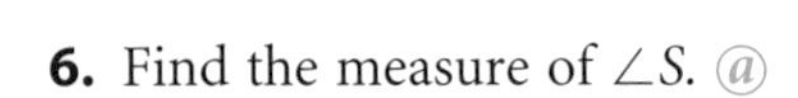

6. Find the measure of $\angle S$. ⓐ

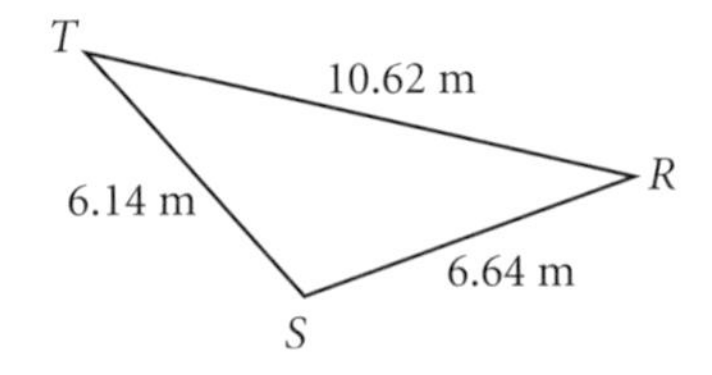

7. APPLICATION Seismic exploration identifies underground phenomena, such as caves, oil pockets, and rock layers, by transmitting sound into the earth and timing the echo of the vibration. From a sounding at point A, a "thumper" truck locates an underground chamber 7 km away. Moving to point B, 5 km from point A, the truck takes a second sounding and finds the chamber is 3 km away from that point. Assume that the underground chamber lies in the same vertical plane as A and B. What more can you say about the location of the underground chamber? ⓗ

Science CONNECTION

Sources of oil are often found by seismic exploration. At one time, seismologists used explosives to generate shock waves, but now they use "thumper" trucks that send controlled vibrations through the ground. Because seismic exploration may be disruptive to a particular environment, oil explorers sometimes use gravitational or magnetic exploration to determine the composition of rock formations based upon Earth's natural gravity or magnetic fields.

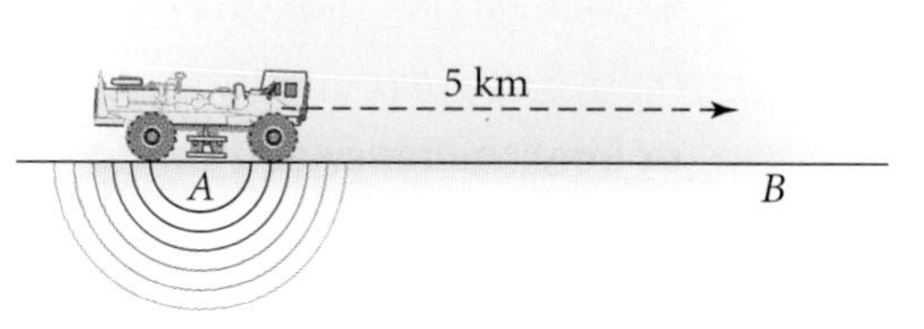

8. A folding chair's legs meet to form a 50° angle. The rear leg is 55 cm long and attaches to the front leg at a point 75 cm from the front leg's foot. How far apart are the legs at the floor? ⓐ

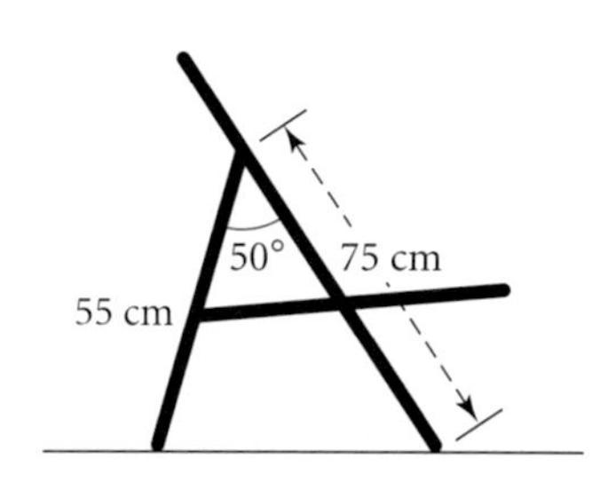

American designer Charles Eames (1907–1978) used mathematics to design this chair, which is suitable for sitting or sprawling.

9. APPLICATION Triangulation is used to locate airplanes, boats, or vehicles that transmit radio signals. The distances to the vehicle are found and the directions calculated by measuring the strength of the signal at three fixed receiving locations. Receiver B is 18 km due east of Receiver A. Receiver C is 20 km due north of Receiver A.

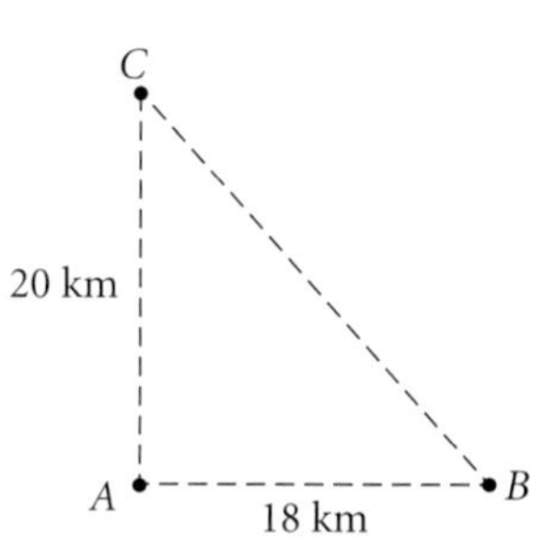

a. The signal from a source vehicle to Receiver A indicates a distance of 15 km. Make a sketch showing the possible locations of the source vehicle.

b. The signal from the source vehicle to Receiver B indicates a distance of 10 km. Add this information to your sketch from 10a, and describe the possible locations of the source vehicle.

c. Receiver C receives a signal from the source vehicle, but due to a malfunction in its equipment, it cannot determine the distance. However, Receiver C has a range of only 20 km. Use your sketch to determine the location of the source vehicle.

d. Determine the angle between the source vehicle, Receiver A, and Receiver C, where Receiver A is the vertex.

Review

10. Find the equation of the parabola that passes through the points $(-2, -20)$, $(2, 0)$, and $(4, -14)$.

11. Here are the batting averages of the National League's Most Valuable Players from 1985 to 2005. (*The New York Times Almanac 2007*)

{.353, .290, .287, .290, .291, .301, .319, .311, .336, .368, .319, .326, .366, .308, .319, .334, .350, .370, .359, .362, .335}

a. Make a box plot of these data.

b. Give the five-number summary.

c. Find the range and interquartile range.

d. Find the mean and standard deviation.

12. Find the total surface area of the figure below. Round your answer to the nearest square centimeter. ⓐ

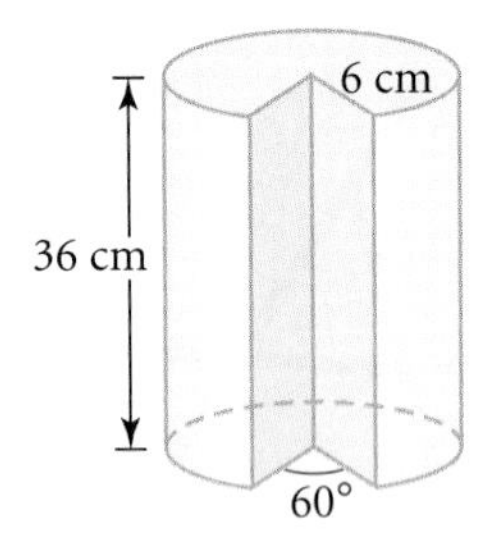

13. Congruent circles A and B are tangent at point C. The radius of each circle is 3 units. Rays RU and ST are tangent at U and S, respectively, and intersect at T. Find ST. ⓗ

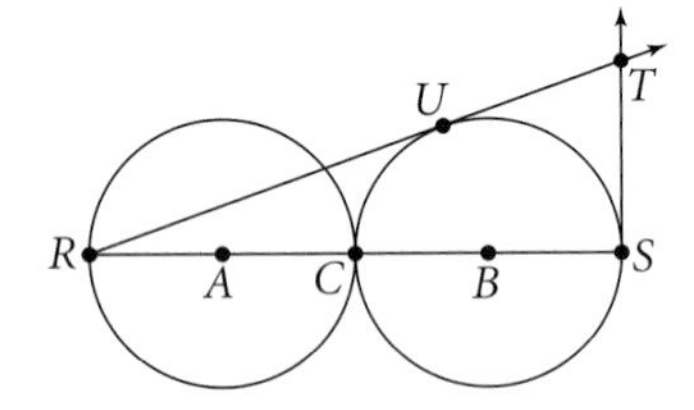

LESSON

12.4

Extending Trigonometry

In Lesson 12.1, the definitions given for the sine, cosine, and tangent ratios applied to acute angles in right triangles. In Lessons 12.2 and 12.3, you solved problems using sines and cosines of angles in oblique triangles. In this lesson you'll extend the definitions of trigonometric ratios to apply to any size angle.

Start by placing the angles on the coordinate plane. When measuring angles on a clock or a compass, you begin at the top and move around the circle to the right (clockwise). For measuring angles on the coordinate plane, however, you begin at the right and move in the opposite direction. You might think of beginning in Quadrant I and moving in order through Quadrants II, III, and IV.

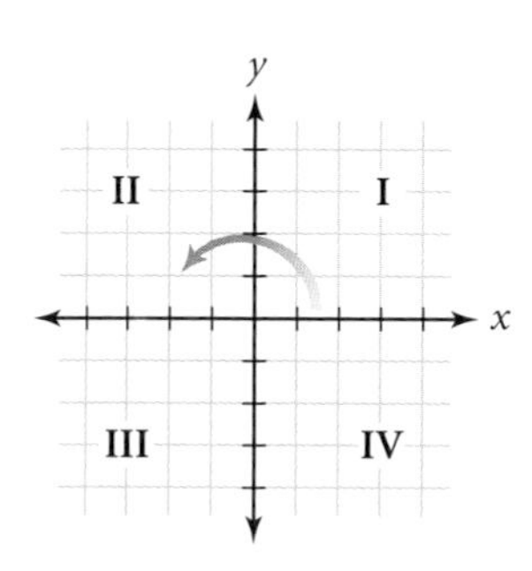

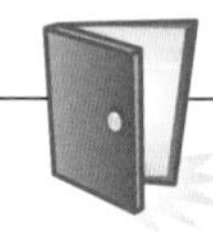

Investigation
Extending Trigonometric Functions

You will need

- graph paper
- a protractor

In this investigation you'll learn how to calculate the sine, cosine, and tangent of non-acute angles on the coordinate plane.

Procedure Note

1. Draw a point on the positive x-axis. Rotate the point counterclockwise about the origin by the given angle measure and draw the image point. (If the angle measure is negative, rotate clockwise.) Then connect the image point to the origin. The angle between the segment and the positive x-axis, in the direction of rotation, represents the amount of rotation.
2. Use your calculator to find the sine, cosine, and tangent of this angle.
3. Estimate the coordinates of the rotated point.
4. Use the distance formula to find the length of the segment.

Step 1 Follow the Procedure Note for each angle measure given below. An example is shown for 120°.

a. 135°

b. 210°

c. 270°

d. 320°

e. −100°

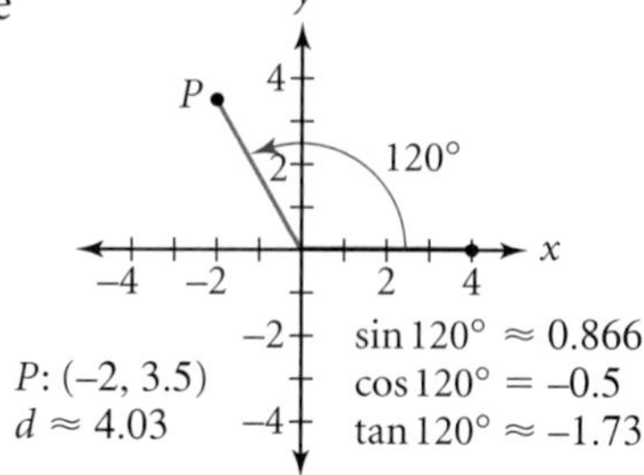

Step 2	Experiment with the estimated x- and y-coordinates and the segment length to find a way to calculate the sine, cosine, and tangent. (Your values are all estimates, so just try to get close.)
Step 3	Plot the point $(-3, 1)$ and draw a segment from it to the origin. Label as A the angle between the segment and the positive x-axis. Use your method from Step 2 to find the values of $\sin A$, $\cos A$, and $\tan A$. What happens when you try using the inverse $\sin^{-1}$ to find the value of A? What happens for $\cos^{-1}$ and $\tan^{-1}$?
Step 4	Now consider the general case of the angle θ between the positive x-axis and a segment connecting the origin to the point (x, y). Give definitions for the values of $\sin\theta$, $\cos\theta$, and $\tan\theta$.

Suppose you know that $\sin B \approx 0.47$. That information isn't enough to determine whether B is about 28°, 151°, or even −208°. You need additional information to determine the measure of angle B.

sin(28.0343)	0.470000053382
sin(151.9657)	0.470000053382
sin(-208.0343)	0.470000053382

In Step 3 of the investigation, the point at $(-3, 1)$ had a sine of $\frac{1}{\sqrt{10}}$, a cosine of $\frac{-3}{\sqrt{10}}$, and a tangent of $\frac{1}{-3}$.

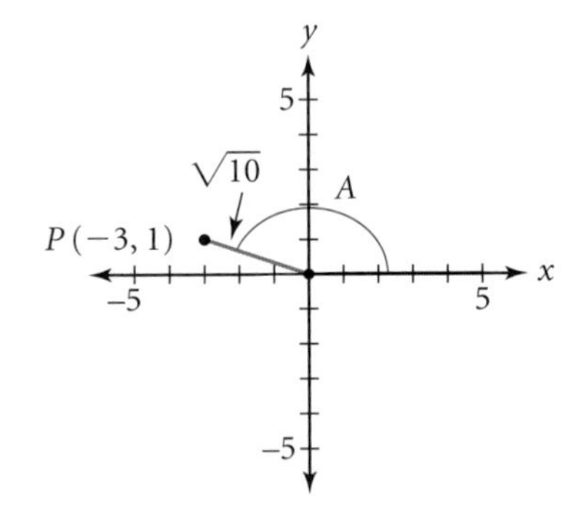

The calculator gives three different angles for the three inverses. Only one corresponds to the angle from Step 3.

$\sin^{-1}\left(\frac{1}{\sqrt{10}}\right)$	18.4349488229
$\cos^{-1}\left(\frac{-3}{\sqrt{10}}\right)$	161.565051177
$\tan^{-1}\left(\frac{1}{-3}\right)$	-18.4349488229

You can use a graph to find the angle. Create a right triangle by drawing a vertical line from the end of the segment to the x-axis. Use right triangle trigonometry to find the measure of the angle with its vertex at the origin. The acute angle in this **reference triangle,** labeled B, is called the **reference angle.** Use the measure of the reference angle to find the angle you are looking for. In this case, angle B has measure 18.435°. Angle A measures $180° - 18.435°$, or 161.565°.

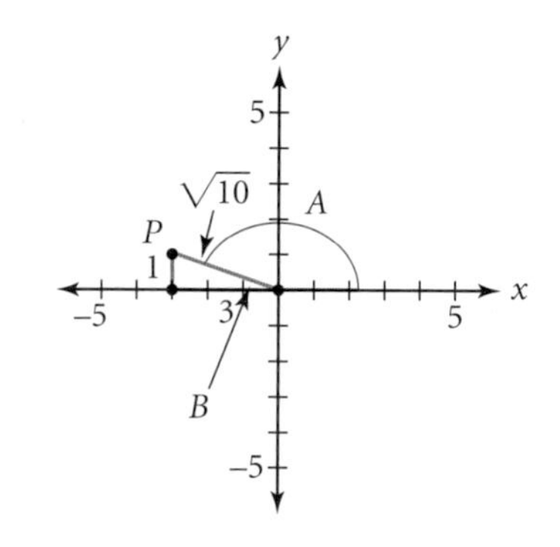

EXAMPLE A | Find sin 300° without a calculator.

▶ Solution

Rotate a point counterclockwise 300° from the positive x-axis. The image point is in Quadrant IV, 60° below the x-axis. The reference angle is 60°. The sine of a 60° angle is $\frac{\sqrt{3}}{2}$. The sine of an angle is the y-value divided by the distance from the origin to the point. Because the y-value is negative in Quadrant IV, $\sin 300° = -\frac{\sqrt{3}}{2}$.

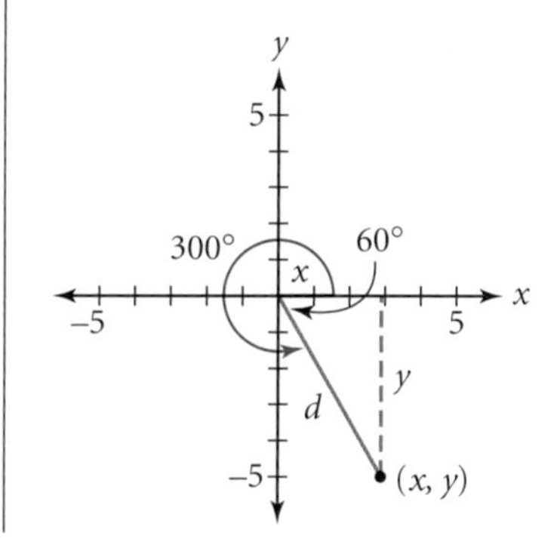

You can generalize this process to find the sine, cosine, and tangent of any angle.

Trigonometric Functions for Real Angle Values

Suppose angle θ is formed by rotating a ray along the positive x-axis about the origin. Point (x, y) on the ray is d units from the origin, where $d = \sqrt{x^2 + y^2}$. Then $\sin\theta = \frac{y}{d}$, $\cos\theta = \frac{x}{d}$, and $\tan\theta = \frac{y}{x}$.

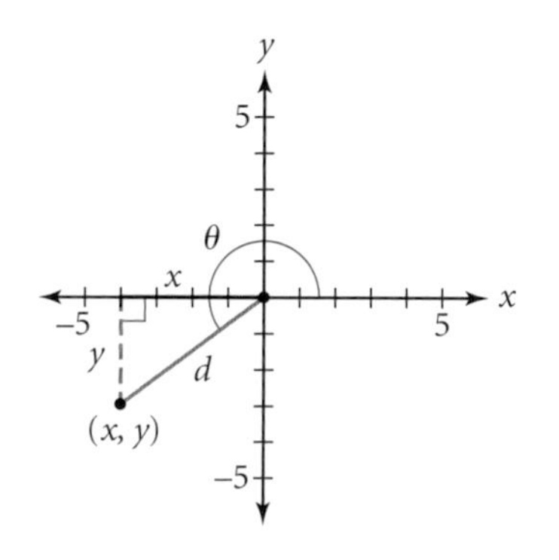

EXAMPLE B What measure describes an angle, measured counterclockwise, from the positive x-axis to the ray from the origin through $(-4, -3)$?

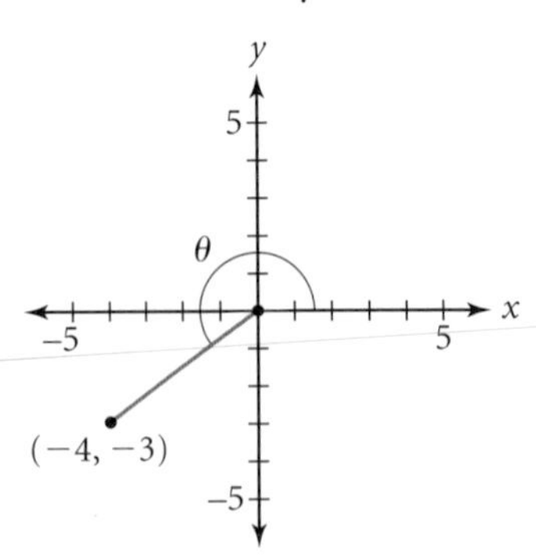

▶ Solution A reference triangle constructed with a perpendicular to the x-axis has sides 3-4-5. The reference angle has measure $\tan^{-1}\left(\frac{3}{4}\right) \approx 36.9°$. The graph shows that θ is more than 180°, so $\theta \approx 180° + 36.9°$, or 216.9°.

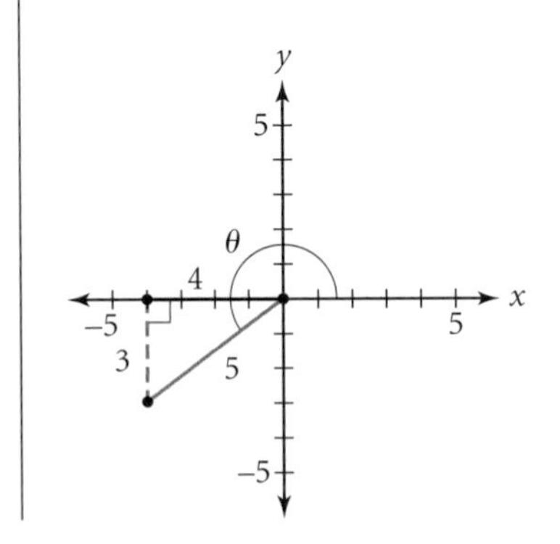

You've seen in this lesson that when you need to find the measure of an angle that is larger than 90° or negative, you will have to decide whether to add to or subtract from 180° or 360°. Your choice will depend on the quadrant of the rotated ray or point and whether the angle measure is positive or negative.

EXERCISES

You will need

A graphing calculator for Exercise **15.**

Practice Your Skills

1. Find the trigonometric value requested for each angle.

a. $\sin\theta$ ⓐ

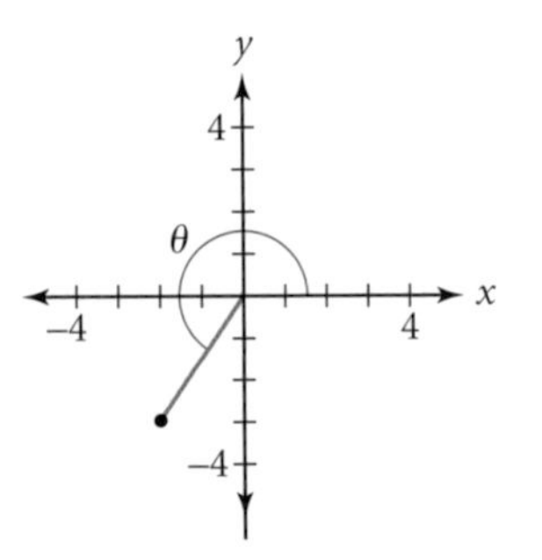

b. $\tan\alpha$

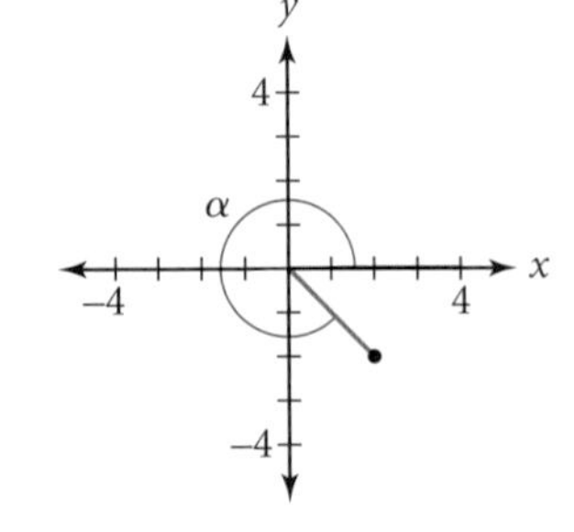

c. $\cos\beta$

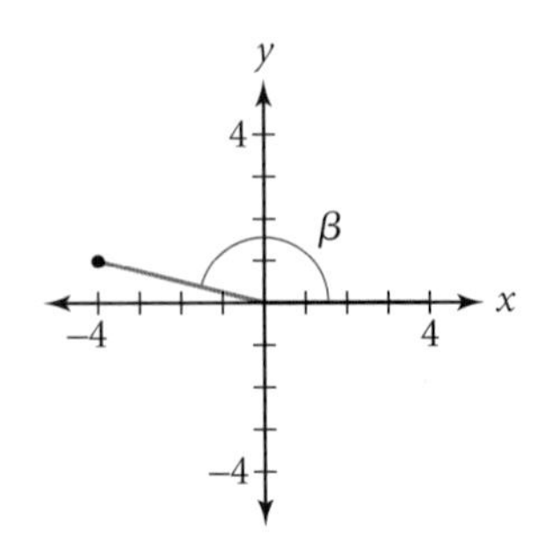

2. For each graph in Exercise 1, determine whether it represents an angle that is ⓐ

i. between 0° and 90°
ii. between 90° and 180°
iii. between 180° and 270°
iv. between 270° and 360°

3. Sketch each angle on the coordinate plane.

a. 280° ⓐ **b.** 101° **c.** 222° **d.** −135°

4. Find the measure of the reference angle for each angle.

a. 280° ⓐ **b.** 101° **c.** 222° **d.** −135°

5. Without using a calculator, determine whether each value is positive or negative.

a. sin 280° ⓐ **b.** cos 280° **c.** sin 101°
d. tan 101° **e.** cos 222° **f.** sin −135°

Recreation
CONNECTION

Skateboarders, snowboarders, and stunt bicyclists all do tricks that involve rotations of greater than 180°. The conservation of angular momentum, the force of gravity, Newton's laws of motions, and other physics concepts can be used to explain how the athletes are able to propel themselves and do aerial flips and twists.

To celebrate the launch of Einstein Year at the Science Museum of London, Cambridge University physicist Helen Czerski worked with Professional BMX rider Ben Wallace to design a new stunt, which they named the "Einstein Flip." Wallace performed the stunt, which involved a 360° backward rotation, for the first time in January 2005.

Reason and Apply

6. Use trigonometry to find the measure of each angle in Exercise 1. (*Note:* Assume the measures are between 0° and 360°.) ⓐ

7. Robyn stands 80 m east of Nathan. Keshon stands directly north of Robyn. Nathan measures the angle between Robyn and Keshon as 28°. How far is Nathan from Keshon? ⓗ

8. Amnah stands 200 yd east of Warren, and Jana stands 100 yd north and 30 yd west of Warren. From Warren, what is the angle between Amnah and Jana? ⓐ

9. Ryan stands 50 m west and 20 m north of Candice, and Kyle stands 40 m east and 90 m south of Candice. From Candice, what is the angle between Ryan and Kyle?

10. A point starts at the coordinates (6, 0) and rotates counterclockwise 140° about the origin. What are its new coordinates? ⓐ

11. A point starts at the coordinates (−8, 2) and rotates counterclockwise 100° about the origin. What are its new coordinates? ⓗ

12. A point has a y-coordinate of −4, and the cosine of the angle from the positive x-axis is −0.7. What is the x-coordinate?

13. *Mini-Investigation* Follow these steps to explore the relationship between the tangent ratio and the slope of a line. For 13a and b, find $\tan\theta$ and the slope of the line.

a.

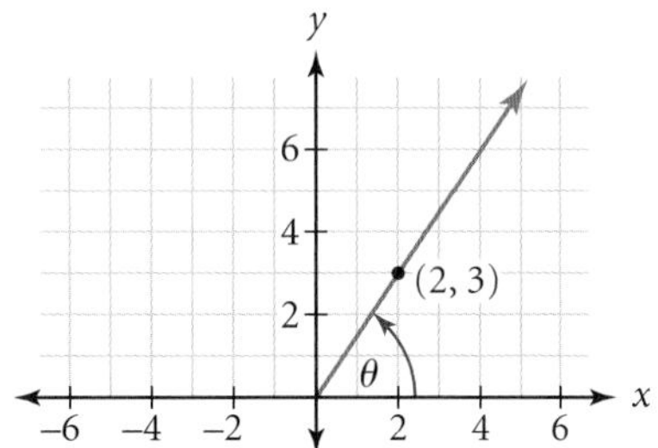

b.

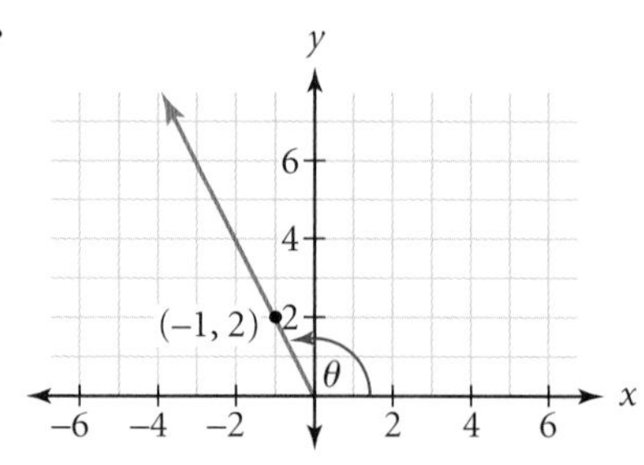

c. What is the relationship between the tangent of an angle in standard position, and the slope of its nonhorizontal side?

d. Find the slope of this line.

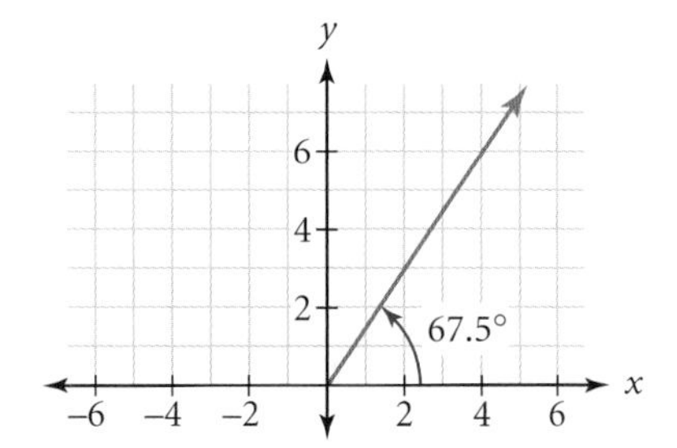

Review

14. Find the equation of the line through (−3, 10) and (6, −5).

15. Consider the function $f(x) = 3 + \sqrt[3]{(x-1)^2}$. ⓐ

a. Find

i. $f(9)$
ii. $f(1)$
iii. $f(0)$
iv. $f(-7)$

b. Find the equation(s) of the inverse of $f(x)$. Is the inverse a function?

c. Describe how you can use your calculations in 15a to check your inverse in 15b.

d. Use your calculator to graph $f(x)$ and its inverse on the same axes.

16. Find a polynomial equation of least degree with integer coefficients that has roots -3 and $\left(\frac{1}{2} - \sqrt{3}\right)$.

17. Write an equation for the ellipse at right.

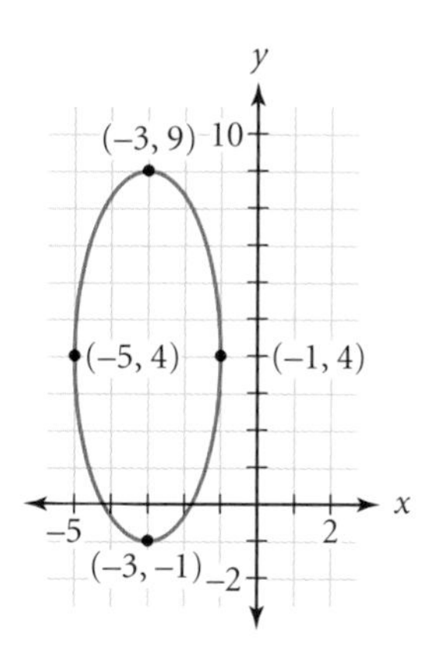

18. APPLICATION The Hear Me Now Phone Company plans to build a cell tower to serve the needs of Pleasant Beach and the beachfront. It decides to locate the cell tower so that Pleasant Beach is 1 mi away at an angle of 60° clockwise from north. The range of the signal from the cell tower is 1.75 mi. The beachfront runs north to south. How far south of Pleasant Beach will customers be able to use their cell phones?

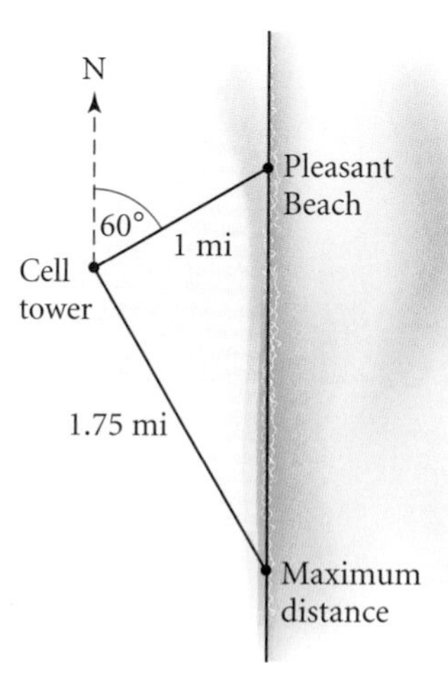

Technology CONNECTION

The placement of cell towers is crucial to providing a variety of cellular services. Mathematical models are used to analyze possible sites. Cell towers are located by individual companies based on their own business plan for the market they serve. For greatest cost efficiency, cellular companies consider the physical design, topography, population density, environmental impact, engineering, and aesthetics of their towers.

This cell phone tower is located in Jonesboro, Georgia.

LESSON 12.5

Introduction to Vectors

Arriving at one point is the starting point to another.
JOHN DEWEY

Many of the quantities you work with in mathematics, such as those representing area, volume, and money in a bank account, are measures, or counts. Other quantities involve both a measure and a direction. For example, distances often have directions associated with them. Other examples include velocity and acceleration.

Directed quantities like these are often represented by vectors. A **vector** can be thought of as a directed line segment. It has a length, or **magnitude,** and a **direction.** A vector is represented by a segment with an arrowhead on one end The end with the arrowhead is called the **head** or **tip,** and the other end is the **tail.**

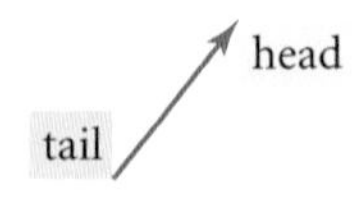

In contrast to vectors, numbers without an associated direction are called **scalars.**

Vectors can be represented in several ways. One way gives the magnitude and the angle the vector makes with the positive x-axis. This is called the **polar form** of the vector. For example, the polar form $\langle 2 \angle 60° \rangle$ represents a vector 2 units long directed at an angle of 60° counterclockwise from the positive x-axis.

Another way to describe a vector is to give the horizontal and vertical change from the tail to the head. This is called the **rectangular form** of the vector. When the tail of a vector is at the origin, then its rectangular form is the same as the coordinates of its head. The rectangular form of the vector at right would be designated as $\langle 1, \sqrt{3} \rangle$.

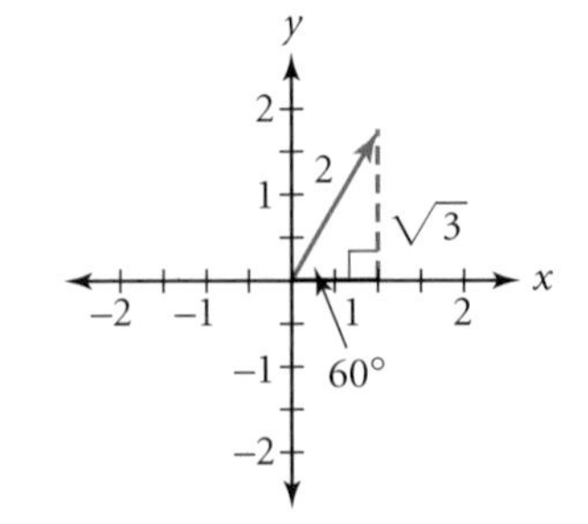

Vectors are **equivalent** if they have the same length and point in the same direction. The location of the vector on the coordinate plane doesn't matter. All of the vectors in the picture at right are equivalent. Each could be described by either $\langle 2 \angle 60° \rangle$ or $\langle 1, \sqrt{3} \rangle$.

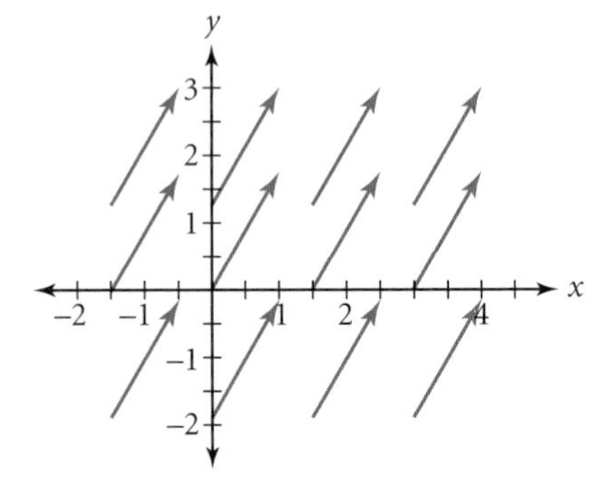

Just as you can add and subtract numbers without direction, or scalars, you can add and subtract vectors. Vector addition can be accomplished geometrically by placing two vectors in a "tip to tail" arrangement. The sum, or **resultant** vector, is the vector that connects the tail of the first vector to the head of the final vector. The figure at right shows the addition $\vec{a} + \vec{b} = \vec{c}$.

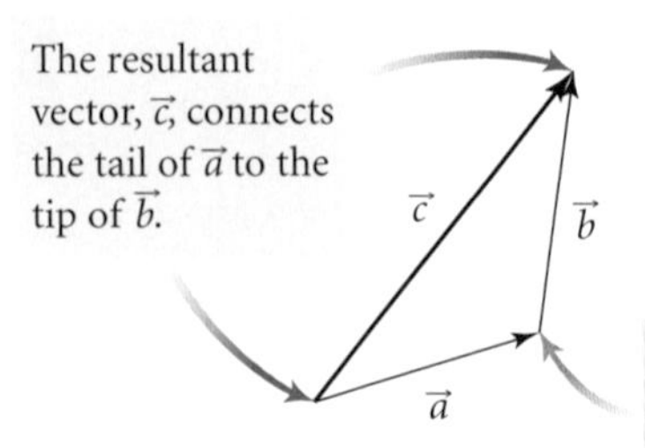

When writing by hand, you use arrows above the letters to refer to vectors. In printing, as in this book, vectors are often designated by boldface type without the arrows, as in **a** + **b** = **c.**

In the investigation you will explore some of the properties of vector addition and subtraction.

Investigation
Vector Addition and Subtraction

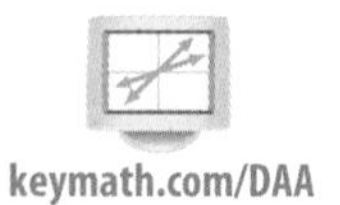
keymath.com/DAA

You will need

- graph paper

You will use these vectors in this investigation.

a = ⟨2, 3⟩ **b** = ⟨4, 1⟩ **d** = ⟨−3, 2⟩ **e** = ⟨3, −1⟩ **f** = ⟨−1, −2⟩

Step 1 On graph paper, draw a set of axes and the vector **a** = ⟨2, 3⟩. Remember to draw an arrowhead at the head, or tip, of the vector.

Step 2 Add the vector **b** = ⟨4, 1⟩ to **a.** Draw **b** so that the tail of **b** starts at the tip of **a.** The tip of **b** should be 4 units to the right and 1 unit up from its tail. Don't forget the arrowhead at the tip.

Step 3 Draw the sum, or resultant vector, **c.** What is its rectangular form?

Step 4 Repeat Steps 1–3 to complete these vector sums:

i. b + **a** **ii. d** + **e** **iii. b** + **f** **iv. a** + **e**

Step 5 Look at the rectangular form of the resultant vectors. Complete the following definition of vector addition for vectors in rectangular form.

If **a** = $\langle a_1, a_2\rangle$ and **b** = $\langle b_1, b_2\rangle$, then the sum **a** + **b** is $\langle a_1, a_2\rangle + \langle b_1, b_2\rangle$ = ⟨ ? , ? ⟩.

Step 6 Subtracting a number is the same as adding its opposite. It's the same for vectors. The opposite of vector **b** is called −**b.** It has the same magnitude as **b,** but it points in the opposite direction. The difference **a** − **b** is the same as the sum **a** + −**b.**

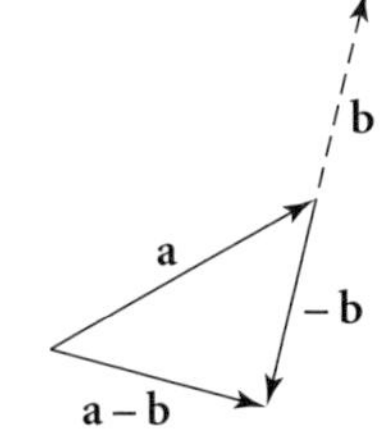

Draw a representation of each difference.

i. a − **b** **ii. b** − **a**
iii. d − **e** **iv. e** − **f**

Step 7 Based on your drawings, complete the following definition of vector subtraction for vectors in rectangular form.

If **a** = $\langle a_1, a_2\rangle$ and **b** = $\langle b_1, b_2\rangle$, then the difference **a** − **b** is $\langle a_1, a_2\rangle - \langle b_1, b_2\rangle$ = ⟨ ? , ? ⟩.

Step 8 Create a conjecture about multiplying a vector by a scalar (number). For example, what would it mean to multiply 2 · **a**? (*Hint:* This is the same as adding **a** + **a.**) Complete the following definition of scalar multiplication.

If **a** = $\langle a_1, a_2\rangle$ and k is a scalar, then the product k · **a** is $k \cdot \langle a_1, a_2\rangle$ = ⟨ ? , ? ⟩.

Step 9 | The magnitude (length) of a vector is symbolized by placing the vector name inside vertical bars, like an absolute-value sign. Find the magnitudes of **a** and **b**, then complete the following definition of the magnitude of a vector in rectangular form.

If $\mathbf{a} = \langle a_1, a_2 \rangle$, then the magnitude of **a**, $|\,\mathbf{a}\,|$, is __?__.

Vectors are useful for representing situations involving motion.

EXAMPLE A

Ernie drives a taxi in a large city in which the streets are laid out in a square grid. One night he starts out from the garage and travels along this route:

- 5 blocks east and 2 blocks north
- 3 blocks west and 7 blocks north
- 6 blocks east and 8 blocks south
- 9 blocks east and 12 blocks north
- 10 blocks west and 4 blocks south

Where is Ernie relative to the garage at the end of this trip?

You could draw all of this on graph paper, but it is easier to think of each of the five trip segments as a vector.

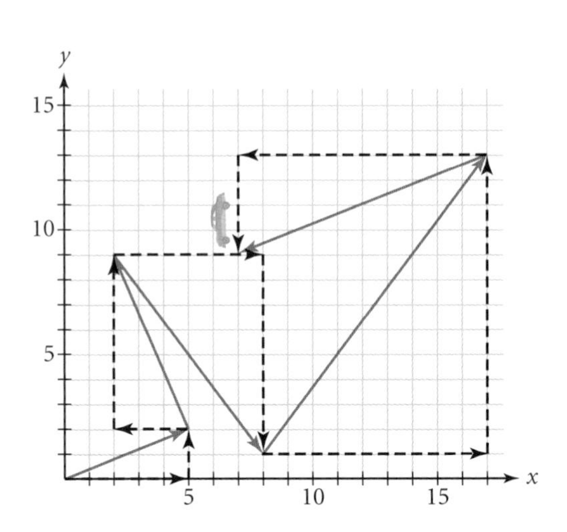

Represent the first trip segment as a vector: Five blocks east and 2 blocks north is the same as the vector $\langle 5, 2 \rangle$.

Similarly, the other trips can be represented by the vectors $\langle -3, 7 \rangle$, $\langle 6, -8 \rangle$, $\langle 9, 12 \rangle$, and $\langle -10, -4 \rangle$. Finding the sum of all five vectors gives the final position:

$$\langle 5 - 3 + 6 + 9 - 10, 2 + 7 - 8 + 12 - 4 \rangle = \langle 7, 9 \rangle.$$

Ernie ends up 7 blocks east and 9 blocks north of the garage.

When traveling in a city with streets and buildings, it may be convenient to consider the rectangular form of the vector. But to represent movement in open space, you may want to use polar form to show the distance and the direction (angle) of the movement. You can use trigonometry to convert from one form to the other.

EXAMPLE B Convert the rectangular form, $\langle 4, 9\rangle$, of a vector to its polar form.

▶ Solution

You need to find both the length, or magnitude, of the vector and its angle with the horizontal. Draw a diagram. The vector is the hypotenuse of a right triangle with legs of lengths 4 and 9. To find its magnitude, use the Pythagorean Theorem.

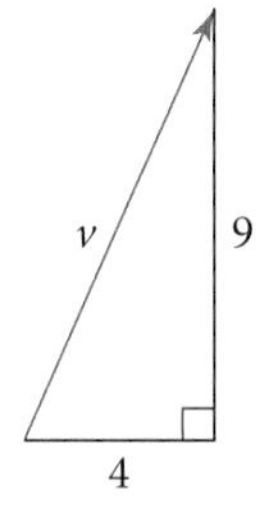

$$4^2 + 9^2 = |v|^2$$
$$16 + 81 = |v|^2$$
$$|v|^2 = 97$$
$$|v| = \sqrt{97} \approx 9.849$$

To find the vector's angle, you can use the inverse tangent: $\tan^{-1}\left(\frac{9}{4}\right) \approx 66.038°$.

So the polar form of the vector is $\langle 9.849 \angle 66.038°\rangle$.

You can use vectors to model the paths of boats and airplanes. In some cases the heading, or direction of motion, is not given by the angle with the horizontal but rather as a **bearing,** or angle clockwise from north. For example, a bearing of 28° refers to a 28° angle measured clockwise from north. Often a distance and a bearing are the most useful representation of a vector in these cases. But if you need to add or subtract vectors, you will need to convert to the rectangular form.

EXAMPLE C A ship leaves port and travels 47 miles on a bearing of 28° to get out of a bay. It then turns and travels 94 miles on a bearing of 137° to reach its destination port. Find the ship's distance and bearing from port.

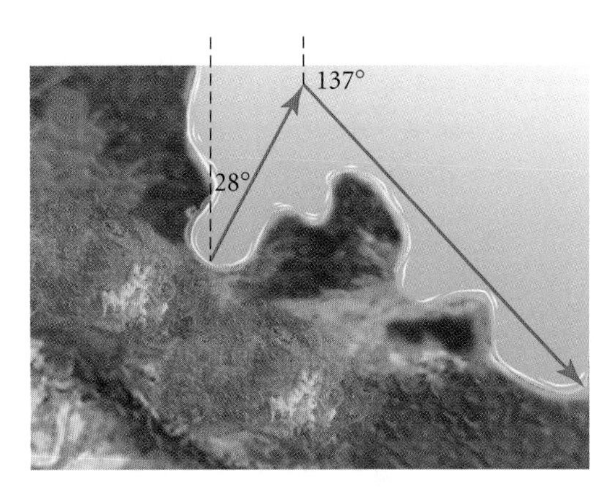

▶ Solution

You can add the two vectors to obtain a single vector from the initial port to the destination, but you will first need to use trigonometry to convert the vectors from polar to rectangular form.

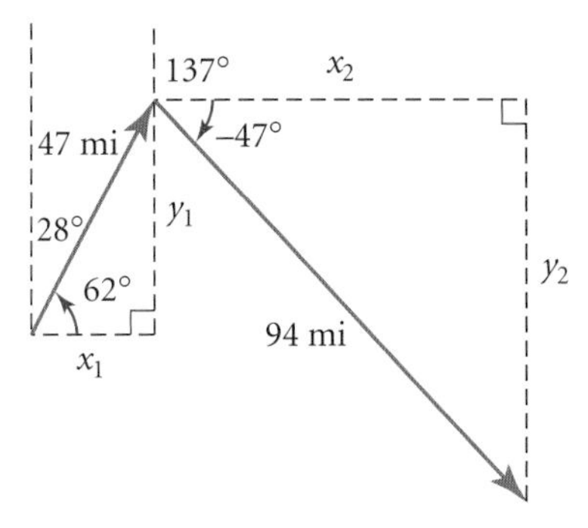

Can you explain where the 62° and the −47° came from?

Use trigonometry to solve for the horizontal and vertical distances.

$$\cos 62° = \frac{x_1}{47} \qquad \sin 62° = \frac{y_1}{47}$$
$$x_1 = 47\cos 62° \qquad y_1 = 47\sin 62°$$
$$x_1 \approx 22.07 \text{ mi} \qquad y_1 \approx 41.50 \text{ mi}$$

$$\cos(-47°) = \frac{x_2}{94} \qquad \sin(-47°) = \frac{y_2}{94}$$

$$x_2 = 94\cos(-47°) \qquad y_2 = 94\sin(-47°)$$

$$x_2 \approx 64.11 \text{ mi} \qquad y_2 \approx -68.75 \text{ mi}$$

So the two vectors are $\langle 22.06, 41.5\rangle$ and $\langle 64.11, -68.75\rangle$.

The rectangular form of the vector between the two ports is the sum of these vectors: $\langle 86.17, -27.25\rangle$.

To represent the resultant vector in the same form as the original vectors, you can first find its length by applying the Pythagorean Theorem: $|d| = \sqrt{(86.17)^2 + (-27.25)^2} \approx 90.38$.

The vector has an angle of $\tan^{-1}\left(\frac{-27.25}{86.17}\right) \approx -17.55°$, so its bearing is 107.55°. The ship is about 90 miles from port at a bearing of about 108°.

Converting between angle measures and bearing can be confusing. You'll find it helpful to draw a reference diagram.

Exercises

Practice Your Skills

1. Let $\mathbf{a} = \langle 2, 5\rangle$, $\mathbf{b} = \langle -4, 0\rangle$, $\mathbf{c} = \langle 6, -7\rangle$. Evaluate each of these expressions.

a. $\mathbf{a} + \mathbf{b}$ (a) **b.** $\mathbf{b} - \mathbf{c}$ **c.** $3\mathbf{a}$ (a)

d. $2\mathbf{b} + 4\mathbf{c}$ **e.** $|\mathbf{a}|$ (a)

2. Write each vector in polar form.

a. $\langle 1, \sqrt{3}\rangle$ (a) **b.** $\langle 4, 0\rangle$

c. $\langle -2, 2\rangle$ (a) **d.** $\langle -\sqrt{3}, -1\rangle$

3. Write each vector in rectangular form.

a. $\langle 5 \angle 30°\rangle$ **b.** $\langle 2 \angle 135°\rangle$ (a)

c. $\langle 6 \angle 180°\rangle$ **d.** $\langle 4 \angle 240°\rangle$ (a)

4. Convert each set of directions to a vector in the form specified.

a. Move 4 blocks east and 5 blocks north in rectangular form. (a)

b. Move 7 miles west and 3 miles north in rectangular form.

c. Move 8 miles at an angle of 30° north of west in polar form. (a)

d. Move 250 km on a bearing of 240° in polar form.

Reason and Apply

5. A plane flies 250 miles on a bearing of 80° and then turns to fly another 185 miles on a bearing of 20°. What are the distance and bearing from the plane's starting point to its final location? (h)

6. Joy set out to visit a new art gallery. After calling for directions, she walked 4 blocks east and then 2 blocks north. She then realized that she should have walked 2 blocks east and 4 blocks north. What path should she follow to get to her destination? Show how to find this answer using vector subtraction. ⓐ

7. Two forces are pulling on an object. Force A has magnitude 50 newtons (N) and pulls at an angle of 40°, and Force B has magnitude 90 N and pulls at an angle of 140° as shown.

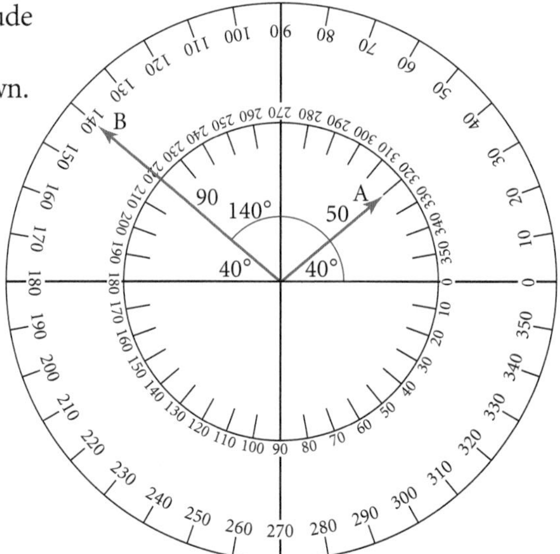

a. Find the rectangular form of the Force A vector. ⓐ

b. Find the rectangular form of the Force B vector.

c. Find the sum of the two force vectors. ⓐ

d. What are the magnitude and direction of the sum vector?

e. What additional force will balance Forces A and B and keep the object in equilibrium? (It should be the same magnitude as the sum of Forces A and B, but in the opposite direction.)

Science CONNECTION

Biomechanics provides an understanding of the internal and external forces acting on the human body during movement. Knowing the role muscles play in generating force and controlling movement is necessary to understanding the limitations of human motion. Information gained from biomechanics helps athletes prepare better for their sports and sporting goods manufacturers produce better equipment. Research in biomechanics also contributes to better treatment and rehabilitation in case of injury. You can learn more about the study and application of biomechanics with the Internet links at www.keymath.com/DAA .

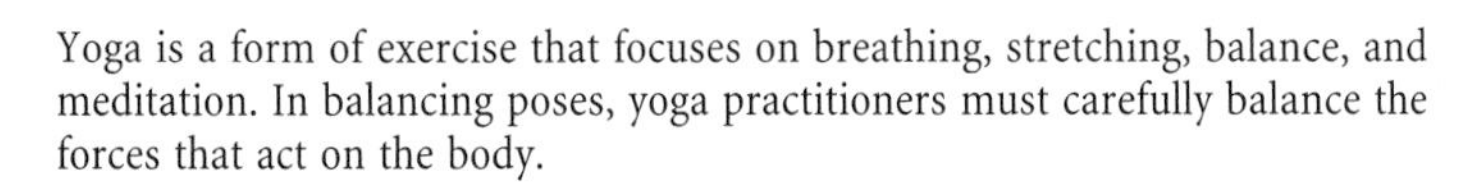

Yoga is a form of exercise that focuses on breathing, stretching, balance, and meditation. In balancing poses, yoga practitioners must carefully balance the forces that act on the body.

8. *Mini-Investigation* Is $|\mathbf{a}| + |\mathbf{b}|$ equal to $|\mathbf{a} + \mathbf{b}|$? Follow the steps below to investigate this question.

a. If $\mathbf{a} = \langle 4, 7\rangle$ and $\mathbf{b} = \langle 3, -1\rangle$, find $|\mathbf{a}|$ and $|\mathbf{b}|$.

b. What is the vector $\mathbf{a} + \mathbf{b}$?

c. What is the magnitude of the resultant vector from 8b?

d. Based on your work in 8a–c, is $|\mathbf{a}| + |\mathbf{b}|$ equal to $|\mathbf{a} + \mathbf{b}|$?

e. Rewrite the equation from 8d as an inequality to make a true statement. The expression you have written is sometimes called the **triangle inequality.** Sketch the vectors and then explain why this is an appropriate name for this property.

History
CONNECTION

A **compass rose** displays the orientation of the cardinal directions, north, south, east, and west, on a map or a nautical chart. An 8-point compass rose displays the intermediate directions northeast, southeast, southwest, and northwest. The 32-point compass rose was developed by Arab navigators in the Middle Ages and showed directions in gradations of 11.25°.

9. APPLICATION A pilot wants to fly directly west from Toledo, Ohio, to Chicago, Illinois. Her plane can fly at 120 mi/h. The wind is blowing at 25 mi/h from the south. The forces of both the plane and the wind contribute to the actual path of the plane. It is often easiest to represent the two forces separately and then add the vectors to find the actual motion. ⓗ

a. If there were no wind, what vector would describe the position of the plane after 2 hours? ⓐ

b. What vector describes the position of something carried solely by the wind after 2 hours? ⓐ

c. Find the sum of the vectors from 9a and b.

d. Where is the plane's position relative to Toledo after 2 hours? State your answer using a distance and an angle.

10. A plane is flying on a bearing of 310° at a speed of 320 mi/h. The wind is blowing directly from the east at a speed of 32 mi/h. ⓐ

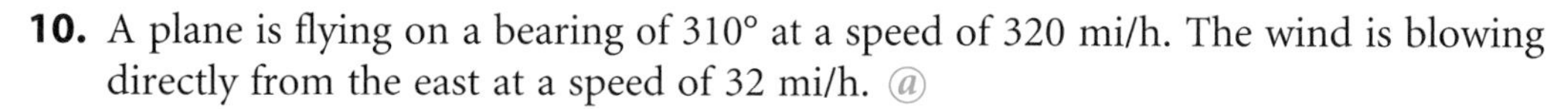

a. The length of the vector representing the position of the plane if there were no wind is given by the expression $320t$, where t represents the length of time the plane has been in the air. Write the vector in rectangular form for the plane after 3 hours.

b. The length of the vector representing the position of something carried solely by the wind is $32t$, where t is the same time used in 10a. Write the vector in rectangular form for something carried by a wind from the east after 3 hours.

c. Find the sum of the vectors from 10a and b.

d. What is the plane's position relative to its starting point after 3 hours? State your answer using a distance and an angle.

11. APPLICATION A plane is headed from Memphis, Tennessee, to Albuquerque, New Mexico, 1000 mi due west. The plane flies at 250 mi/h, and the pilot encounters a 20 mi/h wind blowing from the northwest.

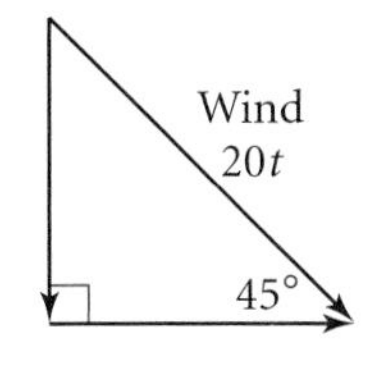

a. Give the rectangular form of the vector from start to destination. Call it vector **c**. @

b. How many hours would this flight take at 250 mi/h?

c. Give the rectangular form of the vector, **b**, for the motion by the wind over this time.

d. Find vector **a** such that **a** + **b** = **c**.

e. Give vector **a** in polar form and explain the real-world meaning of this vector.

Review

12. A rectangle's length is three times the width. Find the angles, to the nearest degree, at which the diagonals intersect.

13. Without graphing, determine whether each quadratic equation has no real roots, one real root, or two real roots. If a root is real, indicate whether it is rational or irrational.

a. $y = 2x^2 - 5x - 3$

b. $y = x^2 + 4x - 1$

c. $y = 3x^2 - 3x + 4$

d. $y = 9x^2 - 12x + 4$

14. Consider this system of equations:

$$\begin{cases} 5x - 3y = -1 \\ 2x + 4y = 5 \end{cases}$$

a. Write the augmented matrix for the system of equations.

b. Use row reduction to write the augmented matrix in row-echelon form. Show each step and indicate the operation you use.

c. Give the solution to the system. Check your answer.

LESSON

12.6 Parametric Equations

Envisioning the end is enough to put the means in motion.

DOROTHEA BRANDE

Sherlock Holmes followed footprints and other clues to track down suspected criminals. As he followed the clues, he knew exactly where the person had been. The path could be drawn on a map, every location described by x- and y-coordinates.

But how could he determine *when* the suspect was at each place? He needed to know how x and y depended on a third variable, *time*. Two variables are often not enough to describe interesting situations fully.

Sidney Paget (1860–1908) created the illustrations of Sherlock Holmes for the original short stories by Arthur Conan Doyle. The stories were published in *The Strand,* a British magazine. Paget pictured Sherlock Holmes in a deerstalker cap and a cape. Both these clothing items and the dark style of Paget's drawings have become strongly associated with the character of Holmes.

You can use **parametric equations** to describe the x- and y-coordinates of a point separately as functions of a third variable, t, called the **parameter.** Parametric equations provide you with more information and better control over which points you plot.

In this example the variable t represents time, and you will write parametric equations to simulate motion on your calculator screen.

EXAMPLE A

Hanna's hot-air balloon is ascending at a rate of 15 ft/s. A wind is blowing continuously from west to east at 24 ft/s. Write parametric equations to model this situation, and decide whether or not the hot-air balloon will clear power lines that are 300 ft to the east and 95 ft tall. Find the time it takes for the balloon to touch or pass over the power lines.

Hot-air balloons are made in all sorts of shapes.

▶ Solution

Create a table of time, ground distance, and height for a few seconds of flight. Set the origin as the initial launching location of the balloon. Let x represent the ground distance traveled to the east in feet, and let y represent the balloon's height above the ground in feet. The table below shows these values for the first 4 s of flight.

Time (s) t	Ground distance (ft) x	Height (ft) y
0	0	0
1	24	15
2	48	30
3	72	45
4	96	60

The parametric equations that model the motion are $x = 24t$ and $y = 15t$. Graph this pair of equations on your calculator. [▶ See **Calculator Note 12C** to learn how to enter and graph parametric equations. ◀] Observe how t controls the x- and y-values.

You can picture the power lines by plotting the point (300, 95). If you trace the graph to a time of 1 s, you will see that the balloon is 24 ft to the east, at a height of 15 ft. At 12.5 s, it has traveled 300 ft to the east and has reached a height of 187.5 ft. Hanna's balloon will not touch the power lines.

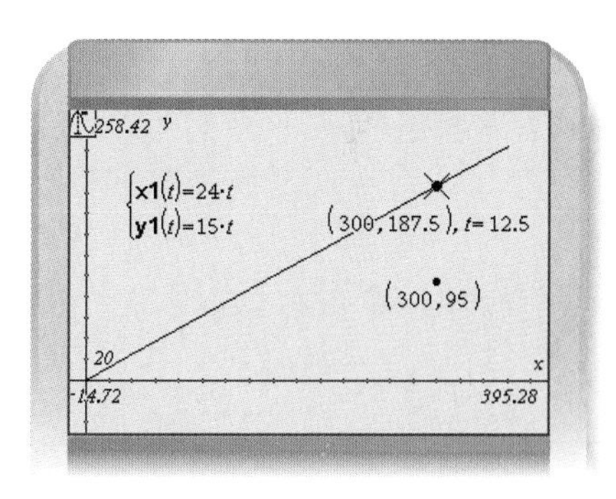

To solve the problem analytically, first find the time when the balloon will be 300 ft to the east. Substitute 300 for x into the equation $x = 24t$ and solve for t: $300 = 24t$, so $t = 12.5$ s. Then substitute this time into the equation that determines the height of Hannah's balloon: $y = 15(12.5) = 187.5$ ft. These answers confirm the results you found by graphing.

As you saw in Example A, modeling motion with parametric equations is much like the graphing you have done in earlier chapters, but you deal with each of the directions independently. This can make complicated situations easier to model.

Many pairs of parametric equations can be written as a single equation using only x and y. If you can eliminate the parameter in parametric equations, then you'll have two different ways to study a relationship.

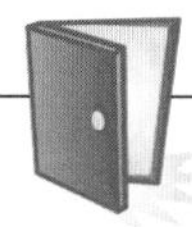

Investigation
Parametric Walk

You will need

- two motion sensors
- masking tape

This investigation involves four participants: a walker, recorder X, recorder Y, and a director.

Procedure Note

1. The walker starts at one end of the segment and walks slowly for 5 s to reach the other end.
2. Recorder X points a motion sensor set for 5 s at the walker and moves along the *y*-axis, keeping even with the walker, thus measuring the *x*-coordinate of the walker's path as a function of time.
3. Simultaneously, recorder Y points a motion sensor set for 5 s at the walker and moves along the *x*-axis, keeping even with the walker, thus measuring the *y*-coordinate of the walker's path as a function of time.
4. The director starts all three participants at the same moment and counts out the seconds.

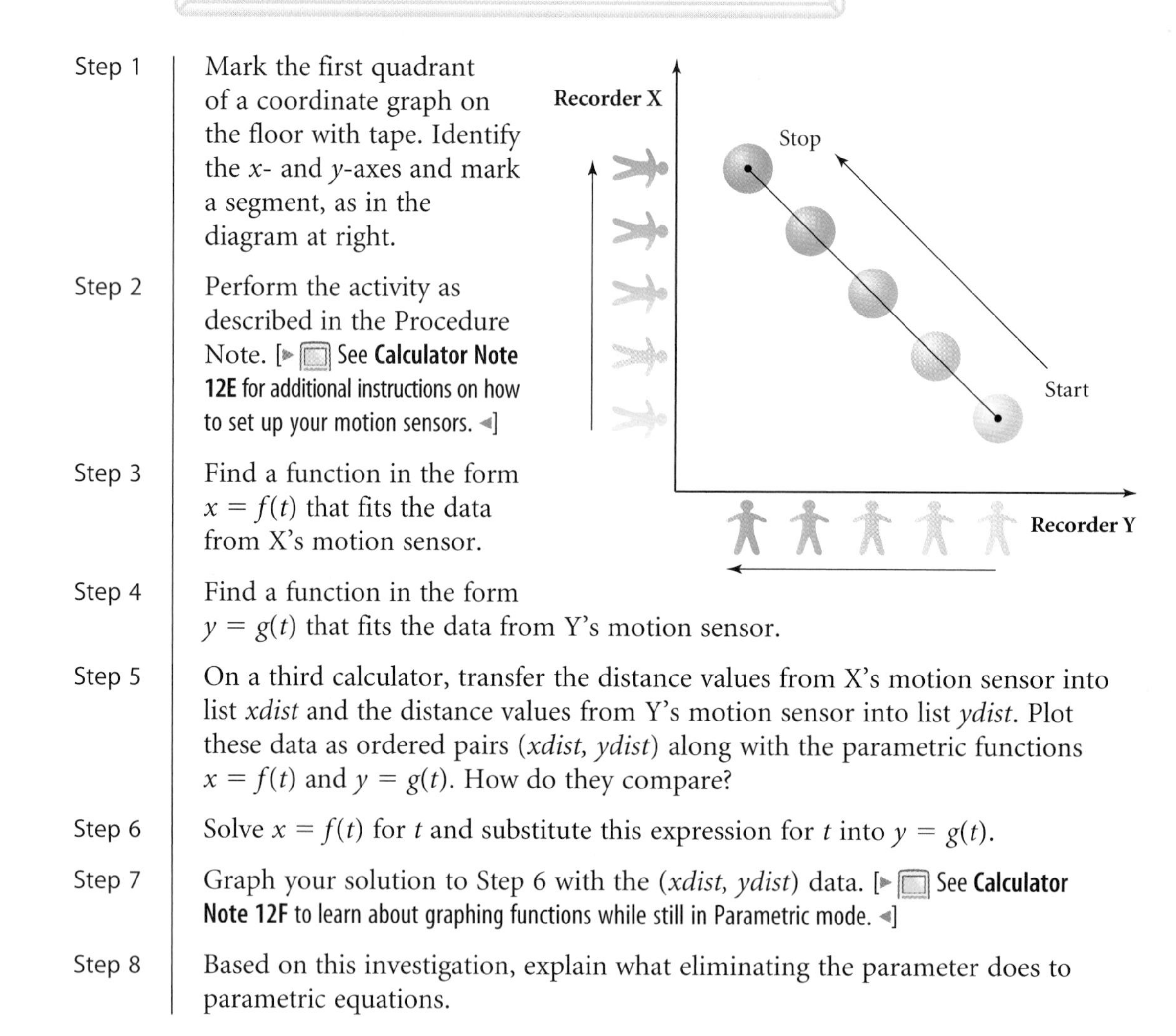

Step 1 Mark the first quadrant of a coordinate graph on the floor with tape. Identify the *x*- and *y*-axes and mark a segment, as in the diagram at right.

Step 2 Perform the activity as described in the Procedure Note. [See **Calculator Note 12E** for additional instructions on how to set up your motion sensors.]

Step 3 Find a function in the form $x = f(t)$ that fits the data from X's motion sensor.

Step 4 Find a function in the form $y = g(t)$ that fits the data from Y's motion sensor.

Step 5 On a third calculator, transfer the distance values from X's motion sensor into list *xdist* and the distance values from Y's motion sensor into list *ydist*. Plot these data as ordered pairs (*xdist, ydist*) along with the parametric functions $x = f(t)$ and $y = g(t)$. How do they compare?

Step 6 Solve $x = f(t)$ for t and substitute this expression for t into $y = g(t)$.

Step 7 Graph your solution to Step 6 with the (*xdist, ydist*) data. [See **Calculator Note 12F** to learn about graphing functions while still in Parametric mode.]

Step 8 Based on this investigation, explain what eliminating the parameter does to parametric equations.

Parametric equations can also model nonlinear movement. Consider a projectile launched at an angle. In Lesson 7.3, you saw that its height at time t is given by a quadratic function, $y = at^2 + v_0t + s_0$, where a is half the downward acceleration due to gravity, v_0 is the initial vertical velocity, and s_0 is the initial height. What if you don't know the initial vertical velocity, but know only the velocity at an angle? Then trigonometry can help you write the vertical and horizontal components of that velocity.

Parametric Equations for Projectile Motion

You can model projectile motion parametrically with these equations, where x is a measure of horizontal position, y is a measure of vertical position, and t is a measure of time.

$$x = v_0t\cos A + x_0$$

$$y = -\frac{1}{2}gt^2 + v_0t\sin A + y_0$$

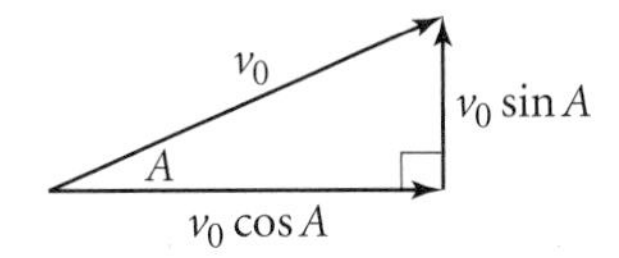

The point (x_0, y_0) is the initial position at time $t = 0$, v_0 is the velocity at time $t = 0$, A is the angle of initial motion measured from the horizontal, and g is the acceleration due to gravity. The value of g on earth is typically 9.8 m/s^2, or 32 ft/s^2.

EXAMPLE B

Carolina hits a baseball so that it initially travels at a speed of 120 ft/s and at an angle of 30° relative to the ground. If her bat contacts the ball at a height of 3 ft above the ground, how far does the ball travel horizontally before it hits the ground?

▸ Solution

Draw a picture and write equations for the x- and y-components of the motion. You'll first write equations that just show the motion based on the initial speed and angle.

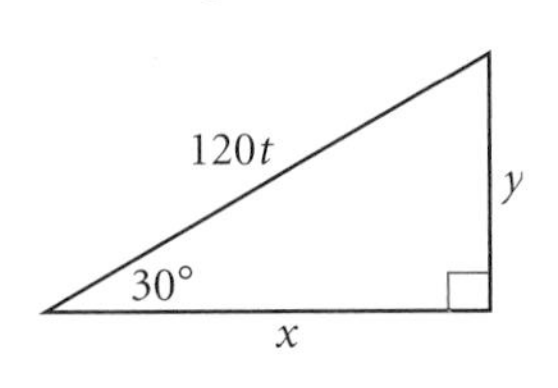

$$\cos 30° = \frac{x}{120t} \qquad \sin 30° = \frac{y}{120t}$$

$$x = 120t\cos 30° \qquad y = 120t\sin 30°$$

Now you'll refine the equations to reflect the effects of gravity and the starting position. The horizontal motion is affected only by the initial speed and angle, so the horizontal position is modeled by $x = 120t\cos 30°$.

The vertical motion is affected by the force of gravity pulling the ball down, as well as by the initial speed, height, and angle. The distances are given in feet, so $-\frac{1}{2}g = -16$. The initial height is 3 ft. Refer to the box on the previous page. The equation that models the vertical position is $y = -16t^2 + 120t\sin 30° + 3$.

To find out when the ball hits the ground, find the time t that makes this vertical position 0.

$-16t^2 + 120t\sin 30° + 3 = 0$ — Original equation.

$-16t^2 + 120t(0.5) + 3 = 0$ — Evaluate sin 30°.

$-16t^2 + 60t + 3 = 0$ — Multiply.

$t = \dfrac{-60 \pm \sqrt{60^2 - 4(-16)(3)}}{2(-16)}$ — Use the quadratic formula. Substitute -16 for a, 60 for b, and 3 for c.

$t \approx -0.049$ or $t \approx 3.799$ — Evaluate.

A negative value for time t doesn't make sense, so use only the positive answer. The ball reaches the ground about 3.8 s after being hit.

To determine how far the ball travels before it hits the ground, substitute this t-value into the parametric equation for x:

$$x = 120(3.799)\cos 30°$$
$$x \approx 395$$

The ball will travel about 395 ft horizontally before it hits the ground.

This example can also be solved by eliminating the parameter first. You will use this method in Exercise 12.

EXERCISES

You will need

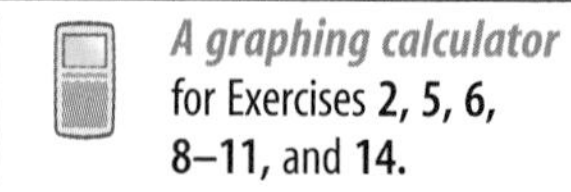

A graphing calculator for Exercises **2, 5, 6, 8–11,** and **14.**

Practice Your Skills

1. Create a table for each equation with $t = \{-2, -1, 0, 1, 2\}$.

a. $x = 3t - 1$, $y = 2t + 1$ ⓐ

b. $x = t + 1$, $y = t^2$

c. $x = t^2$, $y = t + 3$ ⓐ

d. $x = t - 1$, $y = \sqrt{4 - t^2}$

2. Graph each pair of parametric equations on your calculator. Sketch the result and use arrows to indicate the direction of increasing t-values along the graph. For 2a and b, use the window $-10 \le x \le 10$ and $-6 \le y \le 6$ and find a t-value that shows all of the graph. For 2c and d, limit your t-values as indicated.

a. $x = 3t - 1$, $y = 2t + 1$ ⓐ

b. $x = t + 1$, $y = t^2$

c. $x = t^2$, $y = t + 3$, $-2 \le t \le 1$ ⓐ

d. $x = t - 1$, $y = \sqrt{4 - t^2}$, $-2 \le t \le 2$

3. Write a single equation (using only x and y) that is equivalent to each pair of parametric equations.

a. $x = 2t - 3$
$y = t + 2$ ⓐ

b. $x = t^2$
$y = t + 1$

c. $x = \frac{1}{2}t + 1$
$y = \frac{t-2}{3}$ ⓐ

d. $x = t - 3$
$y = 2(t - 1)^2$

4. Find the position at the time given of a projectile in motion described by the equations

$x = -50t\cos 30° + 40$
$y = -490t^2 + 50t\sin 30° + 60$

a. 0 seconds

b. 1 second ⓐ

c. 2 seconds ⓐ

d. 4 seconds

In dog and disc competitions, the Frisbee is one projectile (although it does face wind resistance) and the dog is another. Each object has a horizontal and vertical component to its motion.

5. A ball rolls off the edge of a 12 m tall cliff at a velocity of 2 m/s.

a. Write parametric equations to simulate this motion. ⓗ

b. What equation can you solve to determine when the ball hits the ground?

c. When and where does the ball hit the ground?

d. Describe a graphing window that you can use to model this motion.

Reason and Apply

6. These parametric equations simulate two walkers, with x and y measured in meters and t measured in seconds.

$x = 1.4t$
$y = 3.1$
and
$x = 4.7$
$y = 1.2t$

a. Graph their motion for $0 \leq t \leq 5$.

b. Give real-world meanings for the values of 1.4, 3.1, 4.7, and 1.2 in the equations.

c. Where do the two paths intersect?

d. Do the walkers collide? How do you know?

7. By how much does the ball in Example B clear a 10 ft fence that is 365 ft away if the wind is blowing directly from the fence toward Carolina at 8 mi/h?

8. The graphs of $x = f(t)$ and $y = g(t)$ are shown below. ⓐ

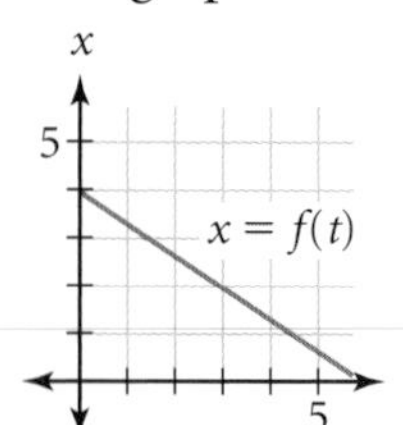

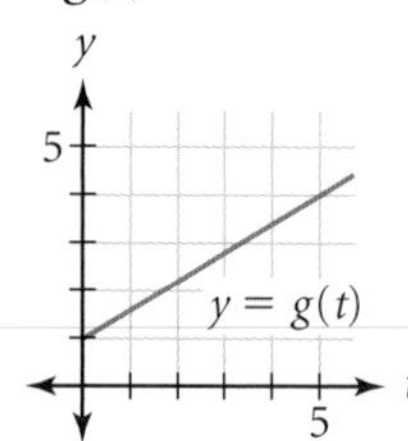

a. Use the graphs to create a graph of y as a function of x.

b. Write parametric equations for $x = f(t)$ and $y = g(t)$ and an equation for y as a function of x. How do the slopes of the graphs compare?

9. APPLICATION A river is 0.3 km wide and flows south at a rate of 7 km/h. You start your trip on the river's west bank, 0.5 km north of the dock, as shown in the diagram at right.

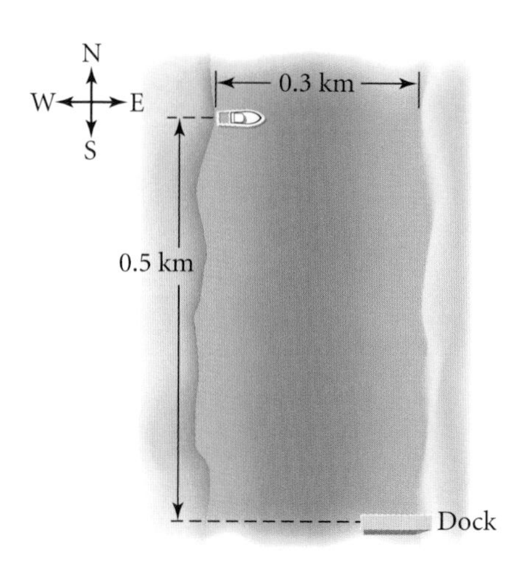

a. If the dock is at the origin, (0, 0), what are the coordinates of the boat's starting location? ⓐ

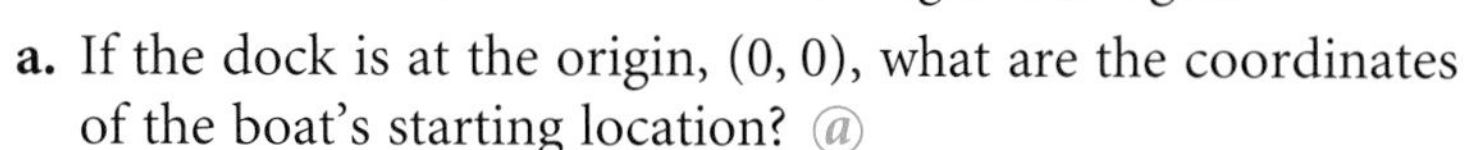

b. Write an equation for x in terms of t that models the boat's horizontal position if you aim the boat directly east traveling at 4 km/h. ⓐ

c. Write an equation for y in terms of t that models the boat's vertical position as a result of the flow of the river.

d. Enter the parametric equations from 9b and c into your calculator, determine a good viewing window and range of t-values, and make a graph to simulate this situation.

e. Determine when and where the boat meets the river's east bank. Does your boat arrive at the dock? ⓗ

f. How far have you traveled?

10. APPLICATION A plane is headed from Memphis, Tennessee, to Albuquerque, New Mexico, 1000 mi due west. The plane flies at 250 mi/h, and the pilot encounters a 20 mi/h wind blowing from the northwest. (That means the direction of the wind makes a 45° angle below the x-axis.)

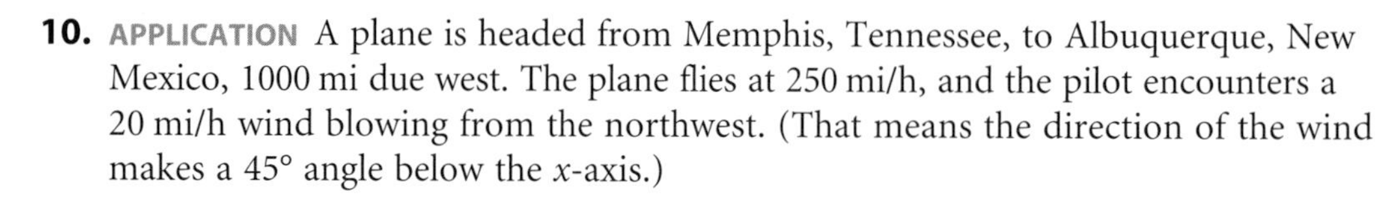

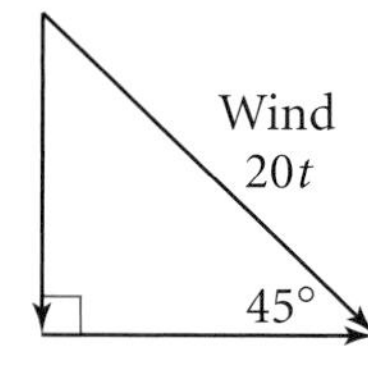

a. Write an equation modeling the southward component of the wind.

b. Write an equation modeling the eastward component of the wind.

c. If the pilot does not compensate for the wind, explain why the final equations for the flight are $x = -250t + 20t\cos 45°$ and $y = -20t\sin 45°$.

d. What graphing window and range of t-values can you use to simulate this flight?

e. Solve the equation $-1000 = -250t + 20t\cos 45°$. What is the real-world meaning of your answer?

f. Use your answer from 10e to find how far south of Albuquerque the plane ended up.

11. Gonzo, the human cannonball, is fired out of a cannon 10 ft above the ground at a speed of 40 ft/s. The cannon is tilted at an angle of 60°. His net hangs 5 ft above the ground. Where does his net need to be positioned so that he will land safely?

Pictured here in Berlin, Germany, a spring-loaded cannon propelled the "Human Cannonball" of the 1920s, Paul Leinert.

12. Use the equations in Example B that modeled the path of the ball that Carolina hit.

a. Find an equation for y in terms of x by eliminating the parameter.

b. Solve your new equation for x when $y = 0$.

c. How does your answer for 12b compare with the solution found in Example B?

Review

13. What is the equation of the image of $y = \frac{2}{3}x - 2$ after a translation right 5 units and up 3 units?

14. Graph the parabola $y = 35 - 4.9(x - 3.2)^2$.

a. What are the coordinates of the vertex? (a)

b. What are the x-intercepts? (h)

c. Where does the parabola intersect the line $y = 15$?

15. Write the equation of the circle with center $(2.6, -4.5)$ and radius 3.6.

16. Find exact values of missing side lengths for each triangle.

a.

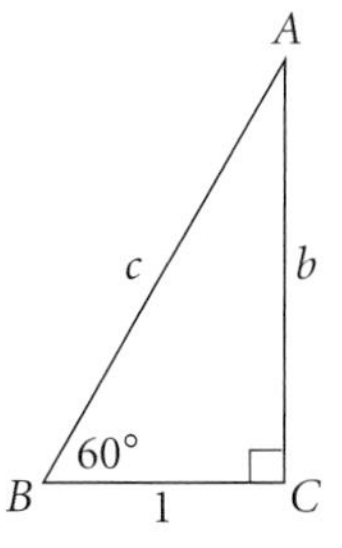

b.

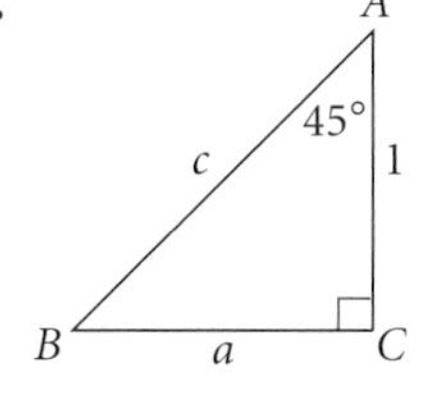

c.

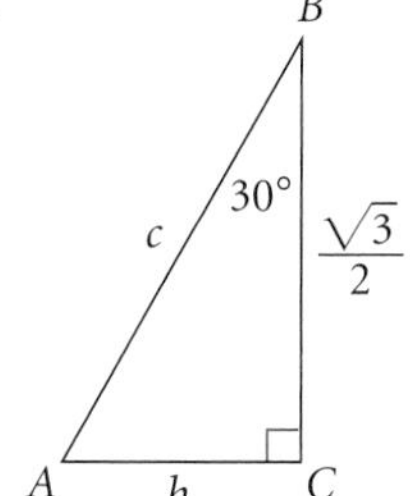

d.

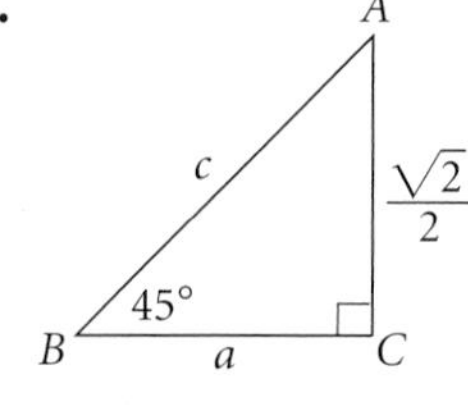

17. Find the height, width, area, and distance specified.

a. total height of the tree

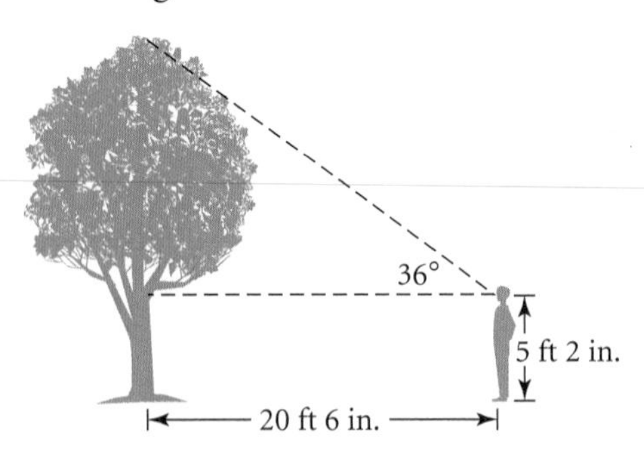

b. width of the lake

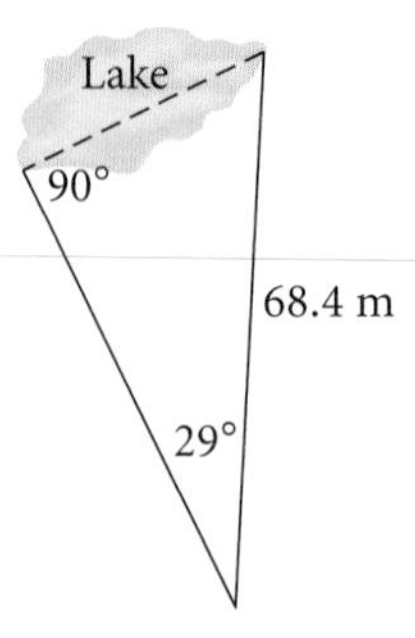

c. area of the triangle

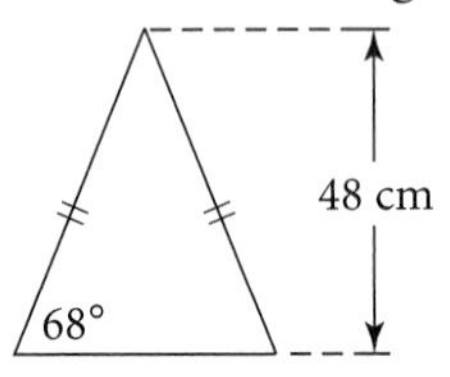

d. distance from the boat to the lighthouse

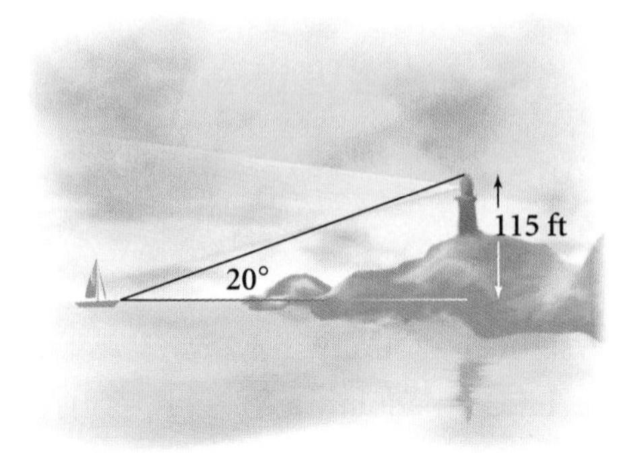

Project

CATAPULT

If you launch a ball at a particular angle and initial velocity, you can determine how far it will travel. Is there another angle at which a ball can be launched, with the same initial velocity, that will cause the ball to travel exactly the same horizontal distance? What angle(s) will cause the ball to travel the farthest distance possible?

Your project should include

- A conjecture about what kinds of angles (if any) will cause a ball to travel equal distances, and evidence to support it.
- Angle(s) that will cause a ball to travel the farthest distance possible.

© The New Yorker Collection from CartoonBank.com/Corbis

CHAPTER 12 REVIEW

The **trigonometric ratios**—sine, cosine, and tangent—relate the side lengths of a triangle to the measure of an angle. The **sine** of an angle is the ratio of the opposite leg to the hypotenuse in a right triangle. The **cosine** of an angle is the ratio of the adjacent leg to the hypotenuse, and the **tangent** of an angle is the ratio of the opposite leg to the adjacent leg. You can use these ratios to find missing side lengths and angles in right triangles. You can use the **Law of Sines** and the **Law of Cosines** to find missing side lengths and angles in triangles that do not contain a right angle.

To use the Law of Sines and the Law of Cosines, it was necessary to extend the definitions for sine and cosine to apply to non-acute angles. You learned to place an angle on the coordinate plane by rotating a point counterclockwise from the x-axis. You used a **reference triangle** and **reference angle** to find the sine and cosine, which are the y-coordinate and x-coordinate of the rotated point, respectively.

You learned about the properties of **vectors,** which allow you to express quantities that involve direction. You learned how to express vectors in both **polar form** and **rectangular form,** and you discovered how to represent addition and subtraction of vectors both geometrically and symbolically. You also explored **parametric equations,** which describe the locations of points by using a third variable, t, called a **parameter.** In many cases this third variable represents time.

EXERCISES

You will need

A graphing calculator for Exercises **11** and **12.**

@ Answers are provided for all exercises in this set.

1. For each triangle, find the measure of the labeled angle or the length of the labeled side.

a.

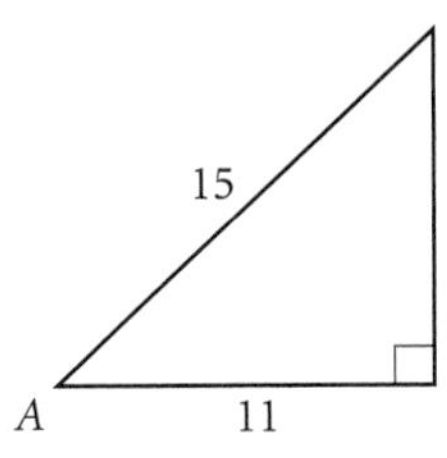

b.

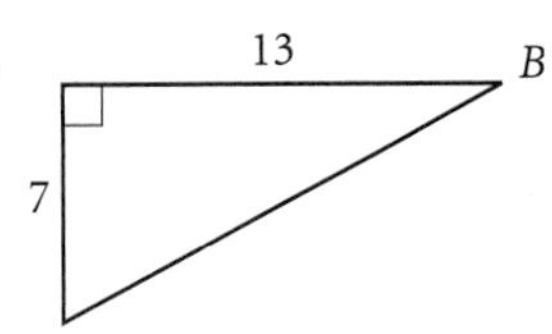

c.

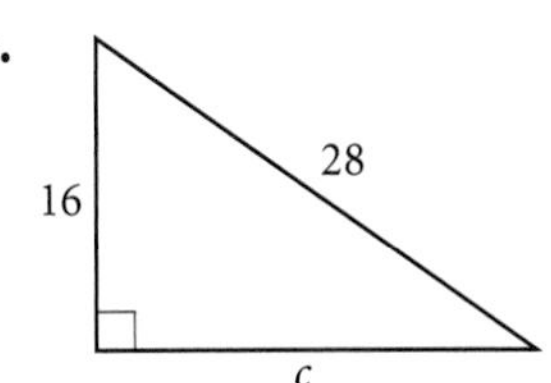

d.

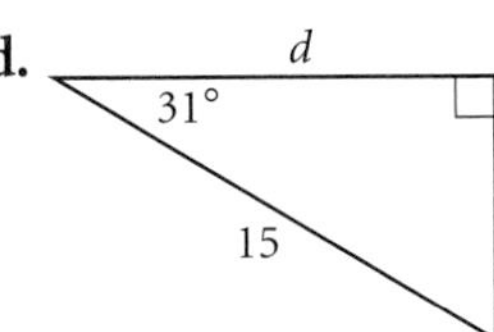

e.

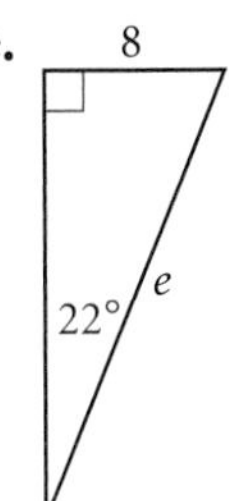

f.

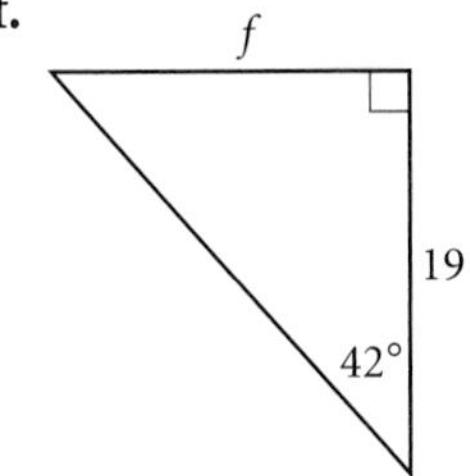

2. Aliya sees a coconut 66 ft up in a coconut palm tree that is 20 ft away from her. What is the angle of elevation from her position to the nut?

3. Find all of the missing side lengths and angle measures.

a.

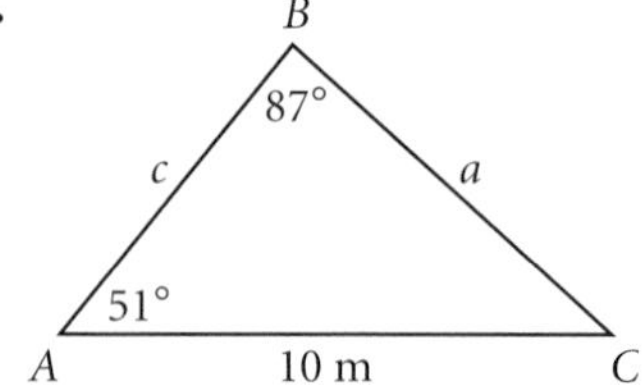

b.

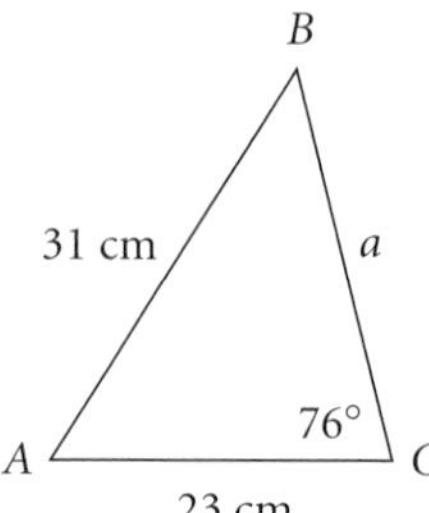

4. A hiker left a north-south road and walked for 2 km on a bearing of 305°. Then she changed direction and walked 5 km back to the road. She arrived directly north of where she started. How far is she from where she started?

5. Find all of the missing side lengths and angle measures.

a.

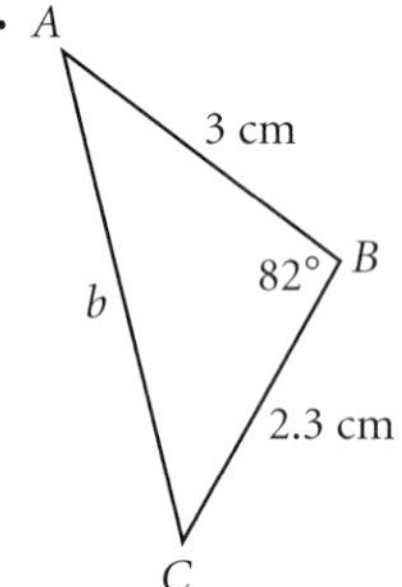

b.

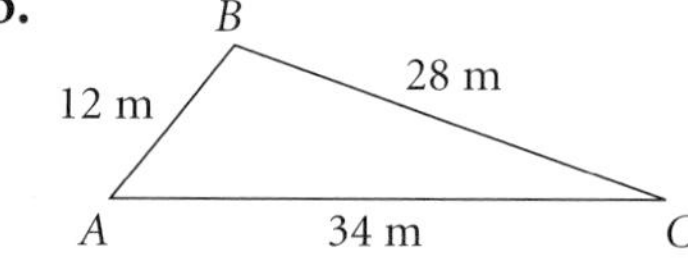

6. Two tankers leave Corpus Christi, Texas, with an angle of 27° between their paths. After 2 hours, the first tanker is 44 miles from Corpus Christi and the second tanker is 36 miles from Corpus Christi. What is the distance between the two tankers?

7. The point (6, 0) is rotated 150° counterclockwise about the origin. What are the coordinates of its image point?

8. A diver runs off a 10 m platform with an initial horizontal velocity of 4 m/s. The edge of the platform is directly above a point 1.5 m from the pool's edge.

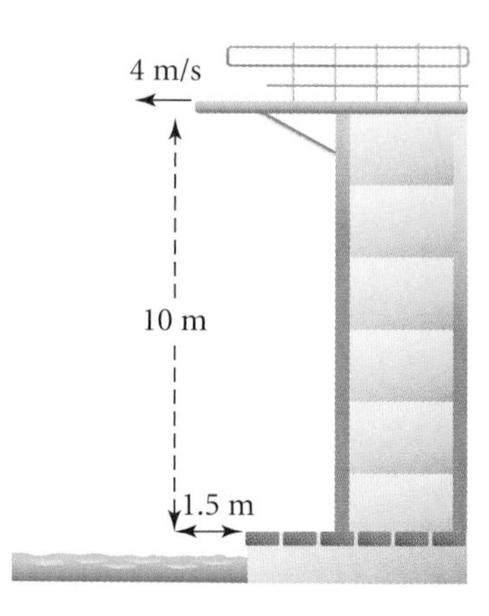

a. Let the origin represent the pool's edge under the platform. What are the coordinates of the edge of the platform?

b. Suppose the diver continues to move horizontally at 4 m/s for 1.5 seconds before hitting the water. What are the coordinates of the point where the diver enters the water?

c. What is the "line of sight" distance between the edge of the platform and the point of entry?

d. Would the diver have traveled the path from 8c, a longer path, or a shorter path? Explain.

Sports CONNECTION

In competitive diving, a running dive must be at least four steps long. The takeoff phase determines the diver's path through the air. Once in the air, the diver has less than 2 seconds to finish the dive, which should end with the diver's body almost perpendicular to the water's surface.

A diver practices at the National Aquatic Center in Beijing, China in 2008.

9. Add the vectors $\langle 47 \angle 30° \rangle$ and $\langle 25, 75 \rangle$, and express the resultant vector in polar form.

10. APPLICATION A pilot is flying to a destination 700 mi away at a bearing of 105°. The cruising speed of the plane is 500 mi/h, and the wind is blowing between 20 mi/h and 30 mi/h at a bearing of 30°. At what bearing should she aim the plane to compensate for the wind?

11. Use the parametric equations $x = -3t + 1$ and $y = \frac{2}{t + 1}$ to answer each question.

a. Find the x- and y-coordinates of the points that correspond to the values of $t = 3$, $t = 0$, and $t = -3$.

b. Find the y-value that corresponds to an x-value of -7.

c. Find the x-value that corresponds to a y-value of 4.

d. Sketch the curve for $-3 \leq t \leq 3$, showing the direction of movement. Trace the graph and explain what happens when $t = -1$.

12. A raft is moving east at 20 m/s and south at 30 m/s due to a strong wind and current.

a. What is the raft's position after 8 s relative to its starting position?

b. What equations simulate this motion?

The Family (1962) by Venezuelan sculptor and painter Marisol Escobar (b 1930) has individual panels that collectively contribute to the whole piece, just as parametric equations combine individual equations to describe one mathematical piece.

TAKE ANOTHER LOOK

1. Select a pair of noncongruent angles that are supplementary (whose measures sum to 180°). Use your calculator to find the sine, cosine, and tangent of each angle measure. What relationships do you notice? Try other supplementary pairs to verify your relationships. Then select a pair of complementary angles (whose measures sum to 90°), and find the sine, cosine, and tangent of each angle measure. What relationships do you notice? Verify these relationships with other complementary pairs. Draw geometric diagrams that prove the relationships you find.

2. In an earlier math course, you probably learned the formula $Area = bh$ for the area of a parallelogram, where b represents the length of the base, and h represents the height of the parallelogram, drawn perpendicular from the base to the opposite side. If you solved the puzzle on p. 697, you found a formula for the area of a triangle that you can use when you know the lengths of two sides and the measure of the included angle. Use the same reasoning to find a formula for the area of a parallelogram given two adjacent side lengths and the measure of one angle.

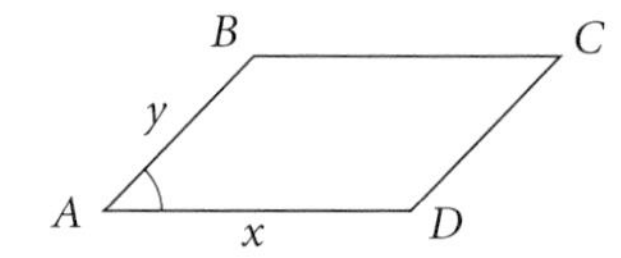

3. What if you know the lengths of three sides of a triangle but don't know any of the angle measures? How can you find the area of a triangle using only the three side lengths? Describe the steps you would use to find the area of $\triangle ABC$.

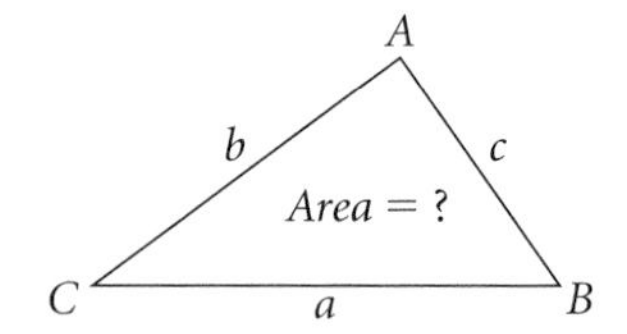

Assessing What You've Learned

WRITE IN YOUR JOURNAL In this chapter you extended the use of sine and cosine to non-right triangles, and even to angles greater than 180° and less than 0°. You also learned about vectors and how trigonometry can be used to convert between rectangular and polar form. Draw diagrams representing each use of trigonometry. Which ideas from this chapter were familiar from previous courses? Which ideas were new? Describe any concepts that are confusing for you, and discuss them with another student or your teacher.

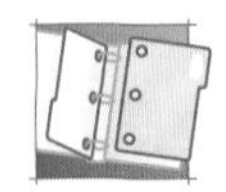

ORGANIZE YOUR NOTEBOOK Make sure your notebook is complete and well organized. Be sure to include all of the definitions and formulas that you have learned in this chapter. You may want to include at least one example of each of the different applications of trigonometry.

WRITE TEST ITEMS Write at least four test items for this chapter. Include items that cover different applications of trigonometry. You may want to include the use of trigonometry for right triangles and oblique triangles, as well as applications of vectors and parametric equations.

CHAPTER

13 Trigonometric Functions

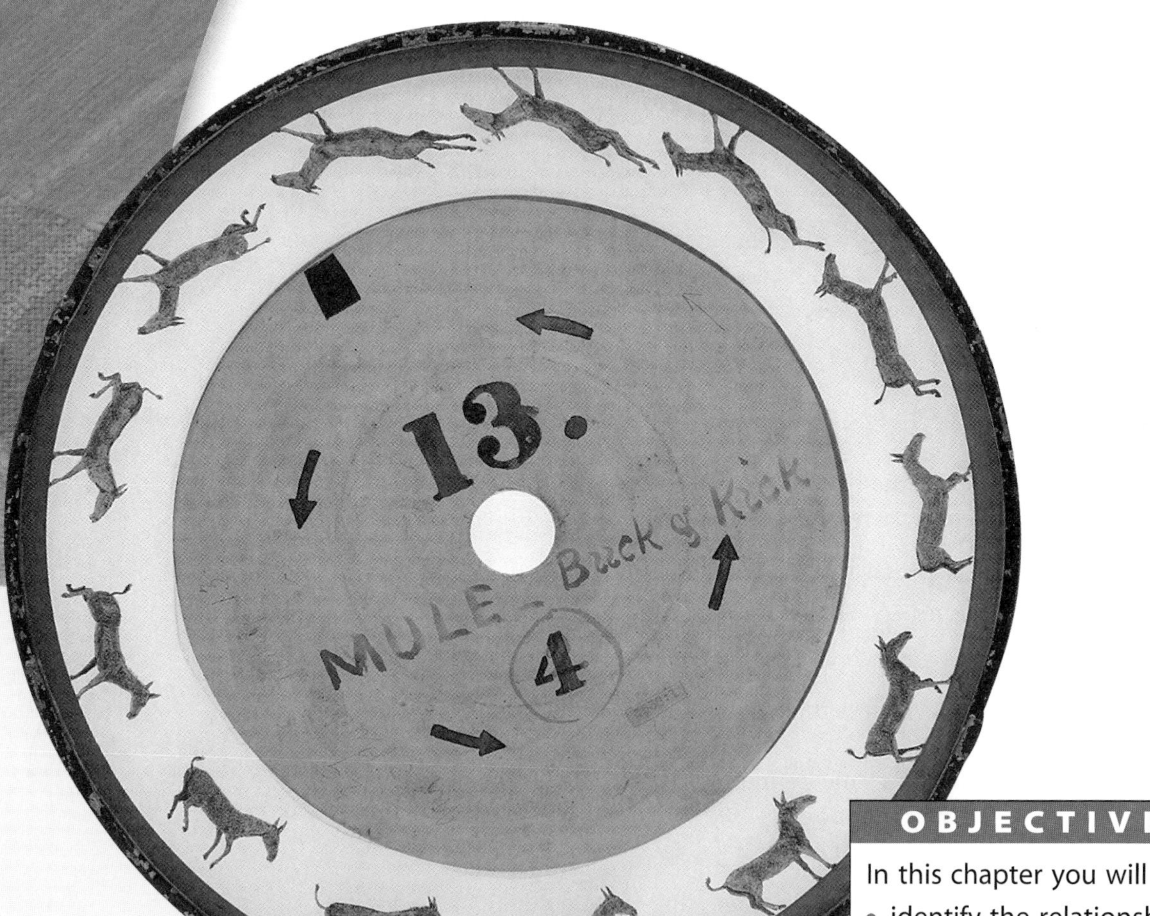

In 1880, English-American photographer Eadweard J. Muybridge (1830–1904) created the zoopraxiscope, an early motion picture machine that projected a series of images on a spinning disk. The series of photographs shown here depicts a mule bucking and kicking. When the disk is spun quickly, it creates the illusion of a cyclical, repetitive motion. Muybridge spent many years studying animal and human movement, and perfecting his method of photographing motion.

Courtesy George Eastman House

OBJECTIVES

In this chapter you will

- identify the relationship between circular motion and the sine and cosine functions
- use a unit circle to find values of sine and cosine for various angles
- learn a new unit of measurement for angles, called radians
- apply your knowledge of transformations to the graphs of trigonometric functions
- model real-world phenomena with trigonometric functions
- study trigonometric identities

Geometry of the Circle

In this lesson you'll review some properties of circles that will be useful as you deepen your knowledge of trigonometry. Recall these geometric terms:

The **circumference** of a circle is the distance around the circle. For a circle with radius r, $C = 2\pi r$.

A **central angle** is an angle formed by two radii.

An **arc** is a piece of the circle.

The **measure of an arc** of a circle is the same as the measure of the central angle that intercepts the arc.

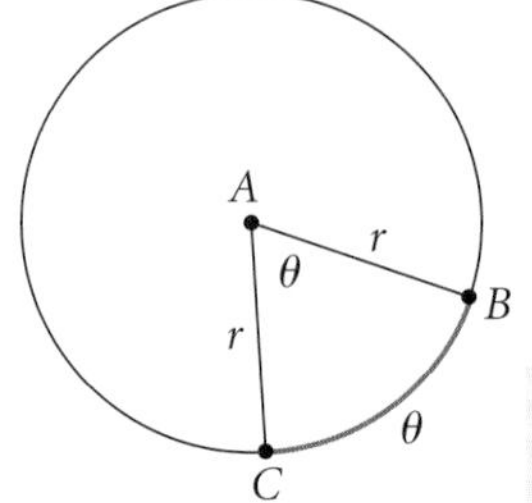

$\angle BAC$ is a central angle.
$m\angle BAC = m\overset{\frown}{BC}$

The **length of an arc** is given in the same units as the radius or circumference.

You can use ratios and proportions to find the lengths and measures of arcs.

EXAMPLE A

Use the information provided in the diagrams to answer the questions below for $\overset{\frown}{DE}$, $\overset{\frown}{FP}$, and $\overset{\frown}{GH}$.

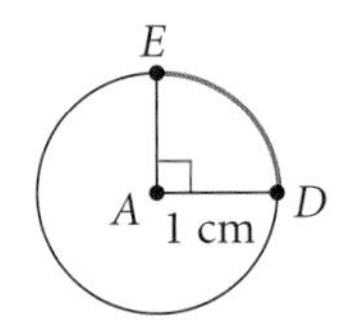

a. What is the measure of each arc?

b. What fraction of the circumference is the length of each arc?

c. What is the length of each arc?

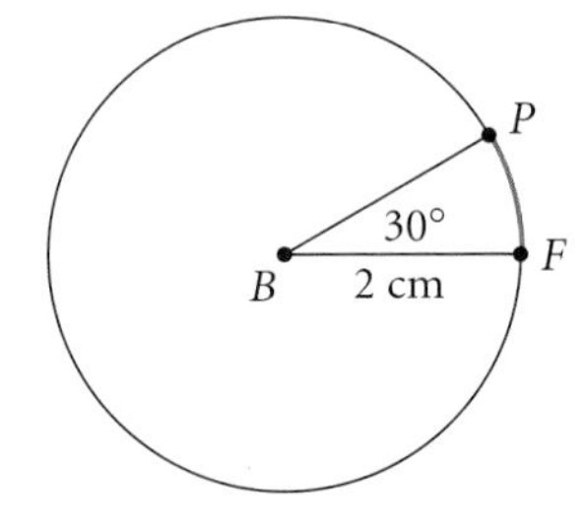

Solution

Refer to the definitions above as you answer each question.

a. The measure of an arc is the same as the measure of the central angle that intercepts the arc. So $m\overset{\frown}{DE} = 90°$, $m\overset{\frown}{FP} = 30°$, and $m\overset{\frown}{GH} = 135°$.

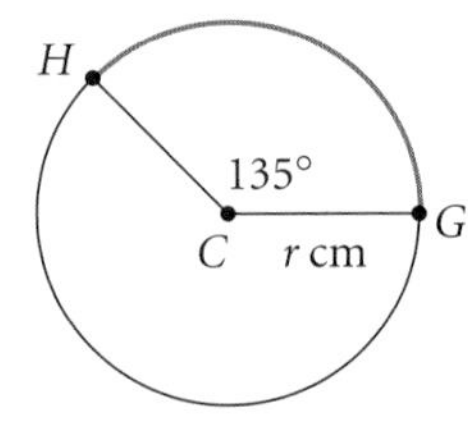

b. Find the ratio of each arc measure to the measure of a revolution about a full circle, which is 360°.

Because $\frac{90°}{360°} = \frac{1}{4}$, the length of $\overset{\frown}{DE}$ is $\frac{1}{4}$ the circumference of Circle A. Because $\frac{30°}{360°} = \frac{1}{12}$, the length of $\overset{\frown}{FP}$ is $\frac{1}{12}$ the circumference of Circle B. The length of $\overset{\frown}{GH}$ is $\frac{135°}{360°}$, or $\frac{3}{8}$ the circumference of Circle C.

c. Arc lengths are fractional parts of the circumference. Use $C = 2\pi r$ to find each circumference. You can find the length of each arc by using the proportion $\frac{\text{arc length}}{\text{circumference}} = \frac{\text{arc measure}}{360°}$.

To find the length of $\overset{\frown}{DE}$ in Circle A, write the proportion

$$\frac{\text{length of } \overset{\frown}{DE}}{2\pi \cdot 1} = \frac{90°}{360°}$$

Multiply both sides by 2π. The length of $\overset{\frown}{DE} = 2\pi \cdot \frac{1}{4} = \frac{\pi}{2}$ cm, or about 1.57 cm. Using the ratio from part b, $\overset{\frown}{FP}$ is $\frac{1}{12}$ the circumference of Circle B. So the length of $\overset{\frown}{FP} = \frac{1}{12} \cdot 2\pi \cdot 2 = \frac{\pi}{3}$ cm, or about 1.05 cm. Similarly, the length of $\overset{\frown}{GH} = \frac{3}{8} \cdot 2\pi r = \frac{3\pi r}{4}$ cm, or about 2.36r cm.

Keep in mind that the *measure* of an arc is not the same as the *length* of an arc. For example, all 90° arcs have the same measure (90°), but they can have different lengths, because their lengths depend on the radius of the circle.

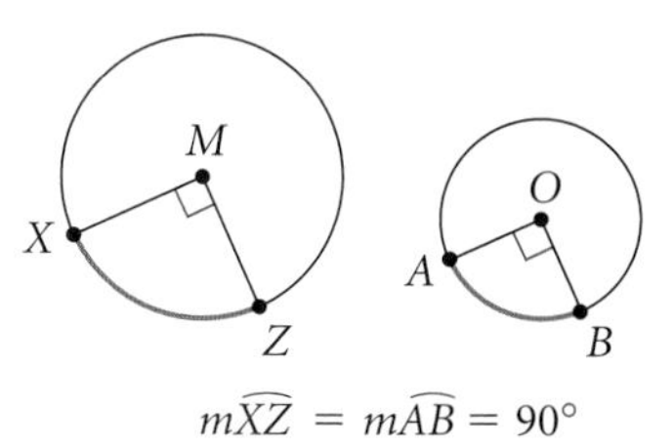

$m\overset{\frown}{XZ} = m\overset{\frown}{AB} = 90°$

EXAMPLE B | Circle X has radius 5 cm and arc $\overset{\frown}{PQ}$ with length 5 cm. What is $m\overset{\frown}{PQ}$?

▶ ***Solution***

The arc length, l, satisfies the proportion $\frac{l}{C} = \frac{m}{360°}$, where m is the measure of $\overset{\frown}{PQ}$ and C is the circumference of Circle X. Substitute 5 for both the arc length and the radius, and solve for the measure.

$$\frac{5}{2\pi \cdot 5} = \frac{m}{360°}$$

$$m = \frac{5}{2\pi \cdot 5} \cdot 360° = \frac{180°}{\pi} \approx 57.3°$$

The measure of $\overset{\frown}{PQ}$ is about 57.3°.

EXERCISES

1. Find the circumference of each circle. Leave answers in terms of π.

a. Circle A has radius 16 ft. ⓐ

b. Circle B has diameter 16 ft.

2. Find the arc measure if the ratio of arc length to circumference is

a. $\frac{1}{2}$ ⓐ **b.** $\frac{7}{8}$ **c.** $\frac{1}{6}$

3. Find the length of the arc intercepted by $\angle AOP$. ⓐ

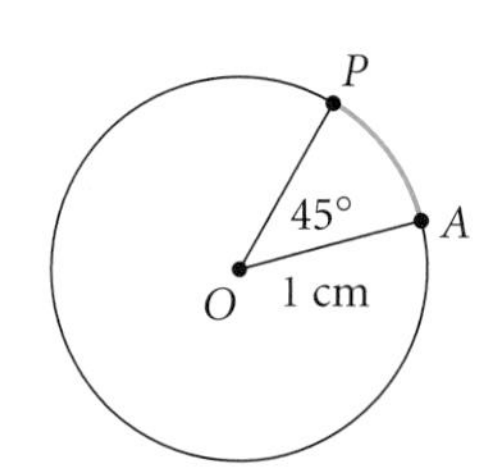

4. Find the measure of an arc in a unit circle (radius = 1) if the arc length is $\frac{3\pi}{2}$.

LESSON

13.1 Defining the Circular Functions

It is by will alone I set my mind in motion.

MENTAT CHANT

Many phenomena are predictable because they are repetitive, or cyclical. The water depths caused by the tides, the motion of a person on a swing, your height above ground as you ride a Ferris wheel, and the number of hours of daylight each day throughout the year are all examples of cyclical patterns. In this lesson you will discover how to model phenomena like these with the sine and cosine functions.

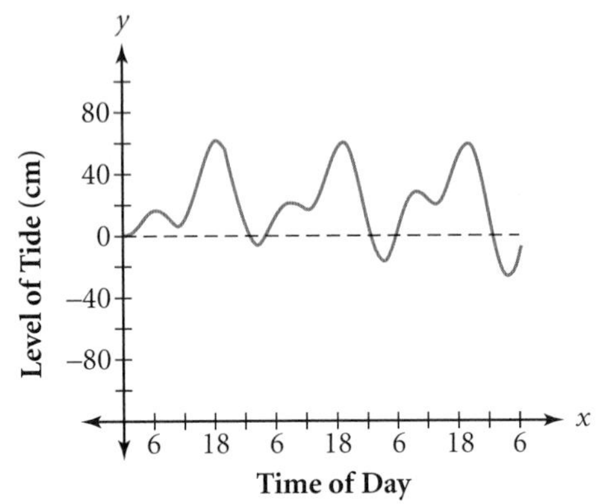

The water levels in the canals of Venice rise and fall periodically, as shown in the graph. During very high tides, the lower-lying areas of Venice can be flooded. This phenomenon is called *Acqua Alta* (high water). Acqua Alta occurs more frequently when the moon is full or new and when there are high winds.

The cathedral of Venice, Italy, Saint Mark's Basilica, is reflected in the flood waters in Saint Mark's Square.

Science CONNECTION

High and low ocean tides repeat in a continuous cycle, with two high tides and two low tides every 24.84 hours. The highest tide is an effect of the gravitational pull of the Moon, which causes a dome of water to travel under the Moon as it orbits Earth. The entire planet Earth experiences the Moon's gravitational pull, but because water is less rigid than land, it flows more easily in response to this force. The lowest tide occurs when this dome of water moves away from the shoreline.

The circle at right has radius r and center at the origin. A central angle of t degrees is shown. You can use your knowledge of right triangle trigonometry to write the equations $\sin t = \frac{y}{r}$ and $\cos t = \frac{x}{r}$. Sometimes it is useful to write these equations as $y = r\sin t$ and $x = r\cos t$.

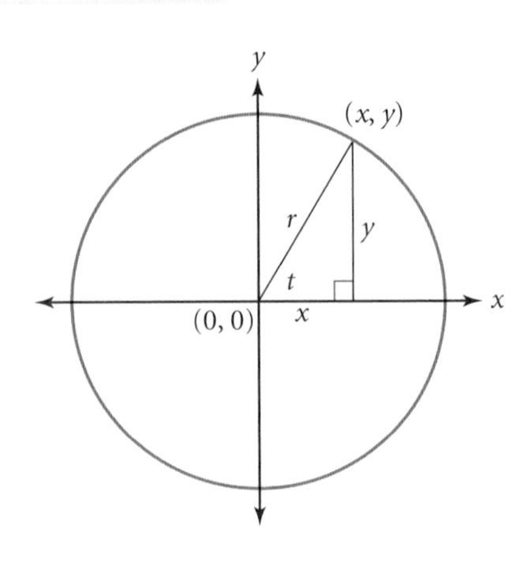

In this investigation you'll see how to use the sine and cosine functions to model circular motion.

Investigation
Paddle Wheel

While swimming along, a frog reaches out and grabs onto the rim of a paddle wheel with radius 1 m. The center of the wheel is at water level. The frog, clinging tightly to the wheel, is immediately lifted from the surface of the river.

Step 1 The wheel spins counterclockwise at a rate of one rotation every 6 minutes. Through how many degrees does the frog rotate each minute? Each second?

Step 2 Create three lists on your calculator. Name the first list *time* and fill it with the values {0, 15, 30, 45, . . . , 900}. Name the second list *hpos* and define it as $hpos = \cos(time)$. Name the third list *vpos* and define it as $vpos = \sin(time)$. Make a scatter plot of (*hpos, vpos*) in a square window and trace the path of the frog.

Step 3 Explain how to find the x- and y-coordinates of any point on the circle. In this context, what is the meaning of those points? What might be more appropriate names for lists *hpos* and *vpos*?

Step 4 Scroll down your lists and describe any patterns you see.

Step 5 Use your lists to answer these questions.

a. What is the frog's location after 1215°, or 1215 s? When, during the first three rotations of the wheel, is the frog at that same location?

b. When is the frog at a height of -0.5 m during the first three rotations?

c. What are the maximum and minimum x- and y-values?

Step 6 Make scatter plots of (*time, hpos*) and (*time, vpos*) on the same screen, using a different symbol for each plot. Use the domain $0 \text{ s} \leq time \leq 360 \text{ s}$. How do the graphs compare? How can you use the graphs to find the frog's position at any time? Why do you think the sine and cosine functions are sometimes called circular functions?

Recall that a circle with radius 1 unit centered at the origin is called the unit circle. While using the unit circle in the investigation, you discovered that values for sine and cosine repeat in a regular pattern. When output values of a function repeat at regular intervals, the function is **periodic.** The **period** of a function is the smallest distance between values of the independent variable before the cycle begins to repeat.

EXAMPLE A | Find the period of the cosine function.

▶ ***Solution***

In the investigation the frog returned to the same position each time the paddle wheel made one complete rotation, or every 360°. It seems reasonable to say that the function $\cos x$ has a period of 360°. You can verify this by looking at a graph of $y = \cos x$.

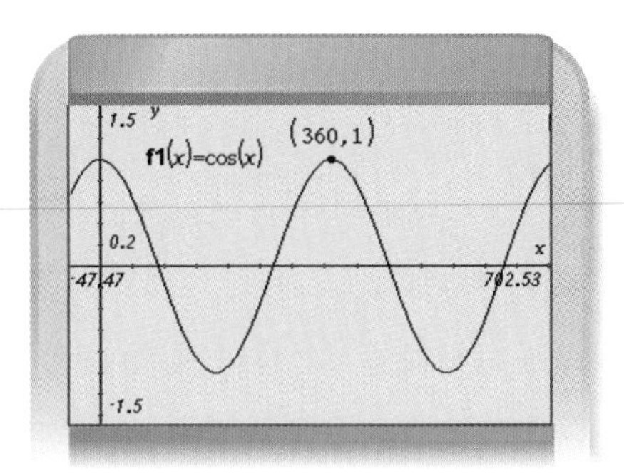

The graph shows that from $x = 0°$ to $x = 360°$, the function completes one full cycle.

Because the period of the cosine function is 360°, you can translate the graph of the function $y = \cos x$ left or right by 360° and it will look the same. Verify this by graphing $y = \cos(x + 360°)$ and $y = \cos(x - 360°)$ on your calculator on the same axes as the parent function $y = \cos x$.

The graph at right of $y = \sin x$, with one cycle highlighted between 0 and 360°, confirms that the sine function also has a period of 360°.

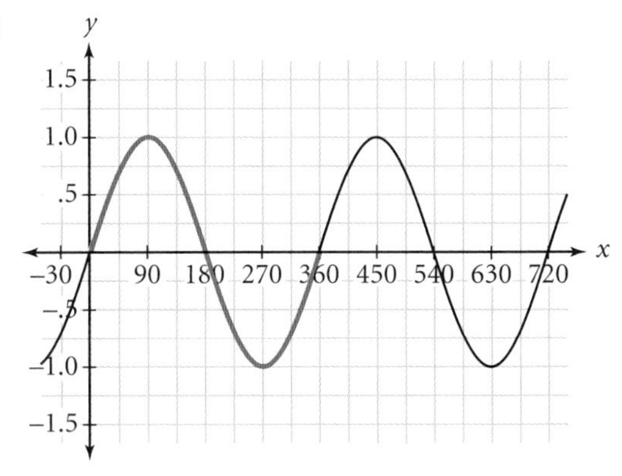

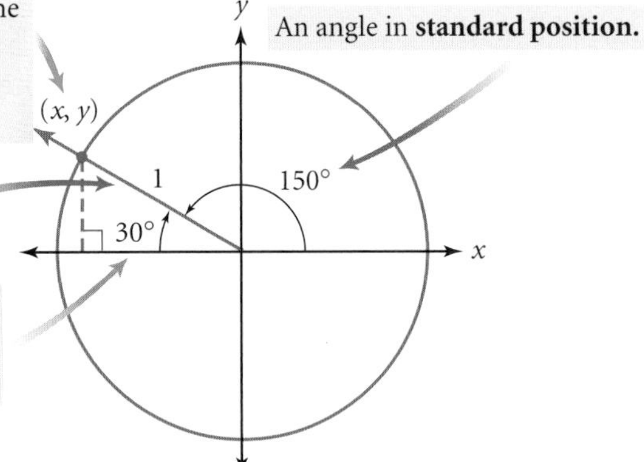

Notice that the domains of the sine and cosine functions include all positive and negative numbers. In a unit circle, angles are measured from the positive x-axis. Positive angles are measured in a counterclockwise direction, and negative angles are measured in a clockwise direction. An angle in **standard position** has one side on the positive x-axis, and the other side is called the **terminal side.**

The x- and y-coordinates of the point where the terminal side touches the unit circle determine the values of the cosine and sine of the angle. Identifying a **reference angle,** the acute angle between the terminal side and the x-axis, and drawing a **reference triangle** can help you find these values. [▶ You can use the **Dynamic Algebra Exploration** at www.keymath.com/DAA to explore the vocabulary of this lesson and deepen your understanding of the sine and cosine functions.◀]

keymath.com/DAA

EXAMPLE B Find the value of the sine or cosine for each angle. Explain your process.

a. $\sin 150°$ **b.** $\cos 150°$ **c.** $\sin 210°$ **d.** $\cos 320°$

▶ Solution

For each angle in a–d, rotate the point (1, 0) counterclockwise about the origin. Draw a ray from the origin through the image point to be the terminal side. Drop a perpendicular line from the image point to the x-axis to create a reference triangle and then identify the reference angle.

a. The y-coordinate of the rotated point on the unit circle is the sine of the angle. For 150°, the reference angle measures 30°. Using your knowledge of 30°-60°-90° triangle side relationships and the fact that the hypotenuse of the triangle is 1, you can find that the lengths of the legs of the reference triangle are $\frac{1}{2}$ and $\frac{\sqrt{3}}{2}$. Thus the coordinates of the image point are $\left(-\frac{\sqrt{3}}{2}, \frac{1}{2}\right)$. Therefore, $y = \sin 150° = \frac{1}{2}$.

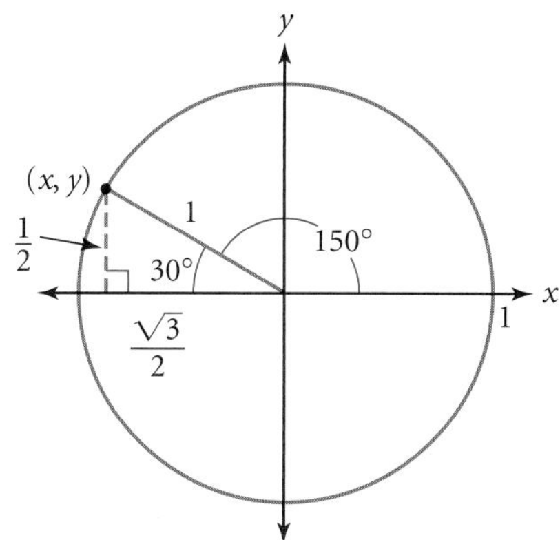

Notice that y remains positive in Quadrant II. Use your calculator to verify that $\sin 150°$ equals 0.5.

b. You can use the same unit circle diagram as above. The cosine is the x-coordinate of the rotated point. The length of the adjacent leg in the reference triangle is $\frac{\sqrt{3}}{2}$. Because the reference triangle is in Quadrant II, the x-coordinate is negative. Therefore, $x = \cos 150° = -\frac{\sqrt{3}}{2}$.

The calculator gives -0.866, which is approximately equal to $-\frac{\sqrt{3}}{2}$, as a decimal approximation of $\cos 150°$.

c. Rotate the point counterclockwise 210°, then draw the reference triangle. The reference angle in this triangle again measures 30°. Because this angle is in Quadrant III, the y-value is negative, so $\sin 210° = -\frac{1}{2}$.

The point $(210, -0.5)$ on the graph of $f(x) = \sin x$ verifies this result.

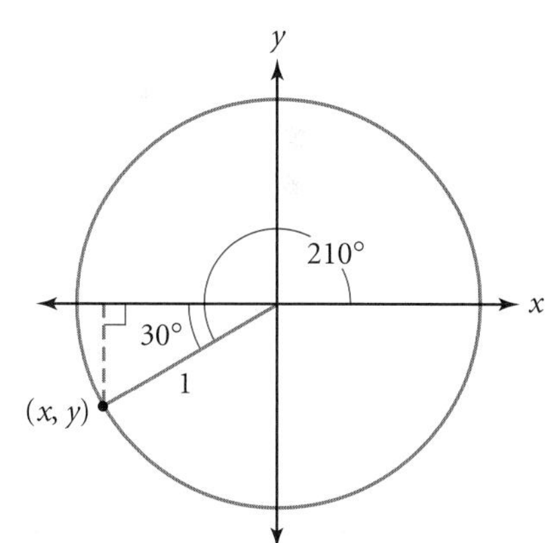

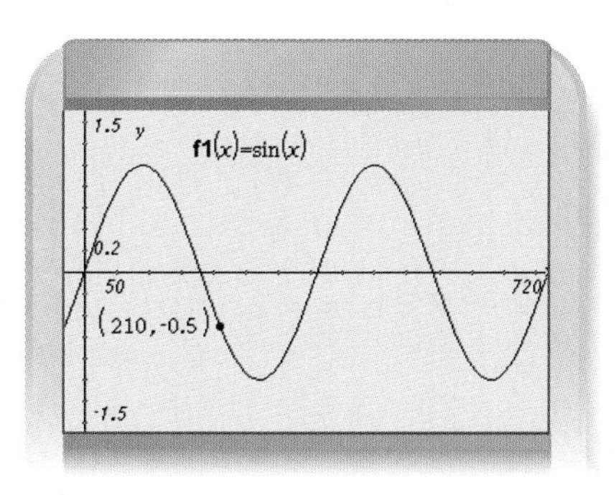

d. If you draw a reference triangle for 320°, you will find a reference angle of 40° in Quadrant IV. In this quadrant, x-values are positive. So, $\cos 320° = \cos 40°$. An angle measuring 40° is not a special angle, so you don't know its exact trigonometric values. According to the calculator, either $\cos 320°$ or $\cos 40°$ is approximately 0.766.

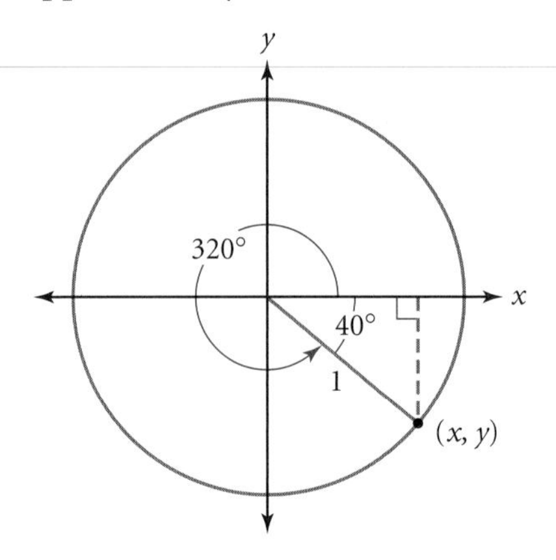

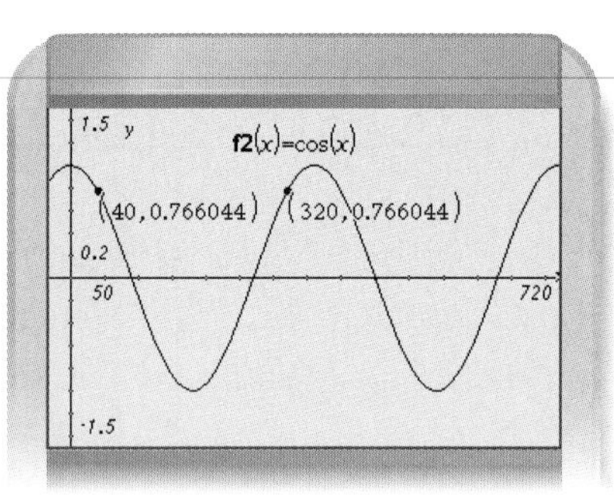

Angles in standard position are **coterminal** if they share the same terminal side. For example, the angles measuring $-145°$, 215°, and 575° are coterminal, as shown. Coterminal angles have the same trigonometric values.

Greek letters like θ (theta) and α (alpha) are frequently used to represent the measures of unknown angles. You'll see both of these variables used in the exercises.

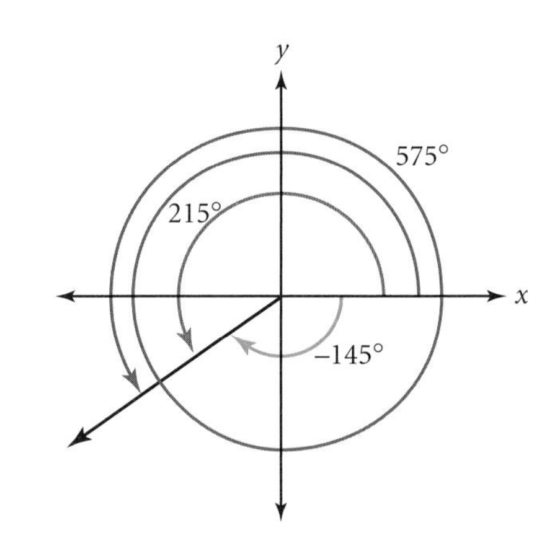

EXERCISES

You will need

A graphing calculator for Exercises **3**, **11**, **13–15**, and **20**.

Practice Your Skills

1. After 300 s, the paddle wheel in the investigation has rotated 300°. ⓐ

a. Draw a reference triangle and find the frog's height at this time.

b. Draw a graph of $y = \sin x$ and highlight the portion of that graph that pictures the frog's height during the first 300 s.

2. Use your calculator to find each value, approximated to four decimal places. Then draw a diagram in a unit circle to represent the value. Name the reference angle.

a. $\sin(-175°)$ ⓐ **b.** $\cos 147°$ **c.** $\sin 280°$ ⓐ **d.** $\cos 310°$ **e.** $\sin(-47°)$

3. Create a sine or cosine graph, and trace to find the value of each expression.

a. $\sin 120°$ **b.** $\sin(-120°)$ **c.** $\cos(-150°)$ **d.** $\cos 150°$

4. The functions $y = \sin x$ and $y = \cos x$ are periodic. How many cycles of each function are pictured?

a.

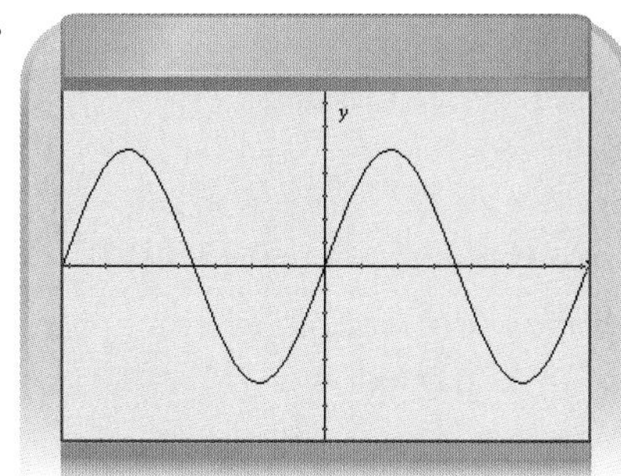

b.

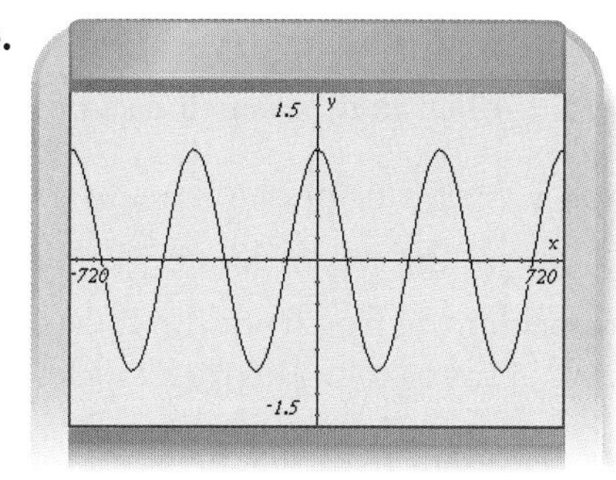

5. Which of the following functions are periodic? For each periodic function, identify the period.

a.

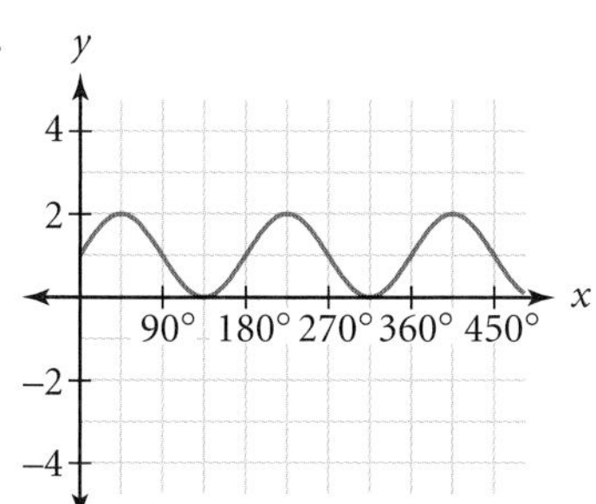

b.

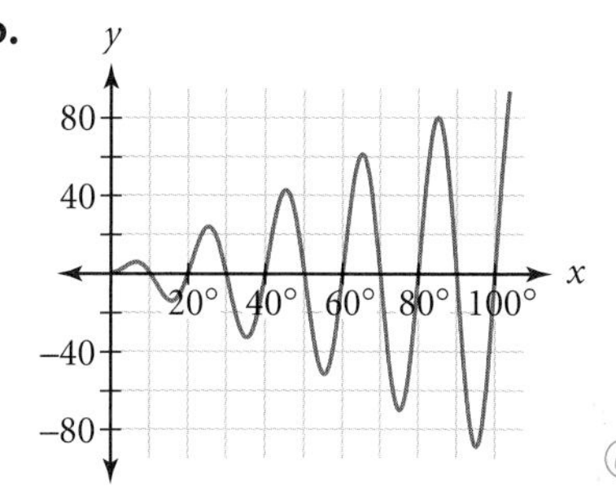

@

c.

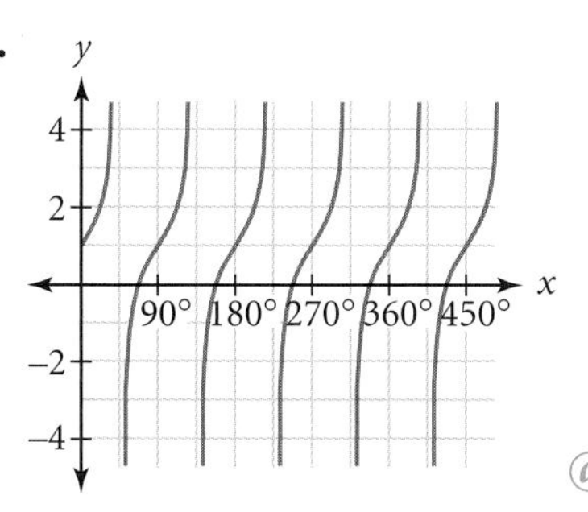

@

d.

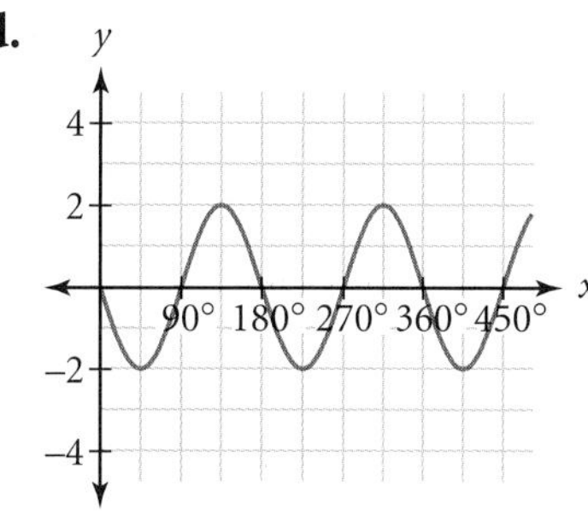

Reason and Apply

6. Identify an angle θ that is coterminal with the given angle. Use domain $0° \leq \theta \leq 360°$.

a. $-25°$ **b.** $-430°$ @ **c.** $435°$ @ **d.** $1195°$

7. For each Quadrant, I–IV, shown at right, identify whether the values of $\cos\theta$ and $\sin\theta$ are positive or negative.

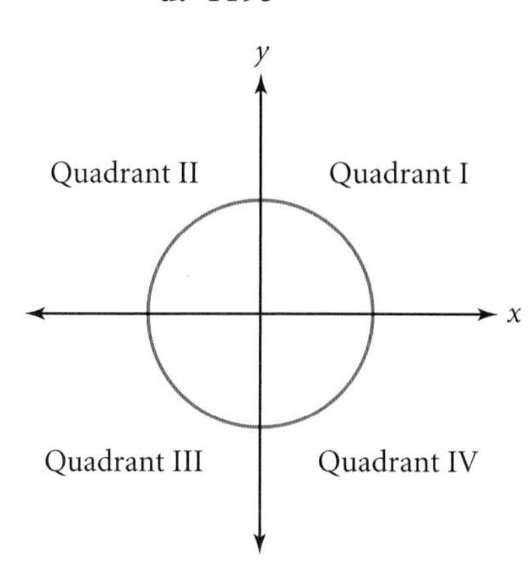

8. Carefully sketch a graph of the function $y = \sin x$ over the domain $-360° \leq x \leq 360°$. Identify all values of x in this interval for which $\sin x = 0$.

9. Carefully sketch a graph of the function $y = \cos x$ over the domain $-360° \leq x \leq 360°$. Identify all values of x in this interval for which $\cos x = 0$.

10. Suppose $\sin\theta = -0.7314$ and $180° \le \theta \le 270°$. ⓐ

a. Locate the point where the terminal side of θ intersects the unit circle.

b. Find θ and $\cos\theta$.

c. What other angle α has the same cosine value? Use domain $0° \le \alpha \le 360°$.

11. Graph $y = \sin x$ using the domain $0 \le x \le 720°$.

a. What are the x-coordinates of all points on the graph such that $\sin x = -0.7314$? ⓗ

b. Consider the frog from the Investigation Paddle Wheel. How do your answers from 11a relate to the frog's location and the amount of time passed?

12. Let θ represent the angle between the x-axis and the ray with endpoint (0, 0) passing through (2, 3).

a. Find $\sin\theta$ and $\cos\theta$. ⓗ

b. Find the coordinates of the point where the terminal side of θ intersects the unit circle. ⓗ

13. APPLICATION For the past several hundred years, astronomers have kept track of the number of sunspots. This table shows the average number of sunspots each day for the years from 1972 to 2006.

Year	Number of sunspots	Year	Number of sunspots	Year	Number of sunspots	Year	Number of sunspots
1972	68.9	1981	140.4	1990	142.6	1999	93.3
1973	38.0	1982	115.9	1991	145.7	2000	119.6
1974	34.5	1983	66.6	1992	94.3	2001	111.0
1975	15.5	1984	45.9	1993	54.6	2002	104.0
1976	12.6	1985	17.9	1994	29.9	2003	63.7
1977	27.5	1986	3.4	1995	17.5	2004	40.4
1978	92.5	1987	29.4	1996	8.6	2005	29.8
1979	155.4	1988	100.2	1997	21.5	2006	15.2
1980	154.6	1989	157.6	1998	64.3		

(www.sidc.oma.be/sunspot-data/)

a. Make a scatter plot of the data and describe any patterns that you notice. ⓐ

b. Estimate the length of a cycle. ⓐ

c. Predict the next period of maximum solar activity (years with larger numbers of sunspots) after 2006.

Science CONNECTION

Sunspots are dark regions on the Sun's surface that are cooler than the surrounding areas. They are caused by magnetic fields on the Sun and seem to follow short-term and long-term cycles.

Sunspot activity affects conditions on Earth. The particles emitted by the Sun during periods of high sunspot activity disrupt radio communications and have an impact on Earth's magnetic field and climate. Some scientists theorize that ice ages are caused by relatively low solar activity over a period of time.

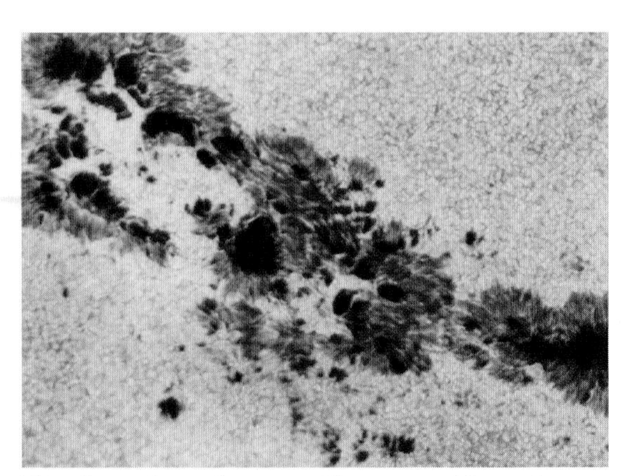

This photo of several sunspots shows powerful eruptions occurring on the Sun's surface.

14. Find each angle θ with the given trigonometric value. Use the domain $0 \le \theta \le 360°$.

a. $\cos\theta = -\frac{\sqrt{3}}{2}$ **b.** $\cos\theta = -\frac{\sqrt{2}}{2}$ **c.** $\sin\theta = -\frac{3}{5}$ **d.** $\sin\theta = 1$

15. ***Mini-Investigation*** Make a table of the values of $\sin\theta$, $\cos\theta$, $\tan\theta$, and $\frac{\sin\theta}{\cos\theta}$, using values of θ at intervals of 30° over the domain $0° \le \theta \le 360°$. What do you notice? Use the definitions of trigonometric ratios to explain your conjecture.

Review

16. Annie is standing on a canyon floor 20 m from the base of a cliff. Looking through her binoculars, she sees the remains of ancient cliff dwellings in the cliff face. Annie holds her binoculars at eye level, 1.5 m above the ground.

a. Write an equation that relates the angle at which she holds the binoculars to the height above ground of the object she sees. ⓐ

b. The top of the cliff is at an angle of 58° above horizontal when viewed from where Annie is standing. How high is the cliff, to the nearest tenth of a meter? ⓐ

c. Lower on the cliff, Annie sees ruins at angles of 36° and 40° from the horizontal. How high are the ruins, to the nearest tenth of a meter?

d. There is a nest of cliff swallows in the cliff face, 10 m above the canyon floor. At what angle should Annie point her binoculars to observe the nest? Round your answer to the nearest degree.

The Cliff Palace in Mesa Verde National Park, Colorado, was constructed by Ancestral Puebloan people around 1200 C.E.

17. Convert to the specified units using ratios. For example, to convert 0.17 meter to inches:

$$0.17\ \cancel{\text{m}} \cdot \frac{100\ \cancel{\text{cm}}}{1\ \cancel{\text{m}}} \cdot \frac{1\text{ in.}}{2.54\ \cancel{\text{cm}}} \approx 6.7\text{ in.}$$

a. 0.500 day to seconds **b.** 3.0 mi/h to ft/s (*Note:* 1 mile = 5280 feet)

18. Find the circumference and area of the circle with equation $2x^2 + 2y^2 - 2x + 7y - 38 = 0$.

19. Rewrite each expression as a single rational expression in factored form. ⓐ

a. $\frac{x+1}{x-4} - \frac{x+2}{x+4} + \frac{4x}{16-x^2}$ **b.** $\frac{2x^2-2}{x^2+3x+2} \cdot \frac{x^2-x-6}{x^2-4x+3}$ **c.** $\frac{1+\frac{a}{3}}{1-\frac{a}{6}}$

20. Write an equation for a rational function, $f(x)$, that has vertical asymptotes $x = -4$ and $x = 1$, horizontal asymptote $y = 2$, and zeros $x = -2$ and $x = 5$. Check your answer by graphing the equation on your calculator.

21. Find an angle in standard position, θ, for a plane flying on these bearings. Use domain $-180° \le \theta \le 180°$. ⓗ

a. 105° **b.** 325° **c.** 180° **d.** 42°

LESSON

13.2

Radian Measure

The measure of our intellectual capacity is the capacity to feel less and less satisfied with our answers to better and better problems.

C. WEST CHURCHMAN

You've learned that in a circle, the measure of an arc in degrees, m, the length of the arc, s, and the circumference, C, of the circle are related by the proportion $\frac{m}{360°} = \frac{s}{C}$. This formula applies when you are measuring the arc in degrees. However, the choice to divide a circle into 360° is rooted in the history of mathematics and is not connected to any fundamental properties of circles. In this lesson you will learn about a different angle measure that is based on an essential property of circles.

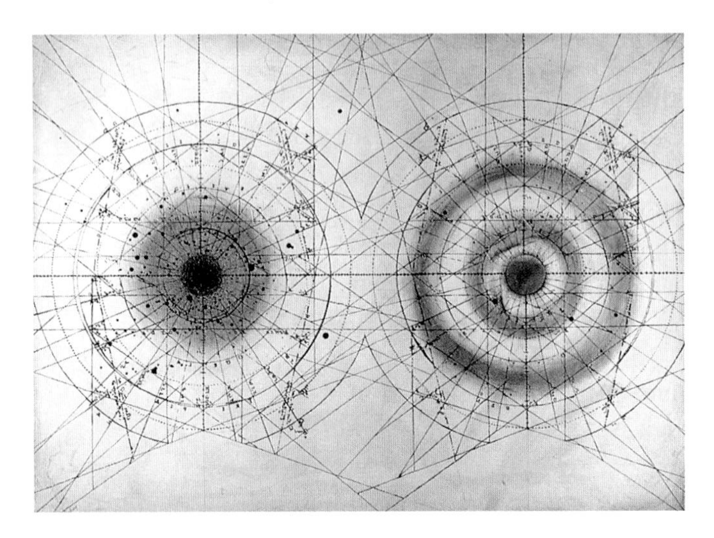

American artist Penny Cerling (b 1946) created this piece, titled *Black Hole and Big Bang* (1999), using pen, ink, and oil on wood. Cerling explores themes of science and nature in many of her works.

History CONNECTION

More than 5000 years ago, the Sumerians in Mesopotamia used a base-60 number system. They may have chosen this system because numbers like 30, 60, and 360 can be evenly divided by many numbers. The Babylonians and Egyptians then borrowed this system and divided the circle into 360 degrees. The Egyptians also devised the symbol for degrees and went on to divide both Earth's equator and north-south great circles into 360 degrees, inventing latitude and longitude lines. The Greek astronomer and mathematican Hipparchus of Rhodes (ca. 190–120 B.C.E.) is credited with introducing the Babylonian division of the circle to Greece and producing a table of chords, the earliest known trigonometric table. Hipparchus is often called the "founder of trigonometry."

Investigation
A Radian Protractor

You will need

- string
- the worksheet A Radian Protractor

Use the semicircle on the worksheet to complete the investigation.

Step 1 Mark the length of the radius of the semicircle on your string.

Step 2 Starting from the right-hand base, use the radius length of string to measure off an arc whose length is the same as the radius.

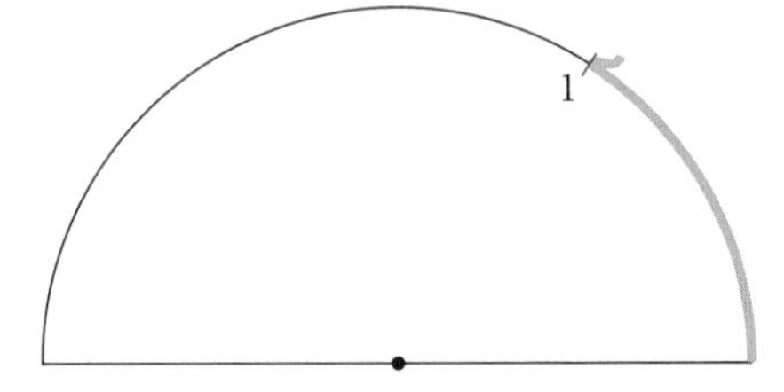

Step 3 The central angle intercepting this arc has a measure of 1 **radian.** Use your radius length of string to mark points that correspond to angles with measures of 2 and 3 radians.

Step 4 Fold your radius length of string in half and use this length to locate points corresponding to angles with measures 0.5, 1.5, and 2.5 radians.

Step 5 If the radius of your circle is r, calculate the length of a semicircular arc. How many radians is this? Mark this radian value on your protractor.

Step 6 What is the length of the arc intercepted by a right angle? How many radians are associated with a right angle? Mark this radian value on your protractor.

Step 7 Find radian values that correspond with these common angle measurements and mark them on your protractor: 30°, 45°, 60°, 120°, 135°, and 150°.

Like degrees, radians are used to measure angles and rotations. The radian measure of an angle can be found by constructing *any* arc intercepted by that angle, with its center at the vertex of the angle, and dividing the length of that arc by its radius.

$$\textit{radian measure} = \frac{\textit{arc length}}{\textit{radius}}$$

arc length

s

r

radius

The circumference of a circle depends on the length of its radius. Using the radian measure definition and the formula $C = 2\pi r$, you will find that the radian measure of a full circle, or 360° rotation, is $\frac{\textit{arc length}}{\textit{radius}} = \frac{2\pi r}{r} = 2\pi$ radians.

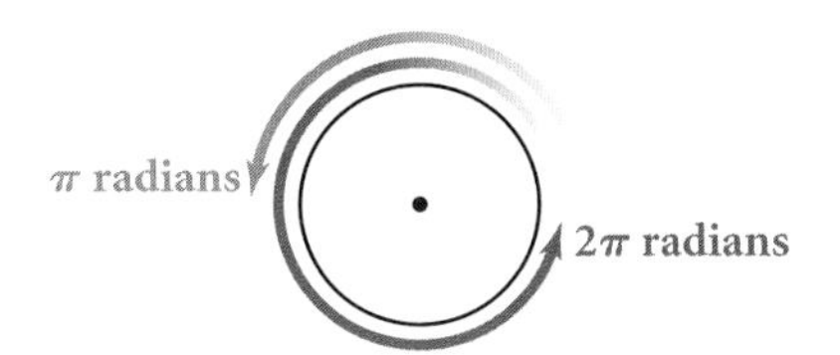

The conversion formula

$$\frac{\textit{angle in degrees}}{360} = \frac{\textit{angle in radians}}{2\pi}$$

is based on the full circle, or one rotation. You can also use an equivalent formula based on a half rotation.

$$\frac{\textit{angle in degrees}}{180} = \frac{\textit{angle in radians}}{\pi}$$

If you solve the last proportion for either the angle in degrees or the angle in radians, you get a formula that you can use to convert from one system to the other.

$$\textit{angle in degrees} = \frac{180}{\pi}(\textit{angle in radians})$$

$$\textit{angle in radians} = \frac{\pi}{180}(\textit{angle in degrees})$$

Radian measures of most common angles are irrational numbers, but you can express them as exact values in terms of π.

Radian measure is based on a fundamental property of a circle: The circumference of any circle is 2π times its radius. For this reason, radian measure is often preferred in advanced mathematics and in physics. You can learn to recognize, compare, and use radians and degrees, just as in the past you have worked with inches and centimeters or Fahrenheit and Celsius.

Degree measures are always labeled with the symbol °. Radian measures are not usually labeled, but they can be labeled for clarity.

EXAMPLE A

Convert degrees to radians or radians to degrees.

a. $\frac{2\pi}{3}$

b. n radians

c. $225°$

d. $n°$

▶ Solution

You can use either conversion formula. Here we will use the equivalence $180° = \pi$ radians.

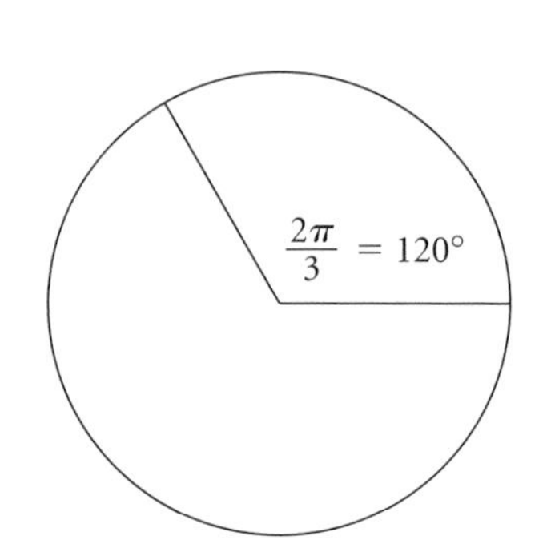

a. You can think of $\frac{2\pi}{3}$ as $\frac{2}{3}(\pi)$, or $\frac{(2\pi)}{3}$. Although no units are given, it is clear that these are radians.

$$\frac{x}{180} = \frac{\frac{2\pi}{3}}{\pi} \text{ or } \frac{x}{180} = \frac{2}{3}$$

So $x = 120°$. Therefore, $\frac{2\pi}{3} = 120°$.

b. $\frac{x}{180} = \frac{n}{\pi}$ or $x = \frac{n}{\pi} \cdot 180$

So n radians $= \left(\frac{180n}{\pi}\right)°$.

c. $\frac{225}{180} = \frac{x}{\pi}$ or $x = \frac{225}{180} \cdot \pi = \frac{5\pi}{4}$

So $225° = \frac{5\pi}{4}$ radians.

d. $\frac{n}{180} = \frac{x}{\pi}$ or $x = \frac{n}{180} \cdot \pi$

So $n° = \frac{n\pi}{180}$ radians.

You can use the relations you found in parts b and d in Example A to convert easily between radians and degrees.

You can also use dimensional analysis to convert between degrees and radians. Dimensional analysis is a procedure based on multiplying by fractions formed of conversion factors to change units. For example, to convert 140° to radian measure, you write it as a fraction over 1 and multiply it by $\frac{\pi \text{ radians}}{180°}$, which is a fraction equal to 1.

$$\frac{140 \cancel{\text{ degrees}}}{1} \cdot \frac{\pi \text{ radians}}{180 \cancel{\text{ degrees}}} = \frac{140\pi \text{ radians}}{180} = \frac{7\pi}{9} \text{ radians}$$

You can use your calculator to check or approximate conversions between radians and degrees. [▶ See **Calculator Note 13B.** ◀]

Earlier you reviewed the proportion $\frac{m}{360°} = \frac{s}{C}$, which relates the arc length, *s*, and arc measure in degrees, *m*. Using radians, you can write the relationship $\frac{s}{2\pi r} = \frac{\theta}{2\pi}$, where *s* is the arc length, *r* is the radius, and θ is the central angle measured in radians. If you solve this equation for *s*, you'll get $s = r\theta$ as a formula for arc length. You can also use radian measure to write a simple formula for the area of a sector.

EXAMPLE B

Circle *O* has diameter 10 m. The measure of central angle *BOC* is 1.4 radians. What is the length of its intercepted arc, $\overset{\frown}{BC}$? What is the area of the shaded sector?

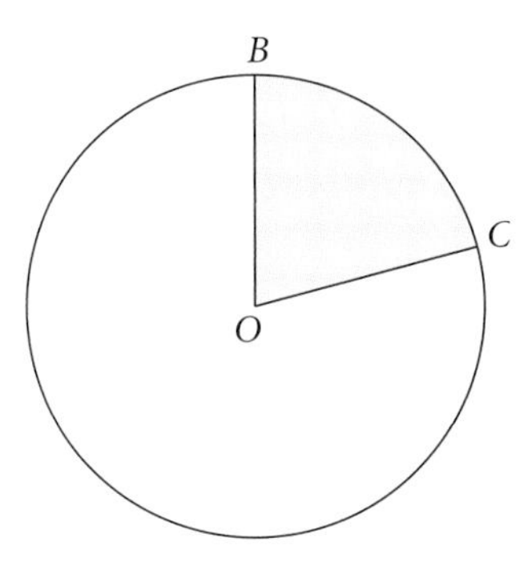

▸ Solution

In the formula $s = r\theta$, substitute 5 for *r* and 1.4 for θ, and solve for *s*. The length of $\overset{\frown}{BC}$ is $5 \cdot 1.4$, or 7 m.

The ratio of the area of the sector to the total area of the circle is the same as the ratio of the measure of $\angle BOC$ to 2π radians.

$$\frac{A_{\text{sector}}}{A_{\text{circle}}} = \frac{1.4}{2\pi}$$

The area of the circle is πr^2, or 25π m^2.

$$\frac{A_{\text{sector}}}{25\pi} = \frac{1.4}{2\pi}$$

Solving the equation, you find that the area of the sector is 17.5 m^2.

As you saw in the example, the area of a sector with radius *r* and central angle θ can be found using the proportion

$$\frac{A_{\text{sector}}}{\pi r^2} = \frac{\theta}{2\pi}$$

You can rewrite this relationship as

$$A_{\text{sector}} = \frac{\pi r^2\theta}{2\pi} = \frac{r^2\theta}{2} = \frac{1}{2}r^2\theta$$

Length of an Arc and Area of a Sector

When a central angle, θ, of a circle with radius *r* is measured in radians, the length of the intercepted arc, *s*, is given by the equation

$$s = r\theta$$

and the area of the intercepted sector, *A*, is given by the equation

$$A = \frac{1}{2}r^2\theta$$

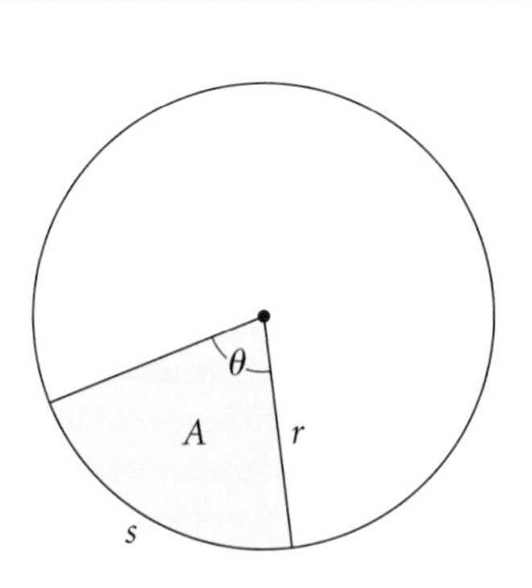

When an object follows a circular path, the distance it travels is the arc length. You can calculate its speed as distance traveled per unit of time. The amount of rotation, or angle traveled per unit of time, is called the **angular speed.**

EXAMPLE C

The Cosmo Clock 21 Ferris wheel at the Cosmo World amusement park in Yokohama, Japan, has a 100 m diameter. This giant Ferris wheel, with 60 gondolas and 8 people per gondola, makes one complete rotation every 15 minutes. The wheel reaches a maximum height of 112.5 m from the ground.

a. Find the speed of a person on this Ferris wheel as it is turning.

b. Find the angular speed of this person.

The Cosmo Clock 21 Ferris wheel, which opened in 1999, can hold as many as 480 passengers at a time.

▸ *Solution*

The speed is the distance traveled per unit of time, measured in units such as meters per second. The angular speed is the rate of rotation, measured in units like radians per second or degrees per second.

a. The person travels one complete circumference every 15 min, or

$$\frac{2\pi r \text{ m}}{15 \text{ min}} = \frac{2\pi \cdot 50 \text{ m}}{15 \text{ min}} \approx 20.94 \frac{\text{m}}{\text{min}}$$

So the person travels at about 21 m/min. You can also use dimensional analysis to express this speed as approximately 1.26 km/h.

$$\frac{20.94 \cancel{\text{m}}}{1 \cancel{\text{min}}} \cdot \frac{60 \cancel{\text{min}}}{1 \text{ h}} \cdot \frac{1 \text{ km}}{1000 \cancel{\text{m}}} \approx \frac{1.26 \text{ km}}{1 \text{ h}}$$

b. The person completes one rotation, or 2π radians, every 15 min.

$$\frac{2\pi \text{ radians}}{15 \text{ min}} \approx 0.42 \text{ radian/min}$$

So the person's angular speed is 0.42 radian/min.

Using dimensional analysis, you can convert between units to express answers in any form.

$$\frac{2\pi \cancel{\text{radians}}}{15 \text{ min}} \cdot \frac{360°}{2\pi \cancel{\text{radians}}} = \frac{24°}{1 \text{ min}}$$

So you can also express the angular speed as 24°/min.

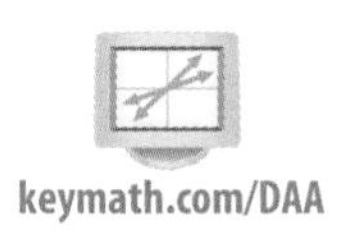
keymath.com/DAA

A central angle has a measure of 1 radian when its intercepted arc is the same length as the radius. Similarly, the number of radians in an angle measure is the number of radii in the arc length of its intercepted arc. [▶ You can use the **Dynamic Algebra Exploration** at www.keymath.com/DAA to explore radians through interactive sketches.◀]

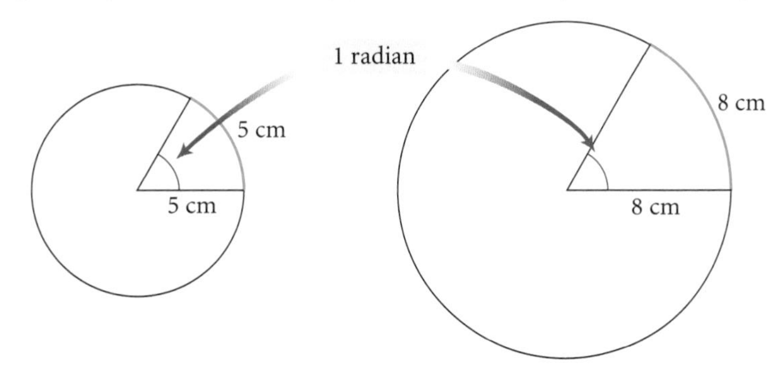

EXERCISES

You will need

A graphing calculator for Exercises **7–9**.

Practice Your Skills

1. Convert between radians and degrees. Give exact answers. (Remember that degree measures are always labeled °, and radians generally are not labeled.)

a. 80° @ **b.** 570° @ **c.** $-\frac{4\pi}{3}$ **d.** $\frac{11\pi}{9}$

e. $-\frac{3\pi}{4}$ **f.** 3π **g.** $-900°$ @ **h.** $\frac{5\pi}{6}$

2. Find the length of the intercepted arc for each central angle.

a. $r = 3$ and $\theta = \frac{2\pi}{3}$ @

b. $r = 2$ and $\theta = 2$

c. $d = 5$ and $\theta = \frac{\pi}{6}$

3. Draw a large copy of this diagram on your paper. Each angle shown has a reference angle of 0°, 30°, 45°, 60°, or 90°.

a. Find the counterclockwise degree rotation of each segment from the positive x-axis. Write your answers in both degrees and radians.

b. Find the exact coordinates of points A–N.

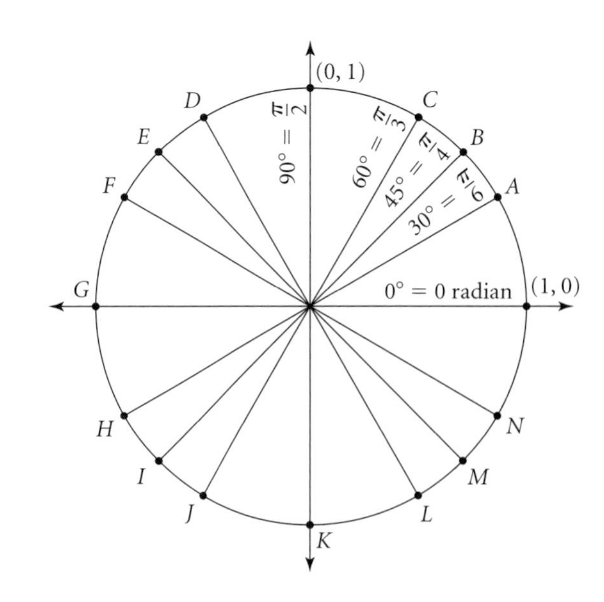

4. One radian is equivalent to how many degrees? One degree is equivalent to how many radians?

Reason and Apply

American actor Harold Lloyd (1893–1971), shown here hanging from a clock in the film *Safety Last* (1923), often performed daring physical feats in his more than 500 comedic films.

5. Are 6 radians more than, less than, or the same as one rotation about a circle? Explain.

6. The minute hand of a clock is 15.2 cm long.

a. What is the distance the tip of the minute hand travels during 40 minutes?

b. At what speed is the tip moving, in cm/min? (a)

c. What is the angular speed of the tip in radians/minute?

7. *Mini-Investigation* Without using a graphing calculator, graph $y = \sin x$ over the domain $0 \leq x \leq 2\pi$.

a. On the x-axis, label all the x-values that are multiples of $\frac{\pi}{6}$.

b. On the x-axis, label all the x-values that are multiples of $\frac{\pi}{4}$.

c. What x-values in this domain correspond to a maximum value of $\sin x$? A minimum value of $\sin x$? $\sin x = 0$?

d. Use your graphing calculator to check your graph. (h)

8. *Mini-Investigation* Without using a graphing calculator, graph $y = \cos x$ over the domain $0 \leq x \leq 2\pi$.

a. On the x-axis, label all the x-values that are multiples of $\frac{\pi}{6}$.

b. On the x-axis, label all the x-values that are multiples of $\frac{\pi}{4}$.

c. What x-values in this domain correspond to a maximum value of $\cos x$? A minimum value of $\cos x$? $\cos x = 0$?

d. Use your graphing calculator to check your graph. (h)

9. *Mini-Investigation* Without using a graphing calculator, graph $y = \frac{\sin x}{\cos x}$ (that is, $y = \tan x$) over the domain $-\pi \leq x \leq \pi$.

a. On the x-axis, label all the x-values that are multiples of $\frac{\pi}{6}$ or $\frac{\pi}{4}$.

b. Name the x-values in this domain where the denominator, $\cos x$, is 0. What can you say about the function at these values?

c. Which x-values in the domain $-\pi \leq x \leq \pi$ produce vertical asymptotes?

d. Use your graphing calculator to verify your graph of $y = \frac{\sin x}{\cos x}$ over the domain $-\pi \leq x \leq \pi$. Confirm that the graph is the same as the graph of $y = \tan x$. (h)

10. Find the value of r in the circle at right.

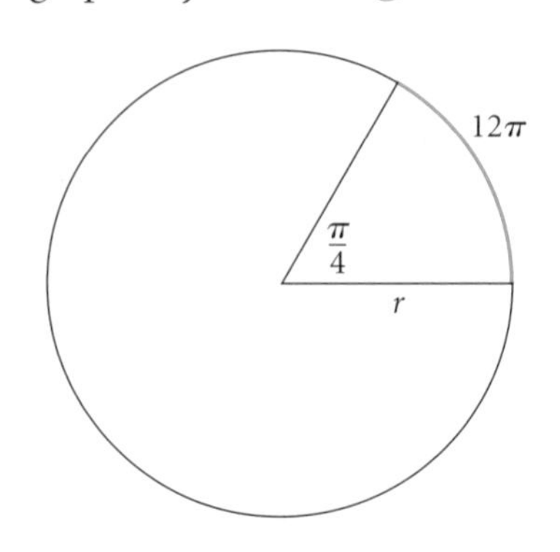

11. Solve for θ. Express your answers in radians.

a. $\cos\theta = -\frac{1}{2}$ and $\pi \le \theta \le \frac{3\pi}{2}$ ⓐ

b. $\cos\theta = \frac{\sqrt{2}}{2}$ and $\frac{3\pi}{2} \le \theta \le 2\pi$

c. $\frac{\sin\theta}{\cos\theta} = \sqrt{3}$ and $0 \le \theta \le \frac{\pi}{2}$

12. Suppose you are biking down a hill at 24 mi/h. What is the angular speed, in radians per second, of your 27-inch-diameter bicycle wheel? ⓐ

Recreation CONNECTION

Fred Rompelberg of the Netherlands broke the world record in 1995 for the fastest recorded bicycle speed, 167.043 mi/h. He rode a specially-designed bicycle behind a race car, which propelled him forward by slipstream, an airstream that reduces air pressure. The record-breaking event took place at Bonneville Salt Flats in Utah.

Dutch cyclist Fred Rompelberg (b 1945)

13. A sector of a circle with radius 8 cm has central angle $\frac{4\pi}{7}$.

a. Use the formula $A = \frac{1}{2}r^2\theta$ to find the area of the sector.

b. Set up a proportion of the area of the sector to the total area of the circle, and a proportion of the central angle of the sector to the total central angle measure.

c. Solve your proportion from 13b to show that the formula you used in 13a is correct.

14. Sitting at your desk, you are approximately 6380 km from the center of Earth. Consider your motion relative to the center of Earth as Earth rotates on its axis.

a. What is your angular speed?

b. What is your speed in km/h?

c. What is your speed in mi/h?

15. APPLICATION The two cities Minneapolis, Minnesota, and Lake Charles, Louisiana, lie on the 93° W longitudinal line. The latitude of Minneapolis is 45° N (45° north of the equator), whereas the latitude of Lake Charles is 30° N. The radius of Earth is approximately 3960 mi. ⓗ

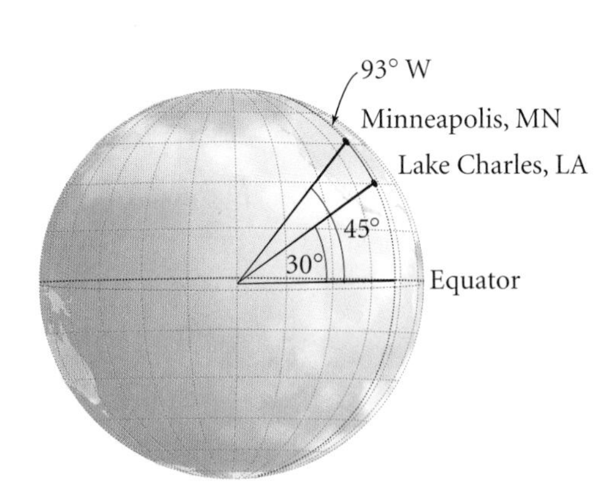

a. Calculate the distance between the two cities.

b. Like hours, degrees can be divided into minutes and seconds for more precision. (There are 60 seconds in a minute and 60 minutes in a degree.) For example, 61°10′ means 61 degrees 10 minutes. Write this measurement as a decimal.

c. If you know the latitudes and longitudes of two cities, you can find the distance in miles between them, D, using this formula:

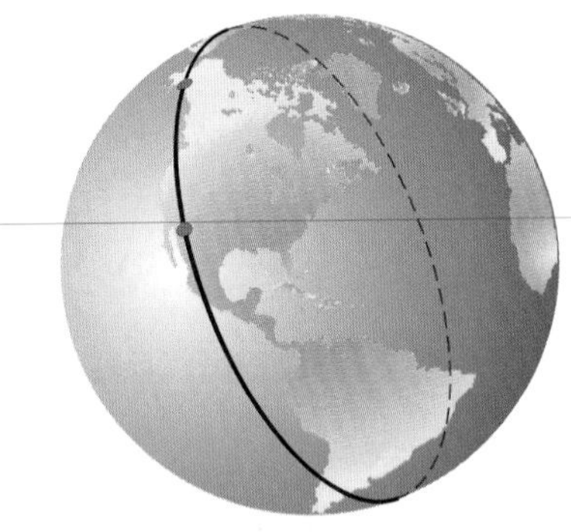

$$D = \frac{\pi \cdot r}{180}\cos^{-1}\left[\sin\phi_A \sin\phi_B + \cos\phi_A \cos\phi_B \cos\left(\theta_A - \theta_B\right)\right]$$

In the formula, r is the radius of Earth in miles, ϕ_A and θ_A are the latitude and longitude of city A in degrees, and ϕ_B and θ_B are the latitude and longitude of city B in degrees. North and east are considered positive angles, and south and west are considered negative.

Using this formula, find the distance between Anchorage, Alaska (61°10′ N, 150°1′ W), and Tucson, Arizona (32°7′ N, 110°56′ W).

History CONNECTION

To find the shortest path between two points on a sphere, you connect the two points with a great arc, an arc of a circle that has the same center as the center of Earth. Pilots fly along great circle routes to save time and fuel. Charles Lindbergh's carefully planned 1927 flight across the Atlantic was along a great circle route that saved about 473 miles compared with flying due east.

Review

16. List the transformations of each graph from its parent function.

a. $y = 2 + (x + 4)^2$

b. $\frac{y}{3} = \left(\frac{x-5}{4}\right)^2$ @

c. $y + 1 = |x - 3|$

d. $y = 3 - 2|x + 1|$ @

17. Write an equation for each graph. @

a.

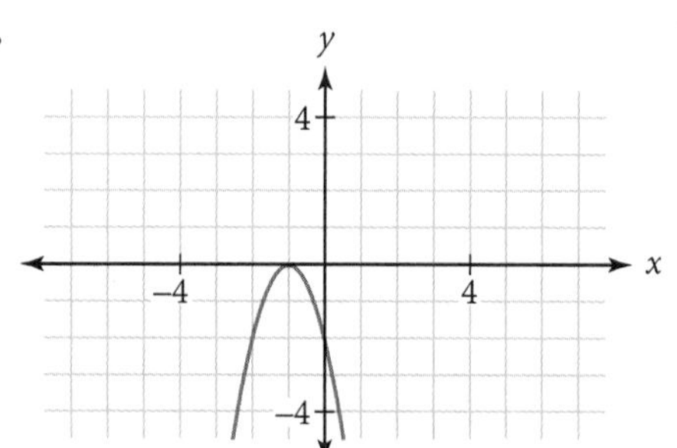

b.

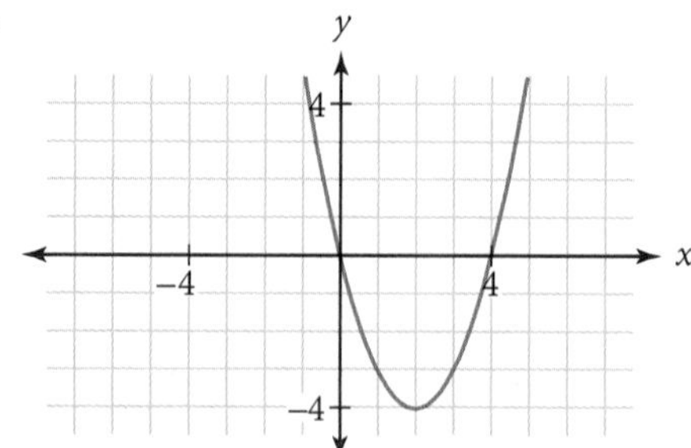

c.

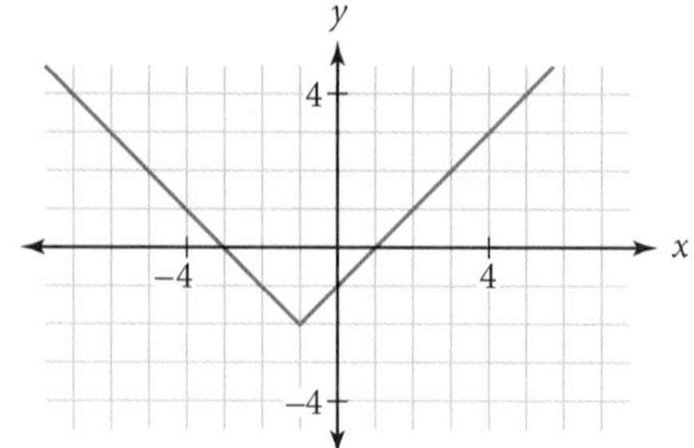

d.

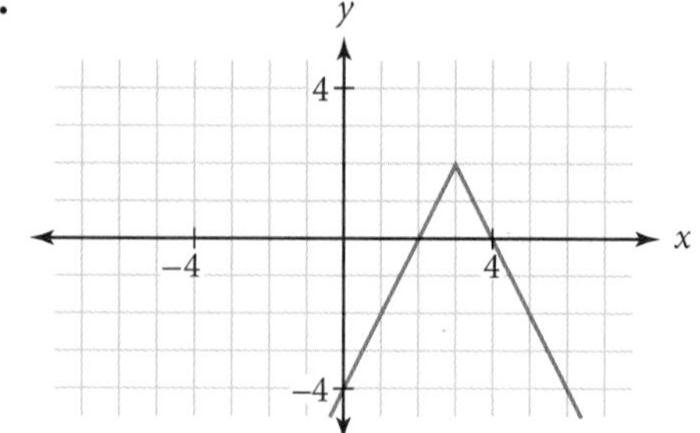

18. Find a second value of θ that gives the same trigonometric value as the angle given. Use domain $0° \leq \theta \leq 360°$.

a. $\sin 23° = \sin\theta$

b. $\sin 216° = \sin\theta$

c. $\cos 342° = \cos\theta$

d. $\cos 246° = \cos\theta$

19. While Yolanda was parked at her back steps, a slug climbed onto the wheel of her go-cart just above where the wheel makes contact with the ground. When Yolanda hopped in and started to pull away, she did not notice the slug until her wheels had rotated $2\frac{1}{3}$ times. Yolanda's wheels have a 12 cm radius.

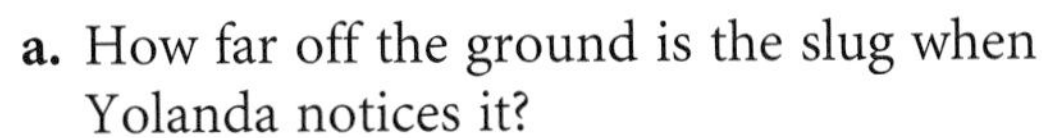

a. How far off the ground is the slug when Yolanda notices it?

b. How far horizontally has the slug moved?

20. Sketch a rectangle with length a and width $2a$. Inscribe an ellipse in the rectangle. What is the eccentricity of the ellipse? Why? ⓐ

21. Circles O and P, shown below, are tangent at A. Explain why $AB = BC$. (*Hint:* One method uses congruence.) ⓗ

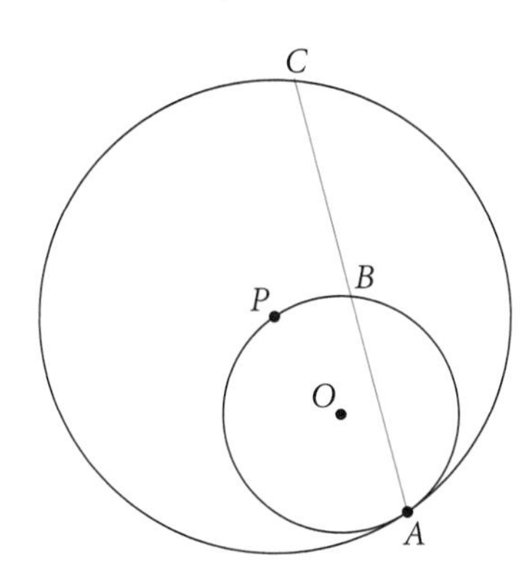

EXPLORATION

Circular Functions

In Lesson 13.1, you plotted the position of a frog on a paddle wheel over time and saw how the graphs of sine and cosine are related to a point rotating around a circle. In this exploration you will construct a dynamic sketch that will allow you to review and further explore the graphs of the sine, cosine, and tangent functions.

Activity

Tracing Parent Graphs

You'll start by creating a coordinate system, constructing a unit circle, and making some measurements.

Step 1 In a new sketch, choose **Edit | Preferences.** Use the Preferences dialog box to set the Angle Units to **radians** and the Precision for slopes and ratios to **thousandths.**

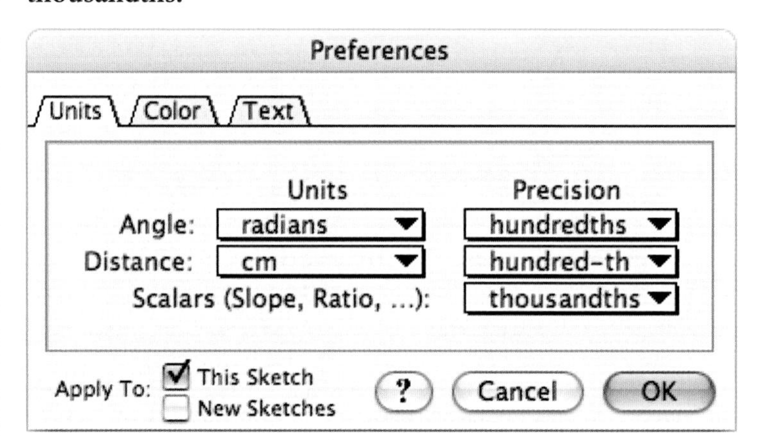

Step 2 Choose **Graph | Show Grid.** Resize the axes by dragging the number on one of the tick marks, and drag the y-axis to the left to show the domain $-2 \leq x \leq 7$.

Step 3 Use the **Label** tool to label the points at the origin and (1, 0) as A and B, respectively.

Step 4 Construct a unit circle. Use the **Arrow** tool to select point A and point B, in order, and choose **Construct | Circle By Center + Point.** Use the **Point** tool to construct a point on the circle and then label the point C.

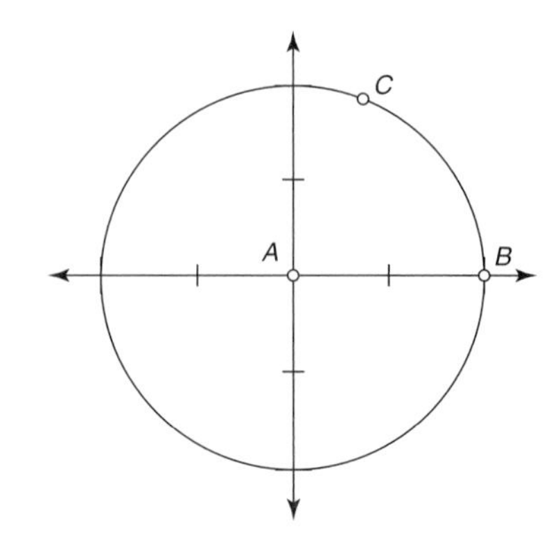

Step 5 Measure the x- and y-coordinates of this new point separately. With only the point selected, choose **Measure | Abscissa (x).** Click in white space, select the point again, and choose **Measure | Ordinate (y).**

Step 6 Choose the **Line** tool and construct a line through the origin and point C. With the line selected, measure its slope by choosing **Measure | Slope.**

Step 7 Now you will construct $\overset{\frown}{BC}$. Click in white space. Then select, in order, the circle, point B, and point C. Choose **Construct | Arc On Circle.** Choose **Display | Line Width | Thick** to make it thick. If you want, you can change the color of the arc by choosing **Display | Color** and selecting a color.

Step 8 With the arc still selected, measure its arc angle. Choose **Measure | Arc Angle.** Drag point C around the circle and observe how all four measurements change. When does each measurement reach its maximum and minimum values?

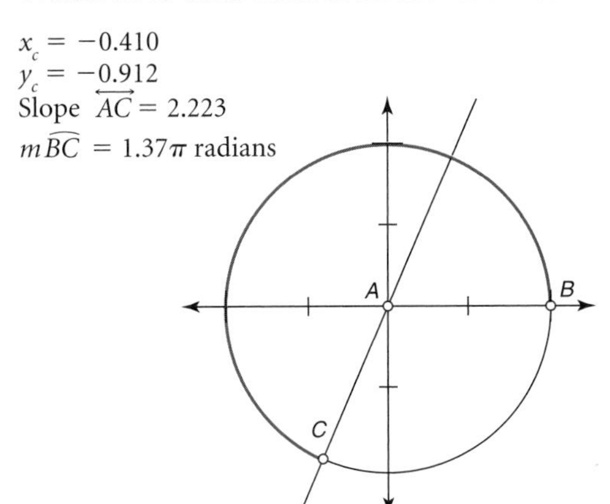

Now you will plot each measurement as a function of the arc angle. Can you predict what the three graphs will look like?

Step 9 Plot the y-coordinate of point C as a function of the arc angle. Select the arc angle measurement, $m\overset{\frown}{BC}$, and the y-coordinate measurement, y_C, in order, and choose **Graph | Plot as (x, y).** With the plotted point selected, choose **Display | Trace Plotted Point.**

Step 10 Drag point C around the circle and examine the trace that appears. Which trigonometric function is this? When you have finished exploring, select the plotted point and turn off the tracing. You might also erase the traces before tracing the next graph.

Step 11 Now plot the x-coordinate of point C as a function of the arc angle. Turn on tracing for this plotted point (you can change the color of the point to change the color of the trace). Drag C around the circle and observe the trace of the plotted point. Which trigonometric function is this?

Step 12 Plot the slope of the line as a function of the arc angle. Turn on tracing, drag C, and observe the result.

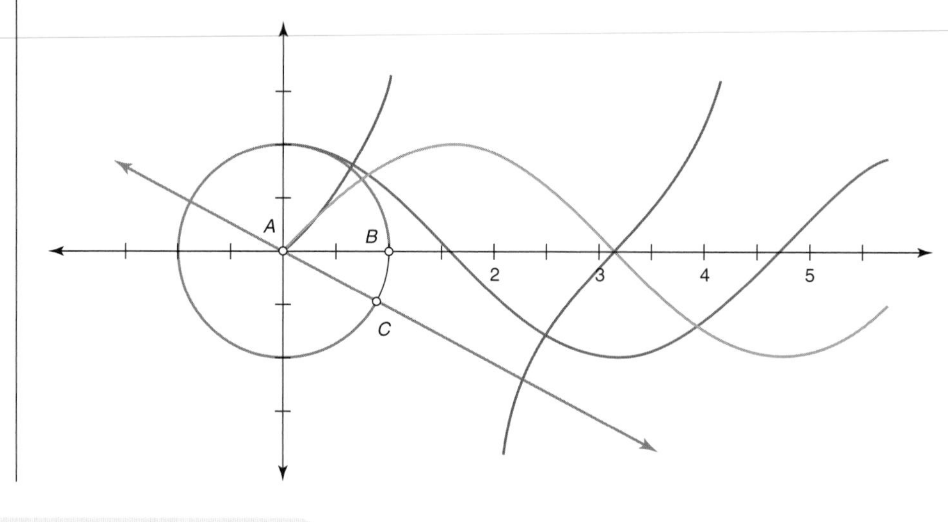

Questions

1. You have seen that the sine and cosine graphs are periodic. How much of a cycle is shown of each graph by the trace? Explain.

2. Does the graph from Step 12 appear to be periodic? Explain.

3. Choose **Measure | Calculate** and calculate the value of y_C / x_C. (Click on each measure on the screen to enter it into the calculator.) Compare this value with the value of the slope while you drag point C. What do you notice? Explain your observations.

4. What arc measures correspond to a slope of 0? To an undefined slope? How are these values represented on the graph of (*arc angle, slope*)?

5. Choose **Display | Erase Traces.** Then choose **Graph | Plot New Function** and graph $y = \tan(x)$. What is the period of the tangent function? Describe other characteristics of the graph that you can observe in the sketch.

LESSON

13.3

Graphing Trigonometric Functions

The best way to predict the future is to invent it.

BENJAMIN PIMENTEL

Graphs of the functions $y = \sin x$ and $y = \cos x$ and transformations of these graphs are collectively called **sine waves** or **sinusoids.** You will find that reflecting, translating, and dilating sinusoidal functions and other **trigonometric functions** is very much like transforming any other function. In this lesson you will explore many real-world situations in which two variables can be modeled by a sinusoidal function in the form $y = k + b\sin\left(\frac{x-h}{a}\right)$, or $\frac{y-k}{b} = \sin\left(\frac{x-h}{a}\right)$.

The wavy terraces of the Hsinbyume Pagoda in Mingum, Myanmar, may represent the seven surrounding hills or the seven seas of the universe.

EXAMPLE A

The graph of one cycle $(0 \leq x \leq 2\pi)$ of $y = \sin x$ is shown at right. Sketch the graph of one cycle of

a. $y = 2 + \sin x$

b. $y = \sin(x - \pi)$

c. $y = 3 + 2\sin(x + \pi)$

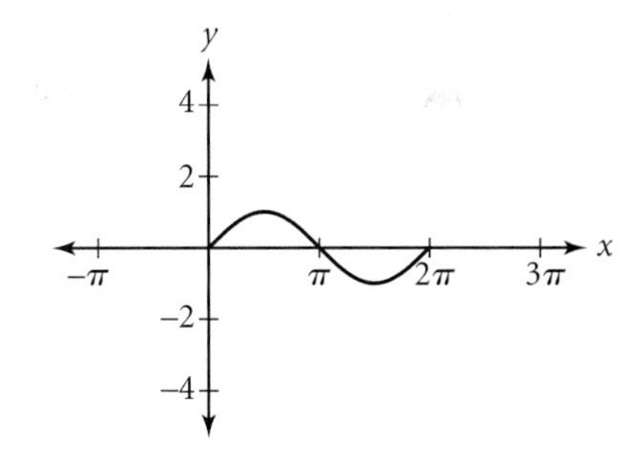

▶ Solution

Use your knowledge of transformations of functions.

a. In relation to the graph of any parent function, $y = f(x)$, the graph of $y = 2 + f(x)$ is a translation up 2 units. The graph of one cycle of $y = 2 + \sin x$ is therefore a translation up 2 units from the graph of $y = \sin x$.

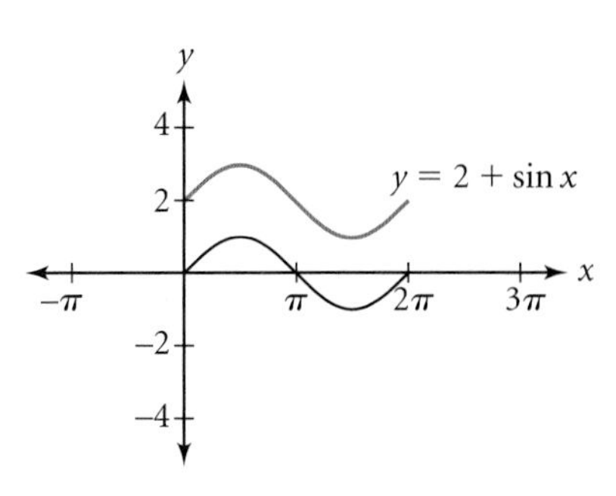

b. When x is replaced by $(x - \pi)$ in any function, the graph translates right π units.

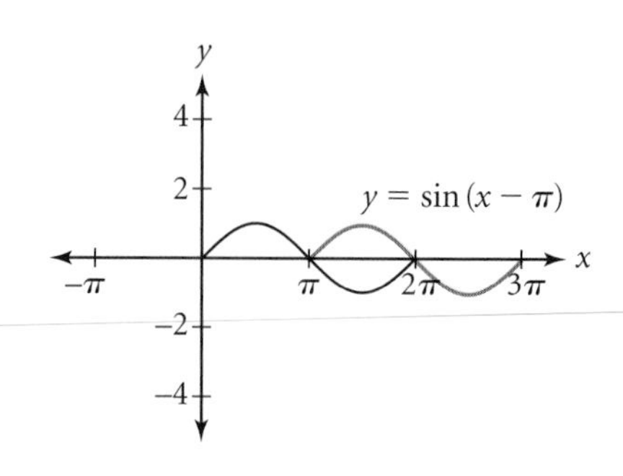

c. The coefficient 2 means the graph $y = \sin x$ must be dilated vertically by a factor of 2. The graph must also be translated left π units and up 3 units.

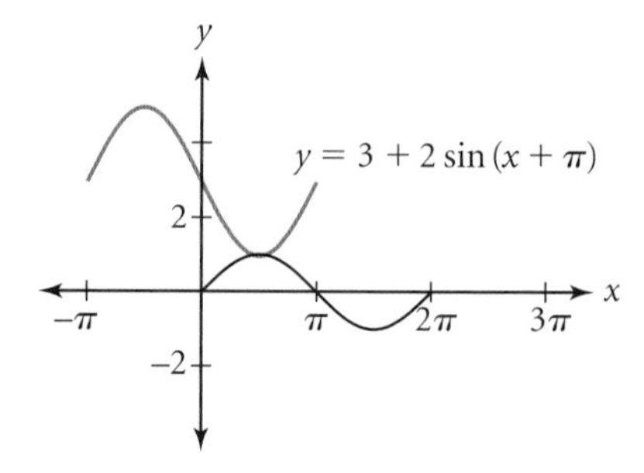

Recall that the period of a function is the smallest distance between values of the independent variable before the cycle begins to repeat. You've discovered that the period of both $y = \sin x$ and $y = \cos x$ is 2π, or 360°. The period of each of the three functions in Example A is also 2π, because they were not dilated horizontally.

The **amplitude** of a sinusoid is half the difference of the maximum and minimum function values, or $\frac{maximum - minimum}{2}$. This is the same as the absolute value of the vertical scale factor, or $|b|$. The amplitude of $y = \sin x$ and $y = \cos x$ is 1 unit.

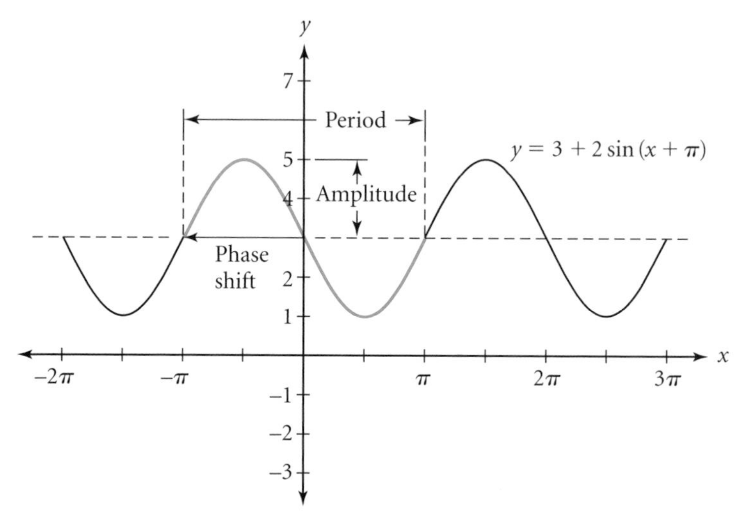

In Example A, the amplitude of $y = 2 + \sin x$ and $y = \sin(x - \pi)$ is 1 unit. The amplitude of $y = 3 + 2\sin(x + \pi)$, shown above, is 2.

The horizontal translation of a sine or cosine graph is called the **phase shift.** The phase shift of $y = 3 + 2\sin(x + \pi)$ is $-\pi$.

Cosine function sinusoids are transformed in the same way as sine function sinusoids. In fact, a cosine function is simply a horizontal translation of a sine function.

Consider the graph of the function $\frac{y-4}{3} = \cos\left(\frac{x-\frac{\pi}{3}}{2}\right)$.

The parent cosine function is shown in black. First perform the dilations. The red image graph is a dilation of the parent graph vertically by a factor of 3 and horizontally by a factor of 2.

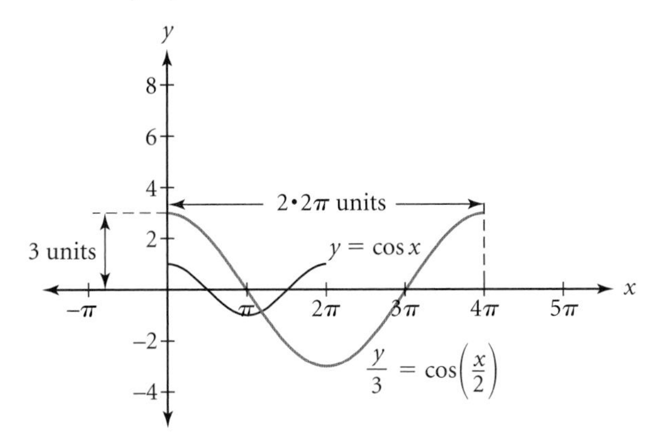

Now apply the translations. Translate the red image graph right $\frac{\pi}{3}$ units and up 4 units.

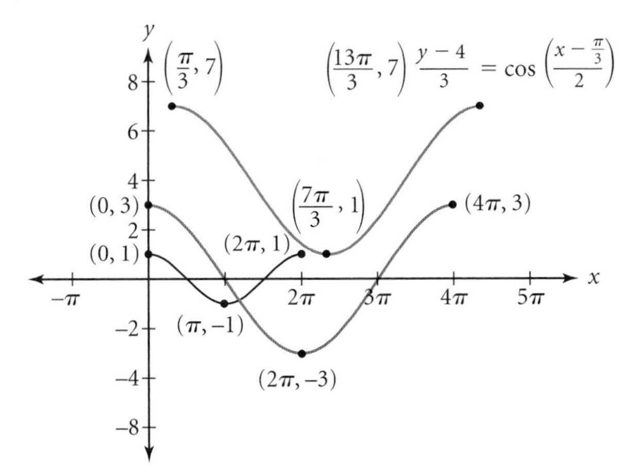

Values of the function $\frac{y-4}{3} = \cos\left(\frac{x-\frac{\pi}{3}}{2}\right)$, graphed in blue, vary from a minimum of 1 to a maximum of 7, so it has amplitude 3. The horizontal scale factor 2 means its period is dilated to 4π. The horizontal translation makes the phase shift $\frac{\pi}{3}$ so that the cycle of this graph starts at $\frac{\pi}{3}$, has a minimum at $2\pi + \frac{\pi}{3}$, or $\frac{7\pi}{3}$, and ends at $4\pi + \frac{\pi}{3}$, or $\frac{13\pi}{3}$.

In the investigation you will transform sinusoidal functions to fit real-world, periodic data.

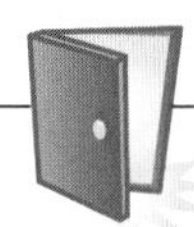

Investigation
The Pendulum II

You will need

- a washer
- string
- a motion sensor

Suspend a washer from two strings so that it hangs 10 to 15 cm from the floor between two tables or desks. Place the motion sensor on the floor about 1 m in front of the washer hanging at rest. Pull the washer back about 20 to 30 cm and let it swing. Collect data points for 2 s of time. Model your data with both a sine function and a cosine function. Give real-world meanings for all numerical values in each equation.

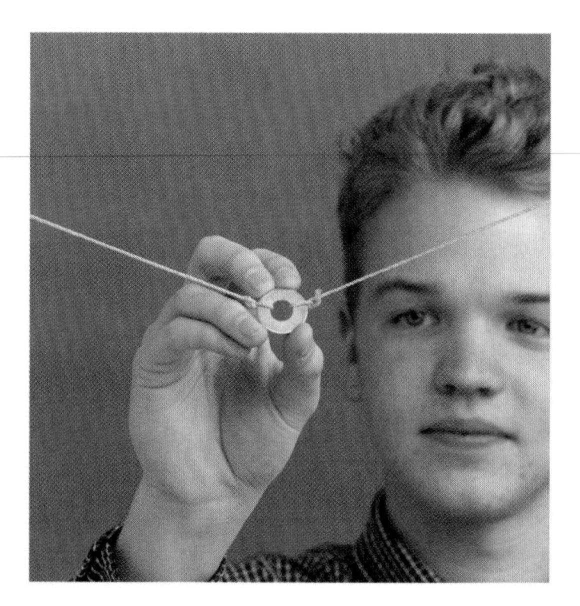

You can use sinusoidal functions to model many kinds of situations.

EXAMPLE B

After flying 300 mi west from Detroit to Chicago, a plane is put in a circular holding pattern above Chicago's O'Hare International Airport. The plane flies an additional 10 mi west past the airport and then starts flying in a circle with diameter 20 mi. The plane completes one circle every 15 min. Model the east-west component of the plane's distance from Detroit as a function of time.

► Solution

To help understand the situation, first sketch a diagram. The plane flies 300 mi to Chicago and then 10 mi past Chicago. At this time, call it 0 min, the plane begins to make a circle every 15 minutes. The diagram helps you see at least five data points, recorded in the table.

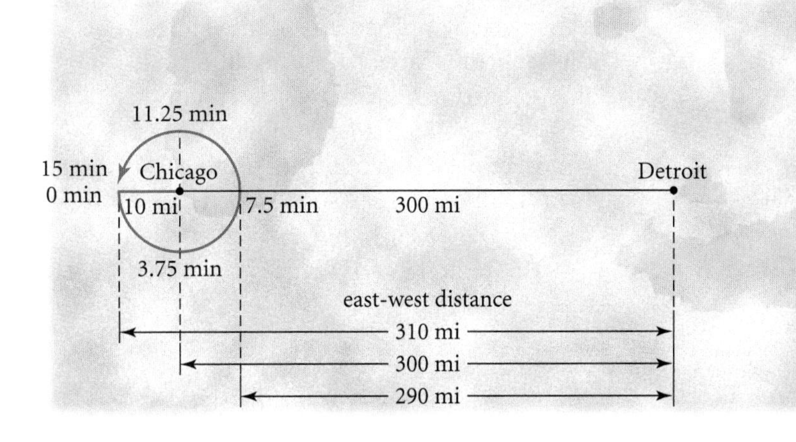

Time from beginning of circle (min) x	East-west distance from Detroit (mi) y
0	310
3.75	300
7.5	290
11.25	300
15	310

A quick plot of these points suggests that a cosine function might be a good model.

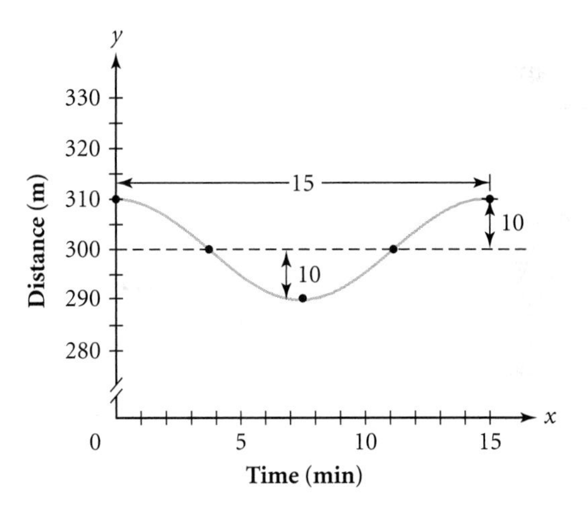

The period of $y = \cos x$ is 2π (from 0 to 2π). The period of this model should be 15 min (from 0 to 15). The horizontal scale factor that stretches 2π to 15 is $a = \frac{15}{2\pi}$.

The amplitude of the model should be 10 mi, so use the vertical scale factor $b = 10$.

The plane's initial flight of 300 mi means that the function must be translated up 300 units, so use $k = 300$.

One possible model is

$$y = 300 + 10\cos\left(\frac{x}{\frac{15}{2\pi}}\right), \text{ or } y = 300 + 10\cos\left(\frac{2\pi x}{15}\right)$$

Although the sine and cosine functions describe many periodic phenomena, such as the motion of a pendulum or the number of hours of daylight each day, there are other periodic functions that you can create from the unit circle.

In right triangle trigonometry, the tangent of angle A is the ratio of the length of the opposite leg to the length of the adjacent leg.

$$\tan A = \frac{a}{b}$$

The definition of tangent can be extended to apply to any angle. Here, the tangent of angle A is the ratio of the y-coordinate to the x-coordinate of a point rotated $A°$ (or radians) counterclockwise about the origin from the positive ray of the x-axis.

$$\tan A = \frac{y\text{-coordinate}}{x\text{-coordinate}}$$

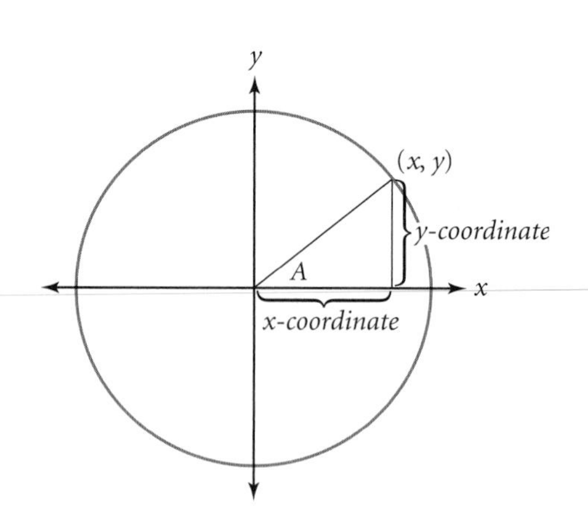

You can transform the graph of the tangent function in the same way that you transform sinusoidal functions. The asymptotes are changed only by horizontal transformations. Use the x-intercepts to help you determine translations. You can use the fact that $\tan\frac{\pi}{4} = 1$ to help you with vertical dilations.

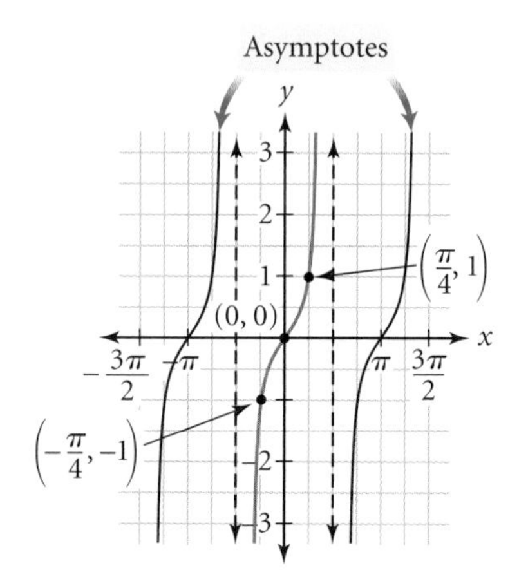

EXAMPLE C

The graph of $y = \tan x$ from $-\pi$ to 2π is shown in black. Find an equation for the red curve, which is a transformation of the graph of $y = \tan x$.

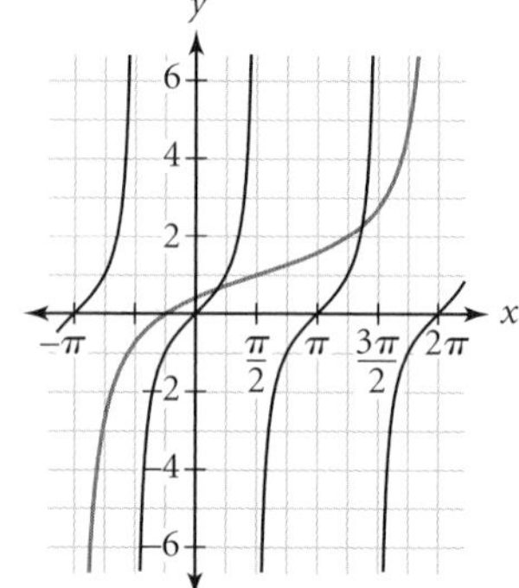

▶ Solution

Notice that the parent tangent curve, $y = \tan x$, has a period of π. The red curve appears to run from $-\pi$ to 2π, a distance of 3π, so the horizontal scale factor is 3.

Notice how the parent curve bends as it passes through the origin, (0, 0). The red curve appears to bend in the same way at $\left(\frac{\pi}{2}, 1\right)$. That is a translation right $\frac{\pi}{2}$ units and up 1 unit. So an equation for the red curve is

$$y = 1 + \tan\left(\frac{x - \frac{\pi}{2}}{3}\right)$$

In Example C, the point $\left(\frac{\pi}{2}, 1\right)$ on the red curve could have been considered the image of $(\pi, 0)$, rather than $(0, 0)$. This would indicate a translation *left* $\frac{\pi}{2}$ units and up 1 unit. So the equation

$$y = 1 + \tan\left(\frac{x + \frac{\pi}{2}}{3}\right)$$

will also model the red curve. Periodic graphs can always be modeled with many different, but equivalent, equations.

EXERCISES

You will need

A graphing calculator for Exercises **4, 6, 7, 11, 13,** and **18.**

Practice Your Skills

1. For 1a–f, write an equation for each sinusoid as a transformation of the graph of either $y = \sin x$ or $y = \cos x$. More than one answer is possible. Describe the amplitude, period, phase shift, and vertical shift of each graph.

a.

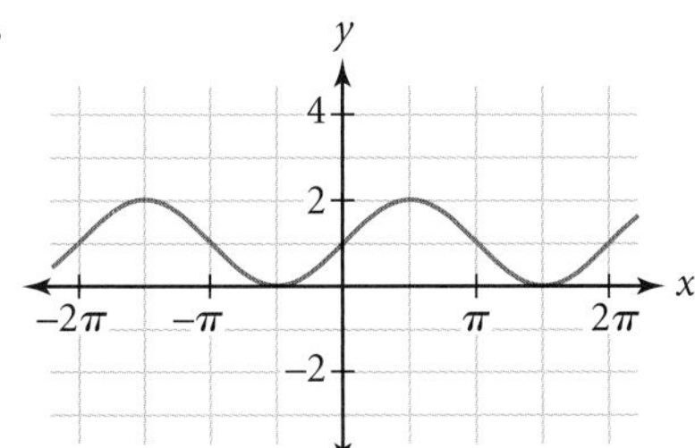

b.

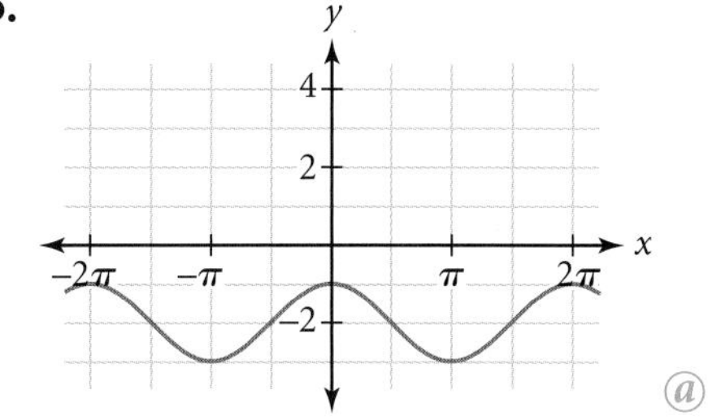

c.

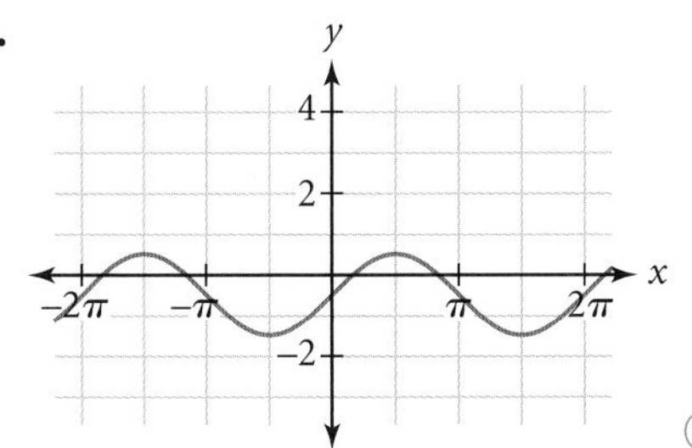

d.

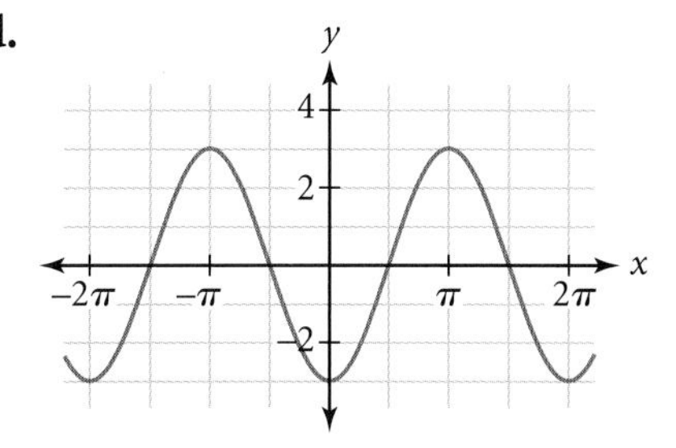

e.

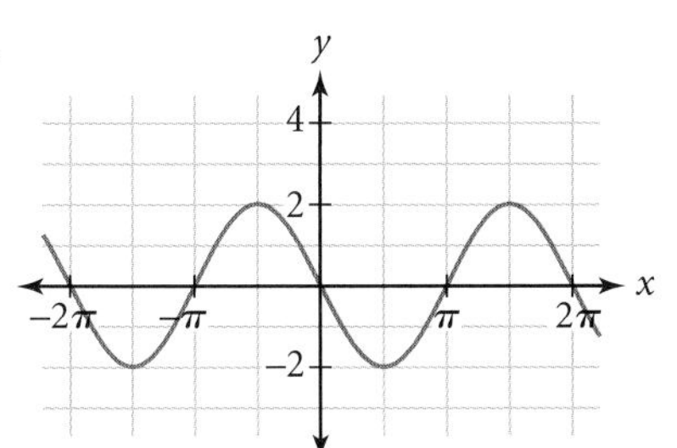

f.

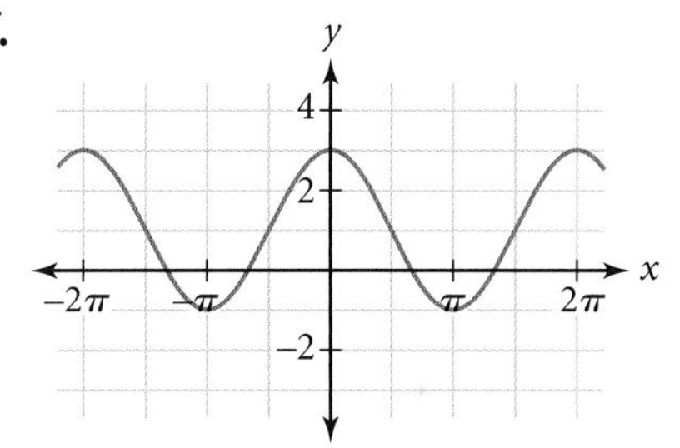

2. For 2a–f, write an equation for each sinusoid as a transformation of the graph of either $y = \sin x$ or $y = \cos x$. More than one answer is possible. Describe the amplitude, period, phase shift, and vertical shift of each graph.

a.

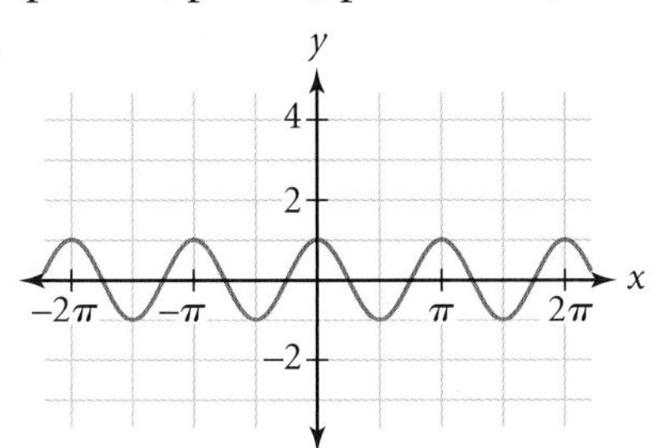

b.

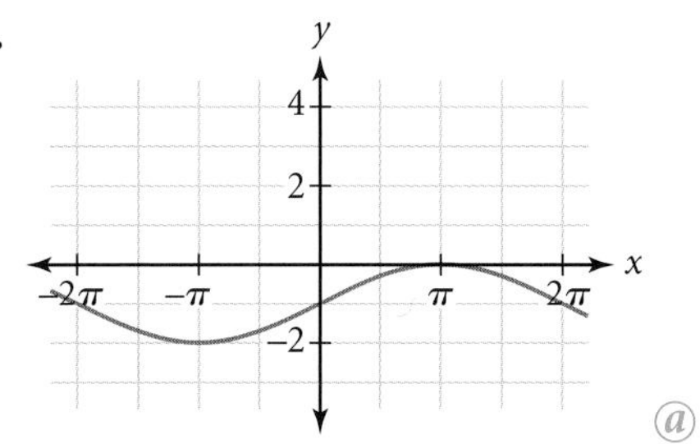

c.

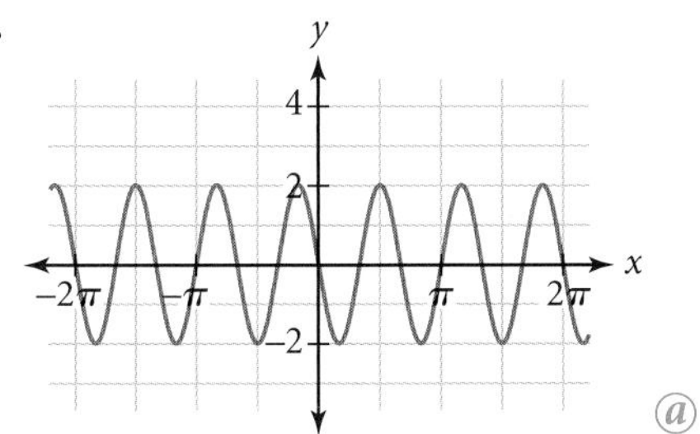

d.

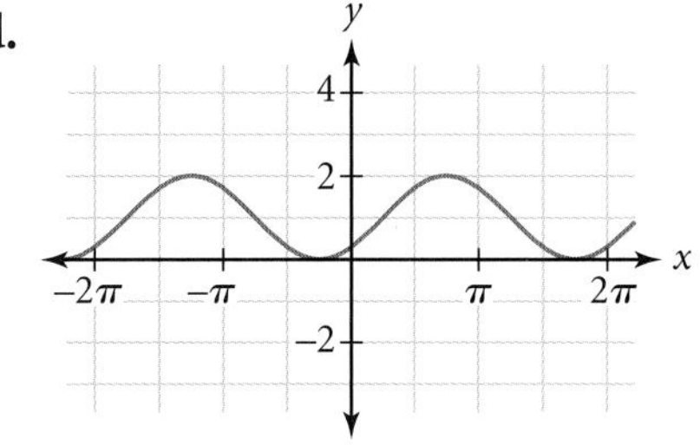

e.

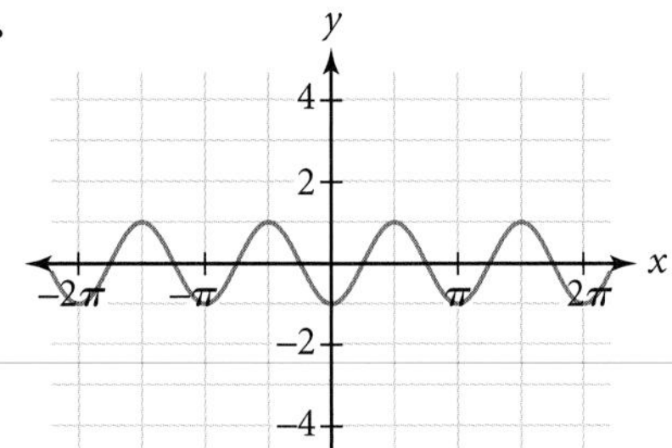

f.

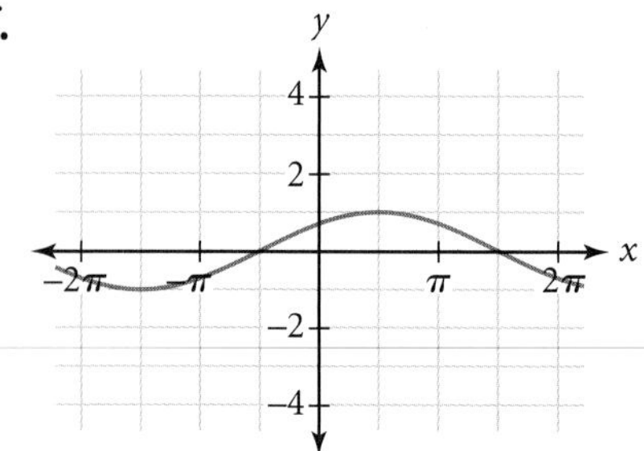

Reason and Apply

3. Consider the graph of $y = k + b\sin\left(\frac{x-h}{a}\right)$.

a. What effect does k have on the graph of $y = k + \sin x$?

b. What effect does b have on the graph of $y = b\sin x$? What is the effect if b is negative?

c. What effect does a have on the graph of $y = \sin\left(\frac{x}{a}\right)$?

d. What effect does h have on the graph of $y = \sin(x - h)$?

4. Sketch the graph of $y = 2\sin\left(\frac{x}{3}\right) - 4$. Use a calculator to check your sketch.

5. Describe a transformation of the graph of $y = \sin x$ to obtain an image that is equivalent to the graph of $y = \cos x$.

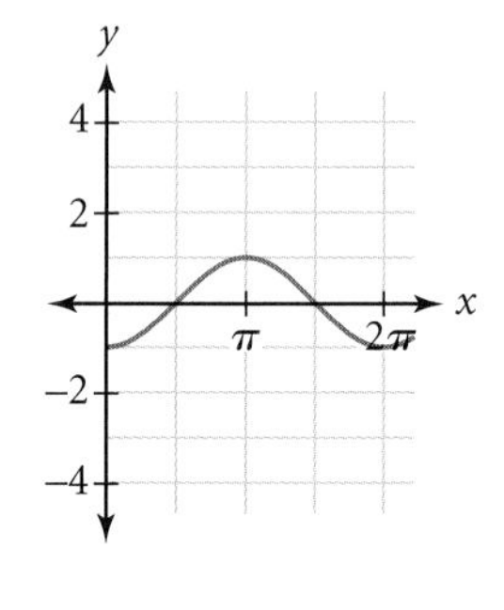

6. Write three different equations for the graph at right. Use a calculator to verify your answers. ⓐ

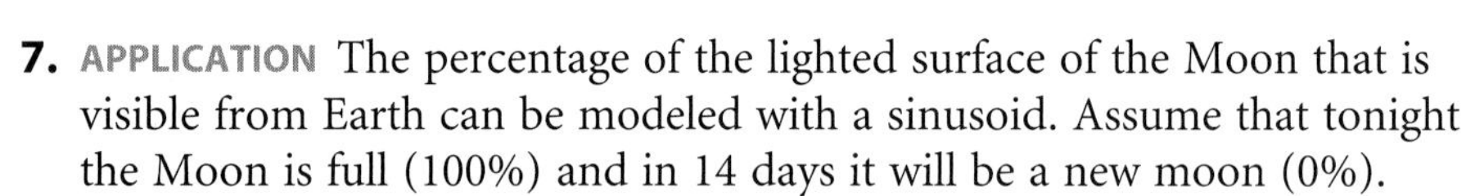

7. APPLICATION The percentage of the lighted surface of the Moon that is visible from Earth can be modeled with a sinusoid. Assume that tonight the Moon is full (100%) and in 14 days it will be a new moon (0%).

a. Define variables and find a sinusoidal function that models this situation. ⓗ

b. What percentage will be visible 23 days after the full moon?

c. What is the first day after a full moon that shows less than 75% of the lit surface?

Science CONNECTION

Half of the Moon's surface is always lit by the Sun. The phases of the Moon, as visible from Earth, result from the Moon's orientation changing as it orbits Earth. A full moon occurs when the Moon is farther away from the Sun than Earth, and the lighted side faces Earth. A new moon is mostly dark because its far side is receiving the sunlight, and only a thin crescent of the lighted side is visible from Earth. For more information about the phases of the Moon, see the links at www.keymath.com/DAA .

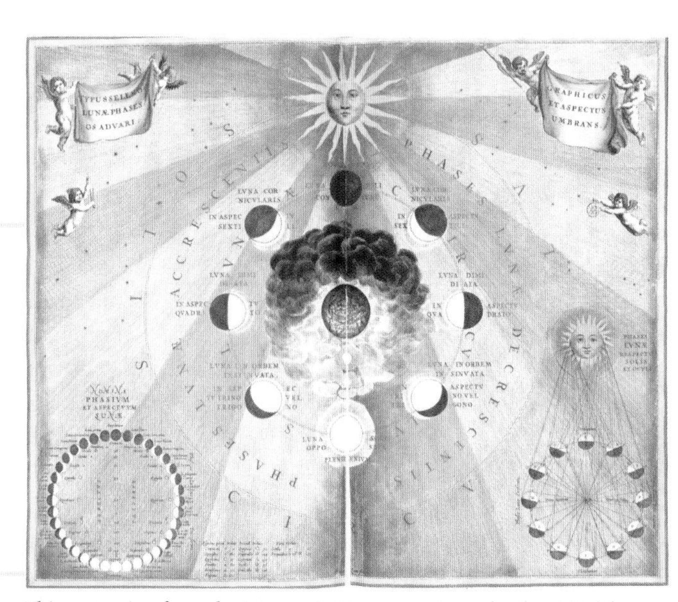

This engraving from the Harmonia Macrocosmica Atlas (ca. 1661) by Dutch-German mathematician Andreas Cellarius (ca. 1596–1665) depicts how astronomers of the 17th century interpreted the phases of the Moon.

8. Estimate the amplitudes and periods of the three graphs pictured.

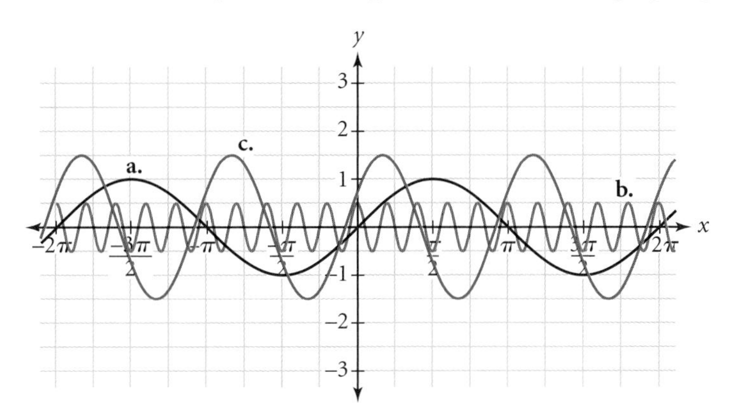

9. For each property, tell whether it applies to sine, cosine, tangent, or none of these functions.

a. The graph is symmetric over the y-axis.

b. The graph is symmetric over the x-axis.

c. The graph is continuous.

d. The domain is the set of real numbers.

e. The range is the set of real numbers.

f. The function contains the point $\left(\frac{\pi}{4}, 1\right)$.

10. Give an equation showing the transformation of the graph of $y = \cos x$ to $y = 1 + 3\sin\left(\frac{x}{2}\right)$.

11. Fluorescent lights do not produce constant illumination as incandescent lights do. Ideally, fluorescent light cycles sinusoidally from dim to bright 60 times per second. At a certain distance from the light, the maximum brightness is measured at 50 watts per square centimeter (W/cm^2) and the minimum brightness at 20 W/cm^2.

a. To collect data on light brightness from three complete cycles, for how much total time should you record data?

b. Sketch a graph of the sinusoidal model for the data collected in 11a if the light climbed to its mean value of 35 W/cm^2 at 0.003 s.

c. Write an equation for the sinusoidal model.

12. Imagine a unit circle in which the point (1, 0) is rotated A radians counterclockwise about the origin from the positive x-axis. Copy this table and record the x-coordinate and y-coordinate for each angle. Then use the definition of tangent to find the slope of the segment that connects the origin to each point.

Angle A	0	$\frac{\pi}{6}$	$\frac{\pi}{4}$	$\frac{\pi}{2}$	$\frac{3\pi}{4}$	π	$\frac{4\pi}{3}$	$\frac{5\pi}{3}$	$\frac{11\pi}{6}$
***x*-coordinate**									
***y*-coordinate**									
Slope or tan A									

13. ***Mini-Investigation*** Graph $y = \tan x$ on your calculator. Set your graphing window to $-2\pi \leq x \leq 2\pi$ and $-5 \leq y \leq 5$.

  a. What happens at $x = \frac{\pi}{2}$? Explain why this is so, and name other values when this occurs.

  b. What is the period of $y = \tan x$?

  c. Explain, in terms of the definition of tangent, why the values of $\tan\frac{\pi}{5}$ and $\tan\frac{6\pi}{5}$ are the same.

  d. Carefully graph two cycles of $y = \tan x$ on paper. Include vertical asymptotes at x-values where the graph is undefined.

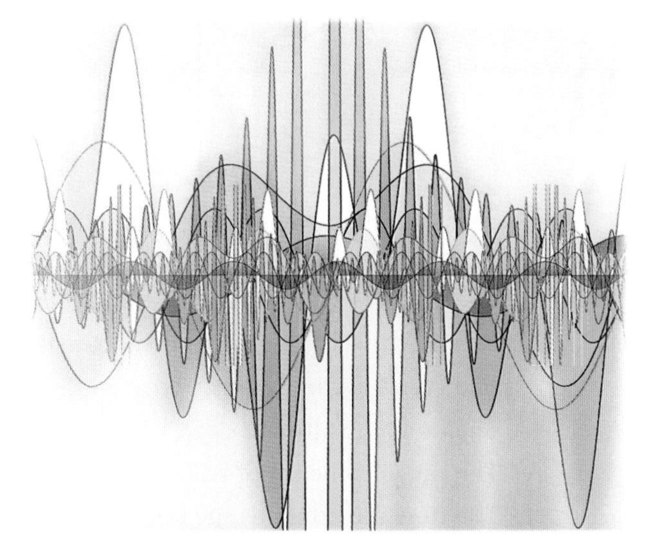

14. Write equations for sinusoids with these characteristics:

  a. a cosine function with amplitude 1.5, period π, and phase shift $-\frac{\pi}{2}$

  b. a sine function with minimum value -5, maximum value -1, and one cycle starting at $x = \frac{\pi}{4}$ and ending at $x = \frac{3\pi}{4}$ ⓐ

  c. a cosine function with period 6π, phase shift π, vertical translation 3, and amplitude 2

Review

15. Find the measure of the labeled angle in each triangle. ⓗ

  a.

40°
7
4.7
A

  b.

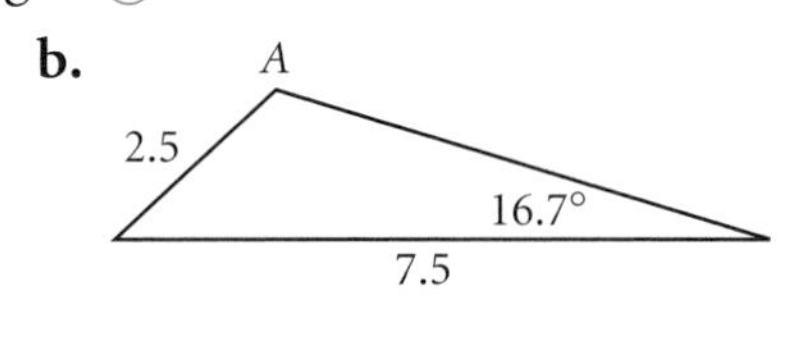

16. Make a table of angle measures from 0° to 360° by 15° increments. Then find the radian measure of each angle. Express the radian measure as a multiple of π.

17. The second hand of a wristwatch is 0.5 cm long. ⓐ

  a. What is the speed, in meters per hour, of the tip of the second hand?

  b. How long would the minute hand of the same watch have to be for its tip to have the same speed as the second hand?

  c. How long would the hour hand of the same watch have to be for its tip to have the same speed as the second hand?

  d. What is the angular speed, in radians per hour, of the three hands in 17a–c?

  e. Make an observation about the speeds you found.

18. Consider these three functions:

i. $f(x) = -\frac{3}{2}x + 6$ @ **ii.** $g(x) = (x + 2)^2 - 4$ **iii.** $h(x) = 1.3^{x+6} - 8$

a. Find the inverse of each function.

b. Graph each function and its inverse.

c. Which of the inverses, if any, are functions?

19. Use what you know about the unit circle to find possible values of θ in each equation. Use domain $0° \leq \theta \leq 360°$ or $0 \leq \theta < 2\pi$.

a. $\cos\theta = \sin 86°$ **b.** $\sin\theta = \cos\frac{19\pi}{12}$

c. $\sin\theta = \cos 123°$ **d.** $\cos\theta = \sin\frac{7\pi}{6}$

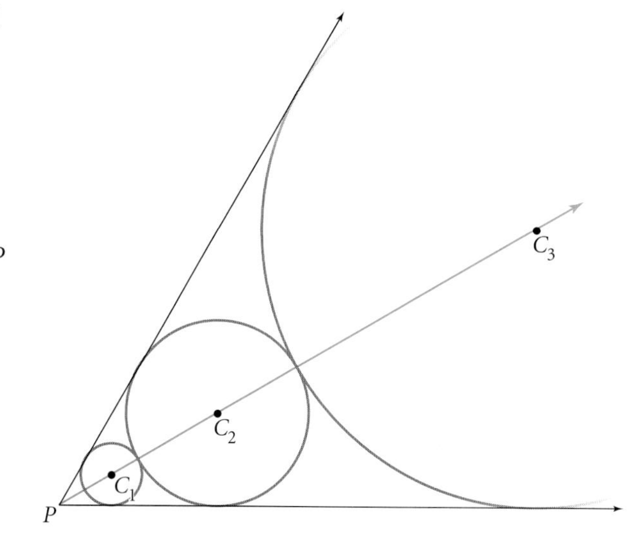

20. Circles C_1, C_2, . . . are tangent to the sides of $\angle P$ and to the adjacent circle(s). The radius of circle C_1 is 6. The measure of $\angle P$ is 60°.

a. What are the radii of C_2 and C_3? (*Hint:* One method uses similar right triangles.)

b. What is the radius of C_n?

Project

DESIGN A PICNIC TABLE

A well-made piece of wood furniture is carefully designed so that the pieces fit together perfectly. If the design is not perfect, assembly will be more difficult, and the furniture may not be stable.

Your task is to design a picnic table and draw the plans for it. Decide on the shape and size of your table. Make scale drawings showing your design. Then make a separate drawing of each piece, labeling all the lengths and all the angle measures.

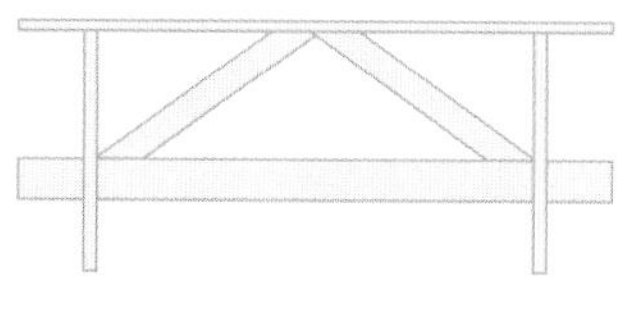

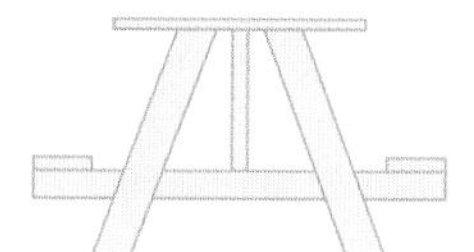

Your project should include

- Scale drawings of your table showing the front view, side view, and top view.
- A drawing of each piece.
- The calculations you made to determine the lengths and angles, clearly labeled and organized.
- A description of your design process, including any problems you encountered and how you solved them.

LESSON 13.4

Inverses of Trigonometric Functions

As you learned in Chapter 12, trigonometric functions of an angle in a right triangle give the ratios of sides when you know an angle. The inverses of these functions give the measure of the angle when you know a ratio of sides.

$$\sin^{-1}\left(\frac{a}{c}\right) = A$$

$$\cos^{-1}\left(\frac{b}{c}\right) = A$$

$$\tan^{-1}\left(\frac{a}{b}\right) = A$$

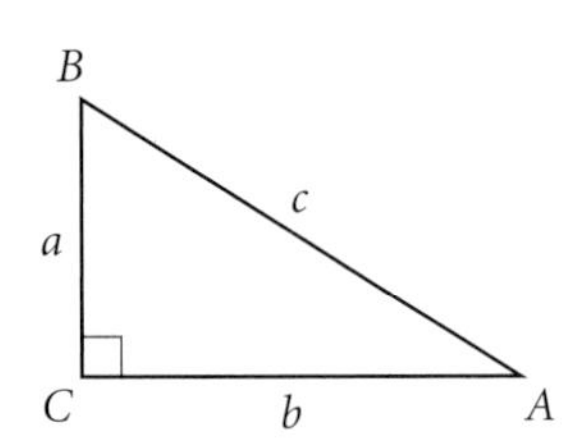

You have learned that the trigonometric ratios apply to any angle measure, not just measures of acute angles. The sine, cosine, and tangent functions have repeating values. So, in a non-right triangle, if you want to find an angle whose sine is 0.8796, there may be more than one answer. For this reason, the calculator answer for an inverse trigonometric function is not always the angle that you're looking for.

EXAMPLE A

Find the measure of $\angle B$.

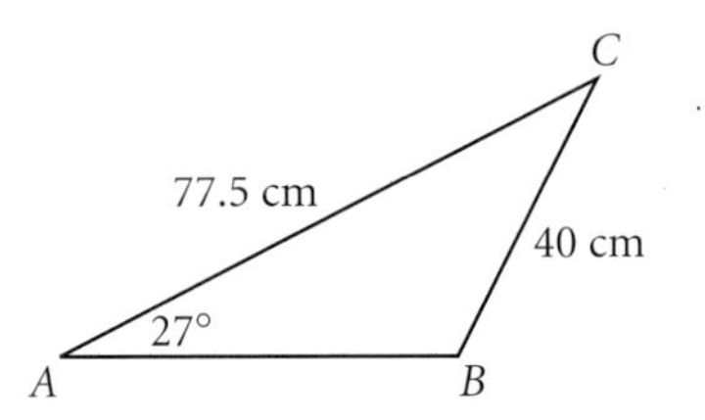

▸ Solution

This is not a right triangle, but you can use the Law of Sines to find $m\angle B$.

$$\frac{\sin 27°}{40} = \frac{\sin B}{77.5}$$

$$\sin B = \frac{77.5 \sin 27°}{40}$$

$$B = \sin^{-1}\left(\frac{77.5 \sin 27°}{40}\right) \approx 61.6°$$

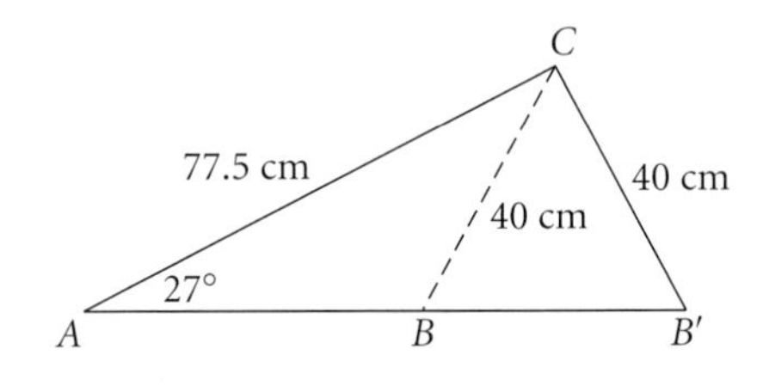

Recall from geometry that two sides and a non-included angle do not determine a specific triangle. This situation is an ambiguous case. So two different triangles can be constructed with the information given in the diagram. In one of these cases, $\angle B$ is acute; in the other, $\angle B$ is obtuse. The calculator has given you the reference angle, or the acute angle. The obtuse angle with the same sine as 61.6° measures 180° − 61.6°, or 118.4°.

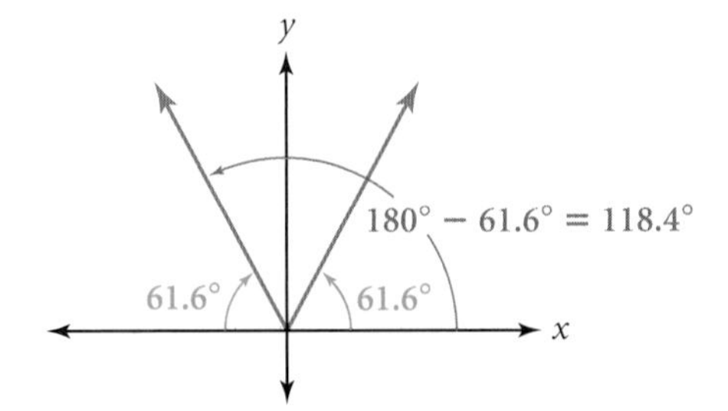

$$\sin 118.4° = \sin 61.6° \approx 0.8796$$

If the first diagram of $\triangle ABC$ is accurate, then $\angle B$ is obtuse, so $B \approx 118.4°$. Because you don't know whether the diagram is accurate, you should give both possible answers: $B \approx 118.4°$ or $B \approx 61.6°$.

Shown at right are the graphs of $y = \sin x$ and $y = \frac{77.5 \sin 27°}{40}$. In the interval $0° \le x \le 180°$, the graphs intersect at 61.6° and again at 118.4°. The two solutions are based on the symmetry of the sine graph—118.4° is the same distance from the x-intercept at 180° as 61.6° is from the x-intercept at 0°.

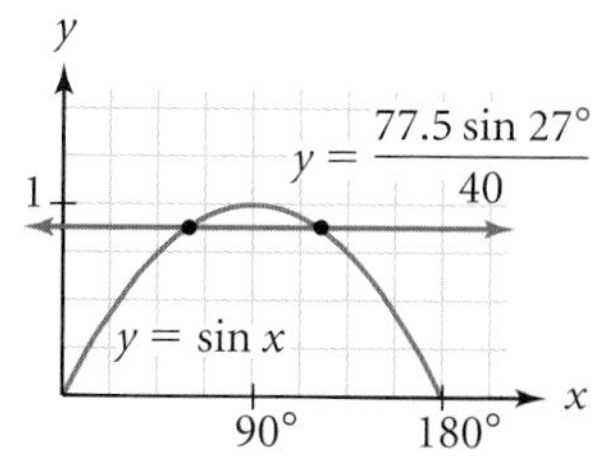

When you evaluate an inverse trigonometric function with your calculator, you get only one answer. A graph can help you make sense of a situation that has more than one solution.

You may recall that the inverse of a relation maps every point (x, y) to a point (y, x). You get the inverse of a relation by exchanging the x- and y-coordinates of all points. That's why a graph and its inverse are reflections of each other across the line $y = x$.

At right are the graphs of the exponential function $y = b^x$ and its inverse, $x = b^y$, which you've learned to call $y = \log_b x$. Notice that each is a function.

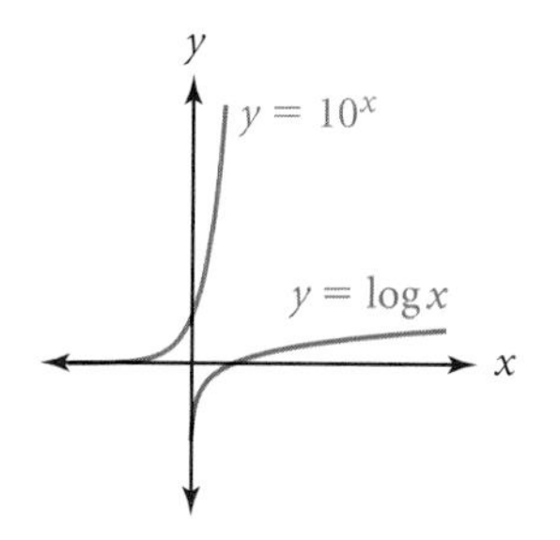

The graphs of the equation $y = x^2$ and its inverse, $x = y^2$, are also shown. Notice that the inverse of $y = x^2$ is not a function. When you solve the equation $4 = y^2$, how many solutions do you find?

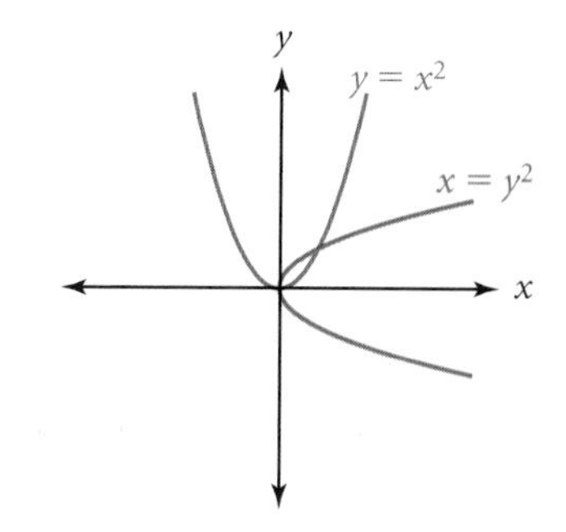

In the investigation you'll explore similar ambiguities with the inverses of the sine and cosine functions.

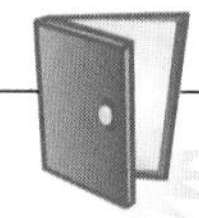

Investigation
Exploring the Inverses

You will need

- graph paper

In this investigation you will explore the graphs of trigonometric functions and their inverses.

Step 1 On graph paper, create a coordinate plane with x- and y-axes ranging from -10 to 10. Mark both axes from -10 to 10 at intervals of 1 unit. Then mark the values of π, 2π, and 3π on each axis. Use the same scale for both axes. That is, the distance from 0 to π on your y-axis should be the same as the distance from 0 to π on your x-axis. Carefully graph $y = \sin x$. Test a few points on your graph to make sure they fit the sine function.

Step 2 Add the line $y = x$ to your graph. Fold your paper along this line and then trace the image of $y = \sin x$ onto your paper.

Verify that this transformation maps the point $\left(\frac{\pi}{2}, 1\right)$ onto $\left(1, \frac{\pi}{2}\right)$, $(\pi, 0)$ onto $(0, \pi)$, and $\left(\frac{3\pi}{2}, -1\right)$ onto $\left(-1, \frac{3\pi}{2}\right)$. In general, every point (x, y) should map onto (y, x).

Step 3 If your original graph is $y = \sin x$, then the equation of the inverse, when x and y are switched, is $x = \sin y$. Is the inverse of $y = \sin x$ a function? Why or why not?

Step 4 Darken the portion of the curve $x = \sin y$ between $y = -\frac{\pi}{2}$ and $y = \frac{\pi}{2}$. Is this portion of the graph a function? Why or why not?

Step 5 Carefully sketch graphs of $y = \cos x$ and its inverse, $x = \cos y$, on axes similar to those you used in Step 1. Then darken a portion of the curve $x = \cos y$ that is a function. What interval did you select for y?

When you solve an equation like $x^2 = 7$ with your calculator, the square root function gives only one value. But there are actually two solutions. This is because the inverse of the parabola $y = x^2$ is not a function.

Similarly, the calculator command $\sin^{-1}$ is a function—it gives only one answer. However, the actual inverse of the sine function is a relation that is not a function. So the equation $\sin x = 0.7$ has many solutions because x can be any value whose sine is 0.7.

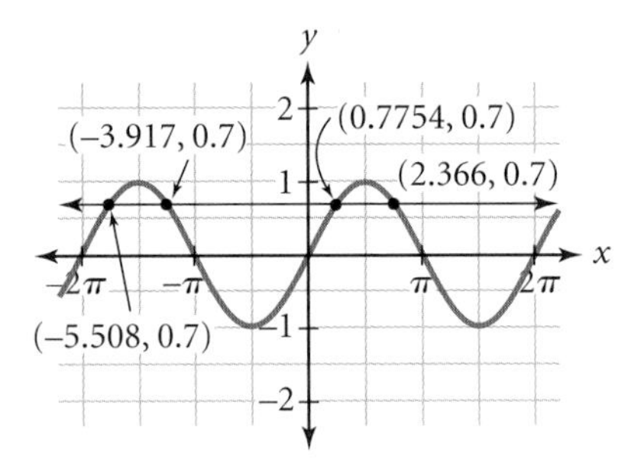

EXAMPLE B

Find the first four positive values of x for which $3 + 2\cos\left(x + \frac{\pi}{2}\right) = 2$.

▶ Solution

Graphically, this is equivalent to finding the first four intersections of the equations $y = 3 + 2\cos\left(x + \frac{\pi}{2}\right)$ and $y = 2$ for positive values of x. Examine this graph.

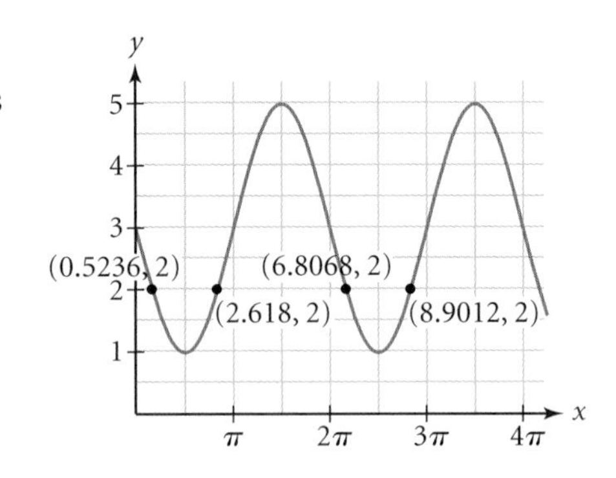

Solving the system of equations symbolically, you will find one answer.

$$3 + 2\cos\left(x + \frac{\pi}{2}\right) = 2$$

$$2\cos\left(x + \frac{\pi}{2}\right) = -1$$

$$\cos\left(x + \frac{\pi}{2}\right) = -\frac{1}{2}$$

$$x + \frac{\pi}{2} = \cos^{-1}\left(-\frac{1}{2}\right)$$

$$x = \cos^{-1}\left(-\frac{1}{2}\right) - \frac{\pi}{2}$$

The unit circle shows that the first positive solution for $\cos x = -\frac{1}{2}$ is $\frac{2\pi}{3}$. Substitute $\frac{2\pi}{3}$ for $\cos^{-1}\left(-\frac{1}{2}\right)$ and continue solving.

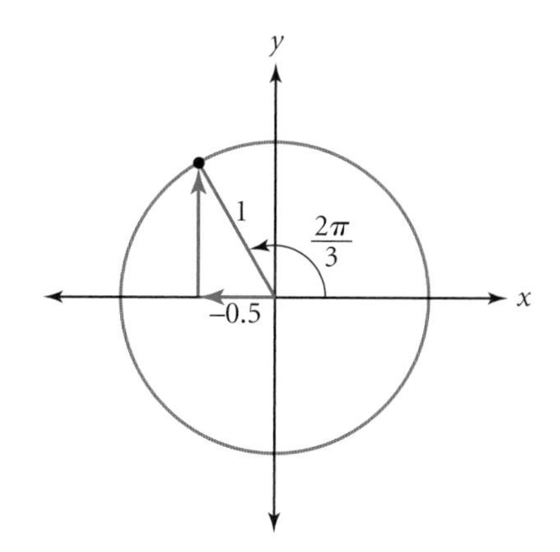

$$x = \cos^{-1}\left(-\frac{1}{2}\right) - \frac{\pi}{2}$$

$$x = \frac{2\pi}{3} - \frac{\pi}{2}$$

$$x = \frac{\pi}{6} \approx 0.5236$$

So one solution is $x = \frac{\pi}{6}$, or approximately 0.5236. Graphs of trigonometric functions follow a pattern and have certain symmetries. You can look at the graph to find the other solutions.

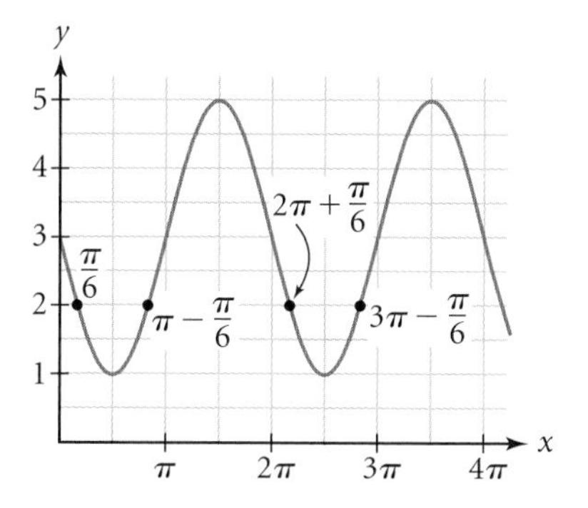

So the next solutions are

$$x = \pi - \frac{\pi}{6} = \frac{5\pi}{6} \approx 2.6180$$

$$x = 2\pi + \frac{\pi}{6} = \frac{13\pi}{6} \approx 6.8068$$

$$x = 3\pi - \frac{\pi}{6} = \frac{17\pi}{6} \approx 8.9012$$

The function $y = \sin^{-1}x$ is the portion of the graph of $x = \sin y$ such that $-\frac{\pi}{2} \leq y \leq \frac{\pi}{2}$ (or $-90° \leq y \leq 90°$). This restriction of the domain of $x = \sin y$ ensures that $y = \sin^{-1}(x)$ is a function.

Similarly, the function $y = \cos^{-1}x$ is the portion of the graph of $x = \cos y$ such that $0 \leq y \leq \pi$ (or $0° \leq y \leq 180°$).

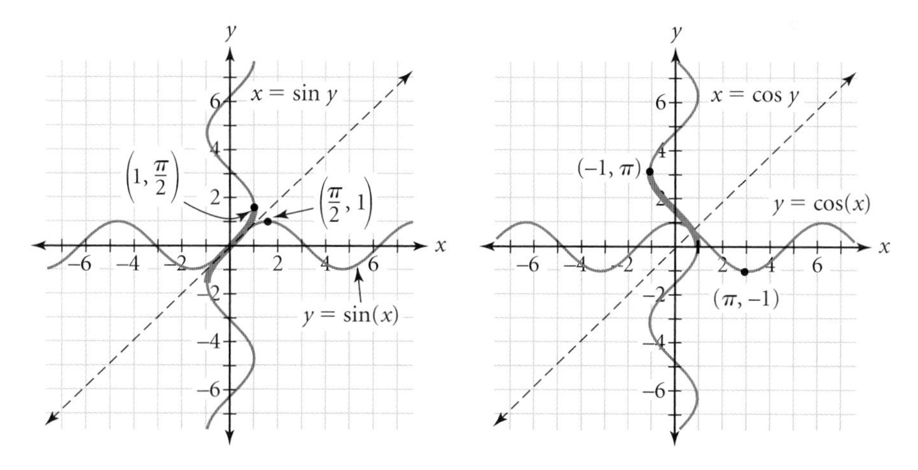

Because inverse cosine is defined as a function, the equation $x = \cos^{-1}0.5$ has only one solution, whereas $\cos x = 0.5$ has infinitely many solutions. As you saw in Example B, some problems may have multiple solutions, but the inverse cosine and inverse sine functions on your calculator will give you only the value within the ranges given above. This value is called the **principal value.**

EXERCISES

You will need

A graphing calculator for Exercises **7, 8, 11,** and **17.**

Practice Your Skills

1. Find the principal value of each expression to the nearest tenth of a degree and then to the nearest hundredth of a radian.

a. $\sin^{-1} 0.4665$ **b.** $\sin^{-1}(-0.2471)$ ⓐ **c.** $\cos^{-1}(-0.8113)$ ⓐ **d.** $\cos^{-1} 0.9805$

2. Find all four values of x between -2π and 2π that satisfy each equation.

a. $\sin x = \sin\frac{\pi}{6}$ ⓐ **b.** $\cos x = \cos\frac{3\pi}{8}$ ⓐ **c.** $\cos x = \cos 0.47$ **d.** $\sin x = \sin 1.47$

3. Illustrate the answers to Exercise 2 by plotting the points on the graph of the sine or cosine curve. ⓗ

4. Illustrate the answers to Exercise 2 by drawing segments on a graph of the unit circle. ⓐ

Reason and Apply

5. Explain why your calculator can't find $\sin^{-1} 1.28$.

6. On the same coordinate axes, create graphs of $y = 2\cos\left(x + \frac{\pi}{4}\right)$ and its inverse.

7. Find values of x that satisfy the conditions given.

a. Find the first two positive solutions of $0.4665 = \sin x$. ⓐ

b. Find the two negative solutions closest to zero of $-0.8113 = \cos x$.

8. How many solutions to the equation $-2.6 = 3\sin x$ occur in the first three positive cycles of the function $y = 3\sin x$? Explain your answer.

9. Find the measure of the largest angle of a triangle with sides 4.66 m, 5.93 m, and 8.54 m. ⓐ

10. In $\triangle ABC$, $AB = 7$ cm, $CA = 3.9$ cm, and $m\angle B = 27°$. Find the two possible measurements for $\angle C$.

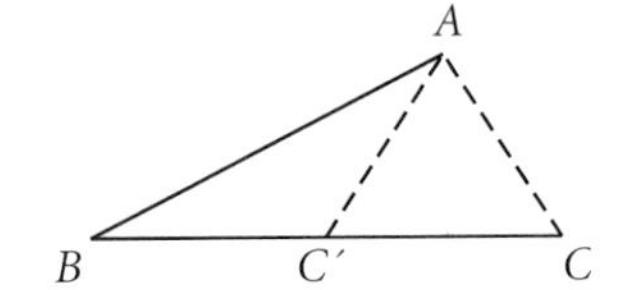

11. *Mini-Investigation* Consider the inverse of the tangent function.

a. Find the values of $\tan^{-1} x$ for several positive and negative values of x.

b. Based on these answers, predict what the graph of $y = \tan^{-1} x$ will look like. Sketch your prediction.

c. How does the graph of $y = \tan^{-1} x$ compare with the graph of $x = \tan y$?

d. Use your calculator to verify your graph of the function $y = \tan^{-1} x$.

12. Shown at right are a constant function and two cycles of a cosine graph over the domain $0° \leq x \leq 720°$. The first intersection point is shown. What are the next three intersection values?

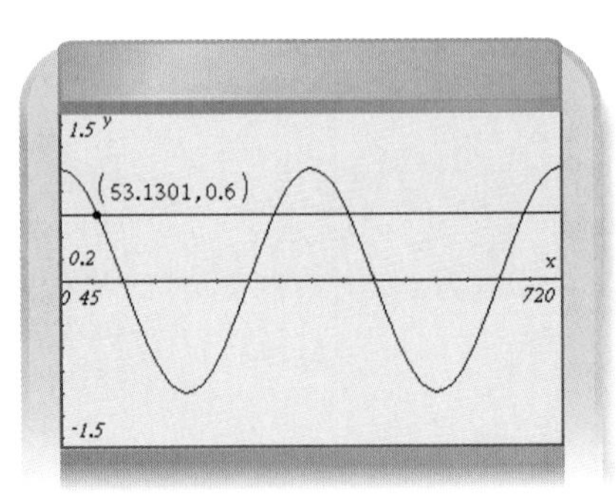

13. The cosine graph at right shows two maximum values, at 50° and 770°. If the first point of intersection with the constant function $y = 0.5$ occurs when $x = 170°$, what is the value of x at the second point of intersection? @

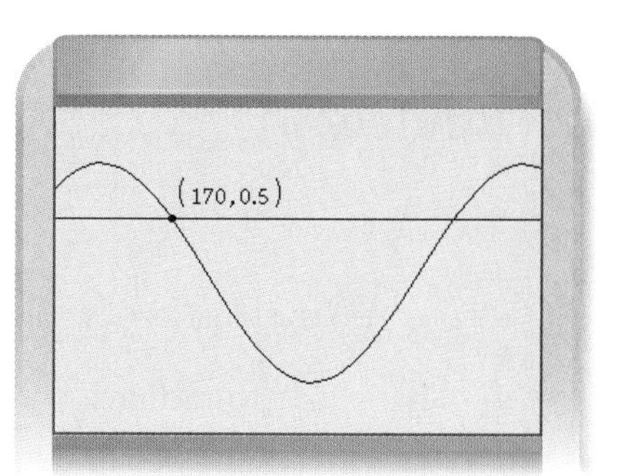

14. Find the exact value of each expression.

a. $\sin\left(\sin^{-1}\frac{4}{5}\right)$ **b.** $\sin\left[\sin^{-1}\left(-\frac{2}{3}\right)\right]$ @

c. $\sin^{-1}\left(\sin\frac{2\pi}{3}\right)$ @ **d.** $\cos^{-1}[\cos(-45°)]$

15. APPLICATION When a beam of light passes through a polarized lens, its intensity is cut in half, or $I_1 = 0.5I_0$. To further reduce the intensity, you can place another polarized lens in front of it. The intensity of the beam after passing through the second lens depends on the angle of the second filter to the first. For instance, if the second lens is polarized in the same direction, it will have little or no additional effect on the beam of light. If the second lens is rotated so that its axis is $\theta°$ from the first lens's axis, then the intensity in watts per square meter (W/m^2) of the transmitted beam, I_2, is

$$I_2 = I_1\cos^2\theta$$

where I_1 is the intensity in watts per square meter of the incoming beam.

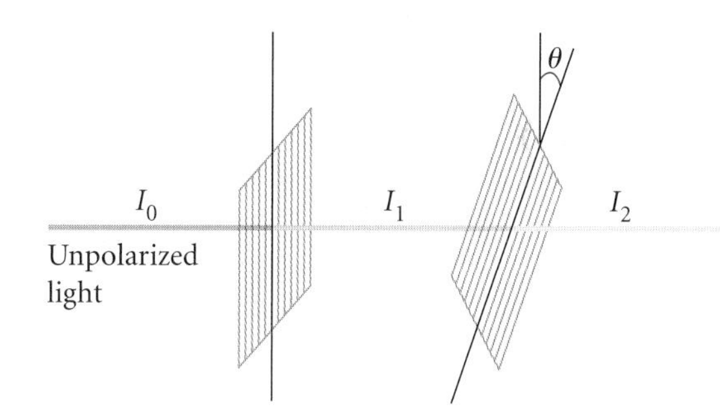

a. A beam of light passes successively through two polarized sheets. The angle between the polarization axes of the filters is 30°. If the intensity of the incoming beam is 16.0×10^{-4} W/m², what is the intensity of the beam after passing through the first filter? The second filter? @

b. A polarized beam of light has intensity 3 W/m². The beam then passes through a second polarized lens and intensity drops to 1.5 W/m². What is the angle between the polarization axes?

c. At what angle should the axes of two polarized lenses be placed to cut the intensity of a beam to 0 W/m²?

Technology CONNECTION

Skiers, boaters, and photographers know that ordinary sunglasses reduce brightness but do not remove glare. Polarized lenses will eliminate the glare from reflective surfaces such as snow, water, sand, and roads. This is because glare is the effect of reflected light being polarized parallel to the reflective surface. Polarized lenses will filter, or block out, this polarized light. You can think of a polarized lens as a set of tiny slits. The slits block light at certain angles while allowing light to pass through select angles.

The photo on the left was taken without a polarized lens and the one on the right was taken with a polarized lens. Notice how the polarized lens reduces glare.

Review

16. Convert

a. $\frac{7\pi}{10}$ radians to degrees

b. $-205°$ to radians

c. 5π radians per hour to degrees per minute

17. Write an equation for each graph.

a. @

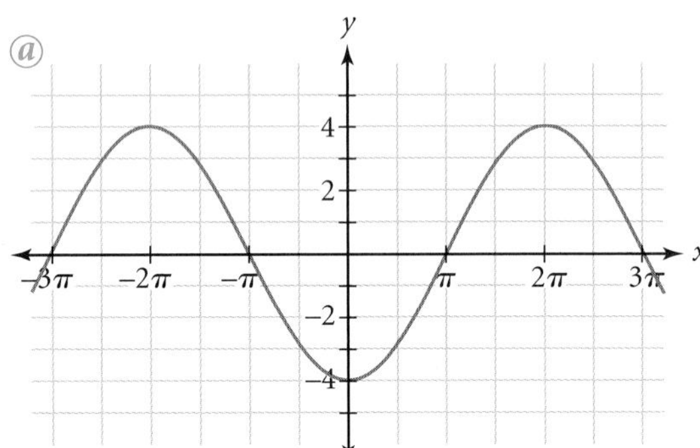

b.

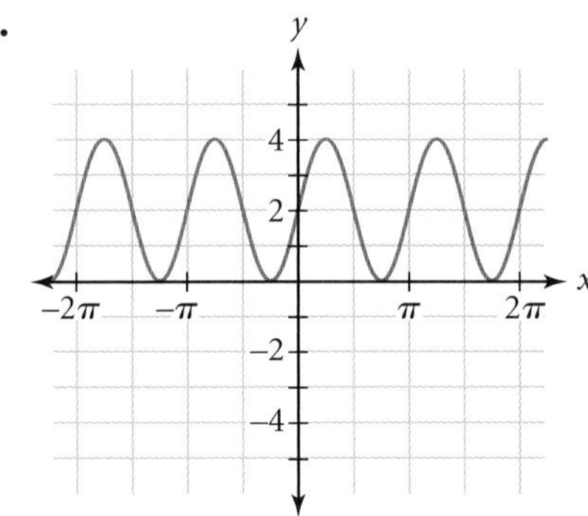

c.

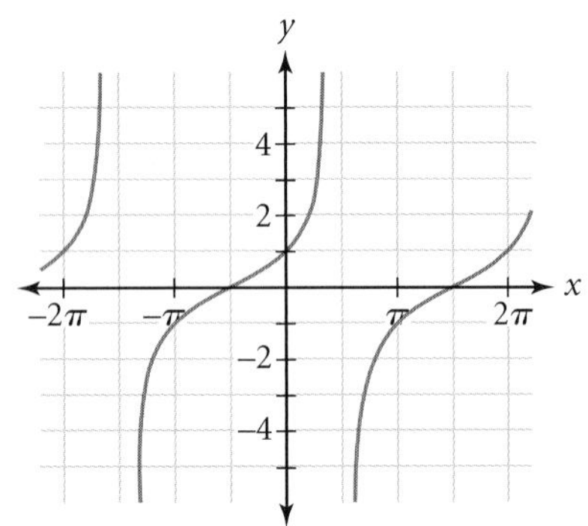

d.

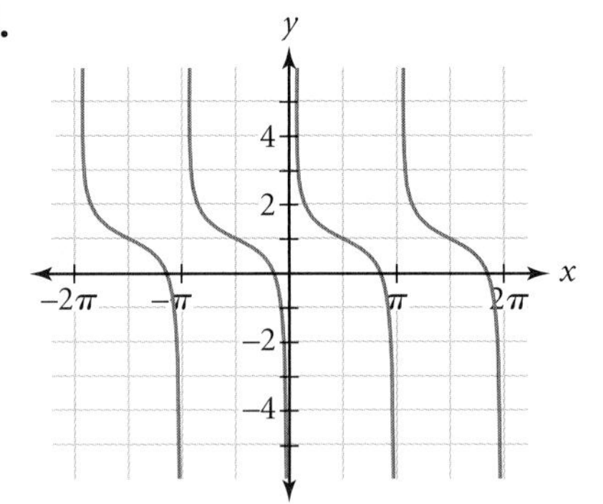

18. Find all roots, real or nonreal. Give exact answers.

a. $2x^2 - 6x + 3 = 0$ @

b. $13x - 2x^2 = 6$

c. $3x^2 + 4x + 4 = 0$

LESSON

13.5

Modeling with Trigonometric Equations

Language exerts hidden power, like the moon on the tides.

RITA MAE BROWN

Tides are caused by gravitational forces, or attractions, between the Moon, Earth, and the Sun as the Moon circles Earth. You can model the height of the ocean level in a seaport with a combination of sinusoidal functions of different phase shifts, periods, and amplitudes.

The tide-predicting machine shown above records the times and heights of high and low tides. Designed in 1910, it combines sine functions to produce a graph that predicts tide levels over a period of time.

This mural is painted in the style of *The Great Wave,* by Japanese artist Hokusai (1760–1849). Like tides, the motion of waves can be modeled with trigonometric functions.

Science CONNECTION

The Moon's gravitational pull causes ocean tides and also tidal bores in rivers. Tidal bores occur during new and full moons when ocean waves rush upstream into a river passage, sometimes at speeds of more than 40 mi/h. The highest recorded river bore was 15 ft high, in China's Fu-ch'un River.

EXAMPLE A

The height of water at the mouth of a certain river varies during the tide cycle. The time in hours since midnight, t, and the height in feet, h, are related by the equation

$$h = 15 + 7.5\cos\left[\frac{2\pi(t-3)}{12}\right]$$

a. What is the length of a period modeled by this equation?

b. When is the first time the height of the water is 11.5 feet?

c. When will the water be at that height again?

▶ Solution

One cycle of the parent cosine function has length 2π.

a. The cosine function in the height equation has been dilated horizontally by a scale factor of $\frac{12}{2\pi}$. So the length of the cycle is $\frac{12}{2\pi} \cdot 2\pi = 12$ hours.

b. To find when the height of the water is 11.5 feet, substitute 11.5 for h.

$$11.5 = 15 + 7.5\cos\left[\frac{2\pi(t-3)}{12}\right] \qquad \text{Substitute 11.5 for } h.$$

$$\frac{-3.5}{7.5} = \cos\left[\frac{\pi(t-3)}{6}\right] \qquad \text{Subtract 15, and divide by 7.5.}$$

$$\cos^{-1}\left(\frac{-3.5}{7.5}\right) = \frac{\pi(t-3)}{6} \qquad \text{Take the inverse cosine of both sides.}$$

$$\frac{6}{\pi} \cdot \cos^{-1}\left(\frac{-3.5}{7.5}\right) + 3 = t \qquad \text{Multiply by } \tfrac{6}{\pi}\text{, and add 3.}$$

$$t \approx 6.93 \text{ hours}$$

So the water height will be 11.5 feet after approximately 6.93 hours, or at 6:56 A.M.

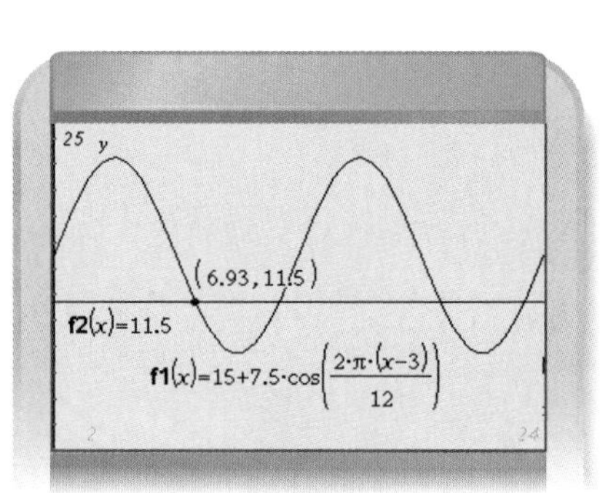

You can also find this answer by graphing the height function and the line $h = 11.5$ and by approximating the point of intersection. The graph verifies your solution. (*Note:* On your calculator, use x in place of t and y in place of h.)

c. The graph in part b shows several different times when the water depth is 11.5 feet. The third occurrence is 12 hours after the first, because each cycle is 12 hours in length. So the third occurrence is at 6:56 P.M. But how can you find the second occurrence? First, consider that the graph is shifted right 3 units, so the cycle begins at $t = 3$. It has a period of 12, so it ends at $t = 15$.

The first height of 11.5 feet occurred when $t \approx 6.93$, or about 3.93 hours after the cycle's start. The next solution will be about 3.93 hours before the end of the cycle, that is, at $15 - 3.93 \approx 11.07$ hours, or 11:04 A.M. This value will also repeat every 12 hours. A graph confirms this approximation.

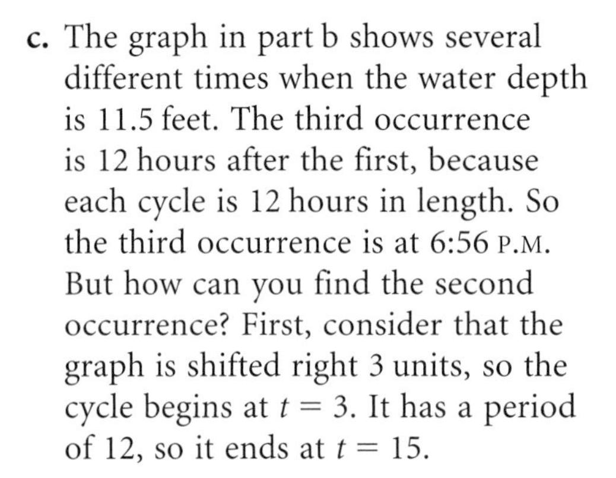

The cycle begins at (3, 22.5). The water is at its highest point, 22.5 feet, at 3 hours after midnight.

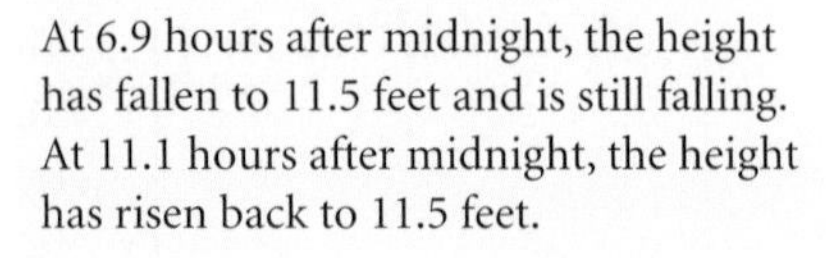

At 6.9 hours after midnight, the height has fallen to 11.5 feet and is still falling. At 11.1 hours after midnight, the height has risen back to 11.5 feet.

You can use both sine and cosine functions to model sinusoidal patterns because they are just horizontal translations of each other. Often it is easier to use a cosine function, because you can identify the maximum value as the start value of a cycle.

When a function has a short cycle, it can be more useful to talk about how many cycles are completed in one unit of time. This value, called the **frequency** of the function, is the reciprocal of the period. For example, if a wave has a period of 0.01 second, then it has a frequency of 100 cycles per second. In Example A, the period is $\frac{1}{2}$ day, so the frequency is 2 cycles per day.

$$frequency = \frac{1}{period}$$

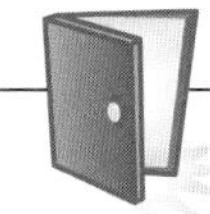

Investigation

A Bounding Spring

You will need

- a motion sensor
- a spring
- a mass of 50 to 100 g
- a support stand

In this experiment you will suspend a mass from a spring. When you pull down on the mass slightly, and release, the mass will move up and down. In reality, the amount of motion gradually decreases, and eventually the mass returns to rest. However, if the initial motion is small, then the decrease in the motion occurs more slowly and can be ignored during the first few seconds.

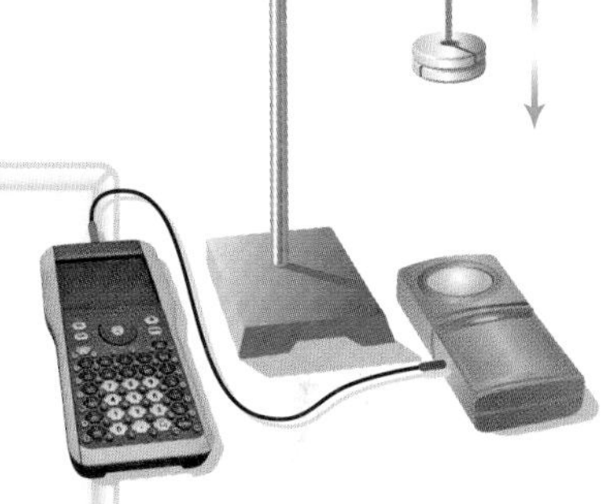

Procedure Note

1. Attach a mass to the bottom of a spring. Position the motion sensor directly below the spring, leaving space to pull down on the mass.
2. Set the motion sensor to collect about 5 s of data. Pull the mass down slightly, and release at the same moment as you begin gathering data.

Step 1 Follow the Procedure Note to collect data on the height of the bouncing spring for a few seconds.

Step 2 Delete values from your lists to limit your data to about four cycles. Identify the phase shift, amplitude, period, frequency, and vertical shift of your function.

Step 3 Write a sine or cosine function that models the data.

Step 4 Answer these questions, based on your equation and your observations.

a. How does each of the numbers in your equation from Step 3 correspond to the motion of the spring?

b. How would your equation change if you moved the motion sensor 1 m farther away?

c. How would your equation change if you pulled the spring slightly lower when you started?

This car is performing in a car dance competition in Los Angeles, California. Springs are often used as shock absorbers.

In the investigation, you modeled cyclical motion with a sine or cosine function. As you read the next example, think about how the process of finding an equation based on given measurements compares to your process in the investigation.

EXAMPLE B

On page 748, you learned that the Cosmo Clock 21 Ferris wheel has a 100 m diameter, takes 15 min to rotate, and reaches a maximum height of 112.5 m.

Find an equation that models the height in meters, h, of a seat on the perimeter of the wheel as a function of time in minutes, t. Then determine when, in the first 30 min, a given seat is 47 m from the ground if the seat is at its maximum height 10 min after the wheel begins rotating.

▸ Solution

The parent sinusoid curve has an amplitude of 1. The diameter of the Ferris wheel is 100 m, so the vertical scale factor, or amplitude, is 50. The period of a parent sinusoid is 2π and the period of the Ferris wheel is 15 min, so the horizontal scale factor is $\frac{15}{2\pi}$. The average value of the parent sinusoid is 0, but the average height of the wheel is 62.5 m, so the vertical translation is 62.5.

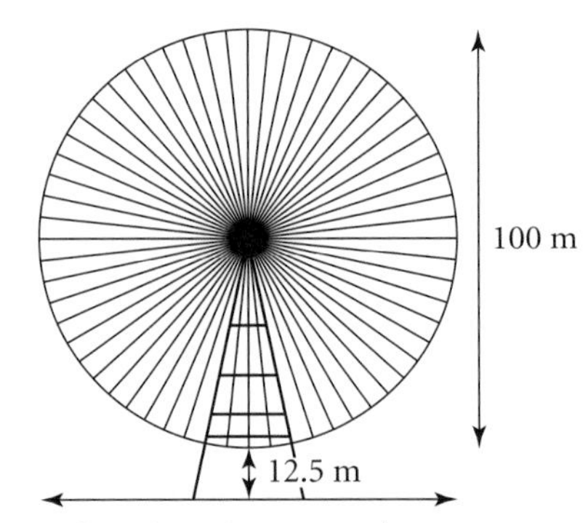

The top of a sinusoid corresponds to the maximum height of a seat. The cosine curve starts at a maximum point, so it will be easiest to use a cosine function. Because the first maximum of the Ferris wheel occurs after 10 min, the phase shift of its equation is 10 to the right. Incorporate these values into a cosine function.

$$h = 50\cos\left[\frac{2\pi(t-10)}{15}\right] + 62.5$$

Now, to find when the height is 47 m, substitute 47 for h.

$$50\cos\left[\frac{2\pi(t-10)}{15}\right] + 62.5 = 47$$ Substitute 47 for h.

$$\cos\left[\frac{2\pi(t-10)}{15}\right] = \frac{47-62.5}{50} = -0.31$$ Subtract 62.5 and divide by 50. Evaluate.

$$\frac{2\pi(t-10)}{15} = \cos^{-1}(-0.31)$$ Take the inverse cosine of both sides.

$$t - 10 = \frac{15}{2\pi}\cdot\cos^{-1}(-0.31)$$ Multiply by $\frac{15}{2\pi}$ on both sides.

$$t = \frac{15}{2\pi}\cdot\cos^{-1}(-0.31) + 10$$ Add 10.

$$t \approx 14.5$$ Approximate the principal value of t.

So the given seat is at a height of 47 m approximately 14.5 min after the wheel starts rotating. But this is not the only time. The period is 15 min, so the seat will also reach 47 m on the second rotation, after 29.5 min. Also, the height is 47 m once on the way up and once on the way down. The seat is at 47 m 4.5 min after its maximum point, which is at 10 min. It will also be at the same height 4.5 min before its maximum point, at 5.5 min, and 15 min after that, at 20.5 min.

So in the first 30 min, the seat reaches a height of 47 m after 5.5, 14.5, 20.5, and 29.5 min. A graph confirms this.

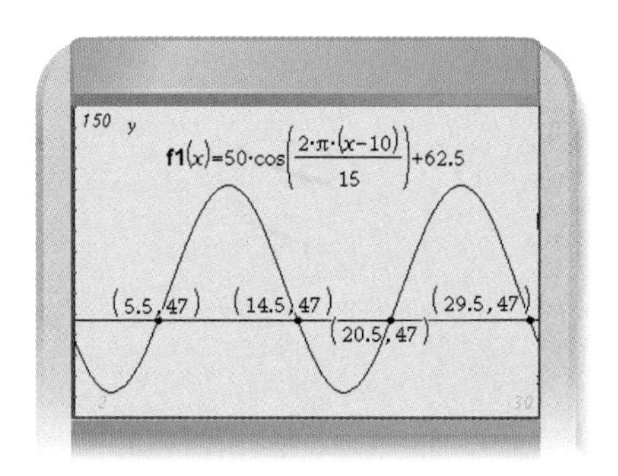

Using sinusoidal models, you can easily find y when given an x-value. As you saw in the examples, the more difficult task is finding x when given a y-value. There will be multiple answers because the graphs are periodic. You should always check the values and number of solutions with a calculator graph.

EXERCISES

You will need

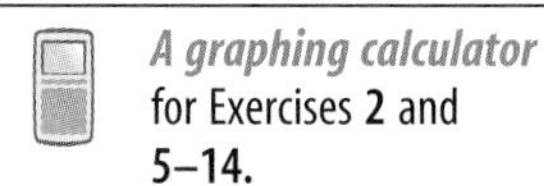
A graphing calculator for Exercises **2** and **5–14.**

Practice Your Skills

1. Find the first four positive solutions. Give exact values in radians.

a. $\cos x = 0.5$ ⓐ

b. $\sin x = -0.5$

2. Find all solutions for $0 \le x < 2\pi$, rounded to the nearest thousandth.

a. $2\sin(x + 1.2) - 4.22 = -4$ ⓐ

b. $7.4\cos(x - 0.8) + 12.3 = 16.4$

3. Consider the graph of the function $h = 5 + 7\sin\left[\frac{2\pi(t-9)}{11}\right]$.

a. What are the vertical translation and average value? ⓗ

b. What are the vertical scale factor, minimum and maximum values, and amplitude? ⓐ

c. What are the horizontal scale factor and period?

d. What are the horizontal translation and phase shift?

4. Consider the graph of the function $h = 18 - 17\cos\left[\frac{2\pi(t+16)}{15}\right]$.

a. What are the vertical translation and average value?

b. What are the vertical scale factor, minimum and maximum values, and amplitude?

c. What are the horizontal scale factor and period?

d. What are the horizontal translation and phase shift?

Reason and Apply

5. A walker moves counterclockwise around a circle with center (1.5, 2) and radius 1.2 m and completes a cycle in 8 s. A recorder walks back and forth along the x-axis, staying even with the walker, with a motion sensor pointed toward the walker. What equation describes the distance as a function of time? Assume the walker starts at point (2.7, 2).

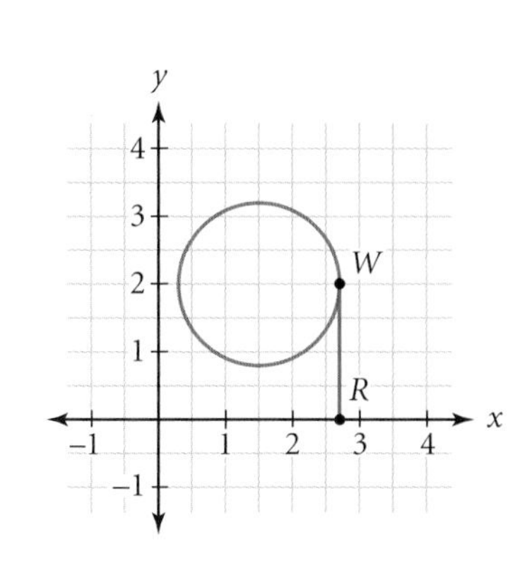

6. A mass attached to a spring is pulled down 3 cm from its resting position and then released. It makes ten complete bounces in 8 s. At what times during the first 2 s was the mass 1.5 cm above its resting position?

7. Household appliances are typically powered by electricity through wall outlets. The voltage provided varies sinusoidally between $-110\sqrt{2}$ volts and $110\sqrt{2}$ volts, with a frequency of 60 cycles per second.

a. Use a sine or cosine function to write an equation for voltage as a function of time. (h)

b. Sketch and label a graph picturing three complete cycles.

8. The time between high and low tide in a river harbor is approximately 7 h. The high-tide depth of 16 ft occurs at noon and the average harbor depth is 11 ft.

a. Write an equation modeling this relationship.

b. If a boat requires a harbor depth of at least 9 ft, find the next two time periods when the boat will not be able to enter the harbor.

9. Two masses are suspended from springs, as shown. The first mass is pulled down 3 cm from its resting position and released. A second mass is pulled down 4 cm from its resting position. It is released just as the first mass passes its resting position on its way up. When released, each mass makes 12 bounces in 8 s.

a. Write a function for the height of each mass. Use the moment the second mass is released as $t = 0$. (a)

b. At what times during the first 2 s will the two masses be at the same height? Solve graphically, and state your answers to the nearest 0.1 s.

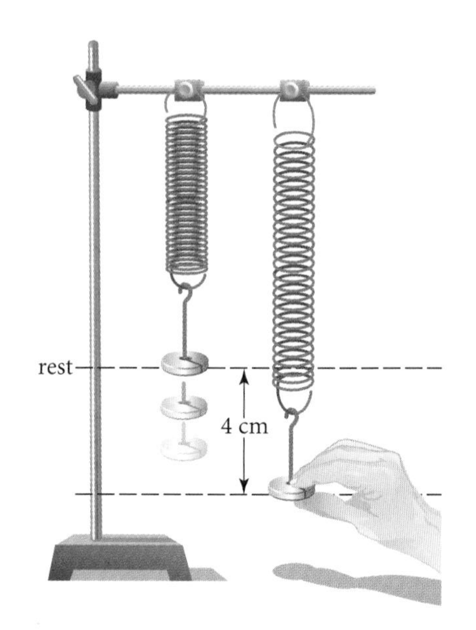

10. APPLICATION An AM radio transmitter generates a radio wave given by a function in the form $f(t) = A \sin 2000\pi nt$. The variable n represents the location on the broadcast dial, $550 \leq n \leq 1600$, and t is the time in seconds.

a. For radio station WINS, located at 1010 on the AM radio dial, what is the period of the function that models its radio waves?

b. What is the frequency of your function from 10a?

c. Find a function that models the radio waves of an AM radio station near you. Find the period and frequency.

Technology CONNECTION

Amplitude modulation, or AM, is one way that a radio transmitter can send information over large distances. By varying the amplitude of a continuous wave, the transmitter adds audio information to the signal. The wave's amplitude is adjusted simply by changing the amount of energy put in. More energy causes a larger amplitude, whereas less energy causes a smaller amplitude. One problem with AM (as opposed to FM, or frequency modulation) is that the signal is affected by electrical fields, such as lighting. The resulting static and clicks decrease sound quality.

A radio transmitter from the early 1900s used electric pulses and vacuum tubes to create radio waves.

11. A point rotates at 3 rev/min counterclockwise around a circle with center (0, 12) and radius 5 m, as shown at right. Write and graph a function of height versus time showing one complete revolution.

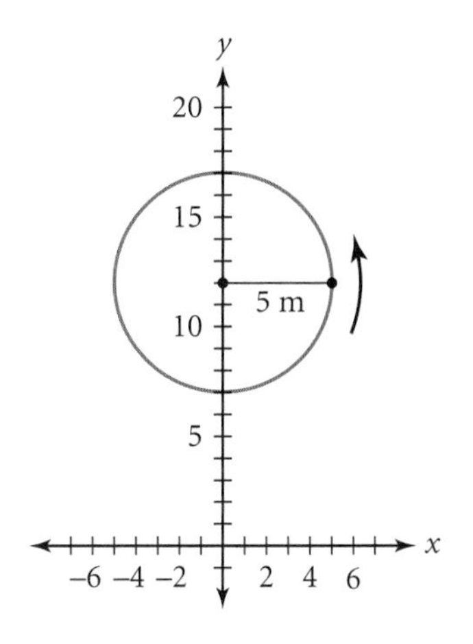

12. The number of hours of daylight, y, on any day of the year in Philadelphia can be modeled using the equation

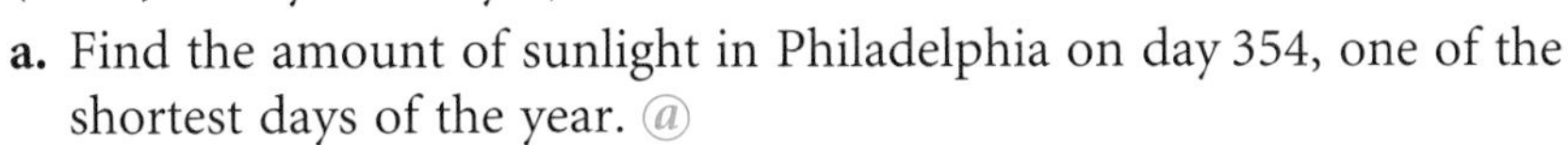

$y = 12 + 2.4\sin\left[\frac{2\pi(x-80)}{365}\right]$, where x represents the day number (with January 1 as day 1).

a. Find the amount of sunlight in Philadelphia on day 354, one of the shortest days of the year. @

b. Find the dates for which the model predicts exactly 12 hours of daylight.

13. A popular amusement park ride is the double Ferris wheel. Each small wheel takes 20 s to make a single rotation. The two-wheel set takes 30 s to rotate once. The dimensions of the ride are given in the diagram.

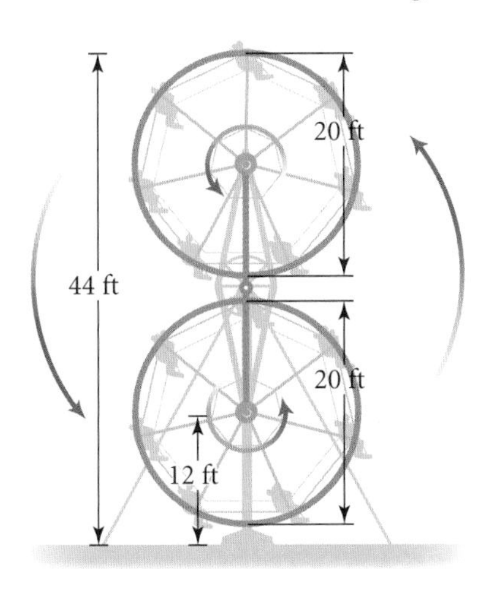

The Sky Wheel, a double Ferris wheel, operated at the Cedar Point theme park in Sandusky, Ohio, from 1962 to 1981.

a. Sandra gets on at the foot of the bottom wheel. Write an equation that will model her height above the center of this wheel as the wheels rotate.

b. The entire ride (the two-wheel set) starts revolving at the same time that the two small wheels begin to rotate. Write an equation that models the height of the center of Sandra's wheel as the entire ride rotates.

c. Because the two motions occur simultaneously, you can add the two equations to write a final equation for Sandra's position. Write this equation.

d. In a 5 min ride, during how many distinct intervals of time is Sandra within 6 ft of the ground?

[► You can use the **Dynamic Algebra Exploration** at www.keymath.com/DAA to explore a dynamic model of a double Ferris wheel.◄]

keymath.com/DAA

Review

14. Solve $\tan\theta = -1.111$ graphically. Use domain $-180° \leq \theta \leq 360°$. Round answers to the nearest degree. ⓐ

15. Which has the larger area, an equilateral triangle with side 5 cm or a sector of a circle with radius 5 cm and arc length 5 cm? Give the area of each shape to the nearest 0.1 cm.

16. Find the equation of the circle with center $(-2, 4)$ and tangent line $2x - 3y - 6 = 0$.

17. Consider the equation $P(x) = 2x^3 - x^2 - 10x + 5$.

a. List all possible rational roots of $P(x)$.

b. Find any actual rational roots.

c. Find exact values for all other roots.

d. Write $P(x)$ in factored form.

Project

SUNRISE, SUNSET

On December 21, 2008, in Anchorage, Alaska, the sun rose at 10:42 A.M. and set at 3:42 P.M., giving 5 h 38 min of daylight. On the same day in San Juan, Puerto Rico, there were 11 h 2 min of daylight. You have seen a model for the number of hours of daylight on any day of the year in Philadelphia. How would this model compare with a model for Anchorage or San Juan? How would it compare with a model for your hometown?

Sunrise in Wrangell-St. Etias National Park, Alaska.

Choose a city and find data on the times for sunrise and sunset for the period between winter solstice (around December 21) and summer solstice (around June 21). Calculate and record the number of daylight hours for one to two days per week for this time period. Assign December 31 as day 0, and let x represent the number of days after December 31. Let y represent the number of hours between sunrise and sunset. Graph your data and find a sinusoidal model.

Your project should include

- A table of the data you used and the source citation for your data.
- The graph of your data.
- An equation for a sinusoidal model of your data.
- A description of how you found the model and an explanation of how different coefficients in the equation relate to the context.
- Data, graphs, and models for other cities in the world and an explanation of how the data and models vary for different locations.

LESSON

13.6

Fundamental Trigonometric Identities

First say to yourself what you would be; and then do what you have to do.

EPICTETUS

In this lesson you will discover several equations that express relationships among trigonometric functions. When an equation is true for all values of the variables for which the expressions are defined, the equation is called an **identity.**

You've already learned that $\tan A = \frac{y\text{-coordinate}}{x\text{-coordinate}}$ of the image point when a point $P(x, y)$ is rotated counterclockwise through $\angle A$ about the origin from the positive x-axis. You've also seen that the y-coordinate is equivalent to $r\sin A$, where r is the distance between the point and the origin, and the x-coordinate is equivalent to $r\cos A$. You can use these relationships to derive an identity that relates $\tan A$, $\sin A$, and $\cos A$.

$\tan A = \frac{y}{x}$ Definition of tangent.

$\tan A = \frac{r\sin A}{r\cos A}$ Substitute $r\sin A$ for y and $r\cos A$ for x.

$\tan A = \frac{\sin A}{\cos A}$ Reduce.

The reciprocals of the tangent, sine, and cosine functions are also trigonometric functions. The reciprocal of tangent is called **cotangent,** abbreviated cot. The reciprocal of sine is called **cosecant,** abbreviated csc. The reciprocal of cosine is called **secant,** abbreviated sec. These definitions lead to six more identities.

History CONNECTION

The reciprocal trigonometric functions were introduced by Muslim astronomers in the 9th and 10th centuries C.E. Before there were calculating machines, these astronomers developed remarkably precise trigonometric tables based on earlier Greek and Indian findings. They used these tables to record planetary motion, to keep time, and to locate their religious center of Mecca. Western Europeans began studying trigonometry when Arabic astronomy handbooks were translated in the 12th century.

Reciprocal Identities

$\csc A = \frac{1}{\sin A}$ or $\sin A = \frac{1}{\csc A}$

$\sec A = \frac{1}{\cos A}$ or $\cos A = \frac{1}{\sec A}$

$\cot A = \frac{1}{\tan A}$ or $\tan A = \frac{1}{\cot A}$

Your calculator probably does not have special keys for secant, cosecant, and cotangent. However, you can use the reciprocal identities to enter these functions into your calculator. For example, to graph $y = \csc x$, you use $y = \frac{1}{\sin x}$. The calculator screen at right shows the graphs of the principal cycles of the parent sine and cosecant functions. [See **Calculator Note 13C** for more information about using secant, cosecant, and cotangent on your calculator.]

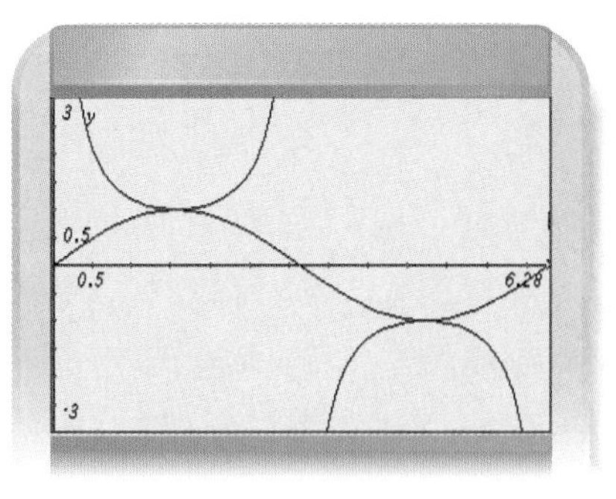

Once you know a few identities, you can use them to prove other identities. One strategy for proving that an equation is an identity is to verify that both sides of the equation are always equivalent. You can do this by writing equivalent expressions for one side of the equation until it is the same as the other side.

EXAMPLE Is the statement $\cot A = \frac{\cos A}{\sin A}$ an identity? Investigate with a graph or table and then write an algebraic proof.

▶ ***Solution***

First, look at the tables for $y = \cot A$ and $y = \frac{\cos A}{\sin A}$. The table shows that the function values appear to be the same. Now use an algebraic proof to verify this conjecture.

x	f1(x):= cot(x)	f2(x):= cos(x)/(si..
0.	#UNDEF	#UNDEF
0.261799	3.73205	3.73205
0.523599	1.73205	1.73205
0.785398	1.	1.
1.0472	0.57735	0.57735

Use definitions and identities that you know in order to show that both sides of the equation are equivalent. Be sure to work on only one side of the equation.

$$\cot A \stackrel{?}{=} \frac{\cos A}{\sin A}$$ Original statement.

$$\frac{1}{\tan A} \stackrel{?}{=} \frac{\cos A}{\sin A}$$ Use the reciprocal identity to replace $\cot A$ with $\frac{1}{\tan A}$.

$$\frac{1}{\frac{\sin A}{\cos A}} \stackrel{?}{=} \frac{\cos A}{\sin A}$$ Replace $\tan A$ with $\frac{\sin A}{\cos A}$.

$$\frac{\cos A}{\sin A} = \frac{\cos A}{\sin A}$$ $1 \div \frac{\sin A}{\cos A}$ is equivalent to $1 \cdot \frac{\cos A}{\sin A}$, or simply $\frac{\cos A}{\sin A}$.

Therefore, $\cot A = \frac{\cos A}{\sin A}$ is an identity. Now that you have proved this identity, you can use it to prove other identities.

In the investigation you will discover a set of trigonometric identities that are collectively called the Pythagorean identities. To prove a new identity, you can use any previously proved identity.

Investigation
Pythagorean Identities

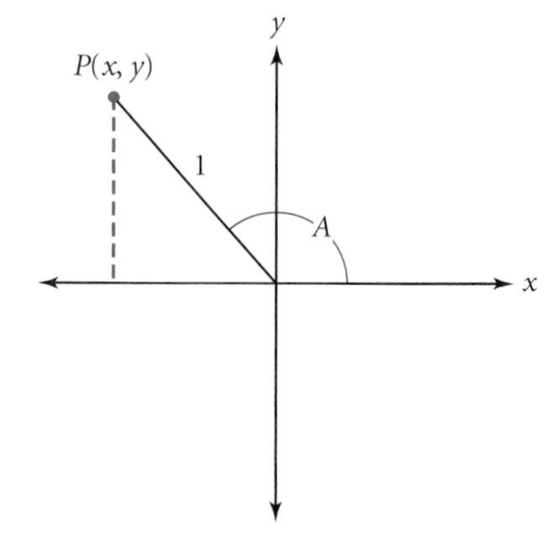

Step 1 Use your calculator to graph the equation $y = \sin^2 x + \cos^2 x$. (You'll probably have to enter this as $y = (\sin x)^2 + (\cos x)^2$.) Does this graph look familiar? Use your graph to write an identity.

Step 2 Use the definitions for $\sin A$, $\cos A$, and the diagram at right to prove your identity.

Step 3 Explain why you think this identity is called a Pythagorean identity.

Step 4 Solve the identity from Step 1 for $\cos^2 x$ to get another identity. Then solve for $\sin^2 x$ to get another variation.

$\cos^2 x =$ ___?___

$\sin^2 x =$ ___?___

Step 5 Divide both sides of the identity from Step 1 by $\cos^2 x$ to develop a new identity. Simplify so that there are no trigonometric functions in the denominator.

Step 6 Verify your identity from Step 5 with a graph or table. Name any domain values for which the identity is undefined.

Step 7 Divide both sides of the identity from Step 1 by $\sin^2 x$ to develop a new identity. Simplify so that there are no trigonometric functions in the denominator.

Step 8 Verify the identity from Step 7 with a graph. Name any domain values for which the identity is undefined.

You have verified identities by setting each side of the equation equal to y and graphing. If the graphs and table values match, you may have an identity. You may have used your calculator in this way to verify the Pythagorean identities in the investigation. If not, try it now on one of the three identities below.

Pythagorean Identities

$\sin^2 A + \cos^2 A = 1$

$1 + \tan^2 A = \sec^2 A$

$1 + \cot^2 A = \csc^2 A$

You should always use an algebraic proof to be certain that you have an actual identity.

EXERCISES

You will need

A graphing calculator for Exercises **2, 6, 9–12, 15,** and **17.**

Practice Your Skills

1. Explain how you can graph $y = \cot x$ on your calculator without using a built-in cotangent function. ⓐ

2. Use graphs or tables to determine which of these equations may be identities.

a. $\cos x = \sin\left(\frac{\pi}{2} - x\right)$

b. $\cos x = \sin\left(x - \frac{\pi}{2}\right)$ ⓐ

c. $(\csc x - \cot x)(\sec x + 1) = 1$

d. $\tan x(\cot x + \tan x) = \sec^2 x$ ⓐ

3. Prove algebraically that the equation in Exercise 2d is an identity. ⓐ

4. Evaluate.

a. $\sec\frac{\pi}{6}$

b. $\csc\frac{5\pi}{6}$

c. $\csc\frac{2\pi}{3}$

d. $\sec\frac{3\pi}{2}$

e. $\cot\frac{5\pi}{3}$

f. $\csc\frac{4\pi}{3}$

Reason and Apply

5. In your own words, explain the difference between a trigonometric equation and a trigonometric identity.

6. A function f is **even** if $f(-x) = f(x)$ for all x-values in its domain. It is **odd** if $f(-x) = -f(x)$ for all x-values in its domain. Determine whether each function is even, odd, or neither.

a. $f(x) = \sin x$ @ **b.** $f(x) = \cos x$ **c.** $f(x) = \tan x$ @

d. $f(x) = \cot x$ **e.** $f(x) = \sec x$ @ **f.** $f(x) = \csc x$

7. In the next lesson, you'll see that $\cos 2A = \cos^2 A - \sin^2 A$ is an identity. Use this identity and the identities from this lesson to prove that (h)

a. $\cos 2A = 1 - 2\sin^2 A$ **b.** $\cos 2A = 2\cos^2 A - 1$

8. Sketch the graph of $y = \frac{1}{f(x)}$ for each function.

a.

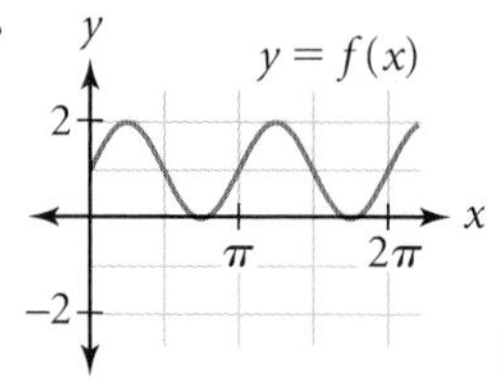

@

b.

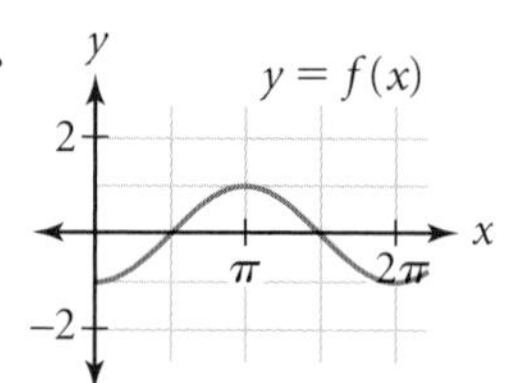

9. Find another function that has the same graph as each function named below. More than one answer is possible.

a. $y = \cos\left(\frac{\pi}{2} - x\right)$ @ **b.** $y = \sin\left(\frac{\pi}{2} - x\right)$ **c.** $y = \tan\left(\frac{\pi}{2} - x\right)$

d. $y = \cos(-x)$ @ **e.** $y = \sin(-x)$ @ **f.** $y = \tan(-x)$

g. $y = \sin(x + 2\pi)$ **h.** $y = \cos\left(\frac{\pi}{2} + x\right)$ **i.** $y = \tan(x + \pi)$

10. Find the first three positive x-values that make each equation true.

a. $\sec x = -2.5$ **b.** $\csc x = 0.4$

11. Sketch a graph for each equation with domain $0 \leq x < 4\pi$. Include any asymptotes and state the x-values at which the asymptotes occur. (h)

a. $y = \csc x$ **b.** $y = \sec x$ **c.** $y = \cot x$

12. Write an equation for each graph. More than one answer is possible. Use your calculator to check your work.

a.

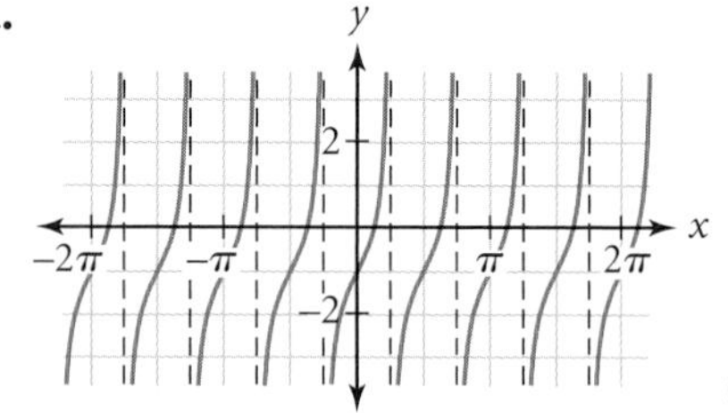

@

b.

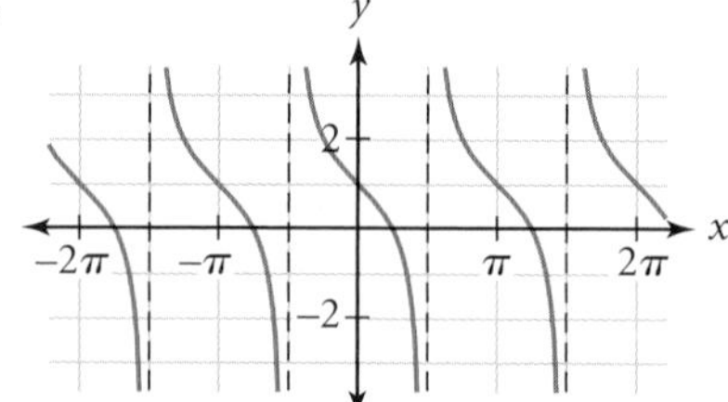

c.

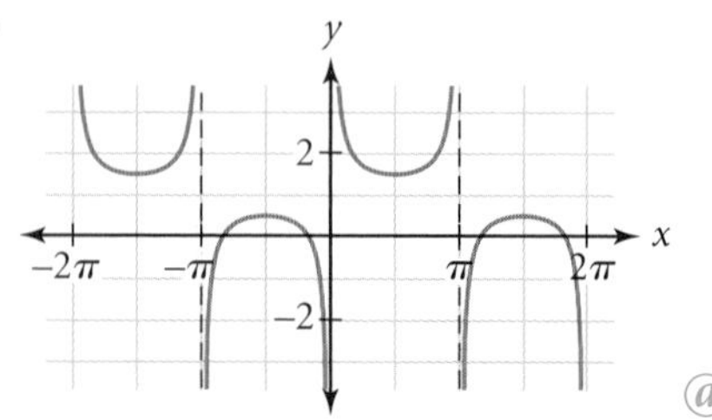

@

d.

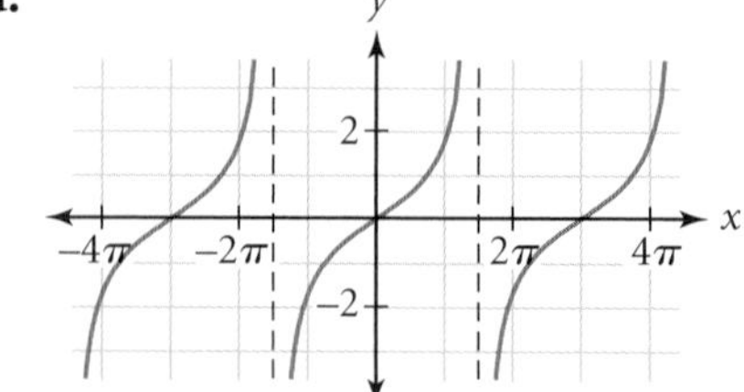

13. Use definitions and identities that you know to prove that the two sides of each equation are equivalent. Be sure to change only one side of the equation.

a. $\dfrac{\sin x}{1+\cos x} = \dfrac{1-\cos x}{\sin x}$ (h)

b. $\dfrac{1-\cos^2 x}{\tan x} = \sin x \cos x$

c. $\dfrac{\sec^2 x - 1}{\sin x} = \tan x \sec x$

14. Use the trigonometric identities to rewrite each expression in a simplified form that contains only one type of trigonometric function. For each expression, give values of θ for which the expression is undefined. Use domain $0 \le \theta < 2\pi$.

a. $2\cos^2\theta + \dfrac{\sin\theta}{\csc\theta} + \sin^2\theta$

b. $(\sec\theta)(2\cos^2\theta) + (\cot\theta)(\sin\theta)$

c. $\sec^2\theta + \dfrac{1}{\cot\theta} + \dfrac{\sin^2\theta - 1}{\cos^2\theta}$

d. $\sin\theta(\cot\theta + \tan\theta)$

Review

15. APPLICATION Tidal changes can be modeled with a sinusoidal curve. This table gives the time and height of the low and high tides for Burntcoat Head, Nova Scotia, on two consecutive days. (h)

High and Low Tides for Burntcoat Head, Nova Scotia

	Low	High	Low	High
	Time (AST)/Height (m)	Time (AST)/Height (m)	Time (AST)/Height (m)	Time (AST)/Height (m)
Day 1	04:13/1.60	10:22/13.61	16:40/1.42	22:49/14.15
Day 2	05:12/1.01	11:19/14.09	17:38/0.94	23:45/14.71

(*www.ldeo.columbia.edu*)

a. Select one time value as the starting time and assign it time 0. Express the other time values in minutes relative to the starting time.

b. Let x represent time in minutes relative to the starting time, and let y represent the tide height in meters. Make a scatter plot of the eight data points.

c. Find mean values for

  i. Height of the low tide.

  ii. Height of the high tide.

  iii. Height of "no tide," or the mean water level.

  iv. Time in minutes of a tide change between high and low tide.

d. Write an equation to model these data.

e. Graph your equation with the scatter plot in order to check the fit.

f. Predict the water height at 12:00 on Day 1.

g. Predict when the high tide(s) occurred on Day 3.

The Bay of Fundy borders Maine and the Canadian provinces of New Brunswick and Nova Scotia. This bay experiences some of the most dramatic tides on Earth, with water depths fluctuating up to 18 m.

16. APPLICATION Juan's parents bought a \$500 savings bond for him when he was born. Interest has compounded monthly at an annual fixed rate of 6.5%.

a. Juan just turned 17, and he is considering using the bond to pay for college. How much is his bond currently worth? @

b. Juan also considers saving the bond and using it to buy a used car after he graduates. If he would need about \$4,000, how long would he have to wait?

17. APPLICATION A pharmacist has 100 mL of a liquid medication that is 60% concentrated. This means that in 100 mL of the medication, 60 mL is pure medicine and 40 mL is water. She alters the concentration when filling a specific prescription. Suppose she alters the medication by adding water.

a. Write a function that gives the concentration of the medication, $d(x)$, as a function of the amount of water added in milliliters.

b. What is the concentration if the pharmacist adds 20 mL of water?

c. How much water should she add if she needs a 30% concentration?

d. Graph $y = d(x)$. Explain the meaning of the asymptote.

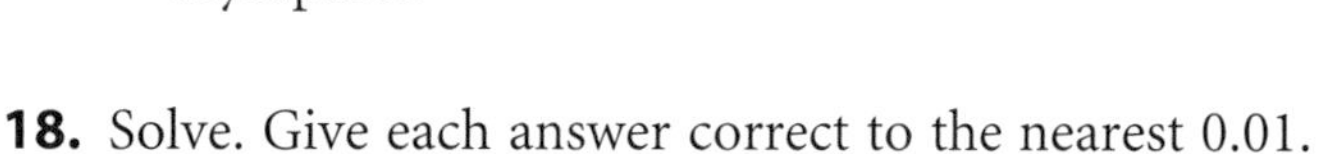

18. Solve. Give each answer correct to the nearest 0.01.

a. $4 + 5^x = 18$

b. $\log_3 15 = \dfrac{\log x}{\log 3}$

c. $120(0.5)^{2x} = 30$

d. $\log_6 100 = x$

e. $2 \log x = 2.5$

f. $\log_5 5^3 = x$

g. $4 \log x = \log 16$

h. $\log(5 + x) - \log 5 = 2$

IMPROVING YOUR VISUAL THINKING SKILLS

An Equation Is Worth a Thousand Words

You have learned to model real-world data with a variety of equations. You've also seen that many types of equations have real-world manifestations. For example, the path of a fountain of water is parabolic because its projectile-motion equation is quadratic.

Look for a photo of a phenomenon that suggests the graph of an equation. Impose coordinate axes, and find the equation that models the photo. If you have geometry software, you can import an electronic version of your photo and graph your function over it.

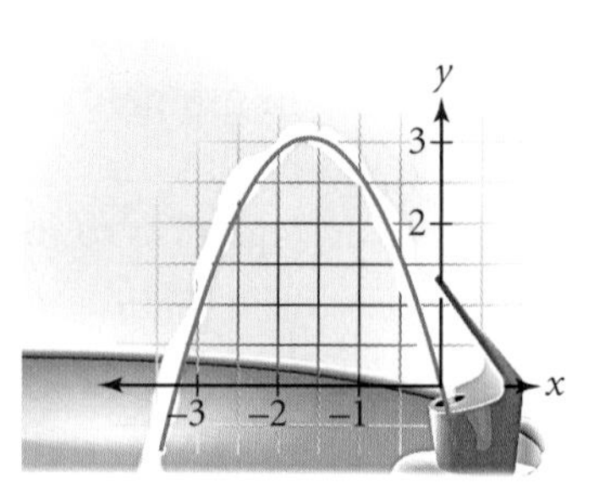

LESSON

13.7 Combining Trigonometric Functions

Music is the pleasure that the human soul experiences from counting without being aware that it is counting.

GOTTFRIED LEIBNIZ

For many applications of trigonometric functions, you'll need to add several functions to get a realistic model. When you add sinusoidal functions, your result may also be a periodic function. Two cycles of the graph of $y = \sin 2x + \sin 3x$ are shown below.

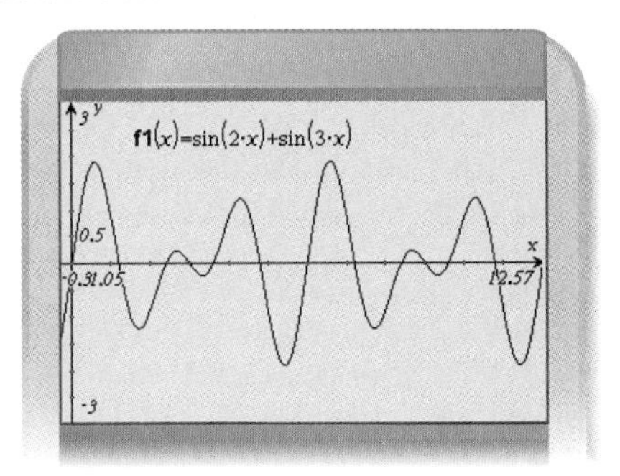

American cellist Yo Yo Ma (b 1955) is one of the world's most renowned classical musicians. He has released more than 75 albums and won more than 15 Grammy awards.

For example, the sound produced by a musical instrument can be modeled by a sum of several sinusoidal functions. Each of these functions represents a single tone, and their sum models the sound that's characteristic of the instrument. As an instrument plays an A note at a frequency of 440 cycles per second, for instance, it also produces several other tones, such as the next higher A at 880 cycles per second. These "overtones" and their relative loudness determine the way the instrument sounds.

The graphs and equations below show a flute and a violin playing the same A above middle C.

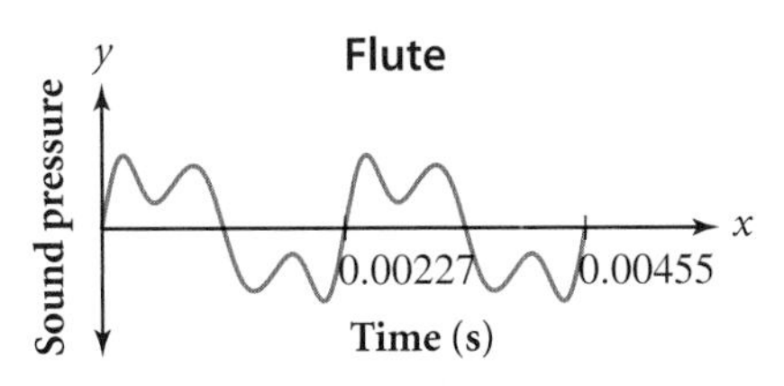

$$y = 16\sin(440 \cdot 2\pi x) + 9\sin(880 \cdot 2\pi x) + 3\sin(1320 \cdot 2\pi x) + 2.5\sin(1760 \cdot 2\pi x) + 1.0\sin(2200 \cdot 2\pi x)$$

The coefficient of a term represents the loudness of the tone. The coefficients in this equation are relatively small compared to the leading coefficient of 16. This is why a flute has a pure sound.

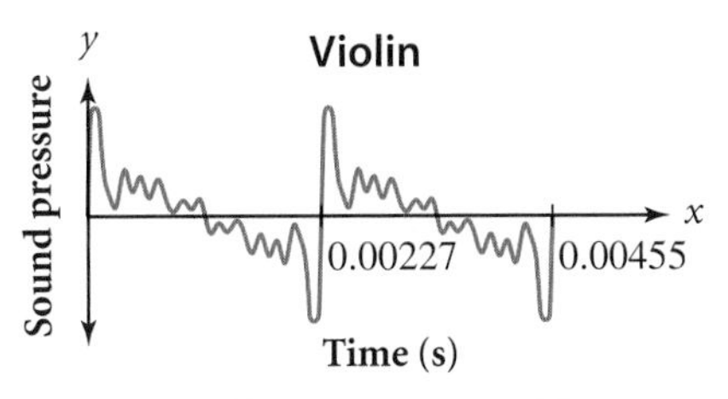

$$y = 19\sin(440 \cdot 2\pi x) + 9\sin(880 \cdot 2\pi x) + 8\sin(1320 \cdot 2\pi x) + 9\sin(1760 \cdot 2\pi x) + 12.5\sin(2200 \cdot 2\pi x) + 10.5\sin(2640 \cdot 2\pi x) + 14\sin(3080 \cdot 2\pi x) + 11\sin(3520 \cdot 2\pi x) + 8\sin(3960 \cdot 2\pi x) + 7\sin(4400 \cdot 2\pi x) + 5.5\sin(4840 \cdot 2\pi x) + 1.0\sin(5280 \cdot 2\pi x) + 4.5\sin(5720 \cdot 2\pi x) + 4.0\sin(6160 \cdot 2\pi x) + 3\sin(6600 \cdot 2\pi x)$$

Some of the coefficients in this equation are quite large relative to the leading coefficient of 19. This causes its graph to be more "bumpy," reflecting a violin's complex sound.

The periods of both graphs on the preceding page are the same and are determined by the coefficient of the argument of the leading term, 440, which represents the primary A tone or fundamental frequency above middle C. The "bumps" in the graphs are caused by the overtones, which are described by the remaining terms in each equation.

The numbers 440, 880, 1320, 1760, and so on are the frequencies of the tones. Each of these frequencies is a multiple of the primary frequency, 440 cycles per second. One cycle is completed in $\frac{1}{440}$ of a second, or approximately 0.00227 s.

When the frequency increases, the period decreases, and you hear a higher sound. When frequency decreases, you hear a lower sound. The amplitude indicates the loudness of a tone.

Each musical instrument has its own typical sound and its own graph for the sound of a particular note. A musician can only slightly affect the sound of the note and the shape of the graph it produces. Try entering the flute and violin equations on page 789 into your calculator to reproduce the graphs. Then change the coefficients of some of the terms and observe how the graph is affected.

Investigation
Sound Wave

You will need

- a microphone probe
- two tuning forks
- musical instrument, *optional*

In this investigation you'll explore the frequency of some tones and combinations of tones.

Procedure Note

1. Set up your calculator and microphone probe to collect sound frequency data. [See **Calculator Note 13D.**]
2. To ring a tuning fork, rap it sharply on a semisoft surface like a book or the heel of your shoe. Hold the fork close to the microphone and begin collecting data. When you use more than one fork, be sure to hold them equidistant from the microphone.

Step 1 Choose a tuning fork and collect data as described in the Procedure Note. Find an equation to fit the data. Repeat this process with a second tuning fork.

Step 2 Using the same two tuning forks that you used in Step 1, ring both forks simultaneously, and collect frequency data. You should see a combination of sinusoids, rather than a simple sinusoid. Model the data with an equation that is the sum of two simple sinusoid equations.

Step 3 Select a musical instrument, perhaps a flute, violin, piano, timpani, or your voice. Play one note (or string) and collect data. You should see a complex wave, probably too complex for you to write an equation. Identify the fundamental frequency. See if you can identify the frequencies of some of the overtones as well.

Technology
CONNECTION

Producing music electronically on a synthesizer involves a series of steps. First a sequencer creates the electronic equivalent of sheet music. This information is sent to the synthesizer using a digital code called MIDI (Musical Instrument Digital Interface). The synthesizer then reproduces each instrument's sound accurately by producing the correct strength of each individual overtone frequency. Some early synthesizers did this by adding together sine waves. Programs called wave editors let you create your own new "instruments" by specifying what their waves will look like.

Relationships modeled by adding sinusoids are not limited to music and sound. These patterns occur in the motion of moons, planets, tides, and satellites, and in any gear-driven mechanism, from a wristwatch to a car.

You can write a horizontally translated sinusoid as a sum of two untranslated curves. For example, $y = \cos(x - 0.6435)$ is equivalent to $y = 0.8\cos x + 0.6\sin x$. (Check this on your calculator.) Here is a proof that $\cos(A - B) = \cos A\cos B + \sin A\sin B$.

The diagram at right shows the terminal sides of $\angle A$ and $\angle B$ with $m\angle A - m\angle B = m\angle C$. Note the coordinates of the intersections of the terminal sides with the unit circle. The distance c between those points can be determined by two equations, one using the distance formula and one using the Law of Cosines.

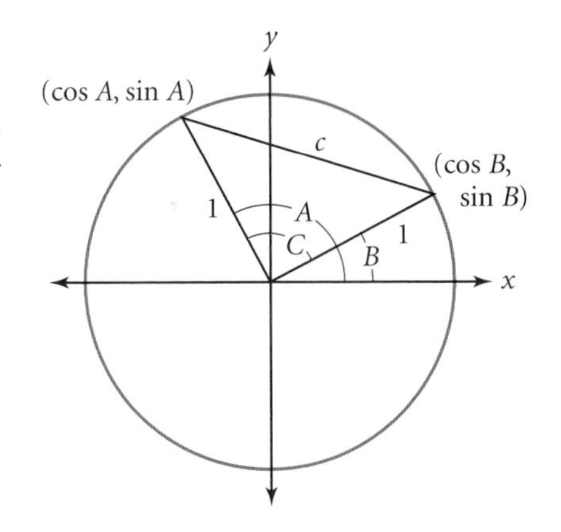

$$c = \sqrt{(\cos A - \cos B)^2 + (\sin A - \sin B)^2}$$

and

$$c = \sqrt{1 + 1 - 2\cos C}.$$

$\sqrt{2 - 2\cos C} = \sqrt{(\cos A - \cos B)^2 + (\sin A - \sin B)^2}$	Set the expressions for c equal to each other.
$2 - 2\cos C = (\cos A - \cos B)^2 + (\sin A - \sin B)^2$	Square both sides.
$2 - 2\cos C = \cos^2 A - 2\cos A\cos B + \cos^2 B + \sin^2 A - 2\sin A\sin B + \sin^2 B$	Expand.
$2 - 2\cos C = \left(\sin^2 A + \cos^2 A\right) + \left(\sin^2 B + \cos^2 B\right) - 2\cos A\cos B - 2\sin A\sin B$	Reorder terms.
$2 - 2\cos C = 1 + 1 - 2\cos A\cos B - 2\sin A\sin B$	Use the Pythagorean identity, $\sin^2 A + \cos^2 A = 1$.
$-2\cos C = -2\cos A\cos B - 2\sin A\sin B$	Subtract 2 from both sides.
$\cos C = \cos A\cos B + \sin A\sin B$	Divide both sides by -2.
$\cos(A - B) = \cos A\cos B + \sin A\sin B$	Substitute $(A - B)$ for C.

So $\cos(A - B) = \cos A \cos B + \sin A \sin B$ is an identity. You can use this identity to find exact cosine values for some new angles, using values you already know.

EXAMPLE A | Find the exact value of $\cos\frac{\pi}{12}$.

▶ Solution

You know exact values of the sine and cosine of 0, $\frac{\pi}{6}$, $\frac{\pi}{4}$, $\frac{\pi}{3}$, and π. So rewrite $\frac{\pi}{12}$ as a difference of these values.

$$\cos\frac{\pi}{12} = \cos\left(\frac{3\pi}{12} - \frac{2\pi}{12}\right) = \cos\left(\frac{\pi}{4} - \frac{\pi}{6}\right)$$

Rewrite $\frac{\pi}{12}$ as a difference of two fractions, and reduce.

$$\cos\frac{\pi}{12} = \cos\frac{\pi}{4} \cdot \cos\frac{\pi}{6} + \sin\frac{\pi}{4} \cdot \sin\frac{\pi}{6}$$

Rewrite $\cos\left(\frac{\pi}{4} - \frac{\pi}{6}\right)$ using the identity $\cos(A - B) = \cos A \cos B + \sin A \sin B$.

$$\cos\frac{\pi}{12} = \frac{\sqrt{2}}{2} \cdot \frac{\sqrt{3}}{2} + \frac{\sqrt{2}}{2} \cdot \frac{1}{2}$$

Substitute exact values for sine and cosine of $\frac{\pi}{4}$ and $\frac{\pi}{6}$.

$$\cos\frac{\pi}{12} = \frac{\sqrt{6} + \sqrt{2}}{4}$$

Combine into one rational expression.

So the exact value of $\cos\frac{\pi}{12}$ is $\frac{\sqrt{6} + \sqrt{2}}{4}$. You can check your work by evaluating $\frac{\sqrt{6} + \sqrt{2}}{4}$ and $\cos\frac{\pi}{12}$ with your calculator.

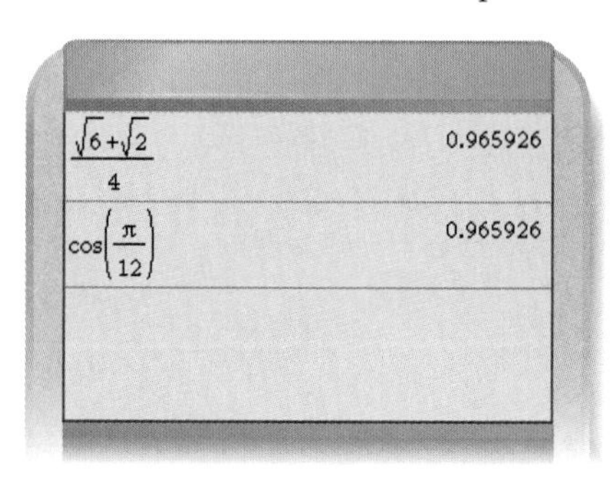

The next example shows you how to develop another identity based on identities that you already know.

EXAMPLE B | Develop an identity for $\cos(A + B)$.

▶ Solution

Notice that $\cos(A + B)$ is similar to $\cos(A - B)$, except the sign of B is changed. Start with the identity $\cos(A - B) = \cos A \cos B + \sin A \sin B$, and replace B with $-B$.

$$\cos[A - (-B)] = \cos A \cos(-B) + \sin A \sin(-B)$$

Replace B with $-B$.

$$\cos(A + B) = \cos A \cos(-B) + \sin A \sin(-B)$$

Rewrite $A - (-B)$ as $A + B$.

Earlier in this chapter, you found that $\cos(-x) = \cos x$ and that $\sin(-x) = -\sin x$.

$$\cos(A + B) = \cos A \cos B - \sin A \sin B$$

Replace $\cos(-B)$ with $\cos B$ and $\sin(-B)$ with $-\sin B$.

So $\cos(A + B) = \cos A \cos B - \sin A \sin B$ is an identity.

There are many other trigonometric identities that can be useful in calculations or for simplifying expressions. As in Example B, you will be asked to prove new identities using existing identities. The box on the next page includes several relationships you have already seen and a few new ones.

Sum and Difference Identities

$$\cos(A - B) = \cos A \cos B + \sin A \sin B$$
$$\cos(A + B) = \cos A \cos B - \sin A \sin B$$
$$\sin(A - B) = \sin A \cos B - \cos A \sin B$$
$$\sin(A + B) = \sin A \cos B + \cos A \sin B$$
$$\tan(A - B) = \frac{\tan A - \tan B}{1 + \tan A \tan B}$$
$$\tan(A + B) = \frac{\tan A + \tan B}{1 - \tan A \tan B}$$

Double-Angle Identities

$$\cos 2A = \cos^2 A - \sin^2 A = 1 - 2\sin^2 A = 2\cos^2 A - 1$$
$$\sin 2A = 2\sin A \cos A$$
$$\tan 2A = \frac{2\tan A}{1 - \tan^2 A}$$

Half-Angle Identities

$$\sin\frac{A}{2} = \pm\sqrt{\frac{1 - \cos A}{2}}$$
$$\cos\frac{A}{2} = \pm\sqrt{\frac{1 + \cos A}{2}}$$
$$\tan\frac{A}{2} = \pm\sqrt{\frac{1 - \cos A}{1 + \cos A}} = \frac{\sin A}{1 + \cos A} = \frac{1 - \cos A}{\sin A}$$

For the half-angle identities, the sign of the answer is determined by the quadrant in which the terminal side of the given angle lies.

In the exercises you will be asked to prove some of these identities. As you do so, remember to work on only one side of the equation.

EXERCISES

You will need

A graphing calculator for Exercises **14–19.**

Practice Your Skills

1. Decide whether each expression is an identity by substituting values for A and B.

a. $\cos(A + B) = \cos A + \cos B$ @ **b.** $\sin(A + B) = \sin A + \sin B$ @

c. $\cos(2A) = 2\cos A$ **d.** $\sin(2A) = 2\sin A$

2. Prove each identity. The sum and difference identities will be helpful.

a. $\cos(2\pi - A) = \cos A$ **b.** $\sin\left(\frac{3\pi}{2} - A\right) = -\cos A$

3. Rewrite each expression with a single sine or cosine.

a. $\cos 1.5 \cos 0.4 + \sin 1.5 \sin 0.4$ @ **b.** $\cos 2.6 \cos 0.2 - \sin 2.6 \sin 0.2$

c. $\sin 3.1 \cos 1.4 - \cos 3.1 \sin 1.4$ @ **d.** $\sin 0.2 \cos 0.5 + \cos 0.2 \sin 0.5$

4. Use identities to find the exact value of each expression.

a. $\sin \frac{-11\pi}{12}$ ⓐ **b.** $\sin \frac{7\pi}{12}$ **c.** $\tan \frac{\pi}{12}$ **d.** $\cos \frac{\pi}{8}$

Reason and Apply

5. Given $\pi \le x \le \frac{3\pi}{2}$ and $\sin x = -\frac{2}{3}$, find the exact value of $\sin 2x$. ⓗ

6. Use the identity for $\cos(A - B)$ and the identities $\sin A = \cos\left(\frac{\pi}{2} - A\right)$ and $\cos A = \sin\left(\frac{\pi}{2} - A\right)$ to prove that

$$\sin(A + B) = \sin A \cos B + \cos A \sin B$$ ⓐ

7. Use the identity for $\sin(A + B)$ from Exercise 6 to prove that

$$\sin(A - B) = \sin A \cos B - \cos A \sin B$$ ⓗ

8. Use the identity for $\sin(A + B)$ to prove the identity $\sin 2A = 2 \sin A \cos A$.

9. Use the identity for $\cos(A + B)$ to prove the identity $\cos 2A = \cos^2 A - \sin^2 A$. ⓗ

10. What is wrong with this statement?

$$\cos(\tan x) = \cos\left(\frac{\sin x}{\cos x}\right) = \sin x$$

11. Show that $\tan(A + B)$ is not equivalent to $\tan A + \tan B$. Then use the identities for $\sin(A + B)$ and $\cos(A + B)$ to develop an identity for $\tan(A + B)$. ⓗ

12. Use your identity from Exercise 11 to develop an identity for $\tan 2A$.

13. You have seen that $\sin^2 A = 1 - \cos^2 A$ and $\cos^2 A = 1 - \sin^2 A$.

a. Use one of the double-angle identities to develop an expression that is equivalent to $\sin^2 A$ but does not contain the term $\cos^2 A$. ⓐ

b. Use another double-angle identity to develop an expression equivalent to $\cos^2 A$ that does not contain $\sin^2 A$.

14. *Mini-Investigation* Set your graphing window to $0 \le x \le 4\pi$ and $-2 \le y \le 2$.

a	1	2	2	3	3	4	4	4
b	2	3	4	4	6	6	8	12
Period								

a. Graph equations in the form $y = \sin ax + \sin bx$, using the a- and b-values listed in the table. Record the period for each pair. ⓐ

b. Explain how to find the period of any function in the form $y = \sin ax + \sin bx$, where a and b are whole numbers.

15. *Mini-Investigation* Set your graphing window to $0 \le x \le 48\pi$ and $-2 \le y \le 2$.

a	2	2	2	4	3
b	4	3	5	6	6
Period					

a. Graph equations in the form $y = \sin\frac{x}{a} + \sin\frac{x}{b}$, using the a- and b-values listed in the table. Record the period for each pair.

b. Explain how to find the period of any function in the form $y = \sin\frac{x}{a} + \sin\frac{x}{b}$, where a and b are integers.

c. Predict the period of $y = \sin\frac{x}{3} + \sin\frac{x}{4} + \sin\frac{x}{8}$. Explain your reasoning.

16. When a tuning fork for middle C is struck, the resulting sound wave has a frequency of 262 cycles per second. The equation $y = \sin(262 \cdot 2\pi x)$ is one possible model for this wave.

a. Identify the period of this wave. Make a graph showing about five complete cycles.

b. Suppose middle C on an out-of-tune piano is played and that the resulting wave has a frequency of 265 cycles per second. Write an equation to model the out-of-tune wave. Identify its period. Then make a graph using the same window as in 16a.

c. A piano tuner plays the out-of-tune C at the same time she uses the tuning fork. The two waves are added to produce a new sound wave. Write the equation that models the sum of the two waves. Graph a 0.5 s interval of this new equation.

Music

CONNECTION

The sound waves from an out-of-tune piano and a tuning fork have slightly different frequencies. When they are played together, the resulting wave will vary in amplitude, getting louder and softer in cycles. These periodic variations are called beats. If the difference between the number of cycles is three (as with 265 − 262), there are three beats per second. The loudness will rise and fall three times per second. Musicians listen for beats to see if their instruments are out of tune. An out-of-tune piano is tuned by adjusting the tension of a string until the beats disappear.

Review

17. Solve. (h)

a. $\sec 144° = x$ **b.** $\csc\frac{24\pi}{9} = x$ (a) **c.** $\cot 3.92 = x$ **d.** $\cot 630° = x$ (a)

18. Find all values of θ that satisfy each equation. Use domain $0° \leq \theta < 360°$.

a. $\tan\theta = 0.5317$ **b.** $\sec\theta = -3.8637$ (a) **c.** $\csc\theta = 1.1126$ (a) **d.** $\cot\theta = -4.3315$

19. A fishing boat rides gently up and down, 10 times per minute, on the ocean waves. The boat rises and falls 1.5 m between each wave crest and trough. Assume the boat is on a crest at time 0 min.

a. Sketch a graph of the boat's height above sea level over time.

b. Use a cosine function to model your graph.

c. Use a sine function to model your graph.

EXPLORATION

Polar Coordinates

You are very familiar with graphing points on a plane with rectangular coordinates. When you graph points in the form (x, y), you need both the x-coordinate and the y-coordinate to identify the exact location of any point. But this is not the only way to locate points on a plane.

In this chapter you have worked with circles centered at the origin. As you move around a circle, you can identify any point on the circle with coordinates in the form (x, y). However, because the radius of the circle remains constant, you can also identify these points by the radius, r, and the angle of rotation from the positive x-axis, θ.

Imagine infinitely many concentric circles covering a plane, all centered at the origin. You can identify any point on the plane with coordinates in the form (r, θ), called **polar coordinates.** Polar equations in the form $r = f(\theta)$ may lead to familiar or surprising results.

This point can be identified as $(-2\sqrt{2}, 2\sqrt{2})$ with rectangular coordinates, or $(4, 135°)$ with polar coordinates.

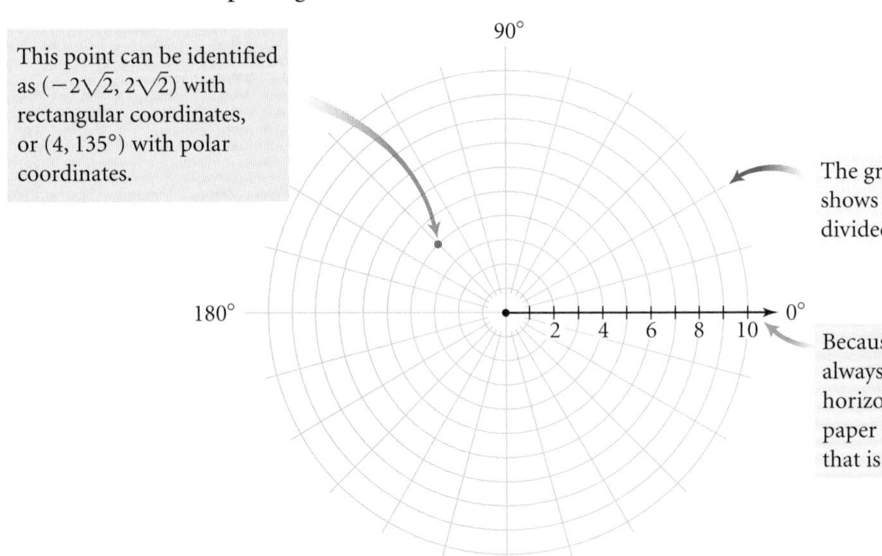

The grid on polar graph paper shows concentric circles, often divided into 15° sectors.

Because polar coordinates always relate to the positive horizontal axis, polar graph paper often shows a single axis that is a ray.

For instance, look at the graph of the polar equation $r = 4$, shown at right.

There is no θ in this equation, so r will always be 4. This is the set of all points 4 units from the origin—a circle with radius 4. Notice how much simpler this equation is than the equation of the same circle using rectangular coordinates!

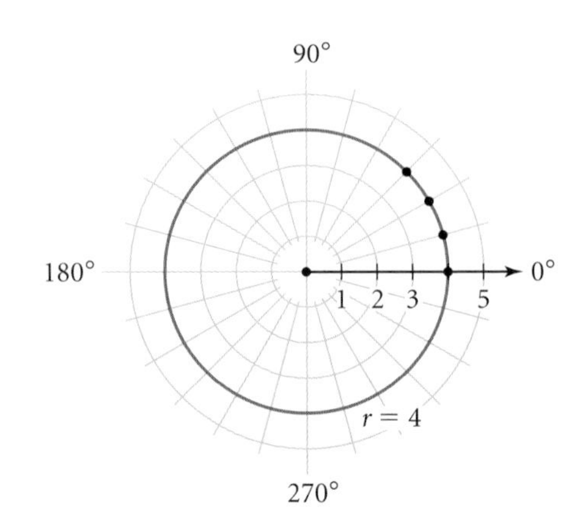

Imagine standing at the origin looking in the direction of the positive horizontal axis. First, rotate counterclockwise by the necessary angle. Then, imagine walking straight out from the origin and placing a point at distance *r*. If *r* is positive, walk forward. If *r* is negative, walk backward.

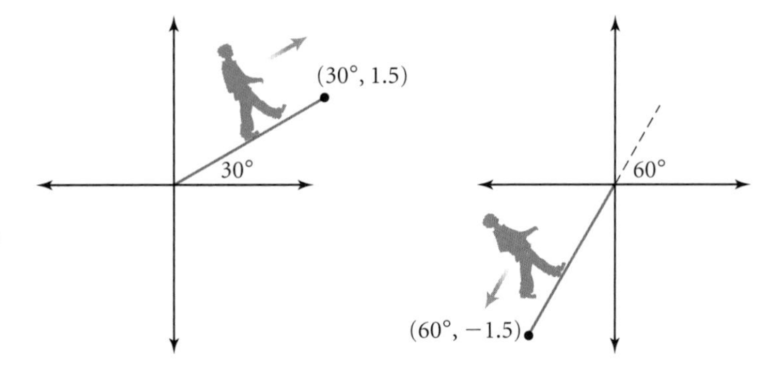

In the activity and questions that follow, you'll explore some of the elegant and complicated-looking graphs that result from polar equations.

Activity
Rose Curves

Use these steps to explore polar equations in the form $r = a\cos n\theta$.

For Steps 1–3, make a table of values, plot the points, and connect them in order with a smooth curve. The results are called rose curves, and each will look like a flower with petals. You can easily check your work with your graphing calculator. [▶ See **Calculator Note 13E** to learn how to graph polar equations on your calculator. ◀]

Step 1 Graph the family of curves $r = a\cos 2\theta$ with $a = \{1, 2, 3, 4, 5, 6\}$. How does the coefficient a affect the graph?

Step 2 Graph the family of curves $r = 3\cos n\theta$ with $n = \{1, 2, 3, 4, 5, 6\}$. Generalize the effect of the coefficient n. Write statements that describe the curves when n is even and when n is odd.

Step 3 Graph the family of curves $r = 3\sin n\theta$ with $n = \{1, 2, 3, 4, 5, 6\}$. Generalize your results. How do these curves differ from the curves in Step 2?

Step 4 Find a way to graph a rose curve with only two petals. Explain why your method works.

Questions

1. Find a connection between the graph of the polar equation $r = a\cos n\theta$ and the graph of the associated rectangular equation $y = a\cos nx$. Explain whether or not you can look at the graph of $y = a\cos nx$ and predict the shape and number of petals in the polar graph.

2. The graphs of polar equations in the forms $r = a(\cos\theta + 1)$, $r = a(\cos\theta - 1)$, $r = a(\sin\theta + 1)$, and $r = a(\sin\theta - 1)$ are called cardioids because they resemble hearts. Graph several curves in the cardioid family. Generalize your results by answering the questions at the top of page 798.

Technology CONNECTION

Cardioid microphones are designed to pick up sound at equal intensity levels from any point on a cardioid around the microphone. These microphones are designed so that the dimple of the curve is directed toward the base of the microphone, so that they minimize sound coming from behind. For this reason, cardioid microphones are especially useful for recording sound from a stage because they pick up little noise from the audience.

a. How do the graphs of $r = a(\cos\theta + 1)$ and $r = a(\cos\theta - 1)$ differ?

b. How do the cardioids created with sine differ from those created with cosine?

c. What is the effect of the coefficient a?

3. Write polar equations to create each graph. For 3b and d, you'll need more than one equation.

a.

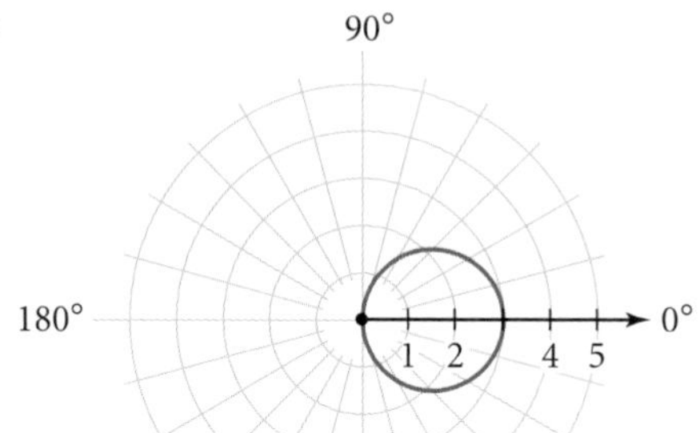

b.

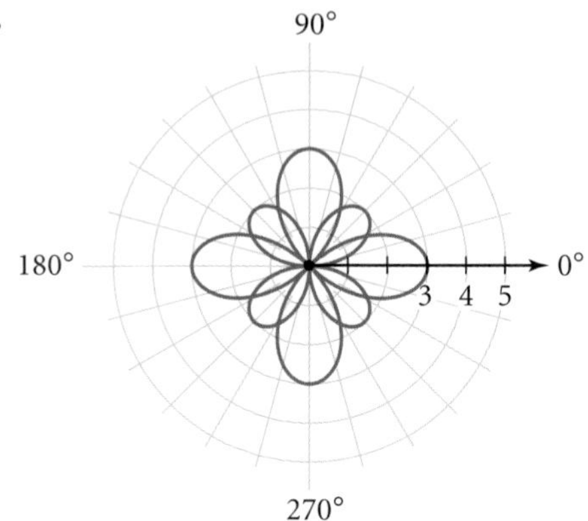

c.

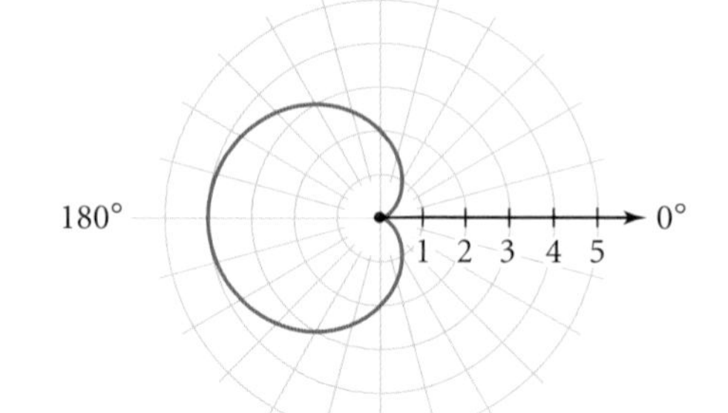

d.

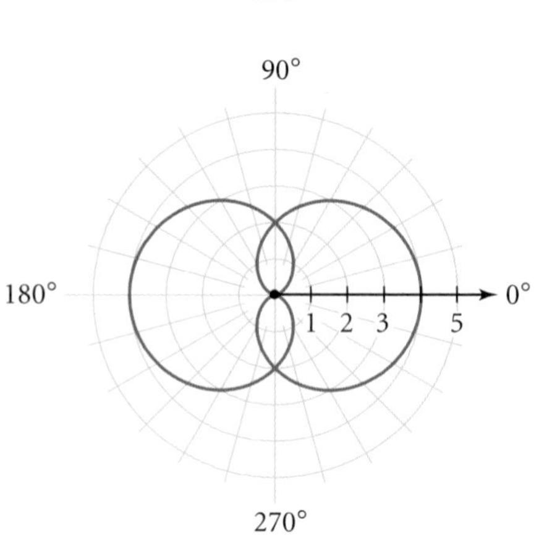

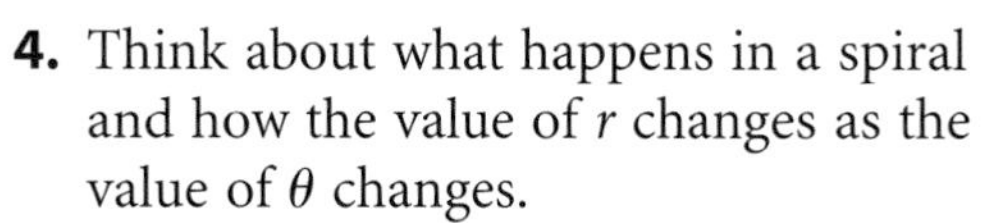

4. Think about what happens in a spiral and how the value of r changes as the value of θ changes.

a. Find an equation that creates a spiral. Check your work by graphing on your calculator with the domain $0° \leq \theta \leq 360°$. What is the general form of the equation that creates a spiral?

b. What happens as you extend the domain of θ to values greater than 360°?

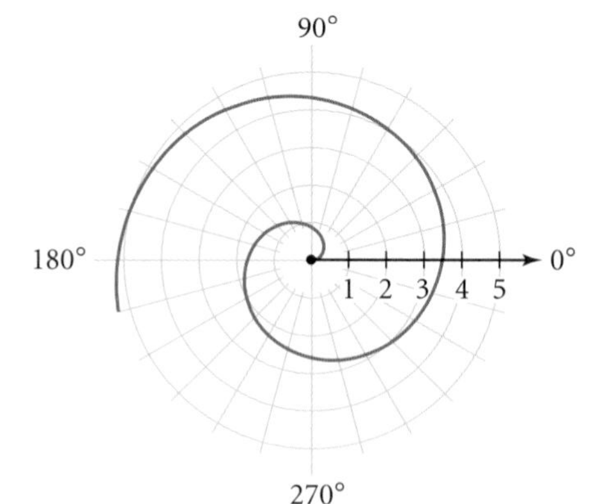

c. What happens if you change the domain of θ to include negative values?

5. In general, are polar equations functions? Explain your reasoning.

CHAPTER 13 REVIEW

In this chapter you learned to measure angles in **radians,** and you then identified relationships among radian measure, arc length, speed, and **angular speed.** You studied circular functions, their graphs, and their applications, and you learned to think of the sine and cosine of an angle as the y- and x-coordinates of a point on the unit circle. This allowed you to identify angles in **standard position** that are **coterminal.** That is, they share the same **terminal side.** Coterminal angles have the same trigonometric values. The remaining **trigonometric functions**—tangent, **cotangent, secant,** and **cosecant**—are also defined either as a ratio involving an x-coordinate and a y-coordinate, or as a ratio of one of these coordinates and the distance from the origin to the point.

Graphs of the trigonometric functions model periodic behavior and have domains that extend in both the positive and negative directions. You worked with many relationships that can be modeled with **sinusoids.** You studied transformations of sinusoidal functions and defined their **amplitude, period, phase shift, vertical translation,** and **frequency.**

You learned the difference between an inverse trigonometric relation and an inverse trigonometric function, and used this distinction to solve equations involving periodic functions. You learned, for example, that $x = \cos y$ provides infinitely many values of y for a given choice of x, whereas the function $y = \cos^{-1}x$ provides exactly one value of y for each value of x, called the **principal value.**

Finally, you discovered several properties and identities involving trigonometric expressions, and you learned how to prove that an equation is an identity.

EXERCISES

You will need

A graphing calculator for Exercises **4, 5, 12, 18, 22,** and **23.**

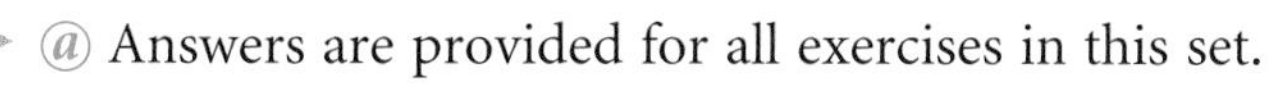

@ Answers are provided for all exercises in this set.

1. For each angle in standard position given, identify the quadrant that the angle's terminal side lies in, and name a coterminal angle. Then convert each angle measure from radians to degrees or vice versa.

a. $60°$ **b.** $\frac{4\pi}{3}$ **c.** $330°$ **d.** $-\frac{\pi}{4}$

2. Find exact values of the sine and cosine of each angle in Exercise 1.

3. State the period of the graph of each equation, and write one other equation that has the same graph.

a. $y = 2\sin\left[3\left(x - \frac{\pi}{6}\right)\right]$ **b.** $y = -3\cos 4x$

c. $y = \sec 2x$ **d.** $y = \tan(-2x) + 1$

4. For the sinusoidal equations in 3a and b, state the amplitude, phase shift, vertical translation, and frequency. Then sketch a graph of one complete cycle.

5. Write an equation for each graph.

a.

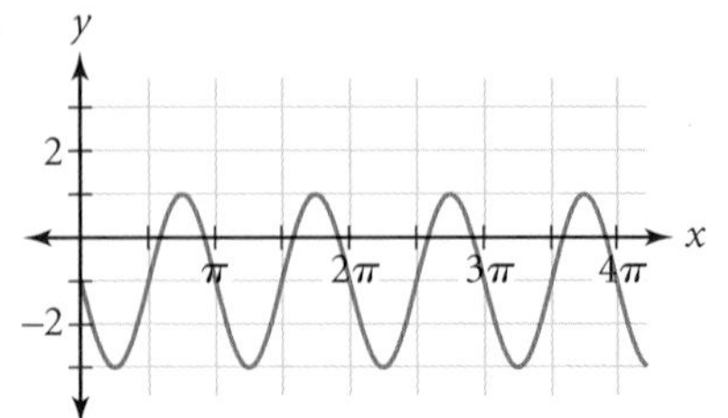

b.

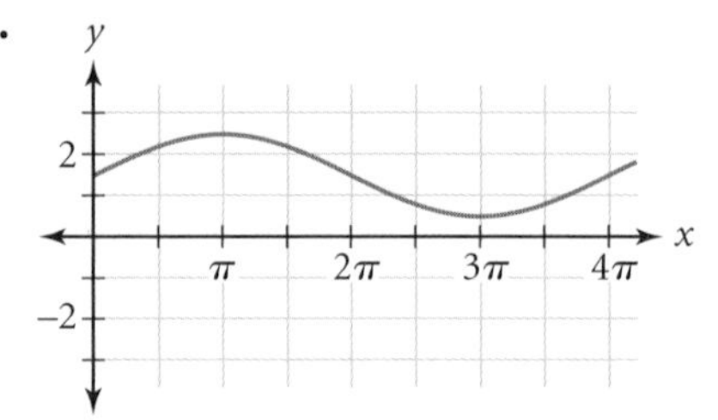

c.

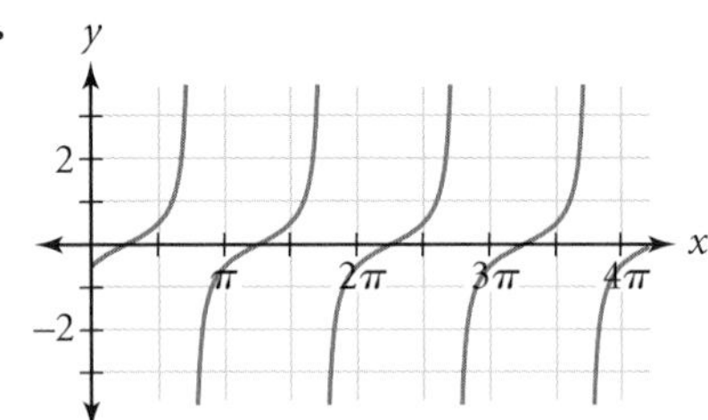

d.

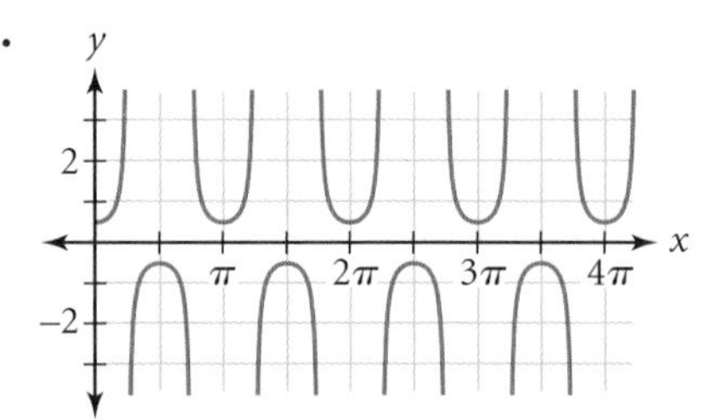

6. Find the area and arc length of a sector of a circle that has radius 3 cm and central angle $\frac{\pi}{4}$. Give exact answers.

7. Identify the domain and range of $\cos y = x$ and $y = \cos^{-1}x$.

8. Find these values without using your calculator. Then verify your answers with your calculator.

a. $\sin\left(\tan^{-1}\frac{3}{4}\right)$ **b.** $\cos\left(\sin^{-1}\frac{3}{5}\right)$ **c.** $\sin\left(\sin^{-1}\frac{8}{17}\right)$

9. Write an equation for a transformation of $y = \sin x$ that has a reflection across the x-axis, amplitude 3, period 8π, and phase shift $\frac{\pi}{2}$.

10. Prove that each of these identities is true. You may use any of the identities that have been proved in this chapter.

a. $\sec A - \sin A \tan A = \cos A$

b. $\frac{1}{\sin^2 A} - \frac{1}{\tan^2 A} = 1$

c. $\frac{\cos B - \tan A \sin B}{\sec A} = \cos(A + B)$

A detail from American sculptor Ruth Asawa's (b 1926) hourglass-shaped baskets shows her focus on circular patterns as a metaphor for family and community circles. The piece is called *Completing the Circle.*

11. A mass hanging from a spring is pulled down 2 cm from its resting position and released. It makes 12 complete bounces in 10 s. Each bounce has the same amplitude. At what times during the first 3 s was the mass 0.5 cm below its resting position?

12. APPLICATION These data give the ocean tide heights each hour on November 17, 2002, at Saint John, New Brunswick, Canada.

a. Create a scatter plot of the data.

b. Write a function to model the data, and graph this function on the scatter plot from 12a.

c. What would you estimate the tide height to have been at 3:00 P.M. on November 19, 2002?

d. A ship was due to arrive at Saint John on November 20, 2002. The water had to be at least 5 m for the ship to enter the harbor safely. Between what times on November 20 could the ship have entered the harbor safely?

Time	Tide height (m)
00:00	5.56
01:00	4.12
02:00	2.71
03:00	1.77
04:00	1.56
05:00	2.09
06:00	3.21
07:00	4.70
08:00	6.18
09:00	7.20
10:00	7.50
11:00	7.09

Time	Tide height (m)
12:00	6.09
13:00	4.65
14:00	3.08
15:00	1.85
16:00	1.33
17:00	1.60
18:00	2.54
19:00	3.96
20:00	5.51
21:00	6.75
22:00	7.33
23:00	7.17

MIXED REVIEW

13. Sketch a graph of each equation, and identify the shape formed.

a. $\frac{x^2}{12} - \frac{(y+3)^2}{9} = 1$

b. $\left(\frac{x-1}{5}\right)^2 + \left(\frac{y-1}{4}\right)^2 = 1$

c. $(y-3)^2 = \frac{x-4}{3}$

d. $-4x^2 - 24x + y^2 + 2y = 39$

14. If the probability of a snowstorm in July is 0.004, and the probability you will score an A in algebra is 0.75, then what is the probability of a snowstorm in July or an A in algebra?

15. Find the sums of each series.

a. Find the sum of the first 12 odd positive integers.

b. Find the sum of the first 20 odd positive integers.

c. Find the sum of the first n odd positive integers. (*Hint:* Try several choices for n until you see a pattern.)

16. APPLICATION A Detroit car rental business has a second outlet in Chicago. The company allows customers to make local rentals or one-way rentals to the other location. At the end of each month, one-eighth of the cars that start the month in Detroit will end up in Chicago, and one-twelfth of the cars that start the month in Chicago will end up in Detroit.

a. Write a transition matrix to represent this situation.

b. If there are 500 cars in each city at the start of operations, what would you expect the distribution to be four months later? In the long run?

17. APPLICATION A store in Yosemite National Park charges \$6.60 for a flashlight. Approximately 200 of them are sold each week. A survey indicates that the sales will decrease by 10 flashlights per week for each \$0.50 increase in price.

a. Write a function that describes the weekly revenue in dollars, y, as a function of selling price in dollars, x.

b. What selling price provides maximum weekly revenue? What is the maximum revenue?

Camping at Glacier Point, Yosemite National Park, California

18. Consider the function $y = \cos x$.

a. Write the equation of the image after the function is reflected across the x-axis, dilated by a vertical scale factor of $\frac{1}{2}$ and a horizontal scale factor of 2, and translated up 6 units.

b. What is the period of the image, in radians? What are the amplitude and phase shift?

c. Graph the function and its image on the same graph.

19. APPLICATION Lily and Philip both go to their doctor, complaining of the same symptoms. The doctor tests them for a rare disease. Data have shown that 20% of the people with these symptoms actually have the disease. The test the doctor uses is correct 90% of the time. Calculate the probabilities in the table below, and explain the meaning of the results.

		Test results	
		Accurate	**Inaccurate**
Patient's condition	Doesn't have the disease		
	Has the disease		

20. Identify each sequence as arithmetic, geometric, or neither. Then write both a recursive and explicit formula to describe the pattern, if possible.

a. 3, 9, 27, 81, 243, . . .

b. $-1, -3, -5, -7, -9, \ldots$

c. 2, 5, 10, 17, 26, . . .

d. $1, -\frac{1}{2}, \frac{1}{4}, -\frac{1}{8}, \frac{1}{16}, \ldots$

21. The circle at right has radius 4 cm, and the measure of central angle ACB is 55°.

a. What is the measure of $\angle ACB$ in radians?

b. What is the length of $\overset{\frown}{AB}$?

c. What is the area of sector ACB?

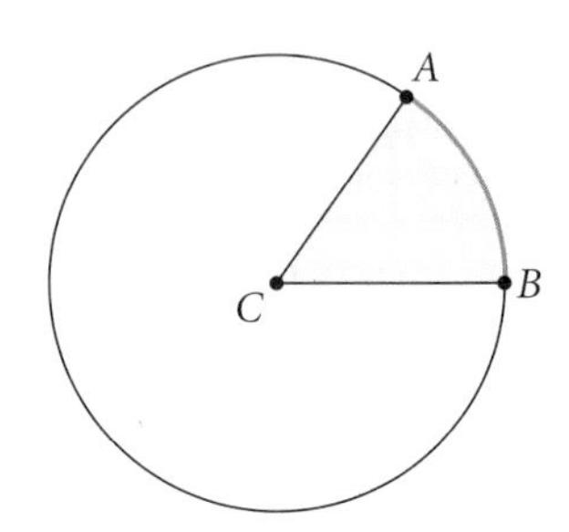

22. The population of Mumbai, India (formerly Bombay) in various years is given in the table below.

Year	1950	1970	1990	2000
Population (in millions)	3.0	6.2	12.3	16.1

(*The New York Times Almanac 2007*)

a. The population roughly doubled in the 20 years between 1950 and 1970 and in the 20 years between 1970 and 1990. What is a good estimate of the growth rate?

b. Find an exponential equation to model Mumbai's population.

c. Use your model to predict the population in 2015.

d. *The New York Times Almanac 2007* predicts that the population in 2015 will be 22.6 million. How does this compare with your prediction? Give a possible explanation for any discrepancy.

Mumbai, India

23. The lengths in feet of the main spans of 40 notable suspension bridges in North America are

{4260, 4200, 3800, 3500, 2800, 2800, 2388, 2310, 2300, 2190, 2150, 2000, 1850, 1801, 1750, 1632, 1600, 1600, 1600, 1596, 1550, 1500, 1495, 1470, 1447, 1400, 1380, 1263, 1207, 1200, 1150, 1108, 1105, 1080, 1060, 1059, 1057, 1050, 1030, 1010}

(*The World Almanac and Book of Facts 2007*)

a. What are the mean, median, and mode of these data?

b. Make a box plot of these data. Describe the shape.

c. What is the standard deviation?

New York's Manhattan Bridge was constructed from 1901 to 1909 over the East River.

24. Solve each system of equations.

a. $\begin{cases} 3x - y = -1 \\ 2x + y = 6 \end{cases}$

b. $\begin{cases} 2x + 4y = -9 \\ x - y = -6 \end{cases}$

25. Consider this series.

$$\frac{1}{10}+\frac{1}{30}+\frac{1}{90}+\frac{1}{270}+\cdots$$

a. What is the sum of the first five terms?

b. What is the sum of the first ten terms?

c. What is the sum of infinitely many terms?

26. Consider the functions $f(x)=\sqrt{2x-3}$ and $g(x)=6x^2$.

a. What are the domain and range of $f(x)$?

b. What are the domain and range of $g(x)$?

c. Find $f(2)$.

d. Find x such that $g(x)=2$.

e. Find $g(f(3))$.

f. Find $f(g(x))$.

27. Two people begin 400 m apart and jog toward each other. One person jogs 2.4 m/s, and the other jogs 1.8 m/s. When they meet, they stop.

a. Write equations to simulate the movement of the joggers.

b. How far does each person run before they meet?

c. How long does it take for them to meet?

28. The heights of all adults in Bigtown are normally distributed with a mean of 167 cm and a standard deviation of 8.5 cm.

a. Sketch a graph of the normal distribution of these heights.

b. Shade the portion of that graph showing the percentage of people who are shorter than 155 cm.

c. What percentage of people are shorter than 155 cm?

TAKE ANOTHER LOOK

1. Consider the geometry-software diagram at right. You have seen that in a unit circle, the length AB has the same value as $\cos\theta$. The lengths AF, GI, AD, AI, and CF correspond to other trigonometric values of θ. Decide which segment length equals each of the values $\sin\theta$, $\tan\theta$, $\sec\theta$, $\csc\theta$, and $\cot\theta$. Justify your answers.

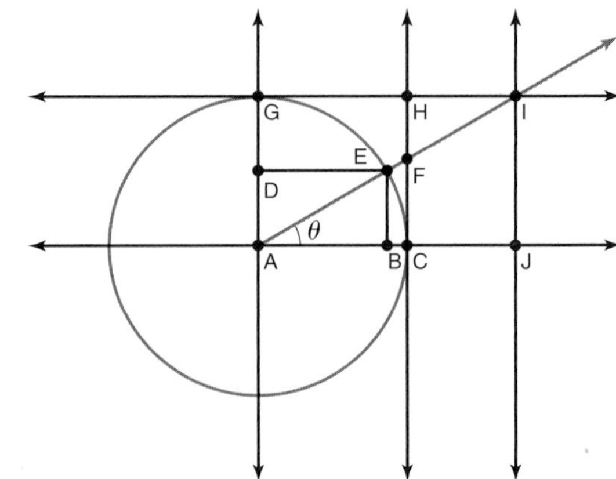

2. You have seen how to find the exact value of $\cos\frac{\pi}{12}$ by rewriting the expression as $\cos\left(\frac{\pi}{4}-\frac{\pi}{6}\right)$ and using a trigonometric identity to expand and evaluate. How can you find the exact value of $\sin\frac{5\pi}{12}$? (*Hint:* Write $\frac{5\pi}{12}$ as a sum or difference of terms.) Find some other exact values of sine, cosine, or tangent using this method.

3. You are now familiar with angle measures in degrees and radians, but have you ever heard of gradians? Research the gradian angle measure. Explain how it compares with radians and degrees, when and where it was used, and any advantages it might have. Can you find any other units that measure angle or slope?

4. The path traced by a fixed point, *P*, on a moving wheel is called a **cycloid** and is shown below. A cycloid can be defined with parametric equations. Use the diagram to derive parametric equations for *x* and *y*, the coordinates of point *P*. Note that the length of arc *CP* equals the length of segment *CO*.

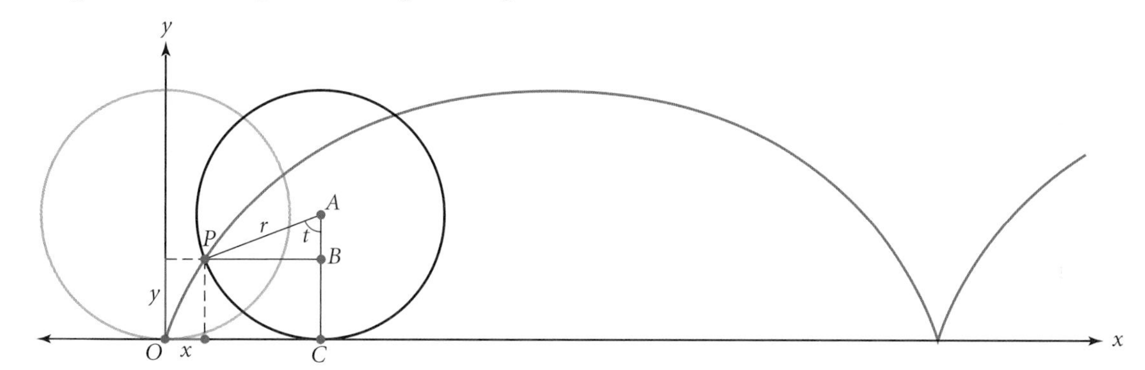

5. In Exercise 13b on page 794, you developed the formula $\cos^2 A = \frac{1 + \cos 2A}{2}$. Use this formula to develop the half-angle formula for cosine. Begin by taking the square root of both sides, then substitute $\frac{\theta}{2}$ for *A*. Use the formula from Exercise 13a for $\sin^2 A$ and a similar process to develop the half-angle formula for sine. Then use the half-angle formulas for sine and cosine to develop the half-angle formula for tangent.

Assessing What You've Learned

PERFORMANCE ASSESSMENT Demonstrate for a friend or family member how to find an equation that models periodic motion data. Be sure to use the words *amplitude, period, frequency, phase shift,* and *vertical translation.* Describe what each of these values tells you about the data.

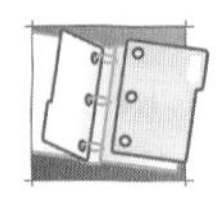

ORGANIZE YOUR NOTEBOOK Check that your notebook is in order. Make sure that you have definitions of all the terminology from this chapter. Include terms related to angles, like *standard position* and *terminal side,* and terms related to sinusoidal graphs, like *amplitude* and *period.* Check that all the trigonometric identities you have learned are in your notes as well.

WRITE IN YOUR JOURNAL Are your understandings of the sine, cosine, and tangent functions different now than they were when you started this chapter? Write a journal entry that describes how your understanding of the trigonometric functions has changed over time.

Selected Hints and Answers

This section contains hints and answers for exercises marked with ⓗ or ⓐ in each set of Exercises.

CHAPTER 0 • CHAPTER  CHAPTER 0 • CHAPTER

LESSON 0.1

1a. Begin with a 10-liter bucket and a 7-liter bucket. Find a way to get exactly 4 liters in the 10-liter bucket.

4a. 1

5. *Hint:* Your strategy could include using objects to act out the problem and/or using pictures to show a sequence of steps leading to a solution.

6a.

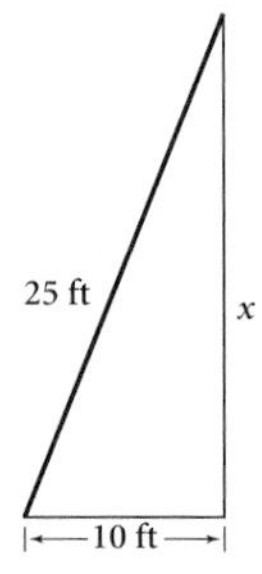

7a. 312 g

7b. 12.5 oz

8a. $a = 12$

9. *Hint:* Try using a sequence of pictures similar to those on page 2.

11a. $x^2 + 11x + 28$

	x	4
x	x^2	$4x$
7	$7x$	28

11b. *Hint:* First rewrite the expression as $(x + 5)(x + 5)$.

15c. $\frac{2}{9}$

LESSON 0.2

1a. Subtract 12 from both sides.

3b. $c = 5.8$

6b. If x represents the first three digits of your phone number, the expression for Step 2 is $80x$.

7b. *Hint:* L represents the price of each large bead. What units are the prices in?

10a. Solve Equation 1 for a to get $a = 5b - 42$. Substitute $5b - 42$ for a in Equation 2 to get $b + 5 = 7[(5b - 42) - 5]$.

11a. *Hint:* $45° - 30° = 15°$. How can you show this with the triangle tools?

12c. *Hint:* Remember, $x \cdot x = x^2$.

13a. 98

15. *Hint:* Try using a sequence of pictures similar to those on page 2. Also be sure to convert all measurements to cups.

LESSON 0.3

1a. approximately 4.3 s

2a. Let e represent the average problem-solving rate in problems per hour for Emily, and let a represent the average rate in problems per hour for Alejandro.

2b. *Hint:* The variable e represents the number of problems that Emily solves in 1 hour.

4c. *Hint:* Remember, subtracting 2 is the same as adding -2. Use this fact when you distribute.

5a. $a = 12.8$

7. *Hint:* How much does Alyse earn when she works overtime? Let r represent the hours worked at the regular rate and t represent the hours worked at the time-and-a-half rate. Write two equations. One equation is $r + t = 35$.

8a. $\frac{12960 \text{ cm}^3}{(18 \text{ cm} \cdot 30 \text{ cm})} = 24 \text{ cm}$

9. *Hint:* The well water is in the bottle.

10a. *Hint:* The area painted is calculated by multiplying *painting rate* · *time*. What fraction expresses Paul's painting rate in $\frac{\text{ft}^2}{\text{min}}$?

12a. r^{12}

CHAPTER 0 REVIEW

1. Sample answer: Fill the 5-liter bucket; pour 3 liters into the 3-liter bucket, leaving 2 liters in the 5-liter bucket; empty the 3-liter bucket; pour the remaining 2 liters from the 5-liter bucket into the 3-liter bucket. Fill the 5-liter bucket and pour 1 liter into the 3-liter bucket, completely filling the 3-liter bucket and leaving 4 liters in the 5-liter bucket.

2a. $x^2 + 7x + 12$

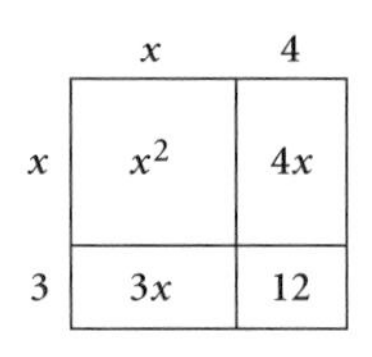

2b. $2x^2 + 6x$

	x	3
$2x$	$2x^2$	$6x$

2c. $x^2 + 4x - 12$

	x	-2
x	x^2	$-2x$
6	$6x$	-12

2d. $2x^2 - 9x + 4$

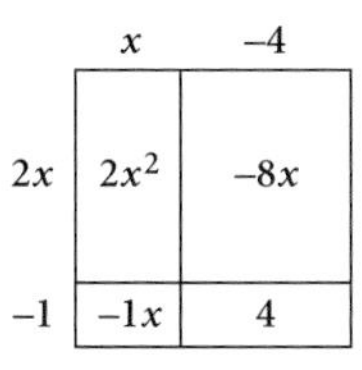

3a. $x = \sqrt{18}$ cm $= 3\sqrt{2}$ cm ≈ 4.2 cm

3b. $y = \sqrt{25}$ in. $= 5$ in.

4a. 168 mi

4b. 9.5 gal

4c. Car B can go 102 mi farther than Car A.

4d. Car A: 24 mi/gal; Car B: 36 mi/gal

4e. For Car A, the slope is $\frac{120}{5} = 24$, which means the car can drive 24 mi/gal of gasoline. For Car B, the slope is $\frac{180}{5} = 36$, which means the car can drive 36 mi/gal.

5a. $x = 13$

5b. $y = -2.5$

6. Let n represent the unknown number.

6a. $2n + 6$

6b. $5(n - 3)$

6c. $2n + 6 = 5(n - 3)$; $n = 7$

7. Let m represent the number of miles driven, and let c represent the total cost in dollars of renting the truck.

7a. $c = 19.95 + 0.35m$

7b. possible answer: \$61.25

7c. \$8.40

8a. Let w represent the mass in grams of a white block, and let r represent the mass in grams of a red block.

8b. $4w + r = 2w + 2r + 40$; $5w + 2r = w + 5r$

8c. $w = 60$; $r = 80$

8d. The mass of a white block is 60 g, and the mass of a red block is 80 g.

9. 17 years old

10. 12 pennies, 8 nickels, 15 dimes, 12 quarters

11a. $h = 0$. Before the ball is hit, it is on the ground.

11b. $h = 32$. Two seconds after being hit, the ball is 32 ft above the ground.

11c. $h = 0$. After 3 s, the ball lands on the ground.

12a. $4x^{-1}$, or $\frac{4}{x}$

12b. $\frac{1}{2}x^{-1}$, or $\frac{1}{2x}$

12c. x^{15}

13a. $y = 2^0 = 1$

13b. $y = 2^3 = 8$

13c. $y = 2^{-2} = \frac{1}{2^2} = \frac{1}{4}$

13d. $x = 5$

14a. $5 \text{ gal} \cdot \frac{4 \text{ qt}}{1 \text{ gal}} \cdot \frac{4 \text{ c}}{1 \text{ qt}} \cdot \frac{8 \text{ oz}}{1 \text{ c}} = 640 \text{ oz}$

14b. $1 \text{ mi} \cdot \frac{5280 \text{ ft}}{1 \text{ mi}} \cdot \frac{12 \text{ in.}}{1 \text{ ft}} \cdot \frac{2.54 \text{ cm}}{1 \text{ in.}} \cdot \frac{1 \text{ m}}{100 \text{ cm}} =$ 1609.344 m

15.

Sales Rep	Mr. Mendoza	Mr. Bell	Mrs. Plum
Client	Mr. Green	Ms. Phoung	Ms. Hunt
Location	Conference room	Convention hall	Lunch room
Time	9:00 A.M.	12:00 noon	3:00 P.M.

CHAPTER 1 • CHAPTER  CHAPTER 1 • CHAPTER

REFRESHING YOUR SKILLS FOR CHAPTER 1

1a. difference = 20, ratio = 1.2

1c. difference = −60, ratio = 0.70

2a. iv

3a. $20(1 + 0.15) = 23$

3c. $300(1 - 0.18) = 246$

4. $30{,}000(1 - 0.15) - 1200 = 25{,}500 - 1200 = \$24{,}300$

LESSON 1.1

2a. iv. −18, −13.7, −9.4, −5.1; arithmetic; $d = 4.3$

2b. ii. 47, 44, 41, 38; arithmetic; $d = -3$

2c. i. 20, 26, 32, 38; arithmetic; $d = 6$

2d. iii. 32, 48, 72, 108; geometric; $r = 1.5$

3.

n	1	2	3	4	5	...	9
u_n	40	36.55	33.1	29.65	26.2	...	12.4

$u_1 = 40$ and $u_n = u_{n-1} - 3.45$ where $n \geq 2$

5d. $u_1 = -6.24$ and $u_n = u_{n-1} + 2.21$ where $n \geq 2$; $u_{20} = 35.75$

6. $u_1 = 4$ and $u_n = u_{n-1} + 5$ where $n \geq 2$; $u_{46} = 229$

8a. 13 min

11a. \$60

13a. *Hint:* How many times do you add the common difference to 35 to get to 51?

14a.

Elapsed time (s)	Distance from motion sensor (m)
0.0	2.0
1.0	3.0
2.0	4.0
3.0	5.0
4.0	4.5
5.0	4.0
6.0	3.5
7.0	3.0

15a. $\frac{70}{100} = \frac{a}{65}$; $a = 45.5$

LESSON 1.2

1a. 1.5; growth; 50% increase

2a. $u_0 = 100$ and $u_n = 1.5u_{n-1}$ where $n \geq 1$; $u_{10} \approx 5766.5$

3a. 2000, 2100, 2205, 2315.25

4A. ii. The graph and rule both indicate decay.

5a. $x(1 + A)$

5b. $(1 - 0.18)A$, or $0.82A$

6a. *Hint:* $\textit{rebound ratio} = \dfrac{\textit{rebound height}}{\textit{previous rebound height}}$

8. \$250,000 was invested at 2.5% annual interest in 2008. $U_{2012} = \$27{,}595.32$

10a. $u_0 = 2000$
$u_n = (1 + 0.085)u_{n-1}$ where $n \geq 1$

11a. \$595.51

11c. *Hint:* You will receive $\frac{6\%}{4}$ of the balance 12 times.

15a. 98.22%

15b. *Hint:* The growth rate for a 30-year period was about 0.98. Use this figure to make an educated guess. You'll revise your estimate in 15d.

18b. 10 s

20a. $x \approx 43.34$

21b. $y = 55$

LESSON 1.3

1a. 31.2, 45.64, 59.358; shifted geometric, increasing

1d. 40, 40, 40; arithmetic or shifted geometric, neither increasing nor decreasing

2b. $b = 1200$ **2c.** $d = 0$

3a. *Hint:* Solve the equation $a = 0.95a + 16$.

4b. $u_0 = 0$ and $u_n = 0.5u_{n-1} + 10$ where $n \geq 1$

5a. The first day, 300 g of chlorine were added. Each day, 15% disappears, and 30 g are added.

6b. $a_0 = 24{,}000$ and $a_n = a_{n-1}\left(1 + \frac{0.034}{12}\right)a - 100$, or $a_n = a_{n-1}(1.00283) - 100$

6c. \$23,871.45. This is the balance on 2/2/09.

8c. $c = 0.88c + 600$

9. $u_0 = 20$ and $u_n = (1 - 0.25)u_{n-1}$ where $n \geq 1$; 11 days ($u_{11} \approx 0.84$ mg)

11a. *Hint:* You might use the Pythagorean Theorem or use the properties of special right triangles. Look for patterns.

14. 23 times

LESSON 1.4

1b. 0 to 19 for n and 0 to 400 for u_n

1d. 0 to 69 for n and 0 to 3037 for u_n

2B. iv. arithmetic

3a. geometric, nonlinear, decreasing

4b. i. sample answer: $u_0 = 200$

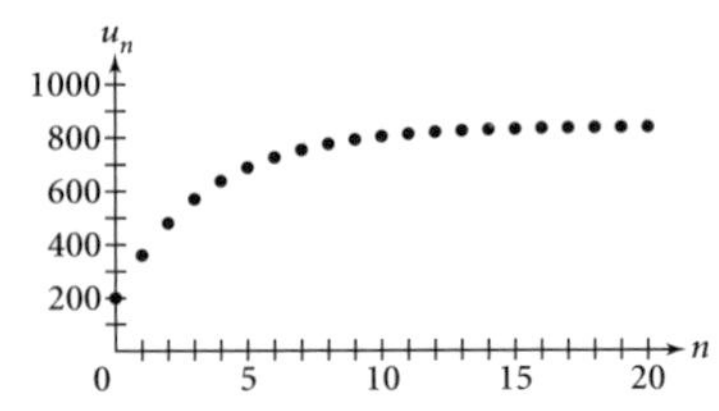

6a. $u_0 = 18$ and $u_n = -0.75u_{n-1}$ where $n \geq 1$

6b.

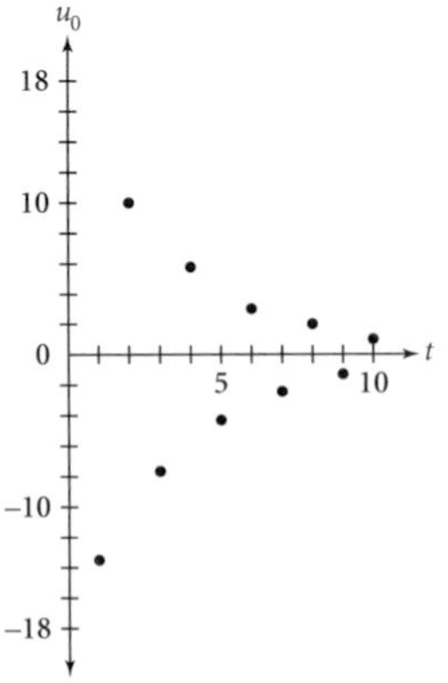

Because every term alternates signs, the graph is different from other graphs you have seen; each point alternates above or below the n-axis. The points

above the n-axis, however, create a familiar geometric pattern, as do the points below the n-axis. If the points below the n-axis were reflected across the n-axis, you would have the graph of $u_0 = 18$ and $u_n = 10.75u_{n-1}$ where $n \geq 1$.

6c. 0

10a. $u_0 = 2$ and $u_n = (1 - 0.15) \cdot u(n-1)$; between six and seven days

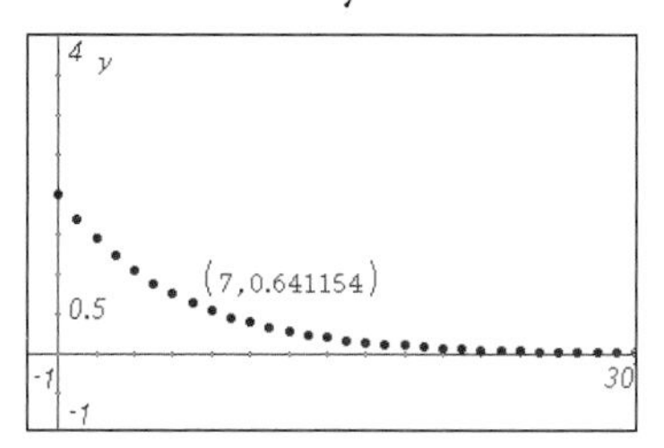

11. *Hint:* If the rate of chirping increases by 20 chirps/min per 5° increase in temperature, how much does the rate of chirping increase per 1° increase in temperature?

12. *Hint:* Complete the table and graph the adjusted data. Notice that the table gives population values every 5 years, but you will need to find the yearly growth rate.

Adjusted year	Adjusted population (thousands)
53	867 − 239 = 628
58	950 − 239 = 711
63	1045 − 239 = 806
68	
73	
78	
83	
88	
93	
98	

15b. $66\frac{2}{3}$

LESSON 1.5

1a. investment, because a deposit is added

1b. \$450

1c. \$50

1d. 3.9%

1e. annually (once a year)

3a. \$130.67

3b. \$157.33

4a. $u_0 = 10{,}000$ and $u_n = u_{n-1}\left(1 + \frac{0.10}{12}\right) - 300$ where $n \geq 1$

5. \$588.09

8a. Both deposit \$1,000 to start and \$1,200 each year. There is no difference.

8b.

Year	Beau	Shaleah
0	\$1,000.00	\$1,000.00
1	\$2,265.00	\$2,303.38
2	\$3,612.23	\$3,694.04
3	\$5,047.02	\$5,177.84
...	...	...

Shaleah's account always has a higher balance. The difference between Beau's and Shaleah's balances gets greater over time.

10a. \$1,990, \$1,979.85, \$1,969.55, \$1,959.09, \$1,948.48, \$1,937.70

11a. *Hint:* The sequence can be defined as $u_0 = 60{,}000$ and $u_n = u_{n-1}\left(1 + \frac{0.096}{12}\right) - p$, where p is the monthly payment; 25 years is 300 months, so guess-and-check to find the value of p that gives $u_{300} = 0$.

13. *Hint:* Find the differences and ratios for two pairs of consecutive terms.

CHAPTER 1 REVIEW

1a. geometric

1b. $a_1 = 256$ and $a_n = 0.75a_{n-1}$ where $n \geq 2$

1c. $a_8 \approx 34.2$

1d. $a_{10} \approx 19.2$

1e. $a_{17} \approx 2.57$

2a. arithmetic

2b. $u_1 = 3$ and $u_n = u_{n-1} + 4$ where $n \geq 2$

2c. $u_{128} = 511$

2d. $u_{40} = 159$

2e. $u_{20} = 79$

3a. −3, −1.5, 0, 1.5, 3; 0 to 6 for n and −4 to 4 for a_n

3b. 2, 4, 10, 28, 82; 0 to 6 for n and 0 to 100 for a_n

4a. $u_0 = 14.7$ and $u_n = (1 - 0.20)u_{n-1}$ where $n \geq 1$

4b.

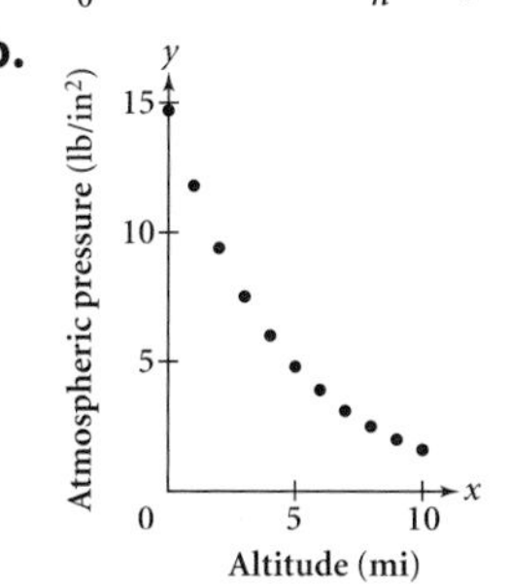

4c. approximately 3.1 lb/in²

4d. 11 mi

Selected Hints and Answers

5A. iv. The recursive formula is that of an increasing arithmetic sequence, so the graph must be increasing and linear.

5B. iii. The recursive formula is that of a growing geometric sequence, so the graph must be increasing and curved.

5C. i. The recursive formula is that of a decaying geometric sequence, so the graph must be decreasing and curved.

5D. ii. The recursive formula is that of a decreasing arithmetic sequence, so the graph must be decreasing and linear.

6. 88 gal

7. approximately 5300; approximately 5200; $u_0 = 5678$ and $u_n = (1 - 0.24)u_{n-1} + 1250$ where $n \geq 1$

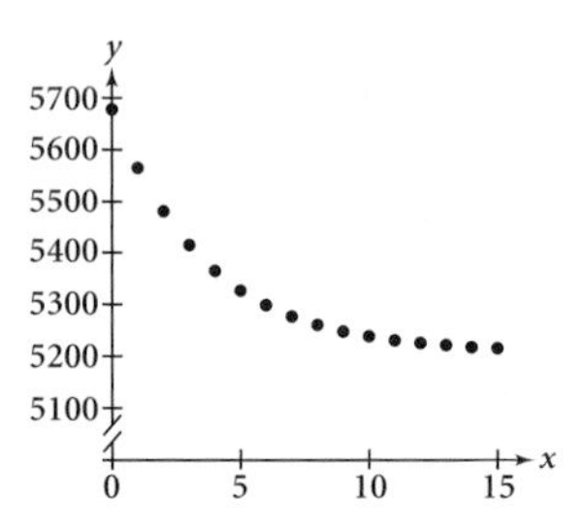

8a. \$657.03 **8b.** \$4,083.21

9. possible answer: $u_{1980} = 70.8$, $u_n = u_{n-1}(1 + 0.08)$ where $n \geq 1981$

10. \$637.95

CHAPTER 2 • CHAPTER **2** CHAPTER 2 • CHAPTER

REFRESHING YOUR SKILLS FOR CHAPTER 2

1a. mean: 29.2 min; median: 28 min; mode: 26 min

1b. *Hint:* When the data set has an even number of values, the median is the mean of the two middle values.

2a. 36 days

3. values: 30, 35, 37, 37, 38, 39, 39, 39, 43, 45, 51

5d. *Hint:* You may recall statistics such as minimum value, maximum value, and range from previous math courses.

LESSON 2.1

1. minimum: 1.25 days; first quartile: 2.5 days; median: 3.25 days; third quartile: 4 days; maximum: 4.75 days

4. Range is approximately 120 for all four plots. (The actual range is 123.) Approximate *IQR* for each plot: Plot A: 60 (64); Plot B: 60(66); Plot C: 10(15); Plot D: 50(51)

5a. 9, 10, 14, 17, 21

6. *Hint:* Make seven blanks and fill them in. Remember that the mean equals the median if the data are perfectly symmetric.

8. *Hint:* Consider the definitions of each of the values in the five-number summary.

9b. 25, 51, 58, 65, 72

9e. *Hint:* If the mean is 60 for 12 years, then what is the total number of runs?

13b. $25 - 1.5(5) = 17.5$ g and $30 + 1.5(5) = 37.5$ g

15c. *Hint:* Compare the range, *IQR,* and how the data are skewed. If you conclude that the data are significantly different, then Rayleigh's conjecture is supported.

19a. $6\sqrt{2} \approx 8.5$

20. *Hint:* Refer to Lesson 1.5 for help calculating compound interest.

LESSON 2.2

1a. 47.0 **1b.** −6, 8, 1, −3 **1c.** 6.1

3. *Hint:* The sum of the deviations is always 0.

4. *Hint:* The sum of all six deviations squared is only 2.

7a. Group A: mean = 184; Group B: mean = 84

7b. Group A: $s \approx 1.5$; Group B: $s \approx 21.6$

9a. possible answer:

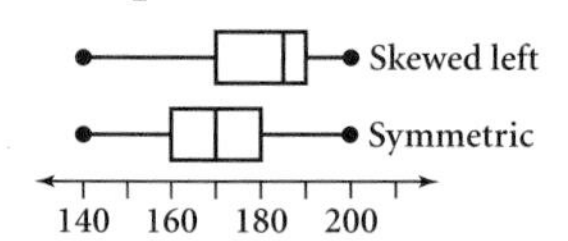

9c. *Hint:* The highest and lowest values for each set must be equal, and the skewed data will have a higher median value.

12a. median = 75 packages; *IQR* = 19 packages

12b. $\bar{x} \approx 80.9$ packages; $s \approx 24.6$ packages

14f. *Hint:* Consider whether your decision should be based on the data with or without outliers included. Decide upon a reasonable first bid, and the maximum you would pay.

17a. $x = 59$

LESSON 2.3

1c. *Hint:* Choose values that reflect the number of backpacks within each bin.

2b. *Hint:* Look for the outlier.

3b. *Hint:* Divide the number of data values below 34 by the total number of data values, 20.

4b. *Hint:* How many different ways are there to roll a 7? A 4? A 2?

6a. SAT scores for all students in one year

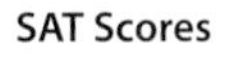

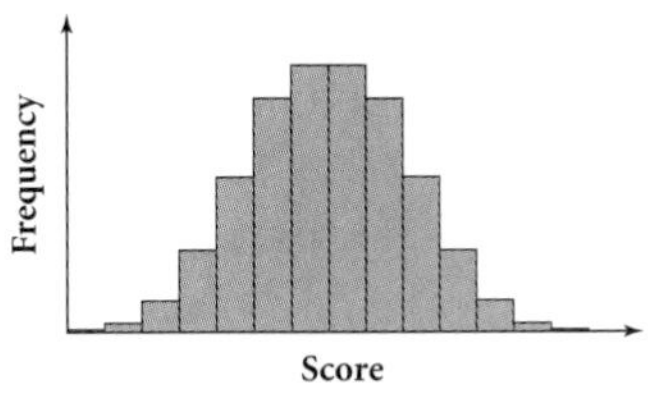

7a.

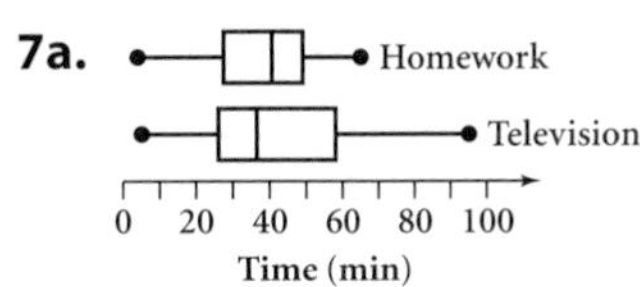

Five-number summaries: homework: 4, 27.5, 40.5, 49, 65; television: 5, 26, 36.5, 58, 95. Television has the greater spread.

9a.

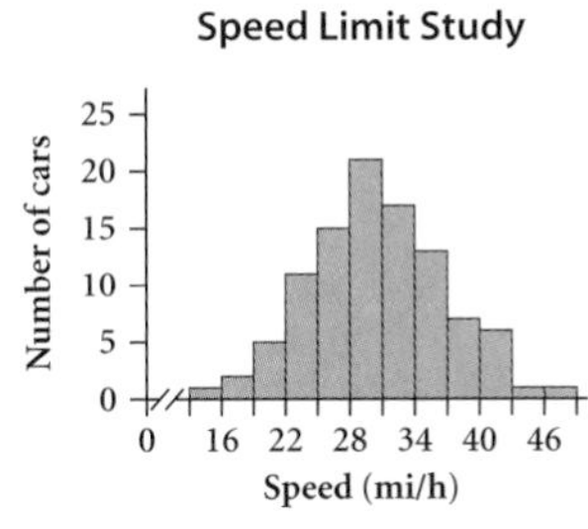

11d. *Hint:* How does translating the data affect the standard deviation and *IQR*?

12a. $10.99

13. *Hint:* Look at Example A in Lesson 0.3 for help with unit conversion.

CHAPTER 2 REVIEW

1. Plot B has the greater standard deviation, because the data have more spread.

2a. Seven values are represented in each whisker of Group A. Six values are represented in each whisker of Group B.

2b. Group B has the greater standard deviation, because the data have more spread.

2c. possible answer:

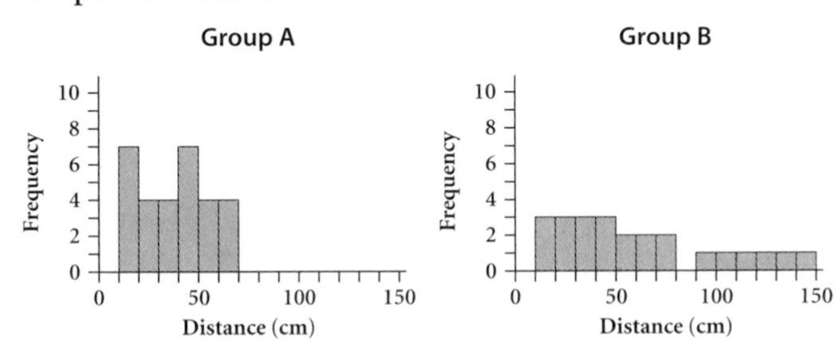

3a. mean: 540.467; median: 349; mode: none

3b. five-number summary: 41, 192, 349, 740, 1599

3c.

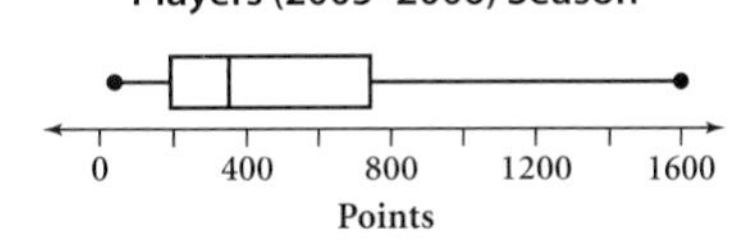

skewed right

3d. $IQR = 548$

3e. Answers will vary. Possible answer: The three top scorers—Ginobili, Parker, and Duncan—are close to double the next lower player; thus they could be outliers. Using the $1.5 \cdot IQR$ rule, only Duncan is an outlier.

4. One strategy is to make Set A skewed and Set B symmetric. Possible answer: Set A: $\{1, 2, 3, 4, 5, 6, 47\}$; $s \approx 16.5$; $IQR = 4$. Set B: $\{1, 5, 7, 9, 11, 13, 17\}$; $s \approx 5.3$; $IQR = 8$.

5. There are 49 values (sum all the frequencies), and the median is between 5 and 10, maybe about 5 or 6. The mean is higher, maybe about 15. Q_1 is in the 0–5 bin, and Q_3 is in the 15–20 bin. So the *IQR* may be about 17 or 18. The range is about 93. These data are highly skewed to the right. There are three outliers, at about 67, 87, and 93. Most songs on the list stay for 2 months or less, but there is a cluster (16%) lasting 6 to 10 months and an occasional hit (6%) lasting more than a year on the chart.

6a. mean ≈ 38.5, median $= 33.5$

6b. mean ≈ 44.13, median $= 42$

6c.

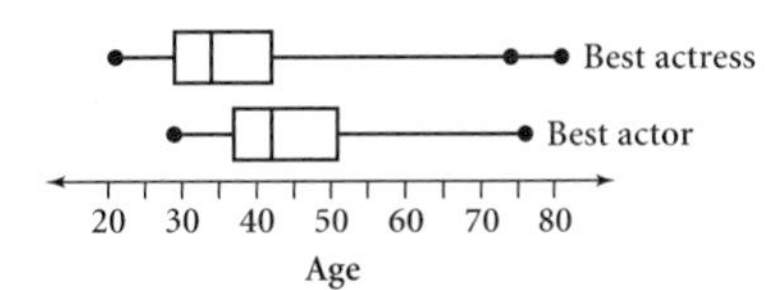

6d.

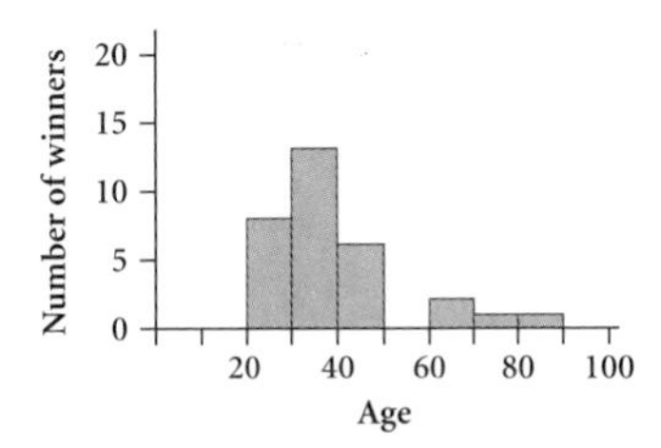

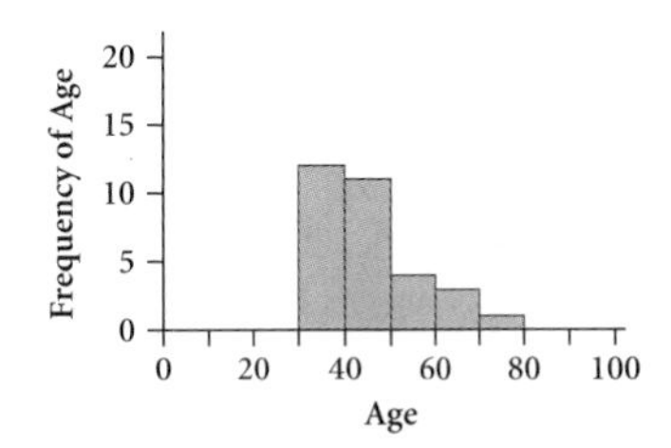

6e. The Best Actress data set should have the greater standard deviation because it has more spread. Best Actress: $s \approx 13.78$; Best Actor: $s \approx 10.45$.

6f. 34th percentile. 34% of the Best Actress winners from 1976 to 2007 were younger than 33 when they won their award.

7. Answers will vary. In general, the theory is supported by the statistics and graphs.

8a. 2005 U.S. Passenger-Car Production

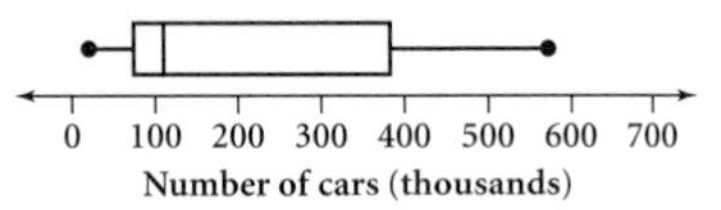

8b. 2005 U.S. Passenger-Car Production

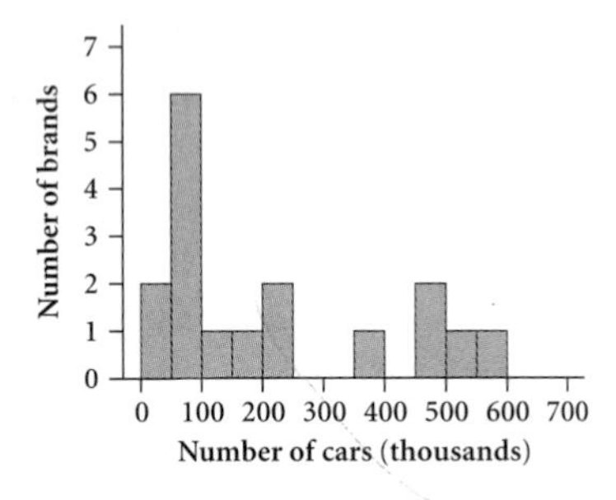

8c. The box plot will be shifted right, and some of the histogram's bins will change frequency as some brands fall into a different bin.

8d. 62nd percentile

8e. 76th percentile

CHAPTER 3 • CHAPTER **3** CHAPTER 3 • CHAPTER

REFRESHING YOUR SKILLS FOR CHAPTER 3

1a. $y = -26$

1c. $t \approx -2.216$

2a. $x = \dfrac{y + 11}{3}$

2c. $s = \dfrac{18.1 + 3.7t}{4.5}$

3a. $y = -5$

LESSON 3.1

1a.

1b. -3; The common difference is the same as the slope.

1c. 18; The y-intercept is the u_0-term of the sequence.

1d. $y = 18 - 3x$

3. $y = 7 + 3x$

5a. 1.7

5d. *Hint:* Plot some points on the line: (0, 12), (1, 12), (2, 12). What is the change in the y-value for each unit change in the x-value?

7a. 190 mi

8a. *Hint:* The point (3 cars, $-\$2050$), corresponds to the term $t_3 = -2050$.

10c. 34 days

11a. *Hint:* The x-values must have a difference of 5.

11d. *Hint:* The linear equation will have a slope of 4. Find the y-intercept.

12c. 304 ft

14a. $x = (1 - 0.4)x + 300$; $x = 750$

LESSON 3.2

1a. $\frac{3}{2} = 1.5$

2c. 15

3a. $y = 14.3$

4c. The equations have the same constant, -2. The lines share the same y-intercept. The lines are perpendicular, and their slopes are reciprocals with opposite signs.

5. *Hint:* Graph the original line and then graph several parallel lines and a perpendicular line. Also, refer to your answers to 4c and d.

7a. Possible answer: Using the 2nd and 7th data points gives a slope of approximately 1.47 volts/battery.

8a. *Hint:* Think about the units of y and x and the units of the slope.

13a. $-10 + 3x$

LESSON 3.3

1a. possible answer: $y = 1 + \frac{2}{3}(x - 4)$

2a. $y = -7 + \frac{2}{3}(x - 5)$

3a. $u_n = 31$

5d. *Hint:* What can you say about the slope and the x- and y-intercepts of a vertical line?

7a. Possible answer: The y-intercept is about 1.7; (5, 4.6); $\hat{y} = 1.7 + 0.58x$

9b. possible answer: $\hat{y} = 0.26x + 0.71$

11. 102

12. *Hint:* Refer to the example in Lesson 1.3 to review graphical and algebraic methods for finding the long-run behavior.

LESSON 3.4

1a. 17, 17, 17

2a. $y \approx 15.7 - 0.6739(x - 8.1)$ or $y \approx 9.5 - 0.6739(x - 17.3)$

4. $y = 72.9 - 1.8x$

6b. (1935, 59.45), (1965, 66.8), (1992.5, 72.15)

7a. $\hat{y} = 3 + 5x$

8b. *Hint:* Think about the units for x and y.

8c. 240.10 s, or 4:00.10. This prediction is 0.7 s slower than Roger Bannister actually ran.

9b. No, it appears that there are two patterns here.

13. *Hint:* What percentage of the students may have scored lower than Ramon? Consider the students that definitely scored lower than 35 and the students that *may* have scored lower than 35 according to the histogram.

LESSON 3.5

1a. -0.2

2. 82.3, 82.9, 74.5, 56.9

4. approximately 0.28

5a. Let x represent age in years, and let y represent height in cm; $\hat{y} = 81.\bar{6} + 5.\bar{6}x$.

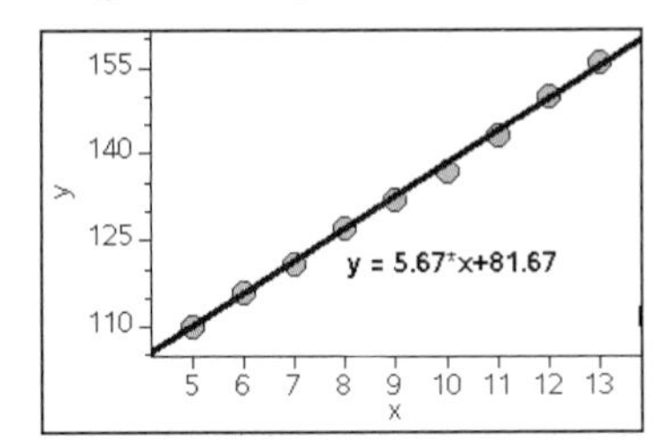

7. *Hint:* Using the data point and its residual, find the coordinates of a point on the line.

8e. i. 27.5°C

12a. $y = 4.7 + 0.6(x - 2)$

12b. *Hint:* You can review the meaning of *direct variation* in Lesson 0.3.

13. *Hint:* The difference between the 2nd and 6th values is 12.

14a. $x = 2$

15b. i. deposited: \$360; interest: \$11.78

LESSON 3.6

1a. $(1.8, -11.6)$

4a. $x = \frac{32}{19} \approx 1.684$

5a. $(4.125, -10.625)$

7a. No. At $x = 25$, the cost line is above the income line.

8c. The median-median lines predict that in 2010, the men's winning time will be 102.3 s, or 1:42.3, and the women's winning time will be 108 s, or 1:48.

9c. *Hint:* Two step functions may not have any intersection or they may intersect over a short interval rather than a single point.

11a. Let l represent length in centimeters, and let w represent width in centimeters; $2l + 2w = 44$, $l = 2 + 2w$; $w = \frac{20}{3}$ cm, $l = \frac{46}{3}$ cm.

11b. *Hint:* An isosceles triangle has two or more sides that are equal in length.

14a. $y = \dfrac{3x - 12}{8} = 0.375x - 1.5$

LESSON 3.7

1b. $h = \dfrac{18 - 2p}{3} = 6 - \frac{2}{3}p$

2a. (2, 21)

3a. $(3.1, -1.8)$

5a. *Hint:* Look for a way to use substitution.

6. *Hint:* The second equation in this system is $F = 3C$.

7a. cost for first camera: $y = 47 + 11.5x$; cost for second camera: $y = 59 + 4.95x$

14a. $y = \dfrac{7 - 3x}{2} = \frac{7}{2} - \frac{3}{2}x = 3.5 - 1.5x$

CHAPTER 3 REVIEW

1. $-\dfrac{975}{19}$

2a. 23.45

2b. possible answer: $y = 23.45x$

2c. possible answer: $y = -\frac{20}{469}x$, or $y \approx -0.0426x$

3a. approximately (19.9, 740.0)

3b. approximately (177.0, 740.0)

4a. $y = 3 - (x + 2)$, or $y = 1 - x$

4b. $y = -8 + 2(x - 0)$, or $y = -8 + 2x$

4c. $y = 13.2 - 0.46(x - 1999)$, or $y = 8.6 - 0.46(x - 2009)$

4d. $y = 7 + 0(x - 2)$, or $y = 7$

5a. Poor fit; there are too many points above the line.

5b. Reasonably good fit; the points are well distributed above and below the line and are not clumped. The line follows the downward trend of the data.

5c. Poor fit; there are an equal number of points above and below the line, but they are clumped to the left and to the right, respectively. The line does not follow the trend of the data.

6a. (10, 28); consistent, independent

6b. every point; same line; consistent, dependent

6c. $\left(\frac{2}{3}, 1\right) = (0.\bar{6}, 1)$; consistent, independent

6d. (1, 0); consistent, independent

6e. $\left(\frac{42}{5}, \frac{53}{10}\right) = (8.4, 5.3)$; consistent, independent

6f. No intersection; the lines are parallel; inconsistent.

7a. The ratio 0.38 represents the slope; that is, for each pound on Earth, a person would weigh 0.38 pound on Mercury.

7b. $m = 0.38(160) = 60.8$. The student's weight on Mercury would be about 60.8 lb.

7c. moon: D, $y_1 = 0.17x$; Mercury: C, $y_2 = 0.38x$; Earth: B, $y_3 = x$; Jupiter: A, $y_4 = 2.54x$

8a.

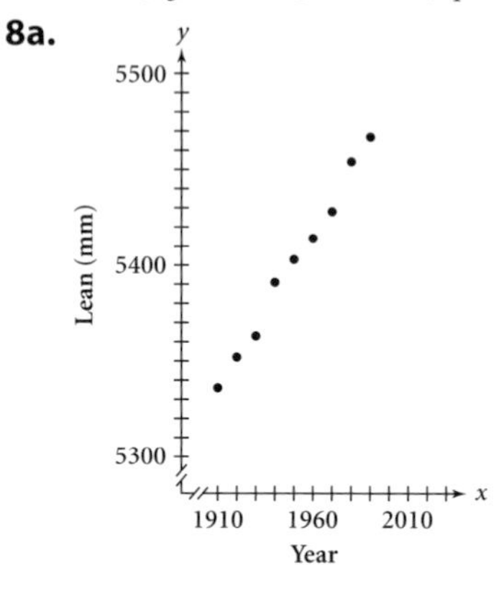

8b. $\hat{y} = 2088 + 1.7x$

8c. 1.7. For every additional year, the tower leans another 1.7 mm.

8d. 5474.4 mm

8e. Approximately 5.3 mm; the prediction in 8d is probably accurate within 5.3 mm. In other words, the actual value probably was between 5469.1 and 5479.7.

8f. Possible answer: $1173 \leq$ domain ≤ 1992 (from year built to year retrofit began); $0 \leq$ range ≤ 5474.4 mm

9. $u_{23} = 16.5$

10a. geometric; curved; 4, 12, 36, 108, 324

10b. shifted geometric; curved; 20, 47, 101, 209, 425

11a. $u_0 = 500$ and $u_n = (1 + 0.059)u_{n-1}$ where $n \geq 1$

11b. 500, 529.50, 560.74

11c. the amount in the account after 3 yr

11d. \$3,718.16

11e. \$14,627.20

12a. Possible answer: $p_{2010} = 6{,}972{,}791{,}646$ and $p_n = (1 + 0.015)p_{n-1}$ where $n \geq 2011$. The sequence is geometric.

12b. 7,511,676,903 people

12c. 2034

12d. Answers will vary. An increasing geometric sequence has no limit. But the model will not work for the distant future because there is a physical limit to how many people will fit on Earth.

13. $u_0 = 25$ mg, $u_1 = 37.5$ mg, $u_2 = 43.75$ mg, $u_3 =$ 46.875 mg, $u_4 = 48.4375$ mg. In the long run, he'll have about 50 mg of the antibiotic in his body.

14.

0 5 10 15 20

14a. skewed left

14b. 12

14c. 6

14d. 50%; 25%; 0%

15a. mean ≈ 4.470, median $= 3.9$, mode $= 3.7, 3.9$, $s \approx 1.755$

15b. Alaska and Arizona lie more than 2 standard deviations above the mean.

15c.

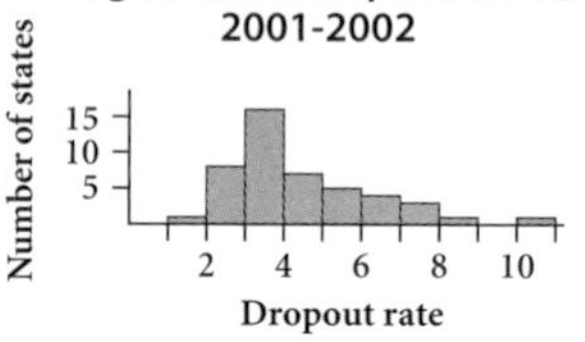

16a. (4.7, 7.1)

16b. $\left(\frac{5}{2}, \frac{11}{6}\right)$

16c. (5.8, 1.4)

17a. fairly linear (a little logistic)

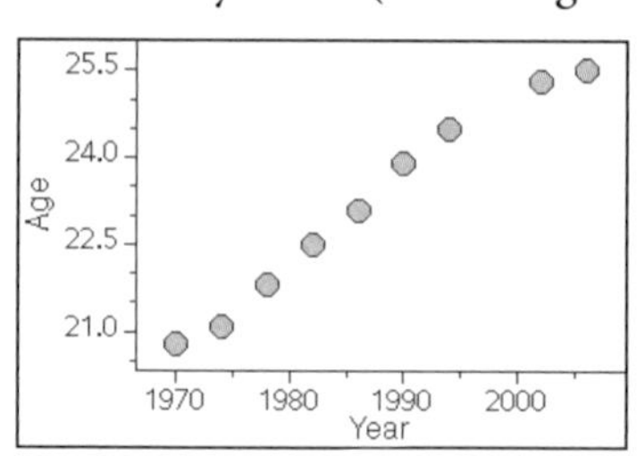

17b. $\hat{y} = -274.9 + 0.15x$

17c. 26.7 years old

17d. Answers depend on year of birth. Possible answers:

Year of birth	Approximate point
1994	(2022, 28.4)
1995	(2024, 28.6)
1996	(2025, 28.8)
1997	(2026, 29.0)
1998	(2027, 29.1)
1999	(2028, 29.3)

This tells me the year in which a woman my age will marry if she marries at the median age and the age she will be that year. If a woman born in 1994 marries at the median age, the model predicts that she will marry at about age 28 in 2022.

17e. Residuals: 0.2, −0.1, 0, 0.1, 0.1, 0.3, 0.3, 0.2, −0.1, −0.5; $s = 0.2622$ yr. The model is within 3 mo of the recorded age.

18a. $u_1 = 6$ and $u_n = u_{n-1} + 7$ where $n \geq 1$

18b. $y = -1 + 7x$

18c. The slope is 7. The slope of the line is the same as the common difference of the sequence.

18d. 223. It's probably easier to use the equation from 18b.

19a. 4.5 **19b.** $y = 7.5 + 4.5x$

19c. The slope of the line is equal to the common difference of the sequence.

CHAPTER 4 • CHAPTER **4** CHAPTER 4 • CHAPTER

REFRESHING YOUR SKILLS FOR CHAPTER 4

1a. Add 7 to each side.

1c. Add −2 to each side or subtract 2 from each side.

2a. $x = 33$ **2c.** $x = 5$ or $x = -3$

LESSON 4.1

1a. A **1d.** B

2b.

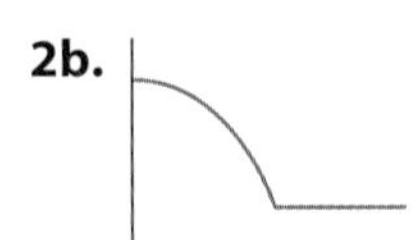

5. *Hint:* Consider the rate and direction of change (increasing, decreasing, constant) of the various segments of the graph.

6b. The car's speed in miles per hour is the independent variable; the braking distance in feet is the dependent variable.

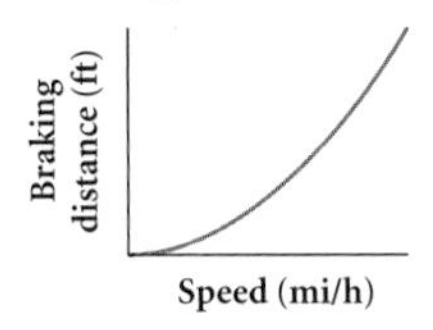

6c. Time in minutes is the independent variable; the drink's temperature in degrees Fahrenheit is the dependent variable.

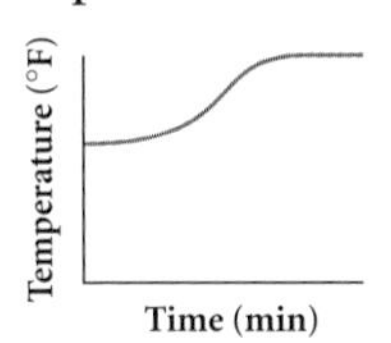

6d. *Hint:* Remember, you are graphing the speed of the acorn, not its height.

7c. Foot length in inches is the independent variable; shoe size is the dependent variable. The graph will be a series of discontinuous horizontal segments, because shoe sizes are discrete.

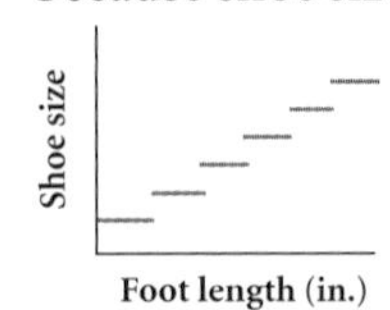

9a. *Hint:* This is not a (*time, distance*) graph.

11a. Let x represent the number of pictures, and let y represent the amount of money (either cost or income) in dollars; $y = 155 + 15x$.

11b. $y = 27x$

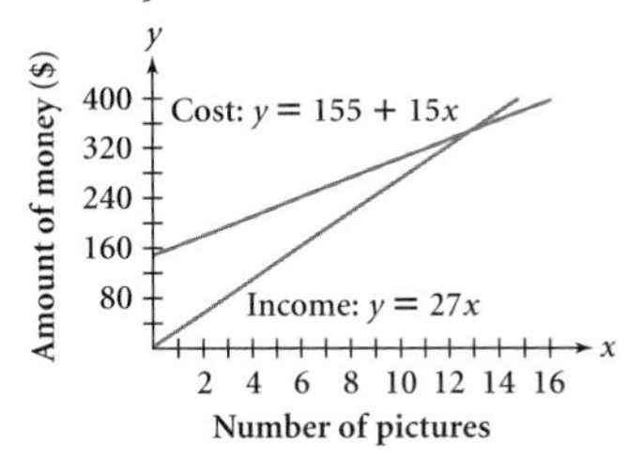

11c. 13 pictures

11d. The income, \$216, is less than the cost, \$275.

12a. \$142,784.22

13a. $3x + 5y = -9$

13b. $6x - 3y = 21$

LESSON 4.2

1c. Function; each x-value has only one y-value.

2d. 11 **2e.** $\frac{11}{3}$

4d. $5 = E$

4g. *Hint:* $\sqrt[3]{\ }$ means the cube root. The cube root of x is the number that you cube to get x. For example, $\sqrt[3]{8} = 2$ because $2^3 = 8$.

4j. $18 = R$

8. domain: $-6 \leq x \leq 5$; range: $-2 \leq y \leq 4$

10d.

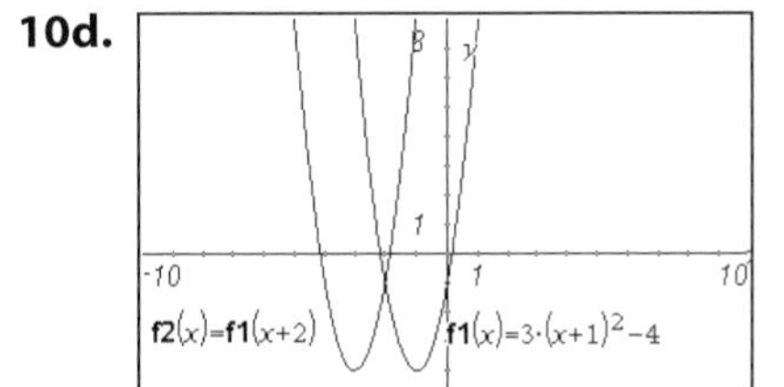

The graphs are the same shape. The graph of $f(x + 2)$ is shifted 2 units to the left of the graph of $f(x)$.

13b. *Hint:* You might use guess-and-check or a graphing calculator.

14a.

Height

Time

15. Sample answer: Eight students fall into each quartile. Assuming that the mean of each quartile is the midpoint of the quartile, the total will be $8(3.075 + 4.500 + 5.875 + 9.150)$, or \$180.80.

18a.

	x	3
x	x^2	$3x$
7	$7x$	21

LESSON 4.3

2. translated right 3 units

3a. $-2(x + 3)$, or $-2x - 6$

3b. $-3 + (-2)(x - 2)$, or $-2x + 1$

4a. $y = -4.4 - 1.1\overline{48}(x - 1.4)$ or
$y = 3.18 - 1.1\overline{48}(x + 5.2)$

5a. $y = -3 + 4.7x$

6a. $y = -2 + f(x)$

8a. *Hint:* Which axis pictures differences in time and which pictures differences in distance?

10b. $y = \frac{c}{b} - \frac{a}{b}x$; y-intercept: $\frac{c}{b}$; slope: $-\frac{a}{b}$

10d. ii. $4x + 3y = -8$ **10d. vi.** $4x + 3y = 10$

12b. $y = \frac{1}{5}x + 65$

12c. *Hint:* Substitute 84 for y in the function from 12b and solve for x.

LESSON 4.4

2a. $y = x^2 - 5$

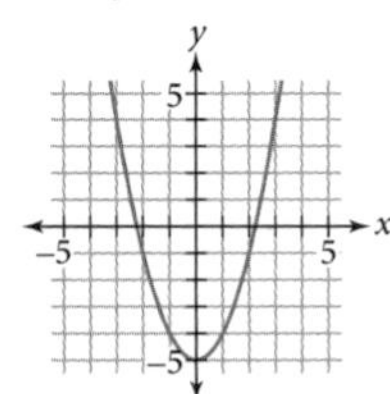

2c. $y = (x - 3)^2$

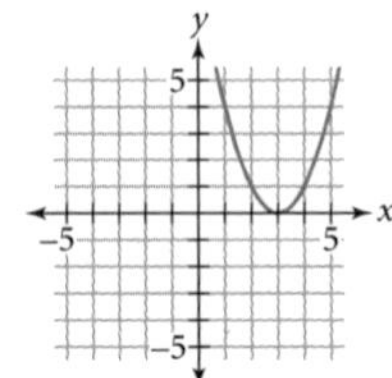

3d. translated horizontally -4 units

5. *Hint:* You can verify your answers with a table or a graph.

5b. $x = 4$ or $x = -4$

7c. $(6, -2), (4, -2), (7, 1), (3, 1)$. If (x, y) are the coordinates of any point on the black parabola, then the coordinates of the corresponding point on the red parabola are $(x + 5, y - 3)$.

7d. Segment b has length 1 unit, and segment c has length 4 units.

9. *Hint:* You might model this situation with points on a circle to represent the teams and segments that show which teams have played each other.

10a. $x = 9$ or $x = 1$

14.

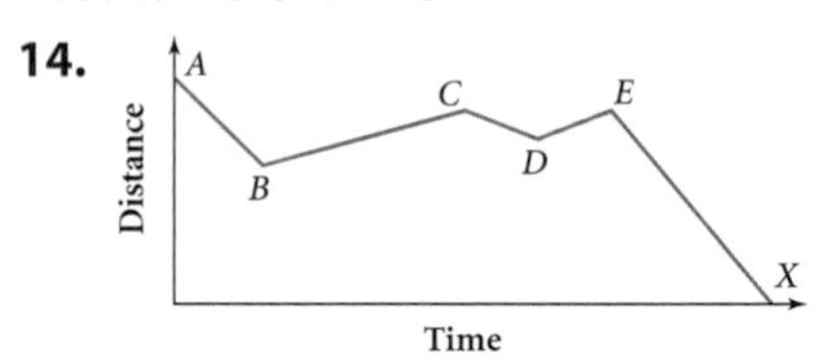

LESSON 4.5

1c. $y = \sqrt{x + 5} + 2$ **1d.** $y = \sqrt{x - 3} + 1$

2a. translated horizontally 3 units

2c. translated vertically 2 units

4a.

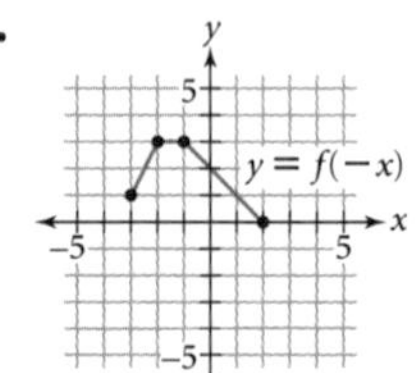

5a. $y = -\sqrt{x}$ **5b.** $y = -\sqrt{x} - 3$

5c. $y = -\sqrt{x + 6} + 5$ **5d.** $y = \sqrt{-x}$

5e. $y = \sqrt{-(x - 2)} - 3$, or $y = \sqrt{-x + 2} - 3$

8b. i. $y = \pm\sqrt{x + 4}$

8b. ii. $y = \pm\sqrt{x} + 2$

9a. *Hint:* Use Chicago time on the horizontal axis and distance on the vertical axis. What is Arthur's distance at the beginning of the trip?

12e. *Hint:* Which piece of the graph is defined for $x = 2$?

13e.

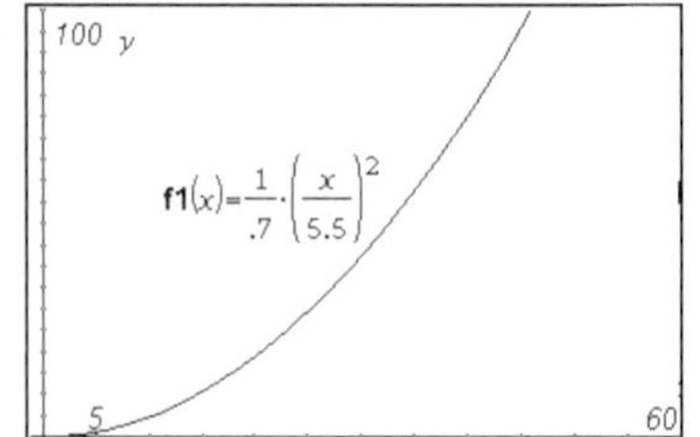

It is a parabola, but the negative half is not used because the distance cannot be negative.

15a. $x = 293$

15d. $x = -13$

17b. $y = \frac{1}{2}(x - 8) + 5$

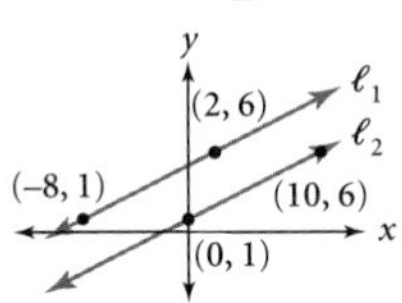

18a. 35, 37.5, 41.5, 49, 73

18b.

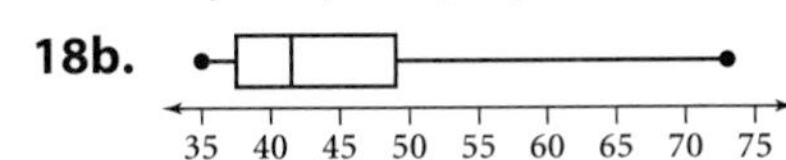

18c. 11.5

18d. 70 and 73

LESSON 4.6

1e. $y = |x| - 1$

1f. $y = |x - 4| + 1$

1m. $y = -(x + 3)^2 + 5$

1n. $y = \pm\sqrt{x - 4} + 3$

1p. $\frac{y}{-2} = \left|\frac{x-3}{3}\right|$, or $y = -2\left|\frac{x-3}{3}\right|$

4. *Hint:* Remember to choose some negative values and some values between -1 and 1.

6. $\hat{y} \approx |x - 18.4|$. The transmitter is located on the road approximately 18.4 mi from where you started.

7a. $(6, -2)$

7b. $(2, -3)$ and $(8, -3)$

9. *Hint:* Refer to Example B of this lesson for help.

11c.

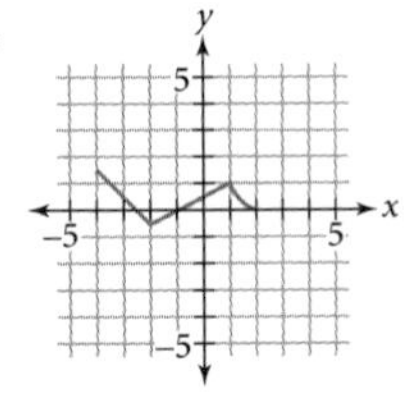

13a. $\bar{x} = 83.75$, $s \approx 7.45$

LESSON 4.7

1.

Reflection	Across x-axis	N/A
Dilation	Horizontal	4
Dilation	Vertical	0.4
Translation	Horizontal	2
Reflection	Across y-axis	N/A

3c.

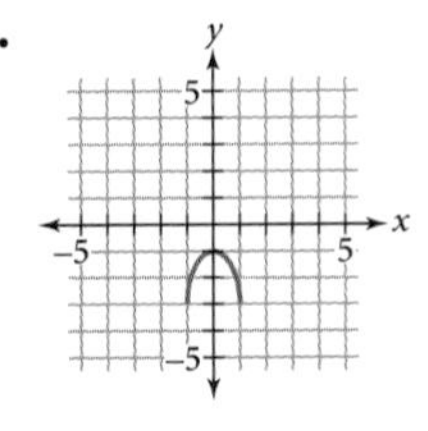

4d. $\frac{y-1}{2} = \sqrt{1 - (x-3)^2}$, or

$y = 2\sqrt{1 - (x-3)^2} + 1$

4e. $\frac{y-3}{-5} = \sqrt{1 - \left(\frac{x+2}{2}\right)^2}$, or

$y = -5\sqrt{1 - \left(\frac{x+2}{2}\right)^2} + 3$

5c. $y = \pm 2\sqrt{1 - x^2}$, or $x^2 + \left(\frac{y}{2}\right)^2 = 1$

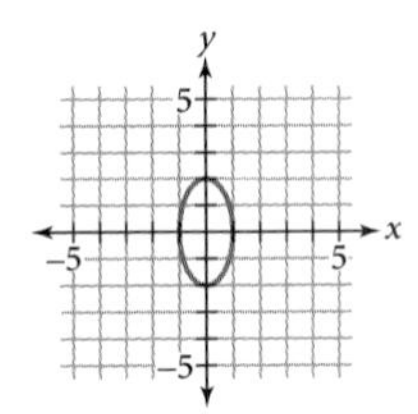

6b. $\frac{x}{3}$

6c. $g(x) = f\left(\frac{x}{3}\right)$

8a. $y = 3\sqrt{1 - \left(\frac{x}{0.5}\right)^2}$ and $y = -3\sqrt{1 - \left(\frac{x}{0.5}\right)^2}$

8b. $y = \pm 3\sqrt{1 - \left(\frac{x}{0.5}\right)^2}$

8c. $y^2 = 9\left[1 - \left(\frac{x}{0.5}\right)^2\right]$, or $\frac{x^2}{0.25} + \frac{y^2}{9} = 1$

9a.

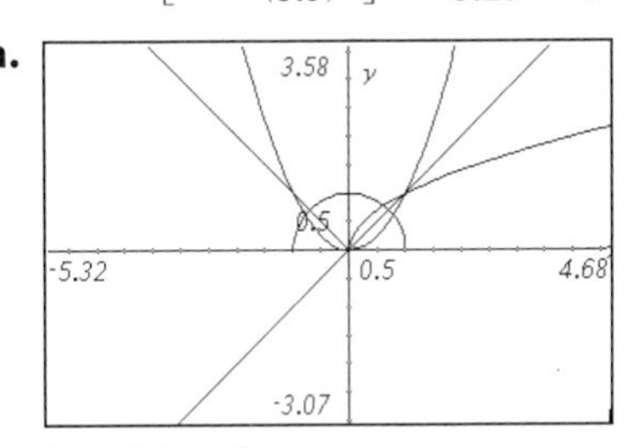

(0, 0) and (1, 1)

9b. *Hint:* Compare the coordinates of the points to the radius of the quarter-circle.

13c. $c = 0.2$ or 3.8

15a, c, d.

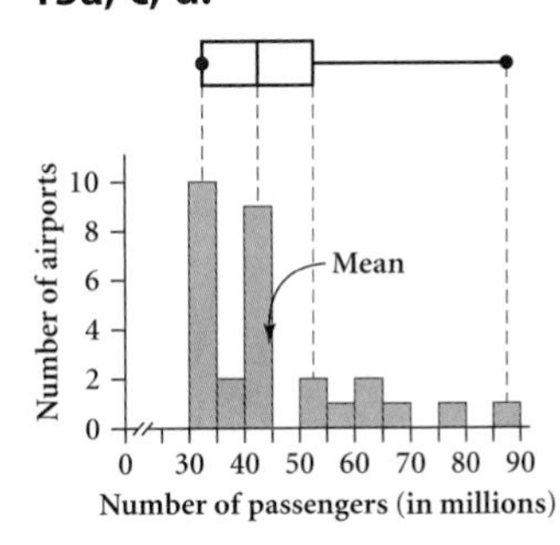

15b. Using the midpoint value for each histogram bin, there were 1340 million, or 1,340,000,000 passengers.

15c. mean = 44.67 million

15d. Five-number summary: 32.5, 32.5, 42.5, 52.5, 87.5; assume that all data occur at midpoints of bins.

LESSON 4.8

1a. 6

1b. 7

2b. 1

4b. composition: $g(f(x))$ where $f(x) = |x + 5|$ and $g(x) = 3 + (x - 3)^2$

6a. 2

7. *Hint:* First create a table of values.

7a. *Hint:* The graphs should be linear.

7b. approximately 41

7c. possible answer: $B = \frac{2}{3}(A - 12) + 13$

8. *Hint:* First sketch your answer conjectures and then confirm with your graphing calculator.

10e. $-x^4 + 8x^3 - 22x^2 + 24x - 5$

11. *Hint:* Use two points to find both parabola and semicircle equations for the curve. Then substitute a third point into your equations and decide which is most accurate.

13a. $x = -5$ or $x = 13$

13b. $x = -1$ or $x = 23$

13c. $x = 64$

13d. $x = \pm\sqrt{1.5} \approx \pm 1.22$

14d. $\hat{y} = 0.2278x$

14e. The ohm rating is the reciprocal of the slope of this line.

CHAPTER 4 REVIEW

1. Sample answer: For a time there are no pops. Then the popping rate slowly increases. When the popping reaches a furious intensity, it seems to level out. Then the number of pops per second drops quickly until the last pop is heard.

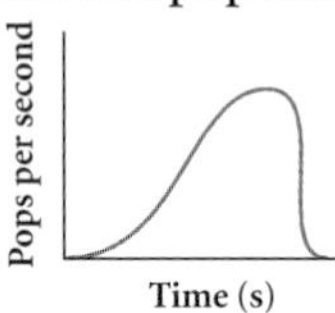

2a. -1

2b. 7

2c. $(x + 3)^2 - 3$

2d. -7

2e. -1

2f. 100

2g. $-2a^2 + 11$

2h. $4a^2 - 28a + 47$

2i. $4a^2 - 32a + 64$

3a.

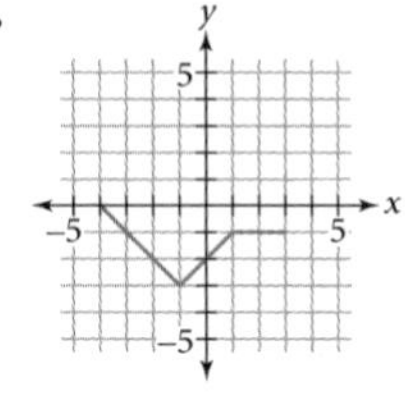

3b.

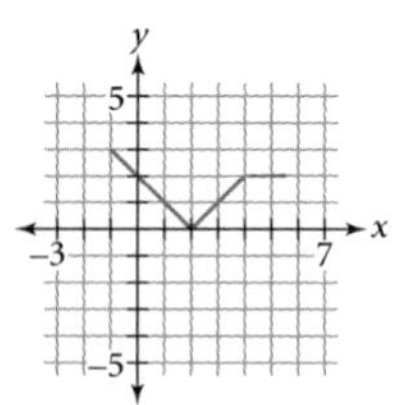

3c.

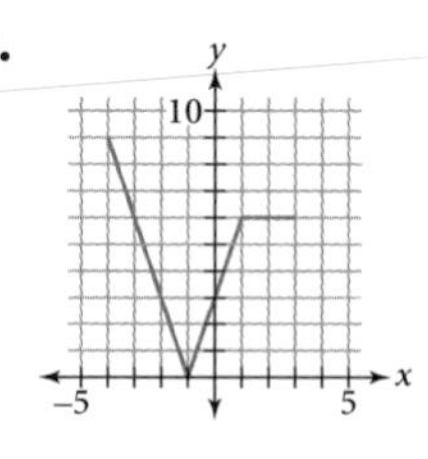

3d.

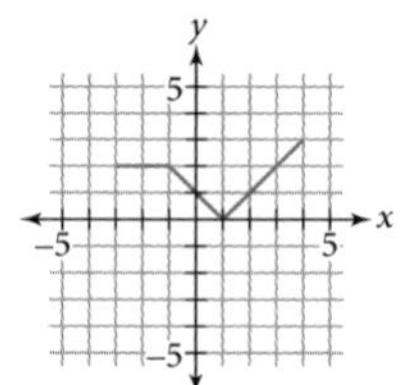

4a. Translate horizontally -2 units and vertically -3 units.

4b. Dilate horizontally by a factor of 2, and then reflect across the x-axis.

4c. Dilate horizontally by a factor of $\frac{1}{2}$, dilate vertically by a factor of 2, translate horizontally 1 unit and vertically 3 units.

5a.

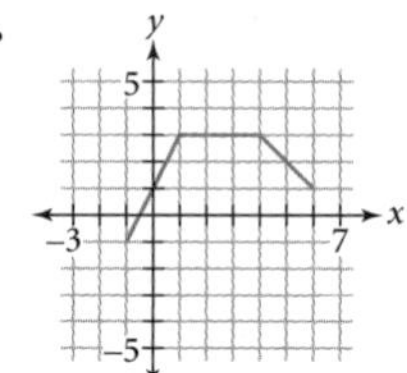

5b.

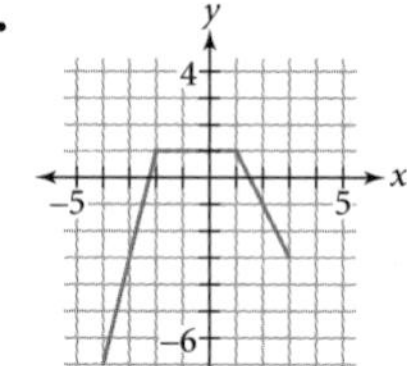

5c.

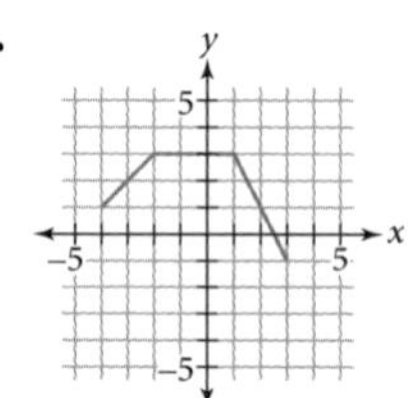

5d.

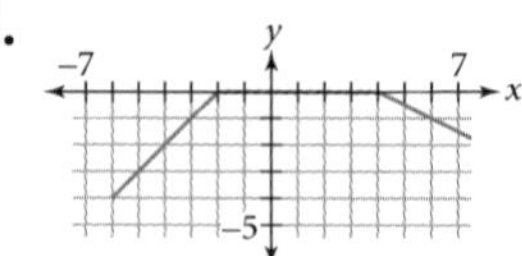

5e.

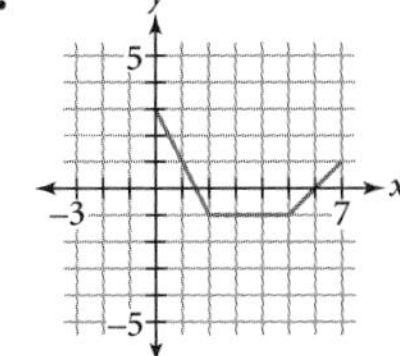

5f.

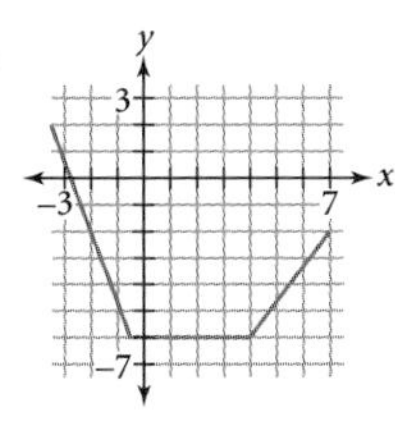

6a. $y = \sqrt{1 - x^2}; y = 3\sqrt{1 - x^2} - 1$

6b. $y = \sqrt{1 - x^2}; y = 2\sqrt{1 - \left(\frac{x}{5}\right)^2} + 3$

6c. $y = \sqrt{1 - x^2}; y = 4\sqrt{1 - \left(\frac{x - 3}{4}\right)^2} - 1$

6d. $y = x^2; y = (x - 2)^2 - 4$

6e. $y = x^2; y = -2(x + 1)^2$

6f. $y = \sqrt{x}; y = -\sqrt{-(x - 2)} - 3$

6g. $y = |x|; y = 0.5|x + 2| - 2$

6h. $y = |x|; y = -2|x - 3| + 2$

7a. $y = \frac{2}{3}x - 2$

7b. $y = \pm\sqrt{x + 3} - 1$

7c. $y = \pm\sqrt{-(x - 2)^2} + 1$

8a. $x = 8.25$

8b. $x = \pm\sqrt{45} \approx \pm 6.7$

8c. $x = 11$ or $x = -5$

8d. no solution

9a.

17,000	16,000	15,000	14,000	13,000	12,000	11,000	10,000
18,700	19,200	19,500	19,600	19,500	19,200	18,700	18,000

9b.

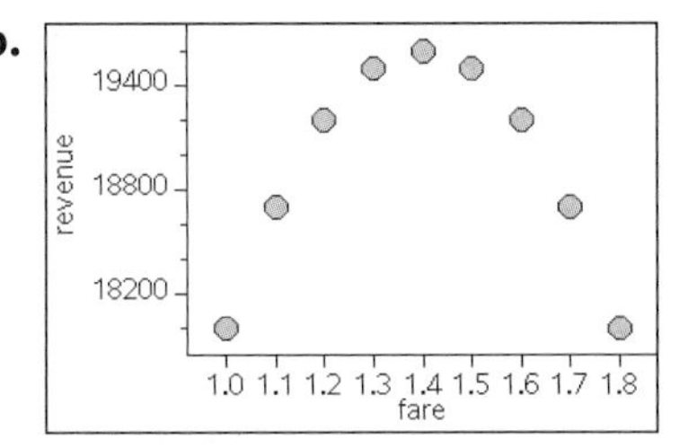

9c. (1.40, 19,600). By charging \$1.40 per ride, the company achieves the maximum revenue, \$19,600.

9d. $\hat{y} = -10{,}000(x - 1.4)^2 + 19{,}600$

9d. i. \$16,000

9d. ii. \$0 or \$2.80

REFRESHING YOUR SKILLS FOR CHAPTER 5

1a. between 3 and 4 (about 3.3)

1c. between 7 and 8 (about 7.4)

2a. $2\sqrt{6}$

2c. $3\sqrt{5}$

3a. iii, B

3c. iv, D

4a. $>$

4c. $<$

LESSON 5.1

1a. $f(5) \approx 3.52$

1c. $h(24) \approx 22.92$

2a. 16, 12, 9; $y = 16(0.75)^x$

3a. $f(0) = 125, f(1) = 75, f(2) = 45$; $u_0 = 125$ and $u_n = 0.6u_{n-1}$ where $n \geq 1$

4c. 0.94; 6% decrease

5a. $u_0 = 1.211$, $u_n = u_{n-1} \cdot 1.015$

5b.

Year	Estimated population (billions)
1995	1.211
1996	1.229
1997	1.248
1998	1.266
1999	1.285
2000	1.305
2001	1.324
2002	1.344

5c. y represents the estimated population x years after 1995; $y = 1.211(1.015)^x$.

5d. The equation predicts that the population of China in 2006 was 1.426 billion. This is larger than the actual value. This means that the population is growing at a slower rate than it was in 1995.

6b. *Hint:* Because you have an explicit equation for the height, you can use decimal values for x.

8. *Hint:* You will need to experiment with transformations of the parent function $\frac{y-k}{a} = 2^{[(x-h)/b]}$, or $y = k + a \cdot 2^{[(x-h)/b]}$. Compare the marked points on the black curve and the red curve to identify translations and stretches.

8a. $y = 2^{x-3}$

8c. $\frac{y}{3} = 2^x$, or $y = 3 \cdot 2^x$

10. *Hint:* You will need to experiment with transformations of the parent function, $\frac{y-k}{a} = 0.5^{[(x-h)/b]}$, or $y = k + a \cdot 0.5^{[(x-h)/b]}$. Use what you learned in Exercise 8.

10a. $y - 5 = 0.5^x$, or $y = 5 + 0.5^x$

10c. $y + 2 = 0.5^{(x-1)}$, $y = -2 + 0.5^{(x-1)}$

11a. $\frac{27}{30} = 0.9$

11b. $f(x) = 30(0.9)^x$

11e. *Hint:* Your equation should contain the number 30.

11f. *Hint:* Think about what x_1, y_1, and b represent.

12. Answers will vary but will be in the form $y = y_1 \cdot 1.8^{x-x_1}$, with (x_1, y_1) being any point from the table.

LESSON 5.2

1a. $\frac{1}{125}$

1c. $-\frac{1}{81}$

1e. $\frac{16}{9}$

2a. a^5

2d. e^3

4a. $x = -2$

4c. $x = -5$

5a. $x \approx 3.27$

5d. *Hint:* Think about the order of operations to determine the first step to undo.

5f. $x = 1$

6c. $10x^7$

6f. $\frac{1}{25}x^{-12}$

7. *Hint:* Is $(2 + 3)^2$ equivalent to $2^2 + 3^3$? Is $(2 + 3)^1$ equivalent to $2^1 + 3^1$? Is $(2 - 2)^3$ equivalent to $2^3 + (-2)^3$? Is $(2 - 2)^2$ equivalent to $2^2 + (-2)^2$?

10c. $4y = x^3$, or $y = \frac{1}{4}x^3$

12a. *Hint:* Write the original information as two ordered pairs, (3, 30) and (6, 5.2).

12b. $30.0r^{x-3} = 5.2r^{x-6}$; $r \approx 0.5576$

12c. *Hint:* The height the ball was dropped from is the same as the height of bounce zero.

13c. *Hint:* First reduce the left side of the equation. Then expand what is left.

14d. $42(0.9476)^{1980-2002} \approx 137.2$; $39.8(0.9476)^{1980-2003} \approx 137.2$; both equations give approximately 137.2 rads.

15b. $x = -4$

15c. $x = 4$

LESSON 5.3

1a. One set of equivalent expressions is a, e, and j.

2d. Power; x is equivalent to x^1.

2g. Exponential; $\frac{12}{3^t}$ is equivalent to $12(3)^{-t}$.

2j. Neither; the function is not a transformation of either x^a or b^x.

3b. $b^{4/5}$, $b^{8/10}$, or $b^{0.8}$

3c. $c^{-1/2}$, or $c^{-0.5}$

4b. $b^{4/5} = 14.3$; raise both sides to the power of $\frac{5}{4}$: $b = 14.3^{5/4} \approx 27.808$.

4c. $c^{-1/2} = 0.55$; raise both sides to the power of -2: $c = 0.55^{-2} \approx 3.306$.

5. *Hint:* Write the given information as two ordered pairs in the form (*number of gels, intensity of light*).

9a. exponential

9b. neither

9c. exponential

9d. power

10. *Hint:* Compare the two points on the black curve with the points on the red curve to identify the translations and stretches.

10a. $y = 3 + (x - 2)^{3/4}$

10b. $y = 1 + [-(x - 5)]^{3/4}$

11b. $x = 180^{1/4} \approx 3.66$

13a. *Hint:* Solve for k.

14a. $27x^9$

14d. $108x^8$

LESSON 5.4

1a. $x = 50^{1/5} \approx 2.187$

1c. no real solution

2a. $x = 625$

2d. $x = 12(-1 + 1.8125^{1/7.8}) \approx 0.951$

3a. $9x^4$

5. *Hint:* Refer to Example B of this lesson.

5a. She must replace y with $y - 7$ and y_1 with $y_1 - 7$; $y - 7 = (y_1 - 7) \cdot b^{x-x_1}$.

6b. *Hint:* Start with a value for a that is less than 1 and a value for b that is a little larger than 1.

7b. 54 ft

11a. 0.0466, or 4.66% per year

11b. 6.6 g

13. $x = -4.5$, $y = 2$, $z = 2.75$

LESSON 5.5

2b. 2

2c. *Hint:* If $g^{-1}(20) = v$, then $g(v) = \underline{\ ?\ }$.

4. *Hint:* Graph the four linear functions and identify the function-inverse pairs. Then do the same for the nonlinear functions.

4b. b and d are inverses

4c. c and g are inverses

5a. $f(7) = 4$; $g(4) = 7$

5b. They might be inverse functions.

6b. *Hint:* Find an equation for $f^{-1}(x)$.

8a. $f(x) = 2x - 3$; $f^{-1}(x) = \frac{x+3}{2}$, or $f^{-1}(x) = \frac{1}{2}x + \frac{3}{2}$

8c. $f(x) = \frac{-x^2 + 3}{2}$, or $f(x) = -\frac{1}{2}x^2 + \frac{3}{2}$; $y = \pm\sqrt{-2x + 3}$ (not a function)

9b. i. $f^{-1}(x) = \frac{x - 32}{1.8}$

9b. ii. $f(f^{-1}(15.75)) = 15.75$

9b. iii. $f^{-1}(f(15.75)) = 15.75$

9b. iv. $f(f^{-1}(x)) = f^{-1}(f(x)) = x$

10a. The equation of the median-median line is $f(x) \approx -0.006546x + 14.75$.

10c. The equation of the median-median line is $g(x) \approx -0.003545x + 58.81$.

10e. *Hint:* Think carefully about the units and what each equation uses as its independent variable.

12. Your friend's score is 1. Sample answers are given for explanations of incorrect answers. Problem 1 is correct. Problem 2 is incorrect: The notation $f^{-1}(x)$ indicates the inverse function related to $f(x)$, not the exponent -1. Problem 3 is incorrect: The expression $9^{-1/5}$ can be rewritten as $\frac{1}{9^{1/5}}$. Problem 4 is incorrect: The expression 0^0 is not defined.

14e. *Hint:* The compositions $g(c(x))$ and $c(g(x))$ should both be equivalent to x.

14f. *Hint:* Use an "average" month of 30 days.

14g. *Hint:* The product of length, width, and height should be equivalent to the volume of water, in cubic inches, saved in a month.

16. $f(x) = 12.6(1.5)^{x-2}$, or $f(x) = 42.525(1.5)^{x-5}$

18. *Hint:* Consider a vertically oriented parabola and a horizontally oriented parabola.

LESSON 5.6

1a. $10^x = 1000$

1c. $7^{1/2} = x$

2a. $x = 3$

2b. $x = 4$

2c. $x = \sqrt{7} \approx 2.65$

2d. $x = 2$

2e. $x = \frac{1}{25}$

2f. $x = 0$

3a. $x = \log_{10} 0.001$; $x = -3$

3c. $x = \log_{35} 8$; $x \approx 0.5849$

7a. sometime in 1977

8a. $y = 100(0.999879)^x$

9. *Hint:* Write the given information as two points: (0, 88.7) and (6, 92.9).

9a. $y \approx 88.7(1.0077)^x$

11a. Median-median line equation using years since 1900 is $\hat{y} = -1121 + 17.1x$.

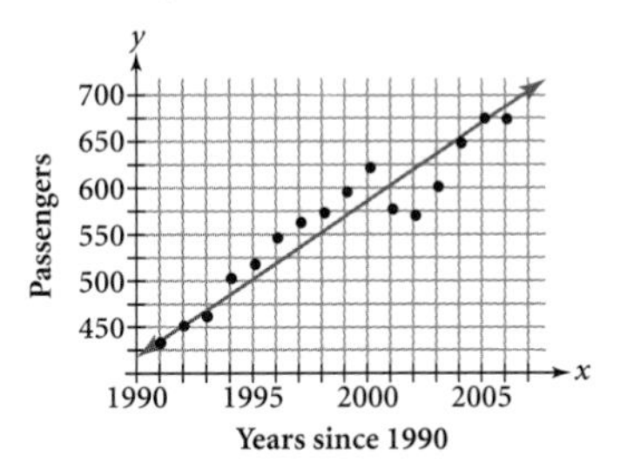

11c. 25.288196. Predictions based on this model will generally be within 25.3 million of the correct number of passengers.

12b. $y = 16.35(2)^x$, where x represents C-note number and y represents frequency in cycles per second

15c. Possible answer: Translate horizontally 1 unit and vertically 4 units. $2(x - 1) - 3(y - 4) = 9$.

LESSON 5.7

1a. log 55

1c. log 4

2a. log 2 + log 11

2b. many possible answers, such as log 26 − log 2

3a. $x \log 5$

3c. $\frac{1}{2} \log 3$

4a. true

4c. true

6a. $y \approx 100(0.999879)^x$

8c. $x \approx 11.174$

8d. $x \approx 42.739$

9a. *Hint:* Write the given information as two data points and then find an equation for an exponential graph through the two points.

9c. $y = 8.91$ lb/in^2

10b. $y = 100(0.965)^x$, with x in minutes

11. *Hint:* If more than one input value results in the same output value, then a function's inverse will not be a function. What does this mean about the graph of the function?

14a. False. If everyone got a grade of 86% or better, one would have to have gotten a much higher grade to be in the 86th percentile.

14d. true

LESSON 5.8

1a. 2.90309

1d. 1.4123

2a. $\log(10^{n+p}) = \log(10^n \cdot 10^p)$

$(n + p)\log 10 = \log 10^n + \log 10^p$

$(n + p)\log 10 = n \log 10 + p \log 10$

$(n + p)\log 10 = (n + p)\log 10$

Because the logarithm of the left side equals the logarithm of the right, the left and right sides are equal. Or, because $\log(10^{n+p}) = \log[(10^n)(10^p)]$, $10^{n+p} = (10^n)(10^p)$.

3. *Hint:* If the annual interest rate is 6.75%, what is the monthly interest rate?

4a. $h = 146(0.9331226)^{T-4}$

5c. $x \approx 72.09$. After 72 days, 6000 games have been sold.

6a. *Hint:* The given value, 10^{-13}, is a value for the variable I. The loudness is a value for the variable D.

6c. $I = 10^{10.7} \cdot 10^{-16} = 10^{-5.3} \approx 5.01 \times 10^{-6}$ W/cm^2

7b. $(\log x, y)$ is a linear graph.

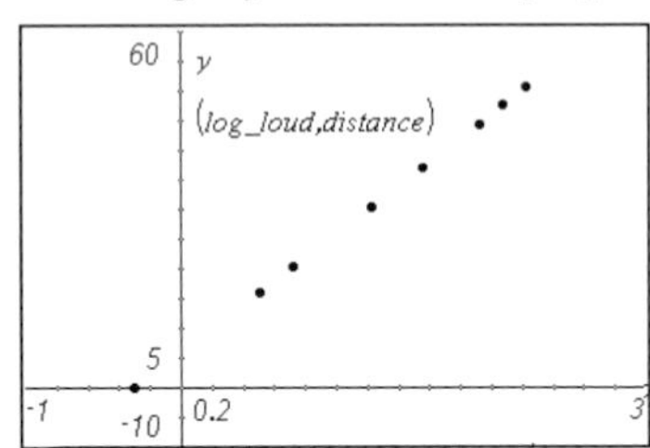

8b. *Hint:* This curve is both reflected and translated.

9a. The data are the most linear when viewed as (log(*height*), log(*distance*)).

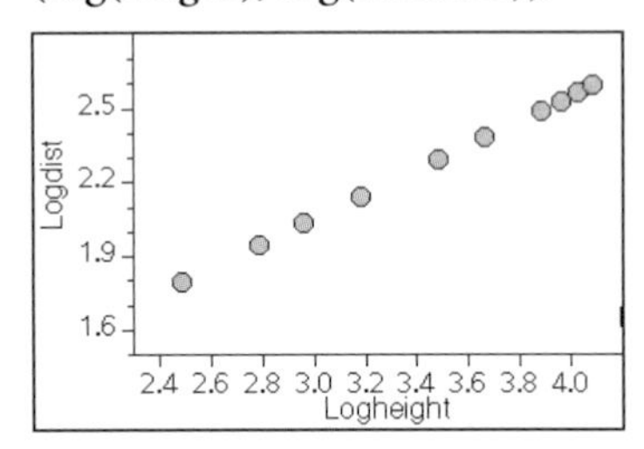

10a. 14.6 qt after 1 day; 13.41 qt after 2 days; $u_0 = 16$, $u_n = u_{n-1}(1 - 0.15) + 1$, $n \geq 1$

11a. $y = 18(\sqrt{2})^{x-4}$, $y = 144(\sqrt{2})^{x-10}$, or $y = 4.5(\sqrt{2})^x$

11b. $y = \dfrac{\log x - \log 18}{\log \sqrt{2}} + 4$, $y = \dfrac{\log x - \log 144}{\log \sqrt{2}} + 10$, or $y = \dfrac{\log x - \log 4.5}{\log \sqrt{2}}$

14a. $x = 3418^{1/5} \approx 5.09$

14b. $x = 256^{1/4} + 5.1 = 9.1$, or $x = 1.1$

14c. $x = \pm\left(\dfrac{55}{7.3}\right)^{1/6} \approx 1.40$

CHAPTER 5 REVIEW

1a. $\frac{1}{16}$ **1b.** $-\frac{1}{3}$ **1c.** 125

1d. 7 **1e.** $\frac{1}{4}$ **1f.** $\frac{27}{64}$

1g. -1 **1h.** 12 **1i.** 0.6

2a. $\log_3 7 = x$, or $x \log 3 = \log 7$

2b. $\log_4 5 = x$, or $x \log 4 = \log 5$

2c. $\log 7^5 = x$

3a. $10^{1.72} = x$ **3b.** $10^{2.4} = x$

3c. $5^{-1.47} = x$ **3d.** $2^5 = x$

4a. $\log xy$ **4b.** $\log z - \log v$

4c. $2.1x^{6.8}$ **4d.** $k \log w$

4e. $x^{1/5}$ **4f.** $\dfrac{\log t}{\log 5}$

5a. $x = \dfrac{\log 28}{\log 4.7} \approx 2.153$

5b. $x = \pm\sqrt{\dfrac{\log 2209}{\log 4.7}} \approx \pm 2.231$

5c. $x = 2.9^{1/1.25} = 2.9^{0.8} \approx 2.344$

5d. $x = 3.1^{47} \approx 1.242 \times 10^{23}$

5e. $x = \left(\dfrac{101}{7}\right)^{1/2.4} \approx 3.041$

5f. $x = \dfrac{\log 18}{\log 1.065} \approx 45.897$

5g. $x = 10^{3.771} \approx 5902$

5h. $x = 47^{5/3} \approx 612$

6a. $x = 0.5(2432^{1/8} - 1) \approx 0.825$

6b. $x = 114^{1/2.7} \approx 5.779$

6c. $x = \dfrac{\log \frac{734}{11.2}}{\log 1.56} \approx 9.406$

6d. $x = 20.2$

6e. $x = \left(\dfrac{147}{12.1}\right)^{1/2.3} - 1 \approx 1.962$

6f. $x = 5.75^2 + 3 \approx 36.063$

7. $x = \dfrac{16 \log \frac{8}{45}}{\log 0.5} \approx 39.9$; about 39.9 h

8a. 1 **8b.** $\dfrac{(x+1)^3 + 2}{4}$

8c. $\frac{1}{2}$ **8d.** 12

9. $y = 5\left(\dfrac{32}{5}\right)^{(x-1)/6}$

10.

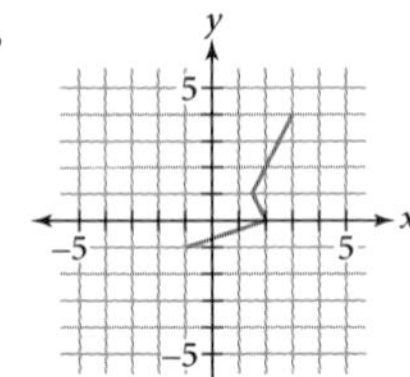

11a. $a = 0.50$ **11b.** $b = \dfrac{2.94}{\log 15} \approx 2.4998$

11c. $\log x = \frac{-0.50}{2.4998} \approx -0.2$; $x \approx 10^{-0.2} \approx 0.63$. The real-world meaning of the x-intercept is that the first 0.63 min of calling is free.

11d. \$4.19 **11e.** about 4 min

12a.

125 y

$f1(x)=100-80\cdot(.750)^x$

x 18

12b. domain: $0 \le x \le 120$; range: $20 \le y \le 100$

12c. Vertically dilate by a factor of 80; reflect across the x-axis; vertically shift by 100.

12d. 55% **12e.** about 4 yr old

13a. approximately 37 sessions

13b. approximately 48 wpm

13c. Sample answer: It takes much longer to improve your typing speed as you reach higher levels. 60 wpm is a good typing speed, and very few people type more than 90 wpm, so $0 \le x \le 90$ is a reasonable domain.

14a. $u_0 = 1, u_n = (u_{n-1}) \cdot 2, n \ge 1$

14b. $y = 2x$

14c.

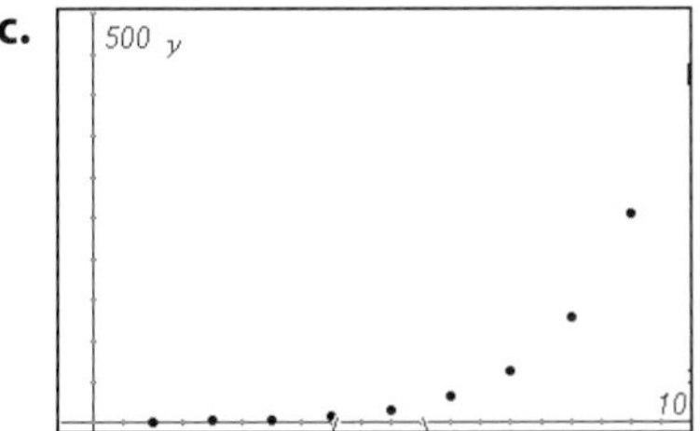

14d. Answers will vary but can include curving upward, increasing, increasing at an increasing rate, discrete.

14e. after 20 cell divisions

14f. after 29 divisions

CHAPTER 6 • CHAPTER  CHAPTER 6 • CHAPTER

REFRESHING YOUR SKILLS FOR CHAPTER 6

1A. d **1C.** a

2a. sometimes true ($x = 5$); commutative property of subtraction; not a property for all real numbers

2c. never true; associative property of division; not a property for all real numbers

3a. $x \ge -2$

−6 −4 −2 0 2 4 6 8

LESSON 6.1

1a.

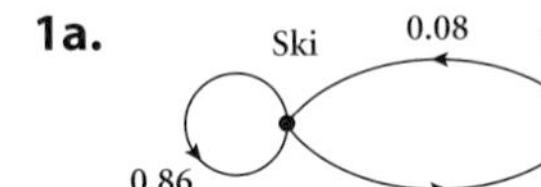

1b. $\begin{bmatrix} 0.86 & 0.14 \\ 0.08 & 0.92 \end{bmatrix}$

2. 0.53; 0.40

4a. $A(-3, 2)$, $B(1, 3)$, and $C(2, -2)$

5b. 18 boys

6a.

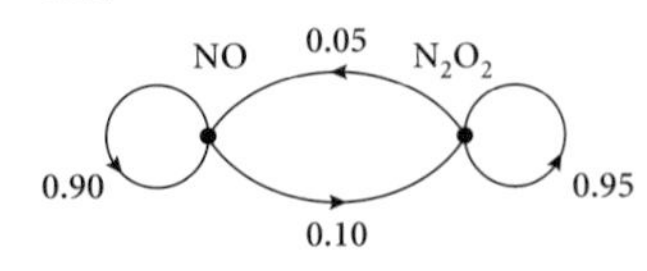

6b. *Hint:* The initial value matrix is $[40 \quad 200]$.

7b. $\begin{bmatrix} 0.99 & 0.01 \\ 0.10 & 0.90 \end{bmatrix}$

8a. 50 waters were sold at the back entrance.

11b. $m_{32} = 1$; there is one round-trip flight between City C and City B.

11c. *Hint:* Count the connections at each vertex.

13. $7.4p + 4.7s = 100$

16a. 45 dB

LESSON 6.2

2a. $x = 7, y = 54$

2b. $c_{11} = 0.815, c_{12} = 0.185, c_{21} = 0.0925, c_{22} = 0.9075$

3a. $\begin{bmatrix} 7 & 3 & 0 \\ -19 & -7 & 8 \\ 5 & 2 & -1 \end{bmatrix}$

3c. $[13 \quad 29]$

3d. not possible because the inside dimensions do not match

5a.

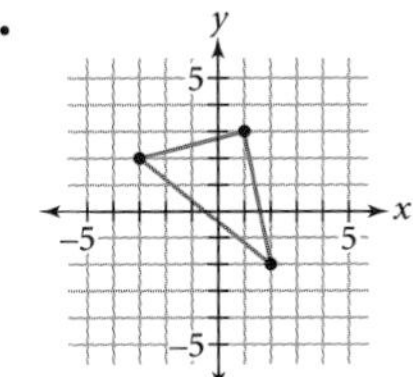

5b. $\begin{bmatrix} 3 & -1 & -2 \\ 2 & 3 & -2 \end{bmatrix}$

7b. $[4800 \quad 4200]$

9a. $a = 3, b = 4$

10. *Hint:* Draw a transition diagram of the spider's movements.

12a. $\begin{bmatrix} 0.50 & 0.45 & 0.05 \\ 0.25 & 0.50 & 0.25 \\ 0.30 & 0.30 & 0.40 \end{bmatrix}$

14a. $a + b$

14b. $3a$

LESSON 6.3

1a. $\begin{bmatrix} 3 & 4 \\ 2 & -5 \end{bmatrix}\begin{bmatrix} x \\ y \end{bmatrix} = \begin{bmatrix} 11 \\ -8 \end{bmatrix}$

1b. $\begin{bmatrix} 1 & 2 & 1 \\ 3 & -4 & 5 \\ -2 & -8 & -3 \end{bmatrix}\begin{bmatrix} x \\ y \\ z \end{bmatrix} = \begin{bmatrix} 0 \\ -11 \\ 1 \end{bmatrix}$

2a. $\begin{bmatrix} 15 & -19 \\ 22 & -27 \end{bmatrix}$

3a. $\begin{bmatrix} 1a + 5c & 1b + 5d \\ 6a + 2c & 6b + 2d \end{bmatrix} = \begin{bmatrix} -7 & 33 \\ 14 & -26 \end{bmatrix}$; $\begin{bmatrix} a & b \\ c & d \end{bmatrix} = \begin{bmatrix} 3 & -7 \\ -2 & 8 \end{bmatrix}$

4a. $\begin{bmatrix} 1 & 0 \\ 0 & 1 \end{bmatrix}$; yes

5a. $\begin{bmatrix} 4 & -3 \\ -5 & 4 \end{bmatrix}$

5b. *Hint:* You will need to solve three systems of three equations. The 3×3 identity matrix is $\begin{bmatrix} 1 & 0 & 0 \\ 0 & 1 & 0 \\ 0 & 0 & 1 \end{bmatrix}$.

6a. $\begin{bmatrix} 8 & 3 \\ 6 & 5 \end{bmatrix}\begin{bmatrix} x \\ y \end{bmatrix} = \begin{bmatrix} 41 \\ 39 \end{bmatrix}$; $x = 4, y = 3$

7b. $28.50

8. *Hint:* Let x and y represent the amount in each account.

9. *Hint:* Create a system of three equations by substituting each pair of coordinates for x and y. Then solve the system for a, b, and c.

15a. possible answer: $2y = 4x + 8$

16a. possible answer: $y = 2x + 6$

LESSON 6.4

1a. $\begin{cases} 2x + 5y = 8 \\ 4x - y = 6 \end{cases}$

2a. $\left[\begin{array}{ccc|c} 1 & 2 & -1 & 1 \\ 2 & -1 & 3 & 2 \\ 2 & 1 & 1 & -1 \end{array}\right]$

3a. $\left[\begin{array}{ccc|c} 1 & -1 & 2 & 3 \\ 0 & 3 & -5 & -2 \\ 2 & 1 & -1 & 2 \end{array}\right]$

4a. $\left[\begin{array}{cc|c} 2 & 5 & 8 \\ 4 & -1 & 6 \end{array}\right]$

4c. $\frac{R_2}{-11} \rightarrow R_2$

5a. $\left[\begin{array}{ccc|c} 1 & 0 & 0 & -31 \\ 0 & 1 & 0 & 24 \\ 0 & 0 & 1 & -4 \end{array}\right]$

5c. cannot be transformed into reduced row-echelon form (dependent system)

7. *Hint:* What do you know about the sum of the angles of any triangle?

8. *Hint:* There are three equations here, one equal to $1,715, one equal to 22, and one equal to the number of full-page ads.

11a. first plan: $14,600; second plan: $13,100

14. $\overline{AB}: y = 6 - \frac{2}{3}(x - 4)$ or $y = 4 + \frac{2}{3}(x - 1)$; $\overline{BC}: y = 4 - \frac{2}{3}(x - 7)$ or $y = 6 - \frac{2}{3}(x - 4)$; $\overline{CD}: y = 1 + 3(x - 6)$ or $y = 4 + 3(x - 7)$; $\overline{DE}: y = 1$; $\overline{AE}: y = 4 - 3(x - 1)$ or $y = 1 - 3(x - 2)$

LESSON 6.5

1a. $y < \frac{10 - 2x}{-5}$, or $y < -2 + 0.4x$

2a.

3a. $y < 2 - 0.5x$

3c. $y > 1 - 0.75x$

5.

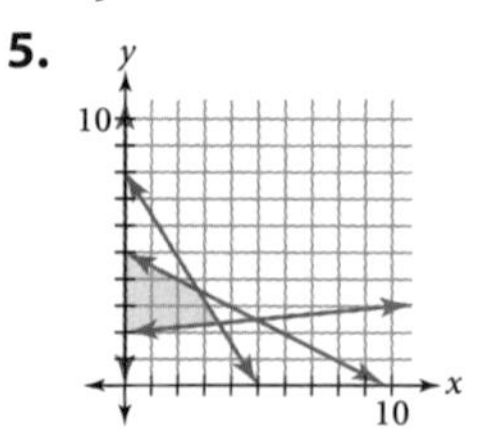

vertices: (0, 2), (0, 5), (2.752, 3.596), (3.529, 2.353)

7.

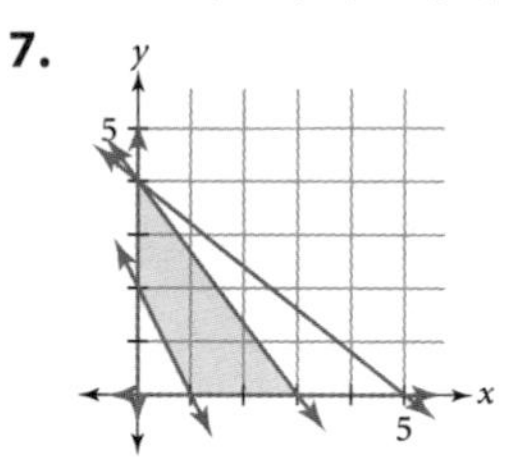

vertices: (0, 4), (3, 0), (1, 0), (0, 2)

9. *Hint:* If the length is x and the height is y, then what are the formulas for area and perimeter?

9a. Let x represent length in inches, and let y represent width in inches.

$\begin{cases} xy \geq 200 \\ xy \leq 300 \\ x + y \geq 33 \\ x + y \leq 40 \end{cases}$

LESSON 6.6

1.

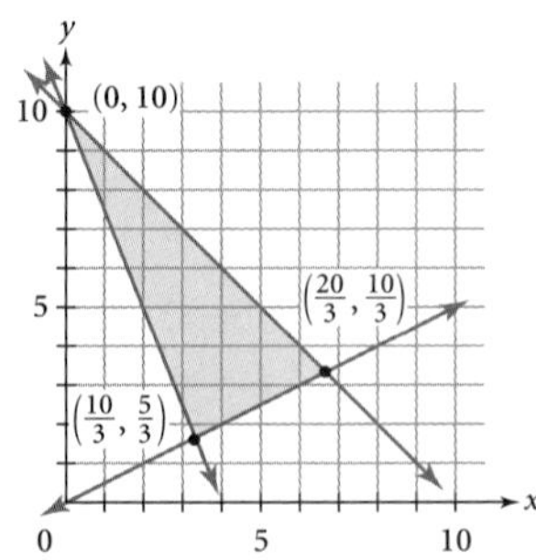

2a. $\left(\frac{20}{3}, \frac{10}{3}\right)$

5a. possible answer:

$$\begin{cases} y \geq 7 \\ y \leq \frac{7}{5}(x-3)+6 \\ y \leq -\frac{7}{12}x + 13 \end{cases}$$

6. *Hint:* One constraint (concerning food) is $2p + 6s \leq 100$.

7. 5 radio minutes and 10 newspaper ads to reach a maximum of 155,000 people. This assumes that people who listen to the radio are independent of people who read the newspaper, which is probably not realistic, and that the shop buys both. This also assumes that each newspaper ad reaches a fresh set of readers (instead of the same subscribers day after day who each see the ad 10 times), and that each radio minute reaches a fresh set of listeners (instead of the same group of people who hear the radio spot 5 separate times). 20 newspaper ads and 0 radio minutes would reach 160,000 people.

8. *Hint:* One constraint (concerning sulfur) is $0.02L + 0.06H \leq 0.04(L + H)$.

10a. Let x represent the length in inches, and let y represent the girth in inches.

$$\begin{cases} x + y \leq 130 \\ x \leq 108 \\ x > 0 \\ y > 0 \end{cases}$$

13. *Hint:* One inequality is $y \leq 5$.

CHAPTER 6 REVIEW

1a. impossible because the dimensions are not the same

1b. $\begin{bmatrix} -4 & 7 \\ 1 & 2 \end{bmatrix}$

1c. $\begin{bmatrix} -12 & 4 & 8 \\ 8 & 12 & -8 \end{bmatrix}$

1d. $\begin{bmatrix} -3 & 1 & 2 \\ -11 & 11 & 6 \end{bmatrix}$

1e. impossible because the inside dimensions do not match

1f. $[-7 \quad -5 \quad 6]$

2a. $\begin{bmatrix} 0.8 & -0.6 \\ 0.2 & -0.4 \end{bmatrix}$

2b. $\begin{bmatrix} -\frac{3}{85} & \frac{16}{85} & -\frac{2}{85} \\ \frac{18}{85} & -\frac{11}{85} & \frac{12}{85} \\ -\frac{32}{85} & \frac{29}{85} & \frac{7}{85} \end{bmatrix}$, or

$\begin{bmatrix} -0.0353 & 0.1882 & -0.0235 \\ 0.2118 & -0.1294 & 0.1412 \\ -0.3765 & 0.3412 & 0.0824 \end{bmatrix}$

2c. does not exist

2d. $\begin{bmatrix} -\frac{5}{56} & \frac{1}{7} & \frac{1}{8} \\ -\frac{3}{56} & \frac{2}{7} & -\frac{1}{8} \\ -\frac{29}{56} & \frac{3}{7} & \frac{1}{8} \end{bmatrix}$

3a. $x = 2.5, y = 7$

3b. $x = 1.22, y = 6.9, z = 3.4$

4a. $x = 2.5, y = 7$

4b. $x = 1.22, y = 6.9, z = 3.4$

5. Inequalities should be equivalent to $y \leq 6 - \frac{2}{5}x$, $y \leq 2(7 - x)$, $y \geq 0$, and $x \geq 0$.

6a.

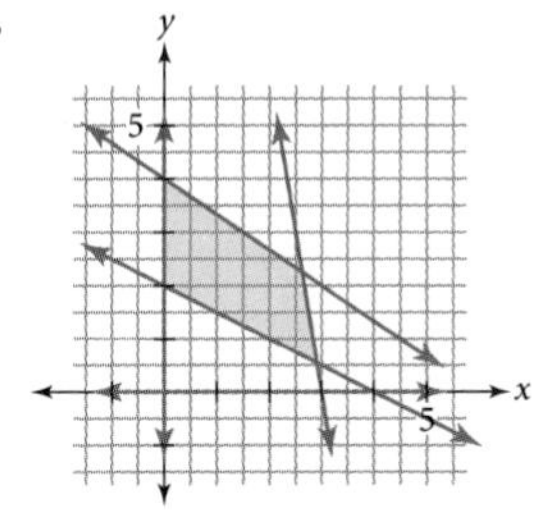

Vertices: (0, 4), (2.625, 2.25), $(2.\overline{90}, 0.\overline{54})$, (0, 2). The maximum occurs at (0, 4): $1.65(0) + 5.2(4) = 20.8$.

6b.

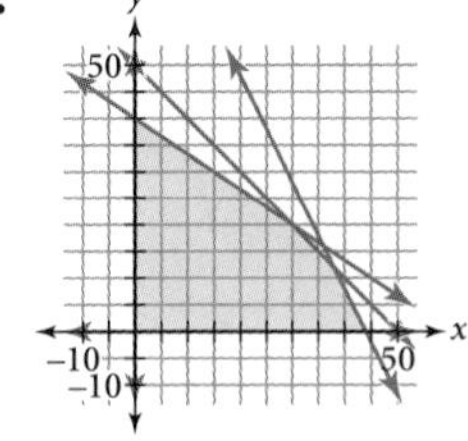

Vertices: (0, 0), (0, 40), (30, 20), (38, 12), (44, 0). The maximum occurs at (30, 20): $6(30) + 7(20) = 320$.

7. about 4.4 yr

8a. $\begin{bmatrix} 0.92 & 0.08 & 0.00 \\ 0.12 & 0.82 & 0.06 \\ 0.00 & 0.15 & 0.85 \end{bmatrix}$

8b. i. Mozart: 81; Picasso: 66; Hemingway: 63

8b. ii. Mozart: 82; Picasso: 70; Hemingway: 58

8b. iii. Mozart: 94; Picasso: 76; Hemingway: 40

9. $y = 4x^2 - 18x + 5$

10a. $-1 \leq x \leq 2$

10b. $x < -5$ or $x > 1$

10c. $x < -2$ or $x > 2$

11. They should make 4 shawls and 2 blankets to make the maximum profit of \$104.

12a. $a < 0; p < 0; d > 0$

12b. $a > 0; p > 0$; d cannot be determined.

12c. $a > 0; p = 0; d < 0$

13a. 3

13b. -5 or approximately 3.5

13c. It is a function because no vertical line crosses the graph in more than one place.

13d. $-6 \leq x \leq 5$

13e. $-2 \leq y \leq 4$

14. 20 students in second period, 18 students in third period, and 24 students in seventh period

15a. Let x represent the year, and let y represent the number of barrels per day in millions.

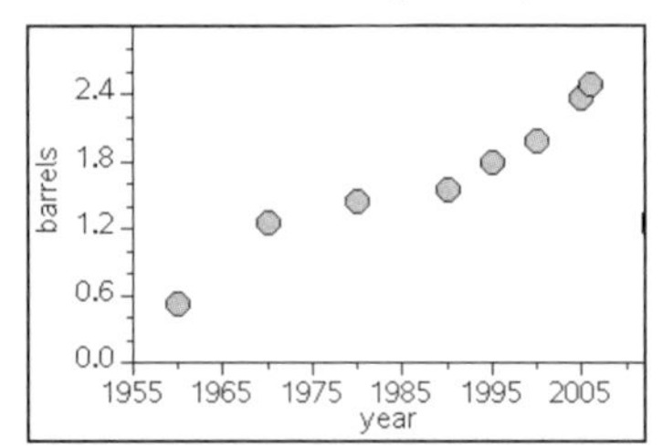

15b. M_1 (1970, 1.26), M_2 (1992.5, 1.675), M_3 (2005, 2.37); $\hat{y} \approx 1.7683 + 0.03171(x - 1989.2)$, or $\hat{y} = -61.317 + 0.03171x$

15c. 2.429 million barrels per day

16a. $x = 245$

16b. $x = 20$

16c. $x = -\frac{1}{2}$

16d. $x = \dfrac{\log \frac{37,000}{15}}{\log 9.4} \approx 3.4858$

16e. $x = 21$

16f. $x = \dfrac{\log 342}{\log 36} \approx 1.6282$

17a. 85%

17b.

Distance (mi)	0	10	20	30	40	50	60	70	...
Percentage of original signal	100	85	72.3	61.4	52.2	44.4	37.7	32.1	...

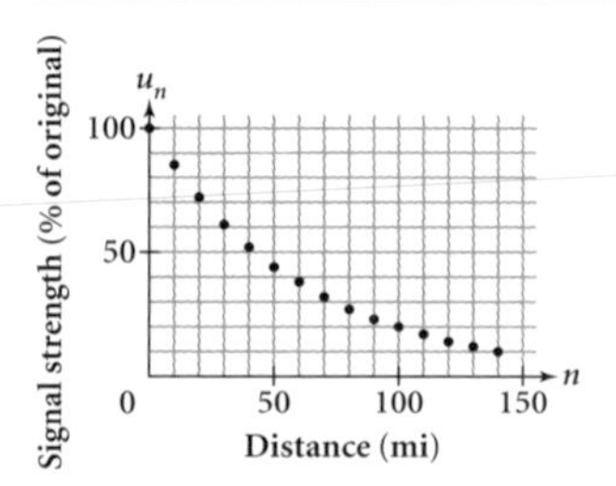

17c. 280 mi

18a. $y = 50(0.72)^{x-4}$ or $y = 25.92(0.72)^{x-6}$

18b. 0.28; decay

18c. approximately 186

18d. 0

19a. sample answer using a bin width of 10:

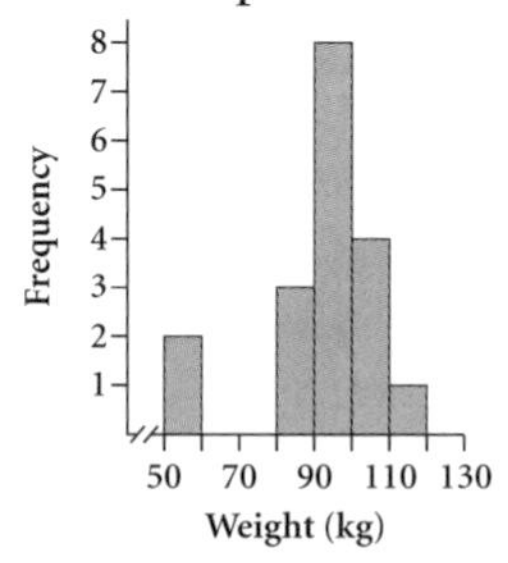

19b. skewed left

19c. 50 kg and 55 kg

19d. 67th percentile

20a. consistent and independent

20b. consistent and dependent

20c. inconsistent

20d. inconsistent

21a. Let x represent the year, and let y represent mean household population. Sample answer using the median-median method: $\hat{y} \approx 45.15 - 0.021286x$.

21b. Sample answer: What was the mean household population in 1965? Approximately 3.325 people.

21c. Sample answer: In what year will the mean household population be less than 2 people? In approximately 2027.

22a. a translation right 5 units and down 2 units

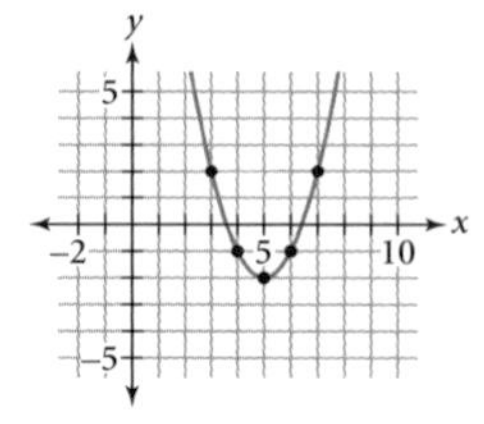

22b. a reflection across the x-axis and a vertical dilation by a factor of 2

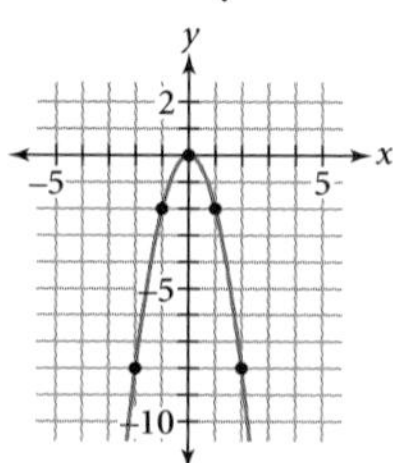

22c. $-1 \cdot [P] = \begin{bmatrix} 2 & 1 & 0 & -1 & -2 \\ -4 & -1 & 0 & -1 & -4 \end{bmatrix}$; This is a reflection across the x-axis and a reflection across the y-axis. However, because the graph is symmetric with respect to the y-axis, a reflection over that axis does not change the graph.

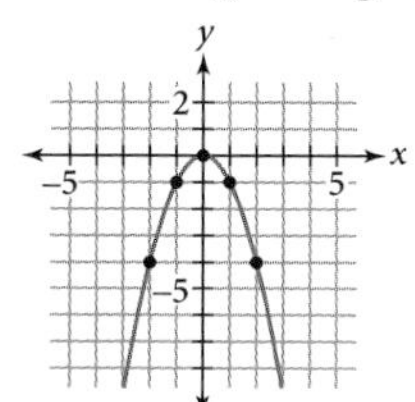

22d. $[P] + \begin{bmatrix} -2 & -2 & -2 & -2 & -2 \\ 3 & 3 & 3 & 3 & 3 \end{bmatrix}$

$= \begin{bmatrix} -4 & -3 & -2 & -1 & 0 \\ 7 & 4 & 3 & 4 & 7 \end{bmatrix}$

CHAPTER 7 • CHAPTER 7 CHAPTER 7 • CHAPTER

REFRESHING YOUR SKILLS FOR CHAPTER 7

1a. $8x + 6$

1c. $-3a^2 - 9a + 7$

2a. $x^2 + 6x + 9$

3a.

	x	5
x	x^2	$5x$
5	$5x$	25

4a. $(x + 5)(x + 5) = x^2 + 10x + 25$

LESSON 7.1

1d. 5

2a. polynomial; 3; $\frac{5}{9}x^3 - 3.5x^2 + 4x - 3$

2c. not a polynomial because $4\sqrt{x^3} = 4x^{3/2}$ has a non-integer exponent

3b. no; {0.007, 0.006, 0.008, 0.010}

5c. 4 points. You have to find the finite differences twice, so you need at least four data points to calculate two D_2 values that can be compared.

5e. *Hint:* Find some pennies and try to arrange each number of pennies into a triangle.

8b. *Hint:* Because measurements are inexact, do not expect the differences to be perfectly constant.

9. $D_1 = \{6, 10, 14, 18, 22, 26\}$; $D_2 = \{4, 4, 4, 4, 4\}$. The second differences are constant, so a quadratic function expresses the relationship. Let x represent the energy level, and let y represent the maximum number of electrons. $y = 2x^2$.

11c. $x = \dfrac{\log 16}{\log 5} \approx 1.7227$

12c. *Hint:* You need to break $8x$ into two terms.

13. $y \geq -\frac{1}{2}(x - 3) + 3$ or $y \geq -\frac{1}{2}(x - 11) - 4$

$y \leq \frac{1}{2}(x + 3) + 3$ or $y \leq \frac{1}{2}(x - 5) + 7$

$y \leq -\frac{11}{6}(x - 5) + 7$ or $y \leq -\frac{11}{6}(x - 11) - 4$

LESSON 7.2

1c. factored form and vertex form

1d. none of these forms

1e. factored form

2b. $(-4, -2)$

3a. -1 and 2

4c. $y = -2x^2 + 20x - 46$

6c. $y = -3(x - 4.5)^2 + 19$

7b. $y = ax^2 - 8ax + 16a$

7d. $y = -0.5x^2 - (0.5r + 2)x - 2r$

7f. $y = ax^2 - a(r + s)x + ars$

9. *Hint:* Substitute the coordinates of the y-intercept into each function to solve for the scale factor, a.

9b. $y = -0.5(x + 2)(x - 3)$

10b. $D_1 = \{9.5, 8.5, 7.5, 6.5\}$; $D_2 = \{-1, -1, -1\}$

10c. *Hint:* You are writing an equation that describes the ordered pairs (*selling price, revenue*).

12b. At approximately 23°C, the rate of photosynthesis is maximized at 100%.

12c. 0°C and 46°C

13d. $9x^2 - 6x + 1$

	$3x$	–1
$3x$	$9x^2$	$-3x$
–1	$-3x$	1

14. *Hint:* Distribute the second factor to each term in the first factor.

15d. *Hint:* Another perfect-square trinomial is $x^2 + 6x + 9$.

LESSON 7.3

1b. $\left(x+\frac{5}{2}\right)^2$

1d. $(x-y)^2$

2b. 12.25, or $\frac{49}{4}$

2d. -3

3b. $y=(x-3.5)^2+3.75$

3d. $y=5(x+0.8)^2-3.2$

4c. $4x^2+6x-3$; $a=4$, $b=6$, $c=-3$

6. *Hint:* Refer to Example C of this lesson.

7. *Hint:* Sketch a graph with time on the horizontal axis and height above ground on the vertical axis. Plot all the points you know.

7a. Let x represent time in seconds, and let y represent height in meters; $y=-4.9(x-1.1)(x-4.7)$, or $y=-4.9x^2+28.42x-25.333$.

10a. $y=-4.9t^2+100t+25$

10b. 25 m; 100 m/s

10c. 10.2 s; 535 m

11a. *Hint:* Make a table showing price and number sold to help you find this function.

11b. $R(p)=-2p^2+100p$

13. $x=2$, $x=-3$, or $x=\frac{1}{2}$

15a. Let x represent the year, and let y represent the number of endangered species.

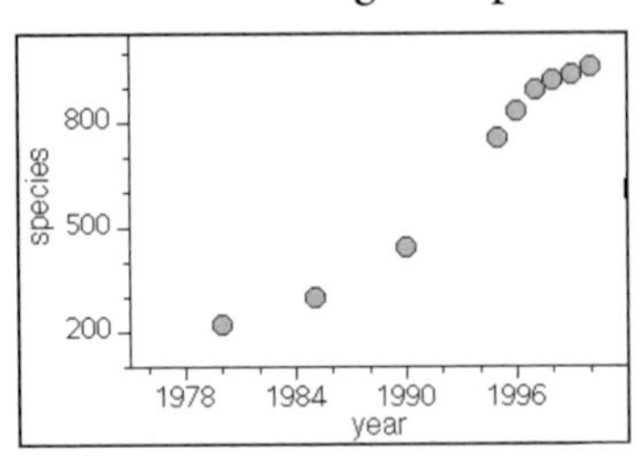

15b. Answers will vary. The median-median line is $\hat{y}\approx 45.64x-90289$.

15c. Answers will vary. Using the equation from 15b, approximately 1225 species in 2005; 3278 species in 2050.

15d. The prediction of about 1225 species is too high. You might lower your prediction for 2050, but there isn't very much data on which to base a prediction.

15e. The more recent data appear to lie on a curve that is leveling off. Possible answers: a logarithmic function, a quadratic function. You should justify your choice of model.

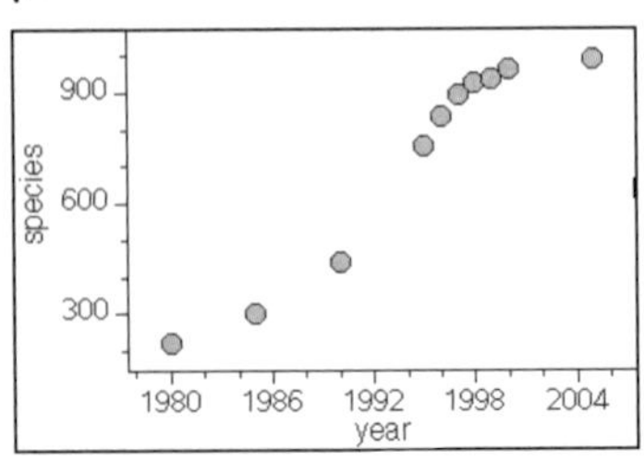

LESSON 7.4

1b. $x^2-5x-13=0$; $a=1$, $b=-5$, $c=-13$

1c. $3x^2+5x+1=0$; $a=3$, $b=5$, $c=1$

2b. -5.898

2d. 8.243

3c. $x=-1$ or $x=-1.4$

4c. $y=5(x+1)(x+1.4)$

6a. $x^2+9x+10=0$

7a. $y=a(x-3)(x+3)$ for $a\neq 0$

8. *Hint:* Think carefully about the value inside the square root, b^2-4ac.

9. *Hint:* When will the quadratic formula result in no real solutions?

11. *Hint:* The plug was pulled at time 0.

11a. $y=-4x^2-6.8x+49.2$

13a. $x^2+14x+49=(x+7)^2$ or $x^2+(-14x)+49=[x+(-7)]^2$

13b. $x^2-10x+25=(x-5)^2$

14c. $y=\pm\sqrt{x+6}-1$

16a. $y=\sqrt{400-x^2}$

16b. approximately 17.32 ft

LESSON 7.5

1b. 7

1c. $4-2i$

2a. $20+22i$

2c. $15+6i$

3d. $-2.35+2.71i$

4. A: $5-4i$; B: $-3i$; C: $-3-i$; D: -6; E: $-2+4i$; F: 1; G: $1+i$

5a.

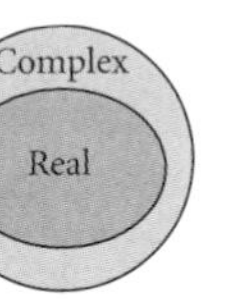

5c.

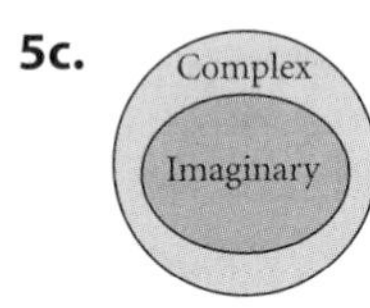

7a. $-i$

7b. 1

9. $0.2+1.6i$

11. *Hint:* Write each function in factored form first.

12. *Hint:* Assume complex zeros occur in conjugate pairs.

13c. The coefficients of the quadratic equations are nonreal.

15b. $0, i, -1+i, -i, -1+i, -i$; alternates between $-1+i$ and $-i$

15d. $0, 0.2+0.2i, 0.2+0.28i, 0.162+0.312i, 0.129+0.301i, 0.126+0.277i$; approaches $0.142+0.279i$

LESSON 7.6

1c. x-intercepts: 3, -2, -5; y-intercept: 60

2b. $y = -0.25(x + 1.5)(x + 6)$

3c. $y = x^3 - 64x$

6b. $6\left(x - \frac{5}{3}\right)\left(x + \frac{1}{2}\right)$, or $(3x - 5)(2x + 1)$

6f. $(x - 8)(x + 8)$

6h. $\left(x - \sqrt{7}\right)\left(x + \sqrt{7}\right)$

9b. $y = 2(x + 5)(x - 3)(x - 6)$

9c. $y = 2(x + 5)(x - 3)(x - 6) + 100$

10a. $(T + t)^2$, or $T^2 + 2Tt + t^2$

	T	t
T	TT	Tt
t	Tt	tt

10b. $(T + t)^2 = 1$, or $T^2 + 2Tt + t^2 = 1$

10c. *Hint:* Tasters include those who have at least one taster gene.

11. No. These points are collinear.

13a. $x = \pm\sqrt{50.4} \approx \pm 17.1$

13b. $x = \pm\sqrt{13} \approx \pm 3.6$

LESSON 7.7

1b. $x = -6, x = -3, x = 2$, and $x = 6$

2b. $(0, 108)$

3b. 4

4b. $y = 0.5(x + 6)(x + 3)(x - 2)(x - 6)$

5b. $y = a(x - 4)^2$ where $a \neq 0$

6b. *Hint:* You may need a large vertical window to see important graph features.

6c. *Hint:* Think about zooming in around each x-intercept.

7a. 4

7b. 5

7c. $y = -x(x + 5)^2(x + 1)(x - 4)$

8b. $y = -2(x + 4)(x - 5)(x + 2)^2$

8d. *Hint:* Remember that imaginary and complex roots come in conjugate pairs.

9. The leading coefficient is equal to the y-intercept divided by the product of the zeros if the degree of the function is even or the y-intercept divided by -1 times the product of the zeros if the degree of the function is odd.

11a. ii. $y = -(x + 5)^2(x + 2)(x - 1)$

11b. ii. $x = -5, x = -5, x = -2$, and $x = 1$

13. *Hint:* A polynomial function of degree n will have at most $n - 1$ extreme values and at most n x-intercepts.

15. $3 - 5\sqrt{2}$; $0 = a\left(x^2 - 6x - 41\right)$ where $a \neq 0$

16c. $36 - 18\sqrt{2}$

18. $S = \frac{20}{\sqrt{75}}\sqrt{\ell}$; approximately 17.9 knots

LESSON 7.8

1a. $3x^2 + 7x + 3$

2a. $3x^3 + 22x^2 + 38x + 15 = (x + 5)(3x^2 + 7x + 3)$

2b. $6x^3 + 11x^2 - 19x + 6 = (3x - 2)\left(2x^2 + 5x - 3\right)$

3a. $a = 12$

3c. $c = 7$

4a. $3x^3 - 11x^2 + 7x - 44 = (x - 4)(3x^2 + x + 11)$

4c. $4x^3 - 8x^2 + 7x - 6 = (x - 1.5)\left(4x^2 - 2x + 4\right)$

6. *Hint:* Try working through the procedure shown on pages 433–434.

6a. $47 = 11 \cdot 4 + 3$

6b. $P(x) = (x - 1)\left(6x^3 + x^2 + 8x - 4\right) + 11$

8a. 4

8b. $x = 1, x = 2, x = -5$, and $x = -1$

8c. $y = (x + 5)(x + 1)(x - 1)(x - 2)$

11a. $x = \pm 2, x = 1$, and $x = \pm 2i$

11b. $x \approx -7.01, x \approx -0.943$, and $x \approx 0.454$

13. *Hint:* The first term of the quotient will be $\frac{2x^3}{x} = 2x^2$. Write this term above $2x^3$ and then multiply to find the term in the lower left cell.

14a. $f(x) = 0.00639x^{3/2}$

14b. $f^{-1}(x) \approx (156x)^{2/3}$

14c. 33 in.

14d. about 177 ft

16a.

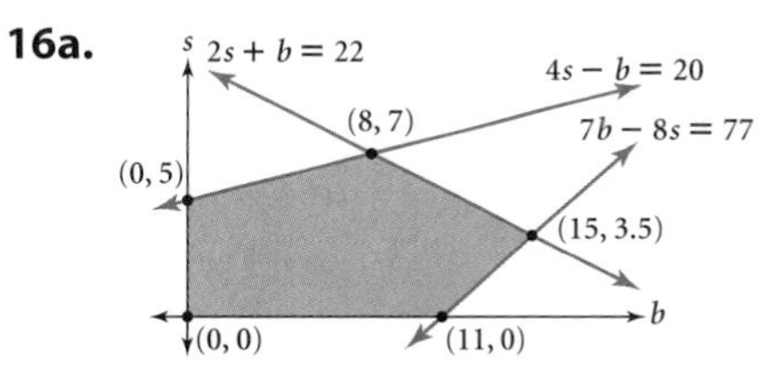

16b. 14 baseball caps and 4 sun hats; \$32

17a. $y = x^2 - 4x - 12$, $y = (x - 6)(x + 2)$; vertex: $(2, -16)$; y-intercept: -12; x-intercepts: 6, -2

17b. $y = 3x^2 + 6x - 24$, $y = 3(x - 2)(x + 4)$; vertex: $(-1, -27)$; y-intercept: -24; x-intercepts: 2, -4

CHAPTER 7 REVIEW

1a. $2(x - 2)(x - 3)$

1b. $(2x + 1)(x + 3)$ or $2(x + 0.5)(x + 3)$

1c. $x(x - 12)(x + 2)$

2a. $x = 9$ or $x = -1$

2b. $x = 0, x = 3$, or $x = -5$

3. 1; 4; 10; $\frac{1}{6}n^3 - \frac{1}{2}n^2 + \frac{1}{3}n$

4a. vertex form; general form: $y = 2x^2 - 8x - 8$; factored form:
$y = 2\left[x - \left(2 + 2\sqrt{2}\right)\right]\left[x - \left(2 - 2\sqrt{2}\right)\right]$

4b. factored form; general form: $y = -3x^2 + 12x + 15$; vertex form: $y = -3(x - 2)^2 + 27$

4c. general form; factored form: $y = (x + 2)(x + 1)$; vertex form: $y = (x + 1.5)^2 - 0.25$

4d. factored form; general form: $y = x^3 + 2x^2 - 11x - 12$; no vertex form for cubic equations

4e. general form; factored form:
$y = 2\left(x - \frac{-5 + \sqrt{73}}{4}\right)\left(x - \frac{-5 - \sqrt{73}}{4}\right)$;
vertex form: $y = 2(x + 1.25)^2 - 9.125$

4f. vertex form; general form:
$y = -x^2 - 14x - 51$; factored form:
$y = -\left[x - \left(-7 + i\sqrt{2}\right)\right]\left[x - \left(-7 - i\sqrt{2}\right)\right]$

5a.

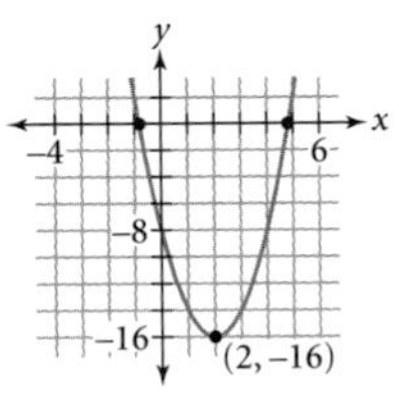

zeros: $x = -0.83$ and $x = 4.83$

5b.

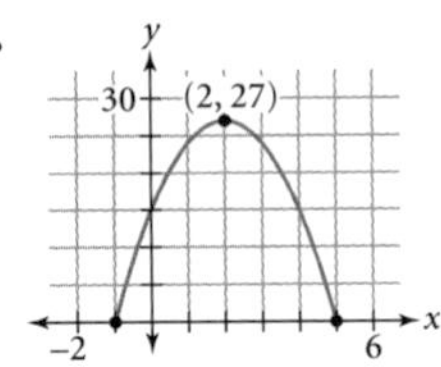

zeros: $x = -1$ and $x = 5$

5c.

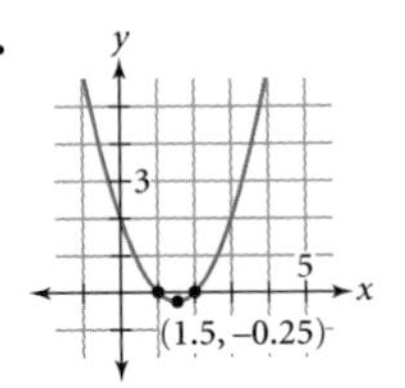

zeros: $x = 1$ and $x = 2$

5d.

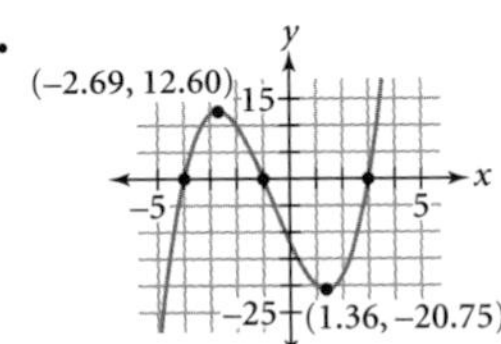

zeros: $x = -4$, $x = -1$, and $x = 3$

5e.

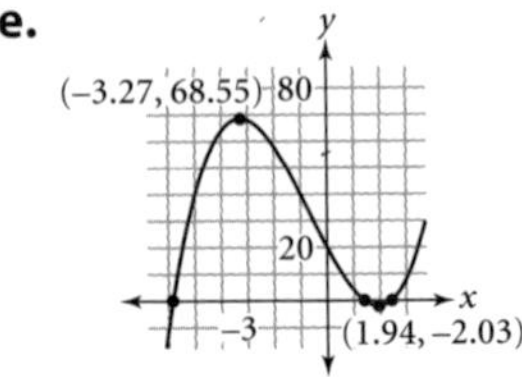

zeros: $x = -5.84$, $x = 1.41$, and $x = 2.43$

5f.

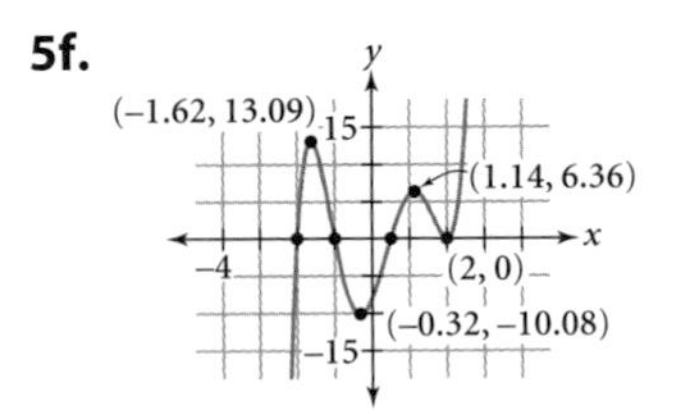

zeros: $x = -2$, $x = -1$, $x = 0.5$, and $x = 2$

6a. $y = 2(x + 1)(x - 4)$

6b. $y = 2(x + 3)^2(x - 1)$

6c. $y = -(x + 2)(x - 3)^3$

6d. $y = 0.5(x + 4)(x - 2)(x - 3i)(x + 3i)$

7. 18 in. by 18 in. by 36 in.

8. approximately 227 m, or 740 ft

9a. $y = 0.5x^2 + 0.5x + 1$

9b. 16 pieces; 56 pieces

10a. $y = (26 - 2x)(21 - 2x)$

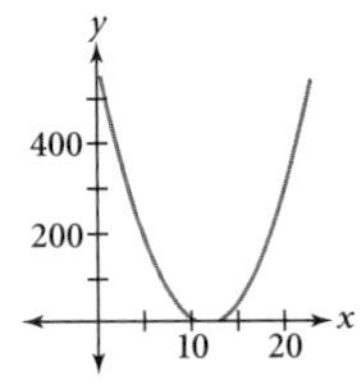

10b. domain: $0 \leq x \leq 10.5$; range: $0 \leq y \leq 546$

10c. $x \approx 3.395$ cm

11a. $\pm 1, \pm 3, \pm 13, \pm 39, \pm\frac{1}{3}, \pm\frac{13}{3}$

11b. $x = -\frac{1}{3}$, $x = 3$, $x = 2 + 3i$, and $x = 2 - 3i$

12a. $0 + 30i$

12b. $-6 - 9i$

12c. $0.4 - 0.2i$

13. $2x^2 + 4x + 3$

CHAPTER 8 • CHAPTER 8 CHAPTER 8 • CHAPTER

REFRESHING YOUR SKILLS FOR CHAPTER 8

1a. 5

2a. Smallville Middletown Bigcity
x $147 - x$
147 mi

3a. $AB = \sqrt{13}$, $AC = \sqrt{17}$, $BC = \sqrt{20}$; $\overline{BC}$ is the longest.

4. $y = 11$ or $y = 3$

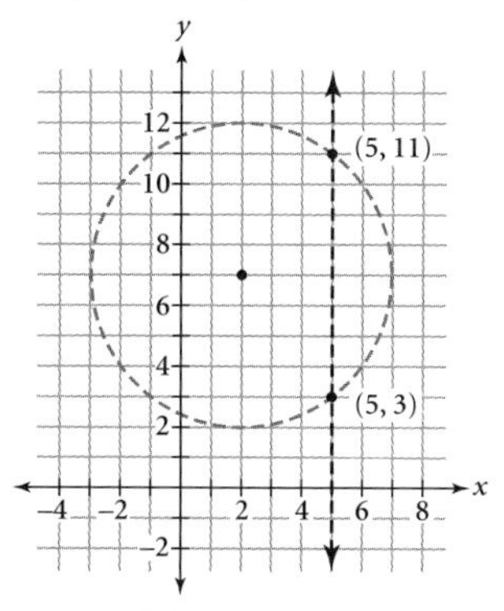

6a. $\sqrt{(x - 2^2) + (1 - 5)^2}$, or $\sqrt{(x - 2)^2 + 16}$

7a. *Hint:* The point (8, 0) satisfies the condition because it is 6 units from (2, 0) and 3 units from (5, 0). What other points can you find?

9a. $\sqrt{x^2 + 4}$ miles through the field and $3 - x$ miles on the road

9b. *Hint:* Refer to Example A of this lesson to see how to express the time.

10a. $y = \sqrt{10^2 + x^2} + \sqrt{(20 - x)^2 + 13^2}$

11a.

Time (s)	Height (ft)
0	24.00
1	23.92
2	23.66
3	23.24
4	22.63
5	21.82
6	20.78
7	19.49
8	17.89
9	15.87
10	13.27
11	9.59
12	0

15a. $(x + 3)^2 = 14$

15c. $(x + 3)^2 + (y - 2)^2 = 17$

16b. median from A to $\overline{BC}$: $y = -0.\overline{36}x + 0.\overline{90}$, or $y = -\frac{4}{11}x + \frac{10}{11}$; median from B to $\overline{AC}$: $y = -1.7x + 6.7$; median from C to $\overline{AB}$: $y = 13x - 57$

17a. $\frac{35}{3x}$

1a. center: (0, 0); radius: 2

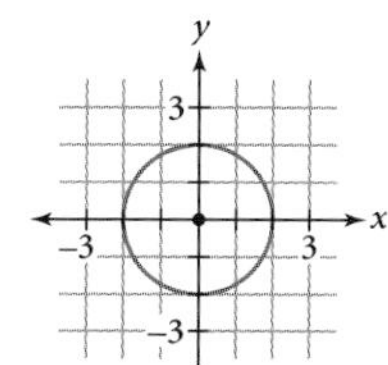

1c. center: (−1, 2); radius: 3

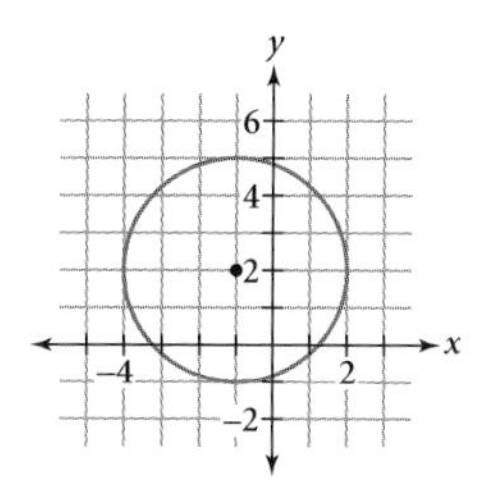

2a. $(x - 3)^2 + y^2 = 25$

2b. $(x + 1)^2 + (y - 2)^2 = 9$

3a. $(x - 2)^2 + (y + 3)^2 = 25$

4a. endpoints: (2, 0), (−2, 0), (0, 4), (0, −4)

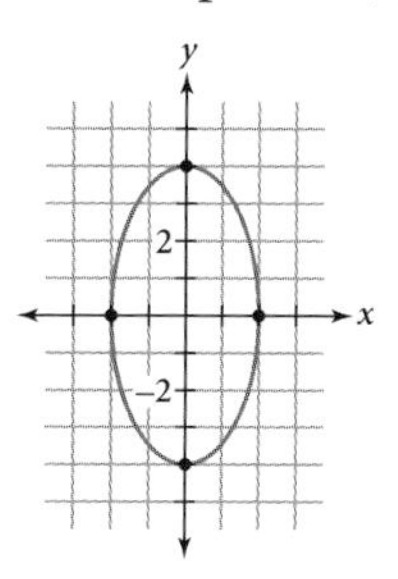

4c. Endpoints: (1, 1), (7, 1), (4, 4), (4, −2). Endpoints may vary because the graph is a circle.

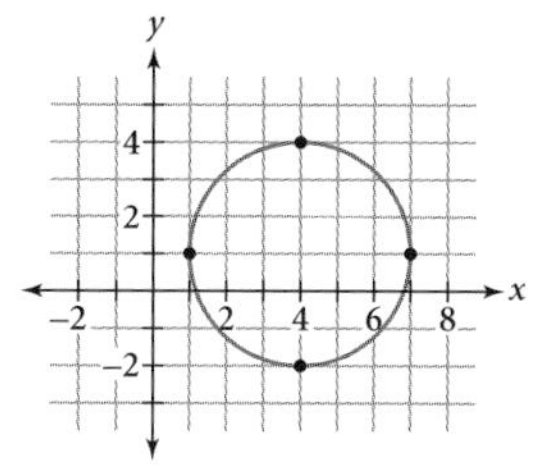

6a. $\left(\frac{x}{6}\right)^2 + \left(\frac{y}{3}\right)^2 = 1$

6c. $\left(\frac{x + 1}{4}\right)^2 + \left(\frac{y - 2}{3}\right)^2 = 1$

7a. $(\sqrt{27}, 0), (-\sqrt{27}, 0)$

7c. $(-1 + \sqrt{7}, 2), (-1 - \sqrt{7}, 2)$

9. *Hint:* Plot the data and estimate values for a and b.

10a. *Hint:* The nails are at the foci.

12a. The whisperers should stand at the two foci, approximately 5.2 m from the center of the room, along the major axis.

14. $y = -(x + 3)^2 + 2$

LESSON 8.3

1a. $(1, 0.5)$ **1b.** $y = 8$

2a. line of symmetry: $x = 0$

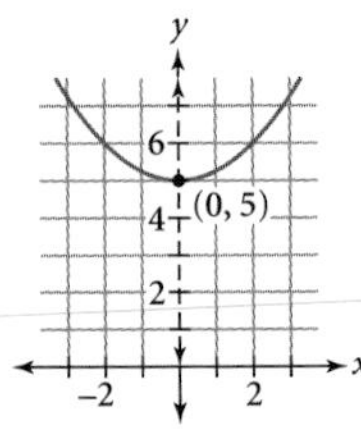

2b. line of symmetry: $y = -2$

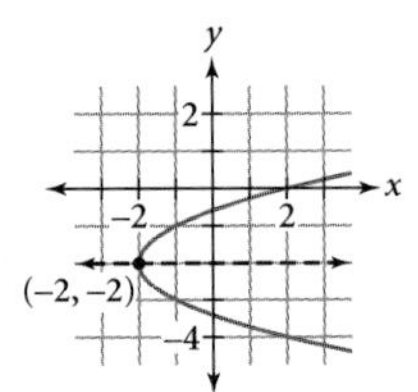

3a. focus: $(0, 6)$; directrix: $y = 4$

3b. focus: $(-1.75, -2)$; directrix: $x = -2.25$

4a. $(y - 2)^2 = x$ **4b.** $\frac{y - 4}{-1} = x^2$

5a. focus: $(0.25, 2)$; directrix: $x = -0.25$

5b. focus: $(0, 3.75)$; directrix: $y = 4.25$

6. *Hint:* Make a sketch and locate the focus.

7. $y = \frac{1}{8}(x - 1)^2 + 1$

8a. $y = \frac{1}{8}x^2 + 1$

10. *Hint:* Make a sketch to scale. The food should be placed at the focus of the parabola.

11a. $\frac{-11}{6x}$

14. $\left(\frac{x - 2}{16}\right)^2 + \left(\frac{y - 1}{\sqrt{192}}\right)^2 = 1$, or

$\left(\frac{x - 2}{16}\right)^2 + \left(\frac{y - 1}{8\sqrt{3}}\right)^2 = 1$

15a. $\pm 1, \pm 2, \pm 5, \pm 10, \pm\frac{1}{2}, \pm\frac{5}{2}$

17b. $x(x - 7)(x + 7)$

LESSON 8.4

1a. vertices: $(-2, 0)$ and $(2, 0)$; asymptotes: $y = \pm 2x$

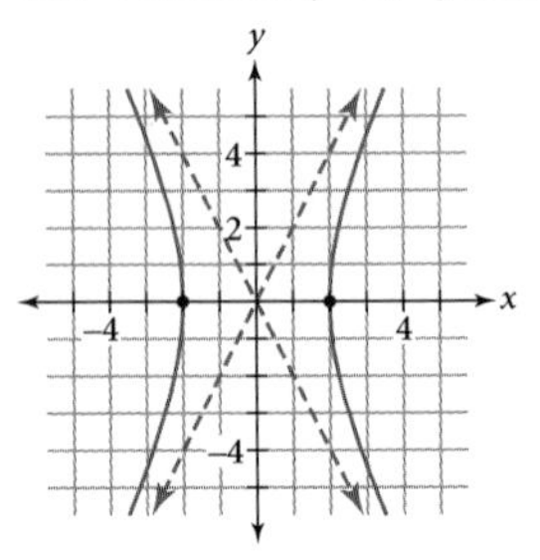

1b. vertices: $(2, -1)$ and $(2, -3)$; asymptotes: $y = \frac{1}{3}x - \frac{8}{3}$ and $y = -\frac{1}{3}x - \frac{4}{3}$

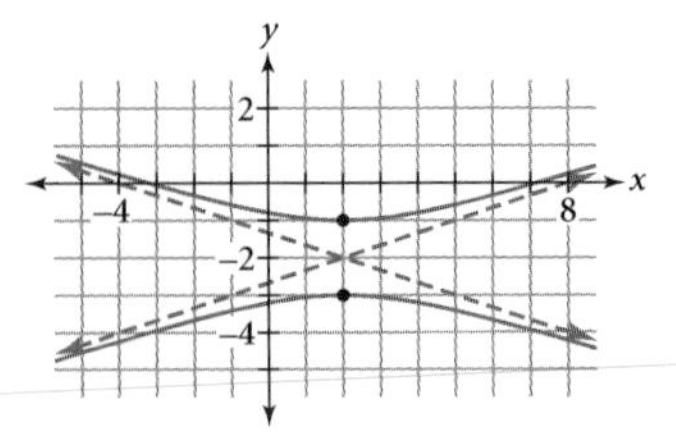

2a. $(\sqrt{20}, 0)$ and $(-\sqrt{20}, 0)$, or $(2\sqrt{5}, 0)$ and $(-2\sqrt{5}, 0)$

2b. $\left(2, -2 + \sqrt{10}\right)$ and $\left(2, -2 - \sqrt{10}\right)$

3a. $\left(\frac{x}{2}\right)^2 - \left(\frac{y}{1}\right)^2 = 1$

3b. $\left(\frac{y + 3}{2}\right)^2 - \left(\frac{x - 3}{2}\right)^2 = 1$

4a. $\left(\sqrt{5}, 0\right)$ and $\left(-\sqrt{5}, 0\right)$

4b. $\left(3, -3 + 2\sqrt{2}\right)$ and $\left(3, -3 - 2\sqrt{2}\right)$

5a. $y = \pm 0.5x$

5b. $y = x - 6$ and $y = -x$

6. $\left(\frac{x - 1}{5}\right)^2 - \left(\frac{y - 1}{\sqrt{11}}\right)^2 = 1$

8a. possible answer: $\left(\frac{y}{3}\right)^2 - \left(\frac{x - 1}{2}\right)^2 = 1$

10a. *Hint:* Plot the data and estimate the location of the center and asymptotes.

12. *Hint:* Sketch the hyperbola and its asymptotes first.

14. $\left(\frac{x - 11{,}900{,}000}{57{,}900{,}000}\right)^2 + \left(\frac{y}{56{,}700{,}000}\right)^2 = 1$

15a. *Hint:* What shape is this curve? What is special about the point where the ball reaches its maximum height?

LESSON 8.5

2a. $x^2 + 14x - 9y + 148 = 0$

2b. $x^2 + 9y^2 - 14x + 198y + 1129 = 0$

4a. $\left(\frac{y - 5}{\sqrt{11}}\right)^2 - \left(\frac{x + 4}{\sqrt{11}}\right)^2 = 1$; hyperbola

4b. $\left(\frac{x - 3}{6}\right)^2 + \left(\frac{y - 8}{\sqrt{72}}\right)^2 = 1$, or

$\left(\frac{x - 3}{6}\right)^2 + \left(\frac{y - 8}{6\sqrt{2}}\right)^2 = 1$; ellipse

5a. $y = \frac{\pm\sqrt{400x^2 + 1600}}{-8}$, or $y = \pm\frac{5}{2}\sqrt{x^2 + 4}$

5b. $y = \frac{-16 \pm \sqrt{160x - 320}}{8}$, or $y = \frac{-4 \pm \sqrt{10x - 20}}{2}$

9a. (2.711, −1.627), (1.510, −2.779), (−1, −3), (−2.954, −1.127)

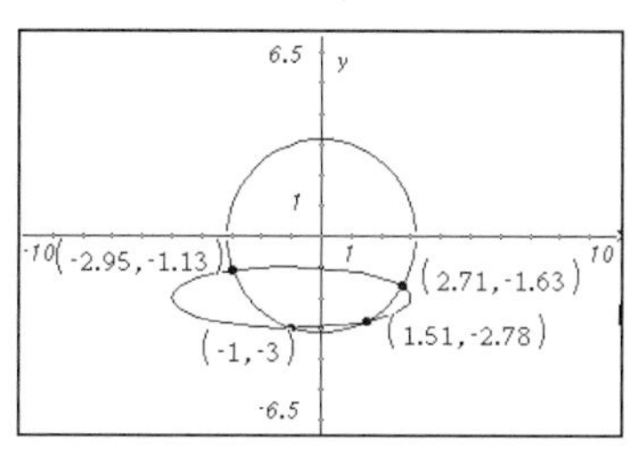

9b. (−5.33, 32), (6.24, 19.5), (−0.443, −5.92), (3.53, −1.62)

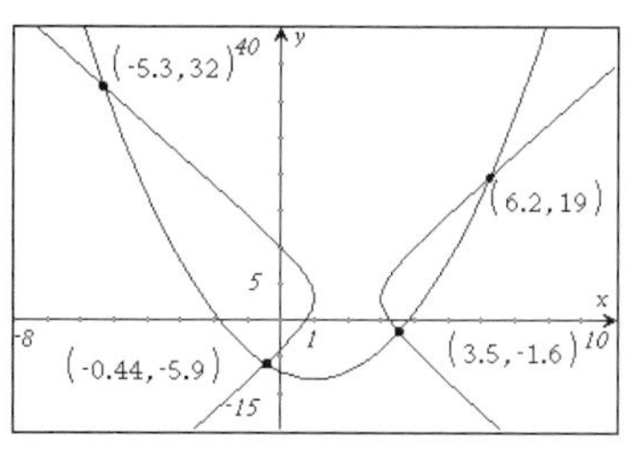

10. *Hint:* Sketch the ellipses and then find one intersection point.

11c. *Hint:* The center of the hyperbola is at (200, 0).

LESSON 8.6

1a. $f(x) = \frac{1}{x} + 2$

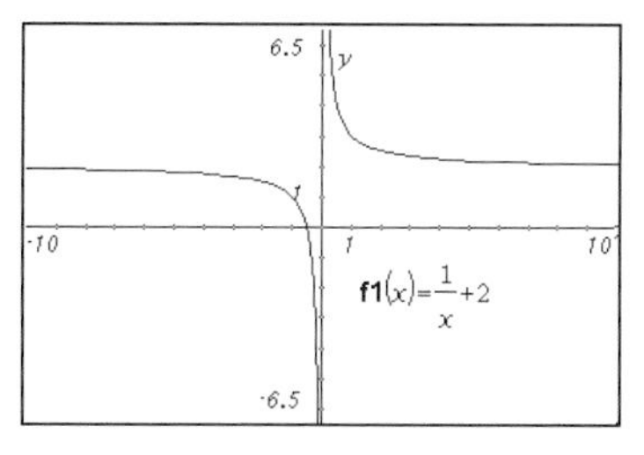

1e. $f(x) = 3\left(\frac{1}{x}\right) + 1$, or $f(x) = \frac{3}{x} + 1$

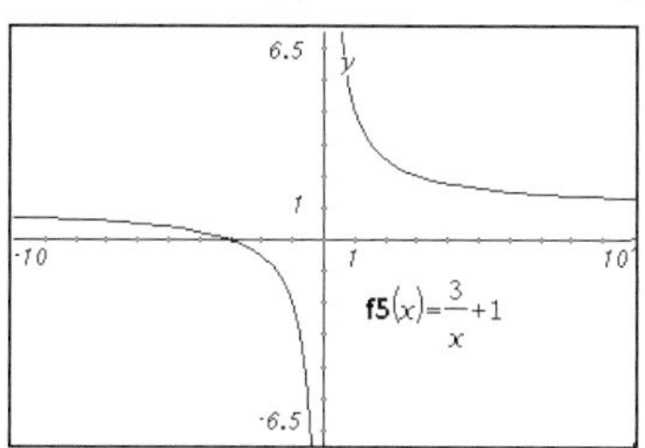

2a. translated left 3 and down 2

3a. $y = \frac{1}{x} + 2$

3c. $y = \frac{1}{x+4} + 3$

4a. $y = 1, x = 0$

4c. $y = -1, x = -2$

6. 12 games

8a. 20.9 mL

10a. *Hint:* Divide the expression as in Example A of this lesson.

12a. *Hint:* The graph should be a rotated hyperbola.

13a. 14.4 ohms

20. *Hint:* Concentric circles have the same center but different radii.

LESSON 8.7

1a. $\frac{(x+3)(x+4)}{(x+2)(x-2)}$

2a. $x = 2$ and $x = -2$

4a. $\frac{3x-10}{x-4}$

5. *Hint:* See Example C of this lesson.

5a. $\frac{7x-7}{x-2}$

6d. *Hint:* Look at a table of values for each graph if your graphing calculator doesn't display the holes.

7a. $y = \frac{x+2}{x+2}$

LESSON 8.8

1a. $\frac{x(x+2)}{(x-2)(x+2)} = \frac{x}{x-2}$

1c. $\frac{3x(x-2)}{(x-4)(x-2)} = \frac{3x}{x-4}$

2a. $(x+3)(x-3)(x-2)$

2c. $(x+2)(x-2)(x+3)$

3a. $\frac{(2x-3)(x+1)}{(x+3)(x-2)(x-3)}$

3b. $\frac{-x^2+6}{(x+2)(x+3)(x-2)}$

4a. $\frac{1}{x-3}$

4b. $\frac{(x-4)(x-2)}{x+4}$

5a. $\frac{2(x-2)}{x+1}$

7a. *Hint:* Factor the denominator.

8a. *Hint:* Look at the denominators.

8b. $\frac{x^2-2}{(x-1)(x-2)}$

10a. $y = \frac{v^2+25v+1250}{25v}$

11b. translation right 3 units and up 1 unit

13a. $370.09

CHAPTER 8 REVIEW

1a.

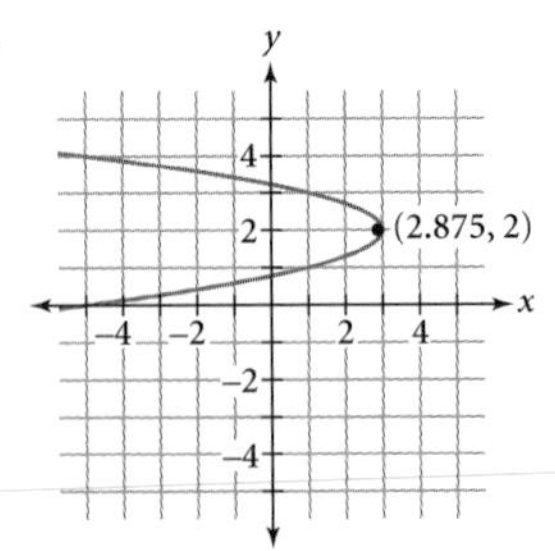

1b.

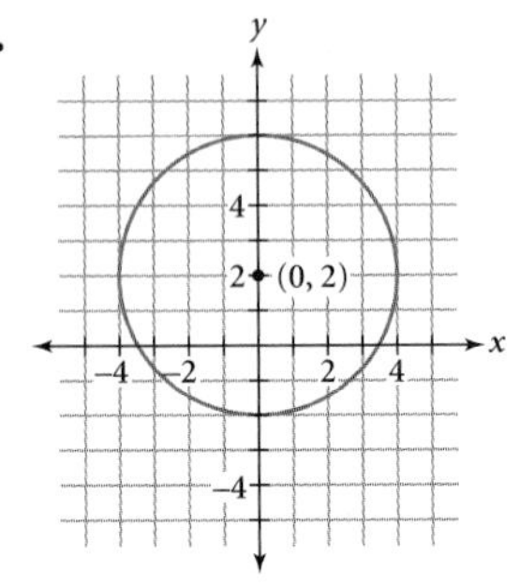

1c.

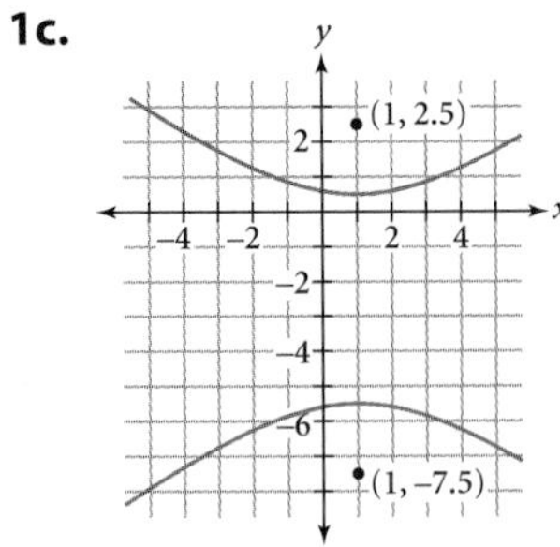

1d.

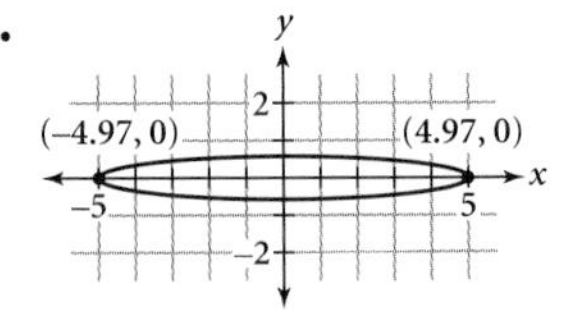

2a. $\left(\frac{x-5}{3}\right)^2+\left(\frac{y+2}{4}\right)^2=1$

2b. $(5,-2)$; $\left(5,-2+\sqrt{7}\right)$ and $\left(5,-2-\sqrt{7}\right)$

2c. $16x^2+9y^2-160x+36y+292=0$

3a. $y=\pm0.5x$

3b. $x^2-4y^2-4=0$

3c. $d=0.5x-\sqrt{\frac{x^2}{4}-1}$

3d. As x-values increase, the curve gets closer to the asymptote.

x	2	10	20	100
d	1	0.101	0.050	0.010

4. $(x+4)^2+(y-1)^2=25$. The graph is a circle with center $(-4, 1)$ and radius 5.

5. $\left(\frac{y-4}{2}\right)^2=x-3$; vertex: $(3, 4)$; focus: $(4, 4)$; directrix: $x=2$

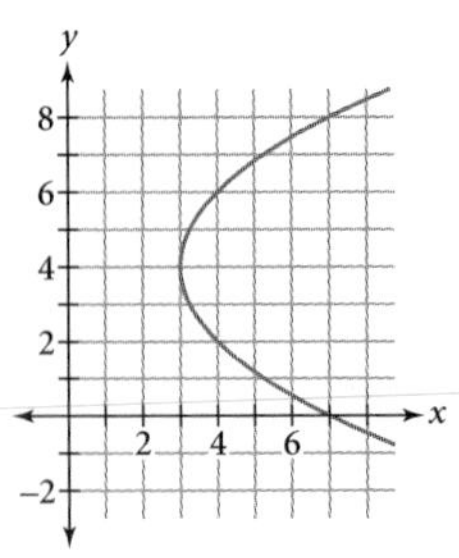

6. 10 oz

7a. $y=1+\frac{1}{x+2}$, or $y=\frac{x+3}{x+2}$

7b. $y=-4+\frac{1}{x}$, or $y=\frac{-4x+1}{x}$

8.

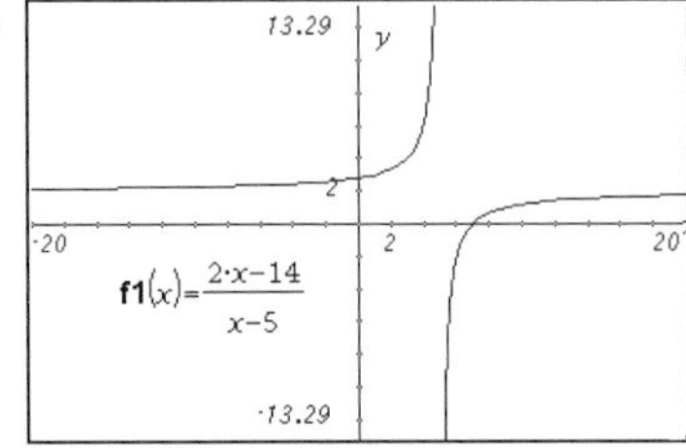

horizontal asymptote: $y=2$; vertical asymptote: $x=5$

9. Multiply the numerator and denominator by the factor $(x+3)$.

$$y=\frac{(2x-14)(x+3)}{(x-5)(x+3)}$$

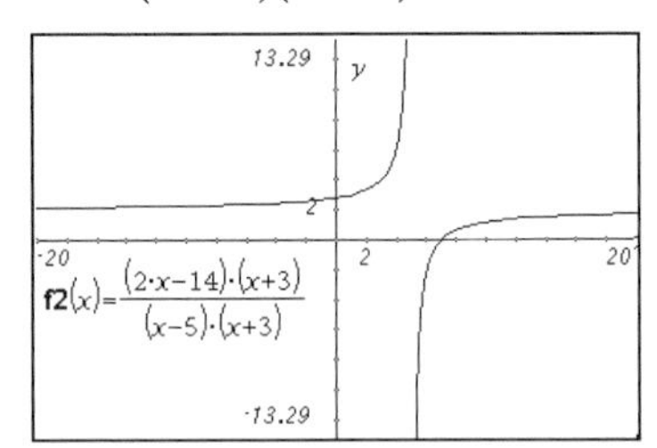

10. 23.3 mi/h and then 43.3 mi/h

11a. $\frac{3x^2+8x+3}{(x-2)(x+1)(x+2)}$

11b. $\frac{3x}{x+1}$

11c. $\frac{(x+1)^2(x-1)}{x(x-2)}$

12. $(0.5, 1.936)$, $(0.5, -1.936)$, $(-2, 0)$

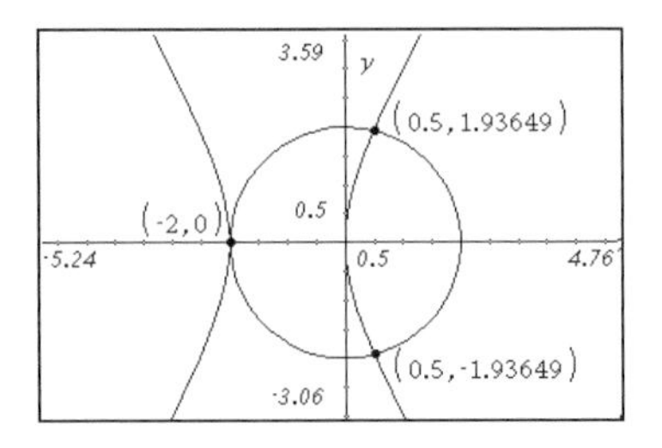

REFRESHING YOUR SKILLS FOR CHAPTER 9

1a. geometric; common ratio = 0.5

1c. Neither; this is the Fibonacci sequence.

2a. $a_1 = 14$ and $a_n = 0.5 \cdot a_{n-1}$ where $n \geq 2$

2b. $a_1 = 47$ and $a_n = a_{n-1} - 6$ where $n \geq 2$

2c. $a_1 = 1, a_2 = 1$, and thereafter $a_n = a_{n-1} + a_{n-2}$ where $n \geq 3$

3a. $u_{10} \approx 115.4968$

4a. The number of bulbs would decrease and level off at 4,000 bulbs after a few years.

LESSON 9.1

2. $S_1 = 2, S_2 = 8, S_3 = 18, S_4 = 32, S_5 = 50$

3b. $-2 + 1 + 6; 5$

4. *Hint:* 6 divides each multiple without a remainder. Find the value of u_{50}.

5. $S_{75} = 5700$

6b. $S_{75} = 5625$

7a. $u_{46} = 229$

10a. 3, 6, 9, 12, 15, 18, 21, 24, 27, 30

10b. $u_1 = 3$ and $u_n = u_{n-1} + 3$ where $n \geq 2$

10d. *Hint:* Write an equation that shows the sum of the cans in the display is 288.

11a. *Hint:* Find the partial sums recursively until you have a total as close to 1000 as possible without going over.

11e. Each of these 1000 values is 2000 more than the corresponding value between 1 and 1000, so add 1000(2000) to the first sum.

14b. $u_n = 9.8n - 4.9$

14d. 490 m

15. $S_x = x^2 + 64x$

17a. 576,443 people

19a. $81, 27, 9, 3, 1, \frac{1}{3}$

19b. $u_1 = 81$ and $u_n = \frac{1}{3}u_{n-1}$ where $n \geq 2$

21a. $y = 2 + \dfrac{\frac{5}{2}}{x - \frac{1}{2}}$

21b. $x = \frac{1}{2}, y = 2$

21c. $\left(\frac{3}{2}, \frac{9}{2}\right)$

LESSON 9.2

2a. $0.47 + 0.0047 + 0.000047 + \cdots$

2b. $u_1 = 0.47, r = 0.01$

2c. $S = \frac{47}{99}$

4. *Hint:* Write an equation using the formula from page 528 and solve for r.

5. $u_1 = 32{,}768$

6a. $S_{10} \approx 209.767$

7a. 96, 24, 6, 1.5, 0.375, 0.09375, 0.0234375, 0.005859375, 0.00146484375, 0.0003662109375

8. *Hint:* Find the distance it falls and the distance it goes up separately. How far does it go up on the first bounce?

9b. $62,500,000

10. *Hint:* Think about summing the distances, where hops to the right are positive and hops to the left are negative.

11a. $\sqrt{2}$ in.

11b. 0.125 in^2

12a. *Hint:* How does the perimeter of one small triangle relate to the perimeter of a small triangle in the previous stage? How does the total perimeter in any stage relate to the perimeter in the previous stage?

12b. *Hint:* What fraction of the shaded area in stage 1 is shaded in stage 2? What fraction of the shaded area in stage 2 is shaded in stage 3?

14A. iv.

15a. $56,625

LESSON 9.3

1a. $u_1 = 12, r = 0.4, n = 8$

1b. $u_1 = 75, r = 1.2, n = 15$

2b. u_{10}

2d. $S_7 = 887.3125$

4b. $u_1 = 3.2, r = 1.5, S_7 = 102.95$

4c. $u_1 = 5.7, d = 2.5, S_{27} = 1031.4$

5a. 3069

5b. 22

7a. $S_{10} = 15.984375$

8a. $25,342.39

12. *Hint:* What type of sequence does each prize represent?

13a. neither

13b. $S_8 \approx 2.717857$

15a. *Hint:* How many games will each team member play?

16. $637.95

18. $f(x) = 3x^3 + 16x^2 + 27x - 26$

CHAPTER 9 REVIEW

1a. $u_{128} = 511$ **1b.** $u_{40} = 159$

1c. $u_{20} = 79$ **1d.** $S_{20} = 820$

2a. $u_{11} \approx 17.490$

2b. $S_{10} \approx 515.687$

2c. $S_{20} \approx 605.881$

2d. The partial sum approaches 625.

3a. 144; 1,728; 20,736; 429,981,696

3b. $u_1 = 12$ and $u_n = 12u_{n-1}$ where $n \geq 2$ and where $n = \frac{d}{5} + 1$, where d is the number of the day

3c. $u_n = 12^n$

3d. approximately 1.2×10^{14}

4a. -6639.7

4b. $S_{67} = \sum_{n=1}^{67} [125.3 - 6.8(n - 1)]$

5a. approximately 56.49 ft

5b. 60 ft

6a. geometric; $r = \frac{1}{2}$

6b. 0.0039 ft; 0.996 ft to the right

6c. approximately 9.5×10^{-7} ft; approximately 1 ft to the right of its starting point

6d. $u_n = 0.5^n$; $S_n = \dfrac{0.5(1 - 0.5^n)}{1 - 0.5}$

6e. the point 1 ft to the right from its starting point

7a. $S_{10} \approx 12.957$; $S_{40} \approx 13.333$

7b. $S_{10} \approx 170.478$; $S_{40} \approx 481{,}571.531$

7c. $S_{10} = 40$; $S_{40} = 160$

7d. for $r = 0.7$

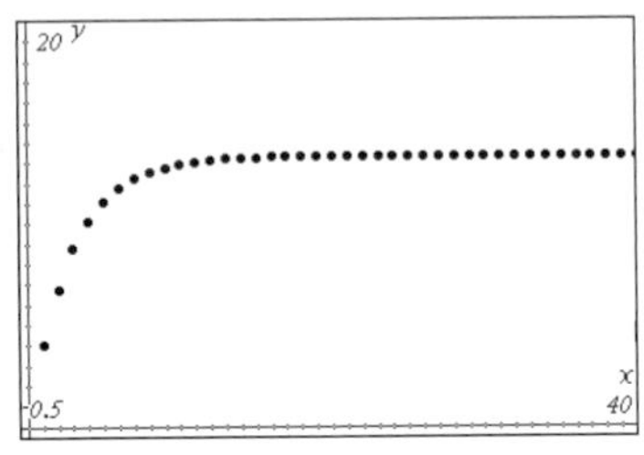

for $r = 1.3$

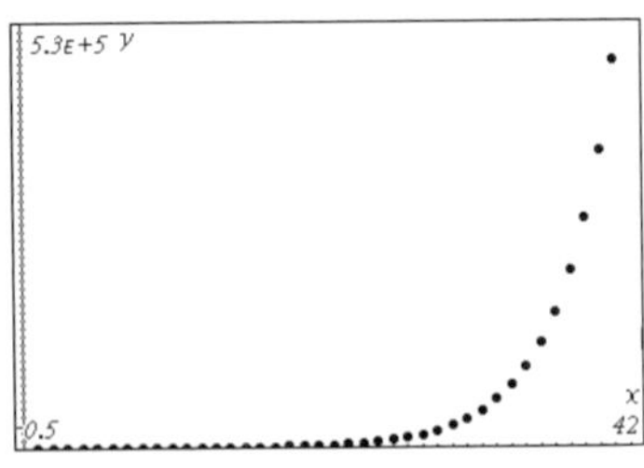

for $r = 1$

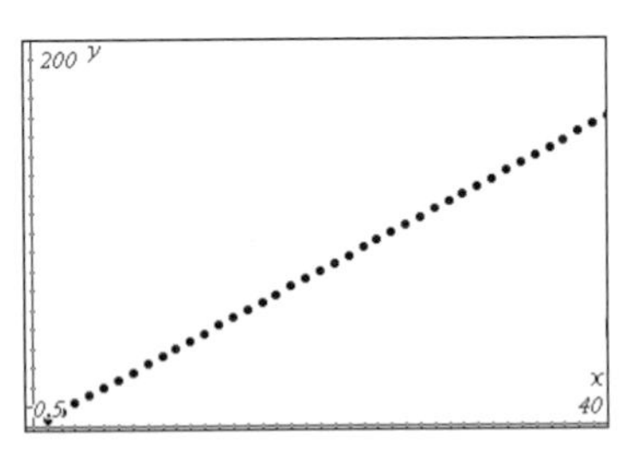

7e. 0.7

8a. 0.8888888888

8b. 0.8888888888888888 **8c.** $\frac{8}{9}$

9a. $\bar{x} = 31.\bar{6}$; median = 33; mode = 34

9b.

0 10 20 30 40 50
Number of students

9c. $s \approx 7.57$. The data are fairly spread out from the mean. If the standard deviation were smaller, the data would be more closely grouped around $31.\bar{6}$ students per class.

10a. $y = 2|x|$ **10b.** $y = 2|x - 4|$

10c. $y = 2|x - 4| - 3$

11a. 91.45 million mi **11b.** 94.55 million mi

11c. 93 million mi **11d.** 1.55 million mi

11e. 92.99 million mi **11f.** $\left(\frac{x}{93}\right)^2 + \left(\frac{y}{92.99}\right)^2 = 1$

12a. Not possible. The number of columns in $[A]$ must match the number of rows in $[B]$.

12b. Not possible. To be added, matrices must have the same dimensions.

12c. $\begin{bmatrix} -3 & 1 \\ 1 & -5 \end{bmatrix}$ **12d.** $\begin{bmatrix} -2 & 6 \\ -10 & 0 \\ -7 & 3 \end{bmatrix}$ **12e.** $\begin{bmatrix} 5 & -3 \\ -1 & 9 \end{bmatrix}$

13a. $\left(\frac{y}{5}\right)^2 - \left(\frac{x}{2}\right)^2 = 1$; hyperbola

13b. $(y + 2)^2 = \dfrac{(x - 2)}{\frac{2}{5}}$; parabola

13c. $(x + 3)^2 + (y - 1)^2 = \frac{1}{4}$; circle

13d. $\left(\dfrac{x - 2}{\sqrt{8}}\right)^2 + \left(\dfrac{y + 2}{\sqrt{4.8}}\right)^2 = 1$, or $\left(\dfrac{x - 2}{2\sqrt{2}}\right)^2 + \left(\dfrac{y + 2}{2\sqrt{1.2}}\right)^2 = 1$; ellipse

14. $y = 0.00115x^2 + 4$

15a. $x \approx 1.64$ **15b.** $x \approx -0.66$

15c. $x = 15$ **15d.** $x \approx 2.57$

15e. $x \approx 17.78$ **15f.** $x = 3$

15g. $x = 2$ **15h.** $x = 495$

15i. $x \approx \pm 4.14$

16a. $-4 + 9i$ **16b.** $-5 - 5i$

16c. $3 - 4i$ **16d.** $\frac{1}{5} - \frac{7}{5}i$

REFRESHING YOUR SKILLS FOR CHAPTER 10

1a. $\frac{5}{10}$, or $\frac{1}{2}$

1c. $\frac{4}{10}$, or $\frac{2}{5}$

2a. $\frac{10}{36}$, or $\frac{5}{18}$

2b. 7 is most likely; probability of 7 is $\frac{6}{36}$, or $\frac{1}{6}$.

3a. $\frac{26}{52}$, or $\frac{1}{2}$

3b. $\frac{8}{52}$, or $\frac{2}{13}$

3c. *Hint:* How many black suits are there?

LESSON 10.1

1a. $\frac{6}{15} = 0.4; \frac{7}{15} \approx 0.4\overline{6}; \frac{2}{15} \approx 0.1\overline{3}$

1b. experimental

2c. $\frac{228}{435} \approx 0.524$

3a. $\frac{4}{14} \approx 0.286$

3e. *Hint:* Think about graphing the line $x + y = 2$.

4c. *Hint:* Use the graph in Example B of this lesson to help you.

7. Answers will vary. Each of these methods has shortcomings.

7. i. Middle numbers (3–7) are more common than getting only 1 or 2 or 8 or 9 heads in one trial of dropping pennies.

7. ii. Very few pencils will be 0 or 1 in. long; students throw away their pencils long before that.

7. iii. This is the best method, although books tend to open to pages that are used more than others.

9a. $4; \frac{4}{36} \approx 0.\overline{1}$

9b. $5; \frac{5}{36} \approx 0.13\overline{8}$

9c. $10; \frac{10}{36} \approx 0.2\overline{7}$

9d. $2; \frac{2}{36} \approx 0.0\overline{5}$

9e. $10; \frac{10}{36} \approx 0.2\overline{7}$

11c. $\frac{44}{144}$

11f. *Hint:* How much area does the line itself occupy?

13a. 270

13c. $\frac{270}{1380} \approx 0.196$

14. *Hint:* Where is each type of face located on the original cube?

16. $x^4 - 4x^3y + 6x^2y^2 - 4xy^3 + y^4$

19a.

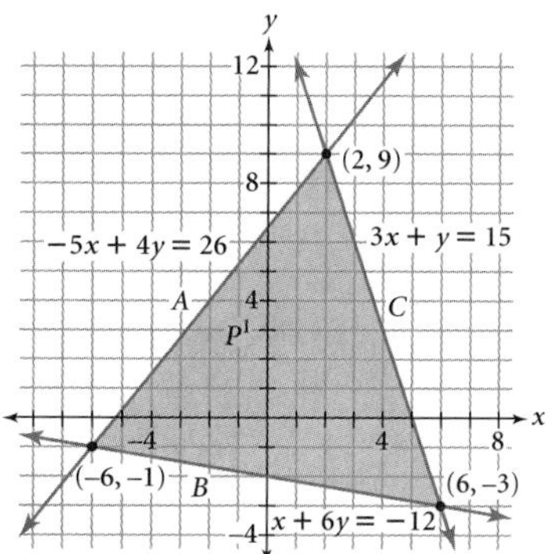

19b. $(2, 9), (-6, -1), (6, -3)$

19c. 68 units^2

20. a parabola with focus $(3, 0)$ and directrix $y = 6$; $y = -\frac{1}{12}x^2 + \frac{1}{2}x + \frac{9}{4}$

21a. Set i should have a larger standard deviation because the values are more spread out.

21b. i. $\bar{x} = 35, s \approx 22.3$

21b. ii. $\bar{x} = 117, s \approx 3.5$

21c. The original values of $\bar{x}$ and s are multiplied by 10.

21c. i. $\bar{x} = 350, s \approx 223.5$

21c. ii. $\bar{x} = 1170, s \approx 35.4$

21d. The original values of $\bar{x}$ are increased by 10, and the original values of s are unchanged.

21d. i. $\bar{x} = 45, s \approx 22.3$

21d. ii. $\bar{x} = 127, s \approx 3.5$

LESSON 10.2

1.

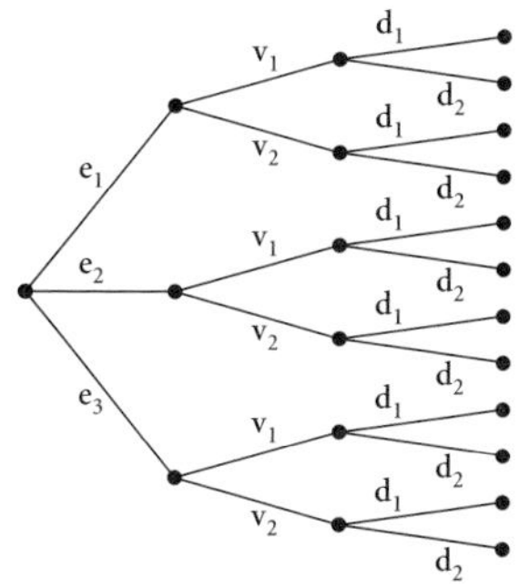

2. $P(\text{a}) = 0.675; P(\text{b}) = 0.075; P(\text{c}) = 0.05; P(\text{d}) = 0.2; 1$

3. *Hint:* 0.42 is the product of 0.6 and the probability of path a.

4a. $\frac{1}{8} = 0.125$

4c. *Hint:* Which paths show Celina and one other student were successful? Which show two successful students?

6a. $\frac{182}{420} = 0.4\overline{3}; \frac{98}{420} = 0.2\overline{3}; \frac{98}{420} = 0.2\overline{3}; \frac{42}{420} = 0.1$

6b. no, because the probabilities of the four paths are not all the same

6c. $\frac{420}{420} = 1$

7a. 24 **7f.** $\frac{12}{24} = 0.5$

9c. 16 **9e.** 1024

11. *Hint:* Guessing randomly is like flipping a coin for each question on the quiz.

11g. 0.3125

12a.

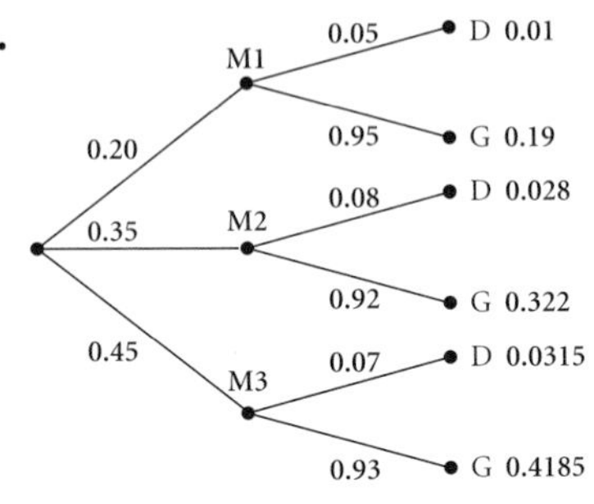

12d. ≈ 0.403

14. *Hint:* This is like flipping a coin with sides boy and girl instead of heads and tails.

16a. $\frac{349}{798} \approx 0.437$

16b. $\frac{512}{1424} \approx 0.360$

16c. The events are dependent, because $P(\text{10th grade} \mid \text{female}) \neq P(\text{10th grade})$. The probability of choosing a 10th grader from the female students is greater than the probability of choosing a 10th grader from all students.

17. 64

18a. $-3 + 2i$

18b. $2 + 24i$ **18c.** $\frac{18}{29} + \frac{16}{29}i$

LESSON 10.3

2a. 0.25 **2b.** 0.12

2c. $\frac{0.15}{0.15 + 0.12} \approx 0.\overline{5}$ **2d.** 0.37

3.

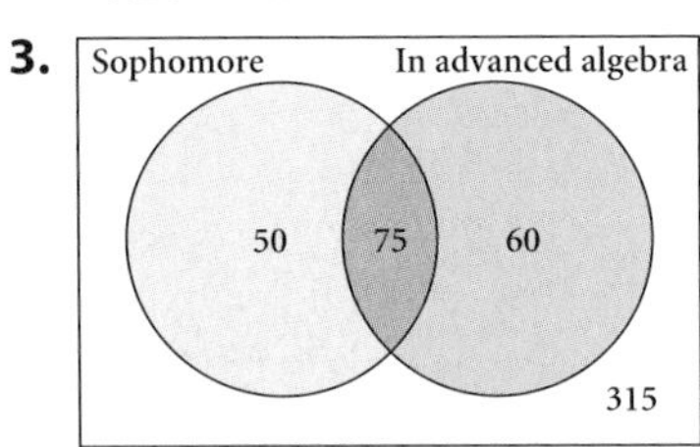

4. *Hint:* Does $P(S) \cdot P(A) = P(S \text{ and } A)$?

5a. yes, because they do not overlap

5b. No. $P(\text{A and B}) = 0$. This would be the same as $P(A) \cdot P(B)$ if they were independent.

6a.

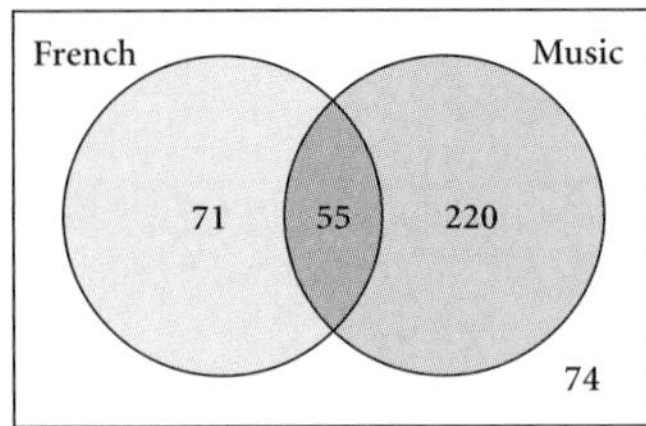

6b. approximately 13%

6c. 74

7a. *Hint:* $P(A \mid B) = \frac{P(A \text{ and } B)}{P(B)}$

8. $0 \leq P(A \text{ and } B) \leq 0.4$, $0.5 \leq P(A \text{ or } B) \leq 0.9$. The first diagram shows $P(A \text{ and } B) = 0$ and $P(A \text{ or } B) = 0.9$. The second diagram shows $P(A \text{ and } B) = 0.4$ and $P(A \text{ or } B) = 0.5$.

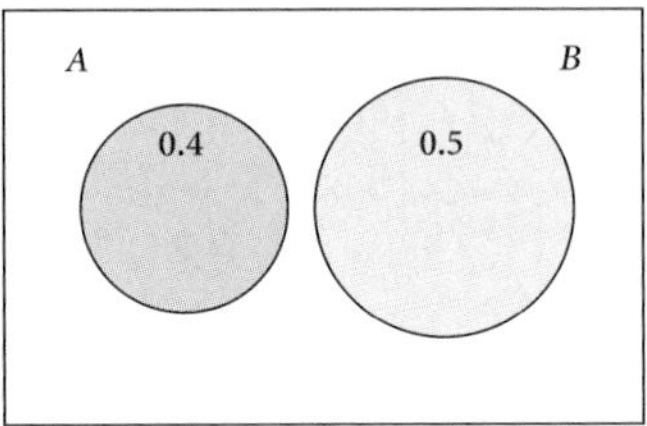

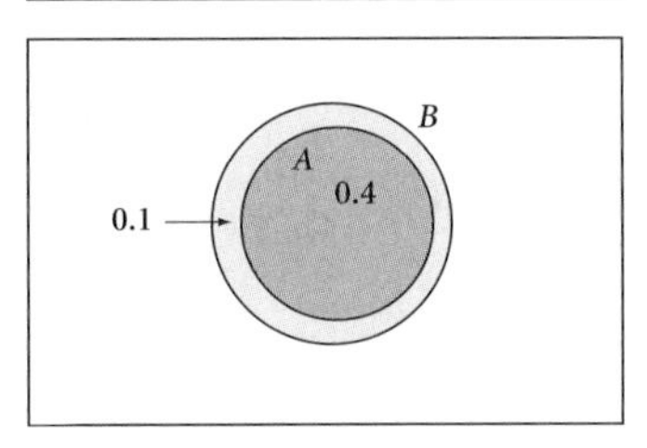

9. *Hint:* Find the new numbers of students in each section before the 20 sophomores move.

10. *Hint:* See the related Science Connection.

10a. yellow

11b. *Hint:* Because all three friends' phone usage is independent, $P(A) \cdot P(B) \cdot P(C) = P(\text{A and B and C})$.

12a. $\frac{280}{1500} = 0.18\overline{6}$ **12b.** $\frac{775}{1500} = 0.51\overline{6}$

12c. $\frac{145}{355} \approx 0.408$ **12d.** $\frac{145}{775} \approx 0.187$

15c. $2xy^2\sqrt{15xy}$

LESSON 10.4

1a. Yes; the number of children will be an integer, and it is based on a random process.

1b. No; the length may be a non-integer.

1c. Yes; there will be an integer number of pieces of mail, and it is based on random processes of who sends mail when.

2. *Hint:* A geometric variable is a count of independent trials before something happens.

4. *Hint:* Make a tree diagram.

5a. 0 **5b.** 0

6d. -0.25

7a. *Hint:* The probability of getting a section A ticket is 0.16 and the ticket is worth $35.

7b. $0.\overline{6}$

8a. Answers will vary.

8b. Sample answer: Assign each of the letters in the word CHAMPION a different number from 1 to 8. Randomly generate numbers between 1 and 8. Count how many digits you must generate until you have at least one of each number.

8c. Answers will vary.

8d. Answers will vary. Theoretically, it should be about 22 boxes.

8e. Answers will vary. Average number of boxes should be about 22 boxes.

9a. 0.2

9c.

Successful hits	Probability
0	0.2
1	0.16
2	0.128
3	0.1024
4	0.08192
5	0.065536

12. *Hint:* Start a tree diagram to help you see the pattern.

13. 1

17. 44

LESSON 10.5

1a. Yes. Different arrangements of scoops are different.

1b. No. We are not counting different arrangements separately.

1c. No. Repetition is not allowed in permutations.

1d. No. Repetition is not allowed in permutations.

2a. 12 **2e.** 14,280 **2f.** $n(n-1)$

3a. 210 **3c.** $\frac{(n+2)!}{2}$

4a. 24 **4c.** 256

5a. 10,000; $27.\overline{7}$ h

6. *Hint:* Can you find consecutive integers with product 720?

8. 60

9a. 40,320

10. *Hint:* Use the information in the table to determine how long it takes to list one permutation.

13b. $\frac{1}{4} = 0.25$ **13e.** $\frac{4}{16} = 0.25$

16a. $\frac{1}{8} = 0.125$

18. *Hint:* Each box is an open cube. The box side lengths form a geometric sequence.

LESSON 10.6

1a. 120 **1d.** 1

2b. 35

3. *Hint:* Use the distributive property to rewrite the left side of the equation. Use a reciprocal trigonometric identity to rewrite $\cot A$, then simplify. Use a Pythagorean identity to complete the proof.

3a. $\frac{{}_7P_2}{2!} = {}_7C_2$

5. *Hint:* What if $n = 7$? Find another good choice for n.

7a. 35 **7b.** $\frac{20}{35} \approx 0.571$

8. *Hint:* What is the difference between permutations and combinations? How does this apply to the lock?

9a. 4 **9b.** 8 **9c.** 16

9d. The sum of all possible combinations of n things is 2^n; $2^5 = 32$.

10b. $\frac{1000}{{}_{47}C_6} \approx 0.000093$

11. *Hint:* When you draw a chord, you are choosing two points out of all the points on the circle.

12a. 26,466,926,850 ways

13a. *Hint:* Assume that one of your four tickets has been awarded first prize. How many ways are there to draw the remaining three tickets from the pool of tickets if the first, second, and third prize tickets are not in the pool?

13c. $1 - \frac{{}_{48}C_4}{{}_{50}C_4} \approx 0.155$

14b. *Hint:* First multiply $(x+y)(x+y)$ and then multiply the result by $(x+y)$.

15. 15 speeds

16a. 0.0194 is the probability that someone is healthy but tests positive.

17. $C < \frac{157}{4}$, or $C < 39.25$

LESSON 10.7

1a. x^{47} **1b.** $5{,}178{,}066{,}751x^{37}y^{10}$

2a. 0.75 **2b.** 0.0625

3a. 0.299 **3b.** 0.795, 0.496

3c. 0.203, 0.502, 0.791

3d. Both the "at most" and "at least" numbers include the case of "exactly." For example, if "exactly" 5 birds (0.165) is subtracted from "at least" 5 birds (0.203), the result (0.038) is the same as $1 - 0.962$ ("at most" 5 birds).

3e. The probability that at least 5 birds survive is 20.3%.

4. What is the probability of exactly 35 successes in 50 trials?

6a. HH, HT, TH, TT

6b. HH, HT, TH, TT

6c. Both diagrams would look the same.

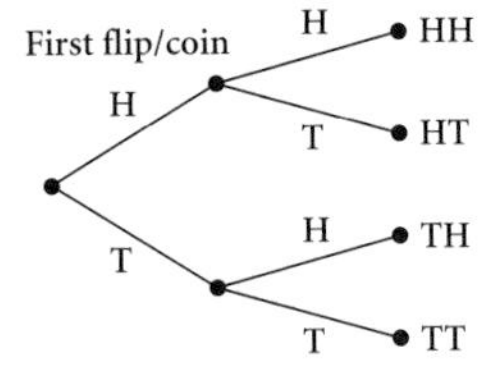

6d. ${}_2C_0 = 1$ is the number of ways of getting 0 tails, ${}_2C_1 = 2$ is the number of ways of getting 1 tail, and ${}_2C_2 = 1$ is the number of ways of getting 2 tails.

6e. There is 1 way of getting 2 heads; there are 2 ways of getting 1 head and 1 tail; and there is 1 way of getting 2 tails.

7a. $x^4 + 4x^3y + 6x^2y^2 + 4xy^3 + y^4$

7c. $8x^3 + 36x^2 + 54x + 27$

8a. ${}_{50}C_{40} \cdot p^{10}q^{40}$ or ${}_{50}C_{10} \cdot p^{10}q^{40}$

9. *Hint:* Think about which term in the binomial expression represents this situation.

9c. $f(x) = {}_{30}C_x(0.97)^{30-x}(0.03)^x$

11a. 0.000257

12. *Hint:* Refer to Example B of this lesson.

14a. 0.025 **14c.** 0.0002

15a. Sum of the first 2 terms: 2, 2, 2, 2; Sum of all the terms: 2, 2.25, ≈ 2.370, ≈ 2.441

16. *Hint:* Consider two cases: First, if the new student is not selected, and second, if the new student is selected.

19a. *Hint:* Graph data in the form (*distance, period*), (log (*distance*), *period*), (*distance*, log (*period*)), and (log (*distance*), log (*period*)), and identify which is the most linear. Find an equation to fit the most linear data you find, then substitute the appropriate variables (*distance, period*, log (*distance*), log (*period*)) for *x* and *y*, and solve for *y*. You should find that $period \approx distance^{1.50} \times 10^{-9.38}$.

19b. *Hint:* Substitute the period and distance values given in the table into your equation from 19a. Errors are most likely due to rounding.

CHAPTER 10 REVIEW

1. Answers will vary depending upon whether you are generating random decimal numbers or random integers. To generate random decimal numbers, you might look at a random-number table and place the decimal point after the first digit in each group of numbers. Alternatively, you could use a calculator command, such as **10*rand(numTrials)** on the TI-Nspire, or **10*rand** on the TI-84 Plus. To generate random integers, you might number 11 chips or slips of paper and randomly select one. Alternatively, you could use a calculator command, such as **randInt(0, 10)**.

2a, b.

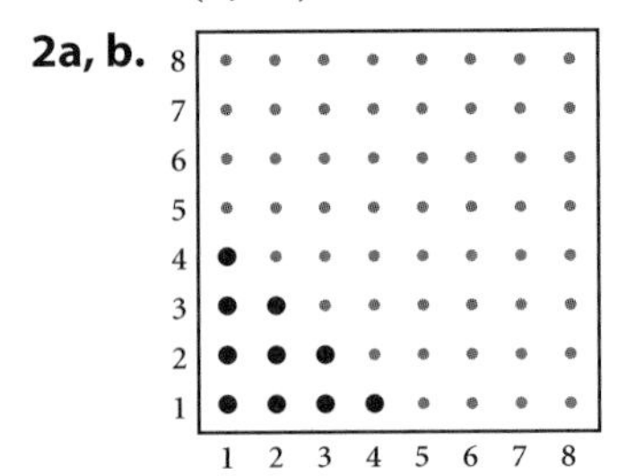

2c. $\frac{10}{64} = 0.15625$ **2d.** $\frac{49}{64} \approx 0.766$

3a. 0.5 **3b.** 17.765 units^2

4a.

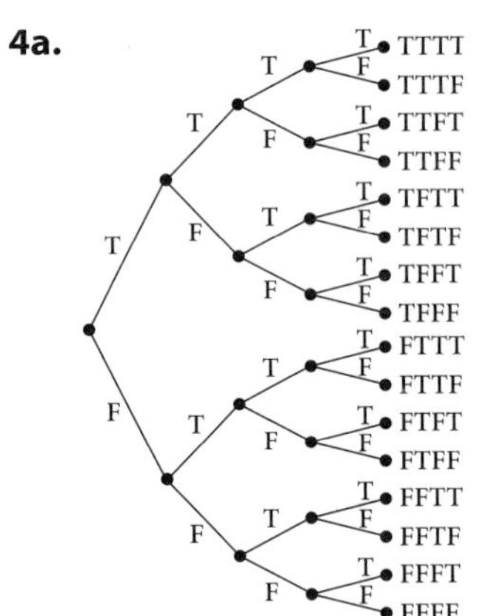

4b. 4

4c. Because the order in which the true and false answers occur doesn't matter, use combinations: ${}_4C_3 = 4$.

4d. $\frac{1}{2} = 0.5$

5a.

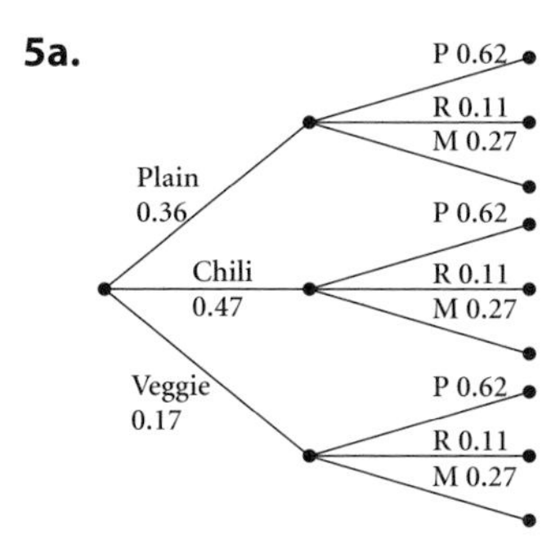

5b. 0.0517

5c. 0.8946

5d. 0.3501

6a. Ice cream: 254; Whipped cream: 118; Total: 23, 55, 122, 172, 372

6b. $\frac{37}{55} = 0.6\overline{72}$ **6c.** $\frac{37}{122} \approx 0.303$

6d. $\frac{18}{254} \approx 0.071$ **6e.** $\frac{118}{372} \approx 0.317$

7. 110.5

8.

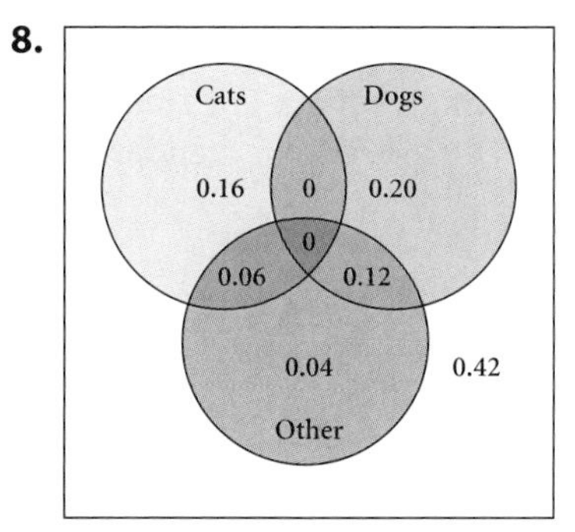

9. approximately 0.044

10a. 1 **10b.**  $\frac{x^{99}}{12^{99}}$ **10c.** $293,930a^{12}b^9$

CHAPTER 11

REFRESHING YOUR SKILLS FOR CHAPTER 11

1b. $s \approx 9.91$

2a. mean $\approx$ 21.208 kg, median = 20 kg, IQR = 5.8 kg

3a. 105 cm: 20th percentile; 115 cm: $93.\overline{3}$ percentile

LESSON 11.1

1a–3a. Answers will vary. Sample answers are given.

1a. This is a survey (more specifically, an interview survey). Nick chose the subjects. They choose whether to respond.

2a. The treatment is that the person has experienced the movie in question. The measure is how much he or she liked the movie.

3a. The subjects who choose this movie are more likely to enjoy this type of movie than if you had selected a random group of people to watch the movie.

6a. Use an experiment to control as many of the factors as you can. Purchase the same size bag of popcorn for each brand, and assume that there are approximately the same number of kernels in each bag. Using one microwave, pop one bag of each brand and count the unpopped kernels in each. Record the deviation from the mean. In other words, if you have five brands and you found 20, 35, 15, 21, and 24 unpopped kernels, then you calculate the mean (23) and record −3, 12, −8, −2, 1. Now do the same with nine different microwaves. The brand with the lowest sum of all these deviations is the winner.

7a. $A = 12$ units^2

8. approximately 0.137

10c. 20th percentile

LESSON 11.2

1. Distribution A: $\frac{1}{8}$

1. *Hint:* Distribution D: You can divide the figure into a trapezoid and a rectangle, or a right triangle and two rectangles.

2. Distribution A: $\frac{9}{16}$

3. Answers will vary. Distribution A: $\sqrt{8} = 2\sqrt{2}$

4. Answers will vary. Distribution A: approximately 2.6

5a.

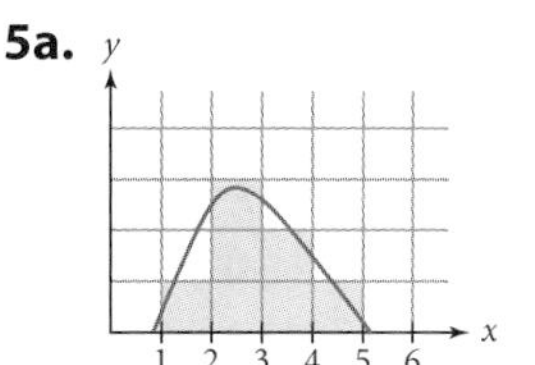

7a. false

9. *Hint:* For 9a, create a random list of 100 numbers from 0 to 1, and store it in a list, A. Enter values of $(A)^2$ in a second list, and graph a histogram of these values. You may want to rerandomize the first list several times, and then generalize the shape of the histogram. See Calculator Notes 1L and 11D for help with these calculator functions. Use a similar process for 9b and c.

11a. Sample answer: The histogram is symmetric. The data set that might produce this graph is the percentile rankings of students' standardized test scores. The range of values on the x-axis is 0% to 100%.

13a.

x	0–3	3–6	6–9	9–12	12–15	15–18
$P(x)$	0.015	0.046	0.062	0.092	0.2	0.138

x	18–21	21–24	24–27	27–30	30–33
$P(x)$	0.123	0.154	0.092	0.062	0.015

14b.

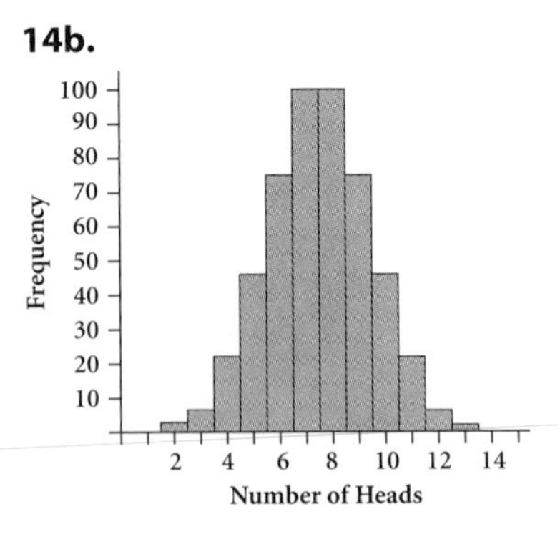

15. 0.022

16a. Big Bird has a more predictable arrival time. The standard deviation for Big Bird is only 1.58 minutes versus 6.85 minutes for Flying Bus.

17. *Hint:* See page 84 to review the meaning of *skewed left* and *skewed right*.

LESSON 11.3

1a. *Hint:* Enter the functions as f_1 and f_2. Then graph or create a table of values, and confirm that they are the same.

1b. $y \approx 0.242$ and $n(x, 0, 1) \approx 0.242$

2a. $\mu = 47, \sigma = 5$

2c. $\mu = 5.5, \sigma = 0.31$

3a. *Hint:* To estimate the standard deviation, look for the inflection points.

4a. $y = \dfrac{1}{2.5\sqrt{2\pi}}\left(\sqrt{e}\right)^{-[(x-18)/2.5]^2}$

6a.

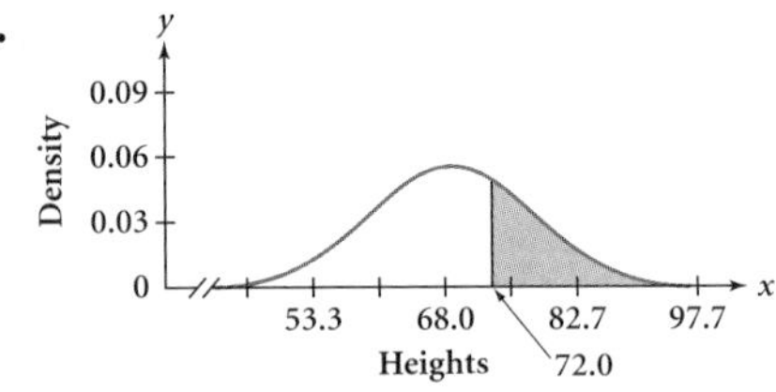

8a. *Hint:* Ninety percent of the area must be above the value 12.

9.

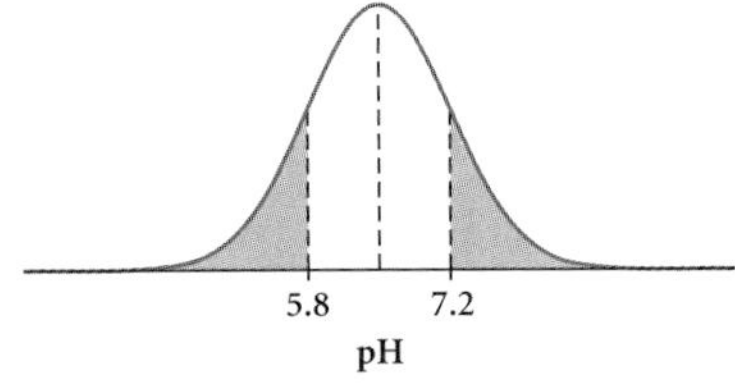

Shade the area greater than or less than 1 standard deviation away from the mean.

10a. $\mu \approx 165; \sigma \approx 5.91$

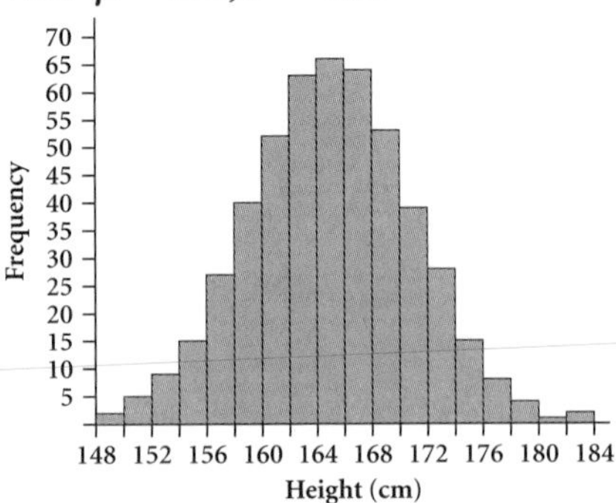

13a. *Hint:* Consider what the mean and standard deviation tell you about the distribution of test scores. Can you be sure which test is more difficult?

13c. *Hint:* Determine how many standard deviations each student's score is from the mean. This will tell you how each student scored relative to other test-takers.

15. *Hint:* Refer to page 606 in Lesson 10.7.

16. *Hint:* For 16a, substitute each point's coordinates into the standard equation of a parabola, $y = ax^2 + bx + c$. You will need to solve a system of three equations to find the *a, b,* and *c* values. For 16b, you will follow a similar procedure using the standard equation of a circle.

LESSON 11.4

2a. $z = 1$

2c. $z \approx 1.57$

3a. 122.6

3c. 131.96

5a. $z = 1.8$

6. *Hint:* Refer to Example B of this lesson.

7a. (3.058, 3.142)

9. *Hint:* Think about where each value appears in the formula and how increasing or decreasing each value affects the result.

12a. 68%

13a. *Hint:* The area under the probability curve must total 1 unit.

13c. 0.1875

15a. *Hint:* Plan 1 can be represented by an arithmetic series.

LESSON 11.5

1b. −0.95

2a. Sample answer:

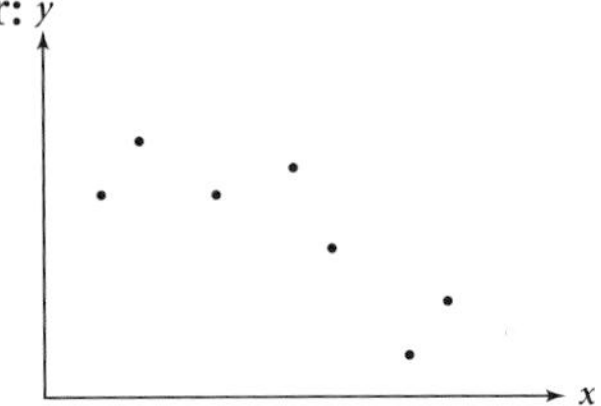

3d. -0.9723

3f. *Hint:* Do the points seem to decrease linearly?

4a. *Hint:* The explanatory variable corresponds to the independent variable.

5a. *Hint:* What lurking variable might be correlated with television ownership?

8a.

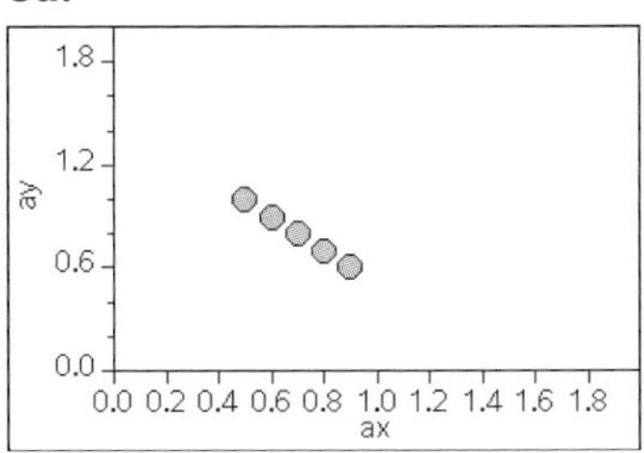

$r = -1$. This value of r implies perfect negative correlation consistent with the plot.

9. *Hint:* The correlation coefficient gives you information about the linearity of the data.

10a. The correlation coefficients for (*year, country*) and (*year, oldies*) are -0.972 and 0.659, respectively.

11. 12

15. *Hint:* The average speed is $\frac{\textit{total distance}}{\textit{total time}}$. If the distance is d, then the total distance is $2d$. To find the total time, use the relationship $t = \frac{d}{r}$.

LESSON 11.6

1a. $\bar{x} = 1975$

1c. $s_x \approx 18.71$

2b. $\hat{y} = 5 - 22.5(x - 0.31)$, or $\hat{y} = 11.975 - 22.5x$

2c. $\hat{y} = 6 - 0.36(x - 88)$, or $\hat{y} = 37.68 - 0.36x$

3a. 0.3166, -0.2292, 0.1251, 0.1794, -1.3663, 0.9880

3c. 0.1002, 0.0525, 0.0157, 0.0322, 1.8667, 0.9761

4a. 31.02%

4c. 46.23%

5a.

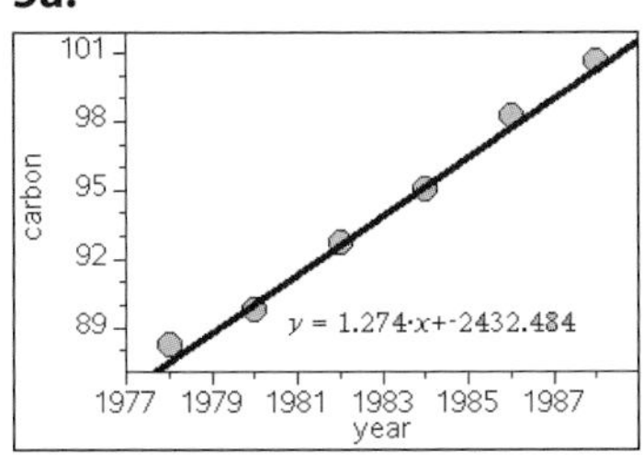

$\hat{y} \approx 1.2741x - 2432.5$

6b. *Hint:* Is the correlation coefficient close to 1?

7a. $\hat{y} \approx 7.56 + 0.4346x$

8. *Hint:* You may want to consider how many points are used to calculate each line of fit, whether each is affected by outliers, and which is easier to calculate by hand.

10a. $y = 1000$ **10b.** $y = \frac{100}{\sqrt{x}}$

12. *Hint:* Use the x-intercepts to write the equation in factored form, then use the y-intercept to solve for the stretch factor.

CHAPTER 11 REVIEW

1. Answers will vary. Sample answers are given.

1a. Observational study; for an experimental study, the dogs would be randomly assigned to one of the two types of food.

1b. Survey; for an observational study, the researcher could have a cashier give out the incorrect amount of change and observe the behavior of a randomly selected group of customers.

2a. $0.5(20)(0.1) = 1$

2b. $20 - \sqrt{150} \approx 7.75$ **2c.** 0.09

2d. $\frac{77}{300} \approx 0.257$

3. $\sqrt{\frac{2}{\pi}}$

4a. $\bar{x} = 10.55$ lb; $s = 2.15$ lb

4b. 6.25 lb to 14.85 lb

5a.

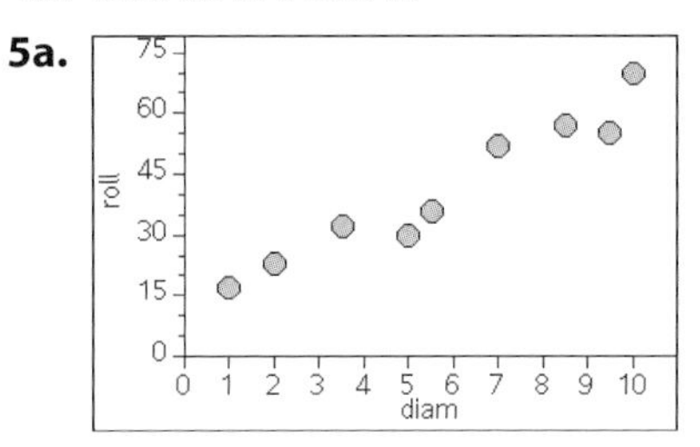

5b. yes; $r \approx 0.965$, indicating a relationship that is close to linear

5c. $\hat{y} \approx 5.322x + 10.585$

5d. The rolling distance increases 5.322 in. for every additional inch of wheel diameter. The skateboard will skid approximately 10.585 in. even if it doesn't have any wheels.

5e. 7.5 in.

6a. approximately 32.3%

6b. approximately 14.2%

6c. between 152 lb and 198 lb

6d. $y = \frac{1}{14\sqrt{2\pi}}(\sqrt{e})^{-[(x-175)/14]^2}$

7. approximately 0.062

8a. Chipley and Panama City ($r \approx 0.859$)

8b. For the graph of (*Chipley, Panama City*), $\hat{y} \approx 0.852x + 0.52$. The slope indicates that for every 0.852 in. of rain in Panama City, there is 1 in. of rain in Chipley.

8c. Crestview and Panama City have the weakest correlation ($r \approx 0.551$). The equation of the least squares line for (*Crestview, Panama City*) is $\hat{y} \approx 0.655x + 1.4$. The y-intercept indicates that the least squares line predicts that in a month with zero rainfall in Crestview, there would be 1.4 in. of rain in Panama City. This is not supported by the data in the table.

CHAPTER 12

REFRESHING YOUR SKILLS FOR CHAPTER 12

1. $\frac{30}{\sqrt{3}}$, or $10\sqrt{3}$ cm

3a. 30°-60°-90° **3c.** neither

LESSON 12.1

1. $\sin A = \frac{k}{j}$; $\sin B = \frac{h}{j}$; $\sin^{-1}\left(\frac{k}{j}\right) = A$; $\sin^{-1}\left(\frac{h}{j}\right) = B$; $\cos B = \frac{k}{j}$; $\cos A = \frac{h}{j}$; $\cos^{-1}\left(\frac{k}{j}\right) = B$; $\cos^{-1}\left(\frac{h}{j}\right) = A$; $\tan A = \frac{k}{h}$; $\tan B = \frac{h}{k}$; $\tan^{-1}\left(\frac{k}{h}\right) = A$; $\tan^{-1}\left(\frac{h}{k}\right) = B$

2a.

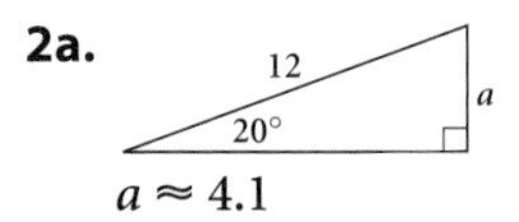

$a \approx 4.1$

2c.

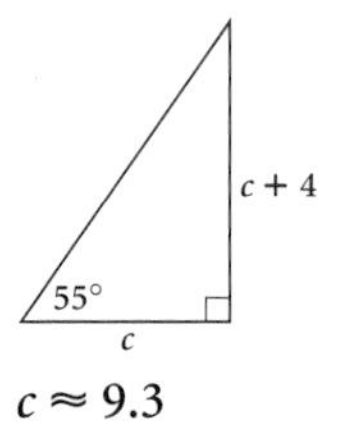

$c \approx 9.3$

3a. $a \approx 17.3$ **3b.** $b \approx 22.8$

4a. $A \approx 17°$

5b. 726 steps

6. *Hint:* Refer to Refreshing Your Skills for Chapter 12 to review the properties of special right triangles.

7a. $BD \approx 25.7$ cm

7c. $AD \approx 70.5$ cm, $DC \approx 19.5$ cm

8. *Hint:* First find the height of the triangle.

11. *Hint:* Make a right triangle by drawing a segment from the point perpendicular to the x-axis.

12a. *Hint:* Pay careful attention to units. You will need to convert km/h to m/s.

12b. 396 km/h

LESSON 12.2

1. 9.7 cm

2. 63.2°

3. $X \approx 50.2°$ and $Z \approx 92.8°$

4. 7.8 cm

6. *Hint:* What is the measure of the third angle?

7a. 12.19 cm

8. *Hint:* You should get two possible answers.

9. *Hint:* The angle in the triangle at the point of the second sighting is 109°.

10. 5.4685 light-years at the first reading and 5.4684 light-years at the second reading

11a. 41° **11c.** 0°

12a. $x = \frac{4 \pm \sqrt{6}}{2}$

LESSON 12.3

1. approximately 6.1 km

2. approximately 35.33°

3a. $A \approx 41.4°$

4a. $AC \approx 6.4$ **4b.** $D = 72°$

5. *Hint:* Refer to the diagram at the beginning of Lesson 12.2.

6. 112.3°

7. *Hint:* You are given the three sides of a triangle. Make a sketch and then solve for the angles to help you describe the location of the chamber.

8. approximately 58 cm

12. 1751 cm^2

13. *Hint:* $\triangle BUR$ is similar to $\triangle TSR$, and both are right triangles.

LESSON 12.4

1a. $\frac{-3}{\sqrt{13}} \approx -0.832$

2a. iii.

3a.

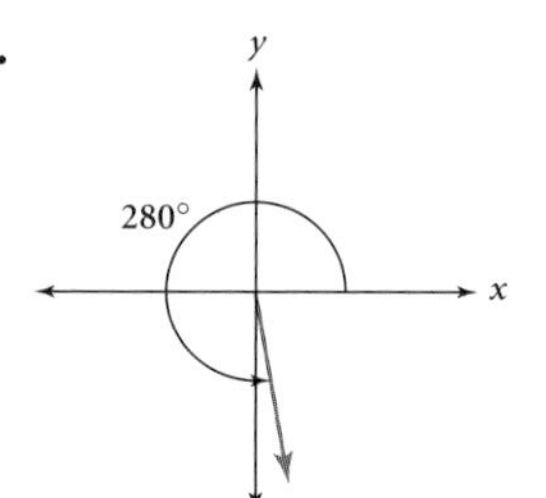

4a. 80°

5a. negative

6a. about 236.3°

7. *Hint:* Place Robyn at the origin and plot the positions of Nathan and Keshon.

8. about 106.7°

10. about (−4.60, 3.86)

11. *Hint:* First calculate the angle of the original point. Add 100° to find the angle of the point after it rotates.

15a. i. 7 **15a. ii.** 3

15a. iii. 4 **15a. iv.** 7

15b. $y = 1 \pm \sqrt{(x-3)^3}$; not a function

15c. Substitute each y-value found in 15a for x into the inverse. Check that the output is equivalent to the original x-value.

15d.

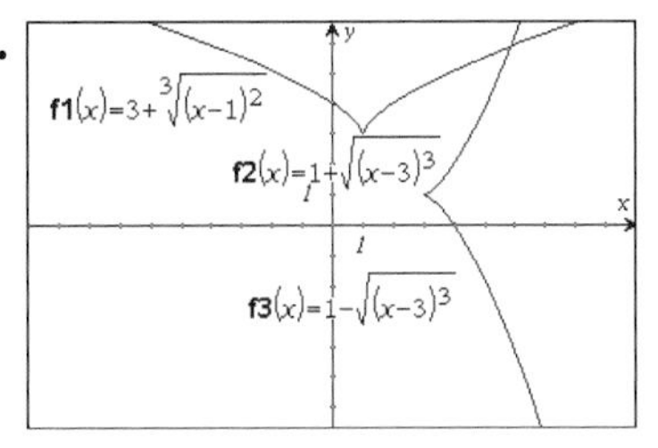

LESSON 12.5

1a. $\langle -2, 5\rangle$ **1c.** $\langle 6, 15\rangle$

1e. approximately 5.385

2a. $\langle 2 \angle 60°\rangle$ **2c.** approximately $\langle 2.828 \angle 135°\rangle$

3b. approximately $\langle -1.41, 1.41\rangle$ **3d.** approximately $\langle -2, -3.46\rangle$

4a. $\langle 4, 5\rangle$ **4c.** $\langle 8 \angle 150°\rangle$

5. *Hint:* Make a drawing and then refer to Example C of this lesson.

6. 2 blocks west and 2 blocks north; $\langle 2, 4\rangle - \langle 4, 2\rangle = \langle -2, 2\rangle$

7a. approximately $\langle 38.30, 32.14\rangle$

7c. approximately $\langle -30.64, 89.99\rangle$

9. *Hint:* Write your vectors in rectangular form.

9a. $\langle -240, 0\rangle$ or $\langle 240 \angle 180°\rangle$

9b. $\langle 0, 50\rangle$ or $\langle 50 \angle 90°\rangle$

10a. $\langle 960t \angle 140°\rangle = \langle -245.13t, 205.69t\rangle = \langle -735.40, 617.08\rangle$

10b. $\langle 32t \angle 180°\rangle = \langle -32t, 0\rangle = \langle -96, 0\rangle$

10c. $\langle -831.40, 617.08\rangle$ **10d.** $\langle 1035.38 \angle 143.42°\rangle$

11a. $\langle -1000, 0\rangle$

LESSON 12.6

1a.

t	x	y
−2	−7	−3
−1	−4	−1
0	−1	1
1	2	3
2	5	5

1c.

t	x	y
−2	4	1
−1	1	2
0	0	3
1	1	4
2	4	5

2a.

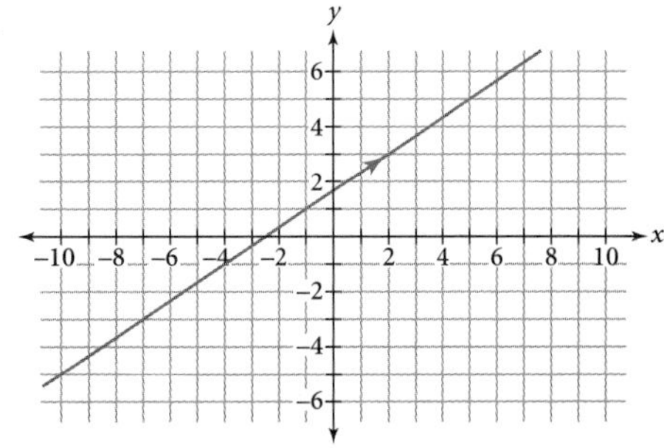

2c.

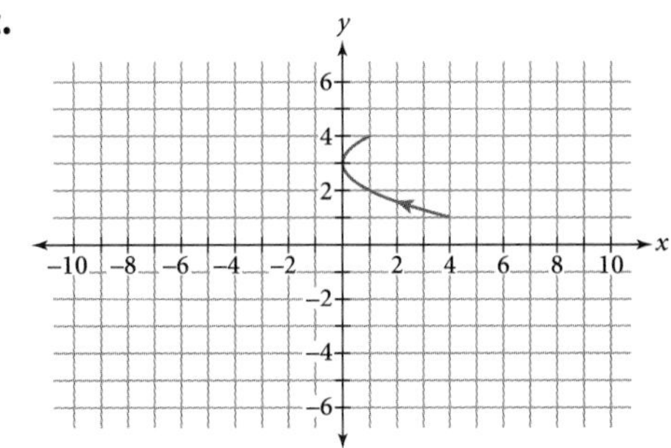

3a. $y = \frac{x+7}{2}$ **3c.** $y = \frac{2x-4}{3}$

4b. left 3.3 cm and down 405 cm

4c. left 46.6 cm and down 1850 cm

5a. *Hint:* One of the equations is $x = 2t$.

8a.

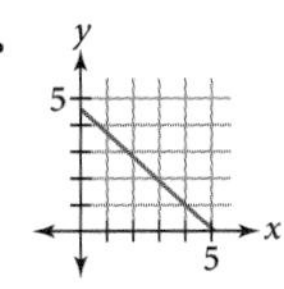

8b. $x = -\frac{2}{3}t + 4$, $y = \frac{3}{5}t + 1$; $y = -\frac{9}{10}x + \frac{23}{5}$. The slope of (x, y) is the ratio of the two slopes of (t, y) and (t, x).

9a. $(-0.3, 0.5)$ **9b.** $x = -0.3 + 4t$

9e. *Hint:* The x-coordinate of the east bank is 0.

14a. $(3.2, 35)$

14b. *Hint:* At these points the y-coordinate is 0.

CHAPTER 12 REVIEW

1a. $A \approx 43°$ **1b.** $B \approx 28°$

1c. $c \approx 23.0$ **1d.** $d \approx 12.9$

1e. $e \approx 21.4$ **1f.** $f \approx 17.1$

2. about 73°

3a. $a \approx 7.8$ m, $c \approx 6.7$ m, $C = 42°$

3b. $A \approx 58°$, $B \approx 46°$, $a \approx 27$ cm

4. about 5.9 km

5a. $A \approx 40°$, $b \approx 3.5$ cm, $C \approx 58°$

5b. $A \approx 51°$, $B \approx 110°$, $C \approx 19°$

6. about 20.2 mi

7. $(-3\sqrt{3}, 3)$

8a. $(1.5, 10)$ **8b.** $(7.5, 0)$

8c. about 11.7 m

8d. A longer path; the path will be parabolic rather than along a straight line.

9. approximately $\langle 118.4 \angle 56.3° \rangle$

10. She should fly on a bearing of 107.77° if the wind averages 25 mi/h.

11a. $t = 3: x = -8, y = 0.5; t = 0: x = 1, y = 2;$ $t = -3: x = 10, y = -1$

11b. $y = \frac{6}{11}$ **11c.** $x = \frac{5}{2}$

11d. When $t = -1$, the y-value is undefined.

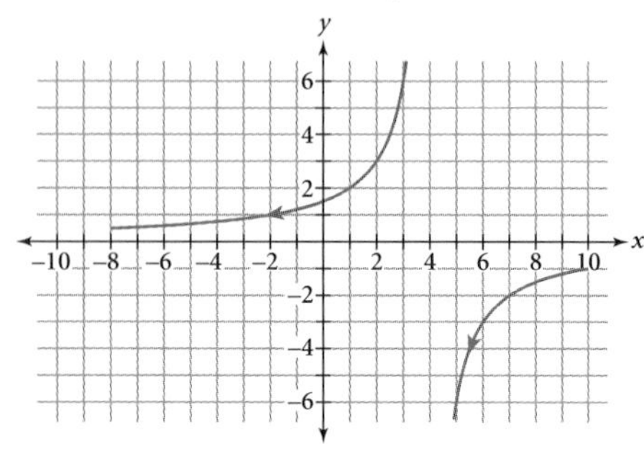

12a. 160 m east and 240 m south

12b. $x = 20t, y = -30t$

CHAPTER 13 • CHAPTER **13** CHAPTER 13 • CHAPTER

REFRESHING YOUR SKILLS FOR CHAPTER 13

1a. $C = 2\pi r = 32\pi$ ft

2a. 180°

3. $\frac{\pi}{4}$ cm, or about 0.79 cm

LESSON 13.1

1a.

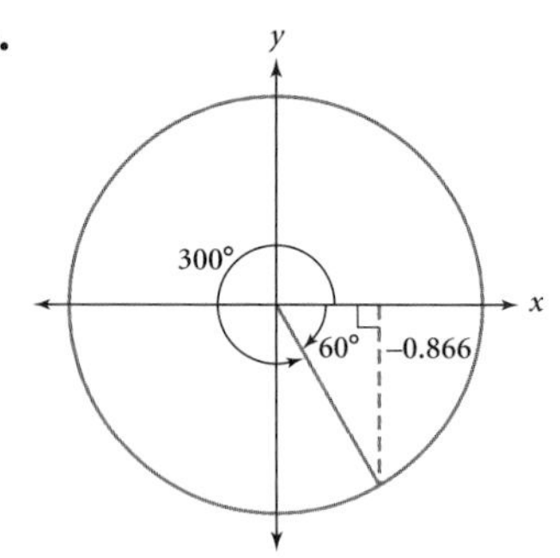

approximately -0.866 m

1b.

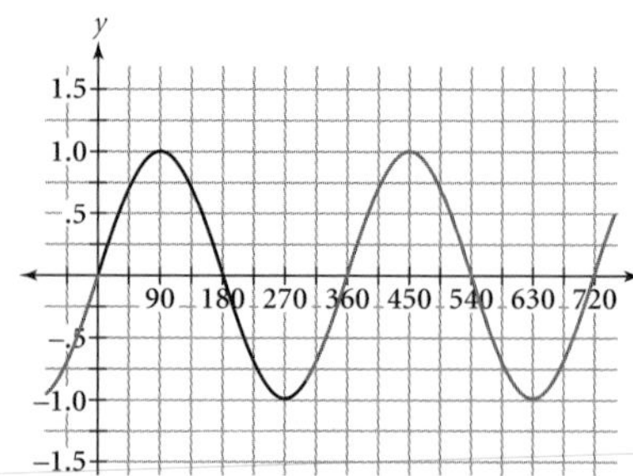

2a. -0.0872; $\sin(-175°) = -\sin 5°$; reference angle 5°

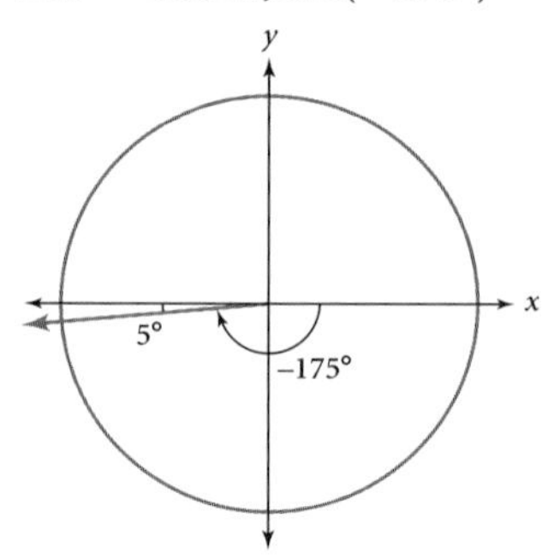

2c. -0.9848; $\sin 280° = -\sin 80°$; reference angle 80°

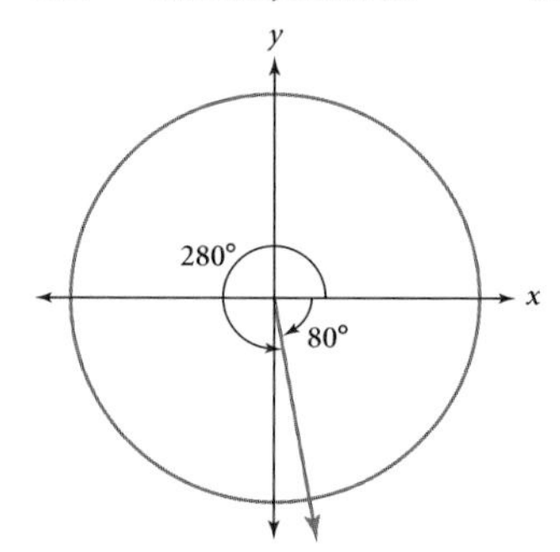

5b. not periodic

5c. periodic, 90°

6b. $\theta = 290°$

6c. $\theta = 75°$

10a.

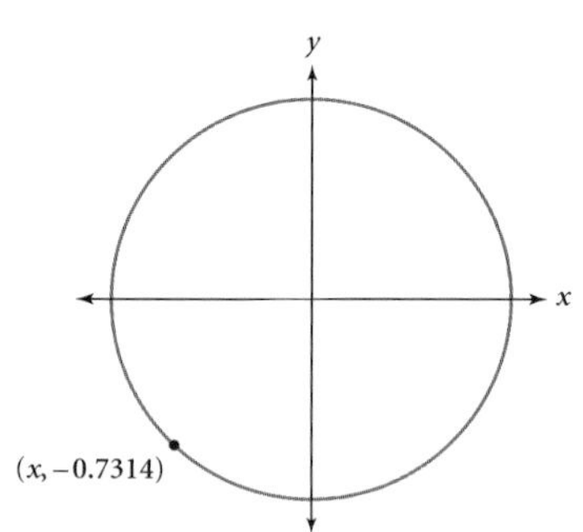

10b. $\theta \approx 227°$; $\cos\theta = -0.682$

10c. $\alpha = 360° - \theta = 133°$

11. *Hint:* Graph a constant function and find the intersection points with $y = \sin(x)$.

12a. *Hint:* Use the Pythagorean Theorem.

12b. *Hint:* Draw similar right triangles.

13a.

13b. 10–11 yr

16a. $h = 1.5 + 20 \tan A$

16b. 33.5 m

19a. $\frac{3}{x-4}, x \neq \pm 4$ **19b.** $2, x \neq -2, -1, 1, 3$

19c. $\frac{2(3+a)}{6-a}, a \neq 6$

21. *Hint:* A bearing is an angle measured clockwise from north.

LESSON 13.2

1a. $\frac{4\pi}{9}$ **1b.** $\frac{19\pi}{6}$ **1g.** -5π

2a. 2π

6b. 1.6 cm/min

7d, 8d, 9d. *Hint:* Make sure your calculator is set to radian mode.

11a. $\frac{4\pi}{3}$

12. 31.3 radians/s

15. *Hint:* Draw an arc (a great arc) on a ball, balloon, or globe.

16b. dilated vertically by a factor of 3 and horizontally by a factor of 4 and translated right 5 units

16d. dilated vertically by a factor of 2, reflected across the x-axis, and translated up 3 units and left 1 unit

17a. $y = -2(x+1)^2$

17b. $y + 4 = (x-2)^2$, or $y = (x-2)^2 - 4$

17c. $y + 2 = |x+1|$, or $y = |x+1| - 2$

17d. $-\frac{y-2}{2} = |x-3|$, or $y = -2|x-3| + 2$

20.

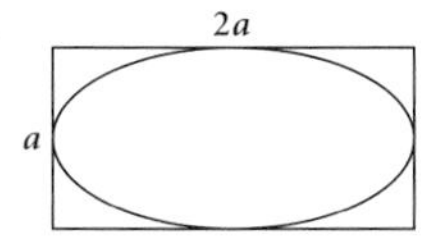

The foci are $\frac{\sqrt{3}}{2}a$ units from the center. Therefore the eccentricity is $\frac{\frac{\sqrt{3}}{2}a}{a} = \frac{\sqrt{3}}{2}$.

21. *Hint:* Construct $\overline{AP}$, $\overline{BP}$, and $\overline{CP}$. $\triangle APC$ is isosceles because $\overline{AP}$ and $\overline{CP}$ are radii of the same circle. $\angle ABP$ measures 90° because the angle is inscribed in a semicircle. Use these facts to prove that $\triangle ABP \cong \triangle CBP$.

LESSON 13.3

1a. $y = \sin x + 1$; amplitude = 1, period = 2π, phase shift = 0, vertical shift = 1

1b. $y = \cos x - 2$; amplitude = 1, period = 2π, phase shift = 0, vertical shift = -2

1c. $y = \sin x - 0.5$; amplitude = 1, period = 2π, phase shift = 0, vertical shift = -0.5

2b. $y = \sin\frac{x}{2} - 1$; amplitude = 1, period = 4π, phase shift = 0, vertical shift = -1

2c. $y = -2\sin 3x$; amplitude = 2, period = $\frac{2\pi}{3}$, phase shift = 0, vertical shift = 0

6. possible answer: $y = \sin\left(x - \frac{\pi}{2}\right)$, $y = -\sin\left(x + \frac{\pi}{2}\right)$, or $y = -\cos x$

7a. *Hint:* What are the independent and dependent variables?

14b. $y = -3 + 2\sin 4\left(x - \frac{\pi}{4}\right)$

15. *Hint:* There are two possible measures for each angle A. The drawings are not necessarily to scale.

17a. 1.885 m/h

17b. 30 cm

17c. 360 cm

17d. second hand: 377 radians/h; minute hand: 6.28 radians/h; hour hand: 0.52 radian/h

17e. The speed of the tip of a clock hand varies directly with the length of the hand. Angular speed is independent of the length of the hand.

18a. i. $y = -\frac{2}{3}x + 4$

LESSON 13.4

1b. $-14.3°$ and -0.25

1c. 144.2° and 2.52

2a. $\frac{\pi}{6}, \frac{5\pi}{6}, -\frac{11\pi}{6}, -\frac{7\pi}{6}$

2b. $\frac{3\pi}{8}, \frac{13\pi}{8}, -\frac{13\pi}{8}, -\frac{3\pi}{8}$

3. *Hint:* Graph $y = \sin x$ or $y = \cos x$ for $-2\pi \leq x \leq 2\pi$. Plot all points on the curve that have a y-value equal to the y-value of the expression on the right side of the equation. Then find the x-value at each of these points.

4a.

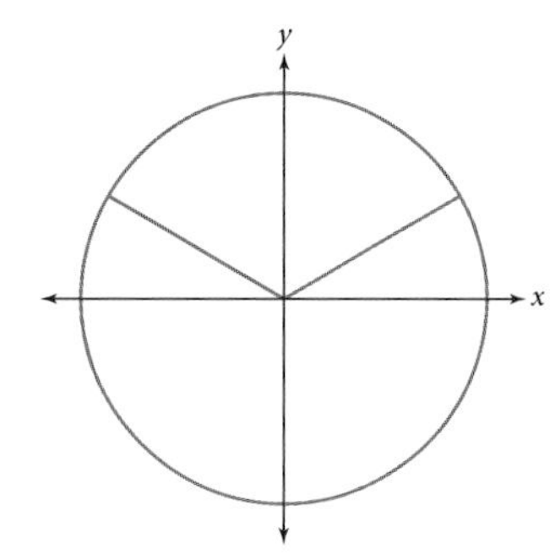

4b.

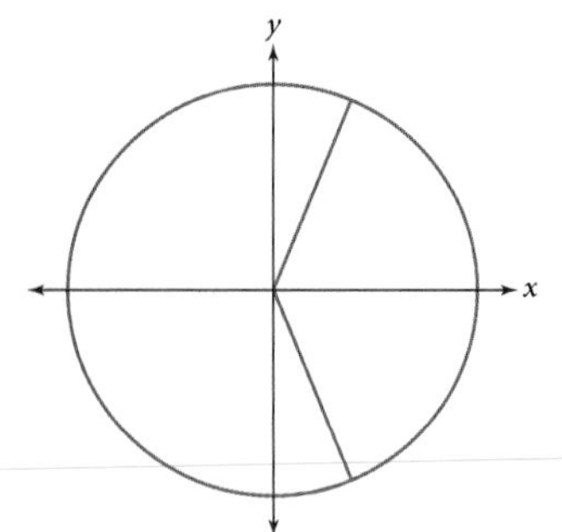

4c.

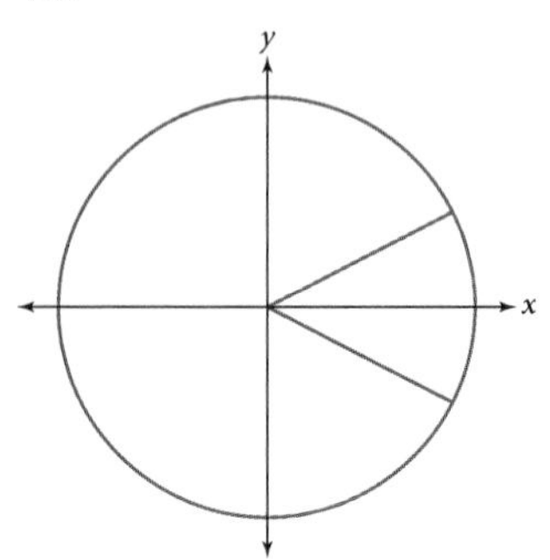

4d.

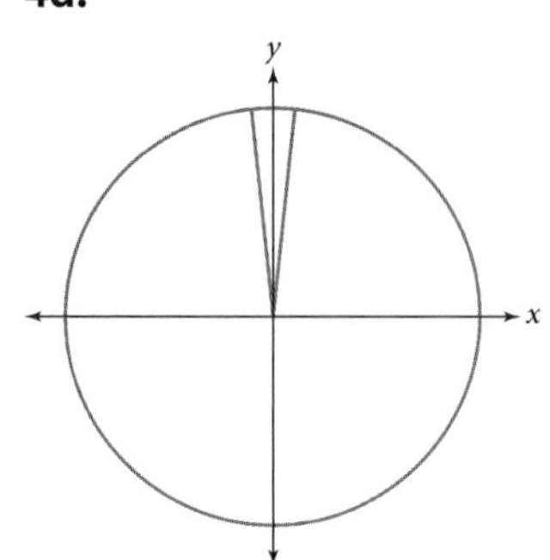

7a. $x \approx 0.485$ or $x \approx 2.656$

9. 106.9°

13. 650°

14b. $-\frac{2}{3}$ **14c.** $\frac{\pi}{3}$

15a. 8.0×10^{-4} W/m^2; 6.0×10^{-4} W/m^2

17a. $y = -4\cos\frac{x}{2}$

18a. $x = \frac{6 \pm \sqrt{12}}{4}$, or $x = \frac{3 \pm \sqrt{3}}{2}$

LESSON 13.5

1a. $x = \left\{\frac{\pi}{3}, \frac{5\pi}{3}, \frac{7\pi}{3}, \frac{11\pi}{3}\right\}$

2a. $x = \{1.831, 5.193\}$

3a. *Hint:* The average value of a sinusoid is the center line.

3b. 7; −2; 12; 7

7a. *Hint:* You can assume a phase shift of 0.

9a. $y_1 = -3\cos\left[\frac{2\pi(t + 0.17)}{\frac{2}{3}}\right]$,

$y_2 = -4\cos\left(\frac{2\pi t}{\frac{2}{3}}\right)$, or $y_1 = 3\sin 3\pi t$, $y_2 = -4\cos 3\pi t$

12a. about 9.6 h

14.

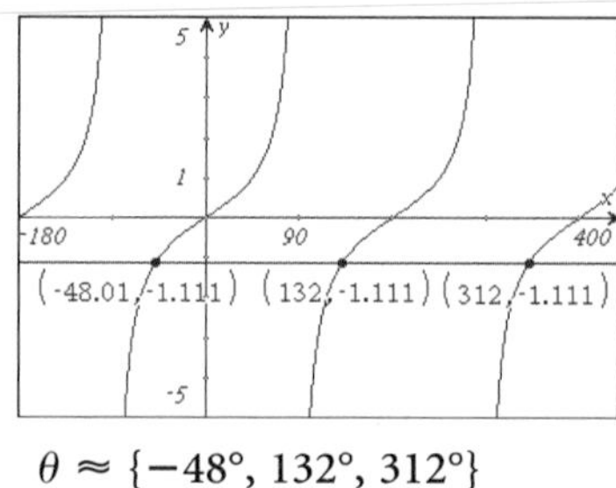

$\theta \approx \{-48°, 132°, 312°\}$

LESSON 13.6

1. Graph $y = \frac{1}{\tan x}$.

2b. not an identity **2d.** may be an identity

3. Proof:

$\tan x(\cot x + \tan x) \stackrel{?}{=} \sec^2 x$	Original equation.
$\tan x \cot x + (\tan x)^2 \stackrel{?}{=} \sec^2 x$	Distribute.
$\tan x \cdot \frac{1}{\tan x} + \tan^2 x \stackrel{?}{=} \sec^2 x$	Reciprocal identity.
$1 + \tan^2 x \stackrel{?}{=} \sec^2 x$	Reduce.
$\sec^2 x = \sec^2 x$	Pythagorean identity.

6a. odd **6c.** odd **6e.** even

7a. *Hint:* Replace $\cos 2A$ with $\cos^2 A - \sin^2 A$. Rewrite $\cos^2 A$ using a Pythagorean identity. Then combine like terms.

7b. *Hint:* Replace $\cos 2A$ with $\cos^2 A - \sin^2 A$. Rewrite $\sin^2 A$ using a Pythagorean identity. Then combine like terms.

8a.

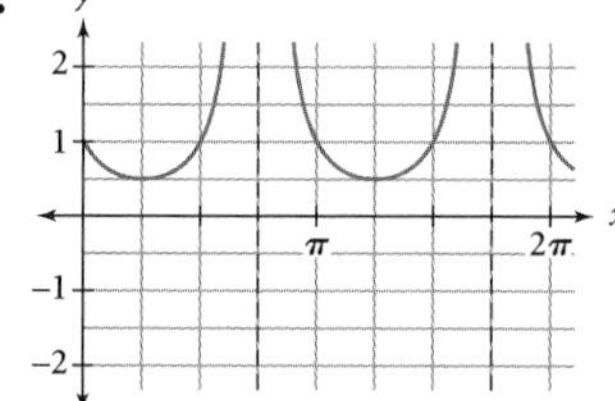

9a. $y = \sin x$ **9d.** $y = \cos x$ **9e.** $y = -\sin x$

11a–c. *Hint:* Use the reciprocal trigonometric identities to graph each equation on your calculator, with window $0 \leq x \leq 4\pi$, $-2 \leq y \leq 2$.

12a. $y = \tan 2x - 1$ **12c.** $y = 0.5\csc x + 1$

13a. *Hint:* Multiply the left side by a fraction equivalent to 1. You will be using the identity $1 - \cos^2 x = \sin^2 x$.

15. *Hint:* AST stands for Atlantic Standard Time.

16a. $1,505.12

LESSON 13.7

1a. not an identity

1b. not an identity

3a. cos 1.1 **3c.** sin 1.7

4a. $\frac{-\sqrt{6}+\sqrt{2}}{4}$

5. *Hint:* Find cos *x*. Then use an identity to find sin 2*x*.

6. proof:

$$\sin(A+B) \stackrel{?}{=} \sin A\cos B + \cos A\sin B$$

$$\cos\left[\frac{\pi}{2}-(A+B)\right] \stackrel{?}{=} \sin A\cos B + \cos A\sin B$$

$$\cos\left[\left(\frac{\pi}{2}-A\right)-B\right] \stackrel{?}{=} \sin A\cos B + \cos A\sin B$$

$$\cos\left(\frac{\pi}{2}-A\right)\cos B + \sin\left(\frac{\pi}{2}-A\right)\sin B \stackrel{?}{=} \sin A\cos B + \cos A\sin B$$

$$\sin A\cos B + \cos A\sin B = \sin A\cos B + \cos A\sin B$$

7. *Hint:* Begin by writing $\sin(A-B)$ as $\sin(A+(-B))$. Then use the sum identity given in Exercise 6. Next, use the identities $\cos(-x)=\cos x$ and $\sin(-x)=-\sin x$ to simplify further.

9. *Hint:* Begin by writing sin 2*A* as $\sin(A+A)$. Then use a sum identity to expand, and simplify by combining like terms.

11. *Hint:* Show that $\tan(A+B) \neq \tan A + \tan B$ by substituting values for *A* and *B* and evaluating. To find an identity for $\tan(A+B)$, first rewrite as $\frac{\sin(A+B)}{\cos(A+B)}$. Then use sum identities to expand. Divide both the numerator and denominator by cos *A* cos *B*, and rewrite each occurrence of $\frac{\sin\theta}{\cos\theta}$ as $\tan\theta$.

13a.
$$\cos 2A = 1 - 2\sin^2 A$$
$$2\sin^2 A = 1 - \cos 2A$$
$$\sin^2 A = \frac{1-\cos 2A}{2}$$

14a. sample graph when $a = 3$ and $b = 4$:

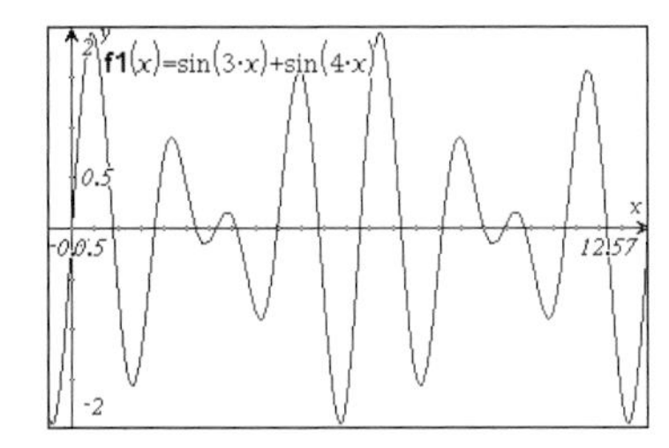

17. *Hint:* Refer to the reciprocal identities on page 783.

17b. $x \approx 1.1547$ **17d.** $x = 0$

18b. 105° or 255° **18c.** 64° or 116°

CHAPTER 13 REVIEW

1a. I; 420°; $\frac{\pi}{3}$

1b. III; $\frac{10\pi}{3}$; 240°

1c. IV; −30°; $\frac{11\pi}{6}$

1d. IV; $\frac{7\pi}{4}$; −45°

2a. $\sin 60° = \frac{\sqrt{3}}{2}$; $\cos 60° = \frac{1}{2}$

2b. $\sin\frac{4\pi}{3} = -\frac{\sqrt{3}}{2}$; $\cos\frac{4\pi}{3} = -\frac{1}{2}$

2c. $\sin 330° = -\frac{1}{2}$; $\cos 330° = \frac{\sqrt{3}}{2}$

2d. $\sin\left(-\frac{\pi}{4}\right) = -\frac{\sqrt{2}}{2}$; $\cos\left(-\frac{\pi}{4}\right) = \frac{\sqrt{2}}{2}$

3. Other equations are possible.

3a. period $= \frac{2\pi}{3}$; $y = -2\cos\left[3\left(x-\frac{2\pi}{3}\right)\right]$

3b. period $= \frac{\pi}{2}$; $y = 3\sin\left[4\left(x-\frac{\pi}{8}\right)\right]$

3c. period $= \pi$; $y = \csc\left[2\left(x+\frac{\pi}{4}\right)\right]$

3d. period $= \frac{\pi}{2}$; $y = \cot\left[2\left(x-\frac{\pi}{4}\right)\right] + 1$

4a. 2; $\frac{\pi}{6}$; 0; $\frac{3}{2\pi}$

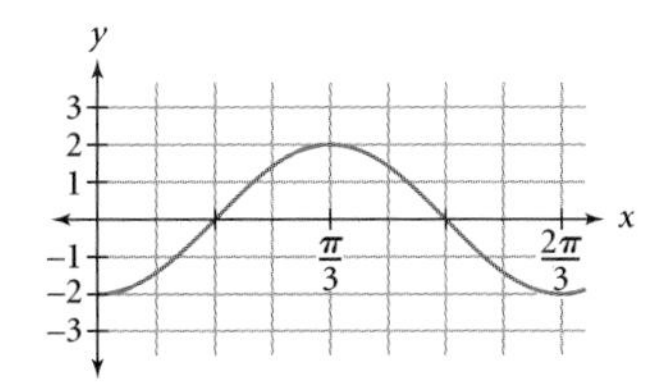

4b. 3; 0; 0; $\frac{2}{\pi}$

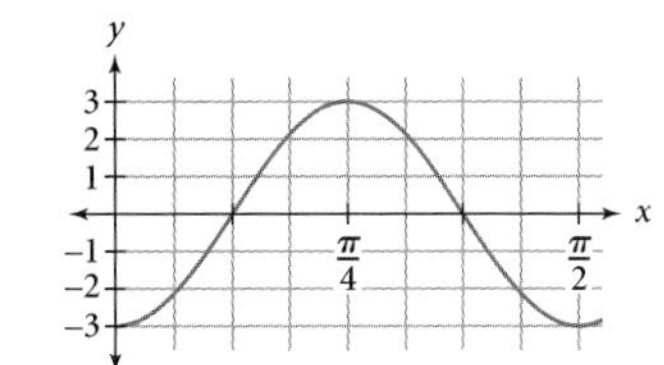

5a. $y = -2\sin 2x - 1$

5b. $y = \sin 0.5x + 1.5$

5c. $y = 0.5\tan\left(x-\frac{\pi}{4}\right)$

5d. $y = 0.5\sec 2x$

6. area $= \frac{9\pi}{8}$ cm²; arc length $= \frac{3\pi}{4}$ cm

7. $\cos y = x$: domain: $-1 \le x \le 1$; range: all real numbers. $y = \cos^{-1}x$: domain: $-1 \le x \le 1$; range: $0 \le y \le \pi$

Selected Hints and Answers

8a. $\frac{3}{5}$ **8b.** $\frac{4}{5}$ **8c.** $\frac{8}{17}$

9. $y = -3\sin\left(\frac{x - \frac{\pi}{2}}{4}\right)$

10a. proof: $\sec A - \sin A \tan A \stackrel{?}{=} \cos A$

$$\frac{1}{\cos A} - \sin A\left(\frac{\sin A}{\cos A}\right) \stackrel{?}{=} \cos A$$

$$\frac{1 - \sin^2 A}{\cos A} \stackrel{?}{=} \cos A$$

$$\frac{\cos^2 A}{\cos A} \stackrel{?}{=} \cos A$$

$$\cos A = \cos A$$

10b. proof: $\frac{1}{\sin^2 A} - \frac{1}{\tan^2 A} \stackrel{?}{=} 1$

$$\frac{1}{\sin^2 A} - \frac{\cos^2 A}{\sin^2 A} \stackrel{?}{=} 1$$

$$\frac{1 - \cos^2 A}{\sin^2 A} \stackrel{?}{=} 1$$

$$\frac{\sin^2 A}{\sin^2 A} \stackrel{?}{=} 1$$

$$1 = 1$$

10c. proof: $\frac{\cos B - \tan A \sin B}{\sec A} \stackrel{?}{=} \cos(A + B)$

$$\frac{\cos B}{\sec A} - \frac{\tan A \sin B}{\sec A} \stackrel{?}{=} \cos(A + B)$$

$$\cos B \cos A - \frac{\sin A}{\cos A} \sin B \cos A \stackrel{?}{=} \cos(A + B)$$

$$\cos B \cos A - \sin A \sin B \stackrel{?}{=} \cos(A + B)$$

$$\cos(A + B) = \cos(A + B)$$

11. 0.175 s, 0.659 s, 1.008 s, 1.492 s, 1.841 s, 2.325 s, 2.675 s

12a. Let x represent time in hours, and let y represent tide height in meters.

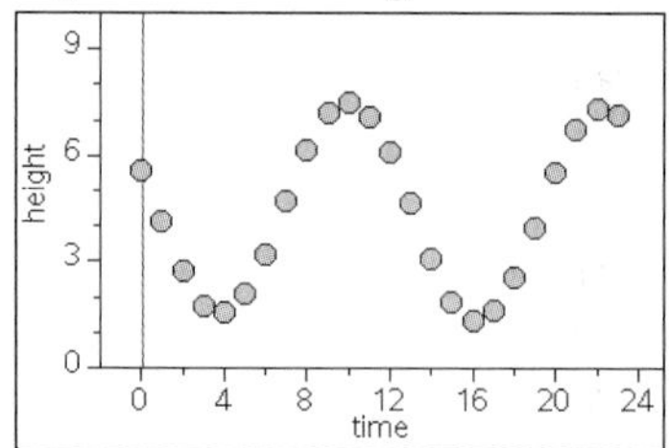

12b. possible answer: $y = 2.985\cos\left[\frac{\pi}{6}(x - 10)\right] + 4.398$

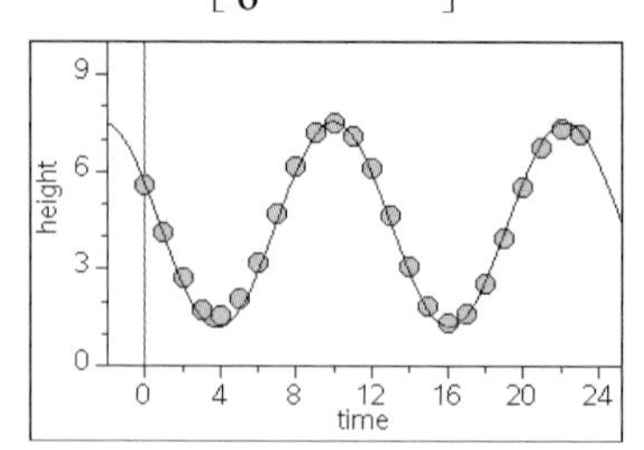

12c. approximately 1.84 m

12d. between 00:00 and 00:37, between 07:22 and 14:10, and between 19:22 and 23:59

13a. hyperbola

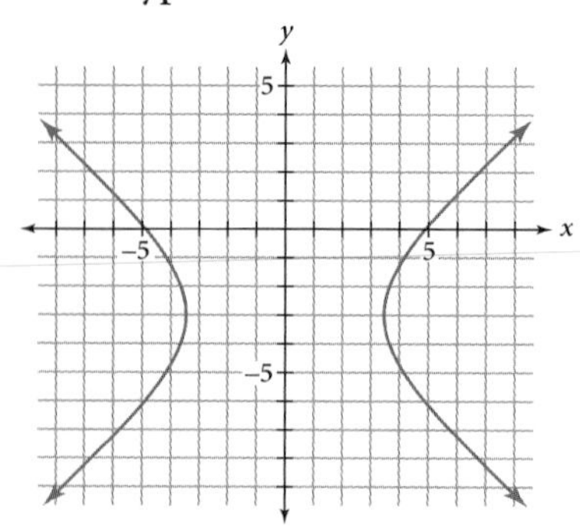

13b. ellipse

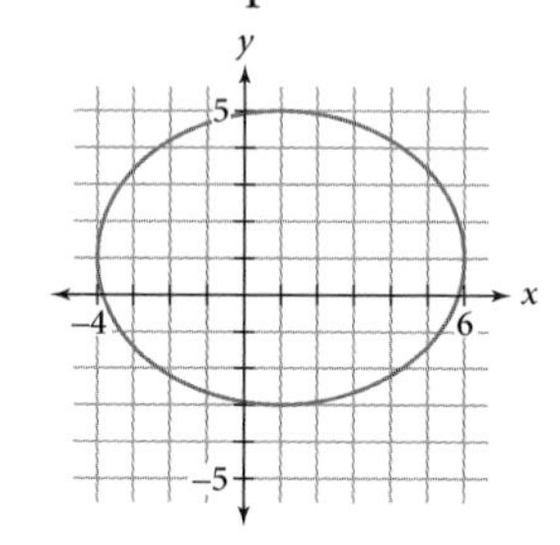

13c. parabola

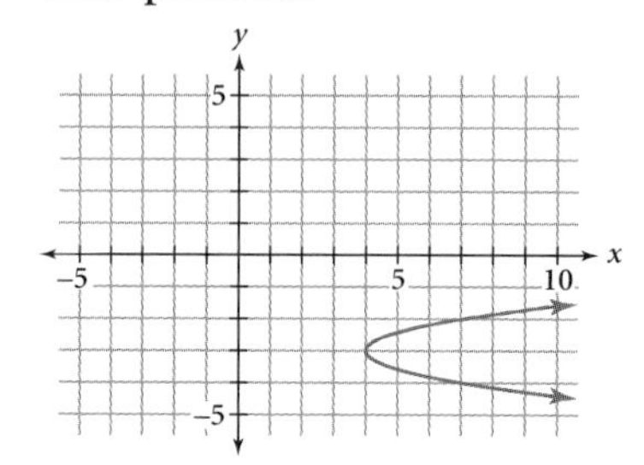

13d. hyperbola

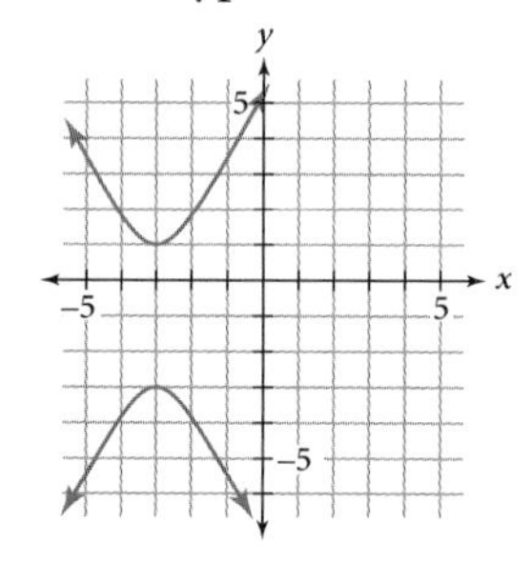

14. 0.751

15a. $S_{12} = 144$ **15b.** $S_{20} = 400$

15c. $S_n = n^2$

16a. $\begin{bmatrix} \frac{7}{8} & \frac{1}{8} \\ \frac{1}{12} & \frac{11}{12} \end{bmatrix}$

16b. approximately 439 cars in Detroit and 561 in Chicago; approximately 400 cars in Detroit and 600 in Chicago

17a. $y = -20x^2 + 332x$ **17b.** \$8.30; \$1,377.80

18a. $y = -\frac{1}{2}\cos\left(\frac{1}{2}x\right) + 6$

18b. period: 4π, amplitude: $\frac{1}{2}$; phase shift: none

18c.

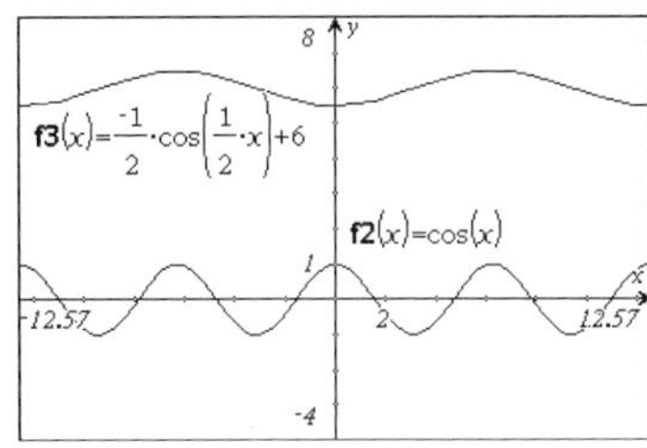

19. Out of 100 people with the symptoms, the test will accurately confirm that 72 do not have the disease while mistakenly suggesting 8 do have the disease. The test will accurately indicate 18 do have the disease and make a mistake by suggesting 2 do not have the disease who actually have the disease.

20a. geometric; $u_1 = 3$ and $u_n = 3u_{n-1}$ where $n \geq 2$; $u_n = 3(3)^{n-1}$ or $u_n = 3^n$

20b. arithmetic; $u_1 = -1$ and $u_n = u_{n-1} - 2$ where $n \geq 2$; $u_n = -2n + 1$

20c. Neither; students are not expected to write a recursive formula; $u_n = n^2 + 1$

20d. geometric; $u_1 = 1$ and $u_n = -\frac{1}{2}u_{n-1}$ where $n \geq 2$; $u_n = \left(-\frac{1}{2}\right)^{n-1}$

21a. $\frac{11\pi}{36}$ **21b.** approximately 3.84 cm

21c. approximately 7.68 cm^2

22a. Answers will vary; possible answer: 3.7% per year.

22b. possible answer: $\hat{y} = 6.0(1 + 0.036)^{x-1970}$

22c. possible answer: 31.8 million

22d. The population predicted by the equation is somewhat higher. Explanations will vary; the growth rate may be declining; the *New York Times Almanac* prediction may be based on more recent data and more data points.

23a. mean: 1818.7 ft; median: 1573 ft; mode: 1600 ft

23b. The data are skewed right.

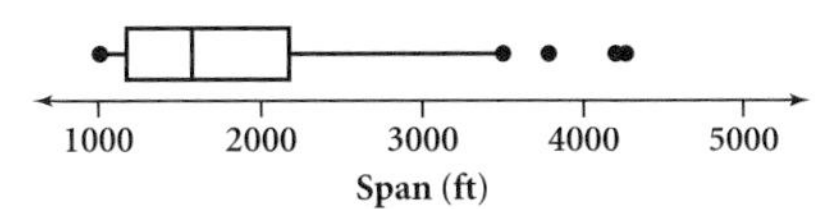

23c. approximately 865.3 ft

24a. $(1, 4)$ **24b.** $(-5.5, 0.5)$, or $\left(-\frac{11}{2}, \frac{1}{2}\right)$

25a. $S_5 = \frac{121}{810} \approx 0.149$ **25b.** $S_{10} \approx 0.150$

25c. $S = \frac{3}{20}$, or 0.15

26a. domain: $x \geq \frac{3}{2}$; range: $y \geq 0$

26b. domain: any real number; range: $y \geq 0$

26c. $f(2) = 1$ **26d.** $x = \pm\sqrt{\frac{1}{3}}$ or $x \approx \pm 0.577$

26e. $g(f(3)) = 18$

26f. $f(g(x)) = \sqrt{12x^2 - 3}$

27a. possible answer: $d = 2.4t$ and $d = 400 - 1.8t$

27b. The faster person runs approximately 229 m; the slower person runs approximately 171 m.

27c. 95.24 *s*

28a and b.

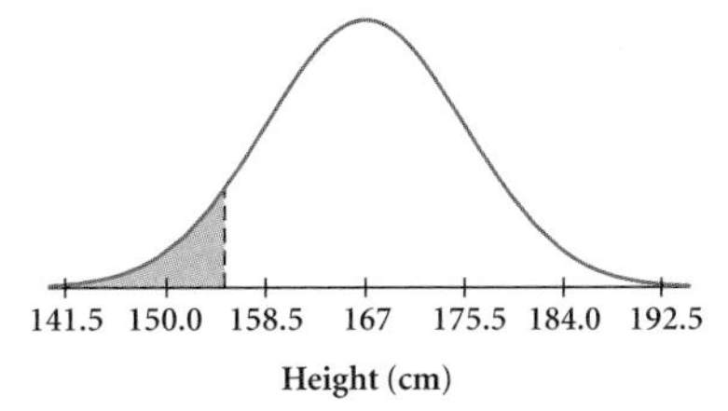

28c. approximately 7.9%

Selected Hints and Answers

Glossary

The number in parentheses at the end of each definition gives the page where each word or phrase is first used in the text. Some words and phrases are introduced more than once, either because they have different applications in different chapters or because they first appeared within features such as Project or Take Another Look; in these cases, there may be multiple page numbers listed.

A

absolute value A number's distance from zero on the number line. The absolute value of a number gives its size, or magnitude, whether the number is positive or negative. The absolute value of a number x is shown as $|x|$. For example, $|-9| = 9$ and $|4| = 4$. (182)

ambiguous case A situation in which more than one possible solution exists. (694)

amplitude Half the difference of the maximum and minimum values of a periodic function. (758)

anecdotal A type of study in which data are collected from informational reports or observations. (618)

angular speed The amount of rotation, or angle traveled, per unit of time. (748)

antilog The inverse function of a logarithm. (293)

arc (of a circle) Two points on a circle and the continuous part of the circle between them. (619)

arc length The portion of the circumference of the circle described by an arc, measured in units of length. (734, 747)

arc measure The measure of the central angle that intercepts an arc, measured in degrees. (734)

arithmetic mean See **mean.**

arithmetic sequence A sequence in which each term after the starting term is equal to the sum of the previous term and a common difference. (33)

arithmetic series A sum of terms of an arithmetic sequence. (519)

association (between variables) A relationship between two variables. In a positive association, an increase in one variable correlates with an increase in the other variable. See also **correlation.** (619)

associative property of addition For any values of a, b, and c, $a + (b + c) = (a + b) + c$. (316)

associative property of multiplication For any values of a, b, and c, $a(bc) = (ab)c$. (315, 316)

asymptote A line that a graph approaches, but does not reach, as the magnitude of the x- or y-values increases without bound. (275)

augmented matrix A matrix that represents a system of equations. The entries include columns for the coefficients of each variable and a final column for the constant terms. (344)

B

base The base of an exponential expression, b^x, is b. The base of a logarithmic expression, $\log_b x$, is b. (259)

bearing An angle measured clockwise from north. (685, 715)

bias A statistical sample of a population may be biased if some of the members of the population are less likely to be included in the sample than others. (82)

bin A column in a histogram that represents a certain interval of possible data values. (103)

binomial A polynomial with two terms. (378)

binomial expansion See **Binomial Theorem** and **expansion.**

Binomial Theorem For any binomial $(p + q)$ and any positive integer n, the binomial expansion is $(p + q)^n = {}_nC_n p^n q^0 + {}_nC_{(n-1)} p^{n-1} q^1 + {}_nC_{(n-2)} p^{n-2} q^2 + \cdots + {}_nC_0 p^0 q^n$. (604)

bivariate sampling The process of collecting data on two variables per case. (655)

box plot A one-variable data display that shows the five-number summary of a data set. (81)

box-and-whisker plot See **box plot.**

causation A relationship in which changes in one variable, called the *explanatory* or *predictor* variable, cause changes in another variable, called the *response* variable. Causation is difficult to prove. (619)

center (of a circle) See **circle.**

center (of an ellipse) The point midway between the foci of an ellipse. (457)

center (of a hyperbola) The point midway between the vertices of a hyperbola. (512)

central angle An angle whose vertex is the center of a circle and whose sides pass through the endpoints of an arc. (734)

circle A locus of points in a plane that are located a constant distance, called the *radius,* from a fixed point, called the *center.* (446, 453)

circumference The perimeter of a circle, which is the distance around the circle. Also, the curved path of the circle itself. (734)

combination An arrangement of choices in which the order is unimportant. (596, 597)

combined variation An equation in which one variable varies directly with one variable and inversely with another variable. For example, in the equation $P = \frac{kT}{V}$, P varies directly with T and inversely with V, and k is the constant of variation. (496)

common base property of equality For $a \neq 1$ and all real values of m and n, if $a^n = a^m$, then $n = m$. (260)

common difference The constant difference between consecutive terms in an arithmetic sequence. (33)

common logarithm A logarithm with base 10, written $\log x$, which is shorthand for $\log_{10} x$. (288)

common ratio The constant ratio between consecutive terms in a geometric sequence. (35)

commutative property of addition For any values of a and b, $a + b = b + a$. (316)

commutative property of multiplication For any values of a and b, $ab = ba$. (314, 316)

compass rose A diagram that shows the orientation of the directions north, south, east, and west, and often intermediate directions such as northeast, southeast, northwest, and southwest. (718)

complements Two events that are mutually exclusive and make up all possible outcomes. (574)

completing the square A method of converting a quadratic equation from general form to vertex form. (395, 398)

complex conjugate A number whose product with a complex number produces a nonzero real number. The complex conjugate of $a + bi$ is $a - bi$. (409)

complex number A number with a real part and an imaginary part. A complex number can be written in the form $a + bi$, where a and b are real numbers and i is the imaginary unit, $\sqrt{-1}$. (409, 410)

complex plane A coordinate plane used for graphing complex numbers, where the horizontal axis is the real axis and the vertical axis is the imaginary axis. (412)

composition of functions The process of using the output of one function as the input of another function. The composition of f and g is written $f(g(x))$. (237)

composition of transformations The resulting transformation when a transformation is applied to a figure and a second transformation is applied to the image of the first. (220)

compound event An event consisting of more than one outcome. (551)

compound interest Interest charged or received based on the sum of the principal and the accrued interest. (42)

conditional probability The probability of a particular dependent event, given the outcome of the event on which it depends. (564)

confidence interval A p% confidence interval is an interval about $\bar{x}$ in which you can be p% confident that the population mean, μ, lies. (648)

congruent Two polygons or other geometric figures are congruent if they are identical in shape and size. (680)

conic section Any curve that can be formed by the intersection of a plane and an infinite double cone. Circles, ellipses, parabolas, and hyperbolas are conic sections. (452)

conjugate pair A pair of expressions in which one is the sum of two terms and the other is the difference of the same terms, such as $x + 3$ and $x - 3$ or $2 + \sqrt{5}$ and $2 - \sqrt{5}$. The complex numbers $a + bi$ and $a - bi$ form a complex conjugate pair. (409)

consistent (system) A system of equations that has at least one solution. (169)

constant of variation (k) The constant ratio in a direct variation or the constant product in an inverse variation. The value of k in the direct variation equation $y = kx$ or the inverse variation equation $y = \frac{k}{x}$. (14)

constraint A limitation in a linear programming problem, represented by an inequality. (353)

continuous random variable A quantitative variable that can take on any value in an interval of real numbers. (624)

convergent series A series in which the terms of the sequence approach zero and the partial sums of the series approach a long-run value as the number of terms increases. (525)

correlation A relationship between two variables. (655)

correlation coefficient (r) A value between -1 and 1 that measures the strength and direction of a linear relationship between two variables. (655, 657)

cosecant The reciprocal of the sine ratio. If A is an acute angle in a right triangle, then the cosecant of angle A is the ratio of the length of the hypotenuse to the length of the opposite leg, or $\csc A = \frac{hyp}{opp}$. See **trigonometric function.** (783)

cosine If A is an acute angle in a right triangle, then the cosine of angle A is the ratio of the length of the adjacent leg to the length of the hypotenuse, or $\cos A = \frac{adj}{hyp}$. See **trigonometric function.** (683)

cotangent The reciprocal of the tangent ratio. If A is an acute angle in a right triangle, then the cotangent of angle A is the ratio of the length of the adjacent leg to the length of the opposite leg, or $\cot A = \frac{adj}{opp}$. See **trigonometric function.** (783)

coterminal Describes angles in standard position that share the same terminal side. (740)

counterexample An example that shows that a given conjecture is not true. (315)

counting principle When there are n_1 ways to make a first choice, n_2 ways to make a second choice, n_3 ways to make a third choice, and so on, the product $n_1 \cdot n_2 \cdot n_3 \cdot \cdots$ represents the total number of different ways in which the entire sequence of choices can be made. (587)

cubic function A polynomial function of degree 3. (416)

curve straightening A technique used to determine whether a relationship is logarithmic, exponential, power, or none of these. See **linearization.** (301)

cycloid The path traced by a fixed point on a circle as the circle rolls along a straight line. (805)

D

decay See **geometric decay.** (42)

degree In a one-variable polynomial, the power of the term that has the greatest exponent. In a multivariable polynomial, the greatest sum of the powers in a single term. (378)

dependent (events) Events are dependent when the probability of occurrence of one event depends on the occurrence of the other. (564)

dependent (system) A system with infinitely many solutions. (169)

dependent variable A variable whose values depend on the values of another variable. (133)

determinant The difference of the products of the entries along the diagonals of a square matrix. For any 2×2 matrix $\begin{bmatrix} a & b \\ c & d \end{bmatrix}$, the determinant is $ad - bc$. (373)

deviation For a one-variable data set, the signed difference between a data value and some standard value, usually the mean. (93)

difference of two squares An expression in the form $a^2 - b^2$, in which one squared number is subtracted from another. A difference of two squares can be factored as $(a + b)(a - b)$. (394)

dilation A transformation that stretches or shrinks a function or graph both horizontally and vertically by the same scale factor. See also **vertical dilation** and **horizontal dilation.** (221, 225, 232)

dimensions (of a matrix) The number of rows and columns in a matrix. A matrix with m rows and n columns has dimensions $m \times n$. (320)

direct variation The relationship between two variables such that the ratio of their values is always the same, or constant. This relationship is written as $y = kx$, where k is the constant of variation. (14)

direction (vector) The orientation of a vector. (712)

directrix See **parabola.**

discontinuity A jump, break, or hole in the graph of a function. (197)

discrete graph A graph made of distinct, nonconnected points. (55)

discrete random variable A random variable that can take on only distinct (not continuous) values. (580)

discriminant The expression under the square root symbol in the quadratic formula. If a quadratic equation is written in the form $ax^2 + bx + c = 0$, then the discriminant is $b^2 - 4ac$. If the discriminant is greater than 0, the quadratic equation has two solutions. If the discriminant equals 0, the equation has one real solution. If the discriminant is less than 0, the equation has no real solutions. (414)

distance formula The distance, d, between points (x_1, y_1) and (x_2, y_2) is given by the formula $d = \sqrt{(x_2 - x_1)^2 + (y_2 - y_1)^2}$. (442)

distributive property For any values of a, b, and c, $a(b + c) = a(b) + a(c)$. (315, 316)

domain The set of input values for a relation. (133)

dot plot A one-variable data display in which each data value is represented by a dot above that value on a horizontal number line. (79)

double root A value r is a double root of an equation $f(x) = 0$ if $(x - r)^2$ is a factor of $f(x)$. (427)

doubling time The time needed for an amount of a substance to double. (254)

E

e A transcendental number related to continuous growth, with a value of approximately 2.718. (307)

eccentricity A measure of how elongated an ellipse is. The ratio of the focal length to the length of the semi-major axis. (458)

element (matrix) See **entry.**

elimination A method for solving a system of equations that involves adding or subtracting multiples of the equations to eliminate a variable. (168)

ellipse A shape produced by dilating a circle horizontally and/or vertically. The shape can be described as a locus of points in a plane for which the sum of the distances to two fixed points, called the foci, is constant. (230, 455)

ellipsoid A three-dimensional shape formed by rotating an ellipse about one of its axes. (459)

end behavior The behavior of a function $y = f(x)$ for x-values that are large in absolute value. (492)

entry Each number in a matrix. The entry identified as a_{ij} is in row i and column j. (320)

equivalent vectors Vectors with the same magnitude and direction. (712)

even function A function that has the y-axis as a line of symmetry. For all values of x in the domain of an even function, $f(-x) = f(x)$. (247, 786)

event A specified outcome or set of outcomes. (551)

expanded form The form of a repeated multiplication expression in which every occurrence of each factor is shown. For example, $4^3 \cdot 5^2 = 4 \cdot 4 \cdot 4 \cdot 5 \cdot 5$. (259)

expansion An expression that is rewritten as a single polynomial. (603)

expected value An average value found by multiplying the value of each possible outcome by its probability, then summing all the products. (580, 581)

experiment A type of study in which a researcher tests a hypothesis by assigning subjects, or experimental units, to specific treatments. (619)

experimental probability A probability calculated based on trials and observations, given by the ratio of the number of occurrences of an event to the total number of trials. (551)

explanatory variable In statistics, the variable used to predict (or explain) the value of the response variable. (657)

explicit formula A formula that gives a direct relationship between two discrete quantities. A formula for a sequence that defines the nth term in relation to n, rather than the previous term(s). (124)

exponent The exponent of an exponential expression, b^x, is x. The exponent tells how many times the base, b, is a factor. (259)

exponential function A function with a variable in the exponent, typically used to model growth or decay. The general form of an exponential function is $y = ab^x$, where the coefficient, a, is the y-intercept and the base, b, is the ratio. (253, 254)

extraneous solution An invalid solution to an equation. Extraneous solutions are sometimes found when both sides of an equation are raised to a power. (219)

extrapolation Estimating a value that is outside the range of all other values given in a data set. (141)

extreme values Maximums and minimums. (423)

F

factored form The form $y = a(x - r_1)(x - r_2) \cdots (x - r_n)$ of a polynomial function, where $a \neq 0$. The values $r_1, r_2, \ldots, r_n$ are the zeros of the function, and a is the vertical scale factor. (388, 419)

factorial For any integer n greater than 1, n factorial, written $n!$, is the product of all the consecutive integers from n decreasing to 1. (589)

Factor Theorem If $P(r) = 0$, then r is a zero and $(x - r)$ is a factor of the polynomial function $y = P(x)$. This theorem is used to confirm that a number is a zero of a function. (431)

fair Describes a coin that is equally likely to land heads or tails. Can also apply to dice and other objects. (549)

family of functions A group of functions with the same parent function. (206)

feasible region The set of points that is the solution to a system of inequalities. (353)

Fibonacci sequence The sequence of numbers 1, 1, 2, 3, 5, 8, . . . , each of which is the sum of the two previous terms. (39)

finite A limited quantity. (518)

finite differences method A method of finding the degree of a polynomial that will model a set of data, by analyzing differences between data values corresponding to equally spaced values of the independent variable. (379)

first quartile (Q_1) The median of the values less than the median of a data set. (82)

five-number summary The minimum, first quartile, median, third quartile, and maximum of a one-variable data set. (82)

focus (plural **foci**) A fixed point or points used to define a conic section. See **ellipse, hyperbola,** and **parabola.**

fractal The geometric result of infinitely many applications of a recursive procedure or calculation. (34, 415)

frequency (of a data set) The number of times a value appears in a data set, or the number of values that fall in a particular interval. (103)

frequency (of a sinusoid) The number of cycles of a periodic function that can be completed in one unit of time. (776)

function A relation for which every value of the independent variable has at most one value of the dependent variable. (190)

function notation A notation that emphasizes the dependent relationship between the variables used in a function. The notation $y = f(x)$ indicates that values of the dependent variable, y, are explicitly defined in terms of the independent variable, x, by the function f. (190)

G

general form (of a polynomial) The form of a polynomial in which the terms are ordered such that the degrees of the terms decrease from left to right. (378)

general form (of a quadratic function) The form $y = ax^2 + bx + c$, where $a \neq 0$. (386)

general quadratic equation An equation in the form $Ax^2 + Bxy + Cy^2 + Dx + Ey + F = 0$, where A, B, and C do not all equal zero. (481)

general term The nth term, u_n, of a sequence. (31)

geometric decay A decay pattern in which amounts decrease by a constant ratio, or percent. In a geometric sequence modeling decay, the common ratio can be represented by $(1 - p)$, where p is the percent change. (42)

geometric growth A growth pattern in which amounts increase by a constant ratio, or percent. In a geometric sequence modeling growth, the common ratio can be represented by $(1 + p)$, where p is the percent change. (42)

geometric probability A probability that is found by calculating a ratio of geometric characteristics, such as lengths or areas. (553)

geometric random variable A random variable that represents the number of trials needed to get the first success in a series of independent trials. The probabilities form a geometric sequence. (580)

geometric sequence A sequence in which each term is equal to the product of the previous term and a common ratio. (35)

geometric series A sum of terms of a geometric sequence. (525)

golden ratio The ratio of two numbers (larger to smaller) whose ratio to each other equals the ratio of their sum to the larger number. Or, the positive number whose square equals the sum of itself and 1. The number $\frac{1+\sqrt{5}}{2}$, or approximately 1.618, often represented with the lowercase Greek letter phi, ϕ. (407)

golden rectangle A rectangle in which the ratio of the length to the width is the golden ratio. (407)

greatest integer function The function $f(x) = [x]$ that returns the largest integer that is less than or equal to a real number, x. (165, 197)

growth See **geometric growth.**

H

half-life The time needed for an amount of a substance to decrease by one-half. (254)

head (or **tip**) The end of a vector with the arrowhead. (712)

histogram A one-variable data display that uses bins to show the distribution of values in a data set. Each bin corresponds to an interval of data values; the height of a bin indicates the number, or frequency, of values in that interval. (103)

hole A missing point in the graph of a relation. (498)

horizontal asymptote See **asymptote.**

horizontal dilation A transformation that increases or decreases the width of a figure or graph. A horizontal dilation by a factor of a multiplies the x-coordinate of every point on a graph by a. (222)

hyperbola A locus of points in a plane for which the difference of the distances to two fixed points, called the foci, is constant. (470)

hyperboloid A three-dimensional shape formed by rotating a hyperbola about the line through its foci or about the perpendicular bisector of the segment connecting the foci. (452, 477)

I

identity An equation that is true for all values of the variables for which the expressions are defined. (783)

identity matrix The square matrix, symbolized by $[I]$, that does not alter the entries of a square matrix $[A]$ under multiplication. Matrix $[I]$ must have the same dimensions as matrix $[A]$, and it has entries of 1's along the main diagonal (from top left to bottom right) and 0's in all other entries. (336)

image A graph of a function or point(s) that is the result of a transformation of an original function or point(s). (200)

imaginary axis See **complex plane.**

imaginary number A number that is the square root of a negative number. An imaginary number can be written in the form bi, where b is a real number $(b \neq 0)$ and i is the imaginary unit, $\sqrt{-1}$. (409, 410)

imaginary unit The imaginary unit, i, is defined by $i^2 = -1$ or $i = \sqrt{-1}$. (409)

inconsistent (system) A system of equations that has no solution. (169)

Glossary

independent (events) Events are independent when the occurrence of one has no influence on the probability of the other. (563)

independent (system) A system of equations that has exactly one solution. (169)

independent variable A variable whose values are not based on the values of another variable. (133)

inequality A statement that one quantity is less than, less than or equal to, greater than, greater than or equal to, or not equal to another quantity. (352)

infinite A quantity that is unending, or without bound. (525)

infinite geometric series A sum of infinitely many terms of a geometric sequence. (525)

inflection point A point where a curve changes between curving downward and curving upward. (638)

intercept form The form $y = a + bx$ of a linear equation, where a is the y-intercept and b is the slope. (131, 198)

interpolation Estimating a value that is within the range of all other values given in a data set. (141)

interquartile range (*IQR*) A measure of spread for a one-variable data set that is the difference between the third quartile and the first quartile. (82)

intersection A set of elements common to two or more sets. In a Venn diagram, it is the overlapping region between two or more sets. (572)

inverse The relationship that reverses the independent and dependent variables of a relation. (280, 282)

inverse matrix The matrix, symbolized by $[A]^{-1}$, that produces an identity matrix when multiplied by $[A]$. (335, 336)

inverse variation A relation in which the product of the independent and dependent variables is constant. An inverse variation relationship can be written in the form $xy = k$, or $y = \frac{k}{x}$. (491)

joint variation An equation in which one variable varies directly with the other variables taken one at a time. In the formula $A = \frac{1}{2}bh$, A varies directly with both b and h, and $\frac{1}{2}$ is the constant of variation. (496)

Law of Cosines For any triangle with angles A, B, and C, and sides of lengths a, b, and c (a is opposite $\angle A$, b is opposite $\angle B$, and c is opposite $\angle C$), $c^2 = a^2 + b^2 - 2ab\cos C$. (699)

Law of Sines For any triangle with angles A, B, and C, and sides of lengths a, b, and c (a is opposite $\angle A$, b is opposite $\angle B$, and c is opposite $\angle C$), these equalities are true: $\frac{\sin A}{a} = \frac{\sin B}{b} = \frac{\sin C}{c}$. (692)

least squares line A line of fit for which the sum of the squares of the residuals is as small as possible. (664)

like terms Monomials with the same base and the same exponents. For example, $3x^2$ and $5x^2$ are like terms, and $2xy^2z$ and $-5xy^2z$ are like terms. (376)

limit A long-run value that a sequence or function approaches. The quantity associated with the point of stability in dynamic systems. (50)

line of fit A line used to model a set of two-variable data. (138)

line of symmetry A line that divides a figure or graph into mirror-image halves. (206)

linear In the shape of a line or represented by a line, or an algebraic expression or equation of degree 1. (55)

linear equation An equation characterized by a constant rate of change. The graph of a linear equation in two variables is a straight line. (124)

linear programming A method of modeling and solving a problem involving constraints that are represented by linear inequalities. (360)

linearization A method of finding an equation to fit data. The x- and/or y-values are transformed until the relation appears linear. Then inverse transformations are applied to the linear model to produce an equation that models the original data. (672)

local maximum A value of a function or graph that is greater than other nearby values. (423)

local minimum A value of a function or graph that is less than other nearby values. (423)

locus A set of points that fit a given condition. (446)

logarithm A value of a logarithmic function, abbreviated log. For $a > 0$ and $b > 0$, $\log_b a = x$ means that $a = b^x$. (287, 296)

logarithm change-of-base property For $a > 0$ and $b > 0$, $\log_a x$ can be rewritten as $\frac{\log_b x}{\log_b a}$. (289, 296)

logarithmic function The logarithmic function $y = \log_b x$ is the inverse of $y = b^x$, where $b > 0$ and $b \neq 1$. (289)

logistic function A function used to model a population that grows and eventually levels off at the maximum capacity supported by the environment. A logistic function has a variable growth rate that changes based on the size of the population. (69)

lurking variable An unmeasured variable that affects the value of the response variable (the variable being studied). (659)

M

magnitude The distance of a number from zero on a number line. Also, the length of a vector. (221, 712)

major axis The longer dimension of an ellipse. Or the line segment with endpoints on the ellipse that has this dimension. (456)

margin of error A measure of accuracy for estimates of a population mean, given by $\frac{z\sigma}{\sqrt{n}}$, where n is the sample size, z is the number of standard deviations from the mean, and σ is the standard deviation. (647)

mathematical induction A type of mathematical proof used to show that a given statement is true for all natural numbers n. A proof by mathematical induction involves first proving the statement is true for $n = 1$, then assuming it is true for $n = k$, and finally proving that it is true for $n = k + 1$. This establishes the truth of the statement for all values of n. (537)

matrix A rectangular array of numbers or expressions, enclosed in brackets. (318)

matrix addition The process of adding two or more matrices. To add matrices, you add corresponding entries. (325, 331)

matrix multiplication The process of multiplying two matrices. The entry c_{ij} in the matrix $[C]$ that is the product of two matrices, $[A]$ and $[B]$, is the sum of the products of corresponding entries in row i of matrix $[A]$ and column j of matrix $[B]$. (327, 331)

maximum The greatest value in a data set or the greatest value of a function or graph. (82, 395, 423)

mean ($\bar{x}$) A measure of central tendency for a one-variable data set, found by dividing the sum of all values by the number of values. See also **population mean.** (79)

measure of central tendency A single number used to summarize a one-variable data set, commonly the mean, median, or mode. (79)

median A measure of central tendency for a one-variable data set that is the middle value, or the mean of the two middle values, when the values are listed in order. See also **population median.** (79)

median-median line A line of fit found by ordering a data set by its x-values, dividing it into three groups, finding three points (M_1, M_2, and M_3) based on the median x-value and the median y-value for each group, and writing the equation that best fits these three points. (145)

minimum The least value in a data set or the least value of a function or graph. (82, 395, 423)

minor axis The shorter dimension of an ellipse. Or the line segment with endpoints on the ellipse that has this dimension. (456)

mode A measure of central tendency for a one-variable data set that is the value(s) that occur most often. See also **population mode.** (79)

model A mathematical representation (sequence, expression, equation, or graph) that closely fits a set of data. (55)

modified box plot A box plot in which any values that are more than 1.5 times the *IQR* from the ends of the box are plotted as separate points. (82)

monomial A polynomial with one term. (376)

multiplicative identity The number 1 is the multiplicative identity because any number multiplied by 1 remains unchanged. (314, 316, 335)

multiplicative inverse Two numbers are multiplicative inverses, or reciprocals, if their product is 1. (314, 316)

mutually exclusive (events) Two outcomes or events are mutually exclusive when they cannot both occur simultaneously. (571)

N

natural logarithm A logarithm with base e, written $\ln x$, which is shorthand for $\log_e x$. (307)

negative exponents For $a > 0$, and all real values of n, the expression a^{-n} is equivalent to $\frac{1}{a^n}$ and $\left(\frac{a}{b}\right)^{-n} = \left(\frac{b}{a}\right)^n$. (260, 296)

nonrigid transformation A transformation that produces an image that is not congruent to the original figure. Dilations are nonrigid transformations (unless the scale factor is 1 or -1). (222)

normal curve The graph of a normal distribution. (634)

normal distribution A symmetric bell-shaped distribution. The equation for a normal distribution with mean μ and standard deviation σ is $y = \frac{1}{\sigma\sqrt{2\pi}}\left(\sqrt{e}\right)^{-[(x-\mu)/\sigma]^2}$. (634, 637)

O

oblique (triangle) A triangle that does not contain a right angle. (690)

observational study A type of study in which the researcher measures variables of interest, but does not assign a treatment to the subjects of the study, and the subjects are not aware of what is being measured. (619)

odd function A function that is symmetric about the origin. For all values of x in the domain of an odd function, $f(-x) = -f(x)$. (247, 786)

one-to-one function A function whose inverse is also a function. (282)

outcome A possible result of one trial of an experiment. (546, 551)

outlier A value that stands apart from the bulk of the data. (82, 91)

P

parabola A locus of points in a plane that are equidistant from a fixed point, called the focus, and a fixed line, called the directrix. (206, 464, 481)

paraboloid A three-dimensional shape formed by rotating a parabola about its line of symmetry. (463)

parameter (in parametric equations) See **parametric equations.**

parameter (statistical) A number, such as the mean or standard deviation, that describes an entire population. (624)

parametric equations A pair of equations used to separately describe the x- and y-coordinates of a point as functions of a third variable, called the parameter. (720)

parent function The most basic form of a function. A parent function can be transformed to create a family of functions. (206)

partial sum A sum of a finite number of terms of a series. (518)

Pascal's triangle A triangular arrangement of numbers containing the coefficients of binomial expansions. The first and last numbers in each row are 1's, and each other number is the sum of the two numbers above it. (602)

percentile rank The percentage of values in a data set that are below a given value. (105)

perfect square A number that is equal to the square of an integer, or a polynomial that is equal to the square of another polynomial. (378)

perfect-square trinomial Any trinomial in the form $a^2 + 2ab + b^2$, whose factored form is a binomial squared, $(a + b)^2$. (384)

period The time it takes for one complete cycle of a cyclical motion to take place. Also, the minimum amount of change of the independent variable needed for a pattern in a periodic function to repeat. (225, 737)

periodic function A function whose graph repeats at regular intervals. (737)

permutation An arrangement of choices in which the order is important. (588, 590)

phase shift The horizontal translation of a periodic graph. (758)

piecewise function A function that consists of two or more functions defined on different intervals. (214)

point-ratio form The form $y = y_1 \cdot b^{x-x_1}$ of an exponential function equation, where the curve passes through the point (x_1, y_1) and has ratio b. (267, 268)

point-slope form The form $y = y_1 + b(x - x_1)$ of a linear equation, where (x_1, y_1) is a point on the line and b is the slope. (139, 198)

polar coordinates A method of representing points in a plane with ordered pairs in the form (r, θ), where r is the distance of the point from the origin and θ is the angle of rotation of the point from the positive x-axis. (796)

polar form (of a vector) A way to represent a vector that describes the magnitude of the vector and the angle it makes with the positive x-axis. (712)

polynomial A sum of terms containing a variable raised to different powers, often written in the form $a_n x^n + a_{n-1}x^{n-1} + \cdots + a_1 x^1 + a_0$, where x is a variable, the exponents are nonnegative integers, and the coefficients are real numbers, and $a_n \neq 0$. (378)

polynomial function A function in which a polynomial expression is set equal to a second variable, such as y or $f(x)$. (378)

population A complete set of people or things being studied. (605, 624)

population mean (μ) In a probability distribution, the sum of each value of x times its probability. (634)

population median For a probability distribution, the median is the number d such that the line $x = d$ divides the area into two parts of equal area. (627)

population mode For a probability distribution, the mode is the value(s) of x at which the graph reaches its maximum value. (627)

population standard deviation (σ) The standard deviation of an entire population. (634)

power function A function that has a variable as the base. The general form of a power function is $y = ax^n$, where a and n are constants. (261)

power of a power property For $a > 0$, and all real values of m and n, $(a^m)^n$ is equivalent to a^{mn}. (260, 296)

power of a product property For $a > 0$, $b > 0$, and all real values of m, $(ab)^m$ is equivalent to $a^m b^m$. (260, 296)

power of a quotient property For $a > 0$, $b > 0$, and all real values of n, $\left(\frac{a}{b}\right)^n$ is equivalent to $\frac{a^n}{b^n}$. (260, 296)

power property of equality For all real values of a, b, and n, if $a = b$, then $a^n = b^n$. (260)

power property of logarithms For $a > 0$, $x > 0$, and $n > 0$, $\log_a x^n$ can be rewritten $n \log_a x$. (296)

prediction interval An interval that predicts the distribution of individual points. For example, a 90% prediction interval means that there is a 90% chance that the next measurement will fall in this interval. (652)

prediction line A line of fit for a data set that can be used to predict the expected value of y for any given x-value. The variable $\hat{y}$ is used in place of y to indicate that a line is a prediction line. (138)

principal The initial monetary balance of a loan, debt, or account. (42)

principal value The one solution to an inverse trigonometric function that is within the range for which the function is defined. (771)

probability distribution A function that will give the probability of an event of a discrete random variable or a function in which the probability of a continuous random variable in the interval (a, b) is the area under the function. (625)

product property of exponents For $a > 0$ and $b > 0$, and all real values of m and n, the product $a^m \cdot a^n$ is equivalent to a^{m+n}. (260, 296)

product property of logarithms For $a > 0$, $x > 0$, and $y > 0$, $\log_a xy$ is equivalent to $\log_a x + \log_a y$. (296)

projectile motion The motion of an object that rises or falls under the influence of gravity. (395, 723)

Q

quadratic formula If a quadratic equation is written in the form $ax^2 + bx + c = 0$, the solutions of the equation are given by the quadratic formula, $x = \frac{-b \pm \sqrt{b^2 - 4ac}}{2a}$. (404)

quadratic function A polynomial function of degree 2. Quadratic functions are in the family with parent function $y = x^2$. (206, 386)

quotient property of exponents For $a > 0$ and $b > 0$, and all real values of m and n, the quotient $\frac{a^m}{a^n}$ is equivalent to a^{m-n}. (260, 296)

quotient property of logarithms For $a > 0$, $x > 0$, and $y > 0$, the expression $\log_a\left(\frac{x}{y}\right)$ can be rewritten as $\log_a x - \log_a y$. (296)

R

radian An angle measure in which one full rotation is 2π radians. One radian is the measure of an arc, or the measure of the central angle that intercepts that arc, such that the arc's length is the same as the circle's radius. (745)

radical A square root symbol. (217)

radius See **circle.**

raised to the power A term used to connect the base and the exponent in an exponential expression. For example, in the expression b^x, the base, b, is raised to the power x. (259)

randomization A characteristic of good data collection in which measurements are taken, subjects are chosen, and/or treatments are applied in a random order. (665)

random number A number that is as likely to occur as any other number within a given set. (550)

random process A process in which no individual outcome is predictable. (548)

random sample A sample in which not only is each person (or thing) equally likely, but all groups of persons (or things) are also equally likely. (82)

random variable A variable that takes on numerical values governed by a chance experiment. (580)

range (of a data set) A measure of spread for a one-variable data set that is the difference between the maximum and the minimum. (82)

range (of a relation) The set of output values of a relation. (133)

rational Describes a number or an expression that can be expressed as a fraction or ratio. (266)

rational exponent An exponent that can be written as a fraction. The expression $a^{m/n}$ can be rewritten as $\left(\sqrt[n]{a}\right)^m$ or $\sqrt[n]{a^m}$, for $a > 0$. (267, 296)

rational function A function that can be written as a quotient, $f(x) = \frac{p(x)}{q(x)}$, where $p(x)$ and $q(x)$ are polynomial expressions and $q(x)$ is of degree 1 or higher. (491)

Rational Root Theorem If the polynomial equation $P(x) = 0$ has rational roots, they are in the form $\frac{p}{q}$, where p is a factor of the constant term and q is a factor of the leading coefficient. (432)

real axis See **complex plane.**

rectangular form (vector) A way to represent a vector in which the x- and y-coordinates of the vector head are used to name the vector (assuming the vector has been translated to the origin). For example, $<2, 5>$ describes a vector with its tail at (0, 0) and its head at (2, 5). (712)

recursion Applying a procedure repeatedly, starting with a number or geometric figure, to produce a sequence of numbers or figures. Each term or stage builds on the previous term or stage. (30)

recursive formula A starting value and a recursive rule for generating a sequence. (31)

recursive rule Defines the nth term of a sequence in relation to the previous term(s). (31)

reduced row-echelon form A matrix form in which each row is reduced to a 1 along the diagonal. All entries above and below the diagonal are 0's. (344)

reference angle The acute angle between the terminal side of an angle in standard position and the x-axis. (707, 738)

reference triangle A right triangle that is drawn connecting the terminal side of an angle in standard position to the x-axis. A reference triangle can be used to determine the trigonometric ratios of an angle. (707, 738)

reflection A transformation that flips a graph across a line, creating a mirror image. (214, 232)

regression analysis The process of finding a model with which to make predictions about one variable based on values of another variable. (664)

relation Any relationship between two variables. (190)

relative frequency histogram A histogram in which the height of each bin shows proportions (or relative frequencies) instead of frequencies. (625)

replication A characteristic of good experimental design in which repeated measurements are made under the same conditions (treatments). (665)

residual For a two-variable data set, the difference between the y-value of a data point and the y-value predicted by the equation of fit. (152)

response variable In statistics, the outcome (dependent) variable that is being studied. (657)

resultant (vector) The sum of two vectors. (712)

rigid transformation A transformation that produces an image that is congruent to the original figure. Translations, reflections, and rotations are rigid transformations. (222)

root mean square error (*s*) A measure of spread for a two-variable data set, similar to standard deviation for a one-variable data set. It is calculated by the formula $s = \sqrt{\frac{\sum_{i=1}^{n}(y_i - \hat{y})^2}{n-2}}$. (154)

roots The solutions of an equation in the form $f(x) = 0$. (387)

row reduction method A method that transforms a matrix into reduced row-echelon form. (344)

S

sample A part of a population selected to represent the entire population. Sampling is the process of selecting and studying a sample from a population in order to make conjectures about the whole population. (605, 624)

scalar A real number, as opposed to a matrix or vector. (712)

scalar multiplication The process of multiplying a matrix by a scalar. To multiply a scalar by a matrix, you multiply the scalar by each value in the matrix. (326)

scale factor A number that determines the amount by which a graph is dilated, either horizontally or vertically. (221)

secant The reciprocal of the cosine ratio. If A is an acute angle in a right triangle, the secant of angle A is the ratio of the length of the hypotenuse to the length of the adjacent leg, or $\sec A = \frac{hyp}{adj}$. See **trigonometric function.** (783)

sequence An ordered list of numbers. (31)

series The indicated sum of terms of a sequence. (518)

shape (of a data set) Describes how the data are distributed relative to the position of a measure of central tendency. (84)

shifted geometric sequence A geometric sequence that includes an added term in the recursive rule. (50)

sigma notation The mathematical notation for summation, as represented by the Greek letter Σ (sigma). For example, $\sum_{n=1}^{4}\frac{1}{n}$ means $\frac{1}{1} + \frac{1}{2} + \frac{1}{3} + \frac{1}{4}$. (97)

significant figures (significant digits) The digits in a measurement that give information about its accuracy. For example, 3.00 has three significant digits, whereas 3 has one significant digit. (90)

similar Geometric figures that are dilations of one another by the same horizonal and vertical scale factors. Two figures are similar if and only if all corresponding angles are congruent and lengths of all corresponding sides, edges, or other one-dimensional measures are proportional. (680)

simple event An event consisting of just one outcome. (551)

simple interest A percentage paid on the principal over a period of time. (42)

simple random sample See **random sample.**

simulation A procedure that uses a chance model to imitate a real situation. (551)

sine If A is an acute angle in a right triangle, then the sine of angle A is the ratio of the length of the opposite leg to the length of the hypotenuse, or $\sin A = \frac{opp}{hyp}$. See **trigonometric function.** (683)

sine wave A graph of a sinusoidal function. See **sinusoid.** (757)

sinusoid A function or graph for which $y = \sin x$ or $y = \cos x$ is the parent function. (757)

skewed (data) Data that are spread out more on one side of the center than on the other side. (84)

slant asymptote A linear asymptote that is not parallel to either the x- or y-axis. (509)

slope The steepness of a line or the rate of change of a linear relationship. If (x_1, y_1) and (x_2, y_2) are two points on a line, then the slope of the line is $\frac{y_2 - y_1}{x_2 - x_1}$, where $x_2 \neq x_1$. (121, 125)

spread The variability in numerical data. (93)

square root function The function that undoes squaring, giving only the positive square root (that is, the positive number that, when multiplied by itself, gives the input). The square root function is written $y = \sqrt{x}$. (213)

standard deviation (s) A measure of spread for a one-variable data set that uses squaring to eliminate the effect of the different signs of the individual deviations. It is the square root of the variance, or $s = \sqrt{\frac{\sum_{i=1}^{n}(x_i - \bar{x})^2}{n-1}}$. See also **population standard deviation.** (97)

standard form (of a conic section) The form of an equation for a conic section that shows the transformations of the parent equation. (454, 455, 466, 474)

standard form (of a linear equation) The form $ax + by = c$ of a linear equation. (203)

standard normal distribution A normal distribution with mean 0 and standard deviation 1. (635)

standard position An angle positioned with one side on the positive x-axis. (738)

standardizing the variable The process of converting data values (x-values) to their images (z-values) when a normal distribution is transformed into the standard normal distribution. (646)

statistic A numerical measure of a data set or sample. (79)

statistics A collection of numerical measures, or the mathematical study of data collection and analysis. (77)

stem-and-leaf plot A one-variable data display in which the left digit(s) of the data values, called the stems, are listed in a column on the left side of the plot, and the remaining digits, called the leaves, are listed in order to the right of the corresponding stem. (112)

step function A function whose graph consists of a series of horizontal line segments. (165, 197)

substitution A method of solving a system of equations that involves solving one of the equations for one variable and substituting the resulting expression into the other equation. (163)

survey A type of study in which a researcher collects information about treatment(s) and results from the subjects. (619)

symmetric (data) Data that are balanced, or nearly so, about the center. (84, 628)

synthetic division An abbreviated form of dividing a polynomial by a linear factor. (433, 434)

system of equations A set of two or more equations with the same variables that are solved or studied simultaneously. (161)

tail The end of a vector that does not have an arrowhead. (712)

tangent If A is an acute angle in a right triangle, then the tangent of angle A is the ratio of the length of the opposite leg to the length of the adjacent leg, or $\tan A = \frac{opp}{adj}$. See **trigonometric function.** (683)

term (algebraic) An algebraic expression that represents only multiplication and division between variables and constants. (376, 378)

term (of a sequence) Each number in a sequence. (31)

terminal side The side of an angle in standard position that is not on the positive x-axis. (738)

theoretical probability A probability calculated by analyzing a situation, rather than by performing an experiment, given by the ratio of the number of different ways an event can occur to the total number of equally likely outcomes possible. (551)

third quartile (Q_3) The median of the values greater than the median of a data set. (82)

transcendental number An irrational number that, when represented as a decimal, has infinitely many digits with no pattern, such as π or e, and is not the solution of a polynomial equation with integer coefficients. (307)

transformation A change in the size or position of a figure or graph. (206, 220)

transition diagram A diagram that shows how something changes from one time to the next. (318)

transition matrix A matrix whose entries are transition probabilities. (318)

translation A transformation that slides a figure or graph to a new position. (198, 200, 232)

tree diagram A diagram whose branches show the possible outcomes of an event, and sometimes probabilities. (560)

triangle inequality theorem For any triangle, the length of a side must be less than or equal to the sum of the other two sides, but greater than or equal to the difference between the two sides. (717)

trigonometric function A periodic function that uses one of the trigonometric ratios to assign values to angles with any measure. (757)

trigonometric ratios The ratios of lengths of sides in a right triangle. The three primary trigonometric ratios are sine, cosine, and tangent. (682)

trigonometry The study of the relationships between the lengths of sides and the measures of angles in triangles. (682)

trinomial A polynomial with three terms. (378)

U

unbiased estimate A simple random sample will produce unbiased estimates because the data are collected from a sample in which every member of the population is equally likely to be selected. (82)

union The set of elements that contains all elements of the sets involved. The word *or* is often associated with the union. (572)

unit circle A circle with radius of one unit. The equation of a unit circle with center (0, 0) is $x^2 + y^2 = 1$. (229)

unit hyperbola The parent equation for a hyperbola, $x^2 - y^2 = 1$ or $y^2 - x^2 = 1$. (471)

V

variance (s^2) A measure of spread for a one-variable data set that uses squaring to eliminate the effect of the different signs of the individual deviations. It is the sum of the squares of the deviations divided by one less than the number of values, or $s^2 = \frac{\sum_{i=1}^{n}(x_i - \bar{x})^2}{n-1}$. (96)

vector A quantity with both magnitude and direction. (712)

velocity A measure of speed and direction. Velocity can be either positive or negative. (129)

Venn diagram A diagram of overlapping circles that shows the relationships among members of different sets. (413, 571)

vertex (of a conic section) The point or points where a conic section intersects the axis of symmetry that contains the focus or foci. (206)

vertex (of a feasible region) A corner of a feasible region in a linear programming problem. (353)

vertex-edge graph A diagram comprised of a set of points, called vertices, along with segments or arcs, called edges, connecting some or all of the points. The essential information in the graph is shown by the way the vertices are connected by the edges. (323, 324)

vertex form The form $y = a(x - h)^2 + k$ of a quadratic function, where $a \neq 0$. The point (h, k) is the vertex of the parabola, and a is the vertical scale factor. (386, 389)

vertical dilation A transformation that increases or decreases the length of a figure or graph. A vertical dilation by a factor of b multiplies the y-coordinate of every point on a graph by b. (222)

Z

zero exponent For all values of a except 0, $a^0 = 1$. (260)

zero-product property If the product of two or more factors equals zero, then at least one of the factors must equal zero. A property used to find the zeros of a function without graphing. (387)

zeros (of a function) The values of the independent variable (x-values) that make the corresponding values of the function ($f(x)$-values) equal to zero. Real zeros correspond to x-intercepts of the graph of a function. See **roots.** (387)

***z*-value** or ***z*-score** The number of standard deviations that a given x-value lies from the mean. (646)

Index

A

B

C

D

N

O

P

Q

R

S

T

U

Z

Photo Credits

Abbreviations: top (*t*), middle (*m*), bottom (*b*), left (*l*), right (*r*)

Cover

Background and center image: NASA; all other images: Ken Karp Photography

Front Matter

v (*t*): Ken Karp Photography; **vi (*t*):** Ken Karp Photography; **vi (*b*):** Ken Karp Photography; **vii (*b*):** Ken Karp Photography; **ix (*t*):** Ken Karp Photography; **x (*t*):** Ken Karp Photography; **xi (*t*):** Exhibit and Photography by Shab Levy; **xii:** Ken Karp Photography;

Chapter 0

1: California Academy of Sciences. Used with permission; **3:** Michael Nicholson/Corbis; **4:** AP-Wide World Photos; **6:** Christie's Images; **7 (*t*):** Getty Images; **7 (*bm*):** Bodleian Library, Oxford, U.K. Copyright (MS Hunt. 214 Title Page); **11:** © Images .com/Corbis; **12:** Christie's Images; **13 (*t*):** Cheryl Fenton; **13 (*b*):** Ken Karp Photography; **14:** Scala/Art Resource, NY; **16 (*t*):** Ken Karp Photography; **16 (*l*):** AFP/Corbis; **16 (*r*):** Reuters NewMedia/Corbis; **18:** Daniel Castor; **19:** Corbis; **20 (*b*):** © The New Yorker Cartoon Collection 2001 Jack Ziegler from *cartoonbank.com*. All Rights Reserved; **21:** Cheryl Fenton; **24:** NASA

Chapter 1

27: *Residual Light* by Anthony Discenza. Photos courtesy of the artist; **28:** © Juergen Sack/*iStockphoto.com*; **30 (*l*):** Sandro Vannini/Corbis; **30 (*r*):** Tom & Dee Ann McCarthy/Corbis; **32:** Owaki-Kulla/Corbis; **37:** Gary D. Landsman/Corbis; **38:** Ken Karp Photography; **39:** The Image Bank/Getty Images; **40:** Bohemia Nomad Picturemakers/Corbis; **41:** Ken Karp Photography; **41:** RF; **43:** Tom Collicott/Masterfile; **45:** Bettmann/ Corbis; **46 (*t*):** Richard T. Nowitz/Science Source/Photo Researchers, Inc.; **46 (*b*):** Archive Photos/PictureQuest; **47:** Bob Krist/Corbis; **48 (*l*):** Duomo/Corbis; **48 (*r*):** © Peter Essick/Aurora/ Getty Images; **49:** Ken Karp Photography; **50:** The Cleveland Memory Project; **51:** Tony Freeman/PhotoEdit/PictureQuest; **52:** RF; **53:** Cheryl Fenton; **55 (*t*):** Cheryl Fenton; **55 (*b*):** NOAA; **56:** © Lawson Wood/CORBIS; **58:** Archivo Iconografico, S. A./ Corbis; **59:** Bill Ross/Corbis; **60:** Getty Images; **61:** Paul Skelcher/ Rainbow; **65:** John Henley/Corbis; **66:** Stuart Westmorland/Index Stock Imagery/PictureQuest; **67:** Réunion des Musées Nationaux/ Art Resource, NY/Artist Rights Society; **69:** Lester Lefkowitz/ Corbis; **72:** Tobias Titz/Getty Images RF; **74:** Galen Rowell/Corbis; **75:** Martha Bates/Stock Boston Inc./PictureQuest

Chapter 2

78: © 2003 Monika Sprüth Gallery/Artists Rights Society (ARS), New York/VG Bild-Kunst, Bonn; Courtesy of Monika Sprüth/ Philomene Magers, Cologne/Munich/Art Resource, NY; **80:** Ken Karp Photography; **81:** Laima Druskis/Stock, Boston Inc./Picture Quest; **83 (*l*):** Gianni Dagli Orti/Corbis; **83 (*r*):** Werner Forman Archive/Museum fur Volkerkunde, Berlin/Art Resource, NY; **84:** Ken Karp Photography; **86:** © AP Photo; **87:** Corbis; **94:** Ken Karp Photography; **95:** Ken Karp Photography; **96:** Photos courtesy of Lorraine Serena/Women Beyond Borders; **97:** NASA; **100:** Bohemian Nomad Picturemakers/Corbis; **104:** © Jim Craigmyle/ Corbis; **106:** Chase Swift/Corbis; **109:** Ken Karp Photography; **110:** © Royalty-Free/Corbis; **111:** Phillip James Corwin/Corbis; **113:** Bettmann/Corbis; **117:** © AP Photo/Eric Gay; **118:** Getty Images

Chapter 3

121: Zaha Hadid Architects; **124:** Pablo Corral/Corbis; **125:** RF; **128:** Alan Schein Photography/Corbis; **129:** Heather Ackroyd & Dan Harvey; **131:** Gale Beery/Index Stock/PictureQuest; **132:** Ken Karp Photography; **135:** Deborah Davis/PhotoEdit/ PictureQuest; **136:** Andre Jenny/Focus Group/PictureQuest; **139:** David Muench/Corbis; **140:** NASA; **141:** Chuck Savage/ Corbis; **143:** Ken Karp Photography; **144:** Chuck Fishman/Contact Press Images/PictureQuest; **147:** *Tourists on the Moon #2* by Yoshio Itagaki, 1998/Courtesy of the artist and Cristinerose Gallery, New York; **150:** Tim Wright/Corbis; **151:** Michael A. Dwyer/Stock Boston Inc./PictureQuest; **155:** Getty Images/RF; **157:** Cheryl Fenton; **158:** Reuters NewMedia Inc./Corbis; **161:** *Green Mass*, I. Pereira, Gift of Leslie Bokor and Leslie Dame, Photograph © 2002 Board of Trustees, National Gallery of Art, Washington, 1950, oil on canvas; **165 (*t*):** AP-Wide World Photos/ Itsuo Inouye; **165 (*bl*):** E. R. Degginger/Bruce Coleman Inc.; **165 (*br*):** E. R. Degginger/Bruce Coleman Inc.; **167:** Courtesy of the National Museum of the American Indian, Smithsonian Institution. Photo by R. A. Whiteside; **171 (*t*):** Ken Karp Photography; **171 (*b*):** E. R. Degginger/Bruce Coleman Inc.; **172 (*m*):** Burstein Collection/Corbis; **172 (*mr*):** Dominique Berretty/Black Star Publishing/PictureQuest; **176:** Jose Fuste Raga/Corbis

Chapter 4

181: Courtesy of the artist and Greenberg Van Doren Gallery, New York; **184:** Lorna Simpson, *Wigs (portfolio)*, 1994, waterless lithograph on felt, 72 in. by 162 in. overall installed/Collection Walker Art Center, Minneapolis/T. B. Walker Acquisition Fund, 1995; **185:** DiMaggio/Kalish/Corbis; **186:** PEANUTS reprinted by permission of United Feature Syndicate, Inc.; **187:** Ellsworth Kelly, *Blue Green Curve*, 1972, Oil on canvas, 87-3/4 × 144-1/4 in. The Museum of Contemporary Art, Los Angeles, The Barry Lowen Collection; **189:** Douglas Kirkland/ Corbis; **191:** Archivo Iconografico, S. A./Corbis; **192:** Getty Images; **194:** David Bentley/Corbis; **195:** Robert Holmes/Corbis; **196:** Chris Chen/2c Management; **198 (*l*):** SIMPARCH, *Free Basin*, birch wood, metal supports, courtesy of SIMPARCH, installation view at the Wexner Center for the Arts, photo by Dick Loesch; **198 (*r*):** SIMPARCH, *Free Basin*, birch wood, metal supports, courtesy of SIMPARCH, installation view at the Wexner Center for the Arts, photo by Roman Sapecki; **199:** Ken Karp Photography; **200 (*m*):** Cheryl Fenton; **201 (*b*):** Danny Lehman/ Corbis; **202:** Cheryl Fenton; **203:** © 2002 Eun-Ha Paek, stills from *L'Faux Episode 7* on *MilkyElephant.com*; **205 (*l*):** AP-Wide World Photos; **205 (*r*):** *Bessie's Blues* by Faith Ringgold © 1997, acrylic on canvas, 76 in. by 79 in., photo courtesy of the artist; **206:** Phil Schermeister/Corbis; **209:** Bo Zaunders/Corbis; **211:** Robert Ginn/Index Stock Imagery/PictureQuest; **212:** © 1998 Jerry Lodriguss; **213:** Ken Karp Photography; **214:** James L. Amos/ Corbis; **218:** Tony Freeman/PhotoEdit/PictureQuest; **224:** Archivo Iconografico, S. A./Corbis; **228:** Bettmann/Corbis; **229:** Tom Nebbia/Corbis; **231:** Lester Lefkowitz/Corbis; **237:** AFP/Corbis; **239:** Ken Karp Photography; **247:** Sense oil on birch panel © 2002 Laura Domela, photo courtesy of the artist

Chapter 5

249: *Tree Mountain—A Living Time Capsule*—11,000 Trees, 11,000 People, 400 Years 1992–1996, Ylojarvi, Finland, (420 meters long by 270 meters wide by 28 meters high) © Agnes Denes. Photo courtesy of Agnes Denes; **253:** Roger Ressmeyer/Corbis; **255:** Buddy Mays/Corbis; **257:** Clementine Gallery/Courtesy Heidi Steiger Collection; **258:** Bettmann/Corbis; **262:** Eric Kamischke; **266:** Paul Almasy/Corbis; **268:** Franz-Marc Frei/Corbis; **269:** © Eliza Snow/ iStockphoto.com; **270:** Neal Preston/Corbis; **272 (*t*):** JPL/NASA; **272 (*b*):** Geoff Tompkinson/Photo Researchers, Inc.; **273:** Ruet Stephane/Corbis Sygma; **274:** © 2002 Eames Office (*www .eamesoffice.com*); **277:** Ken Karp Photography; **278:** NASA; **283:** Alice Arnold/Retna Ltd.; **285:** Getty Images; **291 (*t*):** Tom Bean/Corbis; **291 (*b*):** Corbis; **292:** Getty Images; **293:** Bettmann/ Corbis; **295 (*t*):** Cheryl Fenton; **295 (*b*):** Getty Images;

297: Dr. Oyvind Hammer; **298 (*t*):** Reuters NewMedia Inc./Corbis; **298 (*b*):** © Paul Seheult, Eye Ubiquitous/Corbis; **299:** Getty Images; **300:** Cheryl Fenton; **301:** Tom Stewart/Corbis; **303:** Ken Karp Photography; **304:** Bettmann/Corbis; **305:** Stock Portfolio-Stock Connection/PictureQuest; **306:** Brownie Harris/Corbis; **308 (*l*):** Holt Confer/Grant Heilman Photography; **308 (*r*):** Grant Heilman Photography; **311:** H. Turvey/Photo Researchers, Inc.

Chapter 6

313: *Fragile* by Amy Stacey Curtis. Photos of Fragile © 2000 by Amy Stacey Curtis/*www.amystaceycurtis.com*; **314:** Mark Ralston/AFP Getty Images; **318:** David Stoecklein/Corbis; **319:** Ken Karp Photography; **321:** Karl Weatherly/Corbis; **322:** Corbis; **328:** Ken Karp Photography; **329:** Copyright Robert Silvers 2001 Photomosaics™. Photomosaics™ is the trademark of Runaway Technology, Inc. *www.photomosaic.com* U.S. Patent No. 6,137,498; **332:** Albright-Knox Art Gallery/Corbis; **333:** Getty Images/RF; **334:** Blaine Harrington III/Corbis; **340:** Lester Lefkowitz/Corbis; **342:** © The Nobel Foundation; **344:** © Tate Gallery, London/Art Resource, NY © 2003 C. Herscovici, Brussels/Artists Rights Society (ARS), New York; **349:** Mark Sadan; **351:** Bettmann/Corbis; **354:** Tom Stewart/Corbis; **357 (*t*):** © Christie's Images/Corbis; **357 (*b*):** Melanie Einzig; **358 (*t*):** Graig Orsini/Index Stock Imagery/PictureQuest; **358 (*b*):** NASA; **360:** CNAC/MNAM/Dist. Reunion des Musees Nationaux/Art Resource, NY, © 2003 Artists Rights Society (ARS), New York/ADAGP, Paris; **365 (*t*):** Getty Images/RF; **365 (*b*):** George D. Lepp/Corbis; **368:** © Christie's Images/Corbis; **369:** Anna Clopet/Corbis; **370:** Lowell Georgia/Corbis; **371 (*t*):** Masaaahiro Sano/Corbis; **371 (*b*):** Hulton-Deutsch Collection/Corbis; **372:** Corbis

Chapter 7

375: *Things Fall Apart: 2001,* mixed media installation with vehicle; variable dimensions/San Francisco Museum of Modern Art, Accessions Committee Fund purchase © Sarah Sze; **379:** Getty Images; **381:** Ken Karp Photography; **382 (*l*):** Science Museum/Science & Society Picture Library; **382 (*r*):** Science Museum/Science & Society Picture Library; **384:** Ken Karp Photography; **385:** Kevin Fleming/Corbis; **386:** Mike Souther-Eye Ubiquitous/Corbis; **389 (*t*):** Elizabeth Catlett (American, b 1915), *Singing Their Songs,* 1992, Lithograph on paper (a.p.#6) 15-3/4 in. by 13-3/4 in./National Museum of Women in the Arts, purchased with funds donated in memory of Florence Davis by her family, friends, and the NMWA Women's Committee; **389 (*b*):** Ken Karp Photography; **391:** Ken Karp Photography; **393 (*t*):** Michael Boys/Corbis; **393 (*b*):** Josiah Davidson/PictureQuest; **395:** Scott T. Smith/Corbis; **396:** Ken Karp Photography; **398:** Christie's Images/Corbis; **399:** © Dustin Snipes/Icon SMI/Corbis; **401 (*m*):** Ken Karp Photography; **401 (*bl*):** The Granger Collection, New York City; **401 (*br*):** Bettmann/Corbis; **402 (*bl*):** A. H. Rider/Photo Researchers, Inc.; **402 (*br*):** Nigel J. Dennis/Photo Researchers, Inc.; **407:** © Jonathan Ferguson/Sunforge Studios; **409:** Bettmann/Corbis; **411:** Christie's Images/SuperStock; **415 (*t*):** Mehau Kulyk/Photo Researchers, Inc.; **415 (*m*):** Hank Morgan/Photo Researchers, Inc.; **417:** Cornelia Parker Mass *(Colder Darker Matter),* 1997, charcoal, wire and black string, Collection of Phoenix Art Museum, Gift of Jan and Howard Hendler 2002.1; **418:** Ken Karp Photography; **421:** Children Beyond Borders/VSA Arts (*www.vsarts.org*); **423:** Karim Rashid Inc.; **426:** Courtesy of 303 Gallery, New York-Victoria Miro Gallery, London; **429:** W. Perry Conway/Corbis; **430:** Mark Burnett/Photo Researchers; **439:** Owen Franken/Corbis

Chapter 8

441: Marion Brenner/Maya Lin Studio; **444:** Corbis; **445:** George Hall/Corbis; **448:** Curtis O'Shock/ImageState; **449:** © Royalty-Free/Corbis; **450:** Scott Snibbe; **451:** Courtesy of the artist and Clifford-Smith Gallery, Boston. Photo courtesy of the artist; **452 (*tl*):** Reunion des Musees Nationaux/Art Resource, NY; © L&M Services B. V. Amsterdam 20030605; **452 (*tr*):** Archivo Iconografico, S. A./Corbis; **452 (*bl*):** Lee Snbider/Corbis; **452 (*br*):** St. Louis Science Center; **455:** Ken Karp Photography; **458:** Ken Karp Photography; **459:** Alan Schein Photography/Corbis; **461:** Museum of Science and Industry, Chicago, Illinois; **463 (*t*):** Paul Almasy/Corbis; **463 (*b*):** Michael Gadomski/Photo Researchers; **465:** Robert Landau/Corbis; **468:** Lee Foster/Bruce Coleman Inc.; **469:** Ed Eckstein/Corbis; **470:** © Ronan/iStockphoto.com; **474:** Ken Karp Photography; **477:** W. Cody/Corbis; **481:** Frank Zullo/Photo Researchers, Inc.; **482 (*t*):** Fermi National Accelerator Laboratory; **482 (*b*):** Fermi National Accelerator Laboratory; **484:** Christie's Images/Corbis; **488:** Art © Estate of Alexander Rodchenko RAO, Moscow/VAGA, NY; **489:** U.S. Coast Guard; **490:** Mary Sullivan; **503:** Robert Essel NYC/Corbis; **508:** Alinari/Art Resource, NY © 2003 Artists Rights Society (ARS) New York/Pro Litteris, Zurich; **510:** Cheryl Fenton; **511:** Vadim Makarov

Chapter 9

515: Courtesy of the artist and Lehmann Maupin Gallery, New York; **518:** Binod Joshi/AP-Wide World Photos; **522:** Bettmann/Corbis; **523 (*t*):** Cheryl Fenton; **523 (*b*):** Joseph Sohm-Visions of America/Corbis; **524:** Mark Langford-Index Stock Imagery/PictureQuest; **525:** Ken Karp Photography; **527:** © PictureNet/Corbis; **529:** Lester Lefkowitz/Corbis; **530 (*t*):** George D. Lepp/Corbis; **532:** Ken Karp Photography; **536:** Ali Meyer/Corbis; **537:** Ken Karp Photography; **540:** Lewis Kemper-Index Stock Imagery/PictureQuest; **542:** Julian Calder/Corbis

Chapter 10

545: Smithsonian American Art Museum, Washington, D.C./Art Resource, NY: **548:** Mary Messenger-Stock Boston Inc./PictureQuest; **550 (*l*):** Roger Wood/Corbis; **550 (*r*):** National Museum of India, New Delhi/The Bridgeman Art Library; **553 (*t*):** © Royalty-Free/Corbis; **554:** Royal Asiatic Society, London, UK/The Bridgeman Art Library; **565:** Ken Karp Photography; **568:** Bettmann/Corbis; **571:** Albright-Knox Art Gallery/Corbis; **572:** © Frank Siteman/Index Stock Imagery/Photolibrary.com; **575:** Ken Karp Photography; **577:** RF; **579:** Francois Gohier/Photo Researchers, Inc.; **582:** Robert Young Pelton/Corbis; **583:** Mark C. Burnett-Stock Boston Inc./PictureQuest; **584:** Getty Images; **588:** Chris Rogers-Index Stock Imagery/PictureQuest; **589:** RF; **592:** Ted Horowitz/Corbis; **593:** V & L/Photo Researchers, Inc.; **595:** Ken Karp Photography; **600:** © Royalty-Free/Corbis; **604:** S. Charles Brown-Frank Lane Picture Agency/Corbis; **606:** Bob Daemmrich-Stock Boston Inc./PictureQuest; **608:** Wolfgang Kaehler/Corbis; **610:** © Bettmann/Corbis; **612:** Pablo Corral/Corbis

Chapter 11

615: Naoki Okamoto/SuperStock; **618:** © Corbis/SuperStock; **619:** Ken Karp Photography; **620:** © Yves Forestier/Corbis Sygma; **622:** © Barbara Sauder/iStockphoto.com; **623:** © Bettmann/Corbis; **624 (*t*):** © Kevin Radford/SuperStock; **624 (*b*):** © Paul Ramirez Jonas, courtesy of the LFL Gallery, New York; **631 (*tl*):** Bettmann/Corbis; **631:** Courtesy of John F. Kennedy Presidential Library and Museum, Boston; **631 (*b*):** Ken Karp Photography; **634:** Exhibit and photography by Shab Levy; **637:** © Steve Snyder/iStockphoto.com; **639:** Hulton-Deutsch Collection/Corbis; **640 (*bl*):** Cheryl Fenton; **640 (*br*):** Bettmann/Corbis; **641:** AP-Wide World Photos; **645:** SuperStock; **647:** Cheryl Fenton; **648:** P. Lahalle/Corbis; **650:** Phillip James Corwin/Corbis; **652:** Roger Ball/Corbis; **655:** Jose Carrillo-Stock Boston Inc./PictureQuest; **661:** Galen Rowell/Corbis; **662 (*t*):** AP-Wide World Photos; **662 (*b*):** © Randall Michelson/WireImage/Getty; **664:** © digitalskillet/iStockphoto.com; **665:** Ken Karp Photography; **669:** NASA; **670:** Michael S. Yamashita/Corbis; **675:** Eric Perlman/Corbis; **676:** © Charles Matton, courtesy of Forum Gallery, New York; **678:** Peter Steiner/Corbis

Chapter 12

679: Corey Rich/Coreyography LLC; **682:** Robert Dowling/Corbis; **685:** Roger Ressmeyer/Corbis; **691:** James L. Amos/Corbis; **695:** Chabot Observatory; **696:** George Robinson-Stock Connection/PictureQuest; **698:** Jay Syverson/Corbis; **700:** Ken Karp Photography; **702:** Stone/Getty Images; **703:** Bettmann/Corbis; **709:** © David Bebber/Reuters/Corbis; **711:** Johnny Crawford/The Image Works; **717:** Jose Luis Pelaez, Inc./Corbis; **718 (*t*):** © Matthias Kulka/zefa/Corbis; **718 (*b*):** © Mike Agliolo/Corbis; **720 (*t*):** © Historical Picture Archive/Corbis; **720 (*b*):** Bettmann/Corbis; **725:** William James Warren/Corbis; **727 (*b*):** Underwood & Underwood/Corbis; **728:** © The New Yorker Collection from *CartoonBank.com*/Corbis; **731 (*t*):** © Diego Azubel/epa/Corbis; **731 (*b*):** Digital Image © The Museum of Modern Art/Licensed by Scala/Art Resource, NY

Chapter 13

733: Courtesy of George Eastman House, International Museum of Photography and Film; **736:** © Michael Hanschke/dpa/Corbis; **742:** Courtesy National Solar Observatory/Sacramento Peak; **743:** Lester Lefkowitz/Corbis; **744:** Courtesy of the artist and Hooks-Epstein Gallery, Houston, TX; **748:** © Royalty-Free/Corbis; **750:** Bettmann/Corbis; **752:** Tim Bird/© Royalty-Free/Corbis; **753:** FOXTROT © Bill Amend. Reprinted with permission of UNIVERSAL PRESS SYNDICATE. All rights reserved; **757:** Corbis; **760 (*t*):** Ken Karp Photography; **760 (*b*):** © Royalty-Free/Corbis; **764:** Historical Picture Archive/Corbis; **773:** Ken Karp Photography; **775 (*r*):** NOAA/Center for Operational Oceanographic Products and Services; **775:** Michael S. Yamashita/Corbis; **777:** Getty Images; **780:** Bettmann/Corbis; **781:** Cedar Point Park; **782:** © Frans Lanting/Corbis; **787:** Scott Walking Adventures; **789:** Reuters NewMedia Inc./Corbis; **791:** © Jack Hollingsworth/Corbis; **800:** Collection of the Artist, Courtesy of the Oakland Museum of California, photo by Michael Temperio; **802:** Getty Images; **803 (*t*):** Jonathan T. Wright/Bruce Coleman Inc.; **803 (*b*):** Corbis